Indian Institute of Metals Series

About the Book Series:

The study of metallurgy and materials science is vital for developing advanced materials for diverse applications. In the last decade, the progress in this field has been rapid and extensive, giving us a new array of materials, with a wide range of applications, and a variety of possibilities for design of new materials, processing and characterizing the materials. In order to make this growing volume of knowledge available, an initiative to publish a series of books in Metallurgy and Materials Science was taken during the Diamond Jubilee year of the Indian Institute of Metals (IIM) in the year 2006, and has been published in partnership with Springer since 2016.

This book series publishes different categories of publications: textbooks to satisfy the requirements of students and beginners in the field, monographs on select topics by experts in the field, professional books to cater to the needs of practising engineers, and proceedings of select international conferences organized by IIM after mandatory peer review. The series publishes across all areas of materials sciences and metallurgy. An panel of eminent international and national experts serves as the advisory body in overseeing the selection of topics, important areas to be covered, and the selection of contributing authors.

Raghumani S. Ningthoujam · A. K. Tyagi
Editors

Handbook of Materials Science, Volume 1

Optical Materials

IIM
Metallurgy
Materials Engineering

Springer

Editors
Raghumani S. Ningthoujam
Chemistry Division
Bhabha Atomic Research Centre
Mumbai, Maharashtra, India

A. K. Tyagi
Chemistry Division
Bhabha Atomic Research Centre
Mumbai, Maharashtra, India

ISSN 2509-6400 ISSN 2509-6419 (electronic)
Indian Institute of Metals Series
ISBN 978-981-99-7144-2 ISBN 978-981-99-7145-9 (eBook)
https://doi.org/10.1007/978-981-99-7145-9

This Springer imprint is published by the registered company Springer Nature Singapore Pte Ltd.
The registered company address is: 152 Beach Road, #21-01/04 Gateway East, Singapore 189721, Singapore

Paper in this product is recyclable.

Series Editor's Preface

The *Indian Institute of Metals Series* is an institutional partnership series focusing on metallurgy and materials science and engineering.

About the Indian Institute of Metals

The Indian Institute of Metals (IIM) is a premier professional body (since 1947) representing an eminent and dynamic group of metallurgists and materials scientists and engineers from R&D institutions, academia, and industry, mostly from India. It is a registered professional institute with the primary objective of promoting and advancing the study and practice of the science and technology of metals, alloys, and novel materials. The institute is actively engaged in promoting academia–research and institute–industry interactions.

Genesis and History of the Series

The study of metallurgy and materials science and engineering is vital for developing advanced materials for diverse applications. In the last decade, progress in this field has been rapid and extensive, giving us a new array of materials, with a wide range of applications and a variety of possibilities for processing and characterizing the materials. In order to make this growing volume of knowledge available, an initiative to publish a series of books in metallurgy and materials science and engineering was taken during the Diamond Jubilee year of the Indian Institute of Metals (IIM) in the year 2006. IIM entered into a partnership with Universities Press, Hyderabad, in 2006, and from 2016 the book series is under MoU with M/s Springer Nature, and as part of the IIM book series, a total of 22 books were published till 2022. The books were authored by eminent professionals in academia, industry, and R&D with outstanding background in their respective domains, thus generating unique resources of validated expertise of interest in metallurgy. The international character of the authors' and editors has enabled the books to command national and global readership. This book series includes different categories of publications: textbooks to satisfy the requirements of undergraduates and beginners in the field, monographs on select topics by experts in the field, and proceedings of select international conferences organized by IIM, after mandatory peer review. An eminent panel of international and national experts constitutes the advisory body in overseeing the selection of topics, important areas to be covered, in the books, and the selection of contributing authors.

About "Handbook of Materials Science: Vol. 1 Optical Materials"

The Book on "Handbook of Materials Science: Vol. 1 Optical Materials" with R. S. Ningthoujam and A. K. Tyagi as editors brings out the developments in optical materials with 23 articles highlighting various aspects of optical materials. The book is part of the Handbook of Materials Science comprising of ten volumes on optical materials, magnetic materials, catalytic materials, photocatalytic materials, nuclear materials, sensor materials, materials for health care, energy materials, materials for environmental applications, and soft materials. Synthesis, characterization, properties of optical materials including nanomaterials, recent advances in optical materials, and their applications in various areas are some of the areas deliberated in this volume.

The editors R. S. Ningthoujam and A. K. Tyagi have made excellent efforts to coordinate with professionals with rich experience in the respective filed from various reputed academic and R&D institutions to provide state-of-the-art articles of high relevance. I have no doubt that this book will be a rich treasure for those who are interested in learning the recent developments with respect to synthesis, characterization, and application of optical materials. The IIM-Springer Series gratefully acknowledges both the editors and all the authors of the articles for their excellent efforts in covering a wide range of information on the optical materials which would be of great interest to the readers.

Dr. U. Kamachi Mudali
Editor-in-Chief
Series in Metallurgy and Materials
Engineering & Vice Chancellor
Homi Bhabha National Institute
DAE, Mumbai

Current Series Information

To increase the readership and to ensure wide dissemination among global readers, this new chapter of the series has been initiated with Springer in the year 2016. The goal is to continue publishing high-value content on metallurgy and materials science and engineering, focusing on current trends and applications. So far, eleven important books on state of the art in metallurgy and materials science and engineering have been published and three books were released during IIM-ATM 2022 at Hyderabad. Readers who are interested in writing books for the Series may contact the Series Editor-in-Chief, Dr. U. Kamachi Mudali, Former President of IIM and Vice-Chancellor of Homi Bhabha National Institute (HBNI), DAE, Mumbai, at ukmudali1@gmail.com, vicechancellor@hbni.ac.in or the Springer Editorial Director, Ms. Swati Meherishi at swati.meherishi@springer.com

Preface

The growth of human civilization has always depended on materials. Thus, evolution of materials has been an integral part of human beings right from the primitive times. The importance of materials can be gauged by the fact that various historic ages have been named based on the materials used by the people belonging to a particular era, such as stone age, bronze age, iron age, concrete age, steel age, silicon age, etc. Some of the earlier processes like forging and smelting have played a key role in the growth of materials and their large-scale deployment for the benefit of humankind. However, during the initial phase, the materials were used without much understanding of the factors responsible for their functioning. For example, Romans who were great innovators of glasses and invented the glass window could not predict how important this material would be in future. It is believed that the first metal tool appeared perhaps only 6000 years ago as a result of knowledge of melting and casting processes. Materials that have immensely changed our lives are wood, clay, bronze, iron, steel, cement, polymers, aluminum, glasses, and semiconductors. It shall not be wrong to claim that the twenty-first century is going to witness a tremendous progress in the field of materials science as the need for advanced materials in various areas like engineering, electronics, communications, health care, societal sectors, natural resources management, generation of wealth from waste is going to exhibit rapid growth. Even the demand for materials for high-tech sectors shall witness exponential growth. Net-zero target will also be intricately linked with the advancement in the field of materials science. The quest for developing better, efficient, and cost-effective materials has been a constant endeavor and shall continue forever. It is generally believed that materials are going to play the most important role to achieve the sustainable development goals, in particular for clean energy, clean water, climate change, health care, industries, innovation, infrastructures, etc. In this context, there is an urgent need to develop efficient materials consisting of earth-abundant elements. The solution for some of the challenges such as production of green hydrogen and green synthesis of ammonia will also come from advanced materials. The rational design of newer and efficient materials by interplay of synthesis methodologies, crystallographic structures, morphologies, and dimensionality is going to be of paramount importance.

In the coming years, a diverse range of materials will have to be developed to meet these goals. In view of this, it was felt appropriate to publish a Handbook of Materials Science comprising of 10 volumes on optical materials, magnetic materials, catalytic materials, photocatalytic materials, nuclear materials, sensor materials, materials for health care, energy materials, materials for environmental applications and soft materials.

Volume I (Optical Materials): In general, the materials that are used to manipulate the flow of light are considered as optical materials. The efficiency of a material for a particular optical application is generally related to materials' characteristics, wavelength of light used, surrounding environment as well as intensity of light excitation. Therefore, a basic understanding of the interaction between light and matter is very important. Synthesis, characterization, properties of optical materials including nanomaterials, recent advances in optical materials, and their applications in various areas are some of the areas deliberated in this volume. Some specific topics covered in this volume are classification of optical materials, photoluminescence, down-conversion, quantum cutting, upconversion, core–shell nanostructures, the fundamentals of plasmon–fluorophore interactions with metal nanoparticles that support localized surface plasmons (LSPs) and thin metal films that support propagating surface plasmon polaritons (SPPs), sensing, imaging, single-molecule detection and display technologies, lanthanide-doped materials for optical applications, defect modulated trap engineering of long persistent and mechano-luminescence, phosphors for advanced applications, optical materials for sensing radioactive elements, the chemistry of triphenylamine-based molecules and materials, their applications towards luminophores, sensing, aggregation-induced emission, framework materials, lanthanide ions-doped luminescent nanomaterials for anti-counterfeiting, thermally activated delayed fluorescence (TADF) as a triplet harvesting mechanism in metal-free OLED emitters, luminescent materials for radiation dosimetry, carbon nanotube (1D), graphene (2D), and diamond (3D) materials as prototype examples in nanoscale optics and plasmonics, random laser emission from dye loaded bamboo leaves, chlorophyll and bio-dye, lanthanide-based inorganic–organic and inorganic-organic-plasmonic hybrid materials for white light emitting diode (WLED), light emitting diode (LED), luminescent solar concentrators (LSCs), luminescent thermometers, ultra-violet (UV) light detection, boradipyrromethene (BODIPY) dyes for laser dyes, chemical/bio-sensors, cellular imaging agents, triplet photosensitizers in photodynamic therapy, fluorescent-positron emission tomography (PET) probes, solar energy conversion agents in organic photovoltaics (OPV) and, in photocatalysis, supramolecular interactions of dyes with various hosts (host–guest interaction) and their applications, calcium aluminate-based phosphors for persistent luminescence and radiation dosimetry, silicon-based nanocomposites as photoluminescent materials, graphene and graphene-based nanocomposites in sensing, photocatalysis, photonics, bio-photonics and solar cells, fabrication of scintillator single crystals, quantum dots and its applications, etc.

We are immensely thankful to all the authors for their rich contributions towards this handbook series. Although due efforts have been taken to make the book as error-free as possible, some may have crept in unnoticed. We shall be thankful to the readers for bringing such unintentional errors to our notice. Finally, we sincerely hope that our efforts will be of use to a wide range of researchers working in the field of materials science.

Mumbai, India

Raghumani S. Ningthoujam
A. K. Tyagi

Contents

About the Editors

Dr. Raghumani S. Ningthoujam obtained M.Sc. in Chemistry from Manipur University in 1994. He was awarded Ph.D. in Chemistry from IIT Kanpur in area of superconducting, electron transport, and magnetic properties of nanostructured transition metal nitrides in 2004. He joined Bhabha Atomic Research Centre (BARC), Mumbai, as Scientific Officer (D) in 2006 after completion of Dr. K. S. Krishnan Fellowship. Presently, he is working in area of Luminescent and Magnetic Nanomaterials and their applications in sensors, imaging, diagnosis and therapy. He did Post-doctoral Fellowship at University of Victoria, Canada, in the area of quantum dots. Presently, he is Scientific Officer (G) at Chemistry Division, BARC, Mumbai. He is recognized for Guideship of Ph.D. at Homi Bhabha National Institute (HBNI), Mumbai, and Mumbai University, Mumbai. Many research scholars and students finished their projects under his guidance. He has published about 166 papers in the refereed journals, 6 review articles, and 13 book chapters and edited 3 books. In recognition of his significant contributions to the chemical science, he has been awarded DAE-Scientific and Technical Excellence Award (2012), Young Achiever Award, SSPS (2010), Chemical Research Society of India's Bronze Medal (2022) and Materials Research Society of India's Medal (2022). He has been elected as Fellow, The National Academy of Sciences, India (FNASc), in 2016 and Fellow, Maharashtra Academy of Sciences (FMASc), in 2013.

Dr. A. K. Tyagi obtained his M.Sc. (Chemistry) degree in 1985 from Meerut University, Meerut and joined BARC Training School, Mumbai in the same year. After completing one year orientation course in Nuclear Science and Technology, he joined Chemistry Division, Bhabha Atomic Research Centre (BARC), Mumbai in 1986. Presently, he is Director, Chemistry Group, BARC, Mumbai, Distinguished Scientist, DAE and a Senior Professor of Chemistry at Homi Bhabha National Institute (HBNI), Mumbai. His research interests are in the field chemistry of materials, which includes functional materials, nanomaterials, nuclear materials, energy materials, metastable materials, hybrid materials and structure-property correlation. He has published more than 650 papers in journals, several books and has supervised 35 PhD students.

He was awarded PhD by Mumbai University, Mumbai in 1992. He did postdoctoral research at Max-Planck Institute for Solid State Research (MPI-FKF), Stuttgart, Germany during 1995-96. Subsequently, he regularly visited MPI-FKF, Stuttgart as a visiting scientist. In addition, he has also visited Institute of Superior Technology, Portugal; Institute for Chemical Process and Environmental Technology, Ottawa, Canada; Dalhousie University, Halifax, Canada; Moscow State University, Moscow, Russia; Institute for Materials, Nantes, France; University of Malay, Malaysia; National Institute of Materials Science, Tsukuba, Japan; National University of Singapore, Singapore; Royal Institute of Technology, Stockholm, Sweden; Rice University, Houston, USA; Shanghai, China; University of Valencia, Valencia, Spain; Weizmann Institute of Science, Israel; University of Queensland, Brisbane, Australia; US-Air Force Research Lab, Dayton, USA, Institute for Studies of Nanostructured Materials, Palermo, Italy and iThemba Labs, Cape Town, South Africa.

In recognition to his significant contributions to the field of chemistry of materials, he has been conferred with many prestigious awards, such as Dr. Lakshmi award by the Indian Association of Solid State Chemists and Allied Scientists (2001); Rheometric Scientific-Indian Thermal Analysis Society Award (2002); Medal of Indian Nuclear Society (2003); Materials Research Society of India's Medal (2005); Chemical Research Society of India's Bronze Medal (2006); DAE-Homi Bhabha Science and Technology Award (2006); IANCAS-Dr. Tarun Datta Memorial Award (2007); Rajib Goyal Prize in Chemical Sciences (2007); RD Desai Memorial Award from Indian Chemical Society (2008) and DAE-SRC Outstanding Research Investigator Award (2010); CRSI-Prof. CNR Rao National Prize in Chemical Sciences (2012); ISCB Award for Excellence in Chemical Sciences (2013); MRSI-ICSC Materials Science Senior Award (2014); Coastal Chemical Research Society's Award (2014); Platinum Jubilee Lecture Award in Materials Science from Indian Science Congress Association (2015); Metallurgist of the Year Award (2017), from Ministry of Steel, Government of India; Chemical Research Society of India's Silver Medal (2018); Materials Research Society of India's Prof. CNR Rao Prize in Advanced Materials (2018); JNCASR's National Prize in Solid State and Materials Chemistry (2018) and Acharya PC Ray Memorial Award from Indian Science Congress Association (2020).

He is an elected Fellow of the Indian National Science Academy (FNA), Indian Academy of Sciences (FASc); National Academy of Sciences, India (FNASc); Indian National Academy of Engineering (FNAE); Maharashtra Academy of Sciences (FMASc); Royal Society of Chemistry, UK (FRSC); Asia Pacific Academy of Materials; World Academy of Ceramics, African Academy of Sciences and UNESCO- The World Academy of Sciences (FTWAS).

Chapter 1
Basic Concept of Optical Materials: Classification, Properties and Applications

Raghumani S. Ningthoujam and A. K. Tyagi

1.1 Introduction

Materials used to manipulate the flow of light are considered as optical materials. Upon interaction of light within matter, various phenomena can occur: absorption, scattering, focusing, reflection, refraction, diffraction or splitting of light, emission or ionization, ejection of electron, dissociation of chemical bond, formation of new bond, formation of free radicals, isomerization, initiation or stopping of chemical reaction, or transmission of light. The efficiency of a material for a particular application is generally related to the wavelength of light, the surrounding environment, as well as the intensity of light. Therefore, a basic understanding of the interaction between light and matter is important.

Light is an electromagnetic radiation, which has wave and particle nature. It comprises of cosmic rays, gamma rays, X-rays, UV, visible and IR rays, etc. Electromagnetic wave shows the electric and magnetic components of wave (perpendicular to each other) and propagation of wave through trough and crest (Fig. 1.1) (Kemp 1975).

Material is defined as a substance, which is identified in X, Y and Z coordinates. It can be solid, liquid, gas or plasma.

Few examples of optical materials that can absorb light are SnO_2, ZnO, Y_2O_3, etc. The absorbance (A) is defined as: $A = \varepsilon cl = \log(1/T)$. Where T = transmittance (I/I_0), and I_0 and I are the light intensity before and after interaction with material. ε, c and l are the absorption coefficient, concentration and path length of materials in which light is passing. Here, it is assumed that no reflection takes place. Usually, path

R. S. Ningthoujam (✉) · A. K. Tyagi
Chemistry Division, Bhabha Atomic Research Centre, Mumbai 400085, India
e-mail: rsn@barc.gov.in

R. S. Ningthoujam · A. K. Tyagi (✉)
Homi Bhabha National Institute, Mumbai 400094, India
e-mail: aktyagi@barc.gov.in

R. S. Ningthoujam and A. K. Tyagi (eds.), *Handbook of Materials Science, Volume 1*, Indian Institute of Metals Series, https://doi.org/10.1007/978-981-99-7145-9_1

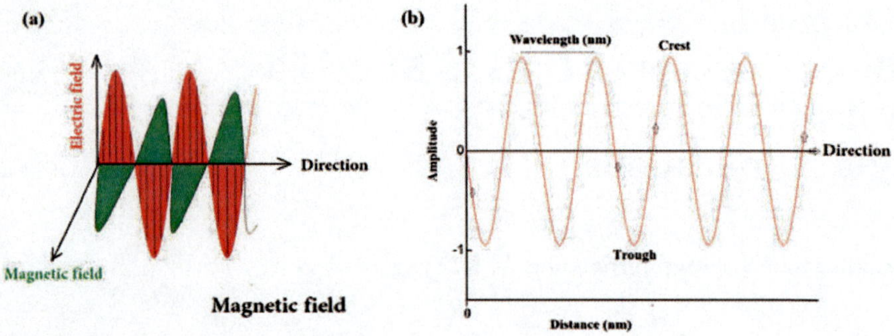

Fig. 1.1 **a** Schematic diagram representing electromagnetic wave, which has the electric and magnetic components of wave (perpendicular to each other) and **b** Propagation of wave showing trough and crest

length is fixed at 1 cm, the absorbance (A) will be proportional to the concentration of material (c = g/ml or g/cm^3 or moles per litre). The optical materials can be glassy, amorphous, crystalline, polymer, liquid or plastics.

1.2 Classifications

On the basis of types of absorption behaviour by materials in UV–visible-NIR-IR regions, optical materials can be classified or grouped as (Curie 1963; Brabec et al. 2003; Blasse and Grabmaier 1994; Simmons et al. 2000):

(i) Metallic materials
(ii) Semiconducting materials
(iii) Insulating materials
(iv) Materials with defects
(v) Materials having metal–ligand or host–guest complexes
(vi) Materials doped with s- and p-block ions
(vii) Materials doped with d-block ions
(viii) Materials doped with f-block ions
(ix) Low dimensional materials
(x) Meta-materials

The element(s) constituting a material can have a characteristic absorption edge in the X-ray regions. From this, it is possible to study oxidation states of ions or atom or chemical bonding or coordination numbers of centre ion. X-ray absorption near edge structure (XANES) and extended X-ray absorption fine structure (EXAFTS) have been used in X-ray absorption spectroscopy. From the absorption of light in UV–visible-NIR-IR regions, band gap of material or charge transfer process or surface plasmon absorption can be predicted.

(i) Metallic materials

Metallic optical materials are noble metals (such as Au, Ag, Cu) and CuS. Metallic nanoparticles can absorb light in UV–visible-NIR region due to surface plasmon resonance (SPR). Au nanoparticles can absorb light at 520 nm, whereas hollow Au nanoparticles absorb light in 600–700 nm (Damani et al. 2022). Ag nanoparticles can absorb light at 390 nm, whereas hollow Ag nanoparticles can absorb light in 500–600 nm (Gautam et al. 2021; Bhavsar et al. 2023). Cu nanoparticles can absorb light at 600–750 nm (Han et al. 2018; Liu et al. 2015). CuS can absorb the light in 800–1000 nm (Chakravarty et al. 2016). The SPR wavelength can also be tuned by changing the shape and size of the nanoparticles. In the form of thin films, these materials can have good reflectivity (Prajwala et al. 2020). Al and Ag films show the high reflectance (93%) in 300–2500 nm solar spectrum. The films can also support propagating surface plasmon.

(ii) Semiconducting materials

Semiconducting optical materials in general are oxides, sulphides, organic molecules and elementals (SnO_2, ZnO, ZnS, PbSe, Si, Ge, organic molecules) having band gap in the range 0.1–4.0 eV. SnO_2, TiO_2, ZnO, Zn_3P_2, Si, Ge and PbSe, can absorb the light at 345 nm (Ningthoujam et al. 2007a, 2008a; Luwang et al. 2010; Ningthoujam and Kulshreshtha 2009), 390 nm (Nair et al. 2022; Rawool et al. 2018), 370 nm (Lanje et al. 2013; Sahu et al. 2012; Ghosh et al. 2011; Singh et al. 2009a), 880 nm (Aleksandar et al. 2019), 1120 eV (Wang et al. 2014), 2133 nm (Mellaerts et al. 2021) and 4770 nm (Ningthoujam et al. 2017), respectively. Organic compounds having conjugated double bonds (anthracene, naphthalene) can absorb the light in 300–600 nm (Kemp 1975; Pan et al. 2017). Fluorescein isothiocyanate (FITC) and 4,6-diamidino-2-phenylindole (DAPI) can absorb the light in visible range (Joshi et al. 2022; Veeranarayanan et al. 2012; Prajapati et al. 2016). These are used as fluorescent labels for bio-imaging. BODIPY (abbreviation for boron-dipyrromethenes, 4,4-difluoro-4-borata-3a-azonia-4a-aza-s-indacene) and its derivatives can absorb the light in the visible-NIR regions depending on the substituted or added molecules/group/part (Bassan et al. 2021). These have been used in many applications such as fluorescent labels for bio-imaging, light-harvesting antennas in photosynthesis, photoluminescent materials, electro-luminescence devices, fluorescent sensors, optoelectronic devices, non-linear optical devices, photodynamic therapy, organic light emitting diodes, etc.

(iii) Insulating materials

Insulating optical materials have wide band gap >5 eV. These can be phosphates, some oxides and fluorides and organic molecules. Few examples are YPO_4 (band gap, E_g~8.6 eV), $NaAF_4$(A = Y, Gd, Lu, E_g~8–10 eV), YF_3(E_g~5.8 eV), which absorb the light in the range of 5–11 eV (Ningthoujam et al. 2015; Huang et al. 2016; Giordano et al. 2021; Vali 2011). The sigma bond or single-double-containing organic compounds (CH_4, $H_2C = CH_2$) can absorb the light in the range of 6–10 eV (Kemp 1975).

(iv) Materials with defects

Materials with defects can absorb light with energy less than the actual band gap of the material. NaCl and KCl with F-centre can absorb light and emit emission (Goldstein 1968; Seitz 1946). ZnO and SnO_2 having interstitial, stoichiometric, surface or lattice defects absorb light and emit light in a broad spectrum (Ningthoujam and Kulshreshtha 2009; Ningthoujam et al. 2008a; Ghosh et al. 2011; Singh et al. 2009a).

(v) Materials having metal–ligand or host–guest complexes

Materials with metal–ligand or host–guest charge transfer (CT) are observed in YVO_4, $CaMoO_4$, UO_2^{2+}, metal-porphyrin, protein-I_2, cellulose-I_2 complex. YVO_4 or $LaVO_4$ or $GdVO_4$(V–O CT) can absorb light in the range of 280–320 nm (Wangkhem et al. 2021; Perala et al. 2021a, 2021b; Okram et al. 2014; Singh et al. 2012; Luwang et al. 2011). $CaMoO_4$ (Mo–O CT) or $SrMoO_4$ or $CaWO_4$ can absorb light in the range of 250–270 nm (Wangkhem et al. 2018; Soni et al. 2018; Singh et al. 2014a, 2014b). UO_2^{2+}(U–O CT) can absorb light at 420 nm (Meinrath et al. 2000; Soni et al. 2021). The metal-porphyrin and haemoglobin can absorb light in the range of 400–600 nm and this absorption peak depends on central metal ion, which can make coordinated bonds with N of porphyrin (Lan et al. 2007; Kumar et al. 2016). The protein-I_2 and cellulose-I_2 have absorption peaks in the range of 300–600 nm (Mizrahi and Domb 2007; Acland 1971; Basu and Nandi 1952). Metal–organic framework systems can also absorb light (Zhou et al. 2012). However, depending on the associated metal ion, absorption band varies. A few examples of host–guest complexes are cyclodextrins (CDs), cucurbit[n]urils (CBn) and crown ethers as host molecules with guest such as dyes/drug molecules (Crini 2014; Schneider et al. 2008). They form supramolecular host-dye complexes, which enhances luminescence quantum yield, solubility, loading of drug, photostability, etc.

(vi) Materials doped with s- and p-block ions

Materials doped with s- and p-block ions include insulating materials doped with s- and p-block ions. Examples are KCl: Tl, CaS: $Bi^{3+}$and host: Bi^{3+} (Curie 1963; Patterson 1960; Krishnamoorthy Sakthivel et al. 2021; Boutinaud 2023; Dongdong et al. 1999). Another example is ortho-ester diazophenylcalix[4]arene, which can show colour changes in solutions of alkali metal ions, alkaline earth metal ions and transition metal ions (Kim et al. 2007).

(vii) Materials doped with d-block ions

Materials doped with d-block ions include semiconducting or insulating materials doped with d-block ions (Kim et al. 2007; Tsujikawa 1971; Batista et al. 2021; Babu et al. 2014). Examples are ZnO: Mn^{2+} or Cr^{3+}. They can absorb light in green to red region. Depending on the crystal field environment, absorption band varies. In general, the d-d transition is broad as well as its position varies with crystal field environment (geometry such as square planar, tetrahedral and octahedral and surrounding ions/atoms such as O/N/S). Organic complexes with Ru(ii), Pt(ii) and

Ir(iii) can exhibit strong absorption in visible region. They have long-lived triplet-excited states. (Zhao et al. 2012).

(viii) Materials doped with f-block ions

Materials doped f-block ions include semiconducting or insulating materials doped with f-block ions. Examples are YPO_4:RE, $ZnGa_2O_4$:RE, SnO_2:RE, TiO_2:RE, ZnO:RE, $ZrO_2:Eu^{3+}$, Y_2O_3:RE, La_2O_3:RE and $NaYF_4$:RE. Here, RE = Ce^{3+}, Eu^{3+}, Gd^{3+}, Pr^{3+}, Sm^{3+}, Ho^{3+}, Dy^{3+}, Tb^{3+}, Er^{3+}, Yb^{3+} (Wei et al. 2014; Ginistrelli et al. 2017; Tian et al. 2012; Ningthoujam et al. 2007b, 2007c, 2008b, 2008c, 2009a, 2009b, 2009c; Gajbhiye et al. 2008; Singh et al. 2008a, 2008b, 2009b, 2009c; Rao et al. 2008; Srinivasu et al. 2009; Yaiphaba et al. 2010; Phaomei et al. 2010; Singh and Ningthoujam 2010; Ningthoujam 2010; Meetei et al. 2012). These can absorb light from UV to visible to NIR regions depending on type of RE. Ce^{3+} can absorb light at 258 and 280 nm due to f-d transition. The f-d transition is an allowed transition, but its peak is broad. Most RE ions show f-f transitions, with weak absorption cross-sections as these are not allowed transitions. The f-f transitions are sharp and the peak position does not change. This is due to screening of 4f electrons by $5s^2$ and $5p^6$ outermost electrons of RE ion. Eu^{3+} ion shows various absorption peaks in UV region (322, 395, 460, 530 nm) and visible region (460, 530 nm). Gd^{3+} ion has absorption peaks at 278 and 310 nm. Other examples are Pr^{3+} (205 nm), Sm^{3+} (363, 377 and 405 nm), Ho^{3+} (369, 450, 1100 nm), Dy^{3+} (350, 370, 390 nm), Tb^{3+} (350, 378 nm), Er^{3+} (370, 519, 542, 651, 800, 980, 1530 nm), Nd^{3+} (808, 885, 1054, 1344, 1858 nm) and Yb^{3+} (980 nm).

(ix) Low-dimensional materials

Low-dimensional materials are the materials, whose dimensions are reduced from 3 to 2D, 1D or 0D. These materials are considered as nanomaterials. In many cases, band gap increases with decrease in dimensions. 0D materials are known as quantum dots (QDs) whose particle size is less than that of exciton Bohr's radius. PbSe bulk (E_g = 0.4 eV) has exciton Bohr's radius of 47 nm (Ningthoujam et al. 2017). PbSe QDs with various size ranges from 3 to 30 nm can show the quantum confinement effects. Band gap of PbSe QDs increases with decrease of particles. Absorption spectrum can show many discrete energy levels as well as the lowest energy level (band gap), whereas its emission shows only one peak, which is slightly longer in wavelength (nm) than that of the lowest energy level. 1D materials are considered as nanorods with length (L) and diameter (D; L>>D). Nanorods of PbSe can show two emission peaks related to the length of nanorods (known as polarizability effect and this occurs at longer wavelength) as well as end of nanorods (related to diameter, i.e., shorter wavelength) (Koh et al. 2010). Different aspect ratios (length to diameter) of rod-shaped materials show interesting absorption and luminescence behaviour (Li et al. 2017). 2D materials are considered as sheet or plane or platelets. PbSe nanosheets with single crystal of 20–60 nm thickness and 0.5–5.0 μm in-plane sizes have been prepared (Wang et al. 2008). PbSe dendrites and cubes grown by a solvothermal method show different photoluminescence behaviours (Kim et al. 2017; Melnychuk and Sionnest 2021). Many semiconductors such as PbS, ZnO, GaN can show low-dimensional properties. Layered (2D) materials or multilayer quantum wells, perovskites, graphene, AM_2 layers (MoS_2, MoW_2), etc. are also examples of low-dimensional materials (Nozik 2001; Iyer et al. 2020; Gutiérrez 2020; Zhang et al. 2016). In general, band gap of semiconductor increases with decrease of size or dimensions (3D to 2D to 1D to 0D). Interestingly, the MoS_2 can be transformed to the 2D materials with various polymorphs (1 T, 2H) in which 1 T is metallic, whereas 2H is semiconducting (Eg = 1.8 eV, direct band gap). The bulk band gap of MoS_2 is 1.20 eV (indirect band gap) (Li and Zhu 2015; Guo et al. 2015).

(x) Meta-materials

Meta-materials are defined as engineered structures having negative refractive index (Gangwar and Paras 2014). There are many arrays of structures smaller than the electromagnetic wavelength of interest. They can manipulate electromagnetic wave in different directions, which is not possible by normal materials. Meta-materials are prepared from nanoparticles, semiconductor, graphene, etc. These have been used in many applications such as sensor detection, smart solar power management, laser guiding, telecommunications, etc.

Classification of optical materials can also be expressed on the basis of types of materials: (i) inorganic glasses, (ii) amorphous materials (inorganic/organic compounds/molecules/polymers), (iii) crystalline materials, (iv) liquid, (v) gases, (vi) dielectric materials, (vii) insulating materials, (viii) semiconducting materials, (ix) anti-reflection coatings, (x) optical fibres, (xi) liquid crystals, etc.

1.3 Properties of Optical Materials

After light absorption by optical materials, excited energy can be converted to heat or light or stored or transferred to another level of energy of surrounding ions or system. In case of heat release, non-radiative process occurs. In case of light release, radiative process occurs. In case of energy storage, it can be stable as long as there is no disturbance. This can be seen in thermoluminescence. In case of energy transfer, amount of transfer depends on the levels of excited energy of surrounding ions of system. This energy transfer can be of radiative or non-radiative or phonon-mediated types.

The optical properties of matter include the followings (Kemp 1975; Curie 1963; Brabec et al. 2003; Blasse and Grabmaier 1994; Simmons et al. 2000; Palik 1997; Donnat and Rauch 1997; Ningthoujam et al. 2012; Kerker 1969; O'Dell 1993; Jain and Singh 2003; Huang et al. 2015; Liang and Wang 2020; Grzybowski and Majewski 2019; Zhao et al. 2022; An et al. 2018; Moon et al. 2019; Rajan 2022; Kubin and Fletcher 1982): (i) Refractive index, (ii) dispersion, (iii) absorption and transmittance, (iv) scattering, (v) turbidity, (vi) reflectance, (vii) albedo, (viii) perceived colour, (ix) fluorescence, (x) photoluminescence, (xi) optical saturation, (xii) dichroism, (xiii) optical bi-stability, (xiv) birefringence, (xv) photo-stability, (xvi) isotropic/

anisotropic, (xvi) non-linear properties. Here, a few properties of optical materials will be discussed briefly.

(i) Refractive index

Refraction is the phenomenon of the light-bending ability of a medium or material when incident light is passing through it. If θ_i and θ_r are angle of incidence and angle of refraction, respectively, it can have a relationship between the refractive index or refraction index or the index of refraction of the two media (n_i and n_r) as:

$$n_r/n_i = \text{Sin}\theta_i/\text{Sin}\theta_r$$

This relationship is known as Snell's law. It is shown in Fig. 1.2. It can also be represented by taking θ_1, n_1 for medium 1 and θ_2, n_2 for medium 2. The refractive index is a unit less quantity. Also, the refractive index is expressed in terms of speed of light in a medium (v) with respect to that in vacuum (c), i.e., $n = v/c$. The refractive indices for materials are generally expressed as a single value when the wavelength of incident rays is 633 nm (Red light). The refractive indices of water and glass are 1.33 and 1.5, respectively.

In Fig. 1.2, the angle of incidence (θ_i) is more than that of refraction (θ_r) when light is passing from medium 1 to medium 2. It means that the refractive index of a

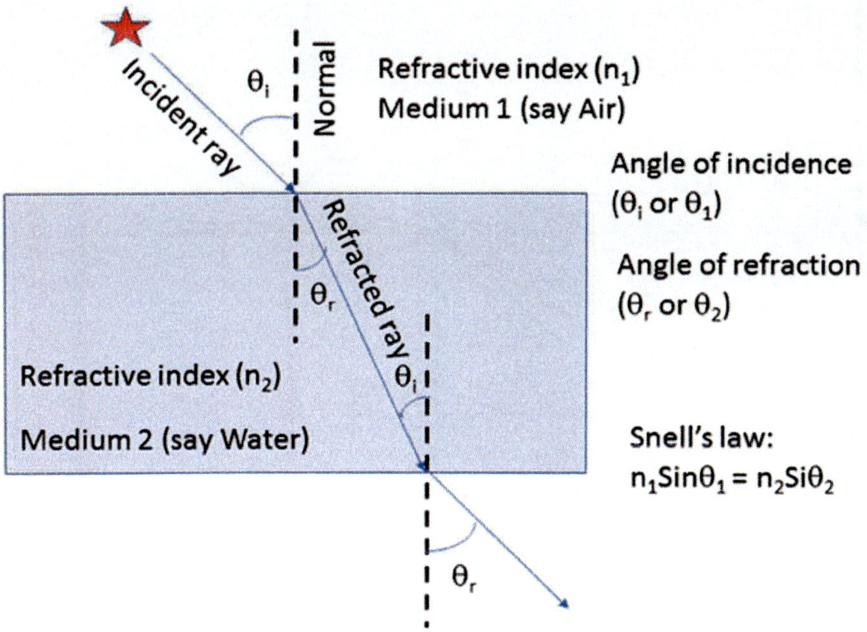

Fig. 1.2 Schematic diagram showing incident light from one medium (say air) to another medium (water)

medium 1 is lower than that of medium 2. On the other hand, when light is passing from medium 2 to medium 1, angle of incident (θ_i) is lower than that of refraction (θ_r). If angle of incidence increases, angle of refraction can become 90°. In this situation, refracted rays will pass through interface or layer of both media. Angle of incidence is known as the critical angle (θ_c). The different materials have their respective θ_c. In general, medium with lower refractive index has the larger critical angle compared to other with higher refractive index. However, the critical angle depends on types of two media (glass to water, glass to air, diamond to air, diamond to glass, diamond to water, etc.). In general, vacuum and air have refractive index of 1. When angle of incident is more that the critical angle for refraction, the light is reflected to same medium. This is known as total internal reflection (Fig. 1.3).

(ii) Dispersion

When white light passes through a triangular glass prism, it splits into seven colours (VIBGYOR). This process is known as dispersion. It is indicated that refractive index is dependent on wavelength of incident ray. Thus, the refractive indices for materials are generally expressed as a single value when the wavelength of incident rays is

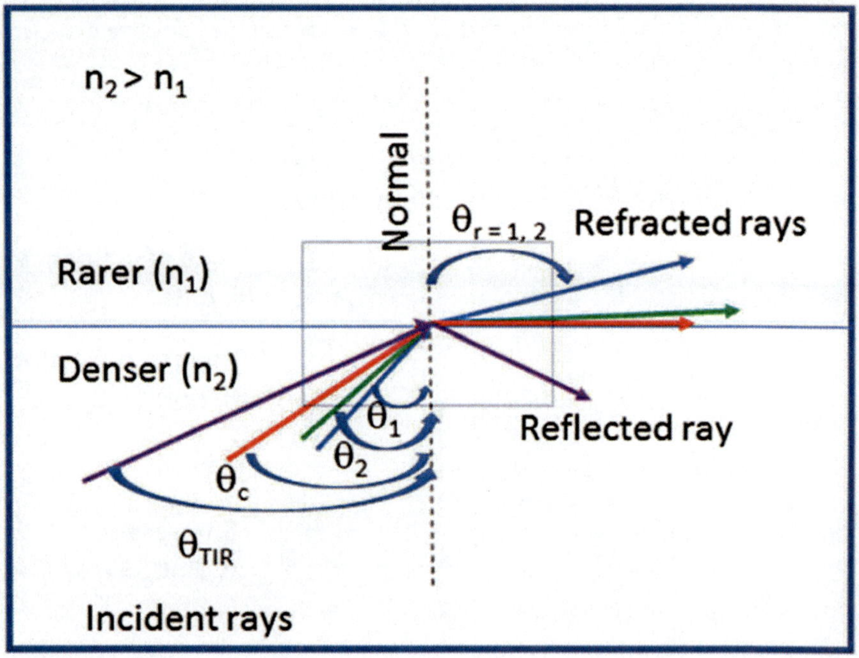

Fig. 1.3 Schematic diagram showing incident light from one denser medium (say water) to another rarer medium (air). With increase of incident angle, angle of refraction increases. At a particular angle of incident rays, the refracted rays go to the interface of two mediums. This angle of incident is called the critical angle. When angle of incident rays is more than the critical angle, the rays come to the denser medium. This is known as total internal reflection

633 nm (red light). The light propagation in absorbing materials can be expressed in terms of a complex-valued refractive index. The real part is accounted for refraction, whereas the imaginary part is accounted for attenuation. Here, attenuation of light means the decrease in its intensity as travels through a medium due to scattering or absorption of light.

(iii) Absorption and Transmission

When light passes through a medium, some part of light intensity is absorbed, some part may be reflected/scattered and some part will be transmitted. The absorbed light is expressed in absorbance (A), which is expressed according to Beer-Lambert' Law:

$$A = \log(I_0/I) = \epsilon cl \tag{1.1}$$

where I_0 and I are the intensities of the light before and after interaction with the sample, respectively (Fig. 1.4). The "c" is the concentration of solute in solvent (mol dm^{-3}). The "l" is the path length of the sample (in cm). The "ε" is the molar absorptivity (molecular extinction coefficient) and its unit is m^2 mol^{-1}.

The transmittance (T) of a sample is expressed as

$$T = I/I_0 \tag{1.2}$$

The relationship between A and T can be

$$A = \log(1/T) \tag{1.3}$$

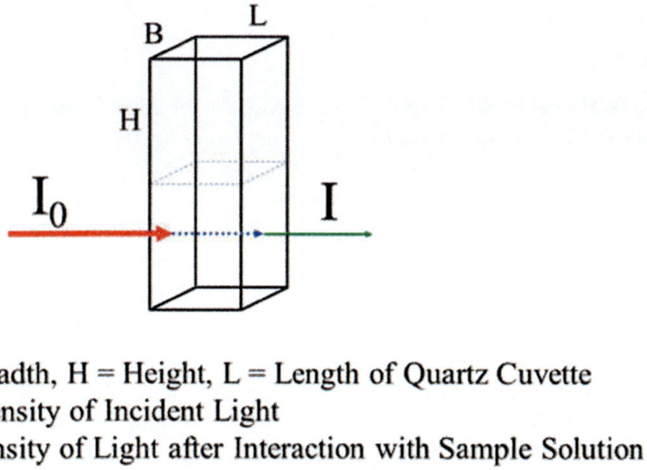

Fig. 1.4 Schematic diagram illustrating passage of light through a liquid/solution in a cuvette

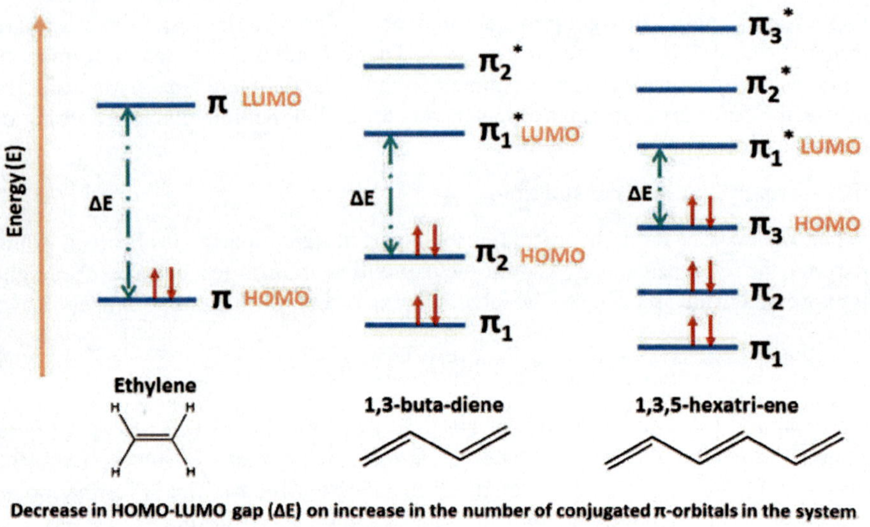

Fig. 1.5 Schematic diagram demonstrating the electronic transitions in $H_2C = CH_2$, $H_2C = CH–CH = CH_2$, $H_2C = CH–CH = CH–CH = CH_2$. Highest occupied molecular orbital = HOMO. Lowest unoccupied molecular orbital = LUMO

The electronic transitions in double bond (sigma, σ and pi, π bonds) containing molecules ($H_2C = CH_2$, $H_2C = CH–CH = CH_2$, $H_2C = CH–CH = CH–CH = CH_2$) take place between the highest occupied molecular orbital (HOMO) and the lowest unoccupied molecular orbital (LUMO). With increase of conjugated π-bonds, gap between LUMO–HOMO decreases. This is schematically shown in Fig. 1.5.

(iv) Scattering

When visible light falls on molecules or particles or atom or ions (charge particles), instead of reflection in a particular direction, the deflections of light in all directions can occur (besides its incident direction). This phenomenon is known as scattering of light. Examples are scattering of sunlight by dust, water molecules, water vapour, etc. The scattering can be divided into two: (1) Elastic and (2) In-elastic scattering. In elastic scattering, there is no loss of energy between incoming (before) and outgoing (after interaction) light. Examples are Rayleigh scattering of light, Mie scattering of light and Thomson scattering. In case of Rayleigh scattering, wavelength of light in visible to NIR is used and its wave interacts with small molecules such as water or organic compounds. The wavelength ($\lambda = 300$–800 nm) of light is much longer than the size of molecules (0.1–2 nm). The probability (P) of scattering is inversely proportional to the fourth power of wavelength (λ).

$$P \propto 1/\lambda^4 \tag{1.4}$$

In case of Mie scattering, size of particles must be comparable with wavelength of light. Examples are the scattering of ultraviolet to visible light ($\lambda = 200$–700 nm) by nanoparticles (5–100 nm). The probability (P) of scattering is inversely proportional to square root of wavelength (λ).

$$P \infty 1/\lambda^2 \tag{1.5}$$

When particle size is much bigger than that of wavelength of excitation, the probability (P) of scattering is independent of wavelength (λ) of excitation. Here, scattering is very high. There, the intensity (I) of scattering by molecules or ions or particles can be related by the following:

$$I = A/\lambda^4 + B/\lambda^2 + C \tag{1.6}$$

In in-elastic scattering, there is energy transfer from incoming light to system/particles/charged ions or vice versa. Thus, a loss or gain of energy takes place in outgoing light. A typical example is Raman scattering. In case of electromagnetic wave in the high energy region say X-rays or γ-rays, after interaction with a charged particle, in-elastic scattering takes either by loss of gain of energy of a charged particle. This is known as Compton effect.

(v) Turbidity

Turbidity is the haziness or cloudiness of a fluid (liquid or air) caused by the scattering of light due to the presence of a large number of particles. In general, such particles are foreign materials, which are invisible to the naked eye. When light passes through a liquid (say water), intensity of scattered light increases with increase in the concentration of suspended particles. Such suspended particles (known as colloidal) get settled down with time. In air, the presence of smoke causes the reduction in air quality. Increase in turbidity in sea causes the decrease of the amount of light reaching to lower depths. Consequently, it inhibits growth of submerged aquatic plants as well as reduces oxygen content in water. In glass or plastic, turbidity decreases the transparency. The turbidity is measured as Nephelometric turbidity unit (NTU).

(vi) Reflectance

When light falls on a surface, some part of incident light will be absorbed, some part will be transmitted and remaining will be reflected or scattered. The total radiant of absorbed (A), transmitted (T) and reflected (R) will be equal to spectral radiant flux (φ) that is falling on the surface.

$$\Phi = \Phi(A) + \Phi(T) + \Phi(R) \tag{1.7}$$

$$1 = A + T + R \tag{1.8}$$

Thus, the reflectance is the ratio of energy/light reflected to the energy/light falling on a substance and this is expressed in terms of percentage. When sunlight falls on a substance having red colour, it reflects only red colour wavelengths, which is visible to our eye. Remaining wavelengths in solar spectrum will be used as absorption as well as transmission. The reflectance characteristics for a material such as leaf, soil, plant, human body in solar spectrum are different in UV–Visible-NIR-IR-Mid-IR range.

(vi) Albedo

The fraction of sunlight that is reflected from the surface of a body is expressed in units known as Albedo. In terms of fraction, Albedo can be from 0 to 1. A black body exhibits Albeto of zero. It is an important idea in astronomy, atmospheric, climatology and environmental management. In terms of surface Albedo, ratio of *radiosity* J_e to the *irradiance* E_e (flux per unit area) received by a surface is an important parameter. Radiosity is the radiant flux leaving a surface per unit area (Watt per square metre) and this can be emitted, reflected and transmitted.

(vii) Perceived colour

The perception of colour is the visual sensation of a part of the field of view on the eye, through which this can be distinguished from another part when observed with a single, unmoving eye. The intensity of light falling on the eye is the deciding factor for the brightness of object. The reflection of a particular wavelength of light, its intensity, radiation of self-emitting light source, night vision and day vision are important parameters in the perception of colour by our eye.

(viii) Photo-stability

Some compounds such as organic dyes or chromophores are unstable in the presence of light and after longer exposure, compounds get decomposed. Similarly, most quantum dots (PbSe, PbS, PbTe, $CsPbI_3$) are unstable to long exposure of light (UV light). Whereas, inorganic compounds such as $NaYF_4$: RE, YF_3: RE and CaF_2: RE are highly stable upon exposure to light.

(ix) Isotropic/Anisotropic/Birefringence

In optically isotropic materials, the optical properties do not depend on the direction of light propagation and polarization. Most samples have isotropic property. In single crystals, the refractive index changes with the propagation direction at the polarization direction (known as birefringent). These materials are considered as optically anisotropic. Anisotropy studies can be done with the help of polarizers.

(x). Photoluminescence

The emission of light from a substance upon absorption with a particular wavelength or energy of photon is known as photoluminescence. Excitation energy is equal to or slightly more than that of energy gap of the excited state and ground state or of two states/levels. Wavelength of the emitted light is either more or less that of the excitation photon. If wavelength of the emitted light is longer than that of the

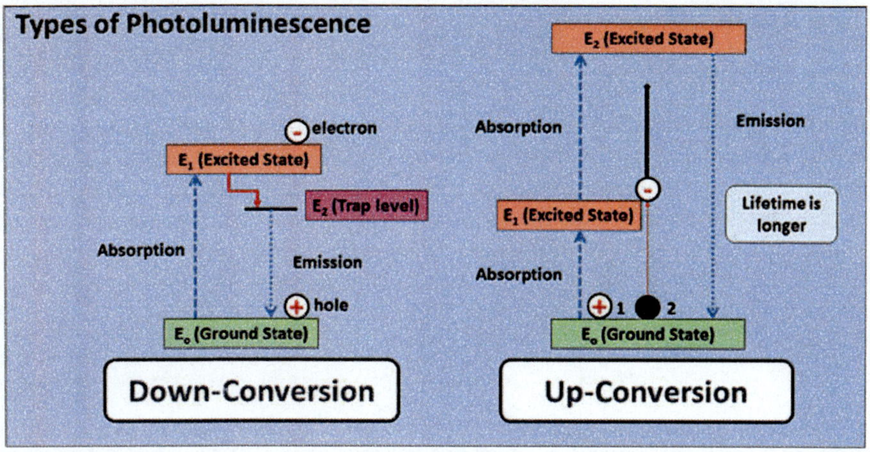

Fig. 1.6 Types of photoluminescence: down-conversion and up-conversion

excitation photon, the process of photoluminescence is known as down-conversion (Fig. 1.6). In opposite case, this is considered as up-conversion. In down-conversion process, an electron leaves ground state (+hole, E_o) upon absorption of resonance energy of light and reaches the excited state (−ve, E_1). The excited electron comes down to the trap level by loss of energy from which e–h pair formation takes place leading to emission of light. In this case, energy of emitted light is less than that of absorbed light. In up-conversion process, electron leaves ground state (+hole, E_o) upon absorption of resonance energy of light and reaches the excited state (−ve, E_1). Absorption of electron takes place within 10^{-15} s. If lifetime of excited electron at a particular level (E_1) is longer (micro-seconds), another electron can hit excited electron upon continuous exposure of light. In this, an excited electron can go to another higher excited level (E_2), from which emission can take place. In turn, energy of an emitted light is more than that of an absorbed light.

During the process of down-conversion or up-conversion, energy transfer process can take place. Quantum efficiency (φ) is the ratio of the emitted light to the absorbed light. The quantum efficiency (φ) values of normal dyes such as rhodamine B can reach 65% in ethanol medium and 31% in water medium. Water acts as quencher. This value is less in case of rare-earth doped particles (Y_2O_3: Eu^{3+}) due to high energy losses. In case of up-conversion process, this value is very low (<1.0%) because of huge loss of excitation energy.

1.4 Applications

Various optical materials (organic molecules, inorganic molecules, polymers, glass, metallic, semiconductors, insulators) can be used in many applications such as photoluminescence-based sensors, scattering-based instrumental response functions (IRF) for lifetime measurement of phosphor samples, anti-reflection coating, wave-guiding, particle size-dependent studies, surface roughness determination, dispersion of light, getting of monochromatic light through diffraction process, quantum-based sensors, fabrication of lasers, random laser, solar cells, imaging, bio-imaging, light emitting diodes, anisotropic study, etc. These aspects will be discussed in detail in this volume.

1.5 Conclusions and an Outlook to the Overall Content of This Volume

In this chapter, the researchers or even beginners can learn definition of optical materials, their classification on the basis of types of absorption or types of materials, their properties such as refractive index, reflection, absorption, scattering, perception of colour, turbidity, photoluminescence, non-linear properties, etc. Finally, their applications in various areas such as anti-reflection coating, wave-guiding, low-dimensional areas, surface roughness determination, sensors and fabrication of lasers, etc. are mentioned.

This volume I: optical materials include the classification of optical materials (bulk to nanomaterials), properties and their applications.

In chap. 2, definition of photoluminescence, its types such as down conversion (down shifting and quantum cutting) and up-conversion (types of up-conversion and photo avalanche). Many examples of phosphors (organic, inorganic, polymers, d/f ions containing compounds) are provided. Variation in luminescence properties in core@shell and core@shell1@shell2, etc. as compared to core counterpart and their applications in light-emitting materials, bio-imaging, drug-loading, light harvesting, solar cells, etc. are mentioned.

In chap. 3, definition of plasmonics and plasmonic materials beyond Au, Ag, metal nanoparticles (localized surface plasmons, LSPs) and thin film of metals (surface plasmon polaritons, SPPs) are provided. Various applications of plasmon-coupled fluorescence in imaging, sensing, single molecule detection and display technologies are mentioned.

In chap. 4, the basics of lanthanides, their spectroscopic properties, different host materials for lanthanides to get luminescence (inorganic phosphors, organic complexes, hybrid materials, metal–organic framework and halide perovskite quantum dots) have been provided. The role of sensitizers and effects of core–shell structure formation, antenna effect, surface plasmon resonance, quantum confinement effect in quantum dots are mentioned. Also, applications of lanthanide-doped materials in the area of light-emitting diodes, optical temperature sensing, optical encryption, laser emission, and optical imaging have been elaborated.

In chap. 5, the role of defect-related trap centres in optical materials is mentioned. It is important in designing long persistent luminescence (LPL), photo-stimulated luminescence (PSL), mechanoluminescence (ML) and thermally stimulated luminescence (TSL). Types of the defects in crystal lattice such as point defects, line defects, planar defects, bulk defects and their characterization techniques such as HRTEM/STEM, XPS, Raman spectroscopy, XAS, EPR spectroscopy, positron annihilation, EELS, thermoluminescence, etc. are mentioned.

In chap. 6, various optical materials for radioactive ions sensing have been provided. Particularly, uranium ion detection is focussed here. The effects of these radioactive ions above toxicity levels on environment have been provided. To develop potential optical materials for sensing, many factors such as cost, use of environmental friendly materials, stability of the probe, selectivity, sensitivity and response time need to be considered.

In chap. 7, triarylamines and its derivatives have been demonstrated as a versatile molecular building block for optoelectronic materials. For example, triphenylamine (TPA), a propeller-shaped, electron-rich molecular motif and its derivatives can be generated by addition of other functional groups on the phenyl group. N atom can be positive and phenyl group can be negative. Depending on additional functional group, their reactivity to other atom/group of atoms will be different. In general, TPA is considered as a donor (D), which can form complex with other acceptors (A) in the forms of DA, DA2, DA3, etc. Applications of triphenylamine-based molecules and materials towards luminophores, sensing, aggregation-induced emission, framework materials are provided.

In chap. 8, many synthesis methods such as co-precipitation, hydrothermal, solvothermal, thermal decomposition, thermolysis, sol–gel, ionic liquid-assisted synthesis, microwave-assisted synthesis and emulsion-based techniques for lanthanide ion-doped optical materials have been provided. These have been fabricated for security ink applications under UV/NIR excitation. Various fabrication printing methods such as screen printing, aerosol jet printing, inkjet printing and photolithography are utilized to make codes and graphics for security purposes.

In chap. 9, metal-free organic light-emitting diodes (OLED) emitters have shown interesting optical properties such as thermally activated delayed fluorescence (TADF) through a triplet harvesting process. There is a thermal equilibrium of the lowest singlet (S_1) and triplet (T_1) states, where the rate constant for reverse intersystem crossing (k_{RISC}) was dependent on the temperature. Their property can be changed by steric hindrance between donor (D) and acceptor (A) groups in D-A-D molecules. The intersystem crossing (ISC) between triplet charge transfer (^3CT) and triplet local exciton (^3LE) states is governed by the extent of vibronic coupling and also by the adiabatic energy difference between them.

In chap. 10, solid-state luminescence dosimetry techniques such as thermoluminescence (TL) and optically stimulated luminescence (OSL) and possible mechanisms for these processes have been provided for the assessment of radiation doses received by the occupational workers. These techniques provide inherent operational simplicity as well as cost-effectiveness. Limitations of TL, examples of phosphors, chronological development of some important TL and OSL phosphors with potential application in dosimetry are mentioned.

In chap. 11, carbon material-based nanoscale optics and plasmonics have been provided. Classification of carbon-based materials has been done on the basis of carbon nanotube (1D), graphene (2D) and single-crystal diamond (3D) materials, which have shown their potential for next-generation optoelectronics and photonic devices. Synthesis, characterization and properties of carbon-based materials are mentioned.

In chap. 12, theoretical and experimental aspects of random laser (RLs) are provided. It provides a lasing device on the basis of random geometrical framework. This includes the followings: Some emerging areas of random lasers, some nature inspired scattering structures for random laser generation and their applications in its speckle free imaging due to its low spatial coherence properties, its footprint in medical sector such as diagnosis methods and blood monitoring, anti-counterfeiting, refractive index sensor, flexible laser display and bit generation, etc.

In chap. 13, one of the boron derivative dyes, boradipyrromethene (BODIPY) is provided for its synthesis procedure, preparation of its derivatives, properties and applications. Various synthetic approaches are provided to tune their optical properties so that these can be appropriate for various applications. Absorption and photoluminescence bands lie in visible range; and its quantum yield varies from one to another derivative. Uses of BODIPY in various areas such as laser dyes, chemical/bio-sensors, cellular imaging agents, triplet photosensitizers in photodynamic therapy, fluorescent-positron emission tomography (PET) probes, solar energy conversion agents in organic photovoltaics (OPV) and, in photo-catalysis and developing self-assembled architectures are mentioned.

In chap. 14, the optical properties of fluorescence dyes decorated functional materials via supramolecular interactions of dyes with various hosts as host–guest chemistry are provided. Various dyes such as BODIPY, hemicyanine dyes, LDS-698 dye and rhodamine B and various hosts such as cyclodextrins (CDs), calix[n]arenes (CXn), cucurbit[n]urils (CBn), crown ethers, pill[n]arenes, cylophanes are mentioned. Also, host-assisted nano-aggregate formation for chromophoric dyes is one of the important topics in contemporary supramolecular research fields, which is provided. Formation of supramolecular polymers and related other supramolecular nanostructures are important. For example, a polymer system comprising of 1-methylpyridium substituted cyanostilbene (Py^+–CNSB–Py^+) derivative as the chromophoric guest is non-fluorescence/weak luminescence, but when it interacts with CB8 as the macrocyclic host, there is a substantial escalation in fluorescence. Diverse applications of dyes decorated functional materials in physics, chemistry, material sciences, biology and environments are provided.

In chap. 15, dimensional engineering of perovskite halides from 3 to 2D or mixture 3D/2D for an efficient and stable solar cells is provided. The perovskite halide materials have the general formula of ABX_3 (where A = $CH_3NH_3^+$ (methyl ammonium ion, MA), FA^+ (Formamidinium ($HC(NH_2)_2^+$) ion), and Cs; B = Pb, Sn, and X = I, Br or Cl). A few examples are $MAPbI_3$, $CsPbI_3$, $FAPbI_3$, which belong to the 3D type of perovskite halide materials, but they are unstable in moisture, heat, oxygen since chemical bonding in compounds ionic. In order to make stable at ambient environment, it is required to coat the surface by organic surfactants/molecules. Better chemical stability can be brought by dimensional engineering from 3 to 2D by substituting A site by long chain organic cation. 2D Ruddlesden-Popper (RP) has the following formula: $L_2A_{n-1}B_nX_{3n+1}$, where L is a long-chain organic cation (spacer), examples include phenyl–ethyl-ammonium (PEA^+) and butylammonium (BA^+). When n = 1, 2D materials are formed, whereas when n = ∝ , 3D materials are formed. Quasi-2D materials are formed for n = 2–5. The synthesis and characterization of 3D or 2D and their photoconversion efficiency (PCE) properties are provided.

In chap. 16, types of luminescence, persistent luminescence, dosimetry and thermoluminescence are mentioned. The basic required properties of thermoluminescence materials used in radiation dosimetry are provided. It is more focused on calcium aluminate-based phosphors. Many examples of TLD phosphors and their dosimetric characteristics are provided. The dopants (s, p, d and f ions) change their properties. The literature on thermoluminescence of calcium aluminate-based phosphors and various preparation methods of calcium aluminate are mentioned.

In chap. 17, tuning of band gap of silicon (Si) from 1.1 to 3.5 eV by reducing particle size to nanometres is provided and their luminescence intensity can be improved. The quantum confinement effects are observed in Si nanoparticles. Various synthesis methods including ionizing radiation-assisted approach for Si nanoparticles are provided. Applications of Si-based nanoparticles in light-emitting diodes (LEDs), lasers, sensors, photovoltaics, bio-imaging, etc. are provided.

In chap. 18, various synthesis methods for graphene and graphene-based nanocomposites are provided. Various synthesis methods such as chemical vapour deposition (CVD), mechanical exfoliation, chemical oxidation and reduction methods, epitaxial growth using metals etc. have been mentioned. Nanocomposites of graphene with metallic nanoparticles, oxides/hydroxides of transition metals and sulphides/nitrides of transition metals are provided. Their applications in area of sensing, photocatalysis, photonics, biophotonics and solar cells are presented.

In chap. 19, Quantum Dots (QDs) as optical materials and the birth of nanotechnology are associated with two events: Nobel Laurate Richard Feynman' famous lecture entitled "There's plenty of room at the bottom" in American Physical Society meeting in 1959 and Norio Taniguchi' talk "On the basic concept of Nanotechnology" at the International Conference on Production Engineering in Tokyo in 1974 are provided. The role of core–shell nanostructures in low-dimensional materials and tuning of absorption and emission wavelengths are mentioned. Various synthesis methods for QDs, characterization and optical properties are provided and their applications in areas of solar cells, light-emitting diodes (LEDs), optical switching and optical limiting, optical gain and lasing, biosensing, gene and drug delivery and photodynamic therapy are mentioned.

In chap. 20, X-ray absorption fine structure (XAFS) for local structure and valence state information of materials is provided. The energy of incident X-ray energy is tuned around the binding energy (E_0) of a component element of sample. The theory, experimental scheme and analytic strategies of XAFS are mentioned.

The uniqueness, versatility and vast domain of XAFS and fitting parameters are provided. Many examples of samples such as alloys, nanocomposites, metallic glass and magneto-electronic systems in which XAFS analysis is provided.

In chap. 21, foundation, perception and challenges for high-resolution molecular secondary ion mass spectrometry (HRM-SIMS) for absolute quantification of materials in low-dimensional structures are provided. SIMS can provide the surface composition and structure. Under ion bombardment on surface of material, secondary ion emission is secondary ion emission is produced from sputtering of materials and is a complex inelastic process of ion-surface interactions. Emission intensity of secondary ions is related to "ionization efficiency", which strongly depends on instantaneous local surface chemistry, known as "Matrix Effect". Many examples are provided with quantitative analysis.

In chap. 22, definition of the scintillators, its classification (inorganic and organic materials based on composition), synthesis procedures for single crystal-based scintillators, mechanism and its applications based on radiation detection and counting (nuclear medicine, home-land security, high energy nuclear physics, nuclear non-proliferation, etc.) are provided. Further, scintillators are grouped into solid, liquid and gas. Types of radiation (charged and neutral) and their interaction process with materials are mentioned. Characteristics of scintillators such as scintillation yield, non-proportionality, energy resolution, decay time and pulse shape discrimination and important scintillator parameters are provided.

In chap. 23, the total reflection X-ray Fluorescence (T-XRF) based on X-ray fluorescence upon the glancing angle of the incident X-rays less than 0.1° is mentioned, which is different from normal energy-dispersive X-ray Fluorescence (ED-XRF), where incident angle of X-ray takes place at 45°. The reduction of matrix effects and compositions of elementals in a sample up to nano-gram or ppb level detection is provided. Principle of external total reflection of X-rays, calibration, quantification, sensitivity, limitations in various atmospheres, and its applications in areas of environmental and geological studies of soil, water samples, and airborne particulates are provided.

References

Acland, J.D.: The interpretation of the serum protein-bound iodine: a review. J. Clin. Pathol. **24**, 187–218 (1971)

Aleksandar, Ž.A., Farkaš, B., Uahengo, V., Leeuw, N.H.D., Dzade, N.Y.: First-principles DFT insights into the structural, elastic, and optoelectronic properties of α- and β-ZnP_2: implications for photovoltaic applications. J. Phys. Condens. Matter **31**, 265501 (2019)

An, R., Zhang, F., Zou, X., Tang, Y., Liang, M., Oshchapovskyy, I., Liu, Y., Honarfar, A., Zhong, Y., Li, C., Geng, H., Chen, J., Canton, S.E., Pullerits, T., Zheng, K.: Photostability and photodegradation processes in colloidal $CsPbI_3$ perovskite quantum dots. ACS Appl. Mater. Interfaces **10**, 39222–39227 (2018)

Babu, B., Manjari, V.P., Aswani, T., Rao, T.G., Stella, R.J., Ravikuma, R.V.S.S.N.: Structural, optical and magnetic properties of Cr^{3+} doped ZnO nanopowder. Indian J. Phys. **88**, 683–690 (2014)

Bassan, E., Gualandi, A., Cozzi, P.G., Ceroni, P.: Design of BODIPY dyes as triplet photosensitizers: electronic properties tailored for solar energy conversion, photoredox catalysis and photodynamic therapy. Chem. Sci. **12**, 6607–6628 (2021)

Basu, S., Nandi, S.P.: Optimum pH and iodine absorption of an enzyme molecule. Nature **169**, 799 (1952)

Batista, E.A., Silva, A.C.A., Lima, T.K.D., Guimaraes, E.V., Silva, R.S.D., Dantas, N.O.: Effect of the location of Mn^{2+} ions in the optical and magnetic properties of ZnO nanocrystals. J. Alloys Comp. **850**, 156611 (2021)

Bhavsar, C., Joshi, R., Fernandes, T., Chandramouleeswaran, S., Khan, T., Hassan, P.A., Momin, M., Ningthoujam, R.S.: Glutathione-capped hollow silver nanoparticles: optimization of surface plasmon resonance, photothermal and in vitro and in vivo biocompatibility studies. ACS Appl. Nano Mater. **6**, 9276 (2023)

Blasse, G., Grabmaier, B.C.: Luminescent Materials. Springer, Berlin (1994)

Boutinaud, P.: Luminescence–structure relationships in solids doped with Bi3+. Phys. Chem. Chem. Phys. **25**, 11027 (2023)

Brabec, C.J., Dyakonov, V., Parisi, J., Sariciftci, N.S.: Organic Photovoltaics: Concepts and Realization. Springer-Verlag Berlin Heidelberg (2003)

Chakravarty, R., Chakraborty, S., Ningthoujam, R.S., Nair, K.V.V., Sharma, K.S., Ballal, A., Guleria, A., Kunwar, A., Sarma, H.D., Vatsa, R.K., Dash, A.: Industrial-scale synthesis of intrinsically radiolabeled ^{64}CuS nanoparticles for use in positron emission tomography (PET) imaging of cancer. Ind. Eng. Chem. Res. **55**, 12407–12419 (2016)

Crini, G.: Review: a history of cyclodextrins. Chem. Rev. **114**, 10940–10975 (2014)

Curie, D.: Luminescence in Crystals, 1st edn. John Wiley & Sons Inc., London (1963)

Damani, M., Desai, N., Singh, B.P., Ningthoujam, R.S., Momin, M., Khan, T.: Synthesis of hollow gold nanoparticles–impact of variables on process optimization. J. Pharm. Sci. **111**, 2907–2916 (2022)

Dongdong, J., Bo-qun, W., Jing, Z., Lu, L.: Luminescence of Bi^{3+} and Eu^{2+} double centers doped in CaS host. Acta Phys. Sin. (overseas Edn) **8**, 813 (1999)

Donnat, P., Rauch, J.: Modelling the dispersion of light. In: Rauch, J., Taylor, M. (eds.) Singularities and Oscillations. The IMA Volumes in Mathematics and its Applications, vol. 91. Springer, New York, NY (1997). https://doi.org/10.1007/978-1-4612-1972-9_2

Gajbhiye, N.S., Ningthoujam, R.S., Ahmed, A., Panda, D.K., Umre, S.S., Sharma, S.J.: Redispersible Li^{+} and Eu^{3+} co-doped CdS nanoparticles: luminescence studies. Pramana J. Phys. **70**, 313 (2008)

Gangwar, K., Paras, G.R.P.S.: Metamaterials: characteristics, process and applications. Adv. Electron. Electr. Eng. **4**, 97–106 (2014)

Gautam, A., Komal, P., Singh, R.S., Gautam, P., Manjari, S.K.V., Ningthoujam, R.S.: Hard core proof of the polyvinyl alcohol as a reducer for the formation of gold nanoparticles. J. Mol. Liquids **334**, 116112 (2021)

Ghosh, M., Ningthoujam, R.S., Vatsa, R.K., Das, D., Nataraju, V., Gadkari, S.C., Gupta, S.K., Bahadur, D.: Role of ambient air on photoluminescence and electrical conductivity of assembly of ZnO nanoparticles. J. Appl. Phys. **110**, 054309 (2011)

Ginistrelli, E.C., Smith, C., Pugliese, D., Lousteau, J., Boetti, N.G., Clarkson, W.A., Poletti, F., Milanese, D.: Nd-doped phosphate glass waveguides: from materials fabrication to power scaling tests. J. Alloy. Compd. **722**, 599–605 (2017)

Giordano, L., Nunes, M.F., Teixeira, V.C., Rodrigues, L.C.V.: Green synthesis of upconverting $NaYF_4$ and $NaGdF_4$ materials and energy levels determination. Braz. Chem. Soc. **32**, 1552–1558 (2021)

Goldstein, F.T.: F centre production by ultraviolet irradiation of KCl and NaCl. Solid State Commun. **6**, 565–567 (1968)

Grzybowski, A., Majewski, K.K.: What is color and how it is perceived? Clin. Dermatol. **37**, 392–401 (2019)

Guo, X., Yang, G., Zhang, J., Xu, X.: Structural, mechanical and electronic properties of in-plane 1T/2H phase interface of MoS_2 heterostructures. AIP Adv. **5**, 097174 (2015)

Gutiérrez, H.R.: Two-dimensional layered materials offering expanded applications in flatland. ACS Appl. Nano Mater. **3**, 6134–6139 (2020)

Han, P., Martens, W., Waclawik, E.R., Sarina, S., Zhu, H.: Metal nanoparticle photocatalysts: synthesis, characterization, and application. Part. Part. Syst. Charact. **35**, 1700489 (2018)

Huang, K., Pan, W., Zhu, J.F., Li, J.C., Gao, N., Liu, C., Ji, J., Yu, E.T., Kang, J.Y.: Asymmetric light reflectance from metal nanoparticle arrays on dielectric surfaces. Sci. Rep. **5**, 18331 (2015)

Huang, B., Dong, H., Wong, K.L., Sun, L.D., Yan, C.H.: Fundamental view of electronic structures of β-$NaYF_4$, β-$NaGdF_4$, and β-$NaLuF_4$. J. Phys. Chem. C **120**, 18858–18870 (2016)

Iyer, P.P., DeCrescent, R.A., Mohtashami, Y., Lheureux, G., Butakov, N.A., Alhassan, A., Weisbuch, C., Nakamura, S., DenBaars, S.P., Schuller, J.A.: Unidirectional luminescence from InGaN/GaN quantum-well metasurfaces. Nat. Photonics **14**, 543–548 (2020)

Jain, S.K., Singh, V.P.: Developments in Water Science, Encyclopedia of Analytical Science 2nd edn, vol. 51, pp. 3–858. Elsevier Science B.V. (2003). https://www.sciencedirect.com/topics/earth-and-planetary-sciences/reflectance

Joshi, R., Shelar, S.B., Srivastava, M., Singh, B.P., Goel, L., Ningthoujam, R.S.: Development of core@Shell γ-Fe_2O_3@Mn_xO_y@SiO_2 nanoparticles for hyperthermia, targeting, and imaging applications. ACS Appl. Bio Mater. **5**, 5386–5393 (2022)

Kemp, W.: Organic Spectroscopy. Macmillan, Pennsylvania State University (1975)

Kerker, M.: The Scattering of Light and Other Electromagnetic Radiation in Physical Chemistry: A Series of Monographs, Academic Press, vol. 16. Published by Elsevier Inc., NC (1969)

Kim, T.H., Kim, S.H., Tan, L.V., Seo, Y.J., Park, S.Y., Kim, H., Kim, J.S.: Transition metal ion selective ortho-ester diazophenylcalix[4]arene. Talanta **71**, 1294–1297 (2007)

Kim, J., Ahn, H.Y., Kim, S.G., Oh, E., Ju, B.K., Choi, W.J., Cho, S.H.: Photoluminescence properties of lead selenide produced by selenization and a solvothermal method. Nanotechnology **28**, 015702 (2017)

Koh, W.K., Bartnik, A.C., Wise, F.W., Murray, C.B.: Synthesis of monodisperse PbSe nanorods: a case for oriented attachment. J. Am. Chem. Soc. **132**, 3909–3913 (2010)

Krishnamoorthy Sakthivel, A.P., Devadoss, I., Rajathi, V.M.A.: Role of Bi^{3+} ions on structural, optical, photoluminescence and electrical performance of $Cd_{0.9-x}Zn_{0.1}Bi_xS$ QDs. Appl. Sci. **3**, 694 (2021)

Kubin, R.F., Fletcher, A.N.: Fluorescence quantum yields of some rhodamine dyes. J. Lumin. **27**, 455–462 (1982)

Kumar, A., Ali, M., Ningthoujam, R.S., Gaikwad, P., Kumar, M., Nath, B.B., Pandey, B.N.: The interaction of actinide and lanthanide ions with hemoglobin and its relevance to human and environmental toxicology. J. Hazard. Mater. **307**, 281–293 (2016)

Lan, M., Zhao, H., Yuan, H., Jiang, C., Zuo, S., Jiang, Y.: Absorption and EPR spectra of some porphyrins and metalloporphyrins. Dyes Pigm. **74**, 357–362 (2007)

Lanje, A.S., Sharma, S.J., Ningthoujam, R.S., Ahn, J.S., Pode, R.B.: Low temperature dielectric studies of zinc oxide (ZnO) nanoparticles prepared by precipitation method. Adv. Powder Technol. **24**, 331–335 (2013)

Li, X., Zhu, H.: Two-dimensional MoS2: properties, preparation, and applications. J. Materiom. **1**, 33–44 (2015)

Li, Y., Du, H., Zhang, J., Liu, Z., Tian, M., Che, R.: Aspect ratio tuned red-shift of photoluminescence emission of PbSe nanorods investigated by electron holography. J. Colloid Interf. Sci. **493**, 385–392 (2017)

Liang, S., Wang, J.: Chapter 6–Broadband albedo, Advanced Remote Sensing (Second Edition) Terrestrial Information Extraction and Applications, pp. 193–250. Academic Press (2020). https://www.sciencedirect.com/topics/earth-and-planetary-sciences/albedo

Liu, P., Wang, H., Li, X., Ruib, M., Zeng, H.: Localized surface plasmon resonance of Cu nanoparticles by laser ablation in liquid media. RSC Adv. **5**, 79738–79745 (2015)

Luwang, M.N., Ningthoujam, R.S., Singh, N.S., Tewari, R., Srivastava, S.K., Vatsa, R.K.: Surface chemistry of surfactant AOT-stabilized SnO_2 nanoparticles and effect of temperature. J. Colloids Interf. Sci. **349**, 27 (2010)

Luwang, M.N., Ningthoujam, R.S., Srivastava, S.K., Vatsa, R.K.: Preparation of white light emitting YVO4:Ln^{3+} and silica-coated YVO_4:Ln^{3+} (Ln^{3+} = Eu3+, Dy3+, Tm3+) nanoparticles by CTAB/n-butanol/hexane/water microemulsion route: energy transfer and site symmetry studies. J. Mater. Chem. **21**, 5326 (2011)

Meetei, S.D., Singh, S.D., Singh, N.S., Sudarsan, V., Ningthoujam, R.S., Tyagi, M., Gadkari, S.C., Tewari, R., Vatsa, R.K.: Crystal structure and photoluminescence correlations in white emitting nanocrystalline ZrO_2:Eu^{3+} phosphor: effect of doping and annealing. J. Lumin. **132**, 537–544 (2012)

Meinrath, G., Lis, S., Stryla, X., Noubactep, C.: Lifetime and fluorescence quantum yield of uranium(VI) species in hydrolyzed solutions. J. Alloy. Compd. **300–301**, 107–112 (2000)

Mellaerts, S., Afanasev, V., Seo, J.W., Houssa, M., Locquet, J.P.: Efficient direct band-gap transition in Germanium by three-dimensional strain. ACS Appl. Mater. Interfaces **13**(26), 30941–30949 (2021)

Melnychuk, C., Sionnest, P.G.: Multicarrier dynamics in quantum dots. Chem. Rev. **121**, 2325–2372 (2021)

Mizrahi, B., Domb, A.J.: Mucoadhesive tablet releasing iodine for treating oral. J. Pharm. Sci. **96**, 3144 (2007)

Moon, H., Lee, C., Lee, W., Kim, J., Chae, H.: Stability of quantum dots, quantum dot films, and quantum dot light-emitting diodes for display applications. Adv. Mater. **31**, 1804294 (2019)

Nair, R.V., Gummaluri, V.S., Matham, M.V., Vijayan, C.: A review on optical bandgap engineering in TiO_2 nanostructures via doping and intrinsic vacancy modulation towards visible light applications. J. Phys. D Appl. Phys. **55**, 313003 (2022)

Ningthoujam, R.S.: Generation of exciton in two semiconductors interface: SnO_2:Eu-Y_2O_3. Chem. Phys. Lett. **497**, 208 (2010)

Ningthoujam, R.S.: Enhancement of luminescence by rare earth ions doping in semiconductor host. In: Rai, S.B., Dwivedi, Y. (eds.) Synthesis, Characterization and Applications of Multifunctional Materials Material Science and Technologies, Chap 7, pp. 145–182. Nova Science Publishers Inc., USA (2012)

Ningthoujam, R.S., Kulshreshtha, S.K.: Nanocrystalline SnO2 from thermal decomposition of tin citrate crystal: luminescence and Raman studies. Mater. Res. Bull. **44**, 57 (2009)

Ningthoujam, R.S., Lahiri, D., Sudarsan, V., Poswal, H.K., Kulshreshtha, S.K., Sharma, S.M., Bhushan, B., Sastry, M.D.: Nature of Vn+ ions in SnO2: EPR and photoluminescence studies. Mater. Res. Bull. **42**, 1293 (2007a)

Ningthoujam, R.S., Sudarsan, V., Godbole, S.V., Kienle, L., Kulshreshtha, S.K., Tyagi, A.K.: SnO_2:Eu^{3+} nanoparticles dispersed in TiO_2 matrix: improved energy transfer between semiconductor host and Eu^{3+} ions for the low temperature synthesized samples. Appl. Phys. Lett. **90**, 173113 (2007b)

Ningthoujam, R.S., Sudarsan, V., Kulshreshtha, S.K.: SnO_2: Eu nanoparticles dispersed in silica: a low temperature synthesis and photoluminescence study. J. Lumin. **127**, 747 (2007c)

Ningthoujam, R.S., Mishra, R., Das, D., Dey, G.K., Kulshreshtha, S.K.: Excess enthalpy and luminescence studies of SnO_2 nanoparticles. J. Nanosci. Nanotech. **8**, 4176 (2008a)

Ningthoujam, R.S., Sudarsan, V., Vinu, A., Srinivasu, P., Ariga, K., Kulshreshtha, S.K., Tyagi, A.K.: Luminescence properties of SnO_2 nanoparticles dispersed in Eu^{3+} doped SiO_2. J. Nanosci. Nanotech. **8**, 1489 (2008b)

Ningthoujam, R.S., Gajbhiye, N.S., Ahmed, A., Umre, S.S., Sharma, S.J.: Re-dispersible Li^+ and Eu^{3+} co-doped nanocrystalline ZnO: luminescence and EPR studies. J. Nanosci. Nanotech. **8**, 3059 (2008c)

Ningthoujam, R.S., Vatsa, R.K., Vinu, A., Ariga, K., Tyagi, A.K.: Room temperature exciton formation in SnO_2 nanocrystals in SiO_2: Eu matrix: quantum dot system, heat-treatment effect. J. Nanosci. Nanotech. **9**, 2634 (2009a)

Ningthoujam, R.S., Shukla, R., Vatsa, R.K., Duppel, V., Kienle, L., Tyagi, A.K.: $Gd_2O_3:Eu^{3+}$ particles prepared by glycine-nitrate combustion: phase, concentration, annealing, and luminescence studies. J. Appl. Phys. **105**, 084304 (2009b)

Ningthoujam, R.S., Sudarsan, V., Vatsa, R.K., Kadam, R.M., Jagannath, G.A.: Photoluminescence studies on Eu doped TiO_2 nanoparticles. J. Alloys Comp. **486**, 864 (2009c)

Ningthoujam, R.S., Sharma, A., Sharma, K.S., Barick, K.C., Hassan, P.A., Vatsa, R.K.: Roles of solvent, annealing and Bi^{3+}co-doping on the crystal structure and luminescence propertiesof $YPO_4:Eu^{3+}$ nanoparticles. RSC Adv. **5**, 68234–68242 (2015)

Ningthoujam, R.S., Gautam, A., Padma, N.: Oleylamine as reducing agent in syntheses of magic-size clusters and monodisperse quantum dots: optical and photoconductivity studies. Phys. Chem. Chem. Phys. **19**, 2294–2303 (2017)

Nozik, A.J.: Spectroscopy and hot electron relaxation dynamics in semiconductor quantum wells and quantum dots. Rev. Phys. Chem. **52**, 193–231 (2001)

O'Dell, J.W.: Determination of turbidity by Nephelometry: Environmental Monitoring Systems, p. 45268. Laboratory Office of Research and Development, US Environmental Protection Agency, Cincinnati, Ohio (1993)

Okram, R., Yaiphaba, N., Ningthoujam, R.S., Singh, N.R.: Is higher ratio of monoclinic to tetragonal in $LaVO_4$ a better luminescence host? Redispersion and polymer film formation. Inorg. Chem. **53**, 7204–7213 (2014)

Palik, E.D.: Chapter 2, Refractive Index in Handbook of Optical Constants of Solids, vol. 5, pp. 5–114. Academic Press (1997)

Pan, Y., Huang, J., Wang, Z.M., Yu, D.W., Yang, B., Ma, Y.G.: Computational investigation on the large energy gap between the triplet excited-states in acenes. RSC Adv. **7**, 26697 (2017)

Patterson, D.A.: Excitation and high temperature absorption of KCl: Tl. Phys. Rev. **119**, 962 (1960)

Perala, R.S., Singh, B.P., Putta, V.N.K., Acharya, R., Ningthoujam, R.S.: Enrichment of crystal field modification via incorporation of alkali K^+ ions in $YVO_4:Ho^{3+}/Yb^{3+}$ nanophosphor and its hybrid with superparamagnetic iron oxide nanoparticles for optical, advanced anticounterfeiting, uranyl detection, and hyperthermia applications. ACS Omega **6**, 19517–19528 (2021a)

Perala, R.S., Joshi, R., Singh, B.P., Putta, V.N.K., Acharya, R., Ningthoujam, R.S.: Brilliant nonlinear optical response of Ho^{3+} and Yb^{3+} activated YVO_4 nanophosphor and its conjugation with Fe_3O_4 for smart anti-counterfeit and hyperthermia applications. ACS Omega **6**, 19471–19483 (2021b)

Phaomei, G., Ningthoujam, R.S., Singh, W.R., Singh, N.S., Luwang, M.N., Tewari, R., Vatsa, R.K.: Low temperature synthesis and luminescence properties of re-dispersible Eu3+ doped $LaPO_4$ nanorods by ethylene glycol route. Opt. Mater. **32**, 616 (2010)

Prajapati, R., Chatterjee, S., Kannaujiya, K.K., Mukherjee, T.K.: Effect of compartmentalization of donor and acceptor on the ultrafast resonance energy transfer from DAPI to silver nanoclusters. Nanoscale **8**, 13006 (2016)

Prajwala, K., Priyanka, G.L., Hasan, M.A., Esther, A.C.M., Sridhara, N., Rajendra, A., Arya, S.B., Dey, A.: Development of reflective co-sputtered nanostructured metallic films. Surface Eng. 400–405 (2020). https://doi.org/10.1080/02670844.2020.1781376

Rajan, M.S.M.: Chapter 1–Photonic crystal fibers for various sensing applications, pp. 3–21. Industrial Applications of Nanocrystals, Micro and Nano Technologies (2022). https://www.sciencedirect.com/topics/materials-science/birefringence

Rao, C.M., Sudarsan, V., Ningthoujam, R.S., Gautam, U.K., Vatsa, R.K., Vinu, A., Tyagi, A.K.: Luminescence studies on low temperature synthesized $ZnGa_2O_4:Ln^{3+}$ (Ln = Tb and Eu) nanoparticles. J. Nanosci. Nanotech. **8**, 5776 (2008)

Rawool, S.A., Pai, M.R., Banerjee, A.M., Arya, A., Ningthoujam, R.S., Tewari, R., Rao, R., Chalke, B., Ayyub, P., Tripathi, A.K., Bharadwaj, S.R.: Pn Heterojunctions in $NiO:TiO_2$ composites with Type-II band alignment assisting sunlight driven photocatalytic H_2 generation. Appl. Catal. B **221**, 443–458 (2018)

Sahu, R.K., Ganguly, K., Mishra, T., Mishra, M., Ningthoujam, R.S., Roy, S.K., Pathak, L.C.: Stabilization of intrinsic defects at high temperatures in ZnO nanoparticles by Ag. J. Colloid Interf. Sci. **366**, 8–15 (2012)

Schneider, H.J., Anatoly, K., Yatsimirsky, A.K.: Selectivity in supramolecular host–guest complexes. Chem Soc. Rev. **37**, 263–277 (2008)

Seitz, F.: Color centers in alkai halide crystals. Rev. Mod. Phys. **18**, 384 (1946)

Simmons, J.H., Potter, K.S.: Optical Materials. Academic Press, New York (2000)

Singh, L.R., Ningthoujam, R.S.: Critical view on energy transfer, site symmetry, improvement in luminescence of Eu^{3+}, Dy^{3+} doped YVO_4 by core-shell formation. J. Appl. Phys. **107**, 104304 (2010)

Singh, L.R., Ningthoujam, R.S., Sudarsan, V., Singh, S.D., Kulshreshtha, S.K.: Probing of surface Eu^{3+} ions present in ZnO: Eu nanoparticles by covering ZnO: Eu core with Y_2O3 shell: luminescence study. J. Lumin. **128**, 1544 (2008a)

Singh, N.S., Ningthoujam, R.S., Devi, L.R., Yaiphaba, N., Sudarsan, V., Singh, S.D., Vatsa, R.K., Tewari, R.: Luminescence study of Eu^{3+} doped GdVO4 nanoparticles: concentration, particle size and core-shell effects. J. Appl. Phys. **104**, 104307 (2008b)

Singh, L.R., Ningthoujam, R.S., Singh, S.D.: Tuning of ultra-violet to green emission by choosing suitable excitation wavelength in ZnO: Quantum dot, nanocrystals and bulk. J. Alloys Comp. **487**, 466 (2009a)

Singh, N.S., Ningthoujam, R.S., Yaiphaba, N., Singh, S.D., Vatsa, R.K.: Lifetimes and quantum yield studies of Dy^{3+} doped $GdVO_4$: concentrations and annealing effect. J. Appl. Phys. **105**, 064303 (2009b)

Singh, L.R., Ningthoujam, R.S., Singh, N.S., Singh, S.D.: Probing Dy^{3+} ions on the surface of nanocrystalline YVO_4: luminescence study. Opt. Mater. **32**, 286 (2009c)

Singh, N.S., Ningthoujam, R.S., Phaomei, G., Singh, G.D., Vinu, A., Vatsa, R.K.: Re-dispersion and film formation of $GdVO_4$:Ln^{3+} ($Ln^{3+} = Dy^{3+}$, Eu^{3+}, Sm^{3+}, Tm^{3+}) nanoparticles: particle size and luminescence studies. Dalton Trans. **41**, 4404 (2012)

Singh, B.P., Parchur, A.K., Ningthoujam, R.S., Ansari, A.A., Singh, P., Rai, S.B.: Enhanced photoluminescence in $CaMoO_4$:Eu^{3+} by Gd^{3+} co-doping. Dalton Trans. **43**, 4779–4789 (2014a)

Singh, B.P., Parchur, A.K., Ningthoujam, R.S., Ansari, A.A., Singh, P., Rai, S.B.: Influence of Gd^{3+} co-doping on structural property of $CaMoO_4$:Eu. Dalton Trans. **43**, 4770–4778 (2014b)

Soni, A.K., Joshi, R., Jangid, K., Tewari, R., Ningthoujam, R.S.: Low temperature synthesized $SrMoO_4$:Eu^{3+} nanophosphors functionalized with ethylene glycol: a comparative study of synthesize route, morphology, luminescence and annealing. Mater. Res. Bull. **103**, 1–12 (2018)

Soni, A.K., Yadav, K.K., Singh, B.P., Joshi, R., Chakraborty, S., Chakravarty, R., Nagaraja, N.K., Singh, D.K., Kain, V., Dash, A., Ningthoujam, R.S.: Smart YPO_4:Er-Yb nanophosphor for optical heating, hyperthermia, security ink, cancer endoradiotherapy and uranyl recovery. ACS Appl. Nano Mater. **4**, 850–860 (2021)

Srinivasu, K., Ningthoujam, R.S., Sudarsan, V., Vatsa, R.K., Tyagi, A.K., Srinivasu, P., Vinu, A.: Eu^{3+} and Dy^{3+} doped YPO_4 nanoparticles: low temperature synthesis and luminescence studies. J. Nanosci. Nanotech. **9**, 3034 (2009)

Tian, Y., Zhang, J., Jing, X., Xu, S.: Optical absorption and near infrared emissions of Nd^{3+} doped fluorophosphate glass. Spectrochim. Acta Part A Mol. Biomol. Spectrosc. **98**, 355–358 (2012)

Tsujikawa, I.: Absorption lines in transition metal complexes. Pure Appl. Chem. **27**, 227–250 (1971)

Vali, R.: Infrared and Raman spectra and the band structure of yttrium trifluoride YF3. Comput. Mater. Sci. **50**, 2391–2396 (2011)

Veeranarayanan, S., Poulose, A.C., Mohamed, S., Aravind, A., Nagaoka, Y., Yoshida, Y., Maekawa, T., Kumar, D.S.: FITC labeled silica nanoparticles as efficient cell tags: uptake and photostability study in endothelial cells. J. Fluoresc. **22**, 537–548 (2012)

Wang, X., Xi, G., Liu, Y., Qian, Y.: Controllable synthesis of PbSe nanostructures and growth mechanisms. Cryst. Growth Des. **8**, 1406–1411 (2008)

Wang, Q., Xu, B., Sun, J., Liu, H., Zhao, Z., Yu, D., Fan, C., He, J.: Direct band gap silicon allotropes. J. Am. Chem. Soc. **136**, 9826–9829 (2014)

Wangkhem, R., Yaba, T., Singh, N.S., Ningthoujam, R.S.: Red emission enhancement from $CaMoO_4:Eu^{3+}$ by co-doping of Bi^{3+} for near UV/blue LED pumped white pcLEDs: energy transfer studies. J. Appl. Phys. **123**, 124303 (2018)

Wangkhem, R., Yaba, T., Shanta Singh, N., Ningthoujam, R.S.: Surface functionalized $GdVO_4:Eu^{3+}$ nanocrystals as turn-off luminescent probe for selective sensing of Cu^{2+} ions: role of pH. Appl. Surf. Sci. **563**, 150350 (2021)

Wei, T., Tian, Y., Chen, F., Cai, M., Zhang, J., Jing, X., Wang, F., Zhang, Q., Xu, S.: Mid-infrared fluorescence, energy transfer process and rate equation analysis in Er^{3+} doped germanate glass. Sci. Rep. **4**, 6060 (2014)

Yaiphaba, N., Ningthoujam, R.S., Singh, N.S., Vatsa, R.K., Singh, N.R.: Probing of inversion symmetry site in Eu^{3+} doped $GdPO_4$ by luminescence study: concentration and annealing effect. J. Lumin. **130**, 174 (2010)

Zhang, W., Wang, Q., Chen, Y., Wang, Z., Wee, A.T.S.: Van der Waals stacked 2D layered materials for optoelectronics. 2D Mater. **3**, 022001 (2016)

Zhao, J., Ji, A., Wu, W., Wu, W., Guo, H., Sun, J., Sun, H., Liu, Y., Li, Q., Huang, L.: Transition metal complexes with strong absorption of visible light and long-lived triplet excited states: from molecular design to applications. RSC Adv. **2**, 1712–1728 (2012)

Zhao, Y., Song, Q., Lin, Y., Chu, F., Wei, Y., Liu, S., Pan, C., Quan, L., Wang, Y.: Improving the photostability of fluorescent dyes by polymer nano-insulating layer. J. Appl. Polym. Sci. **139**, e51625 (2022)

Zhou, H.C., Long, J.R., Yaghi, O.M.: Introduction to metal-organic frameworks. Chem. Rev. **112**, 673–674 (2012)

Chapter 2
Photo-Luminescent Materials: Down-Conversion, Quantum Cutting, Up-Conversion, Photo-Avalanche, Core@Shell Nanostructures

Ruchi Agrawal, Manas Srivastava, and Raghumani S. Ningthoujam

2.1 Introduction

Photo-luminescent materials are materials, which can emit light after absorption of suitable light (Ningthoujam et al. 2012, 2022; Yadav and Ningthoujam 2021; Curie 1963; Blasse and Grabmaier 1994; Brabec et al. 2003; Simmons et al. 2000). Upon excitation, electrons in the ground state get excited to a higher energy state. Once excitation source is removed, the electrons in the higher energy state return to the ground state giving rise to emission. Materials can be of solid, liquid and gas. Solid materials can be amorphous, glass, plastic, single-crystalline, nanocrystalline and polycrystalline. Examples of solid materials are ZnO, SnO_2, TiO_2, YVO_4, $CaMoO_4$, PbSe, InP, PbS, GaN, InGaN, etc. (Singh et al. 2009a; Sahu et al. 2012a; Ghosh et al. 2011; Ningthoujam et al. 2007a, 2009a, 2017; Ganapathi et al. 2022; Luwang et al. 2010; Ningthoujam and Kulshreshtha 2009; Haider et al. 2019; Longo et al. 2011; Chen et al. 2020; Mamiyev and Balayeva 2023; Strite and Morkoç 1992; Sheen et al. 2022). Liquid samples are dyes dissolved in water or alcohol. Examples are Fluorescein isothiocyanate (FITC) dissolved in water (Chaganti et al. 2018; Joshi et al. 2022a), Fluorinated Boron-Dipyrromethene (BODIPY), 4',6-diamidino-2-phenylindole (DAPI) and Curcumin dissolved in water/alcohol (Boens et al. 2012; Omelon et al. 2016; Shetty et al. 2015). In order to change/tune the luminescence intensity or peaks, sometimes, 2 or 3 systems of different phosphor materials or ions are interfaced to form composite or core@shell or substitution (Pan et al. 2018; Shao et al. 2017; Parchur et al. 2012a; Dutta et al. 2012). Also, many ions

R. Agrawal · M. Srivastava · R. S. Ningthoujam (✉)
Chemistry Division, Bhabha Atomic Research Centre, Mumbai 400085, India
e-mail: rsn@barc.gov.in

R. Agrawal · R. S. Ningthoujam
Homi Bhabha National Institute, Anushakti nagar, Mumbai 400094, India

R. S. Ningthoujam and A. K. Tyagi (eds.), *Handbook of Materials Science, Volume 1*, Indian Institute of Metals Series, https://doi.org/10.1007/978-981-99-7145-9_2

are doped to the host to change luminescence or control luminescence behaviour. One example is Tl-doped NaCl used in dosimetry (Takizawa et al. 2021). The 'd' or 'f' block ions are doped to host. Examples of 'd' block ion doped material is host: Mn or host: Cr (Ahmed 2017; Lokesha et al. 2023). The 'f' block ions doped materials include the followings: ZnO_2:Ln^{3+}, SnO_2:Ln^{3+}, ZrO_2:Eu^{3+}, TiO_2:Ln^{3+}, $CaMoO_4$:Ln^{3+}, AVO_4:Ln^{3+} (A = Y, La, Gd, Lu), and APO_4: Ln^{3+}, etc. (Ningthoujam et al. 2007b, c, 2008a, b, 2009a, b, c; Gajbhiye et al. 2008; Singh et al. 2008a, b, 2009b, c, 2010, 2012, 2014a, b, 2015; Rao et al. 2008; Srinivasu et al. 2009; Yaiphaba et al. 2010a, b, c; Phaomei et al. 2010, 2011a, b, 2013; Shukla et al. 2010; Lahiri et al. 2010; Singh and Ningthoujam 2010, 2011; Ningthoujam 2010, 2013; Meetei et al. 2012; Sahu et al. 2012b, 2014; Parchur et al. 2012b, 2014; Parchur and Ningthoujam 2012a; Loitongbam et al. 2013; Prasad et al. 2013; Okram et al. 2014). Here, Ln^{3+} = Ce^{3+}, Eu^{3+}, Gd^{3+}, Pr^{3+}, Sm^{3+}, Ho^{3+}, Dy^{3+}, Tb^{3+}, Er^{3+}, Yb^{3+}. In some cases, 2 or more dopants of different elements are also co-doped to host. There are many phosphors in the form of liquid crystals (Wang et al. 2015). Also, shape and size engineering can bring many photo-luminescent materials for advanced flexible materials in the forms of thin films, single crystals, poly-crystals, amorphous, etc. Perovskite compounds (ABO_3) of pure inorganic, organic, inorganic–organic or lanthanide doped particles (quantum dots, polycrystalline) have been extensively studied for their either excellent luminescence emission, high quantum yields or ferroelectrics (Tai et al. 2019; Ye et al. 2018; Wang et al. 2020). A few examples are MDABCO (N-methyl-N'-diazabicyclo[2.2.2]octonium)–ammonium, $HC(NH_2)_2PbBr_3$, $CsPbX_3$ (X = Cl, Br, I). Organic phosphors are very light as compared to inorganic phosphors.

There are two terms: sensitizer and activator in luminescence (Ningthoujam et al. 2012; Yadav and Ningthoujam 2021). The sensitizer is ion/species, which has high absorption cross-section and its emission is absorbed by another species or ion known as activator through an energy transfer process (radiative or non-radiative process). In this way, activator can produce an improved emission intensity. One example is APO_4:Ce-Tb (A = La, Y, Gd) (Ningthoujam et al. 2012; Yadav and Ningthoujam 2021; Phaomei et al. 2011a; Sahu et al. 2014). Ce^{3+} ions have high absorption cross-section in UV region (250–280 nm) due to allowed transition (f-d transition) and emit a broad emission in 320–360 nm. On the other hand, Tb^{3+} ions have low absorption cross-section in 320–400 nm due to Laporte forbidden (f-f transition). By absorption of UV light at 250 nm by Ce^{3+}, its luminescence intensity at 350 nm is absorbed by Tb^{3+} ions and thus an enhanced luminescence intensity from Tb^{3+} can be obtained. In general, Förster and Dexter energy transfer processes can occur in sensitizer to activator combination (Cravcenco et al. 2020; Lakowicz 2006). Luminescence intensity of activator decreases due to the presence of quencher. This can be brought by the addition of a large number of activators or presence of magnetic ions (Ningthoujam et al. 2012). Increase of activator ions in rare-earth ions doped systems reduces luminescence intensity. This is known as concentration quenching effect.

In the study of luminescence decay, various lifetimes (τ) are observed from the different materials: ns (nanoseconds), μs (microseconds), ms (milliseconds), minutes and hours (Ningthoujam et al. 2012; Curie 1963; Blasse and Grabmaier 1994; Lakowicz 2006). Most semiconductors, insulators or organic phosphors exhibit lifetimes in ns range. The rare-earth doped semiconductors or insulators exhibit lifetimes in μs–ms range. Some insulators doped with Ln^{3+}, Mn^{2+} or Cr^{3+} can exhibit lifetimes in minutes to hours after removal of the excitation source. This is considered as persistence or delay luminescence (Zhuang et al. 2014; Sharma et al. 2023).

In some cases, the excitation energy is less than that of emitted light and the process is known as frequency up-conversion. Here, suitable light in NIR (near infrared) region is used to excite materials and these excited electrons are again brought to the higher excited state through many processes (energy transfer). Examples are $NaYF_4$:Er, Yb, $NaGdF_4$:Er, Yb, and LaF_3:Ho-Yb, etc. (Auzel 2004). Under 980 nm excitation, visible emission is observed. Similarly, organic compounds ($[Ru(dmb)_3]^{2+}$/DPA system; dmb = tris(4,4′-dimethyl-2,2′-bipyridine); DPA = 9,10-diphenyl-anthracene) can show the blue light under green light excitation (i.e., up-conversion) (Singh-Rachford and Castellano 2010).

There are many examples of photo-luminescent materials in the form of organic molecules, polymers, inorganic molecules, host–guest molecules, supermolecules, metal–organic frameworks, antenna-types molecules, hybrid, composite and core@ shell nanoparticles. These luminescent materials are used in many applications such as LEDs (light emitting diodes), power saving, security ink, imaging, cancer detection, biological imaging and therapy, etc. (Ningthoujam et al. 2012, 2022; Yadav and Ningthoujam 2021; Curie 1963; Blasse and Grabmaier 1994; Kumari et al. 2019a, b, 2021, 2023). Such materials can be grouped into 2: down-conversion and up-conversion luminescent materials.

In this chapter, frequency down-conversion and up-conversion luminescent materials will be discussed with many examples. The quantum cutting process is addressed here. In the last section, proximity effects due to core@shell nanostructures are provided. The effects of particle size on luminescence intensity are mentioned by taking a few examples.

2.2 Types of Photoluminescence Process

On the basis of the frequency conversion from excitation to emission, there are 2 types of photoluminescence process: Frequency down-conversion and Frequency up-conversion (Ningthoujam et al. 2012, 2022; Yadav and Ningthoujam 2021; Curie 1963; Blasse and Grabmaier 1994).

2.2.1 Frequency Down-Conversion

Frequency down-conversion optical materials are materials, which are able to convert an absorbed light into an emitted light of lower energy. Here, the excitation energy is more than the emitted energy or wavelength of the excitation energy is less than that

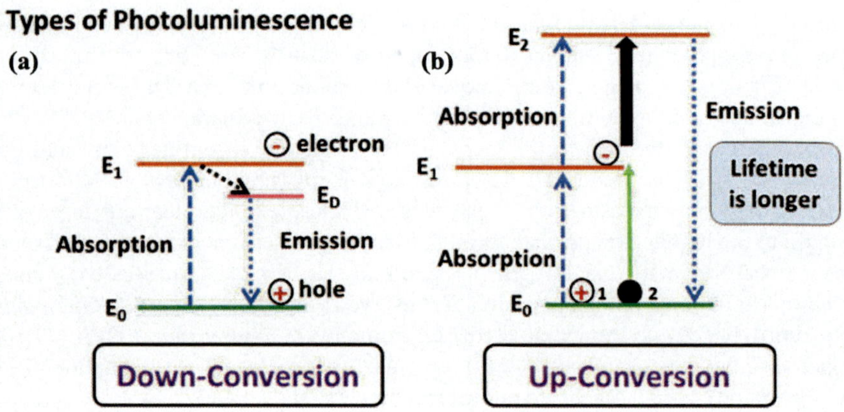

Fig. 2.1 Two types of photoluminescence process: **a** Frequency down-conversion and **b** Frequency up-conversion

of emitted light ($\lambda_{em} > \lambda_{ex}$). The schematic diagram illustrating frequency down-conversion is shown in Fig. 2.1a. At initial stage, ground state electrons absorb light and electrons go from the ground state to the excited state ($E_0 \rightarrow E_1$), hole is formed at the ground state and excited electrons at E_1 can come back to the ground state radiatively or non-radiatively. If there are defects or impurities, excited electrons stay sometime at the level (E_D), which is just below E_1. Then emission occurs. Here, the emitted energy is less than the absorbed energy. Here, 2 types of frequency down-conversion luminescent materials can be seen.

2.2.1.1 Normal (or Down-Shifting)

This was observed in organic compounds, semiconductors, d- and f- block ions and metal–ligand charge transfer systems. A few examples of organic compounds are anthracene, 4′,6-diamidino-2-phenylindole (DAPI), Fluorescein isothiocyanate (FITC), fluorinated boron-dipyrromethene (BODIPY), dyes, curcumin, etc. Metal centre-organic molecule complexes are iron propylene and haemoglobin. Host–guest complexes include supermolecular molecules, in which dye molecules are trapped inside host molecule. Examples of d block containing inorganic compounds are the first-row transition metal complexes, zirconium and hafnium complexes with pyridine dipyrrolide ligands (Wegeberg and Wenger 2021; Leary et al. 2023). Examples of f block containing inorganic compounds are lanthanide complexes and lanthanide–titanium oxo clusters (Hasegawa et al. 2022; Tanner and Pan 2009; Chen et al. 2019; Li and Li 2021). In organic compounds, π- bonds conjugation helps in the reduction of energy gap between HOMO (highest occupied molecular orbitals) and LUMO (lowest unoccupied molecular orbitals) (Brabec et al. 2003; Lakowicz 2006; Kemp 1975). Let us take a few examples. C_2H_6, C_2H_4 and C_2H_2 molecules have C–C, C

= C and C≡C bonds, respectively. Hybridization of C in C_2H_6, C_2H_4 and C_2H_2 molecules form sp^3, sp^2 and sp, respectively. Benzene molecule is having 3 π-bonds and 6 σ-bonds in adjacent bonds to form 6 carbon atoms ring. In resonance, these 3 π-bonds form a ring inside 6 carbon atoms ring. Absorption band shifts to a longer wavelength with an increase of π- bonds conjugation in a series of organic molecules and thus, the wavelength of its emission moves longer. This trend is seen in benzene, naphthalene, and anthracene. Similarly, ZnO absorbs light in 300–360 nm and its emission spectrum can produce a peak at 380 nm (corresponding to absorption edge) along with a broad emission band in 400–600 nm (Singh et al. 2009a; Ghosh et al. 2011).

There is another process known as quantum cutting, which is also a part of frequency down-conversion.

2.2.1.2 Quantum Cutting

Quantum cutting (QC) is one type of down-conversion (DC) process in which two or more low-energy photons are produced from a high-energy photon absorption (Yadav and Ningthoujam 2021; Mir et al. 2020; Zou et al. 2016; Zhang and Huang 2010). The QC process is observed in both singly rare-earth (RE) doped materials as well as in doubly or triply doped rare-earth doped materials. Here, energy transfer occurs through cooperative energy transfer (CET) process. In most cases, emitted light due to QC is observed in NIR range, whereas excitation wavelength is chosen in UV or visible light. In some cases, QC is observed in visible range upon excitation in vacuum UV region. The light emitted due to down-conversion or down-shifting in visible range is linear-dependent on excitation intensity, whereas the light emitted due to QC in NIR range is non-linear-dependent on excitation intensity. Figure 2.2 shows the schematic diagrams of QC mechanism for single ion and double ions in various hosts.

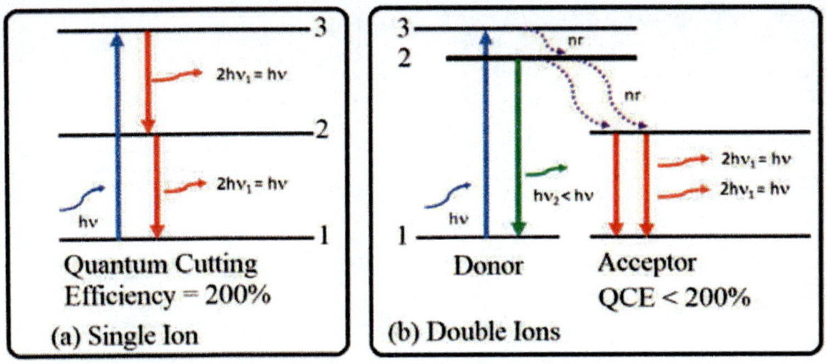

Fig. 2.2 Schematic diagram showing Quantum Cutting (QC) in single ion and double ions doped samples. Here, metastable energy levels help in getting QC

The first theoretical proposal to produce light with quantum yield (QY) more than 100% was reported in 1957 by Dexter (Dexter 1957). The transfer of energy from one sensitizer to two or more activators under different conditions can produce light with QY > 100%. It was realized experimentally in Pr^{3+} doped LaF_3, YF_3 and $NaYF_4$ in 1973 by a group of researchers from The Netherlands and USA (Sommerdijk et al. 1974; Piper et al. 1974). Upon excitation at 210 nm or 185 nm, the peak at 400–410 nm was observed. This transition corresponds to $^1S_0$-$^3P_2$ (4f-4f transition of Pr^{3+}) and is related to quantum cutting (QC) process because half of 400–410 nm is 200–205 nm. Figure 2.3 shows the schematic energy level diagram of Pr^{3+} ($4f^2$ electrons) and 5d levels (Piper et al. 1974). In a weak crystal field environment, 5d level is above $^1S_0$. In such conditions, the emission peak at 400–410 nm is observed under 185 nm excitation (Fig. 2.4) (Piper et al. 1974). If the excitation is done at 213 nm, an emission peak at 400–410 nm is observed, but this emission peak could not be observed under 228 nm excitation (Fig. 2.5) (Sommerdijk et al. 1974). In a strong crystal field environment, 5d levels become below $^1S_0$. Here, 5d levels are split into t_{2g} and e_g levels. Then excited energy is transferred to 5d levels non-radiatively, and the transition from $^1S_0$ becomes weaker. This was observed in CaF_2 and oxide compounds doped with 1 at.% Pr^{3+}. Instead of this, other transitions above 450 nm occur. If energy of excitation is less than $^1S_0$ level, other 4f-4f transitions above 450 nm occur.

In general, the materials having quantum cutting properties contain rare-earth ions, which may act as a part of host or dopants. Systems are Tb^{3+}–Yb^{3+}, Tm^{3+}–Yb^{3+}, Pr^{3+}–Yb^{3+}, Ce^{3+}–Yb^{3+}, Er^{3+}–Yb^{3+}, Ce^{3+}–Tb^{3+}–Yb^{3+}, and Gd^{3+}–Eu^{3+}, etc.

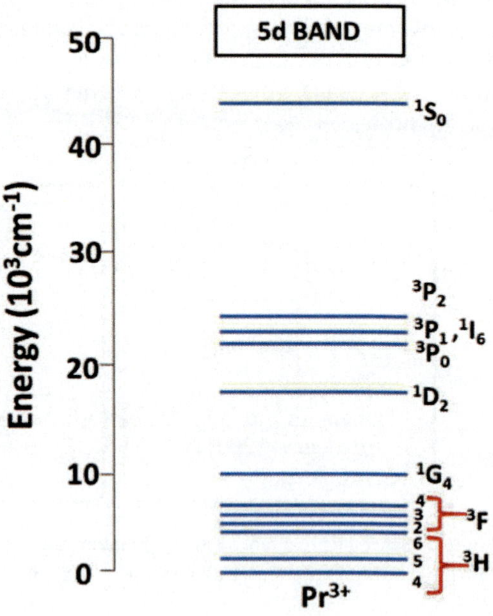

Fig. 2.3 Schematic energy level diagram for Pr^{3+} showing the positions of $4f^2$ levels as well as 4f-5d transition but the position of 5d level is dependent on the strength of ligand neighbouring to Pr^{3+} ion. It is redrawn and adapted from Elsevier Publisher (Piper et al. 1974)

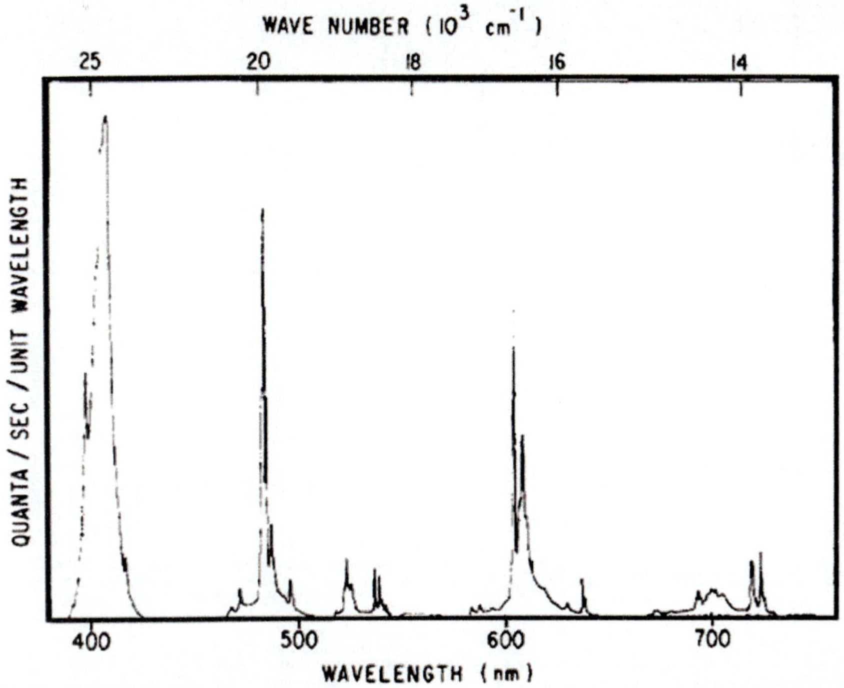

Fig. 2.4 Photoluminescence spectrum for YF_3: 0.1 at.% Pr^{3+} under excitation at 185 nm. It is reproduced with permission from Elsevier Publisher (Piper et al. 1974)

(Yadav and Ningthoujam 2021; Mir et al. 2020; Zou et al. 2016; Zhang and Huang 2010; Huang and Zhang 2010). Nowadays, other ions (from d or p block elements) are added as sensitizers (Bi^{3+}–Yb^{3+}). It is to be remembered that f electrons can have many fixed energy levels at different excited states. From this, possibility of many emission peaks can occur from UV to visible to NIR regions depending on the availability of levels. Photon-cascade emission phenomena can occur through frequency down and up-conversion processes.

In vacuum ultraviolet (VUV) excitations, visible emission peaks in 560–640 nm (maximum peak at 590 nm) are observed from Gd^{3+} doped $LiYF_4$ (singled ion) (Wegh et al. 1997). This is due to QC process. These transitions are assigned to $^6G_j \rightarrow {}^6P_j$ (Here j = spin–orbit coupling that can produce the Stark Splitting). Under VUV, electrons at ground state $^8S_{7/2}$ go to the excited state ^6G_j. Then excited electrons come to the lower energy levels such as ^6D_j, ^6I_j and ^6P_j through radiative or non-radiative processes or mixing. Then radiative transition from ^6G_j to ^6P_j produces visible light. The transition from ^6G_j to $^8S_{7/2}$ at 205 nm, ^6I_j to $^8S_{7/2}$ at 275 nm and ^6P_j to $^8S_{7/2}$ at 310 nm can produce emission. These are explained in Figs. 2.6, 2.7 and 2.8 (Wegh et al. 1997).

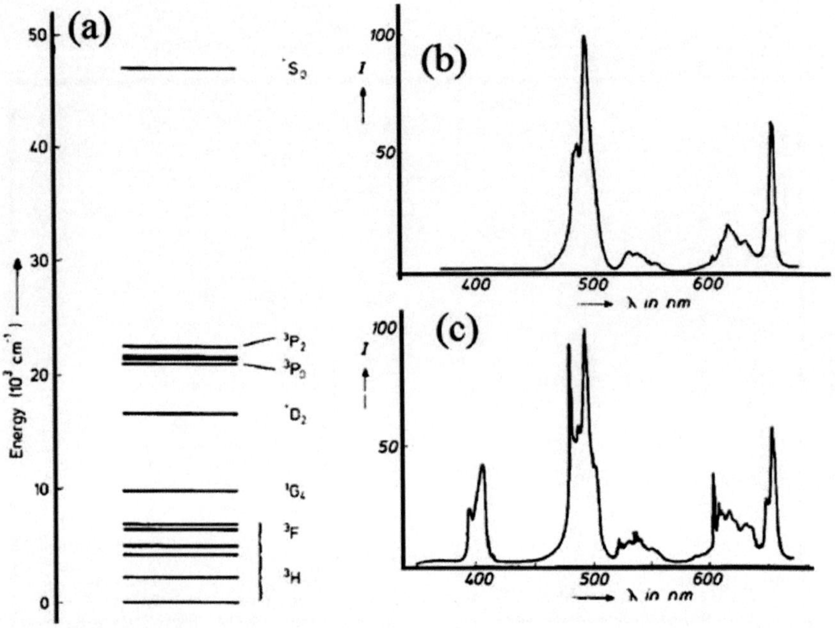

Fig. 2.5 **a** 4f energy level scheme of Pr^{3+}. Photoluminescence spectrum for YF_3: 0.1 at.% Pr^{3+} after excitation at **b** 228.8 (Cd lamp) and **c** 213 nm (Zn lamp). It is reproduced with permission from Elsevier Publisher (Sommerdijk et al. 1974)

Fig. 2.6 Schematic energy level scheme in the 0–50,000 cm^{-1} range for Gd^{3+} in fluoride lattice. It is redrawn and adapted from American Physical Society (APS) Publisher (Wegh et al. 1997)

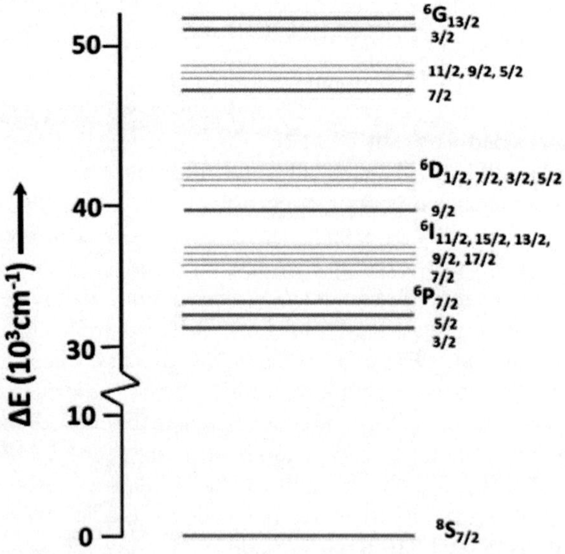

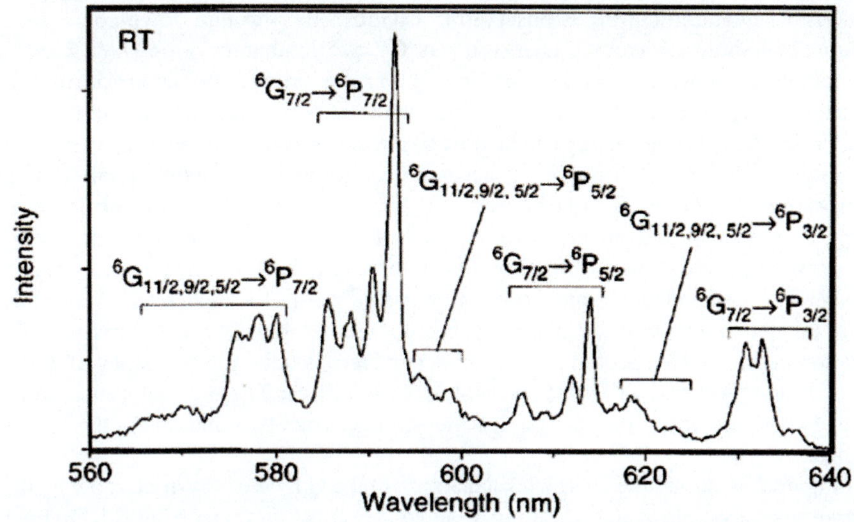

Fig. 2.7 $^6G_j \rightarrow {}^6P_j$ emission spectrum of $LiYF_4$: 5 at.% Gd^{3+} upon excitation at 202.1 nm ($^8S_{7/2} \rightarrow {}^6G_j$) at 300 K. It is reproduced with permission from American Physical Society (APS) Publisher (Wegh et al. 1997)

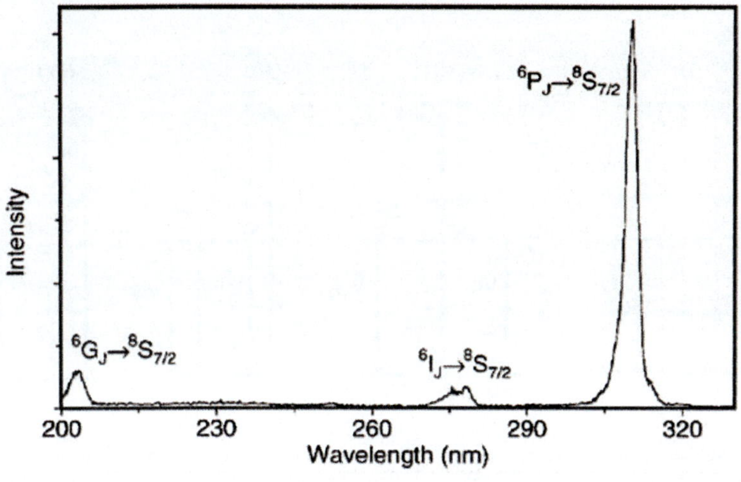

Fig. 2.8 UV emission spectrum of $LiYF_4$:5 at.% Gd^{3+} upon excitation at 194.7 nm ($^8S_{7/2} \rightarrow {}^6G_{13/2}$) at 300 K. It is reproduced with permission from American Physical Society (APS) Publisher (Wegh et al. 1997)

Visible quantum cutting is observed in $LiGdF_4:Eu^{3+}$ through down-conversion. Figure 2.9 shows schematic energy levels for two lanthanide ions (1 & 2 types hypothetically), which can have an energy transfer process for down-conversion process (Wegh et al. 1999). Ion 1 can undergo absorption and emission in one case and in another case, its energy can be transferred to ion 2. (I) A single ion (1) exhibits quantum cutting (QC) through the successive emission of two visible photons after absorption of UV light. **(II)** Quantum cutting through a two-step energy transfer. First step: Energy transfer from ion 1 to ion 2 takes place as a part of the excitation energy by cross-relaxation. Then, ion 2 comes back to the ground state with the emission of one photon in the visible range. Remaining energy of ion 1 is in an excited state and can transfer to a second ion of type 2. From a second ion of type 2, emission of a photon in the visible spectral region can occur. Here, a quantum efficiency of 200% is expected. **(III** and **IV)** Other possible QC processes: There is only one energy transfer process from ion 1 to ion 2, which can generate two photons in the visible range.

Figure 2.10 shows the energy level diagram of the Gd^{3+}-Eu^{3+} system, showing the possibility of visible quantum cutting by a two-step energy transfer from Gd^{3+} to Eu^{3+} (Wegh et al. 1999). In Gd^{3+} -Eu^{3+} system (Eu^{3+} doped $LiGdF_4$), emission intensity coming out from $^5D_0$ is more than a factor of 2 stronger for excitation at 202 nm (^6G_j) than that for excitation at 273 nm (^6I_j). This is due to QC effect. In the case of excitation at 202 nm, excited electrons go to the higher excited state of Gd^{2+} (^6G_j)

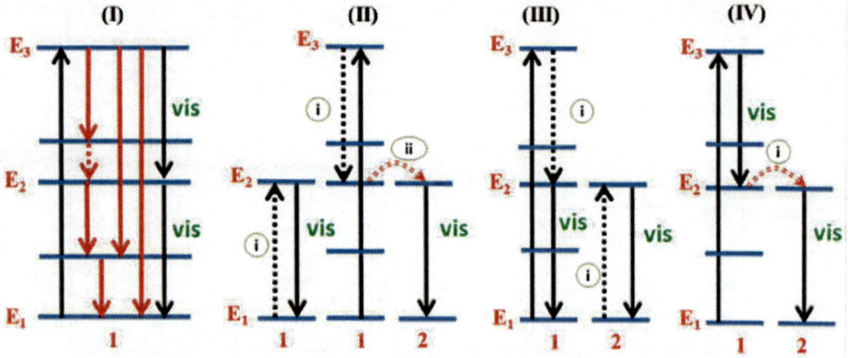

Fig. 2.9 Schematic energy level diagrams for two types of lanthanide ions (1 and 2), which can show possibility for the quantum cutting processes through frequency down-conversion. Possible way **(I)**: A single ion (1) exhibits quantum cutting (QC) through the successive emission of two visible photons after absorption of UV light. Way **(II)**: Quantum cutting through a two-step energy transfer. First step: Energy transfer from ion 1 to ion 2 takes place as a part of the excitation energy by cross-relaxation. Then, ion 2 comes back to the ground state with the emission of one photon in the visible range. Remaining energy of ion 1 is in an excited state and can transfer to a second ion of type 2. From a second ion of type 2, emission of a photon in the visible spectral region can occur. Ways **(III** and **IV)**: There is only one energy transfer process from ion 1 to ion 2, which can generate two photons in the visible range. It is redrawn and adapted from the American Association for the Advancement of Science (AAAS) Publisher (Wegh et al. 1999)

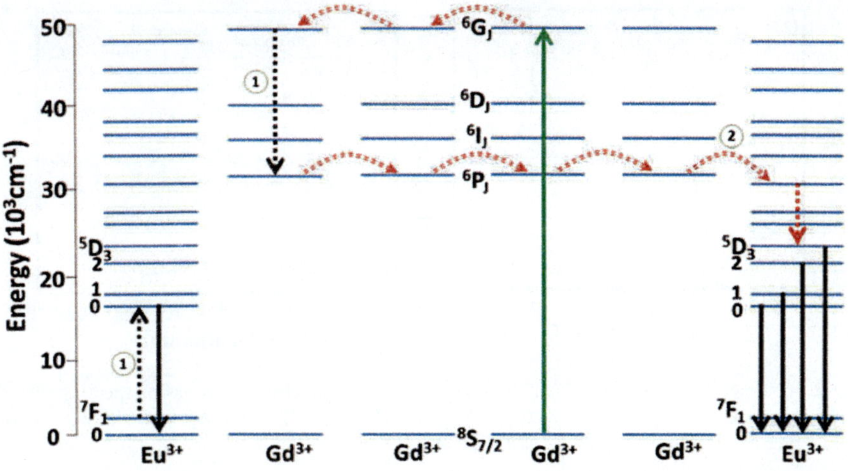

Fig. 2.10 Schematic diagram illustrating energy levels of the Gd^{3+}-Eu^{3+} pair with the possibility of quantum cutting in the visible range through a two-step energy transfer from Gd^{3+} to Eu^{3+}. Here, Gd^{3+} ions absorb light of a single photon in the UV region and it produces visible light. It is redrawn and adapted from the American Association for the Advancement of Science (AAAS) Publisher (Wegh et al. 1999)

and come to a lower energy level (^6P_j). During this transition, electrons in ground state of Eu^{3+} (^7F_j) go to excited state ($^5D_0$). This is known as cross-relaxation. Then visible red emission from $^5D_0$ level occurs. Simultaneously, the remaining excited electrons at ^6P_j of Gd^{3+} are transferred or migrated to the excited states of ^5D_j of Eu^{3+}. The visible red emission occurs. In this way, two photons (left from Eu^{3+} due to cross-relaxation and right from Eu^{3+} due to energy transfer or migration in Fig. 2.10) in visible red region occur. Here, one Gd^{3+} ion acts as sensitizer and 2 Eu^{3+} ions act as activator. In case of excitation at 273 nm (^6I_j), excited electrons go to the higher excited state of Gd^{3+} (^6I_j) and come to a lower energy level (^6P_j) through non-radiative process. Then excited electrons at ^6P_j of Gd^{3+} are transferred or migrated to excited states of ^5D_j of Eu^{3+}. The visible red emission occurs. In this way, one photon in visible red emission occurs.

In Yb^{3+}-Tm^{3+} system (YPO_4: Yb-Tm), an emission peak of 647 nm due to Tm^{3+} occurs after excitation at 475 nm excitation (direct excitation of Tm^{3+}) (Xie et al. 2009). When amount of Yb^{3+} increases, emission intensity of Tm^{3+} decreases, whereas the emission intensity due to Yb^{3+} in 970–1000 nm increases. There is an energy transfer from Tm^{4+} to Yb^{3+}. This was supported by a decrease in the lifetime of 647 nm peak after addition of Yb^{3+}. This is explained by QC process because 2 times of 475 nm gives 950 nm, which is near to 970 nm. After phonon relaxation, emission occurs in 970–1000 nm. This is explained by Figs. 2.11 and 2.12 (Xie et al. 2009). This energy transfer process is cooperative energy transfer.

In case of Bi^{3+}-Yb^{3+} system (Gd_2O_3: Bi-Yb), emission spectrum of Yb^{3+} in 900–1100 nm is observed after excitation through Bi^{3+} (320–390 nm) (Huang and Zhang

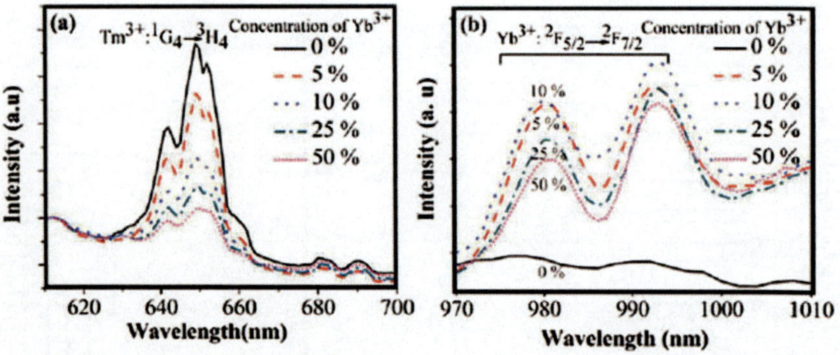

Fig. 2.11 Emission spectra of YPO_4: Yb-Tm after excitation at 475 nm: **a** emission spectra of Tm^{3+} in visible range and **b** emission spectra of Yb^{3+} in NIR range. It is reproduced with permission from American Institute of Physics (AIP) Publisher (Xie et al. 2009)

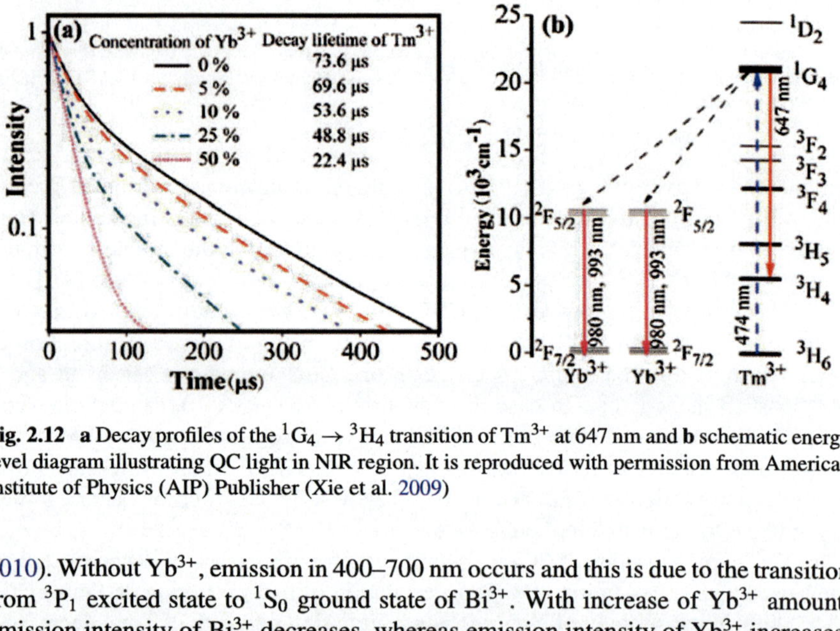

Fig. 2.12 **a** Decay profiles of the $^1G_4 \rightarrow {}^3H_4$ transition of Tm^{3+} at 647 nm and **b** schematic energy level diagram illustrating QC light in NIR region. It is reproduced with permission from American Institute of Physics (AIP) Publisher (Xie et al. 2009)

2010). Without Yb^{3+}, emission in 400–700 nm occurs and this is due to the transition from $^3P_1$ excited state to $^1S_0$ ground state of Bi^{3+}. With increase of Yb^{3+} amount, emission intensity of Bi^{3+} decreases, whereas emission intensity of Yb^{3+} increases. This is related to QC effect. Here, the lifetime of Bi^{3+} after excitation at 347 nm decreases with an increase of Yb^{3+} amount. This transfer of energy from Bi^{3+} to Yb^{3+} is referred as cooperative energy transfer (CET). The schematic diagram as well as the associated emission and decay processes is shown in Figs. 2.13, 2.14 and 2.15 (Huang and Zhang 2010).

The energy transfer efficiency (ETE, η_{ETE}) is calculated using the following:

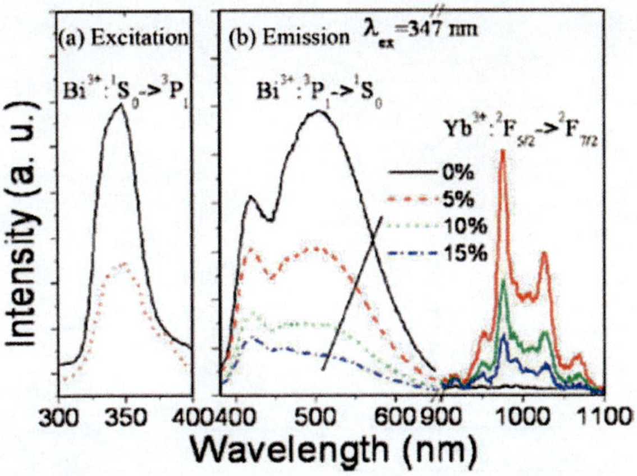

Fig. 2.13 **a** PL excitation spectra of $(Gd_{0.99}Bi_{0.01})_2O_3$ sample (solid line) monitored emission at 506 nm (Bi^{3+} emission peak) and $(Gd_{0.94}Yb_{0.05}Bi_{0.01})_2O_3$ sample (dashed line) monitored emission at 977 nm. **b** PL spectra of $(Gd_{0.99\text{-}x}Yb_xBi_{0.01})_2O_3$ samples in visible—NIR range after excitation at 347 nm (Bi^{3+} excitation/absorption peak). It is reproduced with permission from American Institute of Physics (AIP) Publisher (Huang and Zhang 2010)

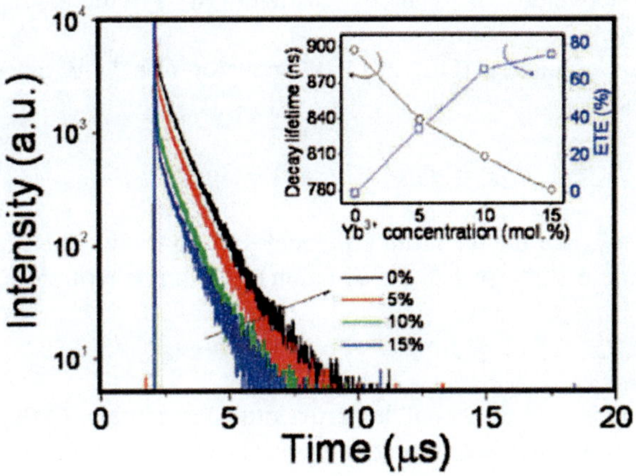

Fig. 2.14 Decay profiles of Bi^{3+} ($^3P_1 \rightarrow {}^1S_0$) emission peak upon excitation at 347 nm in $(Gd_{0.99\text{-}x}Yb_xBi_{0.01})_2O_3$ samples. Inset shows decay lifetime values and energy transfer efficiency from Bi^{3+} to Yb^{3+} at different amounts of Yb^{3+}. It is reproduced with permission from American Institute of Physics (AIP) Publisher (Huang and Zhang 2010)

$$\eta_{ETE} = \eta_{x\%Yb} = 1 - \frac{\int I_{x\%Yb}dt}{\int I_{0\%Yb}dt} \tag{2.1}$$

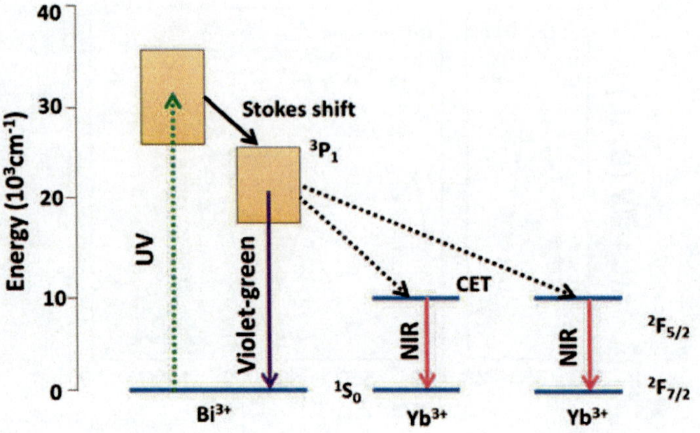

Fig. 2.15 Schematic diagram illustrating energy levels of ions (Bi^{3+} and Yb^{3+}) in Gd_2O_3 host. CET process for QC luminescence in NIR range under UV excitation is provided. It is redrawn and adapted from American Institute of Physics Publisher (Huang and Zhang 2010)

Here, $\int I_{x\%Yb}dt$ *and* $\int I_{0\%Yb}dt$ are luminescence intensities of sensitizer (Bi^{3+}) in the presence of activator (Yb^{3+}) and of sensitizer (Bi^{3+}) in the absence of activator (Yb^{3+}), respectively.

The quantum efficiency (QE, η_{QE}) of activator (Yb^{3+}) is calculated by the following relation:

$$\eta_{QE} = \eta_{Bi}(1 - \eta_{ETE}) + 2\eta_{Yb}\eta_{ETE} \quad (2.2)$$

where, η_{Bi} and η_{Yb} are the quantum efficiencies of Bi^{3+} and Yb^{3+}, respectively. If their values become 100% or 1.0, the equation (2.2) can be written as

$$\eta_{QE} = (1 - \eta_{ETE}) + 2\eta_{ETE} \quad (2.3)$$

Lanthanide doping in metal halide perovskite nanocrystals ($CsPbX_3$:Ln^{3+} (X = Cl, Br, I) also exhibit quantum cutting luminescence (Mir et al. 2020). Near-infrared quantum cutting long persistent luminescence (NQPL) has been observed in phosphor $Ca_2Ga_2GeO_7$:Pr^{3+}, Yb^{3+} (Zou et al. 2016). Many examples of quantum cutting can be found in Ln doped oxides and fluorides (Zhang and Huang 2010).

2.2.2 Frequency Up-Conversion Luminescence

In case of up-conversion luminescence, excitation energy or frequency is less than that of emission ($E_{em} > E_{exc}$ or $\nu_{em} > \nu_{exc}$) (Yadav and Ningthoujam 2021; Ningthoujam et al. 2022; Auzel 2004). It means that wavelength of excitation is more than that of emission ($\lambda_{em} < \lambda_{exc}$). Figure 2.1b shows the up-conversion process schematically. At initial stage, an electron at the ground state goes to the excited state ($E_0 \rightarrow E_1$) by absorption of energy (preferably laser source) with decay lifetimes of a few microseconds. Absorption takes place within 10^{-15} s. Before coming down to the ground state, excited electron at E_1 is hit by another electron, which absorbs energy (if continuous wave laser is used). Now, excited electron at E_1 jumps to higher excited level E_2 ($E_1 \rightarrow E_2$). From higher excited level (E_2), electron comes back to the ground state (E_0). However, its emission energy is more than that of absorbed energy. Sometimes, if excited energy levels have metastable states (like ladder like states in case of lanthanide ions), excited electron can jump to higher energy states ($E_2 \rightarrow E_3, E_3 \rightarrow E_4$, ---) by multi-photon absorption process. Various energy transfer processes can take place within neighbouring ions.

Different types of up-conversion processes can take place depending on materials, excitation source and condition (Fig. 2.16) (Yadav and Ningthoujam 2021; Ningthoujam et al. 2022; Auzel 2004). In most cases, resonance type absorption occurs. There is another process known as photon avalanche, which is similar to up-conversion process, but excitation intensity is more than the critical value as well as non-resonance type absorption occurs in high concentration of activators/dopants present in a host.

The frequency up-conversion mechanisms are useful in the following applications such as (a) infrared (IR) quantum counter detectors, (b) security ink, (c) temperature sensors, (d) three different colour emitting phosphors, (e) compact visible or (f) ultraviolet (UV) solid state lasers, (g) imaging through NIR excitations, (h) energy conservations, (i) light induced isomerism in organic molecules, (j) MRI contrast agents, (k) multi-utility at different excitation sources such UV, Visible, NIR lights, (l) deep tissue photothermal therapy, (m) photodynamic therapy, etc.

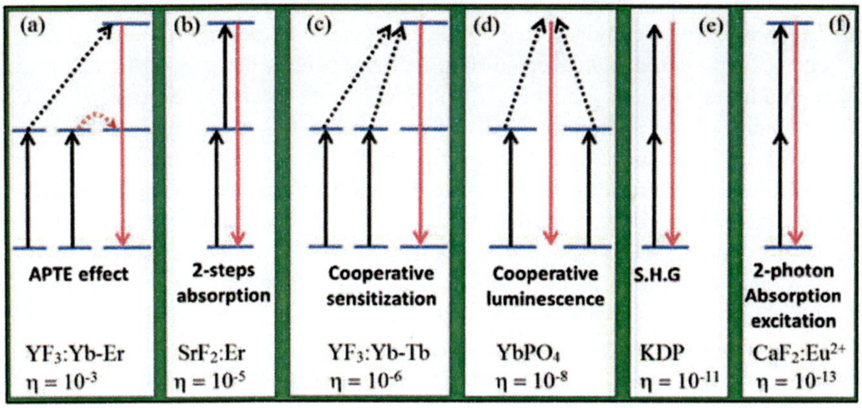

Fig. 2.16 Schematic diagram illustrating various 2-photon up-conversion processes observed in different systems. Here, relative efficiency in respective hosts is provided. It is redrawn and adapted from ACS Publisher (Auzel 2004)

2.2.2.1 Organic Based Frequency Up-Converters

Organic based frequency up-converters have been produced. Here sensitizer-activator combination can bring a light in UV to visible by absorbing suitable light in the whole solar spectrum. This will be useful in light energy harvesting. Here, triplet–triplet annihilation (TTA) mechanism is involved (Singh-Rachford and Castellano 2010). The sensitizer absorbs light from the ground state (GS = $^1S_0{}^s$ to the first excited state ($^1ES^*$ = $^1S_1{}^s$). This excited energy is transferred to its triplet state ($^3ES^*$ = $^3T_1{}^s$) through inter-system-crossing (ISC). The energy of $^3T_1{}^s$ is absorbed by the triplet state (^{3}Acceptor* = $^3T_1{}^a$) of activator through triplet–triplet energy transfer (TTET). At the same time, another molecule of activator has a similar TTET since a continuous source of excitation is provided. Combined 2 triplets of activators ($^3T_1{}^a$–$^3T_1{}^a$) produce light, which is absorbed by activator in such a way that the electrons in ground state of acceptor/activator go to the singlet excited state (^{1}Acceptor*, i.e., from GS ($^1S_0{}^a$) to ES ($^1S_1{}^a$)). This process is known as annihilation (loss or extinction or destruction) or triplet–triplet annihilation. Finally, light is produced from the excited state of activator ($^1S_1{}^a$) to ground state ($^1S_0{}^a$). The frequency of light produced is more than that of absorption by sensitizer. It is shown in Fig. 2.17 (Singh-Rachford and Castellano 2010). To meet the above condition, the energy gap of sensitizer phosphor is less than that of activator phosphor. The triplet excited energy of sensitizer is slightly more than that of triplet excited energy of activator. Energy of 2 triplet excited states of 2 activators is equal to or a slightly more than that of singlet excited state of activator. The difference will be provided by phonon relaxation. It is to be noted that triplet excited state has a relatively longer lifetime (μs to ms), which helps in triplet–triplet annihilation. Absorption of light by sensitizer takes place within 10^{-15} s.

In general, metal complex molecules having metal-to-ligand charge transfer (MLCT) is used as sensitizer. Metals are heavy noble metals such as Pd, Pt and Ru. Ligands have π-conjugated aromatic rings such as porphyrins and phthalocyanines. Such central metal ion in porphyrins enhances the spin–orbit coupling, which produces efficiency of unity for singlet–triplet intersystem crossing (ISC) (Singh-Rachford and Castellano 2010).

At 450 nm laser excitation, $[Ru(dmb)_2(bpy\text{-}An)]^{2+}$ in CH_3CN medium or solution exhibits the up-converted delayed singlet anthracene fluorescence through the bimolecular triplet–triplet annihilation (Fig. 2.18) (Kozlov and Castellano 2004). Excitation wavelength at 450 nm is related to MLCT. Emission occurs UV to visible range. Here, dmb is 4,4'-dimethyl-2,2'-bipyridine. bpy-An is 4-methyl-4'-(9- anthrylethyl)-2,2'-bipyridine. Laser power dependent emission studies show 2 photon absorption processes. Interestingly, solution having equal molar concentration of $[Ru(bpy)_3]^{2+}$ and anthracene (3.5×10^{-5} M) gives threefold enhancement in up-converted emission as compared to $[Ru(dmb)_2(bpy\text{-}An)]^{2+}$. It is suggested that intermolecular interaction (non-covalent interaction) is a better up-converted process as compared to that of intramolecular interaction (covalent interaction).

Without anthracene, $[Ru(bpy)_3]^{2+}$ alone does not produce up-converted emission instead, it produces down-converted emission after excitation at 450 nm. The

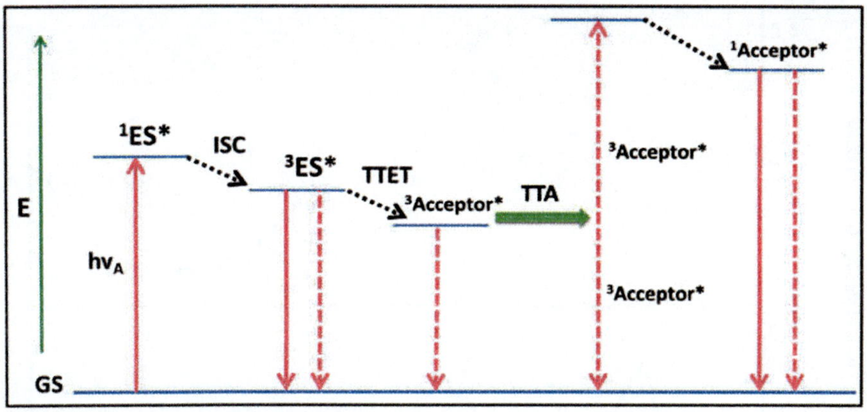

Fig. 2.17 Schematic diagram showing the up-conversion processes from the triplet state of the sensitizer molecule and the triplet state of the acceptor molecule and it leads to the singlet luminescence at higher energy. GS: ground state. ISC: intersystem crossing. ES: excited state. TTA: triplet–triplet annihilation and TTET: triplet–triplet energy transfer. Symbols "1" and "3" indicate the singlet and triplet states, respectively. It is redrawn and adapted from Elsevier Publisher (Singh-Rachford and Castellano 2010)

triplet excited state to singlet ground state transition (emission) occurs at 610 nm in CH_3CN solution and its lifetime is 920 ns. Laser power dependent emission studies show 1 photon absorption process. Anthracene alone does not produce up-converted emission after excitation at 450 nm. Chemical structures of $[Ru(dmb)_2(bpy\text{-}An)]^{2+}$, $[Ru(bpy)_3]^{2+}$ and anthracene are shown here (Fig. 2.19) (Kozlov and Castellano 2004). Figure 2.20 shows the digital photo of up-converted light (blue) after excitation at the green light (laser at 532 nm) from the solution containing $[Ru(dmb)_3]^{2+}$ and 9, 10-diphenyl-anthracene in CH_3CN solvent (Singh-Rachford and Castellano 2010).

2.2.2.2 Transition Metal Ions (d^{n+}) Based Frequency Up-Converters

Transition metal ions (TM = 3d, 4d, 5d) containing parent compounds or doped to other hosts have shown frequency up-converters (Ye et al. 2016). TM ions are Mn^{2+}, Cr^{3+}, Re^{4+}, V^{3+}, Mo^{3+}, Ni^{2+}, and Ti^{2+}, etc. However, they are extensively investigated at cryogenic temperatures except for Mn^{2+} and Cr^{3+} ions. This is due to a large influence from the electron–phonon coupling effect for d electrons of TM ions as compared to f electrons of Ln^{3+} ions. Also, there are large energy gaps between the ground state ($^6A_1$ (Mn^{2+}) or $^4A_2$ (Cr^{3+})) and its first excited states (i.e., emitting levels = $^4T_1$ and $^4T_2$ (Mn^{2+}) or ^2E (Cr^{3+})) in cases of Mn^{2+} and Cr^{3+} ions. The energy gaps of about 10,000–20,000 cm^{-1} are observed and this reduces multi-phonon relaxation and thereby radiative emission at room temperature is observed. Figure 2.21

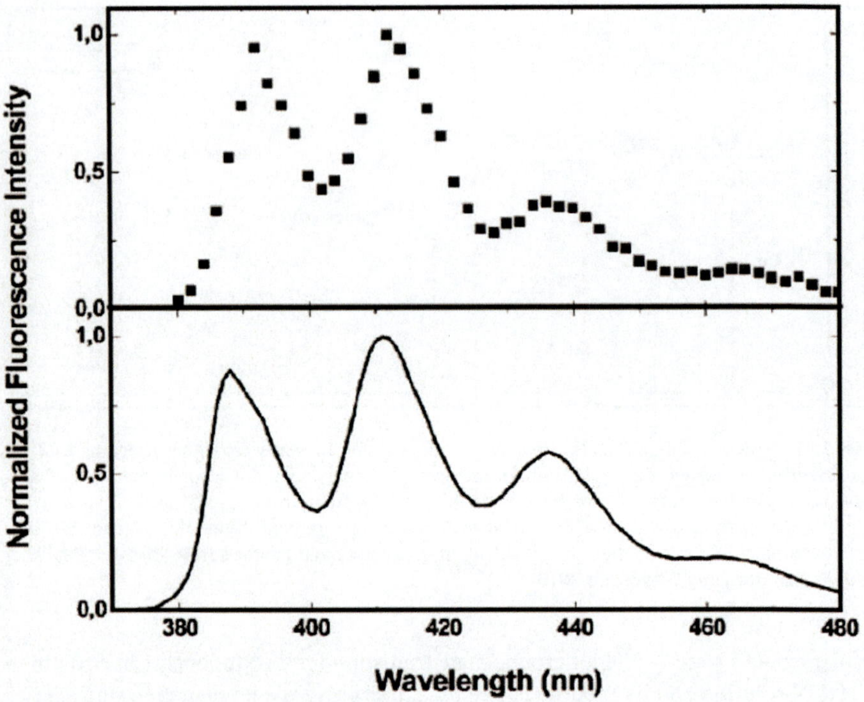

Fig. 2.18 Delayed luminescence spectrum of $[Ru(dmb)_2(bpy\text{-}An)]^{2+}$ (solid squares) in deaerated CH_3CN (3.5×10^{-5} M) measured 9 μs after a 450 nm excitation (power of 60 μJ laser pulse excitation). Residual luminescence from this (black line) using UV light 360 nm lamp excitation. It is reproduced with permission from RSC Publisher (Kozlov and Castellano 2004)

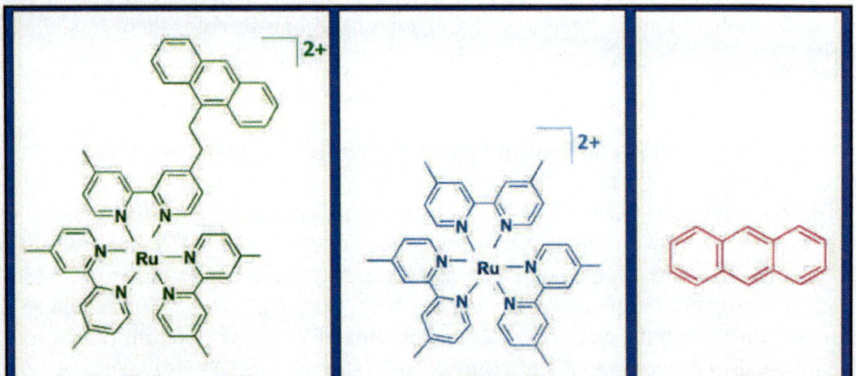

Fig. 2.19 Chemical structures of $[Ru(dmb)_2(bpy\text{-}An)]^{2+}$, $[Ru(bpy)_3]^{2+}$ and anthracene

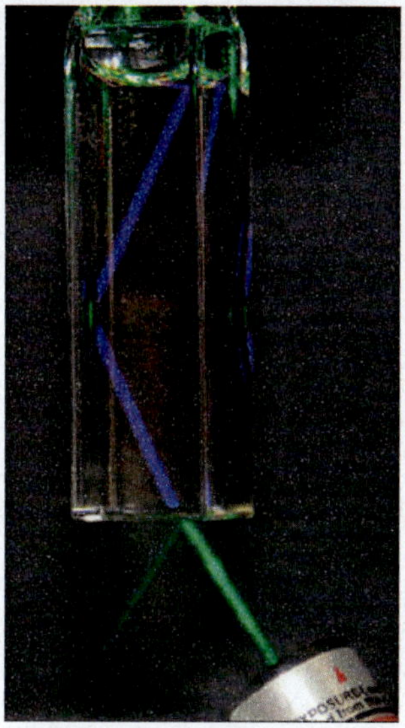

Fig. 2.20 Digital photograph of the up-converted light (blue) after excitation at the green light (laser at 532 nm) from solution containing $[Ru(dmb)_3]^{2+}$ and 9, 10-diphenyl-anthracene in CH_3CN solvent. It is reproduced with permission from Elsevier Publisher (Singh-Rachford and Castellano 2010)

shows Tanabe-Sugano diagram of commonly used d^3, d^5 and d^8 ions (Ye et al. 2016). These ions can show the room temperature up-conversion luminescence.

A broad yellow UC luminescence is observed at room temperature $Ca_3Y(GaO)_3(BO_3)_4$: Yb^{3+}-Mn^{2+} (CYGB: Yb-Mn) and CYGB: Yb phosphors upon 980 nm laser excitation (Fig. 2.22) (Xie et al. 2020). Two UC emission bands are observed at 503 and 600 nm. The 1st peak is related to the cooperative luminescence of Yb^{3+} pairs (CYGB: Yb) and 2nd peak is related to the transition of super-exchange coupled $Yb^{3+} - Mn^{2+}$ pairs. There is no resonance energy transfer between coupled $Yb^{3+} - Mn^{2+}$ pairs. However, UC mechanism is explained on the basis of the ground state absorption (GSA) and followed by the excited state absorption (ESA) in this system (Fig. 2.23). Without Mn^{2+}, $Ca_3Y(GaO)_3(BO_3)_4$:Yb^{3+} shows emission at 503 nm due to emission from virtual level of Yb^{3+} by 2 photon absorption process (Fig. 2.23) (Xie et al. 2020).

2.2.2.3 Lanthanide Ions (Ln^{n+}) Based Frequency Up-Converters

Lanthanides include the elements having $4f^n$ electrons (n = 1–14) and comprise of 15 elements from lanthanum (atomic number 57) to lutetium (atomic number 71) (Ningthoujam et al. 2012). These have similar chemical properties with scandium (atomic number 21) and yttrium (atomic number 39), and are rarely available and also

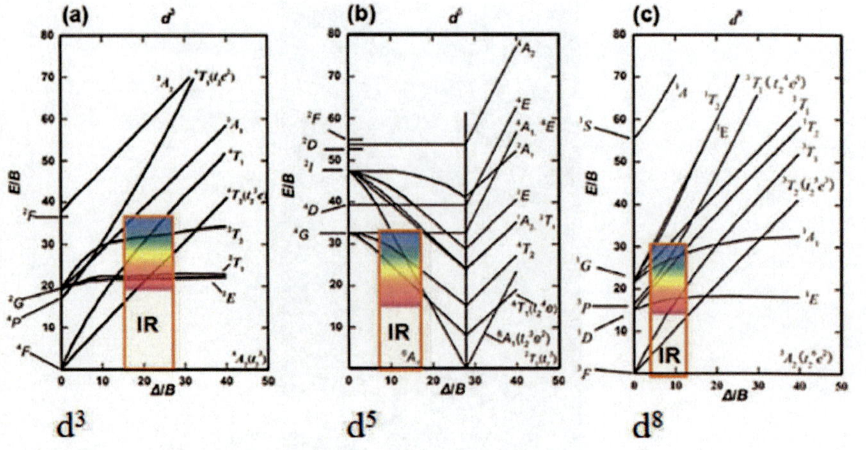

Fig. 2.21 Tanabe-Sugano diagram of a few 3d metal ions (d^3, d^5 and d^8). Energy gap between the excited state and ground state is represented by the histogram. In this histogram, the coloured filled region denoted the energy levels located at visible range, whereas the empty region denoted the energy levels located at infrared (IR) range. It is reproduced under creative common & Wiley Online Library Publisher (Ye et al. 2016)

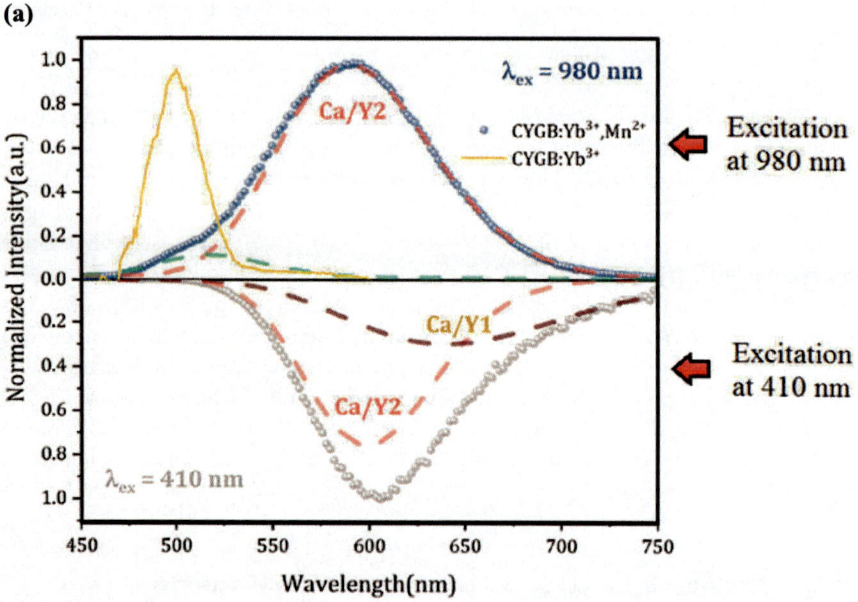

Fig. 2.22 Upper portion: normalized PL spectra (UC) for samples CYGB:0.01Yb^{3+}, 0.45Mn^{2+} and CYGB:0.01Yb^{3+} after excitation at 980 nm. Lower portion: PL spectrum (DC) for sample CYGB:0.01Yb^{3+}, 0.45Mn^{2+} after excitation at 410 nm. Gaussian fitting to spectrum is shown. It is reproduced with permission from ACS Publisher (Xie et al. 2020)

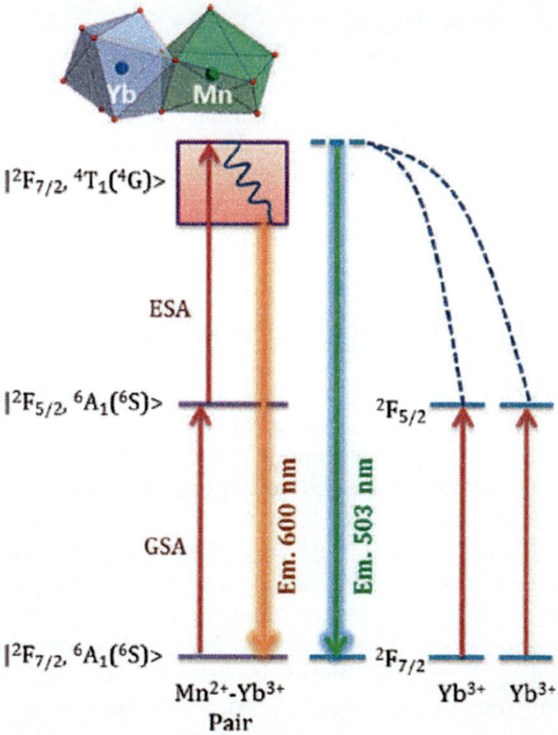

Fig. 2.23 Diagram showing proposed up-conversion of Yb^{3+} and Mn^{2+} in $Ca_3Y(GaO)_3(BO_3)_4$: Yb^{3+}-Mn^{2+} phosphor under 980 nm excitation. It is reproduced with permission from ACS Publisher (Xie et al. 2020)

generally found in the same ore deposits as the lanthanides. These 17 elements are considered as the rare-earth elements. They have a stable oxidation state of + 3 except in a few cases. Ce, Pr, Tb and Dy can have + 3 or + 4 depending on the oxidizing environment. Sm, Eu, Tm and Yb can exhibit + 2 or + 3 depending on the reducing environment. Nd can have + 2 or + 3 or + 4. The ground state electronic configuration of Ln is $1s^2$, $2s^22p^6$, $3s^23p^63d^{10}$, $4s^24p^64d^{10}4f^{n \text{ or } n+1}$, $5s^25p^65d^{1 \text{ or } 0}$, $6s^2$ ($0 \leq n \geq 14$). La, Ce, Gd, and Lu have [Xe] $4f^n$, $5d^1$, $6s^2$ and the remaining Ln elements have [Xe] $4f^{n+1}$, $6s^2$. In case of Ln^{3+}, the ground state electronic configuration is $1s^2$, $2s^22p^6$, $3s^23p^63d^{10}$, $4s^24p^64d^{10}4f^n$, $5s^25p^6$ since 3 electrons from outer orbitals such as $6s^2$ and $5d^1$ or $4f^{n+1}$ are removed.

What types of electronic transition can take place in $4f^n$? $4f^n$ electron can go to higher excited states such as 4f or 5d. The $4f^n$ to $4f^{n-1}5d^1$ transition occurs at vacuum UV ($\lambda_{exc} < 200$ nm) or UV ($\lambda_{exc} = 200$–300 nm) region. However, $4f^n$ to $4f^n$ transition occurs from UV to visible to NIR region. In general, $4f^n$ to $4f^n$ transition produces a sharp peak absorption, whereas, $4f^n$ to $4f^{n-1}5d^1$ transition produces a broad peak absorption because $5d^1$ degeneracy is lifted due to the crystal field from ligands present in around Ln^{3+} ion. $4f^n$ to $4f^n$ transition is forbidden and transition occurs due to relaxation such as mixing of orbitals or phonon relaxation or crystal field environment (Parchur and Ningthoujam 2012b; Srivastava et al. 2022). Ln^{3+}

ions with 4f electrons of n = 1 or 14 (Ce^{3+} or Lu^{3+}) do not undergo f-f transition in UV–visible-NIR range. Gd^{3+} with n = 7 (half field electrons) produces 4f-4f transition in UV region.

Due to the presence of metastable levels in higher excited states (due to spin–orbit coupling), many transitions (absorption and emissions) can take place. Such metastable levels can be used for up-converting process. Examples of up-converters are $NaYF_4$: Er-Yb, $NaYF_4$: Tm-Yb, CaF_2: Er-Yb, LaF_3: Er-Yb, etc. (Auzel 2004; Chen et al. 2004; Borse et al. 2022).

2.2.2.4 Actinide Ions (Ac^{n+}) Based Frequency Up-Converters

Actinide ions based frequency up-conversions are very less due to energy gaps between metastable levels are low and also association of f-d transition. In most cases, excited state absorption (ESA) up-conversion process is more as compared to that of sequential energy transfer or energy transfer (ETU) up-conversion process. Examples of Ac^{3+} ions doped samples are U^{4+} ($5f^2$) ions doped $ThCl_4$ or $ThBr_4$ and $ZnCl_2$ based glass, which have shown up-conversion luminescence (Auzel 2004; Karbowiak et al. 2003; Shalav et al. 2007). Under 950 nm or 1170 nm excitation, green and red emission is observed and this up-conversion process is due to ESA because its emission intensity is linearly dependent on excitation laser power or its emission intensity is linearly dependent on the concentration of U^{4+} ions. $LaCl_3$:U^{3+} ($5f^3$) shows the red emission under green laser excitation. Cm^{3+} ions doped samples have shown up-conversion luminescence.

2.2.2.5 Photon Avalanche

Photon avalanche (PA) is a process of frequency up-conversion in which a sudden increase in intensity of emitted light occurs when non-resonance type excitation is applied (Auzel 2004; Silva et al. 2020; Lee et al. 2021; Peng et al. 2022; Li et al. 2019). Generally, a high concentration of dopant (lanthanides) is associated in host so that cross-relaxation among dopant ions can occur. In such high concentration of dopant ions, when laser power of excitation source is provided above a critical value, huge cross-relaxation among dopant ions occurs and thereby, high intensity of frequency up-converted light is observed. It is a non-linear process and follows $I \propto P^n$ (I is intensity of converted light, P is power of excitation source and n is number of photons absorbed). Avalanched up-conversion phenomenon occurs n>10; normal up-conversion process occurs when n = 2–3 and in between say n = 4–7, looping up-conversion occurs (Lee et al. 2021). In normal energy transfer up-conversion (sequential energy transfer, ETU), dopant concentration of activator ion is small (>2 at.%, for example, $NaYF_4$:2 at.%Er, 10 at.%Yb). Similarly, excited state absorption (ESA) up-conversion materials (for example $NaYF_4$:Er) also require less amount of activators (2 at.% Er). Otherwise, concentration quenching occurs. In ETU and ESA, the excitation wavelength should be close to the resonance with gap between ground state energy and excited state energy, whereas the excitation wavelength of PA generally does not require resonance with gap between ground state energy and excited state energy, but it requires resonance with excited state intermediate levels (Lee et al. 2021; Levy et al. 2016).

For the first time, it was reported that when a $LaCl_3$ or $LaBr_3$ crystal doped with 4.8 at.% Pr^{3+} is exposed to simultaneous IR light at 4.5 μm and green CW laser, a red emission was obtained (Chivian et al. 1979). With increase of laser-pump power radiation, intensity of red emission increases. Above the critical power density of green CW laser 1.2–12.2 W/cm^2, red emission intensity is suddenly increased. Now, the intensity of red emission was independent of NIR light. This is known as Photon Avalanche. This was explained as follows: Pr^{3+} ion has the ground state energy level $^3H_4$ and metastable excited states start from $^3H_5$, $^3H_6$, $^3F_2$, $^3F_3$, $^3F_4$, $^1G_4$, $^1D_2$, $^3P_0$, $^3P_1$, $^1I_6$ to $^3P_2$. The energy levels are shown in Fig. 2.24 (Chivian et al. 1979).

When Pr^{3+} is exposed to IR at 4.5 μm, 4f^2 -electrons in ground state $^3H_4$ go to next higher excited state $^3H_5$. Further excitation with laser at green wavelength, excited electrons go to $^3P_1$ level (higher level). Possible emission transitions such as $^3P_1$–$^3F_2$, $^3P_0$–$^3F_2$, $^3P_0$–$^3H_6$ occur. This is one type of up-conversion process because of simultaneous two absorptions IR and green light. The intensity of red emitted light is linearly dependent on input power of IR or green light when input laser power is less than the critical value (P_c). At or above P_c, intensity of red light increases suddenly and shows non-linear property and without IR source, intensity of red light is very high (Fig. 2.25) (Chivian et al. 1979). This is explained by the non-radiative

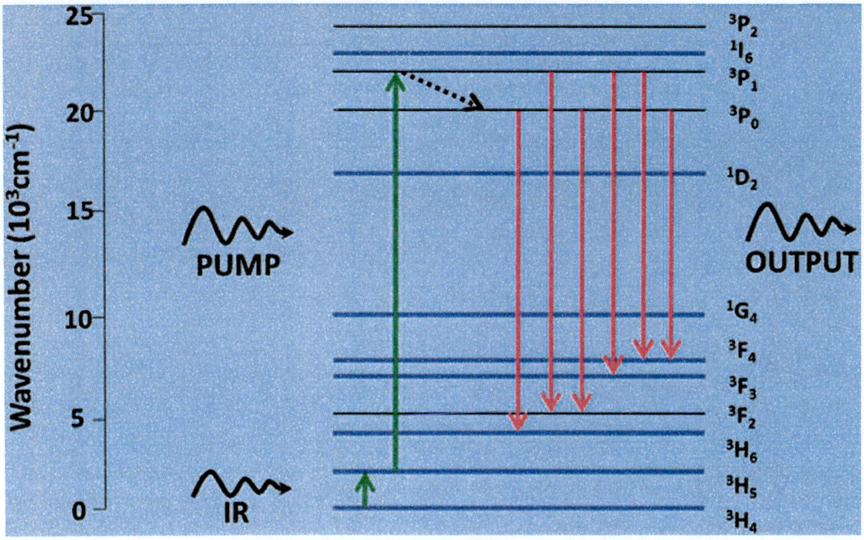

Fig. 2.24 Schematic energy levels of Pr^{3+} in $LaCl_3$ and $LaBr_3$. Here, simultaneous IR light at 4.5 μm and green CW laser are provided. It is redrawn and adapted from AIP Publisher (Chivian et al. 1979)

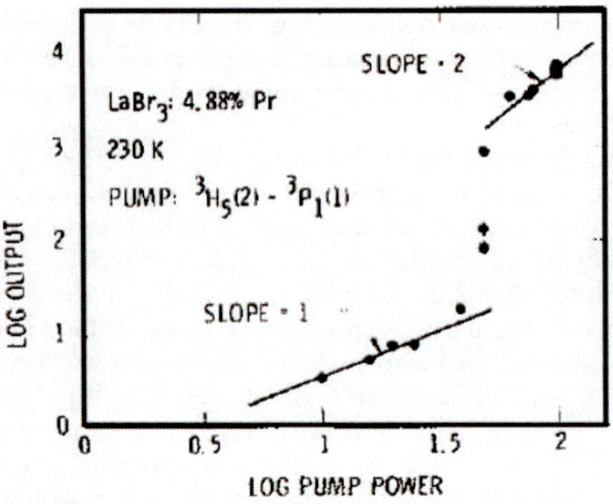

Fig. 2.25 Fluorescence output of Pr^{3+} versus incident pump power. The slope changes near to critical pump power density. It is reproduced with permission from AIP Publisher (Chivian et al. 1979)

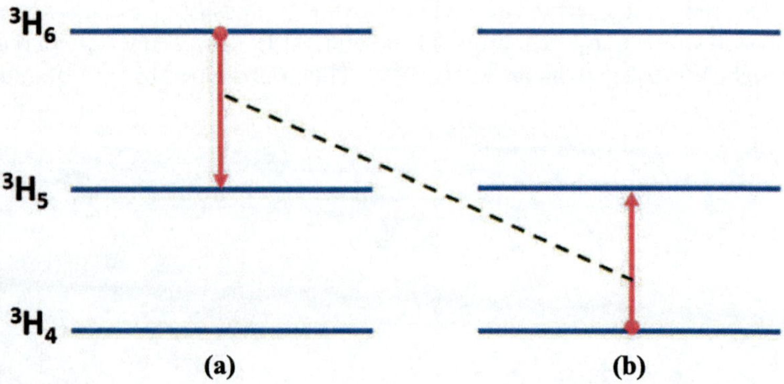

Fig. 2.26 Cross-relaxation among 2 Pr^{3+} ions **a** and **b**, which helps in population of excited electrons at $^3H_5$ level. It is redrawn and adapted from AIP Publisher (Chivian et al. 1979)

decay of $^3H_6$ to $^3H_5$, which is absorbed by another ion at ground state $^3H_4$ and goes to next higher energy state $^3H_5$. Such energy transfer is commonly known as cross-relaxation among two Pr^{3+} ions pair (Fig. 2.26) (Chivian et al. 1979). Above P_c, cross-relaxation process allows high population of electrons at $^3H_5$, which is further lifted to $^3P_1$ and then it emits light at red region ($^3H_6$, $^3F_2$). Again cross-relaxation starts. This produces a loop or cycle or continuous process if the input green laser power is more than P_c.

Photon avalanche (PA) process is an unconventional pumping mechanism, which leads to a strong up-converted emission. Figure 2.27 shows the schematic way of

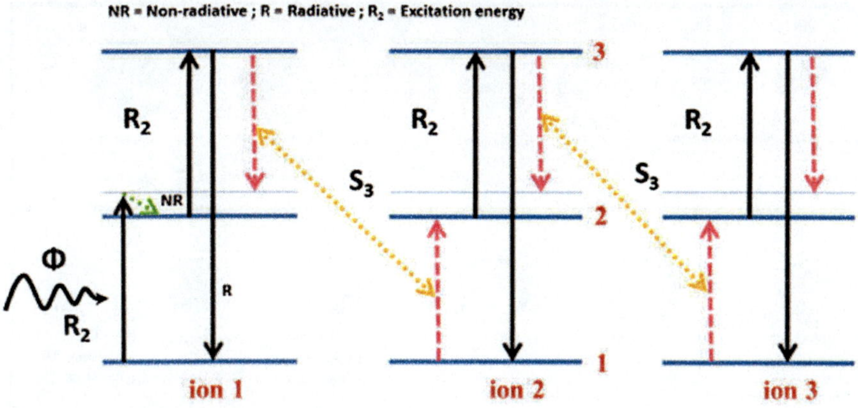

Fig. 2.27 Energy scheme for the simplest PA process in nearby 2 ions (1 and 2) and its energy transfer through cross-relaxation (S_3). It is redrawn and adapted from Ph. D. Thesis (Singh 2018)

PA process (Singh 2018). Pump photon is provided at metastable state (level 2) and excited electrons reach level 3. Some part of the excited electrons at level 3 comes back to level 2 and simultaneously electrons at the ground state go to the metastable level 2. This is known as cross-relaxation (CR) between 2 ions, which is shown by symbol (S_3). In this way, level 2 has a high population of excited electrons. Remaining part of the excited electrons at level 3 emits up-converted light and comes back to ground state level 1. Here, there is no ground state absorption (GSA) using a resonance energy between levels 1 and 2. Many theoretical models can be found in this review article and also many examples are provided (Joubert 1999).

2.3 Core@Shell Nanostructures

In terms of materials chemistry, the core–shell structure is a structure in which core particle is enclosed by a shell of different material/composition (Ningthoujam et al. 2022; Joshi et al. 2022a). Here, a core particle means a group of atoms that form either a perfect lattice or amorphous. Thickness may vary from 1 nm to a few microns. Different layers of shells can be created. It is not necessary to be nano-meter-size and can be micron size. Between core@shell, there will be an interface, which may be a secondary phase different from core and shell in terms of atomic ordering. Similarly, interface arises between layers of shells in a multilayer system. A core particle can have different shapes: sphere, cube, hexagon, prism, octahedron, disc, wire, rod, tube, sheet, star, tetra-pods, irregular, etc. If the size/thickness of at least one axis in x, y and z axis (3D) is nanometer, they can be considered in the nanomaterial category. Types of possible spherical core@shell structures are shown in Fig. 2.28 (Ningthoujam

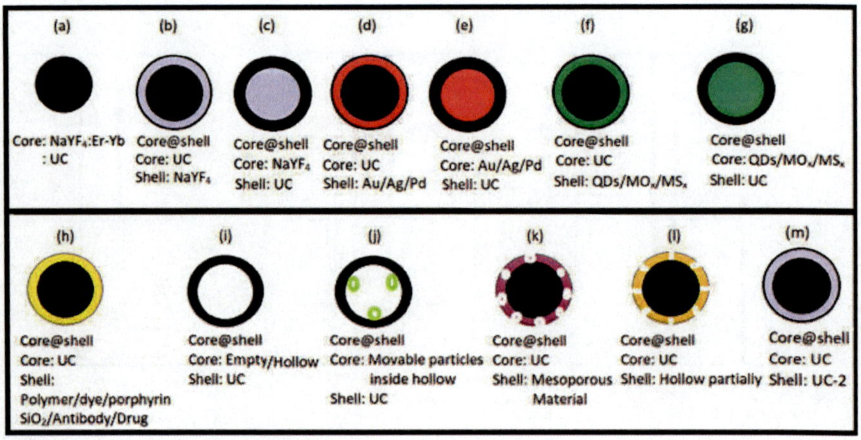

Fig. 2.28 Possible spherical Core@shell nanostructures. It is adapted from reference (Ningthoujam unpublished)

unpublished). In this study, up-converted nanomaterials are most considered. The symbol "UC" or "active UC" means that the nanoparticle is having active rare-earth ions or dopant ions, which are feasible for up-conversion process. Inactive means that there are no dopant ions. Similarly, other core@shell1@shell2, core@shell1@shell2@shell3 structures and so on can be possible (Figs. 2.29, 2.30, 2.31, 2.32, 2.33, 2.34 and 2.35) (Ningthoujam unpublished). Here, shell2 may be same as core material or a different material. In a similar way, shell3 may be same as shell1 or a different material. In nanosize range, it is easy to form core@shell formation kinetically. Nanoparticle is a group of atoms (100 to 10,000) that form a perfect lattice or amorphous phase in the nanosize range and at least one size is less than 50 nm in a particular axis (x, y or z axis). They are classified as 0D, 1D, 2D and 3D particles. Sometimes, many core particles can be enclosed by a particular shell. Such core particles can move on such hollow portions. If sufficient number of atoms is not available to make uniform shell thickness, it forms dispersed atoms over a particle. Interconnectivity between atoms of shell material may not be possible in such situation. In some cases, core particle is formed and same material is coated over to get different shapes and sizes (Oswald ripening route). So far, we assume that there is uniform thickness in the shell. To achieve such uniform thickness of shell, it requires special synthesis routes. In highly crystalline samples of similar crystal structure (hexagonal phase of β-$NaYF_4$, β-$NaGdF_4$ and doped samples), the layer-by-layer epitaxial growth can be performed. Sometimes, metal nanoparticles, SiO_2, mesoporous particles, polymers and any ligands can be used as core or shell (Johnson et al. 2012; Chaudhuri and Paria 2012; Mahdavi et al. 2020; Huseien 2023; Joshi et al. 2022b). In this system, it forms partial epitaxial growth or simply coating over the core. The formation of core@shell can be analysed by the following techniques (Ningthoujam et al. 2022).

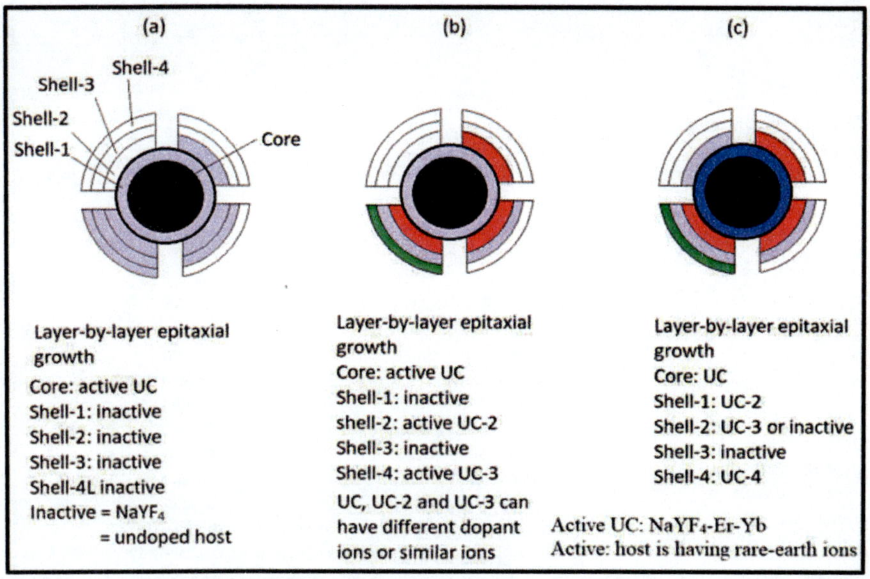

Fig. 2.29 Possible ways for layer-by-layer epitaxial growth of core and shell materials. Here, core is active up-conversing material. It is adapted from reference (Ningthoujam unpublished)

1. Electron energy loss spectroscopy (**EELS**), 2. Line scanning energy dispersive x-ray spectroscopy (**EDAX**), but its resolution is less than that of **EELS**, 3. High-angle annular dark-field imaging (**HAADF**), which is a STEM technique, 4. Secondary ion mass spectrometry (**SIMS**), and 5. The x-ray absorption near edge spectroscopy (**XANES**).

2.3.1 Why Core@Shell Is Required?

What happens when core is coated with shell? The coating of core with other materials has many changes in nanosize particles such as suppression of surface dangling bonds or decrease of surface defects, band-gap manipulation for a particular application, easy to surface functionalization, stability, easy to dispersion in water or oil depending on the presence of surface functional groups, controlled release of drug loaded in core, reduction in consumption of precious core material, targeted therapy, improvement in diagnosis and therapy due to proper coating, manipulation of energy transfer or migration or concentration quenching or electron transfer in luminescence, manipulation in magnetic properties, change in catalytic properties, change in dielectric properties, changes in magnetoresistance, and so on (Ningthoujam et al. 2022; Joshi et al. 2021, 2022a, b; Auzel 2004; Ningthoujam unpublished; Johnson et al. 2012; Chaudhuri and Paria 2012; Mahdavi et al. 2020; Huseien 2023). In d- or f- ions doped core, it is assumed that ions are distributed homogenously in the

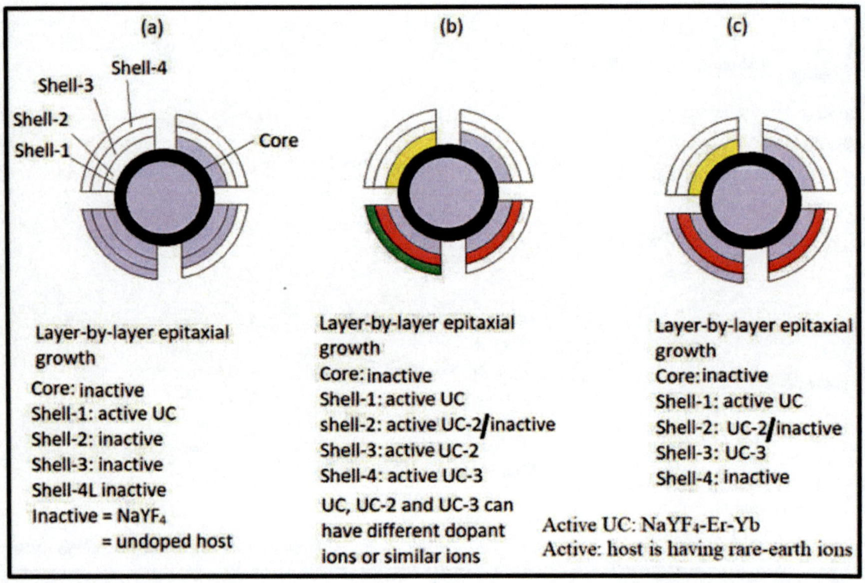

Fig. 2.30 Possible ways for layer-by-layer epitaxial growth of core and shell materials. Here, core is inactive material or undoped host material. It is adapted from reference (Ningthoujam unpublished)

core host lattice. This can be referred as homogenous system. When core is covered with undoped shell or doped shell and further coated with different shells, ions are no longer homogenous in the system (core@shell, core@shell1@shell2, core@shell1@shell2@shell3, etc.). This can be referred as heterogenous system in terms of distribution of d- or f- ions in the system. In homogenous system, concentration quenching due to high doping of d- or f- ions in luminescence is found. However, heterogenous system has unique properties, which are different from core properties. Even high concentrations of dopants can increase luminescence intensity. Mechanisms of energy transfer, energy migration, energy trapping, back energy transfer and cross-relaxation are different in core and core@shell systems.

Let us take homogenous system UCNPs (up-conversion nanoparticles) of $NaYF_4$: 2 at.% Er^{3+}-10 at.% Yb^{3+}, which is the best material so far in terms of efficiency at lower laser power excitation at 980 nm (Ningthoujam et al. 2022; Joshi et al. 2021, 2022a, b; Auzel 2004; Ningthoujam unpublished; Johnson et al. 2012; Chaudhuri and Paria 2012; Mahdavi et al. 2020; Huseien 2023). Now, let us take heterogenous system core@shell UCNPs nanostructures and possible nanostructures with similar atomic concentrations of Na, Y, Er, Yb and F in each system are $NaYF_4$:Er-Yb@$NaYF_4$, $NaYF_4$@$NaYF_4$:Er-Yb, $NaYF_4$:Er@$NaYF_4$-Yb and $NaYF_4$:Yb@$NaYF_4$-Er (Leary et al. 2023). Here, their luminescence intensities are different even when other conditions are fixed in terms of particle size, shape, excitation source and concentration. It means that energy transfer, migration and exchange interaction are different in those heterostructures. Even, if the thickness of shell at the same size

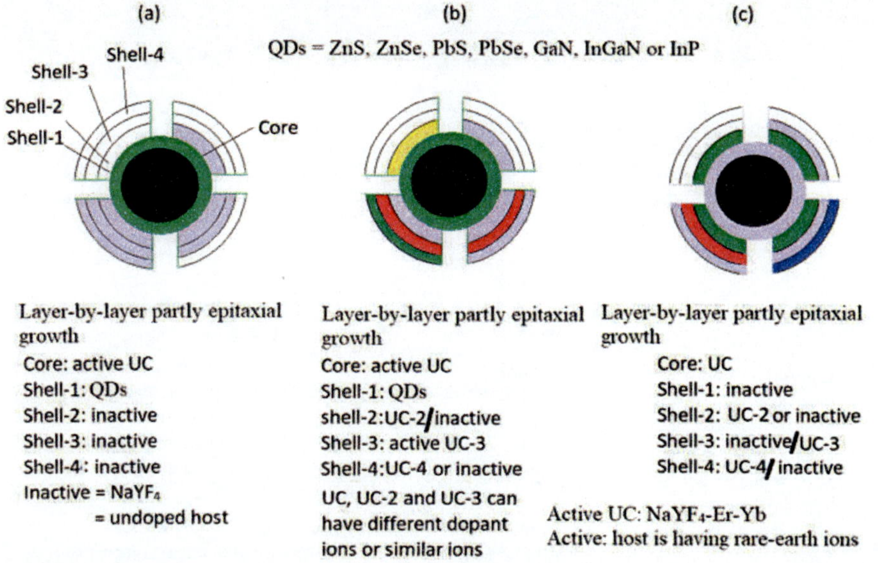

Fig. 2.31 Possible ways for layer-by-layer partly epitaxial growth of core and shell materials. Here, core is active up-conversing material. Here, QDs are associated. It is adapted from reference (Ningthoujam unpublished)

of core varies, their properties will change. If another shell (undoped, doped one) is introduced, a more complicated system arises in terms of study.

These core–shell heterostructures have been used in biomedical and pharmaceutical applications, electronics, catalysis, enhancing photoluminescence, tuning of a particular wavelength for a particular application, sensors, creating photonic crystals, data storage, conversion of one isomer to another isomer in light sensitive organic molecules, etc. In the biomedical field, these have been extensively used in imaging, cell labelling, bio-sensing, targeted drug delivery, controlled drug release through light or pH trigger, radio-isotopes based imaging or ac-frequency and therapy and tissue engineering applications, etc. (Ningthoujam et al. 2022; Joshi et al. 2021, 2022a, b; Auzel 2004; Ningthoujam unpublished; Johnson et al. 2012; Chaudhuri and Paria 2012; Mahdavi et al. 2020; Huseien 2023).

2.4 Classes of Core@Shell Nanostructures

Classes of core@shell nanostructures: Core@shell nanostructures can be classified depending on the types of material used in core and shell: This includes the following (Ningthoujam et al. 2022; Joshi et al. 2021, 2022a, b; Auzel 2004; Ningthoujam

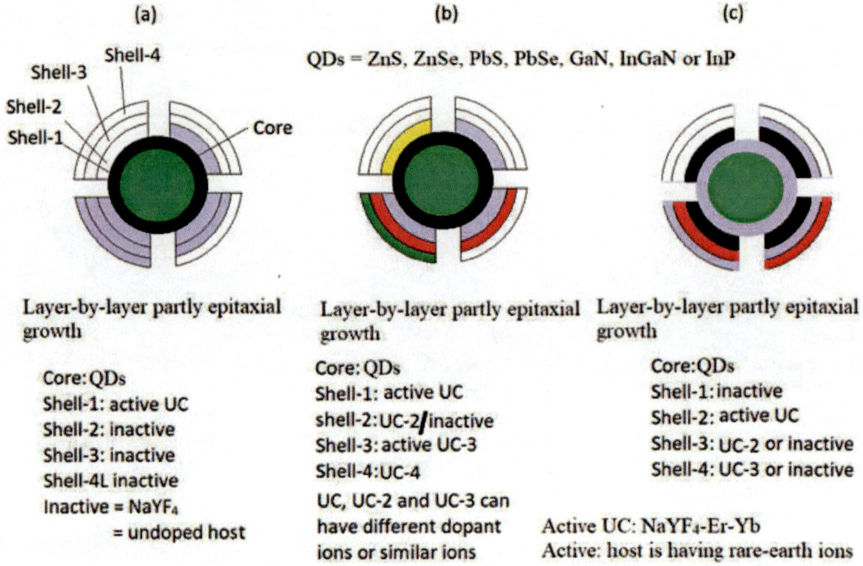

Fig. 2.32 Possible ways for layer-by-layer partly epitaxial growth of core and shell materials. Here, core is inactive such as QDs. It is adapted from reference (Ningthoujam unpublished)

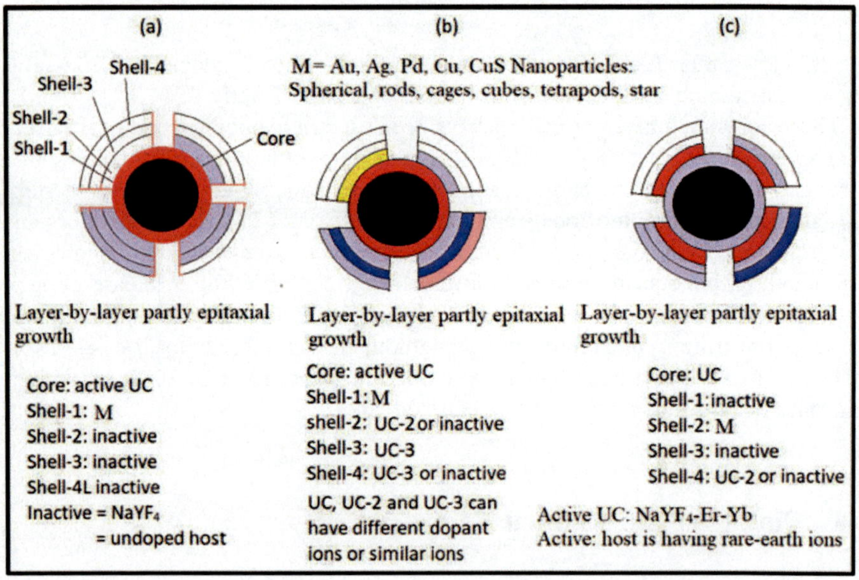

Fig. 2.33 Possible ways for layer-by-layer partly epitaxial growth of core and shell materials. Here, core is active up-conversing material. Here, metallic nanoparticles are associated. It is adapted from reference (Ningthoujam unpublished)

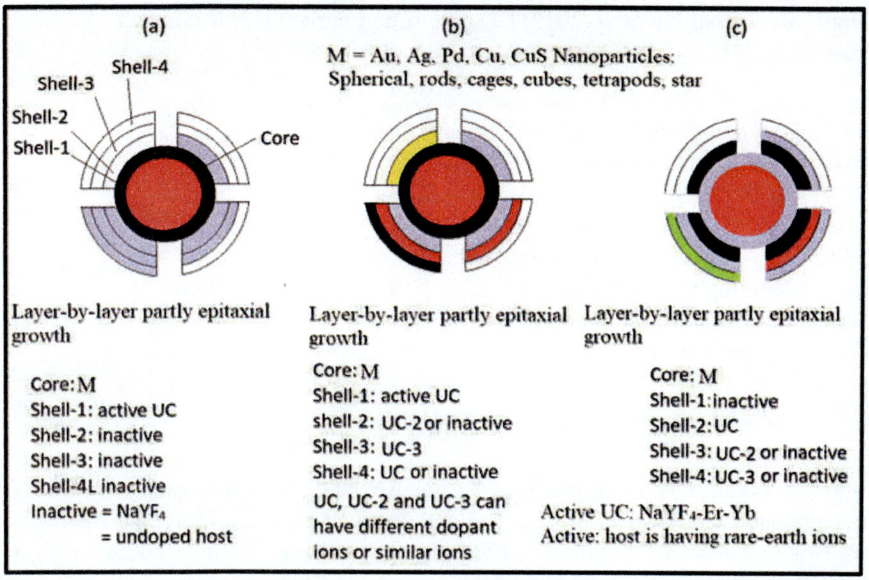

Fig. 2.34 Possible ways for layer-by-layer partly epitaxial growth of core and shell materials. Here, core is inactive such as metallic nanoparticles. It is adapted from reference (Ningthoujam unpublished)

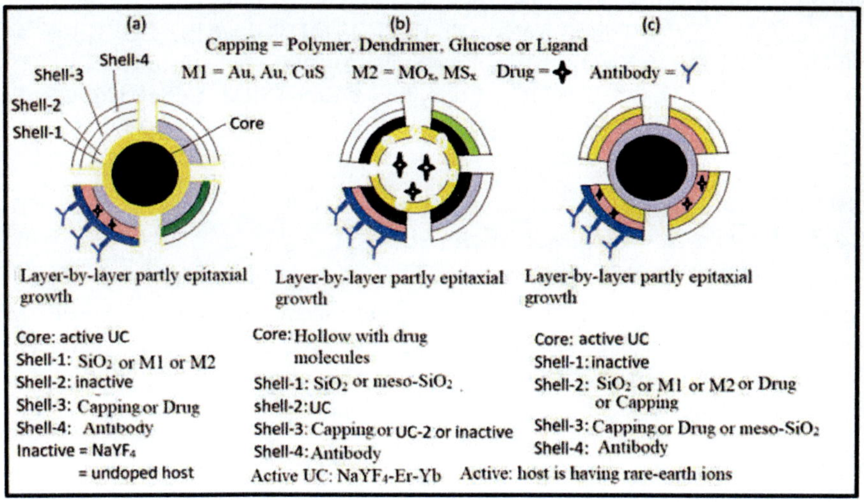

Fig. 2.35 Possible ways for layer-by-layer partly epitaxial growth of core and shell materials. Here, core is active UC or inactive such as metal nanoparticles or metal oxides. Drug and antibody molecules can be loaded. Even radio isotopes can be adsorbed or added. It is adapted from reference (Ningthoujam unpublished)

unpublished; Johnson et al. 2012; Chaudhuri and Paria 2012; Mahdavi et al. 2020; Huseien 2023):

2.4.1 Inorganic@inorganic materials
2.4.2 Inorganic@organic materials
2.4.3 Organic@inorganic materials
2.4.4 Organic@organic materials
2.4.5 Core@multilayers (shell1@shell2, shell1@shell2@shell3 and so on)
2.4.6 Liposome based formulation
2.4.7 Dendrimer based formation
2.4.8 Hollow core@shell materials

2.4.1 Inorganic@Inorganic materials

Inorganic core can be metal particles such as Au, Ni or different metals (Au, Pt, Pd) or semiconductors (C, CdS, CdSe, Fe_2O_3) or insulators SiO_2. Inorganic shells can be semiconductors or insulators (ZnS, LaF_3, β-$NaYF_4$:Yb^{3+}, Er^{3+}, YVO_4-Eu^{3+}). It can be happened with opposite trends.

Examples of inorganic@inorganic materials are Au@Ag, Au@SiO_2, Au@C, Au/CdS, Zn@ZnO, Fe_3O_4@Au, Au/Fe_2O_3, Fe_3O_4@SiO_2, PbS@PbSe, ZnS/CdSe, ZnS:Mn/ZnO, Gd_2O_3:Tm^{3+}@SiO_2, $NaYF_4$:Ln@$NaYF_4$, $NaYF_4$:Ln@$NaGdF_4$, $NaYF_4$:Ln@CaF_2, Fe_3O_4@ $NaYF_4$:Yb-Er, β-$NaYF_4$:Yb^{3+}-Er^{3+}@β-$NaYF_4$, β-$NaYF_4$:Yb-Tm@β-$NaYF_4$:Yb-Er, LaF_3@$Eu_{0.2}La_{0.8}F_3$, $LaPO_4$:Eu^{3+}@$LaPO_4$, YVO_4-Eu^{3+}@YVO_4, YVO_4:Ln@SiO_2, Au@SiO_2@Y_2O_3:Eu^{3+}, Fe_3O_4@SiO_2@Al_2O_3, Fe_3O_4@SiO_2@TiO_2, $CdSe_xTe_{1-x}$@ZnS@SiO2, CdSe@CdS@ZnS, γ-Fe_2O_3@$NaYF_4$:Yb-Er, etc.

2.4.2 Inorganic@Organic Materials

Core is inorganic materials such as metals, semiconductors or insulators. Metals are Au, Ag and Fe. Semiconductors are Fe_3O_4, Fe_2O_3, ZrO_2, PbSe, PbS. Insulators are SiO_2, $NaYF_4$:Ln, CaF_2:Ln. Shell material is organic material. Examples are PEG, PMMA, block polymers, etc.

2.4.3 Organic@Inorganic Materials

Here, organic core is made of organic polymer. A few examples of organic polymer are mentioned here: dextrose, polystyrene, polyurethane, poly(ethylene oxide), poly(vinyl benzyl chloride), poly(vinyl pyrrolidone), surfactants, and different copolymers, such as acrylonitrile–butadiene–styrene, poly- (styreneacrylic acid), and styrenemethyl methacrylate. Inorganic shells will be metals (Au, Ag), semiconductors (c-dots, metal oxide TiO_2, Fe_3O_4, metal chalcogenide ZnS) or insulators (SiO_2).

2.4.4 Organic@Organic Materials

Core particle is made of organic polymer. A few examples of organic polymer are mentioned here: polystyrene (PS), polyphenylene, and poly(D,L-lactic-co-glycolide) (PLGA), poly(N-isopropylacrylamide) (variously abbreviated polymers: PNIPA, PNIPAAm, NIPA, PNIPAA or PNIPAm). Shell particle is organic polymer, such as 1,2-distearoyl-sn-glycero-3- phosphoethanolamine-N-carboxy(poly(ethylene glycol))2000 (DSPE-PEG-COOH), drug, polyethylene oxide (PEO), polyethylene glycol (PEG).

2.4.5 Core@Multilayers (Shell1@Shell2, Shell1@Shell2@Shell3 and So On)

Core particle is made of inorganic or organic material. Shell1 with another material (inorganic or organic). Shell2 will be another material. Shell3 will be another material. They shell layer makes multilayers of different materials. Examples are CdSe@CdS@ZnS, CdSe@ZnSe@ZnS, Au@SiO_2@Y_2O_3:Eu^{3+}, LaF_3:Eu@LnF_3:Eu^{3+}@$La(NO_3)_3$, CdSe@CdTe@ZnSe, CdSe@CdS@$Zn_{0.5}Cd_{0.5}S$@ZnS, $InAs_xP_{1-x}$@InP@ZnSe, InAs@CdSe@ZnSe, CdSe@HgTe@CdTe, CdS@HgS@CdS, CdSe@CdS@ZnTe, CdS@CdSe@CdS, CdSe@ZnS@CdSe, ZnS@CdS@ZnS, CdS@CdSe@CdS, CdSe@ZnS@CdTe, Fe_3O_4@PGMA@PS (here PGMA = poly(glycidyl methacrylate), PS = polystyrene), Fe_3O_4/PEO-PPO-PEO block polymer (known as pluronic P123, which consists of a symmetric triblock copolymer comprising poly(ethylene oxide) (PEO) and poly(propylene oxide) (PPO) in an alternating linear fashion, PEO-PPO-PEO and it has a chemical formula $HO(CH_2CH_2O)_{20}(CH_2CH(CH_3)O)_{70}(CH_2CH_2O)_{20}H$), γ-Fe_2O_3@PEI + PEO-PGA (PEI = poly-(ethylene amine), PEO-PGA = poly-(ethylene oxide)-poly-(glutamic acid)), iron oxide@SiO_2 composite@PS, $CoFe_2O_4$@DTPA-CS (DTPA = Diethylenetriamine pentaacetate, CS = chitosan), Sb_2O_3@PMMA@PVC (polymethyl methacrylate (PMMA) and polyvinyl chloride (PVC) polymer blend, PEG@Si@ Al composite oxide, PS@TiO_2@SiO_2, PS@styrene and MPS hybrid copolymer, etc.

2.4.6 Liposome Based Formulation

Liposome is a spherical vesicle in which the core is having aqueous medium and shell is having lipid bilayers (hydrophobic). Lipid is generally composed of phospholipids and cholesterol. If one lipid bilayer exists, this is known as uni-lamellar (diameter: 30–100 nm or 100–300 nm). If two or more lipid bilayers exist, this is known as multi-lamellar (diameter = 1–5 μm). Between lipid bilayers, aqueous medium exists. Water soluble/dispersible particles or molecules can be trapped into aqueous core, whereas oil soluble/dispersible particles or molecules can be trapped into lipid bilayers. Au or Fe_3O_4 nanoparticles dispersible in water can be prepared in the core of liposome. Oil soluble Au or Fe_3O_4 nanoparticles can be entrapped in a lipid bilayer (shell). Generally, targeted drug delivery and high accumulation of drugs in cells use the concept of liposome-based system. Luminescent nanomaterial can be trapped into aqueous core and oil-soluble lipid layers depending on the types of nanomaterials.

2.4.7 Dendrimer Based Formation

Dendrimer has highly ordered branched polymeric molecules around the core. Generally, core is hydrophobic whereas outer branched polymeric molecules are hydrophilic. However, core can be modified in order to get functional group such as NH_2 or COOH. In this way, nanoparticles (Au, Fe_3O_4, ZnS, ZnO in 2–50 nm) can be entrapped into the core (interaction may be ionic or covalent type). Such branched polymeric molecules are soluble in aqueous medium. Nanoparticles encapsulated dendrimers can be used in drug-delivery systems and also agglomeration among nanoparticles is very less.

2.4.8 Hollow Core@Shell Materials

In case, if core@shell nanoparticles are brought to a particular medium or gas medium or laser medium, either core or shell can be removed. Using SIMS experiment or high-power laser, thickness of shell can be decreased. Finally, the shell can be removed. On the other hand, if a particular solvent, which can be capable for soluble only in core material (e.g., Co or Fe core particles can be dissolved into Co^{2+} or Fe^{2+}/Fe^{3+} in pH < 4), but not affect the shell, hollow core with shell can be formed. Here core is scarified and shell remains. If metal ions are added to such hollow core@shell in aqueous medium, metal ions enter into the hollow core. When a reducing agent is added, reduction of metal ions to metal nanoparticles can occur. Similarly, if oxidizing agent is added, metal ions can be converted to metal oxide nanoparticles. Such nanoparticles formed can form agglomerated particles inside the core. If the number of nanoparticles is less, individual particles can exist and can move in the core. Even, luminescent nanoparticles can be generated in the hollow core region.

2.5 Rate of Electronic Decay, Lifetime, Quantum Yield

During electronic transition from ground state to excited state, excited electrons come back to ground state if excitation source is removed. The excitation rate is equal to decay rate. If N_0 is the number of electrons in the ground state and its excited rate is K_A, the number of excited electrons will be N_1. Decay rate of the excited electrons will be determined by radiative rate (K_R) and non-radiative rate (K_{NR}) (Ningthoujam et al. 2012; Curie 1963; Blasse and Grabmaier 1994; Lakowicz 2006). At a particular time (t), the number of excited electrons can be determined by

$$\mathrm{d}N_1/dt = K_A N_0 - (K_R N_1 + K_{NR} N_1) \tag{2.4}$$

When photon of constant intensity is excited, it can achieve a steady state of equilibrium.

Here, $dN_1/dt = 0$.

It implies that

$$N_1 = K_A N_0/(K_R + K_{NR}) \tag{2.5}$$

.

And amount of decay at a particular time (t) is given by

$$N_1(t) = N_1 exp(-K_A t) \tag{2.6}$$

where $K_A = K_R + K_{NR}$.

$$N_1(t)/N_1 = exp(-K_A t) \tag{2.7}$$

When $K_{NR} = 0, N_1 = K_A N_0/K_R$. In this, lifetime of excited electrons is considered as natural lifetime or intrinsic lifetime (τ_N), which is defined as:

$$\tau_N = 1/K_R \tag{2.8}$$

And $K_A = K_R$.

Intrinsic decay takes place due to collisional interactions among excited electrons.

$$N_1(t)/N_1 = exp(-K_R t_N) \tag{2.9}$$

$$N_1(t)/N_1 = exp(-1) = 0.36 \tag{2.10}$$

The nature lifetime (τ_N) is equal to the time required to decay excited electron population to 36%.

The experiment lifetime (τ_{exp}) is generally less than the natural lifetime (τ_N).

$$\tau_{exp} = 1/K_A = 1/(K_R + K_{NR}) \tag{2.11}$$

The fluorescence quantum yield (QY) is the ratio between the fluorescence intensity and light absorb.

$$\text{QY} = \text{Photon emitted/Photon absorb} \tag{2.12}$$

$$\text{QY} = K_R/(K_R + K_{NR}) \tag{2.13}$$

$$\text{QY} = \tau_{exp}/\tau_N \tag{2.14}$$

Here, the source of non-radiative decay is due to the presence of surrounding heat sink or quencher. Due to the transfer of energy from excited electrons, it gives rise to heat to surrounding and thereby multi-phonon relaxation may take place. If K_{NR} <<< K_R, luminescence intensity is very high. Luminescence intensity of core activator increases significantly after core–shell formation (shell: inactive or undoped host) due to the increase of distance from core activator to quencher (surface dangling bonds, ligands, solvent, etc.; e.g., core: $NaYF_4$:Er-Yb and core@shell: $NaYF_4$:Er-Yb@$NaYF_4$; and YVO_4:Eu and YVO_4:Eu@YVO_4) (Ningthoujam et al. 2009a, 2022).

2.6 Type of Excitation for Electronic Transitions

There are two ways of excitation: resonance type and non-resonance type (Auzel 2004; Singh 2018). In the resonance type excitation, a large number of electrons go to the excited state having a stable level. In case of non-resonance type, a few electrons go to the excited state having a stable level. Most energy got losses due to the difference between excitation energy and stable energy level and this is provided by multi-phonon relaxation.

However, if high-energy flux (say a few kW/cm^2) is provided, situation is different in resonance or non-resonance type process. Sometimes cascades or avalanche phenomenon occurs. Even, continuous or pulse mode excitation changes the process of energy transfer at excited levels.

2.7 Conclusions

Many examples of photo-luminescent materials are provided and their properties such as down-conversion and up-conversion processes are discussed. Many organic compounds, inorganic compounds and organic dyes show the down-conversion process. Many materials showing quantum-cutting process are provided. Different metal complexes, d and f ions doped inorganic compounds show up-conversion process. Possible core@shell nanostructures are mentioned. This helps in the fabrication of devices, light emitting diodes, drug loading, solar cells, energy harvesting, proximity systems, etc. Shape, size and hollowness engineering in nanostructure are mentioned.

References

Ahmed, S.A.: Structural, optical, and magnetic properties of Mn-doped ZnO samples. Res. Phys. **7**, 604–610 (2017)

Auzel, F.: Upconversion and anti-stokes processes with f and d Ions in Solids. Chem. Rev. **104**, 139–173 (2004)

Blasse, G., Grabmaier, B.C.: Luminescent Materials. Springer, Berlin (1994)

Boens, N., Leen, V., Dehaen, W.: Fluorescent indicators based on BODIPY. Chem. Soc. Rev. **41**, 1130–1172 (2012)

Borse, S., Rafique, R., Murthy, Z.V.P., Park, T.J., Kailasa, S.K.: Applications of upconversion nanoparticles in analytical and biomedical sciences: a review. Analyst **147**, 3155–3179 (2022)

Brabec, C.J., Dyakonov, V., Parisi, J., Sariciftci, N.S.: Organic Photovoltaics: Concepts and Realization. Springer-Verlag Berlin Heidelberg (2003)

Chaganti, L.K., Venkatakrishnan, N., Bose, K.: An efficient method for FITC labelling of proteins using tandem affinity purification. Biosci. Rep. **38**, BSR20181764 (2018)

Chaudhuri, R.G., Paria, S.: Core/shell nanoparticles: classes, properties, synthesis mechanisms, characterization, and applications. Chem. Rev. **112**, 2373–2433 (2012)

Chen, G., Qiu, H., Prasad, P.N., Chen, X.: Upconversion nanoparticles: design, nanochemistry, and applications in theranostics. Chem. Rev. **114**, 5161–5214 (2004)

Chen, R., Hong, Z.F., Zhao, Y.R., Zheng, H., Li, G.J., Zhang, Q.C., Kong, X.J., Long, L.S., Zheng, L.S.: Ligand-dependent luminescence properties of lanthanide-titanium oxo clusters. Inorg. Chem. **58**, 15008–15012 (2019)

Chen, B., Li, D., Wang, F.: InP quantum dots: synthesis and lighting applications. Small **16**, 2002454 (2020)

Chivian, J.S., Case, W.E., Eden, D.D.: The photon avalanche: a new phenomenon in Pr^{3+}-based infrared quantum counters. Appl. Phys. Lett. **35**, 124 (1979)

Cravcenco, A., Ye, C., Gräfenstein, J., Börjesson, K.: Interplay between Förster and Dexter energy transfer rates in isomeric donor–bridge–acceptor systems. J. Phys. Chem. A **124**, 7219–7227 (2020)

Curie, D.: Luminescence in Crystals, 1st edn. John Wiley & Sons Inc., London (1963)

da Silva, J.F., da Silva, R.F., Santos, E.P., Maia, L.J.Q., Moura, A.L.: Photon-avalanche-like upconversion in $NdAl_3(BO_3)_4$ nanoparticles excited at 1064 nm. Appl. Phys. Lett. **117**, 151102 (2020)

Dexter, D.L.: Possibility of luminescent quantum yields greater than unity. Phys. Rev. **108**, 630 (1957)

Dutta, D.P., Ningthoujam, R.S., Tyagi, A.K.: Luminescence properties of Sm^{3+} doped YPO_4: effect of solvent, heat-treatment, Ca^{2+}/W^{6+}-co-doping and its hyperthermia application. AIP-Adv. **2**, 042184 (2012)

Gajbhiye, N.S., Ningthoujam, R.S., Ahmed, A., Panda, D.K., Umre, S.S., Sharma, S.J.: Re-dispersible Li^{+} and Eu^{3+} co-doped CdS nanoparticles: luminescence studies. Pramana J. Phys. **70**, 313 (2008)

Ganapathi, S.K., Kaur, M., Samanta, S., Datta, N., Singh, A., Ningthoujam, R.S., Gadkari, S.C., Debnath, A.K.: Influence of MgPc modification on NO_2 sensing characteristics of RGTO grown SnO_2 thin films. Mater. Sci. Semicond. Process. **152**, 107104 (2022)

Ghosh, M., Ningthoujam, R.S., Vatsa, R.K., Das, D., Nataraju, V., Gadkari, S.C., Gupta, S.K., Bahadur, D.: Role of ambient air on photoluminescence and electrical conductivity of assembly of ZnO nanoparticles. J. Appl. Phys. **110**, 054309 (2011)

Haider, A.J., Jameel, Z.N., Al-Hussaini, I.H.M.: Review on: titanium dioxide applications. Energy Procedia **157**, 17–29 (2019)

Hasegawa, M., Ohmagari, H., Tanaka, H., Machida, K.: Luminescence of lanthanide complexes: from fundamental to prospective approaches related to water- and molecular-stimuli. J. Photochem. Photobiol. C **50**, 100484 (2022)

Huang, X.Y., Zhang, Q.Y.: Near-infrared quantum cutting via cooperative energy transfer in Gd_2O_3:Bi^{3+}, Yb^{3+} phosphors. J. Appl. Phys. **107**, 063505 (2010)

Huseien, G.H.: Potential applications of core-shell nanoparticles in construction industry revisited. Appl. Nano **4**, 75–114 (2023)

Johnson, N.J.J., Korinek, A., Dong, C., van Veggel, F.C.J.M.: Self-focusing by ostwald ripening: a strategy for layer-by-layer epitaxial growth on upconverting nanocrystals. J. Am. Chem. Soc. **134**, 11068–11071 (2012)

Joshi, R., Shelar, S.B., Srivastava, M., Singh, B.P., Goel, L., Ningthoujam, R.S.: Development of core@Shell γ-Fe_2O_3@Mn_xO_y@SiO_2 nanoparticles for hyperthermia, targeting, and imaging applications. ACS Appl. Bio Mater. **5**, 5386–5393 (2022a)

Joshi, R., Patra, S., Srivastava, M., Singh, B.P., Chakraborty, A., Shelar, S.B., Chakravarty, R., Ningthoujam, R.S.: Mesoporous $NaGdF_4$/Ho–Yb@m-SiO_2 upconversion nanophosphors as a potent theranostic probe. ACS Appl. Nano Mater. **5**, 12962–12971 (2022b)

Joshi, R., Perala, R.S., Shelar, S.B., Ballal, A., Singh, B.P., Ningthoujam, R.S.: Super bright red upconversion in $NaErF_4$:0.5%Tm@$NaYF_4$:20%Yb nanoparticles for anti-counterfeit and bio-imaging applications. ACS Appl. Mater. Interf. **13**, 3481–3490 (2021)

Joubert, M.F.: Photon avalanche upconversion in rare earth laser materials. Opt. Mater. **11**, 181–203 (1999)

Karbowiak, M., Mech, A., Drożdżyński, J., Edelstein, N.M.: Crystal-field analysis, upconversion, and excited-state dynamics for (U^{4+}, U^{3+}): Ba_2YCl_7 single crystals. Phys. Rev. B **67**, 195108 (2003)

Kemp, W.: Organic Spectroscopy. Macmillan, Pennsylvania State University (1975)

Kozlov, D.V., Castellano, F.N.: Anti-stokes delayed fluorescence from metal–organic bichromophores. Chem. Commun. 2860–2861 (2004)

Kumari, R., Sunil, D., Ningthoujam, R.S.: Naphthalimides in fluorescent imaging of tumor hypoxia-an up-to-date review. Bioorg. Chem. **88**, 102979 (2019a)

Kumari, R., Vasumathy, R., Sunil, D., Ningthoujam, R.S., Pandey, B.N., Kulkarni, S.D., Varadavenkatesan, T., Venkatachalam, G.A., Anil, N.V.K.: Nitronaphthalimide probe for fluorescence imaging of hypoxia in cancer cells. J. Fluoresc. **31**, 1665–1673 (2021)

Kumari, R., Sunil, D., Ningthoujam, R.S., Pandey, B.N., Kulkarni, S.D., Varadavenkatesan, T., Venkatachalam, G.: Dinitro derivative of naphthalimide as a fluorescent probe for tumor hypoxia imaging. Polycyclic Aromat. Compd. **43**, 54–63 (2023)

Kumari, R., Sunil, D., Ningthoujam, R.S., Kumar, N.V.A.: Azodyes as markers for tumor hypoxia imaging and therapy: an up-to-date review. Chemico-Biol. Interact. **307**, 91–104 (2019)

Lahiri, D., Ningthoujam, R.S., Bhattacharyya, D., Sharma, S.M.: Structural understanding of the spectral characteristics of SnO_2:Eu:Y_2O_3, using extended x-ray absorption fine structure. J. Appl. Phys. **107**, 054316 (2010)

Lakowicz, J.R.: Principles of Fluorescence Spectroscopy, 3rd Edn. Springer Science, LLC, Singapore (2006)

Leary, D.C., Zhang, Y., Rodriguez, J.G., Akhmedov, N.G., Petersen, J.L., Dolinar, B.S., Milsmann, C.: Organometallic intermediates in the synthesis of photoluminescent zirconium and hafnium complexes with pyridine dipyrrolide ligands. Organometallics **42**, 1220–1231 (2023)

Lee, C., Xu, E.Z., Liu, Y., Teitelboim, A., Yao, K., Fernandez-Bravo, A., Kotulska, A.M., Nam, S.H., Suh, Y.D., Bednarkiewicz, A., Cohen, B.E., Chan, E.M., Schuck, P.J.: Giant nonlinear optical responses from photon-avalanching nanoparticles. Nature **589**, 230 (2021)

Levy, E.S., Tajon, C.A., Bischof, T.S., Iafrati, J., Fernandez-Bravo, A., Garfield, D.J., Chamanzar, M., Maharbiz, M.M., Sohal, V.S., Schuck, P.J., Cohen, B.E.: Energy-looping nanoparticles: harnessing excited-state absorption for deep-tissue imaging. ACS Nano **10**(9), 8423 (2016)

Li, P., Li, H.: Recent progress in the lanthanide-complexes based luminescent hybrid materials. Coord. Chem. Rev. **441**, 213988 (2021)

Li, Y., Wang, T., Ren, W., Han, J., Yin, Z., Qiu, J., Yang, Z., Song, Z.: BiOCl: Er nanosheets with tunable thickness for photon avalanche phosphors. ACS Appl. Nano Mater. **2**, 7652–7660 (2019)

Loitongbam, R.S., Singh, N.S., Singh, W.R., Ningthoujam, R.S.: Observation of exceptional strong emission transitions ^5D_j (j = 1–3) to ^7F_j (j = 1–3): multicolor from single Eu^{3+} ion doped La_2O_3 nanoparticles. J. Lumin. **134**, 14–23 (2013)

Lokesha, H.S., Mohanty, P., Prinsloo, A.R.E., Sheppard, C.J.: Impact of Cr doping on the structure, optical and magnetic properties of nanocrystalline ZnO particles. J. Alloy. Compd. **960**, 170815 (2023)

Longo, V.M., Cavalcante, L.S., Paris, E.C., Sczancoski, J.C., Pizani, P.S., Li, M.S., Andres, J., Longo, E., Varela, J.A.: Hierarchical assembly of $CaMoO_4$ nano-octahedrons and their photoluminescence properties. J. Phys. Chem. C **115**, 5207–5219 (2011)

Luwang, M.N., Ningthoujam, R.S., Singh, N.S., Tewari, R., Srivastava, S.K., Vatsa, R.K.: Surface chemistry of surfactant AOT-stabilized SnO_2 nanoparticles and effect of temperature. J. Colloids Interface Sci. **349**, 27 (2010)

Mahdavi, Z., Rezvani, H., Moraveji, M.K.: Core–shell nanoparticles used in drug delivery-microfluidics: a review. RSC Adv. **10**, 18280–18295 (2020)

Mamiyev, Z., Balayeva, N.O.: PbS nanostructures: a review of recent advances. Mater. Today Sustain. **21**, 100305 (2023)

Meetei, S.D., Singh, S.D., Singh, N.S., Sudarsan, V., Ningthoujam, R.S., Tyagi, M., Gadkari, S.C., Tewari, R., Vatsa, R.K.: Crystal structure and photoluminescence correlations in white emitting nanocrystalline ZrO_2:Eu^{3+} phosphor: effect of doping and annealing. J. Lumin. **132**, 537–544 (2012)

Mir, W.J., Sheikh, T., Arfin, H., Xia, Z., Nag, A.: Lanthanide doping in metal halide perovskite nanocrystals: spectral shifting, quantum cutting and optoelectronic applications. NPG Asia Mater. **12**, 9 (2020)

Ningthoujam, R.S.: Generation of exciton in two semiconductors interface: SnO_2:Eu-Y_2O_3. Chem. Phys. Lett. **497**, 208 (2010)

Ningthoujam, R.S.: Enhancement of luminescence by rare earth ions doping in semiconductor host. In: Rai, S.B., Dwivedi, Y. (eds.) Synthesis, Characterization and Applications of Multifunctional Materials Material Science and Technologies, Chapter 7, pp. 145–182. Nova Science Publishers Inc., USA (2012)

Ningthoujam, R.S., Kulshreshtha, S.K.: Nanocrystalline SnO_2 from thermal decomposition of tin citrate crystal: luminescence and Raman studies. Mater. Res. Bull. **44**, 57 (2009)

Ningthoujam, R.S., Lahiri, D., Sudarsan, V., Poswal, H.K., Kulshreshtha, S.K., Sharma, S.M., Bhushan, B., Sastry, M.D.: Nature of V^{n+} ions in SnO_2: EPR and photoluminescence studies. Mater. Res. Bull. **42**, 1293 (2007a)

Ningthoujam, R.S., Sudarsan, V., Godbole, S.V., Kienle, L., Kulshreshtha, S.K., Tyagi, A.K.: SnO_2:Eu^{3+} nanoparticles dispersed in TiO_2 matrix: Improved energy transfer between semiconductor host and Eu^{3+} ions for the low temperature synthesized samples. Appl. Phys. Lett. **90**, 173113 (2007b)

Ningthoujam, R.S., Sudarsan, V., Kulshreshtha, S.K.: SnO_2: Eu nanoparticles dispersed in silica: a low temperature synthesis and photoluminescence study. J. Lumin. **127**, 747 (2007c)

Ningthoujam, R.S., Sudarsan, V., Vinu, A., Srinivasu, P., Ariga, K., Kulshreshtha, S.K., Tyagi, A.K.: Luminescence properties of SnO_2 nanoparticles dispersed in Eu^{3+} doped SiO_2. J. Nanosci. Nanotech. **8**, 1489 (2008a)

Ningthoujam, R.S., Gajbhiye, N.S., Ahmed, A., Umre, S.S., Sharma, S.J.: Re-dispersible Li^+ and Eu^{3+} co-doped nanocrystalline ZnO: luminescence and EPR studies. J. Nanosci. Nanotech. **8**, 3059 (2008b)

Ningthoujam, R.S., Singh, L.R., Sudarsan, V., Singh, S.D.: Energy transfer process and optimum emission studies in luminescence of core-shell nanoparticles: YVO_4:Eu-YVO_4 and surface state analysis. J. Alloys Comp. **484**, 782 (2009a)

Ningthoujam, R.S., Vatsa, R.K., Vinu, A., Ariga, K., Tyagi, A.K.: Room temperature exciton formation in SnO_2 nanocrystals in SiO_2: Eu matrix: quantum dot system, heat-treatment effect. J. Nanosci. Nanotech. **9**, 2634 (2009b)

Ningthoujam, R.S., Sudarsan, V., Vatsa, R.K., Kadam, R.M., Jagannath, G.A.: Photoluminescence studies on Eu doped TiO_2 nanoparticles. J. Alloys Comp. **486**, 864 (2009c)

Ningthoujam, R.S., Gautam, A., Padma, N.: Oleylamine as reducing agent in syntheses of magic-size clusters and monodisperse quantum dots: optical and photoconductivity studies. Phys. Chem. Chem. Phys. **19**, 2294–2303 (2017)

Ningthoujam, R.S.: Finding confined water in hexagonal phase of $Bi_{0.05}Eu_{0.05}Y_{0.90}PO_4.xH_2O$ and its impact for identifying the location of luminescence quencher. Pramana J. Phys. **80**, 1055 (2013)

Ningthoujam, R.S., Joshi, R., Srivastava, M.: Frequency upconversion in core@shell nanoparticles. In: Rai, V.K. (ed) Upconverting Nanoparticles: From Fundamentals to Applications. Wiley (2022)

Ningthoujam RS, Core-shell nanostructures, proximity effect and its applications (unpublished)

Okram, R., Yaiphaba, N., Ningthoujam, R.S., Singh, N.R.: Is higher ratio of monoclinic to tetragonal in LaVO4 a better luminescence host? Redispersion and polymer film formation. Inorg. Chem. **53**, 7204–7213 (2014)

Omelon, S., Georgiou, J., Habraken, W.: A cautionary (spectral) tail: red-shifted fluorescence by DAPI–DAPI interactions. Biochem. Soc. Trans. **44**, 46–49 (2016)

Pan, G.H., Wu, H., He, S., Zhang, L., Hao, Z., Zhanga, X., Zhang, J.: Dye-embedded YAG:Ce^{3+}@ SiO_2 composite phosphors toward warm wLEDs through radiative energy transfer: preparation, characterization and luminescence properties. Nanoscale **10**, 22237–22251 (2018)

Parchur, A.K., Ningthoujam, R.S.: Preparation, microstructure and crystal structure studies of Li^+ co-doped YPO_4:Eu^{3+}. RSC Adv. **2**, 10854–10858 (2012a)

Parchur, A.K., Ningthoujam, R.S.: Behaviour of electric and magnetic dipole transitions of Eu^{3+}, ${}^5D_0 \rightarrow {}^7F_0$ and Eu-O charge transfer band in Li^+ co-doped YPO_4:Eu^{3+}. RSC Adv. **2**, 10859–10868 (2012b)

Parchur, A.K., Prasad, A.I., Rai, S.B., Ningthoujam, R.S.: Improvement of blue, white and NIR emissions in YPO_4:Dy^{3+} nanoparticles on co-doping of Li^+ ions. Dalton Trans. Commun. **41**, 13810 (2012a)

Parchur, A.K., Prasad, A.I., Rai, S.B., Tewari, R., Sahu, R.K., Okram, G.S., Singh, R.A., Ningthoujam, R.S.: Observation of intermediate bands in Eu^{3+} doped YPO_4 host: Li^+ ion effect and blue to pink light emitter. AIP Adv. **2**, 032119 (2012b)

Parchur, A.K., Ansari, A.A., Singh, B.P., Hasan, T.N., Syed, N.A., Rai, S.B., Ningthoujam, R.S.: Enhanced luminescence of $CaMoO_4$: Eu by core@shell formation and its hyperthermia study after hybrid formation with F_e3_O4: cytotoxicity assessing on human liver cancer cells and mesanchymal stem cells. Integr. Biol. **6**, 53–64 (2014)

Peng, Y., Xu, L., Peng, J., Wang, T., Wang, Q., Li, Y., Yin, Z., Han, J., Qiu, J., Yang, Z., Song, Z.: Enhanced upconversion luminescence in layered $Bi_2GdO_4Cl:Yb^{3+}/Er^{3+}$ by reducing OVs and its application in photocatalysis. J. Lumin. **252**, 119269 (2022)

Phaomei, G., Ningthoujam, R.S., Singh, W.R., Singh, N.S., Luwang, M.N., Tewari, R., Vatsa, R.K.: Low temperature synthesis and luminescence properties of re-dispersible Eu^{3+} doped $LaPO_4$ nanorods by ethylene glycol route. Opt. Mater. **32**, 616 (2010)

Phaomei, G., Ningthoujam, R.S., Singh, W.R., Loitongbam, R.S., Singh, N.S., Rath, A., Juluri, R.R., Vatsa, R.K.: Luminescence switching behaviour through redox reaction in Ce^{3+} co-doped $LaPO_4:Tb^{3+}$ nanorods: re-dispersible and polymer film. Dalton Trans. **40**, 11571 (2011a)

Phaomei, G., Singh, W.R., Ningthoujam, R.S.: Solvent effect in monoclinic to hexagonal phase transformation in $LaPO_4$:RE (RE = Dy^{3+}, Sm^{3+}) nanoparticles: photoluminescence study. J. Lumin. **131**, 1164 (2011b)

Phaomei, G., Singh, W.R., Singh, N.S., Ningthoujam, R.S.: Luminescence properties of Ce^{3+} co-activated $LaPO_4:Dy^{3+}$ nanorods prepared in different solvents and tunable blue to white light emission from Eu^{3+} co-activated $LaPO_4:Dy^{3+}$, Ce^{3+}. J. Lumin. **134**, 649–656 (2013)

Piper, W.W., DeLuca, J.A., Ham, F.S.: Cascade fluorescent decay in Pr^{3+}-doped fluorides: achievement of a quantum yield greater than unity for emission of visible light. J. Lumin. **8**, 344–348 (1974)

Prasad, A.I., Parchur, A.K., Juluri, R.R., Jadhav, N., Pandey, B.N., Ningthoujam, R.S., Vatsa, R.K.: Bi-functional properties of Fe_3O_4@YPO_4: Eu hybrid nanoparticles: hyperthermia application. Dalton Trans. **42**, 4885–4896 (2013)

Rao, C.M., Sudarsan, V., Ningthoujam, R.S., Gautam, U.K., Vatsa, R.K., Vinu, A., Tyagi, A.K.: Luminescence studies on low temperature synthesized $ZnGa_2O_4:Ln^{3+}$ (Ln = Tb and Eu) nanoparticles. J. Nanosci. Nanotech. **8**, 5776 (2008)

Sahu, R.K., Ganguly, K., Mishra, T., Mishra, M., Ningthoujam, R.S., Roy, S.K., Pathak, L.C.: Stabilization of intrinsic defects at high temperatures in ZnO nanoparticles by Ag. J. Colloid Interf. Sci. **366**, 8–15 (2012a)

Sahu, N.K., Ningthoujam, R.S., Bahadur, D.: Disappearance and recovery of luminescence in $GdPO_4:Eu^{3+}$ nanorods: propose to water/OH^- release under near infrared and gamma irradiations. J. Appl. Phys. **112**, 014306 (2012)

Sahu, N.K., Shanta Singh, N., Ningthoujam, R.S., Bahadur, D.: Ce^{3+} sensitized $GdPO^4:Tb^{3+}$ nanorods: an investigation on energy transfer, luminescence switching and quantum yield. ACS Photonics **1**, 337–346 (2014)

Shalav, A., Richards, B.S., Green, M.A.: Luminescent layers for enhanced silicon solar cell performance: up-conversion. Sol. Energy Mater. Sol. Cells **91**, 829–842 (2007)

Shao, Q., Zhang, G., Ouyang, L., Hu, Y., Donga, Y., Jiang, J.: Emission color tuning of core/shell upconversion nanoparticles through modulation of laser power or temperature. Nanoscale **9**, 12132–12141 (2017)

Sharma, K.S., Melwani, P.K., Yadav, H.D., Joshi, R., Shetake, N.G., Dubey, A.K., Singh, B.P., Phapale, S., Phadnis, P.P., Vatsa, R.K., Ningthoujam, R.S., Pandey, B.N.: Deoxyglucose-conjugated persistent luminescent nanoparticles for theragnostic application in fibrosarcoma tumor model. RSC Adv. **13**, 13240–13251 (2023)

Sheen, M., Ko, Y., Kim, D.U., Kim, J., Byun, J.H., Choi, Y.S., Ha, J., Yeon, K.Y., Kim, D., Jung, J., Choi, J., Kim, R., Yoo, J., Kim, I., Joo, C., Hong, N., Lee, J., Jeon, S.H., Oh, S.O., Lee, J., Ahn, N., Lee, C.: Highly efficient blue InGaN nanoscale light-emitting diodes. Nature **608**, 56–61 (2022)

Shetty, D., Kim, Y.J., Shim, H., Snyder, J.P.: Eliminating the heart from the curcumin molecule: monocarbonyl curcumin mimics (MACs). Molecules **20**, 249–292 (2015)

Shukla, R., Ningthoujam, R.S., Tyagi, A.K., Vatsa, R.K.: Luminescence properties of Dy^{3+} doped Gd_2O_3 nanoparticles prepared by glycine route: annealing effect. Int. J. Nanotechnol. **7**, 843 (2010)

Simmons, J.H., Potter, K.S.: Optical Materials. Academic Press, New York (2000)

Singh, P.: Optical properties of rare earth doped phosphors and hybrid materials (Doctoral thesis). Banaras Hindu University, UP, India (2018)

Singh, L.R., Ningthoujam, R.S.: Critical view on energy transfer, site symmetry, improvement in luminescence of Eu^{3+}, Dy^{3+} doped YVO_4 by core-shell formation. J. Appl. Phys. **107**, 104304 (2010)

Singh, L.R., Ningthoujam, R.S.: Critical view on luminescence properties of Y_2O_3:Eu^{3+} after dispersion in SiO_2. Chem. Phys. Lett. **510**, 120 (2011)

Singh, L.R., Ningthoujam, R.S., Sudarsan, V., Singh, S.D., Kulshreshtha, S.K.: Probing of surface Eu^{3+} ions present in ZnO: Eu nanoparticles by covering ZnO: Eu core with Y_2O3 shell: luminescence study. J. Lumin. **128**, 1544 (2008a)

Singh, N.S., Ningthoujam, R.S., Devi, L.R., Yaiphaba, N., Sudarsan, V., Singh, S.D., Vatsa, R.K., Tewari, R.: Luminescence study of Eu^{3+} doped $GdVO_4$ nanoparticles: concentration, particle size and core-shell effects. J. Appl. Phys. **104**, 104307 (2008b)

Singh, L.R., Ningthoujam, R.S., Singh, S.D.: Tuning of ultra-violet to green emission by choosing suitable excitation wavelength in ZnO: Quantum dot, nanocrystals and bulk. J. Alloys Comp. **487**, 466 (2009a)

Singh, N.S., Ningthoujam, R.S., Yaiphaba, N., Singh, S.D., Vatsa, R.K.: Lifetimes and quantum yield studies of Dy^{3+} doped $GdVO_4$: concentrations and annealing effect. J. Appl. Phys. **105**, 064303 (2009b)

Singh, L.R., Ningthoujam, R.S., Singh, N.S., Singh, S.D.: Probing Dy^{3+} ions on the surface of nanocrystalline YVO_4: luminescence study. Opt. Mater. **32**, 286 (2009c)

Singh, N.S., Ningthoujam, R.S., Singh, S.D., Viswanadh, B., Manoj, N., Vatsa, R.K.: Preparation of highly crystalline blue emitting MVO_4:Tm^{3+} (M = Gd, Y) spherical nanoparticles: effects of activator concentration and annealing temperature on luminescence, lifetime and quantum yield. J. Lumin. **130**, 2452 (2010)

Singh, N.S., Ningthoujam, R.S., Phaomei, G., Singh, G.D., Vinu, A., Vatsa, R.K.: Re-dispersion and film formation of $GdVO_4$:Ln^{3+} ($Ln^{3+} = Dy^{3+}$, Eu^{3+}, Sm^{3+}, Tm^{3+}) nanoparticles: particle size and luminescence studies. Dalton Trans. **41**, 4404 (2012)

Singh, B.P., Parchur, A.K., Ningthoujam, R.S., Ansari, A.A., Singh, P., Rai, S.B.: Influence of Gd^{3+} co-doping on structural property of $CaMoO_4$:Eu. Dalton Trans. **43**, 4770–4778 (2014a)

Singh, L.P., Srivastava, S.K., Mishra, R., Ningthoujam, R.S.: Multifunctional hybrid nanomaterials from water dispersible CaF_2:Eu^{3+}, Mn^{2+} and Fe_3O_4 for luminescence and hyperthermia application. J. Phys. Chem. C **118**, 18087–18096 (2014b)

Singh, L.P., Jadhav, N.V., Sharma, S., Pandey, B.N., Srivastava, S.K., Ningthoujam, R.S.: Hybrid nanomaterials YVO_4:Eu/Fe_3O_4 for optical imaging and hyperthermia in cancer cells. J. Mater. Chem. C **3**, 1965–1975 (2015)

Singh-Rachford, T.N., Castellano, F.N.: Photon upconversion based on sensitized triplet–triplet annihilation. Coord. Chem. Rev. **254**, 2560–2573 (2010)

Sommerdijk, J.L., Bril, A., de Jager, A.W.: Two photon luminescence with ultraviolet excitation of trivalent praseodymium. J. Lumin. **8**, 341–343 (1974)

Srinivasu, K., Ningthoujam, R.S., Sudarsan, V., Vatsa, R.K., Tyagi, A.K., Srinivasu, P., Vinu, A.: Eu^{3+} and Dy^{3+} doped YPO_4 nanoparticles: low temperature synthesis and luminescence studies. J. Nanosci. Nanotech. **9**, 3034 (2009)

Srivastava, M., Devi, L.S., Joshi, R., Singh, B.P., Ningthoujam, R.S.: Effect of signal to noise ratio on luminescence property of YPO_4:Eu. Mater. Lett. X **16**, 100170 (2022)

Strite, S., Morkoç, T.: GaN, AlN, and InN: a review. J. Vac. Sci. Technol., B **10**, 1237–1266 (1992)

Tai, C.L., Hong, W.L., Kuo, Y.T., Chang, C.Y., Niu, M.C., Ochathevar, M.K.P., Hsu, C.L., Horng, S.F., Chao, Y.C.: Ultrastable, deformable, and stretchable luminescent organic-inorganic perovskite nanocrystal-polymer composites for 3D printing and white light-emitting diodes. ACS Appl. Mater. Interfaces **11**, 30176–30184 (2019)

Takizawa, Y., Kamada, K., Yoshino, M., Yamaji, A., Kurosawa, S., Yokota, Y., Sato, H., Toyoda, S., Ohashi, Y., Hanada, T., Kochurikhin, V.V., Yoshikawa, A.: Growth and scintillation properties of Tl-doped CsI/CsCl/NaCl ternary eutectic scintillators. Jpn. J. Appl. Phys. **60**, SBBK01 (2021)

Tanner, P.A., Pan, Z.: Luminescence properties of lanthanide and transition metal ion-doped Ba_2LaNbO_6: detection of $MnO_6{}^{8-}$ and $CrO_6{}^{9-}$ clusters. Inorg. Chem. **48**, 11142–11146 (2009)
Wang, Y., Shi, J., Chen, J., Zhu, W., Baranoff, E.: Recent progress in luminescent liquid crystal materials: design, properties and application for linearly polarised emission. J. Mater. Chem. C **3**, 7993–8005 (2015)
Wang, J., Zhang, J., Zhou, Y., Liu, H., Xue, Q., Li, X., Chueh, C.C., Yip, H.L., Zhu, Z., Jen, A.K.Y.: Highly efficient all-inorganic perovskite solar cells with suppressed non-radiative recombination by a Lewis base. Nat. Commun. **11**, 177 (2020)
Wegeberg, C., Wenger, O.S.: Luminescent first-row transition metal complexes. JACS Au **1**, 1860–1876 (2021)
Wegh, R.T., Donker, H., Meijerink, A., Lamminmaki, R.J., Holsa, J.: Vacuum-ultraviolet spectroscopy and quantum cutting for Gd^{3+} in $LiYF_4$. Phys. Rev. B **56**, 13841 (1997)
Wegh, R.T., Donker, H., Oskam, K.D., Meijerink, A.: Visible quantum cutting in $LiGdF_4$:Eu^{3+} through downconversion. Science **283**, 664 (1999)
Xie, L., Wang, Y., Zhang, H.: Near-infrared quantum cutting in YPO_4: Yb^{3+}, Tm^{3+} via cooperative energy transfer. Appl. Phys. Lett. **94**, 061905 (2009)
Xie, C., Xie, S., Yi, R., Cao, R., Yuan, H., Xiao, F.: Site-selective occupation for broadband upconversion luminescence in $Ca_3Y(GaO)_3(BO_3)_4$:Yb^{3+}, Mn^{2+} phosphors. J. Phys. Chem. C **124**, 6845–6852 (2020)
Yadav, R.S., Ningthoujam, R.S.: Synthesis and characterization of quantum cutting phosphor materials. In: Tyagi, A.K., Ningthoujam, R.S. (eds.) Handbook on Synthesis Strategies for Advanced Material: Volume-III: Materials specific synthesis strategies, Chapter 7, pp. 213–250. Springer Nature, Singapore (2021)
Yaiphaba, N., Ningthoujam, R.S., Singh, N.S., Vatsa, R.K., Singh, N.R.: Probing of inversion symmetry site in Eu^{3+} doped $GdPO_4$ by luminescence study: concentration and annealing effect. J. Lumin. **130**, 174 (2010a)
Yaiphaba, N., Ningthoujam, R.S., Singh, N.S., Vatsa, R.K., Singh, N.R., Dhara, S., Misra, N.L., Tewari, R.: Luminescence, lifetime, and quantum yield studies of redispersible Eu^{3+}-doped $GdPO_4$ crystalline nanoneedles: core-shell and concentration effects. J. Appl. Phys. **107**, 034301 (2010c)
Yaiphaba, N., Ningthoujam, R.S., Singh, N.R., Vatsa, R.K.: Luminescence properties of redispersible Tb^{3+}-Doped $GdPO_4$ nanoparticles prepared by an ethylene glycol route. Eur. J. Inorg. Chem. 2682 (2010)
Ye, S., Song, E.H., Zhang, Q.Y.: Transition metal-involved photon upconversion. Adv. Sci. **3**, 1600302 (2016)
Ye, H.Y., Tang, Y.Y., Li, P.F., Liao, W.Q., Gao, J.X., Hua, X.N., Cai, H., Shi, P.P., You, Y.M., Xiong, R.G.: Metal-free three-dimensional perovskite ferroelectrics. Science **361**, 151–155 (2018)
Zhang, Q.Y., Huang, X.Y.: Recent progress in quantum cutting phosphors. Prog. Mater Sci. **55**, 353–427 (2010)
Zhuang, Y., Katayama, Y., Ueda, J., Tanabe, S.: A brief review on red to near-infrared persistent luminescence in transition-metal-activated phosphors. Opt. Mater. **36**, 1907–1912 (2014)
Zou, Z., Feng, L., Cao, C., Zhang, J., Wang, Y.: Near-infrared quantum cutting long persistent luminescence. Sci. Rep. **6**, 24884 (2016)

Chapter 3
Advances in Plasmonic Substrate-Coupled Fluorescence

Sharmistha Dutta Choudhury

3.1 Introduction

Plasmonics is a burgeoning field of nanoscience that deals with localization and guiding of light by interaction with metallic nanostructures (Schuller et al. 2010; Atwater 2007). The optical response of plasmonic materials depends on their frequency dependent dielectric function, $(\varepsilon(\omega) = \varepsilon'(\omega) + i\varepsilon''(\omega))$, that is commonly characterized by the Drude model (Maier 2007) as,

$$\varepsilon(\omega) = \left(1 - \frac{\omega_p^2}{\omega^2 + \gamma^2}\right) + i\left(\frac{\omega_p^2\gamma}{\omega^3 + \omega\gamma^2}\right) \tag{3.1}$$

Here ω is the angular frequency of light, γ is electron collision frequency and ω_p, is the bulk plasma frequency of the metal ($\omega_p = \sqrt{\frac{Ne^2}{m_{eff}\varepsilon_0}}$, where ε_0 is the permittivity of free space, N is the electron density, m_{eff} is the effective mass of an electron and e is the electron charge). When the frequency of incident light is above ω_p, the electrons are unable to oscillate in response to the incident field. In this condition, the real part of the dielectric constant of the metal ($\varepsilon'(\omega)$) is positive, and light is either transmitted or absorbed due to interband transitions. When the frequency of light is below ω_p, the free conduction electrons oscillate in a phase opposite to that of the incident electromagnetic field. In this case, $\varepsilon'(\omega)$ is negative and light is reflected by the metal (Figs. 3.1a and b) (Maier 2007; Li et al. 2015).

If the metal dimensions are brought down to the nano regime, two types of plasmonic modes can be generated upon interaction with light, depending on the shape

S. Dutta Choudhury (✉)
Radiation & Photochemistry Division, Bhabha Atomic Research Centre, Mumbai 400085, India
e-mail: sharmidc@barc.gov.in

Homi Bhabha National Institute, Anushaktinagar, Mumbai 400094, India

R. S. Ningthoujam and A. K. Tyagi (eds.), *Handbook of Materials Science, Volume 1*, Indian Institute of Metals Series, https://doi.org/10.1007/978-981-99-7145-9_3

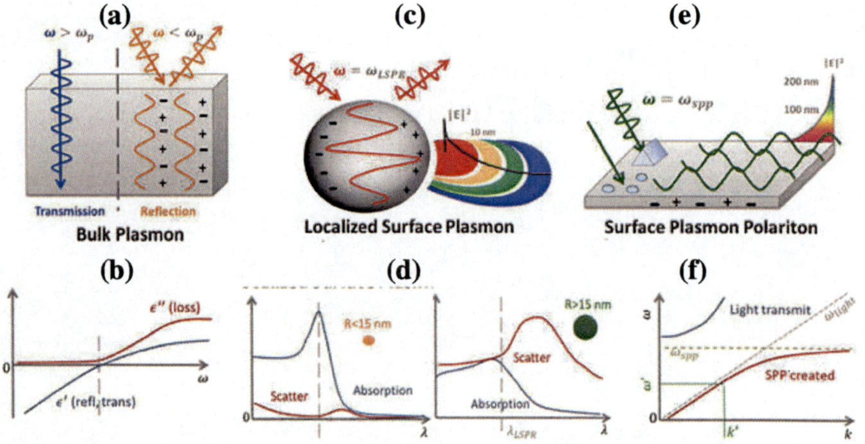

Fig. 3.1 **a** Interaction of light with bulk metal at frequencies below and above its plasma frequency, ω_p, and **b** complex dielectric of the metal. **c** Localised surface plasmons (LSPs) in metal nanoparticle and **d** LSP resonances due to absorption and scattering in particles of different sizes. **e** Propagating surface plasmon polaritons (SPPs) on a thin metal film and **f** the dispersion curve for SPPs. Adapted with permission from Li et al. (2015). Copyright 2015 The Royal Society of Chemistry

and size of the particles. These are: Localized Surface Plasmons (LSPs) and Surface Plasmon Polariton (SPPs).

The LSPs are collective oscillations of surface electrons that arise when the dimensions of metal nanostructures are smaller than the wavelength of light (Fig. 3.1c). LSPs can be directly excited by the incident light. From Mie theory, the extinction coefficient (σ_{ext}) of a metal nanosphere, of volume V in a medium of dielectric constant, $\varepsilon(\omega)_{diel}$, is (Link and El-Sayed 2003):

$$\sigma_{ext} = 9\frac{\omega}{c}(\varepsilon(\omega)_{diel})^{3/2} V \frac{\varepsilon(\omega)''_{metal}}{\left(\varepsilon(\omega)'_{metal} + 2\varepsilon(\omega)_{diel}\right)^2 + \left(\varepsilon(\omega)''_{metal}\right)^2} \tag{3.2}$$

When the real part of the dielectric constant, $\varepsilon(\omega)'_{metal}$, is negative and satisfies the condition, $\varepsilon(\omega)'_{metal} = -2\varepsilon(\omega)_{diel}$, and the imaginary part is small ($\varepsilon(\omega)''_{metal} \approx 0$), the denominator in Eq. 3.2 vanishes, leading to strong plasmon resonance at $\omega_{LSP} = \frac{\omega_p}{\sqrt{1+2\varepsilon(\omega)_{diel}}}$. The extinction coefficient (σ_{ext}) in Eq. 3.2 arises due to both scattering and absorption of light. With increase in size of the nanoparticle, the contribution of the scattering component in σ_{ext} gradually increases (Fig. 3.1d). An important outcome of the confined LSP oscillations is that it leads to intense local electromagnetic fields in the vicinity of the nanoparticle.

Unlike LSPs, the SPPs are propagating charge oscillations on the surface of metallic nanostructures like thin continuous metal films, whose dimensions are greater than the wavelength of light (Fig. 3.1e). The wave vector or momentum of propagating SPPs is expressed as:

$$k_{SPP} = \frac{\omega}{c}\sqrt{\frac{\varepsilon(\omega)_{metal} + \varepsilon(\omega)_{diel}}{\varepsilon(\omega)_{metal}\varepsilon(\omega)_{diel}}} \tag{3.3}$$

The resonance condition for SPPs is, $\omega_{SPP} = \frac{\omega_p}{\sqrt{1+\varepsilon(\omega)_{diel}}}$Maier 2007; Lakowicz 2006a). Evidently, k_{SPP} is greater than the wavevector of the photon in free space ($k_0 = \frac{\omega}{c}$). So the light line lies above the dispersion curve for SPPs (Fig. 3.1f). This means that SPPs cannot be excited directly by incident light. Special conditions requiring gratings, surface corrugations, or prisms are used to excite SPPs by increasing the momentum of the incident light to match with the SPP. The wavevector matching conditions between light and SPPs implies that SPPs at the the metal–dielectric interface can exist only for TM (transverse magnetic; electric field parallel to plane of incidence) or P-polarization of light. The evanescent field of SPPs extends to a distance of about 100–200 nm from the metal film surface.

The unique features of surface plasmon resonances (LSPs and SPPs), specifically, their sensitivity to the surrounding dielectric medium, local field enhancement and creation of evanescent wave on the surface of metal nanostructures, have opened the doors for numerous applications of plasmonics. Surface enhanced Raman spectroscopy, optical sensing, photovoltaics, lasers, quantum computing, are some of the notable areas that have hugely benefitted by contribution from plasmonics (Schuller et al. 2010; Atwater 2007; Li et al. 2015; Yu et al. 2019). The entry of plasmonics in fluorescence spectroscopy is another important development that has seen remarkable progress in recent years.

Fluorescence is the emission of light by radiative relaxation of electronically excited species from their first singlet state to the ground state (Fig. 3.2) (Lakowicz 2006b). It is an indispensable tool in bioanalysis, microscopy, diagnosis and sensing, due to its great versatility, sensitivity, simplicity, and rapid noninvasive measurement technique. The extremely high sensitivity of fluorescence has permitted extraordinary developments in single molecule detection and super-resolution microscopy making it valuable for understanding complex phenomena in great detail.

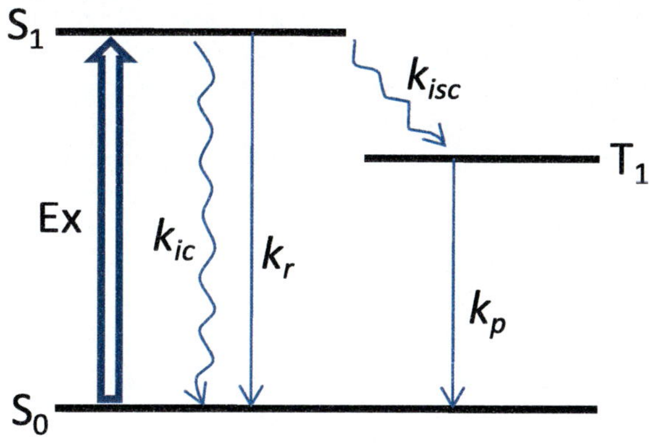

Fig. 3.2 Simplified Jablonski diagram showing the first excited singlet and triplet states and the major photophysical processes in a fluorescent molecule

Fluorescence is characterized by two important parameters; the fluorescence quantum yield (Φ) and the fluorescence lifetime (τ), which are expressed in terms of the radiative decay rate (k_r) and nonradiative decay rate (k_{nr}) (Lakowicz 2006b).

$$\Phi = \frac{k_r}{k_r + k_{nr}} \tag{3.4}$$

$$\tau = \frac{1}{k_r + k_{nr}} \tag{3.5}$$

The nonradiative decay rate ($k_{nr} = k_{ic} + k_{isc}$) includes both intersystem crossing (*isc*) and internal conversion (*ic*). Most of the commonly known fluorophores have aromatic structures or conjugated carbon chains. Apart from molecular fluorophores, many other photoluminescent materials and quantum emitters like, semiconductor quantum dots, metal nanoclusters, upconverting rare-earth phosphors, carbon dots and perovskite nanocrystals have also been developed over the years to meet the growing demands and applications of fluorescence spectroscopy. Despite many developments, classical fluorescence, which relies on measuring the spontaneous emission of light in an optically transparent medium in the far field, is limited by challenges due to weak emission intensities, poor photostability and low collection efficiency.

Under steady state conditions, the emission photon count rate for a single fluorescent molecule (CR) is (Jiao et al. 2017),

$$CR = k_c \phi \frac{\sigma I_e}{1 + I_e/I_s} \tag{3.6}$$

Here, k_c is the light collection efficiency of the optical system, I_e is the excitation rate, σ is the absorption cross-section and I_s is the saturation intensity ($I_s = (k_r + k_{nr})/\left[\sigma(1 + k_{nr}/k_p)\right]$, where k_p is the radiative relaxation of the triplet state). Since classical fluorescence is isotropic in nature, less than 1% of the emitted light can be detected in a standard spectro-fluorometer setup (Lakowicz 2004). Moreover, bulky optical components are required to focus and collect the emitted light, which makes it difficult to progress toward device miniaturization, work with small sampling volumes and meet point-of-care requirements. With the integration of plasmonics in fluorescence, it has become possible to surpass many of the limitations of classical fluorescence and achieve enhanced brightness, better photostability and improved collection efficiency. Plasmonic coupling offers the opportunity to control emission at a fundamental level by modifying the radiative decay rates of fluorophores. Moreover, plasmonics enables emission to be controlled at the point of origin, without any external optics.

3.2 Mechanism and Effects of Fluorophore-Plasmon Interactions

Fluorophores interact with metal nanoparticles that support LSPs and metal nano films that support SPPs, leading to different kinds of effects. While the interaction with LSPs is useful for fluorescence enhancement and tailoring of radiative decay rates (Lakowicz 2006a), the interaction with SPPs is useful for transforming isotropic emission into directional, polarized and wavelength-resolved emission (Lakowicz 2004). The evanescent waves of SPPs are also used to obtain increased excitation fields for improving fluorescence signal in free space (Wang et al. 2010). The different kinds of interactions and effects observed in plasmon-coupled emission are discussed below.

3.2.1 *Interaction with Metal Nanoparticles*

The interaction between fluorophores and metal nanoparticles can be described as a combination of several different effects: (i) local field enhancement in the vicinity of the nanoparticle and hence around the nearby fluorophores (plasmonic antenna effect), which leads to increased excitation rates for the fluorophores; (ii) enhanced coupling efficiency of fluorescence to the far field via LSP resonance, especially the scattering component in σ_{ext}; (iii) increase in radiative decay rate of fluorophores (known as radiative decay engineering, RDE); and (iv) increased nonradiative decay due to energy transfer from fluorophores to the metal surface leading to quenching of fluorescence, a phenomenon that occurs when fluorophores are adjacent to the metal surface (Lakowicz 2006a, 2005; Khatua et al. 2014). When the fluorophore position is more than 5 nm from the metal surface, which can be achieved by using suitable spacer layers, the quenching effect is suppressed, and the plasmon-coupled fluorophore or "plasmophore" radiates with an overall increase in fluorescence intensity due to the synergistic effects of (i), (ii) and (iii). This phenomenon is known as metal-enhanced fluorescence (MEF). The effective fluorescence enhancement, ξ is expressed as, $\xi \approx E_{exc} \times E_{em}$, where E_{exc} corresponds to the excitation enhancement and E_{em} is the overall emission enhancement.

The increase in radiative decay rate for a plasmon-coupled fluorophore can be understood from Fermi's Golden Rule (Eq. 3.7) that relates radiative rate of an excited molecule to the quantum mechanical transition probability, W_{ij}.

$$W_{ij} = \frac{2\pi}{\hbar}\left|\mu_{ij}\right|^2 \rho\left(\nu_{ij}\right) \tag{3.7}$$

Here μ_{ij} is the transition dipole moment connecting the initial (i) and final (j) states and $\rho(\upsilon_{ij})$ is the photonic mode density at the transition frequency. For fluorophores located within nanoscale distances from a plasmonic substrate, the local density of

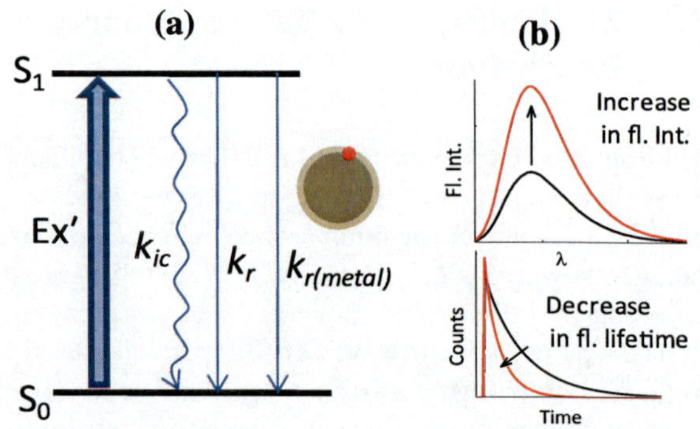

Fig. 3.3 **a** Simplified Jablonski diagram showing the changes in excitation and radiative rate for a plasmon-coupled fluorophore. **b** Schematic representations of the simultaneous increase in intensity and decrease in lifetime

states is substantially high, which leads to increase in the radiative decay rate. An important consequence of this effect is that MEF is characterized by simultaneous increase of quantum yield (Φ_{metal}) and decrease of fluorescence lifetime (τ_{metal}) (Eqs. 3.8 and 3.9) (Lakowicz 2005; Khatua et al. 2014).

$$\Phi_{metal} = \frac{k_r + k_{r(metal)}}{k_r + k_{r(metal)} + k_{nr}} \quad (3.8)$$

$$\tau_{metal} = \frac{1}{k_r + k_{nr} + k_{r(metal)}} \quad (3.9)$$

Here $k_{r(metal)}$ is the additional increase in the radiative rate of fluorophores because of increase in the local density of states. The shortening in fluorescence lifetime is advantageous because photochemical destruction reduces for a molecule that stays for lower time in the excited state. Therefore, the enhanced fluorescence quantum yield and reduced lifetime achieved through plasmonics, greatly increases detection limit, while at the same time reducing photobleaching of fluorophores. The major photophysical changes for a plasmon-coupled fluorophore are shown in Fig. 3.3.

Observation of MEF is critically dependent on the distance between the metal nanoparticle and the fluorophores. By depositing fluorophores at different distances from metal nanostructures using Langmuir–Blodgett monolayers or polyelectrolyte layer-by-layer assembly, it has been shown that fluorescence enhancement is optimum at distances between 5 and 20 nm from the surface, where fluorophores experience enhanced excitation field intensity without being quenched by the metal (Ray et al. 2006; Akbay and Lakowicz 2012). The calculated near-field intensity map and overall fluorescence enhancement, ξ, for a fluorophore at different distances from an Au nanorod are depicted in Figs. 3.4a and 4b, respectively (Khatua et al. 2014).

MEF also depends on the shape and size of metal nanoparticles and the extent of spectral overlap between fluorescence and the LSP resonance. On comparing the fluorescence enhancement by two different Au nanostructures, viz. nanorods and nanoshells, Halas and co-workers found that the quantum yield of IR800 dye (~7%) is increased to ~86% on coupling with nanoshells whereas it increases to ~74% when

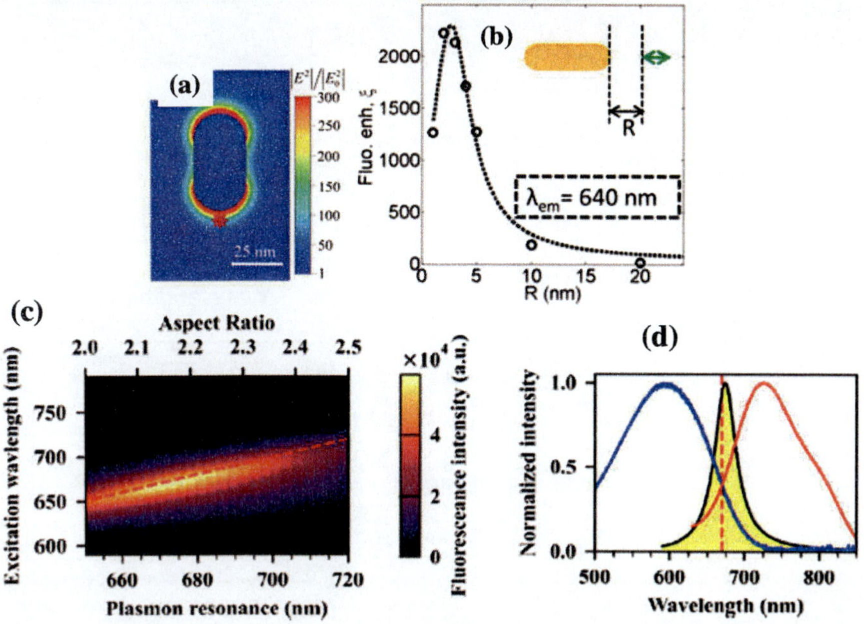

Fig. 3.4 **a** Calculated near-field intensity of an Au nanorod (47 nm long and 25 nm wide, having LSP resonance at 629 nm), and **b** calculated fluorescence enhancement, ξ, as a function of distance between the fluorophore and nanorod. Adapted with permission from Khatua et al. (2014). Lu et al. (2020) Copyright 2014 American Chemical Society. **c** Calculated enhanced emission intensity as a function of excitation wavelength and LSP resonance wavelength of Au nanorods having different aspect ratios and **d** overlap of the scattering spectra of the fabricated optimum sized Au nanorod with the absorption/emission spectra of the coupled naphthalenediimide-based NIR dye. Adapted from. Copyright 2020 American Chemical Society under Creative Commons License CC-BY-NC-ND

coupled with nanorods (Bardhan et al. 2009). This difference is ascribed to greater scattering cross-section of the LSP resonance in nanoshells compared to nanorods, at the emission wavelength of the dye. Thus, a high cross-section for scattering at the emission wavelength is favourable for efficient fluorescent enhancement. Very recently, Lu et. al. have shown that detection of single molecule fluorescence from NIR emissive dyes having quantum yield as low as 10^{-4}, is possible by coupling with Au nanorods, provided their LSP resonance is optimized with the excitation and emission wavelengths of the dye (Lu et al. 2020). Figure 3.4c shows a map of the calculated fluorescence intensity enhancement as a function of the excitation wavelength and LSP resonance for Au nanorods having different aspect ratios (length/ width of nanorod). Based on these calculations, Au nanorods of suitable geometry were fabricated so as to have maximal overlap of the plasmon resonance with the absorption/ emission spectra of the NIR dye (Fig. 3.4d), for obtaining the largest possible enhancement factor.

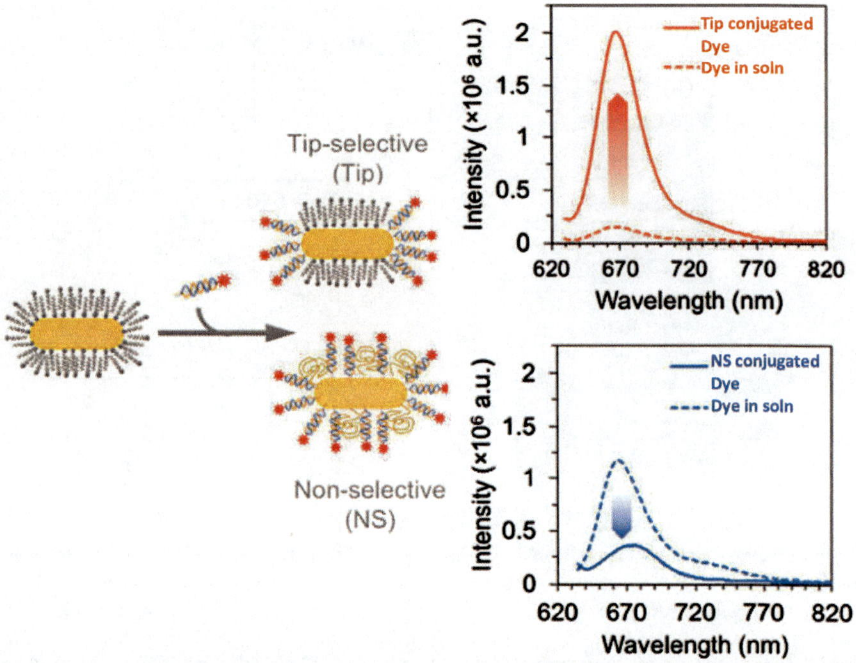

Fig. 3.5 Schematics for fluorophore conjugation specifically at the tip or non-specifically throughout the surface of Au nanorods, along with the emission spectra of the coupled fluorophore, showing enhancement and quenching effects, respectively, in the two cases. Adapted with permission from Botequim et al. (2020). Copyright 2020 The Royal society of Chemistry

The position and orientation of fluorophores on the nanoparticle is another important parameter for MEF, since the local electromagnetic field enhancement is not uniform around the nanoparticle. Higher field enhancements are observed at the tips and sharp edges around nanoparticles (cf. Fig. 3.4a). Recently, Pedro et. al. have used DNA-directed assembly to precisely position fluorescent dyes on the Au nanorod surface (Botequim et al. 2020). Their work revealed that dyes placed on the tips of the nanorods show significantly enhanced emission compared to free dyes in solution, while dyes placed in a nonspecific manner throughout the nanorod surface actually show reduced fluorescence intensity compared to that in solution (Fig. 3.5).

3.2.2 Interaction with Planar Metal Nanofilms

Although light incident directly from the far field on a thin metal film is incapable of creating SPPs on the metal–dielectric interface due to wavevector mismatch between photons and SPPs, interestingly, the emission dipoles of excited fluorophores placed on a metal film are able to create and couple with surface plasmons in the near-field. The coupled emission can subsequently be extracted into free space by wavevector matching, using a medium with high refractive index (glass prism) (Lakowicz 2004; Gryczynski et al. 2004; Su et al. 2021).

In a medium of refractive index, n_p, the wavevector of light increases to $k = k_0 n_p$ (where $k_0 = \omega/c$ is the wavevector in air). The component along the metal surface is given as $k_x = k_0 n_p sin\theta$, where θ is the angle from the surface normal. At a certain angle θ_{sp}, the value of $k_0 n_p sin\theta$ becomes equal to k_{SPP}.

$$k_0 n_p sin\theta_{sp} = k_{SPP} = \frac{\omega}{c}\sqrt{\frac{\varepsilon(\omega)_{metal} + \varepsilon(\omega)_{diel}}{\varepsilon(\omega)_{metal}\varepsilon(\omega)_{diel}}} \tag{3.10}$$

In this situation, the surface plasmon-coupled fluorescence can radiate through the prism as a cone of emission (in Figs. 3.6a and b). This phenomenon is known as surface plasmon-coupled emission (SPCE). Since the optical properties of the metal are frequency dependent, the emission angle θ_{sp} is also sensitive to emission wavelength. The emission at longer wavelength appears at smaller angles than that at shorter wavelength. Furthermore, since SPP can couple only with P-polarized light, the fluorescence emission is also P-polarized, even from randomly oriented fluorophores placed on the metal surface. Thus, SPCE transforms the usual isotropic unpolarized emission into a cone of wavelength-resolved, P-polarized emission.

The ability of SPCE to demarcate different wavelengths directly by their angular separation without using additional dispersive optics is promising for instrument miniaturization. Further, it allows multiple emissive species in a complex mixture to be easily sorted, based on the observation of their emissions at different angles. This has been demonstrated experimentally by Sathish et al. in the spectral resolution of emission from different Rhodamine6G dye aggregates (Sathish et al. 2009). Directional emission is another important advantage of SPCE, because signal collection efficiency can be substantially improved by aligning the detector in a specific direction (Dutta Choudhury et al. 2015a). Furthermore, since the evanescent SPP field extends to a distance of about 200 nm from the metal, SPCE allows signal from fluorophores located near the surface to be exclusively selected without perturbation from background fluorescence. This aspect has been utilised by Matveeva et. al. to carry out fluorescence immunoassays directly in human serum and whole blood without any significant attenuation of the fluorescence signal (Matveeva et al. 2005).

Two types of configurations are generally used in SPCE; the Kretschmann (KR) and the Reverse Kretschmann (RK) (Lakowicz 2004; Gryczynski et al. 2004; Su et al. 2021). In RK configuration, fluorophores are excited directly by incident light from free space and the coupled emission is subsequently collected using a prism, as depicted schematically in Fig. 3.6c. In the KR approach, fluorophores are excited by light incident through prism side. When light with excitation wavelength, λ_{ex} is incident at the correct angle ($\theta_{sp\text{-}ex}$), SPPs can be created and fluorophores within the near-field can be excited by the evanescent field of SPPs. The emitted light can

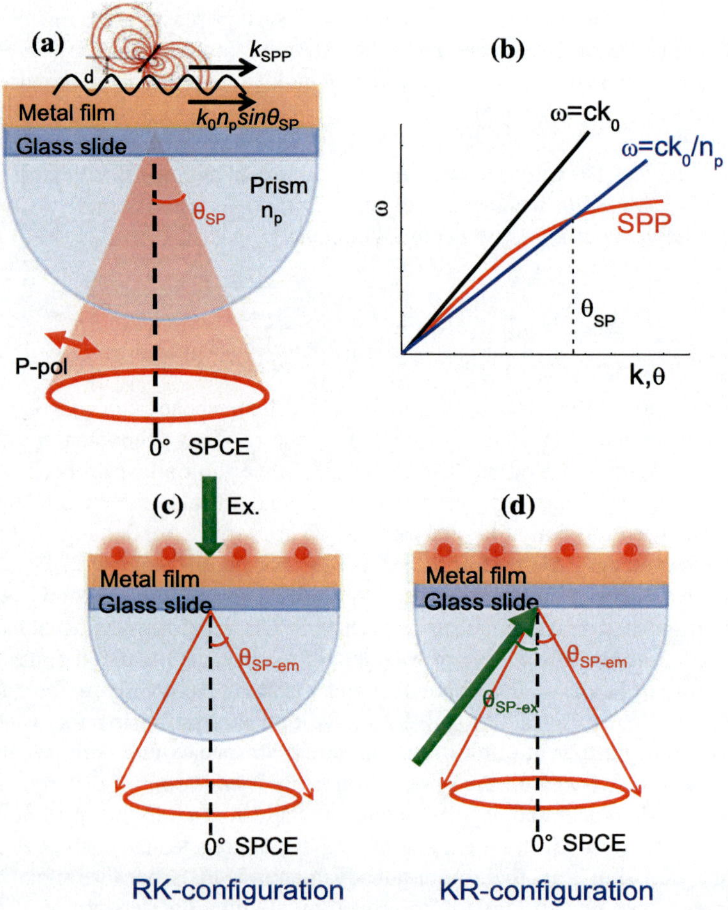

Fig. 3.6 **a** Schematic representation of SPCE, **b** the wavevector matching condition achieved by using a prism, **c** and **d** RK- and KR-configurations for SPCE

subsequently couple back with SPPs and radiate at a different angle ($\theta_{sp\text{-}em}$) corresponding to the emission wavelength, λ_{em} (Fig. 3.6d). Although the KR configuration benefits from enhanced excitation field due to creation of SPPs by the excitation light, most studies of SPCE are based on direct excitation of fluorophores in the RK configuration, because of its simplicity.

The SPCE measurements can be carried out either by angle scanning to generate angular emission patterns (Fig. 3.7a), or by leakage radiation microscopy to directly capture the image of the SPCE ring in the Fourier plane (Fig. 3.7b) (Chen et al. 2013). By changing the orientation of a polarizer placed in front of the detector, interesting variations are observed in the intensity of SPCE due to P-polarized nature of the emission.

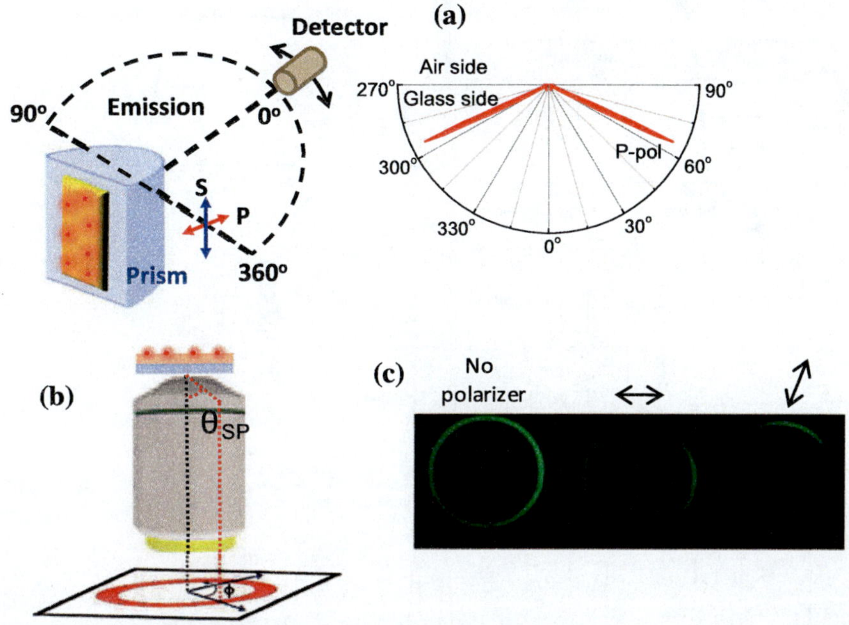

Fig. 3.7 Measurement of SPCE by **a** angle scanning to generate angular emission pattern and **b** leakage radiation microscopy to directly image the SPCE ring. **c** Images of SPCE rings with different polarizer orientations

The SPP modes in metal nanofilm substrates are often interpreted by reflectivity simulations. Light of different wavelengths is considered to be incident on the metal film at different angles through a high refractive index medium. The metal film is highly reflective at all angles above the critical angle, except for P-polarized light at $\theta_{sp\text{-}\lambda}$, when wavevector matching conditions are satisfied for creation of SPPs. Figure 3.8a depicts the calculated reflectivity dispersion map for a substrate composed of Ag film (50 nm thickness) coated with a dielectric layer (50 nm) (Dutta Choudhury et al. 2015a). The dark area is the region of decreased reflectivity due to creation of SPP. As an example, Fig. 3.8b shows that P-polarized light having wavelength of 515 nm has a dip in reflectivity at around 65°, while S-polarized light at this wavelength has a high reflectivity at all incident angles.

Interesting features appear when the dielectric layer thickness is increased in the metal-dielectric substrate, to be comparable with the wavelength of light. Now, in addition to the SPP mode, other optical modes (OMs) having both P- and S-polarizations can be sustained, with their electric fields located in different zones of the substrate (Badugu et al. 2015). The reflectivity dispersion map for such a substrate is depicted in Fig. 3.8c along with the electric field distributions corresponding to

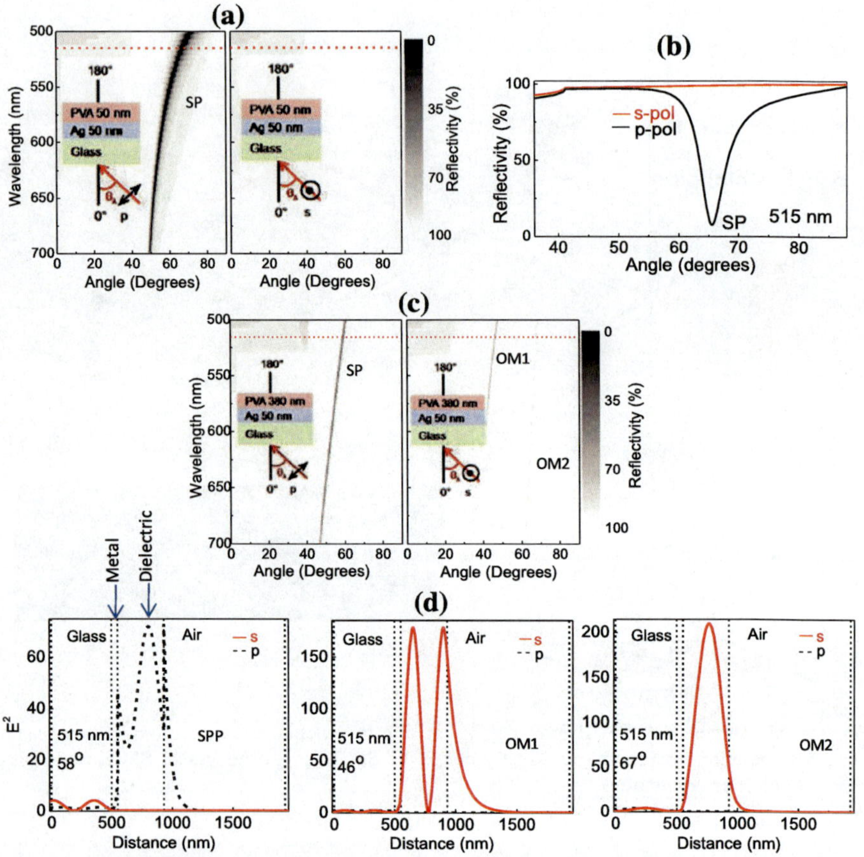

Fig. 3.8 **a** Calculated reflectivity dispersion map for a metal-dielectric substrate with dielectric thickness of 50 nm and **b** reflectivity variation with the incident angle for light having wavelength of 515 nm (corresponding to red dotted line in **a**. **c** Reflectivity dispersion map calculated for substrate with dielectric thickness of 380 nm and showing surface plasmon and OM modes. **d** The electric field intensity distributions of the optical modes in **c**. Adapted with permission from Dutta Choudhury et al. 2015a. Copyright 2015 American Chemical Society

each OM (Fig. 3.8d) (Dutta Choudhury et al. 2015a). The emission from fluorophores placed on this type of plasmon-coupled waveguide structures can interact with each of the OMs in a wavelength-dependent manner, to provide emissions at multiple angles with either S- or P-polarization. In an interesting study, this approach has been used to spectrally resolve and impart selective polarization to the intrinsically unpolarized multiple emission lines of Eu(III) ions (Dutta Choudhury et al. 2014). In another study, Lakowicz and co-workers have shown that the multiple OMs in these substrates can be exploited to selectively image emission from either the surface or the bulk, due to different polarizations and penetration depths of the OMs into the sample space (Wang et al. 2015). This is of immense significance for monitoring weak binding events in surface-based fluorescence assays.

Another approach that combines the well established procedures of fluorescence spectroscopy with SPPs as an enhanced interfacial light source is the surface plasmon field-enhanced fluorescence (SPEF). In this case, the evanescent field of the SPPs are used for excitation of surface bound fluorophores in the KR configuration. However, unlike SPCE, instead of detecting the coupled emission, the free space emission is detected above the metal surface (Fig. 3.9a) (Neumann et al. 2002; Hageneder et al. 2021).

In the SPEF approach, plasmonic chips have been created by using different kinds of spacer molecules, so that fluorophores are placed at appropriate distances from

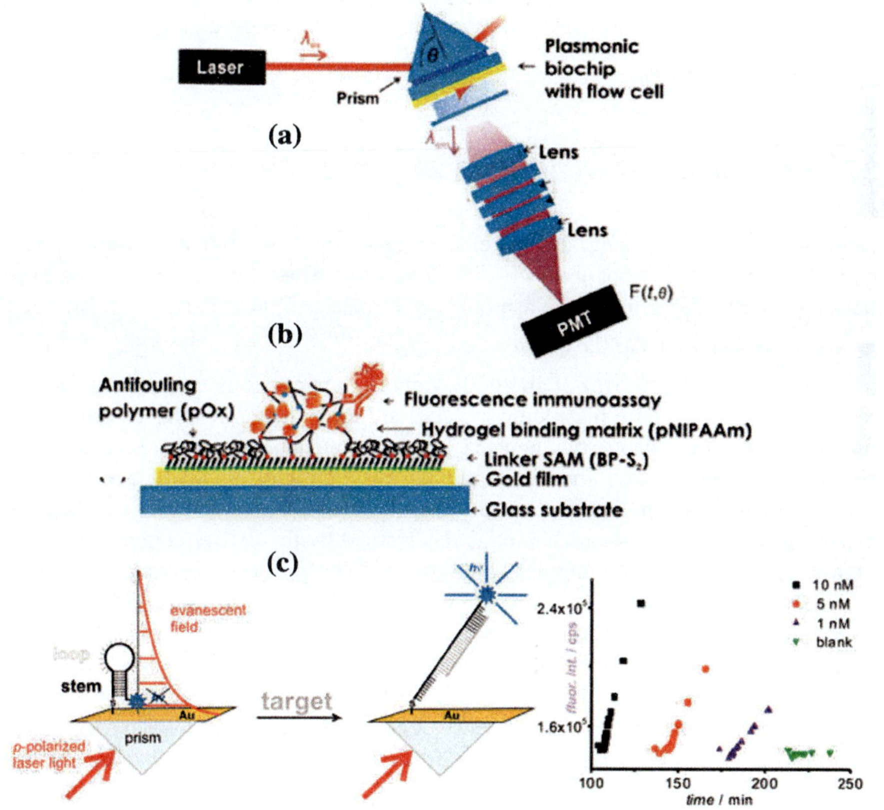

Fig. 3.9 Schematic of the optical setup for SPEF **a** and the plasmonic substrate for fluorescence immunoassay using a hydrogel binding matrix **b**. Adapted from Hageneder et al. (2021). Su et al. (2014) Copyright 2021 under Creative Commons License, published by American Chemical Society. **c** Detection and kinetic quantification of DNA target by SPEF, employing molecular beacon as a self-spacing switchable structure. Adapted with permission from. Copyright 2014 American Chemical Society

the metallic surface to avoid the quenching zone and benefit from excitation field enhancement. In this manner superior fluorescence performance and lower detection limits have been achieved in surface bound assays. Figure 3.9b shows schematics for a plasmonic biochip used for enhanced fluorescence immunoassay of human IgG antibodies for Epstein – Barr virus (EBV) (Hageneder et al. 2021). Here, the poly(N-isopropylacrylamide) (pNIPAAm)-based hydrogel serves as a binding matrix for hosting large amounts of the biomolecule. Figure 3.9c shows another example where SPEF has been used for sensitive detection and kinetic quantification of a DNA target, employing the molecular beacon as a self-spacing switchable structure (Su et al. 2014). Without DNA, the molecular beacon is in closed state, placing the fluorophores in the quenching zone of the plasmonic substrate. On binding to DNA, the molecular beacon attains an elongated double-helix structure, placing fluorophores in the plasmonically enhanced fluorescence zone, thus improving detection of the analyte. The change in fluorescence intensity was followed in real time to monitor the kinetics of binding.

3.3 Plasmonic Materials for Fluorescence Coupling

Generally, plasmonic materials are selected based on their ability to support high quality SPPs or LSP resonances, which is directly governed by the dielectric properties of the material. To quantify plasmonic performances of materials, several metrics have been defined such as the quality factor ($Q = \frac{\omega\left(d\varepsilon(\omega)'/d\omega\right)}{2\varepsilon(\omega)''}$) or the figure-of-merit ($FOM = \frac{-\varepsilon(\omega)'}{\varepsilon(\omega)''}$) (Gutiérrez et al. 2020). Important parameters that are considered in choosing the optimum plasmonic material are the plasmon resonance range, the tunability of optical properties and the losses. In metals, two types of losses are operative: the optical loss and the ohmic loss. The optical loss arises due to inter- and intra-band transitions in the metals. The ohmic loss arises due to the resistance experienced by free electrons, which is determined by the carrier concentration. A high carrier concentration implies lower losses. The plasmon resonance range is the spectral region where plasmons are resonant with incident light. This is again determined by the carrier concentration of the material. However, it can be tuned to a certain extent by controlling size, shape and assembly of the nanostructures. It may be mentioned that the plasmonic properties can also be affected by surface defects and grain boundaries. The presence of grain boundaries can impact the quality factor of the surface plasmon in thin film substrates leading to losses that are greater than theoretically calculated values. This results in reduced coupling or damping of the plasmophore emission. Figure 3.10a shows the typical plasmon spectral range of some common metals according to the shapes of their nanoparticles (Fontaine et al. 2020). Apart from plasmonic properties, other practical considerations such as stability and chemical reactivity of the material, ease of fabrication of their nanostructures and cost effectiveness are also important aspects in deciding the usefulness of a plasmonic material.

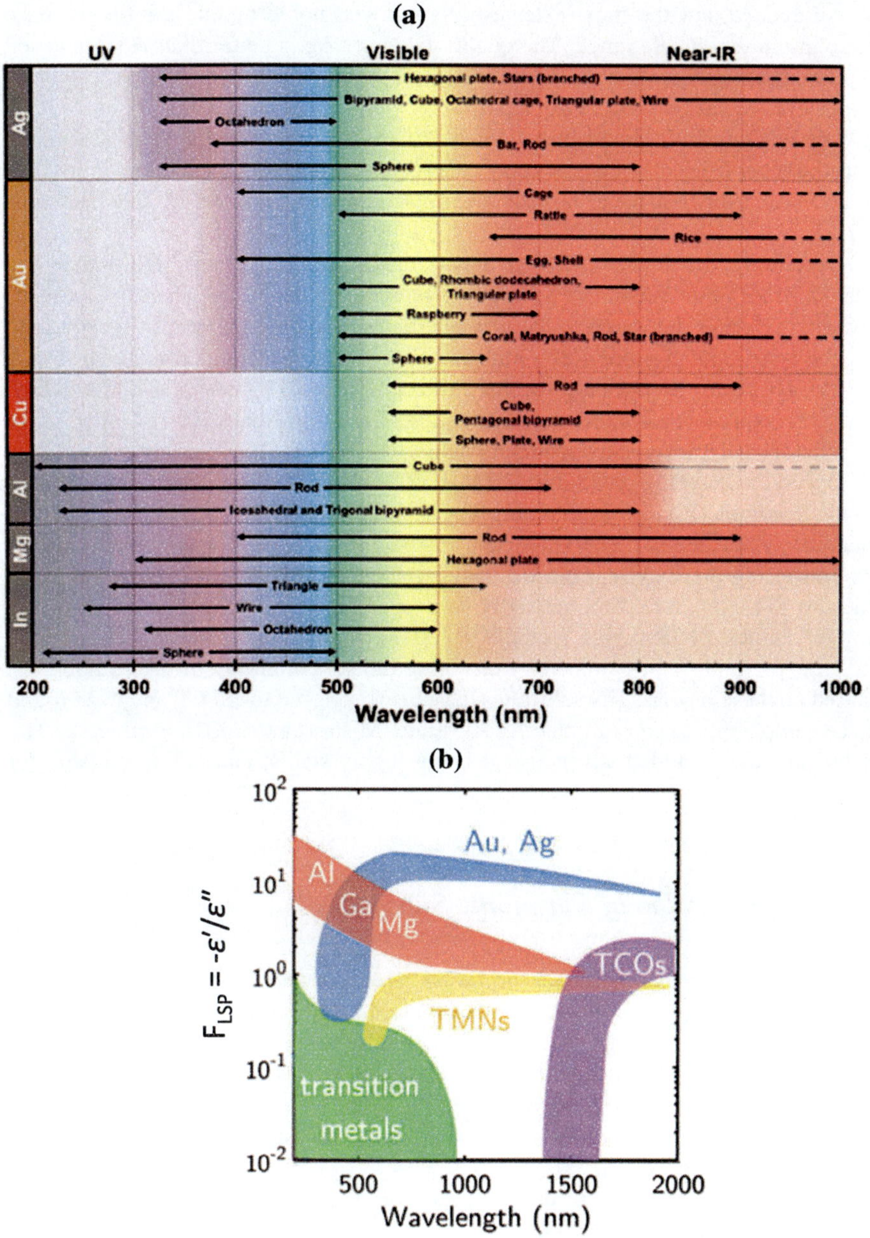

Fig. 3.10 **a** The plasmonic spectral range of some common metals based on the shapes of their nanostructures. Reprinted with permission from Fontaine et al. (2020). Gutiérrez et al. (2020) Copyright 2020 The Royal Society of Chemistry. **b** The plasmonic figure-of-merits for different families of materials, including noble metals, transition metals, transparent conducting oxides (TCOs) and transition metal nitrides (TMNs). Reprinted with permission from. Copyright 2020 AIP Publishing

Au and Ag are the most extensively employed noble metals for fluorescence coupling in the visible range. The quality factor of Ag is better than Au. However, Ag tarnishes easily while Au is more stable and biocompatible, though it is more expensive. Al has reasonably good plasmonic performance in the UV region. Akbay and co-workers have shown that Al and Pt nanostructures can serve as plasmonic substrates for MEF and are promising candidates for label-free detection of proteins and DNA bases that have intrinsically weak fluorescence (Akbay and Lakowicz 2012).

Despite the dominance of metals as plasmonic materials, their high optical and ohmic losses pose challenges for many plasmonic applications. Therefore, intense research is being dedicated to identify alternate plasmonic materials, having good tunability of optical properties. Graphene is a widely investigated material in this regard. However, its plasmon resonance occurs in the THz spectral range, which makes it unsuitable for fluorescence coupling (García de Abajo 2014). Doped semiconductors and transparent conducting oxides (TCOs) like, In-doped tin oxide (ITO), Al-doped ZnO (AZO), or Ga-doped ZnO (GZO) also have their plasmon resonances in the IR region (Fig. 3.10b) (Gutiérrez et al. 2020). Recently, studies have emerged that poorly plasmonic transition metals can be imparted good plasmonic properties by converting them to their nitrides and carbides. Particularly, the transition metal nitrides (TMNs), TiN, ZrN and HfN, are reported to have plasmonic resonances in the visible to NIR region (Naik et al. 2013). Interestingly, the optical response of these materials can be tuned by changing the stoichiometry of the compounds. Based on these reports, TiN substrates have been explored for SPCE and were found to be comparable to conventional Au substrate (Mishra et al. 2022). Further, the TiN substrate has the added advantage of being highly stable, and is thus suitable for reuse.

3.3.1 Fabrication of Plasmonic Substrates

While new materials continue to be investigated, Ag and Au are the most widely established metals for fabrication of plasmonic substrates for fluorescence coupling. The feasibility of making nanostructures with well-controlled morphologies has spurred the use of Ag and Au for MEF in a variety of forms, including colloidal nanoparticles in solution, films deposited on glass/quartz slides, or more complex architectures like nano arrays, plasmonic gratings and bow tie nanoantennas (Fig. 3.11) (Semeniak et al. 2022).

Colloidal nanoparticles of different morphologies like nanospheres, nanorods, nanoshells or nanocubes have been shown to provide large enhancements in fluorescence, especially by optimizing their LSP resonances with the absorption/emission spectra of the coupled fluorophores (Khatua et al. 2014; Bardhan et al. 2009; Lu et al. 2020). Some representative nanoparticles used in fluorescence coupling are shown in Fig. 3.12. Nanorods are especially attractive for MEF as their scattering spectra can be readily controlled by varying the aspect ratios (cf. Fig. 3.4c). Different

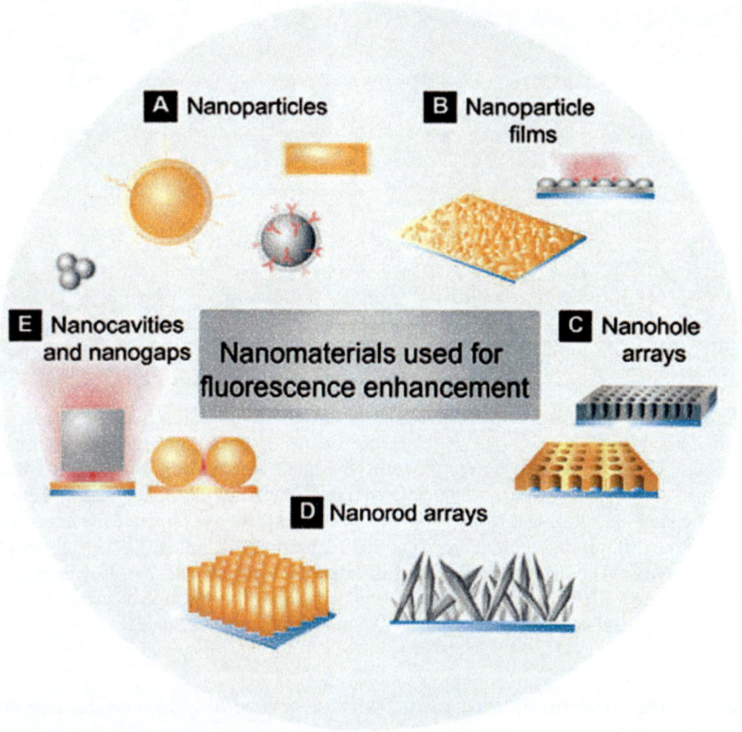

Fig. 3.11 Different kinds of plasmonic substrates used for fluorescence enhancement. Reprinted from Semeniak et al. (2022). Copyright 2022 under Creative Commons License CC BY-NC-ND 4.0, published by Wiley–VCH GmbH

wet chemical synthesis routes have been established for reproducible fabrication of nanoparticles having various shapes and sizes. Nanospheres of Ag and Au are generally obtained by the reduction of their salts ($AgNO_3$/$HAuCl_4$) with sodium citrate at controlled temperatures and reaction times, while nanorods and nanocubes are obtained by seed-mediated growth method (Rycenga et al. 2011). Metal shell-silica core nanoparticles and silica shell-metal core nanoparticles are usually prepared by the Stöber method, by using a silane agent like, 3-aminopropyltrimethoxysilane, and silicate solution or tetraethylorthosilicate as the silicon precursor (Li et al. 2017). Halas and co-workers have shown strong fluorescence enhancement in an interesting nanomatryoshka structure consisting of an Au core, an interstitial silica layer, and followed again by an Au shell layer (Orozco et al. 2014). The shell isolated nanoparticle (SHIN) is another interesting morphology for MEF, where an external silica shell is useful for controlling distance between fluorophores and nanoparticles, while at the same time allowing particle functionalization and stability. Unlike the silica shell-metal core nanoparticles prepared by Stöber method, where the silica shell is porous, the SHINs have compact and thin dielectric shells (You et al. 2020).

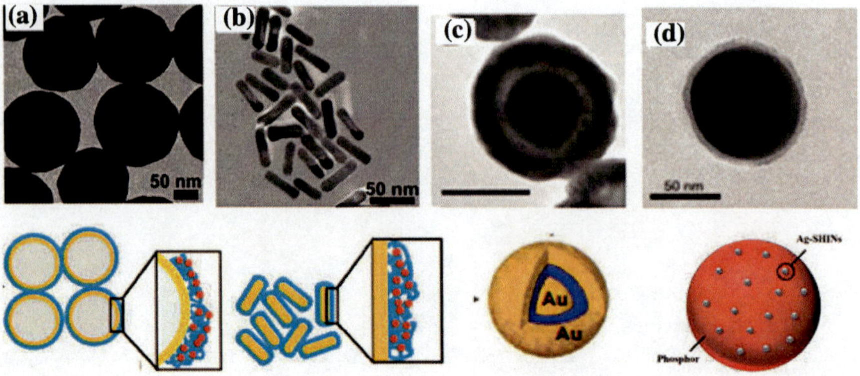

Fig. 3.12 TEM images of **a** Au nanoshells and **b** Au nanorods. Lower panel shows schematic for the conjugation of fluorescent dye on the respective nanoparticle surfaces using human serum albumin protein as spacer and linker for the dye. Reprinted with permission from Bardhan et al. (2009). Orozco et al. (2014)You et al. (2020) Copyright 2009 American Chemical Society. **c** TEM image of an Au nanomatryoshka. Lower panel shows schematic for dye doped in the interstitial SiO_2 layer (blue colour). Reprinted from. Copyright 2014 American Chemical Society under Creative Commons License. **d** TEM image of Ag shell isolated nanoparticle (SHIN). Lower panel shows the conjugation of Ag-SHINs with a red emissive phosphor particle through electrostatic adhesion. Reprinted with permission from. Copyright 2020 American Chemical Society

Metal nanoparticle films with rough surface morphologies can be prepared in different ways (Li et al. 2017). One approach to obtain colloid coated substrates is to first silanize a glass substrate by dipping it in 2-propanol containing (3-mercaptopropyl)trimethoxysilane, followed by immersion in an already prepared Ag or Au sol. The second approach is to deposit nanoparticles in situ by immersing the glass slides directly in the reaction mixture. Silver island films (SIFs) can be prepared in this manner by the well-known silver mirror reaction, where $AgNO_3$ is used to form Tollen's reagent, $Ag(NH_3)_2OH$, which is reduced subsequently to elemental Ag with glucose. The third approach is physical vapour deposition or sputtering methods to deposit a very thin discontinuous layer of Ag/Au on glass substrates. This is followed by an annealing step to generate rough surface morphologies of the metal nanoparticles on the substrates (Dutta Choudhury et al. 2012a). Figure 3.13 shows representative images of Ag nanostructures obtained by chemical reduction and thermal vapour deposition methods.

With developments in lithographic nanofabrication techniques, plasmonic substrates have progressed toward more exotic and complex architectures. A few of these structures are presented in Fig. 3.14, along with brief discussions about their effects on fluorescence.

Using electron beam lithography technique, Kinkhabwala et. al. fabricated Au bow tie nanoantenna structures (Fig. 3.14a) (Kinkhabwala et al. 2009). These nanoantennas enhanced the fluorescence intensity of NIR emissive dyes to a large extent (~1340-fold) when the dyes were optimally situated in the junction between the tips

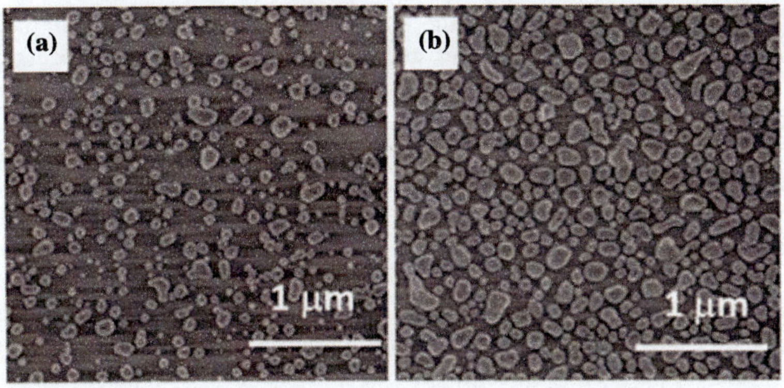

Fig. 3.13 SEM images of silver island films (SIFs) on glass slides obtained by **a** in situ silver mirror reaction and **b** thermal vapour deposition. Reprinted with permission from Dutta Choudhury et al. 2012a. Copyright 2012 American Chemical Society

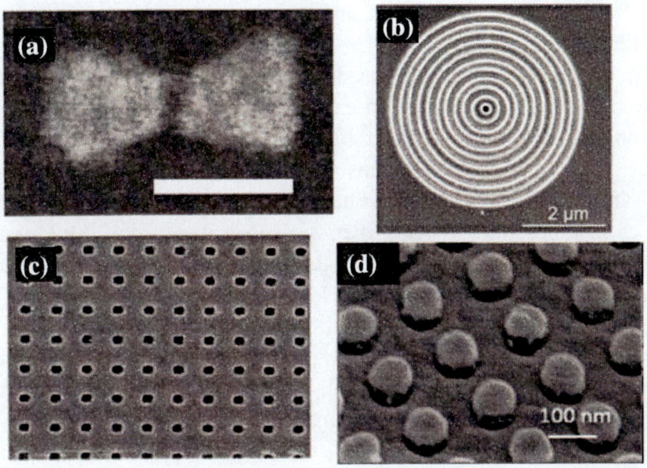

Fig. 3.14 **a** SEM image of Au bow tie nanoantenna fabricated by electron beam lithography. Reprinted with permission from Kinkhabwala et al. (2009). Aouani et al. (2011)Song et al. (2015)Zang et al. (2019) Copyright 2009 Nature Publishing Group. **b** SEM image of a "bulls eyes" structure fabricated by focussed ion beam milling on an Au film. Reprinted with permission from. Copyright 2011 American Chemical Society. **c** SEM image of Au nanoholes fabricated by nanoimprinting. Reprinted with permission from. Copyright 2015 Elsevier. **d** SEM image of 3D nanoantenna structure of Au nanoparticles on SiO_2 pillars fabricated by nanoimprint lithography. Reprinted with permission from. Copyright 2019 Wiley

of two triangles. These positions are the “hot spots” where enhancement of the local electric field is the highest. In another study, Aouani et. al. have used focussed ion beam milling to generate “bulls eyes” structures, which consist of nanoapertures surrounded by periodic corrugations (Fig. 3.14b) (Aouani et al. 2011). For fluorophores located near these nanoapertures, the count rate per molecule was enhanced up to 120-fold. At the same time, emission was coupled into a narrow angular cone perpendicular to the sample. Nanoimprint lithography is another technique that has been used for plasmonic substrate fabrication. Using this method, Song et. al. fabricated Au nanohole arrays (Fig. 3.14c). When applied in an aptamer-based sensor design, these nanohole arrays provided ~20 fold enhancement of quantum dot emission, which resulted in picomolar sensitivity for detection of thrombin (Song et al. 2015). More recently, Zang et al. developed a 3D nanoantenna plasmonic sensor using nanoimprint lithography, for the ultrasensitive detection of Ebola virus antigen (Zang et al. 2019). The 3D nanoantenna structure consisted of SiO_2 nanopillar array with Au nanodisks on top of each pillar, Au nanodots on the sidewalls of the pillars and an Au plane at the base (Fig. 3.14d). This plasmonic sensor platform provided a massive 240,000-fold improvement in sensitivity compared to existing methods for Ebola virus detection.

Apart from plasmonic substrates based on a single metal, bimetallic structures and metal nanocomposites have also been explored to expand the optical range of the plasmon resonance and improve the stability of the substrate. Zhang et al. prepared dielectric core-metal shell particles consisting of two concentric metallic shells, one of them Au and another Ag, by wet chemical synthesis procedures (Zhang et al. 2012). Hao et al. fabricated Al-Ag bimetallic substrates by ion implantation of Ag on an Al film. These substrates were capable of providing broadband fluorescence enhancement from UV to visible regions (Hao et al. 2014). Nanocomposite substrates of Ag and Au have also been prepared through galvanic replacement reaction by immersing a glass slide coated with Ag film into a solution of $HAuCl_4$, which led to the spontaneous replacement of Ag with Au (Fig. 3.15). This nanocomposite material provided $\sim$50-fold enhancement for fluorophores in proximity to the hotspots in the substrate (Dutta Choudhury et al. 2012b). Single molecule fluorescence studies confirmed the increased photostability of dyes coupled to this substrate.

Unlike MEF, the substrates for SPCE are in the form of smooth continuous metal films (Fig. 3.16). In this case, large area substrate fabrication is easily achieved by sputtering or thermal vapour deposition of metallic layers (~50 nm) on glass slides, followed by deposition of a dielectric layer. The dielectric layer can be obtained by thermal vapour deposition (in the case of silica) or by spin coating (in the case of a polymer like polyvinyl alcohol, PVA) (Dutta Choudhury et al. 2015a). The thickness of the layers is controlled by controlling the deposition time, spin coating parameters and concentration of the polymer solution. Different geometries, like metal-dielectric, metal-dielectric-metal or plasmon coupled waveguide structures can be prepared to obtain SPCE or waveguide-coupled emission.

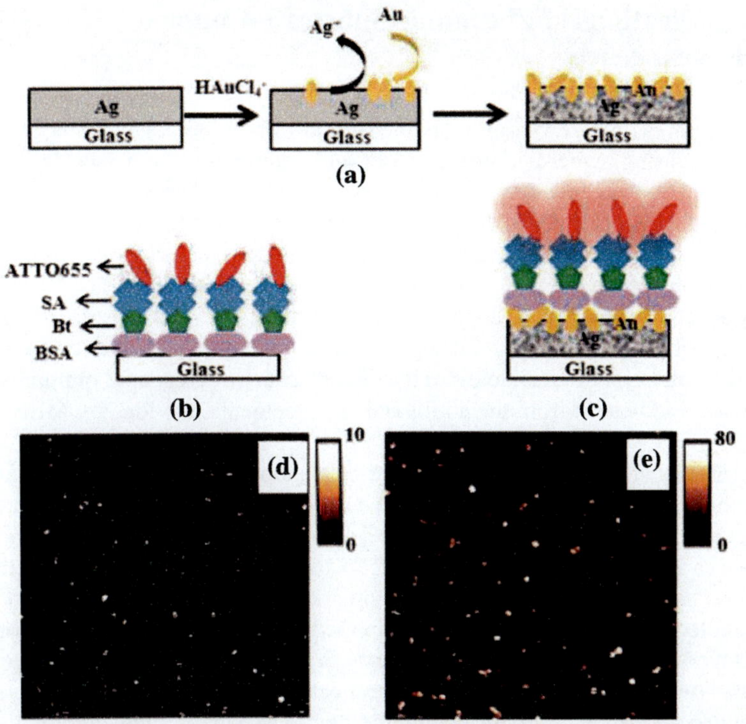

Fig. 3.15 Schematics for fabrication of Au–Ag nanocomposite substrate by galvanic replacement and conjugation of fluorophores (ATTO655) using bovine serum albumin (BSA), biotin (Bt) and streptavidin (SA) assembly **a–c**. Single molecule scanning confocal images of ATTO655 on bare glass **d** and on nanocomposite substrate **e** with respective intensity scales. Adapted with permission from Dutta Choudhury et al. 2012b. Copyright 2012 American Chemical Society

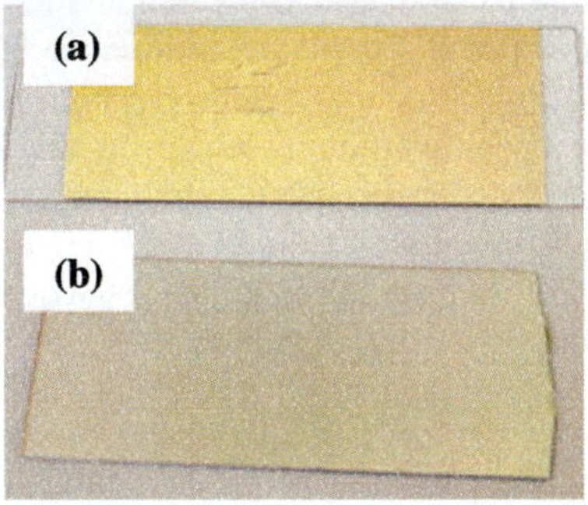

Fig. 3.16 Photographs of continuous thin layer films of **a** Au prepared by thermal vapour deposition and **b** TiN prepared by sputtering

3.4 Applications of Plasmonic Substrate-Coupled Fluorescence

One of the areas where plasmon-coupled fluorescence has made a major impact is optical sensing, where it offers many-fold improvements in sensitivity over conventional fluorescence-based methods. Different kinds of analytes including toxic metals, model biomolecules (DNA, RNA, pyrophosphates, ATP etc.), various disease biomarkers and pathogens, have been detected based on the principles of plasmon-coupled fluorescence (Li et al. 2015; Hageneder et al. 2021; Su et al. 2014; Semeniak et al. 2022; Song et al. 2015; Zang et al. 2019; Wang et al. 2021; Liu et al. 2020). Importantly, plasmonic substrates can be readily incorporated into already established fluorescence technologies like, fluorescence microscopes or fluorescence plate readers to benefit from the additional enhancements provided by MEF, SPCE or SPEF.

Recently, Luan et al. have prepared a flexible nanoparticle embedded substrate that can be applied as an add-on plasmonic patch to obtain uniformly large enhancements (Luan et al. 2018). A schematic of this novel approach used for fluoro-immunoassays is shown in Fig. 3.17. The plasmonic patch was added on a standard 96-well plate with glass bottom to detect early-stage biomarkers for acute kidney injury (kidney injury molecule-1, KIM1). It was also applied in a protein microarray format to detect several protein biomarkers for kidney injury. The plasmonic patch provided more than two orders of magnitude fluorescence enhancement, ~300-fold lower limit of detection and three orders of magnitude higher dynamic range, across all assay formats.

The detection of DNA by the SPEF approach using a molecular beacon has been described earlier (cf. Fig. 3.9b) (Su et al. 2014). An alternative enhanced interfacial molecular beacon strategy for DNA/protein detection using SPCE has been demonstrated by Cao et al. In their approach, graphene oxide has been used as an additional interfacial quencher layer that works in conjunction with the plasmonic metal layer (Cao et al. 2019). In the absence of analyte, the fluorophore labelled DNA probe is in close contact with graphene oxide and the metal surface, leading to strong quenching of fluorescence. On binding to the target analyte, the conformation of the probe changes, placing the fluorophores away from the quenchers (graphene oxide and metal surface) and in the region of plasmonic enhancement (Fig. 3.18a). This results in a better contrast for the quenching-to-enhancing transition, thus improving the sensitivity of detection.

Pang et al. developed a label-free fluorescence sensor for H5N1 influenza virus, based on MEF with core–shell metal nanoparticles (Fig. 3.18b) (Pang et al. 2015). In the presence of the recombinant hemagglutinin (rHA) protein of H5N1 virus, the aptamer sensor attached to the core–shell nanoparticle acquires a G-quadruplex structure, which further binds to the dye Thiazole orange (TO) that is otherwise weekly fluorescent in the free state in solution. This sequence of events brings the dye in close proximity of the metal nanoparticle leading to a large plasmon-coupled enhancement in fluorescence signal.

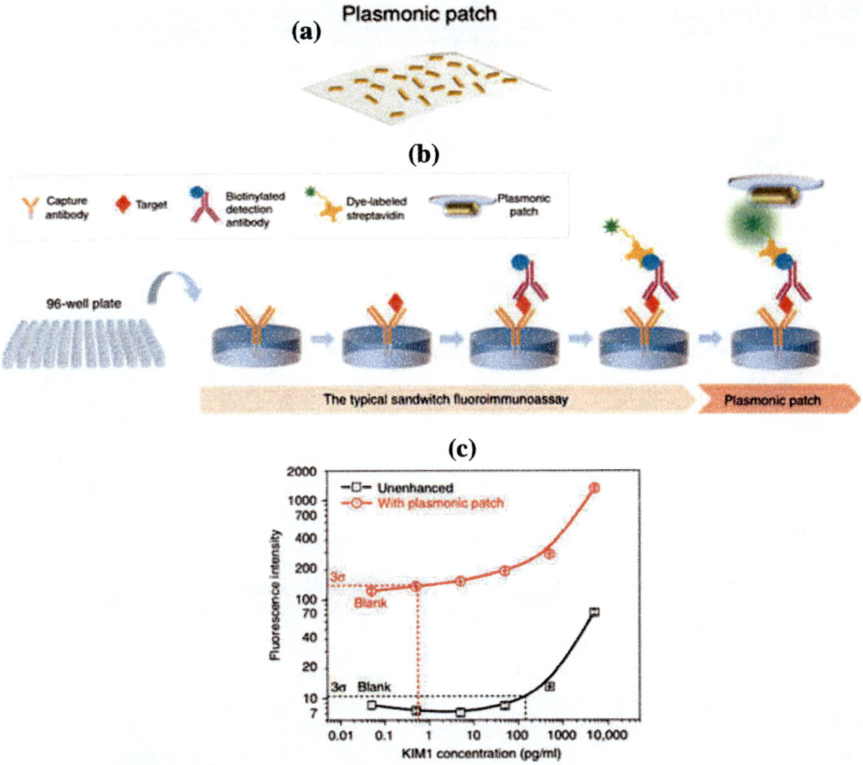

Fig. 3.17 **a** Representation of an add-on plasmonic patch and **b** fluoro-immunoassay using plasmonic patch. **c** Sensor response with and without the plasmonic patch. Adapted from Luan et al. (2018) under Creative Commons License, published by Springer Nature

Recently, Choi et al. have used the MEF strategy to detect cell-free DNA (cfDNA), which are important biomarkers for early stage cancer detection (Choi et al. 2020). Sensing of cfDNA is difficult because of their low abundance in blood and body fluids. Therefore, additional nucleic acid amplification steps are usually required to improve the sensitivity of detection. Choi et al. designed their sensor based on the CRISPR-Cas12a-mediated trans-cleavage reaction, where the activated CRISPR-Cas12a complex selectively degrades single-strand DNA (ssDNA) while leaving double-strand DNA (dsDNA) intact. Two Au nanoparticles (20 and 60 nm) were connected by a short ssDNA strand (2 nm) as well as a long fluorophore labelled dsDNA strand (7 nm). In this state, the fluorescence from the dye is strongly quenched due to its closeness to the 60 nm Au nanoparticle. In the presence of target cfDNA, the CRISPR-Cas12a complex is activated leading to cleavage of ssDNA and release of the fluorophore tagged 20 nm Au nanoparticle. This plasmon-coupled fluorophore is situated at an optimum distance to exhibit MEF, leading to an off-to-on transition in

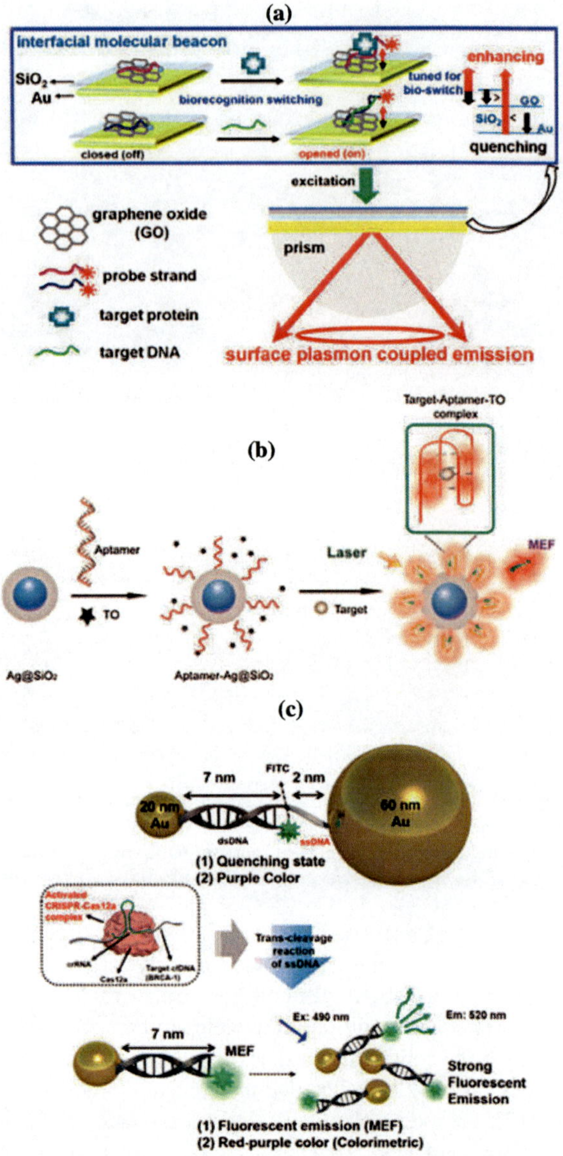

Fig. 3.18 **a** Detection of analytes (DNA and protein) by an interfacial molecular beacon strategy using graphene oxide as a quencher, based on SPCE. Reprinted with permission from Cao et al. (2019). Copyright 2019 American Chemical Society. **b** Detection of rHA protein of H5N1 virus by aptamer sensing, based on MEF. Reprinted with permission from Pang et al. (2015). Copyright 2015 Elsevier. **c** A MEF based approach for detection of cfDNA (cancer biomarker) without involving any nucleic acid amplification step. Reprinted with permission from Choi et al. (2020). Copyright 2021 American Chemical Society

fluorescence signal. The implementation of MEF resulted in the detection of cfDNA at femtomolar concentrations without requiring any amplification steps.

In an interesting application of SPCE, Xu et al. have recently demonstrated a label-free fluorescent nanofilm sensor for in situ monitoring of the growth of metal–organic frameworks (MOF) (Xu et al. 2022). The principle behind this application is that, as the thickness of the dielectric layer above a metal nano film increases, a plasmon-coupled waveguide structure is generated in which emission can be coupled at different angles and with both S- and P-polarizations. The study was carried out in the RK configuration. A thin layer of Rhodamine B dye was first coated on to an Au film, and a zeolite imidazolate MOF film was grown on this substrate. The thickness of the MOF film could be sensitively characterized by the unique angle distribution and polarization of the coupled emission from Rhodamine B (Fig. 3.19).

Single molecule detection is another important application of plasmonic substrate-coupled fluorescence. Several studies have been carried out to optimize the parameters of metal nanostructures (shape, assembly, position of LSP resonances) that can provide the largest enhancement factors to improve the detectability of single fluorescent molecules with low quantum yields (Bardhan et al. 2009; Lu et al. 2020; Dutta Choudhury et al. 2012a; Kinkhabwala et al. 2009). Plasmonic coupling is also useful for fluorescence correlation spectroscopy (FCS), which is an important bioanalytical technique for studying molecular interactions and dynamics based on fluorescence intensity fluctuations in a small detection volume. Plasmonic nanoapertures have been found to reduce the detection volumes so that measurements can be carried out at high concentrations that are more relevant to the realistic biological concentration range. Besides physical confinement within the nanoapertures, the volume reduction also arises because of the creation of a highly fluorescent zone in the vicinity of the nano structures due to MEF effect (Dutta Choudhury et al. 2012a; Winkler et al. 2018). This increases fluorescence count rates and improves detection of single diffusing molecules.

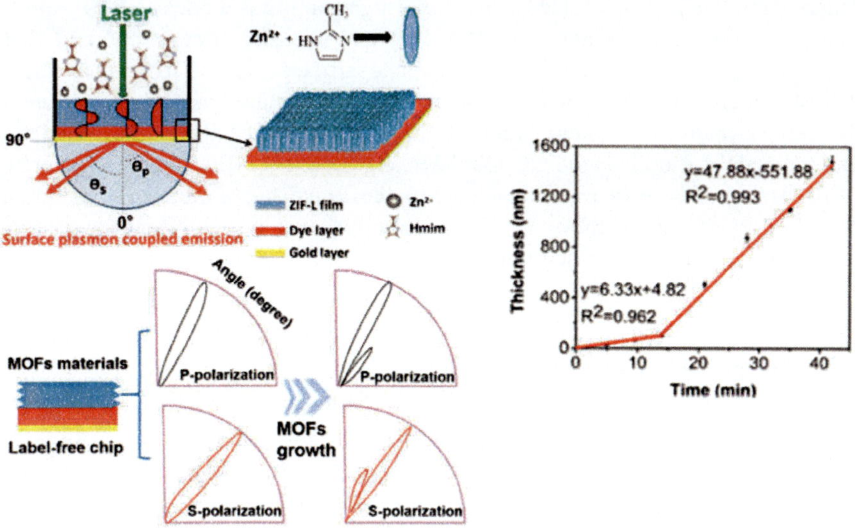

Fig. 3.19 Application of SPCE for in situ monitoring of the growth of metal–organic framework (MOF) layer. Reprinted with permission from Xu et al. (2022). Copyright 2022 American Chemical Society

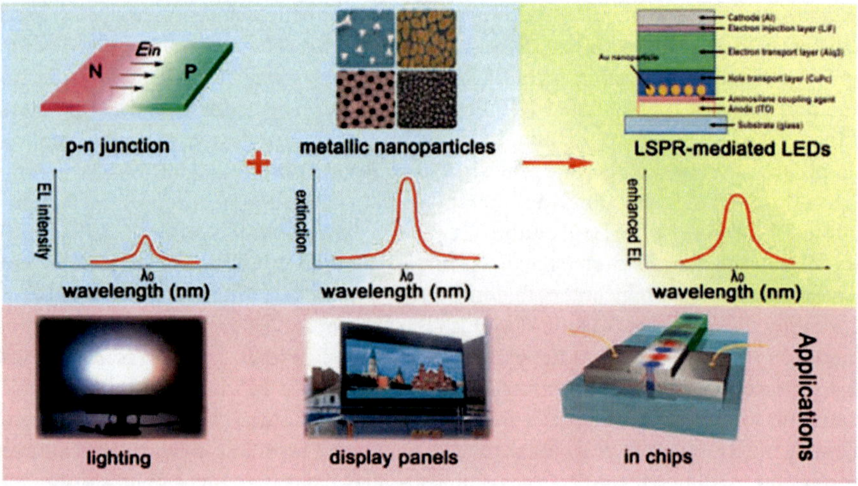

Fig. 3.20 Schematic of plasmon-coupled luminescence in LEDs and its prospective applications in display technologies. Reprinted from Gu et al. (2011). Copyright 2011 under Creative Commons License, published by Springer

Plasmon-coupled fluorescence is also being explored in the field of display technologies. The angle and thickness dependent emission in metal-dielectric-metal and plasmon-coupled waveguide structures is relevant for optical signage applications (Dutta Choudhury et al. 2015b). Plasmonic light emitting diodes (LEDs) are of immense interest for enhancing the efficiency and intensity of conventional LEDs by coupling of excitons with LSPs in metal nanoparticles (Fig. 3.20) (Gu et al. 2011; Okamoto et al. 2017). You et al. have shown that large enhancement in emission of phosphors (which are an integral component of LEDs) can be achieved by coupling with the LSP of shell isolated nanoparticles, Ag-SHINs (cf. Fig. 3.12d) (You et al. 2020). Although there are many challenges in the design and fabrication of optimum structures, plasmonic coupling has the potential to enable high efficiency, energy saving emissive devices in the future.

3.5 Conclusions and Perspectives

The coupling of fluorescence with plasmonics has allowed nanoscale imaging and detection of emission with high efficiency. The unique optical properties of plasmonic nanostructures and a fundamental understanding of metal-fluorophore interactions has generated a wealth of information for developing new generation of fluorescence probes with applications in healthcare and environmental monitoring, as well as in the design of energy saving display technologies.

Surface plasmon resonances can concentrate and magnify the incident electromagnetic field and modulate the fluorescence parameters to enable ultrasensitive detection. This overcomes limitations like, insufficient sensitivity, low specificity, and requirement of amplification steps to multiply and quantify biological species that are in low abundance. The compatibility of plasmonic sensing with portable platforms can accelerate the transformation toward wearable devices and point-of-care kits. We can also look forward to plasmophore coupled nanoscale endoscopic tools that will enable interrogation of small volumes and in vivo applications. The arrival of new plasmonic materials besides metals will facilitate adequate choice of materials and functions for application based structure design. It is anticipated that with progress in the nanofabrication methods, more complex nano architectures will be possible and advanced simulation and machine learning tools will help in assessing the efficacy of a plasmonic structure for fluorescence modulation before its fabrication. The field of plasmon-coupled fluorescence is expanding into diverse areas of material and biological sciences. As the world recovers from the devastating effects of COVID-19, it is quite clear that fast and accurate diagnostic systems are urgently needed to tackle fast-spreading infectious diseases and increase our preparedness for the future. In this regard, plasmon-enhanced fluorescence biosensors hold great promise due to their easy integration with established fluorescence spectroscopy techniques, simple instrumentation and minimal sample pre-treatment.

Acknowledgements The author gratefully acknowledges the support received from Bhabha Atomic Research Centre (BARC), Mumbai and the encouragement from Dr. A. K. Tyagi, Director, Chemistry Group, BARC, Dr. A. Kumar, Head, Radiation & Photochemistry Division, BARC, and Dr. A. C. Bhasikuttan, Head, Molecular Photochemistry Section, BARC. The author also thanks Prof. J. R. Lakowicz for introducing her to the world of plasmonics in fluorescence.

References

Akbay, N., Lakowicz, J.R.: Ray K (2012) Distance-dependent metal-enhanced intrinsic fluorescence of proteins using polyelectrolyte layer-by-layer assembly and aluminum nanoparticles. J. Phys. Chem. C **116**, 10766–10773 (2012)

Aouani, H., Mahboub, O., Bonod, N., Devaux, E., Popov, E., Rigneault, H., Ebbesen, T.W., Wenger, J.: Bright unidirectional fluorescence emission of molecules in a nanoaperture with plasmonic corrugations. Nano Lett. **11**, 637–644 (2011)

Atwater, H.A.: The promise of plasmonics. Sci. Am. **17**, 56–63 (2007)

Badugu, R., Szmacinski, H., Ray, K., Descrovi, E., Ricciardi, S., Zhang, D., Chen, Y., Huo, Y., Lakowicz, J.R.: Fluorescence spectroscopy with metal–dielectric waveguides. J. Phys. Chem. C **119**, 6245–16255 (2015)

Bardhan, R., Grady, N.K., Cole, J.R., Joshi, A., Halas, N.: Fluorescence enhancement by Au nanostructures: nanoshells and nanorods. ACS Nano **3**, 744–752 (2009)

Botequim, D., Silva, I.I.R., Serra, S.G., Melo, E.P., Prazeres, D.M.F., Costa, S.M.B., Paulo, P.M.R.: Fluorescent dye nano-assemblies by thiol attachment directed to the tips of gold nanorods for effective emission enhancement. Nanoscale **12**, 6334–6345 (2020)

Cao, S.H., Weng, Y.H., Xie, K.X., Wang, Z.C., Pan, X.H., Chen, M., Zhai, Y.Y., Xu, L.T., Li, Y.Q.: Surface plasmon coupled fluorescence-enhanced interfacial "molecular beacon" to probe biorecognition switching: an efficient, versatile, and facile signaling biochip. ACS ApplBio Mater **2**, 625–629 (2019)

Chen, Y., Zhang, D., Han, L., Rui, G., Wang, X., Wang, P., Ming, H.: Surface-plasmon-coupled emission microscopy with a polarization converter. Opt. Lett. **38**, 736–738 (2013)

Choi, J.H., Lim, J., Shin, M., Paek, S.H., Choi, J.W.: CRISPR-Cas12a-based nucleic acid amplification-free DNA biosensor via Au nanoparticle-assisted metal-enhanced fluorescence and colorimetric analysis. Nano Lett. **21**, 693−699 (2020)

Dutta Choudhury, S., Ray, K., Lakowicz, J.R.: Silver nanostructures for fluorescence correlation spectroscopy: reduced volumes and increased signal intensities. J. Phys. Chem. Lett. **3**, 2915–2919 (2012a)

Dutta Choudhury, S., Badugu, R., Ray, K., Lakowicz, J.R.: Silver−gold nanocomposite substrates for metal-enhanced fluorescence: ensemble and single-molecule spectroscopic studies. J. Phys. Chem. C **116**, 5042–5048 (2012b)

Dutta Choudhury, S., Badugu, R., Lakowicz, J.R.: Surface-plasmon induced polarized emission from Eu(III)—a class of luminescent lanthanide ions. Chem. Commun. **50**, 9010–9013 (2014)

Dutta Choudhury, S., Badugu, R., Lakowicz, J.R.: Directing fluorescence with plasmonic and photonic structures. Acc. Chem. Res. **48**, 2171–2180 (2015a)

Dutta Choudhury, S., Badugu, R., Ray, K., Lakowicz, J.R.: Directional emission from metal−dielectric−metal structures: effect of mixed metal layers, dye location, and dielectric thickness. J. Phys. Chem. C **119**, 3302–3311 (2015b)

Fontaine, N., Picar-Lafond, A., Asselin, J., Boudreau, D.: Thinking outside the shell: novel sensors designed from plasmon-enhanced fluorescent concentric nanoparticles. Analyst **145**, 5965–5980 (2020)

García de Abajo 2014 García de Abajo, F.J.: Graphene plasmonics: challenges and opportunities. ACS Photon. **1**, 135–152 (2014)

Gryczynski, I., Malicka, J., Gryczynski, Z., Lakowicz, J.R.: Radiative decay engineering 4. Experimental studies of surface plasmon-coupled directional emission. Anal. Biochem. **324**, 170–182 (2004)

Gu, X., Qiu, T., Zhang, W., Chu, P.K.: Light-emitting diodes enhanced by localized surface plasmon resonance. Nanoscale Res. Lett. **6**, 199 (2011)

Gutiérrez, Y., Brown, A.S., Moreno, F., Losurdo, M.: Plasmonics beyond noble metals: exploiting phase and compositional changes for manipulating plasmonic performance. J. Appl. Phys. **128**, 080901 (2020)

Hageneder, S., Jungbluth, V., Soldo, R., Petri, C., Matthias, P., Kreivi, M., Weihaüsel, A., Jonas, U., Dostalek, J.: Responsive hydrogel binding matrix for dual signal amplification in fluorescence affinity biosensors and peptide microarrays. ACS Appl. Mater. Interfaces **13**, 27645–27655 (2021)

Hao, Q., Du, D., Wang, C., Li, W., Huang, H., Li, J., Qiu, T., Chu, P.K.: Plasmon-induced broadband fluorescence enhancement on Al-Ag bimetallic substrates. Sci. Rep. **4**, 6014 (2014)

Jiao, X., Wang, Y., Blair, S.: Plasmonic enhancement of UV fluorescence. In: Geddes, C.D., (ed.), Surface Plasmon Enhanced, Coupled and Controlled Fluorescence, Chap 18, pp. 295–308. John Wiley & Sons, Inc. USA (2017)

Khatua, S., Paulo, P.M.R., Yuan, H., Gupta, A., Zijlstra, P., Orrit, M.: Resonant plasmon enhancement of single-molecule fluorescence by individual gold nanorods. ACS Nano **8**, 4440–4449 (2014)

Kinkhabwala, A., Yu, Z., Fan, S., Avlasevich, Y., Müllen, K., Moerner, W.E.: Large single-molecule fluorescence enhancements produced by a bowtie nanoantenna. Nat. Photonics **3**, 654–657 (2009)

Lakowicz, J.R.: Radiative decay engineering 3. Surface plasmon-coupled directional emission. Anal. Biochem. **324**, 153–169 (2004)

Lakowicz, J.R.: Radiative decay engineering 5: Metal-enhanced fluorescence and plasmon emission. Anal. Biochem. **337**, 171–194 (2005)

Lakowicz, J.R.: Plasmonics in biology and plasmon-controlled fluorescence. Plasmonics **1**, 5–33 (2006a)

Lakowicz, J.R.: Principles of Fluorescence Spectroscopy, 3rd edn. Publisher, Springer, Singapore (2006b)

Li, M., Cushing, S.K., Wu, N.: Plasmon-enhanced optical sensors. Analyst **140**, 193–406 (2015)

Li, J.F., Li, C.Y., Aroca, R.F.: Plasmon-enhanced fluorescence spectroscopy. Chem. Soc. Rev. **46**, 3962–3979 (2017)

Link, S., El-Sayed, M.A.: Optical properties and ultrafast dynamics of metallic nanocrystals. Annu. Rev. Phys. Chem. **54**, 331–366 (2003)

Liu, J., Jalali, M., Mahshid, S., Wachsmann-Hogiu, S.: Are plasmonic optical biosensors ready for use in point-of-need applications? Analyst **145**, 364–384 (2020)

Lu, X., Ye, G., Deep, P., Chiechi, R.C., Orrit, M.: Quantum yield limits for the detection of single-molecule fluorescence enhancement by a gold nanorod. ACS Photon. **7**, 2498–2505 (2020)

Luan, J., Morrissey, J.J., Wang, Z., Derami, H.G., Liu, K.K., Cao, S., Jiang, Q., Wang, C., Kharasch, E.D., Naik, R.R., Singamaneni, S.: Light: science & applications **7**, 29 (2018)

Maier, S.A.: Plasmonics: Fundamentals and Applications. Publisher, Springer, USA (2007)

Matveeva, E.G., Gryczynski, Z., Malicka, J., Lukomska, J., Makowiec, S., Berndt, K.W., Lakowicz, J.R., Gryczynskia, I.: Directional surface plasmon-coupled emission: application for an immunoassay in whole blood. Anal. Biochem. **344**, 161–167 (2005)

Mishra, P., Debnath, A.K., Dutta Choudhury, S.: Titanium nitride as an alternative and reusable plasmonic substrate for fluorescence coupling. Phys. Chem. Chem. Phys. **24**, 6256–6265 (2022)

Naik, G.V., Shalaev, V.M., Boltasseva, A.: Alternative plasmonic materials: beyond gold and silver. Adv. Mater. **25**, 3264–3329 (2013)

Neumann, T., Johansson, M.-L., Kambhampati, D., Knoll, W.: Surface-plasmon fluorescence spectroscopy. Adv. Funct. Mater. **12**, 575–586 (2002)

Okamoto, K., Funato, M., Kawakami, Y., Tamada, K.: High-efficiency light emission by means of exciton–surface-plasmon coupling. J. Photochem. Photobiol. C Photochem. Rev. **32**, 58–77 (2017)

Orozco, C.A., Liu, J.G., Knight, M.W., Wang, Y., Day, J., Nordlander, P., Halas, N.J.: Fluorescence enhancement of molecules inside a gold nanomatryoshka. Nano Lett. **14**, 2926–2933 (2014)

Pang, Y., Rong, Z., Wang, J., Xiao, R., Wang, S.: A fluorescent aptasensor for H5N1 influenza virus detection based-on the core–shell nanoparticles metal-enhanced fluorescence (MEF). Biosens. Bioelect. **66**, 527–532 (2015)

Ray, K., Badugu, R., Lakowicz, J.R.: Distance-dependent metal-enhanced fluorescence from Langmuir Blodgett monolayers of alkyl-NBD derivatives in silver island films. Langmuir **22**, 8374–8378 (2006)

Rycenga, M., Cobley, C.M., Zeng, Z., Li, W., Moran, C.H., Zhang, Q., Qin, D., Xia, Y.: Controlling the synthesis and assembly of silver nanostructures for plasmonic applications. Chem. Rev. **111**, 3669–3712 (2011)

Sathish, S., Kostov, Y., Rao, G.: High-resolution surface plasmon coupled resonant filter for monitoring of fluorescence emission from molecular multiplexes. Appl. Phys. Lett. **94**, 223113 (2009)

Schuller, J.A., Barnard, E.S., Cai, W., Jun, Y.C., White, J.S., Brongersma, M.L.: Plasmonics for extreme light concentration and manipulation. Nat. Mat. **9**, 193–204 (2010)

Semeniak, D., Cruz, D.F., Chilkoti, A., Mikkelsen, M.H.: Plasmonic fluorescence enhancement in diagnostics for clinical tests at point-of-care: a review of recent technologies. Adv. Mater. 2107986 (2022)

Song, H.Y., Wong, T.I., Guo, S., Deng, J., Tan, C., Gorelik, S., Zhou, X.: Nanoimprinted thrombin aptasensor with picomolar sensitivity based on plasmon excited quantum dots. Sens Actuat. B Chem. **221**, 207–216 (2015)

Su, Q., Wesner, D., Schönherr, H., Nöll, G.: Molecular beacon modified sensor chips for oligonucleotide detection with optical readout. Langmuir **30**, 14360–14367 (2014)

Su, Q., Jiang, C., Gou, G., Long, Y.: Surface plasmon-assisted fluorescence enhancing and quenching: from theory to application. ACS Appl. Bio Mater. **4**, 4684–4705 (2021)

Wang, Y., Huang, C.J., Jonas, U., Wei, T., Dostalek, J., Knoll, W.: Biosensor based on hydrogel optical waveguide spectroscopy. Biosens. Bioelectron. **25**, 1663–1668 (2010)

Wang, R., Zhang, D., Zhu, L., Wen, X., Chen, J., Kuang, C., Liu, X., Wang, P., Ming, H., Badugu, R., Lakowicz, J.R.L.: Selectable surface and bulk fluorescence imaging with plasmon-coupled waveguides. J. Phys. Chem. C **119**, 22131–22136 (2015)

Wang, M., Wang, M., Zheng, G., Dai, Z., Ma, Y.: Recent progress in sensing application of metal nanoarchitecture-enhanced fluorescence. Nanoscale Adv **3**, 2448–2465 (2021)

Winkler, P.M., Regmi, R., Flauraud, V., Brugger, J., Rigneault, H., Wenger, J., García-Parajo, M.F.: Optical antenna-based fluorescence correlation spectroscopy to probe the nanoscale dynamics of biological membranes. J. Phys. Chem. Lett. **9**, 110–119 (2018)

Xu, L.T., Chen, M., Weng, Y.H., Xie, K.X., Wang, J., Cao, S.H., Li, Y.Q.: Label-free fluorescent nanofilm sensor based on surface plasmon coupled emission: in situ monitoring the growth of metal−organic frameworks. Anal. Chem. **94**, 6430–6435 (2022)

You, C.Y., Lin, L.H., Wang, J.Y., Zhang, F.L., Radjenovic, P., Yang, Z., Tian, Z.Q., Li, J.F.: Plasmon-enhanced fluorescence of phosphors using shell-isolated nanoparticles for display technologies. ACS Appl. Nano Mater. **3**, 5846–5854 (2020)

Yu, H., Peng, Y., Yang, Y., Li, Z.Y.: Plasmon-enhanced light-matter interactions and applications. Npj Comput Mater **5**, 45 (2019)

Zang, F., Su, Z., Zhou, L., Konduru, K., Kaplan, G., Chou, S.Y.: Ultrasensitive Ebola virus antigen sensing via 3D nanoantenna arrays. Adv. Mater. 1902331 (2019)

Zhang, J., Fu, Y., Mahdavi, F.: Bimetallic nanoshells for metal-enhanced fluorescence with broad band fluorophores. J. Phys. Chem. C **116**, 24224–24232 (2012)

Chapter 4
Lanthanide-Doped Materials for Optical Applications

Priyam Singh, Santosh Kachhap, Manisha Sharma, Prabhakar Singh, and S. K. Singh

4.1 Introduction

Lanthanides (also called rare earth elements) are a family of 15 elements from lanthanum to lutetium, which are located at the top row of the f-block elements lying at the bottom of the periodic table (Gschneidner et al. 1979). Table 4.1 represents the symbol, atomic number, and electronic configuration of the RE^{3+} ions with their respective ground states. The electronic configuration of these elements is $[Xe]4f^n5d^{0,1}6s^2$ where n varies from 0 to 14 when going from lanthanum to lutetium. Xe (xenon) has 54 electrons with electronic configuration as $1s^22s^22p^63s^23p^64s^23d^{10}4p^65s^24d^{10}5p^6$. Because of this electronic similarity, it is quite difficult to separate rare earth elements from each other. The two additional elements of group IIIb viz. Sc and Y, which show similar electronic configurations ($[Ar]3d^14s^2$ for Sc and $[Kr]4d^15s^2$ for Y), also commonly exist in rare earth metal deposits (Boyn 1988). Therefore, the physical and chemical properties of these two elements show similarity with lanthanides. So, the collection of these two elements and 15 lanthanides are called rare earth metals (Judd 1962).

All the lanthanide ions reflect a very stable, common oxidation state of +3. However, +2 (e.g., Sm^{2+}, Yb^{2+} and Eu^{2+}) and +4 (e.g., Ce^{4+} and Tb^{4+}) oxidation states are also possible (Judd 1962; Ofelt 1962). It is not possible to get optical emission of these rare earth elements in their metal form. When the rare earth ions are doped in a suitable host matrix, triple cation corresponding to the $4f^n$ configuration displays unusual characteristics (Chen et al. 2016a; Dhoble et al. 2019). The energy levels corresponding to this $4f^n$ configuration are usually very large in number and are relatively isolated from the $4f^{n-1}5d$ configuration (which lies at much higher energies). The significant aspect of intra f–f transitions is that the absorption or emission spectra of these ions lie in the ultra-violet (UV), visible, and near-infrared (NIR)

P. Singh · S. Kachhap · M. Sharma · P. Singh · S. K. Singh (✉)
Department of Physics, Indian Institute of Technology (BHU), Varanasi 221005, India
e-mail: sunilks.app@itbhu.ac.in

R. S. Ningthoujam and A. K. Tyagi (eds.), *Handbook of Materials Science, Volume 1*, Indian Institute of Metals Series, https://doi.org/10.1007/978-981-99-7145-9_4

Table 4.1 Symbol, electronic configuration, and the ground state of trivalent lanthanide ions

Element	Symbol	Atomic no. (Z)	Electronic configuration (Ln^{3+})	Ground state (Ln^{3+})
Lanthanum	La	57	[Xe] $4f^0$	$^1S_0$
Cerium	Ce	58	[Xe] $4f^1$	$^2F_{5/2}$
Praseodymium	Pr	59	[Xe] $4f^2$	$^3H_4$
Neodymium	Nd	60	[Xe] $4f^3$	$^4I_{9/2}$
Promethium	Pm	61	[Xe] $4f^4$	$^5I_4$
Samarium	Sm	62	[Xe] $4f^5$	$^6H_{5/2}$
Europium	Eu	63	[Xe] $4f^6$	$^7F_0$
Gadolinium	Gd	64	[Xe] $4f^7$	$^8S_{7/2}$
Terbium	Tb	65	[Xe] $4f^8$	$^7F_6$
Dysprosium	Dy	66	[Xe] $4f^9$	$^6H_{15/2}$
Holmium	Ho	67	[Xe] $4f^{10}$	$^5I_8$
Erbium	Er	68	[Xe] $4f^{11}$	$^4I_{15/2}$
Thulium	Tm	69	[Xe] $4f^{12}$	$^3H_6$
Ytterbium	Yb	70	[Xe] $4f^{13}$	$^2F_{7/2}$
Lutetium	Lu	71	[Xe] $4f^{14}$	$^1S_0$

regions (Singh et al. 2016a). Another significant aspect of lanthanides is the existence of sharp and long-lived emission lines (Shahi et al. 2016). The energy states of lanthanide ions in solids are given in Fig. 4.1 (Moore et al. 2009).

The absorption of light by rare earth ions is governed by the odd-parity electric dipole (ED) operator, the electric quadrupole (EQ) operator, and the even-parity magnetic dipole (MD) operator. Therefore, some specific transitions are only allowed as they have to satisfy selection rules. For ED transitions, Laporte's (or parity) rule should be satisfied. As a consequence of Laporte's rule, the sum of the angular momentum must change by an odd integer during any transition. Hence, intra f-f transitions are forbidden. However, when lanthanides doped host matrix is developed, then due to the mixing of the wave functions, relaxation in the parity selection rules takes place. Therefore, intra f-f transitions become partially allowed and are called induced electric dipole transitions (Singh et al. 2016b; Shahi et al. 2017a).

Thus, the fluorescence spectra of lanthanides come through f → f transitions, and the 4f shell is protected by outer $5s^2$ and $5p^6$ shells. Therefore, the effect of ligand/crystal field on the emission spectra of lanthanides is negligible which results in sharp emission spectra. Among all lanthanides, the fluorescence property of Ce^{3+} is different due to the allowed f-d transitions. Instead of sharp lines, its emission is broad. The host matrix (ligand/crystal filed environment) strongly affects the emission peak wavelength (Nargelas et al. 2022). The additional property of the lanthanide is that it possesses dense energy states which facilitate a spectral conversion. Hence, multimode emissions (downconversion (DC), downshifting (DS), and upconversion (UC)) are easily observed in lanthanide-based materials (Singh et al. 2016a; Shahi

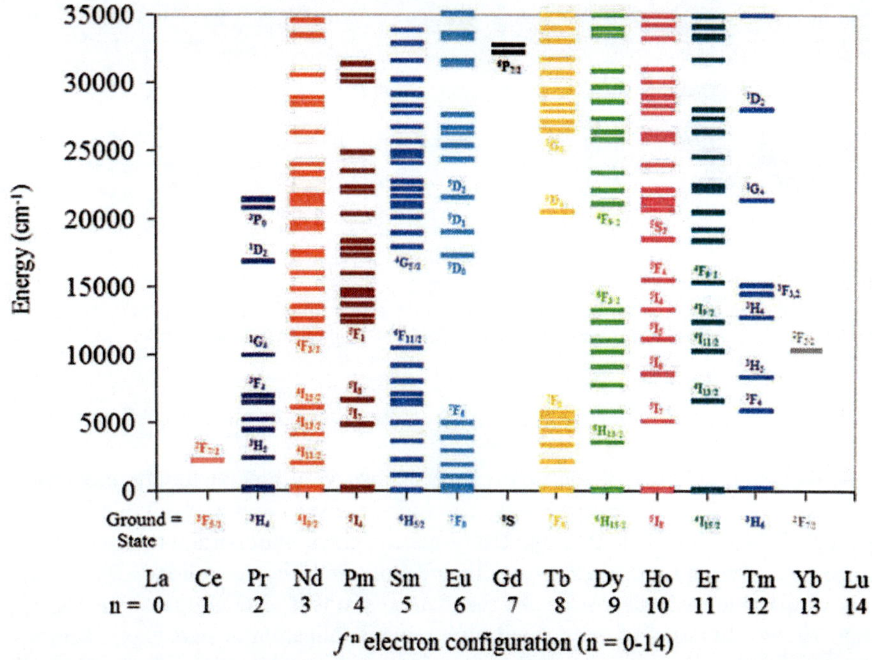

Fig. 4.1 Low-lying energy levels of triply ionized rare earth ions. Reprinted with permission from ref. (Moore et al. 2009). Copyright©2009, American Chemical Society

et al. 2016). This unique property makes them highly suitable for their use in the development of optical materials for different applications (Singh et al. 2021).

4.2 Lanthanide-Activated Luminescent Materials

4.2.1 Lanthanide-Doped Inorganic Phosphors

Phosphors are often rare earth or transition metal-doped inorganic compounds. Thus, in general, inorganic solids with a light emissive property are called phosphors. After systematic scientific research, thousands of phosphors have been prepared so far and extensively used in many areas such as displays, lighting devices, optical switching, cathode-ray-tube (CRT), medical imaging, X-ray fluorescent screens, etc., (Ronda et al. 1998; Gupta et al. 2021). Usually, phosphors are composed of microcrystalline matrix and activator ions. Often a sensitizer ion is also incorporated to induce efficient luminescence from an activator. Activator and sensitizer ions jointly provide efficient emission through the various energy transfer processes. The emission intensity in phosphor material can be readily tuned by changing the activator/sensitizer

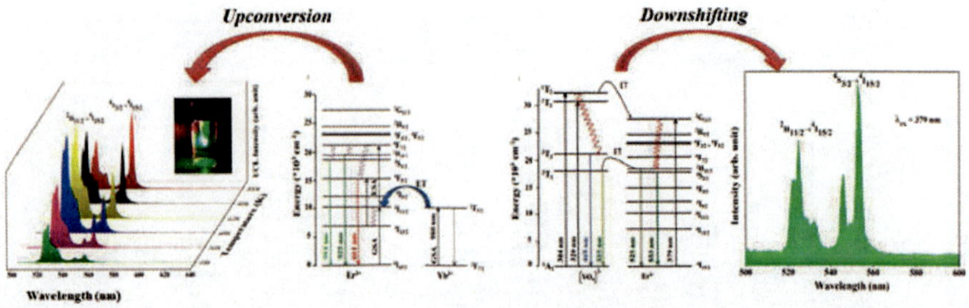

Fig. 4.2 Schematic diagram of the energy transfer process in lanthanide-doped inorganic phosphor. Reprinted with permission from ref. (Sharma et al. 2022). Copyright©2012, Elsevier

concentration without altering the host structure (Singh et al. 2016a; Shahi et al. 2016).

The crystal structure of the host matrix is also very crucial for the electronic transitions of activator/sensitizer ions. Therefore, the selection of a suitable host for getting efficient dual mode (UC and DC) emission plays an important role. The basic properties required for the host matrix include high stability, low phonon frequency, and comparable ionic size of one of the elements to that of the dopant ions (Singh et al. 2016b). In terms of the desired multimode application of phosphors, a series of rare-earth-based oxides, as well as fluorides compounds such as A_2O_3 (A = Y, La, Gd), AF_3 (A = Y, La, Gd), KAF_4 (K = Li, Na K; A = Y, La, Gd), have been chosen as host (Singh et al. 2016b, 2017a; DiMaio et al. 2006). Such phosphors are widely used for different applications (Ronda et al. 1998; Gupta et al. 2021). As shown in Fig. 4.2, the self-activated vanadate phosphor (inorganic host matrix) with relatively high chemical and thermal stability (inorganic host matrix) exhibits dual mode emissions characteristics (Sharma et al. 2022). The two energy transfer process responsible for dual-mode emissions is—(i) ligand to metal energy transfer process: (ii) energy transfer upconversion between lanthanides (sensitizer to activator).

4.2.2 Lanthanide-Activated Organic Complexes

In the case of Ln^{3+} activated inorganic phosphors, the absorption, as well as emission spectra are very narrow and weak, which restricts their practical application. To tackle the problem of low extinction coefficient, lanthanide ions can be combined with an organic ligand (Singh et al. 2017a, 2019). The ligands absorb the incident UV light efficiently and transfer it to the central lanthanide ion in further steps. The mechanism for this process is shown schematically in Fig. 4.3.

The purpose behind combining the lanthanide with organic ligands is to increase its solubility and to prevent the clustering of the ions. Moreover, the clustering of the ions is prevented by the ligands and the hygroscopic nature of the compound decreases. The basic rule behind the luminescence process in organo-lanthanide

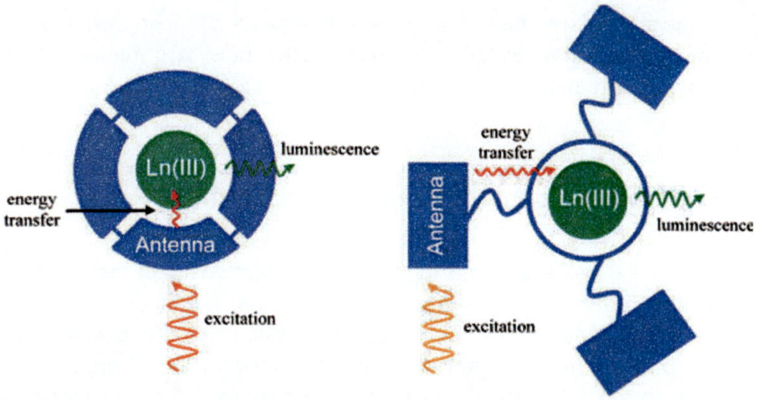

Fig. 4.3 Schematic demonstration of antenna effect in organic complexes. Reprinted with permission from ref. (Moore et al. 2009). Copyright©2009, American Chemical Society

complexes are: (i) absorption of the UV part of electromagnetic (EM) spectrum by the organic ligand; (ii) inter-system crossing (ISC) through which energy from the singlet (S_1) state is transferred to the triplet (T_1) states of the ligand; and (iii) finally, from the triplet state of the ligand to the different excited states of the lanthanide ion (Kido and Okamoto 2002). The mechanism is shown schematically in Fig. 4.4.

In the ground state (S_0), organic ligands absorb UV light and get excited to the first excited singlet state (S_1). The molecule in the excited singlet state then undergoes a fast internal conversion (IC) and thus occupies the lowest vibrational state of the S_1 level. The molecules from the S_1 state can either relax radiatively to the ground state S_0 or it may undergo a non-radiative (NR) intersystem crossing, from the S_1 state to some vibrational level of the triplet state (T_1). A weak phosphorescence (PS) can be observed from the forbidden transition $T_1 \rightarrow S_0$. Alternatively, from the triplet state energy is transferred to an excited state of the lanthanide ion. The excited lanthanide ions undergo a radiative transition to the lower-lying state and confer its characteristic line-like emission spectrum. The energy transfer from the

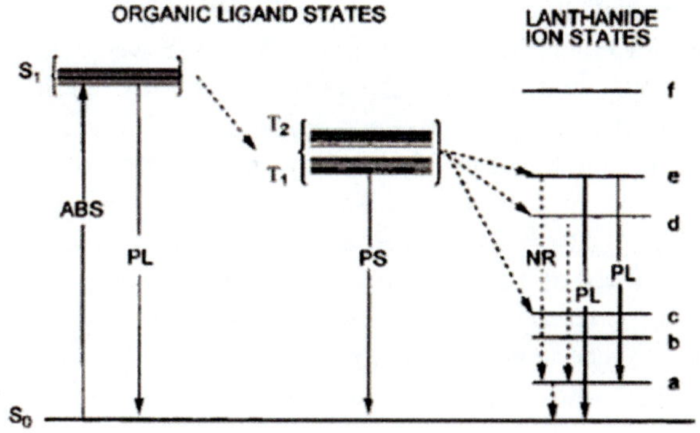

Fig. 4.4 Energy level scheme of photo-physical energy transfer processes in lanthanide complex. Reprinted with permission from ref. (Kido and Okamoto 2002). Copyright©2002, American Chemical Society

excited organic molecule to the rare earth ion is often very efficient. Due to this strong emissions of rare earth ions are observed. The excited rare earth ions may lose their energy through non-radiative relaxation also (Shahi et al. 2017b).

4.2.3 Lanthanide-Activated Composite Materials

It is well known that sensitization is the process of indirectly exciting another luminescent center to improve the absorption cross-section. Aiming to enhance emission efficiency, the synthesis of lanthanide-doped composite/hybrid nanostructure is recently one of the most prominent sensitization approaches. Surface plasmon-enhanced nanoparticles, metal–organic framework nanoparticles, and dye-sensitized lanthanide-activated upconversion nanoparticles (UCNPs) are some examples of hybrid materials (Singh et al. 2017b).

In this context, the sensitization of lanthanide-doped nanomaterials with organic dyes is an efficient technique for boosting luminescence efficiency. Compared to organic dyes, lanthanides have comparatively low absorption cross-sections. One of the largest absorption cross-sections among the lanthanide ions is possessed by the Yb^{3+} ion, which is roughly 1000 times less than the organic dye IR-806 (Beery et al. 2021; Xue et al. 2018). The UCNPs surface is functionalized with organic dye to improve the luminescence efficiency via the so-called "antenna effect". Initially, cyanine dye, IR-806 was used as an antenna for UCNPs. Garfield et al. demonstrated the energy transfer mechanism in cyanine IR-806 dye coupled with $\beta-NaY_{0.78-x}Gd_xYb_{0.2}Er_{0.02}F_4$ ($x = 0$ and 30%) hybrid UCNPs (Garfield et al. 2018). Dye undergoes intersystem crossing (ISC) from the singlet to the triplet state, subsequently, energy transfer can take place from the triplet state to the lanthanide ion resulting in the enhancement of upconversion emission efficiency by 15,000 times for IR-806 sensitized (30%) Gd^{3+} UCNPs. Here, the triplet states of the dye act as an intermediate state to sensitize the lanthanide ions, doped in the nanostructure.

Nasrabadi et al. constructed $NaGdF_4:Yb^{3+}/Er^{3+}@NaGdF_4:Yb^{3+}/Nd^{3+}$ hybrid nanostructure sensitized with six NIR cyanine dyes (Cy 740, Cy 754, Cy 748, Cy 784, Cy 780 and Cy 792) (Nasrabadi et al. 2021). The Cy 754 dye binds strongly with $NaGdF_4:Yb^{3+}$, $Er^{3+}@NaGdF_4$:10% Yb^{3+}, 30% Nd^{3+} exhibits a significant UC enhancement (about 680 times). It was observed that the two dyes Cy 754 and Cy792 firmly bounded with UCNPs enhanced the UC luminescence by 680 and 612 times, respectively. The two other weekly interacting dyes (Cy 740 and Cy 784) enhanced the luminescence by 236 and 75 times, respectively. Chen et al. developed the energy cascade upconversion in $NaYF_4:Yb^{3+}/X^{3+}@NaYF_4:Nd^{3+}/Yb^{3+}$ (X = Tm, Er, and Ho) hybrid nanostructure sensitized with NIR ICG dye (Chen et al. 2016b). They revealed that the spectral overlap of dye with Nd^{3+} and Yb^{3+} that are simultaneously incorporated into the core–shell nanostructure results in Förster energy transfer by the dye to acceptors (Nd^{3+} and Yb^{3+}) incorporated into the shell and subsequently transfer to Yb^{3+} into the core. The upconversion of NIR ICG-sensitized core–shell structures was found 16.4% higher than without dye sensitization (Nasrabadi et al. 2021). Thus, organic dyes enhance lanthanide luminescence through the energy transfer process.

Similarly, surface plasmon-enhanced lanthanide nanoparticles have also been a promising sensitization method for the enhancement of luminescence efficiency. Two potential mechanisms have been used to explain the enhancing effect:—(i) the enhancement of extinction coefficient by the local electromagnetic field, that emerges from the excitation of localized surface plasmon resonance (LSPR) in noble metals (e.g., Au and Ag). (ii) The enhancement of emission efficiency arises from the resonance energy transfer between luminescent centers and surface plasmon resonance of noble metals (Singh et al. 2020a; Xu et al. 2017). Therefore, without disturbing the composition, shape, and size of the nanocrystals, metal-assisted LSPR can significantly change the environment of UCNPs and hence affect their emission and excitation range.

Xu et al. theoretically described the principle of creating the localized EM field in the noble metal (Au or Ag) and their interaction mechanism with the luminescent centers (emitters) (Xu et al. 2017). Their study demonstrated that the amplified EM field caused by SPR is crucial to the electron radiative energy transfer process. The spatial separation between metal nanoparticles and UCNPs plays a crucial role in efficiency enhancement. Singh et al. successfully coupled the silver (Ag) nanoparticles on the surface of the Yb^{3+}/Er^{3+}: La_2O_3 nanostructure through a combustion technique (Singh et al. 2020b). They demonstrated the enhancement in green (32-fold) and red emission (20-fold) for the optimized amount of Yb^{3+}/ Er^{3+}: La_2O_3@Au NPs. Their result revealed that the efficiency enhancement is correlated with a decrease in the decay time of $^4S_{3/2}$ and $^4F_{9/2}$ levels of Er^{3+} with the incorporation of AuNPs. Similarly, Yuan et al. developed the $NaYF_4$: Yb, Er@SiO_2@Ag core–shell nanostructure and observed the effect of SiO_2 shell thickness and size of Ag NPs on the plasmonic enhanced UC emission (Yuan et al. 2012). Jhushan Hu et al. developed the plasmonically enhanced UCNPs bridged with polyethylene glycol for photothermal treatment (Hu et al. 2020). The absorption peak of Au NPs overlaps with the emission of core–shell UCNPs and enhances the efficiency of photo-thermal conversion as shown in Fig. 4.5a. The result revealed that the temperature of the nanocomposite increased by 6 °C in 4 min (as shown in Fig. 4.5b), which is a suitable temperature for killing a cancer cell. Chen et al. designed the Au–Ag/$NaYF_4$: Yb^{3+}, Tm^{3+}/Er^{3+} nanocomposite film by organic self-assembly as well as the template removal method (Chen et al. 2015). Compared to Au and Ag nanoparticles, Au–Ag alloy exhibits high near-field amplification and tunable absorption cross-section tuned with changing ratio of Au/Ag and high structural and chemical stability which make them more advantageous over the pure Ag and Au. Their results revealed that the coupling of surface plasmons of Au–Ag alloy with UCNPs enhanced the luminescence efficiency by ~180 times for the optimized compositions of Au and Ag.

The color tunable and dual-mode luminescence has been studied by Singh et al. in plasmon-enhanced UCNPs@ quantum dots (QDs) (Singh et al. 2020b). The observed tunable color emission obtains from the QDs size variation and spectral overlapping of the absorption band of gold nanoparticle with the UC emission band of UCNPs as shown in Fig. 4.5c. It is also studied that the variation of the red-to-green (R/G) ratio

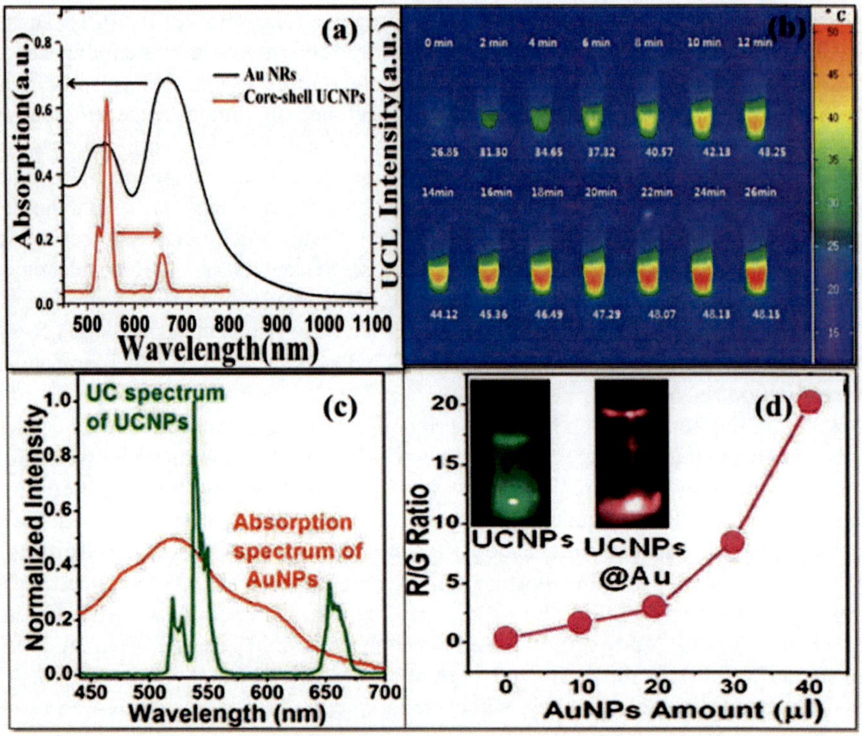

Fig. 4.5 **a** Spectral overlapping of absorption spectra of Au NPs and UCL spectra of core–shell UCNPs. **b** UC emission spectra of dye-sensitized $NaY_{0.78-x}Gd_xYb_{0.20}Er_{0.02}F_4$. Reprinted with permission from ref. (Hu et al. 2020). Copyright©2020, American Chemical Society. **c** Spectral overlap of UC emission spectra of UCNPs and absorption spectra of gold nanoparticles. **d** Concentration-dependent red/green UC emission intensity ratio. Reprinted with permission from ref. (Xu et al. 2017). Copyright©2020, Elsevier B. V.

with the required volume of AuNPs, the maximum calculated R/G ratio is 20.03 for the 40 μl composition of AuNPs as shown in Fig. 4.5d. To study the variation of UC luminescence enhancement factor with the spatial distance between UCNPs and AuNPs. Feng et al. designed the plasmon-enhanced UCNPs@AuNPs encapsulated with polyelectrolyte multilayer as a spacer (Feng et al. 2015). They observed that the enhancement factors depend upon the thickness of the polyelectrolyte spacer and the result revealed a maximum of 22.6-fold enhancement with the optimized spacer thickness of about 8 nm.

4.2.4 Lanthanide-Based Metal–Organic Frameworks

Lanthanide-based metal–organic frameworks (Ln-MOFs) are also promising hybrid nanomaterials that provide a strong basis for addressing the challenges faced by bare lanthanide-based nanostructures. In 2010, Rocha et al. highlighted the multifunctionality of Ln-MOFs as they combine the light emitted by the lanthanides with the inherent properties of the metal–organic complex like microporosity, magnetism, molecular sensing, and chirality (Rocha et al. 2011). A distinct prospect to observe synergistic effects, such as the interaction between photoluminescence (PL) and magnetism, is provided by Ln-MOFs (e.g., modulating PL with a magnetic field). Due to the aforementioned characteristics, MOFs have displayed potential in a wide variety of fields, such as gas adsorption, catalysis, chemical sensing, water treatment, etc. Along with that, the facile functionalization, high drug delivery capacity, low bio-toxicity, excellent biocompatibility, and biodegradability make them suitable for optical imagining, pH-based drug delivery, and anticancer drug carrier also. Xu et al. synthesized the Ln-MOFs with enhanced water and thermal stability by using the hydrothermal method (Xu et al. 2020). When co-doped with Eu^{3+}, the Yb^{3+}/Er^{3+}-MOF exhibited dual-mode emission. The results revealed the variation of DC and UCL intensity as a function of temperature as shown in Fig. 4.6a and c 2D variation of intensities with temperature is demonstrated in Fig. 4.6b and d. Sun et al. successfully designed the lanthanide-organic framework $[Ln(oba)(ox)_{0.5}(H_2O)_2]_n$ (Ln = Y:Yb/Er) (Sun et al. 2009). The presence of low vibrational frequency oxalate ions in the organic complex resulted in the highly intense upconversion emission. The typical two and three-photon processes confer green, blue, and red UC emissions under 980 nm laser excitation.

Li et al. proposed an effective synthesis method for the formation of UCNP@ Fe-MIL-101_NH_2 nanocomposite (Li et al. 2015). To enhance the stability and biocompatibility, the resulting nanocomposite was conjugated with NH_2-PEG-COOH (PEG). The decreasing T_2 relaxation time and also the brightness of T_2 weighted image with increasing concentration of Fe ions suggested that PEG conjugated UCNP@ Fe-MIL-101_NH_2 nanocomposite can be utilized for imaging purposes. Ling et al. have designed the $NaGdF_4$:Yb, Er@$NaGdF_4$, Yb, and Tm@$NaYF_4$ fabricated with Fe-Tetrakis porphine (Fe-TCPP) (Ling et al. 2022). The emission wavelength of $Er^{3+}/$ Tm^{3+} is in resonance with the absorption wavelength of Fe-TCPP, therefore efficient energy transfers from UCNPs to Fe^{3+} lead to the formation of oxygen singlet ($^1O_2$) and hydroxyl radical. The luminescence of UCNPs at 808 nm is not quenched by the MOF absorption that can be used for chemo-dynamic therapy and photodynamic therapy of tumor cells.

Yan et al. reported large porous and stable Eu-Ru$(Phen)_3$-MOF where $Eu(III)_2$ acts as a connecting node and $Ru(phen)_3$ as a linker (Yan et al. 2018). These types of Ln-MOFs are synthesized for the catalysis of CO_2 reduction. It's interesting to observe that the Eu-Ru$(phen)_3$-MOF demonstrates a remarkable rate of 321.9 μmol h^{-1} $mmol_{MOF}{}^{-1}$ for selective CO_2 photoreduction to formate under the

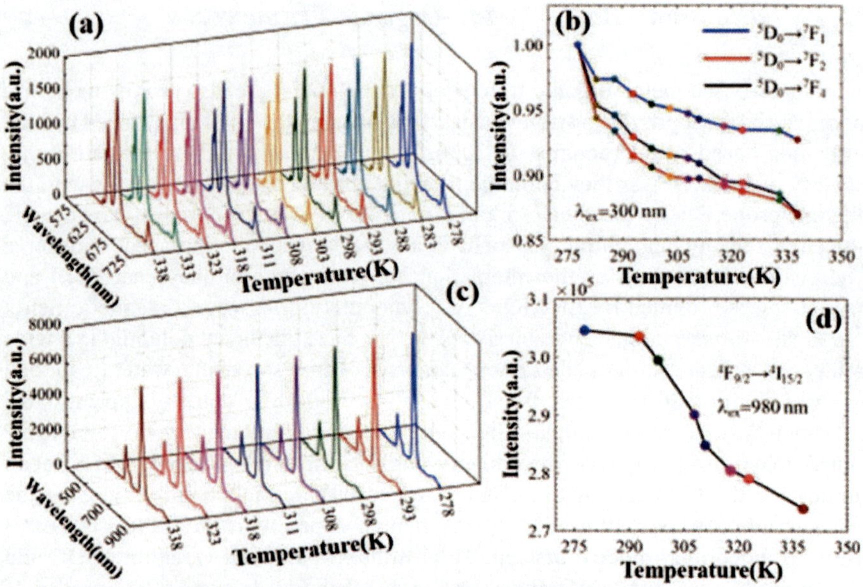

Fig. 4.6 **a** PL emission spectra of Eu^{3+} doped Er^{3+}/Yb^{3+}-MOF recorded at different temperatures. **b** Variation of DC emission peaks with temperature. **c** UC emission spectra of Eu^{3+} doped Er^{3+}/Yb^{3+}-MOF recorded at different temperatures. **d** Variation of DC emission peaks with temperature. Reprinted with permission from ref. (Xu et al. 2020). Copyright©2020, American Chemical Society

influence of visible light. Among all Ln-MOFs, this self-assembled Eu-Ru(phen)$_3$-MOF is notable for being the only one to demonstrate a highly efficient material for selective CO_2 photo-reduction. Additionally, an in-situ electron paramagnetic resonance (EPR) investigation shows unequivocally that the Eu(III)$_2$ clusters turn into active catalytic centers after accepting photo-excited electrons. Chowdhuri et al. prepared NaF_4:Yb^{3+}, Er^{3+}@ ZIF-8/FA nanocomposites which exhibit higher drug encapsulation efficiency (95.54%) and high stability in aqueous solution (Chowdhuri et al. 2016). They demonstrated that the resulting nanocomposite performed as a pH-responsive drug carrier while loaded with 5-Fluorouracil. The releasing rate of 5-Fluorouracil is faster in an acidic medium (pH 5.5) than in a basic medium (pH 7.5).

4.2.5 *Lanthanide-Activated Quantum Dots*

QDs are semiconductor nanocrystals with a particle size between 2 and 10 nm. QDs have attracted the considerable attention of researchers because of their remarkable optical properties such as high extinction coefficient, long PL lifetime, high photostability, and efficient and color-tunable emission, increasing their demand in varieties of research fields and applications such as photovoltaics, photodetectors, diode laser, bio-imaging, etc. (Chen and Liang 2014; Zhang et al. 2021a). In 2019, Song et al. developed the Ag_2Se QDs coupled with core–shell UCNPs to conquer the inadequacy of UCNPs (Song et al. 2019). Here, Ag_2Se QDs with high extinction coefficient act as a sensitizer (antenna) under 808 nm laser excitation and efficiently transfer their energy to Nd^{3+} or Yb^{3+} ions via the resonance energy transfer process, consequently enhancing the UC efficiency of UCNPs. They proposed that after confining the UCNP-QDs with Rose Bengal, this nanocomposite can be used as a photodynamic therapy agent for cancer cell detection. Kumar et al. for the first time designed the Ln^{3+} doped $NaYF_4$@CdS nanocomposite incorporated with AuNPs for plasmonic enhancement (Fig. 4.7a) and demonstrated the efficient UC emission due to the resonance energy transfer from Er^{3+} ions to CdS nanoparticles as shown in Fig. 4.7b (Kumar et al. 2018). The whole nano-system ($NaYF_4$@CdS-Au) was supported by the reduced graphene oxide (RGO) to facilitate the photocatalysis process. Their results demonstrated that increasing the concentration of RGO quenched the UC fluorescence intensity (as shown in Fig. 4.7c), which signifies the successful fabrication of $NaYF_4$@CdS-Au nanoparticles on RGO nanosheets. Here, the absence of an absorption peak of graphene oxide in the nanocomposite signifies that the prepared nanocomposite can more efficiently harness visible light for photocatalysis purposes. Figure 4.7d depicts the photocatalytic degradation of CFX (pharmaceutical drug pollutant) attained by the photocatalysis of $NaYF_4$@CdS-Au-RGO under visible and NIR excitation.

The perovskite material (ABX_3) with a size comparable to the Bohr radius is known as perovskite QDs (PeQDs). PeQDs exhibit some distinctive optical properties such as excellent PL quantum yield (>90%), excessive color purity, narrow emission bandwidth, tunable emission covering the whole visible region, and defect tolerance i.e., defect states created within the band that can quench the chance of non-radiative recombination (Li et al. 2019; Kachhap et al. 2022). Recently, Ruan et al. reported the high stability of ($CsPbBr_3$) perovskite QD in $CsPbBr_3$-$NaYF_4$:Yb^{3+}/Tm^{3+} hybrid nanostructure at high temperature as compared to the conventional QD (bare $CsPbBr_3$) (Ruan and Zhang 2021). The emission spectra consist of blue and cyan emissions under 980 nm excitation and green emission under 375 nm excitation. They have also revealed the fluorescence emission intensity (monitored for excitation at 980 and 375 nm) as a function of temperature. As temperature increases up to 250 °C, no emission peaks are observed corresponding to $CsPbBr_3$ at 522 nm using the 980 nm laser excitation, specifying the low luminescence efficiency and the weak energy transfer rate of α-$NaYF_4$. However, at 300 °C, due to the phase transformation from cubic to hexagonal (β-$NaYF_4$), an intense green emission peak was observed in UC luminescence spectra which corresponds to an efficient energy transfer from β-$NaYF_4$ to $CsPbBr_3$. The photoluminescence quantum yield (PLQY) of the hybrid nanostructure was found to be larger (0.139%) at high temperature (maintained for 60 min at that temperature) when observed under 980 laser excitation. While monitored for 365 nm excitation the result suggested that the PLQY of the hybrid nanostructure is quite higher (21%) than the conventional QD maintained under the same environmental condition (at 300 °C for 60 min), indicating that the

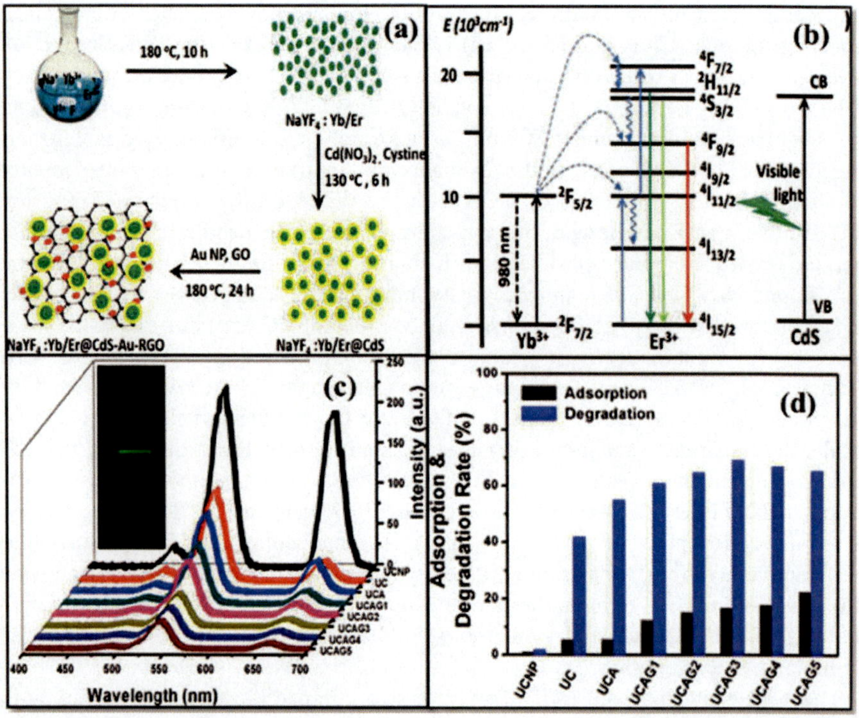

Fig. 4.7 **a** Schematic diagram of the synthesis of $NaYF_4$: Yb/Er@CdS-Au-RGO. **b** Simplified energy level diagram demonstrating the energy transfer mechanism between UCNPs and CdS nanoparticles. **c** UC emission spectra of $NaYF_4$: Yb/Er@CdS-Au-RGO for the different weight ratios of RGO. **d** Histogram of absorption and degradation rate of CFX (pharmaceutical drug pollutant) at different weight ratios of RGO. Reprinted with permission from ref. (Kumar et al. 2018). Copyright©2018, American Chemical Society

hybrid nanostructure nurtures the stability and luminescence of $CsPbBr_3$ QDs even at high temperatures.

In 2018, Zhang et al. reported the NIR-stimulated UC emission in $CsPbX_3$ QDs composite with the UCNPs via radiative energy transfer (Zheng et al. 2018). Their study shows that the UV light emitted by the UCNPs under 980 nm excitation is radiatively reabsorbed by the $CsPbX_3$ QDs, as a result of the large absorption cross-section of QDs in the UV region. Erol et al. observed the white light spectra from the Eu^{3+}/ Dy^{3+} doped $CsPbBr_3$ by optimized Eu^{3+} and Dy^{3+} concentration under 400 nm laser excitation (Erol et al. 2022).

4.3 Application of Lanthanide-Doped Materials

Lanthanide-doped materials such as organic complexes, inorganic phosphors (oxide and fluoride-based), and lanthanide composites have promising optoelectronic properties. The emission wavelength and peak intensity can be tuned either by compositional mapping of the host or by the lanthanide ions. Because of this outstanding behavior, these have been widely used for different applications such as solar cells, light emitting diodes (LEDs) and display devices, optical sensing, optical encoding, optical imaging, and also in laser materials. Out of the numerous versatile applications of phosphors materials, the book chapter only focuses on the optical properties-based applications, which are described in detail hereafter.

4.3.1 In Lighting: Light-Emitting Diodes and Display Devices

Lighting devices are the most fascinating and fast-growing field in the optoelectronic industry to make the world colorful. Synthesizing materials with high color purity for specific applications such as LEDs, display devices, etc., has put the optoelectronic industry in the outstanding stage. Although, a more technical and fundamental deep study is required to achieve an excellent level. The scientific community has been trying to synthesize the materials to achieve outstanding optical behavior by selecting a different host and doping different organic and inorganic ions into them.

For example, a composite of $NaYF_4{:}Yb^{3+}$, $Tm^{3+}@NaYF_4/CsPb(Br_{0.55}I_{0.45})_3$ UCNPs/PeNCs has been designed recently for the white-light-emitting diode (WLEDs) triggered by the NIR light (Naresh et al. 2022). The thin film of UCNPs/PeNCs into polymethylmethacrylate (PMMA)-toluene solution is coupled to the commercial LEDs. The Yb^{3+}, Tm^{3+} based UCNPs emission consists of emission peaks in the blue region while $CsPb(Br_{0.55}I_{0.45})_3$ shows emission in the yellow region at 575 nm as shown in Fig. 4.8d and e. The schematic of white light captured through a UV/IR cut-off lens is indicated in Fig. 4.8a. The appearance of white light without and under a UV/IR cut-off lens is also demonstrated in Fig. 4.8b and c, respectively. The CIE coordinates of the white light generated are (0.318, 0.301) as shown in Fig. 4.8f, which is very close to the standard value of CIE for white light (0.333, 0.333) and color temperature of 6361 K. Similarly, Eu^{3+}-doped $CsPbBr_3/Cs_4PbBr_6$ perovskite composites are also reported for WLEDs application with PLQYs of 87% (Wang et al. 2022a). The designed WLEDs have high efficiency of 43.06 lm/W with a wide color gamut.

The halide perovskite ($CsPbX_3$, X = Cl, Br, I) shows color tunability on partial or complete replacement of the halide ions, suitable for making the color display devices. Taking advantage of the color tunability of $CsPbX_3$, lanthanide ion is incorporated to either enhance the luminescence intensity or stabilize the structure. The Nd^{3+}-ion doped $CsPbBr_3$ is reported for the blue LED with PLQYs of 91% in the

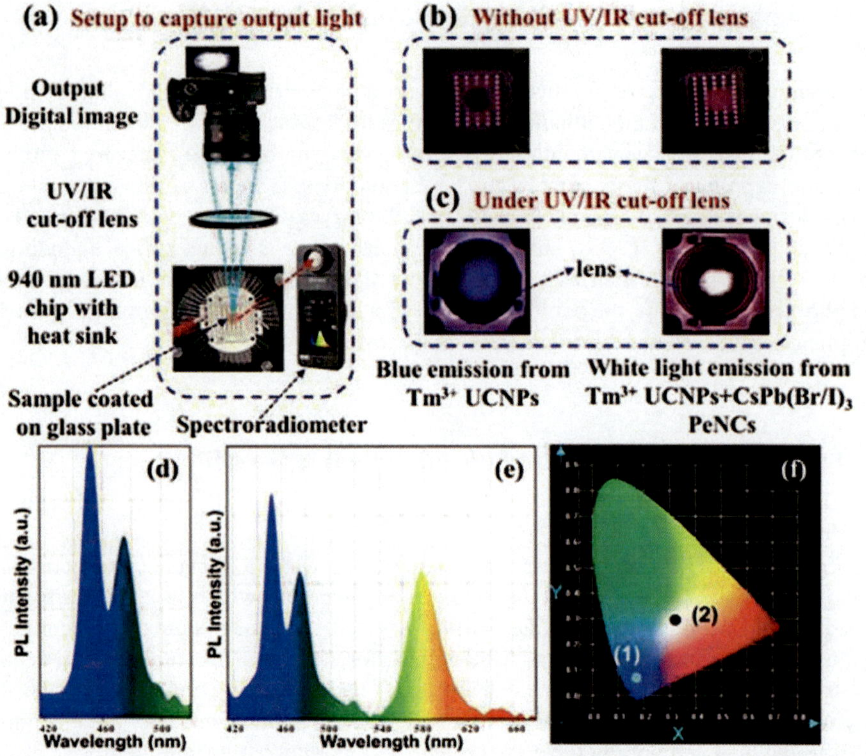

Fig. 4.8 **a** Schematic illustration of NIR triggered UC white light capturing UV/IR cut-off lens, and digital camera, **b** output light without using UV/IR cut-off lens, and **c** the generated UC-blue and white light emissions under UV/IR cut-off lens, **d** UC spectra of the core@shell UCNPs, **e** White light UC spectra of core@shell $NaYF_4$:Yb^{3+}, Tm^{3+}@$NaYF_4$/$CsPb(Br_{0.55}/I_{0.45})_3$, and **f** Color coordinates representation of the UC blue emission (1) and UC white light emission (2). Reprinted with permission from ref. (Naresh et al. 2022). Copyright©2021, Elsevier Ltd.

blue-emitting region (Padhiar et al. 2022). The emission wavelength is also tuned from 515 to 450 nm by varying the concentration of the Nd^{3+}-ion into $CsPbBr_3$ and can be used in making WLEDs. This shows also high stability against the external environment such as water, temperature, and UV exposure. A new double perovskite $NaCaTiTaO_6$:Dy^{3+} with high thermal stability is reported for the WLED application by Guo et al. (Guo et al. 2022). The CIE coordinate value reported is (0.314, 0.323). Eu^{3+}-doped $CaSc_2O_4$ phosphor is used for the red LED and WLED. The CIE coordinate of the red and white LED is (0.5976, 0.3579) and (0.3431, 0.3527), respectively (Chen et al. 2022). Yang et al. reported the tungstate phosphor $CaGdSbWO_8$:Sm^{3+} for WLED application with excellent thermal stability (Yang et al. 2022a). The CIE coordinate of the fabricated WLED is CIE coordinates (0.336, 0.328).

Hence, the lanthanide-doped materials can be suitable candidates for the LEDs and display devices for future perspectives, particularly for industrial applications.

4.3.2 *In Optical Sensing*

Sensing applications based on monitoring the optical parameters of the specimen are nowadays attracting researchers and the scientific community. Luminescence-based optical sensing applications of phosphor materials have great advantages because of their efficacious optical response towards external environments such as temperature (Xia et al. 2022), humidity, gasses, radiations (Shahi et al. 2015), photo-responsive (Parveen et al. 2022), etc. The temperature-dependent PL intensity is given by Yu et al. (2021),

$$I(T) = \frac{I_0}{1 + A\exp(-E_b/kT)} \tag{4.1}$$

where A is the constant, k is the Boltzmann's constant, E_b is the exciton binding energy, and the PL intensity at room temperature is represented by I_0 and at temperature T is I(T).

The fluorescence intensity ratio (FIR) of the thermally coupled energy levels is given mathematically as (Pattnaik and Rai 2022)

$$FIR = \frac{I_1}{I_2} = A\exp\left(-\frac{\Delta E}{kT}\right) \tag{4.2}$$

where, I_1, I_2 are the peak intensities corresponding to two thermally coupled levels, ΔE is the energy difference between thermally coupled states, Boltzmann's constant is represented by k, and T is the absolute temperature. For the thermal sensitivity, the absolute (S_a) and relative (S_r) sensitivity is given as

$$S_a = \frac{d(FIR)}{dT} = FIR \times \left(\frac{\Delta E}{kT^2}\right) \tag{4.3}$$

$$S_r = \frac{1}{FIR} \times \frac{d(FIR)}{dT} = \left(\frac{\Delta E}{kT^2}\right) \tag{4.4}$$

where each symbol has its usual meaning.

Over the last few decades, lanthanide-doped phosphors and organic complexes have been used for non-contact luminescence-based temperature sensing. The Eu^{3+} complex formed by the dibenzoyl methane and 1,10-phenanthroline ($Eu(DBM)_3Phen$) shows an effective temperature response in the range of 50–318 K (Shahi et al. 2015). The dibenzoyl methane acts as a sensitizer and the structure is stabilized by the 1,10-phenanthroline. The temperature-dependent PL study was carried out with the 611 nm emission peak corresponding to the $^5D_0 \rightarrow {}^7F_2$ transition in Eu^{3+} ions.

Similarly, due to the high stability of the oxide perovskite (ABO_3: A, B = metals ions), lanthanide-doped oxide perovskites have also been widely used for temperature sensing applications over the last few decades. For example, Er^{3+}-Nd^{3+}-Yb^{3+}:$CaTiO_3$

shows a noticeable thermal response in the luminescent intensity in the temperature range 303–575 K with S_r value between 0.008–0.011 K^{-1} (Pattnaik and Rai 2021). The molybdate-based phosphor is also a good candidate for the temperature-dependent PL study, because of its high thermal stability. The Er^{3+}-Yb^{3+} co-doped $CaMoO_4$ shows a thermal response in the range of 293 to 473 K with a maximum S_a value of 1.388% K^{-1} at 473 K and S_r value of 1.631% K^{-1} at 293 K (Ran et al. 2022).

The comparative study of the temperature sensitivity in bulk and nanoparticle of Er^{3+}, Yb^{3+} co-doped $CaMoO_4$ is reported in the temperature range 305–575 K (Singh et al. 2022). The maximum value of S_r for the bulk and nano $Ca_{0.79}Er_{0.01}Yb_{0.2}MoO_4$ phosphors are 0.0061 K^{-1} (at 305 K) and 0.0094 K^{-1} (at 299 K), respectively. Whereas, the maximum value of S_a are 0.0150 K^{-1} (at 348 K) and 0.0170 K^{-1} (at 398 K), respectively. These results indicate that the nano-sized $Ca_{0.79}Er_{0.01}Yb_{0.2}MoO_4$ phosphor shows better sensitivity as compared to the bulk counterpart. Similar work is done in $Ca_9Y_{0.647}Li_{0.20}Er_{0.003}Yb_{0.15}(VO_4)_7$ phosphor (Sharma et al. 2022). The obtained maximum S_a value is 0.0125 K^{-1} at 513 K and maximum S_r is 0.012 K^{-1} at 300 K. Wang et al. studied temperature response in Er^{3+}/Yb^{3+} doped $NaLa(WO_4)_2$ (Wang et al. 2022b). The temperature-dependent PL spectra in the temperature range 303–573 K are shown in Fig. 4.9a. The PL intensity decreases linearly with temperature. The FIR from the thermally coupled level is shown in Fig. 4.9b, and Fig. 4.9c shows Ln(FIR) versus 1/T plot. The S_r value is 0.014 K^{-1} at 476 K shown in Fig. 4.9d.

Now a day's lanthanide-doped halide perovskites have also attracted researchers for temperature sensing applications because of their thermal response to luminescence. In all-inorganic halide perovskite, the sharp emission in the visible region with full width at half maxima of 12–42 nm facilitates easy monitoring of even a small variation in the intensity with the temperature (Kachhap et al. 2022). Further, the PL intensity significantly increases with the incorporation of lanthanide ions. These facilitate the dual mode emission in the PL study with dual modes of thermal sensing. In Eu^{3+}-doped $CsPbCl_2Br_1$ QDs, the variation of luminescence intensity with the temperature between 80–440 K was performed (Yu et al. 2021). The maximum value of S_a reported is 0.0315 K^{-1}. Recently, Kachhap et al. summarized the temperature sensitivity of the lanthanide-doped all-inorganic halide perovskites (Kachhap et al. 2022). It is seen that the thermal sensitivity in the lanthanide-doped all-inorganic halide perovskites depends on host compositions as well as the lanthanide ions. The stability of the structure is an important concern and can be tailored by the compositional mapping also. The water stability of the Eu^{3+}-doped $CsPbCl_2Br_1$ QDs glass was tested for 40 days under daylight and UV light (365 nm) exposure (Yu et al. 2021). There is no appreciable change in the intensity. The temperature-dependent PL study is also reported in Yb^{3+}, Er^{3+} co-doped double perovskites $Cs_2AgInCl_6$ and $Cs_2AgBiCl_6$ in the temperature range 100–600 K (Rao et al. 2022). The S_a values reported are 0.0130 and 0.0113 K^{-1}, respectively for the $Cs_2AgInCl_6$ and $Cs_2AgBiCl_6$.

In lanthanide-doped fluoride compounds, thermal studies have also been carried out due to their highly sensitive PL properties under NIR laser excitation. The thermal

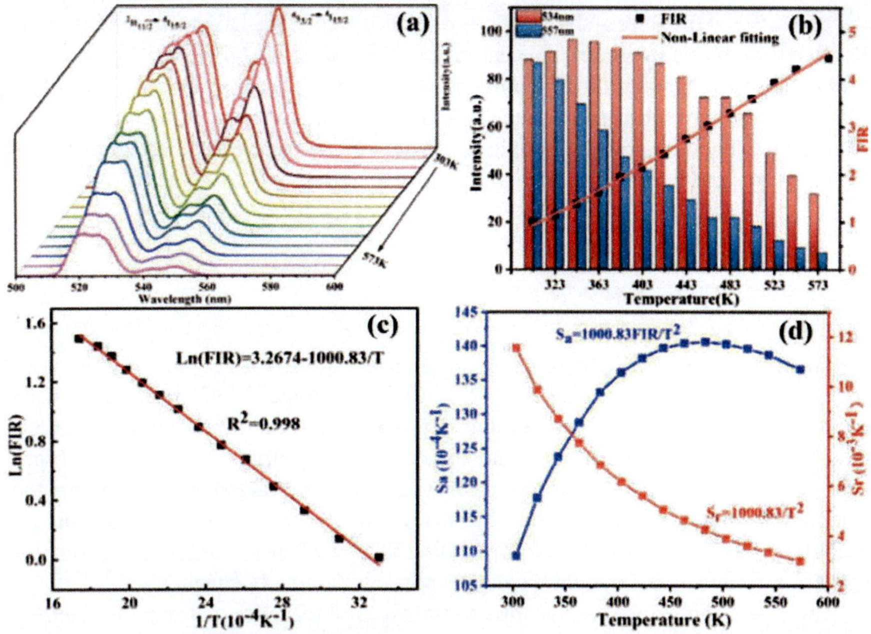

Fig. 4.9 **a** Temperature-dependent UC spectra of Er^{3+}/Yb^{3+}: $NaLa(WO_4)_2$. **b** FIR versus T plot, **c** ln(FIR) versus 1/T plot **d** The variation S_a and S_r with temperature. Reprinted with permission from ref. (Wang et al. 2022b). Copyright©2021, Elsevier B. V.

study of the core/shell $NaYF_4$:Yb^{3+}, Er^{3+}/$NaYF_4$ UCNP is performed in the temperature range of 293 to 358 K (25–85 °C) with the excitation light 975 nm laser (Jurga et al. 2022). The core/shell structure $NaYF_4$:Yb^{3+}, Er^{3+}/$NaYF_4$ UCNP is synthesized using the three different types of reacting precursors (RE chlorides, RE acetates, and RE oleates). The temperature sensitivity of the $NaYF_4$:Yb^{3+}, Er^{3+}/$NaYF_4$ UCNP depends on the precursor type (used in the synthesis process). The S_r value of $NaYF_4$:Yb^{3+}, Er^{3+}/$NaYF_4$ UCNPs synthesized by using RE chlorides, RE acetates, and RE oleates reacting precursors are 1.44%, 1.39%, and 1.28% K^{-1} at 358 K. Ding et al. carried out the thermal sensitivity study in the fluoride β-$NaYF_4$:Yb^{3+}, Er^{3+} core, and β-$NaYF_4$: Yb^{3+}, Er^{3+}@$NaGdF_4$ core–shell structure in the temperature range 100–400 K (Ding et al. 2022). The luminescent intensity of the β-$NaYF_4$: Yb^{3+}, Er^{3+}@$NaGdF_4$ core–shell structure is higher than β-$NaYF_4$:Yb^{3+}, Er^{3+}. For the comparison of the thermal sensitivity, the S_r value is calculated using the luminescent intensity ratio (LIR) of both samples and found that the core–shell structure has a higher S_r value than core only. Also, the LIR of the thermally coupled Stark sub-level (Stark-LIR) technique is used to calculate the S_r value in β-$NaYF_4$:Yb^{3+}, Er^{3+}@$NaGdF_4$ core–shell structure. The S_r value obtained using the LIR techniques is 10.9×10^{-3} K^{-1}, while using Stark-LIR techniques is 17.8×10^{-3} K^{-1} in β-$NaYF_4$:

Yb^{3+}, Er^{3+}@$NaGdF_4$ core–shell structure. Hence, the stark-LIR technique is more effective to improve the thermal sensitivity in core–shell UCNPs.

In conclusion, lanthanide-doped materials are effective for luminescence-based-optical temperature sensing. The sensitivity of the lanthanide-doped materials varies with the host compositions, particle size, doped lanthanide ions (related to the thermally coupled level), etc.

4.3.3 In Optical Encoding

The security issues in duplicating the product and antitheft are one of the major challenges in this era and also for the future. The conventional techniques used for the security check are encoded and duplicated at some level. These problems draw the attention of researchers and the industrial community to develop high-level security technniques. Fluorescent materials have attracted researchers for the anticounterfeiting application widely due to their unique color on external stimulation (Xie et al. 2022; Yang et al. 2022b; Molina-González et al. 2021). However, for high-security levels, the dynamic color response is more advantageous.

Yang et al. synthesized the dynamic responsive fluorescence material by attaching the photochromic molecule spiropyran (SP) into the Ln-MOF (Yang et al. 2022b). The polydimethylsiloxane (PDMS) is doped into the SP@Tb-MOF to make transparent films. Then different numbers or words are encrypted onto this film for the encryption/decryption study. The number "1921" as well as the word "WELCOME" are encrypted on the SP@Tb-MOF/ PDMS film, see Fig. 4.10a.

The dynamic anticounterfeiting level is studied for 60 s. Initially, the encrypted information color turned purple on UV light irradiation. The fluorescence color changes first from yellow-greenish to orange and finally becomes red on irradiation within 60 s. Again, one real information "520" was encrypted from the SP@Tb-MOF/ PDMS film, and fake information "520" was encrypted from the Tb-MOF/PDMS film. Real information "520" encrypted from the SP@Tb-MOF/ PDMS film, shows the change in color on UV light stimulation. Whereas, the pseudo-code doesn't reflect the color change. When the UV light illumination is stopped, the state of the decrypted information again came back to the original encrypted state. This system has a great advantage for complex anti-counterfeiting identification because the emission and colors of the material are regulated through ultraviolet radiation and also have a dynamic fluorescence color response within 60 s.

Rare-earth-activated UCNPs are excellent luminescent materials for optical encryption and decryption. However, this type of UC luminescent (single-mode) material suffers from the risk of data extraction imitation or data extraction by counterfeiting. The applicability of core@shell UCNPs/PeNCs composite was examined for dual-model anti-counterfeiting, which increases the difficulty of counterfeiting the data (Naresh et al. 2022). The information "KMU" was encrypted using the colloidal solutions of $NaYF_4$:Yb^{3+}, Tm^{3+}@$NaYF_4$ UCNPs, and $CsPb(Br_{0.55}/I_{0.45})_3$ PeNCs onto the PMMA surface. The letter "K" is encrypted by $NaYF_4$:Yb^{3+},

Fig. 4.10 **a** Dynamic response of information based on the SP@Tb-MOF/PDMS film. Reprinted with permission from ref. (Yang et al. 2022b). Copyright©2022, American Chemical Society. **b** Images of dual-model anti-counterfeiting under excitation of NIR (980 nm) laser, UV (360 nm) light, and NIR + UV light to encrypt/extract the information. Reprinted with permission from ref. (Naresh et al. 2022). Copyright©2021, Elsevier Ltd. **c** digital photographs for the dynamic anticounterfeiting. Reprinted with permission from ref. (Wang et al. 2022c), Copyright©2022, Royal Society of Chemistry (Open Access, Creative Commons Licenses)

Tm^{3+}@$NaYF_4$ UCNPs, "M" is encrypted using $CsPbBr_3$ and letter "U" using $NaYF_4$:Yb^{3+}, Tm^{3+}@$NaYF_4$/$CsPb(Br_{0.55}/I_{0.45})_3$ shown in Fig. 4.10b. For the decryption of the encrypted anti-counterfeiting information, "KMU" was illuminated under NIR and UV light. Under NIR irradiation, the letter pattern "K" exhibits UC blue emission (due to the presence of Tm^{3+} of the $NaYF_4$:Yb^{3+}, Tm^{3+}@$NaYF_4$ UCNPs structure). The letter 'M' does not emit any color which is made up of $CsPb(Br_{0.55}/I_{0.45})_3$ PeNCs. Where the letter 'U' emits a magenta color from the composition $NaYF_4$:Yb^{3+}, Tm^{3+}@$NaYF_4$/$CsPb(Br_{0.55}/I_{0.45})_3$. On the other hand, under UV (360 nm) light DC emission is observed from the letters 'M' and 'U' that show green and red color, respectively. Whereas the letter 'K' does not show any color emission. However, when the encrypted pattern is irradiated simultaneously with UV and

NIR light the dual color emission from the letter 'U' is observed. The efficiency of the emission intensity monitored for the $CsPb(Br_{0.55}/I_{0.45})_3$ PeNCs under UV light irradiation is higher than the UC emission.

Similarly, the β-$NaYF_4$:RE^{3+} (RE = Sm, Tb, Dy, Pr) was reported as a phosphor with the multi-color emitting anti-counterfeiting application with X-rays exposure (Wang et al. 2022c). The encrypted pattern shows different information on the application of different irradiation times of X-rays. With X-ray irradiation time the encrypted information "Wi-Fi" is only a single dot visible and the curved pattern fully disappeared as indicated in Fig. 4.10c. Similarly, the other encrypted patterns also show dynamic responses to the X-ray radiation. It was also noticed that only a particular part of the encrypted information is decrypted at a particular irradiation time making this store more complex information for the anti-counterfeiting application.

Similarly, the dual mode ink of β-$NaYF_4$:RE^{3+} (RE = Ce, Tb, Eu, Yb, Er, and Tm) phosphors are prepared with polyvinyl alcohol (PVA) and distilled water (DI) (Zhang et al. 2019). The multicolor anti-counterfeiting ink such as β-$NaYF_4$:RE^{3+} phosphors with green (RE = Ce/Tb), red (RE = Ce/Tb/Eu) color DC anti-counterfeiting ink, and β-$NaYF_4$:RE^{3+} phosphors with blue (RE = Yb/Tm), green (Yb/Er) UC anti-counterfeiting ink were prepared. The encrypted information remains invisible in the daylight. However, the encrypted information shows the different colors under a 254 nm UV lamp and 980 nm NIR laser excitation. This can be explored for dual-mode high-security anti-counterfeiting applications.

Although the lanthanides doped organic–inorganic phosphors, complexes, and QDs are the emerging materials for the anti-counterfeiting ink in DC and UC mode, multi-stimuli and dynamic response in the decryption stage are also achieved separately. However, for a high level of security, more than two decryption processes for encrypted data should be fulfilled. These may be achieved either by discovering new host materials or by incorporating the different dopant elements into the reported host which has a promising anti-counterfeit response under different decryption conditions.

4.3.4 *Application in Laser Materials*

The invention of the laser not only resolves the many obstacles of the scientific world, but it also helped the modification of security and military equipment towards the advanced version. Lanthanide-doped materials have been widely used candidates for laser material because of their ladder-like energy levels suitable for selective emission at a particular wavelength. The lanthanide-doped $GdScO_3$ is one of the good candidates for laser materials (Peng et al. 2018; Zhang et al. 2021b; Li et al. 2021). The Dy^{3+} doped $GdScO_3$ single crystal is reported as a yellow laser material shown in (Fig. 4.11a) (Peng et al. 2018). The absorption cross-section found around 453 nm is about 1.8×10^{-21} cm^2. For this absorption band, FWHM is calculated to be about 9 nm (at 300 K). This also shows emission at 577 nm ($^4F_{9/2} \rightarrow {}^6H_{13/2}$

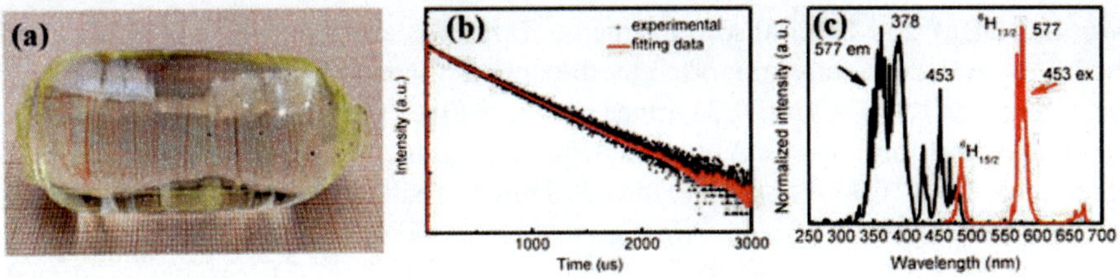

Fig. 4.11 **a** As-grown single crystal of Dy: $GdScO_3$, **b** Fluorescence decay curve of Dy: $GdScO_3$ crystal, **c** Excitation (black) and emission (red) spectra of Dy: $GdScO_3$. Reprinted with permission from ref. (Peng et al. 2018). Copyright©2018, Elsevier B. V.

transition) in Dy^{3+} ion shown in Fig. 4.11c. The stimulated emission cross-section for this transition is 4.1×10^{-21} cm^2 and relatively long fluorescence lifetime (see Fig. 4.11b) with high radiative quantum efficiency. These outstanding properties of Dy^{3+} doped $GdScO_3$ phosphors make it very suitable for yellow laser, pumped by blue GaN LEDs.

Shang et al. summarized the lasing emission in lanthanide-doped UC nanomaterials (Shang et al. 2022). The UC-based lasing action in lanthanide-doped materials is broadly divided into three categories: (a) UC random laser system, (b) whispering gallery mode (WGM)/Fabry–Perot (FP) cavity modulated UC laser, and (c) photon lattice/plasmonic cavity modulated UC micro-lasers. The UC random laser system consists of all the lasing components made up of nanoparticles, while in WGM/FP different microstructures and the cavity is incorporated along with the UNCPs.

Further, the complex formation between lanthanide and organic ligand is also very effective to get long-lived, sharp emissions, because the absorption coefficient of the organic chromophores is 3–5 times higher than the lanthanide ions (Sun et al. 2005). These properties are beneficial for laser emission. The lanthanide complexes such as tris(hexafluoroacetylacetonato) europium(III) 1,2-phenylenebis(diphenylphosphine oxide) $[Eu(hfa)_3(OPPO)_2]$, tris(hexafluoroacetylacetonato) europium(III) 1,1′-biphenyl-2,2′-diylbis(diphenylphosphine oxide) [Eu(hfa)3(BIPHEPO)], tris(hexafluoroacetylacetonato) europium(III) bis(triphenylphosphine oxide) [Eu(hfa)3(TPPO)2], etc., dispersed in the polyphenylsilsesquioxane (PPSQ) polymer result amplified spontaneous emission (ASE) (Shahi et al. 2015). This behavior of lanthanide complexes indicates that these materials can be used for laser application. The PL study of the complex $Eu(DBM)_3Phen$ dispersed in poly (methyl methacrylate) [PMMA] was carried out by 405 nm laser diode excitation. The $Eu(DBM)_3Phen$ shows a sharp emission peak at 611 nm transition from $^5D_0 \rightarrow {}^7F_2$ in Eu^{3+} ion. In the excitation power-dependent PL study, intensity increases linearly up to 25 mW. Beyond this excitation power, the PL intensity increases suddenly in an exponential fashion. The exponential increase in the PL intensity is the indication of the switching from spontaneous to stimulated emission.

Similarly, the ternary $Ln(DBM)_3phen$ complexes (DBM = dibenzoylmethane, phen = 1,10- phenanthroline, and Ln = Nd, Yb) for the laser material are reported by Sun et al. (2005). The the excitation band of the lanthanide complexes overlaps with the DBM and Phen absorption band. This indicates that the sensitization of the Ln^{3+} ion is efficiently supported by the organic ligands DBM and Phen through the antenna effect. The Ln(DBM)$_3$phen complex gives three emission peaks at 880, 1060, and 1340 nm. The 880 nm emission is the transition from the $^4F_{3/2}$ to $^4I_{9/2}$, 1060 nm is from $^4F_{3/2}$ to $^4I_{11/2}$, and the 1340 nm emission peak results from the $^4F_{3/2}$ to $^4I_{13/2}$ transition in Nd^{3+} ion. Among these three emissions, 1060 nm emission is of high intensity and suitable for laser application. The Ln^{3+}/DBM/Phen co-doped luminescent hybrid gels are also reported as the possible laser material (Sun et al. 2005). Both, the ternary Ln(DBM)$_3$Phen complexes and Ln^{3+}/DBM/Phen-co-doped luminescent hybrid gels show efficient NIR luminescence. These are excited by the wavelength that corresponds to the maximum absorption spectra of the ligands. The antenna effect facilitates efficient energy transfer from the ligands to the Ln^{3+} ions. The transition in Nd^{3+} ions from level $^4F_{3/2}$ to $^4I_{11/2}$ is the characteristic transition for the laser application.

In conclusion, lanthanide-doped phosphors, and complexes show appreciable results for laser emission. For future perspective, attention toward stability and tunability of emission peaks with low FWHM is required. Also, deep research is required for lower wavelength laser emissions either by finding new host materials or compositional mapping, tailoring the concentration of the active dopant elements (such as lanthanide).

4.3.5 Application in Optical Imaging

Optical imaging based on various techniques such as photon absorption, fluorescence, and wavelength shift is an emerging area in medical science for biomedical applications. There are various limitations such as photons absorption by the intrinsic tissues, light scattering, etc., which limit the penetration depth and the spatiotemporal resolution. However, recent advancements in optical probes using fluorescent materials and highly sensitive optical detectors have brought a breakthrough in the optical imaging field. For in-vivo and in-vitro probing and imaging, several mechanisms can be used such as voltage-dependent fluorescence and/or absorption of the phosphor/dye, time-dependent spectral shift, etc. Depending upon the response time and intensity of the fluorescence, a designed optical probe can be selected for the precise monitoring of various cells in the species body under study.

The fluorescence imaging technique is achieved by exciting the target fluorescence molecule using the external light of a specific wavelength. The excited fluorescence molecules subsequently release photons of longer/shorter wavelengths depending upon the upconverted/down-converted fluorescence molecule and are used for imaging.

The lanthanide complexes are very useful for in-vivo and in-vitro imaging because of their biocompatibility and stable luminescence characteristics (Wang et al. 2022d; Park et al. 2012). The folic acid (FA)-modified Ba_2LuF_7: Yb^{3+}/ Er^{3+}/Ho^{3+} (abbreviated as BLF@FA) is reported for in vivo tumor detection (Wang et al. 2022d). The

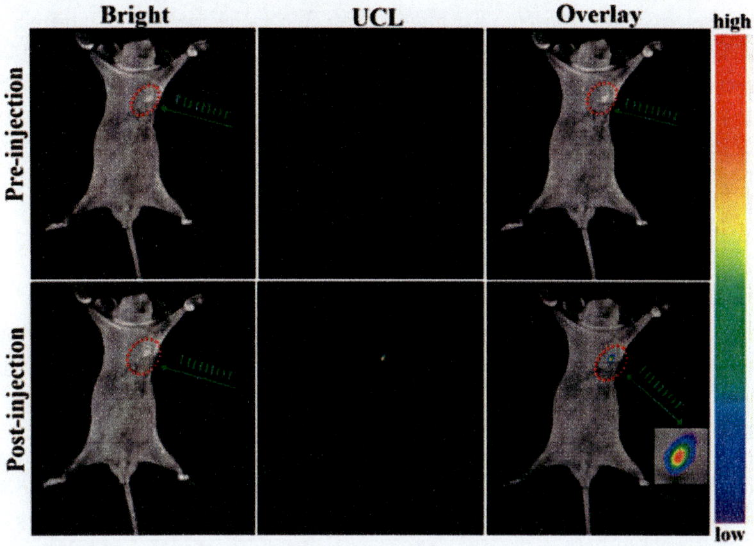

Fig. 4.12 In vivo UCL imaging of tumor-bearing mice with intravenous injection of BLF@FA aqueous suspension. Reprinted with permission from ref. (Wang et al. 2022d). Copyright©2022, Elsevier B. V.

BLF@FA has low toxicity and excellent tumor-targeting- ability. The UCL imaging of tumor-bearing mice pre and post-injection of BLF@FA is shown in Fig. 4.12. From the images shown in Fig. 4.12, it is clear that there is no UCL signal from the tumor site under the NIR light when BLF@FA is absent in the tumor-bearing mice body. However, an intense UCL signal is observed at the tumor site after the injection of BLF@FA under 980 laser excitation.

The UCNPs NaYF4: Yb, Er @NaGdF4 core–shell structure combined with a chlorin e6 (Ce6) is reported for dual-model in photodynamic therapy and in vivo tumor imaging (Park et al. 2012). For imaging studies, the UCNPs are injected into the mice through the tail vein. The fluorescence emission remains absent from the mice body under 980 nm laser irradiation when there is no injection of UCNPs. The liver gives vivid green and red emissions after intravenous injection and is monitored for 1.5 h. The UCNPs are coated with PEG-phospholipid at approximately 70 nm, in which the red signal is more intense than the green signal. These are rapidly accumulated in the liver and spleen. For the in vivo tumor imaging, nude mice bearing a U87MG tumor on the right hind leg, the UCNPs were injected through the tail vein. The UC green and remission are recorded from the mice's bodies at different time intervals. The photosensitizer Ce6 produces a therapeutic effect and also helps UCNPs to accumulate in the tumors. The Ce6 also increases the signal-to-noise ratio of the UCNP-Ce6 effective for tumor imaging. The comparative study indicates that the UCNP–Ce6 provides a clear tumor image than the UCNPs alone. Further, in-vivo bio-imaging has also been done in the NIR-II region using the

808 nm laser excitation in $NaYF_4$: Nd^{3+}@$NaYF_4$, $NaErF_4$@$NaYF_4$, and $NaYbF_4$: Er^{3+}/Ce^{3+}@$NaYbF_4$@$NaYF_4$: Yb^{3+}/Nd^{3+} core–shell structure (Liao et al. 2022).

Hence, lanthanide complexes have good potential for the effective bio-probe for accurate diagnoses such as cancer cell/tumor detection and photodynamic therapy. This may generate new ideas for the manufacturing of bio-imaging probes.

4.3.6 Photocatalytic Application

The population of the world is increasing very rapidly. To fulfill the need of people around the world, industries are producing non-biodegradable products in large amounts. In almost all areas such as the food industry, chemicals are used to grow, rip and keep the product for a longer time. In the manufacturing industry, anti-rusting materials are used for coating the surface of the product to increase its lifetime. For the high demand for electricity to run the industries, hydrothermal power is not sufficient. In this situation, coal is used to a large extent which produces many toxic gasses. These events increase air, water, and soil pollution and the situation may become very dangerous for the generation of the coming future. In this respect photocatalytic is one the best environment-friendly solution to reduce pollution.

The photocatalyst is the kind of nanoparticle that binds with the host matrix and involves in the self-cleaning process. The catalytic action can be initiated by the UV, visible, or NIR light depending upon the type of photocatalyst and host matrix (Wu et al. 2021; Hampel et al. 2021). The photocatalyst generates the electron–hole pairs. These electron–hole pairs bind to the complex pollutant molecule and degrade them to simpler molecules such as hydrogen, carbon dioxide, water, etc. (Yang et al. 2014; Helmy et al. 2022).

4.4 Conclusion and Future Scope

In conclusion, the present book chapter gives the basics of lanthanides and explains their spectroscopic properties at length. The different types of lanthanides doped luminescent materials such as inorganic phosphors, organic complexes, hybrid materials, metal–organic framework, and halide perovskite quantum dots have been discussed with a motive to brief the emission of lanthanides in these hosts. Both the downshifting and upconversion luminescence properties are discussed in detail. A different mechanism for the sensitization process of lanthanides such as co-doping and core–shell structure formation in the case of inorganic phosphors, antenna effect and surface plasmon resonance process in organic ligands, composites, and metal frameworks, quantum confinement effect in Quantum dots, etc., have been discussed. Further, the study covers some of the exotic optical applications of these materials. For example, applications in the area of light-emitting diodes, optical temperature sensing, optical encryption, laser emission, and optical imaging have been elaborated at length. The chapter would be helpful for beginners in the field to understand the basics and to learn the optical emission by lanthanides and its applications in different types of hosts.

Author Contributions The manuscript was written through the contributions of all authors. All authors have approved the final version of the manuscript.

Conflicts of Interest There are no conflicts to declare.

References

Beery, D., Schmidt, T.W., Hanson, K.: Harnessing sunlight via molecular photon upconversion. ACS Appl. Mater. Interfaces **13**(28), 32601–32605 (2021)

Boyn, R.: 4f–4f luminescence of rare-earth centers in II-VI compounds. Phys. Status Solidi B **148**(1), 11–47 (1988)

Chen, Y., Liang, H.: Applications of quantum dots with upconverting luminescence in bioimaging. J. Photochem. Photobiol. B Biol. **135**, 23–32 (2014)

Chen, X., Xu, W., Zhang, L., Bai, X., Cui, S., Zhou, D., Yin, Z., Song, H., Kim, D.: Large upconversion enhancement in the "Islands" Au–Ag Alloy/$NaYF_4$: Yb^{3+}, Tm^{3+}/Er^{3+} composite films, and fingerprint identification. Adv. Funct. Mater. **25**(34), 5462–5471 (2015)

Chen, G., Shao, W., Valiev, R.R., Ohulchanskyy, T.Y., He, G.S., Ågren, H., Prasad, P.N.: Efficient broadband upconversion of near-infrared light in dye-sensitized core/shell nanocrystals. Adv. Opt. Mater. **4**(11), 1760–1766 (2016b)

Chen, B., Wu, M., Liu, Q., He, C., Yang, Y., Liu, Y., Yang, C., Wen, X., Min, X., Huang, Z.: Preparation and photoluminescence properties of CaSc2O4: Eu^{3+} red phosphor for white LEDs. J. Mater. Sci. Mater. Electron. **33**(24), 18939–18951 (2022)

Chen, X., Liu, Y., Tu, D.: Lanthanide-Doped Luminescent Nanomaterials. Springer (2016)

Chowdhuri, A.R., Laha, D., Pal, S., Karmakar, P., Sahu, S.K.: One-pot synthesis of folic acid encapsulated upconversion nanoscale metal organic frameworks for targeting, imaging and pH responsive drug release. Dalt. Trans. **45**(45), 18120–18132 (2016)

Dhoble, S.J., Pawade, V.B., Swart, H.C., Chopra, V.: Spectroscopy of Lanthanide Doped Oxide Materials. Woodhead Publishing (2019)

DiMaio, J.R., Kokuoz, B., Ballato, J.: White light emissions through down-conversion of rare-earth doped LaF 3 nanoparticles. Opt. Express **14**(23), 11412–11417 (2006)

Ding, J., Liao, H., Pang, L., Wang, D., Ye, S.: Improved up-conversion behaviors and temperature sensitivity based on Stark sublevels of Er^{3+} in β-$NaYF_4$: Yb^{3+}, Er^{3+} and β-$NaYF_4$: Yb^{3+}, Er^{3+}@$NaGdF_4$. Opt. Mater. (amst) **128**, 112304 (2022)

Erol, E., Vahedigharehchopogh, N., Ekim, U., Uza, N., Ersundu, M.Ç., Ersundu, A.E.: Ultra-stable Eu^{3+}/Dy^{3+} co-doped $CsPbBr_3$ quantum dot glass nanocomposites with tunable luminescence properties for phosphor-free WLED applications. J. Alloys Compd., **909**, 164650 (2022)

Feng, A.L., You, M.L., Tian, L., Singamaneni, S., Liu, M., Duan, Z., Lu, T.J., Xu, F., Lin, M.: Distance-dependent plasmon-enhanced fluorescence of upconversion nanoparticles using polyelectrolyte multilayers as tunable spacers. Sci. Rep. **5**(1), 1–10 (2015)

Garfield, D.J., Borys, N.J., Hamed, S.M., Torquato, N.A., Tajon, C.A., Tian, B., Shevitski, B., Barnard, E.S., Suh, Y.D., Aloni, S.: Enrichment of molecular antenna triplets amplifies upconverting nanoparticle emission. Nat. Photonics **12**(7), 402–407 (2018)

Gschneidner, K.A., Eyring, L., Roth, T.A.: Handbook on the physics and chemistry of rare earths, vol. 1: metals. J. Electrochem. Soc. **126**(11), 464C (1979)

Guo, J., Li, S., Kong, J., Li, Y., Zhou, L., Lou, L., Lv, Q., Tang, R., Zheng, L., Deng, B.: Synthesis and characterization of a new double perovskite phosphor: $NaCaTiTaO_6$: Dy^{3+} with high thermal stability for w-LEDs application. Opt. Laser Technol. **155**, 108347 (2022)

Gupta, I., Singh, S., Bhagwan, S., Singh, D.: Rare earth (RE) doped phosphors and their emerging applications: a review. Ceram. Int. **47**(14), 19282–19303 (2021)

Hampel, B., Hernádi, K., Baia, L., Pap, Z.: The impact of Au nanoparticles and lanthanide-doped NaYF4 on the photocatalytic activity of titania photocatalyst. Appl. Surf. Sci. **547**, 149123 (2021)

Helmy, R., El-Inany, G.A., Seleem, H.S., Abdel-Salam, M.O., Yoon, T., Saif, M.: A novel Dy^{3+} modified $Zn_2Ti_3O_8$ nanoparticles for efficient hydrogen production photocatalysis. J. Alloys Compd. **907**, 164487 (2022)

Hu, J., Wang, R., Fan, R., Wang, F., Xiong, H., Huang, Z., Liu, L., Fu, H.: Nanocomposites of Au nanorods and core–shell $NaGdF_4$: Yb^{3+}, Er^{3+}@ $NaYF_4$ upconversion nanoparticles for temperature sensing. ACS Appl. Nano Mater. **3**(10), 9679–9685 (2020)

Judd, B.R.: Optical absorption intensities of rare-earth ions. Phys. Rev. **127**(3), 750 (1962)

Jurga, N., Przybylska, D., Kamiński, P., Tymiński, A., Grześkowiak, B.F., Grzyb, T.: Influence of the synthesis route on the spectroscopic, cytotoxic, and temperature-sensing properties of oleate-capped and ligand-free core/shell nanoparticles. J. Colloid Interface Sci. **606**, 1421–1434 (2022)

Kachhap, S., Singh, S., Singh, A.K., Singh, S.K.: Lanthanide-doped inorganic halide perovskites ($CsPbX_3$): novel properties and emerging applications. J. Mater. Chem. C **10**(10), 3647–3676 (2022)

Kido, J., Okamoto, Y.: Organo lanthanide metal complexes for electroluminescent materials. Chem. Rev. **102**(6), 2357–2368 (2002)

Kumar, A., Reddy, K.L., Kumar, S., Kumar, A., Sharma, V., Krishnan, V.: Rational design and development of lanthanide-doped $NaYF_4$@CdS–Au–RGO as quaternary plasmonic photocatalysts for harnessing visible–near-infrared broadband spectrum. ACS Appl. Mater. Interfaces **10**(18), 15565–15581 (2018)

Li, Y., Tang, J., He, L., Liu, Y., Liu, Y., Chen, C., Tang, Z.: Core–shell upconversion nanoparticle@ metal–organic framework nanoprobes for luminescent/magnetic dual-mode targeted imaging. Adv. Mater. **27**(27), 4075–4080 (2015)

Li, Y.-F., Feng, J., Sun, H.-B.: Perovskite quantum dots for light-emitting devices. Nanoscale **11**(41), 19119–19139 (2019)

Li, S.M., Fang, Q.N., Zhang, Y.H., Tao, S.L., Zhang, J.X., Quan, C., Sun, D.L., Zhao, C.C., Hang, Y.: 2 um Ultrabroad spectra and laser operation of Tm:$GdScO_3$ crystal. Opt. Laser Technol. **143**, 107345 (2021)

Liao, J., Yang, L., Wu, S., Yang, Z., Zhou, J., Jin, D., Guan, M.: NIR-II emissive properties of 808 nm-excited lanthanide-doped nanoparticles for multiplexed in vivo imaging. J. Lumin. **242**, 118597 (2022)

Ling, B., Wang, Y., Mi, R., Wang, D., Chen, H., Li, X., Zhang, Y., Wang, L.: Multimodal imaging and synergetic chemodynamic/photodynamic therapy achieved using an $NaGdF_4$, Yb, Er@$NaGdF_4$, Yb, Tm@ $NaYF_4$@ Fe-MOFs nanocomposite. Chem. Asian J. **17**, e2022001 (2022)

Molina-González, J., Arellano-Morales, A., Meza, O., Ramírez-García, G., Desirena, H.: An anti-counterfeiting strategy based on thermochromic pigment activated by highly Yb^{3+} doped photothermal particles. J. Alloys Compd. **850**, 156709 (2021)

Moore, E.G., Samuel, A.P.S., Raymond, K.N.: From antenna to assay: lessons learned in lanthanide luminescence. Acc. Chem. Res. **42**(4), 542–552 (2009)

Naresh, V., Adusumalli, V., Park, Y.I., Lee, N.: NIR triggered NaYF4: Yb3+, Tm^{3+}@ $NaYF_4$/ $CsPb(Br_{1-x}/I_x)_3$ composite for up-converted white-light emission and dual-model anti-counterfeiting applications. Mater. Today Chem. **23**, 100752 (2022)

Nargelas, S., Talochka, Y., Vaitkevičius, A., Dosovitskiy, G., Buzanov, O., Vasil'ev, A., Malinauskas, T., Korzhik, M., Tamulaitis, G.: Influence of matrix composition and its fluctuations on excitation relaxation and emission spectrum of Ce ions in $(Gd_xY_{1-x})\ _3Al_2Ga_3O_{12}$: Ce scintillators. J. Lumin. **242**, 118590 (2022)

Nasrabadi, H.B., Madirov, E., Popescu, R., Štacková, L., Štacko, P., Klán, P., Richards, B.S., Hudry, D., Turshatov, A.: Coordination mechanism of cyanine dyes on the surface of core@ active shell β-$NaGdF_4$: Yb^{3+}, Er^{3+} nanocrystals and its role in enhancing upconversion luminescence. J. Mater. Chem. C **9**(45), 16313–16323 (2021)

Ofelt, G.S.: Intensities of crystal spectra of rare-earth ions. J. Chem. Phys. **37**(3), 511–520 (1962)

Padhiar, M.A., Wang, M., Ji, Y., Yang, Z., Bhatti, A.S.: Tuning optical properties of $CsPbBr_3$ by mixing Nd^{3+} trivalent lanthanide halide cations for blue light emitting devices. Nanotechnology **33**(17), 175202 (2022)

Park, Y.I., Kim, H.M., Kim, J.H., Moon, K.C., Yoo, B., Lee, K.T., Lee, N., Choi, Y., Park, W., Ling, D.: Theranostic probe based on lanthanide-doped nanoparticles for simultaneous in vivo dual-modal imaging and photodynamic therapy. Adv. Mater. **24**(42), 5755–5761 (2012)

Parveen, S., Das, M., Ghosh, S., Giri, P.K.: Experimental and theoretical study of europium-doped organometal halide perovskite nanoplatelets for UV photodetection with high responsivity and fast response. Nanoscale **14**(17), 6402–6416 (2022)

Pattnaik, S., Rai, V.K.: Insight into the spectroscopic and thermometric properties of titanate phosphors via a novel co-excited laser system. Mater. Sci. Eng. B **272**, 115318 (2021)

Pattnaik, S., Rai, V.K.: Tailoring of upconversion luminescence of Al^{3+} engineered titanate phosphor for non-invasive thermometry. Methods Appl. Fluoresc. **10**(3), 34002 (2022)

Peng, F., Liu, W., Zhang, Q., Luo, J., Sun, D., Sun, G., Zhang, D., Wang, X.: Growth, structure, and spectroscopic characteristics of a promising yellow laser crystal Dy: $GdScO_3$. J. Lumin. **201**, 176–181 (2018)

Ran, W., Sun, G., Ma, X., Zhang, Z., Yan, T.: Excellent up conversion luminescence and temperature sensing performance of $CdMoO_4$: Er^{3+}, Yb^{3+} phosphor. Dalt. Trans. (2022)

Rao, Z., Li, Q., Li, Z., Zhou, L., Zhao, X., Gong, X.: Ultra-high-sensitive temperature sensing based on Er^{3+} and Yb^{3+} co-doped lead-free double perovskite microcrystals. J. Phys. Chem. Lett. **13**(16), 3623–3630 (2022)

Rocha, J., Carlos, L.D., Paz, F.A.A., Ananias, D.: Luminescent multifunctional lanthanides-based metal–organic frameworks. Chem. Soc. Rev. **40**(2), 926–940 (2011)

Ronda, C.R., Jüstel, T., Nikol, H.: Rare earth phosphors: fundamentals and applications. J. Alloys Compd. **275**, 669–676 (1998)

Ruan, L., Zhang, Y.: NIR-excitable heterostructured upconversion perovskite nanodots with improved stability. Nat. Commun. **12**(1), 1–10 (2021)

Shahi, P.K., Singh, A.K., Rai, S.B., Ullrich, B.: Lanthanide complexes for temperature sensing, UV light detection, and laser applications. Sens. Actuat. A Phys. **222**, 255–261 (2015)

Shahi, P.K., Singh, P., Rai, S.B., Bahadur, A.: Host-sensitized NIR quantum cutting emission in Nd^{3+} doped $GdNbO_4$ phosphors and effect of Bi^{3+} ion codoping. Inorg. Chem. **55**(4), 1535–1541 (2016)

Shahi, P.K., Singh, P., Singh, A.K., Singh, S.K., Rai, S.B., Prakash, R.: A strategy to achieve efficient dual-mode luminescence in lanthanide-based magnetic hybrid nanostructure and its demonstration for the detection of latent fingerprints. J. Colloid Interface Sci. **491**, 199–206 (2017a)

Shahi, P.K., Singh, P., Rai, S.B.: Sunlight activated lanthanide complex for luminescent solar collector applications: effect of waveguide matrix. J. Phys. D Appl. Phys. **50**(7), 75501 (2017b)

Shang, Y., Chen, T., Ma, T., Hao, S., Lv, W., Jia, D., Yang, C.: Advanced lanthanide doped upconversion nanomaterials for lasing emission. J. Rare Earths **40**(5), 687–695 (2022)

Sharma, M., Singh, P., Singh, S.K., Singh, P.: Li^+ aided self-activated $Ca_9Y_{1-x-y}Er_xYb_y(VO_4)_7$ phosphors for efficient dual-mode emission and temperature sensing application. Opt. Mater. (amst) **133**, 112925 (2022)

Singh, P., Shahi, P.K., Singh, S.K., Rai, S.B.: Photoluminescence, upconversion and quantum-cutting emission in Tm/Tb/Pr and Yb co-doped oxide phosphor: a comparative study. J. Alloys Compd. **681**, 477–485 (2016a)

Singh, P., Shahi, P.K., Rai, A., Bahadur, A., Rai, S.B.: Effect of Li^+ ion sensitization and optical temperature sensing in Gd2O3: Ho^{3+}/Yb^{3+}. Opt. Mater. (amst) **58**, 432–438 (2016b)

Singh, P., Shahi, P.K., Singh, S.K., Singh, A.K., Singh, M.K., Prakash, R., Rai, S.B.: Lanthanide doped ultrafine hybrid nanostructures: multicolour luminescence, upconversion based energy transfer and luminescent solar collector applications. Nanoscale **9**(2), 696–705 (2017a)

Singh, P., Shahi, P.K., Prakash, R., Rai, S.B.: An assembly and interaction of upconversion and plasmonic nanoparticles on organometallic nanofibers: enhanced multicolor upconversion, downshifting emission and the plasmonic effect. Nanotechnology **28**(41), 415701 (2017b)

Singh, P., Singh, P., Prakash, R., Rai, S.B.: Photo-physical studies of ultrasmall upconversion nanoparticles embedded organometallic complexes: probing a dual mode optical sensor for hydrogen peroxide. Opt. Mater. (amst) **98**, 109459 (2019)

Singh, P., Singh, S.K., Singh, P., Prakash, R., Rai, S.B.: Generation of red-NIR bi-modal fluorescence in hybrid nanostructure. Mater. Res. Bull. **122**, 110663 (2020a)

Singh, P., Singh, P., Prakash, R., Rai, S.B., Singh, S.K.: Colour tunability in a bimodal fluorescent hybrid nanostructure UCNPs@ AuNPs@ QDs. Curr. Appl. Phys. **20**(10), 1150–1155 (2020b)

Singh, P., Singh, P., Singh, S.K. (2021) Photon upconversion spectroscopy. Mod. Tech. Spectrosc. Basics Instrument. Appl. **13**, 389

Singh, S., Kachhap, S., Singh, A.K., Pattnaik, S., Singh, S.K.: Temperature sensing using bulk and nanoparticles of $Ca_{0.79}Er_{0.01}Yb_{0.2}MoO_4$ phosphor. Methods Appl. Fluoresc. (2022)

Song, D., Chi, S., Li, X., Wang, C., Li, Z., Liu, Z.: Upconversion system with quantum dots as sensitizer: improved photoluminescence and PDT efficiency. ACS Appl. Mater. Interfaces **11**(44), 41100–41108 (2019)

Sun, L.-N., Zhang, H.-J., Meng, Q.-G., Liu, F.-Y., Fu, L.-S., Peng, C.-Y., Yu, J.-B., Zheng, G.-L., Wang, S.-B.: Near-infrared luminescent hybrid materials doped with lanthanide (Ln) complexes (Ln= Nd, Yb) and their possible laser application. J. Phys. Chem. B **109**(13), 6174–6182 (2005)

Sun, C.-Y., Zheng, X.-J., Chen, X.-B., Li, L.-C., Jin, L.-P.: Assembly and upconversion luminescence of lanthanide–organic frameworks with mixed acid ligands. Inorganica Chim. Acta **362**(2), 325–330 (2009)

Wang, T., Zhao, S., Lei, R., Huang, L., Xu, S.: Eu^{3+} doped and Er^{3+}/Yb^{3+} codoped glass ceramics containing $NaLa(WO_4)_2$ nanoparticles: preparation, optical and temperature sensing properties. J. Non Cryst. Solids **579**, 121379 (2022b)

Wang, B., Wang, Z., Mao, P., Wang, Y.: A multi-color persistent luminescent phosphor β-$NaYF_4$: RE^{3+}(RE= Sm, Tb, Dy, Pr) for dynamic anti-counterfeiting. RSC Adv. **12**(18), 11534–11542 (2022c)

Wang, Y., Ji, C., Tan, Y., Xiang, L., Hou, J.: Construction of multifunctional lanthanide-based nanoparticles Ba2LuF7: Yb/Er/Ho for in vivo dual-modal tumor imaging. Opt. Mater. (amst) **128**, 112369 (2022d)

Wang, S., Xu, Y., Chen, R., Zhu, M., Wang, M., Cao, M., Liu, Y., Ding, H., Zhang, S., Bai, J.: Highly efficient and stable Eu^{3+}-doped $CsPbBr_3/Cs_4PbBr_6$ perovskites for white light-emitting diodes. Coatings **12**(4), 512 (2022)

Wu, Y., Chan, S.Y., Xu, J., Liu, X.: Multiphoton upconversion materials for photocatalysis and environmental remediation. Chem. Asian J. **16**(18), 2596–2609 (2021)

Xia, T., Cao, W., Guan, L., Zhang, J., Jiang, F., Yu, L., Wan, Y.: Three isostructural hexanuclear lanthanide–organic frameworks for sensitive luminescence temperature sensing over a wide range. Dalt. Trans. **51**(14), 5426–5433 (2022)

Xie, Y., Song, Y., Sun, G., Hu, P., Bednarkiewicz, A., Sun, L.: Lanthanide-doped heterostructured nanocomposites toward advanced optical anti-counterfeiting and information storage. Light Sci. Appl. **11**(1), 1–10 (2022)

Xu, W., Chen, X., Song, H.: Upconversion manipulation by local electromagnetic field. Nano Today **17**, 54–78 (2017)

Xu, L., Li, Y., Pan, Q., Wang, D., Li, S., Wang, G., Chen, Y., Zhu, P., Qin, W.: Dual-mode light-emitting lanthanide metal–organic frameworks with high water and thermal stability and their application in white LEDs. ACS Appl. Mater. Interfaces **12**(16), 18934–18943 (2020)

Xue, B., Wang, D., Tu, L., Sun, D., Jing, P., Chang, Y., Zhang, Y., Liu, X., Zuo, J., Song, J.: Ultrastrong absorption meets ultraweak absorption: unraveling the energy-dissipative routes for dye-sensitized upconversion luminescence. J. Phys. Chem. Lett. **9**(16), 4625–4631 (2018)

Yan, Z.-H., Du, M.-H., Liu, J., Jin, S., Wang, C., Zhuang, G.-L., Kong, X.-J., Long, L.-S., Zheng, L.-S.: Photo-generated dinuclear $\{Eu (II)\}_2$ active sites for selective CO_2 reduction in a photosensitizing metal-organic framework. Nat. Commun. **9**(1), 1–9 (2018)

Yang, W., Li, X., Chi, D., Zhang, H., Liu, X.: Lanthanide-doped upconversion materials: emerging applications for photovoltaics and photocatalysis. Nanotechnology **25**(48), 482001 (2014)

Yang, Y., Li, F., Lu, Y., Du, Y., Wang, L., Chen, S., Ouyang, X., Li, Y., Zhao, L., Zhao, J.: CaGdSbWO8: Sm^{3+}: A deep-red tungstate phosphor with excellent thermal stability for horticultural and white lighting applications. J. Lumin. **251**, 119234 (2022a)

Yang, Y., Li, Y., Chen, Y., Wang, Z., He, Z., He, J., Zhao, H.: Dynamic anticounterfeiting through novel photochromic spiropyran-based switch@ Ln-MOF composites. ACS Appl. Mater. Interfaces (2022)

Yu, Y., Shao, G., Ding, L., Zhang, H., Liang, X., Liu, J., Xiang, W.: Ultra-stable Eu^{3+}-doped $CsPbCl_2Br_1$ perovskite quantum dots glass for optical temperature sensing. J. Rare Earths **39**(12), 1497–1505 (2021)

Yuan, P., Lee, Y.H., Gnanasammandhan, M.K., Guan, Z., Zhang, Y., Xu, Q.-H.: Plasmon enhanced upconversion luminescence of $NaYF_4$: Yb, Er@SiO_2@ Ag core–shell nanocomposites for cell imaging. Nanoscale **4**(16), 5132–5137 (2012)

Zhang, D., Ding, M., Dong, B., Zhen, Y., Chang, Q.: Hexagonal sodium yttrium fluoride mesocrystals: one-pot hydrothermal synthesis, formation mechanism and multicolor up-/down-converted luminescence for anti-counterfeiting and fingerprint detection. Ceram. Int. **45**(16), 20307–20315 (2019)

Zhang, Y., Zhu, X., Zhang, Y.: Exploring heterostructured upconversion nanoparticles: from rational engineering to diverse applications. ACS Nano **15**(3), 3709–3735 (2021a)

Zhang, Y., Li, S., Du, X., Guo, J., Gong, Q., Tao, S., Zhang, P., Fang, Q., Pan, S., Zhao, C.: Yb: GdScO 3 crystal for efficient ultrashort pulse lasers. Opt. Lett. **46**(15), 3641–3644 (2021b)

Zheng, W., Huang, P., Gong, Z., Tu, D., Xu, J., Zou, Q., Li, R., You, W., Bünzli, J.-C.G., Chen, X.: Near-infrared-triggered photon upconversion tuning in all-inorganic cesium lead halide perovskite quantum dots. Nat. Commun. **9**(1), 1–9 (2018)

Chapter 5
Defect-Modulated Trap Engineering of Long Persistent and Mechanoluminescence Phosphors for Advanced Applications

Nimai Pathak and Yuanbing Mao

5.1 Introduction

Defect-mediated light emitting materials are gaining unceasing attention as an efficient alternative to the already developed activator-based phosphor materials owing to multiple advantages such as less toxicity, higher stability, easily tunable emission characteristics, and low cost (Loh et al. 2010; Wang et al. 2011). The origins of emissions in this class of materials are very complex to be completely understood. Mounting evidence gathered by using different characterization tools suggests that various defects including vacancies, impurities, radical impurities, and donor–acceptor pairs play a crucial role as color centers similar to activators in doped phosphors. Defect centers create various trap states inside the band gap of host materials through which the electron–hole recombination process occurs. Hence, they impart host materials with light emission at various wavelengths depending on the position of trap states as observed in various oxide-based materials such as ZnO, SnO_2, MgO, $MgAl_2O_4$, $ZnAl_2O_4$, TiO_2, and Al_2O_3 (Zeng et al. 2010; Pathak et al. 2016a, 2016b, 2018; Li et al. 2015; Kar et al. 2011; Ikeda and Uchino 2014). Not only defect-related trap centers act as self-luminescence centers, but their ability to trap charge carriers (electrons and holes) in various activator-doped phosphors also impart them with other luminescence characteristics. For example, defect-related

N. Pathak (✉)
Radiochemistry Division, Bhabha Atomic Research Centre, Mumbai 400085, India
e-mail: nmpathak4@gmail.com; nimai@barc.gov.in

N. Pathak · Y. Mao (✉)
Department of Chemistry, Illinois Institute of Technology, 3101 S Dearborn Street, Chicago, IL 60616, USA
e-mail: ymao17@iit.edu

R. S. Ningthoujam and A. K. Tyagi (eds.), *Handbook of Materials Science, Volume 1*, Indian Institute of Metals Series, https://doi.org/10.1007/978-981-99-7145-9_5

trap centers play pivotal roles in long persistent luminescence (LPL), photostimulated luminescence (PSL), mechanoluminescence (ML), and thermally stimulated luminescence (TSL). Lattice defects in these materials act as the storage of excitation energy which can be exploited for generating ML, PSL, PSL, and TSL.

Lattice defects in a material are due to the deviation from the defined position of an atom in a periodic crystal lattice. Such deviation creates disorder in the lattice bonding and orbital configuration and hence significantly impacts the physical properties created by the free-moving charge carriers. As demonstrated in Fig. 5.1a, these lattice defects create localized electron traps at different energy levels compared to the carrier transport bands. Since the equilibrium nuclear configurations of empty and occupied defect states are different, there will be an energy barrier ΔE between these two states (Jin et al. 2020). Once photo-generated free carriers fall into the localized state they will get trapped energetically (Fig. 5.1b) and their mobility and movement in crystals will be restricted by the additional energy barrier ΔE (Fig. 5.1a). Therefore, the fate of any trapped charge carrier depends on the nature of the traps, which may vary from matrix to matrix. These trapped carriers may escape the localized state and get back into the excited transport levels when some external activation energy is provided either by means of optical or by room-temperature thermal energy. Only those defects-related traps, which are energetically suitably positioned with respect to the conduction band (CB), or valence band (VB) can trap the charge carriers efficiently (Jin et al. 2020) The difference in energy between the transport states and defect levels (ΔE) is called as trap depth. Any defects whose energy level exists above CB or below VB edges cannot act as trap states. As shown in Fig. 5.1c, defects that have a small energy difference to CB/VB edges ($\Delta E \leq k_B T$) are called as shallow trap states while those traps which are energetically positioned toward the middle of the bandgap ($\Delta E > k_B T$) are called deep tarp states. Because of the small trap depth the shallow traps can facilitate both the trapping and de-trapping processes of the charge carriers, unlike the deep traps which sometimes hinder the de-trapping process and result in non-radiative recombination (Fig. 5.1c). The trap's density is generally controlled by its formation energy and the growth temperature during materials synthesis (Jin et al. 2020). Generally higher trap density and large trap depth decrease the carrier's movement and hence it is necessary to engineer defect-traps in such a way that the negative influence of defect centers can be avoided.

As mentioned earlier, defect induced traps play most crucial role for persistent luminescence in LPPs, which possess a special property of glowing for several hours or in few cases for several days once they are excited. LPPs with a longer persistence of luminescence have widespread applications starting from home decoration, safety displays, and dials to various advanced scientific, engineering, and medical fields (Li et al. 2016a; Huang et al. 2021). We have given a simplified pictorial representation of the afterglow mechanism in Fig. 5.2, wherein an electron is first promoted to the excited state from the ground states and then photoionized into the CB with the help of continued excitation energy. Once in the CB, this excited-state electron is free and can be subsequently "trapped" by a trap close to the CB which is either created by the lattice vacancies or the d-orbitals of a rare-earth when used as co-dopant. The thermal energy then helps to release these trapped electrons slowly and get back

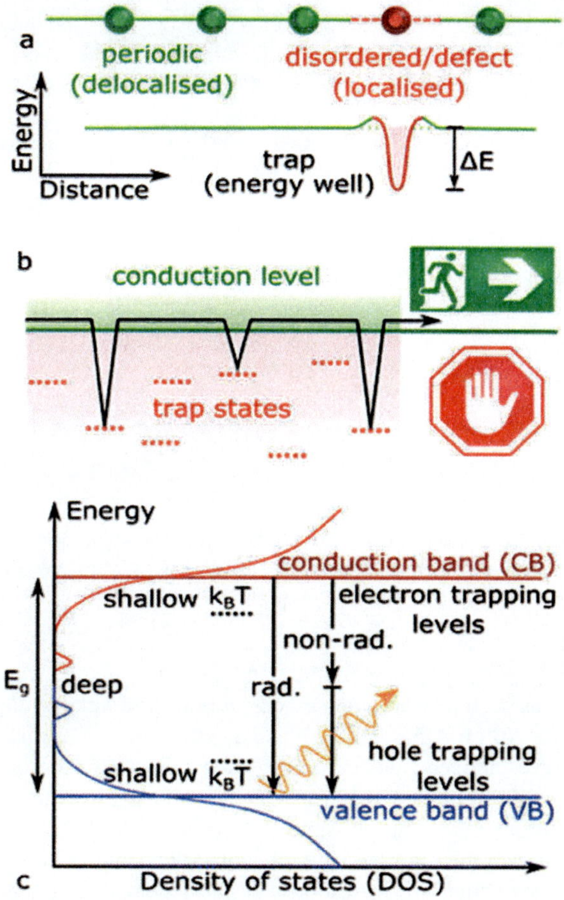

Fig. 5.1 **a** The top of the figure represents charge localization caused by defects while the bottom part shows newly introduced energetic levels. **b** The slowdown of the carrier's mobility because of various trapping and de-trapping events. **c** The density of states of shallow and deep traps inside band gap along with various processes of radiative and non-radiative recombinations. [Reproduced with the permission of the Royal Society of Chemistry (Jin et al. 2020)]

to CB again followed by emission due to electron–hole recombination with long luminescent lifetimes. It is worth mentioning here that trap depth plays a critical role to determine the performance of LPPs. Shallow trap states result in high-intensity light due to quick electron–hole recombination while deep trap states release the charge carriers slowly and enhance the persistency of luminescence. However, as stated earlier that due to higher energy difference deep traps sometimes hinder the de-trapping process and facilitate the non-radiative recombination pathways, which causes less light yield. Therefore, defect-related trap depth engineering plays a crucial role while designing efficient LPPs.

Like LPPs, ML is another luminescence-based technique, which possesses many advanced stress-sensing applications, wherein defect-related trap controlling plays the key role while designing such phosphors. ML is generally observed in a solid material under stress and its intensity is strongly correlated with the applied stress (Zhuang and Xie 2021; Feng and Smet 2018; Zhang et al. 2019). This property makes

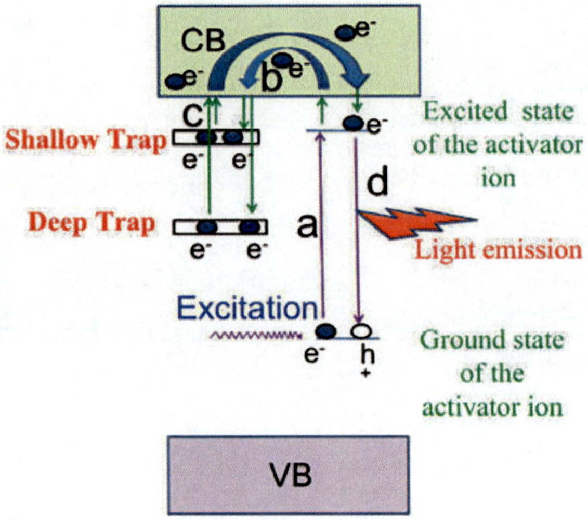

Fig. 5.2 Mechanism of persistent luminescence: **a** The electron is excited from the ground state to the excited state, **b** trapping process of the excited electron into shallow and deep traps, **c** de-trapping process of the electrons, and **d** recombination process. e^- = electron, and h^+ = hole

ML phosphors as potential candidates for many advanced stresses sensing applications such as wearable devices, robots, and flexible electronic devices (Zhuang and Xie 2021). Although the basic mechanism of ML materials is based on mechanical force-induced light emission, such processes are complex in nature and different for different categories of ML phosphors; for example reproducible ML and non-reproducible ML phosphors. Reproducible ML is mostly observed in materials wherein the luminescence phenomena is controlled by the defect-related traps, and they exhibit ML in response to the repeated cycles of non-destructive mechanical force. These ML materials involve two processes: (1) trapping of the charge carriers (electrons/holes) into long-lived trap levels when irradiated by light or another source of energy and (2) de-trapping of the carriers under dynamic loading followed by recombination with the holes/electron which led to the light emissions. The roles of the traps involved in the ML process can be understood in Fig. 5.3.

For example, Sr vacancies can trap the holes in $SrAl_2O_4$:Eu^{2+} ML phosphor are found to have a large impact on the ML intensity (Xu et al. 1999a). As shown in Fig. 5.3a, a hole will escape from Eu^{2+} to the VB and generates Eu^+ upon irradiation with light. The hole will be trapped by a Sr vacancy. Under a mechanical stimulus, the trap releases the hole followed by its recombination with Eu^+ to form an excited Eu^{2+}. The excited Eu^{2+} then comes back to the ground state followed by the emission of a green photon (Xu et al. 1999a). On the other hand Fig. 5.3b illustrates how oxygen vacancies (OVs) act as primary traps for electrons as carriers. While irradiating the materials using UV light, one electron from the Eu^{2+} ion is promoted to the conduction band followed by its trapping by OV. The excited Eu^{2+} ions are then subsequently oxidized to Eu^{3+}. In the next step, when the UV light-irradiated phosphor material

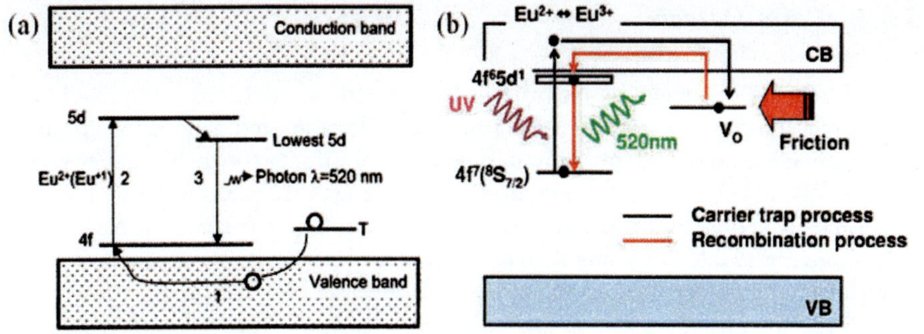

Fig. 5.3 Role of various traps and carriers in the ML process in $SrAl_2O_4:Eu^{2+}$. **a** Hole trap model. [Reproduced with the permission of the American Institute of Physics (Xu et al. 1999a)], **b** Electron trap model. [Reproduced with the permission of The Electrochemical Society (Fu et al. 2007)]

is subjected to a mechanical force, the trapped electrons in OVs will be released followed by recombination with the Eu^{3+}. This will result in the formation of an excited Eu^{2+} followed by the emission of a green photon (Fu et al. 2007).

These defect traps can be engineered in different ways such as exposure to high-energy radiation like X- or γ-ray can generate traps in alkali halide crystals, alkaline earth oxides, and salts (Zhang et al. 2019). In many cases varying composition of reagents, modifying the synthesis method, and using ion-doping techniques such as in various dielectrics like $Sr_3Sn_2O_7:Sm^{3+}$, $CaAl_2Si_2O_8:Eu^{2+}$, $SrAl_2O_4:Eu^{2+}$, and (Ba, Ca)$TiO_3:Pr^{3+}$ also reported generating efficient carriers traps (Zhang et al. 2019). Therefore, before developing any efficient LPP and ML material, a critical understanding of the defect-related traps and the possible ways to tailor and control them is very essential. In this chapter, our focus is to acquaint the reader with the various point defects which create the crucial traps in LPP and ML materials, many advanced characterization tools required to identify them followed by the recent works on trap-controlled LPPs and ML materials and their many potential applications in advanced scientific and engineering field.

5.2 Crystallographic Defects

Any deviation from the well-defined position of an atom in a periodic crystal lattice results in a crystallographic defect. They may be classified as point defects, line defects, and planner defects. Since our main focus of this chapter is on defect-related trap depth engineering based on point defects, we will mostly discuss various point defects with some short descriptions of line defects, planar defects, and bulk defects.

5.2.1 Point Defects

Point defects, also known as intrinsic or thermodynamic defects, are created in a crystalline substance when there are some deviations around a point or an atom from their ideal arrangement. Although there are no limits for their smaller sizes, they involve mostly a few missing atoms. In ionic crystals these point defects are sometimes called as luminescence center, a color center, or F-center. Mostly point defects are represented by using Kröger–Vink notation and a pictorial representation of various point defects in a crystal structure is given in Fig. 5.4.

Vacancy defects:

Such defects are created due to vacancy at the lattice where an atom should be but it is missing (Fig. 5.4c). These defects can be generated by heating the substance and they decrease the volume of the crystals. They can be classified as $\boldsymbol{F}_m{}^n$ center and the $\boldsymbol{V}_{\mathrm{m}}{}^{\mathrm{n}}$ center (*n* is the charge of the ion which is removed from the lattice site and *m* is the number of ***F and V*** centers in the form of aggregate). The $\boldsymbol{F}_m{}^n$ centers

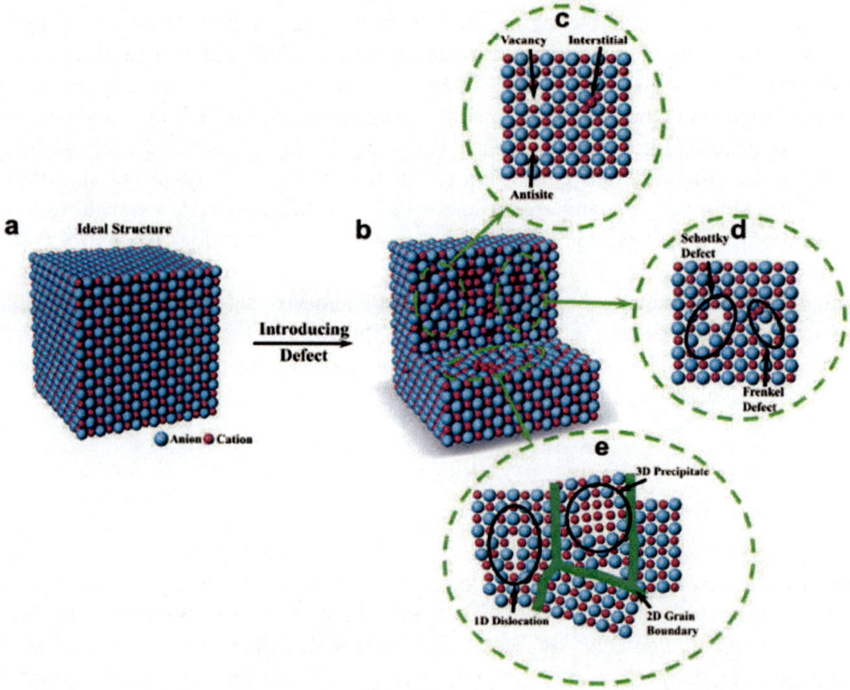

Fig. 5.4 Representation of various crystallographic defects. **a** An ideal structure **b** Defective crystal structure: **c** point defects such as vacancy, antisite, and interstitial defects, **d** defective pairs such as Schottky and Frenkel defect and **e** high dimensional defects. [Reproduced with the permission of the Royal Society of Chemistry (Ran et al. 2018)]

are formed when an electron is trapped at an anion vacancy and they may exist in various aggregate forms of ***F*** centers such as in units of two (called ***M*** centers), three (called ***R*** centers), and four (called ***N*** centers). On the other hand, the $V_m{}^n$ centers are formed when a hole is trapped at a cation vacancy. Here ***n*** is the ion charge that has been removed and ***m*** is the number of ***F and V*** centers that are coupled. The electrons in ***F*** centers are localized by the net positive charge of the missing anion and vice versa for holes in ***V*** centers. These *F and V* centers can be created in simple halide crystals by irradiation with ionizing radiation at low temperatures. To maintain electrical neutrality the cations and anions may also exist as pair, which is known as Schottky defects and mostly found in crystals wherein the cation and anion have almost similar sizes such as NaCl, KCl, CsCl, and AgBr (Zawadzki 2006). They are also called ***P*** centers, i.e., vacancies of both cations and anions. A compound like AgBr shows both Frenkel and Schottky defects (Fig. 5.4d).

Interstitial defects:

When some atoms occupy a site at which there is usually not an atom it is called as an interstitial defect (see Fig. 5.4c) (Zawadzki 2006). Such defects generally have high formation energies, however, small atoms like hydrogen form such defects in palladium without high energy. They increase the density of the substance. They are classified into two types: Self-interstitials and Impurity interstitials. Self-interstitial defects are created by the atoms which are already present in the lattice. They generally occur in low concentrations in metals to minimize the distortion and stress into the tightly packed lattice structure. On the other hand, impurity interstitials are created by the smaller impurity atoms when placed in the open space between the bulk atoms of the lattice structure.

Frenkel defects:

This kind of defects pair is formed when an atom moves into an interstitial site and creates a vacancy at its own lattice site (Fig. 5.4d) (Zawadzki 2006). They are found in ionic substances wherein there is a large difference in the size of the constituent ions such as in ZnS, AgCl, AgBr, and AgI.

Antisite defects:

These defects are formed when different types of atoms exchange their lattice positions (Fig. 5.4c). For example, let us assume a cubic crystal lattice of A and B atoms wherein A type of atoms occupy the corners of a cubic lattice while B atoms occupy the center of the unit cell. Now, in any such unit cell, if the A atom is misplaced at the center, which is the ideal position of B atom, it is called an antisite defect (Tendeloo et al. 2012).

Substitutional or impurity defect:

It is impossible to get a material with 100% purity and these defects are formed when an impurity atom, which is not supposed to be anywhere in the crystal lattice but placed at a regular atomic site. These impurity atoms are generally smaller in size than the ideal lattice atom (ion) and are referred to as off-center ions. The impurity atom may be isovalent to that of the original atom it is replacing, or it may be aliovalent. In the case of aliovalent substitutions there will be a change in the overall charge of the system and to make the system neutral some of the ions will be partially or fully oxidized or reduced, or some new ion vacancies will be formed. For example, some of the lattice sites of Na^+ ions are occupied by Sr^{2+} ions in NaCl crystals due to the presence of a little amount of $SrCl_2$ impurity during the crystallization of molten NaCl (Zawadzki 2006).

5.2.2 *Line Defects*

Line defects are created due to the deviation of an entire row of lattice points from the ideal arrangement (Fig. 5.4e). They are classified into Edge dislocation and Screw dislocation.

5.2.3 *Planar Defects*

If the crystallographic direction of a lattice suddenly changes (Fig. 5.4e) and meets a new one then such defects are formed. Examples of various planner defects are grain boundaries, Stacking faults, and twin boundaries.

5.2.4 *Bulk Defects*

These defects can be three-dimensional pores, cracks, inclusions, or voids made of clusters of vacancies. Sometimes impurities can also form clusters and form precipitates in small regions.

5.3 Defect Characterization

One of the most crucial parts of designing ML and LPP materials is the characterization and identification of these defect-related trap states in detail by using various advanced characterization techniques. These point defects in the host matrix not only change the local structure surrounding the activators but also form the characteristics of a localized electronic trap state inside the optical band gap. An in-depth understanding of these defects will help to tailor the defect modulated traps and subsequently optimize the efficiency of ML and LPP materials. Advanced techniques such as high-resolution transmission electron microscopy (HRTEM) or scanning transmission electron microscopy (STEM) are highly useful to characterize the point defects because of their ability to directly capture the atomic structure with sub-angstrom resolution. Other techniques that provide significant information about various defects in solids include X-ray photoelectron spectroscopy (XPS), Raman spectroscopy, X-ray absorption spectroscopy (XAS), electron paramagnetic resonance (EPR), positron annihilation lifetime spectroscopy (PALS), and electron energy loss spectroscopy (EELS). The thermally stimulated luminescence (TSL) technique is very useful to calculate the trap depth energy. In this section, we have summarized the most used tools to characterize defects.

5.3.1 HRTEM/STEM

HRTEM can provide a direct image at atomic scales while the spherical aberration correctors in STEM enable it to get images in the sub-Å range. The high-angle annular darkfield (HAADF) detector also enables the STEM to get images in a scale that is a square of the atomic number (Tendeloo et al. 2012). Direct atomic resolution images always help understand atomic level defects in crystals significantly. For example, Zhou et al. have reported six different types of point defects in CVD-grown monolayer MoS_2 using STEM-ADF as shown in Fig. 5.5a (Zhou et al. 2013). These vacancies include V_S (mono sulfur vacancy), V_{S2} (disulfur vacancy), *V*MoS3 (vacancy complex of Mo and three nearby sulfur), V_{MoS6} (vacancy complex of Mo and three nearby disulfur pairs), Mo_{S2} (antisite defects created by a Mo atom at S_2 site), and $S2_{Mo}$ (antisite defects created by S_2 column at Mo site).

Similarly, the B vacancies (V_B) in hexagonal boron nitride (h-BN), which shows PL up-conversion, are successfully characterized using STEM-ADF techniques as shown in Fig. 5.5b, wherein the bright and dim spots represent N and B atoms respectively. Figure 5.5c shows the Enlarged STEM-ADF image of a single V_B in monolayer h-BN while Fig. 5.5d represents its simulated STEM-ADF image derived from the DFT-optimized model (Wang et al. 2018a).

5.3.2 XPS

Since defects always reduce the coordination number of the surrounding element, it will also change its bonding energies. This will in turn lead to a slight change in the XPS spectra with respect to intensity or energy position. For example, a prominent peak in the XPS O 1 s spectrum was observed for orthorhombic o-$CoSe_2$-O nanosheets due to the oxygen vacancy, which was absent in the bulk cubic c-$CoSe_2$/DETA sample as shown in Fig. 5.6a (Wang et al. 2018b). Another interesting example where the existence of S vacancies by using the XPS technique was observed is in MoS_2 (Tsai et al. 2017). From the XPS S 2p spectra of pristine MoS_2 (P-MoS_2) and defective MoS_2 (V-MoS_2), it can be clearly seen that desulfurization leads to the creation of S vacancies in P-MoS_2 (Fig. 5.6b).

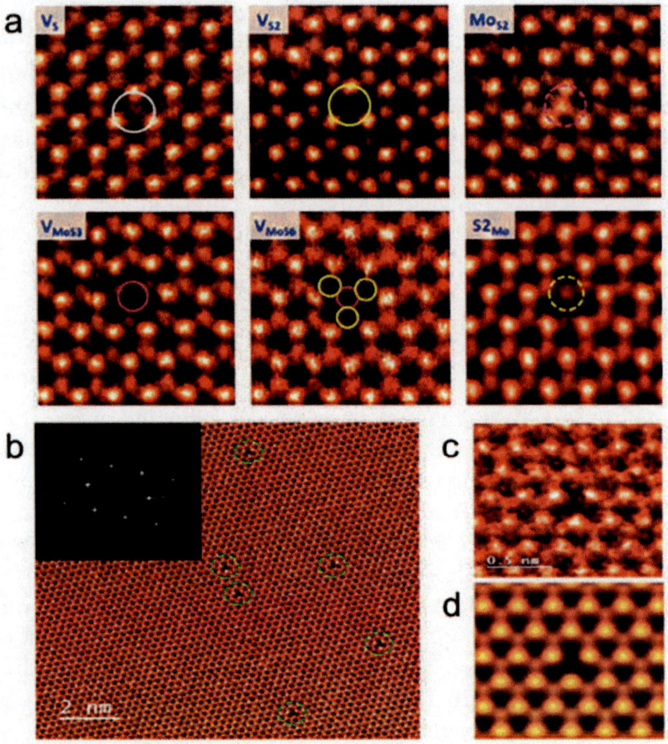

Fig. 5.5 **a** STEM-ADF images of various point defects in monolayer CVD MoS_2. [Reproduced with the permission of the American Chemical Society from (Zhou et al. 2013)], **b** STEM-ADF images of monolayer h-BN with B vacancies (V_B) (Inset: FFT of h-BN), **c** Enlarged and, **d** Simulated STEM-ADF image of a single V_B in monolayer h-BN. [Reproduced with the permission of American Chemical Society (Wang et al. 2018a)]

5.3.3 Raman Spectroscopy

Raman spectroscopy is another useful tool to investigate the defect in solids. Raman scattering is inelastic wherein a monochromatic UV, visible, or NIR laser source is used for excitation. Since crystal defects always influence the vibrational level of the materials, there will be always some energy differences between the incident and the scattered photons as shown in Fig. 5.7a. This will result in a frequency shift or broadening of the Raman bands and sometimes new Raman peaks may also arise due to a change in the local site symmetry caused by the defects.

For example, carbon defects in pristine graphene (G), nitrogen-doped graphene (NG), and defective graphene (DG) were investigated by monitoring the intensity ratio of the Raman D and G bands (ID/IG) as shown in Fig. 5.7b, c (Jia et al. 2016). It has been observed that for pristine graphene (G), the Raman G band is more intense

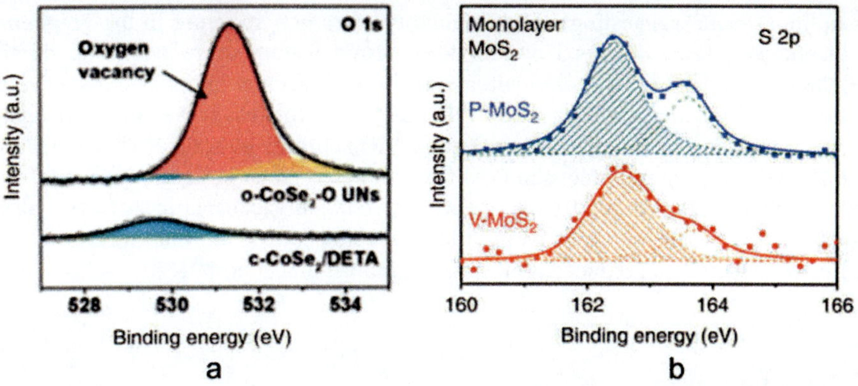

Fig. 5.6 **a** XPS O 1 s spectra of c-$CoSe_2$/DETA and o-$CoSe_2$-O UNs. [Reproduced with the permission of Wiley–VCH (Wang et al. 2018b)], **b** XPS S 2p spectra of P-MoS_2 and V-MoS_2. [Reproduced with the permission of Springer Nature (Tsai et al. 2017)]

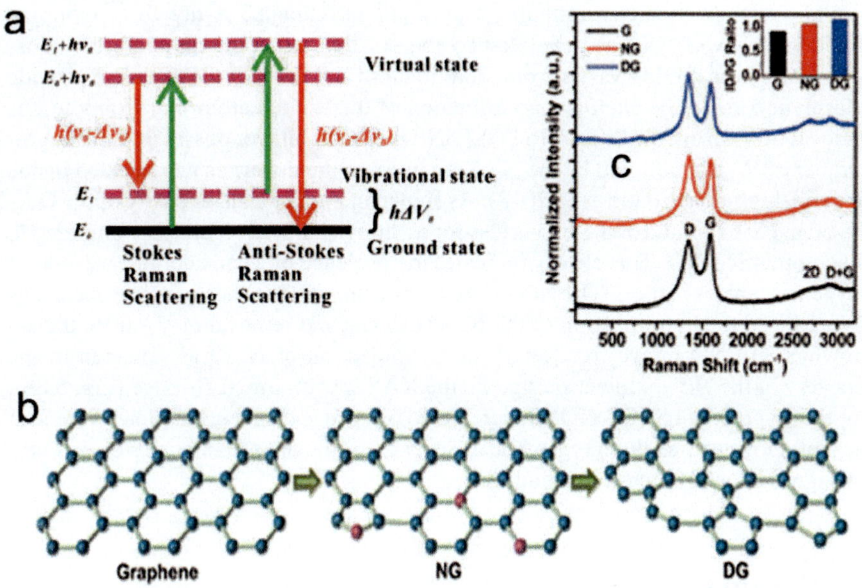

Fig. 5.7 Characterization of defect by Raman spectroscopy: **a** Various transitions in Raman scattering. [Reproduced with the permission of Elsevier (Zhou et al. 2021)]. Characterizations of carbon defects in graphene: **b** The schematic of the formation of NG and DG, **c** Raman patterns of pristine graphene, NG, and DG. [Reproduced with the permission of Wiley–VCH (Jia et al. 2016)]

than the D band, suggesting a high regularity of carbon structure in the graphene with relatively fewer defects. However, when nitrogen atoms were incorporated into G, the hexagonal structures were altered, and some nitrogen impurity defects were introduced. This increased the ratio of ID/IG in NG. Now when the doped nitrogen atoms were also removed from the NG, the ID/IG is further enhanced. This indicates the formation of more defects and disruption of hexagonal structure in the defective graphene (DG). Similarly, the creation of oxygen vacancies in ZnO was also confirmed by Fukushima et al. using Raman spectroscopy when ZnO was heated at different oxidized and reduced atmospheres (Fukushima et al. 2015).

5.3.4 XAS Study

This technique involves the excitation of an electron from the atomic core-orbital to the free or unoccupied levels using an X-ray source. The ejected photoelectrons will then be scattered by the surrounding atoms and an Extended X-ray absorption fine structure (EXAFS) will be generated by the oscillations in the photoelectric cross section. Such scattering varies from atom to atom and the EXAFS data can provide information about the chemical coordination of the excited atom. For example, the removal of V_O from the $SrTiO_3$ (STO)/$LaNiO_3$ (LNO) bilayer possessing the largest number of V_O is confirmed by the XAS study when the materials were heated inside an oxidizing furnace (Guo et al. 2018). As shown in Fig. 5.8a the intensity of the O *K*-pre-edge peak (around 528.5 eV) is less for as-grown STO/LNO bilayer compared to stoichiometric LNO. This clearly indicates the presence of a considerable amount of oxygen vacancies in the STO/LNO as-grown bilayer. Further, these oxygen vacancies also reduced the hybridization of the Ni–O orbitals and removal of V_O from the as-grown STO/LNO bilayer recovered the hybridization of Ni–O as reflected in the recovery of the Ni^{3+} oxidization state in the XAS spectra for Ni L_2-edge (Fig. 5.8b). For the as-prepared STO/LNO bilayer, due to oxygen vacancies, Ni^{2+} ions were also present. However, as the oxygen vacancies were removed the intensity of the Ni^{2+} ion signature was found to be reduced.

5.3.5 EPR Spectroscopy

EPR technique is widely used to characterize any chemical species with unpaired electrons. The basic concept is like that of nuclear magnetic resonance (NMR), but instead of nuclear spin it involves electron spins. Like a proton, the spin of the electron also makes it behave like a tiny bar magnet, and in an external magnetic field with strength H, the electron's magnetic moment may align itself either parallel ($m_s = -1/2$) or antiparallel ($m_s = +1/2$) to the field. Each of the alignments has a specific energy due to the Zeeman effect: $E = m_e g_e Bh$, where g_e is the electron's so-called *g*-factor ($g_e = 2.0023$ for the free electron) and β is the Bohr magneton. Therefore, the energy

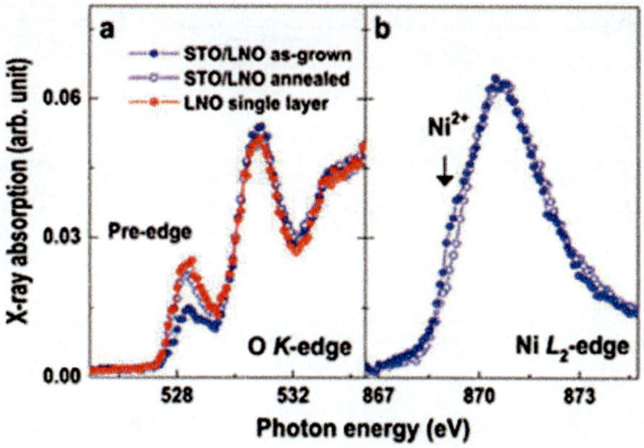

Fig. 5.8 XAS spectra for **a** O K-edge and **b** Ni L_2-edge for an STO/LNO bilayer before and after annealing at 700 °C in atmospheric oxygen. [Reproduced with the permission of Wiley–VCH (Guo et al. 2018)]

difference between the lower and the upper state for an unpaired free electron is $\Delta E = g_e \beta H$ and it is directly proportional to the magnetic field's strength. Defects such as oxygen vacancies (OVs) sometimes trap an electron and behave as unpaired species. In many cases, a hole can also get trapped at a negatively charged cationic vacancy and can be characterized by the EPR technique. For example, Bi^{3+} doped ThO_2 showed OV-induced dilute ferromagnetism because of the presence of unpaired electrons trapped at the OVs. The OVs are created due to the charge difference between Bi^{3+} and Th^{4+} (Guo et al. 2018). EPR spectra of various Bi^{3+} doped ThO_2 compounds revealed a resonance signal around g ~ 2.25 due to the paramagnetic OVs (V_o°) as shown in Fig. 5.9a. At higher concentrations of Bi^{3+} ion, the EPR signal became more intense due to the creation of more and more OVs. Further, the spectra also shifted toward the low-filed region due to the magnetic ordering of the paramagnetic oxygen vacancies. Similar types of paramagnetic OVs are also found to exist in MgO, and ZnO and are characterized by EPR (Pathak et al. 2017; Kaftelen et al. 2012). The EPR technique is also found to be very useful to characterize many other defects. For example, Zhang et al. demonstrated sulfur defects in porous Co_3S_4 nanosheets which are rich in sulfur vacancies (Zhang et al. 2018a). The ultrathin porous Co_3S_4 nanosheets (Co_3S_4 PNS_{vac}) showed a strong EPR signal at $g = 2.003$ in Fig. 5.9b due to the sulfur vacancies while no such signal was observed for the defect-free Co_3S_4 nanosheets (Co_3S_4 NS).

5.3.6 *PALS Study*

PALS is another useful technique to characterize various atomic defects especially the extended free-volume defects composed of multiple vacancies and clusters of vacancies. The basic principle involves the annihilation of a positron through interaction with the electrons inside the free-volume defects followed by the release of

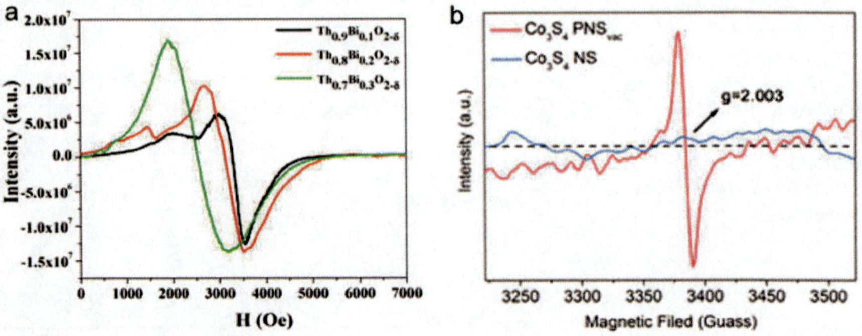

Fig. 5.9 EPR spectra of **a** $Th_{1-x}Bi_xO_{2-\delta}$ solid solutions [Reproduced with the permission of Royal Society of Chemistry (Kanrar et al. 2017)] and **b** Co_3S_4 PNS_{vac} and Co_3S_4 NS [Reproduced with the permission of American Chemical Society (Zhang et al. 2018a)]

two gamma rays of energy (~0.511 meV). Figure 5.10 shows a pictorial representation of a positron from the radioactive ^{22}Na atomic nucleus along with the emission of a neutrino (ν). The time difference between the birth gamma ray (~1.28 meV) and the newly generated gamma rays (~0.511 meV) after annihilation corresponds to the lifetime of the positron. Because of the positive charge, the positrons are repelled by the positively charged vacancies while attracted by the negatively charged vacancies, which are more suitable as positron trap centers.

For example, the PALS study of MgO prepared at an annealing temperature of 600 °C showed two-lifetime components, viz., 237.0 and 445.0 ps, which are attributed to open volume defects. The smaller lifetime component is attributed to the Mg mono-vacancy while the higher positron lifetime component is attributed to a vacancy cluster composed of $2V_{Mg}$ and $2V_O$ or $2(V_{Mg} + V_O)$ wherein the positron gets more time to move around and takes longer time to annihilate (Pathak et al. 2017).

Fig. 5.10 A schematic representation of PALS

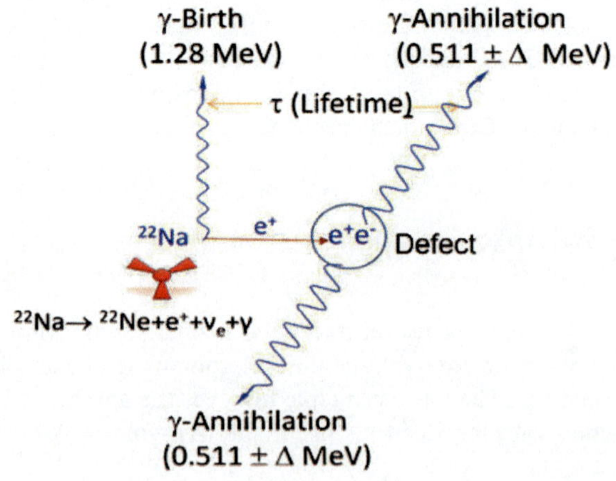

The lifetime of this cluster vacancy was further reduced to 291 ps when annealed at higher temperatures due to the reduction in the size of the vacancy. PALS study was also found to be very helpful to characterize the defects in various halide matrices. We have observed two-lifetime components (~243 and 461 ps) in $KMgF_3$ (Pathak et al. 2020). The smaller component is attributed to Mg-vacancy (mono- or di-) and the large component is attributed to the K-Mg-F vacancy clusters. We also demonstrated that by using a co-dopant ion such as Li^+, the configuration of the vacancy cluster can be changed.

5.3.7 EELS Study

EELS involves the inelastic scattering of electrons. When a known and narrow energy beam of the electron is exposed to the materials, some of the incident electrons will undergo inelastic scattering and lose energy due to various types of reasons including inner shell ionization, phonon excitations, inter- and intra-band transitions, plasmon excitations, Cherenkov radiation. Inner-shell ionizations can provide information about the compositions of various elements along with their electronic structure, and the chemical bonding nature of a material. For example, Liu and co-workers reported the presence of OVs in porous Zn-doped Co_3O_4 sheets by using the EELS technique (Liu et al. 2018). When the EELS spectra were recorded at different positions of the surfaces, the α and β peaks in the O–K edges were found to be sharper for the top/base surfaces than the lateral surface, while the L3 and L2 peaks of the Co–L edges were shifted to higher energy at top surfaces as shown in the EELS spectra in Fig. 5.11b, c. This confirmed the presence of more OVs on the lateral surface of the sheets than that on the top surface.

5.3.8 TL Spectroscopy

TL spectroscopy provides direct information about the number of defect-related traps and their trap depth inside the band gap of the materials. When irradiated with high-energy light or other types of high-energy radiation (X-Rays, gamma, or beta rays), the newly created charged carriers (electrons and holes) will be trapped in the metastable defects-related traps. The depopulation of these trapped electrons from the metastable state can be achieved by heating followed by light emission due to electron–hole recombination (Trojan-Piegza et al. 2008). The defect-related trap's activation energy can be evaluated from the TL spectroscopy. As mentioned earlier, the trap depth of various defect modulated traps is very important while designing any LPP or ML materials (Trojan-Piegza et al. 2008). The TL glow curve is composed of multiple glow peaks originating from various traps at different temperatures due to their different rap depth energy as shown in Fig. 5.12.

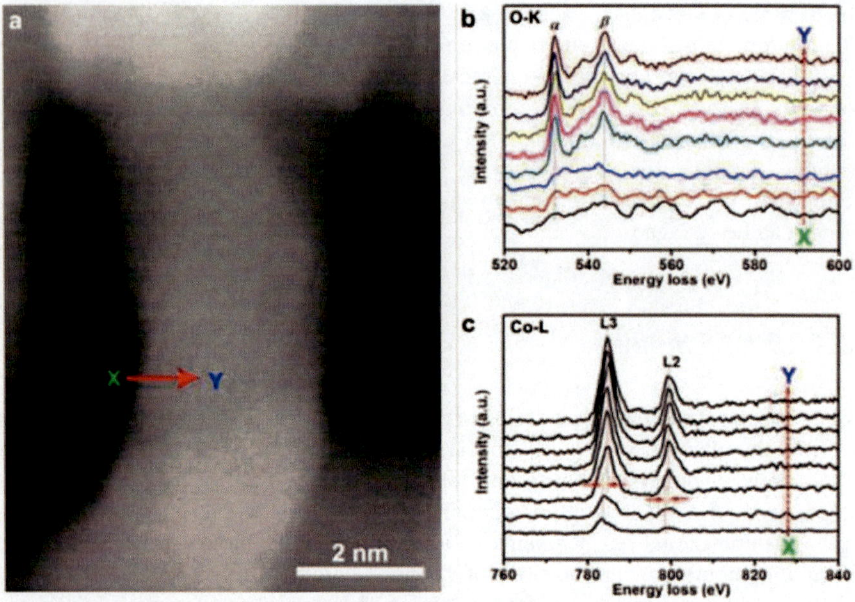

Fig. 5.11 **a** STEM image of porous Zn-doped Co_3O_4 sheets. **b** EELS spectra for the O–K edges and **c** Co–L edges respectively along the arrow (X → Y) in (**a**). X = Lateral surface, Y = Top surface. [Reproduced with the permission of Elsevier (Liu et al. 2018)]

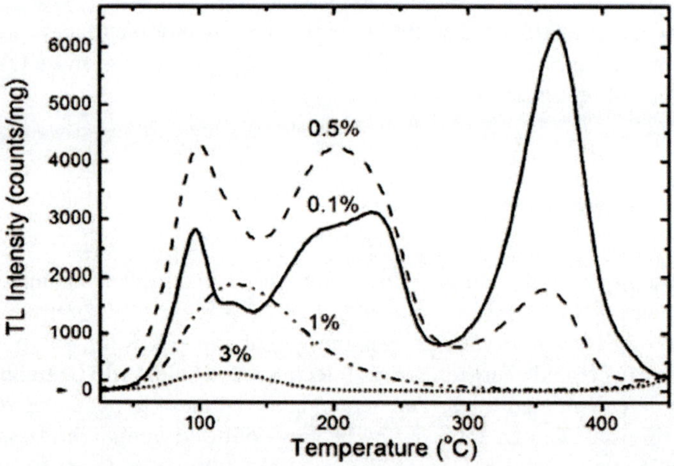

Fig. 5.12 TL glow curves correspond to various traps for vacuum sintered Lu_2O_3:Tb^{3+} compound. [Reproduced with the permission of American Chemical Society (Trojan-Piegza et al. 2008)]

5.4 Defect Engineering for Modulating Trap States

The most crucial task while designing the efficient LPPs and ML materials is the defect-related trap engineering generated from various point defects. One needs to be highly focused on this part before developing those materials. As stated earlier in the introduction part that these traps can be generated in a variety of ways such as by varying the synthesis condition, doping aliovalent ions, or irritating the materials with high-energy radiation (Zhang et al. 2019). For example, heating the material at high temperatures or exposing to excess pressure may create various defects because of thermal dissociation and disorder. Sometimes, by irradiating the samples with an electron beam, high-energy rays (such as gamma and X-rays), nuclear particles (such as alpha particles), or high-intensity lasers also create such defects. In many cases, incorporating a dopant ion into a lattice site also led to the formation of crystal defects. Such substitution can be either equivalent or non-equivalent. In non-equivalent substitution, an aliovalent ion substitutes the lattice ions. For example, if a rare-earth ion with $^{+3}$ charge substitute a divalent host ions (M^{2+}) then due to charge compensation, there will be two different type of point defects, expressed as 2 Re^{3+} + 3 M^{2+} = 2 Re_M + V_M'. Here Re_M will carry one positive while the V_M'' vacancy will carry two negative charges. Now if the Re has a 2+ charge (such as Eu^{2+}) then the substitutional defect will be Re_M' with one negative charge. To balance this negative charge there will be the creation of another anionic vacancy like oxygen vacancy (V_O) (for oxide-based materials) or halide vacancy V_X (for halide compounds), which possess one positive charge. Examples of such defect-related trap engineering can be found in various ML materials such as $Ca_2Al_2SiO_7$:Ce^{3+} (Akiyama et al. 1999), (Ba,Ca)TiO_3:Pr^{3+} (Zhang et al. 2010a), $SrAl_2O_4$:Ce^{3+} (Zhang et al. 2007), $CaYAl_3O_7$:Ce^{3+} (Zhang et al. 2011), $Sr_3Sn_2O_7$:Sm^{3+} (Kamimura et al. 2012), CaZnOS:Er^{3+} (Zhang et al. 2015), mCaO·Nb_2O_5:Pr^{3+} (m = 1, 2 and 3) (Zhang et al. 2016), and $LiNbO_3$:Pr^{3+} (Tu et al. 2017). On the other hand, equivalent substitution creates isoelectronic trap states which are resulted due to the electronegativity differences between the host ions and dopant ions. For example, materials like ZnS:Mn^{2+}, CaZnOS:Mn^{2+}, and BaZnOS:Mn^2 show such defect-related trap states wherein Mn^{2+} substitutes at the Zn^{2+} position (Li et al. 2016b). Because Mn^{2+} has electronegativity lower than that of Zn^{2+} ion it can combine with a hole to act as electron trap centers while Zn^{2+} ions can combine with the electrons to act as hole trap centers. In many cases judicious choice of component materials or by changing synthesis conditions that volatile one element can also generate metal deficiency-related vacancy. For example, Sr vacancies in $SrAl_2O_4$:Eu^{2+} can be created by taking the nonstoichiometric amount of Sr precursor materials while in $Ca_2Al_2SiO_7$:Ce^{3+} such type Ca and Si vacancies can be created by manipulating the compositions of Ca and Si (Xu et al. 1999a; Akiyama et al. 2003). In many cases, OVs are generated by the heat treatment of the compound in a reducing atmosphere (Matsui et al. 2000). Such kind of treatments are also found to be effective to generate and control sulfide vacancies (Gan et al. 2017a).

5.4.1 Defect-Modulated Trap Engineering in LPPs

The three important parameters that determine the efficiency of the persistent luminescence process are the type of traps, their concentrations, and the trap's depth. The type of traps purely depends on the nature of the defect centers, and we will provide a brief discussion about how different types of defect centers create various trap states. Since the efficiency of persistent luminescence is critically dependent on the nature of traps, introducing more effective traps always results in higher efficiency. The trap depth determines the releasing rate of charge carriers, which has an immediate impact on the light yield and its persistency. For example, shallow traps release the carrier at a faster rate and are beneficial for higher light yields. On the contrary, the deep traps release the carriers very slowly and are hardly emptied at room temperature, resulting in less light yield but with a long persistent duration. Therefore, defect-modulated shallow and deep trap engineering plays a key role in optimizing the efficiency of LPPs materials. Here we will discuss the defect-modulated trap engineering by considering three proposed models; namely, the conduction band valence band (CB-VB) model, quantum tunneling model, and oxygen vacancy model.

5.4.1.1 CB-VB Model

This model involves the participation of CB or VB in the excitation process. Here the traps are placed close to CB/VB and the trapping of the excited carriers is accomplished by the spatial localization into the traps. Once a trapped electron or hole is released from a trap, it is free to recombine at a recombination center through a variety of recombination processes, which can be briefly explained by considering different models, e.g., the hole model, electron model, energy-band engineering model, and Quantum tunneling model.

Hole Model

The hole model can be explained by considering the case of $SrAl_2O_4$:Eu^{2+},Dy^{3+} as demonstrated by Matsuzawa et al. in Fig. 5.13a (Yamamoto and Matsuzawa 1997). Photoconductivity measurements showed that holes are the main charge carriers. When Eu^{2+} ions are excited, a hole can escape to the VB and result in an Eu^{+} ion. Being free in the VB the hole can be captured by the Dy^{3+} and forming a Dy^{4+} ion. The room-temperature thermal energy then releases the hole from the Dy^{4+} ion into the VB again. The released hole will then recombine with the Eu^{+} ion, which will return to the ground state followed by the emission of a photon. Therefore, the Dy^{3+} ion acts as a hole trap in this process.

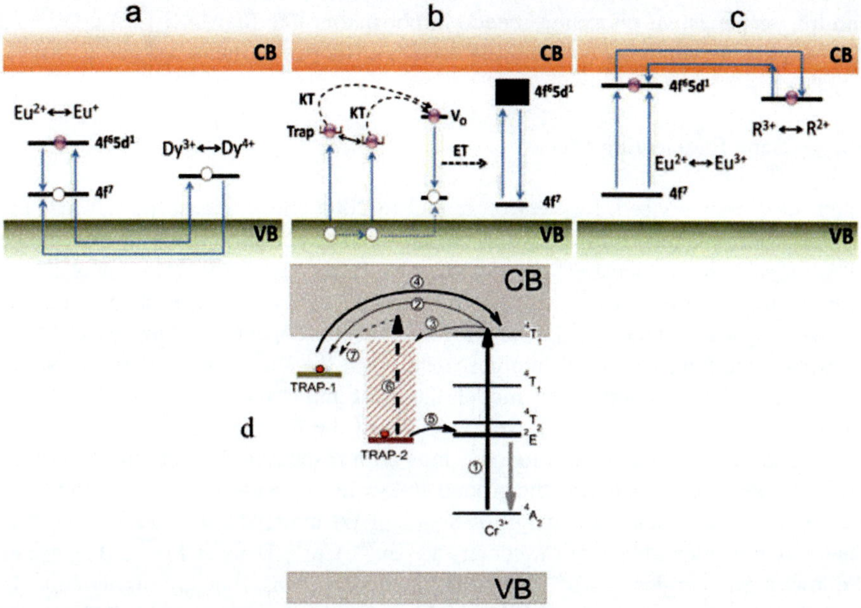

Fig. 5.13 Various afterglow mechanisms: **a** Hole model in $SrAl_2O_4:Eu^{2+},Dy^{3+}$ proposed by Matsuzawa et al. **b** Electron model in $CaAl_2O_4:Eu^{2+},Dy^{3+}$ proposed by Aitasalo et al. **c** Energy band engineering model in aluminate and silicate compounds proposed by Dorenbos et al. (Yamamoto and Matsuzawa 1997; Aitasalo et al. 2003; Dorenbos 2005). [Reproduced with the permission of Royal Society of Chemistry (Li et al. 2016a)] **d** Quantum tunneling model for NIR persistent phosphorescence and photostimulated NIR persistent phosphorescence mechanisms in Cr^{3+}-doped $LiGa_5O_8$. [Reproduced with the permission of Nature Publishing Group (Liu et al. 2013)]

Electron Model

However, the hole model was not found suitable for $CaAl_2O_4:Eu^{2+},Dy^{3+}$ as demonstrated by Aitasalo in 2003 and they proposed that here the electrons were directly excited from the VB into some trap at some unspecified origin as shown in Fig. 5.13b (Aitasalo et al. 2003). From those traps, the electrons are thermally released to the OV-related traps. However, the thermal release of the electrons from the OVs to CB is not possible here because of the higher energy difference. The electron will then recombine with a hole followed by the emission of light. This emitted light can be transferred to the Eu^{2+} ions through an energy transfer process. They did not consider the formation of Eu^{+} and Dy^{4+} as they are chemically unstable ions. This argument that Eu^{+} species were unstable in aluminate is also supported by Holsa et al. who proposed that the co-doping Re^{3+} ion like Dy^{3+} only serve as the trap centers (Holsa et al. 2004). They found that co-doping Sm^{3+} into $SrAl_2O_4$:Eu changes the persistency of luminescence. Further, they also observed different persistency of luminescence when the Sm^{3+} ions were reduced to Sm^{2+}. The fact that doping rare-earth R^{3+} ion only regulates the trap parameters is also supported by the observation of intense and longer persistent phosphorescence in phosphors like $Sr_2MgSi_2O_7$:Eu,Dy, and $SrAl_2O_4$:Eu,Dy wherein Dy^{3+} ion was used as co-dopant (Gluchowski et al. 2015).

Energy Band Engineering Model

Later, Dorenbos revised the electron model in 2005 and proposed that at first the electrons are excited from the $4f^7$ level of divalent europium ions to that of the $4f^65d^1$ level, which is located very close to the CB (see Fig. 5.13c) (Dorenbos 2005). The excited electrons from the excited state can easily be released into the CB followed by its trapping by a trivalent rare-earth co-dopant ion. However, it fails to explain the long persistent luminescence when the Eu^{2+} ion is singly doped in $SrAl_2O_4$. While considering this model, the most important factors are (i) the band gap of the host, (ii) the energy level positions of the Re^{2+} and Re^{3+} ions, and (iii) the trap-level positions of R^{2+} and R^{3+} ions with respect to the bottom of the CB. For example, after considering the ground energy level positions of various Re^{2+} and Re^{3+} ions and the band gap energy of $Sr_2MgSi_2O_7$ materials, it was proposed that the best candidates for afterglow centers are Eu^{2+}, Ce^{3+}, Tb^{3+}, and Pr^{3+}. Because of the lowest 5d energies of Eu^{2+}, Ce^{3+}, Tb^{3+}, and Pr^{3+} ions, they are highly suitable for present-day applications of LPPs which require excitation by visible light or UV light. Further, since the excited 5d levels of these ions are close to the CB, the electron transfer through CB to traps will be more efficient. TL is a very powerful technique to determine trap-level positions. For example, the TL glow curves of YPO4:Ce^{3+},R^{3+} (R = Nd, Sm, Dy, Ho, Er, and Tm) and YPO_4:Pr^{3+},R^{3+} (R = Nd, Dy, Ho, and Er) phosphors are reported by various groups (Lecointre et al. 2011; Bos et al. 2010). Interestingly it has been that the TL peaks of YPO_4:Pr^{3+},R^{3+} and YPO_4:Ce^{3+},R^{3+} compounds are similar. Similar TL peaks were also found with R^{2+} as a co-doped ion for both Pr^{3+} -doped and Ce^{3+}-doped YPO_4 compounds.

5.4.1.2 Quantum Tunneling Model

As stated earlier, deep-level trap states result in the longer persistency of luminescence. The above-mentioned CB–VB models are involved in shallow trap states. The Quantum tunneling model on the other hand explains the afterglow mechanism which involved the deep tarp states. This model is considered when a charge carrier tunnels through a barrier which is not possible classically. For many LPP materials, the tunneling of the electron occurs between the deep trap states and near excited-state levels of activators without the involvement of any CB or VB. For example, various Cr^{3+}–doped zinc gallogermanate showed NIR emission at 650–1000 nm region with a very long afterglow time of 15 days (360 h) (Pan et al. 2012). A similar observation was also found in Cr^{3+}-doped $LiGa_5O_8$ NIR persistent phosphor with luminesce persistent for more than 1000 h (Liu et al. 2013). The material showed an effective afterglow when excited by the UV light, which suggests that UV light is sufficient for the ionization processes required for Cr^{3+} in $Zn_3Ga_2Ge_2O_{10}$ or $LiGa_5O_8$ host followed by the steps 2, 3, and 4 as shown in Fig. 5.13d. However, the materials were interestingly found to generate super long afterglow even with low-energy visible light (400–630 nm) as an excitation source which is below the ionization threshold. This was explained based on the quantum tunneling process (process 5 in Fig. 5.13d). When excited by visible light, the electrons in Cr^{3+} ions might be promoted to higher energy levels that are close to the deep traps. Those deep traps are then filled through the tunneling process from the energetically closed excited levels of Cr^{3+} ions. Similarly, the reverse tunneling processes led to recombination at Cr^{3+} ion which results in weak but super long afterglow.

5.4.1.3 Oxygen Vacancy Model

This model involves OVs acting as the electron trap in oxide-based materials as described earlier in the electron trap model by Aitasalo et al. (2003). Further Clabau et al. also suggested that OVs act as an efficient electron trap in $SrAl_2O_4$:Eu^{2+},Dy^{3+} (Clabau et al. 2005). From TL study it has been observed that although the TL peaks of Eu^{2+} singly doped $SrAl_2O_4$ and Eu^{2+}/Dy^{3+} co-doped $SrAl_2O_4$, are differed in size and location, they are similar in shape. This suggests that the chemical nature of the traps is not much affected by co-doping and OVs are still acting as the main store for the carriers.

5.4.2 *Defect-Modulated Trap Engineering in ML Materials*

Like LPPs materials, engineering efficient defect-related traps in ML materials is very much crucial to improve its performance. ML materials are classified into two groups: one group is based on dielectric host materials and the other group is based on semiconductor host materials. Interestingly, it has been observed that most of the compounds of these groups are piezoelectric in nature which arises due to their non-centrosymmetric crystal structures. It has been proposed that the stress-induced piezoelectric field helps to release the carriers from the traps. Like LPPs, proper control of the defect-related traps holds the key to enhancing the efficiency of ML properties. For example, Zhang et al. demonstrated that the ML intensity of $SrAl_2O_4$:Ce^{3+} phosphor was enhanced by $\approx$100 times when the compound was co-doped with 1.5% Ho^{3+} as shown in Fig. 5.14a (Zhang et al. 2007). This was explained based on the fact that the aliovalent substitution of Ho^{3+} at the Sr^{2+} site creates an abundant amount of shallow traps and thereby increasing the charge storage capacity. Similar co-doping strategy was also found to be effective in other ML materials such as Dy^{3+} in $Ca_2MgSi_2O_7$:Eu^{2+} (Zhang et al. 2008), Sr^{2+} in $CaAl_2Si_2O_8$:Eu^{2+} (Zhang et al. 2010b), Nd^{3+} in CaZnOS:Mn^{2+} (Su et al. 2020), Gd^{3+} in (Ba,Ti)TiO_3:Pr^{3+} (Zhang et al. 2014), Gd^{3+} in $LiNbO_3$:Pr^{3+} (Qiu et al. 2018), La^{3+} in $(Sr,Ca,Ba)_2SnO_4$:Sm^{3+} (Zhao et al. 2016), Zr^{4+} in $SrAl_2O_4$ (Liu et al. 2019), and Ge^{4+} in $Sr_3Sn_2O_7$:Sm^{3+} (Fig. 5.14b–f) (Li et al. 2018). The significant enhancement

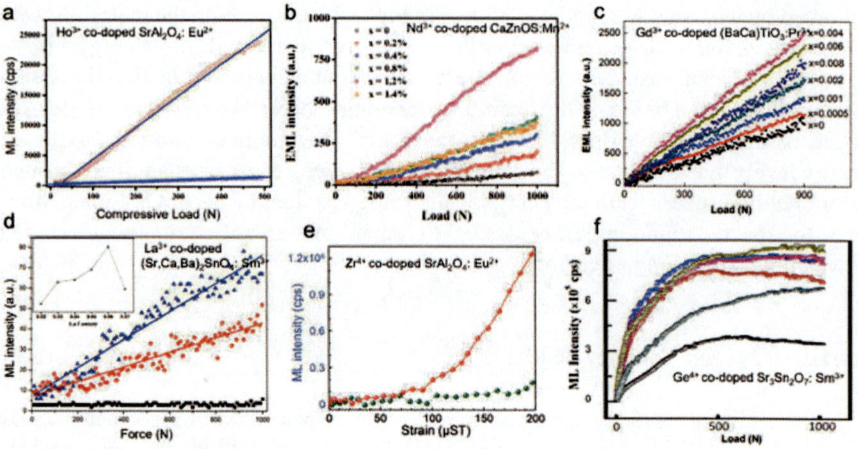

Fig. 5.14 Enhancing the ML intensity by co-doping and controlling the piezoelectricity. **a** ML enhancement in $SrAl_2O_4$:Eu^{2+} by Ho^{3+} co-doping. [Reproduced with the permission of American Institute of Physics (Zhang et al. 2007)] **b** ML enhancement in CaZnOS:Mn^{2+} by Nd^{3+} co-doping. [Reproduced with the permission of Elsevier (Su et al. 2020)] **c** ML enhancement in (Ba, Ca)TiO_3:Pr^{3+} by Gd^{3+} co-doping. [Reproduced with the permission of Optical Society of America (Zhang et al. 2014)] **d** ML enhancement in (Sr, Ca, Ba)$_2SnO_4$:Sm^{3+} by Ln^{3+} doping. [Reproduced with the permission of Elsevier (Zhao et al. 2016)] **e** ML enhancement in $SrAl_2O_4$:Eu^{2+} by Zr^{4+} doping. [Reproduced with the permission of Wiley–VCH (Liu et al. 2019)] **f** ML enhancement in $Sr_3Sn_2O_7$:Sm^{3+} by Ge^{4+} co-doping. [Reproduced with the permission of Elsevier (Li et al. 2018)]

of ML intensity for all these compounds is purely due to the enhancement of trap density. Sometimes varying the heating atmosphere was also found to enhance the trap's density. For example, a ≈100-fold increase in the ML intensity was observed for $ZnAl_2O_4$:Mn^{2+} when the compound was heated in an H_2–N_2 atmosphere at 1300 °C (Matsui et al. 2001). Hydrogenation treatment of ZnS: Cu particle was also found to enhance the ML intensity significantly due to the creation of abundant sulfur vacancies (Gan et al. 2017b). Further, it has been observed that the reddish-orange ML intensity of $Sr_3Sn_2O_7$:Sm^{3+} can be enhanced by 1000 times because of the layered structure with abundant traps and the efficient charge transfer between the host and Sm^{3+} (Fig. 5.16a) (Kamimura et al. 2012). The intensity can be improved further by tailoring the bandgap and trap distribution using co-dopant ions like Si or Ge (Li et al. 2018).

In a piezoelectric host, at first, the charge carriers produced by the excited light are captured by the traps as shown in Fig. 5.15a. When this optically excited material is exposed to low amplitude stress, a local piezoelectric field is produced. This local piezoelectric field will depopulate the traps and facilitate the recombination of carriers in luminescent centers (Fig. 5.15b).

To prove that the piezoelectric field plays a key role in the de-trapping process, Wang et al. fabricated a multilayer structure wherein ZnS:Cu, Mn^{2+} ML material was sandwiched between two PVDF ferroelectric layers (Wang et al. 2019). They

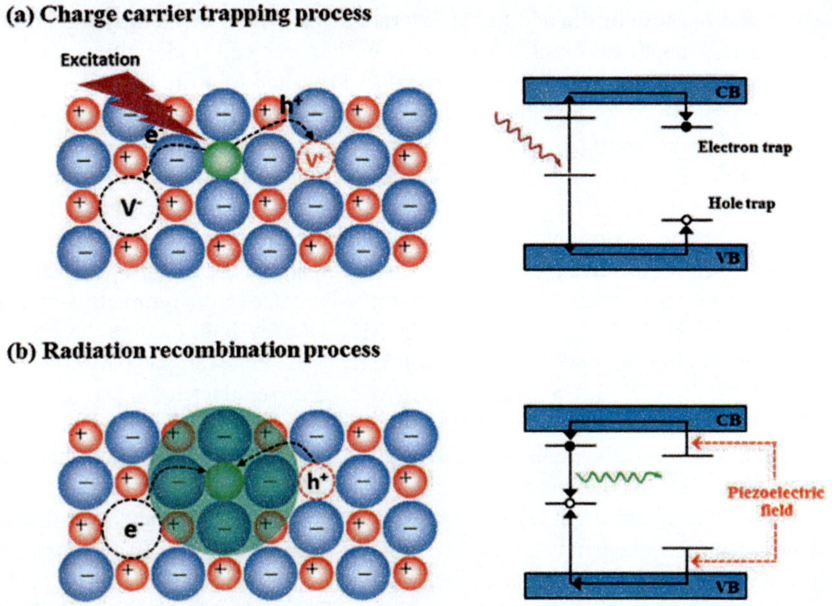

Fig. 5.15 Schematic diagram for ML mechanism in piezoelectric host: Trapping of the carriers and their de-trapping process induced by the local piezoelectric field. [Reproduced with the permission of Elsevier (Zhang et al. 2019)]

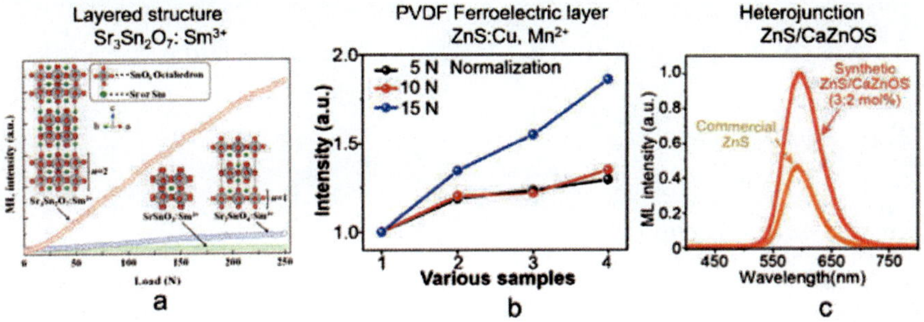

Fig. 5.16 **a** ML enhancement due to the layered crystal structure of $Sr_3Sn_2O_7$:Sm^{3+}. [Reproduced with the permission of American Institute of Physics (Kamimura et al. 2012)] **b** Piezoelectric field induced ML enhancement in ZnS:Cu, Mn^{2+}. [Reproduced with the permission of Elsevier (Wang et al. 2019)] **c** ML enhancement of ZnS/CaZnOS:Mn^{2+} due to the heterojunction structure. [Reproduced with the permission of Wiley–VCH (Peng et al. 2020)]

observed that because of the additional external electric field created by the PVDF layers, the trap depth of ZnS:Cu,Mn^{2+} was further reduced under stress loading which helped to release more charge carriers and resulted in more ML intensity as shown in Fig. 5.16b. Similarly, Peng et al. demonstrated that ML in heterojunctioned ZnS/CaZnOS piezo photonic system is 2 times more intense than that of commercial ZnS:Mn^{2+} as shown in Fig. 5.16c (Peng et al. 2020). This happens because the conduction band offset in the heterojunction interface region facilitates efficient charge transfer and recombination processes. Therefore, controlling both the traps and piezoelectricity in ML materials has a direct impact on the ML properties.

Since the piezoelectricity is originated from the non-centrosymmetric crystal structure; an efficient strategy to control the ML intensity is to change the crystal symmetry by doping an impurity ion with a different size. For example, partial substitution of Sr^{2+} (40mol%) at Ca^{2+} site in $CaAl_2Si_2O_8$:Eu^{2+} enhanced the ML intensity significantly due to the change of the crystal structure to a lower symmetry (Zhang et al. 2010b). The lower symmetry of $CaAl_2Si_2O_8$ crystal's structure upon substitution of Sr^{2+} at Ca^{2+} site happened due to the changes in the cell edge (b and c) and cell volume. This increases the atomic disorder in the crystal lattice and such distortions increase piezoelectricity in the crystal (Zhang et al. 2010b). In many cases, such type of substitution also creates additional traps, which in part contribute to the higher intensity of ML. Few examples are (Ba, Ca)TiO_3:Pr^{3+}, (Sr, M)$MgSi_2O_7$:Eu^{2+} (M = Ba, Sr, Ca), CaZnOS:Mn^{2+},Li^+, (Sr, Ca, Ba)$_2SnO_4$:Sm^{3+},La^{3+} and Sr_3(Sn, Si/Ge)$_2O_7$:Sm^{3+} (Zhang et al. 2019).

5.5 Various Applications of LPPs and ML Materials

5.5.1 Application of LPPs

Displays and safety signs:

Initially, LPPs were mostly being used in luminous Pearl showing green afterglow at night. Later, LPPs found applications in emergency signs, warning signs, luminous paints to mark highways, dial plates of glowing watches, etc. The powder-form LPPs are more favorable for those applications because of their easy mixing into the pigment. Similarly, phosphorescent glasses can also be produced by adding a phosphorescent pigment during their production. These glasses and ceramics will continuously glow in the dark for several hours after switching off the excitation source and found application for decoration and beautification purposes in domestic houses and public buildings. These materials can also be used in safety signs to indicate the escape route during emergency evacuation in the building.

Optical storage, sensors, and detectors:

The traps in LPPs can store the charge carriers (electrons or holes) created during exposure to high-energy radiation. Later on, these trapped charge carriers can be released by the assistance of thermal (TL), optical (photostimulated luminescence or PSL), or mechanical force (ML). The TL, PSL, and ML phenomenon of various LPPs can render their various potential applications in imaging plates, various sensors, and high-energy ray detectors. Exposing the phosphor plates to high-energy rays will "writes" a latent image in the form of trapped electrons and their number is proportional to the amount of radiation dose absorbed. After the initial exposure to high-energy rays, when these materials are further illuminated with a second light source with appropriate wavelength it can "reads" the image because of the visible PSL emission. During this read-out process, most of the trapped electrons will get released and the plate can be further reused. One example of such a phosphor is Cr^{3+}-doped $LiGa_5O_8$ wherein the emission around 716 nm can be repeatedly obtained in a period of more than 1000 h. This phosphor was first irradiated with a UV light (250–360 nm) and then the second illumination was carried out by using light between 380–1000 nm (Liu et al. 2013). The read-out process can also be achieved by using the TL and ML phenomenon. These phosphors have potential applications in erasable and rewritable optical storage, sensors, and detectors.

Biomedical applications:

LPPs with excitation and emission in the biologically transparent NIR window (650–1350 nm) are being used as bio-label for tracing cancer cells (Li et al. 2016a). LPPs have additional advantages over the other labels because their afterglow property permits late time-gated imaging. Therefore, intense research is going on to develop highly sensitive and efficient LPPs that emit light in the first (NIR I, 650–950 nm) and second biologically transparent windows (NIR-II, 1050–1350 nm). For example, Maldiney et al. reported a new generation of bio-labels based on Cr^{3+} doped zinc gallate LPP which can be activated in vivo through living tissues using NIR lights. The signal from persistent luminescence was largely detectable for up to 2 h (Maldiney et al. 2014).

The LPPs also found application as a bio-probe to monitor and identify the variation of biotic habitat. For example, due to the specific reaction of CoOOH and ascorbic acid, the determination and screening of ascorbic acid in living cells and in vivo can be done by using CoOOH-modified LPP nanoparticles (Niu et al. 2014).

Photocatalysis:

TiO_2 has been well known for its photocatalytic properties and for decomposing volatile organic compounds to clear the air under ultraviolet excitation. It has been demonstrated that photocatalytic efficiency can be increased by coupling with LPPs. For example, $CaAl_2O_4$:Eu, Nd LPP with emission around 440 nm and persistent luminescence >10 h can enhance the photocatalytic degradation of acetaldehyde and NO pollutants when coupled with TiO_2 photocatalyst (Li et al. 2012).

Solar cells:

Coupling LPP with a solar cell can also improve solar cell efficiency. For example, $SrAl_2O_4:Eu^{2+},Dy^{3+}$ phosphors can down-convert the 250 nm UV light to 400 nm visible light, which matches the peak absorption region of the most common dye N719. This leads to an overall 13% improvement in the conversion efficiency when the CdS-sensitized solar cells (DSSC) device was coupled with $SrAl_2O_4:Eu^{2+},Dy^{3+}$ LPPs phosphor (He et al. 2013).

5.5.2 Application of ML Phosphors

Since applied stress plays a crucial role in ML phosphors, they have potential application in various stress-sensing technologies. Compared to the conventional stress-sensing technologies presently available in the market, ML phosphor-based sensing technologies are still in the development phase. We have summarized some of their potential applications below.

Electronic Wearable Devices:

These wearable devices are made of ML phosphors-based electronic skin or artificial skin which can mimic the functions of our human skin. Because of the stretchability, self-healing properties, and ability to integrate over large areas, such wearable devices are highly promising for application in health monitoring and prosthetics. The first remote detection of applied stress was done by Xu et al. using the visible light emitted from a composite film containing $SrAl_2O_4:Eu^{2+}$ ML phosphors (Xu et al. 1999b). Following their pioneering work extensive research have been focused on exploring new ML materials for stress-sensing application. Song and co-workers have demonstrated the visualized facial expression by fabricating a skin-attachable stress sensor using SiO_2 nanoparticle-modified $ZnS:Cu/Mn^{2+}@Al_2O_3@PDMS$ (Qian et al. 2018).

Biomechanical and Biomedical Engineering:

ML-based phosphors also have potential application in Biomechanics to understand disease development in biological tissues such as human bones and teeth. For example, $SrAl_2O_4:Eu^{2+},Dy^{3+}$ ML phosphors were used to get the visualization of stress distribution on the surface of the artificial tooth (Jiang et al. 2018). Since the NIR light has excellent penetration capability through the biological tissues, NIR-ML phosphors are the best candidate for biomechanical applications. For example, $SrAl_2O_4:Eu^{2+},Er^{3+}$ phosphor showed ML at $\approx$1530 nm under compressive loading with a linear relationship of ML intensity with the amplitude of the load (Terasawa et al. 2011). Other ML phosphors which showed intense NIR-ML and good penetrability through various other biological tissues are $Sr_3Sn_2O_7:Nd^{3+}$, $CaZnOS:Nd^{3+}$, $LiNbO_3:Nd^{3+}$, and $CaZnOS:Er^{3+}$ (Zhuang and Xie 2021).

Structure Health Diagnosis

ML phosphors-based stress-sensing technology may be helpful for the health diagnosis of many big engineering structures such as bridges, tall buildings, high-pressure vessels, and so on. Like artificial skin, a sensing film can be attached to the surface of a bridge to investigate the real-time stress distribution on the surface (Terasaki and Xu 2013). Similar films can also be used to investigate the stress distribution on the surface of various high-pressure vessels like liquid hydrogen storage cylinders or pipes and thereby providing a real-time health diagnosis (Fujio et al. 2016). They are also highly promising for monitoring the surface stress distribution in aircraft and spacecraft caused by many severe mechanical impacts while flying (Zhuang and Xie 2021).

Electronic Signature and Anticounterfeiting:

They showed potential application in electronic signature system based on ML-based electronic signature skin using different ML phosphors (Zhuang and Xie 2021; Wang et al. 2015). The difficult-to-replicate stress-induced luminescence properties of ML phosphors also make them a very good candidate for anticounterfeiting application. Further, integrating ML with other categories of luminescence properties such as down-conversion, up-conversion, and thermostimulated luminescence can make them more promising candidates for advanced optical anticounterfeiting (Zhang et al. 2018b). Examples of such ML phosphors with multimode emission properties and having potential applications for anticounterfeiting are $LiNbO_3:Pr^{3+}$, $Sr_3Al_2O_6:Eu^{3+}$, $CaZnOS:Mn^{2+}/Ln^{3+}$, and $CaZnOS:Bi^{3+}$ (Zhuang and Xie 2021).

5.6 Summary and Challenges

Here we have tried to summarize full-scale research on LPPs and ML materials starting from their fundamental understandings to many advanced applications. Our specific objective is to present how defect-modulated trap controlling plays a crucial role in designing efficient LPPs and ML materials. We have summarized various advanced characterization techniques for the defect-related tarps and highlighted various ways to control them. In the last a brief overview of their potential application in many advanced scientific and technological fields is also discussed. Despite the substantial progress in the development and exploration of many new classes of ML materials, there are still many challenges that need to be addressed before their routine application.

The first challenge for LPPs and ML materials is brightness and efficiency. Although in the present class of materials defect-related traps play the critical role of storing the charge carriers, in many cases, defect centers provide some non-radiative pathways to the radiative centers and thereby decrease the light yield. Therefore, we must carefully tailor the defect-related trap centers and avoid any defect centers that involve non-radiative pathways. Further, for routine application, the defect-related trap centers should generate a reproducible persistent luminescence and should not disappear while changing the atmospheric condition.

The second challenge of these defect-related trap-controlled LPPs and ML materials is to provide a more convincing mechanism with extensive universality. For example, a concrete mechanism of how stress accelerates the release of the trapped carriers is still missing in the literature. Similarly, for LPPs, it has been proposed that shallow trap states result in the high intensity of light while deep-level trap results in long persistent luminescence. However, a strong correlation between the level of traps and the optimum performance of the persistent luminescence both in terms of light yield and persistency is still missing.

The third and one of the most important challenges of these trap-controlled phosphors is a thorough analysis of the species, concentration, and depth of defect-related traps using multiple techniques. Defects in crystals are a complex subject and their trap's position inside the band gap varies from one matrix to another. Therefore, a similar defect (for example, oxygen vacancy in oxide-based materials) may act as a shallow trap in one matrix but as a deep trap in another matrix due to a change in the band gap. A close collaboration between experimental and theoretical groups is therefore required to establish a strong correlation between the crystal structure and the electronic structure arising due to the activators and the defect traps.

Acknowledgements YM would like to thank the support by IIT startup funds. NP thanks the United States-India Education Foundation (USIEF, India) and the Institute of International Education (IIE, USA) for his Fulbright Nehru Postdoctoral Fellowship (Award# 2681/FNPDR/2021).

References

Aitasalo, T., Deren, P., Holsa, J., Jungner, H., Krupa, J.C., Lastusaari, M., Legendziewicz, J., Niittykoski, J., Strek, W.: J. Solid State Chem. **171**, 114–122 (2003)

Akiyama, M., Xu, C.N., Matsui, H., Nonaka, K., Watanabe, T.: Appl. Phys. Lett. **75**, 2548–2550 (1999)

Akiyama, M., Xu, C.N., Nonaka, K.: J. Electrochem. Soc. **150**, H115–H118 (2003)

Bos, A.J.J., Poolton, N.R.J., Wallinga, J., Bessiere, A., Dorenbos, P.: Radiat. Meas. **45**, 343–346 (2010)

Clabau, F., Rocquefelte, X., Jobic, S., Deniard, P., Whangbo, M.H., Garcia, A., Mercier, T.L.: Chem. Mater. **17**, 3904–3912 (2005)

Dorenbos, P.: J. Lumin. **111**, 89–104 (2005)

Feng, A., Smet, A.P.F.: Materials **11**, 484 (2018)

Fu, X., Yamada, H., Xu, C.N.: J. Electrochem. Soc. **154**, 348–351 (2007)

Fujio, Y., Xu, C.N., Terasawa, Y., Sakata, Y., Yamabe, J., Ueno, N., Terasaki, N., Yoshida, A., Watanabe, S., Murakami, Y.: Int. J. Hydrog. Energy **41**, 1333 (2016)

Fukushima, H., Kozu, T., Shima, H., Funakubo, H., Uchida, H., Katoda, T., Nishida, K.: Evaluation of oxygen vacancy in ZnO using Raman spectroscopy. In: Joint IEEE International Symposium on the Applications of Ferroelectric (ISAF), International Symposium on Integrated Functionalities (ISIF), and Piezoelectric Force Microscopy Workshop (PFM), pp. 28–31 (2015)

Gan, J., Kang, M.G., Meeker, M.A., Khodaparast, G.A., Bodnar, R.J., Mahaney, J.E., et al.: J. Mater. Chem. C **5**, 5387–5394 (2017a)

Gan, J., Kang, M.G., Meeker, M.A., Khodaparast, G.A., Bodnar, R.J., Mahaney, J.E., Maurya, D., Priya, S.: J. Mater. Chem. C **5**, 5387 (2017b)

Gao, R., Kodaimatic, M.S., Yan, D.: Chem. Soc. Rev. **50**, 5564–5589 (2021)

Gluchowski, P., Strek, W., Lastusaari, M., Holsa, J.: Phys. Chem. Chem. Phys. **17**, 17246–17252 (2015)

Guo, E.J., Liu, Y., Sohn, C., Desautels, R.D., Herklotz, A., Liao, Z., Nichols, J., Freeland, J.W., Fitzsimmons, M.R., Lee, H.N.: Adv. Mater. **30**, 1705904 (2018)

He, W., Atabaev, T.S., Kim, H.K., Hwang, Y.H.: J. Phys. Chem. C **117**, 17894–17900 (2013)

Holsa, J., Aitasalo, T., Jungner, H., Lastusaari, M., Niittykoski, J., Spano, G.: J. Alloys Compd. **374**, 56–59 (2004)

Huang, K., Dou, X., Zhang, Y., Gao, X., Lin, J., Qu, J., Li, Y., Huang, P., Han, G.: Adv. Funct. Mater. **31**, 2009920 (2021)

Ikeda, S., Uchino, T.: J. Phys. Chem. C **118**, 4346–4353 (2014)

Jia, Y., Zhang, L., Du, A., Gao, G., Chen, J., Yan, X., Brown, C.L., Yao, X.: Adv. Mater. **28**, 9532–9538 (2016)

Jiang, Y., Wang, F., Zhou, H., Fan, Z., Wu, C., Zhang, J., Liu, B., Wang, Z.: Mater. Sci. Eng. C **92**, 374 (2018)

Jin, H., Debroye, E., Keshavarz, M., Scheblykin, I.G., Roeffaers, M.B.J., Hofkens, J., Steele, J.A.: Mater. Horiz. **7**, 397–410 (2020)

Kaftelen, H., Ocakoglu, K., Thomann, R., Tu, S., Weber, S., Erdem, E.: Phys. Rev. B **86**, 014113 (2012)

Kamimura, S., Yamada, H., Xu, C.N.: Appl. Phys. Lett. **101**, 091113 (2012)

Kanrar, B., Pathak, N., Nayak, C., Prajapat, C.L., Kadam, R.M., Jha, S.N., Bhattacharyya, D., Ravikumar, G., Misra, N.L.: J Mater Chem C **5**, 8836–8846 (2017)

Kar, A., Kundu, S., Patra, A.: J. Phys. Chem. C **115**, 118–124 (2011)

Lecointre, A., Bessiere, A., Bos, A.J.J., Dorenbos, P., Viana, B., Jacquart, S.: J. Phys. Chem. C **115**, 4217–4227 (2011)

Li, H.H., Yin, S., Wang, Y.H., Sato, T.: RSC Adv. **2**, 3234–3236 (2012)

Li, J., Wang, Z., Zhao, A., Wang, J., Song, Y., Sham, T.K.: J. Phys. Chem. C **119**, 17848–17856 (2015)

Li, Y., Gecevicius, M., Qiu, J.: Chem. Soc. Rev. **45**, 2090–2136 (2016a)

Li, L., Wong, K.L., Li, P., Peng, M.: J Mater. Chem. C **4**, 8166–8170 (2016b)

Li, J., Xu, C.N., Tu, D., Chai, X., Wang, X., Liu, L., Kawasaki, E.: Acta Mater. **145**, 462 (2018)

Liu, F., Yan, W.Z., Chuang, Y.J., Zhen, Z.P., Xie, J., Pan, Z.W.: Sci. Rep. **3**, 01554 (2013)

Liu, X., Xi, W., Li, C., Li, X., Shi, J., Shen, Y., He, J., Zhang, L., Xie, L., Sun, X., Wang, P., Luo, J., Liu, L.M., Ding, Y.: Nano Energy **44**, 371–377 (2018)

Liu, L., Xu, C.N., Yoshida, A., Tu, D., Ueno, N., Kainuma, S.: Adv. Mater. Technol. **4**, 1800336 (2019)

Loh, K.P., Bao, Q., Eda, G., Chhowalla, M.: Nat. Chem. **2**, 1015–1024 (2010)

Maldiney, T., Bessiere, A., Seguin, J., Teston, E., Sharma, S.K., Viana, B., Bos, A.J.J., Dorenbos, P., Bessodes, M., Gourier, D., Scherman, D., Richard, C.: Nat. Mater. **13**, 418–426 (2014)

Matsui, H., Xu, C.N., Akiyama, M., Watanabe, T.: Jpn. J. Appl. Phys. **39**, 6582–6586 (2000)

Matsui, H., Xu, C.N., Tateyama, H.: Appl. Phys. Lett. **78**, 1068 (2001)

Niu, J.Y., Wang, X., Lv, J.Z., Li, Y., Tang, B.: TrAC Trends Anal. Chem. 58, 112–119 (2014)

Pan, Z.W., Lu, Y.Y., Liu, F.: Nat. Mater. **11**, 58–63 (2012)

Pathak, N., Ghosh, P.S., Gupta, S.K., Kadam, R.M., Arya, A.: RSC Adv. **6**, 96398–96415 (2016a)

Pathak, N., Ghosh, P.S., Gupta, S.K., Mukherjee, S., Kadam, R.M., Arya, A.: J. Phys. Chem. C **120**, 4016–4031 (2016b)

Pathak, N., Gupta, S.K., Prajapat, C.L., Sharma, S.K., Ghosh, P.S., Kanrar, B., Pujari, P.K., Kadam, R.M.: Phys. Chem. Chem. Phys. **19**, 11975–11989 (2017)

Pathak, N., Ghosh, P.S., Saxena, S., Dutta, D., Yadav, A.K., Bhattacharyya, D., Jha, S.N., Kadam, R.M.: Inorg. Chem. **57**, 3963–3982 (2018)
Pathak, N., Mukherjee, S., Das, D., Dutta, D., Dash, S., Kadam, R.M.: J Mater Chem C **8**, 7149–7161 (2020)
Pei, P., Chen, Y., Sun, C., Fan, Y., Yang, Y., Liu, X., Lu, L., Zhao, M., Zhang, H., Zhao, D., Liu, X., Zhang, F.: Nat. Nanotechnol. **16**, 1011–1018 (2021)
Peng, D., Jiang, Y., Huang, B., Du, Y., Zhao, J., Zhang, X., Ma, R., Golovynskyi, S., Chen, B., Wang, F.: Adv. Mater. **32**, 1907747 (2020)
Qian, X., Cai, Z., Su, M., Li, F., Fang, W., Li, Y., Zhou, X., Li, Q., Feng, X., Li, W., Hu, X., Wang, X., Pan, C., Song, Y.: Adv. Mater. **30**, 1800291 (2018)
Qiu, G., Fang, H., Wang, X., Li, Y.: Ceram. Int. **44**, 15411 (2018)
Ran, C., Xu, J., Gao, W., Huang, C., Dou, S.: Chem. Soc. Rev. **47**, 4581–4610 (2018)
Su, M., Li, P., Zheng, S., Wang, X., Shi, J., Sun, X., Zhang, H.: J. Lumin. **217**, 116777 (2020)
Tendeloo, G.V., Bals, S., Aert, S.V., Verbeeck, J., Dyck, D.V.: Adv. Mater. **24**, 5655–5675 (2012)
Terasaki, N., Xu, C.N.: IEEE Sens. J. **13**, 3999 (2013)
Terasawa, Y., Xu, C.N., Yamada, H., Kubo, M.: IOP Conf. Ser. Mater. Sci. Eng. **18**, 212013 (2011)
Trojan-Piegza, J., Niittykoski, J., Hölsä, J., Zych, E.: Chem. Mater. **20**, 2252–2261 (2008)
Tsai, C., Li, H., Park, S., Park, J., Han, H.S., Nørskov, J.K., Zheng, X., Abild-Pedersen, F.: Nat. Commun. **8**, 15113 (2017)
Tu, D., Xu, C.N., Yoshida, A., Fujihala, M., Hirotsu, J., Zheng, X.G.: Adv. Mater. **29**, 1606914 (2017)
Wang, W.N., Ogi, T., Kaihatsu, Y., Iskandarc, F., Okuyama, K.: J. Mater. Chem. **21**, 5183–5189 (2011)
Wang, X., Zhang, H., Yu, R., Dong, L., Peng, D., Zhang, A., Zhang, Y., Liu, H., Pan, C., Wang, Z.L.: Adv. Mater. **27**, 2324 (2015)
Wang, Q., Zhang, Q., Zhao, X., et al.: Nano Lett. **18**, 6898–6905 (2018a)
Wang, X., Zhuang, L., Jia, Y., Liu, H., Yan, X., Zhang, L., Yang, D., Zhu, Z., Yao, X.: Angew. Chem. Int. Ed. **57**, 16421–16425 (2018b)
Wang, F., Wang, F., Wang, X., Wang, S., Jiang, J., Liu, Q., Hao, X., Han, L., Wang, J., Pan, C., Liu, H., Sang, Y.: Nano Energy **63**, 103861 (2019)
Xu, C.N., Watanabe, T., Akiyama, M., Zheng, X.G.: Appl. Phys. Lett. **74**, 2414–2416 (1999a)
Xu, C.N., Watanabe, T., Akiyama, M., Zheng, X.G.: Appl. Phys. Lett. **74**, 1236 (1999b)
Yamamoto, H., Matsuzawa, T.: J. Lumin. **72–74**, 287–289 (1997)
Zawadzki, M.: Solid State Sci. **8**, 14 (2006)
Zeng, H., Duan, G., Li, Y., Yang, S., Xu, X., Cai, W.: Adv. Funct. Mater. **20**, 561–572 (2010)
Zhang, H., Yamada, H., Terasaki, N., Xu, C.N.: Appl. Phys. Lett. **91**, 081905 (2007)
Zhang, H., Yamada, H., Terasaki, N., Xu, C.N.: J. Electrochem. Soc. **155**, J55 (2008)
Zhang, J.C., Wang, X., Yao, X., Xu, C.N., Yamada, H.: J. Electrochem. Soc. **157**, G269–G273 (2010a)
Zhang, L., Xu, C.N., Yamada, H., Bu, N.: J. Electrochem. Soc. **157**, J50 (2010b)
Zhang, H., Xu, C.N., Terasaki, N., Yamada, H.: Electrochem. Solid-State Lett. **14**, J76-80 (2011)
Zhang, J.C., Wan, Y., Xin, X., Han, W.P., Zhang, H.D., Sun, B., Long, Y.Z., Wang, X.: Opt. Mater. Express **4**, 2300 (2014)
Zhang, H., Peng, D., Wang, W., Dong, L., Pan, C.: J. Phys. Chem. C **119**, 28136–28142 (2015)
Zhang, J.C., Long, Y.Z., Yan, X., Wang, X., Wang, F.: Chem. Mater. **28**, 4052–4057 (2016)
Zhang, C., Shi, Y., Yu, Y., Du, Y., Zhang, B.: ACS Catal. **8**, 8077–8083 (2018a)
Zhang, J.C., Pan, C., Zhu, Y.F., Zhao, L.Z., He, H.W., Liu, X., Qiu, J.: Adv. Mater. **30**, 1804644 (2018b)
Zhang, J.C., Wang, X., Marriott, G., Xu, C.N.: Prog. Mater. Sci. **103**, 678 (2019)
Zhao, H., Chai, X., Wang, X., Li, Y., Yao, X.: J. Alloy. Compd. **656**, 94 (2016)
Zhou, W., Zou, X., Najmaei, S., et al.: Nano Lett. **13**, 2615–2622 (2013)
Zhou, J., Zheng, G., Liu, X., Dong, G., Qiu, J.: Coord. Chem. Rev. **448**, 214178 (2021)
Zhuang, Y., Xie, R.J.: Adv. Mater. **33**, 2005925 (2021)

Chapter 6
Optical Materials for Sensing Radioactive Elements

Pallavi Singhal

Abbreviations

NPs	Nanoparticles
QDs	Quantum Dots
LED	Light-Emitting Diode
UV	Ultraviolet
CDs	Carbon Dots
MPA	Mercaptopropionic Acid
DNA	Deoxyribonucleic Acid
SERS	Surface-Enhanced Raman Scattering
MBs	Magnetic Beads
RLS	Resonance Light Scattering
DI	De-Ionized

6.1 Introduction

Radioactive elements, as the name suggests, are the elements which emit radiation. These elements along with their chemical toxicity also have the associated radiation toxicity. The radiations emitted by the radioactive elements include high energy UV rays, X-rays, γ-rays, α-particles, and β-particles. These radiations have sufficient energy to ionize atoms and molecules and cause adverse effects on the environment and the human body. Depending on the type of radiation, they emit (α, β, γ) the extent of damage is different. Usuall, y radionuclides which emit α radiation cause

P. Singhal (✉)
Environmental Monitoring and Assessment Division, Bhabha Atomic Research Centre, Mumbai 400085, India
e-mail: psinghal@barc.gov.in

R. S. Ningthoujam and A. K. Tyagi (eds.), *Handbook of Materials Science, Volume 1*, Indian Institute of Metals Series, https://doi.org/10.1007/978-981-99-7145-9_6

the maximum damage because of the high energy deposition in short range, while the γ emitting radionuclides cause less damage. Other factors such as half-life of the source and the dose also play a dominant role in deciding the overall effect caused by the radionuclide. Because of the adverse effect of these radiations, numerous devices have been developed using which radiation monitoring and dosimetry are being carried out. In most of these devices, various sensors have been used (Rao et al. 2015; Sizov 2009) but their performance is not up to the mark in the presence of electromagnetic fields or other radiation sources that cater the need to develop other more precise sensors.

Detection of a radioactive element involves two strategies: (1) by detecting it chemically or (2) by detecting the radiation it emits. Chemical detection of a radioactive element involves the detection of element as well as its isotope and the detection by radiation involves measurement of energy and type of radiation it emits (energy spectrometry), since the energy emitted by a radionuclide is a characteristic feature. Energy spectrometry is being carried out for the past several years to identify a radionuclide and numerous instruments such as α-spectrometer, γ-spectrometer (HpGe and NaI(Tl) detectors), ionization chamber, etc., have been used for this purpose. Although these instruments have an exceptional sensitivity, the requirement of tedious sample preparation, large counting time, and high-cost prompt researchers to look for other methods. In most of the cases, the radionuclide needs to be separated from the matrix for measurement. Also, it is difficult to measure low-energy emitting radionuclides because of the instrumental limitations, and the results are not in situ and visible.

Considering the above limitations, optical materials are gaining significant attention in the detection of radioactive elements. Optical materials change their absorption or emission properties on interaction with the analyte. Typically, a whole system consists of an optical material that measures the change in the optical properties, a molecule that will bind both with the material and analyte of interest, and signal transducers. The element will interact with this system and change the optical properties that will be finally measured. The binder molecule here plays a key role that gives the sensor its selectivity and provides information about the presence, concentration, and other physical properties of the analyte. Optical sensors have advantages as compared to other materials because of ease of handling and low cost. Various materials have been used for this purpose and their superior properties such as low cost, in situ, and visual detection make these materials an interesting candidate for sensing applications. In this chapter, we have summarized the work being carried out in this direction and materials developed for the same.

This chapter is dedicated to the work being carried out on different materials for their application as optical sensors to detect radioactive ions. Scheme 6.1 gives an overview of the materials being used. The chapter is divided into several sections based on the type of material used. Herein, we will discuss these materials in detail.

Dyes as the optical material for sensing radioactive elements—Dyes have been used for long as colorant reagents. Initially, these compounds were extracted from plants, insects, fungi, and lichens. However, after the evolution of modern organic

Scheme 6.1 Different optical materials used for radioelement detection

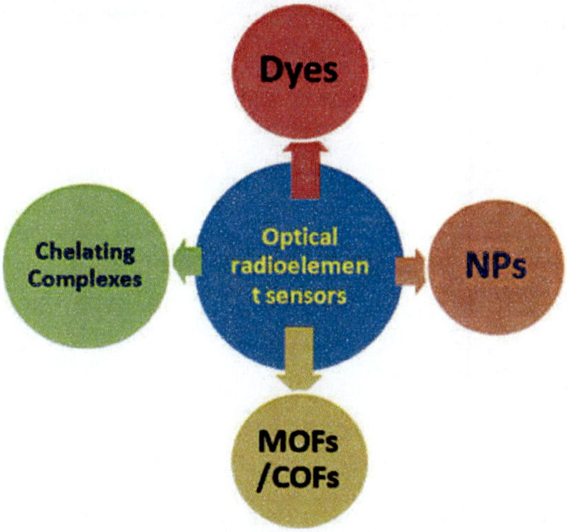

chemistry and expertise in synthesis, a variety of other dyes were designed and prepared depending on the need. Dyes have also been extensively for the detection of radioactive elements because of the color change. Dye molecules interact with uranium and, while interacting, they may either change the structure or form an aggregate. Through this process, a change in color was observed that depends on many factors such as pH, strength of the complex, and concentration of uranium. In this line, Arsenazo III is one of the most studied dyes for the spectrophotometric determination of uranium (Kadam et al. 1981) and many sensors have been developed using this dye as an indicator. Figure 6.1 shows the structure of the Arsenazo III dye and its absorption spectra after interaction with uranium (Orabi 2013). The dye and the uranyl ion form a strong complex and change color from red to violet and this change in color makes the basis for the determination of uranyl ion by this method. Researchers have synthesized membranes, polymers, and composites and immobilized Arsenazo III on it (Baylor and Buchanan 1995; Collins et al. 2002). They have shown that the dye is effective in uranium sensing but the long reaction time of 30–40 min and the interference with other ions (Zn^{2+}, Ca^{2+}, etc.) make use of these sensors practically difficult. Also, a higher concentration of uranium (ppm range) is needed for detection purpose which is more than the prescribed limit in drinking water. To overcome this limitation and detect uranium at <20 ppb levels, Collins et al. (Collins et al. 2002) designed a device that utilizes a renewable reagent system. This system consists of a Nafion microtube in which Arsenazo III is pumped for sampling uranyl ions in groundwater. The uranyl cation formed is detected by a light-emitting diode (LED)-based fiber-optic flow cell. The response time of the device was reported to be 20 min with thorium and calcium as the major interferences. In another study, Serenjeh et al. (2016) used Arsenazo III for uranium detection. They have used a transparent agarose membrane to perform the experiments. Spectrophotometric studies were carried out to study the complexation and it was found that the uranyl

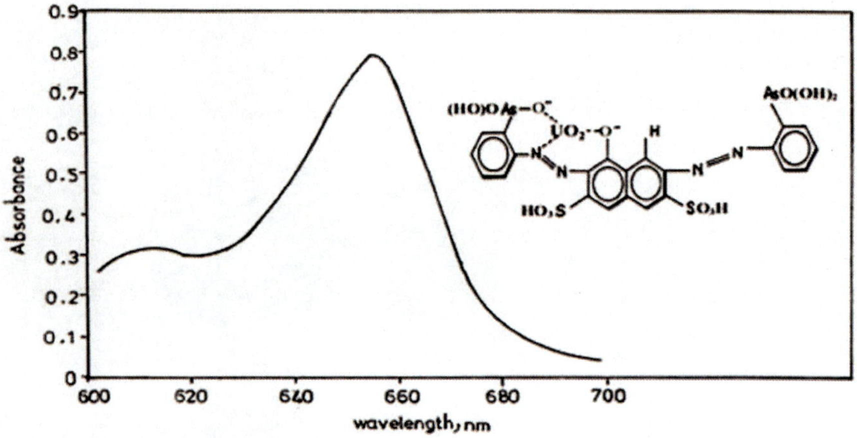

Fig. 6.1 Absorption spectra and structure of Arsenanzo-III uranyl complex. Reproduced with permission from Elsevier Publisher (Orabi 2013) under a creative common license (https://creativecommons.org/licenses/by-nc-nd/3.0/)

ion complex has a much larger stability constant as compared to other metal ions at pH 2 with a distinct change in color change from purple to dark violet. The method does not have interferences from other ions and was successfully applied to determine UO_2^{2+} in spiked water samples.

Another dye that finds promising application in uranium sensing is Alizarin Red S. Alizarin Red S binds selectively with transuranic. Figure 6.2 shows the structure and absorption spectra of dye (Yang and Chan 2009). In this line, Safavi and Bagheri (2005) designed an optical probe in which they have incorporated Alizarin Red S. on triacetylcellulose and tested the membrane for uranium sensing. They have shown that the probe changes color from yellow to violet and the response time was 10 min.

Khalifa (1998) designed a uranium-separating ion exchanger using this dye. The ion exchange was prepared using Alizarin Red S. and anion exchanger Doulite A101. The performance of the resin was evaluated by changing various parameters such as pH, time, and resin capacity and it was observed that the U(VI) sorption occurs over a wide pH range, 2.8–5. The maximum sorption capacity was evaluated to be 0·68 mmol.g^{-1}. Interference studies in the presence of different metal ions were performed and it was observed that Fe(III), Zr(IV), Ti(IV), Cu(II), and Th(IV) were the major interfering ions. Fouad et al. (2019) designed a resin named XAD-2010 using Alizarin Red S for the separation of uranium and thorium in some geological samples. The sorption results follow the Langmuir isotherm sorption capacity of 20.2 mg g^{-1}.

In another study, Zhou et al. (2022) detected uranium using another dye, Nile blue. They done electropolymerization of the dye in modified glassy carbon electrodes and voltammetric techniques to detect uranium. They observed a detection limit of 0.19 μg/L and the sensor was applied for the determination of uranium in natural samples.

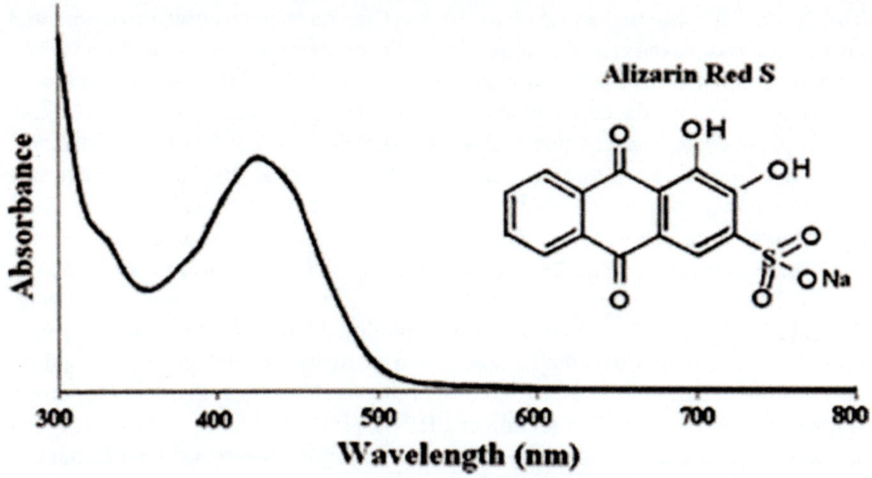

Fig. 6.2 Absorbance spectra and structure of Alizarin Red S dye. Reproduced with permission from Springer Publisher (Yang and Chan 2009)

The above studies suggest that numerous works have been carried out wherein dyes were used for uranium sensing as it was one of the simplest methods and visual detection was possible; however, it has several limitations such as long reaction time, interference from other ions, photo-bleaching and higher detection limit and this prompt researchers to look for other alternatives.

6.2 Nanomaterials as the Optical Sensors for Radioactive Elements

Nanomaterials as the name suggests the materials with at least one dimension in the nanometer range. These materials are used in almost every industry because of their numerous advantages such as high surface area, surface tunability, and shape and size selection. Various optical nanomaterials have been designed for uranium sensing and these materials are further divided into several sections based on the nature of materials used. These materials are described below.

6.2.1 Quantum Dots as Radioactive Elements Sensors

Quantum dots are semiconductor nanomaterials with a size less than their Bohr excitonic radius. They have numerous applications such as in solar energy harvesting, light-emitting diodes, and diagnosis (Singhal et al. 2016, 2017; Singhal and Ghosh

2014, 2015, 2019; Singhal and Pulhani 2019). They have also shown huge potential as metal ion sensors and few researchers have shown their application in the detection of uranium (Singhal et al. 2017; Singhal and Pulhani 2019). Here, we would like to mention that although the application of QDs in sensing was demonstrated but detection mechanism was not very clearly understood. Scheme 6.2 shows the mechanisms by which QDs can lead to metal ion detection (Singhal et al. 2017). It is shown that QDs can interact with metal ions by various pathways and can lead to aggregation, energy transfer, charge transfer, and exchange of its surface ion with foreign ions. By all these processes, either the luminescence intensity will increase or decrease, and this change in intensity is monitored to detect the presence of metal ions. Earlier, Dutta and Kumar (2016) synthesized amine-modified CdS QDs for uranium detection. They have shown that the luminescence intensity of QDs gets quenched on interaction with uranium; however, no proper mechanism was suggested. Our group (Singhal et al. 2017; Singhal and Pulhani 2019) has also worked on this and designed carboxylic functionalized CdSe/CdS core–shell QDs with different shell thicknesses. The shell thickness was determined by the ICP-OES analysis. It was shown that the QDs luminescence is quenched on the addition of uranium and is assigned to the electron transfer from QDs to uranium. We have shown by time-resolved emission spectroscopy that as shell thickness increases, electron transfer rate decreases exponentially and is shown in Fig. 6.3a (Singhal et al. 2017). Stern–Volmer plot shows that the limit of detection for uranium using these dots is 75 ppb and is depicted in Fig. 6.3b.

Since uranium binds with the ligands on the QD surface; it is very important to characterize the QDs surface as it might change the detection efficiency. Our group has carried out this study by synthesizing CdSe/CdS core–shell QDs. Herein, we kept the shell thickness same but changed the number of ligands on the surface. It was shown that the number of ligands plays an important role in binding and can

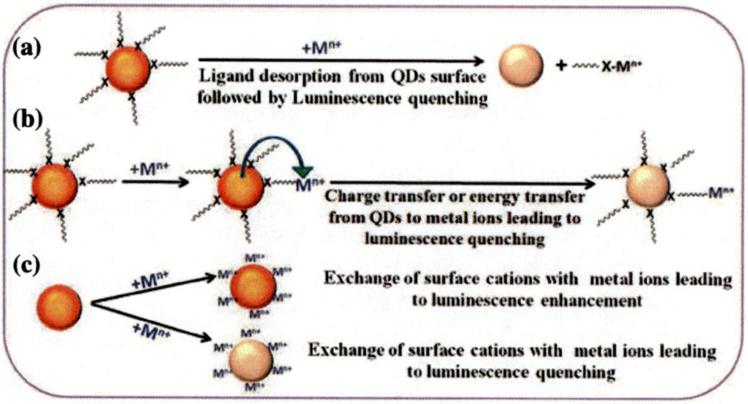

Scheme 6.2 Metal ion interaction with QDs. Reproduced with permission from ACS Publisher (Singhal et al. 2017)

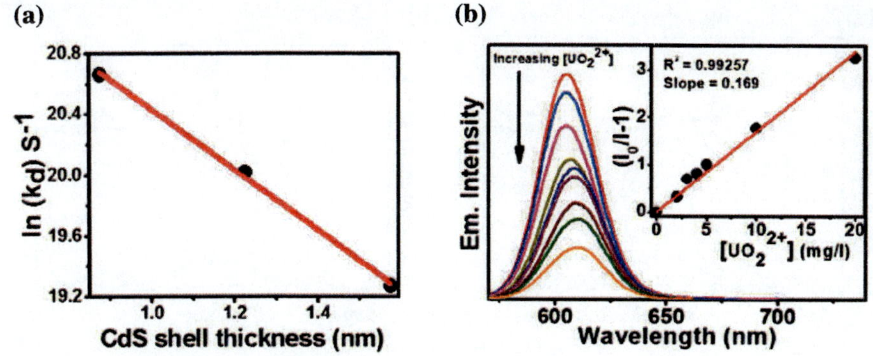

Fig. 6.3 **a** Variation of electron transfer rate with CdS shell thickness. **4b**: Variation in emission intensity of CdSe/CdS Core–Shell QDs with 3.5 monolayers of CdS shell. **b (Inset)**: Stern–Volmer plot. Reproduced with permission from ACS Publisher (Singhal et al. 2017)

change the detection efficiency (Singhal and Pulhani 2019). It is shown in Fig. 6.4 where the change in emission intensity is different on the addition of uranium into QDs. It suggests that in addition to tuning the shell thickness; ligand concentration is also an important factor in deciding the detection efficiency.

Tam et al. (2018) also used CdS QDs for the detection of radiations. They used QDs-doped scintillating polymers, namely, CdS QDs, 2,5-diphenyloxazole (PPO) dye, and dispersed them in polyvinyl toluene (PVT) matrix. A three-stage energy transfer mechanism was observed wherein energy from PVT polymer to PPO and then to CdS QDs was transferred under UV excitation. Interestingly, an enhancement in scintillation efficiency was observed at QDs low loading that further enhances with additional QDs loading. Along with enhancement in scintillation efficiency, a shift in the energy spectrum was also observed. The material was tested with different gamma and beta emitter sources and the energy spectrum had different effects depending on energy. Here, the QDs loading enhanced the performance by more than 3.6 compared to without any QD which is a requirement for a good scintillator. The efficiency to detect high energy beta from Sr/Y-90 decay was evaluated and comes out to be ~9%.

The materials discussed above use Cd-based materials that are toxic and pose a risk to the environment and human health. To make the process environmentally benign, green materials are getting attention from all over the world for numerous applications nowadays (Wang et al. 2015). In this line, carbon-based QDs are getting prominent focus because of their low cost, ease of synthesis, and less toxicity (Wang et al. 2022). They are also used to detect radioactive ions. Wang et al. (2015) have synthesized CDs from the microplasma method and used them for the detection of uranium. They have shown that the blue emission from QDs is quenched with the addition of uranyl ions. The results are shown in Fig. 6.5a. It is observed from Fig. 6.5a that on addition of uranyl ion, the emission is quenched. This quenching in emission on the addition of uranyl ions is plotted in Fig. 6.5b where it is shown that the quenching is linear in the 0–75 ppm concentration range. This quenching

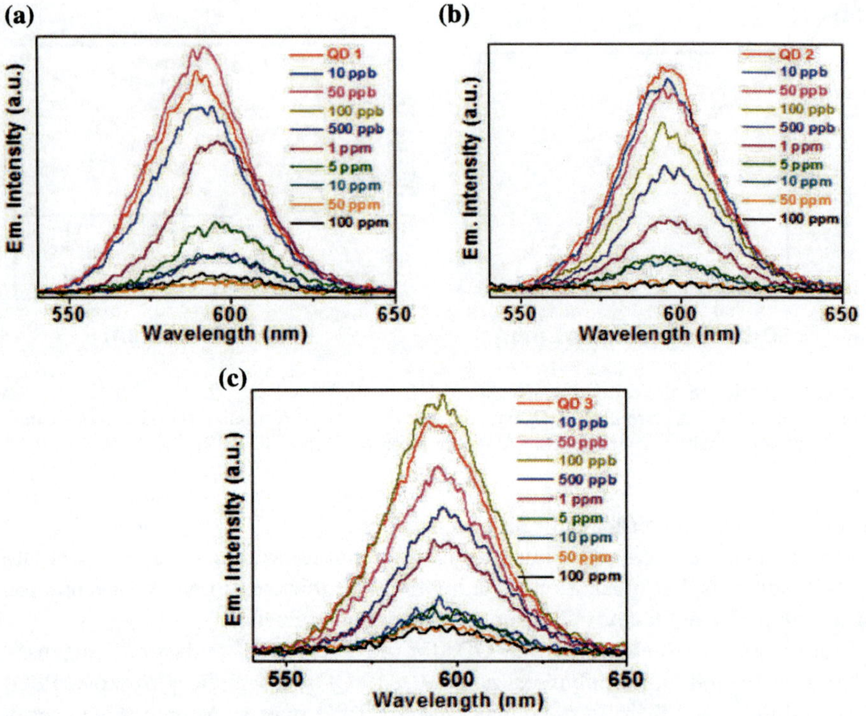

Fig. 6.4 Variation in emission intensity of core–shell QDs on adding different concentrations of uranium. **a** [0.1 μM] QD 1, **b** [0.1 μM] QD 2, and **c** [0.1 μM] QD 3, where the shell thickness was 1 monolayer and the number of ligands was 300, 200, and 114, respectively. Reproduced with permission from John Wiley and Sons Publisher (Singhal and Pulhani 2019)

has been assigned to the interaction between U(VI) ions and the amino groups and hydroxyl groups of CDs. The selectivity of the method was tested and was selective for the majority of the ions and is shown in Fig. 6.5c.

In another study, a similar group has prepared CDs from dopamine and shown that the uranium can be detected with high sensitivity (Wang et al. 2018). The fluorescent polydopamine (FPD) NPs give a detection limit of 2.3 ppm. Time-correlated single-photon counting (TCSPC) and dynamic light scattering (DLS) measurements were carried out to determine the mechanism of interaction. From TCSPC studies, there was no appreciable change in lifetime measurements which suggests static quenching; however, DLS measurements suggest particle aggregation. Based on these observations, it was postulated that the uranyl ion binds chemically with the surface groups of NPs, and because of this binding, aggregation occurs. The results of TCSPC and DLS measurements are shown in Fig. 6.6.

Guin et al. (2018) synthesized CDs and shown that, on the addition of uranyl ions, the emission of CDs is quenched. The quenching in emission has been corroborated by (1) the resonance energy transfer from QDs to uranium and (2) the electron transfer

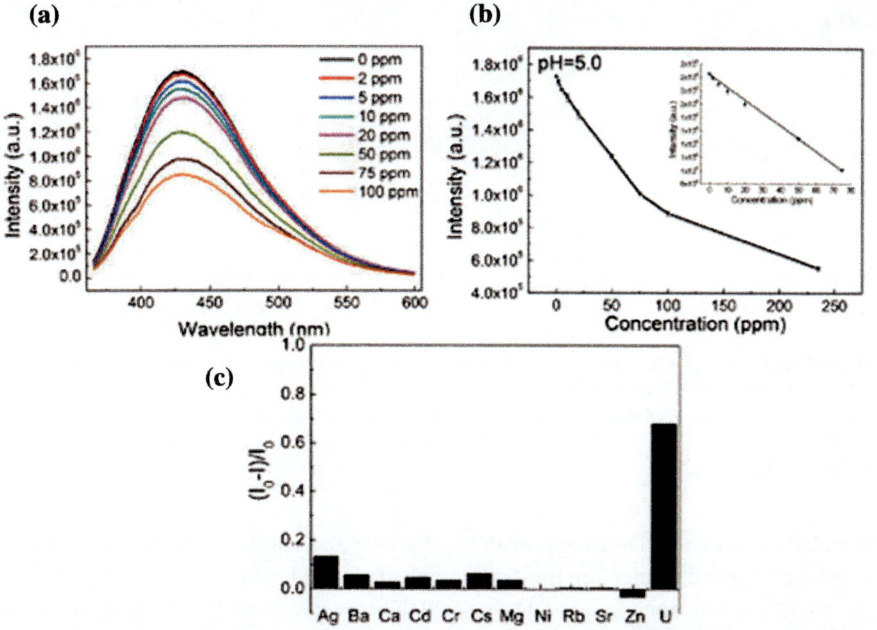

Fig. 6.5 **a** The emission spectra of CDs on the addition of different concentrations of U(VI) (λ_{ex}– 350 nm). **b** The effect of U(VI) addition on emission intensity of CDs. **Insert**: The linear fitting from the curve. **c** Selectivity analysis on the addition of different concentrations of metal ions. Reproduced with permission from RSC Publisher (Wang et al. 2015)

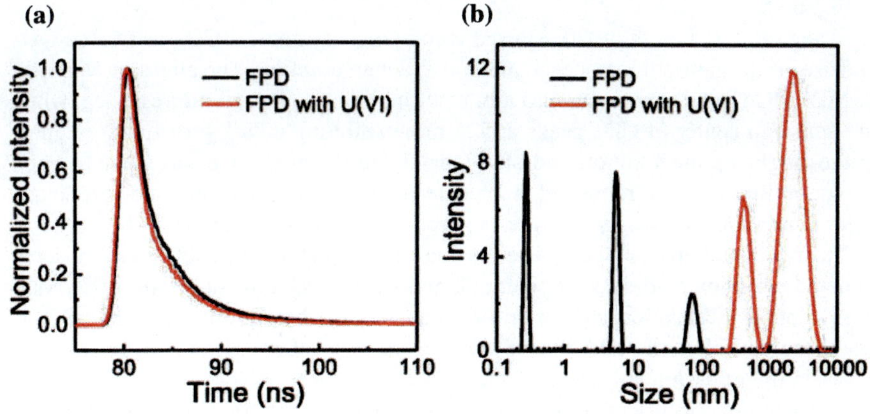

Fig. 6.6 TCSPC and DLS studies of FPD with U(VI). Reproduced with permission from Elsevier Publisher (Wang et al. 2018)

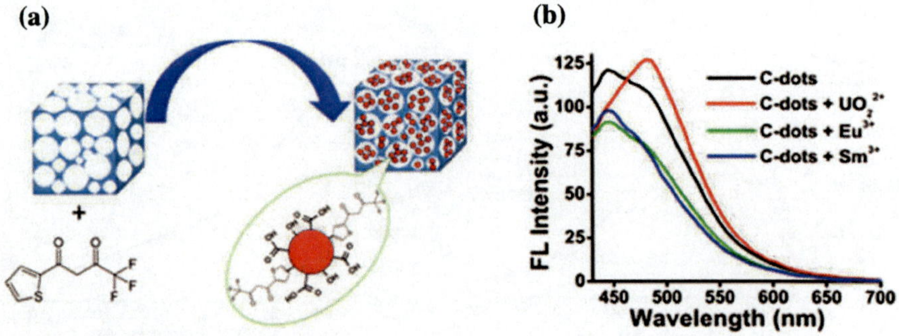

Fig. 6.7 **a** Synthesis mechanism of the aerogel–CDs–TTA complex. **b** Steady stage emission spectra of CDs on the addition of UO_2^{2+}, Sm^{3+}, and Eu^{3+} (λ_{ex}—400 nm). Reproduced from ACS Publisher under Creative Commons public use license (Dolai et al. 2017) (https://pubs.acs.org/doi/10.1021/acsomega.7b01883). Further permissions related to the material excerpted should be directed to the ACS

from CDs to uranium. Dewangan et al. (2019) designed a probe from glucose-derived CDs and achieved a detection limit of 1.5 ppb at pH 7. Dolai et al. (2017) designed a thenoyltrifluoroacetone (TTA)-CDs/aerogel fluorescent sensor for the determination of UO_2^{2+}, Sm^{3+}, and Eu^{3+}. The synthesis mechanism is shown in Fig. 6.7a. They have shown that the binding with the UO_2^{2+} changes the electron density on CDs surface because of the coordination of UO_2^{2+} and TTA which in turn shifts the fluorescence of CDs from 445 to 480 nm. Interestingly with Sm^{3+} and Eu^{3}, no such shift in emission was observed but only quenching in emission was observed. The results are shown in Fig. 6.7b.

Wang et al. (Chen et al. 2018) used a composite system of CDs and CdTe QDs and used radiometric fluorescence method to detect uranium. The emission intensity of MPA@CdTe QDs peaks at 640 nm, resulting in a strong red fluorescence, while the emission center of CDs peaks at 525 nm, exhibiting a dark green fluorescence. Interestingly, on the addition of uranyl ions, the red fluorescence was quenched and the green fluorescence remained stable. Because of this, the resultant fluorescence color continuously changed from red to green as shown in Fig. 6.8a.

To understand this quenching in emission and binding mechanism, optical absorption and emission studies, zeta potential, and DLS measurements were carried out. From optical absorption studies, no change in the absorption spectrum of QDs was observed after the addition of uranyl ion. Also, there is no spectral overlap between the emission spectra of CdTe QDs and absorption spectra of uranyl. This excludes the possibility of luminescence quenching by energy transfer mechanism. DLS measurement and zeta potential studies suggest no change in particle size which excludes the possibility of luminescence quenching via aggregation. Therefore the only possibility which arises is the complex formation between the uranyl ions and the surface functional groups of MPA, because of which luminescence gets quenched through photo-induced electron transfer. No binding of uranyl ion with CDs was observed. The detection limit of the method was calculated to be 4 nM. Using this

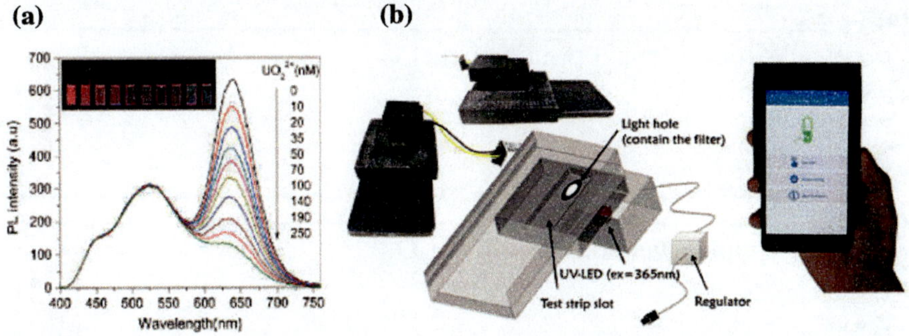

Fig. 6.8 **a** Ratiometric fluorescent spectra of CDs/CdTe QDs composite system (λ_{ex}—365 nm) on the addition of different concentrations of uranyl ions (0–250 nM). The inset photograph shows the corresponding results under UV light excitation. **b** Photograph and structure of the device. Reproduced with permission from ACS Publisher (Chen et al. 2018)

technology, they also developed an App for uranium detection in the 1–50 μM concentration range that can be operated in a smartphone. The results are shown in Figs. 6.8b and 6.9.

CDs have also been used for the detection of iodine. Iodine is a nuclear fission product and ^{129}I particularly is a matter of concern because of its long half-life of ~10^7 years. To detect it, Wang and Wu (2015) designed a fluorescent turn on–off–on probe based on CDs and Hg^{2+}.

The CD fluorescence was quenched on the addition of Hg^{2+} and assigned to the electron transfer from QDs to Hg^{2+}. On the addition of iodine, the fluorescence recovered because of the formation of Hg–I complex and was proportional to the amount of iodine added. Based on the Stern–Volmer plot the detection limit of the

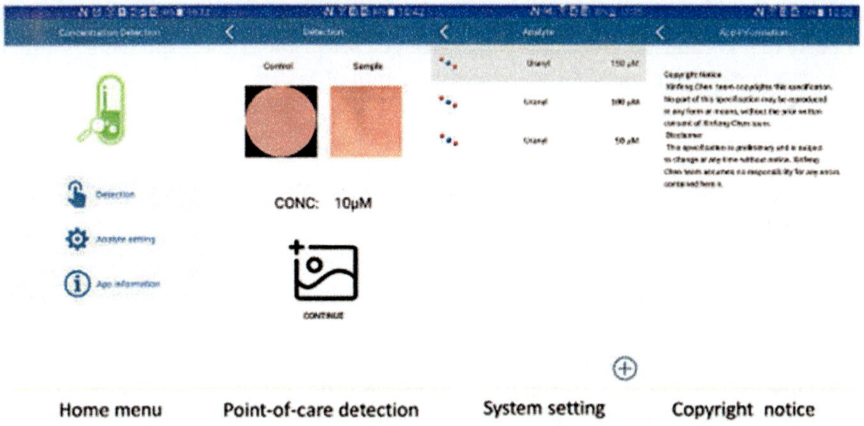

Fig. 6.9 Modules of the "Concentration Detection" app, showing its various components. Reproduced with permission from ACS Publisher (Chen et al. 2018)

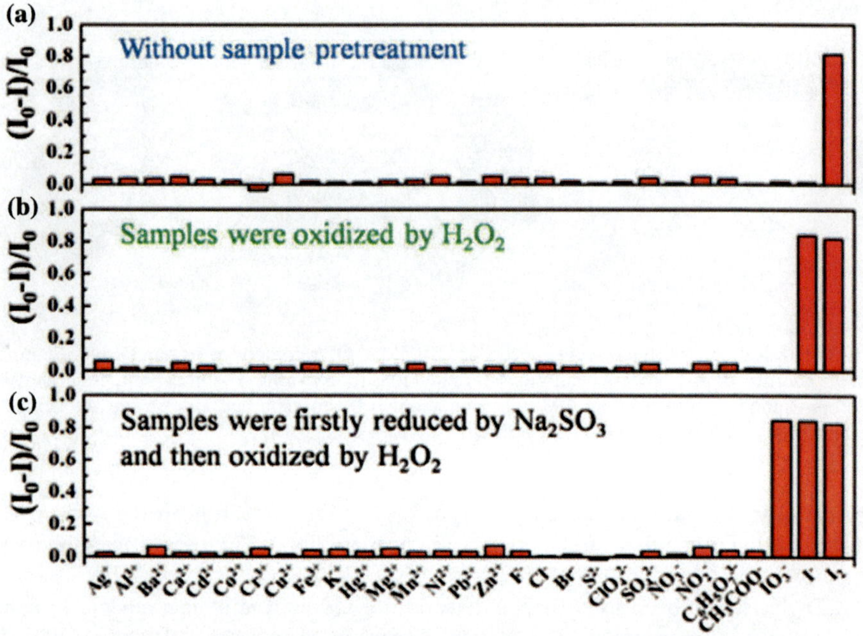

Fig. 6.10 The effect of cation and anion interferences on iodine sensor performance under different conditions Reproduced with permission from CSIRO Publisher (Wang and Wu 2015)

method was 46 nmol/L. Wang et al. (2016) have synthesized the CDs having both hydrophilic and hydrophobic nature. On addition of iodine, the fluorescence was quenched achieving a detection limit of 3.5 nmol/L. It was observed that the sensor was sensitive for the detection of I_2 only and both I^- and IO_3^- can be converted to I_2 for their detection. This is shown in Fig. 6.10. It was observed that the amphiphilic property of CDs and the electron transfer from CDs to I_2 both are responsible for the luminescence quenching of CDs and the detection of I_2.

6.2.2 Uranium Sensing Using a Composite of NPs and DNA

Nowadays, numerous studies has been carried out on DNA as a sensor for uranium because of its selectivity (Wu et al. 2013). Since DNA molecules are very specific and selective to a particular metal ion, binding of uranium ions with DNA can induce structural as well as conformational changes that can result in variable signal output. In this line, researchers are associating various optically active molecules with DNA where the optical property will serve as a sensor, and the selectivity of DNA for uranium will make the probe specific. Wu et al. (2013) reported a DNAzyme-based gold NPs (AuNPs) sensor. In this case, the NPs were functionalized with a DNA

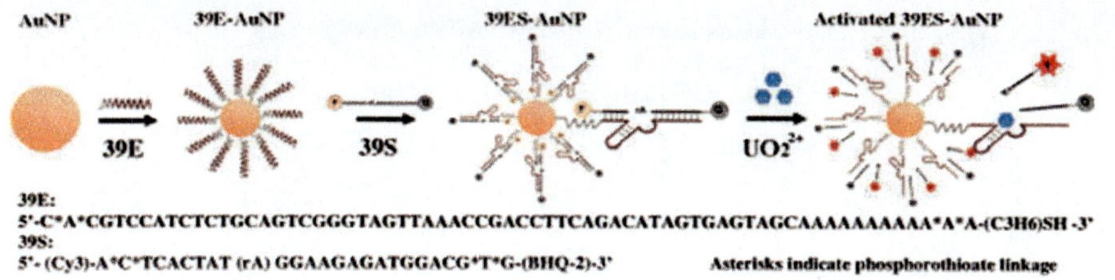

Fig. 6.11 Design of a fluorescent DNA enzyme sensor for the selective detection of UO_2^{2+} ions. Reproduced with permission from ACS Publisher (Wu et al. 2013)

enzyme called 39E, the structure of which is shown in Fig. 6.11. This enzyme binds with the NPs and quencher at 5′ and 3'ends, respectively. Herein, the binding of uranyl ion with the DNA enzyme results in the release of fluorophore because of which fluorescence signal appears that was missing in the absence of uranyl ion. The intensity of the signal is dictated by the concentration of uranyl ion and because of high sensitivity of DNA ultratrace level detection with high specificity is possible. This is shown in Fig. 6.11.

Xiong et al. (2020) presented a probe based on DNA for the fluorescent detection of UO_2^{2+}. Herein, the DNA enzyme breaks in the presence of uranyl ion and this cleavage results in either fluorescence enhancement or decrement. To utilize this strategy, AuNPs and fluorophore were fixed at the DNA enzyme. In the presence of target UO_2^{2+}, DNA enzyme breaks down resulting in the release of fluorescent molecules. Based on the sensitivity of the probe, the detection of the method was pM. Zhu et al. (2019) developed an approach for uranyl ion detection using E-DNA, cleaved substrate strand (S-DNA) and SYBR green I (SG). They also used DNA cleavage technique in the presence of target ion and using the method achieved the detection limit of 200 pM. Wang et al. (2019) used a different strategy. Instead of labeling the dyes at the end, they have embedded the dyes in the middle of DNA. It was found that in the presence of UO_2^{2+}, fluorescence signal was significantly quenched due to the base-stacking interaction. The method is shown in Fig. 6.12.

Fu et al. (Zhang et al. 2015) also presented a fluorescent biosensor for UO_2^{2+} sensing. They coupled molybdenum disulfide (MoS_2) nanosheets with DNAzyme wherein the DNA worked as a selective binder for UO_2^{2+} and MoS_2 worked as the fluorescence probe. When UO_2^{2+} interacts with the system, the DNA cleavage occurs and, because of this cleavage, the released DNA gets adsorbed on the surface of MoS_2 nanosheets, resulting in a decrease in fluorescence signal and achieving a detection limit of 2.14 nM. Yang et al. (2021) reported another sensor for UO_2^{2+} monitoring. The sensor worked on the principle of ratiometric fluorescence. Herein, the sensor contained a DNA enzyme probe (39E-DNA and 39S-DNA) and the split G-quadruplex probes (39F-R and 39F-L). 39S-DNA helps to form a complex between 39F-R and 39F-L. When UO_2^{2+} interacts with the system, it induces the cleavage of 39S-DNA resulting in split 39F-R and 39F-L. Because of this, the fluorescence intensity decreases and this change was linear with an increase in UO_2^{2+} concentration. The results were fitted into the Stern–Volmer plot and detection limit of 0.08 μg/

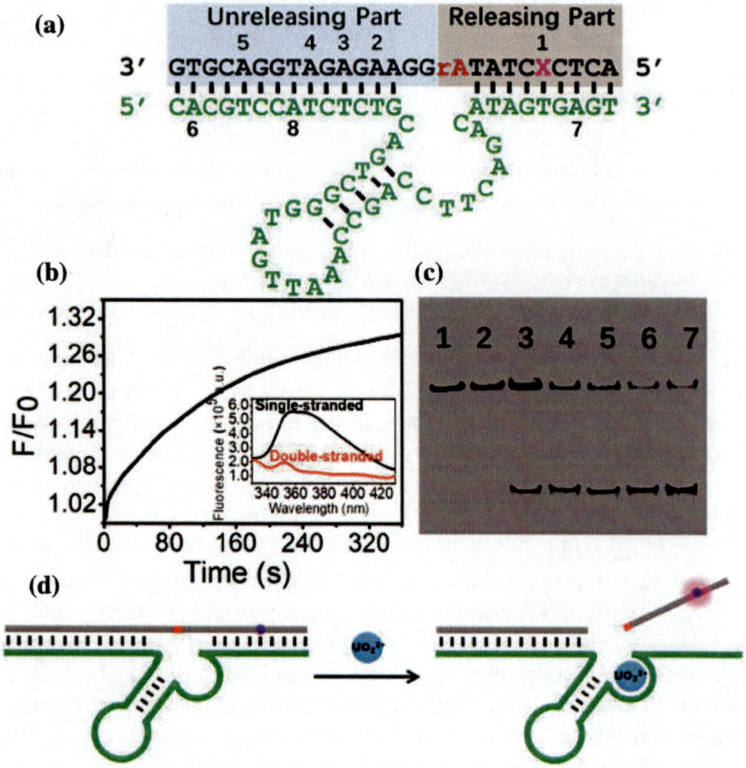

Fig. 6.12 Interaction of DNA enzyme with UO_2^{2+} and corresponding change in fluorescence emission intensity. Reproduced with permission from Elsevier Publisher (Wang et al. 2019)

L was achieved. Fu group (Zhang et al. 2017) reported a colorimetric biosensor for uranyl ion detection using magnetic beads (MBs) and HCR. Initially, the MB surface was immobilized with DNA enzyme. When UO_2^{2+} interacts, the enzyme beaks which results in the release of ssDNA. This released ssDNA acts as a trigger to capture a large amount of horseradish peroxidase (HRP) on MBs surface which, on further interacting with TMB-H_2O_2 solution, results in a color change from colorless to blue. The interaction mechanism is shown in Fig. 6.13.

In another study (Huang et al. 2016), a smart hydrogel based on Au NPs was prepared and used as a sensor for UO_2^{2+} ion detection. The designed sensor was rapid, portable, and sensitive. 50 μL of UO_2^{2+} was added to the hydrogel. Reaction conditions such as incubation at 25 °C and shaking at 150 rpm for 2 h were followed. Interestingly, after the reaction time, it was observed that the hydrogel changed its color and became colorless, while a red color in supernatant solution was observed. It was suggested that the UO_2^{2+} interacts with internal AuNPs because of which it comes out of hydrogel leaving its colorless. The mechanism is shown in Fig. 6.14.

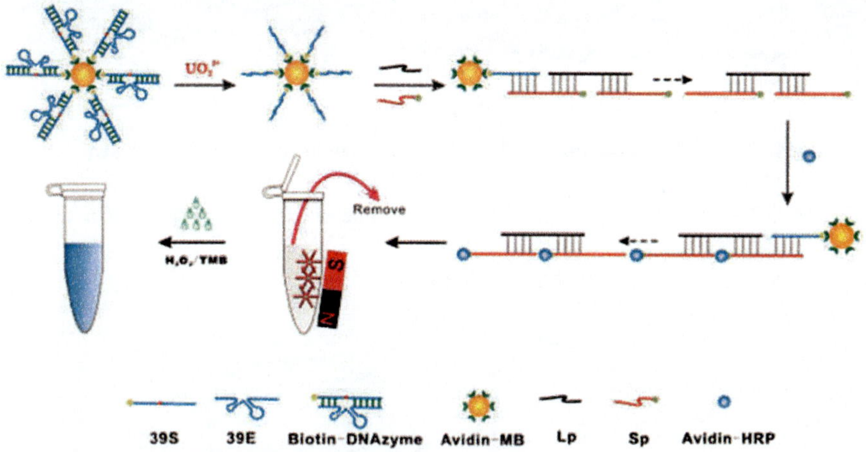

Fig. 6.13 Schematic of the working principle of colorimetric biosensor. Reproduced with permission from Elsevier Publisher (Zhang et al. 2017)

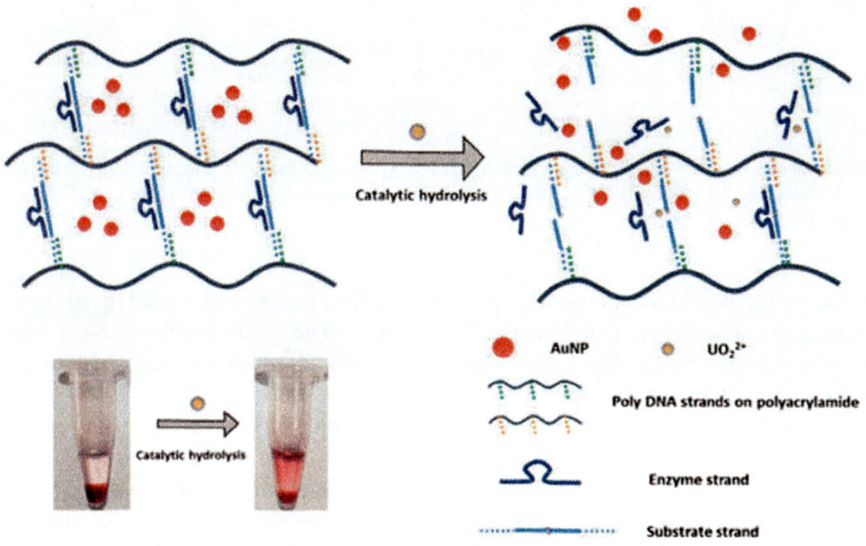

Fig. 6.14 Working principle of the uranyl ion responsive hydrogel. Reproduced with permission from Elsevier Publisher (Huang et al. 2016)

In another study, Zhou et al. (2013) proposed a resonance light scattering (RLS) method for uranium detection. The technique is fast, simple, and has high sensitivity. They have used a labeled DNA and AuNPs composite system for uranium sensing. The schematic of the method is shown in Fig. 6.15. In the presence of uranyl ion, change in the aggregation state of AuNPs occurs because of the cleavage of the

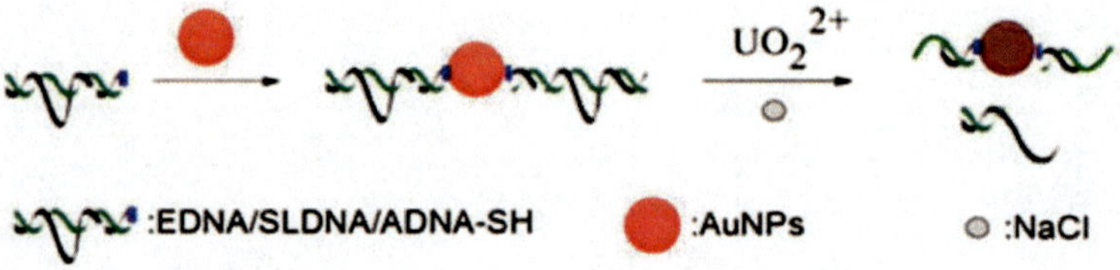

Fig. 6.15 Schematic of detection of UO_2^{2+} ion. Reproduced with permission from Elsevier Publisher (Zhou et al. 2013)

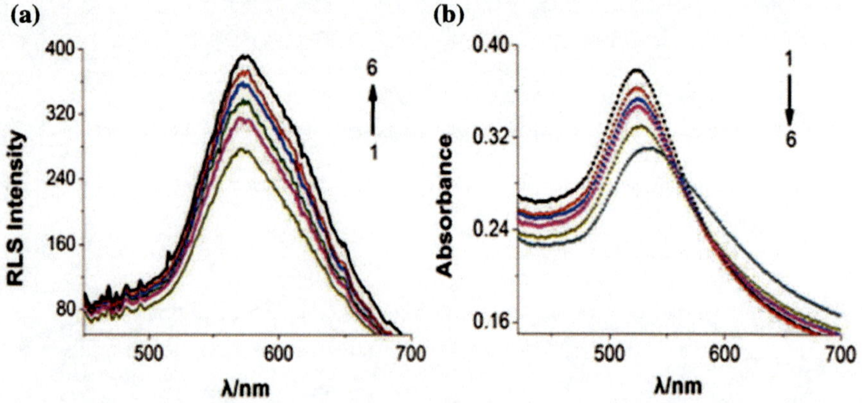

Fig. 6.16 RLS (**a**) and absorption spectra (**b**) of DNAzyme–AuNPs–UO_2^{2+} system. $C_{UO2}{}^{2+}$ (1–6)/ ($\times 10^{-9}$ mol L^{-1}): 0.0, 15.0, 25.0, 75.0, 100.0, 200.0. Reproduced with permission from Elsevier Publisher (Zhou et al. 2013)

substrate strand of DNAzyme. This releases a shorter duplex and results in enhancement in RLS intensity. This is shown in Fig. 6.16. The probe has been tested with interfering ions such as Cu^{2+}, Fe^{2+}, Fe^{3+}, Hg^{2+}, and Tb^{3+} wherein no interferences were observed.

In another study, Manochehry et al. (2018) developed a colorimetric DNAzyme-based biosensor for uranyl ion detection using a litmus test. In this system, the DNA complex was immobilized on the surface of a magnetic bead. The DNA enzyme binds with the uranyl ion, and after this interaction, the cleavage occurs. In this cleaved portion, urea and phenol red were added. During this process, hydrolysis of urea occurs that releases ammonia which changes the pH of the system and finally the color of phenol red changes.

Uranium sensing using chelating complexes—Many complexes have been designed to detect uranium selectively and efficiently. In this line, Wu et al. (2020) designed a sensor named W1H with the help of a rational design strategy. The designed sensor changes color from red to blue and detects uranium selectively and with a limit in the nM range. The method was selective but Co^{2+} and Cu^{2+} interfere with the analysis. The synthesis procedure of the sensor is shown in Scheme 6.3.

In another study, Akl synthesized a Schiff base derivative for uranium sensing and incorporated it in a membrane (Akl 2022). A careful study of the membrane constituents on detection efficiency was performed and an optimal composition for optimal performance was designed. Interaction mechanism of the uranium with the synthesized Schiff base is shown in Fig. 6.17.

Uranium sensing using metal–organic framework and covalent organic framework—Among various sensors of uranium and actinides, metal—organic frameworks (MOFs) have been widely explored due to their fascinating characteristics (Liu et al. 2022). Many researchers have demonstrated uranium detection using MOFs. In this line, Liu et al. (2022) synthesized $[Cd_3L_2(bipy)(H_2O)_2]{\cdot}H_2O$ and observed a quenching in emission luminescence. The detection limit was calculated using Stern–Volmer plot and observed to be 0.9 μg/L and resonance energy transfer from MOF to uranium was assigned as the possible mechanism for luminescence quenching (Liu

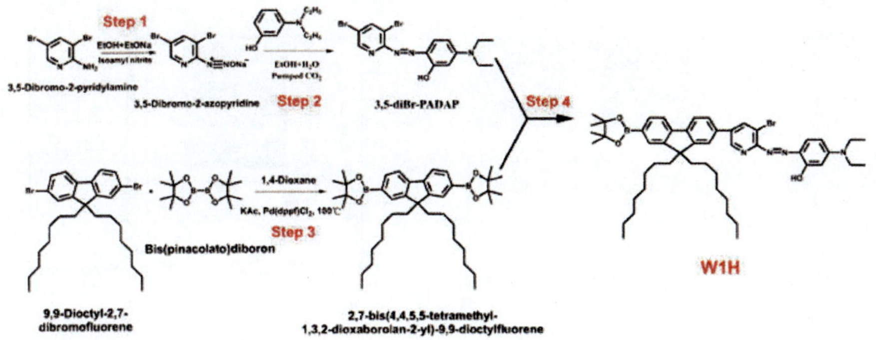

Scheme 6.3 Synthesis procedure of WIH ligand for uranium sensing. Reproduced with permission from Elsevier Publisher (Wu et al. 2020)

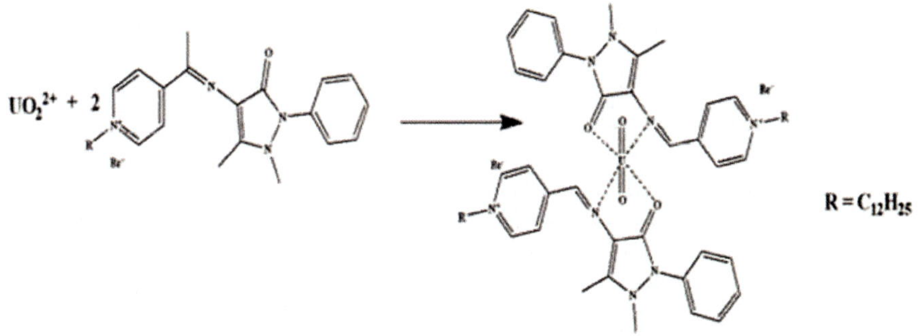

Fig. 6.17 Interaction mechanism of the uranium with synthesized Schiff base. Reproduced from RSC Publisher under Creative Commons Licence (Akl 2022)

et al. 2022). In another study, Ye et al. (2017) synthesized the first efficient dual-channel response luminescence 3D MOF; $[Tb(BPDC)_2]\cdot(CH_3)_2NH_2$ (DUT-101) for the detection of UO_2^{2+} ions. DUT-101 exhibits an intense green emission because of the transition between Tb^{3+} levels and has been assigned to the energy transfer from the ligand to Tb^{3+} ions. This has been confirmed by monitoring the emission of the ligand only without Tb^{3+} ions wherein no green emission was observed but an emission at 408 nm occurred due to the transition between ligand levels. On addition of UO_2^{2+} ions, the green emission gets quenched and the emission intensity at 408 nm increases. This was assigned to the binding of DUT-101 with uranyl ion which inhibits both electron flow and energy transfer from ligands to Tb^{3+} centers. This results in fast luminescence quenching and has been confirmed by the XRD, TG, and EDX analyses. This has been shown in Fig. 6.18.

In another study, Yang et al. (2013) synthesized BTC-capped Tb-MOF-76 and used it for the detection of uranium. Initially, the energy transfer from BTC to Tb^{3+} ions occurs resulting in the green emission. Interestingly, on addition of uranyl ion,

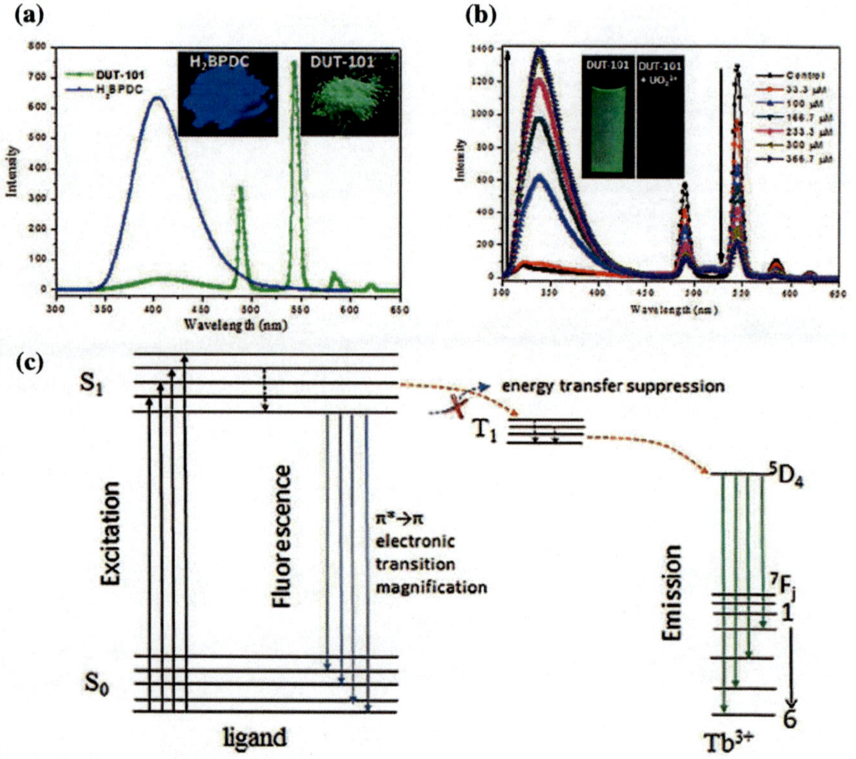

Fig. 6.18 Emission spectra of DUT-101 and H_2BPDC ligand in the presence and absence of UO_2^{2+} and mechanism of the interaction depiction. Reproduced with permission from John Wiley & Sons Publisher (Ye et al. 2017)

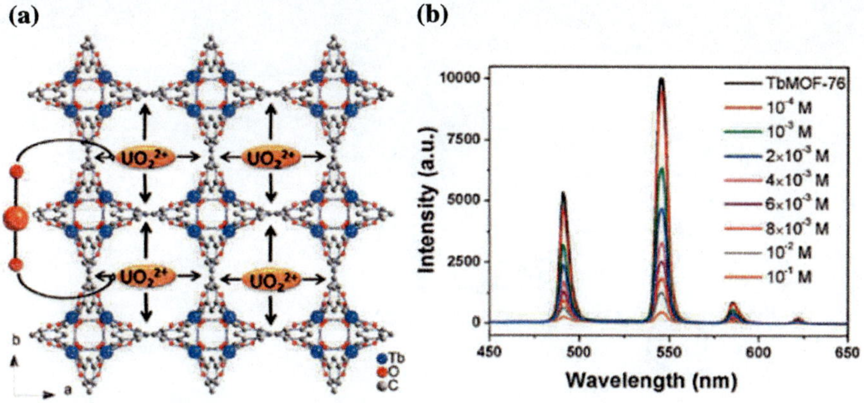

Fig. 6.19 **a** Interaction mechanism of UO_2^{2+} ions with MOF-76 and **b** corresponding change in emission spectra. Reproduced with permission from RSC Publisher (Yang et al. 2013)

this emission quenched. It was postulated that the UO_2^{2+} ions interact with the micropores or the surface of MOF-76. This interaction reduces the efficiency of energy transfer process resulting in a decrease in the emission intensity. The interaction mechanism is shown in Fig. 6.19.

Guo et al. (2019) synthesized Fe_3O_4-CMC @ ZIF-8 @ CDs nanocomposite and used it for UO_2^{2+} detection. They observed that the fluorescence intensity decreaseds gradually with increase in UO_2^{2+} concentration and 100 mg/L was the detection limit of the method. Qin et al. (2020) synthesized a fluorescent zinc-based MOF (HNU-50) for the detection and extraction of UO_2^{2+}. It was shown that the amide group and two uncoordinated carboxyl oxygen atoms in HNU-50 are potential binding sites for uranium binding and this in turn was confirmed by the IR and XPS studies. The mechanism of interaction is shown in Scheme 6.4. On addition of uranyl ion to HNU-50 quenching in luminescence was observed and is because of spectra overlap between emission spectra of HNU-50 and absorption spectra of uranyl ion. The method was found to be selective for most of the ions and a detection limit of 1.2×10^{-8} M was observed. It shows that many studies have been carried out using MOF as sensors. The field is in its nascent stage and many issues such as stability of the probe, selectivity, and better detection limit needs to be achieved before it can be realized into a commercial product.

Covalent organic frameworks (COFs) are another class of interesting materials that have been used for uranium detection. Xu et al. (2022) synthesized a sulfone-based COF (TAS-COF). The synthesis was done following the Schiff base condensation reaction. Interestingly, this COF has high oxidation potential. This COF while interacting with colorless 3,3′,5,5′-tetramethylbenzydine (TMB) solution under visible light oxidizes it, turning it into a blue color solution. When UO_2^{2+} interacts with this oxidized species, because of the binding with the imines, a clear color fading of the sensing system was observed. This is shown in Scheme 6.5.

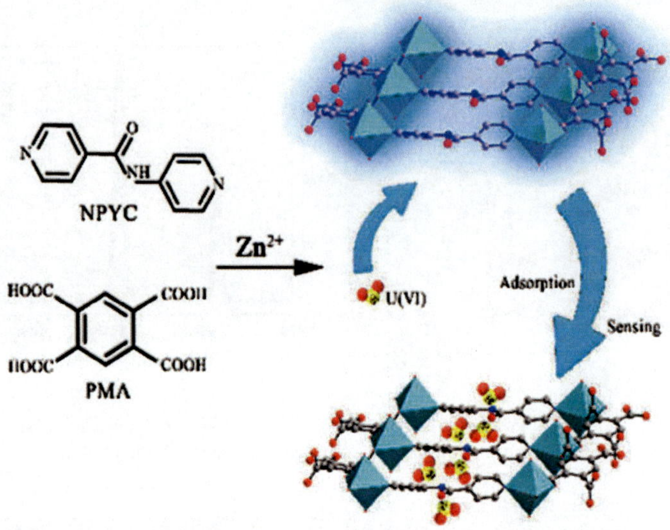

Scheme 6.4 Mechanism of interaction of uranyl ion with HNU-50. Reproduced with permission from ACS Publisher (Qin et al. 2020)

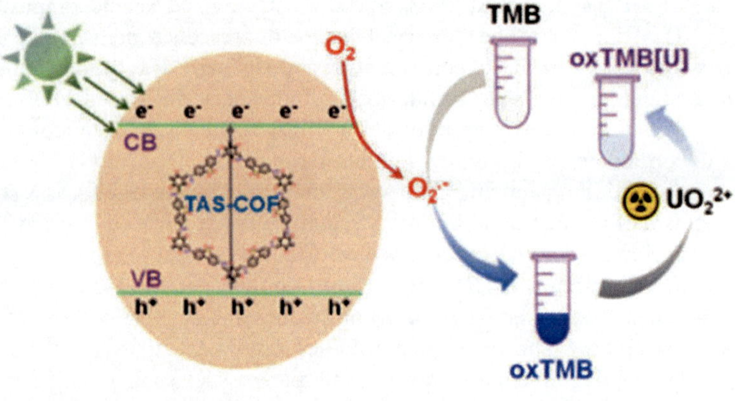

Scheme 6.5 Illustration for detection of UO_2^{2+} using TAS-COF. Reproduced from MDPI Publisher under Creative Commons Licence (Xu et al. 2022)

6.3 Conclusions

Numerous materials have been developed for uranium sensing as discussed above in this chapter. In order to compare the performance, we have tabulated some of these materials with their corresponding detection limit given in Table 6.1. It is interesting to observe from Table 6.1 that many of the synthesized materials have low detection

Table 6.1 Different optical materials used for uranium defection with their detection limits

Material	Detection limit	References
Arsenazo III pumped into a nafion microtube	<20 ppb	Collins et al. (2002)
Nile blue on modified glassy carbon electrode	0.19 ppb	Zhou et al. (2022)
CdSe/CdS core–shell QDs	75 ppb	Singhal et al. (2017)
Glucose-derived CDs	1.5 ppb	Dewangan et al. (2019)
Composite system of CDs and CdTe QDs	4 nM	Chen et al. (2018)
Composite of AuNPs and DNA tweezer probe	25 pM	Xiong et al. (2020)
G-quadruplex-assisted enzyme strand	200 pM	Zhu et al. (2019)
DNAzyme and molybdenum disulfide nanosheets	2.14 nM	Zhang et al. (2015)
Composite of DNAzyme (39E-DNA and 39S-DNA) and the split G-quadruplex probes	0.08 ppb	Yang et al. (2021)
Schiff base derivative incorporated onto a membrane	3.90×10^{-7} mol L^{-1}	Akl (2022)
$[Cd_3L_2(bipy)(H_2O)_2]\cdot H_2O$	0.9 ppb	Liu et al. (2022)
Fe_3O_4-CMC @ ZIF-8 @ CDs nanocomposite	100 ppm	Guo et al. (2019)
Zinc-based MOF (HNU-50)	1.2×10^{-8} M	Qin et al. (2020)

limit, much lower than the permissible limit in drinking water, making these probes a great sensor for uranium monitoring. However, there are many factors that need to be taken into account before realizing the potential of these probes commercially.

The cost of the method—It is an important factor to consider when scaling up the product is considered. As discussed above and can be seen from Table 6.1 also that many probes have shown superior properties for uranium detection in terms of their detection limits, they involve costly reagents such as DNA. In that case, widespread use of these probes is questionable. Also, in many of the cases, the synthesis involved sophisticated equipments and high cost. This increases the overall cost of the process and makes use of these probes difficult in a competitive market.

(i) **Use of environmentally friendly materials**—It is another important point to be considered since these probes will be directly used in the environment. Many probes have been developed that use toxic chemicals and elements such as Cd and Pd. Hence, the use of these probes in environment is not advisable. In view of this, many materials have been developed which are environmentally friendly such as CDs and can be synthesized from low-cost materials. Efforts should be made in the direction to develop these materials and use them.

(ii) **Stability of the probe**—This is another important merit to be considered before deciding the performance of a probe. There are many methods that have been developed but the majority of them are not stable in real environmental conditions. For example, uranium detection in tap water is completely different from the detection in river and seawater. In all these water bodies, the concentration of uranium, competing ions, ionic strength, speciation, and the prevailing pH conditions are different. Many of the probes do not work under such conditions and collapse before detecting uranyl ion. They can only be used on lab scale with pure water. Therefore, to use a probe commercially it is important to check the environmental stability of the probe.

(iii) **Selectivity, sensitivity, and response time**—These are the other parameters that decide the merit of a probe. Uranium detection in a water body involves many other competing ions that may interact with the probe and give signal. In that case, the probe gives a false signal and uranium detection by the method is not reliable. Similarly, many probes detect uranium at high concentration and it is desirable to achieve the lowest detection limit because of its high toxicity so that a remedial action can be taken at a low concentration. Response time of the probe is another important measure and it should be as prompt as possible to take an action immediately.

With this, we conclude that optical materials have an advantage of visual detection and many methods have been developed for the same still the probe that can be used in real environmental conditions is yet to emerge and work is going on in that direction.

Acknowledgements The author acknowledges Dr. Vandana Pulhani, Dr. A. Vinodkumar, and Dr. D. K. Aswal of the Health Safety and Environment Group for their continuous support.

References

Akl, Z.F.: Rapid electrochemical sensor for uranium(vi) assessment in aqueous media. RSC Adv. **12**(31), 20147–20155 (2022)

Baylor, L.C., Buchanan, B.R.: Reflectance probe for Uranium(VI) based on the colorimetric indicator arsenazo III. Appl. Spectrosc. **49**(5), 679–681 (1995)

Chen, X., Mei, Q., Yu, L., Ge, H., Yue, J., Zhang, K., et al.: Rapid and on-site detection of uranyl ions via ratiometric fluorescence signals based on a smartphone platform. ACS Appl. Mater. Interfaces. **10**(49), 42225–42232 (2018)

Collins, G.E., Lu, Q., Abubeker, S., Vajs, E.: Remote fiber-optic flow cell for the detection of Uranium(VI) in groundwater. Appl. Spectrosc. **56**(4), 464–468 (2002)

Dewangan, P.K., Khan, F., Shrivas, K., Sahu, V.: Determination of uranium in environmental sample by nanosensor graphene quantum dots. J. Radioanal. Nucl. Chem. **320**(3), 757–763 (2019)

Dolai, S., Bhunia, S.K., Zeiri, L., Paz-Tal, O., Jelinek, R.: Thenoyltrifluoroacetone (TTA)–carbon Dot/Aerogel fluorescent sensor for lanthanide and actinide ions. ACS Omega **2**(12), 9288–9295 (2017)

Dutta, R.K., Kumar, A.: Highly sensitive and selective method for detecting ultratrace levels of aqueous uranyl ions by strongly photoluminescent-responsive amine-modified cadmium sulfide quantum dots. Anal. Chem. **88**(18), 9071–9078 (2016)

Fouad, H., Abu Elenein, S., Orabi, A., Abdulmoteleb, S.: A new extractant impregnated resin for separation of traces of uranium and thorium followed by their spectrophotometric determination in some geological samples. SN Appl. Sci. **1**(4), 309 (2019)

Guin, S.K., Ambolikar, A.S., Paul Guin, J., Neogy, S.: Exploring the excellent photophysical and electrochemical properties of graphene quantum dots for complementary sensing of uranium. Sens. Actuators B Chem. **272**, 559–573 (2018)

Guo, X., Liu, Q., Liu, J., Zhang, H., Yu, J., Chen, R., et al.: Magnetic metal-organic frameworks/carbon dots as a multifunctional platform for detection and removal of uranium. Appl. Surf. Sci. **491**, 640–649 (2019)

Huang, Y., Fang, L., Zhu, Z., Ma, Y., Zhou, L., Chen, X., et al.: Design and synthesis of target-responsive hydrogel for portable visual quantitative detection of uranium with a microfluidic distance-based readout device. Biosens. Bioelectron. **85**, 496–502 (2016)

Kadam, B.V., Maiti, B., Sathe, R.M.: Selective spectrophotometric method for the determination of uranium(VI). Analyst **106**(1263), 724–726 (1981)

Khalifa, M.E.: Selective separation of uranium using alizarin red S (ARS)-modified anion-exchange resin or by flotation of U-ARS chelate. Sep. Sci. Technol. **33**(14), 2123–2141 (1998)

Liu, H., Fu, T., Mao, Y.: Metal-organic framework-based materials for adsorption and detection of Uranium(VI) from aqueous solution. ACS Omega **7**(17), 14430–14456 (2022)

Manochehry, S., McConnell, E.M., Tram, K.Q., Macri, J., Li, Y.: Colorimetric detection of uranyl using a litmus test. Front. Chem. **6** (2018)

Orabi, A.H.: Determination of uranium after separation using solvent extraction from slightly nitric acid solution and spectrophotometric detection. J. Radiat. Res. Appl. Sci. **6**(2), 1–10 (2013)

Qin, X., Yang, W., Yang, Y., Gu, D., Guo, D., Pan, Q.: A zinc metal-organic framework for concurrent adsorption and detection of uranium. Inorg. Chem. **59**(14), 9857–9865 (2020)

Rao, N.S.V., Sen, S., Prins, N.J., Cooper, D.A., Ledoux, R.J., Costales, J.B., et al.: Network algorithms for detection of radiation sources. Nucl. Instrum. Methods Phys. Res. Sect. A **784**, 326–331 (2015)

Safavi, A., Bagheri, M.: A novel optical sensor for uranium determination. Anal. Chim. Acta **530**(1), 55–60 (2005)

Serenjeh, F.N., Hashemi, P., Ghiasvand, A.R., Rasolzadeh, F.: A new optical sensor for selective quantitation of uranium by the immobilization of arsenazo III on an agarose membrane. Anal. Methods **8**(21), 4181–4187 (2016)

Singhal, P., Ghosh, H.N.: Ultrafast hole/electron transfer dynamics in a CdSe quantum dot sensitized by pyrogallol red: a super-sensitization system. J. Phys. Chem. c. **118**(30), 16358–16365 (2014)

Singhal, P., Ghosh, H.N.: Hot charge carriers in quantum dots: generation, relaxation, extraction, and applications. ChemNanoMat. **5**(8), 985–999 (2019)

Singhal, P., Pulhani, V.: Effect of ligand concentration, dilution, and excitation wavelength on the emission properties of CdSe/CdS core shell quantum dots and their implication on detection of uranium. ChemistrySelect **4**(15), 4528–4537 (2019)

Singhal, P., Ghorpade, P.V., Shankarling, G.S., Singhal, N., Jha, S.K., Tripathi, R.M., et al.: Exciton delocalization and hot hole extraction in CdSe QDs and CdSe/ZnS type 1 core shell QDs sensitized with newly synthesized thiols. Nanoscale **8**(4), 1823–1833 (2016)

Singhal, P., Jha, S.K., Vats, B.G., Ghosh, H.N.: Electron-transfer-mediated uranium detection using quasi-type II core-shell quantum dots: insight into mechanistic pathways. Langmuir **33**(33), 8114–8122 (2017)

Singhal, P., Ghosh, H.N.: Hot-hole extraction from quantum dot to molecular adsorbate. Chem. A Eur. J. **21**(11), 4405–12 (2015)

Sizov, F.: THz radiation sensors. Opto Electron. Rev. **18**, 10–36 (2009)

Tam, A.K., Boyraz, O., Unangst, J., Nazareta, P., Schreuder, M., Nilsson, M.: Quantum-dot doped polymeric scintillation material for radiation detection. Radiat. Meas. **111**, 27–34 (2018)

Wang, Q., Wu, Y.: A carbon nanodots-based fluorescent turn-on probe for iodide. Aust. J. Chem. **68**(10), 1479–1484 (2015)

Wang, Z., Lu, Y., Yuan, H., Ren, Z., Xu, C., Chen, J.: Microplasma-assisted rapid synthesis of luminescent nitrogen-doped carbon dots and their application in pH sensing and uranium detection. Nanoscale **7**(48), 20743–20748 (2015)

Wang, M., Zheng, B., Yang, F., Du, J., Guo, Y., Dai, J., et al.: Synthesis of "amphiphilic" carbon dots and their application for the analysis of iodine species (I2, I− and IO3−) in highly saline water. Analyst **141**(8), 2508–2514 (2016)

Wang, Z., Xu, C., Lu, Y., Wei, G., Ye, G., Sun, T., et al.: Microplasma electrochemistry controlled rapid preparation of fluorescent polydopamine nanoparticles and their application in uranium detection. Chem. Eng. J. **344**, 480–486 (2018)

Wang, X., Zeng, R., Chu, S., Tang, W., Lin, N., Fu, J., et al.: A quencher-free DNAzyme beacon for fluorescently sensing uranyl ions via embedding 2-aminopurine. Biosens. Bioelectron. **135**, 166–172 (2019)

Wang, Z., Zhang, L., Zhang, K., Lu, Y., Chen, J., Wang, S., et al.: Application of carbon dots and their composite materials for the detection and removal of radioactive ions: a review. Chemosphere **287**, 132313 (2022)

Wu, P., Hwang, K., Lan, T., Lu, Y.: A DNAzyme-gold nanoparticle probe for uranyl ion in living cells. J. Am. Chem. Soc. **135**(14), 5254–5257 (2013)

Wu, X., Yin, Q., Huang, Q., Mao, Y., Hu, Q., Wang, H.: Rational designing an azo colorimetric sensor with high selectivity and sensitivity for uranium environmental monitoring. Anal. Chim. Acta **1140**, 153–167 (2020)

Xiong, Z., Wang, Q., Zhang, J., Yun, W., Wang, X., Ha, X., et al.: A simple and programmed DNA tweezer probes for one-step and amplified detection of UO22+. Spectrochim. Acta Part A Mol. Biomol. Spectrosc. **229**, 118017 (2020)

Xu, Y., Wei, J., Chen, X.: Visible light-responsive sulfone-based covalent organic framework as metal-free nanoenzyme for visual colorimetric determination of uranium. Chemosensors. **10**(7), 248 (2022)

Yang, G., Chan, S.-W.: Photocatalytic reduction of chromium(VI) in aqueous solution using dye-sensitized nanoscale ZnO under visible light irradiation. J. Nanopart. Res. **11**, 221–230 (2009)

Yang, W., Bai, Z.-Q., Shi, W.-Q., Yuan, L.-Y., Tian, T., Chai, Z.-F., et al.: MOF-76: from a luminescent probe to highly efficient UVI sorption material. Chem. Commun. **49**(88), 10415–10417 (2013)

Yang, Y., Yang, H., Wan, Y., Zhou, W., Deng, S., He, Y., et al.: Temperature-robust and ratiometric G-quadruplex proximate DNAzyme assay for robustly monitoring of uranium pollution and its microbial biosorbents screening. J. Hazard. Mater. **413**, 125383 (2021)

Ye, J., Bogale, R.F., Shi, Y., Chen, Y., Liu, X., Zhang, S. et al.: A water-stable dual-channel luminescence sensor for UO22+ ions based on an anionic terbium(III) metal–organic framework. Chem. A Eur. J. **23**(32), 7657–7662 (2017).

Zhang, H., Ruan, Y., Lin, L., Lin, M., Zeng, X., Xi, Z., et al.: A turn-off fluorescent biosensor for the rapid and sensitive detection of uranyl ion based on molybdenum disulfide nanosheets and specific DNAzyme. Spectrochim. Acta Part A Mol. Biomol. Spectrosc. **146**, 1–6 (2015)

Zhang, H., Cheng, X., Chen, L., Mo, F., Xu, L., Fu, F.: Magnetic beads-based DNA hybridization chain reaction amplification and DNAzyme recognition for colorimetric detection of uranyl ion in seafood. Anal. Chim. Acta **956**, 63–69 (2017)

Zhou, B., Shi, L.-F., Wang, Y.-S., Yang, H.-X., Xue, J.-H., Liu, L., et al.: Resonance light scattering determination of uranyl based on labeled DNAzyme–gold nanoparticle system. Spectrochim. Acta Part A Mol. Biomol. Spectrosc. **110**, 419–424 (2013)

Zhou, Z., Zhou, Y., Liang, X., Luo, J., Liu, S., Ma, J.: Electrochemical sensor for uranium monitoring in natural water based on poly Nile blue modified glassy carbon electrode. J. Solid State Electrochem. **26**(5), 1139–1149 (2022)

Zhu, P., Zhang, Y., Xu, S., Zhang, X.: G-quadruplex-assisted enzyme strand recycling for amplified label-free fluorescent detection of UO22+. Chin. Chem. Lett. **30**(1), 58–62 (2019)

Chapter 7
Triarylamines—A Versatile Molecular Building Block for Optoelectronic Materials

Muniappan Kalipriyadharshini, Arunachalam Raman, Joseph Ajantha, J. B. Gowthaman, and Shanmugam Easwaramoorthi

7.1 Introduction

Molecular libraries are an integral part of the research which gives various cheminformatics approaches to the chemists to retrieve the synthetic methodologies, properties, and applications. Cheminformatics, in the coming years, becomes increasingly important in the collections of small molecules which would mainly help in organic synthesis, bioactivity, etc. Though no universal molecular library exists, several searchable databases do exist and most of them were meant for the compounds of importance in biology. For organic molecules, one of the prominent research focuses is the development of new molecules, synthetic methodologies, and catalysts aiming at the pharmaceutically important molecules, wherein the largest number of molecules belong to this class. On the other hand, an increasingly important area of organic compounds in the past two decades would be functional organic materials. "Functional organic materials" broadly represent the diverse π-conjugated organic compounds having different chemical and physical properties such as dyes, semiconductors for organic electronics, molecular motors, machines, fluorescent probes, photoswitches, nonlinear optical materials, and sensors. Numerous compounds with extended π-conjugation as small molecules, oligomers, and polymers were developed in structurally diverse configurations and were constructed using several molecular building blocks. In general, the molecular building blocks are aromatics, polycyclic aromatic hydrocarbons, heterocycles, and various electron-rich and -deficient substituents. Indeed, the compounds with the combination of electron-rich (donor, D) and -deficient (acceptor, A) motifs connected through intervening π-bridge are

M. Kalipriyadharshini · A. Raman · J. Ajantha · J. B. Gowthaman · S. Easwaramoorthi (✉)
Inorganic and Physical Chemistry Laboratory, CSIR-Central Leather Research Institute, Adyar, Chennai 600020, India
e-mail: moorthi@clri.res.in

R. S. Ningthoujam and A. K. Tyagi (eds.), *Handbook of Materials Science, Volume 1*, Indian Institute of Metals Series, https://doi.org/10.1007/978-981-99-7145-9_7

found to be attractive in providing the desired optoelectronic properties. The push–pull effect in the molecule induces the intramolecular charge transfer interactions, alters the electronic energy levels, modulates the redox properties, reduces the electronic energy gap, and so on. Not restricted to these, a little more complex donor–acceptor compounds in DA_2, DA_3, AD_2, and AD_3 were also synthesized because of their improved optoelectronic and nonlinear optical properties. In this regard, one of the structural motifs that is capable of forming donor–acceptor systems in various configurations is triphenylamine, a widely exploited structural building block by several research groups to make numerous compounds toward specific applications.

Triphenylamine (TPA), central nitrogen connected with the three-benzene ring through sp^2 hybridized N–C bond and have the lone pair electron on nitrogen $2p_z$ orbital. Though all the N–C bonds are coplanar, the steric hindrance of three phenyl rings makes them twisted to give TPA, a "propeller"-shaped molecule as shown in Fig. 7.1. The propeller-shaped structure prevents intermolecular aggregation and leads to the formation of isotropic amorphous films, an essential requirement for a molecule to be used in organic electronic devices (Zhang et al. 2018). The N lone-pair electron being delocalized over three phenyl groups resulted in the phenyl group acting as an acceptor bearing a positive charge on the nitrogen with a negative charge on the phenyl ring as depicted in Fig. 7.1. Notably, all three phenyl groups have an equal probability to bear the negative charge, which is highly influenced by the substituents present in the phenyl ring and will be discussed later. The delocalization of N lone-pair prevents nitrogen protonation and TPA is generally a nonbasic system. Derivatization of TPA is achieved through the substitution of one or more phenyl rings in *ortho*, *meta*, and *para* positions. The impact of the substitution on the electronic properties of the TPA core could be understood by analyzing the electronic structure of isolated TPA. The valency photoelectron spectra of TPA measured in the gas phase and its comparison with the theoretical total and partial density of states reveals that the highest occupied molecular orbital (HOMO) of TPA has about 33.34% contribution from p_z orbital of the N atom (Zhang et al. 2018). For carbon 2p orbitals, *ortho* and *para* C, respectively, have 31.61 and 20.08%, whereas the negligible contribution from the meta carbon ($s + p$ total 2.75%) highlights that the electron donation predominantly originates from N and the substituent at *ortho* and *para* position could generate pronounced impact on optoelectronic properties when compared to the *meta* position (Zhang et al. 2018). Incidentally, the majority of the literature report on TPA derivatives belongs to *para* position as the substituents at the *ortho* position create steric repulsions and adversely affect the electron-donating property of N. Nevertheless, the strategy to use *ortho* positions in TPA has wisely been used by connecting the two phenyl rings by a methylene bridge to make azatriangulenes, which possess enhanced π-conjugation aided by the enforced coplanarity (Krug et al. 2020; Ward et al. 2019).

TPA absorbs in the UV spectral region and any utility of these compounds is only after substitution of different functional groups by extending the π-conjugation pathway. The general trend for the TPA-based chromophores is the substitution is at the *para* position and may be either one, or two, or all the phenyl rings. Notably, the number of substitutions does have a paramount role in depicting the electronic

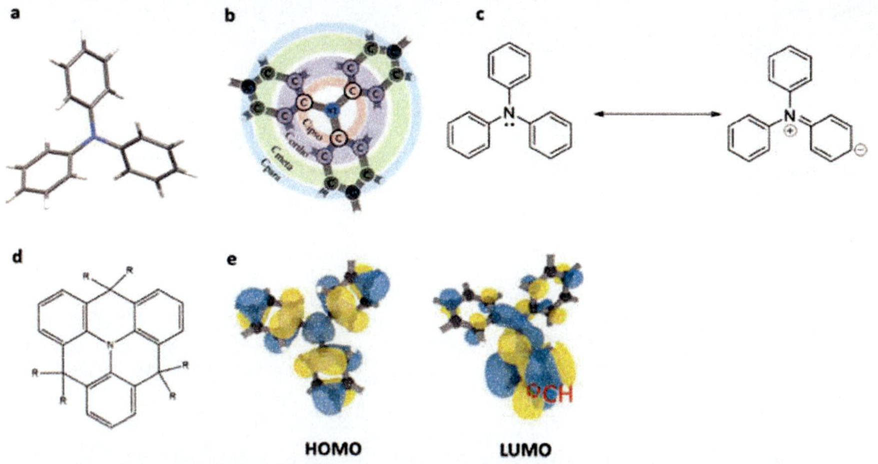

Fig. 7.1 **a** Single-crystal X-ray structure of TPA (CCDC—1,319,035), **b** For TPA, following the nomenclature of benzene-derived compounds, the carbon atoms directly bound to the N atom are denoted as Cipso. The carbon atoms binding to the *Cipso* are denoted *Cortho* followed further away from the central N atom by *Cmeta* and, outermost, by *Cpara*, **c** Resonant electronic structures for the N lone-pair electrons of TPA, **d** Structure of azatriangulenes with fused phenyl rings, and **e** HOMO and LUMO of TPA (**a**-adopted with permission from International Union of Crystallography, **b** and **d** are adopted with permission from reference 1, Copyright American Chemical Society)

properties. To simplify, consider TPA as a donor motif (D) and electron-deficient systems as acceptor (A) motifs. If one of the phenyl rings is substituted with the acceptor motif, the resultant configuration would be the simple donor–acceptor (D-π-A) configuration. It is a well-known fact that the existence of electron-donating and -withdrawing groups connected through intervening π-conjugation always resulted in intramolecular charge transfer process (ICT) to form a charge transferred state, D^+-π-A^-. This process accompanies the changes in the dipole moment between the ground (μ_g) and the excited state (μ_e), experimentally detected using solvent polarity-dependent spectral shift in fluorescence, fluorescence quantum yield, and lifetime measurements. Considering the molecular symmetry of TPA at the ground and excited states as shown in Fig. 7.2, the C_2 rotational symmetry axis passes through one of the phenyl's *para* positions, and C–N bond remains intact. It has been found that TPA with more than one acceptor unit improves the nonlinear optical properties and solar cell efficiencies, thus leading to new molecular designs with diverse structural and functional properties. Those molecules could be classified based on the number of acceptor groups present in the system as shown in Fig. 7.2. For instance, the simplest representation of such configurations A-π-D-π-A and A-π-D(-π-A)$_2$, respectively, denoted as quadrupolar and octupolar chromophores and generally called multipolar chromophores. A well-defined, molecularly engineered multipolar or multichromophoric arrays poses phenomenal improvement in photonic, photovoltaic, and photocatalytic processes. A fundamental understanding of the processes that happens upon photoexcitation such as charge transfer, exciton

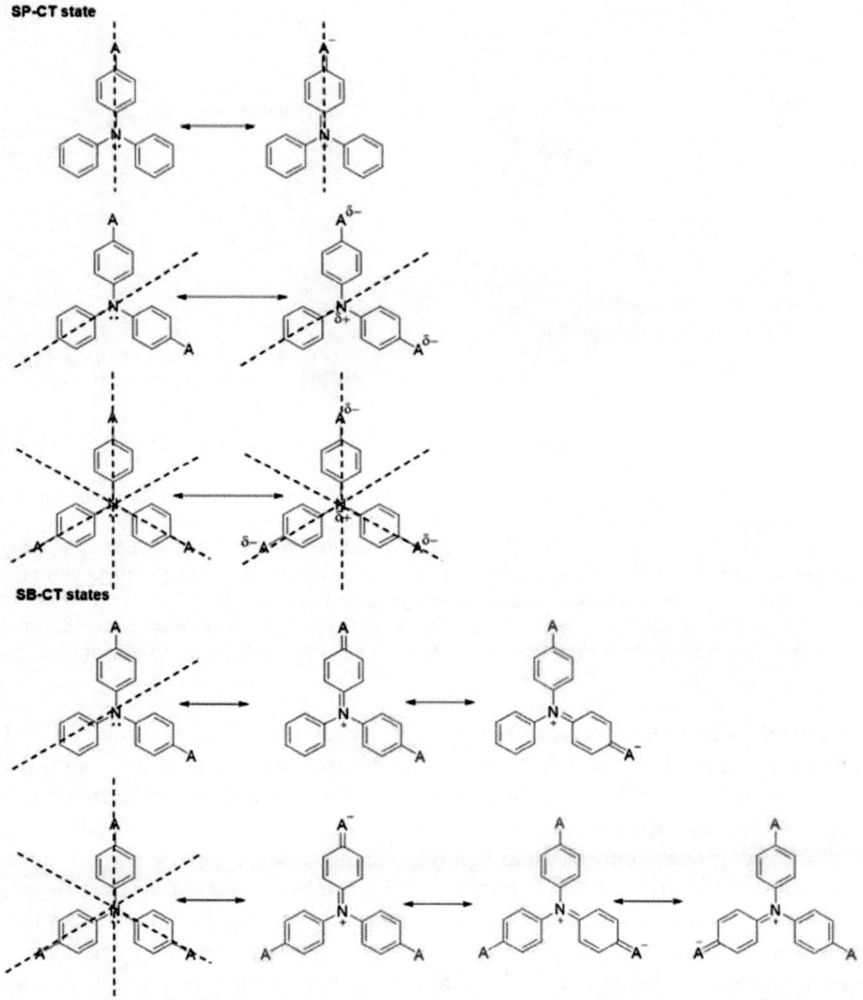

Fig. 7.2 Structure of TPA with different acceptor groups. The dotted line indicates the possible C_2 rotational axis of TPA. The phenyl rings marked in blue are the spectator groups, which are not part of TPA. (SP-CT Symmetry protected CT state; SB-CT Symmetry broken CT state)

localization or delocalization, and impact of the surrounding environment is essential to realize the potential of these kinds of molecules. One of the crucial parameters for their efficiency is highlighted as intramolecular charge transfer processes, nature and stability of the charge transferred state and (de)localization of the charges are considered to be crucial factors.

As discussed previously, the ICT formation of dipolar systems is straightforward, can experimentally be measured, and quantified with various spectroscopic techniques. But for the quadrupolar and octupolar molecules, there exists more than one pathway wherein the entire multichromophoric array might be involved in the charge transfer process in whole or part. For instance, the quadrupolar system with A-π-D-π-A configuration would form $^{\delta-}$A-π-D^{+}-π-A$^{\delta-}$ having both the acceptor involved in ICT with the charge being shared equally. This ICT leads to the dipole moment changes equally by both the acceptors, but in opposite direction, hence the net dipole moment change would be zero. Similarly, for octupolar system, one of the probable ICT states would bear partial negative charges on all the acceptors, thus lead to the no or negligible dipole moment change with respect to ground state upon photoexcitation. These charge transferred states are, respectively, called quadrupolar and octupolar CT states, which could be understood experimentally by observing no solvatochromism in fluorescence due to the absence of ground and excited state dipole moment change. Importantly, the molecular symmetry axis of TPA in the ground state remains intact even at the excited state; hence, it is represented as symmetry preserved CT state (SP-CT). On the other hand, upon photoexcitation the CT process occurs between one of the donor–acceptor pairs and the remaining acceptor would be the spectator group and would not be part of the CT state, phenyl rings marked in blue in Fig. 7.2. Therefore, the radiative excited state deactivation processes might originate from the sub-chromophoric units and not from the entire molecule. Indeed, there is minimum of one acceptor that would not be involved in the CT state but have an equal probability to become part of the sub-chromophore upon excitation. Hence, a minimum of two and three canonicals, CT states are possible, respectively, for quadrupolar and octupolar molecules and all of them are isoenergetic and have equal probability to be part of the CT state. It is understandable that these CT states accompany the dipole moment change and can be experimentally determined by the solvatofluorochromism. A comparison between the rotational symmetry of the ground and the CT states shown in Fig. 7.2 implies that the original symmetry has been reduced after ICT processes. For instance, for quadrupolar molecules, the C_2 rotational axis passing through C–H of unsubstituted phenyl and N–C bond in the ground state disappears after CT state formation, as one of the acceptors is identical with the ground state, while the other one forms quinonoid kind of CT state. Thus, the original symmetry is broken, and the CT states formed by this method are called symmetry broken charge transferred state (SB-CT). A careful evaluation of the light absorption and subsequent excited state processes would be informative to unravel the formation of SB-CT states. Absorption of photon happens in the femtosecond time scale and the charge delocalization is believed to be throughout the molecule. Then the molecules undergo faster relaxation to form the localization of exciton on one of the arms and the fast hopping of excitation energy between the arms could lead to a faster relaxation process than the dipolar analogous compounds. These processes are expected to be faster than the solvent relaxation. On the other hand, solvent relaxation would also lead to the localization of the CT state on one of the branches, thus undergoing SB. Several attempts were made by different research groups to spectroscopically identify the SB process using the femtosecond transient absorption measurements. While the electronic transient absorption spectroscopy does not provide any conclusive evidence, the vibrational spectroscopy in the timescale of femtosecond unraveled the reaction course in polar and nonpolar solvent environments.

The beauty of the TPA could be form D–A compounds in dipolar, quadrupolar, and octupolar configurations, thus offering to establish a systematic structure–property relationship. One of the important features that molecules with ICT character offer is polarization of charge from one end to other end of the molecules, which might result in the atoms adjusting to a newer environment of electron density distribution, i.e., alters the bond length. Unfortunately, the excited state process being the dynamic process, experiments were carried out in the solution state, and experimental identification became tedious in the absence of suitable time-resolved spectroscopic techniques. But we believe that analyzing the single crystal X-structures of dipolar, quadrupolar, and octupolar chromophores would offer preliminary guidelines about the molecular properties. One such important property is the bond length alternation (BLA) values, the difference between longer and shorter bonds of a carbon chain which is significantly influenced by hybridization between the orbitals of separate systems and charge transfer (Rusznyák et al. 2005). The BLA, δr values are calculated using the formula given in Eq. 7.1, and the typical bond lengths a, a', b, b', c, c' are denoted in Fig. 7.3.

$$\mathrm{dr} = \left\{\frac{(a+a')}{2} - \frac{(b+b')}{2}\right\} + \left\{\frac{(c+c)}{2} - \frac{(b+b')}{2}\right\} \tag{7.1}$$

The best example is benzene, where all the C–C bond length is 1.39 Å, which lies between the single and double bond lengths of 1.54 Å and 1. 34 Å, respectively. The δr value of 0 for benzene indicates all C–C bonds are equal and any substitution that would perturb the bond length alternation δr to become nonzero, for instance, δr value of 0.08–0.10 is an indication for the full quinonoid form which can be correlated to the extent of ground state charge transfer in the molecule (Thamaraiselvi et al. 2020). A reasonable approximation could be arrived for the relationship between BLA and electronic properties of the molecule which are useful in accessing the nonlinear optical (NLO) properties, second-order polarizability β or third-order

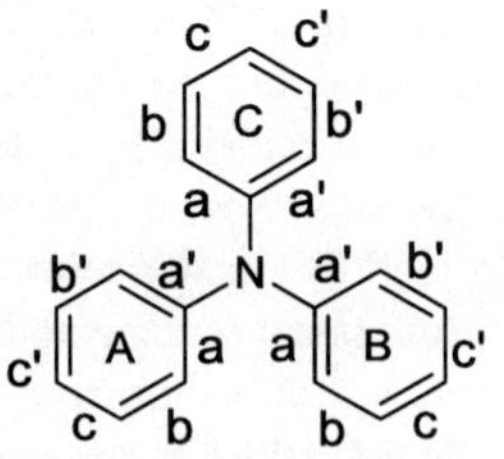

Fig. 7.3 Notation of bonds which are used to calculate the bond length alternation values. Rings "A", "B", and "C" denote the individual phenyl, wherein the mono-substituted TPA, "A" phenyl will be the one with substituent, for disubstituted, the two phenyl groups with substituent are denoted as "A" and "B"; "A", "B", and "C" are equal for tri-substituted TPA

polarizability γ. However, there are concerns raised by some studies that BLA might mislead in accessing NLO properties as the values obtained from crystal data do not include the environmental effects such as solvent–solute, solute–solute interactions, etc. Hence, it has been suggested to simulate the same environmental conditions used to study the NLO properties and BLA calculation, thus most of the studies are based on the computational methods. Nevertheless, values from crystal data are realistic and provide valuable information on the relationship between BLA and electronic properties, thus the BLA values are calculated from single-crystal X-ray structures retrieved from the literatures and Cambridge Crystallographic Data Centre (CCDC). The references are initially retrieved from SciFinder using the keywords "triphenylamine" and refined with "crystal structure". The most relevant structures and the BLA values along with the respective CCDC number are summarized in Table 7.1.

A careful analysis of the BLA listed for TPA in Table 7.1 highlights that even the simple TPA exhibit non-zero BLA with slightly different values for all the three phenyl rings A, B, and C, which would have originated from the propeller-shaped TPA with N p_z would have been aligned with one of the phenyl rings. The *para* substitution of cyano or CHO group in one of the phenyl rings leads δr value significantly different for all the phenyl rings, but complete removal of the *para* hydrogen with the CN leads to a closer δr value of all the phenyl rings. Over several structures analyzed, no conclusive pattern has been arrived in terms of BLA in many cases. For instance, in some of the molecules, the phenyl-bearing acceptor moiety shows a lower BLA value compared to the other, particularly when the substituents comprise of aromatic ring linked through extended π-conjugation. Another observation has been that for TPA with one of the phenyls having a substituent, the remaining two unsubstituted phenyls do have different BLA values, but higher than one with the substituent. This feature would have arisen from the extended conjugation between the phenyl and the substituent. Nevertheless, no systematic relation with respect to BLA and ground state charge transfer character is arrived, the perturbations in δr values are an indication of the influence of the substituent and subsequently the electronic properties.

Unfortunately, the absence of crystal structures of many of the structurally related TPA derivatives renders it difficult in deducing the structure–property relationship based on BLA. Thus, many groups attempted to obtain the BLA values based on the computational methods, though the computed values were seemingly overestimated compared to their corresponding crystal data, employing same methodology to obtain δr value for all the molecules is reasonable for rationalizing the ground state charge transfer character. Accordingly, a systematic approach of increasing the number of acceptor groups to have molecules in dipolar, quadrupolar, and octupolar configurations, revealed that dipolar analogous have higher BLA compared to quadrupolar and octupolar configurations. This feature suggests the existence of more quinonoid character in the dipolar molecule than the multipolar molecules in the ground state. Hence, if quadrupolar and octupolar chromophores undergo SB to form dipolar CT state, the nature, and stability of the state may not be the same as that of their simple dipolar analogues. Because the acceptors which are not part of the SB-CT state would try to destabilize it and consequently impact the stability, which could be understood from the electronic and photophysical properties. The electronic energy

Table 7.1 Bond length alternation values calculated from single crystal X-ray structures of TPA derivatives accessed from CCDC. The values are obtained using Eq. 7.1 and the notation is indicated in Fig. 7.3. The last column indicates the Cambridge Crystallographic Data Centre (CCDC) number of the compound

S.No.	Substituent			δr (Å)			CCDC No
	R_1	R_2	R_3	A	B	C	
1	H	H	H	0.0105	0.007	0.009	1,319,035
2	*p*-CN	H	H	0.0055	0.0165	0.013	963,776
3	*p*-CN	*p*-CN	*p*-CN	0.004	0.003	0.004	963,777
4	*p*-CHO	H	H	0.0105	0.0095	0.0115	994,493
5	p-NO2	H	H	−0.0195	−0.011	−0.017	1,812,891
6	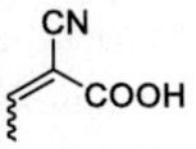	H	H	0.001	0.0095	0.0185	814,763
7	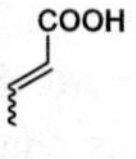	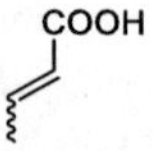	H	0.0015	0.0015	0.016	1,440,441
8	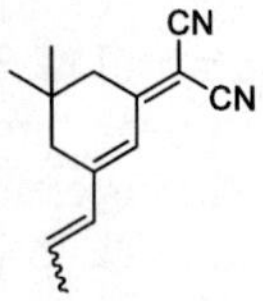	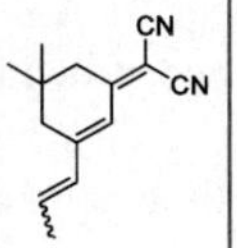	H	0.017	0	0.009	1,040,856
9	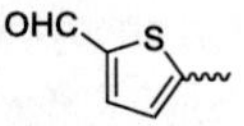	p-Br	p-Br	0.018	0.005	0.0075	1,956,541
10	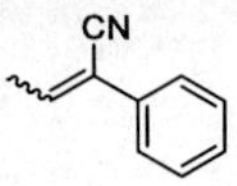	H	H	0.0065	0.0065	0.01	968,916
11	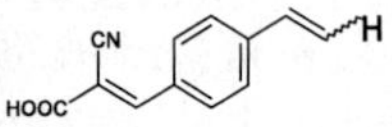	H	H	0.0025	0.013	0.0155	793,942
12	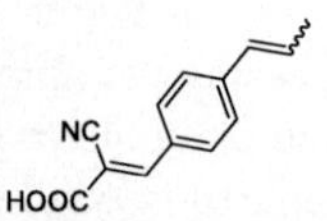	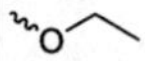	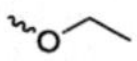	0.008	0.0025	0	814,762
13	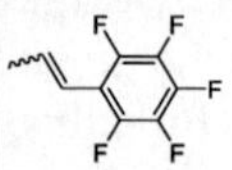	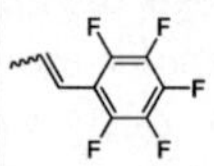	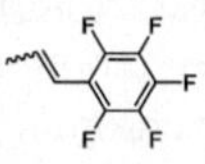	0.003	0.0025	0.0065	833,959

(continued)

Table 7.1 (continued)

S.No.	Substituent			δr (Å)			CCDC No
	R_1	R_2	R_3	A	B	C	
14	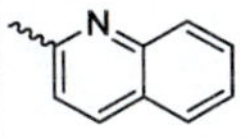	H	H	0.005	0.011	0.0095	1,476,946
15	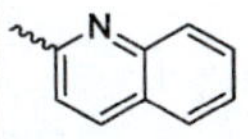	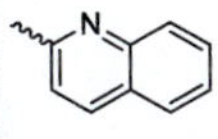	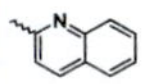	0	0.002	0.009	1,476,949
16	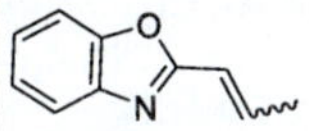	H	H	0.005	0.0055	0.0115	973,275

level of TPA could be tuned with the different electron-deficient substituents such as NO_2, CHO, CN, and aromatic motifs as well as the number of the substituents. A series of dipolar, quadrupolar, and octupolar TPA with NO_2, CHO, and CN as shown in Fig. 7.4 exerts solvent polarity-dependent fluorescence, a characteristic feature of dipolar or SB-CT state, but the magnitude of spectral shift was found to be dipolar > quadrupolar > octupolar. A quantitative correlation using the Lippert–Mataga plot of Stokes shift versus solvent polarity parameter, orientation polarizability wherein the larger the Stokes shift or slope value is an indication of higher CT character. Although no systematic relationship has been derived, it can be assumed that significant differences among the structurally similar dipolar CT state, from the higher slope value for dipolar than their respective quadrupolar and octupolar counterparts suggests, the significant difference among the identical dipolar CT states. Among the substituents, the one with a strong electron-withdrawing nature induces more CT character, a possible structure–property relationship can be derived using the Hammett substituent constant (σ) and slope value of the Lippert–Mataga plot as shown in Fig. 7.5 (Thamaraiselvi et al. 2020; Easwaramoorthi et al. 2014).

A mechanistic understanding of the possible excited state events for the quadrupolar and octupolar TPA derivatives is essential to unravel the ICT behavior of multipolar chromophores. As discussed previously, an essential state model for dipolar and octupolar chromophores, respectively, have two and three isoenergetic dipolar CT states with equal probabilities for their formation. The potential energy diagram of the electronic states depicted in Fig. 7.5b indicates that the driving force for the interconversion between the one canonical form to the other would be N–C bond rotation of the phenyl ring to reach coplanarity with the N p_z orbital, subsequently, it destabilizes the existing localized exciton on one of the arms, thus moving the exciton to another arm. Any factors that influence the rotational motion of N–C bond such as solution viscosity, temperature, solvent polarity, and local environment of the multipolar molecule are critical in depicting the stability of the CT state, hence could be used as a probe to monitor several phenomena in biological and

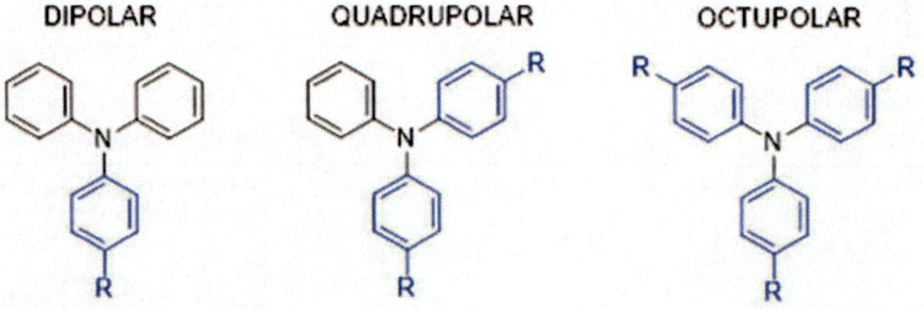

DIPOLAR	QUADRUPOLAR	OCTUPOLAR
1 R= H	**1** R= H	**1** R= H
2 R= CN	**3** R= CN	**4** R= CN
5 R= NO_2	**6** R= NO_2	**7** R= NO_2
8 R= CHO	**9** R= CHO	**10** R= CHO

Fig. 7.4 Structure of TPA derivatives in dipolar, quadrupolar, and octupolar donor–acceptor configuration

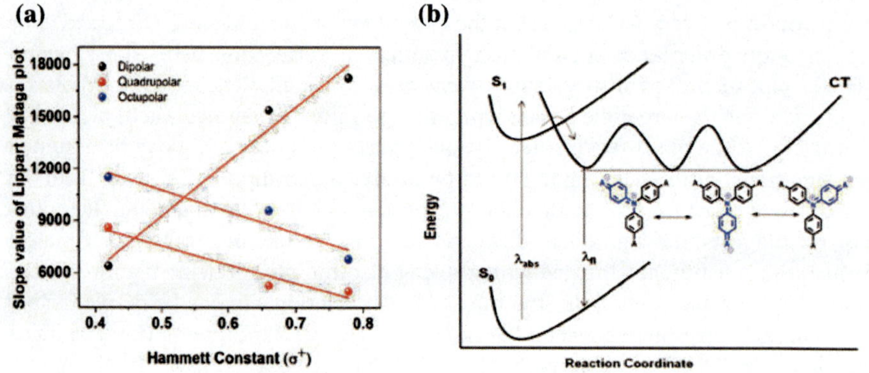

Fig. 7.5 **a** A plot of Hammett constant versus slope value of the Lippert–Mataga plot of dipolar, quadrupolar, and octupolar TPA derivatives **b** Potential energy diagram of feasible dipolar CT states of octupolar (DA_3) TPA

environmental samples. For example, N–C rotation slows down with increased solution viscosity thereby destabilization of SB-CT states of quadrupolar and octupolar chromophores is minimized, thus leads to enhanced fluorescence intensity. Unfortunately, all the SB-CT states are isoenergetic no discrimination between the states and also interconversion between them could not be probed. Solvent polarity plays an important role in the excited state of the SB process. The absorption process in any solvent occurs along the vertical excitation from the ground state having all the coordinates fixed at the equilibrium position, thus absorption spectra are not affected by the excited state symmetry breaking. On the other hand, the steady-state fluorescence originates from the relaxed excited state which will be localized on one of the SB-CT states in polar solvents (Terenziani et al. 2006, 2008). And, an increase in solvent polarity further lowers the energy as understood from the strong solvatochromism in fluorescence. In nonpolar solvents, the ratio between the height of the barrier separating the two or three equivalent minima of SB-CT and rotational motions of N–C decides the chromophore's behavior, higher barrier of the system would be localized on one of the arms with the negligible tunneling, and symmetry is broken effectively. If the barrier is low with respect to the N–C rotational motion, tunneling is fast and effectively restores original symmetry. Both the processes could have different spectroscopic signatures, and as for as TPA derivatives are concerned there is symmetry breaking even in the nonpolar solvent as understood from the linear relationship with the orientation polarizability of the solvent, pointing out the changes in the excited state dipole moments.

7.2 Sensors

Developing molecules for chemical sensing is a timeless allure. New methods empowered by the novel materials or molecules to work in simple test tubes to hand-held devices are definitely the disruptive technologies to get away from the expensive instrumentation and can be handled by anyone. The most important part of any sensor is the sampling, but unfortunately, predominant studies report the sampling method deviates from the practical situation and then renders the practical implementation. Nevertheless, those studies unravel the fundamental molecular engineering aspects of sensor design, thus aiding the improvements in the workability in realistic conditions, sensitivity, and selectivity. Since, predominant contaminants or biomarkers exists at ultra-low concentration, higher sensitivity with stronger signal are an intrinsic advantage. Chemical sensors always encounter the cross-reactivity with other species, but recent developments highlight the unique pattern being observed for the cross-reactions thus still making them employable and also aiding the construction of sensor arrays aiming multiple targets on a single platform. On material aspects, the motifs which exhibit excellent responsiveness in terms of redox, colorimetric, and fluorescence properties which reflect the proximity changes around it have been used as a transducer. TPA as a building block is not exceptional in this regard. Diverse structural features inclusive of small molecules to framework materials were explored for sensing from ionic species to neutral molecules.

The tuneable optoelectronic properties of TPA with judicious choice of the substituent are promising features toward sensor applications. The generic molecular design for the TPA-based sensor is given in Fig. 7.6, wherein the receptor unit specific for the particular analyte is linked to the core covalently through π-conjugation. Modulations in the electronic properties after analyte binding reflected

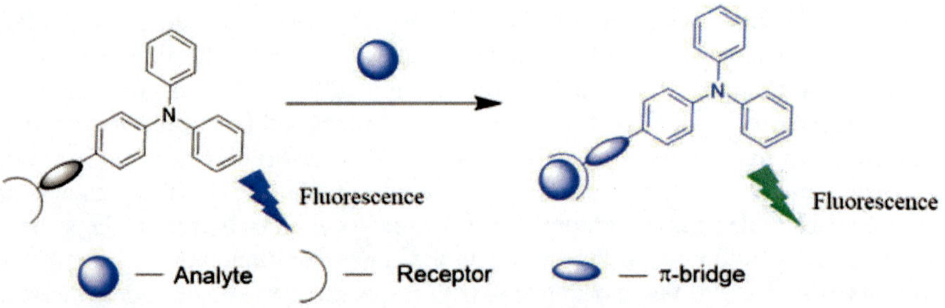

Fig. 7.6 A generic TPA derivative as a sensor with the receptor for analyte binding. Sensing could be detected by colorimetric, fluorescence, and redox properties changes

in terms of changes in color, fluorescence, and redox behavior act as a signaling pathway. Difluoroboron β-diketonate substituted triphenylamine complexes 11 and 12 as shown in Fig. 7.7 exerts intense red luminescence and forms one-dimensional nanofibril by adjusting the concentration, solvent/non-solvent ratio, and temperature. The emission becomes quenched instantaneously and selectively when exposed to gaseous amines such as n-butylamine, dibutylamine, tributylamine, triethylamine, cyclohexylamine, hydrazine, aniline, and N,N-dimethylaniline. While a complete quenching was observed for 11, the intense red emission becomes yellow for 12 in the presence of amines. Mechanistically, in both the cases amines interact with the boron through B…N interaction that leads to quenching, disturbing the interaction by purging the air leads to fluorescence recovery, thus it could be used multiple times. The higher sensitivity and low response time were attributed to the higher surface-to-volume ratio of nanofibrils (Zhai et al. 2016; Zhang et al. 2012). Another systematic approach by Cheng et al. by introducing one, two, and three benzothiadiazole rings on the TPA to afford 13, 14, and 15 the intensely fluorescent compounds which also undergo AIEE to form emissive aggregate in a thin film state (Shi et al. 2012). These probes are sensitive to aniline vapor by turn-off fluorescence by virtue of photoinduced electron transfer and the quenching efficiency is seemingly impacted by the number of benzothiadiazole units. Mechanistically, the HOMO level of 13, 14, and 15 are slightly lower than the aniline, thus the photoinduced electron transfer from aniline to the photoexcited probes leads to fluorescence quenching. On the other hand, the effect of quenching by 4-methylaniline is comparatively less, owing to its higher HOMO energy level and also other aliphatic amines have lower HOMO levels than the probes, thus do not quench the fluorescence from the compounds 13, 14, and 15. Interestingly, sensitivity towards aniline was poor in thin film state attributed to the lower contact area and inferior vapor penetration. A morphological control with aligned ZnO nanorods along with the thin film of 13, 14, and 15 eliminates the aggregation and enhances the quenching efficiency owing to high surface-to-volume ratio and 3D morphology. TPA motif could be used as a central structural core or terminal donor, a tripodal chromophore 16 has been synthesized using triaminoguanidinium cation as a central unit and the TPA has been linked through imine linkage (Xu and Wang 2021). 16 form an organogel through intermolecular π–π interaction and is

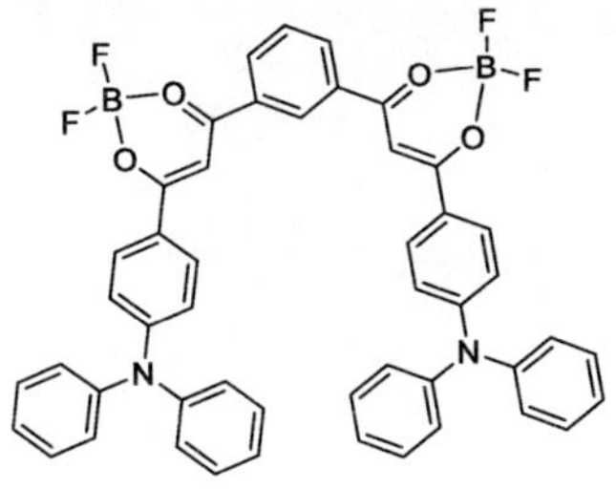

11

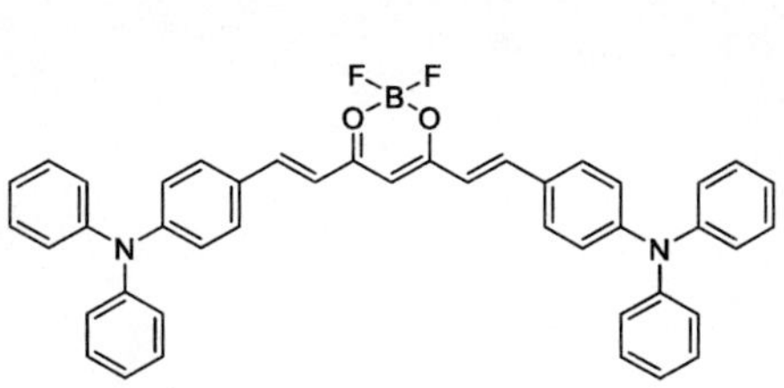

12

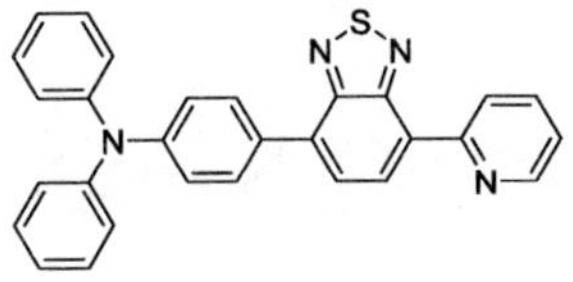

13

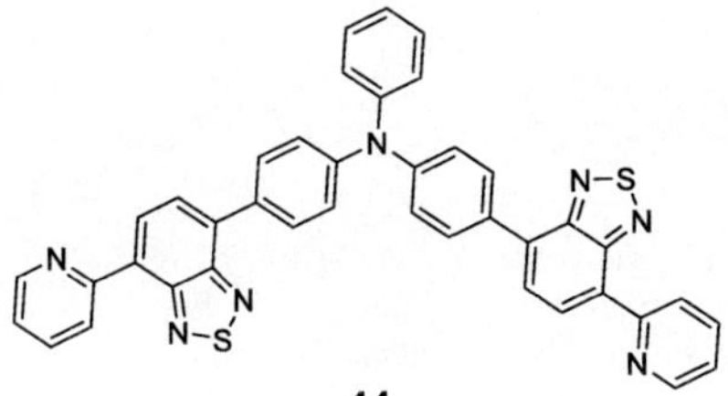

14

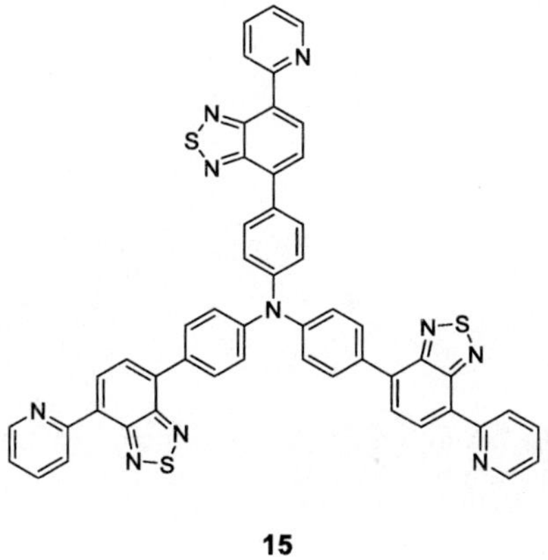

15

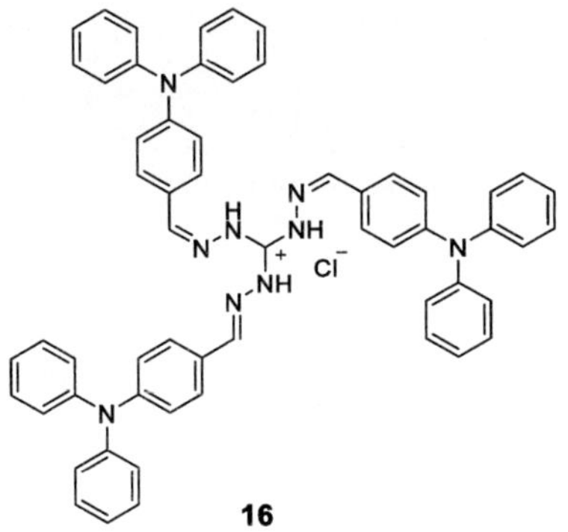

16

Fig. 7.7 Structure of TPA derivatives used for sensing different amines

responsive to Cu^{2+} ions and the gel becomes unstable in the presence of copper ions while heating, but other metal ions do not have any impact. The sensor also detects triethylamine and trifluoroacetic acid reversibly through both color change as well as reversible gel–sol phase transition.

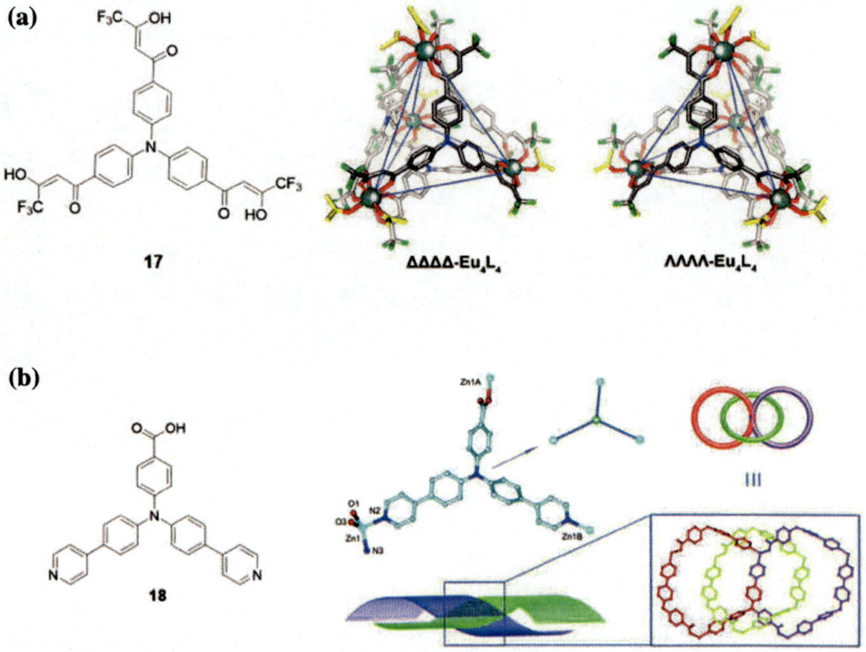

Fig. 7.8 **a** Crystallographic structures of ΔΔΔΔ-Eu_4L_4 (left) and ΛΛΛΛ-Eu_4L_4 (right). The structure of the ligand L is also given. **b** Asymmetric unit of 18 (symmetry codes: A = 1.5 − x, − y, 0.5 + z; B = 1.5 − x, 1 − y, 0.5 + z). Insert shows the 3-connected bpba linker. The hydrogen atoms are omitted for clarity. **b** Schematic of the [2 + 2] polycatenated structure in 1. Inset: highlighting the interlocked hexagonal loop by a pair of edge-shared rings. (The crystal structures in "a" and "b" are adopted from references 13 and 14 with permission, Copyright The American Chemical Society (**a**) and Royal Society of Chemistry)

Inorganic–organic hybrid materials with rare earth metal ions as a coordination center with the functionalized TPA ligands generate the three-dimensional architectures with C_2- or C_3-symmetric ligands. The higher order structures create the porous architecture with the recognition or reaction sites to interact with small molecules, thus act as a sensor for several molecules. The selectivity, in this case, has been due to the pore size of the framework materials, reactivity toward a specific target, and coordination ability of the target with the metal center. In this line, TPA substituted with the 4,4′,4″-Tri(4,4,4-trifluoro-1,3-dioxobutyl)triphenylamine, 17 forms Eu_4L_4 tetrahedral cage having self-assembled by four Eu(III) ions as vertices, and four C_3-symmetric tris-β-diketones as faces as shown in Fig. 7.8 (Yao et al. 2020). The β– diketones with CF_3 make them acceptor units, and TPA is the donor unit, thus four exerts fluorescence originated from the ICT interactions. A spin-coated thin film of 17 exerts a turn-on fluorescence response in the presence of volatile amines owing to the weak nucleophilic interaction and the selectivity and reversibility arise from the ICT character of the ligand and cage-like structure. A two-dimensional, inorganic–organic hybrid system comprises of zinc and 4-(bis(4-(pyridin-4-yl)phenyl)amino)benzoic acid (18), which features 2D + 2D parallel polycatenation structure shown in Fig. 7.8 was developed through the mild hydrothermal process by Zheng et al. (Ji et al. 2018). Photoluminescence of the 2D structure is intense when compared to the free ligand, when it is exposed to the vapor phase amines, the fluorescence is quenched with higher selectivity achieved for 4-aminophenol. The detection limit was down to 0.10 ppm, with a rapid response time of 30 s. Indeed, this real-time vapor sensor could be regenerated by simply heating the sensor at 60 °C in a vacuum oven for a minute, thus making them employable for continuous, real-time monitoring.

Nitroaromatics are a class of organic compounds having an extensive usage as explosives and their widespread deployment necessitates the need for the portable sensor, which is otherwise detected using non-field deployable, X-ray scanning and ion mobility spectrometry. Thus, several chromophoric units have been deployed for sensing the various nitroaromatics in different physical states using colorimetric, fluorescence, and redox techniques. Vamvounis et al. synthesized triarylamine cored dendrimers having phenyl branching unit with thiophene, 19 and bithiophene, 20 as dendrons and dodecyl alkyl chain as surface groups (Vamvounis et al. 2013). The structures of the dendrimer are given in Fig. 7.9. It has been demonstrated that the emissive species could be altered from core to dendron, simply by replacing the thiophene with the bithiophene. The sensor works as turn-off sensors as the fluorescence quenching of trinitrotoluene and its analogous 4-nitrotoluene, 2,4-dinitrotoluene, 1,4-dinitrobenzene and the plastic explosives 2,3-dimethyl-2,3-dinitrobutane. Interestingly, when the emissive species is localized on the center, i.e., TPA core, the sensing ability of nitroaromatics by 19 is slightly less than the tribiphenylamine, suggesting that the dendrons on the dendrimer 19 are affecting the interaction between the sensor and analyte. On the contrary, the quenching efficiency of dendrimer, 20 is comparable with its dendron revealing that chromophore–analyte interactions are not affected when the emissive species is on the dendrimer surface. The higher order, organized, hybrid structures with TPA as a structural building block have been demonstrated to be a better framework material for sensing applications, including nitroaromatics as analytes. The tripodal, TPA derivative as ligand aids the construction of metal–organic frameworks (MOF) with metals such as zinc, lanthanum, etc., and the resultant MOFs are having unique porous architecture dictated by the ligand as well as metal ion. Tricarboxytriphenylamine, 21 is versatile in constructing various framework structures in combination with other ligands and their stoichiometry. For instance, zinc and pimelic acid (pim) in $\{[Zn(\mathbf{21})pim_{0.5}]2H_2O{\cdot}NO_3\}_n$ is a two-fold interpenetrating 3D framework containing four chiral (10,3)-a subnets and $\{[Zn(21)pim]3H_2O\}_n$ is merely a 2D + 2D → 2D entangled network (Yao et al. 2019). The electronic properties of these materials arise from the intraligand charge transfer process which are susceptible to undergo photoinduced electron transfer with nitroaromatic compounds and resonance energy transfer with ofloxacin, a fluoroquinolone drug used to treat bacterial infection. Ligand 21, in combination with N, N-dimethylacetamide forms a three-dimensional MOF $[(Zn_4O)(H_2O)_2(\mathbf{21})_2]{\cdot}8DMA$ having tetragonal channels and nanosized square channels as shown in Fig. 7.9 (Xie et al. 2021). The blue emission from the MOF has been

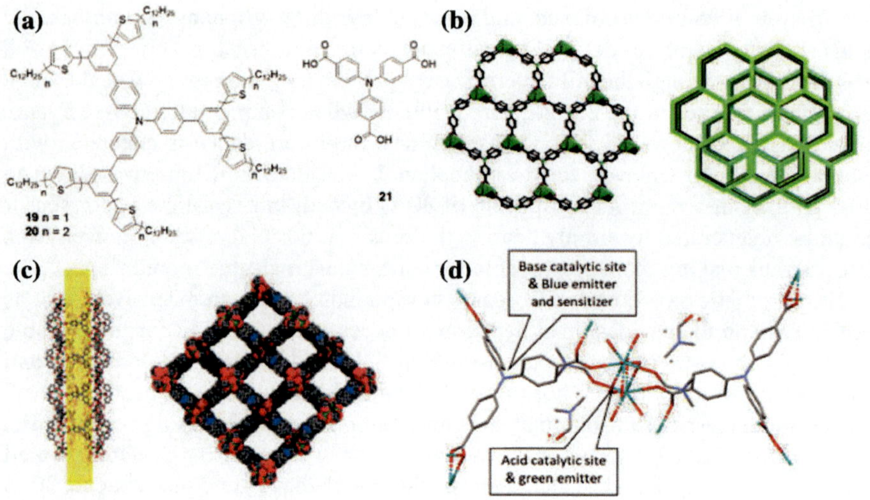

Fig. 7.9 Structures of the dendrimer 19, 20 and the ligand tricarboxytriphenylamine employed to construct the MOFs. **a** the 2D structure along the c axis and **b** the 3D supramolecular architecture of 21 in zinc-based MOF (adopted from reference 16) **c** ball-and-stick representation of a 1D square channel and the space-filling 3D representation of MOF with 21 and DMA viewed from the [101] direction (adopted from reference 17) **d** The crystal structure of terbium with ligand 21 (adopted from reference 18). The figures used with permission, Copyright Royal Society of Chemistry (**b** and **c**) and John Wiley and Sons (**d**)

quenched by the nitroaromatics, specifically picric acid has greater selectivity over its other analogous and the sensor could be regenerated that lead to the usage of multiple cycles. Unlike the previous case, the mechanism has been attributed to the energy transfer from the electron-donating framework to electron-deficient nitroaromatics. When zinc is replaced with terbium, ligand 21 forms Tb-21, MOF by solvothermal method with emission at 540 nm and 435 nm corresponding to lanthanide and TPA-based emission(Wu et al. 2011). While terbium sites act as Lewis acid and TPA sites act as Lewis bases on MOF's internal surfaces (Fig. 7.9), therefore facilitating both Knoevenagel and cyanosilylation reactions in a size-selective fashion. Further, the absorption through porous site-selective interactions between the aldehyde and Tb^{3+} ions resulted in the ratiometric fluorescence response, hence could be employed both as sensor and catalyst.

An unconventional sensing methodology, i.e., photoelectrochemical sensing for cysteine was evaluated using the basic principles of dyes-sensitized solar cells (Wu et al. 2014). The photosensitizer, three anchored on the TiO_2 surface through the carboxyl group undergoes ultrafast charge injection to the conduction band of the semiconductor to produce the photocurrent; any process that hinders the photocurrent generation could be detected instantly. The acrylic moiety is prone to undergo Michael's addition reaction thereby to form an sp^3 hybridized carbon, which would destroy the π-conjugation to the carboxylic group and decrease the photocurrent response. A reaction of cysteine with the acrylic moiety prevents the end-to-end π-conjugation in 3, thus decreasing the charge injection into the conduction band of TiO_2 and photocurrent response. Indeed, other amino acids, bithiols do not interfere in the presence of cysteine owing to the selectivity and steric factors. A systematic study using TPA-rhodanine acetic acid in dipolar (23), quadrupolar (24), and octupolar (25) configurations reported that these molecules were highly selective to detect mercury and silver ions in aqueous solution. Interestingly, for mercury, all of them form a 1:1 complex with similar binding constant values, though 24 and 25 have additional receptor sites available as further metal binding. This feature suggests that the free binding site alone will not be enough for the binding, the changes in the electronic properties of sensor–Hg^{2+} complex would have prevented further binding; thus, a negative cooperative effect was observed (Thamaraiselvi et al. 2019). The detection limit for these metal ions as in the order of ppb. Not limited to the chemical entities, fluorescence sensors for the changes in the physical parameters such as pressure and temperature have also been reported. For instance, to measure the internal temperature of any organ and biological cell, conventional thermometers could not be used. Compound 26, constructed with TPA as a central core donor and terminal donor having ethyl rhodanine as an electron acceptor, D–D–A configuration with ICT characters was developed (Ajantha et al. 2021). The intervening biphenyl configuration between the two TPA units renders the ineffective π-conjugation throughout the molecule, thus 42C has two emissive species, i.e., the terminal TPA and TPA-ethyl rhodanine moiety, wherein the former emits from its local excited state and the emission from the latter has an influence from the ICT characteristics, which cover all three primary color emissions, that is, red, green, and blue (RGB). These multiple emissions from the sub-chromophoric units undergo ratiometric changes when the sample is subjected to various temperatures. At cryogenic temperatures, the lowest energy emission maximum red-shifted from 528 to 564 nm gradually but the higher energy emission remains almost intact, thus serves as a reference point. A plot of temperature with the spectral maximum provided a sensitivity of 0.3 nm K^{-1}. Further, the ratiometric changes in the fluorescence are linear between 180 and 330 K with a temperature coefficient of 0.19 K^{-1}. Thus, 26 serves as a multiparameter temperature sensor without need for the external reference. Mechanistically, the changes in the π-conjugation due to the arrest of the C–C bond, and the subsequent impact on the ICT emissions are responsible for the temperature-dependent phenomena (Fig. 7.10).

7.3 Aggregation Induced Emission (AIE)

Most of the donor–acceptor compounds are meant for optoelectronic applications and are expected to work both in solution, solid/thin film state. Unfortunately, when the studies are extended to solid state, compounds lose their luminescent properties owing to aggregation caused quenching (ACQ) arising from strong intermolecular interaction such as π–π stacking thus limiting their applications in organic light

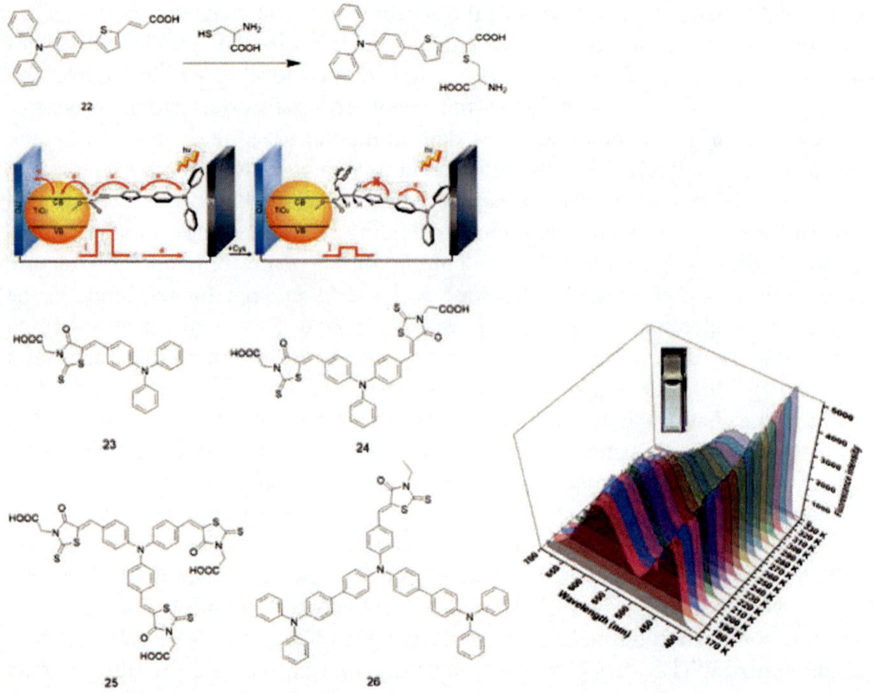

Fig. 7.10 Structures of TPA derivatives used as sensors for cysteine, mercury, silver, and temperature. The scheme at the top describes the photoelectrochemical sensor principles of 22 (Adopted from reference 19 with permission, Copyright American Chemical Society). The spectra given at the bottom are the fluorescence spectra of 26 in methyl tetrahydrofuran solvent in different temperatures (Adopted from reference 21 with permission, Copyright John Wiley and Sons)

emitting devices (OLED), organic field effect transistors (OFET), luminophores and bio-imaging. The stronger intermolecular interactions are prevalent in many chromophores as almost all of them are constructed with the planar π-conjugated aromatics as a building block. A varying degree of ACQ has been reported when they exist as aggregate or cluster. The excited state relaxation of luminophores which are closely spaced owing to the π-π interactions opens up various non-radiative decay pathways, thus ending up in emission quenching (Bünau 1970). In 2001, a new concept of aggregation-induced emission (AIE) has been evolved by Professor Tang wherein the 1-Methyl-1,2,3,4,5-pentaphenylsilole in the dilution solution exerts week emission but acquires intense emission in the aggregated or solid state (Mei et al. 1940; Luo et al. 2001). This phenomenon is exactly opposite to the prevalently known ACQ process and understandably the stronger intermolecular interactions facilitate the emission process. A brief recall of the excited state deactivation channels and the influencing factors would help to understand the basics of ACQ and AIE processes. After photoexcitation, the molecule is excited to the Franck–Condon state, according to the Franck–Condon principle the light absorption process is a vertical transition and

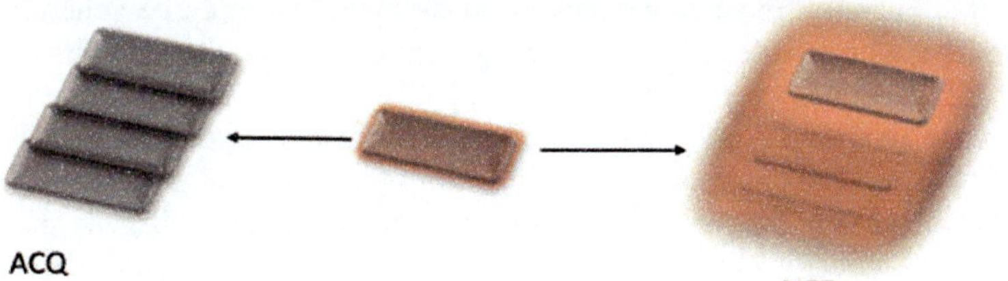

Fig. 7.11 Schematic representation of ACQ and AIEE

the molecular structure does not alter during the absorption process. Since Franck–Condon state is highly unstable, the molecule should relax to a geometry with lower energy. Various modes of structural relaxations such as E/Z isomerization, rotational motion of the single bond, twisted intramolecular charge transfer, and excited-state intramolecular proton transfer (ESIPT) are dynamic, energy consuming, and occur at the excited state would affect both radiative and non-radiative decay pathways which are highly interdependent. The larger proportion of the non-radiative decay channels leads to the molecule to be non-fluorescent or weakly fluorescent. The interplay between the radiative and non-radiative decay channels could be controlled by several external factors which could be rationalized based on the structure of the luminophore (Fig. 7.11).

In general, the AIEgens are non-planar, conformationally flexible, propeller-like conformation, highly twisted conformation which exerts dominant, dynamic, excited state structural changes having weaker or no emission in the dilute solution. The mechanistic insights for the origin of the AIE effect are broadly under three categories viz. restriction of intramolecular rotations (RIR), restriction of intramolecular motions (RIM), and restriction of intramolecular vibrations (RIV) (Mei et al. 2014). For RIM, the best example is hexaphenylsilole (HPS), a six-phenyl group decorated with the five-membered silole core through single bonds, which renders them conformationally flexible, propeller-like shape with twisted conformation, steric repulsion between the adjacent phenyl rings avoids dense packing and essentially no π–π interaction in the solids state. However, the existence of strong intermolecular C–H π interactions between the adjacent HPS synergistically constraints the HPS in the crystal lattices, thereby suppressing the rotation of the phenyl ring in solid and aggregated state to become a popular skeleton for AIEgens. Molecules such as 10,10',11,11'-tetrahydro-bi-5H-dibenzo[a,d]cycloheptene (THBA) which exerts six normal modes of vibrations becomes three in the THBA cluster, thus leading to the arrest of excited state non-radiative decay channels, leading to AIE effect owing to RIV mechanism. When RIR and RIV are mutually responsible for AIE, then it could be classified under RIM, where motion (M) comprises both vibrational and rotational motions. It should be noted that RIR and RIV pathways are not mutually exclusive, but could contribute together to restrict the intramolecular non-radiative decay channels by virtue of structural rigidification to facilitate AIE phenomena.

The requirement of molecular skeleton for the AIEgens could be generalized as the molecules which have sufficient room for the intramolecular motions to dissipate the excited state energy through non-radiative decay channels in the isolated molecule and would have the twisted structures in the aggregates without the π-π interactions. Ironically, extended π-conjugation is an essential structural requirement of almost all luminogens. Thus, molecular engineering of AIEgens should be comprised of a judicious choice of different building blocks that aid AIE without leading to ACQ. Accordingly, many AIEgens were developed by integrating AIE and ACQ motifs to cover the entire visible–NIR spectra or functionalization specific to particular processes. Apart from classical examples tetraphenyleneethylene (TPE) and HPS, one of the predominantly explored structural motifs would be the TPA for AIEgens. The excited state energy dissipating freely rotating three phenyl, propeller-shaped non-planar structure that reduces the opportunity for the intermolecular π-π interactions, provision to have D-A configuration to alter the electronic energy levels cumulatively offer the TPA as versatile building blocks to construct the AIEgens. The utility of TPA in AIEgens could be roughly classified based on the structure to be two, one of the approaches has been that it is used as a terminal group along with the other AIE motifs to alter the electronic properties and the other one would be that it plays a central role to construct the D–A chromophores in various configurations. Experimentally, the AIE phenomenon has been demonstrated by measuring the fluorescence of the probe in solvent mixtures of different proportions. Both the solvents should be miscible, but the probe should be soluble in one of the solvents and insoluble in the other solvent. The solvent which solubilizes the probe is referred to as good solvent and the insoluble solvent is called as bad solvent. Generally, when a solute particle is introduced to a solvent, the solvent molecules surround the solute and orient themselves around it, forming a solvation shell. The orientation of the solvent molecules is such that the polar or charged groups of the solvent molecules are facing the solute while the non-polar groups are facing away from it. This arrangement allows the solute to be stabilized and remain in a state of dissolution. When the proportion of the bad solvent increases in the solution, since it is miscible with the other solvent, both the solvent molecules become evenly distributed throughout the resulting solution. The solvation spheres of each solvent overlap and the solute molecules interact with both solvents simultaneously, which would lead to changes in the solubility of certain solutes. Since the probe has a different solubility in these solvents, when the proportion of the good solvent in its solvation sphere has been replaced with the bad solvent, the solubility of the probe is reduced in the solvent mixture. To adjust the reduced solubility under a mixed solvent environment, the molecules tend to aggregate to form higher order structures with distinctly different intermolecular interactions and fluorescent properties. Tetrahydrofuran and DMSO are the widely used good solvents and the water has been the bad solvent. The aggregates are predominantly in water and therefore extends their applicability in biological medium as well.

The classical AIEgen, HPS upon substitution of TPA at 2,6 position, **27** induces ICT from TPA to silole, the yellow emissive AIEgen shows the solid-state emission quantum yield of 42.5% which is non-emissive in dilute solution. The multilayer OLED devices constructed with the aggregate exhibit the highest performance and also it could act as hole transport layers (Mei et al. 2012). Introducing the electron-deficient benzothiadiazole between the TPA and HPS shifts the emission to the deep red region with a maxima of 635 nm and a solid state quantum yield of 9% (Du and Wang 2011) with the spectra spread over the NIR region. In fact, the work on another classical AIEgen, TPE, 28 has been extensively modified with TPA in diverse combinations to afford the simplest D-A chromophore to polymers with a distinctly different electronic properties. TPA and methoxy substituted TPA directly linked with the triphenylethylene, 29, 30 exerts ICT interaction owing to the D–A configuration, which are the simplest examples for substituted TPE. Both 29 and 30 are weakly fluorescent in the solution state and become strongly emissive in the aggregated and solid state with quantum yields of 98.3 and 91.1%, respectively (Lin et al. 2019). In dilute solutions, weak emission is due to the rotation of all the phenyl rings, which dissipates energy via non-radiative decay channels. Restriction of intramolecular motions in the aggregated state leads to emission enhancement. The single crystal data shown in Fig. 7.12 of 29, 30 highlights the C-H···π hydrogen bonds with a distance of 2.766 Å could exist between the protons of TPA with the TPA of neighboring molecules. These C-H···π hydrogen bonds efficiently lock the RIM, thereby responsible for the intense solid-state emission (Lin et al. 2019). An electrofluorochromic device was also constructed using 30 which shows a shorter switching response time (Zhao et al. 2015). When all the phenyl groups of TPE are replaced with TPA, the solid-state quantum yield of almost unity has been achieved. The multiple TPAs offered better charge carrier mobility in OLED devices and the emission spectra are sensitive to the solid-state packing; for instance, in the crystalline face, it emits green fluorescence at 501 nm, and their grounded powder emits greenish yellow at 531 nm. This phenomenon, referred as mechanochromism arises from the changes in the intrinsic solid-state emission as a result of stressing, deforming, or breaking of solids. The molecules 31–35 are the TPE substituted with TPA, but the torsion angle between both the units is increased owing to the biphenyl configuration which renders the relatively poor end-to-end π-conjugation. In addition, the substitution at *ortho*, *meta*, and *para* positions also has an impact on the electronic properties, and all the molecules are known to be blue emitters, except 31, where both the substituents are at *para* positions (Huang et al. 2014). A core structural change of TPE by replacing the phenyl with the heterocycle, thiophene affords 2,2-(2,2-diphenylethene-1,1-diyl)dithiophene (DPDT), which has further been functionalized with TPA through benzothiadiazole π-bridge. Molecules 36–39 pose symmetrically and asymmetrically substituted DPDT with one side donor TPA and the other side D–A component TPA/TPE with benzothiadiazole as shown in Fig. 7.13. Asymmetric structures 36 and 38 seemingly efficient to show better solid-state fluorescence quantum yields and higher AIE in comparison to their respective symmetric counterparts. The initial addition of water to the THF solution of these compounds reduces the fluorescence intensity because of ICT, after 50% water fraction, the aggregation process started. These probes could be used as two-photon and one-photon excited fluorescent probes for live cell imaging (Chang et al. 2016).

AIEgens constructed on the TPA core have diverse structural features ranging from the simple functionalization at the *para* position of the phenyl ring to extended

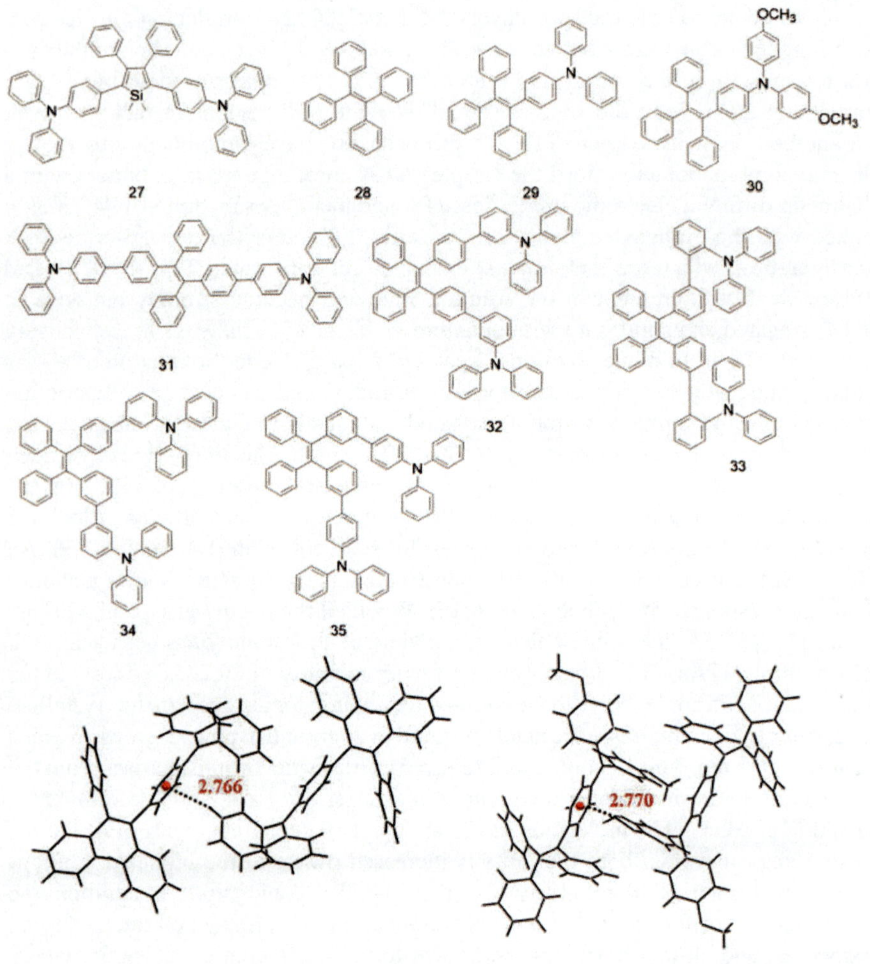

Fig. 7.12 Structures of AIEgens of TPA derivatives of classical AIEgens silole and TPE. The bottom image indicates the crystal structures of 29 and 30 demonstrating the C-H$\cdots\pi$ hydrogen bonds. (Crystal data adopted with permission from Reference 28, Copyright American Chemical Society)

conjugation with several electron-rich or -deficient functional motifs. The structural diversity offers to generate a great number of AIEgen molecular libraries with wider, visible-to-NIR spectral coverage and also its utility from bioprobes to charge transport materials in organic electronic devices. Nitro substitution in chromophores facilitates intersystem crossing (ISC) from an excited singlet state to a nonemissive triplet state thereby quenching the emission. Nitro-substituted TPA, 40, 41 are not exceptional and are weakly emissive in dilute acetonitrile solution. But, an amorphous, aggregated state in acetonitrile–water solution turns 40 and 41 into a yellow

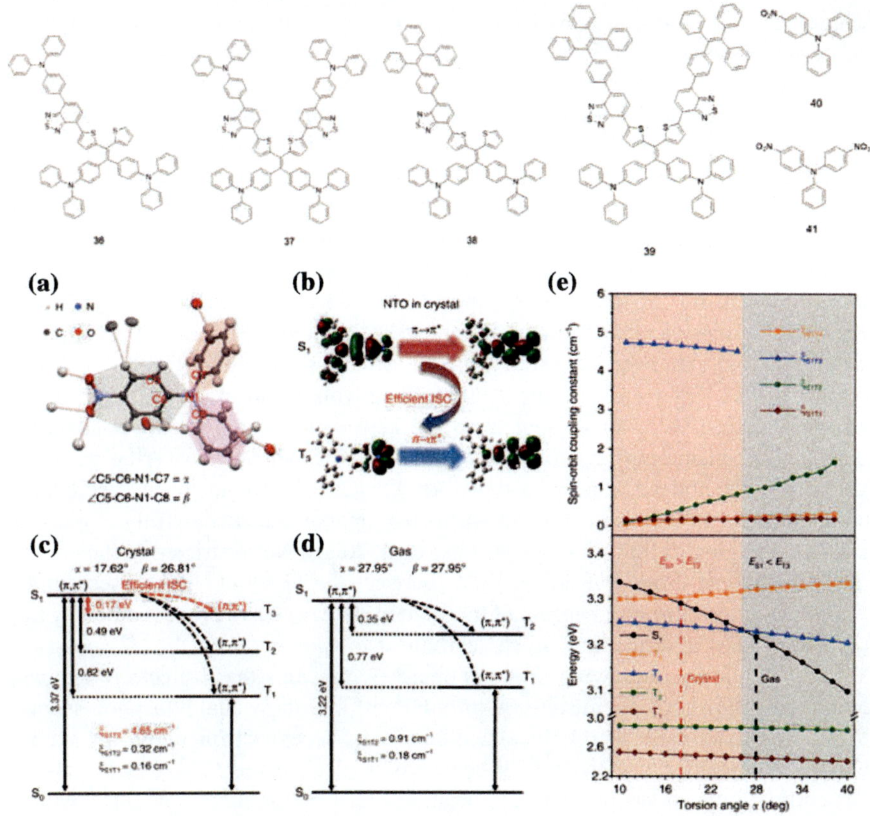

Fig. 7.13 Structures of AIEgens of TPA derivatives with benzothiadiazole and nitro-substituted TPA. Theoretical calculations on the emission mechanism of 40. **a** Crystal structure and intermolecular interactions between 40 and its adjacent molecules. The torsion angle between nitrophenyl group and other two phenyl groups is assigned to α and β, respectively. **b** Nature transition orbitals of the lowest singlet excited state (S_1) and triplet excited state (T_3) of 40 in crystal. **c, d** Calculated energy gaps, electronic configuration character, and related spin–orbit coupling constants in crystalline and free gas states. **e** Potential energy curves of S_1 and T_n and spin–orbit coupling constants (ξ) of related transitions from lowest S_0 to T_n as a function of the torsion angle α (Adopted with permission from Reference 32, Copyright Springer Nature under Creative Commons Licence)

emitting species, apparently indicating that ISC facilitated by the nitro group is overcome by the aggregation leading to the solid-state luminescence (Zhao et al. 2018). Interestingly, both the compounds are non-emissive in crystalline state as the nitrophenyl group facilitates the ISC channels to form a nonemissive triplet state. Nevertheless, mechanical stimulus manipulates the crystalline morphology and changes the molecular conformations, thus AIE effect dominates the photophysical properties to recover the emission. It has been demonstrated that the torsion angles obtained from single crystal, and the calculated gaseous and amorphous states reveal that the amorphous state as shown in Fig. 7.13 is considered to be closer to the gaseous state with similar molecular conformation. Consequently, the computed energy gap between the triplet and singlet first excited state, ΔE_{ST} value corresponds to 0.17 and 0.35 eV for crystalline and amorphous state, respectively, thus the higher ΔE_{ST} values originated from the enforced conformational change in the amorphous state become responsible for overpowering ISC with AIE. This on–off methodology offers these molecules employability as intelligent materials in depicting sensitive mechanoresponsive materials (Zhao et al. 2018). A hybrid luminogen of TPA with salicylaldehyde with other electron-donating substituents infer the effect of AIE on another important excited state deactivation process, excited state intramolecular proton transfer (ESIPT). Compounds 42 and 43 are susceptible to ESIPT but exerts weak emission characteristic of enol state. In an aggregate form, an approximately 36-fold enhanced emission with the Stokes shift value of about 140 nm was noted. The larger Stokes shift is due to the involvement of the ESIPT process and the emission being originated from the keto form, which becomes advantageous when considering its utility as bioprobes as it avoids the overlap with the excitation light. Indeed, this biocompatible AIEgen was found to be specifically staining the lipid droplets in cell in a wash-free manner and successfully employed in in vivo imaging of lipid-rich tissues (Yin et al. 2020). Not restricted to the conjugated substituents, an sp^3 hybridized linker between the TPA and acceptor seemingly provides an interesting engineering of the excited state radiative process aided by the aggregation. For instance, having sp^3 carbon or oxygen between the electron-rich TPA and electron-withdrawing groups effectively prevents the π-electron delocalization, thus the aggregate in solution or polymer films show dual phosphorescence at 77 K originated from largely localized donor and acceptor triplet emitting states; at room temperature, a merged RTP band has been observed (Wang et al. 2021). TPA-carboxylic acid, 44 prone to aggregation aided by the metal ion, Zn^{2+} which electrostatically interacts with the carboxylate to form Zn-44 complex, which further self-assemble to form vesicles which could accommodate the drugs for controlled delivery. The TPA confined in the vesicle membrane depicts, AIE along with the two-photon excited emission, thus capable to be used in theranostic applications, cell imaging, drug carriers, and metal ion detection (Wei et al. 2018).

The prominent molecular engineering approach for TPA-based DA and AIEgens is the extension of π-conjugation with aromatics as π-bridge. Structures of some of the representative AIEgens are given in Fig. 7.14. Compounds 45–47 pose different acceptor groups linked covalently through intervening phenyl bridge extending the π-electron delocalization to induce the ICT character understood by the emission maximum shifted to 668 nm and solvatochromism in fluorescence. The fluorescence quantum yield of the aggregates in THF–water solvent mixture was enhanced to 72% by the RIM mechanism. Being with DA configuration, the two-photon absorption cross section up to 592 GM has been achieved (Liu et al. 2019a). A structurally similar compound, 48 was also investigated for the detailed photophysical studies and demonstrated that a large Stokes shift of about 202 nm was observed owing to the TICT process. These molecules were found to be biocompatible, posses high brightness, selectivity in labeling the lipid droplets, and excellent photostability. Two-photon imaging of lipid droplets with higher three dimension resolution, deeper tissue

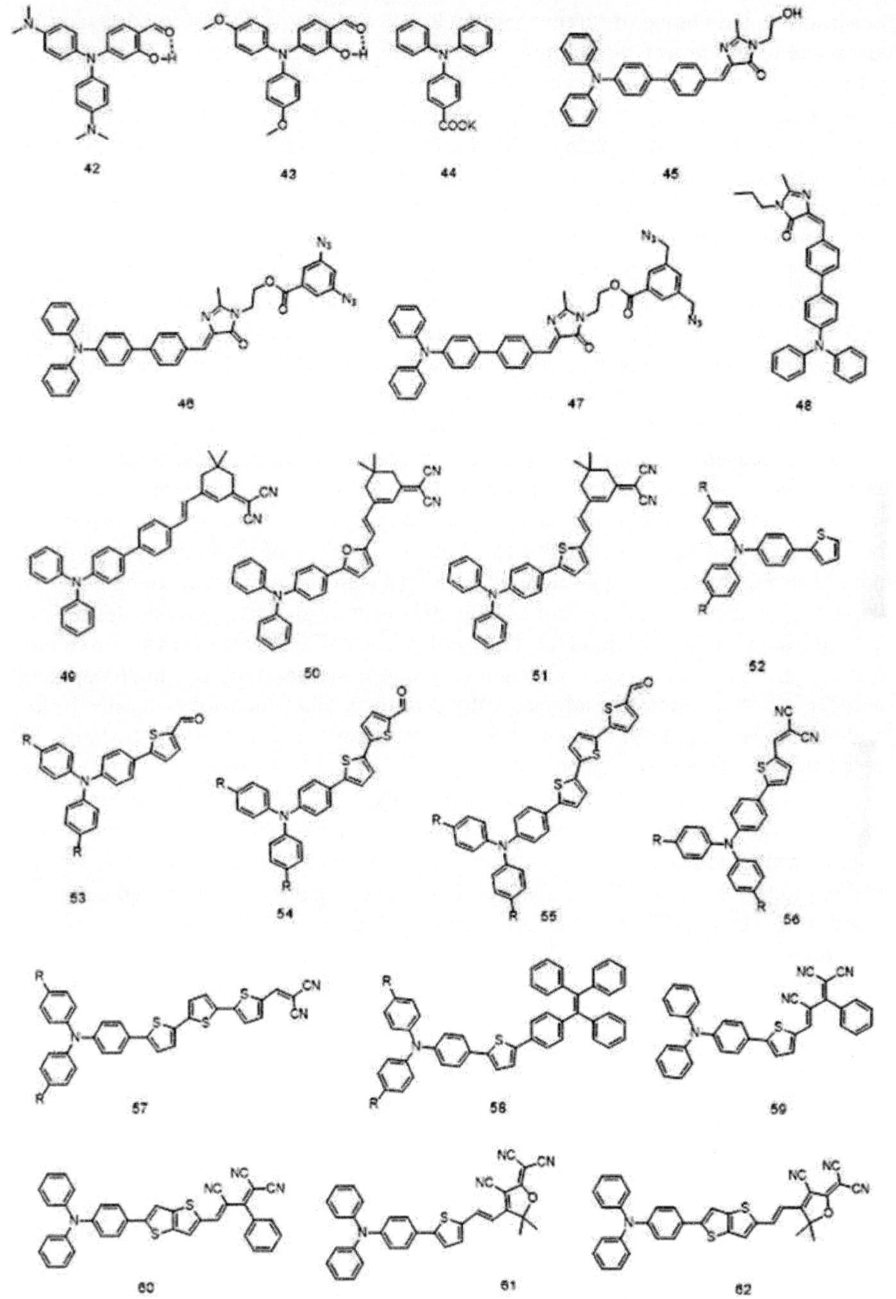

Fig. 7.14 Structure of D–A TPA compounds

penetration, and choice of monitoring the localized polarity in mixed bulk samples are some of the promising features for imaging the cell organelles or biosensing (Jiang et al. 2017). The aromatic linker between the TPA and the acceptor unit has a significant role in depicting the photophysical properties. A systematic study with the replacement of linker phenyl with thiophene and furan, 49–51 in fact alters the effective end-to-end π-delocalization. For instance, the linker phenyl ring forms a biphenyl configuration with a definite dihedral angle owing to the steric hindrance between the ortho hydrogens of the adjacent phenyl ring. But, the heterocycles thiophene and furan avoids the biphenyl configuration and facilitates the effective electron delocalization, which could be understood from red-shifted absorption from 465 to 510 nm while going from benzene to furan. These luminogens are weakly fluorescent in methanol becomes strongly emissive with red-shifted spectral maximum when in aggregate state in methanol–toluene solvent mixture, attributed to the RIM mechanism and the mixing of non-bonding electrons of S and O with the $\pi \rightarrow \pi^*$ transition, subsequent alternations in ICT interactions respectively. Near IR emission centered around 700 nm and above has been noticed for aggregate as well as in solid state that provide the superior resolution for the fluorescence imaging in biological objects (Zhang et al. 2021a). Thiophene is one of the important motifs in optoelectronic materials, a systematic variation with the number of thiophene and its impact on the AIE and electronic properties is unraveled using molecules 52–58. The emission maximum found for these molecules at 402 nm (violet) 482 nm (blue), 531 nm (green), 580 nm (yellow), 612 nm (orange), 649 nm (red), 667 nm (deep red), and 724 nm (NIR) regions, suggesting the extremely wide emission color tunability, which is ascribed to both of their varied π-conjugation and DA effect. It should be noted that increase in emission with respect to bad solvent fraction is not always linear and it depends upon the solvent combination and the excited state process of the molecules. For example, some of the molecules wield decreased fluorescence at initial stage which could be attributed to the other nonradiative excited state deactivation process such as TICT, which could also accompany the red-shifted emission maximum. Thus, TICT and AIE are competitive processes, and the dominance of both processes could be controlled externally by the surrounding environment. The single crystal also infers the twisted conformation of TPA with the intermolecular distance >3.2 Å between the two parallel planes, avoiding intermolecular π–π interactions and abundant C–H…O, C–H…C, and S…C interactions. All these factors are cumulatively responsible for the restricted intramolecular motions thereby leading to solid-state emission. These AIEgens are specifically employed for lipid droplet-specific cell imaging and also generate reactive oxygen species when exposed to white light leading to the photodynamic ablation of cancer cells (Xu et al. 2019). The molecules 59–62 reveal the impact of acceptor groups and planarization of the π–bridge, specifically the replacement of thiophene with thieno[3,2-b]thiophene imparts increased planarity leading to enhanced molar absorptivity, ε. A neat DMSO solution or the aqueous solution of the aggregates when exposed to 660 nm laser irradiation exhibits increased temperature, but the magnitude of temperature is seemingly higher for neat DMSO solution than the aggregate. This feature highlights that the absorbed light energy is dissipated as thermal energy with more efficiency by isolated molecule than the aggregate. The temperature rise could be controlled by the laser energy and repeated for several cycles, thus the AIEgens could be used as photothermal agents for multifunctional phototherapeutics toward cancer therapy (Wang et al. 2022a).

Cationization of the luminogens is one of the better approaches to improve the ICT character as the combination of electron-rich and cation would offer much higher charge separation than the neutral electron-deficient systems. Furthermore, it also provides enhanced solubility in water, stabilization through ion–pair interactions, and targeted binding in specific organelles which are essential for biological applications. Cationization of AIEgens is aimed to target the specific organelle, such as mitochondria or cancer cells, and can not only be used as fluorescent label, could also be used as a photosensitizer to initiate photodynamic or photothermal therapies. The excited state of the AIEgen might undergo either direct electron transfer or energy transfer process to generate localized reactive oxygen species, thus leading to targeted therapies. Combination of dual-cationic triphenylphosphonium alkylated pyridinium substituted with triphenylamine, 63 have a red emission and AIE characteristics; have a selective targeted binding towards mitochondria or lysosomes of cancer cells. Not limited, 63 exerts dark and photo-toxicity and generates white light-triggered reactive oxygen species depending upon the phenyl substitution showing superior killing of cancer cells and inhibiting tumor growth through synergistic effect (Ma et al. 2021). Interestingly, the electronic properties of TPA-based pyridine can also be altered by organic halides used to form pyridinium ion. Specifically, when benzophenone substituted alkyl bromide is used to form pyridinium ion, 100 the resulting molecule is a typical AIEgen with DA system that forms different aggregates characteristic of various treatment methods. Green, yellow, and orange emissions from crystals, amorphous, or their mixer were noticed, which could be controlled by external stimuli of fuming, crystallization, sonication, and grinding, thus 64 could be a mechanochromic luminophore (Yin et al. 2021). These materials could find applications in pressure sensing, surface damage, rewritable materials, and finger print recordings. Compound 65 having dipyridinium cation with hydroxammic acid chelating group is designed so that the pyridinium ions exert electrostatic binding with nucleic acids and the latter acts a chelating group for zinc ions in the active center of histone deacetylases (HDACs). The AIE effect induces a strong fluorescence from 65 in biological medium and can generate reactive oxygen species through photosensitization process thereby efficiently damaging the nucleus DNA and making it an excellent inhibitor of telomerase activity (Wang et al. 2022b). Another AIEgen, triphenylamine thiophen pyridinium, 66 was used as a photosensitizer for the treatment of bacterial endophthalmitis, an acute eye infection and potentially irreversible blinding ocular disease (Fig. 7.15).

66 can selectively bind to bacteria over normal ocular cells owing to the combined presence of cationic charge with hydrophobicity, subsequently kill the bacteria through light-triggered reactive oxygen species. In addition, the efficient antibacterial behavior of 66 triggered an early immune response to the intraocular infection and was seemingly better than the commercially used rose Bengal (Li et al. xxxx). The structurally similar TPA-pyridinium ions 67, and 68–71 have also been employed to

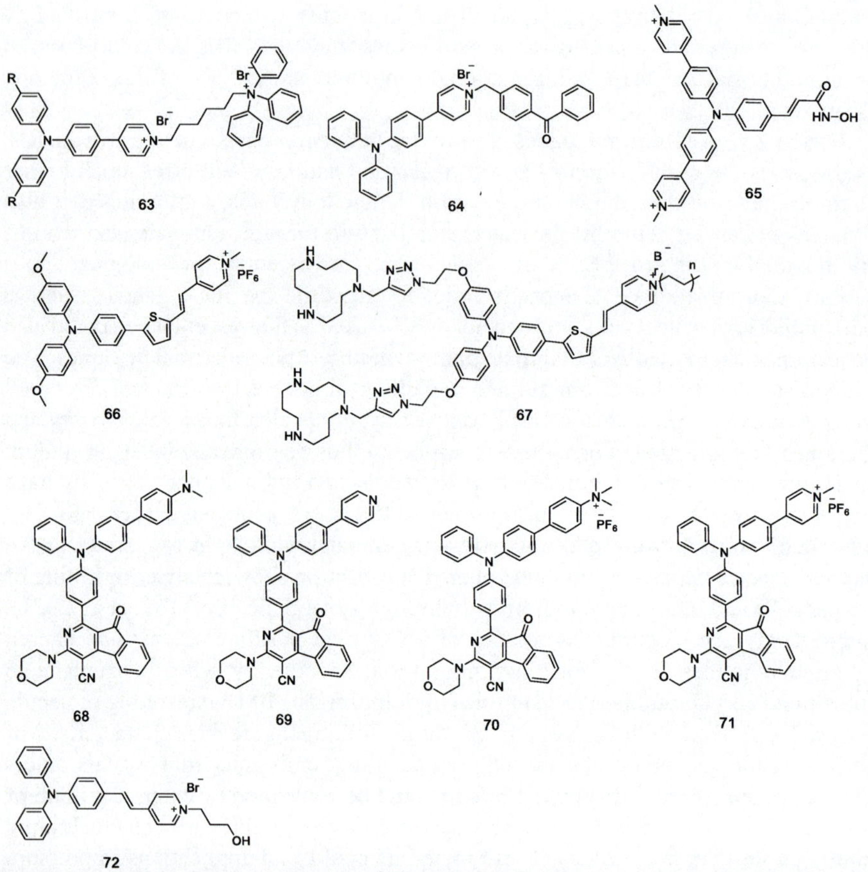

Fig. 7.15 Structure of cationic TPA AIEgens

demonstrate the photodynamic therapy and gene delivery (Tang et al. 2021; Liu et al. 2019b). Cationization significantly improves the photodynamic therapy and reactive oxygen species generation to a greater extent in addition to the specific targeting of mitochondria. The water-soluble AIEgen, 72 is nonfluorescent in pure water and became intensely red fluorescent with the addition of poor solvent, ethyl acetate. The fluorescence quantum yield has been increased by 129 times in aggregate form when compared to water, and the aggregates were tested for solution phase visualization of latent finger prints (LFP) on walls, bricks, and paper. When an aqueous solution of 72 was sprayed on the surfaces having LFP, the molecules will adhere on the lipid secretions in finger prints owing to the hydrophobic–hydrophobic interaction of the lipid. Indeed, AIEgen 72, could provide levels 1–3 details of the fingerprint and the solution possesses relatively better photo and thermal stability along with the good shelf-life (Wang et al. 2020).

Molecular engineering of TPA in combination with several planar π-bridges such as benzothiadiazole and dibenzo[a,c]phenazine in different donor–acceptor configurations are an interesting approach having ACQ and AIE motifs to develop novel materials with unique photophysical properties. Benzothiadiazole is a well-known electron deficient system and in combination with TPA would effectively reduce π–π stacking and facilitate the emission in organic electronic devices. 73–76 with different combinations of TPA, benzothiadiazole, and anthracene exhibit strong electroluminescence in orange to red color with low turn-on voltage in non-doped OLED devices. 73, could also work both as an emitting layer and hole-transport layer with an efficiency comparable to the device with the standard molecule, N,N′-diphenyl-N,N′-bis (1-naphthyl) (1,1′-biphenyl)-4,4′diamine (NPB) (Zhang et al. 2021b). An NIR-emissive TPA-benzothiadiazole covalently linked with luminol through π-conjugation affords 77, which could be converted to dots using Pluronic F127 surfactant. The combination of these chromophores is based on the principle of combining NIR emission with chemiluminescence, wherein the latter has high penetration with external photoexcitation, thus promising for deep tissue imaging. Electron-deficient benzothiadiazole and TPA ensure the NIR chemiluminescence emission in aqueous solution and could be used for the detection of singlet oxygen in biological systems. Indeed, NIR chemiluminescence emission is capable of penetrating through 3-cm-thick pork ham as well as distinguishing tumor cells from normal cells (Liu et al. 2020).

7.4 Summary and Perspective

This article outlines the summary of the chemistry of triphenylamine-based molecules and materials, their applications toward luminophores, sensing, aggregation-induced emission, and framework materials. The bond length alternation of phenyl rings of TPA core and its impact upon substitution with electron deficient or donating groups highlights the intramolecular charge transfer even in the ground state. In fact, the number of substituents does have an impact on the ICT strength. Though more than thirty crystal structures of TPA derivatives from the CCDC database have been analyzed, unfortunately, no systematic correlation could be arrived, which may also be due to the non-availability of crystal data of closely related structures. Development of dipolar, quadrupolar, and octupolar chromophores, respectively, with DA, DA_2, and DA_3 highlights that the emission wavelength could be shifted to near-infrared region with simple substitutions and also have a potential to be used as two-photon fluorescent probes. Fundamentally, when the substitution is at *para*-position both quadrupolar and octupolar chromophores undergo excited state symmetry breaking to form a symmetry-reduced ICT state, similar to their, simple dipolar analogous. Nevertheless, all the ICT states of these compounds are seemingly similar, still the excited state properties are distinctly different which could be spectrally identifiable. Unfortunately, even the femtosecond time scale transient absorption techniques were unable to through the light on the excited state symmetry breaking process, and still further systematic studies are required to unravel the excited state processes. Applications of these probes towards sensing were demonstrated using several chromophores for targeting several probes including the toxic species in the environments, intermediates produced in living systems during biological processes, and explosives. Mechanistically, the impact of intramolecular charge transfer processes of TPA derivative upon analytes interaction, which is directly related with the radiative excited state process is the sensory tool in most of the cases. For instance, for explosives generally the nitroaromatics and electron-rich amines, the turn-off fluorescent sensing owing to the quenching of TPA derivatives excited state has been witnessed. Further, the complexation ability of TPA derivative as a ligand or the reactive nature to construct the framework materials such as metal–organic framework and covalent organic framework with two- or three-dimensional, organized, stable, and porous framework, which provides ample opportunities for materials applications. Few examples discussed in this article also highlight the combination of optoelectronic properties of the ligand and organized porous structure in achieving the selectivity of specific analytes over their structurally similar analogous compounds. Another important field of developing aggregation-induced emission enhancement is an important area in realizing the solid state or aggregate state luminescent applications of TPA-based chromophores, which are otherwise elusive in realizing real-life luminescent-based products. The propeller-shaped structure, wobbly crystal packing aided by the C..H π-interactions, arresting the rotational or vibrational motions in the aggregate state is responsible for the intense luminescence. Perturbations in the intermolecular interactions either by means of mechanical forces, pressure, or infusion of chemical vapor also resulted in luminescent changes, thus leading to the development of mechanochromism and vapochromism phenomena. Applications of AIEgens towards sensing, fabrication of organic electronic devices, and two-photon fluorescent probes are also discussed.

Though large quantum work has been done with TPA-based molecules towards materials for organic photovoltaics, charge transport materials, light emitters for organic light-emitting diodes, sensors, AIEE, and framework materials, still the TPA motif is found to show novel properties by adopting diverse chemistry. Fundamentally, the ICT interactions are crucial for almost all the applications, but still the formation and stability of the dipolar ICT state in multipolar molecules are elusive. Further, TPA core could be altered to create a wide choice of ligands to construct the framework materials, mechanistic understanding of kinetics, stability, and nucleation of the ordered structure formation needs to be studied to develop higher order structures. Thus, we believe that TPA motif could still be further explored in various applications using the fundamental understanding that has been developed so far.

Acknowledgements We thank CSIR-CLRI, MLP-10 project for the financial support. We also acknowledge SERB funding where part of the research work has been carried out in our laboratory.

Author Contributions The manuscript was written through the contributions of all authors. All authors have given approval to the final version of the manuscript.

References

Ajantha, J., Yuvaraj, P., Karuppusamy, M., Easwaramoorthi, S.: Chem.—A Eur. J. **27**, 11319–11325 (2021)

Bünau, G.V.: Berichte der Bunsengesellschaft für physikalische Chemie **74**, 1294–1295 (1970)

Chang, Z.-F., Jing, L.-M., Chen, B., Zhang, M., Cai, X., Liu, J.-J., Ye, Y.-C., Lou, X., Zhao, Z., Liu, B., Wang, J.-L., Tang, B.Z.: Chem. Sci. **7**, 4527–4536 (2016)

Du, X., Wang, Z.Y.: Chem. Commun. **47**, 4276–4278 (2011)

Easwaramoorthi, S., Thamaraiselvi, P., Duraimurugan, K., Beneto, A.J., Siva, A., Nair, B.U.: Chem. Commun. **50**, 6902–6905 (2014)

Huang, J., Jiang, Y., Yang, J., Tang, R., Xie, N., Li, Q., Kwok, H.S., Tang, B.Z., Li, Z.: J. Mater. Chem. C **2**, 2028–2036 (2014)

Ji, N.-N., Shi, Z.-Q., Hu, H.-L., Zheng, H.-G.: Dalton T **47**, 7222–7228 (2018)

Jiang, M., Gu, X., Lam, J.W.Y., Zhang, Y., Kwok, R.T.K., Wong, K.S., Tang, B.Z.: Chem. Sci. **8**, 5440–5446 (2017)

Krug, M., Wagner, M., Schaub, T.A., Zhang, W.-S., Schüßlbauer, C.M., Ascherl, J.D.R., Münich, P.W., Schröder, R.R., Gröhn, F., Dral, P.O., Barbatti, M., Guldi, D.M., Kivala, M.: Angew. Chem. Int. Ed. **59**, 16233–16240 (2020)

Li, T., Wu, Y., Cai, W., Wang, D., Ren, C., Shen, T., Yu, D., Qiang, S., Hu, C., Zhao, Z., Yu, J., Peng, C., Tang, B.Z.: Adv. Sci., n/a 2202485

Lin, H.-T., Huang, C.-L., Liou, G.-S.: ACS Appl. Mater. Inter. **11**, 11684–11690 (2019)

Liu, M.-X., Ma, L.-L., Liu, X.-Y., Liu, J.-Y., Lu, Z.-L., Liu, R., He, L.: ACS Appl. Mater. Inter. **11**, 42975–42987 (2019a)

Liu, Z., Zou, H., Zhao, Z., Zhang, P., Shan, G.-G., Kwok, R.T.K., Lam, J.W.Y., Zheng, L., Tang, B.Z.: ACS Nano **13**, 11283–11293 (2019b)

Liu, C., Wang, X., Liu, J., Yue, Q., Chen, S., Lam, J.W.Y., Luo, L., Tang, B.Z.: Adv. Mater. **32**, 2004685 (2020)

Luo, J., Xie, Z., Lam, J.W., Cheng, L., Chen, H., Qiu, C., Kwok, H.S., Zhan, X., Liu, Y., Zhu, D.: Chem Commun. 1740–1741 (2001)

Ma, Y., Zhuang, Z., Xing, L., Li, J., Yang, Z., Ji, S., Hu, R., Zhao, Z., Huo, Y., Tang, B.Z.: Adv. Func. Mater. **31**, 2106988 (2021)

Mei, J., Leung, N.L.C., Kwok, R.T.K., Lam, J.W.Y., Tang, B.Z.: Chem. Rev. **115**(2015), 11718–11721 (1940)

Mei, J., Wang, J., Sun, J.Z., Zhao, H., Yuan, W., Deng, C., Chen, S., Sung, H.H.Y., Lu, P., Qin, A., Kwok, H.S., Ma, Y., Williams, I.D., Tang, B.Z.: Chem. Sci. **3**, 549–558 (2012)

Mei, J., Hong, Y., Lam, J.W.Y., Qin, A., Tang, Y., Tang, B.Z.: Adv. Mater. **26**, 5429–5479 (2014)

Rusznyák, Á., Zólyomi, V., Kürti, J., Yang, S., Kertesz, M.: Phys. Rev. B **72**, 155420 (2005)

Shi, L., He, C., Zhu, D., He, Q., Li, Y., Chen, Y., Sun, Y., Fu, Y., Wen, D., Cao, H., Cheng, J.: J. Mater. Chem. **22**, 11629–11635 (2012)

Tang, F., Liu, J.-Y., Wu, C.-Y., Liang, Y.-X., Lu, Z.-L., Ding, A.-X., Xu, M.-D.: ACS Appl. Mater. Inter. **13**, 23384–23395 (2021)

Terenziani, F., Painelli, A., Katan, C., Charlot, M., Blanchard-Desce, M.: J. Am. Chem. Soc. **128**, 15742–15755 (2006)

Terenziani, F., Sissa, C., Painelli, A.: J. Phys. Chem. B **112**, 5079–5087 (2008)

Thamaraiselvi, P., Duraipandy, N., Kiran, M.S., Easwaramoorthi, S.: ACS Sustain. Chem. Eng. **7**, 9865–9874 (2019)

Thamaraiselvi, P., Varathan, E., Subramanian, V., Easwaramoorthi, S.: Dyes Pigm. **172**, 107838 (2020)

Vamvounis, G., Shaw, P.E., Burn, P.L.: J. Mater. Chem. C **1**, 1322–1329 (2013)

Wang, T., Hu, Z., Nie, X., Huang, L., Hui, M., Sun, X., Zhang, G.: Nat. Commun. **12**, 1364 (2021)

Wang, H., Wang, Y., Zheng, Z., Yang, F., Ding, X., Wu, A.: J. Mater. Chem. B **10**, 1418–1426 (2022a)

Wang, K.-N., Liu, L.-Y., Mao, D., Hou, M.-X., Tan, C.-P., Mao, Z.-W., Liu, B.: Angew. Chem. Int. Ed. **61**, e202114600 (2022b)

Ward, J.S., Kukhta, N.A., dos Santos, P.L., Congrave, D.G., Batsanov, A.S., Monkman, A.P., Bryce, M.R.: Chem. Mater. **31**, 6684–6695 (2019)

Wei, Y., Wang, L., Huang, J., Zhao, J., Yan, Y.: ACS Appl. Nano Mater. **1**, 1819–1827 (2018)

Wu, P., Wang, J., Li, Y., He, C., Xie, Z., Duan, C.: Adv. Func. Mater. **21**, 2788–2794 (2011)

Wu, S., Song, H., Song, J., He, C., Ni, J., Zhao, Y., Wang, X.: Anal. Chem. **86**, 5922–5928 (2014)

Xie, W., Jiang, W., Xu, G.-J., Zhang, S.-R., Xu, Y.-H., Su, Z.-M.: CrystEngComm **23**, 3901–3906 (2021)

Xu, W., Lee, M.M.S., Zhang, Z., Sung, H.H.Y., Williams, I.D., Kwok, R.T.K., Lam, J.W.Y., Wang, D., Tang, B.Z.: Chem. Sci. **10**, 3494–3501 (2019)

Xu, J., Wang, B.: Dyes Pigm. **194**, 109643 (2021)

Yao, X.-Q., Xiao, G.-B., Xie, H., Qin, D.-D., Ma, H.-C., Liu, J.-C., Yan, P.-J.: CrystEngComm **21**, 2559–2570 (2019)

Yao, Y., Zhou, Y., Zhu, T., Gao, T., Li, H., Yan, P.: ACS Appl. Mater. Inter. **12**, 15338–15347 (2020)

Yin, W., Li, Y., Li, N., Yang, W., An, H., Gao, J., Bi, Y., Zhao, N.: Adv. Opt. Mater. **8**, 1902027 (2020)

Yin, W., Yang, Z., Zhang, S., Yang, Y., Zhao, L., Li, Z., Zhang, B., Zhang, S., Han, B., Ma, H.: Mater. Chem. Front. **5**, 2849–2859 (2021)

Zhang, X., Liu, X., Lu, R., Zhang, H., Gong, P.: J. Mater. Chem. **22**, 1167–1172 (2012)

Zhang, T., Brumboiu, I.E., Grazioli, C., Guarnaccio, A., Coreno, M., de Simone, M., Santagata, A., Rensmo, H., Brena, B., Lanzilotto, V., Puglia, C.: The J. Phys. Chem. C **122**, 17706–17717 (2018)

Zhang, F., Liu, Y., Yang, B., Guan, P., Chai, J., Wen, G., Liu, B.: J. Mater. Chem. B **9**, 2417–2427 (2021a)

Zhang, D., Yang, T., Xu, H., Miao, Y., Chen, R., Shinar, R., Shinar, J., Wang, H., Xu, B., Yu, J.: J. Mater. Chem. C **9**, 4921–4926 (2021b)

Zhai, L., Liu, M., Xue, P., Sun, J., Gong, P., Zhang, Z., Sun, J., Lu, R.: J. Mater. Chem. C **4**, 7939–7947 (2016)

Zhao, L., Lin, Y., Liu, T., Li, H., Xiong, Y., Yuan, W.Z., Sung, H.H.Y., Williams, I.D., Zhang, Y., Tang, B.Z.: J. Mater. Chem. C **3**, 4903–4909 (2015)

Zhao, W., He, Z., Peng, Q., Lam, J.W.Y., Ma, H., Qiu, Z., Chen, Y., Zhao, Z., Shuai, Z., Dong, Y., Tang, B.Z.: Nat. Commun. **9**, 3044 (2018)

Chapter 8
Lanthanide Ions-Doped Luminescent Nanomaterials for Anticounterfeiting

Jyoti Yadav, Satish Kumar Samal, and Boddu S. Naidu

8.1 Introduction

Counterfeiting is a widespread problem all around the world. Though governments and business agencies have made numerous efforts to prevent counterfeiting by passing laws, counterfeiting goods in the apparel, medicine, food, and electronic fields continue to find its place in our daily lives, posing a serious threat to property safety, social stability, and even human health. The greatest approach to combat counterfeiting is to prevent it for which a variety of modern security measures are constantly being investigated (Kumar et al. 2017). As per the past reports, the worldwide monetary losses brought about by counterfeiting might arrive at 1.82 trillion USD in 2020 (Liu et al. 2010). Anticounterfeiting techniques, for example, luminescent barcodes, watermarks, and two-dimensional codes have been explored to battle counterfeiting, but these are easy to replicate. Figure 8.1 provides the number of yearly publications on anticounterfeiting materials (Abdollahi et al. 2020).

The luminous patterns can work as promising warning indications for high-level security and the protection of important information. Conventional fluorescent materials, for instance, organic dye molecules and semiconductor quantum dots (QDs) can generate tunable colors, but poor stability, photobleaching, and high toxicity limit their use in security applications (Yao et al. 2019). Due to the extraordinary optical characteristics of luminescent materials, such as high emission intensity, multicolor emission, prolonged lifespan, and numerous emissions in response to heat, light, mechanical stimuli, etc., luminescence anti-counterfeiting has attracted a great deal of interest (Yu et al. 2021).

Luminescent materials, often known as phosphors, are characterized as solids that can emit light upon the application of specific form of energy (Campos-Cuerva et al.

J. Yadav · S. K. Samal · B. S. Naidu (✉)
Energy and Environment Unit, Institute of Nano Science and Technology (INST), Sector-81, Knowledge City, Mohali Punjab-140306, India
e-mail: naidu245@gmail.com; sanyasinaidu@inst.ac.in

R. S. Ningthoujam and A. K. Tyagi (eds.), *Handbook of Materials Science, Volume 1*,
Indian Institute of Metals Series, https://doi.org/10.1007/978-981-99-7145-9_8

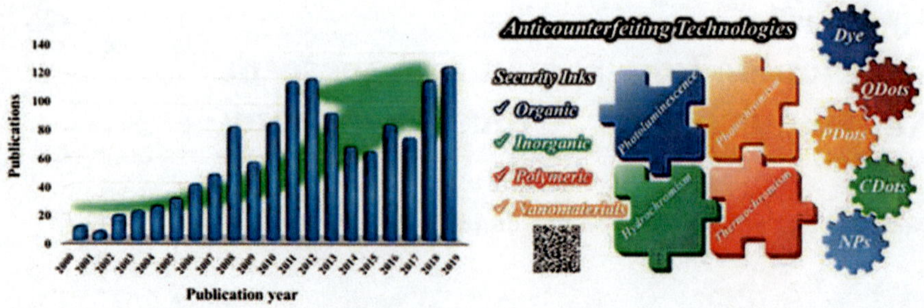

Fig. 8.1 The number of yearly publications on anticounterfeiting materials during the previous 20 years was found by searching for "anticounterfeiting" in the titles of papers and innovations that were published on scholars' websites and different technologies used for anticounterfeiting. Reprinted with permission from ref. (Abdollahi et al. 2020) Copyright 2020, American Chemical Society

2016). Among all phosphors, lanthanide-based materials are significant and desirable choices. Lanthanide ions (Ln^{3+}) are frequently utilized as emitting species in various phosphors because of their wide range of f-orbital configurations and can display strong fluorescence emission via intra *4f* or *4f-5d* transitions (Campos-Cuerva et al. 2016). Since the *f-f* transitions of the Ln^{3+} are disallowed, direct excitation of the same is not preferred. From this perspective, a doping technique is frequently used to create Ln^{3+}-doped materials with desired photoluminescence properties. This process entails incorporating a small amount of ions into the host matrix to produce hybrid materials. The host lattice with Ln^{3+} offers the emitter a stable microenvironment. An effective energy transfer (ET) from host or co-dopant to Ln^{3+} ions can occur when they have suitable energy levels. In addition, Ln^{3+}-doped inorganic nanomaterials have the advantages of a large Stokes shift, long luminescence decay lifetime, a sharp emission spectrum, low toxicity, and high chemical/photochemical stability when compared to metal complex compounds, carbon dots, halide perovskites, quantum dots (QDs), and organic dyemolecules (Gordon et al. 2008; Nichkova et al. 2022).

8.1.1 Basic Concepts of Luminescence

The term "luminescence" comes from the Latin word *lumen*, which means "light". The current understanding defines luminescence as the spontaneous emission of light from an electrically or vibrationally stimulated entity that is not in thermal equilibrium with its surroundings. (Valeur et al. 2011). Based on the form of excitation energy, luminescent materials can be further categorized into several types such as chemiluminescence, bioluminescence, photoluminescence, electroluminescence, triboluminescence, crystaloluminescence, radioluminescence, thermoluminescence, etc. The name of the luminescence is based on the type of excitation energy and

Table 8.1 Various kinds of luminescent materials and corresponding excitation energy along with examples

Type of luminescence		Excitation source	Example
Chemical	Chemiluminescence	Chemical reaction	Firefly lucerin
Biological	Bioluminescence	Biochemical reaction	Glow-worm
Physical	Photoluminescence	Light	$NaYF_4$:Eu
	Electroluminescence	Electric current	ZnS:Cu
	Mechanoluminescence	Mechanical action	Quartz glass
	Crystaloluminescence	Crystallization	NaCI
	Radioluminescence	Ionizing radiation	$NaMgF_3$:Sm
	Thermoluminescence	Heat	$CaTi0_3$:Eu

their examples are shown in Table 8.1. Among these luminescent materials, photoluminescent materials are well used in several fields such as photocatalysis, anti-counterfeiting, solar cells, sensors, latent fingerprint visualization, bioassays, drug delivery, photodynamic therapy, etc. There are several kinds of photoluminescent materials such as quantum dots, halide perovskites, carbon dots, metal clusters, lanthanide-doped materials. In this chapter, the main focus is on lanthanide-doped luminescent materials due to their excellent properties (Kumar et al. 2016).

8.1.2 Lanthanide-Doped Luminescent Nanomaterials

The lanthanides, also known as "rare earths" are the 4f-block elements (La to Lu). They share many similarities, such as ionic radius and oxidation state (III). Because of their 4f electrons, these materials consist of rich energy levels as shown in Fig. 8.2, which leads to showing excellent optical properties such as long luminescence lifetimes (up to several milliseconds), strong emission peaks, narrow full width half maxima, and upconversion luminescence over a range of wavelengths from near-infrared to visible. Due to the fast developments in nanotechnology, especially the progress of new design strategies for materials, great interest has developed in the nanoscale controlled synthesis and luminescence properties of Ln^{3+}-doped materials for various applications. In comparison to their bulk counterparts, nanocrystalline and nanostructured materials, which are commonly characterized as polycrystalline materials with a particle size of 100 nm or less, may display unique optical, electrical, magnetic, and thermodynamic properties. Due to the electronic structure of Ln^{3+} ions in solids, they can emit light from UV to infrared (IR). Based on the luminescence emission mechanism, lanthanide-doped luminescent nanomaterials are divided into three categories: downshifting, upconversion, and quantum cutting luminescent nanomaterials. The luminescence mechanism in these categories is discussed in the following sections.

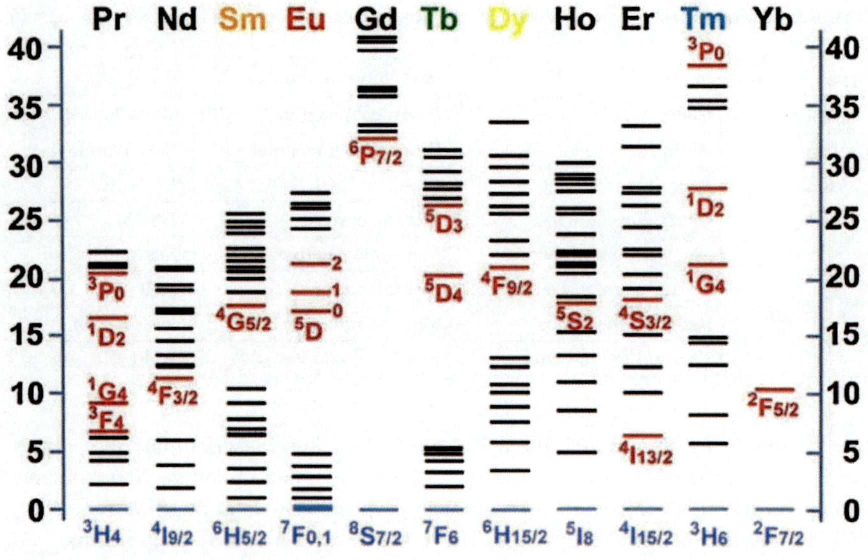

Fig. 8.2 Energy levels diagram for the lanthanide aquo ions. The values of the energy levels are given as multiples of 1000 cm^{-1}. The red lines are the main luminescent levels, while the blue lines are the ground state level. Reprinted with permission from ref. SeethaLekshmi et al. (2017) Copyright 2017, Elsevier Publication

8.1.2.1 Downshifting Luminescence

In the downshifting luminescence (DSL) process, a low-energy photon is emitted upon absorption of a high-energy photon as shown in Fig. 8.3. The corresponding wavenumber shift between the maxima of the emission and absorption bands (E_3–E_2) is identified as the Stokes shift. In DSL, the host lattice is commonly doped with trivalent lanthanide ions. Nowadays, various kinds of rare earth-based host materials like fluoride, phosphates, vanadates, oxides, and oxy-sulfide are reported with excellent luminescence properties. Taking into consideration Eu^{3+}-doped YVO_4 nano/microcrystals, there is a proficient energy transfer from $VO_4{}^{3-}$ groups to Eu^{3+} ions. The most commonly employed activators for DSL are Dy^{3+}, Sm^{3+}, Tb^{3+}, Eu^{3+}, and all of which show numerous emission lines in the visible range when exposed to UV light (Jia et al. 2009). Basically, low concentrations ($\leq$ 5 mol percent) of the activators are doped in the host material to minimize the quenching of luminescence due to the concentration of dopants without changing the crystal structure of the host material. Apart from the dopant concentration, synthesis conditions, calcination temperature, particle size, and the host lattice also have an impact on the PL intensity (Luwang et al. 2011).

To enhance the effective luminescence of Ln^{3+} (Ln = Sm, Dy, Tb, Eu) ions, the Gd^{3+} and Ce^{3+} ions are often co-doped as sensitizers. There are various reports on Ce^{3+}– and Gd^{3+}–sensitized downshifting luminescent nanomaterials. Regardless of

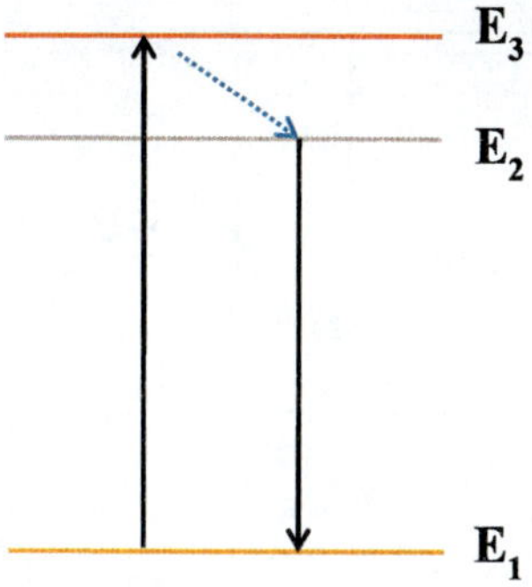

Fig. 8.3 Diagrammatic representation of the downshifting luminescence process

the single-doping approach, codoping of activators with larger absorption coefficient sensitizers could prompt an efficient emission by using ET from sensitizer to activator. For the desired luminescence characteristics in a typical lanthanide luminescent nanosystem, both sensitizers as light absorption centers and activators as light emitting centers are essential and significant. The only sensitizers that are thought to be the most effective across all sensitization systems are Yb^{3+} and Nd^{3+}. As a result, the matching excitation wavelengths are constrained to be between 980 and 808 nm. It is essential and advantageous to investigate more sensitization units that have bigger absorption cross sections, higher energy transfer efficiency, and independent excitation in order to enrich excitation wavelengths and increase luminescence intensity.

8.1.2.2 Upconversion Luminescence

The upconversion (UC) emission process was first suggested by N. Bloembergen (Ge et al. 2017). In upconversion, two or more low-energy photons get absorbed in order to emit a high-energy photon (Ge et al. 2017). Often, the low energy-absorbed photons are from the NIR region and high energy-emitted photons are in the UV or visible region. UC luminescence has numerous advantages over DS luminescence, such as the absence of background emission, long tissue penetration length without damaging, etc. In general, to produce higher energy photons, upconversion employs sequential absorption of numerous photons using a long excited state lifetime and actual ladder-like energy levels of Ln^{3+} ions placed in a reasonable host material (Kang et al. 2020).

Numerous forms of upconversion nanoparticles (UCNPs) have so far been produced using a variety of synthetic techniques. Figure 8.4a–f displays the schematic designs of the upconversion process. UC methods can be classified into the following groups: energy migration-mediated upconversion (EMU), cooperative sensitization upconversion (CSU), photon avalanche (PA), energy transfer upconversion (ETU), and excited state absorption (ESA) (Zhou et al. 2020). ESA (Fig. 8.4a) is the most fundamental method of UC luminescence. ESA is the easiest single-center UC

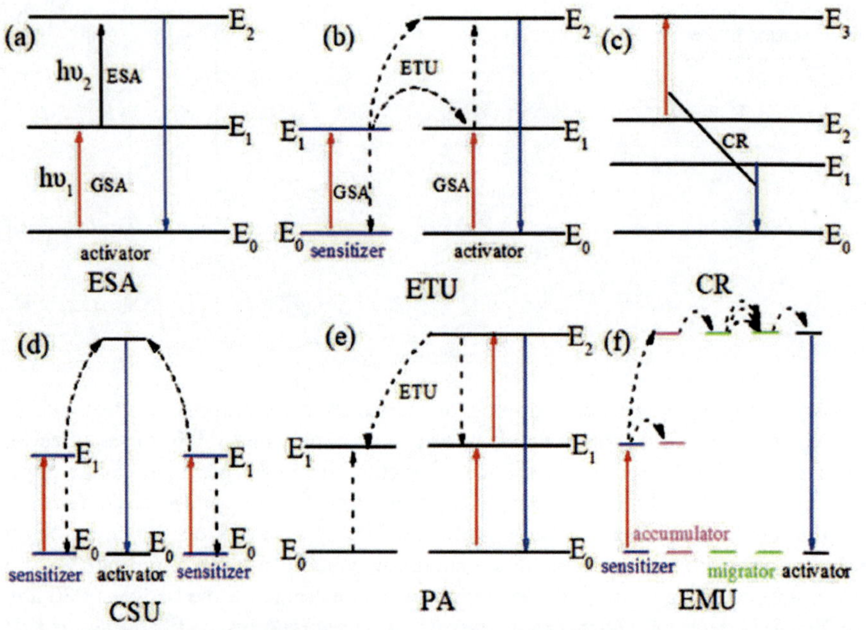

Fig. 8.4 Diagrammatic representation of the upconversion luminescence process of Ln^{3+}-doped materials. Terms mentioned here are ground state (E_0), excited states (E_1, E_2, and E_3), ground state absorption (GSA), excited state absorption (ESA), energy migration-mediated upconversion (EMU), photon avalanche (PA), cooperative sensitization upconversion (CSU), cross-relaxation (CR), energy transfer upconversion (ETU). Reprinted with permission from ref. (Zhou et al. 2020) Copyright 2020, Elsevier Publication

process. By using ground state absorbance (GSA), the atom moves from E_0 state to E_1 state and then by absorbing a second photon, it rises to E_2 state. ESA is the least effective, so ETU has been a far more popular method of providing UC luminescence (Fig. 8.4b) (Zhou et al. 2020). ETU consists mainly of an activator and a sensitizer, both of them are excited to the E_1 state by GSA when exposed to the proper source of excitation. A variation of energy transfer (ET) is cross-relaxation (CR), as seen in Fig. 8.4c. When two kinds of ions are excited simultaneously, one of them gives energy to the other to excite it to a high energy state while relaxing to a lower energy level without emitting radiation. The primary outcome of ion–ion interaction is the CR process, and the efficiency of this method is directly correlated with the amount of dopant present. Although it is the primary cause of the well-known "concentration quenching mechanism" of emission, it can also be purposefully employed to modify the color output in UCNPs or build a reliable PA mechanism. CSU is comparable to ETU, that uses two sensitizers and one activator (Fig. 8.4d). Pump light excites two sensitizers, which may be the same or different, to the E_1 state. The energy is then transferred to the activator, which excites it to the E_2 state and emits UC photons. As a result of having quasi-virtual pair states during transitions, which must be explained in quantum physics with higher order disturbances. CSU's effectiveness is often orders of magnitude smaller than that of the ESA or ETU. The pairing of ESA and CR is called photon avalanche (PA) (Fig. 8.4e). The occurrence of CR between nearby ions causes a rapid, avalanche-like spike in the number of electrons in the excited state. Basically, PA is a periodic method that entails the ESA process of excitation light and the effective CR that produces feedback. EMU is distinctive from the previous method (Fig. 8.4f). EMU is usually applied in core–shell Ln^{3+}-doped UCNPs, where a more efficient emission may result from the active shell. Liu's group (Su et al. 2012) initially reported on this upconversion mechanism. This phenomenon involves the integration of four different types of luminescence components, i.e., a sensitizer, an accumulator, a migrator, and an activator into a single nanoparticle at various layers. Following laser irradiation, the nanoparticle's core undergoes the EMU process, which is followed by a gradual ET from the accumulator to the activator, resulting in upconversion luminescence (Zhou et al. 2015).

8.1.2.3 Quantum Cutting Luminescence

Quantum cutting (QC) luminescence is an opposite process to the upconversion luminescence. In QCL, two or more low-energy photons are emitted by absorbing one high-energy photon. Many ions, including Pr^{3+}, Nd^{3+}, Er^{3+}, Ho^{3+}, Tm^{3+}, Tb^{3+}, Ce^{3+}, Eu^{3+}, Bi^{3+}, and Eu^{2+} have been employed as donor ions in typical QC luminescence materials because of their ladder-like energy levels, which make photon absorption and consequent ET steps easier. Due to its NIR absorption and emission because of the $^2F_{5/2}$-$^2F_{7/2}$ transition, Yb^{3+} ion is typically utilized as an acceptor ion (Zou et al. 2016). Currently, the most effective QC luminescence has been obtained in certain fluorides because the QC luminescence process demands low phonon frequency host to prevent non-radiative losses (Fusari et al. 2008). With a quantum efficiency of more than 100%, the quantum cutting materials open up new possibilities for the improvement of robust security applications, cutting-edge display systems, and solid-state lighting (Huang and Yu 2009). Dexter developed a theoretical foundation for the quantum cutting phenomena (Dexter 1957). There are three different ways that the quantum cutting process can be carried out using (a) host energy levels; (b) single ions; and (c) ion pairs (Ronda 2002). For process (a), host material absorbs the photons having at least double the energy of its band gap to create an e–h pair. These excited e–h pairs generate two e–h pairs having energy nearly equal to the band gap of the host through the Auger interaction process. In process (b), two conditions must be satisfied: (1) to prevent multi-phonon relaxation, the energy gap between neighboring levels has to be significant; and (2) visible emission branching ratio must be large. There are several studies that go into great detail about the single-ion quantum cutting procedure. Quantum cutting effect is first noticed in Pr^{3+}-doped YF_3, Sommerdijk et al. (1974a, b). A few findings have so far been published in the literature and the production of lanthanide fluoride nanoparticles is not an easy operation (Wegh et al. 1999; Lorbeer et al. 1999). Numerous ways to make nanofluorides have been previously investigated. However, extra care must be taken to obtain quantum cutting materials that are completely oxide-free because oxygen impurities can cause non-radiative relaxation via Ln-oxygen charge transfer states, which can then result in emission from these levels (Ghosh and Mudring 2016). Compared to other quantum cutters, ion pairs-based quantum cutting (two or three ions) has attracted a lot of attention recently.

In route (c), one of the pair ions has the capacity to exhibit the quantum cutting phenomena through interaction with the other ion, some of the energy is transmitted to the later ion. These QC nanomaterials are interesting since they split a single high-energy photon into two visible photons with lower energies, and they also increase energy efficiency. As a result, technically, nanomaterials can achieve a quantum efficiency of about 200%. On the other hand, technologies for designing nanoscale particles using room-temperature ionic liquid (RTIL) have gained appeal and popularity. In ambient conditions, RTILs typically have low vapor pressure, but their physicochemical characteristics can be modified by adjusting the length of the alkyl chain, carefully choosing the cation–anion pair, IL concentration, etc. Thus, these solvents are sometimes referred to as "green and designer" solvents. Curiously, temperature and IL are being shown to have an impact on QC (Wegh et al. 1999). Various possible transitions of trivalent lanthanide ions which lead to the DS, UC, and QC luminescence are summarized in Table 8.2. Emission wavelengths and emission intensity of the corresponding transitions are also given in the table.

8.2 Preparation of Ln^{3+}-Doped Nanomaterials

The structure, size, and shape of host nanomaterials doped with lanthanides have a significant impact on their luminescence properties. Thus, the desired optical characteristics of luminescent nanoparticles are largely dictated by the process of synthesis. To explore more of their potential applications, it is essential to design and synthesize nanocrystals in the required shape and size. Various synthetic methods have tremendous influences on the morphologies and structures of nanocrystals. In the past couple of years, various efforts have been made to create various straightforward procedures for the synthesis of photoluminescent nanocrystals with adjustable size and shape. However, compared to the more traditional approaches like solid-state reactions, which are frequently utilized for the development of luminescent nanocrystals, soft chemical methods provide a lot of benefits, viz., (i) For preventing agglomeration, soft chemical route is the most efficient way, where ligands are introduced on the surface of the materials via dispersion in solvents of nonpolar or polar nature. (ii) The synthesis circumstances, which include the raw materials, pH, temperature, solvent used, concentration, additive, and reaction time, can change the shape and size of nanocrystals. (iii) Soft chemical reactions enable the creation of luminescent nanocrystals at large scales, low temperatures, and low cost. Due to these important factors, soft chemical methods have been shown to be the most effective for

Table 8.2 An overview of the fundamental transitions in the emission spectra of typical Ln^{3+} ions used in the downshifting (DS), upconversion (UC), and quantum cutting (QC) photoluminescence processes (Huang et al. 2012)

Ln	Transitions	Emission wavelength (nm)	Energy (cm^{-1})	Intensity	Remarks
Yb	$^2F_{5/2} \rightarrow {}^2F_{7/2}$	980	10,204	Strong	UC, QC and DS
Er	$^4G_{11/2} \rightarrow {}^4I_{15/2}$	380	26,316	Weak	UC and QC
	$^2P_{3/2} \rightarrow {}^4I_{13/2}$	408	24,510	Weak	UC
	$^2P_{3/2} \rightarrow {}^4I_{11/2}$	480	20,833	Weak	UC
	$^2H_{11/2} \rightarrow {}^4I_{15/2}$	525	19,048	Weak	UC and QC
	$^4S_{3/2} \rightarrow {}^4I_{15/2}$	545	18,349	Strong	UC and QC
	$^4S_{3/2} \rightarrow {}^4I_{13/2}$	850	11,765	Weak	UC and QC
	$^4F_{9/2} \rightarrow {}^4I_{15/2}$	665	15,038	Strong	UC and QC
	$^4I_{9/2} \rightarrow {}^4I_{15/2}$	800	12,500	Strong	UC and QC
	$^4I_{11/2} \rightarrow {}^4I_{15/2}$	980	10,204	Strong	UC and QC
	$^4I_{13/2} \rightarrow {}^4I_{15/2}$	1540	6494	Strong	QC
Tm	$^1D_2 \rightarrow {}^3H_4$	360	27,778	Medium	UC and QC
	$^1D_2 \rightarrow {}^3F_4$	450	22,222	Medium	UC and QC
	$^1G_4 \rightarrow {}^3H_6$	475	21,053	Strong	UC and QC
	$^1G_4 \rightarrow {}^3H_4$	650	15,385	Strong	UC and QC
	$^1G_4 \rightarrow {}^3H_5$	786	12,723	Weak	UC and QC
	$^3H_4 \rightarrow {}^3H_6$	800	12,500	Strong	UC and QC
	$^3F_4 \rightarrow {}^3H_6$	1800	5556	Weak	QC
Ho	$^5S_2, {}^5F_4 \rightarrow {}^5I_8$	540	18,519	Strong	UC and QC
	$^5S_2, {}^5F_4 \rightarrow {}^5I_7$	749	13,351	Weak	UC and QC
	$^5S_2, {}^5F_4 \rightarrow {}^5I_6$	1012	9881	Weak	QC
	$^5F_5 \rightarrow {}^5I_8$	644	15,528	Medium	UC and QC
	$^5F_5 \rightarrow {}^5I_7$	966	10,352	Weak	QC
	$^5I_6 \rightarrow {}^5I_8$	1180	8475	Strong	QC
Pr	$^3P_2 \rightarrow {}^3H_4$	440	22,727	Weak	QC
	$^3P_1 \rightarrow {}^3H_4$	470	21,277	Weak	QC
	$^3P_1 \rightarrow {}^1G_4$	872	11,468	Medium	QC
	$^3P_0 \rightarrow {}^3H_4$	480	20,833	Strong	UC, and QC
	$^3P_0 \rightarrow {}^3H_5$	545	18,349	Weak	UC and QC
	$^3P_0 \rightarrow {}^3H_6$	606	16,502	Medium	UC and QC
	$^3P_0 \rightarrow {}^3F_2$	640	15,625	Weak	UC and QC
	$^1D_2 \rightarrow {}^3F_4$	1037	9643	Medium	QC

(continued)

Table 8.2 (continued)

Ln	Transitions	Emission wavelength (nm)	Energy (cm^{-1})	Intensity	Remarks
Nd	$^4D_{3/2} \rightarrow {}^4I_{9/2}$	395	28,169	Weak	UC and QC
	$^2P_{3/2} \rightarrow {}^4I_{9/2}$	388	26,316	Weak	UC and QC
	$^2P_{3/2} \rightarrow {}^4I_{11/2}$	410	24,390	Strong	UC and QC
	$^2P_{3/2} \rightarrow {}^4I_{13/2}$	452	22,124	Strong	UC and QC
	$^4G_{7/2} \rightarrow {}^4I_{9/2}$	545	18,394	Weak to strong	UC, and QC
	$^4G_{7/2} \rightarrow {}^4I_{11/2}$	587	17,036	Weak to strong	UC and QC
	$^4G_{7/2} \rightarrow {}^4I_{13}$	655	15,267	Weak to strong	UC and QC
	$^4F_{3/2} \rightarrow {}^4Ig/2$	886	11,287	Weak to strong	QC and DS
	$^4F_{3/2} \rightarrow {}^4I_{11/2}$	1064	9399	Strong	QC and DS
	$^4F_{3/2} \rightarrow {}^4I_{13/2}$	1340	7463	Weak	QC and DS
Sm	$^4G_{5/2} \rightarrow {}^6H_{5/2}$	564	17,730	Medium	DS
	$^4G_{5/2} \rightarrow {}^6H_{7/2}$	601	16,639	Strong	DS
	$^4G_{5/2} \rightarrow {}^6H_{9/2}$	644	15,528	Medium	DS
Eu	$^5D_0 \rightarrow {}^7F_{0,1,2,3,4}$	570–720	13,889–17,544	Strong	UC and DS
Tb	$^5D_4 \rightarrow {}^7F_{6,5,4,3}$	480–650	15,385–20,833	Strong	UC, QC and DS
Dy	$^4F_{9/2} \rightarrow {}^6H_{15/2}$	486	20,576	Medium	QC and DS
	$^4F_{9/2} \rightarrow {}^6H_{13/2}$	575	17,391	Strong	QC and DS
	$^4F_{9/2} \rightarrow {}^6H_{11/2}$	664	15,060	Weak	QC and DS

the shape and size-controlled production of luminescent nanomaterials. Some soft chemical techniques are discussed below.

8.2.1 Co-precipitation Method

Co-precipitation is one of the most effective and conventional approaches for the preparation of Ln^{3+}-doped nanomaterials. Due to its comparatively moderate reaction conditions, affordable equipment requirements, straightforward protocols, and quick reaction time, this approach is predicted to be more practical and advantageous (Wang and Liu 2009). For the synthesis of doped nanomaterials, complex oxides, sulfides, phosphates, fluorides, molybdates, tungstates, and solid solutions, the co-precipitation method is highly suitable due to the simultaneous precipitation of constituent ions. Nanomaterials of $NaBi_{0.9-x}La_xEu_{0.1}(MoO_4)_2$ ($0.0 \leq x \leq 0.9$) were prepared at room temperature by Pushpendra, et al. and the process is shown in Fig. 8.5. Van et al. synthesized Ln^{3+}-doped LaF_3 (Ln = Ho, Nd, Er, and Eu) (Stouwdam and Veggel 2002) nanomaterials using the co-precipitation process. This method is further expanded to the fabrication of Ln ions-doped $YbPO_4$ and $LuPO_4$ nanocrystals as well as Ln-doped $NaYF_4$ crystals (Lingeshwar Reddy et al. 2018). By tuning the reaction conditions, there is a likeliness for the regulation of the size and form of the nanomaterials produced using this process. However, the luminescence intensity of these nanomaterials is modest and can be improved by heating them at high temperatures, which removes the stabilizing ligand and decreases their hydrophilicity.

The cubic-phased $NaYF_4$ crystals that are produced mainly by the co-precipitation method have less intense UC emission and require post-processing by subjecting to high-temperature calcination for producing hexagonal phase nanocrystals that are highly UCL efficient (Lingeshwar Reddy et al. 2018). Co-precipitation methodology is much simpler and more practical than other methods, since it does not require complicated procedures, expensive setup, or harsh reaction conditions. To use these nanoparticles for the application of luminescent security ink, some relevant surface changes are also required. The co-precipitation technique also has considerable industrial uses because of its environmental friendliness, high yield, low cost, and applicability to synthetic materials. Nanomaterials of $Bi_{0.95-x}Gd_xEu_{0.05}PO_4$ solid solutions prepared by co-precipitation at 185 °C are shown in Fig. 8.6A. Microstructural analysis of $GdPO_4:Eu^{3+}$ nanomaterials made using the coprecipitation method

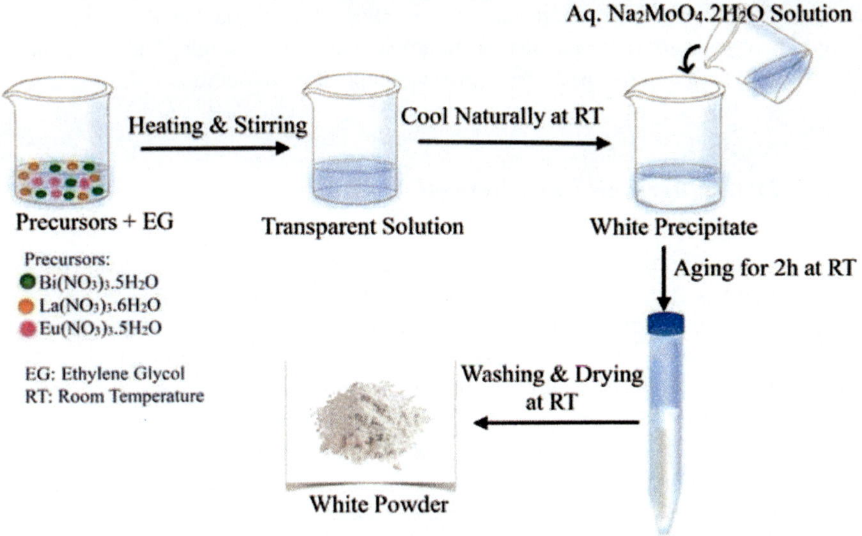

Fig. 8.5 $NaBi_{0.9-x}La_xEu_{0.1}(MoO_4)_2$ ($0.0 \leq x \leq 0.9$) nanomaterials synthesized via coprecipitation. Reproduced with permission of ref. (Pushpendra et al. 2019) Copyright 2019, ACS Publication

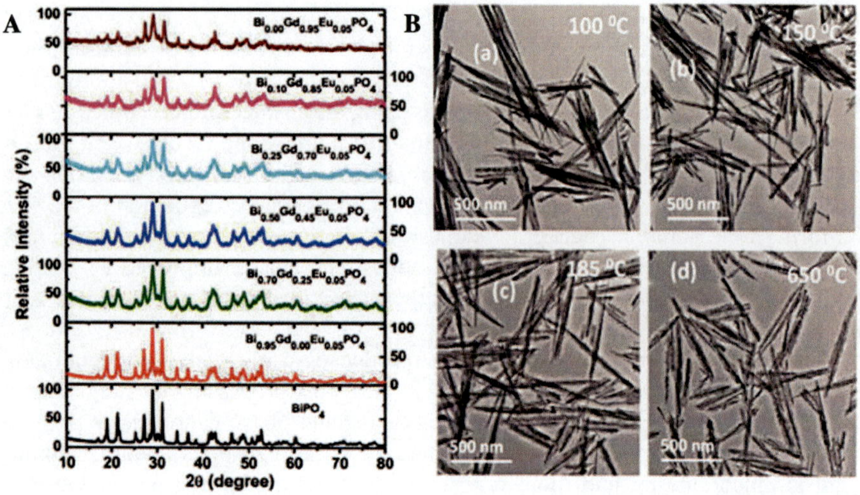

Fig. 8.6 **A** PXRD patterns of $Bi_{0.95-x}Gd_xEu_{0.05}PO_4$ solid solutions and **B** TEM images of $GdPO4:Eu^{3+}$ nanomaterials prepared at different temperatures. Reproduced with permission of ref. (Pushpendra et al. 2021a, b) Copyright 2021, Elsevier Publication and American Chemical Society

is displayed in Fig. 8.6B. By changing the reaction temperature from 100 to 185 °C, the crystal structure of these nanomaterials changes from hexagonal to monoclinic. All of these materials have distinct rod-shaped morphologies with increasing length and width of nanorods by increasing temperature. The shape remains same even though the crystal structure changes from hexagonal to monoclinic. The monoclinic nanorods are stable in terms of structure and shape after calcination at 650 °C.

8.2.2 *Hydrothermal/Solvothermal Method*

The hydro/solvothermal technique is one of the most frequently used methods for the preparation of nanomaterials, and it depends upon the reaction of precursors in liquids at high pressure and temperature. In hydro/solvothermal synthesis, solid precursors are dissolved in liquids and subjected to high pressure in specialized sealed reaction vessels known as autoclaves. In comparison to other synthetic approaches, this technique allows the production of highly crystalline nanomaterials under very mild conditions. The reactants solubility is greatly increased and the reaction for the production of nanoscaled particles is sped up when high pressure and temperature are applied during the solvo/hydrothermal synthesis technique, which is typically carried out above the critical point. However, this process uses a lower temperature than thermal breakdown does.

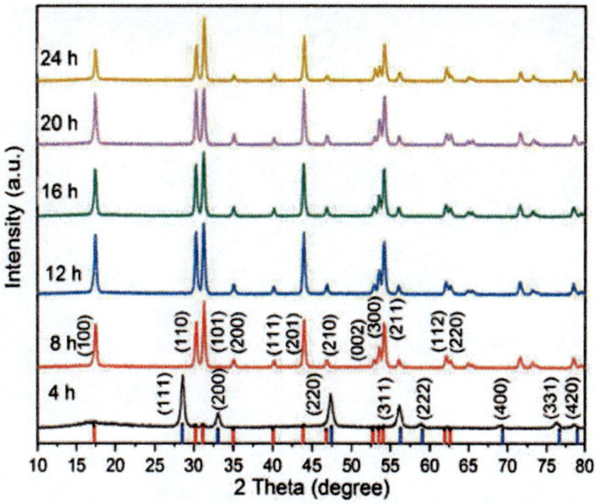

Fig. 8.7 PXRD patterns of $NaYbF_4$ materials synthesized at 200 °C using Na_3Cit with diverse reaction times. The standard JCPDS files of hexagonal (#27–1427 (red)) and cubic $NaYbF_4$ (#77–2043 (blue)) are also given. Reproduced with permission of ref. (Luo et al. 2022) Copyright 2022, Elsevier Publishing

This method is widely used since it is simple to utilize and requires little experimental expertise. To simultaneously control the morphology, size, and crystalline phases along with the desired surface functional groups, certain organic solvents and/or surfactants, such as oleic acid (Yang et al. 2013), polyethyleneimine (Yang et al. 2013; Wang et al. 2013), oleylamine, ethylene glycol, cetyltrimethyl ammonium bromide, etc., are typically added along with the precursors in a hydro/solvothermal synthesis method. Although the hydrothermal approach has been used to produce nanomaterials, there are still several issues that need to be resolved. First of all, since most hydrothermal reaction circumstances allow for prolonged reaction times and high reaction temperatures, the experiments are inherently dangerous. Second, the majority of materials formed via the hydrothermal process are micron-sized crystals. The PXRD patterns of hydro/solvothermally synthesized $NaYbF_4$ nanomaterials are depicted in Fig. 8.7. Initially, it forms a cubic phase and it converts into hexagonal phase by increasing the reaction time.

The $NaYbF_4$ forms irregular particles at a relatively short amount of reaction time (Fig. 8.8a). Figure 8.8b shows that after 8 h of reaction time, non-uniform microdiscs with distinct hexagonal cross sections start developing. With time, it becomes increasingly apparent that the hexagonal prism top/bottom facet is not perfectly flat and has a slightly concave center (Fig. 8.8c). In Fig. 8.8d, after 16 h, the product is made up of mostly homogenous microprisms and undergoes a small overall morphological change. The resulting sample comprises hexagonal microrods with projecting centers after 20 h (Fig. 8.8e). Finally, when the reaction time is increased to 24 h, uniformly shaped polyhedral microcrystals are produced (Fig. 8.8f) (Luo et al. 2022).

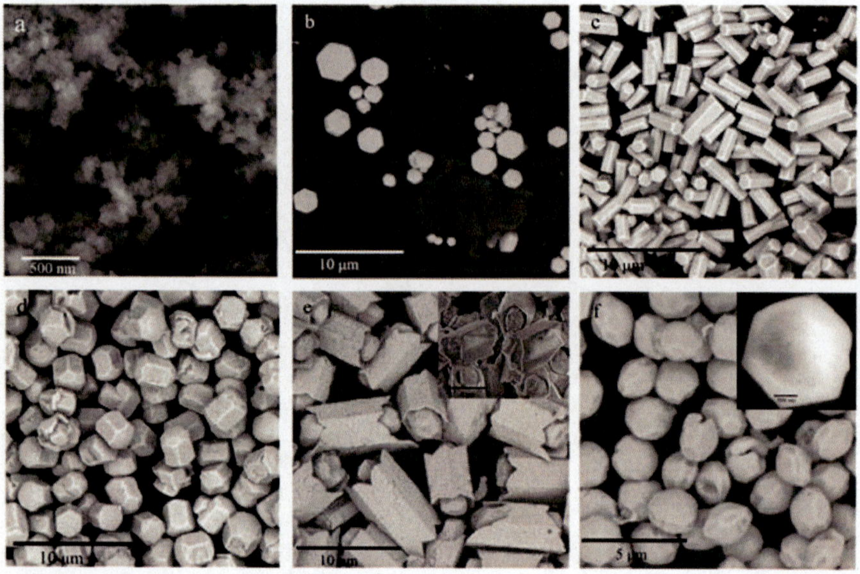

Fig. 8.8 SEM images of $NaYbF_4$ materials synthesized at 200 °C using Na_3Cit with diverse reaction times **a** 4 h, **b** 8 h, **c** 12 h, **d** 16 h, **e** 20 h, and **f** 24 h. Reproduced with permission of ref. (Luo et al. 2022) Copyright 2022, Elsevier Publishing

8.2.3 Thermal Decomposition Method

Another technique that is frequently used is thermal decomposition, which involves the decomposition of organometallic precursor chemicals that disintegrate at a high temperature in organic solvents with high boiling points such as 1-octadecene, oleic acid, tri-n-octyl phosphine oxide, and oleylamine for obtaining monodispersed nanomaterials. Typically, organic acid salts of RE such as acetate, oleate, acetylacetonate, and trifluoroacetate are employed as the organic precursors in a thermal breakdown to generate the corresponding metal and fluoride ions. In addition to the high boiling solvents, the thermal decomposition process also uses surfactants. The surfactant covers nanocrystals' surface, making them highly dispersive in organic solvents and preventing nanoparticle agglomeration. Both oleic acid and oleylamine, which have an organic functional group and a lengthy chain of hydrocarbon, act as surfactants and organic solvents (Liu et al. 2013). By carefully adjusting reaction conditions such as reaction time, amount of metal precursors, type of solvents, and reaction temperature, it is possible to develop superior nanomaterials with variable sizes, good crystallinity, and extraordinary optical properties (Liu et al. 2013). The thermal decomposition process is categorized into the following groups based on the precursors to make the topic easier to follow.

8.2.3.1 Trifluoroacetate Precursor

The first report on the synthesis of oleic acid-capped lanthanum fluoride triangular nanoplates appeared in 2005 by Yan and coworkers using thermal decomposition of a $La(CF_3COO)_3$ single-source precursor (SSP) in a mixed solvent of oleic acid and 1-octadecene (Zhang et al. 2005). Since, the $La(CF_3COO)_3$ precursor supplies both the La^{3+} ions and the fluoride ions, the formation of LaF_3 from SSP is noticeably simple to regulate. As a result, the combination of metal and anions as well as the breakdown of precursors both take place simultaneously. The effect of concentration variation of both anion and cation from distinct sources can be avoided. As a result, trifluoroacetate precursors can be used to create high-quality metal fluoride nanomaterials (Gai et al. 2014). After this first report, numerous publications on the thermal decomposition-based synthesis of Ln^{3+}-doped nanomaterials with a variety of host lattices have been reported. This methodology has gained increasing attention, particularly for the preparation of RE-based core–shell nanomaterials. Synthesis and morphology of $NaGdF_4$:Yb/Tm@$NaGdF_4$:Ce/Mn@$NaYF_4$ nanostructures produced by this method are depicted in Fig. 8.9.

8.2.3.2 Acetate, Acetylacetonate, and Oleate Precursors

Due to the generation of toxic fluoride species resulting from the thermal breakdown of trifluoroacetate, Chen and colleagues presented a non-toxic thermal decomposition method for producing Ln^{3+} doped $REYF_4$ nanomaterials using affordable RE oleate and NaF as precursors (Liu et al. 2009; Wei et al. 2006).

It is interesting to note that by lowering the ratios of NaF to RE oleate from 20 to 4 in oleic acid and 1-octadecene mixture solvents, the shape and size of $NaYF_4$:Yb, Er/Tm nanomaterials can be modified from nanoplates (~200 nm) to nanospheres (~18 nm) (Gai et al. 2014).

8.2.4 *Sol–gel*

Sol–gel method is a wet chemical synthesis route where the colloidal suspension of the precursors is used to produce a variety of nanomaterials with complex chemical compositions, excellent properties with uniform particle size distribution in a straightforward manner. The creation of pure-phase multicomponent nanocrystals with great crystallinity requires further heating at a high temperature. Despite the lengthy procedures involved, the main advantage of the sol–gel technology is the ability to create pure ceramic material near to room temperature. The primary use of the sol–gel process is to synthesize various coatings and films (Kumar et al. 2019; Bandi et al. 2012). The sol–gel technique includes the following crucial steps: hydrolysis, condensation, gelation, aging, drying, densification, and crystallization.

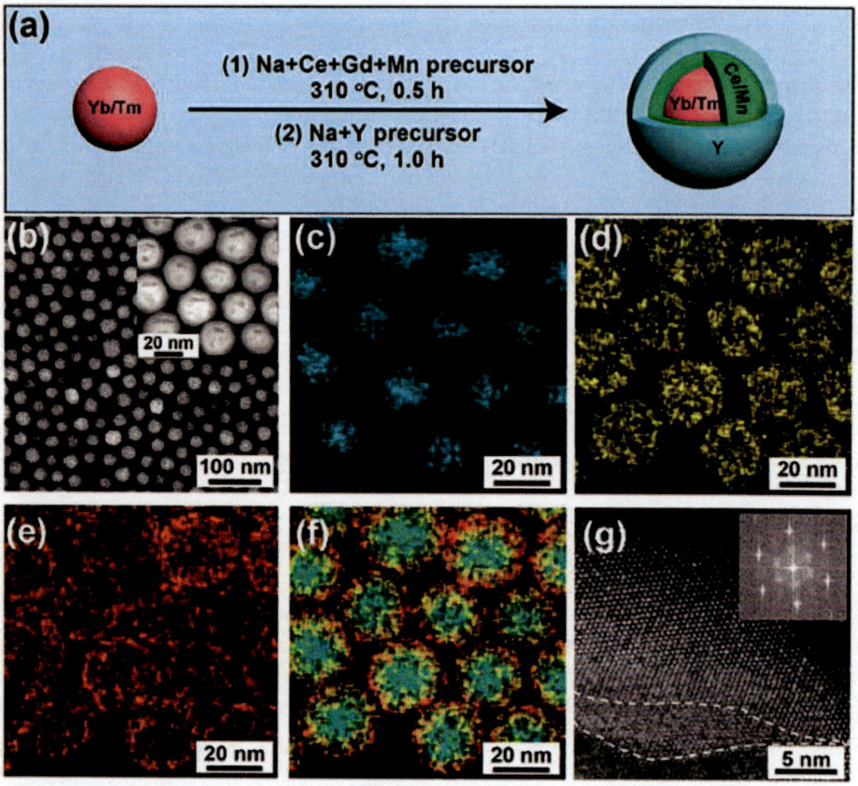

Fig. 8.9 **a** Schematic representation of synthesis **b** TEM image **c–f** Elemental mapping **g** HRTEM image of $NaGdF_4$:Yb/Tm@$NaGdF_4$:Ce/Mn@$NaYF_4$ multilayer nanomaterials. Reproduced with permission of ref. (Liu et al. 2019) Copyright 2019, ACS Publication

P. N. Prasad's group has also reported the production of lanthanide-based nanomaterials through an emulsion sol–gel procedure is also known as sol–gel emulsion (Patra et al. 2003). Figure 8.10 shows XRD and SEM images of precursor gels for Ho, Tm, Yb, and Lu, aluminum that had been calcined at 800 °C for five hours. As can be observed, the calcination results include agglomerated hazy amorphous particles in addition to crystallites of various shapes.

8.2.5 Ionic Liquid-Assisted Synthesis

The synthesis of fluoride-based inorganic nanoparticles benefits greatly from ionic liquid (IL)-based synthesis. ILs are now significant in the fields of electrochemistry, materials science, and chemistry. Because of their adaptable qualities, they have recently attracted a lot of attention for numerous applications. Most ionic liquids can

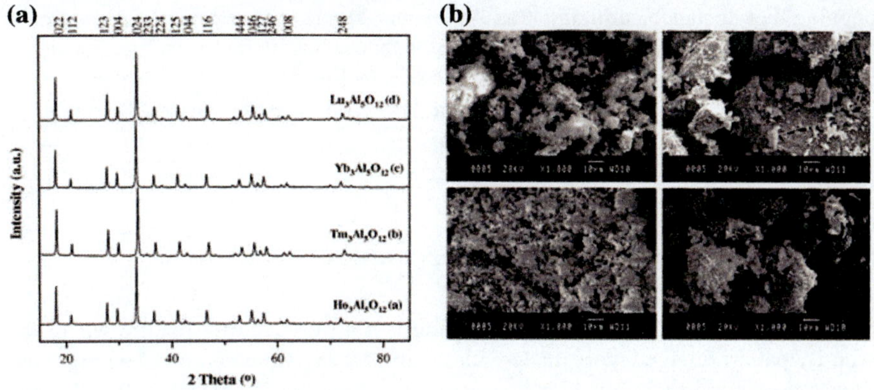

Fig. 8.10 **a** XRD patterns **b** SEM images of $Ho_3Al_5O_{12}$, $Tm_3Al_5O_{12}$, $Yb_3Al_5O_{12}$, and $Lu_3Al_5O_{12}$. Reproduced with permission of ref. (Dubnikova et al. 2011) Copyright 2019, Elsiver Publishing

be used in synthesis as a reaction medium, nanoparticle stabilizing agent, templating agent, reaction partner, and more due to their low vapor pressure, lack of flammability, and ability to serve as all of these additional roles. Ionic liquids are frequently referred to as "designer" solvents because of their remarkable qualities, which allow chemical and physical properties to be practically modified in accordance with the requirements of reaction circumstances. These ionic liquids' main flaw is their extreme sensitivity to ambient moisture and protonic contaminants (Sharma and Ghosh 2021).

8.2.6 Microwave-Assisted Synthesis

Microwave (MW)-assisted synthesis depends on dielectric heating, and it requires dipolar and/or ionic molecules. Solid-phase nanomaterial synthesis with MW assistance is a developing area for the advancement of material science in a variety of fields. In most cases, microwave radiation of an appropriate wavelength cooperates with the reaction material and produces the heat concurrently without leading to any significant thermal destruction in the structure. In the solid phase system, this method typically entails direct MW irradiation of the solid substrates in the MW oven. Chemical changes are brought about in this MW oven by the interaction of the solid precursors with the reflected MW from the cavity walls (Kumar et al. 2020).

8.2.7 Emulsion-Based Techniques

The ability to build "nanoreactors" with customizable dimensions by adjusting the water-to-surface ratio made both the conventional and the reverse micelle approach popular. For instance, utilizing traditional and single microemulsion techniques, monodisperse amorphous spheres, and homogeneous hexagonal and triangular single crystals of YF_3 were produced. Using a reverse micelle technique, Ghosh et al. created $NaYF_4$ nanocrystals with a well-defined and controlled crystal phase (Sharma et al. 2017).

8.3 Various Types of Printing Techniques

Photoluminescence-based anticounterfeiting is a covert security feature where the security pattern/image/letters made with the ink are invisible in normal daylight but show color upon irradiation with a particular wavelength of light (UV/NIR). Surface modification techniques are used to make stable dispersion of luminescent nanoparticles in hydrophilic or hydrophobic solvents before they are widely used by various printing processes (Lupo et al. 2013). Additionally, a variety of printing processes are available to meet the requirements of generating anticounterfeit markings at various levels. They may generally be divided into direct printing and template-based printing and each technique has a different resolution and printing speed. The type of products that need to be covered, nature of surface, cost, complexity, and desired resolution all have a role in selecting the printing method. Various methods used to print the anticounterfeiting patterns are shown in Fig. 8.11. All these printing methods are discussed below in detail.

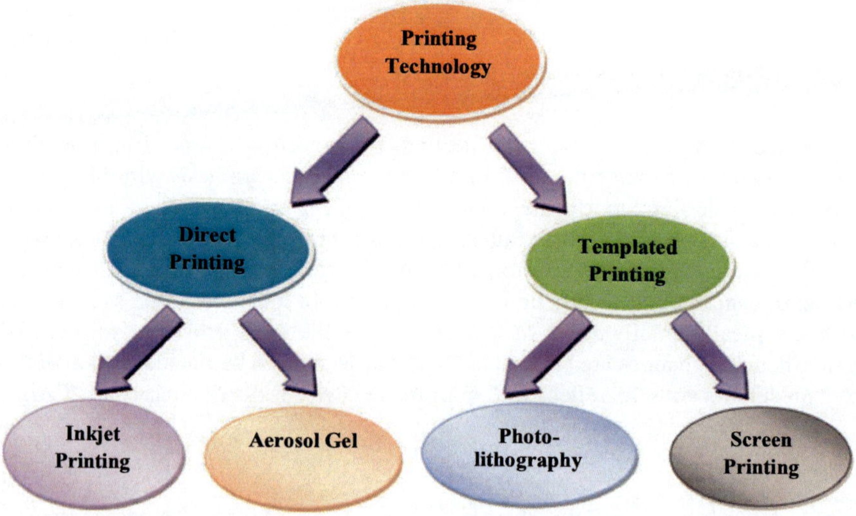

Fig. 8.11 Various types of printing techniques used in anticounterfeiting

8.3.1 Inkjet Printing

In the inkjet printing process, a predetermined amount of ink is ejected from nozzles during through an abrupt, quasi-adiabatic drop in the chamber's volume caused by piezoelectric action. A liquid-filled container contracts when an external voltage is applied. A liquid drop ejects from the nozzle as a result of the abrupt reduction creating a shockwave in the liquid. Recent review studies have discussed and partially explored this mechanism (Kumar et al. 2016). The ejected drop encounters the substrate and travels toward it under the influence of gravity and air resistance till contact is made. From that point, it spreads under the effects of kinematic momentum and surface tension-assisted flow. When polymer solutions are printed using an inkjet printer, non-Newtonian fluid behavior is typically observed. Usually, an elongating filament holds a droplet of polymer solution to the tip for many hundred microseconds (Tekin et al. 2007).

Inkjet printer's practicality for generating security codes can be attributed to their alluring features, such as:

(i) Its suitability for use with substrates with big surfaces.
(ii) The potential for entirely additive operation. Inks are applied based on their requirement.

8.3.2 Aerosol Gel

Aerodynamic focusing is used in the aerosol jet printing process to deposit colloidal dispersion and/or chemical precursor solutions with high resolution.

The fundamental system comprises two essential parts:

(i) A component that atomizes raw liquid ingredients.
(ii) The other module for depositing the droplets and concentrating the aerosol.

The QR (Quick Response) code printing procedure by an aerosol jet printer is demonstrated in Fig. 8.12.

8.3.3 Photolithography

The primary technique used to create printed circuit boards (PCBs) and microprocessors is photolithography. In an essential microfabrication method, photolithography is utilized to pattern substrates for contemporary electronics, sensors, and microfluidics. The interface of a wafer is coated with a light-sensitive polymer known as a photoresist in this precise type of bespoke surface fabrication. In the production of contemporary semiconductors, photolithography employs layers of photoresist to image the mask on a silicon wafer using optical radiation.

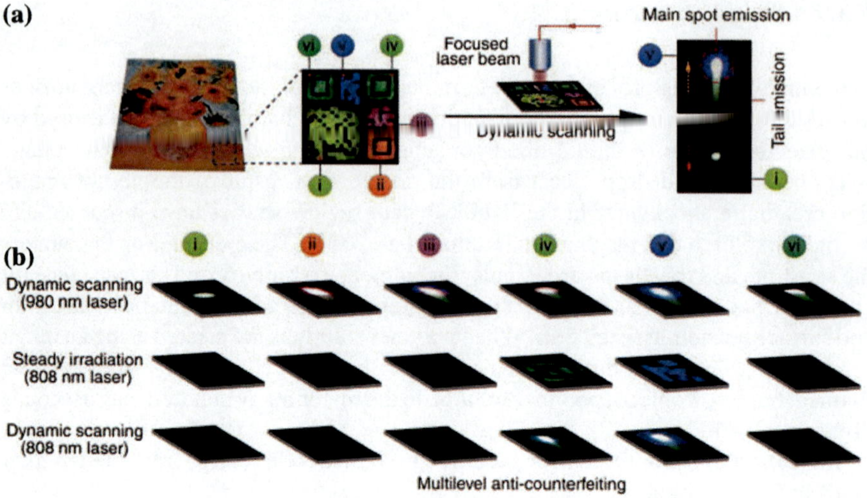

Fig. 8.12 Multilevel anticounterfeiting by using binary temporal upconversion codes of Mn^{2+}-activated nanoparticles. Reproduced with permission of ref. (Liu et al. 2017b) Copyright 2017, under Creative Commons Licence

8.3.4 Screen Printing

In the process of screen printing, ink is transferred to a surface having a mesh, except for locations where a blocking stencil has made the surface impervious to the ink. The open mesh holes are filled with ink and then spread it across the screen with the help of squeegee. Then, to make contact with the substrate along a line of contact, a reverse stroke on screen will help. Figure 8.12a shows 2D designs generally formed by using nanoparticles with various compositions. In contrast to rapid screening of the patterns with a focused beam of lasers (64 W cm^2), continuous irradiation with a 980 nm laser (6 W cm^2) causes the patterns to have multicolor properties. In Fig. 8.12bi–vi under various irradiation scenarios, the emission characteristics of the designs were captured (Liu et al. 2017a).

8.4 Lanthanide-Doped Nanomaterials for Anticounterfeiting

Numerous anticounterfeiting techniques, such as holograms, luminescence, two-dimensional codes, and barcodes have been investigated to stop counterfeiting. There are several conventional anticounterfeiting methods, but they are simple to copy and difficult to use. Since luminescence materials have excellent optical properties such as huge emission intensities, long emission lifetimes, multiple emission colors, and a variety of emission methods including mechanoluminescence, chemiluminescence, and photoluminescence, they have become increasingly popular as anticounterfeiting techniques. Currently, anticounterfeiting patterns made with lanthanide-doped nanomaterial-based ink is the most effective luminescence anticounterfeiting technique. According to the anticounterfeiting level, the luminescence anticounterfeiting strategies can be divided into three categories: single mode, double mode, and multimode anticounterfeiting (Yu et al. 2021).

8.4.1 Single-Mode Luminescence Anticounterfeiting

Single-mode luminescence, which has anticounterfeiting labels that change only once in response to outside stimuli, is the most straightforward photoluminescence anticounterfeiting technique. The single-mode luminescence anticounterfeiting can be further subdivided into downshifting (DS), upconversion (UC), and chemiluminescence anticounterfeiting methods based on the variation in luminescence modes. Figure 8.13 shows the downshifting luminescence-based anticounterfeiting label "SUST" made by the electrospinning method, which is transparent in daylight and turns into green color under UV light.

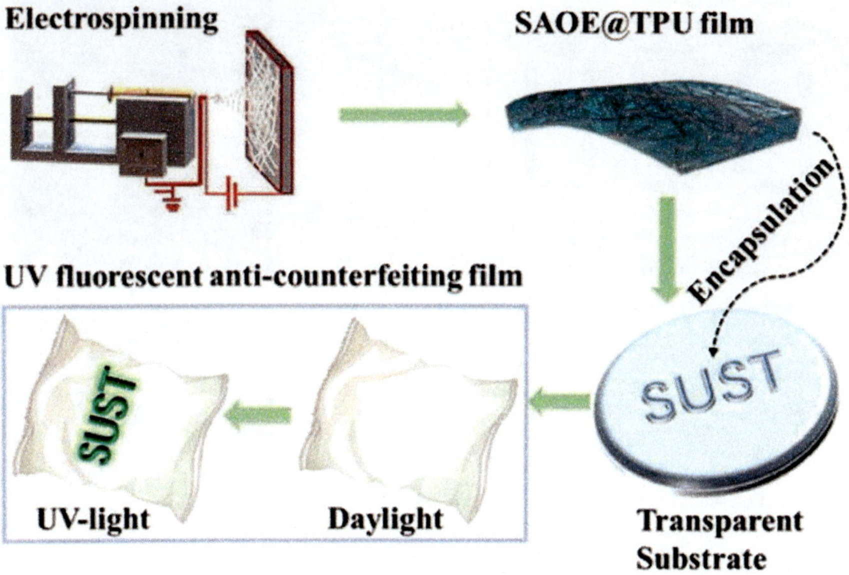

Fig. 8.13 Single-mode anticounterfeiting depending on up- and downshifting photoluminescence. Reproduced with permission of ref. (Gao et al. 2021) Copyright 2021, Elsiver Publishing

8.4.2 Dual-Mode Luminescence Anticounterfeiting

Dual-mode photoluminescence is more sophisticated than single-level photoluminescence anticounterfeiting because, in double mode anticounterfeiting labels vary twice in response to external stimuli and luminescence lifetime regulation, responding to raising the anticounterfeiting level. Based on means of decoding anticounterfeiting tags, the dual mode luminescence anticounterfeiting can be divided into three parts: (1) regulation via excitation wavelength, (2) co-regulation by excitation wavelength and photoluminescence lifetime, and (3) co-operatively regulated by excitation wavelength and additional variables (e.g., heat, mechanical force, and chemical reagents) (Yu et al. 2021). Dual-mode anticounterfeiting pattern made with Ln^{3+}-doped $Na_2CaGe_2O_6$ (NCG) is shown in Fig. 8.14. Under 980 nm excitation, the pattern looks like "381" with three colors, blue, green, and yellow, observed. Additionally, the red "3" and "1" as well as the green "2" are visible in the presence of a 254 nm UV light. Due to the continuous green emission of Tb^{3+} ions, the pattern will retain a green "2" after the UV laser is removed, while other digital designs will dissipate (Jin et al. 2022). Therefore, these phosphors with different luminescence properties offer the potential for more sophisticated dual-mode anticounterfeiting.

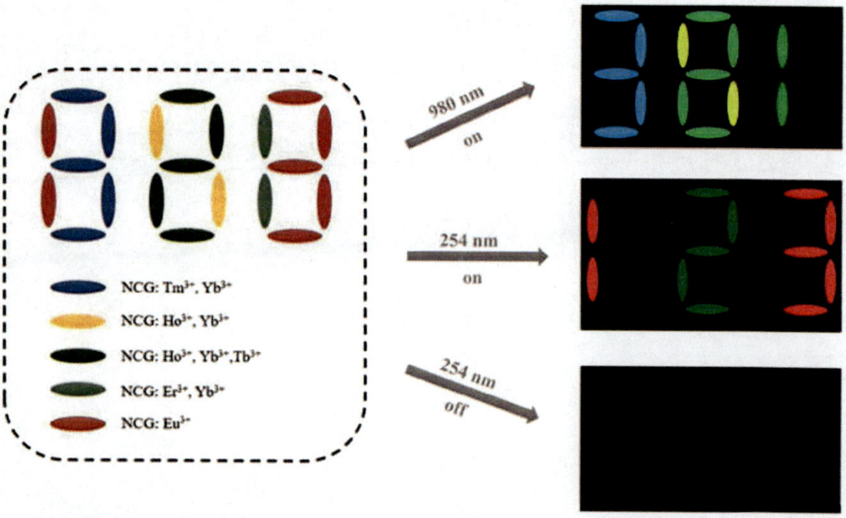

Fig. 8.14 Dual-level anticounterfeiting luminescence with excitation light-dependent co-regulation. Reproduced with permission of ref. Jin et al. (2022) Copyright 2022, Elsevier Publishing

8.4.3 Multimode Luminescence Anticounterfeiting

Anticounterfeiting based on multimode luminescence is the most sophisticated method of anticounterfeiting in comparison to other modes of luminescence anti-counterfeiting. Based on regulation factors, multimode luminescence anticounterfeiting can be categorized into four types: (1) regulation via excitation light, (2) co-regulation by excitation light and photoluminescence lifetime, (3) co-regulation by excitation light and chemical reagents, and (4) excitation from heat and light stimulation co-regulation (Wei et al. 2006). Figure 8.15 comprises the multiple anti-counterfeiting image: a leaf made of NCG:Pb^{2+},Mn^{3+},Yb^{3+} phosphor, a flower design formed with NCG:Pb^{2+},Tb^{3+} phosphor, and a butterfly pattern built of NCG:Pb^{2+},Y^{3+} phosphors (Shi et al. 2019). The image printed on a typical cotton fabric is nearly invisible under natural light, however, it is obviously apparent to the touch. The green–blue leaf, the red-violet flower, and the white-blue butterfly can be easily seen under 254 nm UV light. The next color change is performed by turning off the UV light, which causes the blue fluorescence to suddenly dissipate and be replaced by a green leaf, a red flower, and a deep-blue luminous butterfly. The final color variation occurs under 295 nm excitation, which causes all components of the photo to switch to the deep-blue color. It is obvious that the variety of luminescent patterns can drive numerous innovative anticounterfeiting designs (Shi et al. 2019).

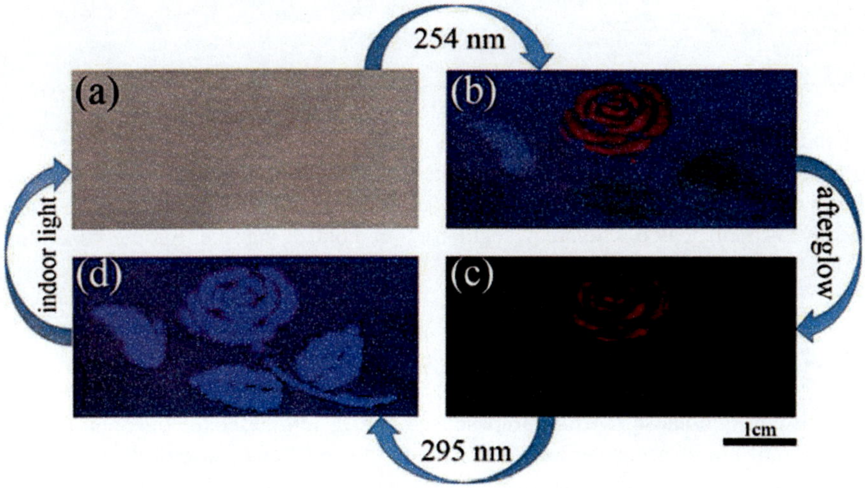

Fig. 8.15 Multimode luminescence anticounterfeiting co-regulated with different excitation wavelengths. Reproduced with permission of ref. Shi et al. (2019) Copyright 2019, American Chemical Society

8.4.3.1 Regulation by Excitation Light

By changing the wavelength of excitation light, irradiation mode, irradiation period, and excitation power, it is easy to achieve multilevel luminescence anticounterfeiting. This effective anticounterfeiting method does not require the use of additional external stimuli. When activated by photons with different wavelengths, the anticounterfeit label produces fluorescence in a variety of hues, achieving multilevel luminescence anticounterfeiting. The luminescence of anticounterfeiting labels can also be controlled by excitation power in a similar way to that of excitation wavelength.

8.4.3.2 Co-regulation by Luminescence Lifetime and Excitation Light

By adjusting the excitation wavelength, irradiation mode, and irradiation time, it is possible to achieve anti-counterfeiting that is also co-regulated by the photoluminescence lifetime and excitation wavelength. On the other hand, due to the addition of the time dimension, only one excitation wavelength may also achieve multimode photoluminescence anticounterfeiting. Excitation light also affects anticounterfeiting and it is possible to extend luminescence lifetime by adjusting the excitation wavelength, irradiation period, and irradiation mode. The irradiation time is also another useful control factor in multimode photoluminescence anticounterfeiting, which can control both the fluorescence and phosphorescence of materials.

8.4.3.3 Co-regulation by Excitation Light and Chemical Reagents

Chemical reagents can also cause several modifications to labels that are intended to prevent counterfeiting. The effectiveness of these anticounterfeiting techniques shows that water is a powerful stimuli component that can take over the role of other sophisticated chemical regulators in controlling the luminescence of anticounterfeiting tags.

8.4.3.4 Co-regulation by Heat and Light Stimulation

Heat stimuli can cause thermoluminescence (ThL), luminescence quenching, and alteration of luminescence color. The change in the heat can cause of discoloration luminescence materials in addition to causing luminescence enhancement. This makes heat stimulation unquestionably an essential stimuli factor in multimode luminescence anticounterfeiting. Heat stimulus helps the anticounterfeit label's fluorescence and color return. This anticounterfeiting tactic's efficiency was demonstrated by the numerous anticounterfeiting label variations that were co-regulated by heat stimuli and excitation light.

8.5 Conclusions

The widespread practice of forging crucial documents, including money, has turned into a serious menace to society. This article examines current anticounterfeiting application developments that are critically needed by the scientific community, businesses, government agencies, and consumers to safeguard their valuable documents. Beginning with a historical overview of luminous nanomaterials, fundamental ideas about the mechanism of luminescence, and a detailed explanation of the complete process from the creation of desirable luminous nanoparticles, ink, and materials synthesis and using various printing methods for security measures before exploring anticounterfeiting applications. In this article, the selection of luminous nanoparticles utilized in the creation of several types of security inks as well as their applications are discussed. The four common printing methods, i.e., screen printing, aerosol jet printing, inkjet printing, and photolithography that are utilized to print security codes and graphics were thoroughly discussed. The main aim of this article is to offer several approaches for the use and selection of different types of luminous nanoparticles, as well as how to make ink for high-level security printing to prevent the forging of papers, banknotes, medicines, and other branded goods.

Acknowledgements Science and Engineering Research Board (SERB), India (File Number: CRG/2021/007668), is acknowledged for the financial support. JY thanks CSIR, India, for the fellowship.

References

Abdollahi, A., Roghani-Mamaqani, H., Razavi, B., Salami-Kalajahi, M.: Photoluminescent and chromic nanomaterials for anticounterfeiting technologies: recent advances and future challenges. ACS Nano **14**, 14417–14492 (2020)

Bandi, V.R., Grandhe, B.K., Jang, K., Lee, H.S., Shin, D.S., Yi, S.S.J.J.: Citric based sol–gel synthesis and luminescence characteristics of $CaLa_2ZnO_5$:Eu^{3+}phosphors for blue LED excited white LEDs. J. Alloys Compd. **512**, 264–269 (2012)

Campos-Cuerva, C., Zieba, M., Sebastian, V., Martínez, G., Sese, J., Irusta, S., Contamina, V., Arruebo, M.S.J.: Screen-printed nanoparticles as anti-counterfeiting tags. Nanotechnology **27**, 0957–4484 (2016)

Dexter, D.L.: Possibility of luminescent quantum yields greater than unity. Phys. Rev. **108**, 630 (1957)

Dubnikova, N., Garskaite, E., Beganskiene, A., Kareiva, A.: Sol–gel synthesis and characterization of sub-microsized lanthanide (Ho, Tm, Yb, Lu) aluminium garnets. Opt Mater (amst) **33**, 1179–1184 (2011)

Fusari, F., Lagatsky, A.A., Richards, B., Jha, A., Sibbett, W.B.C.: Spectroscopic and lasing performance of Tm^{3+}- doped bulk TZN and TZNG tellurite glasses operating around 1.9 μm. Opt. Express **16**, 19146–19151 (2008)

Gai, S., Li, C., Yang, P., Lin, J.: Recent progress in rare earth micro/nanocrystals: soft chemical synthesis, luminescent properties, and biomedical applications. Chem. Rev. **114**, 2343–2389 (2014)

Gao, W., Ge, W., Shi, J., Tian, Y., Zhu, J.L.Y.: Stretchable, flexible, and transparent $SrAl_2O_4$:Eu^{2+}@TPU ultraviolet stimulated anti-counterfeiting film. Chem. Eng. J. **405**, 126949 (2021)

Ge, X., Liu, J., Sun, L.: Controlled optical characteristics of lanthanide doped upconversion nanoparticles for emerging applications. Dalton Trans. **46**, 16729–16737 (2017)

Gordon, W.O., Carter, J.A., Tissue, B.M.: Long-lifetime luminescence of lanthanide-doped gadolinium oxide nanoparticles for immunoassays **108**, 339–342 (2008)

Ghosh, P., Mudring, A.V.: Phase selective synthesis of quantum cutting nanophosphors and the observation of a spontaneous room temperature phase transition. Nanoscale **8**, 8160–8169 (2016)

Huang, X.Y., Yu, D.C.Z.Q.: Enhanced near-infrared quantum cutting in $GdBO_3$: Tb^{3+}, Yb^{3+} phosphors by Ce^{3+} codoping. J. Appl. Phys. **106**, 113521 (2009)

Huang, X., Han, S., Huang, W., Liu, X.: Enhancing solar cell efficiency: the search for luminescent materials as spectral converters. Chem. Soc. Rev. **42**, 173–201 (2012)

Jia, G., Liu, K., Zheng, Y., Song, Y.Y.H.: Facile synthesis and luminescence properties of highly uniform MF/YVO_4:Ln^{3+} (Ln) Eu, Dy, and Sm) composite microspheres. Cryst. Growth Des. **9**, 3702–3706 (2009)

Jin, X., Wang, Z., Wei, Y.F.Z.: Dual-mode multicolor luminescence based on lanthanide-doped $Na_2CaGe_2O_6$ phosphor for anticounterfeiting application. J. Lumin. 118937 (2022)

Kang, D., Jeon, E., Kim, S., Lee, J.: Lanthanide-doped upconversion nanomaterials: recent advances and applications. The Korean BioChip Soc. Springer **14**, 124–135 (2020)

Kumar, P., Singh, S., Gupta, B.K.: Future prospects of luminescent nanomaterial based security inks: from synthesis to anti-counterfeiting applications. Nanoscale **8**, 14297–14340 (2016)

Kumar, P., Nagpal, K., Kumar Gupta, B.: Unclonable security codes designed from multicolor luminescent lanthanide-doped Y_2O_3 nanorods for anticounterfeiting. ACS Appl. Mater. Interfaces **9**, 14301–14308 (2017)

Kumar, K.N., Vijayalakshmi, L., Choi, J.K.J.: Efficient red-luminescence of $CaLa_2ZnO_5$ phosphors co-doped by Ce^{3+} and Eu^{3+} ions. J. Alloys Compd. **787**, 711–719 (2019)

Kumar, A., Kuang, Y., Liang, Z., Sun, X.: Microwave chemistry, recent advancements, and eco-friendly microwave-assisted synthesis of nanoarchitectures and their applications: a review. Mater. Today Nano **11**, 100076 (2020)

Lingeshwar Reddy, K., Balaji, R., Kumar, A.K.V.: Lanthanide doped near infrared active upconversion nanophosphors: fundamental concepts, synthesis strategies, and technological applications. Small **14**, 1801304 (2018)

Liu, C., Wang, H., Li, X., Chen, D.: Monodisperse, size-tunable and highly efficient b-$NaYF_4$:Yb, Er(Tm) up-conversion luminescent nanospheres: controllable synthesis and their surface modifications †. J. Mater. Chem. **19**, 3546–3553 (2009)

Liu, Y., Zheng, Y., Zhu, Y., Ma, F., Zheng, X., Yang, K., Zheng, X., Xu, Z., Ju, S., Zheng, Y.G.T.: Unclonable perovskite fluorescent dots with fingerprint pattern for multilevel anticounterfeiting (2010)

Liu, Y., Tu, D., Zhu, H., Chen, X.: Lanthanide-doped luminescent nanoprobes: controlled synthesis, optical spectroscopy, and bioapplications. Chem. Soc. Rev. **42**, 6924–6958 (2013)

Liu, X., Wang, Y., Li, X., Yi, Z., Deng, R., Liang, L., Xie, X., Loong, D.T., Song, S., Fan, D.A.A.H.X.L.: Binary temporal upconversion codes of Mn^{2+}-activated nanoparticles for multilevel anti-counterfeiting. Nat Commun **8**, 1–7 (2017)

Liu, X., Wang, Y., Li, X., Yi, Z., Deng, R., Liang, L., Xie, X., Loong, D.T., Song, S., Fan, D.A.A.: Binary temporal upconversion codes of Mn^{2+}-activated nanoparticles for multilevel anti-counterfeiting. Nat. Commun. **8**, 1–7 (2017a)

Liu, X., Ji, Q., Hu, Q., Chen, L., Chen, M., Sun, J., Wan, Y., Sun, Q.B.G.: Dual-mode long-lived luminescence of Mn^{2+}-doped nanoparticles for multilevel anticounterfeiting. ACS Appl. Mater. Interfaces **11**, 30146–30153 (2019)

Lorbeer, C., Cybinska, J., Mudring, A.-V.: Facile preparation of quantum cutting GdF_3 : Eu^{3+} nanoparticles from ionic liquidsw. In: Science (1979). Accessed 11 Mar 2023 (1999)

Luo, X., Zhao, W., Chen, Q., Zhi, M.: Controllable hydrothermal synthesis of $NaYbF_4$ and YbF_3 with diverse morphologies and emission enhancement in β-$NaYbF_4$: Er^{3+} upconversion microcrystals via solvents codoping and Na^+ dosage. J. Alloys Compd. **897**, 162672 (2022)

Lupo, D., Clemens, W., Breitung, S.H.K.: OE-A roadmap for organic and printed electronics. applications of organic and printed electronics. In: Applications of Organic and Printed Electronics. Accessed 27 Sep 2022 (2013)

Luwang, M.N., Ningthoujam, R.S., Srivastava, S.K., Vatsa, R.K.: Disappearance and recovery of luminescence in Bi^{3+}, Eu^{3+} codoped YPO_4 nanoparticles due to the presence of water molecules up to 800 °C. J. Am. Chem. Soc. **133**, 2998–3004 (2011)

Nichkova, M., Dosev, D., Gee, S.J., Hammock, B.D.K.I.: Microarray immunoassay for phenoxybenzoic acid using polymer encapsulated Eu:Gd_2O_3 nanoparticles as fluorescent labels. Anal. Chem. **77**, 6864–6873 (2022)

Patra, A., Friend, C.S., Kapoor, R., Prasad, P.N.: Fluorescence upconversion properties of Er^{3+}-doped TiO_2 and $BaTiO_3$ nanocrystallites. Chem. Mater. **15**, 3650–3655 (2003)

Pushpendra, K.R.K., Achary, S.N., Naidu, B.S.: $NaBi_{0.9}Eu_{0.1}(MoO_4)_2$ nanomaterials: tailoring the band gap and luminescence by La^{3+} substitution for light-emitting diodes. ACS Appl. Nano Mater. **2**, 5527–5537 (2019)

Pushpendra, S.S., Srinidhi, S., Kunchala, R.K., Kalia, R., Achary, S.N.N.B.: Structural and excitation-dependent photoluminescence properties of $Bi_{0.95-x}Gd_xEu_{0.05}PO_4 (0 \leq x \leq 0.95)$ solid solutions and their anticounterfeiting applications. Cryst. Growth Des. **21**, 4619–4631 (2021a)

Pushpendra, S.I., Srinidhi, S., Singh, S., Kalia, R., Kunchala, R.K., Mudavath, S.L.N.B.: Downshifting and upconversion dual mode emission from lanthanide doped $GdPO_4$ nanorods for unclonable anti-counterfeiting. Mater. Today Commun. **26**, 102144 (2021b)

Ronda, C.: Luminescent materials with quantum efficiency larger than 1, status and prospects. J. Lumin. **100**, 301–305 (2002)

SeethaLekshmi, S., Ramya, A.R., Reddy, M.L.P., Varughese, S.: Lanthanide complex-derived white-light emitting solids: a survey on design strategies. J. Photochem. Photobiol., C **33**, 109–131 (2017)

Sharma, R.K., Mudring, A.V., Ghosh, P.: Recent trends in binary and ternary rare-earth fluoride nanophosphors: how structural and physical properties influence optical behavior. J. Lumin. **189**, 44–63 (2017)

Sharma, R.K., Ghosh, P.: Lanthanide-doped luminescent nanophosphors via ionic liquids. Front. Chem. **9**, 580 (2021)

Su, Q., Han, S., Xie, X.: The effect of surface coating on energy migration-mediated upconversion. J. Am. Chem. Soc. **134**, 20849–20857 (2012)

Stouwdam, J.W., Van Veggel, F.C.J.M.: Near-infrared emission of redispersible Er3+, Nd3+, and Ho3+ doped LaF3 nanoparticles. Nano Lett. **2**, 733–737 (2002)

Sommerdijk, J.L., Bril, A., de Jager, A.W.: Two photon luminescence with ultraviolet excitation of trivalent praseodymium. J. Lumin. **8**, 341–343 (1974a)

Sommerdijk, J.L., Bril, A., de Jager, A.W.: Luminescence of Pr^{3+}-activated fluorides. J. Lumin. **9**, 288–296 (1974b)

Shi, C., Shen, X., Zhu, Y., Li, X., Pang, Z.G.M.: Excitation wavelength-dependent dual-mode luminescence emission for dynamic multicolor anticounterfeiting. ACS Appl. Mater. Interfaces **11**, 18548–18554 (2019)

Tekin, E., Smith, P.J., Schubert, U.S.: Inkjet printing as a deposition and patterning tool for polymers and inorganic particles. Soft Matter **4**, 703–713 (2007)

Valeur, B., Ario, M., Berberan-Santos, N.: A brief history of fluorescence and phosphorescence before the emergence of quantum theory. J. Chem. Educ. **88**, 731–738 (2011)

Wang, F., Liu, X.: Recent advances in the chemistry of lanthanide-doped upconversion nanocrystals. Chem. Soc. Rev. **38**, 976–989 (2009)

Wang, F., Zhang, Y., Fan, X., Wang, M.: Facile synthesis of water-soluble LaF3: Ln3+ nanocrystals. J. Mater. Chem. **16**, 1031–1034 (2013)

Wegh, R.T., Donker, H., Oskam, K.D., AM,: Visible quantum cutting in $LiGdF_4:Eu^{3+}$ through downconversion. Science **1979**, 663–666 (1999)

Wei, Y., Lu, F., Zhang, X., Chen, D.: Synthesis of oil-dispersible hexagonal-phase and hexagonal-shaped $NaYF_4$:Yb, Er nanoplates. Chem. Mater. **18**, 5733–5737 (2006)

Yang, D., Dai, Y., Ma, P., Kang, X., Cheng, Z., Li, C.L.J.: One-step synthesis of small-sized and water-soluble $NaREF_4$ upconversion nanoparticles for in vitro cell imaging and drug delivery. Chem.–A Eur. J. **19**, 2685–2694 (2013)

Yao, W., Tian, Q., Wu, W.: Tunable emissions of upconversion fluorescence for security applications. Adv. Opt. Mater. **7**, 1801171 (2019)

Yu, X., Zhang, H., Yu, J.: Luminescence anti-counterfeiting: from elementary to advanced. Aggregate **2**, 20–34 (2021)

Zhang, Y.W., Sun, X., Si, R., You, L.P.Y.C.: Single-crystalline and monodisperse LaF_3 triangular nanoplates from a single-source precursor. J. Am. Chem. Soc. **127**, 3260–3261 (2005)

Zhou, B., Shi, B., Jin, D.L.X.: Controlling upconversion nanocrystals for emerging applications. Nat. Nanotechnol. **10**, 924–936 (2015)

Zhou, B., Shi, B., Jin, D.L.X.: Optical temperature sensing of up-conversion luminescent materials: fundamentals and progress. Elsevier Ltd. (2020)

Zou, Z., Feng, L., Cao, C., Zhang, J.W.Y.: Near-infrared quantum cutting long persistent luminescence. Sci. Rep. **6**, 1–7 (2016)

Chapter 9
Thermally Activated Delayed Fluorescence in Metal-Free Small Organic Materials: Understanding and Applications in OLEDs

Biki Kumar Behera and Neeraj Agarwal

9.1 Introduction

The first example of thermally activated delayed fluorescence (TADF) was observed in the solid state by Perrin (1919) and in solution by Lewis (1941) (Randall John Turton 1939; Lewis and Lipkin 1941). A few groups were still active in TADF research; however, in 2012, it gained resurgence due to its applications in organic electronics as shown by Adachi's research group (Uoyama et al. 2012). They reported the TADF mechanism to harvest the triplet excitons produced in the emitting layer *without heavy metal* in organic light-emitting diodes (OLED). Owing to its application in OLED, TADF has been studied extensively to enhance efficiency. According to the spin statistic theorem, direct charge recombination in electroluminescent devices produces singlet and triplet excitons in 1:3 ratios, respectively. The triplet states were considered to be non-emissive at room temperature. Therefore, harvesting non-emissive triplet excitons became the center of attraction for researchers in academia as well as in many industries (Dias et al. 2013). Conventional fluorescent material-based OLEDs have shown excellent stability; however, they only harvest on singlet excitons. On the other hand, heavy metal-based emitters (phosphors) can harvest both singlet and triplet excitons; nevertheless, heavy metal (such as Pt, Ir, and Rh)-based phosphorescent emitters are expensive, hazardous to the environment, and thus are not the first choice for commercial usage. Pure organic emitters exhibiting TADF with external quantum efficiency (EQE) comparable to the best-performing phosphorescent emitters are now known so, TADF-based electroluminescent devices have been the center of focus. Also, the electronic and photophysical properties of

B. K. Behera · N. Agarwal (✉)
Centre for Excellence in Basic Sciences, UM-DAE, University of Mumbai, Kalina Campus, Santacruz (E), Mumbai 400098, India
e-mail: na@cbs.ac.in

R. S. Ningthoujam and A. K. Tyagi (eds.), *Handbook of Materials Science, Volume 1*, Indian Institute of Metals Series, https://doi.org/10.1007/978-981-99-7145-9_9

these TADF molecules can be altered by structural design. A lot of research has been carried out to optimize the molecular designing of organic emitters showing efficient TADF. Apart from TADF, other photophysical mechanisms, such as triplet–triplet annihilation (TTA) and upper triplet crossing which is formally known as "Hot Exciton", have also been shown to harvest the dark triplet states. TTA is also called P-type delayed fluorescence as it was first observed in pyrene and phenanthrene solution. The mechanism of TTA was formulated by Parker and Hatchard (Parker and Hatchard 1961). They observed that the intensity of delayed fluorescence in pyrene and phenanthrene was proportional to the square of the intensity of absorption. The square-law dependence indicates the delayed fluorescence observed in aromatic hydrocarbons is a biphotonic process

$$I_{\mathrm{PD}} \propto I_{\alpha}^{2}$$

Φ_{PD} = Intensity of delayed fluorescence/intensity of absorption $= I_a{}^2/I_a = I_a$.

The mechanism of the observed biphotonic delayed fluorescence process involves the encounter of two triplets with energy transfer between them producing an intermediate (X). In pyrene solution, the intermediate is considered to be dimeric, since the characteristic emission of excimer was also observed along with P-type delayed fluorescence. The intermediate excited state species dissociates to produce a singlet excited state and a ground state. Further, the singlet state undergoes radiative decay. The emission rate constant of P-type delayed fluorescence was observed to be governed by the triplet state production.

$$T + T \rightarrow (X) \rightarrow S_1 + S_0 \rightarrow S_0 + h\nu_f$$

The spin-exchange mechanism involved in the singlet production from two triplet states is shown to be spin allowed

$$T(\uparrow\uparrow) + T(\downarrow\downarrow) \rightarrow S_0(\uparrow\downarrow) + S_1(\uparrow\downarrow)$$

The maximum theoretical yield that can be obtained from TTA as the triplet harvesting mechanism is 62.5%, whereas the maximum theoretical yield that can be obtained from TADF is 100%. TADF molecules require small singlet–triplet splitting (ΔE_{ST}); however, it has been observed that devices with TADF molecules having high ΔE_{ST} also exhibit high EQE (Liu et al. 2019). In these devices, hot exciton transfer ($T_n \rightarrow S_n$) RISC predominates over the cold exciton *RISC* ($T_1 \rightarrow S_1$). Excitons formed in the higher excited state (hot excitons) undergo rapid internal conversion to the lowest excited states (S_1 and T_1) before the radiative decay as per *Kasha's rule.* When the rate of internal conversion (IC) and intersystem crossing (ISC) are comparable then the ratio of singlet to triplet states changes due to the ISC and RISC processes. This can result in a change in fluorescence efficiency (Pan et al. 2014). Wang et al. reported that for a molecule having high S_1-T_1 and T_2-T_1 energy gaps, the $T_2 \rightarrow S_1$ reverse intersystem crossing (RISC) competes with $T_2 \rightarrow T_1$ internal conversion (IC). Unlike phosphorescence and TADF-based emitters, hot exciton-based emitters do not undergo efficiency roll-off. Hence, a hot exciton-based upper triplet transfer mechanism can be used as an alternate triplet exciton harvesting route (Liu et al. 2019). The detailed photophysical processes in TADF molecules were studied by Dias et al. (2017). They demonstrated the dynamic nature of the reverse intersystem crossing (RISC) process in TADF. Apart from molecular designing of TADF molecules, solvent effect or interaction with a host also plays a significant role in determining the efficiency of emitters (Etherington et al. 2016; Méhes et al. 2014; Sun et al. 2017).

9.2 Fundamentals of TADF

Photoexcitation of the ground state fluorophore mostly leads to singlet exciton formation which rapidly undergoes IC to the lowest singlet excited state (S_1), and further it undergoes radiative and/or non-radiative decay. In a conventional TADF molecule, the ΔE_{ST} is small, and consequently the T_1 state can be up-converted to the S_1 state by thermal activation, which can further undergo fluorescence emission to the ground state (Figs. 9.1 and 9.2). Spectral properties of phosphorescence and fluorescence are quite different but the emission spectra of TADF or sometimes also known as delayed fluorescence (DF) and prompt fluorescence (PF) coincide exactly with a larger lifetime for DF, implying that both processes involve emission from S_1 state (Dias et al. 2017).

The complex mechanism of TADF is best perceived when studied using time-resolved methods following the photoluminescence measurements. Time-resolved spectra of TADF molecules usually show a fast decay component corresponding

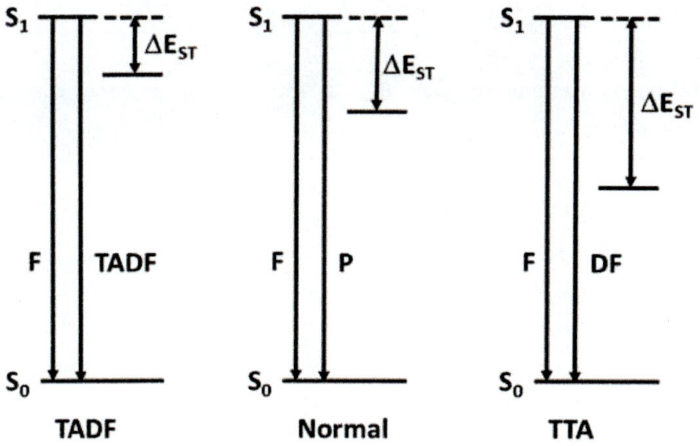

Fig. 9.1 Depiction of the diverse photophysical process involved between lowest singlet and triplet excited states, where F, P, TADF, and TTA denote fluorescence, phosphorescence, thermally activated delayed fluorescence, and triplet–triplet annihilation, respectively

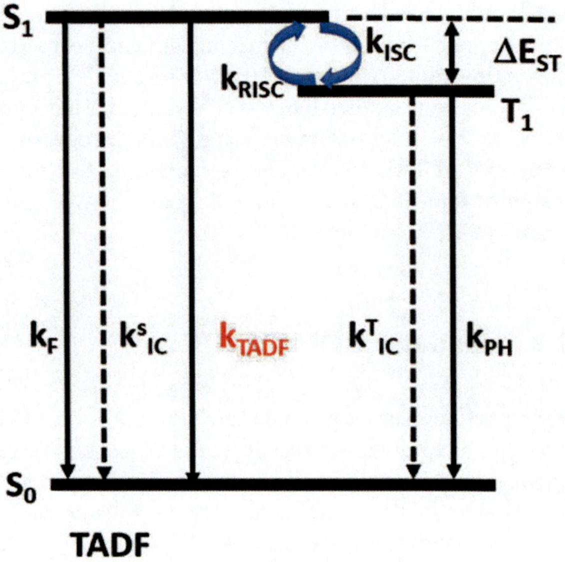

Fig. 9.2 Representation of TADF kinetic mechanism and electronic energy level involved in it

to the prompt fluorescence and a long lifetime component that corresponds to the delayed fluorescence. A high yield of triplet formation (Φ_{ISC}) and a high yield of singlet formation from RISC (Φ_{RISC}) are often required to observe strong TADF emission. These conditions are possible when the rate of RISC is much higher than the combined rate of phosphorescence (k_{Ph}) and vibrational decay of triplet states ($k^{T}{}_{IC}$) involved in TADF (Dias et al. 2017).

$$\Phi_{\text{rISC}} = \frac{k_{\text{rISC}}}{k_{\text{rISC}} + k_{\text{IC}}^{T} + k_{\text{Ph}}} \approx 1$$

The total emission quantum yield of a TADF emitter can be expressed by the following equation:

$$\Phi_F = \Phi_{\text{PF}} + \Phi_{\text{DF}} = \sum_{i=0}^{n} \Phi_{\text{PF}} (\Phi_{\text{ISC}} \Phi_{\text{rISC}})^i$$

The total fluorescence quantum yield approaches unity when the ratio of the quantum yield of PF to the quantum yield of DF became four or greater than four $\left(\frac{\Phi_{\text{DF}}}{\Phi_{\text{PF}}} \approx 4\right)$ where ΔE_{ST} was found to be less than 150 meV (Dias et al. 2016). In such a simplified system, the yield of triplet production was given by the following equation:

$$\Phi_{\text{ISC}} = \frac{\frac{\Phi_{\text{DF}}}{\Phi_{\text{PF}}}}{1 + \frac{\Phi_{\text{DF}}}{\Phi_{\text{PF}}}} = \frac{\Phi_{\text{DF}}}{\Phi_{\text{PF}} + \Phi_{\text{DF}}}$$

The triplet yield (Φ_{ISC}) of a strong TADF emitter can be calculated directly using the above equation so it is very important to discuss different strategies to determine the (Φ_{DF}/Φ_{PF}) ratio from their photophysical measurements. The prompt and delayed components of a TADF molecule can be measured directly from a single time-resolved spectrum using a degassed sample. The fluorescence decay curve shows two exponential decays from PF and DF as per the following equation (Dias et al. 2017):

$$I(t)_{Fl} = A_{PF}\exp\left(-\frac{t}{\tau_{PF}}\right) + A_{DF}\exp\left(\frac{t}{\tau_{DF}}\right)$$

$\frac{\Phi_{DF}}{\Phi_{PF}}$ can be calculated from the integral of PF and DF components.

$$\frac{\Phi_{DF}}{\Phi_{PF}} = \frac{A_{DF}\tau_{DF}}{A_{PF}\tau_{PF}}$$

The second approach includes the determination of (Φ_{DF}/Φ_{PF}) from steady-state measurement. Since molecular oxygen (triplet) strongly quenches the triplet state of the fluorophore, the delayed component gets suppressed in the presence of oxygen in a non-degassed solution. In such cases, emission spectra will only contain the prompt fluorescence component. Whereas the emission spectra obtained in a degassed solution will contain both the PF and DF components. The proportionality constant for both Φ_{PF} and $\Phi_{PF} + \Phi_{DF}$ will be the same since emission in both the solutions originates from the same excited state (S_1) (Dias et al. 2016). The $\frac{\Phi_{DF}}{\Phi_{PF}}$ can be calculated by taking the ratio of integrated fluorescence spectra as per the following equation:

$$\frac{\int I_{DF}^{deg}(\lambda)d\lambda}{\int I_{PF}^{O_2}(\lambda)d\lambda} = \frac{\Phi_{PF} + \Phi_{DF}}{\Phi_{PF}} = 1 + \frac{\Phi_{DF}}{\Phi_{PF}}$$

The lifetime (τ_{PF}) and prompt fluorescence quantum yield (Φ_{PF}) can be determined in a degassed solution or solid film. The radiative rate constant (k_f) can be known by the determination of Φ_{PF} and τ_{PF} as per the following equation:

$$k_f = \frac{\Phi_{PF}}{\tau_{PF}}$$

If k_f is much faster than the rate of internal conversion k_{IC}, the internal quantum efficiency (η_{int}) of a TADF molecule can be expressed as

$$\eta_{int} = \eta_{r_1 S}\Phi_{PF} + \eta_{r_1 S}\Phi_{TADF} + \eta_{r_1 T}\frac{\Phi_{TADF}}{\Phi_{ISC}}$$

where η_{r1S} and η_{r1T} are the singlet and triplet production efficiency, respectively, and Φ_{ISC} is the triplet production quantum yield (Dias et al. 2017). For a TADF emitter to have a high value of η_{int} both Φ_{PF} and $\frac{\Phi_{DF}}{\Phi_{PF}}$ need to be improved.

$$\tau_{PF} = \frac{1}{K_f + K_{IC} + K_{ISC}}$$

The relationship between k_f and absorption coefficient of fluorescent molecules with large HOMO–LUMO overlap was given by *Strickler* and *Berg* as per the following equation (Strickler and Berg 1962):

$$k_f = 2.88 \times 10^{-9} n^2 \langle \overline{v_f^{-3}} \rangle^{-1} \int \varepsilon(v_a) d \ln(v_a)$$

$$\langle v_f^{-3} \rangle^{-1} = \frac{\int f(v_f) dv_f}{\int f(v_f) v_f^{-3} dv_f}$$

where υ_a is the absorption wavenumber, ε (υ_a) denotes the molar absorption coefficient at $\upsilon_{a,}$ n is the refractive index, υ_f is the fluorescence wave number, and $f(\upsilon_f)$ is the fluorescence spectra. The relationship between transition dipole moment (Q) and oscillator strength (F) associated with the absorption was determined as

$$F = 4.32 * 10^{-9} n^{-1} \int \varepsilon(v_a) dv_a = \frac{8\pi^2 m_e c(v_f)}{3he^2} |Q|^2$$

where m_e is the mass of the electron, υ_f is the average fluorescence wavenumber, and c is the speed of light. It was also reported that for molecules with small HOMO–LUMO overlap, the calculated value of k_f from the absorption spectra was found to be the same as the value of k_f calculated from the experimental value of τ_F and Φ_{PF}. In such cases, the oscillator strength and transition dipole moment of fluorescence were the same as that of the oscillator strength and transition dipole moment of absorption. Hence, high Φ_{PF} can be achieved by high k_f, which can be achieved by improving F and Q. Also, when the RISC mechanism follows the Arrhenius model and with the assumption that k_f and k_{ISC} are much higher than k_{IC}, then $\frac{\Phi_{TADF}}{\Phi_{ISC}}$ can be expressed as

$$\frac{\Phi_{TADF}}{\Phi_{ISC}} = \frac{\Phi_{PF}\Phi_{rISC}}{1 - \Phi_{ISC}\Phi_{rISC}} = \frac{\Phi_{PF}}{\frac{k_{rISC+k_{nr}}}{k_{rISC}} - \Phi_{ISC}}$$

$$= \frac{1}{\frac{k_{nr}}{k_{rISC}\Phi_{PF}} - 1} = \frac{1}{1 + \frac{k_{nr}}{\Phi_{PF} A \exp\left(-E\frac{ST}{kT}\right)}}$$

where k_{nr} is the rate constant of non-radiative decay of T_1 state to ground state S_0. It is evident that a small ΔE_{ST} is essential to have a high value of $\frac{\Phi_{TADF}}{\Phi_{ISC}}$. So, molecular designing of TADF should follow the requirement of small ΔE_{ST} and large Q and F to have high internal quantum efficiency (Strickler and Berg 1962).

9.3 Mechanism of Intersystem Crossing

In the mechanism of TADF a lower triplet excited state is up-converted by thermal energy and further undergoes intersystem crossing to a radiative singlet state. Intersystem crossing is a non-radiative decay pathway between two electronic states with different spin multiplicities ($S_j \rightarrow T_k$ or $T_k \rightarrow S_p$). A molecule of N atoms has 3N-6 modes of vibrational degrees of freedom, and consequently, the potential energy will be described by 3N-6 poly-dimensional hyperfaces. Radiation-less transition happens at the isoenergetic points where energy for both curves will be the same. In the theory of non-radiative transition, the states are coupled by the nuclear kinetic energy operator $\hat{J}_{N,}$ which is effective only between Born Oppenheimer states, where electronic and nuclear motions are decoupled. Under the framework of non-radiative transition theory, the probability of energy transfer between two states under the influence of the perturbation $\hat{J}_N$ is given by

$$\text{Probability} \sim \left[\int \Psi_f \hat{J} \Psi_i d\tau_e\right]^2$$

$$\sim \left[\int \Psi_f \hat{J} \Psi_i d\tau_e \int \chi^{v'} \chi^{v''} d\tau_v \int S_f S_i d\tau_S\right]^2$$

where $d\tau_e$, $d\tau_\upsilon$, $d\tau_s$ denote the configuration space for electronic, vibrational, and spin motions, respectively. In the above expression, the electronic wave function is only affected by the perturbation. The spin integral component determines the allowedness of the transition. The spin integral term is one when the transition happens between states of the same spin manifold and zero otherwise. However, it may have a nonzero value due to spin–orbit interaction. The middle integral term is the Frank–Condon overlap integral which defines the efficiency of the transition along with the symmetric restrictions and spin multiplicity rules.

9.3.1 The Coupling Elements

Previously, it was believed that in organic compounds the spin-forbidden non-radiative transitions are slow compared to their spin-allowed counterparts. However, recent findings pointed out that the ISC process in organic compounds can compete with IC (Penfold et al. 2018). Importantly, it has been established that the photophysics of fluorophore depends upon the coupling between electronic structure, nuclear motion, and spin magnetic moment. The electronic Hamiltonian in Pauli's approximation is expressed as the sum of the Electronic Hamiltonian ($\hat{H}_{EL}$) and Spin–Orbit Hamiltonian ($\hat{H}_{SO}$) (Penfold et al. 2018).

$$\hat{H}_{\text{EL}} = \hat{H}_{\text{ES}} + \hat{H}_{\text{SO}}$$

The spin-free electrostatic energy was contained in $\hat{H}_{ES}$ whereas the spin–orbit operator, $\hat{H}_{SO}$ couples the electronic orbitals and spin degrees of freedom. $\hat{H}_{EL}$ and $\hat{H}_{SO}$ both depend upon nuclear coordinate and can be expressed as

$$\hat{H}_{EL} = \hat{H}_{ES}(Q_0) + \hat{H}_{SO}(Q_0) + \sum_{\alpha}\left[\frac{\partial \hat{H}_{ES}}{\partial Q_\alpha} + \frac{\partial \hat{H}_{SO}}{\partial Q_\alpha}\right]_{Q_0}$$

where Q_0 is the Frank–Condon geometry and α is the nuclear degrees of freedom. The coupling element of the vibronic mixing of states with different spin multiplicities depends upon the vibrational degrees of freedom (Penfold et al. 2018).

9.3.2 Spin–Orbit Coupling

The interaction between electron spin and its orbital motion around the nucleus is known as Spin–Orbit Coupling (SOC). It is a relativistic effect, which allows the mixing of electronic states with different spin multiplicities by the mixing of orbital and spin degrees of freedom (Penfold et al. 2018). The SOC operator can be expressed in the most common *Breit-Pauli* form as

$$\hat{H}_{BP}^{SO} = \frac{1}{2m_e^2C^2}\sum_I\sum_i \frac{Z_I}{r_{Ii}^3}\left(\hat{r}_{iI} \times \hat{P}_i\right)\cdot \hat{s}_i - \frac{1}{2m_e^2C^2}\sum\nolimits_i\sum\nolimits_{j\neq i}\frac{1}{r_j^3}\left(\hat{r}_{ij} \times \hat{P}_i\right)\cdot\left(\hat{s}_i + 2\hat{s}_j\right)$$

where $\hat{r}_{iI} \times \widehat{P}_i$ denotes the angular momentum of ith electron with respect to the nucleus I, $\hat{r}_{ij} \times \widehat{P}_i$ denotes the angular momentum of ith electron with respect to jth electron, and $\hat{s}_i$ denotes the spin operator of ith electron. In the above equation, the first term is a one-electron term that arises due to the interaction of the spin magnetic moment of an electron with the induced magnetic moment by its orbital motion in the electric field produced by the nucleus. The second term contains two parts, (i) the spin-same-orbit coupling and (ii) spin-other-orbit coupling. Spin-same-orbit coupling arises from the interaction of the spin angular momentum of an electron with its orbital magnetic moment while the spin-other-orbit coupling part arises due to the interaction of the spin magnetic moment of one electron with the orbital magnetic moment of other electrons. The spin-other-orbit coupling term provides screening to the one-e006Cectron term. The one-electron term increases rapidly with the atomic number, so the contribution of one-electron term is dominant in heavier elements (Pyykkö 2012). The SOC operator can be expressed in a one-center approximation as

$$\widehat{H}_{\text{SO}}^{\text{eff}} = \frac{1}{2m_e^2c^2}\sum_I\sum_i \frac{Z_I^{eff}}{r_{Ii}^3}\widehat{l}_i.\widehat{s}_i$$

where the $\widehat{l}_i$ is the angular momentum and $\hat{s}_i$ is the spin operator of the ith electron and Z_i^{eff} denotes the effective nuclear charge due to the screening effect of electrons (Koseki et al. 1992). More importantly, we must consider the fact that the size of SOC depends upon the integral over two states as

$$\left[\hat{H}_{\text{SO}}\right]_{ij} = \left\langle \Psi_{S_i} | \hat{H}_{\text{SO}} | \Psi_{T_j} \right\rangle.$$

9.3.3 *Spin-Vibronic Coupling*

The static spin–orbit coupling described in the above section is limited only to some specific nuclear configuration, but in reality, it is not constant and depends on the vibrational degrees of freedom. In addition to SOC and non-adiabatic coupling, we have to also consider the scenario where the states are coupled by both spin–orbit coupling and vibrational coupling simultaneously (Penfold et al. 2018). Singlet–triplet coupling by spin-vibronic coupling mechanism is outlined by Albrecht (Albrecht 1963). The full spin–orbit interaction up to the second order can be expressed as

$$\begin{aligned}
\hat{H}_{\text{SO}} &= \langle \psi_{S_i} \langle \hat{H}_{\text{SO}} \rangle \psi_{T_j} \rangle + \sum_\alpha \frac{\delta \langle \psi_{S_i} \left| \hat{H}_{\text{SO}} \right| \psi_{T_j} \rangle}{\delta Q_\alpha} Q_\alpha \\
&+ \frac{1}{2}\sum_\alpha\sum_\beta \frac{\delta^2 \langle \psi_{S_i} \left| \hat{H}_{\text{SO}} \right| \psi_{T_j} \rangle}{\delta Q_\alpha \delta Q_\beta} Q_\alpha Q_\beta \\
&+ \sum_n \frac{\langle \psi_{S_i} \left| \hat{H}_{\text{SO}} \right| \psi_{T_n} \rangle \langle \psi_{T_n} \left| \hat{H}_{\text{SO}} \right| \psi_{T_i} \rangle}{E_{T_2} - E_{S_1}} \\
&+ \sum_n \frac{\langle \psi_{S_i} \left| \hat{\tau}_N \right| \psi_{S_m} \rangle \langle \psi_{S_i} \left| \hat{H}_{\text{SO}} \right| \psi_{T_1} \rangle}{E_{S_1} - E_{T_1}}
\end{aligned}$$

(ψ—Molecular wave function for a particular state, $\hat{\tau}_N$—vibronic mixing operator, $\hat{H}_{\text{SO}}$—spin–orbit coupling). The first term in the above equation denotes the direct spin–orbit coupling. It is driven by the electronic character of the involved states. The second term denotes vibrational spin–orbit coupling, whose magnitude depends upon the motion along a particular nuclear degree of freedom. The third term denotes the first-order correction of vibrational spin–orbit coupling. The fourth term denotes spin-vibronic coupling with vibronic coupling in triplet manifolds and the last term denotes spin-vibronic coupling with vibronic coupling in the singlet manifold (Albrecht 1963).

9.4 Simple Kinetic Picture of TADF

For simplicity purpose, the kinetics of TADF emitter with η_{EL} of near 100% is considered. By ignoring the non-radiative channel, we can assume the emission efficiency of the T_1 to S_1 (RISC) transition to be 100% which further simplifies the kinetics picture of the TADF emitter. Considering the thermal equilibrium between singlet and triplet states we can substitute the k_{ISC} (rate of intersystem crossing) and k_{RISC} (rate of reverse intersystem crossing) as k_1 and k_{-1}, respectively. The zero field splitting (ZFS) energy is very small as compared to the singlet–triplet energy difference (ΔE_{ST}), hence the triplet sublevels can be considered as degenerated. k_{T1} is neglected as compared to k_{S1}, k_{ISC}, and k_{TADF}. Using a three-level model, the TADF emission process can be visualized as per the above-mentioned assumption (Aizawa et al. 2020)

$$T_1 \underset{k_{-1}}{\overset{k_1}{\longleftrightarrow}} S_1 \overset{k_{s_1}}{\rightarrow} S_0 + h\upsilon$$

$$\frac{dh\nu}{dt} = \frac{k_1 k_{S_1}}{k_{-1} + k_{S_1}}[T_1]$$

Case 1: When $k_{S1} >> k_{ISC}$: This is the simplest case where the rate of fluorescence decay is much higher than the rate of intersystem crossing. When the rate of T_1 phosphorescence is much smaller than the rate of k_{ISC} there will be no interconversion of singlet and triplet states and thus S_1 and T_1 states decay independently. This situation may predominate in fluorescence-based OLED and in devices optimized to harvest triplet exciton via triplet–triplet annihilation (Ravinson and Thompson 2020). In these, S_1 state undergoes fluorescence decay, and the triplet state undergoes non-radiative decay.

Case 2: $k_{ISC} >> k_{S1}$: Rapid ISC (high triplet yield) is a common situation in transition metal-based emitters. Singlet and triplet states intermix due to the direct spin–orbit coupling factors (Ravinson and Thompson 2020). Further assuming both k_1 and $k_{-1} >> k_{S1}$ and can be simplified with a pre-equilibrium approximation (Aizawa et al. 2020)

$$\frac{dh\upsilon}{dt} = \frac{k_1 k_{s_1}}{k_{-1}}[T_1] \Rightarrow \tau_{TADF} = \frac{\tau_{S_1}}{K_{eq}}$$

It is evident that K_{eq} (k_1/k_{-1}) is more relevant than the individual k_1 and k_{-1}. In materials having very high fluorescence quantum yield, i.e., $\Phi_{PL} \approx 1$ (non-radiative decay pathways can be neglected), the value of rate constant can be directly measured from the initial emission decay time. The ratio of emission intensity at the earliest time (S_1 emission) to emission intensity at a long time (>500 ns) is equal to K_{eq} (Aizawa et al. 2020). The k_{S1} can be determined by knowing the value of K_{eq} and τ_{TADF} (Aizawa et al. 2020). Furthermore, ΔE_{ST} can be calculated by K_{eq} as per the following equation (Ravinson and Thompson 2020):

$$K_{eq} = \frac{k_1}{k_{-1}} = \frac{1}{3}\exp\left(-\frac{\Delta E_{ST}}{K_B T}\right) \Rightarrow \Delta E_{ST} = -k_B T.\ln(3K_{eq})$$

Case 3: $k_{ISC} \sim k_{S1}$: When k_{ISC} and k_{S1} are comparable, a more complicated situation arises. This situation is predominant in organic TADF emitters. The principle approach involves the minimization of ΔE_{ST} to enhance the rate of ISC. Various strategies were adopted to minimize the ΔE_{ST}, which is further discussed in subsequent sections. The idea of maximizing the ISC rate by minimizing the ΔE_{ST} can be rationalized by the following equation:

$$k_{TADF} \propto \sum_n \frac{\left(T_n\middle|\hat{H}_{SOC}\middle|S_1\right)}{E_{S_1} - E_{Tn}}.\langle S_0|\mu|S_1\rangle$$

where $\hat{H}_{SOC}$ is the spin–orbit coupling operator and μ is the dipole moment operator. In the above equation, only direct spin–orbit coupling between singlet and triplet states and non-adiabatic mixing of T_1 and S_1 states are not considered. To simplify the kinetics of these types of emitters, the non-radiative decay and slow rate of phosphorescence are neglected which is consistent with efficient organic TADF emitters.

$$\tau_{TADF} = \frac{k_{-1} + k_{S_1}}{k_{-1}k_{S_1}} = 3e^{\frac{\Delta E_{ST}}{k_B T}}\left(\frac{1}{k_{S_1}} + \frac{1}{k_{-1}}\right)$$

The simple treatment in the above equation illustrates that decreasing ΔE_{ST} has a positive impact on TADF emission.

9.5 Emission Property of TADF in Organic Molecules

The necessary condition of small ΔE_{ST} for effective TADF can be achieved by separating or minimizing the overlap between HOMO and LUMO. Physical separation of donor and acceptor moieties in a single molecule is reported to be a common way of doing it (Xu et al. 2021). It gives rise to charge transfer of electronic transition, which affects the absorption and emission properties of organic donor–acceptor molecules.

TADF molecules have distinctive absorption and emission spectra along with a broad featureless band corresponding to ground state *CT* absorption (Lee and Kim 2016). Numerous factors affect the absorption and emission properties of TADF molecules which are discussed briefly in the following sections.

9.5.1 Solvatochromism in TADF

In 1878, *Kundt* in *Zurich* studied the absorption of different fluorophores including chlorophyll, fuchsin, quinizarin, aniline green, egg yolk, and cyanine in different solvents of varying polarity and observed the bathochromic shift in absorption spectra. It was assigned to the dispersion interaction of the solvent and the absorbing species in the solution (Reichardt 2007). Later, in 1922, *Hantzsch* termed the reversible change of absorption and emission spectra of molecules in different solvents as the *Solvatochromism effect*. The electronic structures of the ground and excited states of molecules are different and as a result the solvation energy also differs depending upon their interaction with surrounding solvent molecules resulting in the shifting of absorption spectra which further results in the color change (Zuehlsdorff and Isborn 2019).

According to Frank–Condon's principle, solvent reorientation takes longer time compared to the time taken for electronic excitation, and consequently the excited state will experience the same ground state solvent coordination sphere (Zhang et al. 2020). If the excited state of absorbing species is more polar than the ground state then it will be stabilized in a polar environment, and as a result, the emission peak will shift to a longer wavelength. It is known as bathochromic shift or positive solvatochromism. In contrast, a less polar excited state will be destabilized in a polar environment which leads to the blue shift of emission spectra and is known as *hypsochromic shift* or *negative solvatochromism* (Zhang et al. 2020). The $\pi \rightarrow \pi^*$ transition of $C = O$ in the polar solvent will show a *positive solvatochromic effect* as the excited state is more polarized than the ground state whereas $n \rightarrow \pi^*$ transition in the polar solvent will undergo a *negative solvatochromic effect* (Edwards and Alexander 2017). In a less polar solvent, emission spectra display the contribution from both ^3LE and ^1CT states. Further, the time-resolved spectra showed slow charge transfer characteristics as an indication of weak electron coupling between donor and acceptor ^1CT states, whereas in more polar environments a predominant *CT* emission was observed in the steady-state photoluminescence spectra. It was proposed that fast electron transfer and stabilization of separated charge character of excited states in the polar environment resulted in the red-shifted emission spectra (Santos et al. 2016).

9.5.2 *Temperature Dependence of TADF*

Rate constant for E-type delayed fluorescence exhibits temperature dependence as shown by the following equation:

$$k_{\mathrm{TADF}} = \frac{1}{3}k_F \exp\left[\frac{-\Delta E_{\mathrm{ST}}}{k_B T}\right]$$

In this, ΔE_{ST} is the activation energy which corresponds to the S_1-T_1 energy difference. The small ΔE_{ST} can be overcome by providing thermal energy to vibrationally promote the exciton from the relaxed triplet to an isoenergetic point to cross back to the radiative singlet state. Since triplet state density increases due to the reduced thermal activation of T_1 state to S_1 state, neat film photoluminescence spectra of TADF molecules display a decrease in their *PL* quantum efficiency around temperatures below 100 K. Below 100 K, the triplet state intensity increases due to a lack of RISC resulting in low PL efficiency of TADF at low temperature. At this low temperature, TTA annihilation may act as a dominant non-radiative decay pathway along with guest–host interactions (Lee et al. 2017).

Solid-state time-resolved emission of TADF materials consists of sharp decay and a slow long tail (Niwa et al. 2014). The former was attributed to prompt fluorescence while the latter was to the delayed fluorescence. The origin of the delayed fluorescence can be checked by taking the emission spectra at different temperatures. For TADF originated mechanism, fluorescence intensity increases with the temperature (Niwa et al. 2014). Prompt fluorescence slightly increases with a decrease in temperature due to suppression of non-radiative decay pathways while delayed fluorescence increases as RISC becomes the rate-determining step (Uoyama et al. 2012).

9.6 Effect of Regio and Conformational Isomerization in TADF Emission

The substitution of donor, acceptor, and their relative position in the TADF molecules has a prominent effect on their photophysical properties (Xie et al. 2020). The charge transfer properties of an organic TADF molecule with para-substituted carbazole on a central acceptor were reported to be different from the meta-substituted conformer (Zhang et al. 2019). *Para*-substituted carbazole was shown to have high oscillator strength and high PL quantum yield, whereas the *meta*-connected system showed efficient separation of electron and hole densities consequently exhibiting small ΔE_{ST}, thus higher TADF contribution. The molecular conformation of fluorophores plays a very important role in TADF emission as it intrinsically controls the CT properties of the excited state (Zhang et al. 2019). Wang et al. (2018) showed that in pre-twisted molecules the more twisted conformation needs to have different excited state conformation to minimize the frontier orbital overlap, consequently reducing

ΔE_{ST}. The ΔE_{ST} for D-A-type monomeric molecules were found to be less compared to their dimer counterparts. It was attributed to the possibility of monomers undergoing more conformational changes in their excited state with respect to the dimer's excited state. The fluorophores with two stable conformations exhibit dual emissions corresponding to each conformation. Absorption spectra of these fluorophores showed high-energy intense absorption corresponding to *the LE* state and a low-energy weak absorption band corresponding to the *CT* state. The former is due to the near-planar conformation while the latter corresponds to near-orthogonal conformation. Planar conformation with high HOMO–LUMO overlap only exhibits high-energy fluorescence while near-orthogonal HOMO–LUMO exhibits TADF emission. The S_1 and T_1 states of near-planar conformation can lie above the S_1 and T_1 states of near-orthogonal conformer; as a result, the previously wasted triplet exciton in near-planar conformation can be transferred to near-orthogonal conformer avoiding exciton loss (Wang et al. 2018).

Different photophysical processes, including TADF and RTP, may be enhanced by various conformers that dominate in the solids, as a result of studying the connection between the detailed photo-physics of MCL materials and specific conformer-enriched phases (Data et al. 2019). Well-separated HOMO–LUMO electronic configuration in equatorial–equatorial conformation leads to small ΔE_{ST} compared to the axial–axial conformation. Since the HOMO–LUMO overlap in the axial–axial conformer is larger, the corresponding exchange interaction (*J*) destabilizes the ^1CT state, and consequently, it lies much above the ^3LE state. So, thermally activated delayed ISC could not be observed. Whereas in the case of equatorial–equatorial conformation, both PF and TADF could be observed due to the small energy difference between ^1CT and ^3LE states (see Fig. 9.3) (Data et al. 2019).

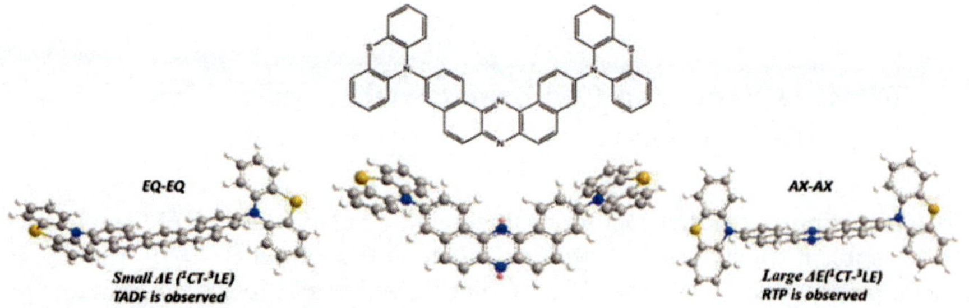

Fig. 9.3 Schematic representation of the effect of different conformers on the ΔE_{ST}. This is redrawn and is adapted from RSC Publisher (Data et al. 2019)

9.7 Solid Host for TADF

TADF molecules were intensively studied for their potential application in electroluminescent devices like OLEDs. In these devices, the emitter molecules were dispersed in a host material to avoid aggregation-induced exciton quenching. So, it is important to discuss the photophysical characterization of the TADF molecules dispersed in solid hosts. In this section, the guest–host interaction and its effects on the electrical and optical properties of the TADF emitter are briefly reviewed. The combination of host and emitter should decrease the RISC energy barrier for efficient TADF property (Zhang et al. 2020; Li et al. 2022). A fundamental requirement of choosing host material aims for the minimum interaction with the ground state emitter to minimize their local exciton property (Xu et al. 2021). The triplet energy of the host material should be higher than the triplet energy of the emitter to avoid unwanted exciton quenching (Li et al. 2022; Méhes et al. 2020).

The TADF character of molecules is linked to their *CT* state and these states are associated with larger dipole moments, their electronic states show a strong dependence on the solvent. Large dipolar interaction between solute and solvents re-orients the solvents, which leads to the bathochromic spectral shift observed in *CT* emission with the increase of solvent polarity (Fig. 9.4) (Ward et al. 2016). Solid-state solvation effects, analogous to solvent effects, are very important in the context of OLED, where active emitting molecules are doped into an amorphous organic thin film (Etherington et al. 2016). In solid state, solvent molecules are sterically constrained in their respective position, which gives rise to local dipolar interactions and leads to the expected spectral shift (Deng et al. 2019). When local dipolar interactions are completely random, no shift is observed in their absorption and emission spectra. Broadening of emission spectra is expected due to the large dispersion in the absorption and emission energies (Etherington et al. 2016). Based on a static state polarization model, Wang and Zhang proposed that the large S_1 state dipole moment and small electron–hole separation lead to reduced ΔE_{ST}, which is helpful for solid-state solvation enhanced thermally activated delayed fluorescence (SSSE-TADF). Apart from the static polarity effect, the dynamic effects in solid-state TADF emission are also observed (Zhang et al. 2020; Lin et al. 2014).

Lin et al. (2014) explored the light-emitting mechanism theoretically using the first principle calculations. The excited state dynamics of molecules in solvent and thin film showed different behaviors corresponding to their reorganizational energy. They projected the reorganizational energy on different vibrational modes and reported that molecules where the low-frequency twisting motions (bending modes) mostly contribute to the reorganizational energy get suppressed easily in solid state which decreases the available non-radiative decay pathways and PLQY was improved in thin film compared to in solvent. Molecules, where the reorganization energy comes from high-frequency vibrational modes (C–C bond stretching mode), cannot be effectively hindered in solid state, resulting in a little difference of PLQY in the thin film compared to solvent.

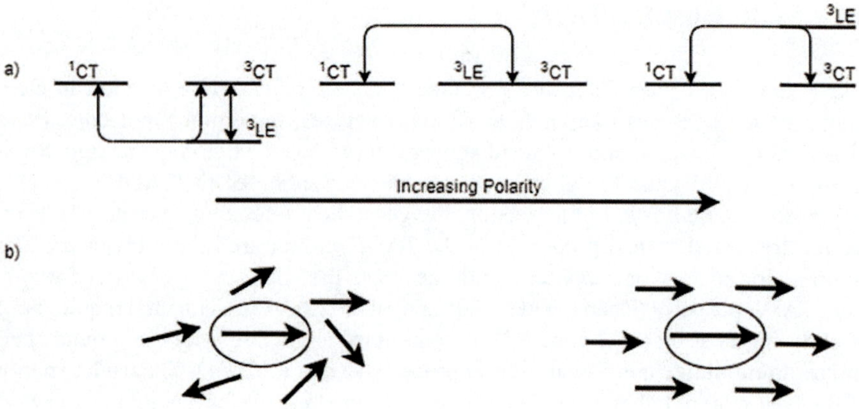

Fig. 9.4 **a** A schematic representation of static solid-state solvation effect depicting the stabilization of CT state due to the polarity of the host. **b** Schematic depiction of the origin of the dynamic stabilization of the CT state, where the solvent dipoles reorient to stabilize the CT excited state. This is adapted from RSC publisher under Creative Commons Licence (Ward et al. 2016)

Recently Sun et al. (2017) reported the reduction of ΔE_{ST} due to the hybrid nature of singlet and triplet excited states. Northey et al. (2017) reported that rigidity and the ground state dipole moment of D-A-D-type TADF emitter molecules along with the dipole moment of the host play crucial roles in determining the magnitude of SSSE. They also proposed that conformational heterogeneity is a responsible factor for the observed dynamic effect associated with SSSE. Time-dependent solvation dynamics studies rule out the guest–host interaction as the dominant effect in the solvation dynamics. Alternatively, they proposed that the charge redistribution of the emitter at the picosecond time scale stabilizes the excited states. Further, the dispersion effect of *CT* states stabilizes the *CT* state in the nanosecond time scale (Etherington et al. 2016).

9.8 Designing of TADF Molecules

Reverse intersystem crossing requires strong coupled T_1 and S_1 states in thermal equilibrium. Assuming, the electronic transition from HOMO to LUMO is localized on the donor and acceptor, respectively. The thermal energy barrier (ΔE_{ST}) can be calculated in a two-state model as

$$E_{S_1} = h_H + h_L + J_{\mathrm{HL}} + K_{\mathrm{HL}}$$

$$E_{T_1} = h_H + h_L + J_{\mathrm{HL}} - K_{\mathrm{HL}}$$

$$\Delta E_{ST} = 2K_{HL}$$

where h_H, h_L is the one-electron energy of HOMO and LUMO, respectively, J_{HL} is the columbic interaction term between electrons at HOMO and LUMO, and K_{HL} is the exchange energy term (Chen et al. 2015a).

$$J_{HL} = \iint d\vec{r_1}\, d\vec{r_2}\, H(\vec{r_1}) H(\vec{r_1}) \frac{1}{r_{12}} L(\vec{r_2}) L(\vec{r_2})$$

$$K_{HL} = \iint d\vec{r_1}\, d\vec{r_2}\, H(\vec{r_1}) L(\vec{r_1}) \frac{1}{r_{12}} H(\vec{r_2}) L(\vec{r_2})$$

It is important to note that the exchange interaction energy (K_{HL}) is directly proportional to the spatial overlap of HOMO and LUMO (Chen et al. 2015a). From the above equation, if the electronic transition is assumed to be purely HOMO–LUMO transition then spatially separating them can result in vanishingly small ΔE_{ST}, but the excitation corresponds to complex electronic configuration, so more factors must be considered to calculate the value of ΔE_{ST}.

While designing TADF molecules, factors affecting delayed and prompt fluorescence have to be considered. To observe room temperature TADF, a small ΔE_{ST} and high radiative decay rate of the S_1 state are necessary. This can be achieved by minimizing the orbital exchange interaction and increasing the oscillator strength and it is not a straightforward process to optimize these contrasting properties. Small ΔE_{ST} can be obtained by minimizing the HOMO–LUMO overlap (Xu et al. 2021). This can be achieved by (i) separating donor and acceptor moieties, (ii) maximizing the D-A dihedral angle, (iii) introducing steric hindrance between donor and acceptor, and (iv) twisted molecular structure with spiro-type D-A junction. Similarly, conditions for high radiative rate are (i) large overlap density between S_0 and S_1 electronic states, (ii) large delocalization of molecular orbitals, (iii) increasing molecular rigidity, and (iv) increasing the π conjugation length (Xu et al. 2021). Efficient blue TADF emitters have been achieved by controlling the conjugation length along with the redox potential of both the donor and acceptor (Zhang et al. 2012).

9.9 Theories of TADF

Initially the mechanism of TADF was based on the thermal equilibrium of the lowest singlet and triplet states and was explained by a simple Arrhenius-type equation.

$$k_{TADF} = \frac{1}{3} k_F \exp\left[\frac{-\Delta E_{ST}}{k_B T}\right]$$

This model assumes thermal equilibrium between the lowest singlet and triplet states and the rate constant of the RISC process (k_{RISC}) is predominantly dependent on temperature. This model is used to describe TADF where detailed knowledge of vibrational energy level density is absent. The equation shows that the ΔE_{ST} is the most critical factor determining k_{RISC}. A more rigorous approach, namely, Fermi's Golden rule employing first-order perturbation theory can be used to determine k_{RISC}:

$$k_{\mathrm{RISC}} = \frac{2\pi}{hZ}\left|\left\langle \Psi_f | \hat{H}_{\mathrm{SOC}} | \Psi_i \right\rangle\right|^2_{q_o} \times \sum_{jk} \exp^{-\beta E_j} \left|\left\langle v_{fk} | v_{ij} \right\rangle\right|^2 \delta\left(E_{ij} - E_{fk}\right)$$

where $\sum_{jk} \exp^{-\beta E_j}$ denotes the vibrational partition function in the initial electronic state and the coupling of electronic states with different spin multiplicities is effectively achieved by the spin–orbit coupling operator ($\hat{H}_{\mathrm{SOC}}$). Generally, electronic SOC dominates the coupling interaction but spin–spin coupling interaction comes into play in case of weak interaction. Effective decoupling of electronic and nuclear motion (Frank–Condon approximation) and coupling between the two states involved in the excited state kinetics should be small as compared to their energy difference. Intrinsic molecular factors along with adiabatic energy difference, the magnitude of spin–orbital coupling matrix element (SOCME), and environmental factors including temperature and specific solvent interactions are the determining factors affecting the efficiency of ISC and RISC process in TADF emission (Dias et al. 2017).

Due to the importance of *CT* states in TADF, it was assumed that the excited states involved in the RISC process were to be *CT* (^1CT *and* ^3CT) states. Spin–orbit coupling between the two intramolecular *CT* states was shown to be zero (Lim et al. 1981). The SOC operator contains both spatial angular quantum numbers and spin magnetic quantum numbers. Since the RISC process involves a change in spin multiplicity, coupling between states with the same spatial occupation is not allowed. Consequently, SOC between singlet and triplet *CT* states is formally zero. Independent tuning of the singlet and triplet excited state was reported to solve this issue (Dias et al. 2017). As a result, the singlet and triplet states involved in the RISC process could be different and spin–orbit coupling between them would be allowed (Gibson et al. 2016). Chen et al. (2015b) proposed that the non-adiabatic effect between low-lying excited states is responsible for the efficient T_1-S_1 up-conversion process in organic TADF molecules. Recent work of Ward et al. (2016) showed that the interplay of TADF and room temperature phosphorescence (RTP) was observed by tuning the steric hindrance between the donor and acceptor group in a series of D-A-D molecules. Moreover, restricting the rotation of the donor–acceptor bond can potentially suppress the TADF emission indicating the dynamic nature of the RISC process in TADF. This indicates the TADF mechanism possibly be promoted by molecular vibration along a specific vibrational degree of freedom (Ward et al. 2016). Compounds **1** and **2** showed strong TADF emission whereas **3** and **4** showed predominant phosphorescence emission (Fig. 9.5).

Theoretical calculation by Chen et al. (2015b) showed that the calculated rate of k_{rISC} is not able to be explained by the spin–orbital coupling between ^1CT and ^3LE states. They explained this deviation as the absence of non-adiabatic coupling

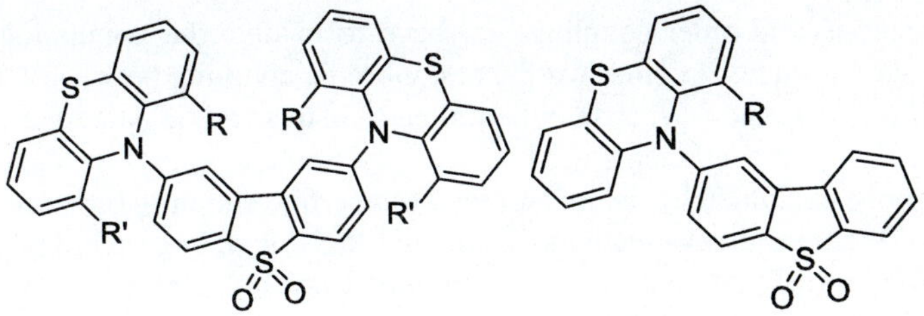

R, R'	R
1. H, H	2. H
3. *i*-Pr, H	4. *i*-Pr

Fig. 9.5 Chemical structures of compounds **1–4**

between ^{3}CT and low-lying localized triplet states (^{3}LE). Using the multi-reference quantum chemistry method, Merien et al. (2016) showed that indeed the calculated rate of RISC using only SOC coupling was not in agreement with the observed high values of K_{RISC}. Instead, they proposed that the mixing of multiple excited states is involved in the RISC process through vibronic coupling along with SOC. Ogiwara et al. (2015) probed the excited state population of ^{3}LE and ^{1}CT states by electron paramagnetic resonance spectroscopy and proposed that the rate of *RISC* is in agreement with the EPR signal of the mixed excited state (^{3}LE and ^{1}CT). They proposed that in addition to SOC-induced *ISC* ($^{3}LE \rightarrow {}^{1}CT$), hyperfine coupling-induced ISC ($^{3}LE \rightarrow {}^{1}CT$) also plays a significant role in the ISC/RISC process in TADF.

Gibson et al. (2016) performed the quantum dynamic simulation using the vibronic coupling Hamiltonian to give a dynamic picture of the ISC and RISC process and the effect of vibronic coupling in k_{ISC} and k_{RISC}. Interestingly, they found that by removing the vibronic coupling between ^{3}CT and ^{3}LE from the Hamiltonian, the population transfer to the ^{3}LE state from the ^{3}CT state becomes negligible, whereas removing the hyperfine coupling between ^{3}CT and ^{1}CT state has very little effect on the kinetics of ISC and RISC processes. They concluded that vibronic coupling plays a critical role in the ISC process of TADF molecules (Northey et al. 2017). This could not be described in the limits of the first-order perturbation theory. A more general description includes second-order perturbation theory as follows:

$$k_{\mathrm{RISC}} = \frac{2\pi}{\hbar Z}\left| \langle 1_{\psi_f} \middle| \widehat{H}_{\mathrm{SOC}} \middle| 3_{\psi_i} \rangle + \frac{\langle 1_{\psi_f} \middle| \widehat{H}_{\mathrm{SOC}} \middle| 3_{\psi_n} \rangle \langle 3_{\psi_n} \middle| \widehat{H}_{\mathrm{vib}} \middle| 3_{\psi_i} \rangle}{\delta\left(3_{E_n} - 3_{E_i}\right)} \right|^2 \delta\left(3_{E_i} - 1_{E_f}\right)$$

The direct second-order coupling will have to involve the population transfer between two *CT* States. In other words, first vibronic coupling between *CT* and *LE* states will populate the ^{3}CT state, which then transfers to ^{1}CT state via hyperfine coupling (HFC) but coefficients of HFC are very small and the observed values of k_{rISC} cannot be explained by the magnitude of hyperfine coupling between ^{1}CT and ^{3}CT states. Whereas the two-step mechanism of RISC with large vibronic coupling between ^{3}LE and ^{3}CT populates the ^{3}LE state with a much faster time scale compared to the RISC process, which further transfers to ^{1}CT state through efficient SOC between ^{3}LE and ^{1}CT. Consequently, the ^{3}CT and ^{1}CT states are coupled through the second-order equation using ^{3}LE as an intermediate state (Dias et al. 2017).

$$k_{\mathrm{RISC}} = \frac{2\pi}{\hbar Z}\left|\langle 3\psi_{CT}\left|\widehat{H}_{\mathrm{SOC}}\right|3\psi_{LE}\rangle\right|^{2} \times \delta(3E_{LE} - 3E_{CT})$$

$$k_{\mathrm{RISC}} = \frac{2\pi}{\hbar Z}\left|\frac{\langle 1\psi_{CT}\left|\widehat{H}_{\mathrm{SOC}}\right|3\psi_{LE}\rangle\langle 3\psi_{LE}\left|\widehat{H}_{\mathrm{vib}}\right|3\psi_{CT}\rangle}{\delta(3E_{LE} - 3E_{CT})}\right|^{2} \times \delta(3E_{CT} - 1E_{CT})$$

The intersystem crossing process between ^{3}CT and ^{3}LE depends on the extent of vibronic coupling and the adiabatic energy difference between them. Good vibrational overlap between ^{3}CT and ^{3}LE states and efficient SOC between them are the necessary conditions for an efficient RISC process in TADF molecules (Northey et al. 2017). This explains the result obtained by Ogiwara et al. (2015) where they demonstrated that the energy gap of ^{1}CT and ^{3}CT and ^{3}CT and ^{3}LE states must be tuned to observe efficient TADF.

9.10 Organic TADF Molecules (Monomolecular System)

Generally, organic TADF molecules have an intramolecular D-A system with a twisted structure or large steric hindrance between the donor and acceptor moieties. These molecules exhibit small ΔE_{ST} due to the distribution of HOMO and LUMO on the electron-donating and electron-accepting units, respectively (Yang et al. 2017). This helps in reducing the HOMO and LUMO spatial overlap. D-A molecules having strong electron-donating ability increases the CT nature of electronic states, so aromatic amines such as carbazole, phenoxazine, phenothiazine, and diphenylamine, and their derivatives were used in designing TADF materials. In this section, we will briefly review some of the reported TADF molecules.

Organic molecules with cyano-substituted aromatic acceptor units have been widely employed as electron acceptors. A series of highly efficient organic emitters with dicyanobenzene as electron acceptor and carbazole as electron donor was reported by Uoyama et al. (2012). Steric hindrances between donor and acceptor distort the carbazole unit from the dicyanobenzene plane consequently localizing the HOMO and LUMO on the donor and acceptor, respectively, resulting in a small exchange interaction (*K*), leading to a small ΔE_{ST}. Moreover, they also reported high photoluminescence efficiency as well as color tuning by varying the relative position and number of substituents on the donor or acceptor. They also reported a high EQE of OLED based on **2CzPN** as the emitter. For this, high EQE (~13.6%) at a low current density of (J = 0.01 $mAcm^{-2}$) and EQE of 3.6% at a current density of 10 $mAcm^{-2}$ are observed. These results indicate efficient EML exciton confinement with charge recombination at low current density, also it suggests the exciton quenching at high current density. At high current densities, EQE is dominated by singlet–triplet annihilation and triplet–triplet annihilation due to the long triplet lifetime of **2CzPN**. Monkman et al. (2005) studied the details of TADF photo-physics and their effects on OLED efficiency. They proposed that molecules with high EQE possess small energy gaps between ^{1}CT and the ^{3}CT states. They proposed the involvement of additional ISC channels by hyperfine coupling in the RISC process of TADF. Excited-state dynamics of **2CzPN** and **4CzPN** molecules reported by Uoyama et al. (2012), Hosokai et al. (2016) revealed that in some cases the two-state model (S_1 and T_1) is not enough to explain the observed rate of the RISC process. **DCzIPN** has been synthesized and characterized by Cho et al. (2015a). It showed blue-shifted emission due to the weak ICT transition. A dual-core TADF molecule (**DDCzIPN**) having high absorption coefficient resulted in a high EQE (18.9%) of the TADF device. In this, steric hindrance between four CN units increased the rigidity of the molecule, consequently shortening the delayed emission (Cho et al. 2015b). The effect of the twisting angle of D-A-type molecules was studied by Li et al. (2014), they reported that a larger dihedral angle (60°) of **26IPNDCz** results in small ΔE_{ST} (0.06 eV) as compared to **35IPNDCz** which has the dihedral angle of 50° and correspondingly the ΔE_{ST} to be 0.14 eV. Kretzschmar al. (2015) studied the effect of halogen substituent on carbazole units (**4CzIPN-2Cl, 4CzIPN-2Br, 4CzIPN-2I**). They reported the effect of halogen atoms on ISC and RISC and found it proportional to the atomic weight of halogen atoms.

Lee et al. (2015) reported a molecule **BTCz-2CN** with benzothienocarbazole as an electron-donating unit and reported better TADF performance compared to **2CzPN**. Taneda et al. (2015) reported an ΔE_{ST} of 0.103 eV for a trigonal molecule with three phenylamine units. Steric hindrance between the alternating benzene substituent in **3DPA3CN** resulted in the decrease of vibrational deactivation (Taneda et al. 2015). An efficient blue TADF emitter **CPC** was designed by Liu et al. (2015) with pyridine as the central unit. The results were comparable to the best blue phosphor-based emitter. The introduction of a fluorine atom to the cyano-substituted aromatic acceptor produces a soluble blue emitter. Compounds **3CzFCN and 4CzFCN** showed very low ΔE_{ST} (0.06 eV) and the solution-processed TADF emitter **4CzFCN** showed comparable EQE to vacuum-deposited blue emitter (Cho et al. 2015c). The effect of *tert*-butyl substitution on the TADF performance of emitters was studied by Zhang et al. (2016) They proposed that the *tert*-butyl group increases the solubility and efficiency of the TADF emitter. In compounds **5CzBN, 4CzBN, 4TCzBN, and 5TCzBN**, the tert-butyl group acts as a shield in **5TCzBN**, protecting the light-emitting core, consequently improving its stability.

The effect of the linker between donor and acceptor was studied by Park et al. (2016). They reported efficient TADF emitters with compounds **Cz-VPN**, **Ac-VPN**, and **Ac-CNP** having angular linked D-A structure combining phthalonitrile with various donor units. TADF molecules with donor–linker–acceptor (D-L-A) framework with carbazole donor and cyano-substituted benzene acceptor were studied by Vikramaditya et al. (2016). They reported compounds **CTBN, CEBN** with thiophene (conjugated) and ethylene (non-conjugated) linker, respectively. Furthermore, they showed that the ΔE_{ST} of **CTBN** (non-conjugated linker) was smaller than the ΔE_{ST} of **CEBN** (conjugated linker). The difference in the efficiency of **Ac-VPN** and **CEBN** was proposed to be the result of reduced conjugation of donor and acceptor units. The transmission of electronic information between orbitals of the donor and acceptor is smaller in D-L-A molecules with non-conjugate linker than in conjugated linker. Endo et al. (2011) reported a TADF molecule **PIC-TRZ** containing an indolocarbazole unit as the donor and an electron-deficient triazine unit as the acceptor. In **PIC-TRZ**, the steric interaction between the biphenyl group and the bulky indolocarbazole unit localizes the HOMO and LUMO on the indolocarbazole and biphenyl triazole unit resulting in a small ΔE_{ST} of 0.08 eV. A novel triazine derivative **PIC-TRZ2** exhibiting ΔE_{ST} of 0.003 eV was reported by Sato et al. (2013). A highly efficient compound **(CC2TA)** was synthesized by Youn Lee et al. (2012) by replacing the indolocarbazole unit with a bicarbazole unit and the resulted ΔE_{ST} was found to be exceedingly small (0.06 eV).

Wang et al. (2015) reported the first-ever near-infrared TADF molecule **TTA-DCPP** with a V-shaped D-π-A-π-D configuration. A small ΔE_{ST} value of 0.13 eV and a high glass-transition temperature of 508 °C were observed for **TTA-DCPP**. D-A-D-type molecules (**DPA-DPS, tDPA-DPS, tDCZ-DPS**) containing diphenyl sulfoxide unit as acceptor were reported by Zhang et al. (2012) ΔE_{ST} values of corresponding molecules as 0.54, 0.45, and 0.32 eV, respectively, were reported. It was shown that the lower value of ΔE_{ST} of **tDCZ-DPS** was due to the stabilization of *CT* State. The introduction of *tert*-butyl on the diphenylamine unit of **tDPA-DPS** enhanced its electron-withdrawing ability consequently stabilizing the *CT* state. The carbazole unit affected ^{1}CT lesser than ^{3}LE and reduces the ΔE_{ST} value for **tDCZ-DPS**. Highly electronegative oxygen of the sulfonyl group gives the electron-withdrawing properties of diphenyl sulfoxide and the limited conjugation to the compound. The tetrahedral geometry of the sulfonyl group on diphenyl sulfoxide makes it a good candidate for electron–acceptor units in D-A-D molecules.

Data et al. (2016) reported a series of U-shaped D-A-D TADF molecules t-**BuCZ-DBPHZ, MeODP-DBPHZ, POZ-DBPHZ** based on dibenzo-[a,j] phenazine (DBPHZ) acceptor unit. A small range of ΔE_{ST} (0.02–0.2 eV) was observed with efficient TADF properties. They revealed the efficient spin flip in the RISC process and reasoned it to the coupling of a low-lying ^{3}LE state with the ^{1}CT state, which provides the spin–orbit charge transfer (SOCT). They also showed that in the polar medium the stabilization of the ^{1}CT state with respect to the ^{3}LE state results in the reduction of ΔE_{ST}, hence enhancing the efficiency of TADF (Table 9.1).

Siddiqui et al. (2018) synthesized two donor–acceptor–donor (D-A-D) materials, **Acr-CBz** and **Acr-CBz-CF3**, for use in optoelectronic applications. These

Table 9.1 PL emission peak (λ_{PL}), PLQY, singlet–triplet energy gap (ΔE_{ST}), EL emission peak of different organic TADF molecules

Compound	Name	λ_{PL} nm	PLQY %	ΔE_{ST} eV	λ_{EL} nm	Refs.
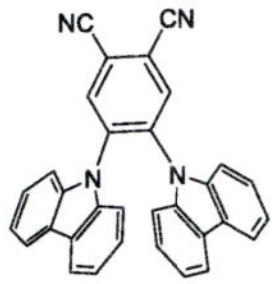	**2CzPN**	473	46.5, 89	0.34	470	Uoyama et al. (2012)
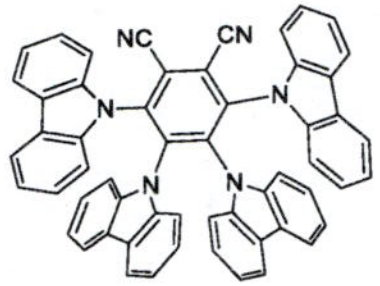	**4CzPN**	525	74.4	0.18	538	Uoyama et al. (2012), Hosokai et al. (2016)
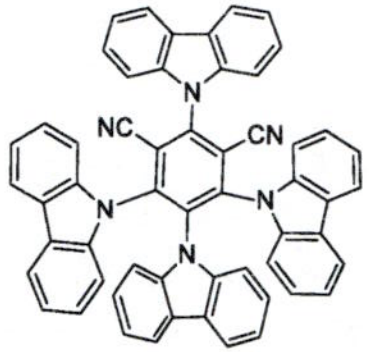	**4CzIPN**	507	93.8	0.08	507	Uoyama et al. (2012)
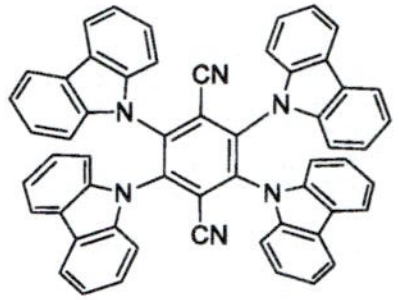	**4CzTPN**	535	71.6	–	580	Uoyama et al. (2012)

(continued)

Table 9.1 (continued)

Compound	Name	λ_{PL} nm	PLQY %	ΔE_{ST} eV	λ_{EL} nm	Refs.
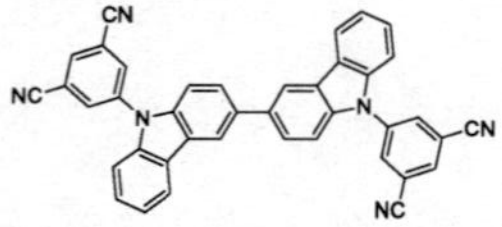	**35IPNDCz**	447	35	0.14	494	Li et al. (2014)
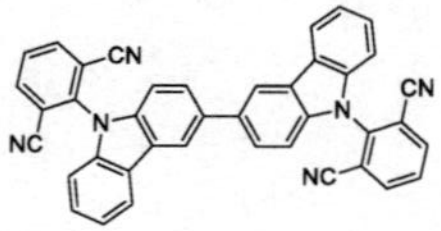	**26IPNDCz**	477	91	0.06	497	Li et al. (2014)
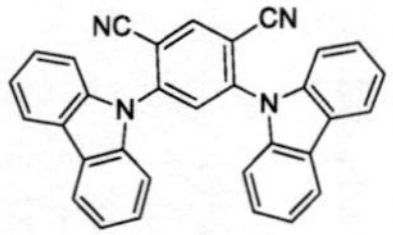	**DCzIPN**	470	50	0.05	487	Cho et al. (2015a)
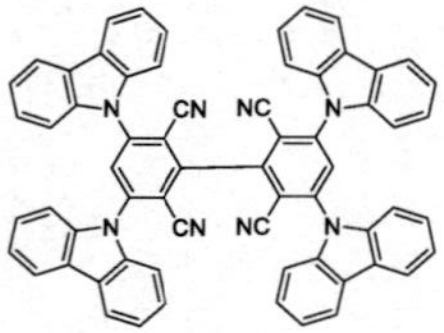	**DDCzIPN**	488	72	0.13	501	Cho et al. (2015b)

(continued)

Table 9.1 (continued)

Compound	Name	λ_{PL} nm	PLQY %	ΔE_{ST} eV	λ_{EL} nm	Refs.
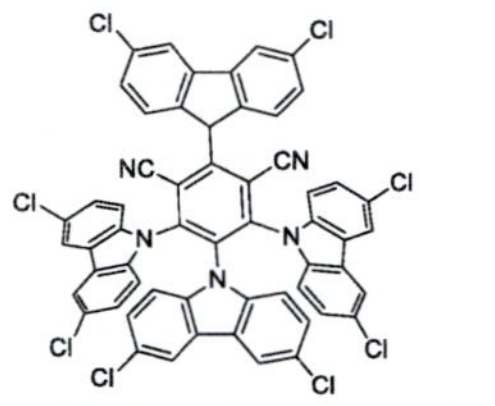	**4CzIPN-2Cl**	518	76	–	–	Kretzschmar et al. (2015)
	**4CzIPN-2Br**	521	67	0.04	–	Kretzschmar et al. (2015)
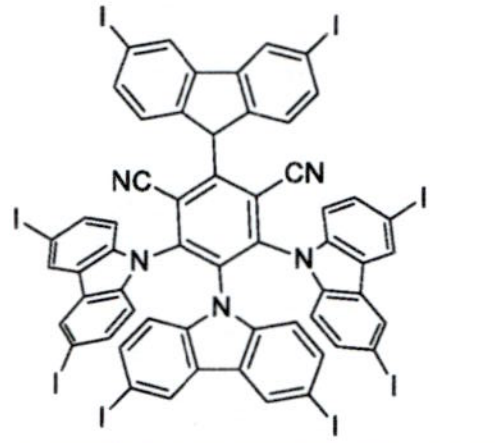	**4CzIPN-2I**	533	70	–	–	Kretzschmar et al. (2015)

(continued)

Table 9.1 (continued)

Compound	Name	λ_{PL} nm	PLQY %	ΔE_{ST} eV	λ_{EL} nm	Refs.
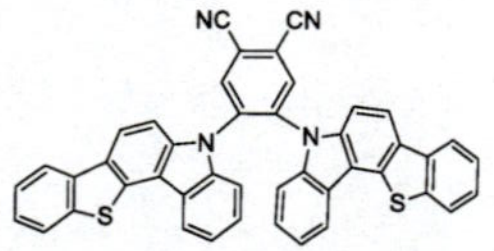	**BTCz-2CN**	328	49	0.13	486	Lee et al. (2015)
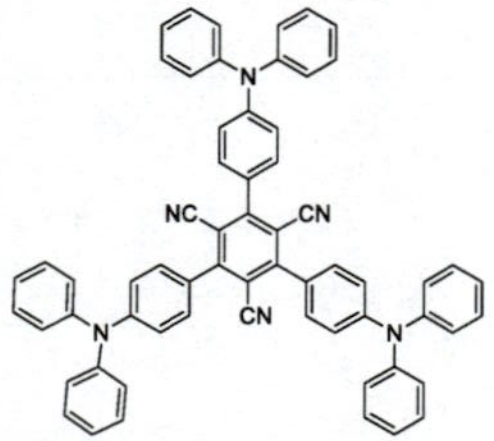	**3DPA3CN**	506	82	0.103	533	Taneda et al. (2015)
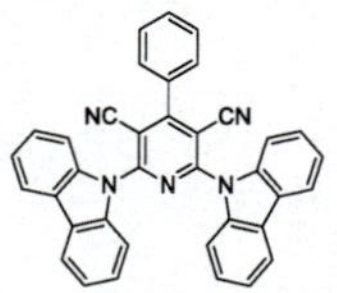	**CPC**	448	49.7	0.04	475–512	Liu et al. (2015)
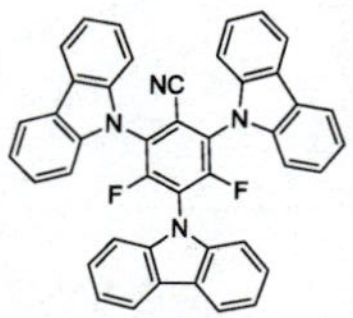	**3CzFCN**	443	76,74	0.06	463	Cho et al. (2015c)

(continued)

Table 9.1 (continued)

Compound	Name	λ_{PL} nm	PLQY %	ΔE_{ST} eV	λ_{EL} nm	Refs.
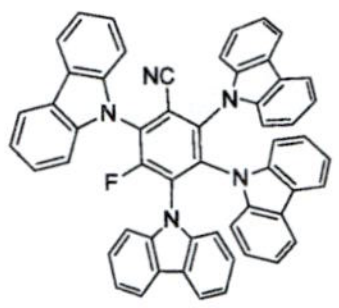	**4CzFCN**	453	81,100	0.06	471	Cho et al. (2015c)
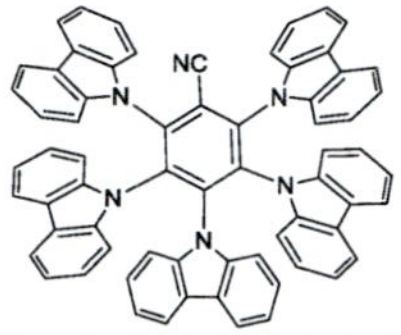	**5CzBN**	464	70	0.22	490	Zhang et al. (2016)
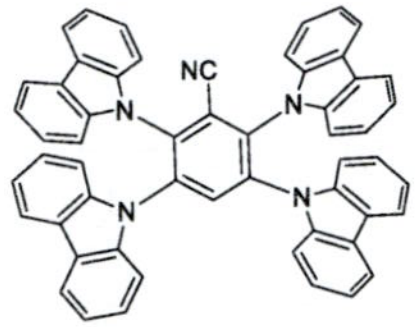	**4CzBN**	442	49	0.30	458	Zhang et al. (2016)
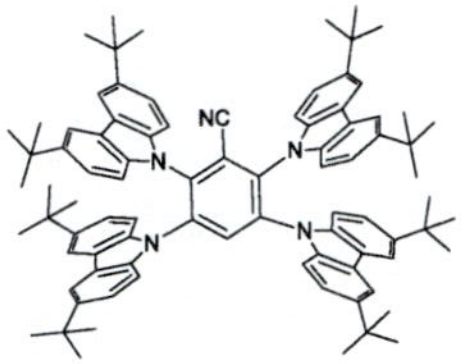	**4TCzBN**	456	68	0.24	463	Zhang et al. (2016)

(continued)

Table 9.1 (continued)

Compound	Name	λ_{PL} nm	PLQY %	ΔE_{ST} eV	λ_{EL} nm	Refs.
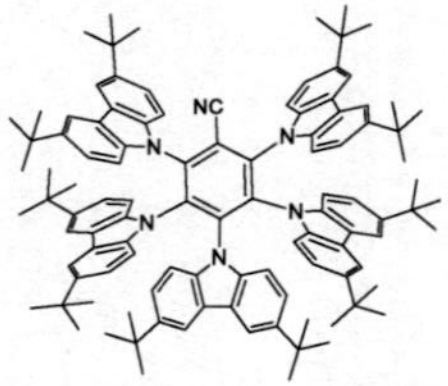	**5TCzBN**	480	86	0.17	490	Zhang et al. (2016)
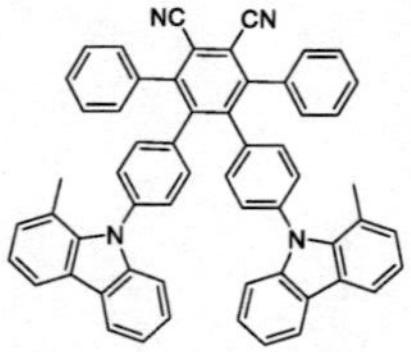	**Cz-VPN**	450	63	0.36	465	Park et al. (2016)
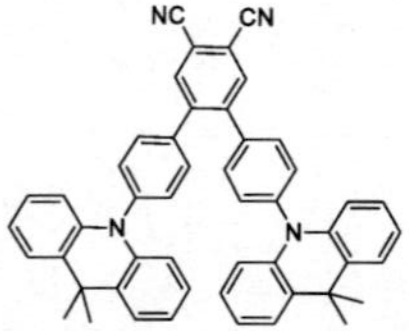	**Ac-VPN**	499	86	0.20	504	Park et al. (2016)

(continued)

Table 9.1 (continued)

Compound	Name	λ_{PL} nm	PLQY %	ΔE_{ST} eV	λ_{EL} nm	Refs.
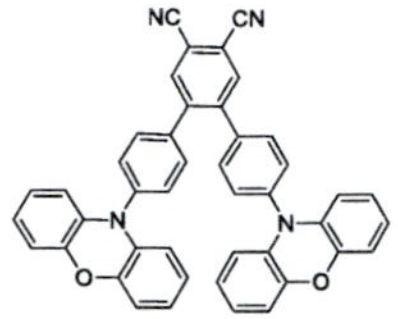	**Ac-CNP**	540	77	0.08	537	Park et al. (2016)
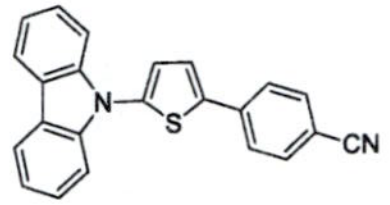	**CTBN**	–	–	–	–	Vikramaditya et al. (2016)
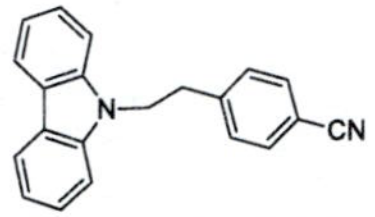	**CEBN**	–	–	–	–	Vikramaditya et al. (2016)
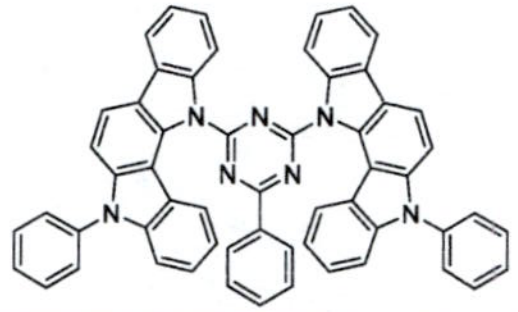	**PIC-TRZ**	466,492	35,39	0.080	506	Endo et al. (2011)

(continued)

Table 9.1 (continued)

Compound	Name	λ_{PL} nm	PLQY %	ΔE_{ST} eV	λ_{EL} nm	Refs.
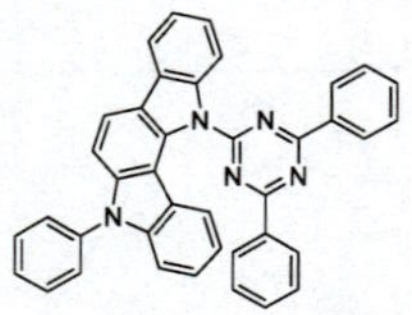	**PIC-TRZ2**	500	59	0.003	500	Sato et al. (2013)
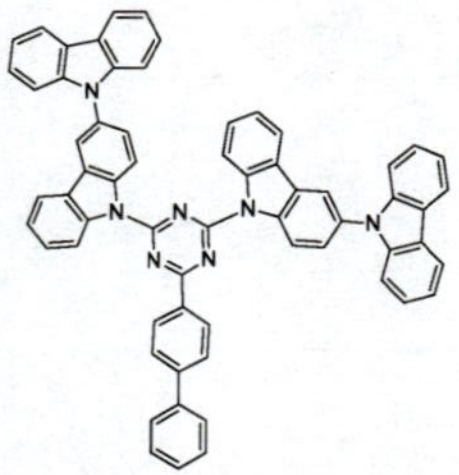	**CC2TA**		62	0.06	485	Youn Lee et al. (2012)
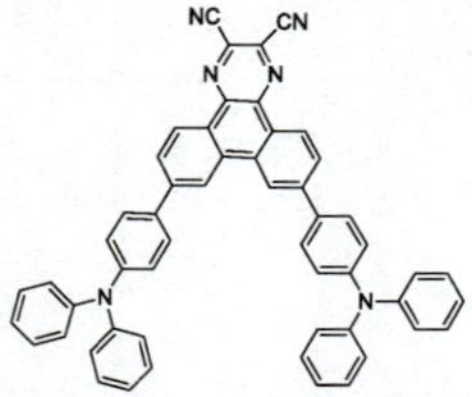	**TTA-DCPP**	588, 708, 645	84, 14, 50	0.13	668, 710	Wang et al. (2015)

(continued)

Table 9.1 (continued)

Compound	Name	λ_{PL} nm	PLQY %	ΔE_{ST} eV	λ_{EL} nm	Refs.
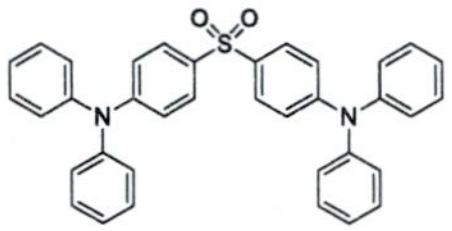	**DPA-DPS**	402, 421	57, 60	0.54	420	Zhang et al. (2012)
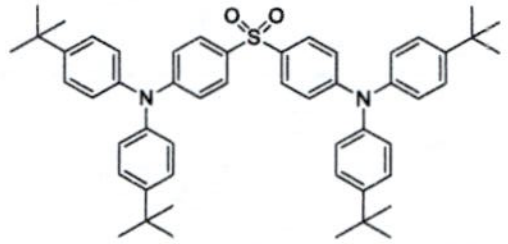	**tDPA-DPS**	419, 430	65, 66	0.45	425	Zhang et al. (2012)
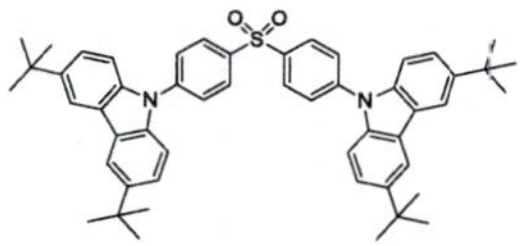	**tDCz-DPS**	404, 423	69, 80	0.32	400	Zhang et al. (2012)
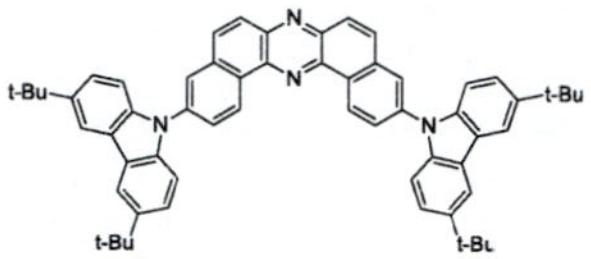	**t-BuCz-DBPHZ**	509	0.31	0.33	–	Data et al. (2016)
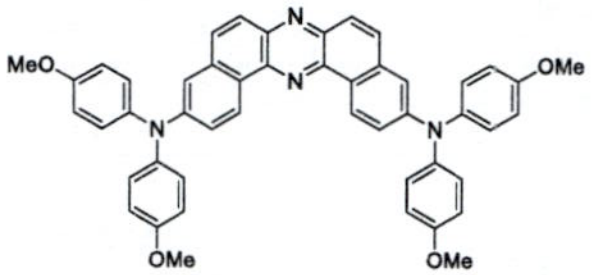	**MeODP-DBPHZ**	592	0.58	0.19	–	Data et al. (2016)

(continued)

Table 9.1 (continued)

Compound	Name	λ_{PL} nm	PLQY %	ΔE_{ST} eV	λ_{EL} nm	Refs.
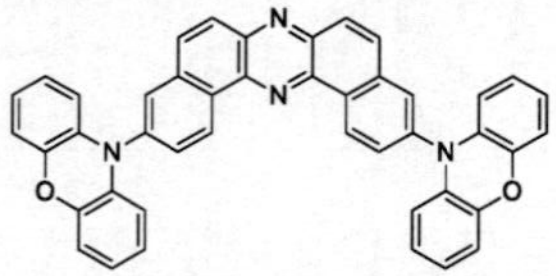	**POZ-DBPHZ**	595	0.79	0.02	–	Data et al. (2016)
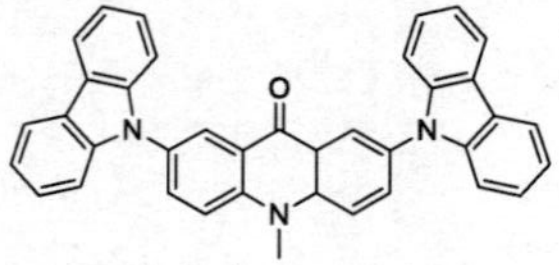	**Acr-CBz**	501	44	0.17	505	Siddiqui et al. (2018)
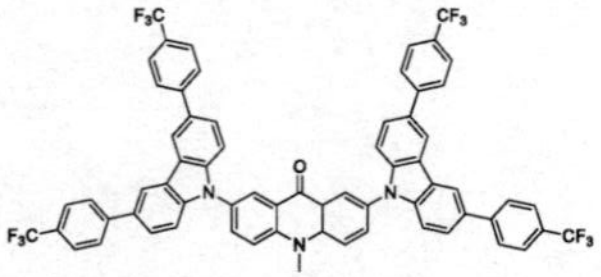	**Acr-CBz -CF_3**	502	40	0.15	490	Siddiqui et al. (2018)
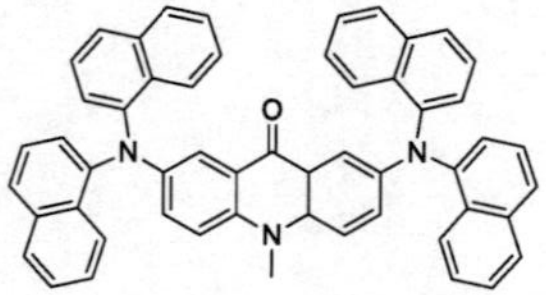 **Acr-Naph**	**Acr-Naph**	550	9	0.3	550	Awasthi et al. (2019)

(continued)

Table 9.1 (continued)

Compound	Name	λ_{PL} nm	PLQY %	ΔE_{ST} eV	λ_{EL} nm	Refs.
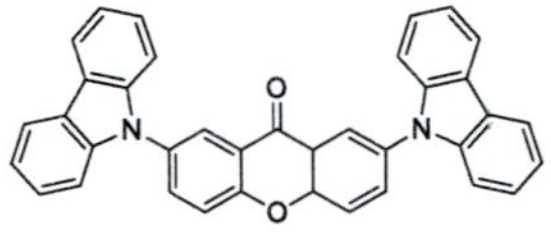 **Xan-CBz**	**Xan-Cbz**	470	22	0.32	465	Siddiqui et al. (2019)
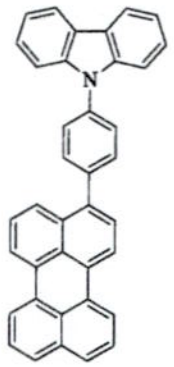	**P-Ph-N-CBz**	–	79	–	557	Dixit et al. (2022)
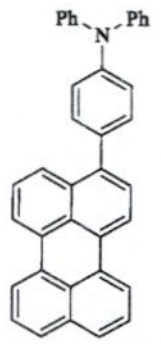	**P-Ph-N-BP**	–	63	–	568	Dixit et al. (2022)

(continued)

Table 9.1 (continued)

Compound	Name	λ_{PL} nm	PLQY %	ΔE_{ST} eV	λ_{EL} nm	Refs.
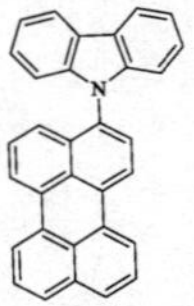	**P-N-CBz**	–	–	–	–	Dixit et al. (2022)
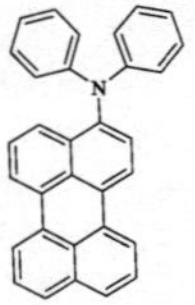	**P-N-BP**	–	–	–	–	Dixit et al. (2022)

compounds consist of acridone as the acceptor unit and carbazole as donor, which were synthesized. In **Acr-CBz**, carbazole was substituted at positions 2 and 7 of acridone, while **Acr-CBz-CF3** contained 3,6-trifluoromethylphenyl carbazole. The researchers investigated the steady-state and time-dependent emission properties of these materials to determine their potential for thermally activated delayed fluorescence (TADF). The singlet–triplet energy gap (ΔE_{ST}) was found to be as low as 0.17 eV for **Acr-CBz** and 0.15 eV for **Acr-CBz-CF3**, making them favorable TADF materials. These compounds were effective green TADF emitters in OLED devices, and interestingly TADF properties were observed in devices based on undoped **Acr-CBz** and **Acr-CBz-CF3**, without the use of a host matrix, which is a departure from the most commonly reported TADF emitters. In addition, the researchers observed an exciplex emission at 465 nm in blends of **Acr-CBz** and **Acr-CBz-CF3** with polyvinyl carbazole (PVK) in a 1:7 (w/w) ratio. OLEDs that used the blend of **Acr-CBz** and **Acr-CBz-CF3** with PVK as the active layer exhibited intense electroluminescence at 465 nm, which was consistent with the exciplex photoluminescence. This study demonstrated that acridone-carbazole derivatives, specifically **Acr-CBz** and **Acr-CBz-CF3**, can exhibit variable electroluminescence as undoped TADF green emitters and blue exciplex emitters when doped in PVK.

Awasthi et al. (2019) investigated the potential of a donor–acceptor–donor (D-A-D) material for use in TADF and OLEDs applications. They synthesized and characterized a compound called **Acr-Naph**, which is based on acridone as the acceptor and naphthylamine as the donor. **Acr-Naph** is a fluorescent compound that emits in the green region (550 nm), with an estimated energy gap between the ground and the lowest excited singlet (S1) state of 2.55 eV. The energy gap between the CT singlet and triplet states (ΔE_{ST}) was found to be approximately 0.3 eV. A delay of 100 μs was observed in the emission at 550 nm, which corresponded to the delayed fluorescence in **Acr-Naph**. The TADF lifetime was determined to be 176 μs. OLEDs based on **Acr-Naph** were fabricated, and their intensity was found to be comparable to that of many known TADF emitters, with a maximum intensity of nearly 17,000 Cd/m^2 at 25 mA/cm^2.

A recent study reported the TADF properties of xanthone derivatives (**Xan-Cbz**) by Siddiqui et al. (2019). Blue TADF emission at 470 nm was observed in neat thin films and further analyzed through delayed fluorescence and lifetime measurements. The study also revealed that a blend of **Xan-Cbz** with NPD produces an exciplex emission at 525 nm in a thin film. OLEDs based on **Xan-Cbz** were fabricated with different device geometries. One device geometry (ITO/PEDOT:PSS/NPD/Xan-Cbz/Bphen/LiF-Al) achieved a luminance of 1.96 $\times$ 104 Cd/m^2 at a current density of 50 mA/cm^2 and VON of ~ 6 V. The electroluminescence showed features of both the neat emission (470 nm) of **Xan-Cbz** and its exciplex (525 nm) with NPD. Additionally, color tuning was observed as a function of applied voltage, and the ratio of light intensity (I525/I470) of neat and exciplex emission decreased with increasing voltage.

Dixit et al. (2022) synthesized and analyzed the photophysical and electroluminescent properties of donor–acceptor dyads containing secondary amine (carbazole and diphenyl amine) linked to perylene through nitrogen, with and without phenyl.

Their DFT studies revealed an almost orthogonal geometry between perylene and carbazole for **P-N-CBZ**. The emission spectra of **P-Ph-N-CBZ** and **P-N-CBZ** in dichloromethane had distinct structures, whereas **P-Ph-N-BP** and **P-N-BP** had broad and featureless spectra, indicating weaker electronic coupling in phenyl-linked dyads. **P-Ph-N-CBZ** and **P-Ph-N-BP** were demonstrated as active emitting materials in OLEDs, with a high luminance of 4.3 × 103 Cd/m^2 at a current density of 100 mA/cm^2 and a maximum EQE of 4.2% and a low turn-on voltage of ~4 V.

9.11 Conclusion

Since the potential of TADF in the harvesting of non-emissive triplet excitons in purely organic emitters was discovered, it has been a hot area of research in both academia and industry. The rate of RISC (k_{rISC}) based on the two-state model of TADF showed that for efficient thermal up-conversion of triplet exciton small ΔE_{ST} is necessary. The detailed mechanism of RISC presented by Fermi's golden rule signifies the effect of the Franck–Condon overlap integral and the SOC operator along with the nature of the excited state involved in the light-emitting mechanism of TADF. The discrepancies between the calculated and the observed values of k_{rISC} were attributed to the Non-Born Oppenheimer effects. Further, the work of Gibson et al. showed that vibronic coupling between the relevant excited states ($^3CT\,and\,^3LE$) involved in TADF plays a significant role in determining k_{RISC}. Since the inclusion of vibronic coupling in the expression of k_{RISC} could not be explained by first-order perturbation theory, a more elaborate description using second-order perturbation theory was employed and it was shown that the relative ordering of excited states corresponds to different mechanisms of the RISC process, which can be controlled by the polarity of the host. So, the choice of host is critical for the performance of a TADF emitter. Also, the designing of organic TADF molecules should look for the tuning of energy gaps between ^1CT and ^3CT and ^3CT and ^3LE states instead of ΔE_{ST}. Further investigation is needed to improve the optoelectronic performance of TADF chromophores and to achieve efficient TADF emission in blue and red regions. Most importantly the synthesis and characterization of large organic TADF molecules are required for their potential use in cheap and large-area lighting.

References

Aizawa, N., Harabuchi, Y., Maeda, S., Pu, Y.J.: Kinetic prediction of reverse intersystem crossing in organicdonor–acceptor molecules. Nat. Commun. **11**, 1–6 (2020)

Albrecht, A.C.: Vibronic—spin-orbit perturbations and the assignment of the lowest triplet state of benzene. J. Chem. Phys. **38**(2), 354–365 (1963)

Awasthi, A.A., Gupta, N., Siddiqui, Q.T., Parab, P., Palit, D.K., Bose, S., Agarwal, N.: Synthesis of acridone-naphthylamine derivative and its thermally-activated delayed fluorescence studies for application in OLEDs. J. Chem. Sci. **131**, 1–8 (2019)

Chen, T., Zheng, L., Yuan, J., An, Z., Chen, R., Tao, Y., Li, H., Xie, X., Huang, W.: Understanding the control of singlet-triplet splitting for organic exciton manipulating: a combined theoretical and experimental approach. Sci. Rep. **5**, 1–11 (2015a)

Chen, X.K., Zhang, S.F., Fan, J.X., Ren, A.M.: Nature of highly efficient thermally activated delayed fluorescence in organic light-emitting diode emitters: nonadiabatic effect between excited states. J. Phys. Chem. C **119**, 9728–9733 (2015b)

Cho, Y.J., Yook, K.S., Lee, J.Y.: Cool and warm hybrid white organic light-emitting diode with blue delayed fluorescent emitter both as blue emitter and triplet host. Sci. Rep. **5**, 1–7 (2015a)

Cho, Y.J., Jeon, S.K., Chin, B.D., Yu, E., Lee, J.Y.: The design of dual emitting cores for green thermally activated delayed fluorescent materials. Angew. Chem. Int. Ed. **54**, 5201–5204 (2015b)

Cho, Y.J., Chin, B.D., Jeon, S.K., Lee, J.Y.: 20% external quantum efficiency in solution-processed blue thermally activated delayed fluorescent devices. Adv. Funct. Mater. **25**, 6786–6792 (2015c)

Data, P., Pander, P., Okazaki, M., Takeda, Y., Minakata, S., Monkman, A.P.: Dibenzo[a j]Phenazine-cored donor-acceptor-donor compounds as green-to-red/NIR thermally activated delayed fluorescence organic light emitters. Angew. Chemie Int. Ed. **55**, 5739–5744 (2016)

Data, P., Okazaki, M., Minakata, S., Takeda, Y.: Thermally activated delayed fluorescence: vs. room temperature phosphorescence by conformation control of organic single molecules. J. Mater. Chem. C **7**, 6616–6621 (2019)

Deng, C., Zhang, L., Wang, D., Tsuboi, T., Zhang, Q.: Exciton-and polaron-induced reversible dipole reorientation in amorphous organic semiconductor films. Adv. Opt. Mater. **7**, 1801644 (2019)

Dias, F.B., Santos, J., Graves, D.R., Data, P., Nobuyasu, R.S., Fox, M.A., Batsanov, A.S., Palmeira, T., Berberan-Santos, M.N., Bryce, M.R., Monkman, A.P.: The role of local triplet excited states and d-a relative orientation in thermally activated delayed fluorescence: photophysics and devices. Adv. Sci. **3**, 1–10 (2016)

Dias, F.B., Penfold, T.J., Monkman, A.P.: Photophysics of thermally activated delayed fluorescence molecules. Methods Appl. Fluoresc. **5**, 12001 (2017)

Dias, F.B., Bourdakos, K.N., Jankus, V., Moss, K.C., Kamtekar, K.T., Bhalla, V., Santos, J., Bryce, M.R., Monkman, A.P.: Triplet harvesting with 100% efficiency by way of thermally activated delayed fluorescence in charge transfer OLED emitters. Adv. Mater. **25**, 3707 (2013)

Dixit, S.J.N., Gupta, C.V., Naidu, G.S., Bose, S., Agarwal, N.: *Peri*-N-amine-perylenes with and without phenyl bridge: photophysical studies and their OLED applications. J. Photochem. Photobio. a: Chem. **426**, 113710 (2022)

Edwards, A.A., Alexander, B.D.: UV-Visible absorption spectroscopy organic applications. In: Encyclopedia of spectroscopy and spectrometry, pp. 511–519 (2017)

Endo, A., Sato, K., Yoshimura, K., Kai, T., Kawada, A., Miyazaki, H., Adachi, C.: Efficient up-conversion of triplet excitons into a singlet state and its application for organic light emitting diodes. Appl. Phys. Lett. **98**, 2011–2014 (2011)

Etherington, M., Gibson, J., Higginbotham, H., et al.: Revealing the spin–vibronic coupling mechanism of thermally activated delayed fluorescence. Nat. Commun. **7**, 13680 (2016)

Gibson, J., Monkman, A.P., Penfold, T.J.: The importance of vibronic coupling for efficient reverse intersystem crossing in thermally activated delayed fluorescence molecules. ChemPhysChem **1**, 2956–2961 (2016)

Hosokai, T., Matsuzaki, H., Furube, A., Tokumaru, K., Nakanotani, H., Yahiro, M., Adachi, C.: Revealing the excited-state dynamics of thermally activated delayed fluorescence molecules by using transient absorption spectroscopy 786–789 (2016)

Koseki, S., Schmidt, M.W., Gordon, M.S.: Calculations of one-electron spin-orbit coupling constants in diatomic molecules. J. Phys. Chem. **96**, 10768–10772 (1992)

Kretzschmar, A., Patze, C., Schwaebel, S.T., Bunz, U.H.F.: Development of thermally activated delayed fluorescence materials with shortened emissive lifetimes. J. Org. Chem. **80**, 9126–9131 (2015)

Lee, K., Kim, D.: Local-excitation versus charge-transfer characters in the triplet state: theoretical insight into the singlet-triplet energy differences of carbazolyl-phthalonitrile-based thermally activated delayed fluorescence materials. J. Phys. Chem. C **120**, 28330–28336 (2016)
Lee, D.R., Hwang, S.H., Jeon, S.K., Lee, C.W., Lee, J.Y.: Benzofurocarbazole and Benzothienocarbazole as donors for improved quantum efficiency in blue thermally activated delayed fluorescent devices. Chem. Commun. **51**, 8105–8107 (2015)
Lee, S.M., Ju, B.K., Lee, C.J.: Temperature dependence of the driving properties for a green thermally activated delayed fluorescence device with a mixed host. Thin Solid Films **660**, 166–170 (2017)
Lewis, G.N., Lipkin, D.: Reversible photochemical processes in rigid media. a study of the phosphorescent state. J. Am. Chem. Soc. **63**, 3005–3018 (1941)
Li, B., Nomura, H., Miyazaki, H., Zhang, Q., Yoshida, K., Suzuma, Y., Orita, A., Otera, J., Adachi, C.: Dicarbazolyldicyanobenzenes as thermally activated delayed fluorescence emitters: effect of substitution position on photoluminescent and electroluminescent properties. Chem. Lett. **43**(3), 319–321 (2014)
Li, N., Ni, F., Lv, X., Huang, Z., Cao, X., Yang, C.: Host-dopant interaction between organic thermally activated delayed fluorescence emitter and host material: insight into the excited state. Adv. Optical Mater. **10**, 2101343 (2022)
Lim, B.T., Okajima, S., Chandra, A.K., Lim, E.C.: Radiationless transitions in electron donor-acceptor complexes: selection rules for $S_1 \rightarrow T$ intersystem crossing and efficiency of $S_1 \rightarrow S_0$ internal conversion. Chem. Phys. Lett. **79**(1), 22–27 (1981)
Lin, L., Fan, J., Cai, L., Wang, C.K.: Excited state dynamics of new-type thermally activated delayed fluorescence emitters: theoretical view of light-emitting mechanism. Mol. Phys. **116**, 19–28 (2014)
Liu, W., Zheng, C.J., Wang, K., Chen, Z., Chen, D.Y., Li, F., Ou, X.M., Dong, Y.P., Zhang, X.H.: Novel carbazol-pyridine-carbonitrile derivative as excellent blue thermally activated delayed fluorescence emitter for highly efficient organic light-emitting devices. ACS Appl. Mater. Interfaces **7**, 18930–18936 (2015)
Liu, J., Li, Z., Hu, T., Wei, X., Wang, R., Hu, X., Liu, Y., Yi, Y., Yamada-Takamura, Y., Wang, Y., Wang, P.: Experimental evidence for "Hot Exciton" thermally activated delayed fluorescence emitters. Adv. Opt. Mater. **7**, 1–9 (2019)
Marian, C.M.: Mechanism of the triplet-to-singlet up-conversion in the assistant dopant ACRXTN. J. Phys. Chem. C **120**, 3715–3721 (2016)
Méhes, G., Goushi, K., Potscavage, W.J., Adachi, C.: Influence of host matrix on thermally activated delayed fluorescence: effects on emission lifetime, photoluminescence quantum yield, and device performance. Org. Electron. **15**(9), 2027–2037 (2014)
Méhes, G., Sandanayaka, A.S.D., Ribierre, J.C., Goushi, K.: Physics and design principles of OLED devices. In: Adachi, C., Hattori, R., Kaji, H., Tsujimura, T. (eds.) Handbook of Organic Light-Emitting Diodes. Springer, Tokyo (2020). https://doi.org/10.1007/978-4-431-55761-6_49-1
Niwa, A., Kobayashi, T., Nagase, T., Goushi, K., Adachi, C., Naito, H.: Temperature dependence of photoluminescence properties in a thermally activated delayed fluorescence emitter. Appl. Phys. Lett. **104**, 213303 (2014)
Northey, T., Stacey, J., Penfold, T.J.: The role of solid-state solvation on the charge transfer state of a thermally activated delayed fluorescence emitter. J. Mater. Chem. C **5**, 11001–11009 (2017)
Ogiwara, T., Wakikawa, Y., Ikoma, T.: Mechanism of intersystem crossing of thermally activated delayed fluorescence molecules. J. Phys. Chem. A **119**, 3415–3418 (2015)
Pan, Y., Li, W., Zhang, S., Yao, L., Gu, C., Xu, H., Yang, B., Ma, Y.: High yields of singlet excitons in organic electroluminescence through two paths of cold and hot excitons. Adv. Opt. Mater. **2**, 510–515 (2014)
Park, I.S., Lee, S.Y., Adachi, C., Yasuda, T.: Full-color delayed fluorescence materials based on wedge-shaped phthalonitriles and dicyanopyrazines: systematic design tunable photophysical properties and OLED performance. Adv. Funct. Mater. **26**, 1813–1821 (2016)

Parker, C.A., Hatchard, C.G.: Triplet-singlet emission in fluid solutions. Phosphorescence of Eosin. Trans. Faraday Soc. **57**, 1894–1904 (1961)

Penfold, T.J., Gindensperger, E., Daniel, C., Marian, C.M.: Spin-vibronic mechanism for intersystem crossing. Chem. Rev. **118**, 6975–7025 (2018)

Pyykkö, P.: Relativistic effects in chemistry: more common than you thought. Annu. Rev. Phys. Chem. **63**, 45–64 (2012)

Randall John Turton: The fluorescence of compounds containing manganese. Proc. r. Soc. Lond. A **170**, 272–293 (1939)

Ravinson, D.S., Thompson, M.E.: Thermally assisted delayed fluorescence (TADF): fluorescence delayed is fluorescence denied. Mater. Horiz. **7**(5), 1210–1217 (2020)

Reichardt, C.: Solvents and solvent effects: an introduction. Org. Process Res. Dev. **11**, 105–113 (2007)

Rothe, C., Monkman, A.: Regarding the origin of the delayed fluorescence of conjugated polymers. J. Chem. Phys. **123**, 1–6 (2005)

Santos, P.L., Ward, J.S., Data, P., Batsanov, A.S., Bryce, M.R., Dias, F.B., Monkman, A.P.: Engineering the singlet-triplet energy splitting in a TADF molecule. J. Mater. Chem. C **4**, 3815–1824 (2016)

Sato, K., Shizu, K., Yoshimura, K., Kawada, A., Miyazaki, H., Adachi, C.: Organic luminescent molecule with energetically equivalent singlet and triplet excited states for organic light-emitting diodes. Phys. Rev. Lett. **110**, 1–5 (2013)

Siddiqui, Q.T., Awasthi, A.A., Bhui, P., Muneer, M., Chandrakumar, K.R.S., Bose, S., Agarwal, N.: Thermally activated delayed fluorescence (Green) in undoped film and exciplex emission (Blue) in acridone-carbazole derivatives for OLEDs. J. Phys. Chem. C **123**, 1003–1014 (2018)

Siddiqui, Q.T., Awasthi, A.A., Bhui, P., Parab, P., Muneer, M., Bose, S., Agarwal, N.: TADF and exciplex emission in a xanthone–carbazole derivative and tuning of its electroluminescence with applied voltage. RSC Adv. **9**, 40248–40254 (2019)

Strickler, S.J., Berg, R.A.: Relationship between absorption intensity and fluorescence lifetime of molecules. J. Chem. Phys. **37**, 814–822 (1962)

Sun, H., Hu, Z., Zhong, C., Chen, X., Sun, Z., Bredas, J.: Impact of dielectric constant on the singlet-triplet gap in thermally activated delayed fluorescence (TADF) materials impact of dielectric constant on the singlet-triplet gap in thermally activated delayed fluorescence materials. J. Phys. Chem. Lett. **8**, 2393–2398 (2017)

Taneda, M., Shizu, K., Tanaka, H., Adachi, C.: High efficiency thermally activated delayed fluorescence based on 1 3 5-Tris(4-(Diphenylamino)Phenyl)-2 4 6-tricyanobenzene. Chem. Commun. **51**, 5028–5031 (2015)

Uoyama, H., Goushi, K., Shizu, K., Nomura, H., Adachi, C.: Highly efficient organic light-emitting diodes from delayed fluorescence. Nature **492**, 234–238 (2012)

Vikramaditya, T., Saisudhakar, M., Sumithra, K.: Computational study on thermally activated delayed fluorescence of donor-linker-acceptor network molecules. RSC Adv. **6**, 37203–37211 (2016)

Wang, S., Yan, X., Cheng, Z., Zhang, H., Liu, Y., Wang, Y.: Highly efficient near-infrared delayed fluorescence organic light emitting diodes using a phenanthrene-based charge-transfer compound. Angew. Chemie Int. Ed. **54**, 13068–13072 (2015)

Wang, K., Shi, Y.Z., Zheng, C.J., Liu, W., Liang, K., Li, X., Zhang, M., Lin, H., Tao, S.L., Lee, C.S., Ou, X.M., Zhang, X.H.: Control of dual conformations: developing thermally activated delayed fluorescence emitters for highly efficient single-emitter white organic light-emitting diodes. ACS Appl. Mater. Interfaces **10**, 31515–31525 (2018)

Ward, J.S., Nobuyasu, R.S., Batsanov, A.S., Data, P., Monkman, A.P., Dias, F.B., Bryce, M.R.: The interplay of thermally activated delayed fluorescence (TADF) and room temperature organic phosphorescence in sterically-constrained donor-acceptor charge-transfer molecules. Chem. Commun. **52**, 2612–2615 (2016)

Xie, F., Zhou, J., Li, Y.: Effects of the relative position and number of donors and acceptors on the properties of TADF materials. J. Mater. Chem. C **8**(28), 9476–9494 (2020)

Xu, Y., Xu, P., Hu, D., Ma, Y.: Recent progress in hot exciton materials for organic light-emitting diodes. Chem. Soc. Rev. **50**, 1030–1069 (2021)

Yang, Z., Mao, Z., Xie, Z., Zhang, Y., Liu, S., Zhao, J., Xu, J., Chi, Z., Aldred, M.P.: Recent advances in organic thermally activated delayed fluorescence materials. Chem. Soc. Rev. **46**, 915–1016 (2017)

Youn Lee, S., Yasuda, T., Nomura, H., Adachi, C.: High-efficiency organic light-emitting diodes utilizing thermally activated delayed fluorescence from triazine-based donor-acceptor hybrid molecules. Appl. Phys. Lett. **101**, 9–13 (2012)

Zhang, Q., Li, J., Shizu, K., Huang, S., Hirata, S., Miyazaki, H., Adachi, C.: Design of efficient thermally activated delayed fluorescence materials for pure blue organic light emitting diodes. J. Am. Chem. Soc. **134**, 14706–14709 (2012)

Zhang, D., Cai, M., Zhang, Y., Zhang, D., Duan, L.: Sterically shielded blue thermally activated delayed fluorescence emitters with improved efficiency and stability. Mater. Horizons **3**, 145–151 (2016)

Zhang, W., Song, H., Kong, J., Kuang, Z., Li, M., Guo, Q., Chen, C.F., Xia, A.: Importance of conformational change in excited states for efficient thermally activated delayed fluorescence. J. Phys. Chem. C **123**, 19322–19332 (2019)

Zhang, X., Shi, Y., Cai, L., Zhou, Y., Wang, C.K., Lin, L.: Solvent effect on the photophysical properties of thermally activated delayed fluorescence molecules. Spectrochim. Acta Part A Mol. Biomol. Spectrosc. **225**, 117473 (2020)

Zuehlsdorff, T.J., Isborn, C.M.: Modeling absorption spectra of molecules in solution. Int. J. Quantum Chem. **119**, e25719 (2019)

Chapter 10
Luminescent Materials for Radiation Dosimetry

N. S. Rawat, B. S. Dhabekar, and B. K. Sapra

10.1 Motivation and Background

Nuclear technology has widespread peaceful applications in our daily life, the most important being the production of green and clean electricity. The applications in health care are well established, specifically for the treatment of most types of cancers. Diagnostic techniques like positron emission tomography (PET), single-photon emission computed tomography (SPECT), and computed tomography (CT) scans are among the various applications of nuclear technology. Other industrial applications include wide use of radioactive sources in smoke detectors, nucleonic gauges (for thickness measurements, level detectors, etc.), well logging, etc. Radiation (energetic electrons from accelerators) is also used for coloration of precious stones. In agriculture, ionizing radiations are used to produce the improved seeds (in terms of high yield, better resistance against deceases, shape, size, etc.) by genetic mutation; radioactive tracers are used to optimize the use of water, fertilizers, and pesticides. Food irradiation increases the shelf life of various food items like onions, potatoes, mangoes, spices, meat products, etc. The nuclear technology is therefore progressing by leaps and bounds with the advent of new facilities like synchrotron, particle accelerators, nuclear reactors, production of radiation sources, etc. However, indiscriminate use of ionizing radiation without adhering to proper safety procedures may lead to excessive radiation exposure of radiation workers, which may increase the risk of developing cancer (a stochastic effect) later in life, the probability of which increases with increase in level of radiation exposure. Consequently, the monitoring of occupational workers and ensuring their compliance with dose limits as stipulated

N. S. Rawat (✉) · B. S. Dhabekar · B. K. Sapra
Radiological Physics and Advisory Division, Bhabha Atomic Research Center, Mumbai 400085, India
e-mail: naru@barc.gov.in

Homi Bhabha National Institute, Mumbai, India

R. S. Ningthoujam and A. K. Tyagi (eds.), *Handbook of Materials Science, Volume 1*, Indian Institute of Metals Series, https://doi.org/10.1007/978-981-99-7145-9_10

by regulatory authorities is mandatory. Solid-state luminescence dosimetry techniques like **Thermoluminescence** (TL) and **Optically Stimulated Luminescence** (OSL) have emerged as the most viable option for assessment of radiation doses received by the occupational workers. These techniques offer inherent operational simplicity and cost-effectiveness. This article attempts to highlight the efforts made by Bhabha Atomic Research Center (BARC), in the recent few years, to augment the existing dosimetric techniques and to develop new modalities for varied applications.

10.2 Introduction

Robert Boyle in 1663 was the first to record the TL response scientifically by heating a diamond, giving a glimmering light (Boyle 1663). However, Wiedemann and Schmidt (1895) of Germany were probably the first to report experimental and radiation-induced TL in a wide variety of synthetic and natural materials, immediately before W. C. Roentgen discovered X-rays. They showed that certain bodies emit light after exposure to cathode rays or the electric spark, when they are heated to a temperature much below that is required to cause incandescence (Rutherford 1913). The study of radiation-induced TL received the much needed impetus from Curie (1904) when she wrote her doctoral thesis which mentioned the TL properties of fluorite exposed to radium. In the late 1940s and early 1950s, Farrington Daniels and his team at the University of Wisconsin (USA) made significant advancements in the use of Thermoluminescence (TL) as a radiation dosimetry tool. Their pioneering work focused on studying Lithium fluoride (LiF) as a TL material (Daniels et al. 1953). Initially, LiF was employed as a dosimeter to measure radiation doses following nuclear weapon tests. As more scientific groups began exploring TL dosimetry (TLD), the idea gained momentum.

Cameron and Kenney (1963), Cameron et al. (1968) discovered that the desirable TL properties in lithium fluoride were associated with complex defects resulting from the presence of Ti and Mg. Their research eventually led to the patenting of TLD-100 by Harshaw Chemical Company in 1963. The first application of TL material to estimate an individual's dose occurred through a unique collaboration between Daniels and Marshal Brucer at the Oak Ridge Institute of Nuclear Studies. The crystal of LiF was ingested by one of Brucer's cancer patients undergoing ^{131}I treatment. After a few days, the crystal was retrieved and analyzed, estimating the delivered dose to be 60 Roentgen (Frame 2005).

Luminescence is a phenomenon in which a material releases its stored excitation energy in the form of UV or visible light. Unlike incandescence, luminescence does not rely on high temperatures and typically does not generate noticeable heat. This characteristic led to the term "cold light" being used to describe these emissions.

Thermoluminescence (TL) or more accurately thermally stimulated luminescence (TSL) is one of the many luminescence processes, viz., photoluminescence, electroluminescence, cathodoluminescence, etc. TL refers to the emission of light, beyond the thermal radiation, by an insulator or semiconductor material that has previously been exposed to ionizing radiation. This emission occurs when the TL material is heated, a process known as stimulation. The TL phenomenon exhibits high sensitivity to the presence of defect centers, which play a crucial role in the TL process. In a study conducted by Townsend and Kelly (1993), it was quantified and estimated that the TL technique is capable of detecting defect levels as low as 109 in the phosphor material. This highlights the remarkable sensitivity of TL as a method for detecting and quantifying defects in materials. The exceptional sensitivity of the TL process allows us to accurately determine extremely low radiation doses, reaching levels as minute as (μGy).

During TL readout, under favorable conditions, the intensity of light emitted from a material is directly proportional to the absorbed dose caused by ionizing radiation. Therefore, by experimentally measuring the quantity of emitted light, it becomes possible to estimate the dose, which represents the energy absorbed per unit mass in the material. This fundamental principle forms the basis of dosimetry.

10.3 TL and OSL for Radiation Dosimetry

10.3.1 Thermoluminescence (TL)

In general, the phenomenon of TL is exhibited by insulators, semiconductors, or organic materials. However, for radiation dosimetry, the wide bandgap materials (insulators) are preferred, either in single crystal or polycrystalline form. Figure 10.1a, b shows the schematic picture of an ideal crystal of an insulator and the corresponding band structure. The well-defined band structure is a consequence of long-range order in the ideal crystal. Such an ideal crystal is never found because of the intrinsic defects. Two defects are always present in the real crystal. The first one is the finite size of the crystal due to which its periodicity is broken at the surface thus leading to imperfection. Also, at the temperatures above 0 K, the lattice vibration leads to the imperfections like interstitials and vacancies to minimize the Gibbs free energy. Apart from these two defects, the real crystals always have vacancies, interstitials, grain boundaries, stacking faults, etc. During the synthesis of any TL/OSL phosphor, defects are deliberately introduced in the crystal to create electron and hole traps that render sensitivity to the material making it amenable for its use in radiation dosimetry. Therefore, perfection of any phosphor material is neither achievable nor many a times desirable. Figure 10.2a, b shows such a crystal and corresponding band structure. It must be noted that the electron traps are empty and hole traps are filled with electrons (represented by blue circle). The TL process consists of following three-stages:

Excitation by Ionizing Radiation: During irradiation with ionizing radiation (excitation) such as gamma rays, electrons from valence or any other inner shells receive sufficient energy and are excited/pumped to the conduction band, creating a hole (absence of electron) in the valence or any other inner shell band. Excitation from

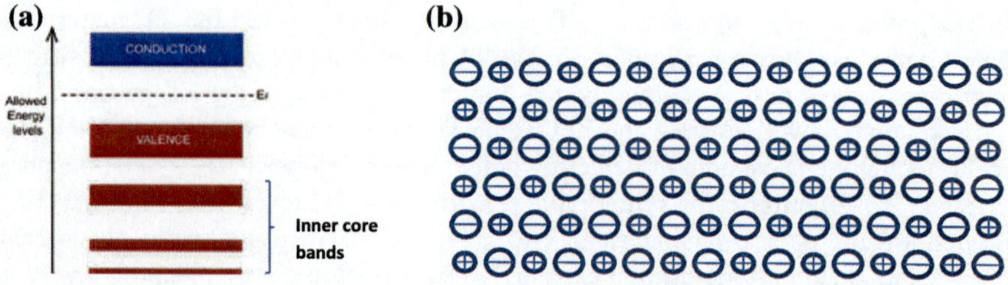

Fig. 10.1 **a** Band structure of an ideal crystal, **b** Schematics of an ideal crystal of an ionic solid

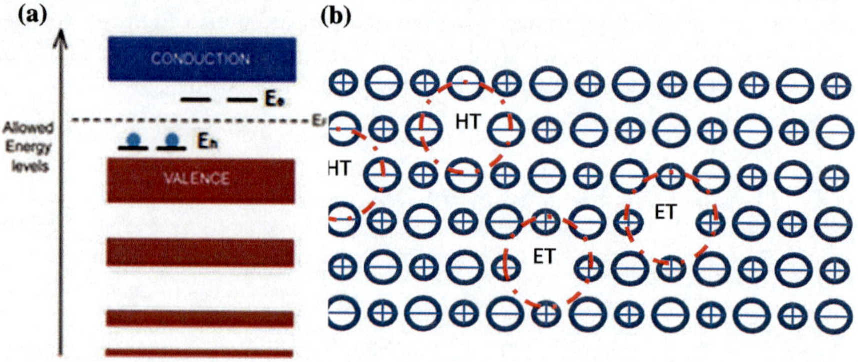

Fig. 10.2 **a** Band structure of TL/OSL phosphor, **b** Crystal structure of a TL/OSL phosphor

the inner shells is an important aspect of TL/OSL process because it is due to this excitation, the response of non-tissue equivalent phosphors is more than one, particularly at low photon energies, where photoelectric effect is dominant. Both the valence band and the conduction band contain mobile entities such as holes and electrons. Typically, the recombination of electrons in the conduction band and holes in the valence band leads to the phenomenon of radioluminescence. However, in exceptional circumstances, electrons in the conduction band can become trapped at specific electron traps (ET) (Fig. 10.3a) and the corresponding holes (represented by green circle) are trapped at hole traps (HT).

Latency Period: If the energy difference between the electron trap and the bottom of the conduction band (E_e) and hole trap and top of the valence band (E_h) is sufficiently large, these trapped electrons and holes are stable at the respective sites (Fig. 10.3a, b). This stage is called as latency period or energy storage period. These trapped entities will remain at these sites until they are stimulated either by heat (which leads to TL) or by light (OSL). The life time of the trapped entities may vary from a few seconds to millions of years depending upon the trap depth (E_e or E_h).

Readout: During TL readout, the temperature of the TL phosphor is gradually increased (generally linearly). If the trapped electrons acquire sufficient energy

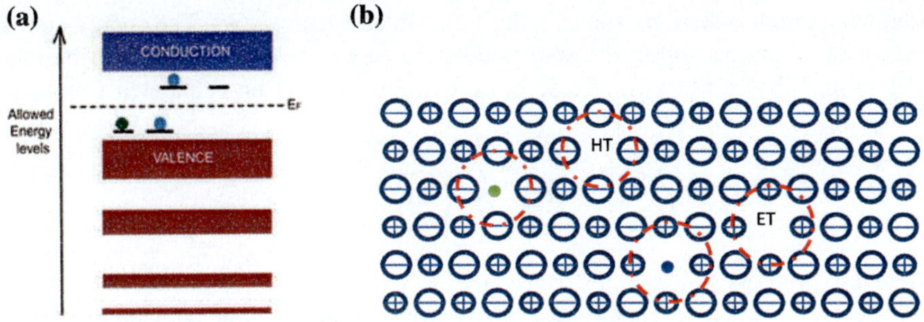

Fig. 10.3 **a** Band structure of TL/OSL phosphor after irradiation, **b** Crystal structure of a TL/OSL phosphor after irradiation

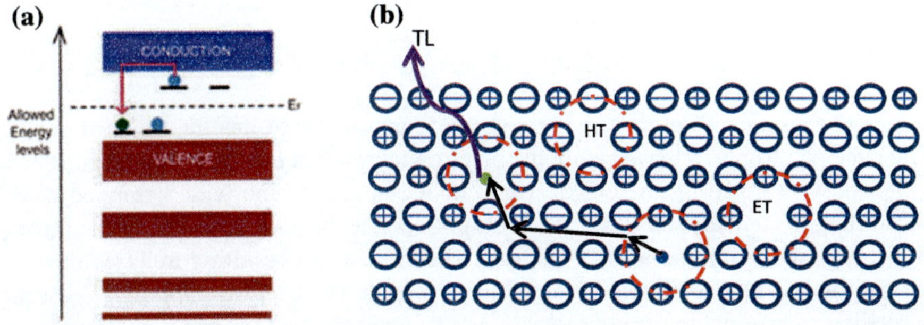

Fig. 10.4 **a** Band structure of TL/OSL phosphor during readout, **b** Crystal structure of TL/OSL phosphor during readout

they are stimulated to the conduction band. Once in the conduction band, these electrons are free to move and can potentially combine with the trapped holes (Fig. 10.4a, b). The energy released during recombination is converted in the form of photons referred as TL. In case of OSL, the stimulation agent is light (usually blue). After readout, the lattice is restored in its original state and hence ready for reuse. It is an important aspect of TL/OSL-based dosimetry.

10.3.1.1 The TL Glow Curve

During TL readout, the intensity or brightness of the emitted light is typically plotted against temperature, resulting in a TL glow curve. The simplified and phenomenological mathematical model describing TL was initially formulated by Randall and Wilkins (1945a, b). This model considers a phosphor with a single type of defects that produce a solitary electron trap level possessing a specific activation energy or trap depth, denoted as **E**. At a given time **t** and temperature **T** (measured in Kelvin) following initial radiation exposure, this trap contains **n** electrons. The energy distribution of electrons within the trap is determined by the Boltzmann distribution. The probability of electron release or de-trapping (**p**) can be calculated using the following Arrhenius equation:

$$p = s\exp\left(\frac{-E}{kT}\right) \tag{10.1}$$

Here, ***k*** represents Boltzmann's constant,

The symbol ***s*** represents the attempt-to-escape factor or frequency factor (often it ranges from ~10^{12}–10^{14} s^{-1} depending on nature of defect and host lattice).

The number of electrons released from this trap will be the product of number of trapped electrons (***n***) and probability of release of an electron (***p***) from the trap, i.e.,

$$-\left(\frac{dn}{dt}\right) = np = ns\exp\left(\frac{-E}{kT}\right) \tag{10.2}$$

where the number of electrons trapped (***n***) at any instance has the units of cm^{-3}. Once the electron is released from the electron trap, it has two possible paths. It can either be re-trapped by the electron trap or it can recombine with a trapped hole. According to Randall and Wilkins' assumption, once the electrons are released from traps, they will combine with holes trapped at hole traps, resulting in TL with zero probability of re-trapping. As a result, the rate of electron de-trapping/release is directly proportional to the concentration of trapped charges (**n**) and is referred to as "first-order" kinetics. In such a scenario, the intensity of the TL glow, represented as ***I(t),*** is proportional to the rate at which trapped electrons are released and can be mathematically expressed as follows:

$$I(t) = -C\left(\frac{dn}{dt}\right) = Cns\exp\left(\frac{-E}{kT}\right) \tag{10.3}$$

where the constant *C* is related to TL efficiency and instrument parameters.

For linear heating profile the temperature varies as $T\,(K) = T_0 + \beta \cdot t\,(s)$, where β is heating rate and T is absolute temperature. The above equation then modifies to.

$$I(T) = -C\left(\frac{dn}{dT}\right) = -\frac{Cns}{\beta}\exp\left(\frac{-E}{kT}\right) \tag{10.4}$$

Figure 10.5a shows theoretically simulated TL glow curve for E = 1 eV, s = 10^{12} s^{-1}, and β = 0.5 K/s. The TL peak occurs at around 200 °C. The presence of a peak indicates that two mutually opposing parameters dictate the TL phenomenon. According to Eq. 10.3, the intensity of TL is influenced by two factors: the quantity of trapped electrons (**n**) at a specific temperature and the probability of electron release (**p**) from the trap. With increasing temperature, the trapped electrons are gradually released (stimulated) and move to the conduction band. Hence, ***n*** decreases with

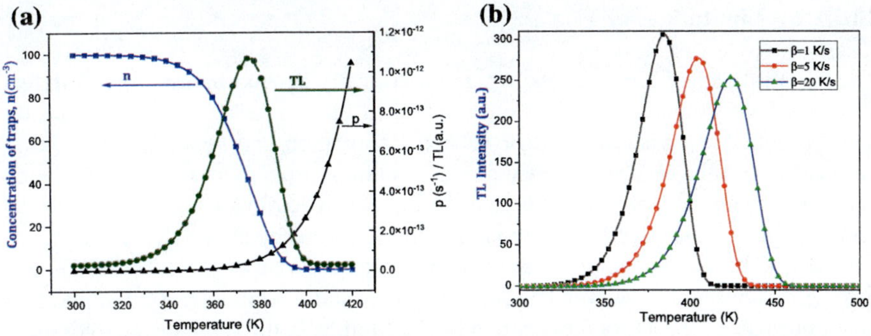

Fig. 10.5 **a** Simulated curves for first-order kinetics and trap parameters E = 1 eV, s = 10^{12} s^{-1}, β = 0.5 K/s showing (i) depletion of trapped charge carriers (n), (ii) TL glow peak, and (iii) Probability of release of trapped charge carries (p) with temperature, **b** Variation of TL intensity with heating rate. With increase in heating rate (β), the TL peak as well as entire curve shift toward higher temperature

temperature. It is not difficult to comprehend that the second factor ***p*** increases with increase in temperature. The variation of ***n*** and ***p*** with ***T*** is shown in Fig. 10.5a.

Figure 10.5b shows the simulated TL glow curves for different heating rates. It is evident that the TL glow peak shifts toward higher temperature region with increase in heating rate. The phenomenon of the shift of a TL glow curve with heating rate is widely recognized and easily observed, as described by the model proposed by Randall and Wilkins (1945a). It is noticed that the TL glow peak moves toward higher temperatures, but occurs earlier in the time domain, as the heating rate increases. Initially, this shift in the TL glow peak with a change in heating rate may not be readily apparent. However, the empirical explanation of this observation was proposed by Kitis et al. (1993). To understand the shift qualitatively, let us assume two heating rates β_1 and β_2 such that $\beta_2 > \beta_1$. Now, at lower heating rate β_1, the phosphor experiences an instantaneous temperature say T_1, for longer duration compared to β_2. As a result, at lower heating rate, higher number of charge carriers (electrons or hole) are stimulated and thermally released at arbitrary temperature T_1. The extent of stimulation depends on the lifetime of trapped charge carriers at given temperature. The phosphor therefore absorbs more energy from the heating system at lower heating rate as compared with that of higher heating rates. Hence, to release the same number of charge carriers thermally at β_2, a higher temperature T_2 is necessary. Consequently, the entire TL glow peak shifts toward the higher temperature region as the heating rate increases. This shift is influenced by both the lifetime of charge carriers and the duration spent at corresponding temperature. An alternative approach related to effect of black-body radiation along with its mathematical formalism has also been suggested by Rawat et al. (2014a). This approach qualitatively and quantitatively explains shift in TL peak with increasing/decreasing heating rate.

10.3.1.2 Limitations of TL

The TL attributes of any phosphor material critically depend on the type of defect structure. The temperature-dependent stability of these defects decides the faithful and reproducible extraction of dose-related information and hence the dosimetric performance of the phosphor. Few materials exhibit sharp reduction in TL efficiency at elevated temperatures. This reduction is attributed to dominant non-radiative relaxation process (Rawat 2012). The reduction in TL efficiency at elevated temperature is called as thermal quenching that hampers the application of the TL material as a dosimetric phosphor. However, heating of a sample and associated rise in temperature being intrinsic to the TL process is inevitable. Moreover, the reuse or redeployment of TL phosphor in radiation dosimetry necessitates the heat treatment of TL phosphor. This is required for resetting of the phosphor or zeroing of the dose before its redeployment. The heat treatment is not completely a non-destructive process and depending on the material may upset the defect structure and remarkably alter the sensitivity and performance of the phosphor (Rawat 2012). Therefore, it is crucial to maintain a precise and consistent heating profile during the TL process. Additionally, repeated heating of a phosphor can negatively impact its chemical integrity. These challenges necessitate a suitable solution in terms of a stimulation technique that is based on a more user-friendly and convenient method, such as light stimulation.

10.3.2 Optically Stimulated Luminescence (OSL)

OSL refers to the emission of light that occurs when crystalline solids (insulators or semiconductors), which were previously excited by ionizing radiation, are stimulated with a different wavelength of light. The initial excitation places the entire crystalline material in a metastable state, wherein electrons and holes become trapped at corresponding defect levels (Rawat 2012). When stimulated by visible photons (light), the trapped electrons and holes are released during the OSL process. This leads to the recombination of electron and hole pairs and subsequent excitation of luminescence centers within the crystalline solid. OSL involves the emission of photons as the excited luminescence centers undergo radiative relaxation back to the ground state.

Earlier, OSL was not a popular technique and thus not used extensively unlike TL technique in radiation dosimetry, as there was dearth of highly sensitive, low effective atomic number luminescent materials with acceptable or negligible post-irradiation fading characteristics. However, the introduction of α-Al_2O_3:C phosphor about 30 years ago, as a highly sensitive dosimetric material, has attracted many researchers to work on its applications in OSL-based radiation dosimetry. Bøtter-Jensen and McKeever (1996) has reviewed the potential of OSL for radiation dosimetry.

Some nationally accredited service providers in the USA are utilizing OSL dosimetry systems based on Al_2O_3:C (Botter-Jensen et al. 2003). This system meeting American National Standards Institute (ANSI) criteria has largely replaced

TL-based dosimetry systems. OSL-based system has also been introduced for personnel monitoring in some other countries (Pradhan et al. 2008). It is estimated that more than 25% of about 5 million badges worldwide are in fact OSL dosimeters (Lee and Jai 2001). The National Council on Radiation Protection and Measurements (NCRP) has approved the use of Al_2O_3:C-based OSL dosimeters for dosimetry of astronauts and the habitable areas of spacecraft (Yukihara et al. 2006a; Yukihara and McKeever 2011). Additionally, OSL technique has emerged as a real-time in vivo dosimetry tool for applications such as mammography, radiotherapy, and CT scans. For the past three decades, OSL phosphors like BaFBr:Eu, etc. are extensively used for storage plates, along with other applications like two-dimensional imaging (Pradhan et al. 2008).

10.3.2.1 Merits of OSL Technique

OSL technique is all optical in nature and therefore obviates the need of heating. As OSL measurements are carried out at room temperature, the problem pertaining to thermal quenching, if any, does not arise unlike in TL. OSL is essentially a non-destructive technique. However, TL is quasi-destructive as it involves heating that may disturb the defect structure in material thus resulting in change of sensitivity and usage for a limited number of cycles when compared to OSL. Due to heating requirement, the TL dosimeter cannot be encapsulated in an inexpensive plastic unlike OSL dosimeter. Further, the TL dosimeter needs to withstand frequent administration of harsh temperature cycle that prevents fabrication of thin TL dosimeters. Therefore, thickness of TL dosimeter is usually high enough to ensure geometrical integrity and prevent any buckling. This makes TL-based beta dosimetry a challenging task and introduces uncertainty in measurement of beta doses. The problem can be overcome by fabricating thin OSL dosimeters required for beta dosimetry. OSL offers inherent simplicity and ease of instrumentation. By controlling stimulation intensity and time, the OSL signal can be obtained in relatively lesser time, which offers fast readout and hence high throughput can be achieved. The scenario of high throughput is essential in case of routine personnel monitoring wherein large number of OSLD badges need to be processed. Further, restricting OSL readout for few seconds paves the way for multiple assessment of doses in continuous wave as well as pulsed mode (Rawat et al. 2015). The detection of doses as low as sub-μGy can be achieved using OSL technique (Rawat et al. 2014b, 2017). Two-dimensional dose imaging with high spatial resolution can be attained using OSL film that can be used to assess the legitimacy and genuine nature of the over-exposure cases in routine personnel monitoring. Remote optical fiber-based dosimetry in real time is another advantage of OSL dosimetry for medical applications.

Al_2O_3:C-based OSL dosimeter offers high sensitivity, low background, high spatial resolution, and availability in various dimensions and shapes. Additionally, it offers the capability of measuring absorbed doses in real time for both electron and photon field, and temperature independence for the ease of calibration and use (Pradhan et al. 2008). Although OSL offers several advantages, a few limitations of

OSL include dark room conditions for handling. The OSLD badge also needs to be light tight else the OSL signal will get erased to certain extent.

10.3.3 Salient Features of Luminescent Material for Dosimetry

TL/OSL phosphors for dosimetric applications should possess the following attributes (Rawat 2012):

- High TL/OSL efficiency, indicating a significant light output in relation to the absorbed dose per unit mass.
- Long-term stability of trapped charge carriers, ensuring for the retention of the stored signal at room temperature even after irradiation.
- Long-term stability of trapped charge carriers (post-irradiation) and therefore the stored signal at room temperature.
- Very good shelf life of the phosphor, i.e., the TL/OSL sensitivity of the phosphor should not exhibit any reduction when used after long duration.
- Chemical stability, enabling resistance to harsh climatic conditions including pressure, temperature, and humidity.
- TL/OSL emission spectrum of the phosphor should match with spectral response of readily available photomultiplier tube/detector system. For instance, it is desirable to have phosphors with characteristic light emission occurring in the high efficiency region (preferably 300–500 nm) of photomultiplier tube. Similarly, TL phosphors with negligible or low luminescence emission in infrared region upon heating are suitable for radiation dosimetry. This is desirable as the luminescence signal can easily be segregated/filtered from background resulting from black-body emission, which lies mostly in infrared region during heating.
- Near-tissue equivalent response is preferable. In other words, effective atomic number (Z) of the phosphor should be close to that of tissue (Z_{tissue} ~ 7.4). This condition ensures that the measured dose is close to the dose absorbed by the human body, particularly for low-energy photon (<200 keV) region where photoelectric effect is dominant.
- It should have linear response over a wide dynamic dose range. This offers the possibility of calibrating the dosimeters at a single dose value, making the calibration process simpler.

In addition to the above, a few other desirable features for TL-based phosphors are

- Simple glow curve structure with single glow peak around 200 °C. If the peak temperature is too low (~100 °C), the TL signal is unstable as it will fade at room temperature. If it is too high (~ 400 °C), there will be issues pertaining to high infrared (IR) background.
- TL material should exhibit negligible influence of room light.

- TL phosphor should not exhibit any thermal quenching.
- Reusable with simple annealing procedure, with no alteration/loss in sensitivity with reuse as well as repeated annealing.

Generally, aforementioned requirements should be fulfilled for a desirable TL/OSL dosimeter. However, as often the case, not all attributes can simultaneously be achieved.

10.4 Widely Used Luminescent Materials for Radiation Dosimetry

Ever since Farrington Daniels and his research group at the University of Wisconsin (USA) initially proposed the use of TL as a radiation dosimetry technique, a significant number of TL/OSL phosphors have been synthesized (Bhatt and Kulkarni 2014). Figure 10.6 provides a chronological overview of some noteworthy TL/OSL phosphors that have been developed. These phosphors are synthesized using various techniques, including: (i) co-precipitation technique, (ii) solid-state diffusion method, (iii) recrystallization from solution, (iv) melt process technique, (v) growing single crystals from melt, (vi) Solution Combustion Synthesis (SCS) method, and (vii) sol–gel method. (International Atomic Energy Agency 2004). Figure 10.6 gives a chronological account of development of important TL/OSL phosphors.

10.4.1 $CaSO_4$:Dy

$CaSO_4$:Dy is the backbone of TLD-based country-wide personnel monitoring program in India. Personnel monitoring is measurement of radiation doses received by an individual from external and/or internal radiation, which also includes keeping records of doses that have been determined. Individual monitoring is employed to assess the efficacy of radiation control measures within the workplace. It serves the purpose of detecting any alterations occurring in the workplace, validating or supplementing stationary workplace monitoring, identifying work methods that minimize radiation exposure, and providing crucial information in case of accidental exposure (International Atomic Energy Agency 2004; Pradhan et al. n.d.).

In India, the official centralized Personnel Monitoring Service (PMS) was started in 1952 by BARC using the Film Badge (Pradhan et al. n.d.). Although the film badge paved the way for personnel monitoring in India, it faced several limitations like fogging caused by high temperature and light, complicated chemical process for read out and non-reusability. This leads to the indigenous development of $CaSO_4$:Dy-based TLD system, which is a better dosimetry technique.

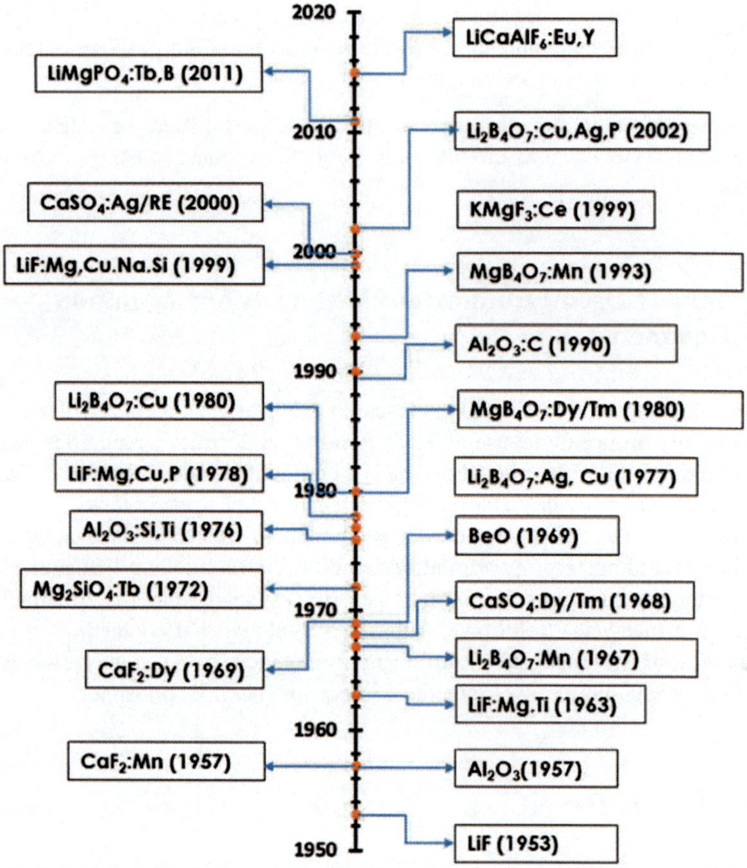

Fig. 10.6 Chronological development of some important TL and OSL phosphors with potential application in dosimetry

Yamashita et al. (1968, 1971) reported the synthesis of highly sensitive Dy- and Tm-doped $CaSO_4$ TL phosphor by recrystallization method. Analytical-reagent-grade $CaSO_4.2H_2O$ and reagent-grade rare oxides were dissolved in hot sulfuric acid kept around 300 °C. This saturated solution of $CaSO_4$ is boiled off to get small crystallites of the material. The sensitivity of the phosphor material prepared by this technique is about 35 times that of LiF:Mg,Ti. Although this method is straightforward, it is not the best for preparing the phosphor in large quantities, wherein evaporating large quantity of conc. H_2SO_4 leads to problems of corrosion and pollution. In the study conducted by Rao et al. (1980), it was observed that the process of acid evaporation in an open-air environment leads to inconsistencies in TL sensitivity between different batches, primarily due to contamination. A novel method was established to standardize the TL phosphor production process. This method involves distilling a saturated solution using air as the carrier gas, enabling the recovery of approximately

98% of H_2SO_4. More than 90% of the recovered H_2SO_4 is of 35–36 N concentration, which is then recycled for the preparation of new batches. The resulting crystalline powder undergoes annealing at 700 °C for 2 h to eliminate any residual traces of H_2SO_4 and enhance the TL characteristics for future reuse (Sunta 1984). This modified technique is employed for the commercial production of the $CaSO_4$:Dy phosphor (600 g per batch) (Bhatt and Kulkarni 2014). BARC also developed an indigenous technique to fabricate PTFE (polytetraflouroethylene)-based $CaSO_4$:Dy TLD disks for routine application in dosimetry. The process of phosphor preparation in powder form, preparation of PTFE-based disks, TLD badges, and dosimetric procedures were standardized which led to transfer of technology to the private sector. This made the decentralization of the PMS feasible and economically viable. Presently, 16 TLD Laboratories (government and private) located in various parts of the country are providing Personnel Monitoring Services (PMS) using the indigenously developed TLD Badge System. This development initiated the replacement of Film Badge by TLD badge starting from 1975 in a phased manner (Vohra et al. 1980; Pradhan 1981; Kher et al. 1983; Lakshmanan et al. 1986).

A TLD badge comprises a TLD card inserted into a cassette that includes appropriate metallic filters. It consists of three PTFE-based $CaSO_4$:Dy disks mechanically clipped on a nickel-plated aluminum card (Fig. 10.7). The TLD cards are read out in an indigenously developed semi-automatic TLD reader system (Fig. 10.8), which can handle 50 cards in a single run. The readout process is very simple as compared to that of film badge. TLDs offer a wider dose measurement range compared to film badges. They can be read onsite, eliminating the need for sending them away for developing. Moreover, TLDs can easily be reused. Thus, the introduction and advent of TLDs completely revamped the personnel monitoring program of India.

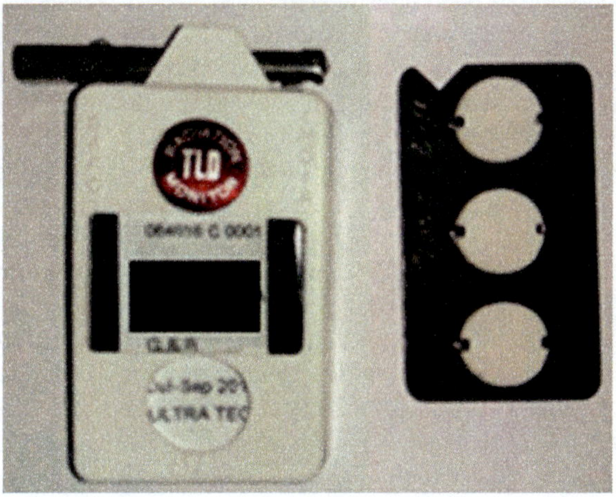

Fig. 10.7 TLD Badge and TLD Card used for country-wide personnel monitoring in India

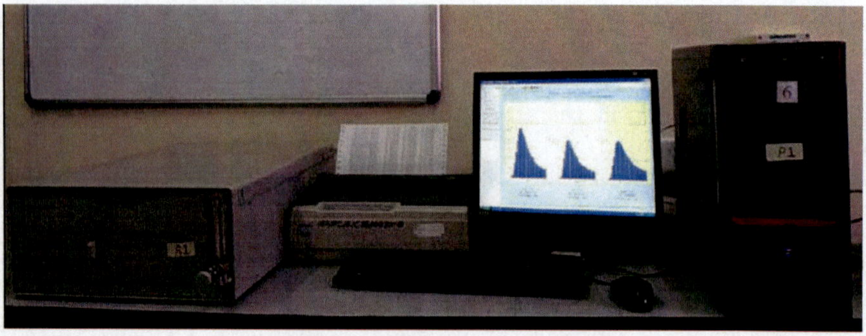

Fig. 10.8 TLD Badge Reader along with PC and other accessories

10.4.2 LiMgPO₄: Tb, B (LMP)

The OSL is now popular as a luminescent dosimetric technique due to various advantages over TL technique. However, not many commercial OSL phosphors are available with exception of artificially grown Al_2O_3:C and BeO. Due to scarcity of dosimetry-grade OSL phosphors, efforts were made to synthesize other OSL materials. In BARC, $LiAlO_2$:Tb/Ce (Dhabekar et al. 2008; Mittani et al. 2008), $ZnAl_2O_4$:Tb (Sanjeev et al. 2008), $MgAl_2O_4$:Tb (Alagu et al. 2009) phosphors were synthesized. However, $LiMgPO_4$:Tb,B phosphor (LMP) proved to be a possible dosimetric phosphor. The effective atomic number of LMP is Z_{eff} ~ 11.44, which is comparable to that of Al_2O_3:C Z_{eff} ~ 11.28 (Bhushan et al. 2011). LMP was prepared by solid-state reaction between lithium hydroxide, magnesium nitrate, and ammonium dihydrogen phosphate in air at 750 °C. The dopants were taken in the form of terbium oxide and boric acid. After confirming the $LiMgPO_4$ phase formation with X-ray diffraction (XRD), LMP was characterized for its application in dosimetry.

The material exhibits a broadband around 240 nm in its photoluminescence (PL) excitation spectrum. This band corresponds to 4f-5d transitions in Tb^{3+}. Additionally, sharp lines beyond 300 nm are observed due to forbidden f-f transitions in Tb^{3+}. The PL and OSL emission spectrums of LMP, as reported by Bhushan et al. (2011) and Rawat et al. (2017), respectively, display multiple lines at approximately 380, 417, 440, 490, 545, and 585 nm. These lines correspond to various transitions, such as $^5D_3$–$^7F_6$, $^5D_3$–$^7F_5$, $^5D_3$–$^7F_4$, $^5D_4$–$^7F_6$, $^5D_4$–$^7F_5$, and $^5D_4$ -$^7F_4$, indicating the presence of Tb^{3+} as a luminescent center in LMP. Of particular interest is the prominent 380 nm OSL emission peak, which can be efficiently detected by the RISO reader system utilizing a U-340 Hoya detection filter and a bi-alkali PMT. The OSL sensitivity of LMP was found to be approximately 1.8 times that of Al_2O_3:C when integrating OSL counts for the initial 1 s (Bhushan et al. 2011). The minimum detectable dose for LMP was optimized to around 0.5 μGy (Rawat et al. 2017).

Furthermore, the PTFE-based disks prepared using LMP were found to be reusable for up to 50 cycles. One unique characteristic of LMP is its linear dose versus

OSL response, ranging from 1 mGy to 1 kGy. It is noteworthy that commonly used TL phosphors such as LiF:Mg,Ti, LiF:Mg,Cu,P, and $CaSO_4$:Dy, as well as generally employed OSL phosphors like Al_2O_3:C or BeO, do not exhibit such a wide linearity. LMP thus is an important phosphor, which has tremendous potential in food irradiation dosimetry.

In India, dosimetry for food irradiation generally involves the use of chemical dosimeters like Fricke and Ceric cerous sulfate. However, these dosimeters require careful handling during irradiation. For example, in Fricke dosimeters, insufficiently clean glassware can lead to false background absorbance, especially at shorter wavelengths (Fricke et al. 1966). The alanine/EPR system is a powerful technique suitable for a wide range of radiation doses encountered during food irradiation. However, due to the high cost of EPR spectrometers and the need for expertise and highly trained personnel, this technique is not yet commonly used (Prakasan et al. 2014). Hence, there is a need for an easily available, simple, fast, and cost-effective dosimetry system that is insensitive to environmental conditions such as temperature and humidity, for routine food irradiation dosimetry in the dose range of 0.01–1 kGy. OSL-based dosimetry can serve as an alternative to chemical dosimeters for food irradiation dosimetry. It offers advantages such as fast and automatic readout, resilience in harsh environments, and ease of handling. The LMP-based OSL dosimetry system possesses all these desirable attributes, making it a promising candidate for dosimetry during phytosanitary applications of radiation in the low-dose range of 0.01–1 kGy. This includes the irradiation of fruits like mangoes, onions, potatoes, garlic, and others (Menon et al. 2019). However, the fading of the OSL signal up to 30% for an extended period of time poses limitation, particularly if the time of irradiation is unknown (Menon et al. 2018). In food irradiation dosimetry, the time of irradiation is normally known and the correction factor for OSL fading can be applied to evaluate the doses accurately. LMP exhibits three TL peaks at ~130, ~160, and ~250 °C (Fig. 10.9) when the TL is recorded up to 350 °C (Menon et al. 2018). Although the first peak fades quickly within 1–2 days, the second TL peak around 160 °C is more stable and is responsible for extended fading of OSL signal in LMP. Thus to have better fading characteristics, the intensity of the second TL peak should be minimized while keeping the intensity of the third peak (~250 °C) almost same. Efforts were made to reduce the intensity of the second peak by modifying the concentration of dopants. It was found that the second TL peak is minimum in $LiMgPO_4$:B (LMP-N) with peak3/peak2 ratio of 2.7 as compared to 0.63 for LMP (Menon et al. 2018). This modified version of the phosphor, $LiMgPO_4$:B (LMP-N), retains all virtues of LMP like high OSL sensitivity, linearity up to 1 kGy. It is also observed that LMP-N exhibits better fading property as compared to that of LMP. The OSL signal in LMP stabilizes (due to fading) at about 70% of initial OSL intensity, while in LMP-N it stabilizes at 85%. Thus by utilizing appropriate procedures for fading correction, LMP-N can be used for various dosimetric applications for a wide range of doses, particularly in low-dose range of food irradiation dosimetry.

Apart from dosimetric characterization, efforts were also made to understand the nature of TL and OSL of LMP. It was confirmed that the OSL signal in this phosphor arises from five optically sensitive traps with different photoionization cross sections

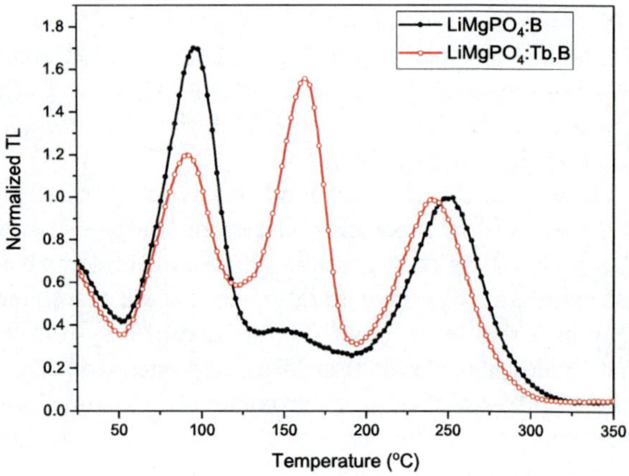

Fig. 10.9 Normalized TL glow curves of $LiMgPO_4$:B and $LiMgPO_4$:Tb,B

and same traps are responsible for both TL and OSL in this phosphor (Singh et al. 2012). The OSL contributions from first, second, and third TL peak were reported to be about 33%, 45%, and 22%, respectively (Munish et al. 2011).

10.4.3 $Li_2B_4O_7$:Cu, Ag

Lithium Borate, i.e., $Li_2B_4O_7$ (LBO) being tissue equivalent (Z_{eff} ~ 8) is a very useful TL material. Further, this material has very intense emission for UV excitation (Senguttuvan et al. 2002).

LBO doped with Cu has been found to be light sensitive, and exposure to sunlight leads to significant fading of the TL signal (Takenaga et al. 1980; Lakshmanan et al. 1981). Although the reduction in TL signal due to sunlight exposure is considered a disadvantage, it can be advantageous when using doped LBO as an OSL material for radiation dosimetry. Single crystals of LBO:Cu grown using the Czochralski technique do not exhibit a significant OSL signal when stimulated by 470 nm light (Babita et al. 2010). However, Czochralski-grown single crystals of $Li_2B_4O_7$:Cu, Ag show a considerable OSL signal when stimulated by blue light (470 nm). In the growth process, commercially available polycrystalline powder material of $Li_2B_4O_7$ (99.99 + % pure) is used as a starting charge, and high-purity CuO and Ag2O (0.25 wt% each) are mixed with the $Li_2B_4O_7$ charge. Babita et al. (Babita et al. 2010) reported first-order kinetics for all the TL peaks exhibited by LBO:Cu, Ag. The emission and excitation bands, peaked at 370 nm and 250 nm, respectively, are attributed to the spin and parity forbidden $3d^94s \rightarrow 3d^{10}$ transitions. LBO:Ag crystal shows an emission band at 267 nm and an excitation band at 210 nm, with a mean decay time of 10 μs. The luminescence in this region is attributed to the $4d^95s \rightarrow 4d^{10}$ transition of Ag^+ ions. The emission spectrum of Ag^+ overlaps with the excitation region of Cu + , resulting in the predominant emission of Cu^+ states at 370 nm in

LBO:Cu, Ag. Consequently, Ag acts as a sensitizer when co-doped with Cu, leading to an overall increased emission. Rawat et al. (2012) conducted OSL investigations on LBO:Cu, Ag and reported it for the first time. Using the continuous wave (CW)-OSL method, the Minimum Measurable Dose (MMD) in this material was found to be approximately 10 mSv at room temperature. Furthermore, a correlation study between CW-OSL and TL suggests that the third TL peak is responsible for the fast OSL component, allowing for achieving an MMD as low as a few mGy (Rawat et al. 2012). The shallow traps delay the occurrence of OSL signal and thus extends the acquisition time. The interference of these traps can be suppressed and minimized by simultaneous application of elevated temperature and hence MMD and sensitivity of LBO:Cu,Ag can be improved.

10.4.4 Al_2O_3:C

A highly promising phosphor TLD-500, i.e., α-Al_2O_3:C was originally developed as a TL phosphor by Urals Polytechnic Institute, Russia (Akselrod et al. 1990). This material exhibited excellent TL sensitivity. However, the fact that the storage of this phosphor post-irradiation required protection from exposure to sunlight/ambient light led a research group at Oklahoma State University to exploit the application of Al_2O_3: C as a highly sensitive OSL phosphor for dosimetry of ionizing radiation. This ceramic material possess extreme thermal, chemical, and mechanical stability. Additionally, it possess a large photoionization cross section in the broadband of stimulation wavelength ranging from 450 to 555 nm (Rawat 2012). The intrinsically high cross section and sensitivity during OSL readout ensures that stimulation of even a tiny fraction of all the trapped charges can yield sufficient luminescence output allowing fast and multiple readout of the sample. Owing to these properties of Al_2O_3:C, currently more than 1.5 million OSL dosimeter badges are routinely used for personnel and environmental monitoring worldwide (Rawat 2012; Pradhan et al. 2008).

Conventionally, the synthesis of α-Al_2O_3:C is carried out using Czochralski-based crystal growth technique. The utilization of a "single crystal" approach is anticipated to present certain challenges due to the specific temperature and growth rate requirements for crystal formation. During the crystal growth process, vacancies in the aluminum oxide melt tend to migrate forward, resulting in an increased concentration of defects along the growth axis. The region of the crystal that solidifies last typically exhibits the highest concentration of F centers. Consequently, it becomes challenging to control the level of carbon incorporation into the lattice and achieve uniformity in both the nature and concentration of defects using this synthesis method. This lack of control leads to undesired variations in the dosimetric properties across different regions of the crystal (Rawat 2012; Gimadova et al. 1990). Another problem associated with crystal growth route is that it demands expensive equipments and is time-consuming, particularly during large-scale phosphor development. These limitations of diverse nature have prompted and propelled the successful development of an indigenous, novel, and cost-effective pathway, i.e., melting of polycrystalline Al_2O_3 in vacuum and reducing atmosphere, that has led to the synthesis of this highly sensitive, multifunctional α-Al_2O_3:C phosphor (Rawat 2012; Muthe et al. 2016). This highly economical technique of melt processing for synthesis of Al_2O_3:C is very effective in generating defects that are dosimetrically relevant.

Melt Processing technique for large-scale production of α-Al_2O_3:C: This method involves stacking of sintered pellets of polycrystalline alumina weighing ~ 50 gm, which is placed in a graphite crucible (that acts as a source of carbon for doping in alumina pellets). Subsequently, these pellets are melted in high-temperature vacuum induction furnace, as depicted in Fig. 10.10, at 2000 °C, 10^{-6} Torr vacuum in a controlled manner (Muthe et al. 2016). The furnace is allowed to cool down to room temperature after reaching 2000 °C. The material thus formed is pulverized and subsequently annealed at 650 °C for 1 h to remove the stress and luminescence signal introduced during crushing and pulverization.

The TL and OSL sensitivity variation of different batches of Al_2O_3:C, thus synthesized, is within ±30% and its sensitivity is comparable to that of commercially available α-Al_2O_3:C from Landauer Inc. USA. It exhibits negligible fading in TL/OSL domain when stored for a duration of 2 months, compared to TL/OSL output measured after 24 h of irradiation. The material showed linear response in the range 1 mGy to 5 Gy, and the MMD was found to be ~20 μGy (within 3σ).

Fig. 10.10 Inductively heated vacuum melting system

10.4.5 $LiCaAlF_6$: Eu, Y (LCAF)

$LiCaAlF_6$ is a scintillating inorganic phosphor, which has application in neutron dosimetry (Okada et al. 2017). TL and OSL properties of the phosphor were investigated by doping it with various dopants. The TL properties of Ce-doped $LiCaAlF_6$ were reported by Yerpude et al. (2016) while the OSL properties of Eu-doped $LiCaAlF_6$ phosphor were reported by More et al. (2015) with OSL sensitivity of approximately one-third to that of $LiMgPO_4$:Tb,B. Rahangdale et al. (2016, 2015) reported the effect of co-doping on the OSL/TL properties of $LiCaAlF_6$. The OSL sensitivity of $LiCaAlF_6$:Eu,Y (LCAF) synthesized by them was found to be about eight times that of commercial Al_2O_3:C (Landauer) and detection threshold of 4.7 μGy. Because of these promising features, the phosphor was further characterized at BARC in a more extensive way. For synthesis of LCAF, the analytical reagent-grade lithium nitrate, calcium hydroxide, and aluminum nitrate were dissolved in the stoichiometric ratios. Hydrofluoric acid (HF) was added drop-wise in the solution kept in a PTFE container to obtain the white precipitate of $LiCaAlF_6$, which was then washed multiple times with distilled water to remove the excess HF in the solution, if any. The precipitate was dried on a hot plate at about 80 °C. Appropriate quantities of europium oxide and yttrium oxide were added to the dried precipitate of $LiCaAlF_6$. The mixture was then melted at about 860 °C for about 10 min in an air furnace and was quenched to room temperature. The particles in the range of 53–150 microns were selected for OSL measurements by crushing it and the mechanical sieving. The annealing of the selected powder was carried out at 400 °C for 1 h to remove the unwanted signal generated by mechanical grinding. The phase purity of the as-prepared samples was confirmed by X-ray diffractometer. The concentration of Eu was optimized to 0.1 mol % and that of Y to 0.5 mol % (Bhushan et al. 2017).

Photoluminescence studies show that when LCAF is excited by 300 nm, its emission band lies around 370 nm. Belsare et al. (2015) attributed the emission at 370 nm in $LiCaAlF_6$:Eu to the transition from the lowest level of $4f^6 5d^1$ configuration to $^8S_{7/2}$ level of $4f^7$configuration of Eu^{2+} ion. The excitation spectrum of LCAF consists of multiple overlapping bands in 290–325 nm range due to the transitions from $^8S_{7/2}$ level of $4f^7$configuration to the levels of $4f^6 5d^1$ configuration of Eu^{2+} ions. Apart from the band around 370 nm, sharp emission lines at 594 nm and 613 nm arising from forbidden f-f transitions from $^5D_o \rightarrow {}^7F_1$ (594 nm) and $^5D_o \rightarrow {}^7F_2$ (613 nm) of Eu^{3+} were observed. The excitation spectrum of sharp Eu^{3+} lines consists of multiple lines in 300–400 nm range due to $^7F_o \rightarrow {}^5L_6$ (392 nm), $^7F_o \rightarrow {}^5G$ (380 nm), $^7F_o \rightarrow {}^5D_4$ (365) nm, $^7F_o \rightarrow {}^5H$ (325–335 nm), and $^7F_o \rightarrow {}^5F$ (300–305 nm) transitions (Belsare et al. 2015). The OSL emission spectrum of LCAF was recorded after exposing it to about 100 Gy of Co-60. The OSL emission spectrum with 520 and 680 nm stimulation exhibits broad 370 nm band due to transition in Eu^{2+} ion (Bhushan et al. 2017). No sharp lines corresponding to f-f transition in Eu^{3+} are observed in the OSL emission spectrum. The 370 nm band corresponding to Eu^{2+} transition is observed in both PL and OSL emission spectra of LCAF. This implies that Eu^{2+} ion acts as a luminescent center in LCAF. The OSL sensitivity of LCAF is about 20 times that of commercially available Al_2O_3:C when compared on RISO reader using Hoya U-340 filter, with 0.5 μGy of MMD (Bhushan et al. 2017). One of the many possible factors responsible for high OSL sensitivity is the existence of 370 nm OSL emission due to divalent europium ion. This is because the Hoya filter allows most of the emission to reach the PMT and this emission in this wavelength also matches with the spectral response of the Bialkali PMT present in the RISO reader system. The phosphor is reusable for the studied 50 cycles with no change in the OSL efficiency. The OSL versus dose response shows linear response in the dose range of 1 mGy to 100 Gy when irradiated with Co-60. Fading studies indicate a decrease of ~6% of OSL signal till 9 days after irradiation, thereafter reaching stabilization (Bhushan et al. 2017). Rawat et al. (Rawat et al. 2020) carried out temperature dependence of OSL on LCAF. "It is found that the simultaneous application of thermal and optical stimulation results in more efficient and faster depletion of signal in comparison to the sum of thermal and optical stimulations alone when carried out in isolation. Thus, simultaneous application of temperature and optical stimulation lead to increased synergistic effect in terms of acquisition of luminescence signal" (Rawat et al. 2020).

10.5 Scope of Future Work

The scope of future work in luminescence dosimetry can be summed up in terms of following existing challenges and pertinent solution for the same.

10.5.1 Materials for Neutron Dosimetry

Nuclear reactors and high-energy accelerators are intense sources of neutrons. The neutron energy spectrum resulting from the fission of ^{235}U has wide energies extending from few keV to greater than 10 meV. These neutrons have average energy of about 2 meV. In a research reactor, neutron beams emerging through the ports are degraded in energy, as they pass through reactor core, coolant system as well as structural materials (Turner 2007). Therefore, personnel dosimetry at nuclear facilities has become challenging and increasingly rigorous. It is necessary to study and investigate the neutron response of dosimetric materials in the aforementioned energy range using neutron beams that are well characterized. A few direct-reading dosimeters exist but they are cost-prohibitive. A neutron dosimeter typically comprises albedo dosimeter (that predominantly measures moderated and body-reflected neutrons). The TLD Badge used in conjunction with it measures mixed gamma/neutron exposures. The albedo neutron dosimetry has limitation, as it requires prior knowledge and information about neutron energy spectrum of concern. The spectrum changes even in the vicinity of human body due to scattering (Muthe et al. 2016; Okada et al. 2017; Yerpude et al. 2016; More et al. 2015; Rahangdale et al. 2016, 2015;

Bhushan et al. 2017; Belsare et al. 2015; Rawat et al. 2020; Turner 2007; Piesch and Burgkhardt 1985; Gibson 1988; Haninger and Fehrenbacher 2007).

Despite very high sensitivity of Al_2O_3:C, the shortcoming associated with this material is that it is insensitive to neutrons. It is due to the elemental composition of Al_2O_3:C that offers negligible or practically insignificant thermal neutron absorption cross section. This shortcoming can be addressed to certain extent by physical mixing or addition of neutron converters (like ^{10}B, ^{6}Li, ^{157}Gd, etc.) to the OSL material. However, the addition of neutron converters may result in alteration of effective atomic number of OSL material, the limited range of secondary charged particles (proton, alpha, tritium, etc.) due to self-absorption which can further hamper and deteriorate the sensitivity of the material (Rawat 2012).

This limitation can be tackled by adopting alternate approach that involves opting for a host lattice like lithium borate whose constituent elements possess large absorption cross section for neutrons. The secondary charged particles generated after interaction with neutrons will deposit their entire or substantial amount of energy inside the host material. This enhances efficiency of detection, and therefore improvement in sensitivity (Rawat 2012). However, the sensitivity is still not enough as natural isotopic abundances of ^{6}Li and ^{10}B isotopes that render high thermal neutron absorption cross section are merely ~7.5% and ~19.5%, respectively. Thus, for improvement in sensitivity, it is required to make Li- and B-based OSL phosphors that are enriched in terms of ^{6}Li and ^{10}B. This being an expensive process, the availability of enriched ^{6}Li and ^{10}B for specialized and strategic applications is an issue.

10.5.2 High LET Radiation

External dosimetry for the personnel stationed on the earth is simple. The low Linear Energy Transfer (LET, L) (<5 keV/μm) radiation normally encountered (photons and betas) can be monitored with TLDs/OSLDs, while high LET (>10 keV/μm) radiation (fast neutrons) are monitored with Plastic Neutron Track Detectors (PNTD), e.g., CR-39. However, the situation for astronauts/cosmonauts is completely different. Exposure to radiation occurs through three primary sources for astronauts: galactic cosmic rays (GCR), solar particle events (SPEs), and charged particles trapped in the Earth's magnetic field (Earth's Radiation Belts, ERB) (Yukihara and McKeever 2011). In addition, secondary particles generated by nuclear interactions between the energetic primary particles and the astronaut's surroundings, as well as the astronaut's own body, contribute to individual doses. This makes the assessment of the total absorbed dose in Gray for each astronaut challenging. Furthermore, to estimate the health hazard posed by space radiation, the absorbed dose needs to be multiplied by a quality factor (Q) to calculate the weighted dose in Sieverts (Sv), considering the significant variation in Linear Energy Transfer (LET) of space radiation. Hence, precise information on the Q is necessary.

In general, the efficiency of TLD/OSLDs decreases with increase in LET of radiation for particles (Yukihara and McKeever 2011; Yukihara et al. 2006b). Thus, TLDs/

OSLDs should be used jointly with the detector that can evaluate doses from high LET radiations with L values up to $\sim 10^3$ keV/μm, such as the PNTDs. Ideally, TLD/OSLD should respond (ideally flat response) only to low LET radiation (L < 10 keV/μm) and zero response to high LET radiation (L < 10 keV/μm) and vice versa for the PNTD (respond to high L with no response to low L radiation). However, these two dosimeters have overlapping responses, with TLDs/OSLDs having some sensitivity up to very high LET radiations, and PNTDs having some response for low LET radiation (up to ~5 keV/μm).

TL/OSL dosimeters are typically calibrated based on the absorbed dose to water using gamma rays. However, it is important to note that the energy deposited in the TLD/OSLD may not be identical to the energy deposited in water under the same exposure conditions. Therefore, a thorough calibration of the dosimeters is crucial. Considering these factors, space dosimetry poses significant challenges that need to be addressed for accurate measurements.

In a few TL materials, some glow peaks are LET dependent, while others are independent of LET. The varying LET efficiencies of the main glow curve peaks in these materials can be advantageous in establishing an LET calibration curve. This enables the determination of an "average" LET in mixed radiation fields. TLD materials like LiF:Mg,Ti (in combination of TLD-100, TLD-600, and TLD-700) (Vana et al. 1996) and CaF_2:Tm (Hajek et al. 2008) are reportedly used for measurements of doses using this high-temperature peak ratio (HTR) method. Similarly, the CW-OSL curve tends to get steeper with increase in LET. Thus, the CW-OSL curve too can provide some information about LET of the incident radiation. To study these effects, indigenously synthesized $LiCaAlF_6$:Eu,Y and Al_2O_3:C TL/OSL phosphors in the form of disks were characterized at the BARC-TIFR pelletron facility under the proton beam with energy ranging from 8 to 20 meV. Identification of LET-dependent parameter and dose response up to 20 mGy was investigated. It was found that the CW-OSL curve becomes steeper with increase in LET. Two LET-dependent CW-OSL-based parameters, ratio of initial to total OSL counts and the mean life of the CW-OSL decay curve were identified. The dose linearity was also established up to 20 mGy for 12 and 20 meV proton beams. However, further characterization of these phosphors in other heavy ion beams is required to cover a wider LET spectrum.

10.5.3 Tissue Equivalent Material

Al_2O_3:C is an excellent OSL phosphor owing to its sensitivity, and other desirable properties. However, its effective atomic number (Z_{eff}) is 11.2 which is above the effective atomic number for tissue. There are very few OSL phosphors that are close to tissue in terms of their Z_{eff}. BeO and $Li_2B_4O_7$:Cu, Ag are few of those. However, the problem with BeO is that it is toxic and poses difficulty in handling, particularly when in powdered form. Similarly, $Li_2B_4O_7$:Cu, Ag is hygroscopic and has far less sensitivity. There is still a scarcity of tissue equivalent OSL material and efforts need to be put in this direction.

10.5.4 High-Dose Dosimetry

The high-dose radiation technologies have applications in radiation-induced polymerization, preservation of agriculture product, and sterilization of medical products (Machi 1995). All these technologies use gamma emitters like ^{137}Cs, ^{60}Co or electron accelerators providing radiation doses as high as several kilo gray (Machi 1996). For quality control and better effectiveness of radiation processing, there is a need to measure associated doses due to ionizing radiation within stipulated uncertainty. Hence, radiation dosimetry holds great significance in radiation processing, ensuring it is conducted with precision, adherence to guidelines, and optimized procedures. Moreover, high-dose measurements are necessary for nuclear power plants and liquid sludge hygienization facilities. The widespread application of radiation in various fields has driven extensive research toward creating an appropriate dosimetry system. In pursuit of this goal, the development of lithium aluminate ($LiAlO_2$) as a radiation detection material has been a dedicated endeavor for over a decade, as evidenced by Twardak et al. (2014a, 2014b) and Bhushan et al. (2008). Gupta et al. (2017) synthesized $LiAlO_2$-based phosphors using solution combustion method. They reported linear dose response in TL domain for $LiAlO_2$:Ce^{3+}, $LiAlO_2$:Dy^{3+}, and $LiAlO_2$: Eu^{2+} phosphors up to 3 kGy, 3.5 kGy, and 10 kGy, respectively. However, issues pertaining to sensitivity and fading remain unresolved. Bakshi et al. (2009) reported the TL/OSL properties of indigenously synthesized CaF_2:Mn which exhibits linear TL response from 50 Gy to 3 kGy. They also claim that in the range of 500–1000 Gy there was a little over response (~10%) which may be considered to be acceptable in dosimetry applications. Apart from synthesizing phosphors for high-dose response, certain techniques can also be devised that can extend linearity of TL/OSL phosphor on either side (lower as well as higher doses). One such technique is thermally assisted OSL (TA-OSL) using which Soni et al. (Soni et al. 2013) has reported the improvement in MMD extended dose linearity by more than one order of magnitude in case of Al_2O_3:C.

10.6 Conclusions

$CaSO_4$:Dy is the back bone and country-wide personnel monitoring program in India. The TLD-based dosimetry system, which includes synthesis of $CaSO_4$:Dy phosphor, PTFE-embedded CaSO4:Dy disks, and the TL reader system, is complctely indigenous and is serving the nation successfully since its inception. However, due to several advantages offered by OSL technique, there is a paradigm shift worldwide from TL- to OSL-based dosimetry. This motivation led to the synthesis of several dosimetry-grade OSL phosphors like LMP, LCAF, LBO, and A12O3:C. Since ionizing radiation has immense applications, there is a huge penetration of radiation-related technology in daily life. For quality control and better effectiveness of radiation processing, there is a need to measure associated doses due to ionizing radiation within stipulated uncertainty. Luminescent materials offer viable solution in this regard. Thus, TL/OSL phosphors have potential application both in low-dose range like personnel monitoring, environmental monitoring, diagnostic radiology, space dosimetry, etc., and in high-dose range like radiation therapy, food irradiation dosimetry, etc. TL/OSL techniques have application in accident dosimetry and in archaeological and geological dating too. Synthesis and characterization of dosimetry-grade TL/OSL is still a vibrant and exciting field in material science, which is ever growing due to several challenges like phosphors for neutron dosimetry, sensitive tissue equivalent material, dosimetry in space, etc.

References

Akselrod, M.S., Kortov, V.S, Kravetsky, D.J., Gotlib, V.I.: Highly sensitive thermoluminescent anion-defect α-Al_2O_3:C single crystal detectors. Radiat. Prot. Dosim. **33**, 119–22 (1990)

Alagu, R.E., Sanjeev, M., Bhushan, D., Rawat, N.S., Gundu Rao, T.K.: J. Lumin. **129**, 829 (2009)

Babita, T., Rawat, N.S., Desai, D.G., Singh, S.G., Tyagi, M., Ratna, P., Gadkari, S.C., Kulkarni, M.S.: J. Lumin. **130**, 2076 (2010)

Bakshi, A.K., Bhushan, D., Rawat, N.S., Singh, S.G., Joshi, V.J., Vijay, K.: Nuclear Instrum. Methods Phys. Res. B **267**, 548–553 (2009)

Belsare, P.D., Joshi, C.P., Moharil, S.V., Omanwar, S.K.: Int. J. Adv. Sci. Eng. Technol. (Special Issue-1) (2015). ISSN: 2321–9009

Bhatt, B.C., Kulkarni, M.S.: Defect and Diffusion Forum Vol. 347, pp 179–227 (2014)

Bhushan, D., Alagu, R.E., Sanjeev, M., Gundu Rao, T.K., Kher, R.K., Bhatt, B.C.: J. Phys. D: Appl. Phys. **41**, 115414 (2008)

Bhushan, D., Menon, S.N., Alagu Raja, E., Bakshi, A.K., Singh, A.K., Chougaonkar, M.P., Mayya, Y.S.: Nucl. Instr. Methods Phys. Res. B **269**, 1844–1848 (2011)

Bhushan, D., Rawat, N.S., Gaikwad, N., Sonal, K., Koul, D.K.: Radiat. Measur. **107**, 7–13 (2017)

Botter-Jensen, L., McKeever, S.W.S.: Radiat. Prot. Dosim. **65**(1–4), 273 (1996)

Botter-Jensen, L., McKeever, S.W.S., Wintle, A.G.: Optically Stimulated Luminescence. BV Publication, Elsevier Science (2003)

Boyle, R.: Register of the Roy. Soc. **1663**, 213 (1663)

Cameron, J.R., Kenney, G.N.: Radiat. Res. **19**, 199 (1963)

Cameron, J.R., Suntharalingham, N., Kenney, G.N.: Thermoluminescence Dosimetry. Univ. of Wisconsin Press, Madison (1968)

Curie, M.: Radioactive Substances (English translation of doctoral thesis presented to the Faculty of Science, Paris), p. 1961. Greenwood Press, Westpoint (1904)

Daniels, F., Boyed, C.A., Saunders, D.F.: Science **117**, 343 (1953)

Dhabekar, B., Raja, E.A., Menon, S., Gundu Rao, T.K., Kher, R.K., Bhatt, B.C.: Identification of defect centres using TSL, PL, OSL and ESR studies in $LiAlO_2$ based phosphors. J. Phys. D Appl. Phys. **41**, 115414 (2008)

Frame, P.W.: Health Phys. **88**, 613 (2005)

Fricke, F.S., Hart, E.J.: Chemical Dosimetry. In: Attix, F.H., Roesch, W.C. (eds.) Radiation dosimetry, vol. 2, 2nd edn., pp. 167–239. Academic Press, New York (1966)

Gibson, J.A.B.: Radiat. Prot. Dosim. **23**, 109 (1988)

Gimadova, T.L., Bessonova, T.S., Tales, I.A., Avvkumova, L.A., Bodyachevsky, S.V.: Radiat. Prot. Dosim. **33**, 47 (1990)

Gupta, K.K., Kadam, R.M., Dhoble, N.S., Lochab, S.P., Dhoble, S.J.: A comparative investigation of Ce3+ /Dy3+ and Eu2+ doped LiAlO 2 phosphors for high dose radiation dosimetry: explanation of defect recombination mechanism using PL, TL and EPR study. J. Lumin. **188**, 81–95 (2017)

Hajek, M., Berger, T., Bergmann, R., Vana, N., Uchihori, Y., Yasuda, N., Kitamura, H.: Radiat. Meas. **43**, 1135–1139 (2008)

Haninger, T., Fehrenbacher, G.: Radiat. Prot. Dosim. **125**, 361 (2007)

International Atomic Energy Agency, Vienna: 2004. "Practical Radiation Technical Manual—Individual Monitoring".

Kher, R.K., et al.: Experiences in the Use of TLD badge for personal monitoring. Bull. Radiat. Protect. **6**(3&4), 83–88 (1983)

Kitis, G., Spiropulu, M., Papadopoulos, J., Charalambous, S.: Nucl. Instrum. Methods Phys. Res. **73**, 367–372 (1993)

Lakshmanan, A.R., et al.: Application of thermoluminescence dosimeters for personnel monitoring in India. Radiat. Prot. Dosimetry. **17**(1–4), 49–52 (1986)

Lakshmanan, A.R., Bhuwan, C., Bhatt, R.C.: Radiat. Prot. Dosim. 1, 191 (1981)

Lee, S.Y., Jai, L.K.: Development of a personal dosimetry system based on optically stimulated luminescence of Al_2O_3: C for mixed radiation fields. Appl. Radiat. Isotopes **54**, 675–685 (2001)

Machi, S.: Radiation technology for sustainable development. Radiat. Phys. Chem. **46**, 399–410 (1995)

Machi, S.: New trends of radiation processing application. Radiat. Phys. Chem. **47**, 333–336 (1996)

Menon, S.N., Dhabekar, B.S., Sonal, K., Koul, D.K.: Fading studies in LiMgPO 4 :Tb,B and synthesis of new LiMgPO 4 based phosphor with better fading characteristics. Nuclear Inst. Methods Phys. Res. B **436**, 45–50 (2018)

Menon, S.N., Singh, A.K., Sonal, K., Mhatre, S., Bhaskar, S., Bhushan, D.: J. Food Process. Preservation **43**, 13891 (2019)

Mittani, J.C., Prokic, M., Yukihara, E.G.: Radiat. Meas. **43**, 323 (2008)

More, Y.K., Wankhede, S.P., Moharil, S.V., Kumar, M., Chougaonkar, M.P.: Optically stimulated luminescence in LiCaAlF6:Eu2+ phosphor. Luminescence **30**, 878–882 (2015)

Munish, K., Bhushan, S., Menon, S.N., Chougaonkar, M.P., Mayya, Y.S.: Nuclear Inst. Methods Phys. Res. B **269**, 1849–1854 (2011)

Muthe, K.P., Kulkarni, M.S., Soni, A., Ajay, S., Rawat, N.S., Mishra, D.R., Pradeep, R., Bhattacharya, S., Sharma, D.N., Gupta, S.K.: Method for large scale synthesis of optically stimulated luminescence grade polycrystalline ceramic material. United States Patent 9523033 B (2016)

Okada, G., Fukuda, K., Kawaguchi, N., Yanagida, T.: Characterizations of $LiCaAlF_6$:Eu^{2+} ceramics as neutron scintillator: Primitive experimental studies and future prospective. Radiat. Meas. **106**, 134–139 (2017)

Piesch, E., Burgkhardt, B.: Albedo neutron dosimetry. Radiat. Prot. Dosim. **10**, 175 (1985)

Pradhan, A.S.: Thermoluminescence dosimetry and its applications. Radiat. Prot. Dosimetry. **1**(3), 153–167 (1981)

Pradhan, A.S., Lee, J.I., Kim, J.L.: Recent developments of optically stimulated luminescence materials and techniques for radiation dosimetry and clinical applications. J. Med. Phys. **33**, 85–99 (2008)

Pradhan, A.S., Adtani, M.M., Varadharajan, G., Bakshi, A.K., Srivastava Kshama and Bihari, R.R.: BARC report, BARC/2002/E/025. "Hand book on the use of TLD badge based on CaSO4:Dy Teflon TLD discs for individual monitoring"

Prakasan, V., Sanyal, B., Chawla, S.P., Chander, R., Sharma, A.: Cyanocobalamin solutions as potential dosimeters in low-dose food irradiations. Appl. Radiat. Isot. **86**, 97–101 (2014)

Rahangdale, S.R., Palikundwar, U.A., Wankhede, S.P., Dhabekar, B., Kadam, S., Moharil, S.V.: Effect of co-doping on luminescence of $LiCaAlF_6$: Eu phosphor. J. Lumin. **167**, 80–84 (2015)

Rahangdale, S.R., Palikundwar, U.A., Wankhede, S.P., Dhabekar, B., Kadam, S., Moharil, S.V.: Luminescence in $LiCaAlF_6$:Eu, La phosphor. J. Lumin. **178**, 446–450 (2016)

Randall, J.T., Wilkins, M.H.F.: Proc. Roy. Soc. London **184**, 365 (1945a)

Randall, J.T., Wilkins, M.H.F.: Proc. Roy. Soc. London **184**, 390 (1945b)

Rao, G.S., Iyer, R.K., Gokhale, Y.W., Gupta, S.K., Deshpande, S.G., Gupta, S.S.: Preparation of CaSO4:Dy phosphor, Bhabha Atomic Research Centre, Report No. BARC/I-591 (1980)

Rawat, N.S.: Characterization of TL/OSL phosphors and study of photo-ionization cross-sections of meta-stable trap levels-An Experimental and Theoretical approach. Ph.D thesis, Homi Bhabha National Institute (HBNI) (2012)

Rawat, N.S., Kulkarni, M.S., Mishra, D.R., Bhatt, B.C., Babu, D.: An attempt to correlate shift in thermoluminescence peak with heating rate and black body radiation. Radiat. Prot. Environ. **37**, 63–67 (2014a)

Rawat, N.S., Dhabekar, B., Kulkarni, M.S., Muthe, K.P., Mishra, D.R., Soni, A., Gupta, S.K., Babu, D.A.R.: Optimization Of CW-OSL parameters for improved dose detection threshold in Al_2O_3:C. Radiat. Meas. **71**, 212–216 (2014b)

Rawat, N.S., Kulkarni, M.S., Tyagi, M., Ratna, P., Mishra, D.R., Singh, S.G., Tiwari, B., Soni, A., Gadkari, S.C., Gupta, S.K.: TL and OSL studies on lithium borate single crystals doped with Cu and Ag. J. Luminescence **132**, 1969–1975 (2012)

Rawat, N.S., Jagriti, S., Dhabekar, B.S., Koul, D.K., Sapra, B.K.: Nuclear Inst. Methods Phys. Res. B **479**, 171–179 (2020)

Rawat, N.S., Bhushan, D., Muthe, K.P., Koul, D.K., Datta, D.: Detection of sub micro Gray dose levels using OSL phosphor LiMgPO4:Tb,B. Nucl. Instrum. Methods B **397**, 27–32 (2017)

Rawat, N.S., Muthe, K.P., Bhushan, D., Gaikwad, N.P., Kulkarni, M.S.: On the feasibility of multiple assessment of dose using CW-OSL technique in Al2O3:C. Radiat. Measur. **82**, 74–82 (2015)

Rutherford, E.: Radioactive Substances and their Radiations. Cambridge University Press, Cambridge, England (1913)

Sanjeev, M., Bhushan, D., Alagu, R.E., More, S.P., Gundu Rao, T.K., Kher, R.K.: J. Lumin. **128**, 1673 (2008)

Senguttuvan, N., Ishii, M., Shimoyama, M., Kobayashi, M., Tsutsui, N., Nikl, M., Dusek, M., Shimizu, H.M., Oku, T., Adachi, T., Sakai, K., Suzuki, J.: Nucl. Instrum. Methods Phys. Res. A **486**, 264 (2002)

Singh, A.K., Menon, S.N., Bhushan, D., Sonal, K., Chougaonkar, M.P., Mayya, Y.S.: Nucl. Instrum. Methods Phys. Res. B **274**, 177–181 (2012)

Soni, A., Mishra, D.R., Bhatt, B.C., Gupta, S.K., Rawat, N.S., Kulkarni, M.S., Sharma, D.N.: Thermally assisted OSL: a potent tool for improvement in dose threshold and extension of dose range of Al_2O_3: C. Geochronometria **40**(4), 258–265 (2013)

Sunta, C.M.: A review of thermoluminescence of calcium fluoride, calcium sulphate and calcium carbonate. Radiat. Prot. Dosim. **8**, 25–44 (1984)

Takenaga, M., Yamamato, O., Yamashita, T.: Nucl. Instrum. Methods **175**, 77 (1980)

Townsend, P.D., Kelly, J.C.: Colour Centres and Imperfections in Insulators and Semiconductors, p. 70. Sussex University Press, London (1993)

Turner James, E.: Atoms, Radiation, and Radiation Protection. 3rd Edition (2007)

Twardak, A., Bilski, P., Marczewska, B., Gieszczyk, W.: Analysis of TL and OSL kinetics of lithium aluminate. Radiat. Meas. **71**, 143–147 (2014a)

Twardak, A., Bilski, P., Marczewska, B., Lee, J.I., Kim, J.L., Gieszczyk, W., Mrozik, A., Wróbel, D.: Properties of lithium aluminate for application as an OSL dosimeter. Radiat. Phys. Chem. **104**, 76–79 (2014b)

Vana, N., Schoner, W., Fugger, M., Akatov, Y.: Absorbed dose measurement and LET determination with TLDs in space. Radiat. Prot. Dosim. **66**, 145 (1996)

Vohra, K.G., et al.: A Personnel dosimetry TLD badge based on CaSO4: Dy Teflon TLD Discs. Health Phys. **38**(2), 193–196 (1980)

Wiedemann, E., Schmidt, G.C.: Ann. Phys. Chem. Neue Folge **54**, 604 (1895)

Yamashita, T., Nada, N., Onishi, H., Kitamura, S.: Calcium sulphate activated by thulium or dysprosium for thermoluminescent dosimetry. Health Phys. **21**, 295–300 (1971)

Yamashita, T., Nada, N., Onishi, H., Kitamura, S.: Calcium sulphate activated by rare earth. In: Auxier, J.A., Becker, K., Robinson, E.M. (eds.), Proceedings Second International Conference on Luminescence Dosimetry, Tennessee, USA, Sept. 23–26, 1968, CONF 680920, pp. 4–17 (1968)

Yerpude, M.M., Dhoble, N.S., Lochab, S.P., Dhoble S.J.: Comparison of thermoluminescence characteristics in γ-ray and C^{5+} ion beam-irradiated $LiCaAlF_6$:Ce phosphor Luminescence **31**, 1115–1124 (2016)

Yukihara, E.G., McKeever, S.W.S.: Optically Stimulated Luminescence. Publication, A John Wiley and Sons Ltd (2011)

Yukihara, E.G., Sawakuchi, G.O., Guduru, S., McKeever, S.W.S., Gaza, R., Benton, E.R.: Application of the optically stimulated luminescence (OSL) technique in space dosimetry. Radiat. Measur. **41**, 1126–1135 (2006a)

Yukihara, E.G., Sawakuch, G.O., Guduru, S., McKeever, S.W.S., Gaza, R., Benton, E.R., Yasud, N., Uchihori, Y., Kitamura, H.: Radiat. Meas. **41**, 1126–1135 (2006b)

Chapter 11
Carbon Material-Based Nanoscale Optics and Plasmonics

Padmnabh Rai and Vivek Kumar Shukla

11.1 Introduction

High-speed and encrypted communication rely on the integration of photonic and nanoscale electronic devices in the chip-scale circuitry (Ozbay 2006; Wang et al. 2020, 2018). Scaling down of electronic devices has brought their ultimate speed and it is limited by short-channel and thermal effects in electronic interconnects (Ozbay 2006). Optical interconnects possess large data-carrying capacity and may offer interesting solution to this problem. Optical alternative may be attractive for next-generation devices which comprise the architecture of electronic computing units connected with high-speed links (Wang et al. 2020, 2018; Zia et al. 2006). However, their integration is hampered due to large size disparity between electronic and dielectric photonic components (Zia et al. 2006). The size of dielectric photonic components is limited by the fundamental laws of diffraction (about half a wavelength of light) (Novotny and Hecht 2006). This size mismatch between electronic and photonic components presents major challenge to interface these technologies. Plasmonics is emerging as an alternate technology, which deals with propagation and manipulation of electromagnetic wave on dielectric waveguide of size at sub-wavelength regime (Zia et al. 2006; Novotny and Hecht 2006; Barnes et al. 2003).

Carbon-based materials (carbon nanotube, graphene, and single-crystal diamond) have emerged as a potential material for developing electronic and optoelectronic devices on the same platform (Rai et al. 2018; Bonaccorso et al. 2010; Aharonovich et al. 2011a; Avouris et al. 2008). Carbon hybridizes into sp, sp^2, and sp^3 states of atomic orbitals and gives a variety of allotropes such as graphite, graphene, carbon nanotube (CNT), fullerene, and diamonds. Graphene, a two-dimensional (2D) atomic

P. Rai (✉) · V. K. Shukla
School of Physical Sciences, UM-DAE Centre for Excellence in Basic Sciences, University of Mumbai, Kalina, Santacruz (E), Mumbai 400098, India
e-mail: padmnabh.rai@cbs.ac.in

R. S. Ningthoujam and A. K. Tyagi (eds.), *Handbook of Materials Science, Volume 1*, Indian Institute of Metals Series, https://doi.org/10.1007/978-981-99-7145-9_11

layer of carbon atoms (sp^2), has shown its application as tunable plasmonic and optoelectronic devices (Bonaccorso et al. 2010). CNT is a rolled graphene sheet in tubular form (1D) with distorted sp^2 hybridization. Carbon nanotube field effect transistor (CNTFET) has been applied as logic gate, light-emitting and detecting devices by manipulating the gate voltage in the same transistor (Avouris et al. 2008). A nanoscale optical source based of CNTFET launches surface plasmon polariton (SPP) in plasmonic circuitry, which opens the opportunity to modulate optical signal at GHz frequency in solid-state devices (Rai et al. 2013). Diamond is a sp^3 hybridized form of carbon atoms arranged in tetrahedral configuration. Due to its high bandgap (5.48 eV), the various color centers or atomic impurities in diamond matrix (NV, SiV, and Cr) can be treated as individual qubits for encrypted data transmission (Aharonovich et al. 2011a; Aharonovich and Neu 2014). Diamond-based qubits are quite stable and having long life time at room temperature. The nitrogen vacancy (NV) center, particularly the NV^-, is the most explored optical center in diamond and it is crucial for applications in quantum information science and optical sensors (Aharonovich and Neu 2014; Balasubramanian et al. 2008; Aharonovich et al. 2011b).

11.2 Classification of Carbon Materials

Carbon-based materials, such as carbon nanotube (1D), graphene (2D), and single-crystal diamond (3D) materials, have displayed their potential for next-generation optoelectronics and photonic devices. Their theoretical developments are discussed as follows.

11.2.1 Graphene

Graphene is a mono-layer derivative of graphite material. It is composed of a closely packed single layer of carbon atoms forming a 2D honeycomb lattice with its thickness ~0.35 nm. The graphene honeycomb lattice is depicted in Fig. 11.1a. Its unit cell includes two atoms, called A or B sub-lattices as each of these two atoms form hexagonal structure. The honeycomb structure of graphene does not belong to Bravais lattices because the arrangement of atoms looks different when it is viewed from A and B sites. The primitive vectors of single-layer graphene are indicated by $\overrightarrow{a_1}$ and $\overrightarrow{a_2}$, , where the lattice constant (a) is ~0.246 nm and the nearest-neighbor distance (a_{cc}) between two carbon atoms in the graphene sheet is a/$\sqrt{3}$ (~0.142 nm). The reciprocal lattice structure and corresponding basis vectors, $\overrightarrow{b_1}$ and $\overrightarrow{b_2}$, are shown in Fig. 11.1b. The real and reciprocal space primitive vectors, respectively, are related to each other as $\overrightarrow{a_i}.\overrightarrow{b_j} = 2\pi\delta_{ij}$, where δ_{ij} is Kronecker's delta defined as $\delta_{ij} = 1(i = j)$ or $\delta_{ij} = 0(i \neq j)$. The band structure of graphite is highly anisotropic, which governs its electrical properties. The out-of-plane carbon bond

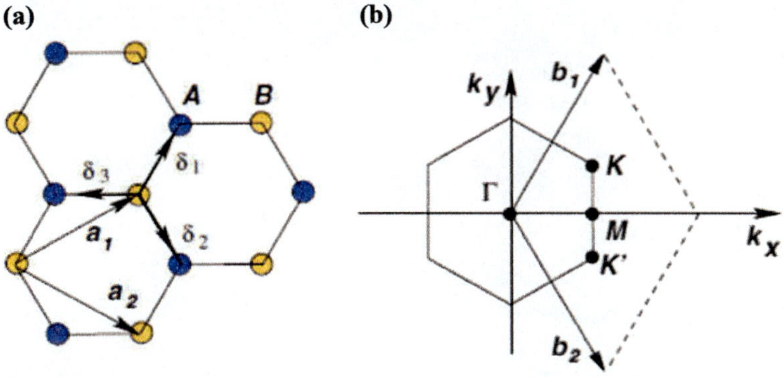

Fig. 11.1 **a** The honeycomb lattice structure of graphene. The unit cell, rhombus, defined by vectors, $\overrightarrow{a_1}$ and $\overrightarrow{a_2}$, containing two atoms belongs to sub-lattices *A* and *B*. **b** Reciprocal lattice structure of graphene. The points K and K′ denote the position of Dirac points and $\overrightarrow{b_1}$ and $\overrightarrow{b_2}$ represent the reciprocal lattice vectors. It is adapted with permission from APS publisher (Castro Neto et al. 2009)

is the result of weak van der Waals interaction and arises by delocalized π-orbitals, whereas in-plane carbon–carbon bonds have a strong covalent bond known as σ-bonds as a result of sp^2 hybridization. Due to the overlap between the π-orbitals on adjacent atoms, electron mobility within planes is high whereas the mobility perpendicular to the planes is relatively low. The electronic band structure of graphene can be determined by employing the tight-binding model. The Schrödinger equation for tight-binding Hamiltonian is given as

$$\hat{H}\psi(r) = E\psi(r) \tag{11.1}$$

The wave function $\Psi(r)$ can be written as the linear superposition of atomic orbital functions, $\Psi(r) = \frac{1}{\sqrt{N}}\sum_l^N c_l e^{ik\cdot R_l}\varphi_l(r)$, where N is the number of unit cells, $\varphi_l(r) = \varphi(r - R_l)$ is the ground state of 2 p_z electron orbital of an isolated atom located at lattice point R_l, and the index l runs over all carbon atom in graphene lattice including identical sub-lattice points A and B (Fig. 11.1a). The non-trivial eigenvalues corresponding to solution of Eq. (11.1) is obtained (Castro Neto et al. 2009) as

$$E(k) = \varepsilon \pm \gamma_0 \sqrt{3 + 2\cos\left(\sqrt{3}k_y a\right) + 4\cos\left(\frac{\sqrt{3}}{2}k_y a\right)cos\left(\frac{3}{2}k_x a\right)} \tag{11.2}$$

where +ve and –ve signs indicate the conduction (π^*-band) and valance band (π-band), respectively, and γ_0 is overlap integral. Figure 11.2 shows energy dispersion relation of graphene as described by Eq. (11.2). It depicts the constant energy contour of conduction and valence bands of 2D graphene sheet. In the first Brillouin zone

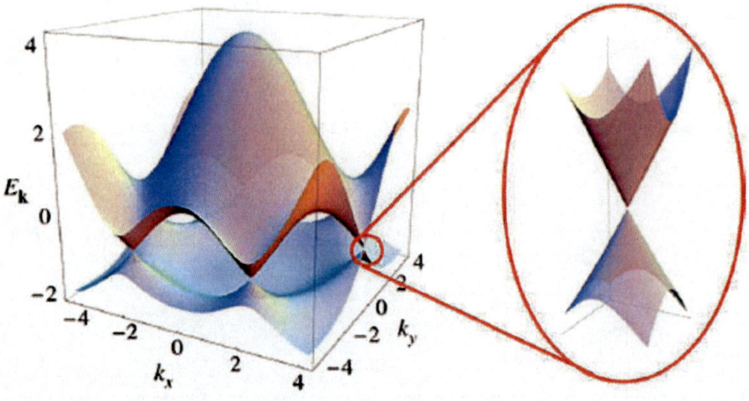

Fig. 11.2 Electronic dispersion relations for 2D honeycomb crystal lattice of graphene under the tight-binding approximation and demonstration of six Dirac points. It is adapted with permission from APS publisher (Castro Neto et al. 2009)

of graphene, the conduction and valence bands touch each other at six Fermi points (K), known as Dirac points. At Γ point ($k = 0$), the eigenenergy is $E = \varepsilon \pm 3\gamma_0$, where $E = \varepsilon - 3\gamma_0$ corresponds to valence band for symmetric bonding orbitals and $E = \varepsilon + 3\gamma_0$ belongs to conduction band for antisymmetric bonding orbitals.

11.2.2 Carbon Nanotube

Carbon nanotube can be understood as hollow cylinder obtained by rolling of 2D graphene sheet (Fig. 11.3). The carbon atom in a nanotube is sp^2 hybridized. A single-walled carbon nanotube (SWNT) consists of graphene sheet that is rolled into a cylinder over a vector called "chiral vector" ($\overrightarrow{C_h}$), whereas a multi-walled carbon nanotube (MWNT) consists of multiple coaxial SWNTs. The geometrical structure of SWNT is defined by the chiral vector ($\overrightarrow{C_h}$), which is expressed as a linear combination of real space primitive vectors $\overrightarrow{a_1}$ and $\overrightarrow{a_2}$ of graphene:

$$\overrightarrow{C_h} = n\overrightarrow{a_1} + m\overrightarrow{a_2} \tag{11.3}$$

where n, m are integers known as chiral index and $0 \leq |m| < n$. In the x, y-coordinates the real space primitive vectors of the hexagonal graphene lattice are $\overrightarrow{a_1} = \frac{3}{2}\, a_{cc}\hat{i} + \frac{\sqrt{3}}{2}\, a_{cc}\hat{j}$ and $\overrightarrow{a_2} = \frac{3}{2}\, a_{cc}\hat{i} - \frac{\sqrt{3}}{2} a_{cc}\hat{j}$, respectively. The inner product between $\overrightarrow{a_1}$ and $\overrightarrow{a_2}$ yields $\overrightarrow{a_1} \cdot \overrightarrow{a_1} = \overrightarrow{a_2} \cdot \overrightarrow{a_2} = a^2$, $\overrightarrow{a_1} \cdot \overrightarrow{a_2} = \frac{a^2}{2}$. The radius of a SWNT is expressed in terms of chiral vector, $|\overrightarrow{C_h}| = 2\pi r$.

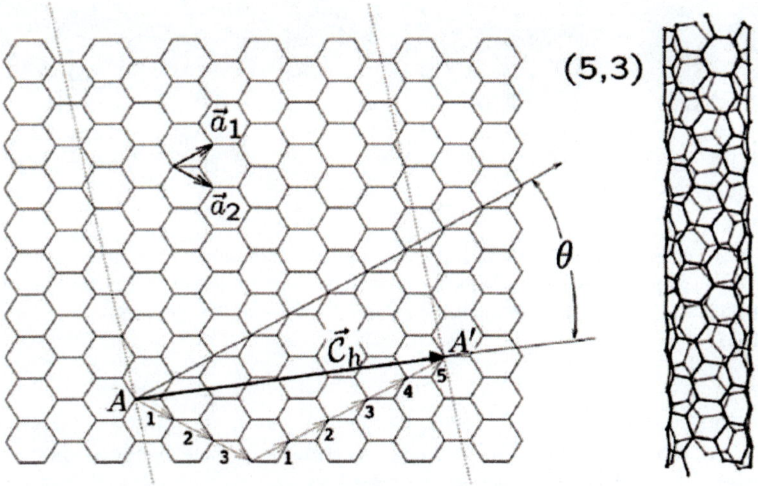

Fig. 11.3 The geometrical structure of SWNT generated by rolling of 2D graphene sheet in tubular form. It is adapted with permission from APS publisher (Charlier et al. 2007)

$$r = \frac{a_{cc}}{2\pi}\sqrt{3\left(m^2 + n^2 + mn\right)} \tag{11.4}$$

The chiral angle is defined as the angle between chiral vector and the primitive vector $\overrightarrow{a_1}$.

$$\theta = tan^{-1}\left(\frac{\sqrt{3}n}{2m + n}\right) \tag{11.5}$$

The set of integers (n, m) defines a unique way of rolling the graphene sheet to form a carbon nanotube. Depending on the rolling direction, SWNT is categorized into three groups such as armchair ($n = m$ and $\theta = 30°$), zigzag ($m = 0$ and $\theta = 0°$), and chiral ($0° < \theta < 30°$) nanotube.

The electronic dispersion relation of a CNT (1D) is determined from the dispersion relation of graphene (2D) by imposing periodic boundary conditions in the circumferential direction ($\overrightarrow{C_h}$). The allowed wave vectors $\overrightarrow{k}$ are obtained from the quantum condition, $\overrightarrow{C_h} \cdot \overrightarrow{k} = 2\,\pi q$, where q is an integer. The permissible modes of k values are represented by a discrete number of parallel lines that emerge as a result of quantization in the reciprocal space of graphene (Dresselhaus et al. 2005). These discrete lines are called sub-bands of a CNT (Fig. 11.4).

If these sub-bands pass through the Fermi points, then nanotube is metallic otherwise it will be semiconducting. The condition for metallic nanotubes is $\overrightarrow{C_h} \cdot \overrightarrow{k_F} = 2\,\pi q$.

Conductivity of CNT is determined by its index (n, m), when $(n - m)/3$ is integral multiple of 3, then it will be metallic otherwise semiconducting. Thus, an armchair

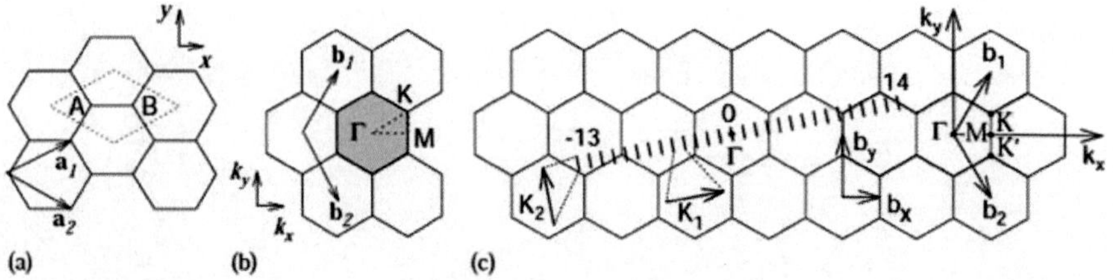

Fig. 11.4 **a** The unit cell structure of 2D graphene, **b** Reciprocal lattice structure of graphene and **c** Reciprocal lattice diagram for (4,2) indexed nanotube. The parallel lines represent the allowed values for the wave vector $\overrightarrow{k}$. It is adapted with permission from Elsevier publisher (Dresselhaus et al. 2005)

(n, n) nanotube is truly metallic whereas zigzag $(n, 0)$ or chiral (n, m) could be metallic or semiconducting. The bandgap of semiconducting nanotube is inversely proportional to its diameter:

$$Eg = \frac{2\gamma_0 a_{cc}}{d_t} \tag{11.6}$$

where d_t is the diameter of nanotubes and γ_0(2.7 eV) is the overlap integral for graphite.

11.2.3 Diamond

Diamond is a wide bandgap ($E_g = 5.48$ eV) semiconducting (Group IV) material with a lattice parameter of 0.357 nm. Its lattice structure can be regarded as two interpenetrating face centered cubic lattices, which are displaced from each other along the body diagonal by 1/4 length of the unit cell. The distance between the C–C atoms in the diamond lattice is 0.154 nm, which is 66% of Si–Si atoms distance in silicon unit cell. The closeness of C atoms results in wide energy gap between the bonding and antibonding orbitals and the strong covalent bond (C–C) makes diamond the hardest material.

Diamond lattice may also incorporate other atomic impurities that alter the its electronic and optical properties, including nitrogen, boron, silicon, hydrogen, and many others (Aharonovich and Neu 2014; Aharonovich et al. 2011b). For instance, even though a pure diamond is a perfect insulator, the incorporation of boron changes its conductivity and transforms it into a p-type semiconductor (Groot-Berning et al. 2014). Nitrogen is the most abundant gas in the atmosphere and it is the most frequent contaminant found in diamond matrix. It may incorporate as substitutional atom within diamond lattice by replacing one carbon atom, denoted as N*s* and usually called the P1 center (Talbot-Ponsonby et al. 1997). The defect, which has no optical activity, serves as electron donor in lattice system. If a nitrogen impurity is associated to a vacancy created in the diamond lattice, the defect becomes optically active and is called N-V color center (Fig. 11.5a). NV center possesses C3v symmetry, which implies 120° symmetrical rotation along the bond connecting the nitrogen atom and vacancy. The neutral NV defect, often known as NV°, is formed by five unpaired electrons, which is paramagnetic in nature. Additionally, the defect has the ability to capture one or more electrons, forming the negatively charged NV^- center due to which the NV^- center has an electron spin S = 1. In both charge states, NV center has an atom-like structure which emits a broadband photoluminescence (PL) signal with a zero-phonon line (ZPL) at $\lambda_{ZPL} = 575$ nm (2.156 eV) and $\lambda_{ZPL} = 637$ nm (1.945 eV) for NV° and NV^- states, respectively. The electronic state of NV^- splits into two energy levels due to spin–spin interaction with zero field splitting D = 2.87 GHz between the singlet state of spin projection (along the NV center quantization axis) $m_s = 0$ of lower energy and the degenerate doublet states $m_s = \pm 1$ (Fig. 11.5b).

The parallel spin states ($m_s = \pm 1$) have an energy higher than the antiparallel spin states ($m_s = 0$) due to quadrupole coupling and nuclear mixing of the hyperfine states. After being optically excited, the NV defect can relax either by the same radiative transition that produces a broadband red PL or through a separate pathway involving non-radiative intersystem crossing (ISC) to a singlet state. The PL emission of NV centers is perfectly photostable at room temperature which enables the development of highly robust single-photon sources operating at ambient conditions (Aharonovich et al. 2011b; Le-Sage et al. 2013).

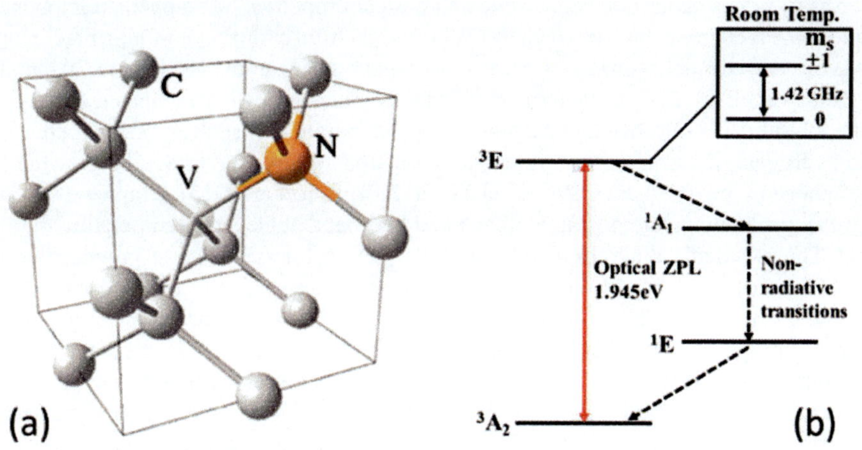

Fig. 11.5 **a** Depiction of NV center in tetrahedral diamond matrix and **b** Energy level diagram corresponds to NV^- center. It is adapted with permission from IOP Publishing (Aharonovich et al. 2011b)

11.3 Synthesis of Carbon Materials

Chemical vapor deposition (CVD) plays a significant role in growth of CNT, graphene, and single-crystal diamond (SCD). Considerable research has been done on CVD process for synthesizing carbon-based materials. Single- and multi-walled carbon nanotube has been grown in various architectural structures for electronic and optoelectronic applications (Wei et al. 2002; Harutyunyan et al. 2009; Rai et al. 2008a; Hazra et al. 2009). Planar graphene (2D) grown on Cu and SiO_2 substrate plays vital role in optoelectronics and plasmonics (Zhu et al. 2010; Avouris and Dimitrakopoulos 2012; Pandey et al. 2012). Homo- and hetero-epitaxial growths of large area SCD are being explored for high-temperature electronic and quantum devices (Aharonovich et al. 2011a; Schreck et al. 2017; Schwander et al. 2011; Tyagi et al. 2006; Mohapatra et al. 2011). The CVD growth of carbon-based materials is continued for the requirement of large area with precise control in its electronic and optical properties.

11.3.1 Synthesis of Graphene

Graphene is synthesized by various methods including mechanical exfoliation and CVD (Pandey et al. 2012; Novoselov et al. 2004). However, CVD is more versatile in terms of precise control of electronic and optical properties of graphene. Graphene and few-layer graphene are grown by CVD method from carbon-containing precursor gas on catalytic metal surface by surface segregation of C atom dissolved in Cu metal (Pandey et al. 2012). The catalytic metals for graphene growth are selected based on the solubility of C atom into metal matrix. Various parameters such as ratio of methane to hydrogen, duration, annealing temperature, and cooling rate affect the growth of graphene in low-pressure CVD (Cunha et al. 2014; Liu et al. 2015). Graphene on Cu foil is grown by decomposing methane and hydrogen at elevated temperature ~900 °C. The schematic diagram of thermal CVD set-up for synthesis of graphene and CNT is shown in Fig. 11.6a. Prior to growth, the copper foil (1 cm × 1 cm) was immersed for 30 s in HNO_3 (5.4%) and then cleaned with deionized water, acetone, and isopropyl alcohol, respectively, followed by hot air drying. Further, the CVD chamber was evacuated to a base pressure of 10^{-2} mbar after loading the copper foil. The furnace temperature was then increased to 900 °C while maintaining the flow of H_2 (10 SCCM) and Ar (100 SCCM) through the CVD chamber. Once the copper foil is heated up, it was annealed for 10 min under reducing ambient (150 SCCM, H_2 and 50 SCCM, Ar). After annealing, methane, H_2, and Ar were introduced into the chamber to initiate graphene growth for 15 min. After deposition, the substrate was quickly allowed to cool by moving the quartz tube away from the hot zone while keeping the flow rate constant unless the furnace temperature fell below 200 $°C$. The sudden cooling decreases the graphene nucleation sites and thereby hinders multilayer graphene growth.

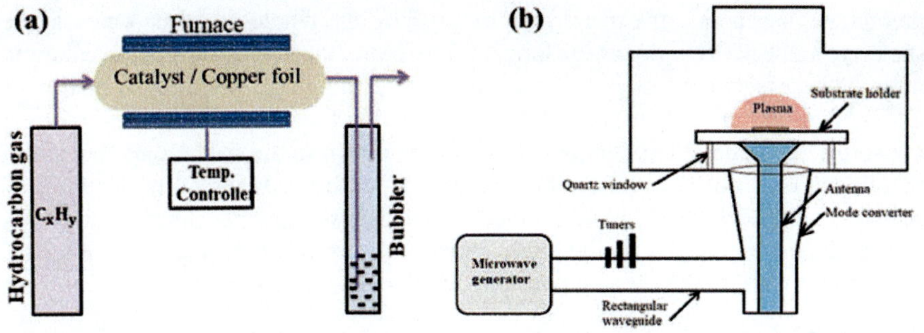

Fig. 11.6 **a** The schematic diagram of thermal CVD set-up employed for growth of graphene and carbon nanotube and **b** A multi-mode microwave plasma CVD reactor for synthesis of diamond and low-dimensional carbon-based materials

11.3.2 Synthesis of Carbon Nanotube

Like graphene, carbon nanotube is also grown by CVD process. Other methods are also applicable to synthesize nanotubes, such as arc discharge and laser ablation (Mann 2006). CNT can be grown in a variety of forms, such as powder, films, aligned or entangled, straight or coiled nanotubes, or a desired architecture of nanotubes on predefined sites of a patterned substrate (Rai et al. 2008a, b; Mann 2006). Nanotube growth involves decomposition of hydrocarbon (~15 min) on catalytic material (Fe, Co, Ni) at elevated temperature (800–1000 °C) in thermal or microwave plasma CVD reactor (Fig. 11.6) (Rai et al. 2008a, b; Misra et al. 2006a, b; 2007). The size of the catalyst particle controls the diameter of the nanotube (Misra et al. 2006a, b, 2007). The catalyst precursor may be used in any form (such as solid, liquid, or gas) and is placed inside the reactor or fed from outside (Rai et al. 2008a, b). Pyrolysis of the catalyst vapor at a suitable temperature liberates metal nanoparticles in situ, known as floating catalyst method or the catalyst-coated substrates can be placed in the hot zone of the furnace to catalyze the CNT growth. Hydrocarbon vapor initially breaks down into carbon and hydrogen species when it comes into touch with the "hot" metal nanoparticles; the hydrogen flies away while the carbon gets dissolved into the metal. The dissolved carbon in the catalyst metal precipitates out and crystallizes in the form of a cylindrical tubular network with no dangling bonds after exceeding the carbon-solubility limit in the metal at that temperature (Mann 2006). Being an exothermic process, hydrocarbon decomposition releases some heat to the metal's exposed zone whereas carbon crystallization takes heat endothermically from the metal's precipitation zone. The process continues due to the precise temperature gradient within the metal particle.

The highly pure CNT is grown by either "tip-growth" or "bottom-growth" model depending on whether the catalyst–substrate interaction is weak or strong, respectively (Mann 2006). In both the models, hydrocarbon decomposes on the top of the metal, and carbon diffuses down through the metal but in the former method, CNT precipitates out across the metal bottom, pushing the whole metal particle off the substrate and CNTs continues to elongate. Therefore, the metal stops being catalytic after the excess carbon has completely covered it, which also stops the formation of CNTs. Whereas in the latter case, CNT precipitation fails to push the metal particle up, so the precipitation is compelled to emerge out from the metal's apex. Carbon first crystallizes out as hemispherical dome and further extends up in the form of seamless graphitic cylinder. As dissolved carbon diffuses upward, subsequent hydrocarbon deposition occurs on the lower peripheral surface of the metal. As a result, CNT grows up with the catalyst particle rooted on its base.

11.3.3 Synthesis of Diamond

Microwave plasma CVD is emerging as the most promising method for producing large area (cm), high-quality diamonds at a reasonable cost using low-pressure microwave plasma (Schreck et al. 2017; Tyagi et al. 2006; Mohapatra et al. 2011). The advancement in the synthesis of diamond using CVD technique leads to grow diamond in the bulk form or coatings on a variety of substrates (Tyagi et al. 2006; Misra et al. 2006a, b). In this technique, carbon atoms that originate from the dissociation of a carbon-containing precursor gas are deposited on a substrate. The substrate is either bulk diamond (natural or lab grown) or a non-diamond substrate (Tyagi et al. 2006; Misra et al. 2006a, b). There are many CVD techniques utilized for synthesis of high-quality diamond materials such as hot filament, thermal, combustion-flame-assisted, microwave plasma, and direct current arc plasma (Schwander and Partes 2011; Sussmann 2009). Microwave plasma chemical vapor deposition (MPCVD) offers a clean environment to regulate and minimize the incorporation of contaminants during the growth process of diamond (Schreck et al. 2017; Tyagi et al. 2006; Tallaire et al. 2020). In this deposition method, plasma is generated by the microwave heating (Fig. 11.6b). The role of plasma is to generate atomic hydrogen and appropriate carbon precursors for the synthesis of diamond. Atomic hydrogen is produced by electron impact dissociation of molecular hydrogen. The carbon precursors, such as methane, aliphatic or aromatic hydrocarbons, alcohols, ketones, amines, ethers, or carbon monoxide, serve as the starting point for the growth of diamond. Methane is utilized extensively in diamond growth process, because of its high purity and tetrahedral structure (sp^3 bonding), which is identical to diamond structure. In MPCVD, the plasma is created in resonant cavity by high-power (6–12 kW) microwave with frequency (2.45 GHz or 915 MHz). In growth process of diamond, free radicals particularly CH_3 and atomic hydrogen play crucial role, which determine the quality and growth rate.

11.4 Characterization of Carbon Materials

Various characterization techniques such as electron microscopy, electron spectroscopy, X-ray diffraction, and Raman and photoluminescence spectroscopies are employed to understand the structural, optical, and electronic properties of carbon-based materials, respectively. Raman and photoluminescence spectroscopies are most frequently used techniques to distinguish the various phases of carbon materials and its optical properties (Rai et al. 2018; Dresselhaus et al. 2005).

11.4.1 Raman Spectroscopy of Carbon Materials

Raman spectroscopy offers a rapid and non-destructive method to characterize and estimate the quality of the carbon-based materials. It is a versatile tool to identify and characterize the chemical and physical properties of graphitic materials such as graphene and CNTs (Cancado et al. 2004; Ferrari and Robertson 2000). The characteristics of Raman spectra of carbon-based materials exhibiting sp^2 (CNT and graphene) and sp^3 (diamond) hybridization are shown in Fig. 11.7.

Mono-layer graphene is the simplest crystal structure with the maximum degree of symmetry and, thus, the simplest Raman spectra in the family of sp^2 carbon systems. The G- and G'-bands of graphene were used to address strain, charge transfer, and disorder effects produced by doping via gate electrode and various types of substrates. By measuring the dependence of the Raman spectra on the number of scattering in graphene layers, Raman imaging is used to define the number of layers in different

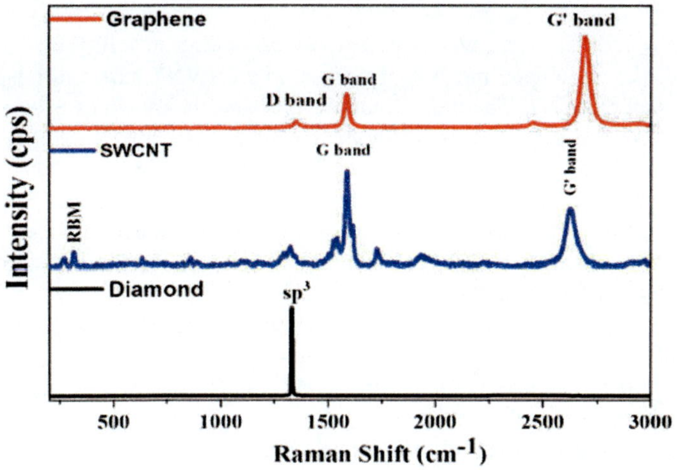

Fig. 11.7 Raman spectra of carbon-based materials such as **a** graphene, **b** carbon nanotube, and **c** diamond

locations of a given graphene flake (Ferrari and Basko 2013; Wang et al. 2008). Figure 11.7a shows the presence of the three characteristics Raman bands (D-, G-, and G'-bands) of graphene. The G-band originates from doubly degenerate phonon mode (E_{2g} symmetry) and appears at ~1580 cm^{-1} whereas the D-band originates from backscattering of phonons by disorder in graphene such as edges and defects and occurs at ~1350 cm^{-1}. Depending on the crystallinity of the material, the relative intensities of these vibrations vary. The intensity ratio of D- to G-bands (I_D/I_G) reflects the purity of the sample. Apart from the G- and D-bands, there is an intense G'-band that corresponds to a second-order process by involving two phonons near K point and occurs at ~2700 cm^{-1}, which exhibits a strong signature of graphene. The G'-band is an overtone of the D-band and does not come from a defect but rather from the backscattering of a phonon by an electron (Pandey et al. 2012; Ferrari and Bosko 2013).

The typical Raman spectrum of CNT recorded in the range between 100 and 3000 cm^{-1} is shown in Fig. 11.7b. The spectral feature of SWNTs is broadly divided into three parts such as radial breathing modes (RBM) (100–500 cm^{-1}), disorder-induced D-band (~1350 cm^{-1}), and tangential vibrational mode G-band (~1580 cm^{-1}). The electronic conductivity of CNT (metallic or semiconducting) is determined by the nature of G-band. The RBM peak of Raman spectrum is used to estimate the diameter of SWNT or inner diameter of double-walled carbon nanotubes. In RBM of nanotube, carbon atoms vibrate in radial direction of cylindrical symmetry. Due to involvement of one phonon in radial vibration of carbon atoms, it is called first-order Raman scattering. The diameter of SWNT in a bunch of nanotubes is determined by RBM frequency (Dresselhaus et al. 2005):

$$\omega_{\mathrm{RBM}}\left(\mathrm{cm}^{-1}\right) = \frac{224}{\mathrm{d_t(mm)}} + \mathrm{A} \tag{11.7}$$

where ω_{RBM} (cm^{-1}) is the wave number corresponding to RBM, d_t (nm) is the diameter of the nanotube, and A = 14 cm^{-1} when SWNT exists in bunches and zero for isolated SWNT. The defect induced (D-band) is known as a second-order Raman scattering because it involves both an elastic scattering by the defect center and a phonon. The G-band corresponds to Raman peak ∼1585 cm^{-1}, and is an in-plane bond stretching mode of the C–C bonds in the hexagonal lattice. All the sp^2 hybridized carbon compounds exhibit the G-band. The G'-band (or 2D-band) appears at ∼2700 cm^{-1} which is another peak that may be found in the Raman spectra of the majority of sp^2 carbon nanotube.

The Raman spectrum of diamond is shown in Fig. 11.7c and the characteristics of Raman peak corresponding to sp^3 bond in diamond appear at 1332 cm^{-1}. The sharp (narrow FWHM) peak at 1332 cm^{-1} corresponds to SCD and precise measurements of any shift of Raman peak relative to unstressed location provide information about the stresses present in diamond. The strain present in the sample is significantly affected by introducing dopants and defects into diamond matrix, which results in small shifts in Raman band. The stress in the diamond film is estimated by using dynamical calculation, $\sigma = k\,(\upsilon_m - \upsilon_0)$ GPa, where k = 0.34 GPa.cm is calculated empirically for diamond and (υ_m-υ_0) corresponds to the shift in sp^3 peak (Di Liscia et al. 2013). In polycrystalline diamond, both sp^3 and sp^2 carbon contributions may be observed in the spectra, where the intensity ratio of the diamond peak 1332 cm^{-1} to the G-band 1575 cm^{-1} is employed to evaluate the crystallite size.

11.4.2 Photoluminescence of Carbon Materials

Photoluminescence is a commonly used spectroscopy tool to determine the bandgap and defect level present in semiconductor materials. Semiconducting carbon-based materials such as carbon nanotube and diamond show PL emission. Graphene belongs to semi-metals and does not show PL. Fluorescence emissions in graphene can be obtained by generating bandgap within it. Numerous experimental approaches have been utilized to introduce bandgaps in the graphene, such as cutting graphene sheets into small pieces or modifying the π-electrical network to form quantum-confined sp^2 "islands" in a graphene sheet, both of which appear to involve the creation or exploitation of structural defects (Bachilo et al. 2002).

Due to unique photo-physical properties, SWNT served as building blocks for biosensors, functional materials, and devices. Direct bandgaps in semiconducting CNTs display favorable optical emission properties. Figure 11.8 shows the PL spectrum of SWNT and single-crystal diamond in the IR region at the room temperature. Semiconducting SWNT exhibits distinct and structure-dependent optical transitions in PL emission (Hartmann et al. 2012). Compared with theoretical calculations, one can assign the most prominent peak 980 nm to (6,5) indexed SWNT. The chirality of (6, 5) is easily recognized by its luminescence spectra and can be excited efficiently near their second-order resonance E_{22} using low-noise lasers (Hartmann et al. 2012).

Nitrogen impurities and other color centers in diamond are easily reflected in PL spectra of diamond (Fig. 11.8) at room temperature. The signature of the nitrogen vacancy centers, created by the zero-phonon line (ZPL) at 575 nm and 637 nm, corresponds to neutral charge state (NV^0) and negatively charged state (NV^-), respectively (Aharonovich et al. 2011b). The broad PL background also appears along with color centers in diamond. A sharp intense peak at 521 nm corresponds to the first-order Raman line which confirms high-quality sp^3 structure. It occurs due to inelastic scattering of the laser light by the diamond lattice. The second-order Raman feature at 554 nm is also observed in the given PL spectrum. A temperature-dependent ZPL at 503 nm originating from the 3H center is also present, which is usually ascribed to some form of self-interstitial defects in diamond. It has been confirmed that the 3H center, which is a negatively charged center present in diamond structure, acts as a donor (Aharonovich et al. 2011b).

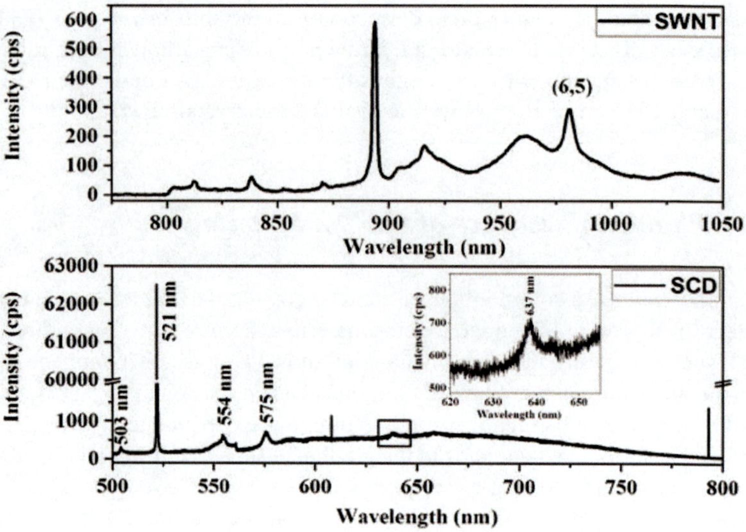

Fig. 11.8 Photoluminescence of **a** single-walled carbon nanotube and **b** single-crystal diamond

11.5 Carbon Material-Based Nanoscale Optics

The manipulation and propagation of electromagnetic field at sub-wavelength regime is the current requirement of next-generation communication devices (Zia et al. 2006; Novotny and Hecht 2006). Surface plasmon is coupled electromagnetic wave with charge oscillation on metal surface that propagates along metal–dielectric interface. Such SPP waves are supported by sub-wavelength-sized metallic waveguides in photonic circuitry. The electrically and optically excited carbon-based nanoscale devices have been demonstrated as nanoscale optical source (Bonaccorso et al. 2010; Aharonovich et al. 2011a; Avouris et al. 2008; Rai et al. 2013). Therefore, carbon-based nanoscale electronic and optoelectronic devices offer a chip-scale platform to integrate the plasmonic circuitry based on these materials.

11.5.1 Nanoscale Optics

The optics at nanometer scale deals the propagation and manipulation of electromagnetic waves without any limitation of fundamental law of diffraction ($\lambda/2$, where λ is wavelength). The propagation of light in free space is governed by the dispersion relation,$\omega = ck$, where ω, c, and k are angular frequency, speed of propagation, and wavevector of electromagnetic wave, respectively. Uncertainty principle states that the product of uncertainty in special position (Δx) and component of momentum ($\Delta \hbar k_x$) of a microscopic particle in the same direction cannot become smaller than

$\hbar/2$. The uncertainty relation for photons is expressed as

$$\Delta x.\Delta\hbar k_x \geq \hbar/2 \tag{11.8}$$

Equation (11.8) can further be written as

$$\Delta x \geq 1/2\Delta k_x \tag{11.9}$$

The maximum possible spread in the x-component of wavevector, k_x, is the length of wavevector of a photon $k = 2\pi/\lambda$, where $k = \sqrt{k_x^2 + k_y^2 + k_z^2}$ and λ is the wavelength of a photon. This leads to the expression:

$$\Delta x \geq \lambda/4\pi \tag{11.10}$$

Thus, the spatial confinement is limited by the spread of wavevector component in the given direction and one can get the resolution better than Rayleigh diffraction criterion (λ/2).

An electromagnetic wave that propagates along metal–dielectric interface and coupled with electron oscillations of metal is known as SPP. Both energy and momentum conservation should be fulfilled in order to excite SPP waves. A propagating SPP along metal–dielectric interface is shown in Fig. 11.9a, where p-polarized light excites the surface plasmon at metal–dielectric interface. The energy–momentum conservation for SPP can be understood by its dispersion relation (Fig. 11.9b). The SPP momentum is obtained (Novotny and Hecht 2006) as

$$k_{SPP} = k_0\sqrt{\frac{\varepsilon_1\varepsilon_2}{\varepsilon_1 + \varepsilon_2}} \tag{11.11}$$

where k_{SPP} and k_0 are the surface plasmon and free space momentum of light, respectively, and ϵ_1 and ϵ_2 are real dielectric constant of metal and dielectric, respectively.

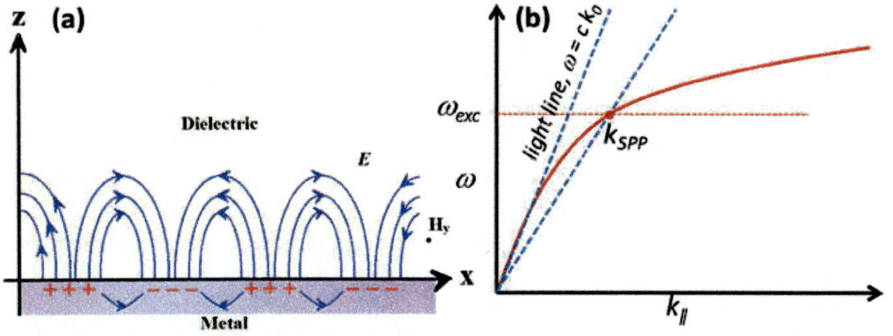

Fig. 11.9 **a** The SPP wave is propagating along the metal surface, which was launched by electromagnetic wave with transverse magnetic mode. **b** The dispersion relation of SPP wave

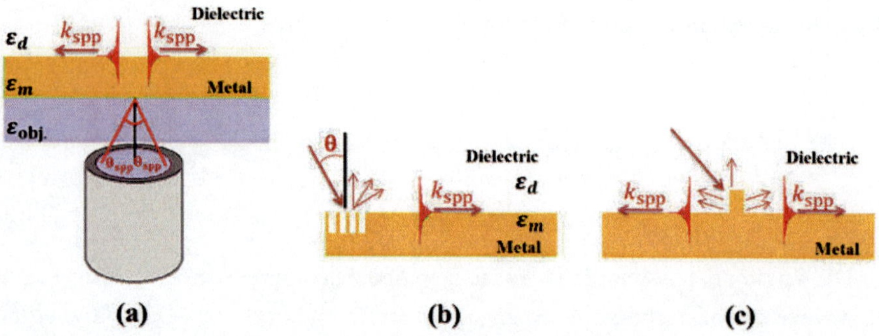

Fig. 11.10 The SPP excitation on the interface of metal–dielectric by **a** high numerical objective in Kretschmann configuration, **b** grating coupler, and **c** metal nanoparticle placed over the metal surface

It is evident from Fig. 11.9b that for a given energy $\hbar\omega$ the wavevector k_x of surface plasmon is larger than wavevector k_0 of light in free space. The increased momentum of SPP is due to strong coupling between light and surface charges, where light field must push the electron along the metal surface. To excite the SPP on metal surface, one must increase the momentum of light by introducing a dielectric with refractive index ($n > 1$) at metal surface (Fig. 11.9b). It is obvious from Fig. 11.9b that the light line is tilted by a factor of n in this case because $\omega = ck/n$. The additional momentum in free space light, which is required to excite the SPPs on metal–dielectric interface, is obtained by Kretschmann method and specially designed optical grating (Novotny and Hecht 2006).

In Kretschmann configuration, a thin metal film is deposited over the base of prism and SPP is excited at the metal–air interface by creating an evanescent wave at glass–metal interface, which must penetrate the metal layer. Alternatively, Kretschmann configuration can also be achieved by coupling the light to metal–dielectric interface through high numerical aperture objective placed below the metal film as shown in Fig. 11.10a.

The grating couplers and metal nanoparticles are alternate experimental approaches to excite SPP on metal–dielectric interface (Fig. 11.10b, c). In grating coupler method, the increase of necessary momentum to match the momentum of SPP is provided by adding the reciprocal vector of the grating to the free space wavevector. If a is the special periodicity of grating structure, then the new parallel momentum can be expressed as $k_{x'} = k_x + 2\pi n/a$, with $2\pi n/a$ being a periodicity of reciprocal lattice vector. In case of metal nanostructure, when light is incident on metal nanoparticle, then there is huge distribution of momentum in the vicinity of nanoparticle and metal film junction. Some of the momentums are matched with SPP momentum and able to excite the SPP on the metal film surface.

Confocal scanning optical microscopy is employed for high accuracy measurement in the field of nanoscale optics and plasmonics (Fig. 11.11). It employs far-field illumination and detection schemes to understand the optical and spectroscopic properties of individual nanoscale material or molecule. Various nanoscale optic-related experiments, such as surface-enhanced Raman scattering or PL, single-photon emission (SPE) from quantum emitters (QE), and free space light coupling into surface plasmon, employee the principle of confocal technique (Novotny and Hecht 2006; Rai and Dubey 2018). In such instruments, the input laser light is spatially filtered by pinhole or sending it through a single-mode optical fiber, which is a requirement of perfect Gaussian beam profile. The spatially filtered laser light is allowed to pass through a lens to convert into collimated beam with large diameter, so that it overfills the back-aperture of microscope objective used to focus the light onto the sample plane.

The lateral spot size (Δx) of laser beam on the sample depends on numerical aperture (NA) of microscope objective and the wavelength of light (λ) used for illumination, which is limited by diffraction of laser light at the entrance aperture of objective.

$$\Delta x = 0.61\frac{\lambda}{NA} \tag{11.12}$$

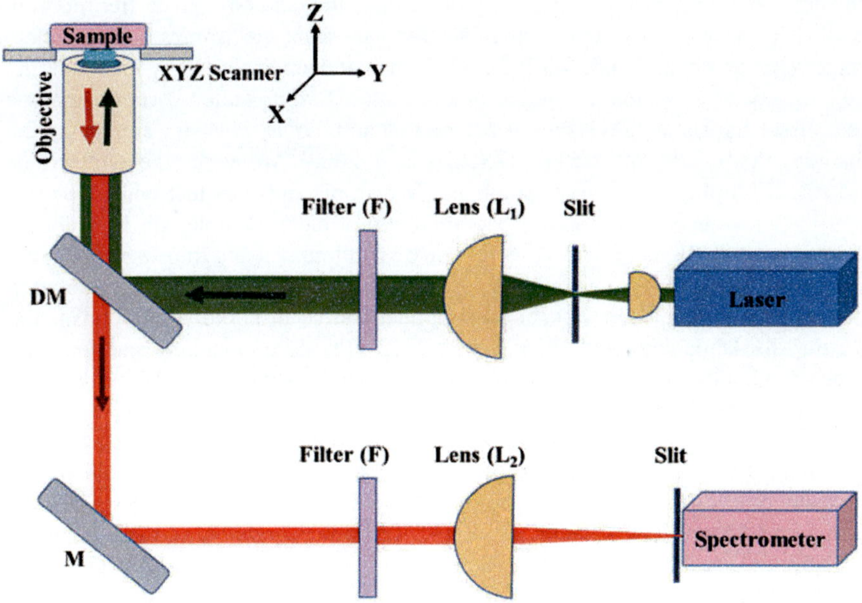

Fig. 11.11 Schematic diagram of confocal scanning optical microscopy set-up. DM and M stand for dichroic mirror and mirror, respectively. The dichroic mirror transmits the light in restricted spectral range, which is further filtered and passed through the pinhole before the entrance slit of the spectrometer

For $NA = 1.49$, the spot size for laser light (λ= 532 nm) is ~218 nm, which is less than Rayleigh criterion limit ($\lambda/2$).

When the incident laser light interacts with the sample, it produces reflected and scattered light at wavelength shifted with respect to the excitation. The same microscope objective is utilized to illuminate the sample and collect the scattered light. The Rayleigh light is cut-off before entering into detector of the spectrometer by using dichroic mirror and notch filter in combination. In order to improve the signal-to-noise ratio of the scattered light, a pinhole is placed in front of the slit of the spectrometer. The pinhole arrangements allow the light emerging from the focal point on the sample, while the laterally displaced or out-of-focal plane light beams are blocked by the aperture. This is called confocal detection technique (Novotny and Hecht 2006). Confocal detection technique provides better resolution and contrast of confocal scanned images and thus it can be employed for detection of individual molecules or quantum emitters.

11.5.2 Carbon Nanotube Plasmonics

The quasi-one-dimensional structure with direct bandgap and high carrier mobility of SWNT makes it a suitable candidate for nanoscale electronics and optoelectronics (Avouris et al. 2008; Rai et al. 2013). Its bandgap is altered by manipulating the chirality of nanotube as evident in Eq. (11.6). Single-walled carbon nanotube field effect transistor (SWNTFET) behaves as ambipolar transistor simply by the changing the polarity of gate bias (Avouris et al. 2008). Moreover, the gate voltage of SWNTFET plays a vital role in tuning the electronic and optoelectronic properties of SWNT (Avouris et al. 2008). Recombination of electron–hole pair in nanotube channel produces light emission, which can be manipulated and translated along the channel by gate voltage. The light emission source produced by electrical excitation method can be employed as nanoscale optical source in photonic and plasmonic circuitry for launching the SPP (Rai et al. 2013). Plasmon launching and detection in plasmonic circuitry can also be achieved by simply using PL of SWNT by laser excitation method (Hartmann et al. 2012; Rai et al. 2012).

The directional excitation of propagating SPP on a thin metal film by PL of individual SWNT is shown in Fig. 11.12 (Hartmann et al. 2012). Inverted confocal optical microscope with an oil immersion objective (NA = 1.4) and linearly polarized laser ($\lambda = 565$ nm) for excitation was employed for real space PL and SPP imaging. The Fourier space image of SPP was acquired by same configuration with additional lens to collimate the light coming due to leakage radiation from thin metal film (Hartmann et al. 2012). For this experiment, (6,5) indexed enriched SWNT was spin coated on SiO_x spacer layer (35 nm) on top of thin gold film (25 nm) deposited over cover-glass slip. Subsequently, certain area of the sample was scanned by confocal imaging. The luminescent spot as shown in Fig. 11.12a is marked as (6,5) indexed SWNT, which was confirmed by PL and Raman spectroscopies. The prominent characteristic of SPP is confirmed by Fourier plane imaging of the emitted radiation.

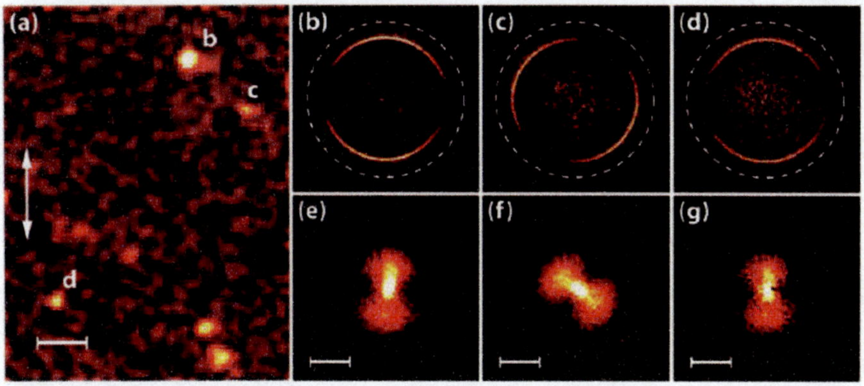

Fig. 11.12 **a** Confocal PL image of individual SWNT on a thin SiO_x (35 nm) spacer layer on top of a thin gold film (25 nm). The scale bar is 2.0 μm and laser polarization is indicated by in-plane double-headed arrow. The Fourier space (k-space) image of PL spots marked as b, c, and d in **a** are shown in (**b–d**). The two sharp arcs represent the excitation of SPP perpendicular to nanotube axis. **e–g** The extended real space images of SPP with scale bar 4.0 μm. It is adapted with permission from ACS publisher (Hartmann et al. 2012)

The half ring-shaped Fourier image (Fig. 11.12 (b-d)) confirms the signature of plasmon radiating into the metal substrate. The half ring pattern of Fourier image arises due to directional excitation of SPP.

The elongated distribution of emission is shown in Fig. 11.12e–g corresponding to real space image recorded at the same location which indicates the nature of directional propagation of SPP. The phenomena of SWNT photoluminescence coupling to surface plasmon are explained by radiating dipole into thin metal surface (Hartmann et al. 2012). The critical angle for effective index and propagation length of propagating plasmon are 44.9° and 4.2 μm, respectively, which is in good agreement with theoretical calculation.

In the reverse process, surface plasmon can also excite the photoluminescence and Raman scattering in nanotube (Rai et al. 2012). In this process, the SPP was excited remotely (few microns away from nanotube) from the individually marked nanotube on the plasmonic waveguide, where SPP gets coupled into PL and Raman scattering of the nanotube (Fig. 11.13). The coupling efficiency depends on the angle between plasmon propagation vector (k_{SPP}) and axis of the nanotube, lateral distance of nanotube from the excitation point and transverse distance of the nanotube from the line of propagation of SPP. The plasmonic wave guide of Au metal was fabricated on ITO-coated cover-glass slip by electron-beam lithography and lift-off process. The length, width, and thickness of the waveguide strips are 10–30 μm, 5 μm, and 35 nm, respectively. On the top of the waveguide, a thin oxide (SiO_x) layer (10 nm) was deposited to suppress the photoluminescence quenching by metal film. A solution of (6,5) enriched SWNT was spin coated on the Au waveguide, which is shown in Fig. 11.13a. A surface plasmon jet was launched by *p*-polarized laser with excitation wavelength of 785 nm (Fig. 11.13b). An optical image shown in

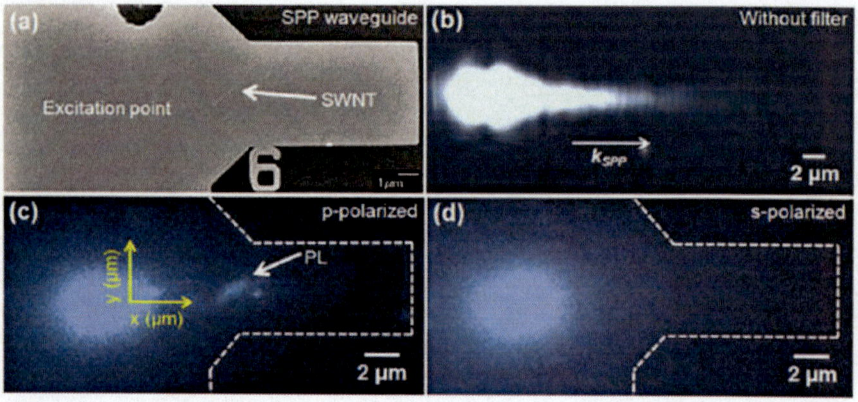

Fig. 11.13 **a** Scanning electron microscopy image of plasmonic waveguide, where nanotube is spin coated on it and marked by white arrow, **b** Optical image of propagating surface plasmon jets along k_{SPP} shown by white arrow, **c** Coupling of SPP into PL/Raman scattering of nanotube for *p*-polarized light, and **d** There is no light at the location of nanotube when the polarization of incident light changed from *p*- to *s*-polarization. It is adapted with permission from The Optical Society (Rai et al. 2012)

Fig. 11.13c was recorded by a combination notch and long pass optical filter, which cut off the excitation wavelength 785 nm.

A light spot can be seen at the location of nanotube, it may be produced by the PL/Raman scattering from the nanotube. It was further confirmed by changing the polarization of incident light from *p* to *s*, for which, there is no plasmon propagating along the waveguide and light at the location of the nanotube was absent (Fig. 11.13d). The SPP coupling to optical excitation of SWNT is maximal when plasmon momentum is parallel to nanotube axis. Plasmon excitation was found weaker when lateral and transverse distance of nanotube from plasmon excitation point was beyond 7.2 μm (plasmon propagation length) and 0.7 μm (FWHM of transverse Gaussian profile of plasmon intensity), respectively. The SPP optical excitation of nanotube is mediated by dipole–dipole interaction between the dipole moment of optically excited nanotube and electric field of surface plasmon wave (Rai et al. 2012). This excitation method can be employed for remote optical sensing of soft matter in the integrated phonic circuitry, where the possibility of photo-bleaching can be minimized.

In plasmonic circuitry, conventional optical method of plasmon excitation cannot be employed because of bulky laser system involved in the process. Electrical excitation of surface plasmon is the necessity of next-generation chip-scale technology for faster communication speed (Rai et al. 2013). Nanoscale optical source generated by electroluminescence (EL) process in CNTFET can be employed for surface plasmon launching in plasmonic waveguide created in electronic circuitry (Rai et al. 2013). The EL process in CNTFET is generated by various electron–hole pair recombination mechanism, such as impact excitation, recombination of carriers during ambipolar transport, phonon-assisted process, and recombination from thermally populated

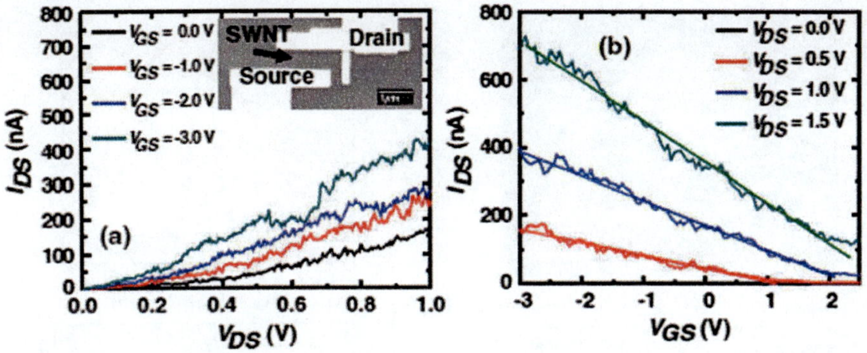

Fig. 11.14 **a** The output and **b** transfer characteristics of SWNTFET. The SEM image of SWNTFET is shown in the inset of (**a**). It is adapted with permission from APS publisher (Rai et al. 2013)

electronic states (Avouris et al. 2008; Rai et al. 2013). For optical and plasmonic applications, SWNTFET was fabricated on ITO-coated cover-glass slip by electron-beam lithography and lift-off process. The electrical gating and plasmonic waveguide were ensured by depositing 5–10 nm of Ti or Pd followed by 45 nm of Au by electron-beam evaporation. A 250-nm-thick layer of SiO_2 was deposited by thermal evaporation method on ITO-coated cover-glass slip to ensure the gate electrode for the CNT channel. The output (I_{DS}-V_{DS}) and transfer (I_{DS}-V_{GS}) characteristics of SWNTFET are shown in Fig. 11.14a, b, respectively. The inset of Fig. 11.14a shows SEM image of SWNTFET. The EL in CNTFET is produced by impact ionization process, which is shown in Fig. 11.15a. The EL spots seem to spread along entire nanotube channel of SWNFET.

To identify the transition mechanism for EL in SWNT, the EL spectra was compared with PL spectra of same batch SWNT (Fig. 11.15b). The emission peaks for EL and PL were found at 995 nm and 982 nm, respectively, which corresponds to E_{11} transition of (6,5) SWNT. The small red-shift in EL spectra is caused due to drain-induced doping in the nanotube. The intensity of EL process depends on bias voltages (Fig. 11.15c, d), which shows that impact ionization process is responsible mechanism for EL in SWNTFET.

Long-channel SWNTFET devices were made with extended electrode for launching surface plasmon by electrical excitation method. The SWNTFET device is made operational at $V_{DS} = 10$ V and $V_{GS} = -5$ V. Scattering of light is observed from the edges of electrode in addition to EL spot at nanotube location (Fig. 11.16a). The far scattering of light from the edges of electrode can only be possible if electroluminescence gets coupled into surface plasmon. The further confirmation of plasmon properties was confirmed momentum space mapping of scattered light (Fig. 11.16b), which is signature of propagating plasmon bound at metal–glass interface.

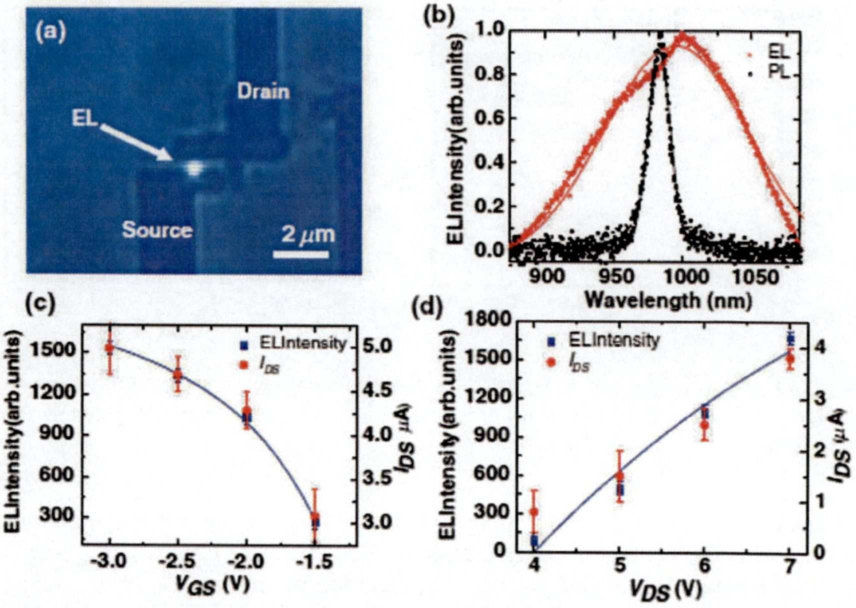

Fig. 11.15 **a** Electroluminescence from SWNTFET, **b** normalized PL and EL spectrums and **c–d** plot of EL intensity and drain current versus bias voltages. It is adapted with permission from APS publisher (Rai et al. 2013)

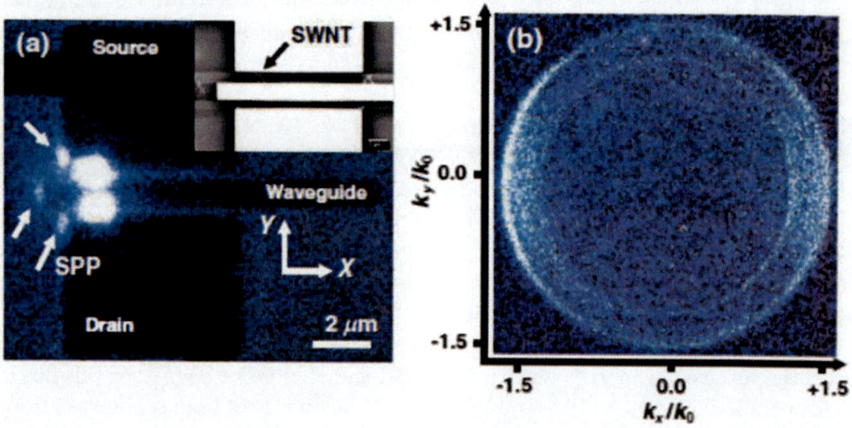

Fig. 11.16 **a** Electroluminescence coupled surface plasmon launching in electrode geometry of SWNTFET and **b** Fourier space image of the corresponding EL-coupled SPP. It is adapted with permission from APS publisher (Rai et al. 2013)

11.5.3 Graphene Plasmonics

Graphene is an attractive 2D planar material from the prospective of its tunable interbond transition in optical response by single particle excitation (electron) (Ju et al. 2011). The linear dispersion relation of graphene demonstrates that its electron behaves like massless Dirac fermions. Graphene plasmon can be understood in terms of collective oscillation of its massless electrons and the plasmon frequency is modulated by its electron density through bias voltage. The physical origin of plasmon can be understood in terms of forced oscillation of collective charge in the presence of external electric field (Fig. 11.17a). When electrons move to screen an externally present electric field (due to electromagnetic wave) then electrons move slightly too far. They are then pulled back by the disturbed charge and overshoot the equilibrium position, which is setting up a damped oscillation. The plasmon dispersion in graphene can be understood in long-wavelength limit. The plasmon frequency in doped graphene is expressed in long-wavelength limit (Grigorenko et al. 2012):

$$w(k) = \sqrt{8E_F\sigma k/\hbar\varepsilon} \tag{11.13}$$

where $E_F \propto n^{1/2}$ for massless electrons (Fermi energy), $\sigma = \pi e^2/2h$ (optical conductivity of graphene, e and h are electronic charge and Plank's constant, respectively), $k = 0.1k_F$, and ε is dielectric constant of the material. Thus, the plasmon frequency varies as $n^{1/4}$, where n is the carrier concentration of graphene. For a typical doping in graphene, $n = 10^{11}$ cm^{-2}, the Fermi energy is ~37 meV and the plasmon energy of graphene on SiO_2 is ~16 meV for $k = 0.6 \times 10^5$ cm^{-1} (Grigorenko et al. 2012). Graphene plasmon frequency lies in terahertz regime and plasmon resonance-induced absorption in it is observed at low temperature (~4 K). The tunable light–matter interaction together with excellent electronic properties of graphene nano-ribbons makes it suitable for terahertz metamaterial application.

Graphene plasmonic can be probed experimentally by various techniques, such as electron energy loss spectroscopy, scanning tunnelling spectroscopy, and optical measurement. The schematic for tuning the graphene plasmon frequency by electrical gating is shown in Fig. 11.17b, where the device is made on graphene micro-ribbons. In this device, the graphene plasmon resonance varies with bias voltage. Furthermore, the plasmon frequency is also scaled down as $1/\sqrt{w}$, where, w is width of graphene nano-ribbons (Fig. 11.17c). The direct interaction of graphene plasmon with infrared light is achieved by stacking of graphene micro-disk (Fig. 11.17d). By this method the plasmon resonance frequency can be tuned by number of layers, disk diameter, and gating. It was found that collective oscillation of massless fermions in graphene is quantum mechanical phenomena, in which resonance frequency is scaled down as $n^{1/4}$.

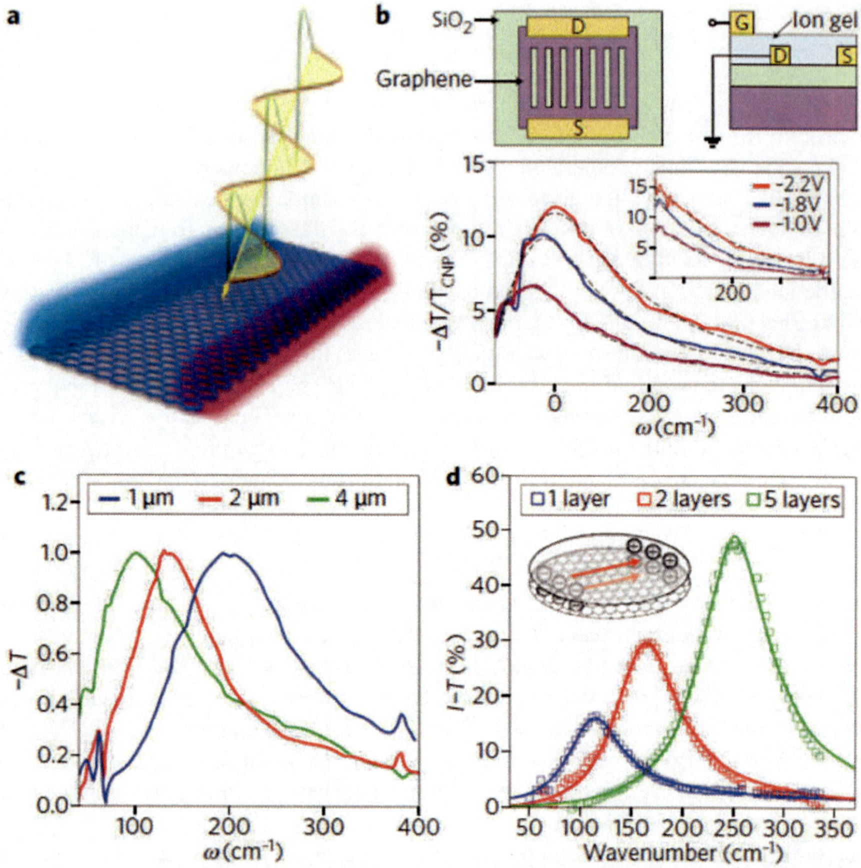

Fig. 11.17 **a** Schematic of graphene plasmon resonance excitation, **b** Plasmon resonances in gated graphene micro-ribbons, **c** variation of resonance frequency with width of graphene micro-ribbons, and **d** extinction measurement in stacked graphene for different number of layers. It is adapted with permission from Springer Nature publisher (Ju et al. 2011, Grigorenko et al. 2012)

11.5.4 Diamond Quantum Emitters

The NV^- centers in diamond matrix are emerging as potential single-photon emission, because it behaves like individual atom in solid-state system. Quantum information processing (QIP) relies on QE present in solid-state system and should have the following properties: (i) emit one and only one photon per excitation pulse, (ii) photostable emission (blinking and bleaching free behaviors), (iii) narrow PL linewidth (emission centered on ZPL), (iv) short excited state lifetime (ps $< \tau <$ ns), (v) fully polarized excitation and emission channel, and (vi) two-level system without metastable state (Aharonovich et al. 2011b).

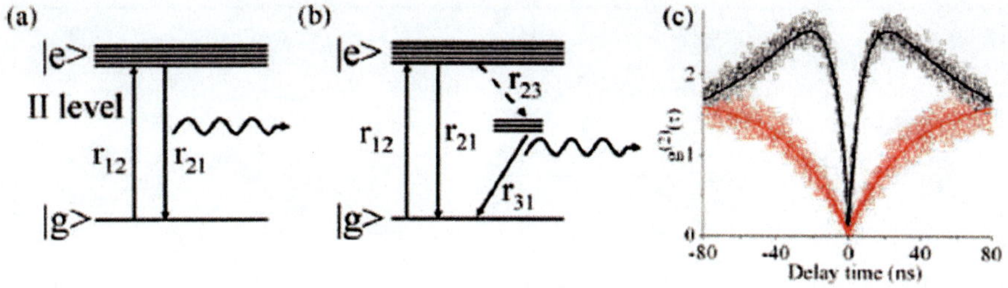

Fig. 11.18 Schematic diagram of **a** two-level and **b** three-level system, where $|g>$ and $|e>$ denote the ground and excited states, respectively. **c** auto-correlation plot for SPE from diamond QE. It is adapted with permission from IOP Publishing (Aharonovich et al. 2011b)

The light sources are distinguished by their photon statics such as Poissonian, sub-Poissonian, or non-correlated statistics. It can be studied by measuring the second-order auto-correlation function ($g^2(\tau)$) of emitted photons (Novotny and Hecht 2006). The auto-correlation function measures the probability of detecting a photon at a delay time τ, if it is given the probability of detecting a photon at time t. The most intuitive pictures of ideal SPE can be described by two-level system but in this case the photons reaching to the detector in bunches and leading to coincidences at t = 0 due to thermal noise and coherency of the sources (Fig. 11.18a). The realistic view of QE is three-level system, where additional level modifies the correlation between photons (Fig. 11.18b). The probability of getting two photons delayed by time τ is obtained by using normalized second-order correlation function (Novotny and Hecht 2006):

$$g^{(2)}(\tau) = \frac{\langle I(t)I(t+\tau)\rangle}{\langle I(t)\rangle^2} \tag{11.14}$$

where the symbol <> denotes the time average and $I(t)$ represents the number of coincidences at time t. Ideally its value will go to zero for $\tau = 0$ delay but due to stray light, APD dark counts, and contribution from neutral NV centers, one can get the value in the range $0 < g^{(2)}(\tau = 0) \leq 0.5$ for single-photon emitters. Figure 11.18c shows the quantum behavior (anti-bunching curve) of light emitted from individual NV center emitter. It is related to the fact that the lifetime of metastable state is longer than the excited state lifetime of the center, where the photons trapped in this state create dark periods between the normal photon emissions from the excited state. However, the high probability of emission at a given delay time indicates the bunching behavior of photons for which $g^{(2)}(\tau) > 1.0$.

The Hanbury Brown–Twiss (HBT) interferometer is employed to measure the photon statistics of light (Fig. 11.19). The interferometer set-up includes two avalanche photodiodes APD1 and APD2 for investigating the fluorescence intensity correlation. SPE from an individual NV center was collected using an APD-based single-photon counting module coupled to time-to-amplitude converter (TAC). The

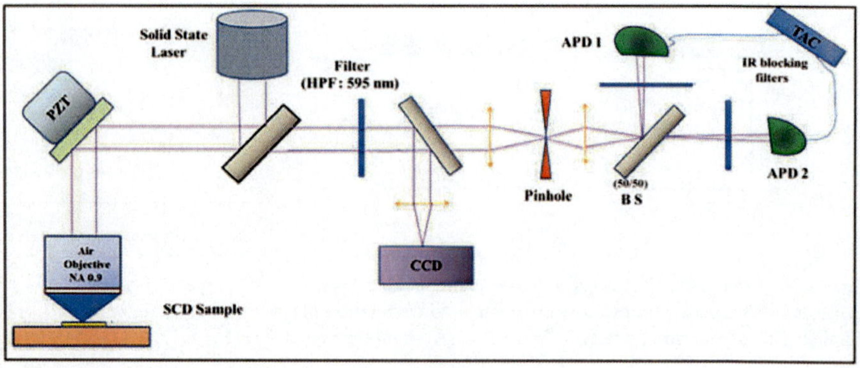

Fig. 11.19 The confocal optical scanning set-up for fluorescent imaging of NV centers present in diamond samples. The two APDs connected through TAC represent HBT interferometer

photon counting and auto-correlation measurements were performed using a time-correlated single-photon counting (TCSPC) module. In order to prevent optical cross exchange, IR filters and pinholes are placed in front of each detector. Light collected from the NV center passes though the 50/50 beam splitter and detected with an APD. TAC helps to convert the time delay between the output pulses from the two APDs into voltage amplitude that is linearly proportional to the delay. A multichannel analyzer records the number of events for each time delay to get the photon statistics.

Other than creating NV centers during growth process, various methods are available to create NV centers in diamond matrix, such as laser writing, ion irradiation, and lithography process. Well-precise spatial distribution of NV centers is required for photonic application. Laser writing is a controlled process to develop spatially distributed NV centers in diamond (Chen et al. 2017). SPE from NV centers created by laser writing process is shown in Fig. 11.20. The spatially distributed square matrix of NV centers (Fig. 11.20) is created by femtosecond laser with wavelength 790 nm, pulse duration 300 fs, and energy range ~16–60 nJ. These vacancies are identified by $g^2(\tau)$ measurement in HBT interferometer. The $g^{(2)}(0) < 0.32$ attributes to the presence of one color center, whereas $0.32 < g^{(2)}(0) < 0.65$ and $0.65 < g^{(2)}(0) < 0.90$ correspond to the presence of two and three color centers, respectively.

The quantum computing device based on NV centers requires integrating NV centers in photonic circuity. Nano-photonics provides the platform where quantum emitters can be placed inside the cavity without dropping the signal due to diffraction loss at sub-wavelength regimes.

A NV center present in nano-diamond is integrated in photonic cavity (Fig. 11.21a), which is addressed through Ta_2O_5 waveguides. The structure was achieved through multi-step lithography process. In this structure, the NV center was excited through port-2, single-photon counting was acquired through port-3, and spectroscopic studies were performed through port-1 (Schrinner et al. 2020). The multi-port characterization process enables efficient coupling of light into quantum emitter in photonic waveguide. A PL spectrum is shown in Fig. 11.21b which

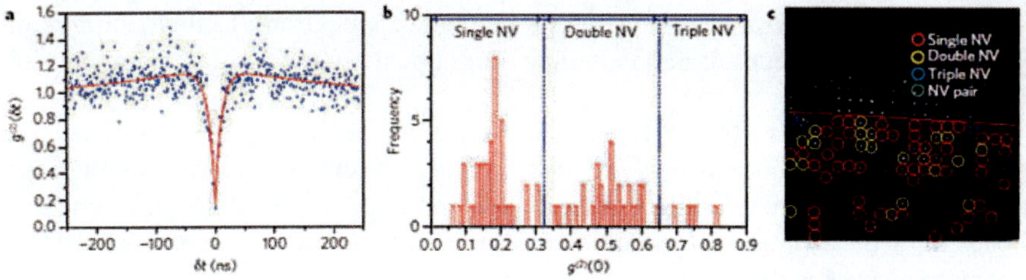

Fig. 11.20 **a** $g^2(\tau)$ plot is showing single-photon emission from NV centers, **b** distribution of single, double, and triple NV centers and **c** spatial distribution of multi-NV centers created by laser irradiation. It is adapted with permission from Springer Nature (Chen et al. 2017)

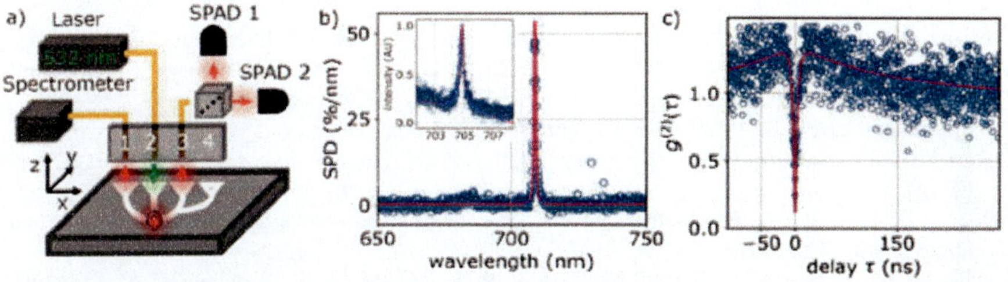

Fig. 11.21 **a** Set-up for addressing NV centers in photonic cavity through waveguide, **b** Spectral power distribution of emission from NV center, and **c** $g^2(\tau)$ plot for identifying the quantum behavior of light. It is adapted with permission from ACS publisher (Schrinner et al. 2020)

confirms the ZPL of NV center present in photonic cavity. The photons emitted from these optical centers show quantum behavior of light (Fig. 11.21c).

11.6 Conclusions

Since past three decades, growth, and electronic and optical properties of carbon-based materials such as carbon nanotube (1D), graphene (2D), and single-crystal diamond (3D) have been explored due to its quantum structures. The carbon-based materials with desired physical properties can be grown by CVD techniques. Due to reduced dimensionality, these materials may become the platform for next-generation optoelectronic devices, where electronic and optoelectronic devices are integrated.

The direct bandgap in semiconducting SWNT made it possible to use SWNTFET as light detection and light-emitting devices. Nanoscale optical source produced by electrical excitation method in SWNTFET launches surface plasmon in plasmonic circuitry. These devices can be applied for frequency modulation of optical signal at circuit level.

Graphene plasmon shows very attractive features such as extremely high localization, strong confinement, and strong light–matter interaction. The tunable inter-band optical transition in graphene is useful for terahertz applications. Light manipulation with intrinsic graphene plasmons in quantum optical regime promises a revolution in optical computing.

The color centers (NV, SiV, etc.) in single-crystal diamond act as a source of single-photon emission in solid-state system. Diamond quantum emitters are highly stable and show longer life time at room temperature with high brightness. It becomes the frontiers of quantum information processing and next-generation quantum technology for twenty-first century.

References

Aharonovich, I., Greentree, A.D., Prawer, S.: Diamond photonics. Nat. Photonics **5**, 397 (2011a)

Aharonovich, I., Castelletto, S., Simpson, D.A., Su, C.-H., Greentree, A.D., Prawer, S.: Diamond-based single-photon emitters. Report Progress Phys. **74**, 076501 (2011b)

Aharonovich, I., Neu, E.: Diamond nanophotonics. Adv. Opt. Mater., 1–18 (2014)

Avouris, P., Dimitrakopoulos, C.: Graphene: synthesis and applications. Materialstoday **15**(3), 86 (2012)

Avouris, P., Freitag, M., Perebeinos, V.: Carbon-nanotube photonics and optoelectronics. Nat. Photonics **2**, 341 (2008)

Bachilo, S.M., Strano, M.S., Kittrell, C., Hauge, R.H., Smalley, R.E., Weisman, R.B.: Structure-assigned optical spectra of single-walled carbon nanotubes. Science **298**, 2361 (2002)

Balasubramanian, G., Chan, I.Y., Kolesov, R., Al-Hmoud, M., Tisler, J., Shin, C., Kim, C., Wojcik, A., Hemmer, P.R., Krueger, A., Hanke, T., Leitenstorfer, A., Bratschitsch, R., Jelezko, F., Wrachtrup, J.: Nanoscale imaging magnetometry with diamond spins under ambient conditions. Nature **455**(7213), 648–651 (2008)

Barnes, W.L., Dereux, A., Ebbesen, T.W.: Surface plasmon subwavelength optics. Nature **424**, 824 (2003)

Bonaccorso, F., Sun, Z., Hasan, T., Ferrari, A.C.: Graphene photonics and optoelectronics. Nat. Photonics **4**, 611 (2010)

Cançado, L.G., Pimenta, M.A., Neves, R.A., Ribeiro, G.M., Enoki, T., Kobayashi, Y., Takai, K., Fukui, K., Dresselhaus, M.S., Saito, R., Jorio, A.: Anisotropy of the Raman spectra of nanographite ribbons. Phys. Rev. Lett. **93**, 047403 (2004)

Castro Neto, A.H., Guinea, F., Peres, N.M.R., Novoselov, K.S., Geim, A.K.: The electronic properties of graphene. Rev. Modern Phys. **81**, 109 (2009)

Charlier, J.-C., Blasé, X., Roche, S.: Electronic and transport properties of nanotubes. Rev. Modern Phys. **79**, 677 (2007)

Chen, Y.-C., Salter, P.S., Knauer, S., Weng, L., Frangeskou, A.C., Stephen, C.J., Ishmael, S.N., Dolan, P.R., Johnson, S., Green, B.L., Morley, G.W., Newton, M.E., Rarity, J.G., Booth, M.J., Smith, J.M.: Laser writing of coherent colour centres in diamond. Nat. Photonics **11**, 77 (2017)

Cunha, T.H.R., Ek-Weis, J., Lacerda, R.G., Ferlauto, A.S.: Graphene chemical vapor deposition at very low pressure: the impact of substrate surface self-diffusion in domain shape. Appl. Phys. Lett. **105**, 073104 (2014)

Di Liscia, E.J., Álvarez, F., Burgos, E., Halac, E.B., Huck, H., Reinoso, M.: Stress analysis on single-crystal diamonds by Raman spectroscopy 3D mapping. Mater. Sci. Appl. **4**, 191–197 (2013)

Dresselhaus, M.S., Dresselhaus, G., Saito, R., Jorio, A.: Raman spectroscopy of carbon nanotubes. Phys. Rep. **409**, 47 (2005)

Ferrari, A.C., Basko, D.M.: Raman spectroscopy as a versatile tool for studying the properties of graphene. Nat. Nanotechnol. **8**, 235 (2013)

Ferrari, A.C., Robertson, J.: Interpretation of Raman spectra of disordered and amorphous carbon. Phys. Rev. B **61**, 14095 (2000)

Grigorenko, A.N., Polini, M., Novoselov, K.S.: Graphene plasmonics. Nat. Photonics **6**, 749 (2012)

Groot-Berning, K., Raatz, N., Dobrinets, I., Lesik, M., Spinicelli, P., Tallaire, A., Achard, J., Jacques, V., Roch, J.-F., Zaitsev, A.M., Meijer, J., Pezzagna, S.: Passive charge state control of nitrogen-vacancy centres in diamond using phosphorous and boron doping. Physica Status Solidi (a) **211**, 2268 (2014)

Hartmann, N., Piredda, G., Berthelot, J., Colas-des-Francs, G., Bouhelier, A., Hartschuh, A.: Launching propagating surface plasmon Polaritons by a single carbon nanotube dipolar emitter. Nano Lett. **12**, 177 (2012)

Harutyunyan, A.R., Chen, G., Paronyan, T.M., Pigos, E.M., Kuznetsov, O.A., Hewaparakrama, K., Kim, S.M., Zakharov, D., Stach, E.A., Sumanasekera, G.U.: Preferential growth of single-walled carbon nanotubes with metallic conductivity. Science **326**, 116 (2009)

Hazra, K.S., Rai, P., Mohapatra, D.R., Kulshrestha, N., Bajpai, R., Roy, S., Misra, D.S.: Dramatic enhancement of the emission current density from carbon nanotube based nanosize tips with extremely low onset fields. ACS Nano **3**(9), 2617 (2009)

Ju, L., Geng, B., Horng, J., Girit, C., Martin, M., Hao, Z., Bechtel, H.A., Liang, X., Zettl, A., Shen, Y.R., Wang, F.: Graphene plasmonics for tunable terahertz metamaterials. Nat. Nanotechnol. **6**, 630 (2011)

Le-Sage, D., Arai, K., Glenn, D.R., DeVience, S.J., Pham, L.M., Rahn-Lee, L., Lukin, M.D., Yacoby, A., Komeili, A., Walsworth, R.L.: Optical magnetic imaging of living cells. Nature **496**, 486 (2013)

Liu, J., Huang, Z., Lai, F., Lin, L., Xu, Y., Zuo, C., Zheng, W., Qu, Y.: Controllable growth of the graphene from millimeter-sized monolayer to multilayer on Cu by chemical vapor deposition. Nanoscale Res. Lett. **10**, 455 (2015)

Mann, D.: Synthesis of carbon nanotubes. In: O'Connell, M.J. (ed.) Carbon Nanotubes Properties and Applications, chap 2, pp. 19–49. Taylor & Francis Group (2006)

Misra, A., Tyagi, P.K., Rai, P., Misra, D.S.: Reorientation of the crystalline planes in confined single crystal nickel nanorods induced by heavy ion irradiation. Appl. Phys. Lett. **89**, 091907 (2006a)

Misra, A., Tyagi, P.K., Yadav, B.S., Rai, P., Misra, D.S.: Hexagonal diamond synthesis on h-GaN strained films. Appl. Phys. Lett. **89**, 071911 (2006b)

Misra, A., Tyagi, P.K., Rai, P., Mahopatra, D.R., Ghatak, J., Satyam, P.V., Avasthi, D.K., Misra, D.S.: Axial buckling and compressive behavior of nickel-encapsulated multiwalled carbon nanotubes. Phys. Rev. B **76**, 014108 (2007)

Mohapatra, D.R., Jain, L., Rai, P., Hazra, K.S., Samajdar, I., Misra, D.S.: Development of crystallographic texture and in-grain misorientation in CVD-produced single and polycrystalline diamond. Chem. Vap. Deposition **17**, 107 (2011)

Novoselov, K.S., Geim, A.K., Morozov, S.V., Jiang, D., Zhang, Y., Dubonos, S.V., Grigorieva, I.V., Firsov, A.A.: Electric field effect in atomically thin carbon films. Science **306**, 666 (2004)

Novotny, L., Hecht, B.: Principles of Nano-Optics. Cambridge University Press, UK (2006)

Ozbay, E.: Plasmonics: merging photonics and electronics at nanoscale dimensions. Science **311**, 189 (2006)

Pandey, S., Rai, P., Patole, S., Gunes, F., Kwon, G.-D., Yoo, J.-B., Nikolaev, P., Arepalli, S.: Improved electron field emission from morphologically disordered monolayer graphene. Appl. Phys. Lett. **100**, 043104 (2012)

Rai, P., Dubey, S.K.: Raman Spectroscopy: a potential characterization tool for carbon materials. In: Sharma, S.K. (ed.) Handbook of Materials Characterization, Chap 11, pp. 405–433. Springer International Publishing AG, Springer Nature (2018)

Rai, P., Mohapatra, D.R., Hazra, K.S., Misra, D.S., Tiwari, S.P.: Nanotip formation on a carbon nanotube pillar array for field emission application. Appl. Phys. Lett. **93**, 131921 (2008a)

Rai, P., Mohapatra, D.R., Hazra, K.S., Misra, D.S., Ghatak, J., Satyam, P.V.: Appearance of radial breathing modes in Raman spectra of multi-walled carbon nanotubes upon laser illumination. Chem. Phys. Lett. **455**, 83 (2008b)

Rai, P., Hartmann, N., Berthelot, J., Colas-des-Francs, G., Hartschuh, A., Bouhelier, A.: In-plane remote photoluminescence excitation of carbon nanotube by propagating surface plasmon. Opt. Lett. **37**(22), 4711 (2012)

Rai, P., Hartmann, N., Berthelot, J., Arocas, J., Colas des Francs, G., Hartschuh, .A, Bouhelier, A.: Electrical excitation of surface plasmons by an individual carbon nanotube transistor. Phys. Rev. Lett. **111**, 026804 (2013)

Schreck, M., Gsell, S., Brescia, R., Fischer, M.: Ion bombardment induced buried lateral growth: the key mechanism for the synthesis of single crystal diamond wafers. Sci. Rep. **7**(44462), 1 (2017)

Schrinner, P.P.J., Olthaus, J., Reiter, D.E., Schuck, C.: Integration of diamond-based quantum emitters with Nanophotonic circuits. Nano Letter **20**, 8170 (2020)

Schwander, M., Partes, K.: A review of diamond synthesis by CVD processes. Diam. Relat. Mater. **20**, 1287–1301 (2011)

Sussmann, R.S.: CVD Diamond for Electronic Devices and Sensors. Wiley (2009)

Talbot-Ponsonby, D., Newton, M., Baker, J.: Electron paramagnetic resonance imaging of the distribution of the single substitutional nitrogen impurity through polycrystalline diamond samples grown by chemical vapor deposition. J. Appl. Phys. **82**(3), 1201 (1997)

Tallaire, A., Brinza, O., Huillery, P., Delord, T., Pellet-Mary, C., Staacke, R., Abel, B., Pezzagna, S., Meijer, J., Touati, N., Binet, L., Ferrier, A., Goldner, P., Hetet, G., Achard, J.: High NV density in a pink CVD diamond grown with N2O addition. Carbon **170**, 421 (2020)

Tyagi, P.K., Misra, A., Narayanan Unni, K.N., Rai, P., Singh, M.K., Palnitkar, U., Misra, D.S., Le Normand, F., Roy, M., Kulshreshtha, S.K.: Step growth in single crystal diamond grown by microwave plasma chemical vapor deposition. Diam. Relat. Mater. **15**, 304 (2006)

Wang, X., Zhi, L., Müllen, K.: Transparent, conductive graphene electrodes for dye-sensitized solar cells. Nano Lett. **8**, 323–327 (2008)

Wang, J., Sciarrino, F., Laing, A., Thompson, M.G.: Integrated photonic quantum technologies. Nat. Photonics **14**(5), 273–284 (2020)

Wang, J., Paesani, S., Ding, Y., Santagati, R., Skrzypczyk, P., Salavrakos, A., Tura, J., Augusiak, R., Mančinska. L., Bacco, D., Bonneau, D., Silverstone, J.W., Gong, Q., Acín, A., Rottwitt, K., Oxenløwe, L.K., O'Brien, J.L., Laing, A., Thompson, M.G.: Multidimensional quantum entanglement with large-scale integrated optics. Science **360**(6386), 285−291 (2018)

Wei, B.Q., Vajtai, R., Jung, Y., Ward, J., Zhang, R., Ramanath, G., Ajayan, P.M.: Organized assembly of carbon nanotubes. Nature **416**, 495 (2002)

Zhu, B.Y., Murali, S., Cai, W., Li, X., Suk, J.W., Potts, J.R., Ruoff, R.S.: Graphene and graphene oxide: synthesis, properties, and applications. Adv. Mater. **22**, 3906 (2010)

Zia, R., Schuller, J.A., Chandran, A., Brongersma, M.L.: Plasmonics: the next chip-scale technology. Materialstoday **9**(7–8), 20 (2006)

Chapter 12
Developments of Random Laser: Fundamentals and Applications

Arindam Dey, Ashim Pramanik, Subrata Biswas, Partha Kumbhakar, and Pathik Kumbhakar

12.1 Introduction

Thanks to the recent technological advances in the field of photonic devices and photonics based systems in which laser source is one integral and main workhorse. Since the emergence of the working laser from the realm of the Science fiction movies in the late 1950's, where it has been shown as a weapon, the laser has found its way into a wide range of the modern applications, including in sophisticated eye and heart surgery to precise cutting of metal sheets in automobile industry (Noginov 2005; Wilson and Hawkes 1987; Townes 2002; Schawlow 1969; Rodberg and Thaler 1985; Einstein 1905; Boltzmann 1884; Anderson 1958; Feynman et al. 1989; Wiersma et al. 1995; Letokhov 1968; Markushev et al. 1986; Cao et al. 1999; Wu et al. 2004; Livdan and Lisyansky 1996; Chen et al. 2013; Cao 2003). As it is well known that laser light exhibit some interesting properties, including high monochromaticity, high directionality, coherence and high intensity. These properties of laser light are intricately related to its construction geometry, properties of the active medium and employment of the process of stimulated emission for achievement of the emission of amplified light (Townes 2002; Schawlow 1969; Rodberg and Thaler 1985; Einstein 1905). Therefore, in laser systems, randomness is strictly undesirable as it emits light through the process of optical amplification based on the stimulated emission of electromagnetic radiation (Wilson and Hawkes 1987; Townes 2002). The term stimulated emission was first coined by the Einstein (Townes 2002) and according to him if an electron is in a higher energy level and a photon comes along with an energy equal to the difference between the electron's energy and a lower energy the photon will stimulate the electron to fall into the lower energy state, thereby emitting

A. Dey · A. Pramanik · S. Biswas · P. Kumbhakar · P. Kumbhakar (✉)
Nanoscience Laboratory, Department of Physics, National Institute of Technology Durgapur, Durgapur 713209, India
e-mail: pkumbhakar.phy@nitdgp.ac.in

R. S. Ningthoujam and A. K. Tyagi (eds.), *Handbook of Materials Science, Volume 1*, Indian Institute of Metals Series, https://doi.org/10.1007/978-981-99-7145-9_12

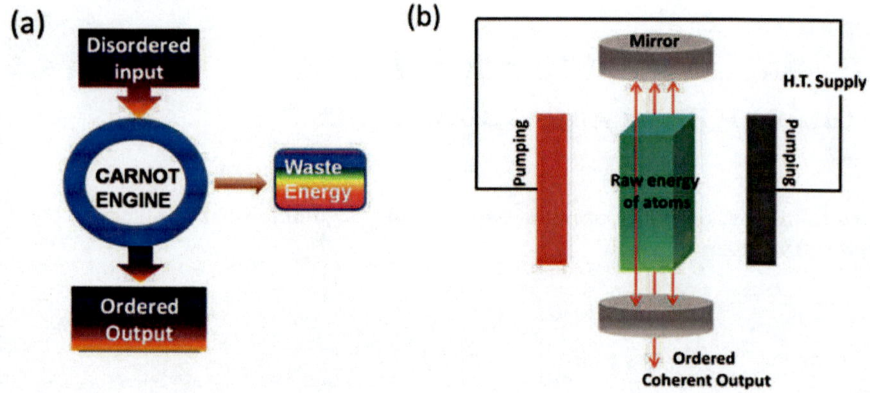

Fig. 12.1 **a** A Schematic of Carnot engine where a raw input energy is converted to more ordered output. **b** A schematic of the conventional laser where raw energy of the atom is converted into intense coherent output

a new photon. The most interesting fact is that the emitted photon will have the same frequency and phase as those of the original photon.

One can compare laser with a Carnot engine (Schawlow 1969) where a raw input energy can be converted to more ordered output energy (shown in the Fig. 12.1a). Henceforth all such laser sources which contain formal optical cavity (i.e., a pair of mirrors) is to be called as conventional laser.

In a conventional laser source, a complex system is needed to produce highly monochromatic and coherent light. However, in random laser (RL), this complexity can be avoided as it does not need any sophisticated amplification cavity and amplification of the input can be obtained from the scattering of random system rather than reflection from mirrors (Noginov 2005). So, there exist some distinctions between conventional laser and RL in terms of their operating principles and their cavity structure. In a conventional laser amplifying of light takes place due to back and forth motion of propagating radiation in the gain medium due to the reflections in the front and back mirrors of optical cavity. And thus, a highly collimated and coherent beam of light is produced. In contrast, in RL, no mirror is required and it operates by multiple scattering of light within the disordered medium. In RLs, the active medium is typically consisting of powdered particles, solution of film containing light-emitting particles or molecules and when they are excited by an external energy source, such as a pump laser or an electrical current, the light emitted by the particles undergoes multiple scattering events within the disordered medium. This randomness in the scattering events leads to a broad spectral emission with large fluctuations in its intensity distribution. However, another difference between conventional lasers and RLs is the size and cost of their respective cavity structures. A conventional laser requires highly precise and expensive mirrors to create a stable cavity structure, whereas a RL can be made by using low-cost and easily available gain materials and nanostructured materials as scatterer.

It may be noted here that conventional laser sources are demonstrated to be very much useful for various technological applications, but still there exist a lot of scope for innovation. Consequently, a lot of stimulus has been given in the recent past by various researchers for the improvement in properties and performances of RLs by employing various luminescent materials as gain media and by using various types of nanostructured materials for achieving efficient scattering (Cao 2003) but still a lot of improvements are needed in understanding the principle of operations of RLs, in the design and developments of innovative and flexible RL systems with low lasing threshold tunable light emission. However, to fill the gap to some extent, in this chapter for seeking the attention of the readers on some recent advances in the fields of development of RLs for various applications is represented. In this chapter, a brief theoretical background of RLs is presented as a mini review along with some recent relevant works, particularly focused on improvement of lasing threshold of RL by structural engineering of gain/scatterer media, by employment of innovative external feedback system, and by employment of nature extricated gain/scatterer molecules/systems etc. It is believed that the present chapter will act as a catalysts/primer for new researchers and may open new possibilities for sustainable and bio-inspired development of RLs for their novel applications in future RL based imaging, sensing, and other photonic devices.

12.1.1 A Brief Theoretical Background

The basic component of a laser contains a gain medium, a mechanism to energize it, and some mechanism to provide optical feedback as shown in the Fig. 12.1b. In most of the cases a pair of highly polished parallel mirrors (Schawlow 1969) provides the optical feedback. Light enters into the gain medium from a highly luminous (pumping) source and then travels through the gain medium and then bounces back and forth by the mirrors and simultaneously gets amplified by the stimulated emission from the gain medium. However, there are several conditions that must be fulfilled if the laser is to operate successfully. These involve the gain and loss of the cavity and the frequency. The cavity condition is so chosen that the total gain in the cavity is larger than the losses, the system reaches a threshold and laser oscillation takes place. The emission characteristics of lasing modes i.e., the directionality and its frequency is determined by the cavity. In conventional laser any randomness is undesirable because it drives the system away from the lasing condition (Noginov 2005; Wilson and Hawkes 1987; Townes 2002; Schawlow 1969).

In secondary classes in school, it is often learned why the sky appears blue? It is due to the scattering (Rodberg and Thaler 1985) of the sunlight by the molecules. Thus, scattering plays a very important role in daily lives. Almost all objects which appear opaque scatter light that means they reflect the light that illuminates them in all directions. Scattering of classical wave in random medium has been studied for centuries. The Brownian motion (Einstein 1905) which is a microscopic manifestation of the diffusion process (wave propagation) in the macroscopic level was first

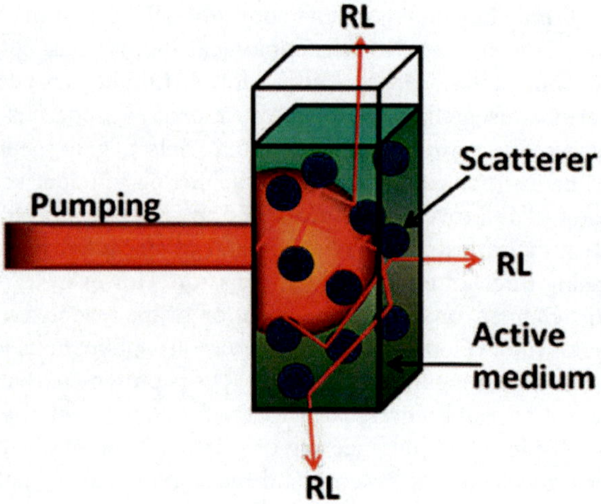

Fig. 12.2 A Schematic of random lasing action with active medium and scatterer

observed by Robert Brown in year 1827 while looking through a microscope at particles trapped in cavities inside pollen grains in water. In the nineteenth century and the beginning of the twentieth century, Brownian motion was studied by Langevin and Einstein (Einstein 1905). The diffusion equation has been obtained and with the diffusion coefficient, the random walk of classical particles could be quantitatively studied. The classical non-equilibrium transport problem has been studied by Boltzmann who developed the famous Boltzmann equation (Boltzmann 1884). In the Boltzmann equation, the particle is a classical one, so when quantum effects are important, one needs to reconsider the Boltzmann formulation. The reformulation of the Boltzmann theory has been done by Anderson in 1958 (Anderson 1958), when he first introduced the localization referring to a dramatic change in the propagation of an electron when it is subjected to a spatially random potential.

The transmission of light in a strongly scattering domain can be analogous to photons travelling in a random medium. The energy density follows a diffusion equation over length scales longer than the scattering mean free path (l_{sc}) associated with this random walk, comparable to the Fick law (Feynman et al. 1989) for particle transport. When compared to ballistic propagation in free space, the dwell period associated with the transmission of light intensity within the medium is significantly enhanced. Multiple scattering of light converts the disordered material into a virtual cavity. This can be used to observe a lasing threshold in the presence of gain and stimulated amplification of spontaneous emission (SE), like in the conventional laser. Such a laser without mirrors is called a "random laser" (Wiersma et al. 1995).

A schematic of RL is illustrated in the Fig. 12.2 with an active medium and scatterer. Here, the beam-like emission and the high degree of spatial coherence characteristic of standard lasers is lost as the cavity is not well defined, and many modes with uncorrelated phases can lase simultaneously. However, the complexity of the virtual disordered cavity turns out to be an advantage as it offers new degrees of freedom to control freely laser emission.

Letokhov (1968) investigated multiple scattering of light in an amplified scattering medium in 1968, using a model established for diffusive light transport. He contended that for any diffusion process with amplification, when the total gain is proportionate to the volume and the losses are proportional to the whole surface, the process is complete. It is therefore clear that there is a critical volume (Letokhov 1968) at which gain exceeds loss and the intensity diverges. If the gain varies with wavelength, this model predicts that the emission spectrum narrows over a specific threshold, with a maximum intensity at the maximum gain wavelength. Moreover, relaxation oscillations and lasing spikes may be detected in a diffusive model. Several features of random laser as predicted by Letokhov (Noginov 2005) are observed experimentally if interference effect between the scattered waves are not considered. However, by considering the interference effect between scattered light, RLs can be divided into two kinds, (i) RL with incoherent feedback i.e., non-resonant feedback and RL with coherent feedback i.e., resonant feedback (Markushev et al. 1986). In case of weakly scattering regime, a photon entering into the disordered medium undergoes a random walk. Moreover, if gain is introduced in the system, then the photon can initiate stimulated emission in the system before leaving the system. In such a system it is difficult to get the required feedback for the lasing oscillation. In case of the strong scattering regime after multiple scattering photons may reach to the same point from where it has started its journey. Thus, it forms a closed loop and in the presence of the gain the amplification in the closed loop path overcomes the loss and lasing oscillation starts. The back scattered photons interfere and determine the lasing frequencies. This fulfils the requirement of the field feedback for laser oscillation to occur and produced random lasing which is termed as coherent random lasing (Cao et al. 1999). Further, there are a few length scales which describe the scattering of photon in disordered dielectrics viz. scattering mean free path (l_s) which is the distance travelled by the photon before being scattered again. The transport mean free path (l_t) (Wu et al. 2004) is defined as the average distance a wave travels before its direction of propagation is randomized. Another two important length scales are gain length (l_g) (Wu et al. 2004) and amplification length (l_{amp}) (Wu et al. 2004). The l_g is described as the length over which the intensity is amplified by a factor of e and l_{amp} is the distance between the staring and end point of the l_g. If L is the smallest dimension of a 3D random medium there are three regimes for light transport (i) ballistic regime, $L \sim l_t$; (ii) diffusive regime, $L >> l_t >> \lambda$; (iii) and Anderson localization occurs at, $kl_t \sim 1$ (Wu et al. 2004; Livdan and Lisyansky 1996), where λ is the incident wavelength of light and k $(=\frac{2\pi}{\lambda})$ is the corresponding wave vector. The l_t in diffusive regime can be obtained by the following relation (Chen et al. 2013),

$$l_t = 1/n_p\sigma_{sca} = L/\ln(I_0/I), \tag{12.1}$$

where, n_p is the number of particles per unit volume, σ_{sca} is the scattering cross section, I_0 is the incident laser pump intensity, I is the transmitted pump intensity and L is the thickness of the sample. When the wavelength of the incident laser radiation is greater than the particle size (r) one approximation of the scattering cross section (σ_{sca}) using the Rayleigh scattering formula can be done, which is given by Wu et al. (2004),

$$\sigma_{\text{sca}} = \frac{8\pi r^2}{3} s^4 \left(\frac{n^2 - 1}{n^2 + 2}\right)^2, \tag{12.2}$$

where, s is the structure factor ($2\pi r/\lambda$) and n is the refractive index of the medium. From Eq. (12.2), l_s which is almost equal to l_t can be approximated by the relation (Wu et al. 2004),

$$l_s \cong l_t = \frac{\lambda}{4 f \pi^4} \left(\frac{\lambda}{r}\right)^4 \left(\frac{n^2 + 2}{n^2 - 1}\right)^2, \tag{12.3}$$

where, $f = \frac{4\pi\rho r^3}{3}$ is called the filling factor and ρ is the density of the material. As discussed earlier in case of the incoherent RL the only energy or intensity feedback is present i.e., no field or phase feedback hold. In such a system the photon transport equation can be written as (Cao 2003),

$$\frac{\partial \varphi(r, t)}{\partial t} = D\nabla^2 \varphi(r, t) + \frac{v}{l_g} \varphi(r, t). \tag{12.4}$$

where, $\varphi(r, t)$ is the photon energy density; $D = vl_t/3$, is the diffusion co-efficient, v is the photon velocity. The solution of the Eq. (12.4) can be approximated as,

$$\varphi(r, t) = \sum_m a_m \varnothing_m(r) e^{-\left(Db_m^2 - \frac{v}{l_g}\right)t}, \tag{12.5}$$

Here, $\varnothing_m(r)$ and b_n are the Eigen value and Eigen function, respectively and a_m is a coefficient and satisfy the relation,

$$\nabla^2 \varnothing_m(r) + b_m^2 \varnothing_m(r) = 0. \tag{12.6}$$

Now at above threshold, Eq. (12.6) can be written as,

$$Db_1^2 - \frac{v}{l_g} = 0. \tag{12.7}$$

where, b_1 is the lowest eigenvalue and can be approximated to ~$1/L$. So, with this approximation and using the relation $D = vl_t/3$ one can rewrite the Eq. (12.7) as (Cao 2003).

$$L_{crit}^3 = V_{\text{crit}} = \left(\frac{l_t l_g}{3}\right)^{\frac{3}{2}}. \tag{12.8}$$

Here, L_{crit} and V_{crit} are the critical length and volume, respectively. When the gain volume of the system exceeds the critical volume as given by Eq. (12.8), the l_s becomes greater than the l_g and gain overcomes the loss and system begins to lase above threshold.

12.2 Some Emerging Areas of Random Lasers

12.2.1 Role of Structural Engineering and External Feedback in Controlling Performance of Random Lasers

The walk of photons in the disordered environment is an interesting topic of recent photonic research. Photon transport processes for RL emission exclusively depend on the types of scattering elements. In earlier reports, opaque or translucent materials have been used for RL generation (Zhang et al. 2013). For a strongly scattering system, photons return to the same coherence volume via multiple light scattering. Therefore, even with weak pumping, sufficient optical feedback for RL emission is achievable (Wu et al. 2006; Andreasen and Cao 2011). However, attaining lasing conditions in weakly scattering systems (Zhang et al. 2013) is tedious, and sometime requires a repetitive feedback and elongated pump strip. Also, spectral and spatial overlap in between the lasing modes is another hurdle in weakly scattering regime. As a solution for demonstrating RL emission in a weakly scattering regime, inhomogeneous gain medium is used by previous researchers for providing optical feedback (Andreasen and Cao 2011; Frolov et al. 1999; Caixeiro et al. 2016). Also, RL threshold can be reduced by using highly reflective scatterers (Kim et al. 1999; Yashchuk et al. 2008). Although employment of reflective scatterers may cause extra loss of pump energy, hence it may lead to the pump volume quenching. Contextually, intentional doping of microbubbles or sub-millimetre air voids inside a disorder active medium helps in lengthening of the effective path length of emitted photons to promote the photon multiplication process (Yashchuk et al. 2008). Likewise, RL emission from polymer films embedded with microbubbles and CdSe/ZnS core–shell quantum dots (QDs) as gain medium has been reported, as shown in Fig. 12.3a–c (Hu et al. 2020) where microbubbles serve as the centres for total internal reflection to amplify the photons. Not only in solid state polymer matrix, but also the microbubbles can help in light localization in the liquid disorder system. For example, application of nano-micro bubbles as SPASER (surface Plasmon amplification by stimulated emission of radiation) element for generating super bright stimulated emission is already known (Noginov et al. 2009). Although, the generation of microbubbles within a liquid medium is experimentally tricky and sometime it requires an appropriate type

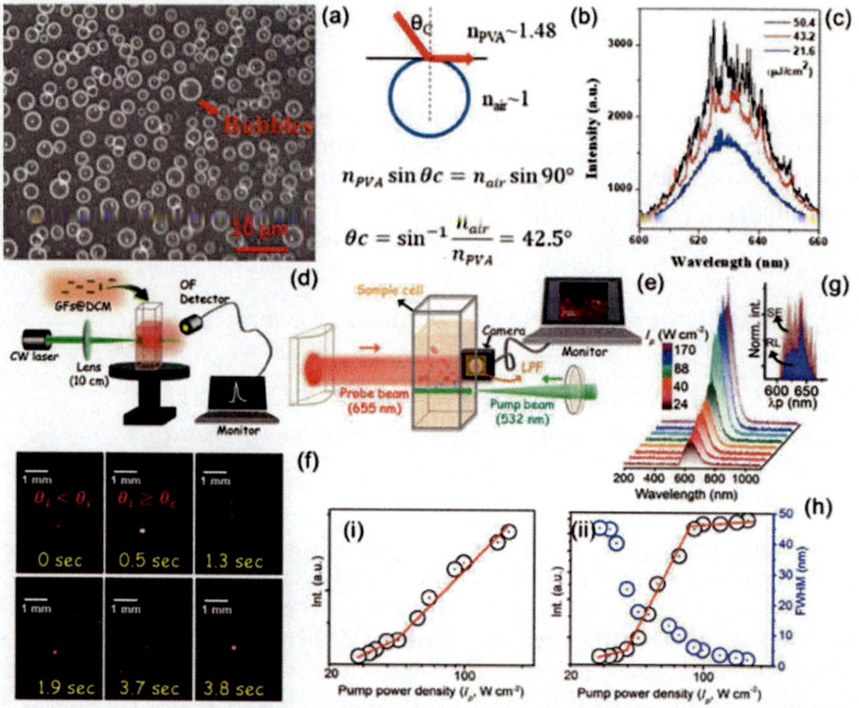

Fig. 12.3 **a** FESEM image of the CdSe/ZnS core–shell QDs embedded polyvinyl alcohol (PVA) film filled with air bubbles. **b** Mechanism of total internal reflection due to the presence of air bubble. **c** Emission spectra of the air bubble incorporated RL device at different pump intensities. (Hu et al. 2020, Copyright ACS, reproduced with permission) The schematic of the experimental setup for **d** RL generation, and **e** pump-probe photography. **f** Identification of microbubble inside liquid disorder active medium. (i) When $\theta_i < \theta_c$, the microbubble appears to be faint due to the refraction of incident light), and (ii) when $\theta_i \geq \theta_c$, the microbubble appears as bright spot due to the total internal reflection. **g** Emission spectra of DCM in the presence of graphene flakes as passive scatterer (number density $N_{SC} = 7 \times 10^{16}$ nos./ml), when it is pumped using a CW laser of 532 nm wavelength with various pump intensities. **h** Variation of emission intensity and FWHM with input pump intensity for (i) bare DCM and (ii) DCM in the presence of graphene flakes. (Pramanik et al. 2021a, Copyright Willey, reproduced with permission)

of catalyst material having a refractive index similar to that of the gain medium (Okamoto and Yoshitome 2017). However, quite interestingly to generate RL emission in a weakly scattering regime, recently utilization of two-dimensional (2D) graphene flakes (GFs) as the passive scatterer and thermo-catalyst for microbubble generation inside liquid suspension of dye molecules is done (Fig. 12.3d–h). As depicted in Fig. 12.3f, the existence of microbubbles as well as their role in RL emission generation is confirmed through pump-probe photography experiment. The phenomenon of total internal reflection is clearly observed, which helps in localization and amplification of photons inside the lasing medium. Thus, RL emission at

638.4 nm at a low lasing threshold, linewidth, and high quality factor (Q-factor) of ~84 W cm^{-2}, 2 nm, and 1127, respectively is demonstrated. The performance of the GFs for RL generation in weakly scattering regime has been reported to be superior over those of spherical shaped TiO_2 and Ag nanoparticles (Pramanik et al. 2021a). The existence of localized bubbles under intense laser excitation has also been proposed by other researchers to achieve the necessary effective feedback during the RL emission from Graphdiyne (Jiang et al. 2020).

Also, structural engineering to provide external feedback is common for achieving RL emission under optical pumping by continuous wave (CW) laser. A key issue for CW pumped lasing is to avoid the accumulation of long-lived excited states in the gain volume, causing induced absorption (Bornemann et al. 2006). Bravo et al. have reported CW pumped upconverting single mode lasing at room temperature from Yb^{3+}/Er^{3+}-co-doped upconverting nanoparticles conformally coated on Ag nanopillar arrays (Fernandez-Bravo et al. 2019). Zhang et al., have used $(PEA)_2PbI_4$ thin flakes as the gain medium to achieve a CW vertical-cavity surface-emitting single mode laser operation with an ultra-low threshold of 5.7 W cm^{-2} (Zhang et al. 2021).

12.2.2 Engineering Random Laser Action with Structural Variation in Gain/Scatterer Media

Already, a number of research studies have been carried out over the past decades to overcome the difficulties related to low photon confinement in the diffusive regime under pulse optical pumping. In this regard, alternations of photon transposition in a disordered medium by means of controlled structural deformation/reorganization in the geometry of RL devices are reported extensively (Dey et al. 2022; Lee et al. 2019; Shen et al. 2014; Tong et al. 2019; Sznitko et al. 2014; Sun et al. 2015). Experimental results shown in Fig. 12.4a–c demonstrate that the cavity path length of the emitted photons can be controlled strongly by altering the curvature of the flexible polymer film containing the scatterer and gain molecule (Lee et al. 2019). Therefore, reduction in RL threshold as well as tuning in spectral position is possible to achieve. Furthermore, it has been observed that engineering over the device architecture is an effective method for producing RL emission with controllable properties. Chip-scale RL action by employing the long-range random nanoscale wrinkles polydimethylsiloxane (PDMS) film has been reported (Shen et al. 2014). Prior to the discussion on RL emission properties, authors have given a detail about the fabrication of the wrinkled RL device, as shown in Fig. 12.4d–g. The optical amplification is achieved via multiple scattering of photons at each of the PDMS wrinkle-dye interfaces. The unique sinusoidal Bragg-grating-like random structure helps in lowering the RL threshold. Additionally, randomness in the wrinkle structure has further offered a tunable RL emission characteristic. Meanwhile, tuning of RL emission wavelength from 561 to 578 nm by change in the excitation area along the thickness gradient of wedge-shaped dye doped polymer film resonator is demonstrated by Tong et al. (2019), as shown in Fig. 12.4h–j. Notably, as the thickness of the wedge-shaped dye doped polymer film is increased, the number of lasing modes is reported to be increased. Therefore, not only wavelength tuning but also mode selective RL emission can be designed by varying the thickness of the disorder active medium. As thickness of the active medium is increased, the effective number density of the scatterer under a fixed area of optical pumping decreased, and hence the gain photons get enough room for forming coherent loops via multiple scattering. Therefore, more number of coherent RL modes may appear. Similar type of mode tunable RL emission generation by altering the effective number density of scatterers under a fixed pump area by stretching flexible active medium has also been reported earlier (Sun et al. 2015), as depicted in Fig. 12.4k–l.

12.2.3 Employment of External Feedback (ExFB) Approach for Controlling Threshold Pump Intensity of RL Emission

It has been shown that, when $L >> l_t$ i.e., at a diffusive regime, pump volume amplification inside the random active medium could be an effective way to achieve tunable RL emission and reduced pump threshold (Ling et al. 2001; Soest et al. 1999). In a recent work, indigenously designed external feedback (ExFB) based RL system as shown in Fig. 12.5a, consisting of a quartz substrate as a front-facing mirror (Pramanik et al. 2021b) is demonstrated. The substrate has been placed just after the DCM@ZnO-MCs doped polymer film which acted as the active medium. A piece of paper having typical thickness of 120–130 μm has been used as the spacer in between the smooth surface of the Si substrate (acted as a mirror) and the polymer film. A CW diode laser of 532 nm is used as a pumping source in the RL experiment. To capture the emission spectra, a fiber optic detector (Avaspec) is placed at right angles to the direction of incidence of the pump beam. In the previously reported work (Pramanik et al. 2021b), the technique of pump volume amplification to reduce the RL threshold is achieved by the following mechanisms.

(i) *Changing the excitation spot area (A)*

In general, with the increase in the value of the excitation spot area A the enhanced amplification of the pump volume takes place and therefore RL emission at a low pump threshold can easily be achieved. The capability of reduction of lasing threshold through pump volume amplification in DCM@ZnO-MCs in a polyvinyl alcohol (PVA) host is checked and the reduction in threshold intensity (I_{Th}) has been observed with an increase in A (Pramanik et al. 2021b). Theoretically; this dependence can be realized according to the random resonator statistical framework proposed by Apalkov and Raikh (Wang et al. 2019; Dhanker et al. 2014) and I_{Th} can be written as,

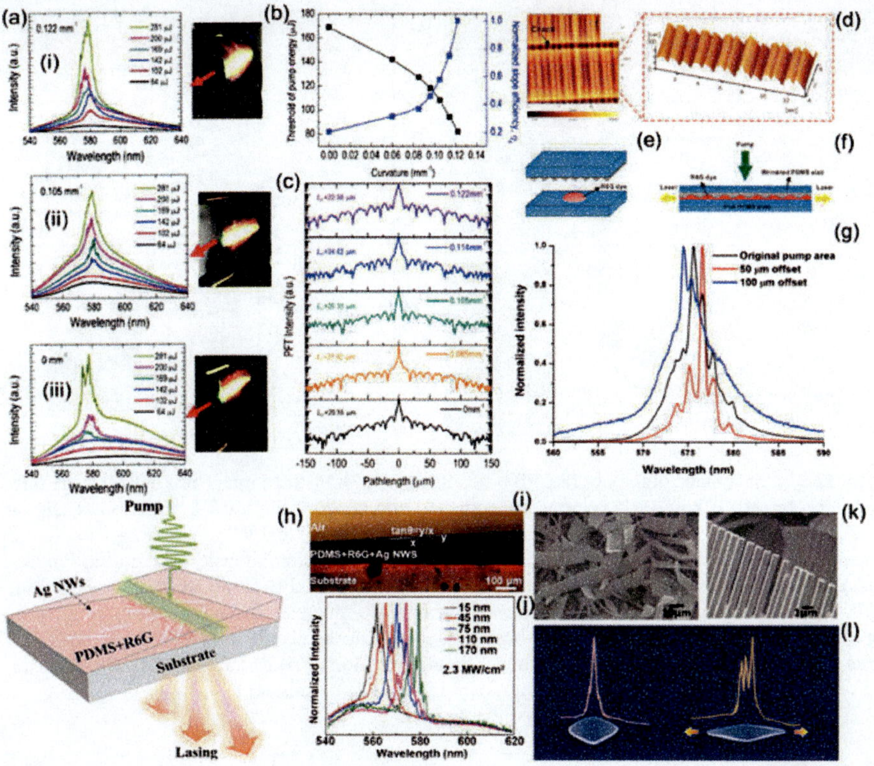

Fig. 12.4 **a** Emission spectra for the ZnO based flexible RL with the increase in the pump energy at a bending curvature (k) = (**i**) 0.122, (**ii**) 0.105, and (**iii**) 0 mm^{-1}; Corresponding digital photographs of the flexible RL at different k values. **b** Characteristic behaviours of the RL threshold and the slope efficiency as a function of k for the ZnO based flexible RL device. **c** The corresponding power Fourier transform (PFT) spectra to derive the cavity path length $L_c = \pi \text{p}_\text{m}/\text{nm}$; (Lee et al. (2019) Copyright RSC, reproduced with permission) Where, p_m is the peak of the Fourier plot, n is the refractive index of the gain medium, and m is the order of the Fourier harmonic. **d** AFM image of PDMS surface wrinkle patterns. **e** Preparation of the sandwich RL device using the dye solution in ethylene glycol is drop casted on to the surface of a flat PDMS and then confined with a wrinkled PDMS. **f** The front-view diagram of the sandwiched RL device. **g** Emission spectra generated from the wrinkled PDMS based RL device, when pumped at different positions. (Shen et al. 2014, Copyright AIP, reproduced with permission). **h** Schematic and **i** optical microscopic image of wedge shaped Plasmonic RL device. **j** Emission spectra generated from the wedge shape Plasmonic RL device, when pumped at different positions. (Tong et al. 2019, Copyright Elsevier reproduced with permission) **k** SEM images of ZnO nanobrushes in low (left) and high (right) magnifications. (**il**) Tuning of RL mode in the stretchable elastomer substrate consisting of ZnO nanobrushes (Sun et al. 2015, Copyright ACS, reproduced with permission)

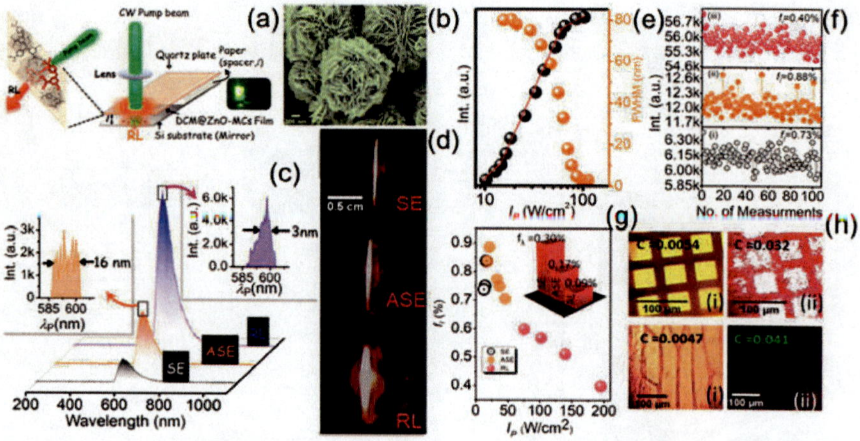

Fig. 12.5 **a** Schematic of the external FB based RL system. **b** FESEM image of ZnO-MCs. **c** Emission spectra of FB based RL systems are collected during SE, ASE and RL emission. **d** Digital images of the active medium are captured taken during SE, ASE and RL emission. **e** Variation of output intensity and FWHM of the RL system with I_p. **f** Emission intensity of the lasing modes as a function of number of measurements carried out for the ExFB based RL system, during (i) SE, (ii) ASE, and (iii) RL emission. The value of intensity fluctuation coefficient ($f_I = \sigma_I/I_{\text{mean}}$ **g** Variation of f_λ with I_P for the ExFB based RL system. Microscopic images of the **h** TEM grid and **e** Bio cell membrane (Allium cell) under the illumination of (i) RL, and (ii) CW He–Ne laser. (Pramanik et al. 2021b, copyright Elsevier, reproduced with permission)

$$I_{\text{Th}} \propto \exp\left[-\left(\frac{\ln\left(A/A_0\right)}{G}\right)^{1/\lambda}\right] \tag{12.9}$$

where, A_0 is the two-dimensional area occupied by typical quasimode, and G is a parameter reflecting the disorder strength of the random medium. The value of λ is ≥ 1, and it is dependent on the degree of disorder. In the limit of $\lambda \to 1$, Eq. (12.9) can be reduced to a simple power-law (Wang et al. 2019; Dhanker et al. 2014).

$$I_{\text{Th}} \propto \left(A/A_0\right)^{-1/G} \tag{12.10}$$

In this case, the experimental data has been fitted with Eq. (12.10), and the disorder strength of the active medium (G) is extracted to be 0.76. In the experiment ~5 times decrease in the CW RL threshold is also observed with ~3 times enhancement in A. The value of $1/G$ reported in Wang et al. (2019) is 0.7 and similar kind of observation is also reported by Dhanker et al., (Dhanker et al. 2014) for achieving random lasing in organo-lead halide perovskite microcrystal networks at a low RL threshold.

(ii) *Pump light reinjection in presence of external feedback*

The characteristics of the light emitted from the designed RL systems which are collected at different input pump intensities (I_p) can be seen from Fig. 12.5c. The gain saturation in RL emission is observed due to mode competition and the sharing of pumping energy among the adjacent modes. Moreover, the formation of multiple fluctuating modes during amplified spontaneous emission (ASE), and a decrement in the number of lasing modes thereafter during the RL emission due to mode competition, can be seen from Fig. 12.5c (left and right inset). In this context, the narrowing in the linewidth (FWHM) is also observed during the transition from ASE to RL as shown in Fig. 12.5e. Also, a change in the nature of the gain volume has been observed during the transition from spontaneous emission (SE) to ASE and then to the RL mode of emission, as shown in Fig. 12.5d. Moreover, Fig. 12.5e shows the variations of the value of integrated emission intensity and FWHM of the collected spectra as a function of I_P. The "*S*" like behavior in the log–log plot is also a signature of clear transformation from SE to ASE and finally to RL action followed by gain saturation (Pramanik et al. 2021b). Furthermore, the loss of the pump photons as well as their reinjection has been optimized by changing the effective cavity length (L_{eff}). Hence, with a L_{eff} of 220 μm, a single-mode RL emission at 598 nm is obtained at a pump threshold intensity of 58.8 W/cm^2, Q-factor of 1090 and efficiency (η) of 47%. Also, a significant reduction of ~60% in I_{th} in the FB based RL system compared to that of the bare polymer films (PF) have been observed.

Previously, the transition from SE to RL (Pramanik et al. 2021b) has been identified by comparing the relative statistical fluctuation in the intensity of the emitted radiation at different I_P values. Herein, similar studies have been also carried out and the results are showcased in Fig. 12.5f. The statistical measurement of emission intensity (in every 1 s within a time window of 2 min) of the FB based RL has been carried out at different I_{Ps} as shown in Fig. 12.5f, and in all cases, the mean intensity is found to be increased, as is expected with increasing gain in the system (Pramanik et al. 2021b). However, generation of multiple fluctuating modes (Pramanik et al. 2021b), leads to a highest value of f_I just above the ASE threshold. Whereas, as shown in Fig. 12.5g, a decrease in the value of f_I on further increase in I_P suggests the mode competition via sharing of pumping energy. As a consequence, saturation of one particular lasing mode intensity during the RL emission is observed. Further, the RL system has been judiciously used for collecting images of *Allium cepa* cells (Pramanik et al. 2021b) with ~ 88% reduction in the value of speckle contrast (*C*) compared to that of a conventional CW laser and with other ordinary lights. Also, as shown in Fig. 12.5h, the value of *C* for the ExFB assisted RL illumination is demonstrated to be ~3 times lower compared to that of the RL light obtained from bare dye doped polymer film, indicating the low spatial coherency of the former light source (Fig. 12.6).

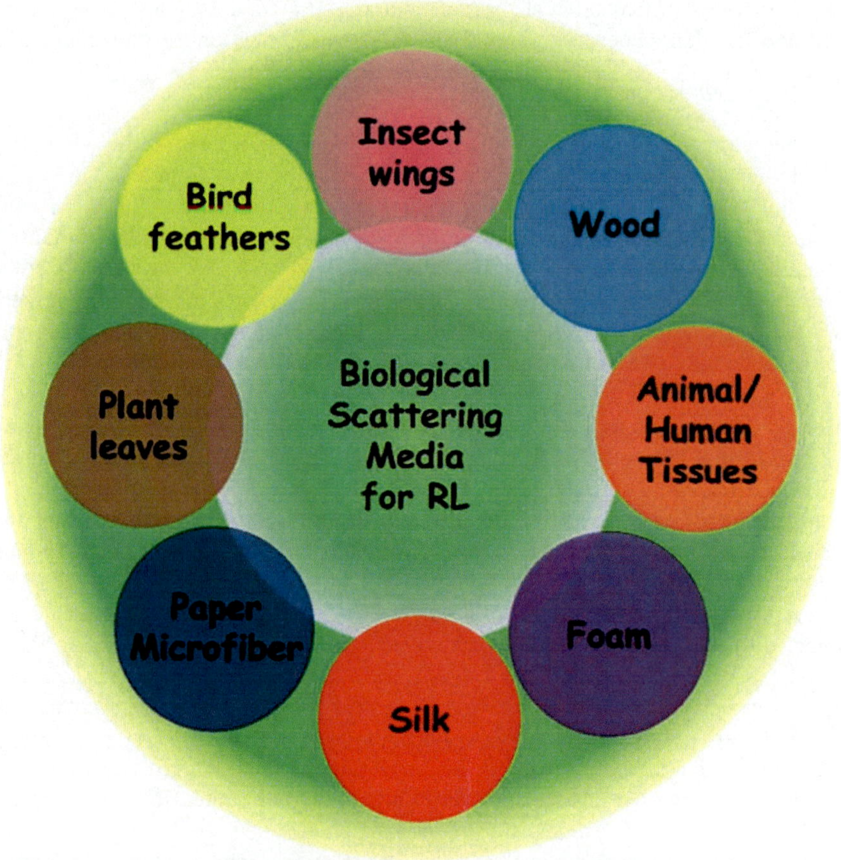

Fig. 12.6 An overview of natural and bio-inspired frameworks for generating RL emission

12.2.4 Some Nature Inspired Scattering Structures for Random Laser Generation

Consumption of naturally procured and bio-degradable constituents is increasing in a day-to-day manner. The traditional malignant commodities used in laboratory-based research activities have some limitations due to their toxicity, non-biodegradability, impact on environmental parameters and as a result employment of these materials have massive drawback for in-vivo research applications. Therefore, the applications of natural derivatives for general research methods are the one-step solution with respect to those harmful vendible objects. Recently, a lot of research studies are carried out with environment-friendly substances, for their successful employment in potentially beneficial domains of research (Moura et al. 2017; Tadepalli et al. 2017; Syurik et al. 2017). Biological tissues are a major type of scattering materials, and they have been demonstrated to be suitable for generating RL radiation. Significantly, in ex-vivo dye infiltrated human tissue, emission spectrum alterations in malignant tissue compared to healthy tissue have been discovered, presenting the intriguing prospect to utilize random laser physics to create novel sensors for detecting biological tissue disorders. Ex-vivo dye penetrated human tissue from the same person was utilized to map the malignant and healthy sections. The finding is related to the extremely active field of diffuse light biomedical imaging and takes use of the fact that various tissues and bone types have various optical structures and mean free paths. Thus, use of the gain medium which is bio-compatible and non-hazardous can make a huge difference in in-vivo bio-imaging of the human tissue. Hence, biocompatible and non-hazardous gain/scattering media for lasing generation, in particular for RL, is becoming a spotlight area in day-by-day manner (Pramanik et al. 2022).

12.2.5 *Nature Extricated Gain/Scatterer Frameworks for RL Applications*

Usually, nature extracted or biologically inspired gain/scatterer media in RL have huge advantages over commercially available objects, as they are less harmful, easily procurable, biodegradable. In this section, discussions about some nature inspired, biodegradable gain/scatterer configurations used for constructing various RL devices and their fruitful applications in daily life is carried out.

Recently, spatial coherency of RL emission has been calculated in dye doped transparent wood (TW) (Koivurova et al. 2018). TWs having high haze value (~95%) and high transmittance (~90%) has complex hierarchical structure, consisting of several components of hemicellulose, cellulose, lignin etc. Figure 12.7a shows the construction of TW based RL arrangement with corresponding pumping scheme as shown in Fig. 12.7b. The typical quasi-RL emission spectrum with incoherent feedback is shown in Fig. 12.7c. The spatial coherency is measured by means of double grating interferometer in the reported R6G dye doped TW based RL device, where root mean square value of degree of spatial coherence ($\overline{\Upsilon}$) is defined as,

$$\overline{\Upsilon}^2 = \frac{\iint I_0(x-\Delta x)I_0(x+\Delta x)|\gamma_0(x-\Delta x, x+\Delta x)|^2 dx d\Delta x}{\iint I_0(x-\Delta x)I_0(x+\Delta x) dx d\Delta x} \tag{12.11}$$

The measured $\overline{\Upsilon}$ is shown in Fig. 12.7d. The overall degree of spatial coherence is estimated to be as low as $\overline{\Upsilon} = 0.16 \pm 0.01$, which is an indication of possible usage of the apparatus in speckle free imaging devices.

In 2012, Zhang et al. have reported coherent RL emission from exotic nanopillar structures in *pomponia imperatoria* cicada wing (Zhang et al. 2012). Figure 12.8a shows the digital image of a *pomponia imperatoria* cicada and in Fig. 12.8b–c the SEM images of cicada wings consisting of nanopillars, at different magnifications are shown. The dye doped polymer film consisting of cicada wings can constitute

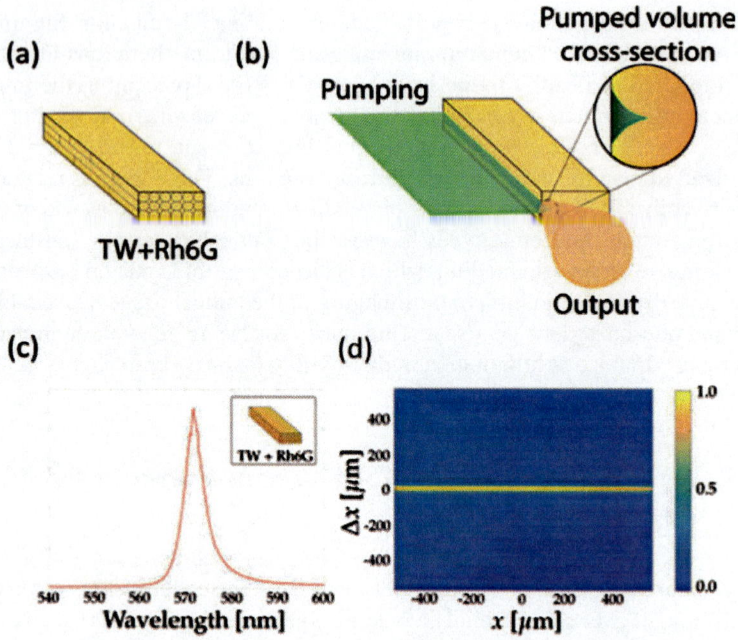

Fig. 12.7 **a** Schematic representation of the TW-R6G sample, **b** Pumping mechanism on the TW-R6G sample, **c** Typical emission spectra from TW-R6G sample, **d** Measured spatial coherency of TW-R6G sample (Koivurova 2018, Copyright Optica, reproduced with permission)

coherent emission at ~605 nm with a threshold intensity value of 70.4 W/cm^2. Basically, the nanopillars in the film provide sustainable feedback of amplified light by recurrent multiple scattering, and thereby creating coherent loops. As a result, RL emission has been achieved upon surpassing the overall loss with the net gain, as shown in Fig. 12.8d, having mode line-width of <0.55 nm.

Another distinctive RL emission has been realized in RhB as the gain medium in polymeric films with diatom frustule as the scatterer (Lamastra et al. 2014). Porous nanostructures of diatomite, produced by solvent casting technique, as shown in Fig. 12.8e provides intensity feedback of light for generation of incoherent RL emission. As shown in Fig. 12.8f, the lowest RL threshold intensity of 308 kW/cm^2 has been optimized by deliberately choosing the number of frustules (Lamastra et al. 2014).

The silk based scatterer for RL framework is another highly intense research area and silk protein-based bio-microfiber disordered structures has been made from *Bombyx Morie* moth (Xie et al. 2020). They used a simple electrospinning technique for preparation of the disordered structure and achieved coherent RL emission with 33 nm wavelength tunability with RL threshold energy of 2.68 mJ. Further, by fabricating inverse photonic glass design in the silk based RL system, measurement of intrinsic pH of NaOH solution has been demonstrated by Caixeiro et al. (2016). A

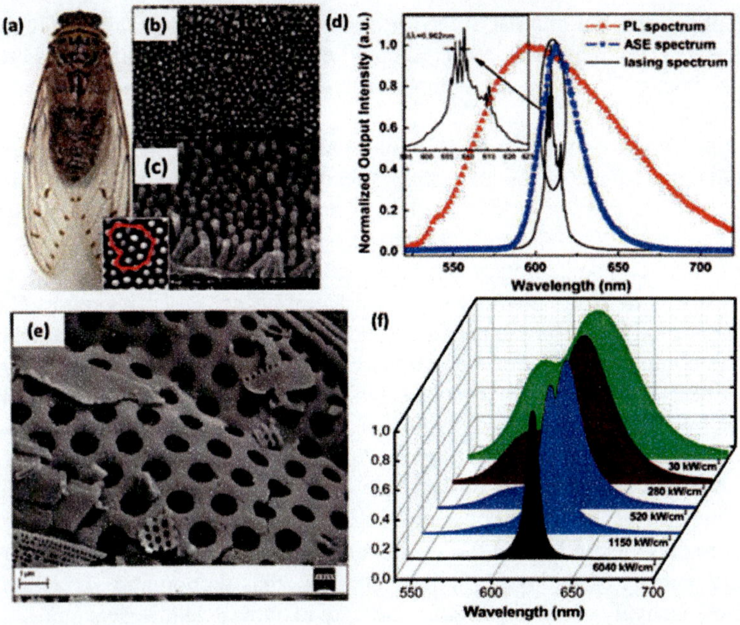

Fig. 12.8 **a** A digital image of pomponia imperatoria cicada, **b–c** SEM images of cicada wing nanopillars at different magnification. **d** Emission spectra from cicada wing nanopillar based DCM:SU8 films (Zhang 2012 Copyright Elsevier, reproduced with permission). **e** SEM images of a diatomite, **f** Normalised emission spectra from PMMA/RhB/DE20 sample at different excitation power density (Lamastra et al. 2014, Copyright RSC, reproduced with permission)

nonlinear increase of FWHM from ~15–20 nm at pH = 7 to ~55 nm at pH = 13 attributes to deprotonation of the dye molecule.

12.2.6 *Generation of Random Lasers in Hibiscus Rosa-Sinensis Leaf Extract (HRLe) and Clitoria Ternatea (CT) Dye*

In general, the commercially available organic dyes are frequently used as gain media for RL emission under pulse pumping owing to their high photoluminescence quantum yield (PLQY). But these commercial dyes are not bio-compatible in most of the cases; thus, they restrict the in-vivo bio-imaging application of RL emission. In this regard, some natural dye containing chlorophyll (Chl) or other natural organic ingredients such as Hibiscus rosa-sinensis leaves extract (HRLe), may fulfil the requirements fabricating bio compatible RL devices (Hindman et al. 1977; Leupold et al. 1990; Chen et al. 2016; Navarro and Werts 2013; Biswas and Kumbhakar 2017). Although laser emission generated under low power CW pumping is essential for some applications, but in most of the previous reports Chl lasers are designed under pulsed lasers (Hindman et al. 1977; Leupold et al. 1990). And stimulated emission from Chl under CW laser excitation is sparsely reported (Hindman et al. 1977; Leupold et al. 1990; Chen et al. 2016).

Hibiscus rosa-sinensis is a perennial green plant that is widely cultivated throughout the tropics and subtropics. The dark green colour of *Hibiscus rosa-sinensis leaves* indicates its use as rich source of Chl (Zhang and Forrest 2011; Hao and Schatz 2004). On the other hand, size and shape dependent Plasmonic response of Au and Ag nanoparticles have created enormous research interest in the past (Navarro and Werts 2013; Biswas and Kumbhakar 2017). As reported earlier through resonant scattering experiments, scattering efficiency of Ag nanoparticles is more in compared to that of Au nanoparticles of similar dimensions, as Ag has a lower value of imaginary dielectric constant compared to that of Au nanoparticles. In a recent work, Navarro and Werts (2013) have experimentally compared resonant light scattering from different Au, Ag and Au–Ag alloy nanoparticles. In this regard, similar to PLQY of fluorescent objects, they have formulated the resonant light scattering efficiency (ϕ_{sca}). The light scattering efficiency is defined as, $\phi_{\text{sca}} = \sigma_{\text{sca}}/\sigma_{\text{ext}}$ and it is mainly dependent on the excitation wavelength, which is unlike the case for PLQY. Here, σ_{sca} and σ_{ext} are the scattering cross section and extinction cross section, respectively. At optical frequencies, Ag nanoparticle interacts more strongly than that of Au nanoparticles (Hao and Schatz 2004) and thus the Ag nanoparticles have higher scattering efficiency.

Meanwhile, the absorption spectrum of triangular nanostructures of silver (TNS) synthesized matches very close to the emission wavelength of 632.8 nm of the pump CW He–Ne laser. Moreover, it has been found that the quenching in the PL intensity of the HRLe takes place in the presence of TNS. Thus, a judicious choice of gain medium and scatterer is essentially needed to generate CW laser pumped emission. Quite interestingly, demonstration of a low power CW He–Ne laser pumped RL emission at ~674 nm in HRLe in the presence of TNS as the plasmonic scatterer (Biswas and Kumbhakar 2017) is reported. Additionally, ~2.54 times reduction in the RL threshold is achieved with the increase in number density of TNS. For comparison, a thin film of PVA doped with commercial MB dye and TNS has been prepared and RL experiment is performed, similar to that with the liquid system, and a systematic reduction in RL threshold with number density of TNS is observed. Notably, in both cases, the variation of RL threshold with number density of TNS follows a power law. Figure 12.9a shows the emission spectra of bare HRLe, which are obtained under various pump laser intensities (*IP*). Whereas, Fig. 12.9b–f show the same but in the presence of TNS at different number densities starting from 5×10^{14} to 8×10^{15} nos./mL (Biswas and Kumbhakar 2017).

It may be noted that irrespective of the pump laser intensity and number density of TNS, the HRLe exhibits a central PL emission band at 674 nm along with two other shoulder peaks at 720 and 648 nm. The lasing threshold for RL generation in HRLe in the presence of the TNS scatterer of a given concentration has been determined by plotting the output power emitted at 674 nm with the variation of the IP and the results are shown in Fig. 12.10a–f.

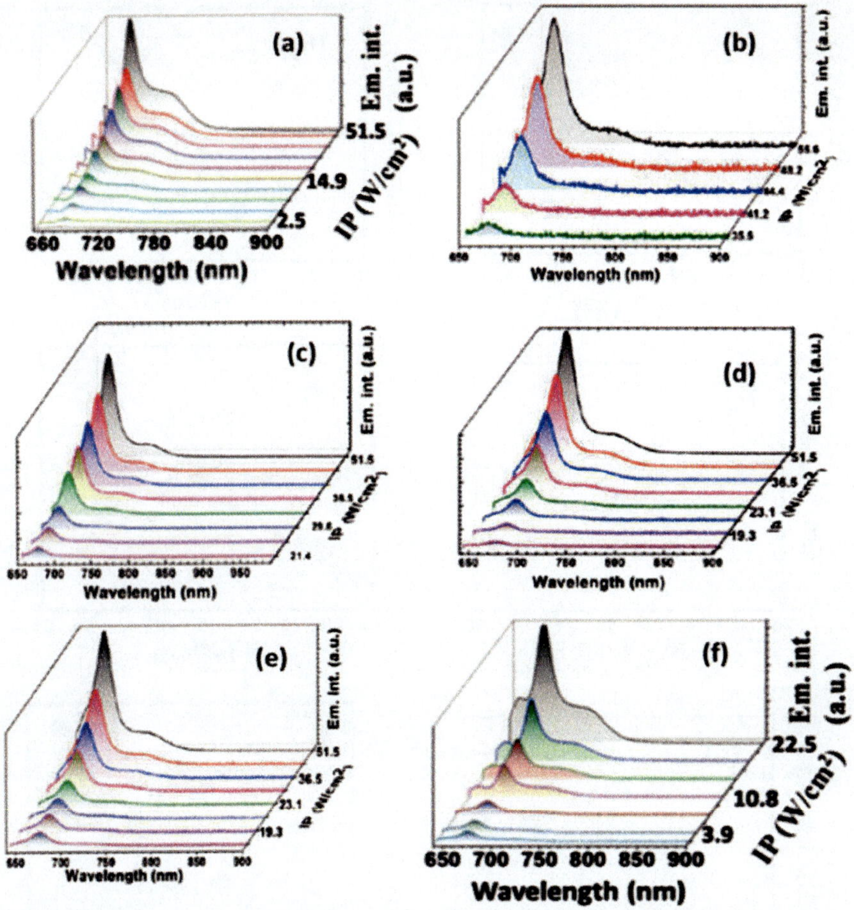

Fig. 12.9 **a** Emission spectra of HRLe without TNS scatterer. **b–f** Emission spectra of HRLe for four different values of N of **b** 5×10^{14}, **c** 1×10^{15}, **d** 2×10^{15}, **e** 4×10^{15} and **f** 8×10^{15} nos./mL, when it is pumped by a He–He laser of 632.8 nm wavelength with various pump intensities (Biswas and Kumbhakar 2017, Copyright RSC, reproduced with permission)

12.2.7 Generation of Random Laser in Clitoria Ternatea Dye

The application of flower-like ZnO nanostructure for the development of RL in different gain media, such as commercially available MB dye, bio-compatible CT flowers extract, and their mixture has been reported earlier. In (Kumbhakar et al. 2019), the demonstration of RL action at 660 nm in butterfly pea (*Clitoria ternatea*; CT) flower extract (used as gain medium) has been made. In (Kumbhakar et al. 2019), flower-like ZnO NPs have been used as passive scatterer and a low power CW He–Ne laser radiation of 632.8 nm has been used as excitation source of light.

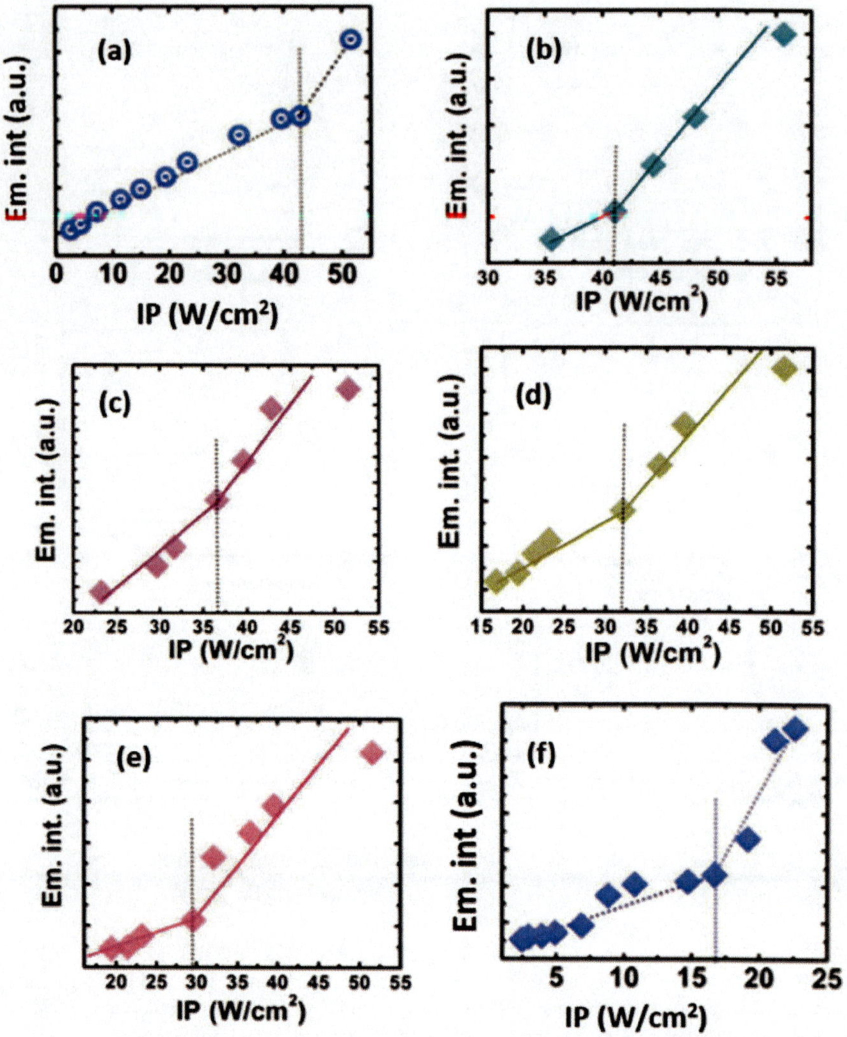

Fig. 12.10 Variations of emission intensity with pump intensity for RL generation in HRLe for different concentrations of TNS **a** 0 **b** 5×10^{14} **c** 1×10^{15} **d** 2×10^{15} **e** 4×10^{15} **f** 8×10^{15} nos/mL of TNS. (Biswas and Kumbhakar 2017, Copyright RSC, reproduced with permission)

For achieving low lasing threshold in these gain media, binary scatterers, such as Ag and ZnO NPs have also been employed as passive scatterer. The resonance energy transfer (RET) process between the two gain media to achieve tunable light emission in RL experiments has been utilized.

The extraction of pigments from CT dye has been carried out by cutting them into small pieces and dipping into a beaker with a particular volume of DI water.

After 24 h, a blue coloured solution is extracted from the mixture by using a filter paper (pore diameter ~50 μm), and then it kept in a separate container (Kumbhakar et al. 2019). As depicted through the fourier transform infrared (FTIR) spectrum in Fig. 12.11**d**, the blue colour solution contains several organic compounds such as flavonoids, anthocyanins, polyphenolic compounds, monoglycosides, diglycosides etc. (Kumbhakar et al. 2019).

Thereafter, the Plasmonic scatterer i.e., TNS have been introduced with the random system consisting of CT dye and ZnO nanorods. The significance of TNS on a random system is judged, by careful mixing of different concentrations of ZnO and TNS in liquid as well as polymer matrix. Figure 12.12a–b show the RL emission spectra of polymer films obtained under different pump powers, in the absence and presence of Ag NPs, but for the same number density of ZnO NPs (Kumbhakar et al. 2019).

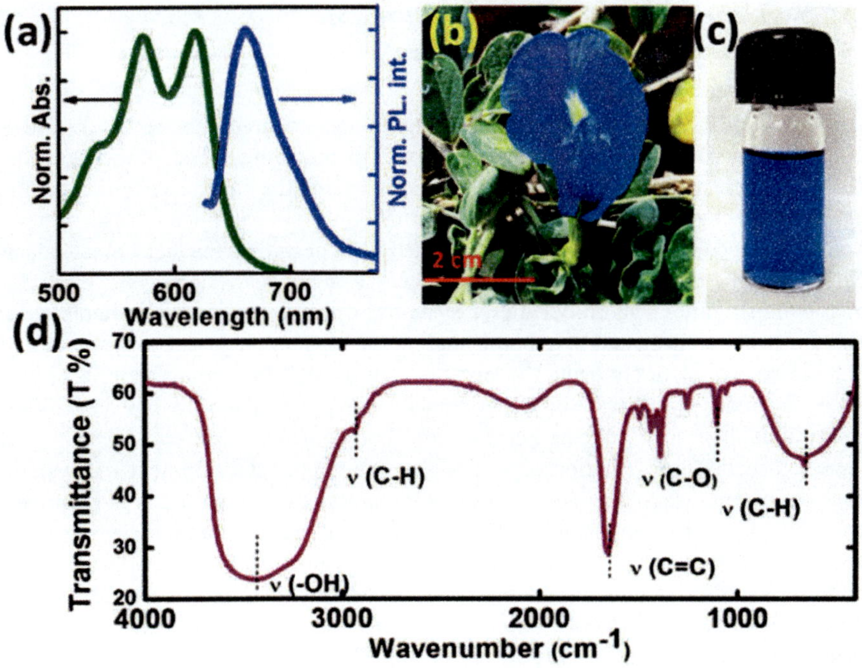

Fig. 12.11 **a** Optical absorption and PL emission spectra of the as extracted CT dye. Digital photographs of CT flower and its extract is shown in (**b**) and (**c**), respectively. **d** FTIR spectra of CT dye. (Kumbhakar et al. 2019, Copyright RSC, reproduced with permission)

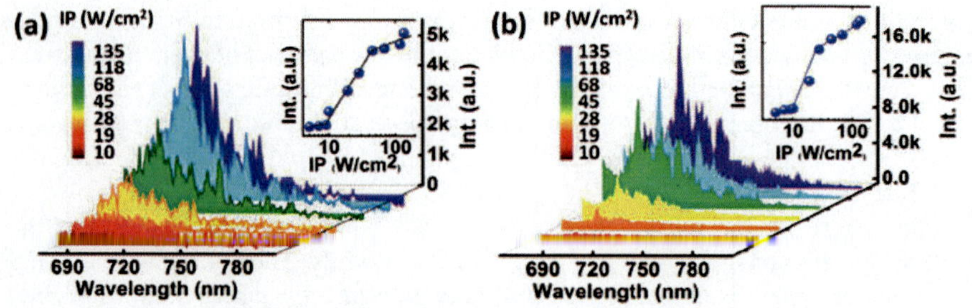

Fig. 12.12 The emission spectra of CT_MB medium in the presence of ZnO scatterer ($n_{ZnO} = 7 \times 10^{12}$) at different IP values of **a** without Ag NPs **b** with Ag NPs ($n_{ZnO} = 20 \times 10^{12}$ and $C_{Ag} = 36 \times 10^{15}$ ml^{-1}), respectively; inset shows the threshold behaviour of random system for RL generation in MB dye. (Kumbhakar et al. 2019, Copyright RSC, reproduced with permission)

12.2.8 Generation of RL from Natural Micro-pillar Based Scatterer in Bambusa Tulda Leaves

Exotic photonic structures found in different nature procured substances are often found to be jaw-dropping due to their complex and astonishing optical properties. The recent research studies from various researchers around the world have enriched the knowledge such that nowadays the conventional bulky and complicated laser system is being replaced by natural and biodegradable RLs. The optical feedback mechanism within this type of scatterer media occurs due to random scattering events from the micro/nanostructured natural scatterer systems, when mixed with a suitable gain medium, when illuminated by a pump source. Low lasing threshold intensity along with the versatility of choosing the suitable gain medium/scattering materials have made the natural scatterer based RL system a convenient and worthy replacement of conventional lasers for imaging applications of any object.

In this section, presentation of work on an indigenous development RL system made up of *bambusa tulda* (Indian bamboo) leaves (Dey et al. 2021) is reported. An exotic RL arrangement has been made by coating the bamboo leaves with Rhodamine-B (RhB) dye. The front surface of those bamboo leaves is made of natural micro-pillars with variable sizes and distances between themselves. Figure 12.13a shows the image of a fresh bamboo leaf, whereas, the FESEM images of the upper surface of a bamboo leaf is shown in Fig. 12.13b, which shows the micro pillar structures. The interactive 3D surface plot of the FESEM image of the marked area of the upper surface is shown in Fig. 12.13c which will make it easy to visualize. These bamboo leaves are coated with specific concentrations of Rhodamine-B dye and it has been pumped by a 532 nm CW laser. At low input intensities of the pump laser, the broad spectra corresponding to the SE of RhB dye is obtained. However, as the input intensity is further increased, emergence of spikes on top of the SE background is found. The occurrence of such spikes is the scattering of amplified light around those natural micro-pillars and this type of RL emission occurs at ~ 582 nm

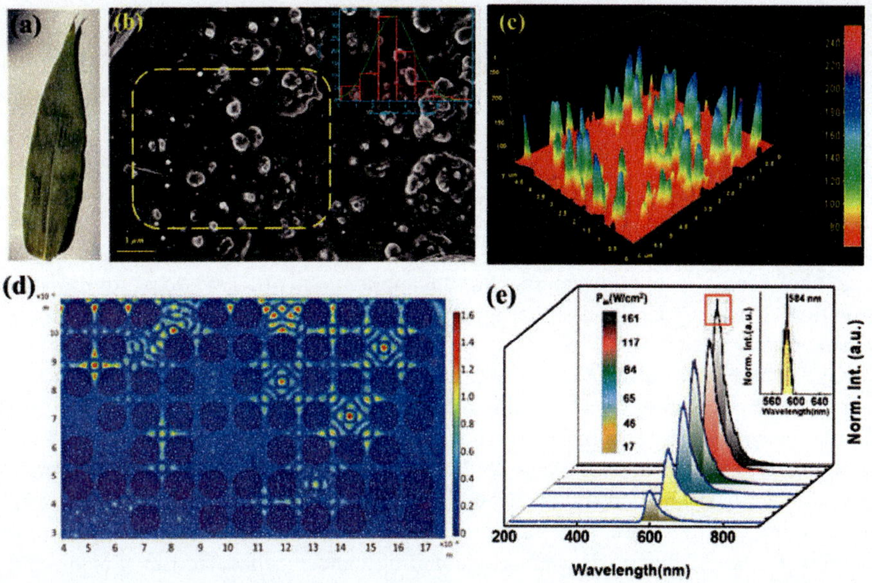

Fig. 12.13 **a** Digital image of a fresh bambusa tulda leaf, **b** FESEM image of a clean leaf (Inset shows the histogram plot of average micro pillar distances), **c** 3D surface plot of the yellow marked area in Fig. 12.13b, shows 3D nature of the micro pillars of bamboo leaves, **d** COMSOL Multiphysics simulation of electric field distribution around the micro-pillars due to scattering of amplified light, **e** Normalized emission spectra from 2 months old S3 sample (Inset shows the zoomed portion of the marked spectra showing a RL mode at ~584 nm appearing at high P_{in}) (Dey et al. 2021, Copyright Optica, reproduced with permission)

is achieved in RhB dye coated bamboo leaves. The threshold input intensity ($^{Th}P_{in}$) of such a lasing action is calculated to be ~132 W/cm^2.

Furthermore, the scattered electric field distribution of the amplified light around those micro-pillars can be numerically simulated by the commercially available COMSOL Multiphysics software. Such field is schematically shown in Fig. 12.13d. Furthermore, this incoherent RL emission is used for speckle free or noiseless imaging technique for better illumination of any object because of the very low degree of coherency. The speckle contrast value for RL illumination was close to zero with a value of 0.03. Hence, the present report (Dey et al. 2021) not only describes simple and exclusive method for synthesizing environment friendly, non-hazardous, plant-extracted RL system via a simple-route technique, but the interplay of scattering by the 'micro-pillar and amplified light' duo can open up huge opportunity to set new platforms for handy, low-cost and non-complex CW laser assisted RL system for bio-imaging applications. Therefore, the synthesized RL system made of dye coated *bambusa tulda* leaves is a promising candidate for advanced photonics application in the future and it can surely be a breakthrough in the field of RL based imaging apparatus (Dey et al. 2021).

12.3 Applications of RL

As mentioned earlier that due to the recent advancement in materials science and due to the improvement in understanding of the principle of operation of RL a lot of researchers are working on it and a plethora of new application opportunities are opening up and a thriving route for RLs additional basic research and applications in the future are anticipated. The practical uses of RL have extended in the vast realm of optical, medical, and commercial and defense sectors. Here, as mentioned in brief about some recent advancement in the applications of RL systems for drawing attention of the readers and some applications of RL in various fields are shown schematically in Fig. 12.14.

One of the major utilizations of RLs are its speckle free imaging due to its low spatial coherence properties (Redding et al. 2012). RLs have been used as a source of imaging employing various nanostructures (Liu et al. 2019; Yang et al. 2019; Bian et al. 2020). The speckle contrast (C) with a low value of 0.0241 has been obtained in Liu et al. (2019), which shows that the generated RL illumination is highly useful for imaging. Furthermore, RLs have also made their footprint in the medical sector as well. Different diagnosis methods including cancer cell detection (Polson and Vardeny 2004; Zhang et al. 2019), blood monitoring (Mendicuti et al. 2021), human

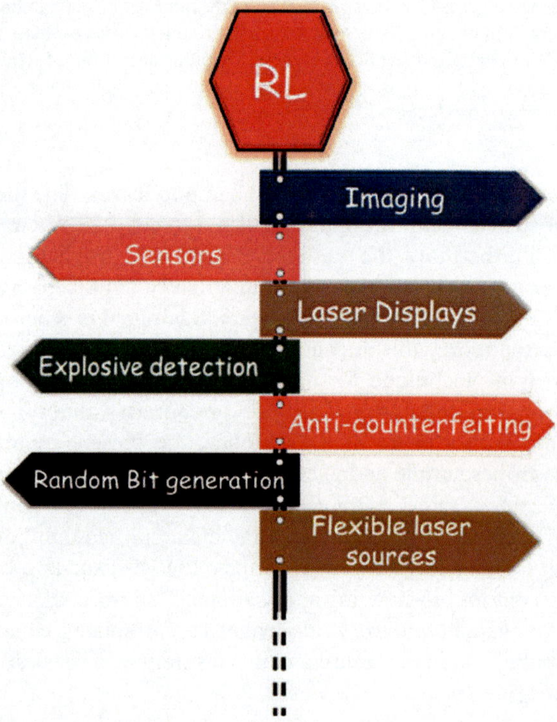

Fig. 12.14 A schematic representation on different applications of RL

wearables (Ge et al. 2021) etc. have been reported, so far, by using different RL systems. However, further implications of RLs include anti-counterfeiting (Ni et al. 2021; Su et al. 2020), refractive index sensor based on fiber RL (Shi et al. 2020), flexible laser display (Hou et al. 2021), bit generation (Sznitko et al. 2021) etc. For instance, the polymer gel containing R6G dye and Ag nanoparticle has been used as a smart ink for detection of friend or foe i.e., anti-counterfeiting (Su et al. 2020). Similar to luminescent quantum dots, array $CsPbBr_3$ NCs/SiO_2 of RLs has been already applied for fabrication of flexible display planes (Hou et al. 2021). However, excitation wavelength dependent biochromatic emission of a dye molecule such as the 3-(2,2-dicyanoethenyl)-1-phenyl-4,5-dihydro-1H-pyrazole (DCNP) can be utilized for generation of RL based optical random numbers (Sznitko et al. 2021). Notably, Ismail et al., have developed a technique for sensing of dopamine by controlling the aggregation of gold nanoparticles inside the RL media (Ismail and Liu 2016).

12.4 Conclusions and Future Scope

In this chapter, a mini review on various theoretical and experimental aspects of RLs has been presented. In opposition with the tricky and fine textures present in conventional lasers, RLs provides us the privilege to construct a lasing device based on any random geometrical framework. Various engineered techniques of fine tuning the emission regimes; with possible scopes for incoherent/coherent RL emissions have been broadly discussed. The demonstration of generation of CW laser induced microbubbles in presence of the GFs may lead to the application of 2D materials as the nano-generator of microbubbles for RL generation in a weakly scattering regime. On the other hand, use of ExFB in the dye and semiconductor scatterer doped polymer film can be an inspiration for fabricating low threshold RL device under CW pumping. Additionally, here the speckle free imaging application of RL using a low-cost experimental technique is reviewed. Also, discussions on possible way out for greener approaches for constructing biodegradable and environment friendly gain/scatterer media are presented and it has been found that bio-pigment extract can be used as the gain medium to achieve visible RL illuminations by using some commercially available CW laser as the pump source. It has been shown that the lasing threshold of RL emission can be reduced considerably in the presence of triangular shaped nanoparticles of silver, i.e. TNS, due to its high scattering cross section and through SPR enhanced photon multiplication mechanism. The Plasmonic TNSs are shown to be utilized here as a suitable Plasmonic scatterer to generate RL emission from a commercially available methylene blue dye and also from bio-extracted CT dye. As the pigment extracted from bio sources are nontoxic and bio-compatible, random lasing from such material can be used safely in bio-imaging purposes in future. This craftsmanship of lower threshold intensity for RLs along with tunability of emission regimes shall establish new pavements for futuristic, scalable, and biocompatible RL devices for sensing and imaging applications in future.

Acknowledgements Authors are thankful to Dept. of Science & Technology and Biotechnology, Govt. of W.B., India for the project grant No. 332 (Sanc.)/ST/P/S&T/16G-24/ 2018 dt. March 06, 2019 for the partial financial support. The partial financial support from CSIR, India via Grant No. 03(1472)/19/EMR-II dt. August 05, 2019 is gratefully acknowledged. Arindam Dey acknowledges Ministry of Education, Govt. of India and NIT Durgapur for the maintenance fellowship.

Declaration The manuscript was written through contributions from all authors. All authors have given approval to the final version of the manuscript.

References

Anderson, P.W.: Phys. Rev. **109**, 1492–1505 (1958)
Andreasen, J., Cao, H.: Opt. Express **19**, 3418–3433 (2011)
Bian, Y., Shi, X., Hu, M., Wang, Z.: Nanoscale **12**, 3166–3173 (2020)
Biswas, S., Kumbhakar, P.: Nanoscale **9**, 18812–18818 (2017)
Boltzmann, L.: Wiedemann's Annalen **22**, 291–294 (1884)
Bornemann, R., Lemmer, U., Thiel, E.: Opt. Lett. **31**, 1669–1671 (2006)
Caixeiro, S., Gaio, M., Marelli, B., Omenetto, F.G., Sapienza, R.: Adv. Opt. Mater. **4**, 998–1003 (2016)
Cao, H.: Waves Random Media **13**, R1–R39 (2003)
Cao, H., Zhao, Y.G., Ho, S.T., Seelig, E.W., Wang, Q.H., Chang, R.P.H.: Phys. Rev. Lett. **82**, 2278 (1999)
Chen, S., Shi, J., Kong, X., Wang, Z., Liu, D.: Laser Phys. Lett. **10**, 55006 (2013)
Chen, Y., Chen, Q., Fan, X.: Lab Chip **16**, 2228–2235 (2016)
Dey, A., Pramanik, A., Kumbhakar, P., Biswas, S., Pal, S.K., Ghosh, S.K., Kumbhakar, P.: OSA Contin **4**, 1712–1722 (2021)
Dey, A., Pramanik, A., Biswas, S., Chatterjee, U., Kumbhakar, P.: J. Lumin. **250**, 119252 (2022)
Dhanker, R., Brigeman, A.N., Larsen, A.V., Stewart, R.J., Asbury, J.B., Giebink, N.C.: Appl. Phys. Lett. **105**, 151112 (2014)
Einstein, A.: Ann d Phys **17**, 549 (1905)
Fernandez-Bravo, A., Wang, D., Barnard, E.S., Teitelboim, A., Tajon, C., Guan, J., Schatz, G.C., Cohen, B.E., Chan, E.M., Schuck, P.J., Odom, T.W.: Nat. Mat. **18**, 1172–1176 (2019)
Feynman, R.P., Leighton, R.B., Sands, M.: The Feynman Lectures on Physics. The California Institute of Technology (1989)
Frolov, S.V., Vardeny, Z.V., Yoshino, K., Zakhidov, A., Baughman, R.H.: Phys. Rev. B **59**, R5284 (1999)
Ge, K., Guo, D., Ma, X., Xu, Z., Hayat, A., Li, S., Zhai, T.: Nanomaterials **11**, 1809 (2021)
Hao, E., Schatz, G.C.: J. Chem. Phys. **120**, 357–366 (2004)
Hindman, J.C., Kugel, R., Svirmickas, A., Katz, J.J.: Proc. Natl. Acad. Sci. u.s.a. **74**, 5–9 (1977)
Hou, Y., Zhou, Z., Zhang, C., Tang, J., Fan, Y., Xu, F.F., Zhao, Y.S.: Sci. China Mater. **64**, 2805–2812 (2021)
Hu, H.W., Haider, G., Liao, Y.M., Roy, P.K., Lin, H.I., Lin, S.Y., Chen, Y.F.: ACS Omega **5**, 18551–18556 (2020)
Ismail, W., Liu, G., Zhang, K., Ewa, G., Dawes, J.: Opt. Express **24**, A85–A91 (2016)
Jiang, X., Zhao, X., Bao, W., Shi, R., Zhao, J., Kang, J., Xia, X., Chen, H., Li, H., Xu, J., Zhang, H.: ACS Appl. Nano Mater. **3**, 4990–4996 (2020)
Kim, W., Safonov, V.P., Shalaev, V.M., Armstrong, R.L.: Phys. Rev. Lett. **82**, 4811 (1999)
Koivurova, M., Vasileva, E., Li, Y., Berglund, L., Popov, S.: Opt. Express **26**, 13474–13482 (2018)
Kumbhakar, P., Biswas, S., Kumbhakar, P.: RSC Adv. **9**, 37705–37713 (2019)
Lamastra, F.R., De Angelis, R., Antonucci, A., Salvatori, D., Prosposito, P., Casalboni, M., Congestri, R., Melino, S., Nanni, F.: RSC Adv. **4**, 61809–61816 (2014)

Lee, Y.J., Yeh, T.W., Yang, Z.P., Yao, Y.C., Chang, C.Y., Tsai, M.T., Sheu, J.K.: Nanoscale **11**, 3534–3545 (2019)
Letokhov, V.S.: J. Exp. Theor. Phys. **26**, 835–840 (1968)
Leupold, D., Struck, A., Stiel, H., Teuchner, K., Oberltider, S., Scheer, H.: Chem. Phys. Lett. **170**, 478–484 (1990)
Ling, Y., Cao, H., Burin, A.L., Ratner, M.A., Liu, X., Chang, R.P.H.: Phys. Rev. A **64**, 063808 (2001)
Liu, Y., Yang, W., Xiao, S., Zhang, N., Fan, Y., Qu, G., Song, Q.: ACS Nano **13**, 10653–10661 (2019)
Livdan, D., Lisyansky, A.A.: J. Opt. Soc. Am. A **13**, 844–850 (1996)
Markushev, V.M., Zolin, V.F., Briskina, Ch.M.: Zh. Prikl. Spektrosk. **45**, 847–850 (1986)
Mendicuti, E., Käferlein, O., García-Segundo, C.: Opt. Lett. **46**, 274–277 (2021)
Moura, A.L., Pincheira, P.I., Maia, L.J., Gomes, A.S., de Araújo, C.B.: J. Lumin. **181**, 44–48 (2017)
Navarro, J.R.G., Werts, M.H.V.: Analyst **138**, 583–592 (2013)
Ni, Y., Wan, H., Liang, W., Zhang, S., Xu, X., Li, L., Shao, Y., Ruan, S., Zhang, W.: Nanoscale **13**, 16872–16878 (2021)
Noginov, M.A.: Solid-State Random Lasers, Springer Series in Optical Sciences, Springer, Berlin (2005)
Noginov, M.A., Zhu, G., Belgrave, A.M., Bakker, R., Shalaev, V.M., Narimanov, E.E., Stout, S., Herz, E., Suteewong, T., Wiesner, U.: Nature **460**, 1110–1112 (2009)
Okamoto, T., Yoshitome, R.: J. Opt. Soc. Am. B **34**, 1497–1502 (2017)
Polson, R.C., Vardeny, Z.V.: Appl. Phys. Lett. **85**, 1289–1291 (2004)
Pramanik, A., Biswas, S., Dey, A., Kumbhakar, P.: Adv. Photon Res. **2**, 2100063 (2021a)
Pramanik, A., Biswas, S., Kumbhakar, P., Kumbhakar, P.: J. Lumin. **230**, 117720 (2021b)
Pramanik, A., Mondal, K., Biswas, S., Pal, S.K., Ghosh, S.K., Ganguly, T., Kumbhakar, P.: Appl. Phys. B **128**, 167 (2022)
Redding, B., Choma, M.A., Cao, H.: Nat. Photon **6**, 355–359 (2012)
Rodberg, L.S., Thaler, R.M.: Introduction to the Quantum Theory of Scattering, 1st edn. Academic Press Inc., 111 Fifth Avenue, New York NY 10003, USA (1985)
Schawlow, L.: Lasers and Light. W. H. Freeman and Co (1969)
Shen, Z., Wu, L., Zhu, S., Zheng, Y., Chen, X.: Appl. Phys. Lett. **105**, 021106 (2014)
Shi, X., Ge, K., Tong, J.H., Zhai, T.: Opt. Express **28**, 12233–12242 (2020)
Su, C.Y., Hou, C.F., Hsu, Y.T., Lin, H.Y., Liao, Y.M., Lin, T.Y., Chen, Y.F.: ACS Appl. Mater. Interfaces **12**, 49122–49129 (2020)
Sun, T.M., Wang, C.S., Liao, C.S., Lin, S.Y., Perumal, P., Chiang, C.W., Chen, Y.F.: ACS Nano **9**, 12436–12441 (2015)
Syurik, J., Siddique, R.H., Dollmann, A., Gomard, G., Schneider, M., Worgull, M., Wiegand, G., Hölscher, H.: Sci. Rep. **7**, 1–11 (2017)
Sznitko, L., Cyprych, K., Szukalski, A., Miniewicz, A., Mysliwiec, J.: Laser Phys. Lett. **11**, 045801 (2014)
Sznitko, L., Chtouki, T., Sahraoui, B., we sliwiec, J.: (2021) ACS Photonics **8**, 1630–1638
Tadepalli, S., Slocik, J.M., Gupta, M.K., Naik, R.R., Singamaneni, S.: Chem. Rev. **117**, 12705–12763 (2017)
Tong, J., Shi, X., Li, S., Chen, C., Zhai, T., Zhang, X.: Org. Electron. **75**, 105337 (2019)
Townes, C.H.: How the Laser Came to Happen: Adventures of a Scientist. Oxford University Press (2002)
Van Soest, G., Tomita, M., Lagendijk, A.: Opt. Lett. **24**, 306–308 (1999)
Wang, Y.C., Li, H., Hong, Y.H., Hong, K.B., Chen, F.C., Hsu, C.H., Lee, R.K., Conti, C., Kao, T.S., Lu, T.C.: ACS Nano **13**, 5421–5429 (2019)
Wiersma, D.S., Albada, M.P., Lagendijk, A.: Nature **373**, 203–204 (1995)
Wilson, J., Hawkes, J.F.B.: Lasers: Principal and Applications. Prentice Hall Publication (1987)
Wu, X.H., Yamilov, A., Noh, H., Cao, H.J.: Opt. Soc. Am. B **21**, 159–167 (2004)

Wu, X., Fang, W., Yamilov, A., Chabanov, A.A., Asatryan, A.A., Botten, L.C., Cao, H.: Phys. Rev. A **74**, 053812 (2006)

Xie, Z., Xie, K., Hu, T., Ma, J., Zhang, J., Ma, R., Cheng, X., Li, J., Hu, Z.: Opt. Express **28**, 5179–5188 (2020)

Yang, T.H., Chen, C.W., Jau, H.C., Feng, T.M., Wu, C.W., Wang, C.T., Lin, T.H.: Appl. Phys. Lett. **114**, 191105 (2019)

Yashchuk, V.P., Prygodiuk, O., Koreniuk, V., Tikhonov, E., Bezrodny, V.: Appl. Phys. B **92**, 593–597 (2008)

Zhang, Y., Forrest, S.R.: Phys. Rev. B **84**, 241301–241304 (2011)

Zhang, D., Kostovski, G., Karnutsch, C., Mitchell, A.: Org. Electron. **13**, 2342–2345 (2012)

Zhang, J., Xu, L., Wang, H., Huang, F., Sun, X., Zhao, H., Chen, X.: Appl. Phys. Lett. **102**, 021109 (2013)

Zhang, D., Wang, Y., Tang, J., Mu, H.: J. Appl. Phys. **125**, 203102 (2019)

Zhang, H., Hu, Y., Wen, W., Du, B., Wu, L., Chen, Y., Feng, S., Zou, C., Shang, J., Fan, H.J., Yu, T.: APL Mater. **9**, 071106 (2021)

Chapter 13
BODIPY: A Unique Dye for Versatile Optical Applications

Soumyaditya Mula

13.1 Introduction

Fluorescent dyes are known for more than one and half a century, and still have the prime attraction to the scientists from multidisciplinary arena. They are the key elements in recent development of personal diagnostics, luminescent based organo-electronic devices and many other hi-tech applications. Thus, demand for the next-generation emissive dyes with tunable optical and electrical properties has increased. Countless classes of fluorescent organic dyes known so far vary in absorptivity and absorbance range, fluorescence colour and intensity, photo and chemical stability, triplet conversion rate etc. These are used for widespread hi-tech applications, such as laser dyes, chemical and bio-sensing, organic photovoltaics, OLEDs, cellular imaging as well as for other biological applications etc. (Valeur and Berberan-Santos 2012). Few of the most popular dyes are Rhodamine, pyrene, perylene, squaraine, cyanine, coumarin, boron-dipyrromethene (BODIPY), diketopyrrolo pyrrole (DPP), etc. (Fig. 13.1). Among these BODIPY (boradipyrromethene, 4,4-difluoro-4-bora-3a,4a-diaza-*s*-indacene) dyes have superior optical and redox properties. The first member of this class of compound was reported by Treibs and Kreuzer in 1968 (Treibs and Kreuzer 1968), although relatively little attention was given to the discovery until the end of the 1980s. After that, their potential applications were explored and the BODIPY dyes have been established as one of the most versatile fluorophores.

Structurally, BODIPY is a boron complex of dipyrromethene unit which can be considered as a "rigidified" monomethine cyanine dye (Fig. 13.1). This structural rigidification results unusually high fluorescence yields of the BODIPY core (Ulrich et al. 2008). The whole organic backbone is conjugated and extension of the conjugation is possible via attachment of different groups at various positions of one or both

S. Mula (✉)
Bio-Organic Division, Bhabha Atomic Research Centre, Mumbai 400085, India
e-mail: smula@barc.gov.in

Homi Bhabha National Institute, Anushakti Nagar, Mumbai 400094, India

R. S. Ningthoujam and A. K. Tyagi (eds.), *Handbook of Materials Science, Volume 1*,
Indian Institute of Metals Series, https://doi.org/10.1007/978-981-99-7145-9_13

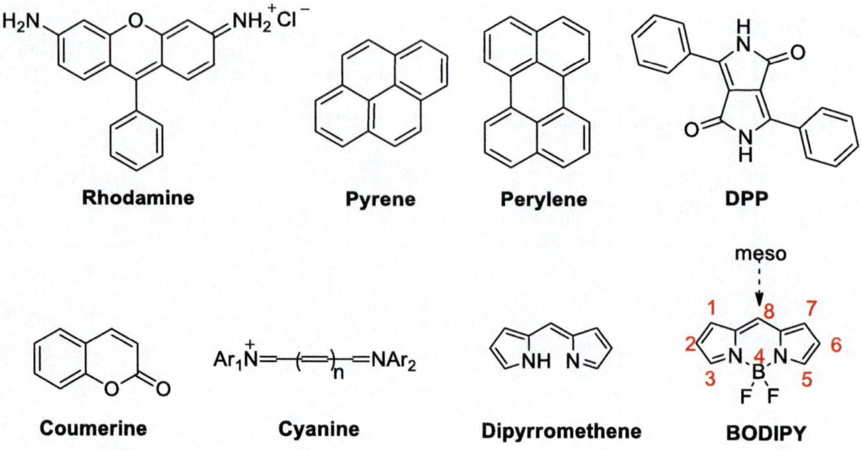

Fig. 13.1 Molecular structures of various organic dyes

pyrrole moieties. This helps to tune the fluorescence colour as well as the intensity of the moiety, which leads to countless numbers of tailored dyes for multiple applications (Mula et al. 2015; Gupta et al. 2013). Other properties such as high absorption coefficient, low triplet-state formation, good solubility in various organic solvents, excellent thermal and photochemical stability, and chemical robustness have made BODIPYs more attractive to the scientific fraternity (Mula et al. 2008).

13.2 Synthesis of BODIPY Dyes

Dipyrromethenes are synthesized by condensation of substituted/unsubstituted pyrroles with acid chlorides (aromatic/aliphatic)/acid anhydrides, or with acetals/aromatic aldehydes followed by oxidation using DDQ or p-chloranil. Then, BF_2-BODIPYs are synthesized by the complexation of corresponding dipyrromethenes with $BF_3.Et_2O$ in the presence of trialkylamine base (Et_3N or *i*-Pr_2NEt). Similarly, the dipyrromethenes can be reacted with BPh_3 to get the corresponding BPh_2-BODIPYs (Scheme 1) (Ulrich et al. 2008; Loudet and Burgess 2007; Mula et al. 2009). It is important to mention here that the *meso*-aryl/alkyl group can easily be chosen during the dipyrromethene synthesis which gives the possibility of synthesizing an array of BODIPY dyes with fluorescence in the UV–Vis region for different applications as discussed *vide infra.*

Aza-BODIPYs (4-bora-3a,4a,8-triazaindacene dyes) are very interesting as they have red shifted fluorescence in the far-red and near-IR region as compared to the conventional BODIPYs discussed above. Aza-dipyrromethenes are synthesized by reacting nitromethane with chalcone, followed by condensation with an ammonium salt (Scheme 2) (Ulrich et al. 2008; Loudet and Burgess 2007). Finally, complexation

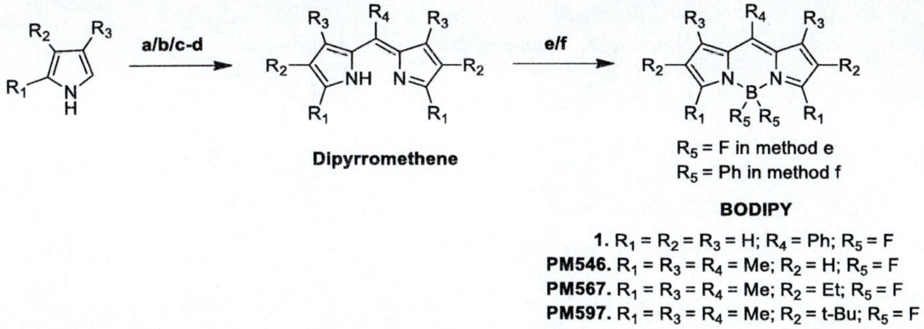

Scheme 1. Synthetic procedures for BODIPY dyes: **a** acid chloride/CH_2Cl_2/25 °C; **b** acid anhydride//25 °C; **c** acetal or aldehyde/TFA/CH_2Cl_2/25 °C; **d** *p*-chloranil or DDQ/CH_2Cl_2/25 °C; **e** Et_3N or *i*-$Pr_2NEt/BF_3.OEt_2/CH_2Cl_2$/25 °C; **f** BPh_3/toluene/90 °C

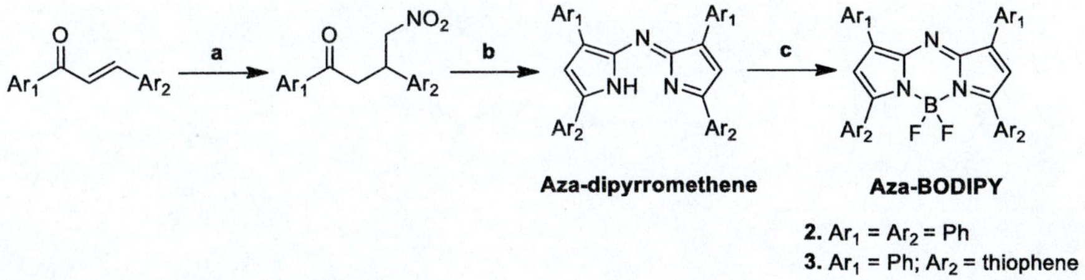

Scheme 2. Synthetic procedures for aza-BODIPY dyes: **a** CH_3NO_2, $HNEt_2$, MeOH, Δ; **b** NH_4OAc; **c** $BF_3{\cdot}OEt_2$, *i*-Pr_2NEt, rt

of aza-dipyrromethene with $BF_3{\cdot}OEt_2$ in the presence of *i*-Pr_2NEt furnished the corresponding BF_2-aza-BODIPYs.

13.3 Chemistry of BODIPY Core

13.3.1 Electrophilic Substitution at the Pyrrole Moiety

2,6-positions of the BODIPY core are the most reactive sites for the electrophilic substitutions. Treibs and Kreuzer first synthesized 2,6-disulfonic acid BODIPY dye by electrophilic substitution using chlorosulfonic acid (Treibs and Kreuzer 1968). This methodology was exploited later extensively to synthesize 2,6-dibromo/iodo/nitro/formyl BODIPYs which were subsequently used for further synthetic modifications (Scheme 3a) (Ulrich et al. 2008; Gupta et al. 2013; Loudet and Burgess 2007). Importantly, B-F bonds remain unaffected during these reactions and only 2,6-substituted products are obtained. Thus, this is an important synthetic methodology for regio-selective syntheses of functional BODIPYs. For example, water-solubility

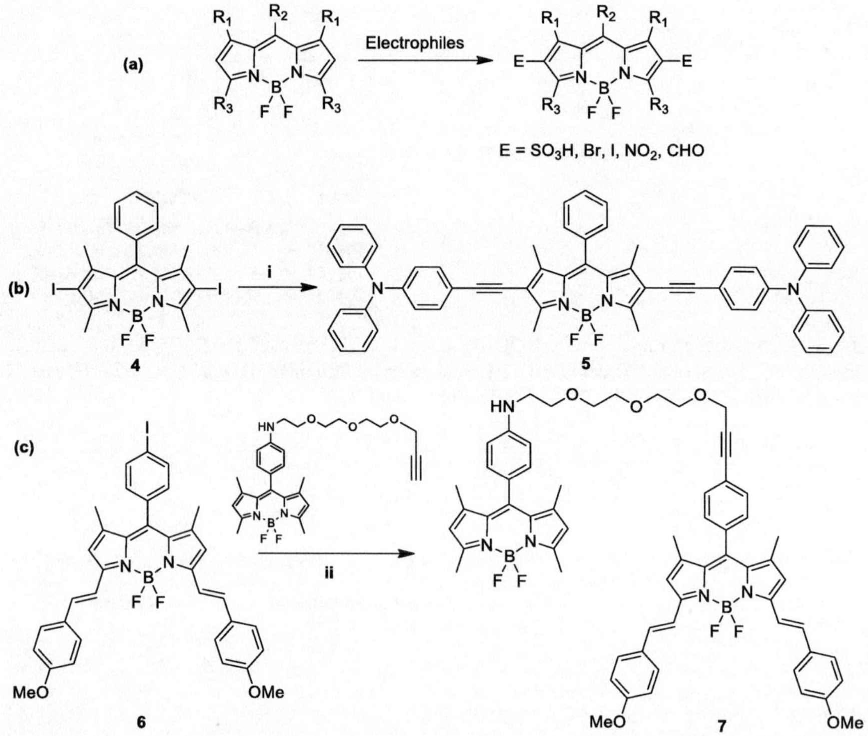

Scheme 3. **a** General procedure for eletrophilic substitution at 2,6 positions of BODIPY dyes: Electrophiles: $ClSO_3H$ or NBS or NIS or HNO_3 or $POCl_3$/DMF; **b** Synthesis of alkenyl BODIPY via Sonogashira coupling: (i) alkyne (2.5 equiv), $Pd(PPh_3)_4$, CuI, THF/NEt_3 (5:1), 60 °C; **c** Synthesis of BODIPY dyad: (ii) [$Pd(PPh_3)_4$] (6 mol%), benzene, Et_3N, 50 °C

of the hydrophobic BODIPY dyes were increased by incorporation of the sulfonate groups keeping the absorption and emission profiles unchanged, whereas substitutions of nitro, bromo or iodo groups reduce the fluorescence drastically as compared to their parent dyes. The reduced fluorescence quantum yields of the bromo and iodo substituted BODIPYs is due to enhanced intersystem crossing (ISC) facilitated via the heavy-atom effect (Kamkaew et al. 2013), whereas nitro-BODIPYs are low/non-fluorescent due to Photoinduced Charge Transfer (PCT) discussed *vide infra.*

13.3.2 Metal-Mediated Cross-Coupling Reaction

Halogenated BODIPYs prepared from electrophilic substitutions discussed above or synthesized from halogen substituted pyrroles are very good precursors for different types of palladium-catalyzed coupling reactions such as Sonogashira, Suzuki, Heck and Stille coupling (Scheme 3b-c). Interestingly, the B-F bonds remain unaffected during these cross-coupling reactions. These methodologies are extensively used to extent the conjugation length of the BODIPY chromophores and to build sophisticated dye structures with red shifted absorptions and emissions (Lu et al. 2014).

Additionally, 8-bromo/iodoaryl BODIPYs are used in similar kind of palladium-catalyzed cross-coupling reactions to connect with another aryl/heteroaryl fluorophore. The resultant multi component systems are used as efficient intermolecular electron and energy transfer molecular assemblies for hi-tech opto-electronic applications (Scheme 3c) (Mula et al. 2010).

13.3.3 Condensation at the Active Pyrrole-Methyl Groups

BODIPY dyes bearing methyl groups at 1, 3, 5, 7 and 8 positions are susceptible for the Knoevenagel type condensation reactions. These pyrrole methyl groups have higher acidity, thus deprotonate easily under mild conditions and the corresponding nucleophiles can be condensed with variety of aromatic aldehydes to synthesize styryl BODIPY dyes (Scheme 4). The reaction condition can be modulated to synthesize different regio-isomers and multi styryl BODIPY dyes with red shifted absorption and emission spectra. This methodology has been used extensively to develop functional BODIPYs emitting in the UV–Vis to NIR region for various applications (Lu et al. 2014; Shivran et al. 2011; Buyukcakir et al. 2009).

13.3.4 Modifications at the Boron Center

Substitution reactions at the boron center of the BODIPY dyes open the new arena of functionalized BODIPY dyes. Fluorine atoms at the B-center can be replaced using organometallic reagents (organolithium or Grignard reagents) to synthesize B-alkyl/aryl/ethynylaryl/ethynyl BODIPYs (Scheme 5) (Bodio and Goze 2019; Gupta et al. 2017; Jagtap et al. 2013). Due to the tetrahedral geometry of the B-center the substituents at the B-center do not have any direct electronic conjugation with the dipyrrin core. Thus, in contrast to the substitutions at the dipyrrin core, B-substitutions do not affect much the absorption and fluorescence wavelengths, but change the 3D structures of the molecules which helps to modify the other properties of the BODIPY dyes such as fluorescence quantum yield, redox properties, photo and chemical stability, solubility, aggregation behavior etc. The modifications at the B-center are highly useful for the preparing new dyads and cascade-type dyes molecules as discussed *vide infra.*

The boron substitutions reaction conditions with Grignard reagents were modified further to increase the yield and to reduce the reaction time. A typical reaction of BODIPY dye with EtMgBr furnished BEt_2-BODIPY with 67% yield in

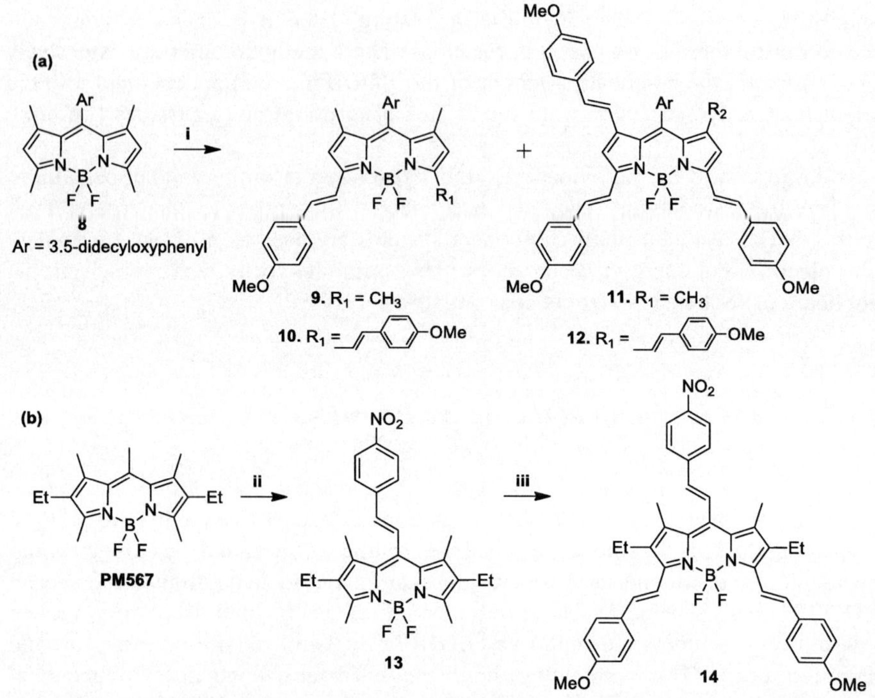

Scheme 4. General procedure for Knoevenagel type condensation reactions of BODIPY dyes: (i) *p*-anisaldehyde/AcOH/piperidine/PhH/Δ; (ii) *p*-nitrobenzaldehyde/AcOH/piperidine/toluene/Δ; (iii) *p*-anisaldehyde/AcOH/piperidine/toluene/Δ

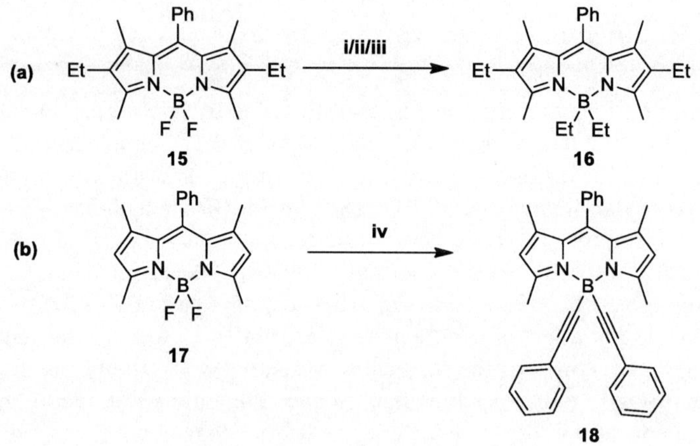

Scheme 5. General methods of different substitutions reactions at boron center of BODIPY dyes: (i) $EtMgBr/Et_2O$/3 h/22 °C; (ii) $BCl_3/EtMgBr/Et_2O$/ 3 h/ 22 °C; (iii) Et_2AlCl/CH_2Cl_2/25 °C/5 min; (iv) Ph-C≡C-Li or Ph-C≡C-H/EtMgBr

3 h (Scheme 5a). But the reaction becomes more facile with the use of Lewis acid reagents. For example, in combination with the BCl_3 and EtMgBr, the yield of the BEt_2-BODIPY increased to 98% (Scheme 5a) (Lundrigan et al. 2012). Further, with the use of Lewis acid Et_2AlCl, the reaction completes within 5 min to give high yield (78%) of the desired product, BEt_2-BODIPY (Scheme 5a) (More et al. 2014).

13.4 Photophysical Properties

BODIPY dyes are known for their outstanding photophysical properties which can be tuned by the chemical modification of the dye core. High extinction coefficient (ε_{max}) and high quantum yield of fluorescence (Φ_{fl}) are the trademark of the BODIPY dyes. The naked BODIPY without any substitutions at the pyrrole rings, **1** (Scheme 1) shows absorption spectra with the longest absorption peak (λ_{abs}) at 503 nm for S0-S1 transition and very low fluorescence with λ_{em} at 521 nm (Table 13.1). Both the absorption and emission spectra shift bathochromically as the substitutions increase around the pyrrole ring of the BODIPY core. For example, PM546 and PM567 (Scheme 1) have λ_{abs} of 493 nm and 518 nm respectively and they show bright greenish yellow fluorescence with λ_{em} of 504 nm and 534 nm respectively (Table 13.1). Boron substitutions do not change the absorption and fluorescence spectral position due to the tetrahedral structure of the boron center. Styryl BODIPY dyes show highly bathochromic shift in absorption and fluorescence spectra, and the bathochromic shift increases with the number of attached styryl groups. For example, λ_{abs} of mono, bis, tri, tetra-styryl BODIPYs (**9–12**) (Scheme 4) show >100 nm bathochromic shift. Mono and bis-styryl BODIPYs (**9, 10**) show orange and red fluorescence whereas dyes **11** and **12** emit in the NIR region. Aza-BODIPYs also show red shifted absorbance and fluorescence. Tetraphenyl aza-BODIPY, **2** (Scheme 2) shows red fluorescence whereas both the absorbance and fluorescence of aza-BODIPY **3** (Scheme 2) are in the NIR region (Table 13.1). Absorbance and fluorescence properties of other BODIPYs are discussed later during the discussion about their applications.

13.5 Applications of BODIPY Dyes

13.5.1 Laser Dye

Apart from the high fluorescence quantum yields and high extinction coefficients of the BODIPY dyes, their low intersystem crossing (ISC) rate and low triplet excitation coefficients over the laser spectral region made them highly efficient laser dyes (Mula et al. 2008). Few BODIPY dyes are known to perform better in terms of lasing efficiency and photochemical stability as compared to

Table 13.1 Photophysical parameters of the BODIPY dyes

Dye	λ_{abs} (nm)	ε_{max} (M^{-1} cm^{-1})	λ_{em} (nm)	Φ_{fl}
1[a]	503	54,000	521	0.05
PM546[b]	493	79,000	504	0.99
PM567[c]	518	71,000	534	0.84
2[d]	643	76,900	673	0.22
3[d]	710	108,600	732	0.46
9[e]	572	58,900	585	0.92
10[e]	645	116,400	660	0.37
11[e]	665	96,900	682	0.35
12[e]	689	127,900	710	0.34

[a] Data reported in toluene (Zhang et al. 2012)
[b] Data reported in MeOH (Gupta et al. 2013)
[c] Data reported in EtOH (Mula et al. 2008)
[d] Data reported in ACN (Wagner and Lindsey 1996)
[e] Data reported in $CHCl_3$ (Buyukcakir et al. 2009)

widely used Rhodamine 6G (Rh6G). Boyer and Pavlopoulos are the pioneers in exploring the lasing properties of the BODIPY dyes. Later on, during the late 1980s and early 1990s, Boyer and co-workers did extensive work to establish BODIPY dyes as highly efficient laser dyes in green yellow to the red spectral region (Shah et al. 1990; Boyer et al. 1993). Various BODIPY analogues were synthesized by changing the substituents at C-2, 6 and 8 (*meso*) positions of the dipyrromethene core (Fig. 13.2). Among these, pyrromethene 567 (PM567, 1,3,5,7,8-pentamethyl-2,6-diethylpyrromethene-difluoroborate complex), pyrromethene 546 (PM546, 1,3,5,7,8-pentamethylpyrromethene-difluoroborate complex), pyrromethene 597 (PM597, 1,3,5,7,8-pentamethyl-2,6-di-t-butylpyrromethene-difluoroborate complex) and pyrromethene 556 (PM556, disodium-1,3,5,7,8-pentamethylpyrromethene-2.6-disulfonate-difluoroborate complex) are highly efficient and most popular laser dyes in the green yellow region which are commercially available (https://www.photonicsolutions.co.uk). Among these dyes, PM567, PM546 and PM597 are used in organic solvents (mostly in ethanol), but PM556 is water soluble, thus useful in water based laser systems.

Despite very good optical properties, there are two major drawbacks of the BODIPY dyes such as their small Stokes shifts and low photochemical stability. The low Stokes shifts effectively enhances ground state absorption (GSA) which eventually decreases their lasing efficiency. On the other hand, faster phtodegradation reduces the laser operation life which is a big hurdle for the long-term operation of BODIPY liquid dye lasers. This becomes more severe problem in case of high average power and high repetition rate dye lasers (Mula et al. 2008).

It is suggested that the excited triplet state of the BODIPY dyes produced on photoexcitation transfers the energy to surrounding triplet oxygen ($^3O_2$) to generate singlet oxygen ($^1O_2$) which reacts at the C-8 olefin moiety of the dyes, leading to

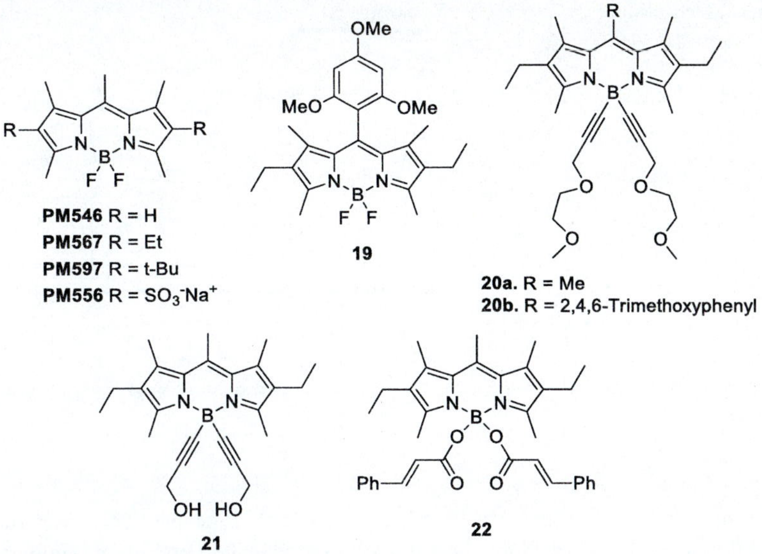

Fig. 13.2 Chemical structures of BODIPY laser dyes

their degradation (Mula et al. 2008). The entire $^1O_2$ generation process and photochemical decomposition of BODIPY dyes is schematically depicted in Fig. 13.3. The dye decomposition can be prevented by using a triplet quencher like benzoquinone (BQ) as well as trace amounts (1 wt% doping level) of antioxidants, such as DABCO, Tin770 and TBP, which supports the proposed dye degradation mechanism. Especially, the highest efficacy of the singlet oxygen-quencher, DABCO, amongst the additives strongly suggests $^1O_2$ as the major causative agent for the dye decomposition (Ray et al. 2007).

BODIPY dyes dispersed in solid matrix such as glass/polymer were also developed to enhance their laser output. In some of these matrices, BODIPY dyes showed better photostability as compared to liquid dye lasers. This is probably due to the unavailability of the oxygen in the solid matrix which reduces the chemical breakdown of the BODIPY dye structures.

Attempts were taken to improve photostability of the BODIPY dyes by their structural modifications. One such effort showed that substitution at the *meso* position of pyrromethene 567 (PM567) laser dye by aryl group increased the photostability (Mula et al. 2008). Especially, *meso*-trimethoxyphenyl BODIPY (**19**) (Fig. 13.2) was twofold more photostable with similar lasing efficiency as compared to PM567 at a significantly lower concentration. The dye **19** generates less $^1O_2$ due to low reactivity of the triplet dye with triplet oxygen. Also due to more steric crowding at *meso*-position, its reaction probability with the $^1O_2$ is lower as compared to PM567. All these enhance its photostability.

In other reports, boron substitutions were shown to enhance the photostability of the BODIPY dyes. For example, dye **20a** and **20b** (Fig. 13.2) with 2,5-dioxaoct-7-yne

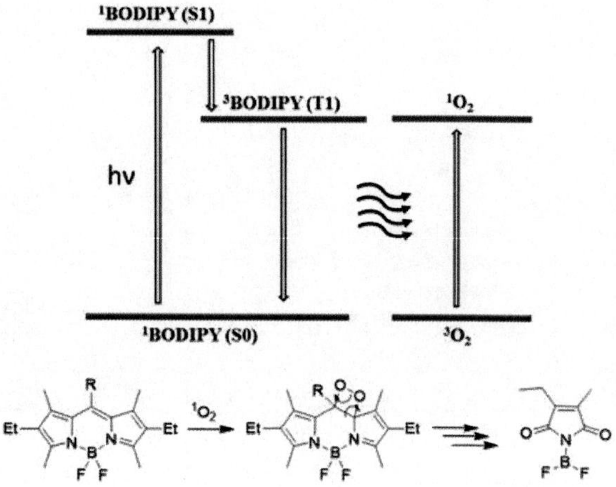

Fig. 13.3 Mechanism of singlet oxygen generation and photochemical decomposition of the BODIPY dyes

substitutions at the boron center of PM567 and BODIPY **19** respectively, were of high lasing efficient with higher photostability as compared to their precursors (Jagtap et al. 2013). Detailed theoretical calculations and pulse radiolysis investigation established that due to the boron substitutions, $^1O_2$ generation capacity of these dyes and also their rate of reaction with $^1O_2$ reduced which effectively increased their lasing life time. Further, the effect of solvent polarity change in the lasing efficiency and photostability of BODIPY dyes were investigated. The BODIPY dye **21** (Fig. 13.2) with propargyl alcohol substitutions at the B-center showed high lasing efficiency with higher photochemical stability as compared to PM567 in both polar (ethanol) and non-polar (1,4-dioxane) solvents (Gupta et al. 2017). The photostability of the dye **21** in 1,4-dioxane is comparable to that of Rhodamine 6G in ethanol, considered to be the benchmark for lasing photostability. The high photostability of dye **21** in the non-polar solvent 1,4-dioxane is mainly due to its low reactivity with $^1O_2$ which was confirmed by methylene blue-induced photo-degradation results, and this was further rationalized by theoretical calculations. Thus, combination of the substitutions at the B-center and use of non-polar solvents as lasing medium are useful to enhance the photostability of the BODIPY dyes immensely keeping their lasing properties unaffected.

Ray et al. developed series of COO-BODIPY laser dyes with outstanding lasing efficiencies and photostabilities (Ray et al. 2020). For example, dye **22** (Fig. 13.2) showed ~68% lasing efficiency as compared to 48% lasing efficiency of the well-known commercial laser dye, PM567 pumped under identical experimental conditions. The dye is highly photostable, no decrease in lasing efficiency was observed until 100,000 pumping pulses whereas 80% drop in lasing emission was found in case of PM567 under the same pumping period and experimental conditions.

BODIPY PM650 (Fig. 13.4), a methylated BODIPY dye with the *meso*-cyano group is a commercially available well known red-emitting BODIPY laser dye (λ_{peak} = 656.5 nm) (Belmonte-Vázquez et al. 2019). The electron withdrawing *meso*-cyano

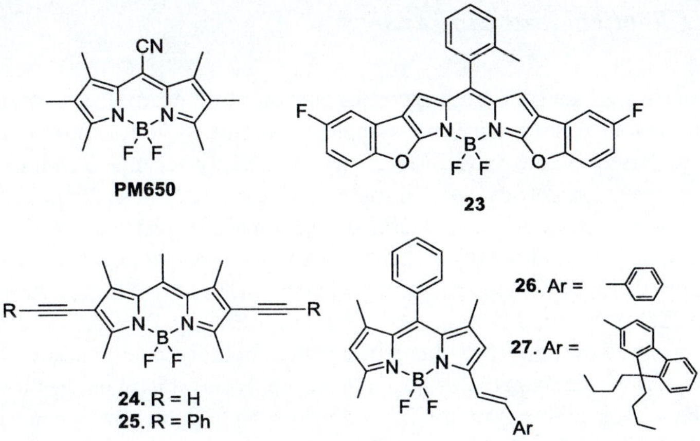

Fig. 13.4 Chemical structures of red shifted BODIPY laser dyes

group lowers its reactivity with the singlet oxygen, responsible for the photodegradation mechanism which eventually makes it highly photostable (tolerance of ≤6.3 GJ/mol). Belmonte-Vázquez et al. synthesized a family of benzofuran-fused BODIPY dyes with bright fluorescence and high lasing efficiencies (>40%) toward the red edge of the visible spectrum (Belmonte-Vázquez et al. 2019). Among these dye **23** has better (43%) lasing efficiency as compared to PM650 (31%) in similar conditions, and its photostability (tolerance of ≤7.4 GJ/mol) surpasses that of the reference dye PM650.

Using the metal mediated cross-coupling reaction discussed *vide supra*, extended π-conjugated dyes, 2,6-diacetylenyl and 2,6-bis(phenylacetylenyl)-BODIPY derivatives (**24** and **25**) were synthesized for lasing applications in the red spectral region (Fig. 13.4) (Maity et al. 2017). Dye **24** and **25** showed stable lasing efficiencies of 41% (tuning range: 561–580 nm) and 36% (tuning range: 602–617 nm) respectively under continuous transverse pumping at 532 nm for 180 and 110 min respectively.

Zhang et al. synthesized a series of highly fluorescent red shifted BODIPY dyes by increasing π-conjugation length as well as tuning Intramolecular Charge Transfer (ICT) effect (Zhang et al. 2011). These dyes showed stable laser emission with high efficiencies in the green to near IR spectra region (570–725 nm) (Zhang et al. 2011). Among these dyes, dye **26** and **27** (Fig. 13.4) showed 46% (λ_{lasing}: 559 nm, pumped at 532 nm) and 57% (λ_{lasing}: 596 nm, pumped at 570 nm) lasing efficiencies respectively.

Thus, tremendous molecular engineering was done to synthesize functional BODIPY laser dyes applicable in the UV–Vis to NIR region. Different strategies were also adopted to enhance their photostability to increase lasing life time.

13.5.2 Chemical and Bio-sensors

Fluorescence based sensors are highly popular due to their high sensitivity, selectivity, rapid response, simplicity, low sample requirements, non-destructiveness etc. Organic dye based chemo- and bio-sensors are mostly relying on modulation of their fluorescence properties upon trapping the analyte. These kinds of efficient chemo- and bio-sensors are very useful in analytical chemistry as well as in clinical, medical, and environmental applications. Fluorescence changes in these sensors occur mainly *via* Photoinduced Electron Transfer (PET) or by Photoinduced Charge Transfer (PCT) or by the Fluorescence Resonance Energy Transfer (FRET) mechanism. Thus, the key design of these molecules includes chemical attachment of some pre-designed site capable of trapping the analytes which changes the fluorescence of the sensor. In this manner, recognition of the analyte can enhance (turn-on) or decrease (turn-off) the fluorescence, but in conventional design, turn-on fluorescence of the sensor in the presence of an analyte is preferred. Change in fluorescence colour of the sensor (ratiometric sensing) after trapping the analyte is also very interesting. Potential of the BODIPY dyes as the molecular sensor was first explored by Daub and Rurack and after that numerous examples of BODIPY-based fluorescent chemo- and bio-sensors were reported (Boens et al. 2012).

BODIPY dyes containing *p*-(N,N-dialkyl)aniline subunit were used extensively as pH sensors. These dyes are low/non-fluorescent due to PET from anilino group to BODIPY core which can be prevented by protonation of the anilino moiety and thus the dyes become fluorescent after protonation due to prevention of PET. For example, *meso p*-(N,N-dialkyl)aniline substituted BODIPY **28** (Fig. 13.5) acts as highly efficient "turn-on" type pH sensor in the greenish-yellow spectral region (Boens et al. 2012). Subsequently, red shifted 3,5-distyryl-BODIPY dye (**29**) (Fig. 13.5) were synthesized via Knoevenagel type condensation reactions discussed vide supra. Dye **29** was low fluorescent ($\lambda_{abs} = 700$ nm, $\lambda_{em} = 753$ nm, $\Phi_{fl} = 0.18$) in the neutral state but in the presence of acid, both the absorption and emission bands hypsochromically shifted and it lit up ($\lambda_{abs} = 620$ nm, $\lambda_{em} = 630$ nm, $\Phi_{fl} = 0.68$) (Ziessel et al. 2009). It was covalently attached with the porous polyacrylate beads and efficiently used as the solid-state colorimetric and fluorescent sensor for analysis of HCl in a gas stream and aqueous solution.

Ratiometric pH sensor based on BODIPY dyes are also known. Imino-BODIPY dye **30** (Fig. 13.5) showed dramatic colour change from pink to yellow in the presence of acid due to the cleavage of its acid-labile imine group. This also resulted in change in fluorescence from orange to green in the pH range of 1.8 to 7.4 (Gupta et al. 2013).

Several anilino substituted BODIPY dyes were also exploited for detection and quantification of various metal ions as well as other important analytes. For example, *meso*-ortho phenylenediamine BODIPY (**31**) is non-fluorescent due to photo-induced electron transfers of amino lone pairs to the BODIPY core. But both the amine residues are highly reactive towards nitric oxide (NO) to form dye **32** which is highly fluorescent (Fig. 13.6). This fluorescence enhancement of dye **31** in the presence of NO was used for efficient detection of in situ generated NO (Gabe et al. 2004). Nitric

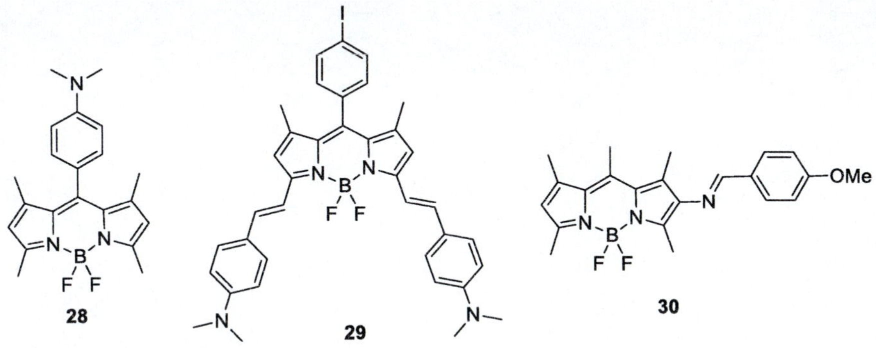

Fig. 13.5 BODIPY based pH sensors

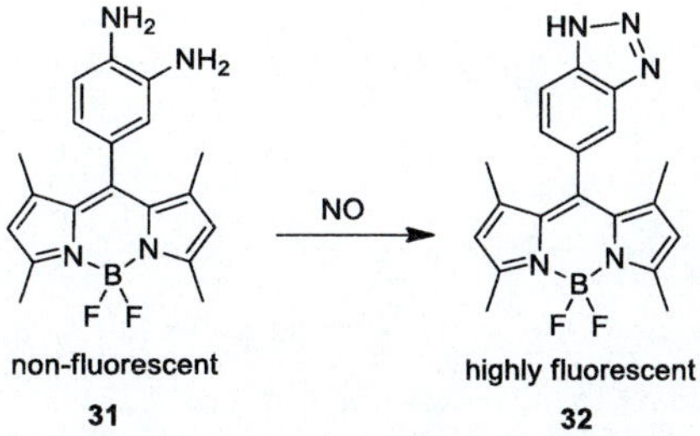

Fig. 13.6 Reaction of *meso*-ortho phenylenediamine BODIPY (**31**) with NO

oxide imaging in living cells using other BODIPY dyes is also discussed *vide infra* in the cellular imaging section.

Adopting the similar strategy, anilino-BODIPY dyes with coordinating ligands were used as sensor for different alkali and transition metal ions. For example, BODIPY dye **33** (Fig. 13.7) with the 13-phenyl-1,4,7,10-tetraoxa-13-azacyclopentadecane chelator is non-fluorescent, but after complexing with different alkali (Li^+, Na^+) and alkaline-earth (Mg^{2+}, Ca^{2+}, Sr^{2+}, Ba^{2+}) metal ions, it became highly fluorescent showing its capability as the metal sensor (Kollmannsberger et al. 1998). The metal specificity was tuned with a different chelator. BODIPY dye **34** (Fig. 13.7) with the *meso*-16-phenyl-1,4,7,10,13-pentaoxa-16-azacyclooctadecane chelating group binds efficiently with Pb^{2+} ion showing turn-on fluorescence (Mbatia et al. 2010). Chang et al. used a mixed N/O/S receptor based ligand to synthesize water soluble BODIPY dye **35** (Fig. 13.7) which showed highly selective turn-on fluorescence in the presence of Ni^{2+}. Dye **35** is capable of detecting intracellular Ni^{2+} and did not show any fluorescence response with the other biologically abundant metal ions (Dodani et al. 2009).

Different other strategies were also used to develop BODIPY based metal sensors. BODIPY-phenanthroline conjugate **36** (λ_{em} = 535 nm) and BODIPY-acetylacetone conjugate **37** (λ_{em} = 570.0 nm) (Fig. 13.7) were shown to have selective Cu^{2+} sensing ability where both the dyes showed turn-off fluorescence responses due

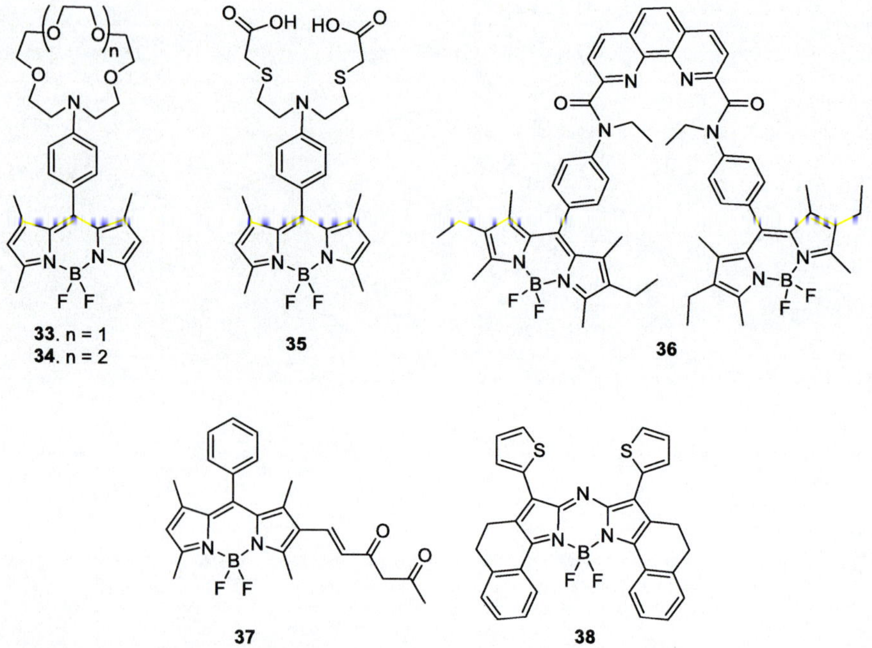

Fig. 13.7 Metal sensitive BODIPY dyes

to BODIPY to metal charge transfers (More et al. 2017; Gorai et al. 2022). Xiao et al. developed a highly selective near-infrared fluorescent probe for Hg^{2+} based on thiophene containing aza-BODIPY **38** ($\lambda_{em} = 782$ nm) (Fig. 13.7) (Jiang et al. 2016). Hg^{2+} ion coordinated with the C-8 N and S atoms of 1,7-thienyl groups inhibiting the free rotations of the thienyl groups. This caused intramolecular charge transfer from thienyl groups to the aza-BODIPY moiety leading to fluorescence quenching. The probe was also shown to detect Hg^{2+} in living cells. Zn^{2+} ion sensing using BODIPY dye and its application is discussed *vide infra* in the cellular imaging section.

BODIPY based efficient bio-sensors are also well-known. BODIPY based fluorescence techniques were developed for detection and quantification of proteins useful for clinical applications. Chang et al. synthesized a mega Stokes shift BODIPY-triazole dye (**39**) (Fig. 13.8) capable of very specific binding with human serum albumin (HSA) as compared to other serum albumins showing 220-fold fluorescence enhancement. Successful quantification of HSA in urine samples within the concentration limits of micro-albuminuria using dye **39** showed its potential in clinical application ability (Er et al. 2013).

BODIPY dyes are also extensively used for the detection of protein fibrils which are highly important for the diagnosis and treatment of amyloid related neurological diseases. Detection of self-assembly of diphenylalanine into nanofibers is very important to get the mechanistic inside of the β-amyloid aggregation which is the central cause for the Alzheimer's disease. Quan et al. reported that the green BODIPY dye

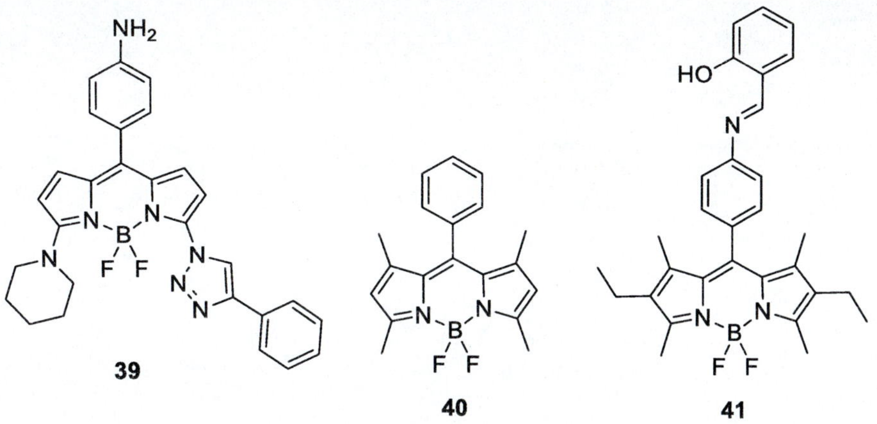

Fig. 13.8 BODIPY dyes used for detection of protein and protein fibrils

40 (Fig. 13.8) showed very high turn-on fluorescence with nanofibers of diphenylalanine, thus can be useful for early detection as well as drug development of Alzheimer's disease (Quan et al. 2019).

Nath et al. also developed a turn-on fluorescent based NIR-emitting BODIPY glycoside probe for detection of matured insulin fibrils (Mora et al. 2021). Most importantly the probe can also detect oligomers formation from the native protein. The probe can be useful for the in vivo imaging of protein oligomers and matured fibrils. In a separate report they also investigated the structure activity relationship of a BODIPY-salicylaldimine Schiff base (**41**) (Fig. 13.8) and its corresponding boron complex dye for their sensitivity towards amyloid fibrils from hen-egg white lysozyme (Sen et al. 2022). The dye **41** showed high fluorescence enhancement as compared to its boron complex upon binding with lysozyme fibrils.

13.5.3 Cellular Imaging

Outstanding fluorescence properties along with the easy cell permeability of the BODIPY dyes (due to its hydrophobicity) made them highly useful candidates for developing cellular imaging agents. Due to their relatively high lipophilicity, they tend to accumulate in the subcellular membranes. Additionally, several molecular engineering techniques were done to develop organelle specific imaging agents.

For example, BODIPY dye **42** (Fig. 13.9) showed preferential localization in the endoplasmic reticulum (ER) confirmed by co-staining experiments (Jiao et al. 2010). The commercially available ER-Tracker™ Red (BODIPY™ TR Glibenclamide) is a highly selective ER staining dye. In this, green-fluorescent BODIPY™ TR dye is attached with glibenclamide which is very specific to bind with the sulphonylurea receptors of ATP-sensitive K^+ channels present in ER.

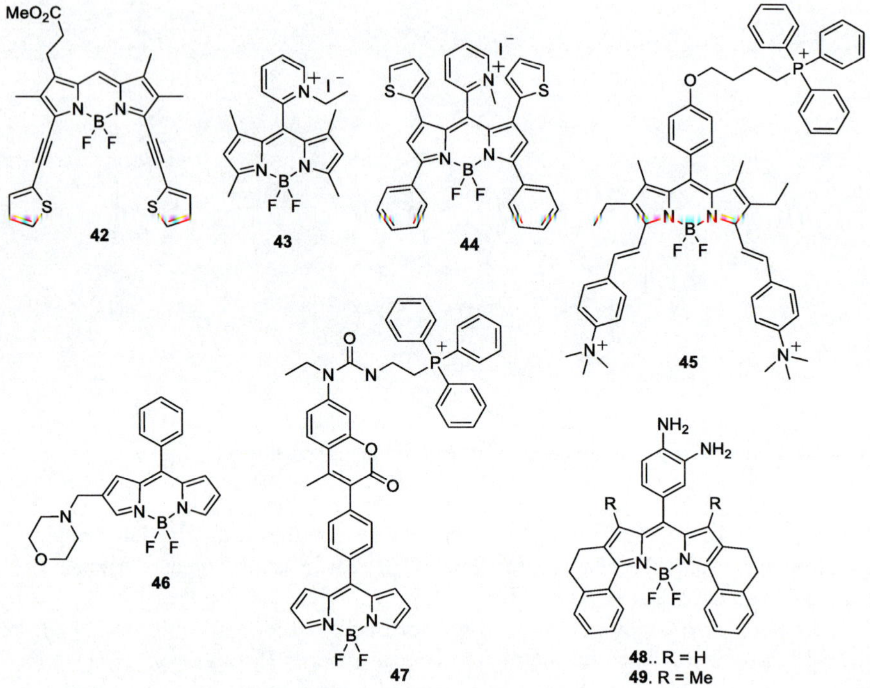

Fig. 13.9 BODIPY dyes for cellular imaging

Several mitochondria specific BODIPY dyes were also developed. As mitochondria is negatively charged, the positively charged dyes tend to accumulate in it. In general, BODIPY dyes are neutral, thus cationic moieties such as ammonium, pyridinium, phosphonium moieties were chemically inserted in the BODIPY core to synthesize positively charged BODIPY dyes for mitochondria imaging. For example, the pyridinium BODIPY dyes **43** (Zhang et al. 2013) and **44** (Jiang et al. 2013) (Fig. 13.9) are water soluble and photostable mitochondria targeting dyes useful for imaging. Jiang et al. developed a red emitting distyryl BODIPY dye (**45**) (Fig. 13.9) as an excellent alternative to the commercially available Mito Tracker Red (Wang et al. 2021). The ammonium groups increased its water solubility and the phosphonium group enhanced its mitochondria targeting ability. The dye penetrates the cellular and mitochondrial membranes and accumulates in mitochondria in high densities. Additionally, it showed low photobleaching and phototoxicity as compared to Mito Tracker Red which established it as an excellent mitochondria imaging dye.

Several BODIPY based viscosity probes were developed to visualize intracellular viscosity changes. Molecular rotor based BODIPY dyes are used for this purpose as these dyes showed restricted bond rotation in viscous medium which changed their fluorescence properties. For example, dye **46** (Fig. 13.9) was synthesized to probe lysosomal viscosity changes (Wang et al. 2013). The attached morpholine moiety made it lysosome specific dye and, reduction in PET from morpholine to BODIPY core in lysosomal acidic environment as well as viscosity driven restricted rotation of the *meso*-phenyl group turn-on its fluorescence.

The dye **47** (Fig. 13.9) was developed for the mitochondrial viscosity measurement. The dye was designed in such a way that the attached phosphonium moiety made it mitochondria targeting and the *meso*-coumarinylphenyl formed molecular rotor structure which showed viscosity responsive turn-on fluorescence of BODIPY core at 516 nm (Yang et al. 2013). The coumarin fluorescence (427 nm) remained unchanged with viscosity change, thus the intensity ratio of BODIPY to coumarin emission peaks showed linear response with increase in viscosity. Using this dye mitochondrial viscosity in living HeLa cells was determined as 62 cP which increased to 110 cP upon treatment with an ionophore like monensin or nystatin.

Nitric oxide (NO), an endogenously produced gaseous signaling molecule can be visualized in cells by using a highly reactive dye. Reactivity towards NO and fluorescence turn-on behaviour of o-phenylenediamine-BODIPY moiety was discussed *vide supra.* A similar strategy was adopted for imaging of NO in cells and tissues. The NIR-fluorescent BODIPYs **48** and **49** (Fig. 13.9) were highly sensitive NO probes showing >400-fold increase in fluorescence intensity in the presence of NO with very low detection limit i.e. 2.1 nM and 0.6 nM respectively. Both the dyes were cell-permeable and capable of NO imaging in living cells without any cytotoxicity (Zhang et al. 2014).

13.5.4 Photodynamic Therapy

Cancer is the leading cause of death, accounting for millions of deaths every year worldwide. Thus, research and treatment of cancer is an urgent topic where almost all the developed countries are spending huge amount of money every year. Most popular tumor therapies are surgery, chemotherapy and radiotherapy. Compared to these, photodynamic therapy (PDT) is relatively a new treatment modality for malignant tumors which has many unique advantages such as (a) useful for the site/organ unsuitable for surgery; (b) suitable for people (infirm, the elderly etc.) not fit for surgery, chemotherapy and radiotherapy; (c) safe and effective reusability; (d) does not suppress the immune system; (5) useable after or at the same time in a synergistic way with surgery, chemotherapy and radiotherapy (Brown et al. 2004).

Photodynamic therapeutic agents are the triplet photosensitizers (PSs) in which intersystem crossing (ISC) is very efficient. Thus, triplet photosensitizers absorb light to excite to their singlet excited state and then eventually go to the triplet state via efficient ISC. Then the radiative decay of the triplet state i.e. phosphorescence occurs which has a long lifetime (microseconds and above). This radiative energy excites the ground sate oxygen (triplet oxygen) to generate singlet oxygen ($^1O_2$) (Fig. 13.3) which is highly reactive and subsequently generates different reactive oxygen species (ROS). ROS including $^1O_2$ are responsible for the killing of the cancer cells where the triplet photosensitizers are accumulated.

Currently, porphyrin derivatives, such as porfimer sodium (Photofrin), protoporphyrin IX, and temoporfin etc. are the most studied photosensitizers for PDT applications (Ethirajan et al. 2011). Although they have their own shortcomings such as, small Stoke shift, low molar absorptivity ($<2 \times 10^4$ cm^{-1} M^{-1} in the range of 650–900 nm), high aggregation tendency due to strong π-π stacking which can reduce the quantum yield of $^1O_2$, synthetic inaccessibility which restricts the production in large amount by conventional organic synthetic strategy etc. Other non-porphyrin photosensitizers such as phthalocyanine, squarine dye and perylene diimide were reported for PDT study, however poor chemical and photostability and, aggregation in a polar environment restrict their practical applications (Wainwright 1996).

In the past decade, BODIPY class of dyes have been established as a promising PDT agent due to their exceptional photophysical properties (Kamkaew et al. 2013; Turksoy et al. 2019). The intersystem crossing (ISC) can be facilitated by the attachment of heavy atoms which eventually increases singlet oxygen yield. Therefore, it is a widely adopted strategy for the development of efficient PDT molecules. Thus, similar strategy was also used to design BODIPY based PDT molecules. Chemically, all the positions of the BODIPY core are reactive for heavy atom (Br/I) addition without disrupting the planarity of the dye. Additionally, cellular localization of these photosensitizers also can be ascertained by the fluorescent imaging technique.

O'shea et al. synthesized 2,6-dibromo aza-BODIPY chromophore (**50, 51**) which showed efficient singlet oxygen generation at lower concentrations as compared to their parent BODIPY dyes (Fig. 13.10) (Killoran et al. 2002; Gorman et al. 2004). Yogo et al. synthesized the simple 2,6-diiodo BODIPY dye **52** (Fig. 13.10) which showed 1.34 times greater $^1O_2$ generation as compared to well-known PDT photosensitizer Rose Bengal (Yogo et al. 2005).

Further, different iodo-substituted BODIPY dyes were synthesized to check the effects of halogen substitution patterns on the photosensitizing ability of BODIPY dyes which eventually helped to increase the heavy atom effect (Ortiz et al. 2012). The study showed that the substitution of iodine at 2 and 6 positions has the highest impact on singlet oxygen generation (**53, 54**). Iodine substitution at 3,5-positions (**55, 56**) did not show any distinct increase in singlet oxygen generation efficiencies (Fig. 13.10).

Thus, 3,5-positions of BODIPY core were used to extend π-conjugation for the synthesis of near-infrared (NIR) absorbing dyes useful for practical application in the body's therapeutic window (650–900 nm). Further, other functionalities were also added to induce water solubility, cancer cell affinity etc. For example, Atilgan et al. developed 2,6-dibromo/iodo-3,5-distyryl derivatives with or without polyethylene glycol (PEG) side chains (**57–59**) (Fig. 13.10) (Atilgan et al. 2006). The halogens were used to generate $^1O_2$, styryl units extended conjugation to the NIR region (650–680 nm) and water-soluble PEG groups enhance their solubility in aqueous solutions, cell permeability and tumor targeting properties. Beside all these, PEG helps in alleviating aggregate formation in aqueous condition which is the main reason for inefficient formation and potential quenching of the triplet state of the dye and singlet oxygen. Compound **59** ($IC_{50} = 11$ $ngmL^{-1}$) was the most potent candidate showing even higher photocytotoxicity as compared to the most used photosensitizer

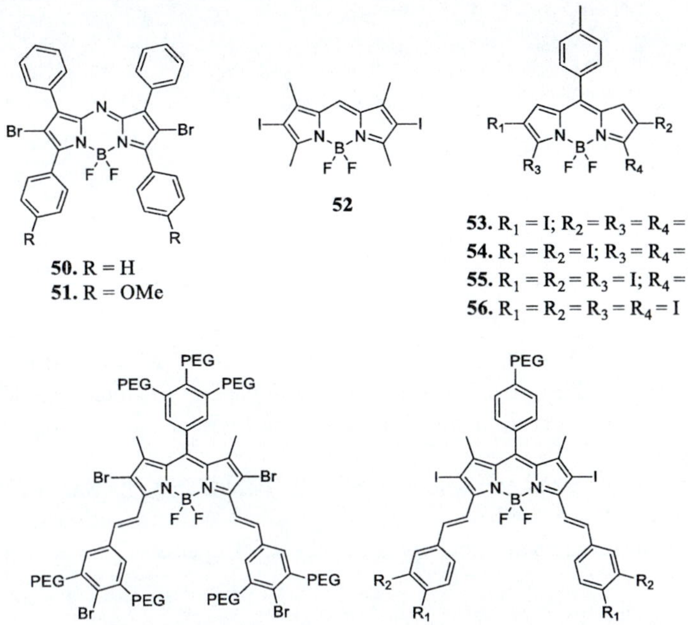

50. R = H
51. R = OMe

52

53. $R_1 = I; R_2 = R_3 = R_4 = H$
54. $R_1 = R_2 = I; R_3 = R_4 = H$
55. $R_1 = R_2 = R_3 = I; R_4 = H$
56. $R_1 = R_2 = R_3 = R_4 = I$

PEG = $O(CH_2CH_2O)_3CH_3$
57

58. $R_1 = OCH_3; R_2 = H$
59. $R_1 = R_2 = O(CH_2CH_2O)_3CH_3$

Fig. 13.10 Halogenated BODIPY dyes for PDT applications

porfimer sodium ($IC_{50} = 4600$ ngmL^{-1}) under the same condition. This is probably due to its low aggregation and high cellular uptake in the biological environment due to attached triethylene glycol chains.

It is well established that the attachment of heavy atoms enhances dark toxicity, decreases triplet state lifetime and, changes bio-distribution and pharmacokinetics of the PDT agents. Thus, heavy-atom-free organic triplet photosensitizers (PSs) are advantages for practical applications. Several heavy atom free BODIPY dyes are also reported to be used as PDT photosensitizers. Shivran et al. synthesed water-soluble BODIPY glycosides with different fluorescence colour and studied their efficiencies as PDT agents (Shivran et al. 2016). Amongst them glycosylated mono-styryl-BODIPY dye **60** ($\lambda_{abs} = 573.8$ nm and $\lambda_{em} = 590$ nm) (Fig. 13.11) showed best PDT activities against the A549 cell line and importantly it is non-toxic to normal lung cells.

Orthogonal BODIPY dimers (Bis-BODIPYs) are another class of heavy atom free BODIPY dyes for potential singlet oxygen photosensitization (Cakmak et al. 2011). As compared to their monomer ($\lambda_{abs} = 530$ nm), the dimer **61** (Fig. 13.11) showed split band absorption maxima (490 and 560 nm) and large Stokes shift (>80 nm) with decreased florescence quantum yield. High ISC from singlet to triplet excited state resulted in high quantum yield of singlet oxygen production (0.4 in toluene and 0.5

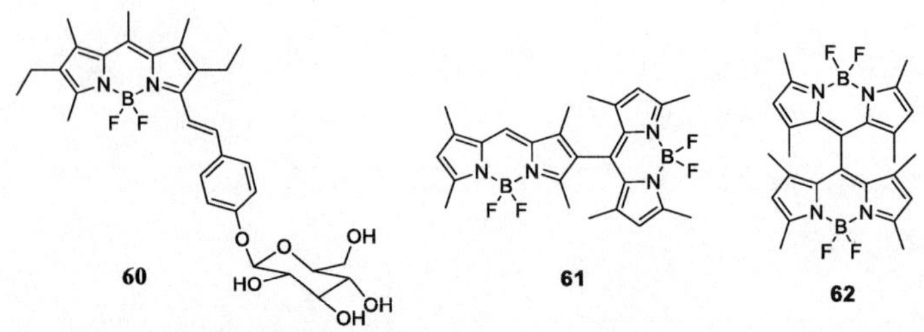

Fig. 13.11 Heavy atom free BODIPY dyes for PDT applications

in dichloromethane) whereas the monomer produces very low $^1O_2$ (<0.1). On the other hand, dimers **62** (Fig. 13.11) showed lower singlet oxygen quantum yield with significant florescence emission and thus promising for dual use as therapeutic and imaging agents.

Thus, both halogenated non-halogenated BODIPY PDT agents were developed. Some of the dyes showed high photosensitizing ability. Further experimentations are required to establish them as clinically applicable PDT agents.

13.5.5 Fluorescent-Positron Emission Tomography Probes

High resolution molecular imaging techniques are important for diagnostic and therapeutic purpose i.e., in vivo imaging of specific biological pathways at the molecular and cellular level. Various high-end imaging techniques are available such as Computed Tomography (CT), Single Photon Emission Computed Tomography (SPECT), Positron Emission Tomography (PET), bioluminescence, fluorescence and Magnetic Resonance Imaging (MRI) etc. But all these molecular imaging techniques have their own advantages and limitations in spatial and temporal resolution, depth penetration, sensitivity and cost (Tsien 2003). Thus, fusion of two or more techniques can be very useful to obtain synergistic effect avoiding these limitations of each individual techniques. PET-CT, PET-MRI are the examples of such dual imaging techniques. Similarly, PET-fluorescence dual modality imaging technique could be of highly beneficial for clinical applications. PET is a sophisticated imaging technique useful for noninvasive diagnosis of cancer via locating the in vivo distribution of radio labeled biomolecules. In contrast, fluorescence imaging is a superior technique useful for intra operative tumor detection (Nguyen et al. 2010). Thus, due to these complementary natures of PET and optical imaging techniques, PET/fluorescence dual modality imaging technique might be greatly useful for both diagnosis and therapeutic purpose of cancers. Noninvasive PET scans will be useful for locating the lesion (diagnosis), and intra operative fluorescence image-guided surgery will be helpful for the surgeons to identify the PET-detected lesions or smaller metastasis (therapy).

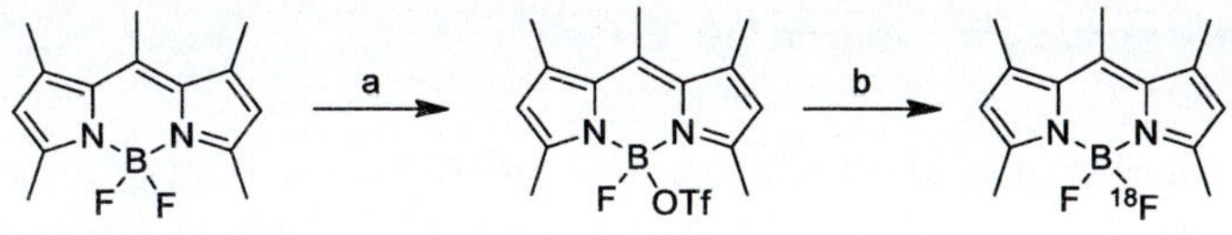

PM546 63 64

Scheme 6. Synthesis of ^{18}F-BODIPY dye: (a) TMSOTf, 20 °C, 1 min; (b) $^{18}F^-$, 20 °C, 1 min

BODIPY dyes are advantageous over other well-known fluorescent dyes for intracellular imaging within living cells as discussed vide supra. Due to their neutral nature, they efficiently penetrate the cell membranes Additionally, ^{19}F/^{18}F exchange is possible at the BF_2 group present in the central core to develop hybrid ^{18}F PET/optical imaging agents.

Few methods are reported to synthesize ^{18}F-BODIPY, among them the easiest method was described by Mazitschek et al. First, activated BODIPY triflate (**63**) was synthesized (1 min, near quantitative yields) followed by addition of free $^{18}F^-$ which converts it to ^{18}F-BODIPY (**64**) almost instantly (Scheme 6) (Hendricks et al. 2012). The method is easily reversible, rapid, and efficient in incorporation of 18-fluoride in the BODIPY core.

Hyunjung Kim et al. synthesized a red shifted ^{18}F-BODIPY dye **65** ($\lambda_{abs} = 580$ nm and $\lambda_{em} = 590$ nm) (Fig. 13.12) useful for PET/fluorescence imaging of brain (Kim et al. 2019). PET/optical imaging data showed that it was suitable for brain penetration with desirable brain pharmacokinetics of the radio ligand and thus can be widely used as a prosthetic group for the brain hybrid PET/optical imaging agent. Further ^{18}F-BODIPY dye (**66**) (Fig. 13.12) was developed by Giuseppe Carlucci et al. for dual-modality optical/PET imaging of PARP1 in Glioblastoma (Carlucci et al. 2015). The fluorescent component of **66** enables optical imaging with cellular resolution, while the radiolabeled component helps whole-body deep-tissue imaging of malignant growth. Thus, BODIPY dyes showed high promise as fluorescent-PET probes for the development of PET/fluorescence dual modality imaging technique.

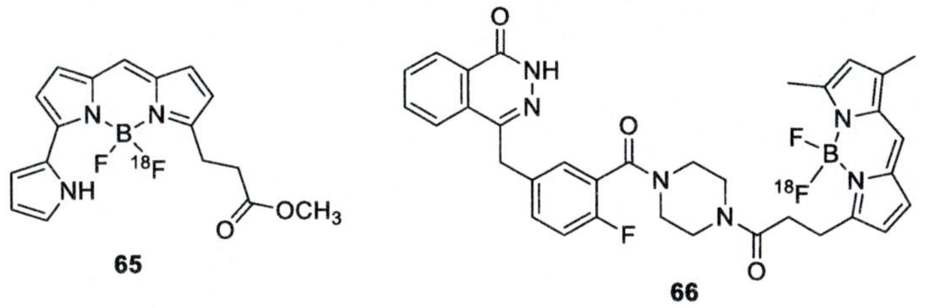

65 66

Fig. 13.12 ^{18}F-BODIPY dyes for PET/optical dual imaging

13.5.6 Organic Photovoltaics (OPV)

Development of organic small molecules for efficient solar energy conversion into chemical energy is a major contemporary challenge for chemists. In the photosynthesis processes, natural photosynthetic organisms act as efficient antennas to absorb photons in the visible spectrum and then funnel energy to the reaction center for photosynthesis. Mimicking this, several BODIPY based light-harvesting systems were developed. High extinction coefficient, wavelength tuning from UV–Vis to NIR regions, good electron mobility etc. are the positive elements which made BODIPY to be considered for futuristic organic photovoltaics.

Ziessel et al. first developed synthetic strategies for polyaromatic appended BODIPY dyes with the aim that the energies absorb by the ancillary light absorbers will channel to the BODIPY core. In this way, several dual-dye systems (**67–70**) (Fig. 13.13) were synthesized and efficient energy transfer (90%) were shown from the attached aromatic polycycles to the BODIPY core (Ulrich et al. 2008).

Further, this concept was extended to develop highly sophisticated energy transfer cassette **71** (Fig. 13.14) where dyads of pyrene and yellow fluorescent BODIPYs are connected with NIR fluorescent BODIPY dye via 1,4-phenylene-diethynylene moieties (Harriman et al. 2009). In **71**, excitation of the pyrene moiety at 370 nm resulted in bright red emission at 670 nm ($\Phi_{fl} = 51\%$) which confirms highly efficient energy transfer from blue fluorescent pyrene to green fluorescent BODIPY to ultimately red fluorescent BODIPY fluorophore. This type of light-harvesting system with wide absorption band are extremely useful for developing organic solar cells.

Leclerc et al. developed strategies to synthesize ethynylene and vinylene bridged dumbbell-shaped triazatruxene-BODIPY conjugates for bulk heterojunction (BHJ) solar cells (Bulut et al. 2017). Both the dyes were used as electron donor in devices formed in blending with phenyl-C71-butyric acid methyl ester ($PC_{71}BM$) as the electron acceptor. The ethylene bridged triazatruxene-BODIPY conjugate **72** (Fig. 13.15)

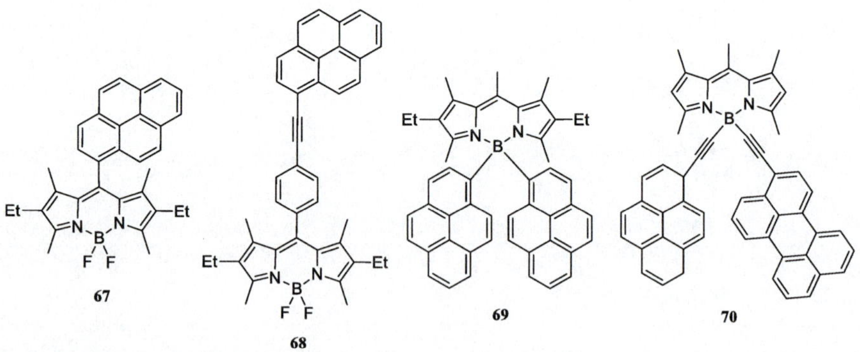

Fig. 13.13 BODIPY dyes attached with secondary polycycle chromophores as ancillary light absorber

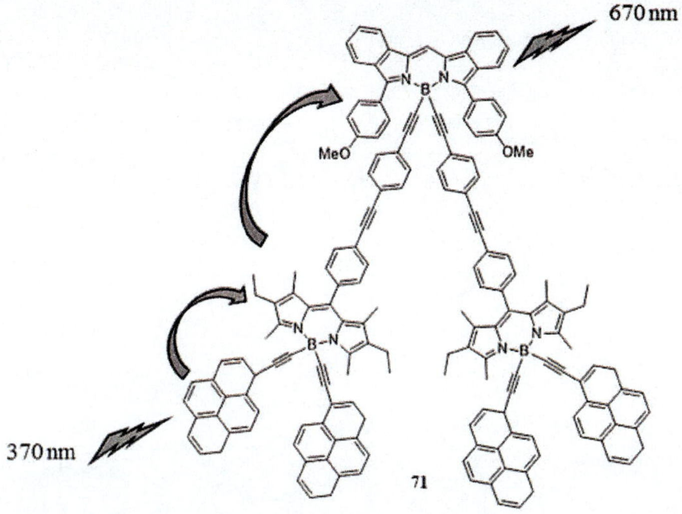

Fig. 13.14 Light-harvesting system containing different coloured BODIPYs and pyrenes

showed high power conversion efficiency of 5.8%. Singh et al. synthesized dithiafulvalene decorated BODIPY core (BODIPY-DTF, **73**) (Fig. 13.15) with efficient electron donor capacity. The OPV devices fabricated by blending BODIPY-DTF and $PC_{71}BM$ as electron acceptor showed highly efficient (7.2%) solar energy conversion (Srinivasa Rao et al. 2017). Several other BODIPY based small molecules showed excellent potential for being used as artificial light harvesters (Bessette and Hanan 2014).

13.5.7 Self-assembled Architectures

Soft materials and nanoparticles prepared from self-assembly of monomers show diversified property useful for numerous applications. Gels, dendrimers, liquid crystals, micelles, liposomes, microcapsules etc. are examples of different soft materials. These kinds of soft materials were also prepared by self-assembly of functional BODIPY based compounds for chemical, biological and optical applications. Ziessel et al. did the pioneering work in this area by incorporating the pre-designed BODIPY core into different supramolecular assemblies to develop coloured liquid crystals and organogels.

For example, 1:2 complex of yellow fluorescent anionic dye (**74**) and red fluorescent cationic dye (**75**) formed liquid crystals on heating in room temperature to above 150 °C (Fig. 13.16) (Olivier et al. 2012). Highly efficient energy transfer from the yellow fluorescent dye to the red fluorescent dye was observed in the mesomorphic state, thus it can be used as the light harvesting array on solar cell applications.

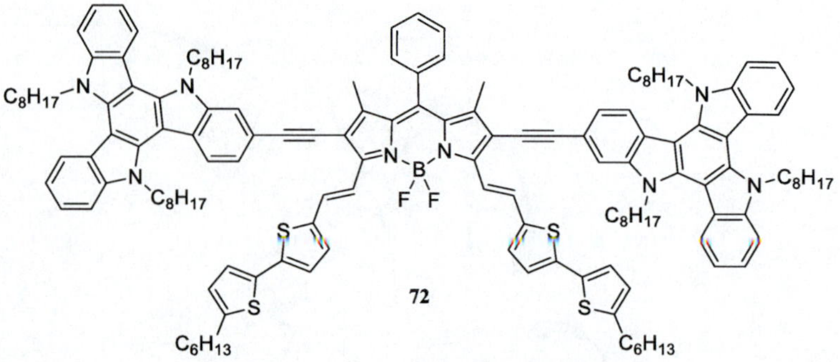

72

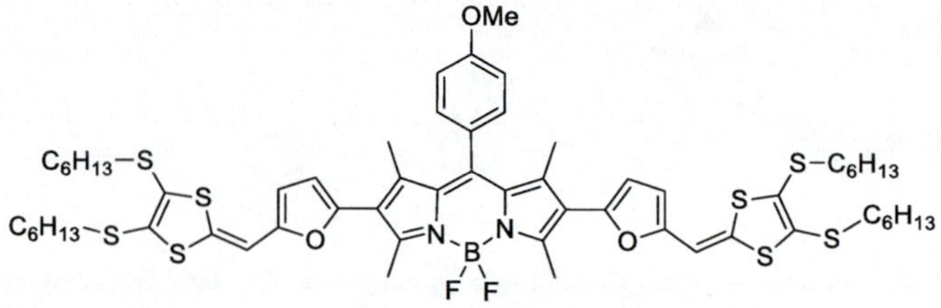

73

Fig. 13.15 BODIPY dyes for solar energy conversion

In another report, green and red emitting dyes were coupled with different generations of dendrimers to develop differently coloured liquid crystals (**76–79**) (Fig. 13.16) (Mula et al. 2015). The supramolecular assembly differs with the dendrimers generation, such as smectic A and C phases were observed with first-generation dendrimers whereas second- and third generation dendrimers showed nematic and/or smectic A phases.

Self-assembled BODIPY nanoparticles were shown to have great potential as sensors for important biological analytes. Amphiphilic BODIPY-*O*-glycoside dye **80** (Fig. 13.17) was reported to be self-assembled in aqueous solution to form non-fluorescent nanoparticles (NPs) (Shivran et al. 2021). These NPs showed high fluorescence enhancement in the presence of serum albumin, thus useful for selective detection and quantification of serum albumin. Human serum albumin (HSA) detection in urine samples were also shown using these NPs. In another report, self-assembled non-fluorescent nanoparticles (NPs) of BODIPY dye **81** (Fig. 13.17) showed selective turn-on fluorescence in the presence of Zn^{2+}. This was used in detection and quantification of Zn^{2+} in human hair (Jia et al. 2016).

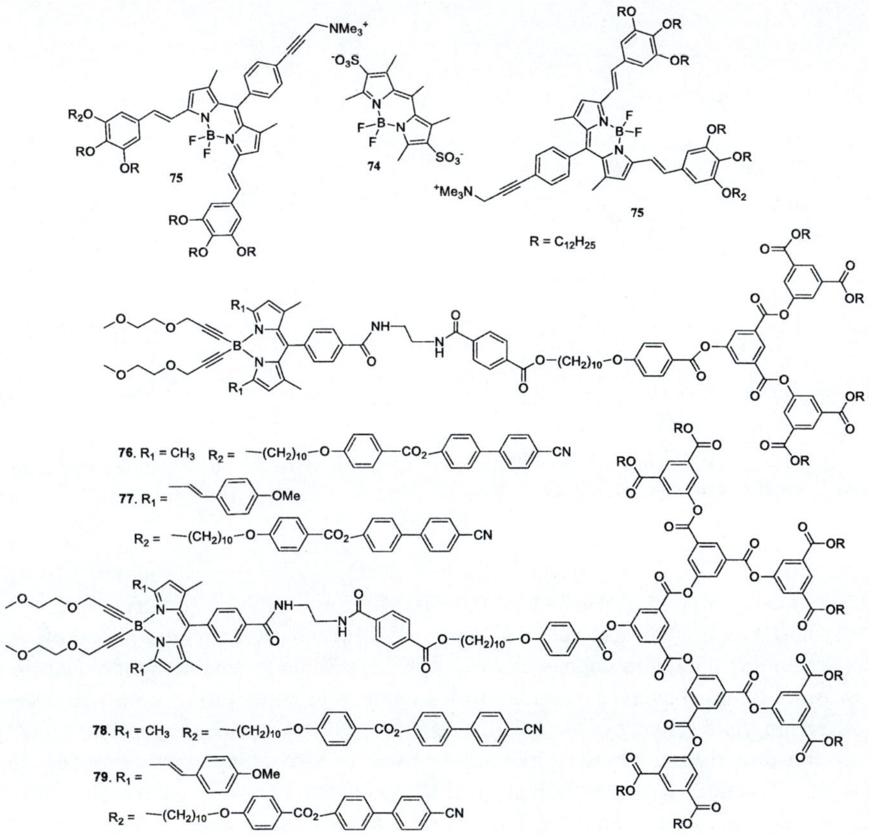

Fig. 13.16 Chemical structures of BODIPY based compounds showing liquid crystal properties

13.5.8 Photocatalysis

BODIPY dyes are also established as an efficient photocatalyst in various important organic reactions as well as chemical processes. Triplet state of the dye is responsible for the catalytic behaviour of the BODIPY dyes. During de-excitation of the triplet state, singlet oxygens are generated (Fig. 13.3) which are shown to be efficiently catalyzed organic reactions. For example, reaction of sulfides with singlet oxygen produce sulfoxides which are important intermediates for the total synthesis of biologically relevant molecules. BODIPYs are successfully used for photooxidation of sulfides into sulfoxides. BODIPY dye **40** was used to oxidize thioanisole (**82**) to corresponding methyl phenyl sulfoxide (**83**) quantitatively which took 24 h to complete the reaction. But the corresponding diiodo compound **84** took only 3 h to complete the reaction (Scheme 7) (Li et al. 2013). Dye **84** has higher ISC (triplet

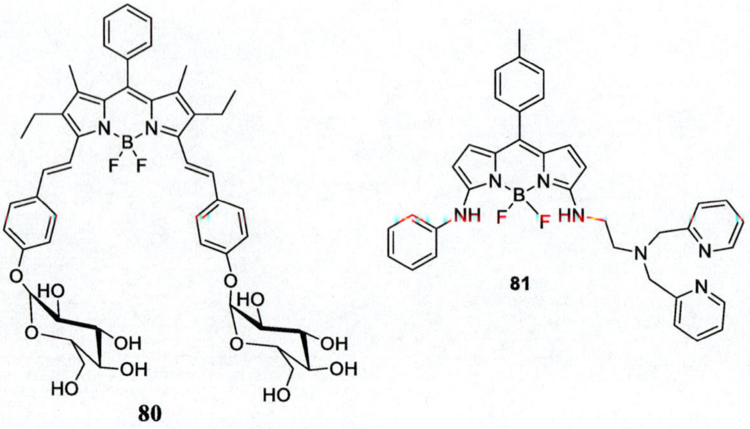

Fig. 13.17 Chemical structures of BODIPY dyes form self-assembled nanoparticles useful for sensing important biological analytes

quantum yield 0.83%) i.e. it generates higher $^1O_2$ generation as compared to **40** (triplet quantum yield 0.01%) which increases its photocatalytic activity.

BODIPY dye **84** is also very effective in photocatalytic oxidation naphthols to corresponding naphthoquinones. For example, dye **84** under visible light excitation (35 W Xe lamp) converts 1-naphthol to 1,4-naphthoquinone (**86**) in 1.5 h with 74% conversion yield (Huang et al. 2013). The 1,4-naphthoquinone was further reacted with substituted anilines and copper acetate to synthesize aminonaphthoquinones in a one pot reaction. The adduction product **87** was obtained in very good yield (75%) (Scheme 7). The yield of aminonaphthoquinone **87** was higher when BODIPY dye **84** was used as a photocatalyst as compared to well-known photocatalyst, tetraphenyl-porphyrin (TPP). The higher photocatalytic activity of **84** is probably its due to its stronger visible light-harvesting ability.

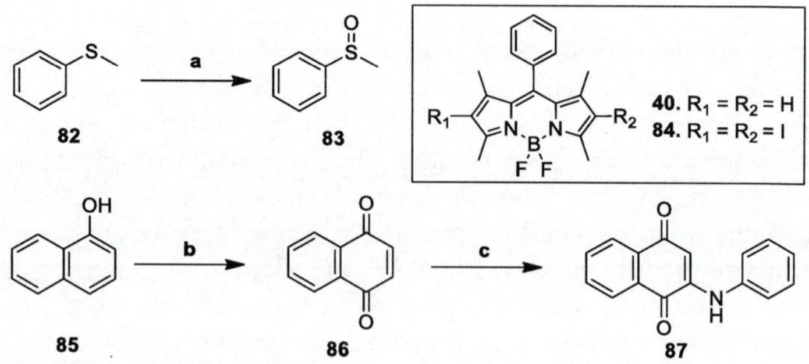

Scheme 7. a 40/84, visible light, air, CH_3OH; **b 84** (2 mol %), visible light, air, CH_2Cl_2/CH_3OH, 20 °C; **c** aniline, copper acetate, acetic acid, 65 °C, 3 h

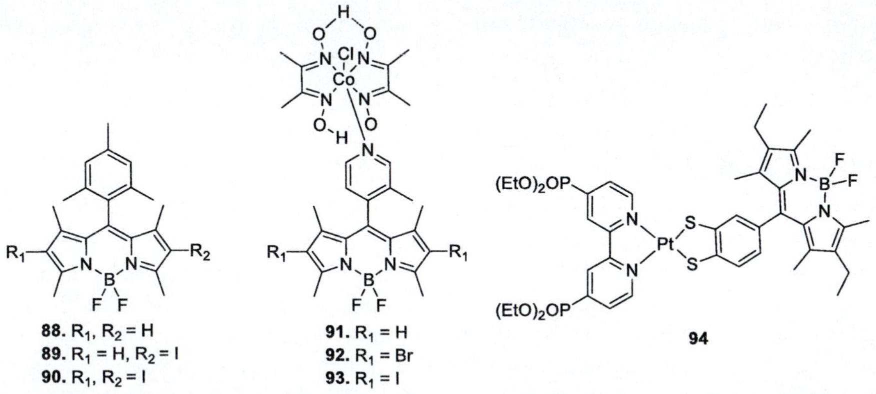

Fig. 13.18 Chemical structures of BODIPY based photocatalyst for hydrogen generation from water

BODIPY dyes are very efficient in catalyzing the hydrogen generation from water using visible light as energy source and converting it into chemical energy. In this system, BODIPY dyes act as photosensitisers (PSs) to harvest light energy and then transfer this energy into electrochemical energy by transferring electrons to a water reduction catalyst (WRC) to catalyze the proton reduction step.

Beweries et al. showed application of BODIPY dyes (**88–90**) (Fig. 13.18) as photosensitizers useful for light-driven hydrogen production in a multicomponent catalyst system comprising of BODIPY dye, bis(dichlorotriphenylphosphinppalladium) $[74]_2$, H_2O and trimethylamine (Dura et al. 2015). The dye **88** is capable of H_2 generation and its activity was further increased by iodination at the BODIPY dyes. Iodination enhances the triplet conversion as well as increases triplet lifetime of **89** and **90** which are essential for the intermolecular electron transfer.

Weare et al. showed the photocatalytic activity of BODIPY—cobaloxime complexes (**91–93**) (Fig. 13.18) for light-driven hydrogen evolution by proton reduction (Bartelmess et al. 2014). The triplet state of the BODIPY chromophores is required to transfer the electron to cobaloxime catalyst to show the catalytic activity. Thus, non-halogenated BODIPY—cobaloxime complex (**91**) was ineffective in hydrogen evolution. As the halogenation enhances the triplet state of BODIPY dyes, the hydrogen evolution efficiency of the complexes also increased with halogenation of BODIPY (**92, 93**). Also the diiodo BODIPY—cobaloxime complex (**93**) was more efficient than the corresponding dibromo BODIPY—cobaloxime complex (**92**).

Eisenberg et al. developed dyad **94** comprising of BODIPY and platinum diimine dithiolate (PtN_2S_2) charge transfer (CT) chromophore which is very efficient in light driven hydrogen generation from aqueous protons (Fig. 13.18) (Zheng et al. 2015). This was attached with the TiO_2 nanoparticle through the attached $PO(OEt)_2$ groups for rapid electron photoinjection into the semiconductor. The dye on photoexciataion at 530 nm in water with ascorbic acid as electron donor, exhibited excellent H_2

production by generating ~40,000 turnover numbers of H_2 over 12 d (with respect to dye).

13.6 Conclusions

Thus, BODIPY is a class of dyes with outstanding optical as well as electrical properties. Synthetic versatility of these dyes showed that their properties can be modulated as per the requirement of different applications. All these attract scientists from almost every discipline of science to utilize these dyes for multiple applications in chemistry, biology, physics, materials science etc. Rational design of BODIPY dyes tuning their properties to develop custom made materials for different applications are discussed in this chapter. Several other applications of these dyes could not be incorporated here. The popularity of the BODIPY dyes among the scientific community clearly established them a unique dye for versatile optical applications.

But, still there are some lacuna/unfulfillment in different fields which need to be addressed in future. Several BODIPY dyes with high lasing efficiency are available from green-yellow-red to far red region. But low photostability is still a bottleneck for their continuous lasing applications. More focused studies are required to enhance their lasing lifetime. Efforts also need to develop water soluble BODIPY laser dyes for easy and safe operation. Similarly, development of high intensity and photostable BODIPY-based probes for single-molecule imaging and super-resolution microscopy is also required to visualize and analyze the localization and dynamics of individual proteins and biomolecules. Water solubility and aggregation of the BODIPY dyes in polar environment still remain a major problem for their applications in cellular imaging and photodynamic therapy of cancers. Despite many highly efficient BODIPY based triplet photosensitizers, they are still far from clinical applications as PDT agents. Priority research should be done to develop cancer cell specific BODIPY dyes targeting the over-expressed cell-surface receptors in tumor cells. Similarly, BODIPY based fluorescent-PET probes are still in the initial stages which need to develop. This can change the present cancer diagnosis and therapeutic scenario.

Numerous chemo/bio-sensor are reported, but water solubility, specificity of the sensors can be further improved for precise applications. Although abundant cation (metal) sensors are reported but sensors to detect quantify physiologically important anions in an aqueous environment are still lacking. BODIPY based photovoltaics showed good promise for future energy conversion devices. Dedicated research is needed to develop commercial small molecules based organic photovoltaics to fulfill the dream of the material chemists. Future development in all these areas will definitely happen as BODIPY class of dyes have all the required properties for these advancement which will further solidify their uniqueness in the world of fluorescent dyes.

References

Atilgan, S., Ekmekci, Z., Dogan, A.L., Guc, D., Akkaya, E.U.: Chem. Commun. 4398–4400 (2006)

Bartelmess, J., Francis, A.J., El Roz, K.A., Castellano, F.N., Weare, W.W., Sommer, R.D.: Inorg. Chem. **53**, 4527–4534 (2014)

Belmonte-Vázquez, J.L., Avellanal-Zaballa, E., Enríquez-Palacios, E., Cerdán, L., Esnal, I., Bañuelos, J., Villegas-Gómez, C., Arbeloa, I.L., Peña-Cabrera, E.: J. Org. Chem. **84**, 2523–2541 (2019)

Bessette, A., Hanan, G.S.: Chem. Soc. Rev. **43**, 3342–3405 (2014)

Bodio, E., Goze, C.: Dyes Pigm. **160**, 700–710 (2019)

Boens, N., Leen, V., Dehaen, W.: Chem. Soc. Rev. **41**, 1130–1172 (2012)

Boyer, J.H., Haag, A.M., Sathyamoorthi, G., Soong, M.-L., Thangaraj, K., Pavlopoulos, T.G.: Heteroat. Chem. **4**, 39–49 (1993)

Brown, S.B., Brown, E.A., Walker, I.: Lancet Oncol. **5**, 497–508 (2004)

Bulut, I., Huaulmé, Q., Mirloup, A., Chávez, P., Fall, S., Hébraud, A., Méry, S., Heinrich, B., Heiser, T., Lévêque, P., Leclerc, N.: Chemsuschem **10**, 1878–1882 (2017)

Buyukcakir, O., Bozdemir, O.A., Kolemen, S., Erbas, S., Akkaya, E.U.: Org. Lett. **11**, 4644–4647 (2009)

Cakmak, Y., Kolemen, S., Duman, S., Dede, Y., Dolen, Y., Kilic, B., Kostereli, Z., Yildirim, L.T., Dogan, A.L., Guc, D., Akkaya, E.U.: Angew. Chem. Int. Ed. **50**, 11937–11941 (2011)

Carlucci, G., Carney, B., Brand, C., Kossatz, S., Irwin, C.P., Carlin, S.D., Keliher, E.J., Weber, W., Reiner, T.: Mol. Imaging Biol. **17**, 848–855 (2015)

Dodani, S.C., He, Q., Chang, C.J.: J. Am. Chem. Soc. **131**, 18020–18021 (2009)

Dura, L., Ahrens, J., Pohl, M.-M., Höfler, S., Bröring, M., Beweries, T.: Chem. Eur. J. **21**, 13549–13552 (2015)

Er, J.C., Tang, M.K., Chia, C.G., Liew, H., Vendrell, M., Chang, Y.-T.: Chem. Sci. **4**, 2168–2176 (2013)

Ethirajan, M., Chen, Y., Joshi, P., Pandey, R.K.: Chem. Soc. Rev. **40**, 340–362 (2011)

Gabe, Y., Urano, Y., Kikuchi, K., Kojima, H., Nagano, T.: J. Am. Chem. Soc. **126**, 3357–3367 (2004)

Gorai, S., Ghosh, A., Chakraborty, S., Retailleau, P., Ghanty, T.K., Patro, B.S., Mula, S.: Dyes Pigm. **203**, 110343 (2022)

Gorman, A., Killoran, J., O'Shea, C., Kenna, T., Gallagher, W.M., O'Shea, D.F.: J. Am. Chem. Soc. **126**, 10619–10631 (2004)

Gupta, M., Mula, S., Tyagi, M., Ghanty, T.K., Murudkar, S., Ray, A.K., Chattopadhyay, S.: Chem. Eur. J. **19**, 17766–17772 (2013)

Gupta, M., Mula, S., Ghanty, T.K., Naik, D.B., Ray, A.K., Sharma, A., Chattopadhyay, S.: J. Photochem. Photobiol Chem. **349**, 162–170 (2017)

Harriman, A., Mallon, L.J., Elliot, K.J., Haefele, A., Ulrich, G., Ziessel, R.: J. Am. Chem. Soc. **131**, 13375–13386 (2009)

Hendricks, J.A., Keliher, E.J., Wan, D., Hilderbrand, S.A., Weissleder, R., Mazitschek, R.: Angew. Chem. Int. Ed. **51**, 4603–4606 (2012)

https://www.photonicsolutions.co.uk

Huang, L., Zhao, J., Guo, S., Zhang, C., Ma, J.: J. Org. Chem. **78**, 5627–5637 (2013)

Jagtap, K.K., Shivran, N., Mula, S., Naik, D.B., Sarkar, S.K., Mukherjee, T., Maity, D.K., Ray, A.K.: Chem. Eur. J. **19**, 702–708 (2013)

Jia, M.-Y., Wang, Y., Liu, Y., Niu, L.-Y., Feng, L.: Biosens. Bioelectron. **85**, 515–521 (2016)

Jiang, N., Fan, J., Liu, T., Cao, J., Qiao, B., Wang, J., Gao, P., Peng, X.: Chem. Commun. **49**, 10620–10622 (2013)

Jiang, X.-D., Zhao, J., Li, Q., Sun, C.-L., Guan, J., Sun, G.-T., Xiao, L.-J.: Dyes Pigm. **125**, 136–141 (2016)

Jiao, L., Yu, C., Uppal, T., Liu, M., Li, Y., Zhou, Y., Hao, E., Hub, X., Vicente, M.G.H.: Org. Biomol. Chem. **8**, 2517–2519 (2010)

Kamkaew, A., Lim, S.H., Lee, H.B., Kiew, L.V., Chung, L.Y., Burgess, K.: Chem. Soc. Rev. **42**, 77–88 (2013)
Killoran, J., Allen, L., Gallagher, J.F., Gallagher, W.M., O′Shea, D.F.: Chem. Commun. 1862–1863 (2002)
Kim, H., Kim, K., Son, S.-H., Choi, J.Y., Lee, K.-H., Kim, B.-T., Byun, Y., Choe, Y.S.: ACS Chem. Neurosci. **10**, 1445–1451 (2019)
Kollmannsberger, M., Rurack, K., Resch-Genger, U., Daub, J.: J. Phys. Chem. A **102**, 10211–10220 (1998)
Li, W., Li, L., Xiao, H., Qi, R., Huang, Y., Xie, Z., Jing, X., Zhang, H.: RSC Adv. **3**, 13417–13421 (2013)
Loudet, A., Burgess, K.: Chem. Rev. **107**, 4891–4932 (2007)
Lu, H., Mack, J., Yang, Y., Shen, Z.: Chem. Soc. Rev. **43**, 4778–4823 (2014)
Lundrigan, T., Crawford, S.M., Cameron, T.S., Thompson, A.: Chem. Commun. **48**, 1003–1005 (2012)
Maity, A., Sarkar, A., Sil, A., BN, S.B., Patra, S.K.: New J. Chem. **41**, 2296–2308 (2017)
Mbatia, H.W., Kennedy, D.P., Camire, C.E., Incarvito, C.D., Burdette, S.C.: Eur. J. Inorg. Chem. **2010**, 5069–5078 (2010)
Mora, A.K., Murudkar, S., Shivran, N., Mula, S., Chattopadhyay, S., Nath, S.: Int. J. Biol. Macromol. **166**, 1121–1130 (2021)
More, A.B., Mula, S., Thakare, S., Sekar, N., Ray, A.K., Chattopadhyay, S.: J. Org. Chem. **79**, 10981–10987 (2014)
More, A.B., Mula, S., Thakare, S., Chakraborty, S., Ray, A.K., Sekar, N., Chattopadhyay, S.: J. Lumin. **190**, 476–484 (2017)
Mula, S., Ray, A.K., Banerjee, M., Chaudhuri, T., Dasgupta, K., Chattopadhyay, S.: J. Org. Chem. **73**, 2146–2154 (2008)
Mula, S., Ulrich, G., Ziessel, R.: Tetrahedron Lett. **50**, 6383–6388 (2009)
Mula, S., Elliott, K., Harriman, A., Ziessel, R.: J. Phys. Chem. A **114**, 10515–10522 (2010)
Mula, S., Frein, S., Russo, V., Ulrich, G., Ziessel, R., Barberá, J., Deschenaux, R.: Chem. Mater. **27**, 2332–2342 (2015)
Nguyen, Q.T., Olson, E.S., Aguilera, T.A., Jiang, T., Scadeng, M., Ellies, L.G., Tsien, R.Y.: Proc. Natl. Acad. Sci. U S A **107**, 4317–4322 (2010)
Olivier, J.-H., Barberá, J., Bahaidarah, E., Harriman, A., Ziessel, R.: J. Am. Chem. Soc. **134**, 6100–6103 (2012)
Ortiz, M.J., Agarrabeitia, A.R., Duran-Sampedro, G., Bañuelos Prieto, J., Lopez, T.A., Massad, W.A., Montejano, H.A., García, N.A., Arbeloa, I.L.: Tetrahedron **68**, 1153–1162 (2012)
Quan, L., Gu, J., Lin, W., Wei, Y., Lin, Y., Liu, L., Ding, H., Pan, C., Xie, Z., Wu, T.: Chem. Commun. **55**, 8564–8566 (2019)
Ray, A.K., Kundu, S., Sasikumar, S., Rao, C.S., Mula, S., Sinha, S., Dasgupta, K.: Appl. Phys. B **87**, 483–488 (2007)
Ray, C., Schad, C., Moreno, F., Maroto, B.L., Bañuelos, J., Arbeloa, T., García-Moreno, I., Villafuerte, C., Muller, G., de la Moya, S.: J. Org. Chem. **85**, 4594–4601 (2020)
Sen, A., Mora, A.K., Koli, M., Mula, S., Kundu, S., Nath, S.: Int. J. Biol. Macromol. **220**, 901–909 (2022)
Shah, M., Thangaraj, K., Soong, M.-L., Wolford, L.T., Boyer, J.H., Politzer, I.R., Pavlopoulos, T.G.: Heteroat. Chem. **1**, 389–399 (1990)
Shivran, N., Mula, S., Ghanty, T.K., Chattopadhyay, S.: Org. Lett. **13**, 5870–5873 (2011)
Shivran, N., Tyagi, M., Mula, S., Gupta, P., Saha, B., Patro, B.S., Chattopadhyay, S.: Eur. J. Med. Chem. **122**, 352–365 (2016)
Shivran, N., Koli, M.R., Chakraborty, G., Srivastava, A.P., Chattopadhyay, S., Mula, S.: Org. Biomol. Chem. **19**, 7920–7929 (2021)
Srinivasa Rao, R., Bagui, A., Hanumantha Rao, G., Gupta, V., Singh, S.P.: Chem. Commun. **53**, 6953–6956 (2017)
Treibs, A., Kreuzer, F.-H.: Justus Liebigs Ann. Chem. **718**, 208–223 (1968)

Tsien, R.Y.: Nat. Rev. Mol. Cell Biol. Suppl. Ss16–21 (2003)
Turksoy, A., Yildiz, D., Akkaya, E.U.: Coord. Chem. Rev. **379**, 47–64 (2019)
Ulrich, G., Ziessel, R., Harriman, A.: Angew. Chem. Int. Ed. **47**, 1184–1201 (2008)
Valeur, B., Berberan-Santos, M.N.: Molecular Fluorescence: Principles and Applications, 2nd edn. Wiley-VCH Verlag GmbH & Co, KGaA (2012)
Wagner, R.W., Lindsey, J.S.: Pure Appl. Chem. **68**, 1373–1380 (1996)
Wainwright, M.: Chem. Soc. Rev. **25**, 351–359 (1996)
Wang, L., Xiao, Y., Tian, W., Deng, L.: J. Am. Chem. Soc. **135**, 2903–2906 (2013)
Wang, J.-L., Zhang, L., Gao, L.-X., Chen, J.-L., Zhou, T., Liu, Y., Jiang, F.-L.: J Mater Chem B **9**, 8639–8645 (2021)
Yang, Z., He, Y., Lee, J.-H., Park, N., Suh, M., Chae, W.-S., Cao, J., Peng, X., Jung, H., Kang, C., Kim, J.S.: J. Am. Chem. Soc. **135**, 9181–9185 (2013)
Yogo, T., Urano, Y., Ishitsuka, Y., Maniwa, F., Nagano, T.: J. Am. Chem. Soc. **127**, 12162–12163 (2005)
Zhang, X., Yu, H., Xiao, Y.: J. Org. Chem. **77**, 669–673 (2012)
Zhang, S., Wu, T., Fan, J., Li, Z., Jiang, N., Wang, J., Dou, B., Sun, S., Song, F., Peng, X.: Org. Biomol. Chem. **11**, 555–558 (2013)
Zhang, H.-X., Chen, J.-B., Guo, X.-F., Wang, H., Zhang, H.-S.: Anal. Chem. **86**, 3115–3123 (2014)
Zhang, D., Martín, V., García-Moreno, I., Costela, A., Pérez-Ojeda, M.E., Xiao, Y.: Chem. Chem. Phys. **13**, 13026–13033 (2011)
Zheng, B., Sabatini, R.P., Fu, W.F., Eum, M.S., Brennessel, W.W., Wang, L., McCamant, D.W., Eisenberg, R.: Proc. Natl. Acad. Sci. U S A **112**, E3987–E3996 (2015)
Ziessel, R., Ulrich, G., Harriman, A., Alamiry, M.A.H., Stewart, B., Retailleau, P.: Chem. Eur. J. **15**, 1359–1369 (2009)

Chapter 14
Dye-Decorated Functional Materials

Goutam Chakraborty, Padma Nilaya Jonnalgadda, and Haridas Pal

Abbreviations

ACQ	Aggregation-caused-quenching
AIE	Aggregation-induced emission
AMP	Adenosine mono-phosphate
ADP	Adenosine di-phosphate
ATP	Adenosine tri-phosphate
ALP	Alkaline phosphatase
AuO	Auramine-O
BODIPY	4,4-Difluoro-4-bora-3a,4a-diaza-s-indacene
BBR	Berberine
BSA	Bovine serum albumin
BSPOTPE	1,2-Bis(4-(3-sulfonatopropoxyl)phenyl)-1,2-diphenylethene
BTASE	N,N′-Bis[4-[[3-(trimethylammonio)-thyl]oxy]salicylidene]-ethylenediamine
CBn	Cucurbit[n]uril
CD	Cyclodextrin
Ct-DNA	Calf-thymus deoxyribonucleic acid
CXn	Calix[n]arene
DLS	Dynamic light scattering
DHXP	3-Dihydro-1H-xanthene-6-ol
DHNp	2,6-Dihydroxynaphthalene
DASPE	2-(4-(Dimethylamino)styryl)-1-ethylpyridinium

G. Chakraborty (✉) · P. N. Jonnalgadda
Laser and Plasma Technology Division, BARC, Mumbai 400085, India
e-mail: gchak@barc.gov.in

P. N. Jonnalgadda · H. Pal (✉)
Homi Bhabha National Institute, Anushaktinagar, Mumbai 400094, India
e-mail: hpal@barc.gov.in

R. S. Ningthoujam and A. K. Tyagi (eds.), *Handbook of Materials Science, Volume 1*,
Indian Institute of Metals Series, https://doi.org/10.1007/978-981-99-7145-9_14

HEPES	N-2-hydroxyethylpiperazine-N′-2-ethanesulfonic acid
HOMO	Highest occupied molecular orbital
Hp	Heparin
IR	Infra-red
ICT	Intramolecular charge transfer
LUMO	Lowest unoccupied molecular orbital
LCG	Lucigenin
MB	Methylene blue
MV	Methyl viologen
Nd-YAG	Neodymium dopped yttrium aluminum garnet
NMR	Nuclear magnetic resonance
Pr	Protamine
PMA	Pyrene methyl ammonium
PSS	Polystyrene sulfonate
PQ	Paraquat
PSV	Poly(sodium vinylsulfonate)
PTS	Pyrene tetrasulfonate
QN	Quinine
SCE	Saturated calomel electrode
SEM	Scanning electron microscopy
TC	3,3′-Diethylthiacyanine
TEM	Transmission electron microscopy
TFPB	Tetrakis(4-fluorophenyl)borate
ThT	Thioflavin T
TPE	Tetraphenyl ethylene
TP	Tetraphenylthiophene
TPB	Tetraphenylborate
TO	Thiazole orange
Tris-HCl	Tris(hydroxymethyl)aminomethane (THAM) hydrochloride
UV	Ultra-violet

14.1 Introduction

Organic dye molecules generally consist of two functional units, one is the chromophore unit, the main color-producing moiety, and the other is the auxichrome, which is capable of either intensifying or reducing the color of the dye molecule. During absorption and emission processes, when an electron undergoes a transition from electronic ground state (HOMO) to electronic excited state (LUMO) and vice versa, the chromophoric molecule displays its characteristic color. Appearance of this color can take place anywhere between far UV to near infra-red region, depending upon the energy spacing between HOMO and LUMO states of the concerned dye molecule. Interestingly, optical properties (absorption and emission) of a dye molecule are possible to be tuned quite effectively by modifying the energies of HOMO and LUMO states, particularly their energy gap, by introducing different substituents in the molecule, especially the auxichromes, which can expand the opportunities of any specific class of the dyes for their wider applications. Optical properties of various chromophoric dyes can also be modulated quite significantly by changing the solvent environments suitably. Other option is to involve supramolecular dye-host interactions that can alter microenvironment of the dye, modulating its optical properties. The supramolecular modulations especially offer the chromophoric dyes to enhance their sensing abilities for various analytes, especially those having different biological relevance. In the forthcoming sections, various methodologies have been systematically discussed, which are applied usefully in modulating the optical properties of different chromophoric dyes for their better applications.

14.2 Incorporation of Suitable Substituent at Proper Position of a Chromophoric Moiety

Incorporation of substituents at the chromophoric moiety of organic dyes can be achieved through various chemical reactions, using suitable reagents and appropriate reaction conditions. These substituents can modulate the electronic states of the chromophoric molecules involving various physiochemical processes like resonance, intramolecular charge transfer, electron delocalization, hydrogen transfer, and so on (Thakare et al. 2018; Díaz et al. 2009). Accordingly, such substituted organic dyes can show notable changes in their optical properties as compared to their unsubstituted counterparts. This strategy has been adopted for various class of organic dyes to improve their absorption and emission features, along with their other attributes like photo-stability, emission quantum yield, excited-state relaxations, lasing efficiency, etc. (Thakare et al. 2018; Díiaz et al. 2009). In this regard, one of the most notable examples is the derivatization of the pyrromethene class of dyes, commonly known as BODIPY dyes. These dye molecules are one of the most extensively used chromophoric systems in dye laser applications, due to their sharp absorption and emission profiles, high fluorescence quantum yield, low singlet to triplet conversion efficiency, etc. Further, these dyes are also used quite extensively in bio-sensing and bio-imaging applications, due to their excellent bio-compatibility (Batat et al. 2011; Boens et al. 2012). In the absence of any substituents, the basic fluorinated boron-dipyrromethene chromophore, usually referred as the BODIPY core, absorbs and emits light in the green region of visible spectrum (Bañuelos et al. 2012). Various substituent groups can, however, be introduced to the basic BODIPY core, causing suitable modifications in absorption and fluorescence characteristics for the resultant derivatized BODIPY molecules, as may be desirable depending upon the applications aimed for such systems. Typical examples of such spectral shifts introduced by various substituents are clearly indicated for the derivatized BODIPY dyes shown in Fig. 14.1.

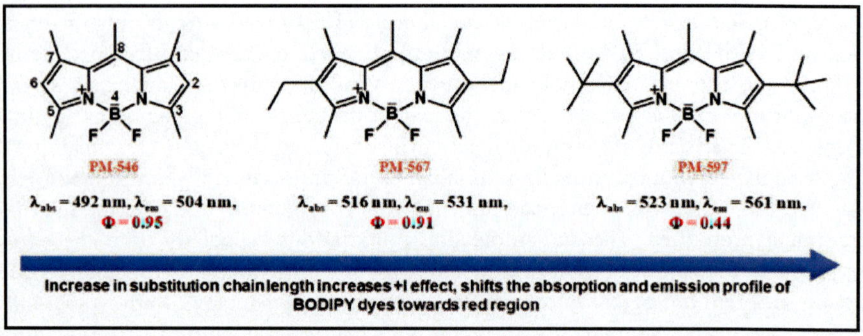

Fig. 14.1 Derivatized BODIPY dyes having different alkyl substituents. Characteristic absorption (λ_{abs}) and emission (λ_{em}) peak positions and fluorescence quantum yield (ϕ) values are also noted along with

As indicated from Fig. 14.1, the presence of alkyl substituents like ethyl and tertiary butyl groups, at the 2 and 6 positions of two pyrrole rings of BODIPY core, leads to substantial bathochromic shift for absorption and emission profiles of the concerned pyrromethene derivatives. This happens mainly due to +I effect of the alkyl substituents, which significantly modulates the electron delocalization in the BODIPY core and thereby reduces the energy gap among the HOMO–LUMO states for the chromophoric moiety (Bañuelos et al. 2012). In another example, as shown in Fig. 14.2, with the introduction of the styryl substituents at the electron deficient 3 and/or 5 positions of BODIPY core, absorption as well as fluorescence spectra undergo significant shifts in the red region. The extended resonance offered by the styryl groups results in stabilization of the LUMO states of the dyes to larger extents. This reduces the energy difference between HOMO and LUMO states significantly, thereby shifting the spectra towards the red region. There are many other examples of this kind of substituent-induced modifications in the optical properties of various classes of organic dyes, which are however, beyond the scope of the current chapter.

14.3 Solvent-Mediated Tuning of the Optical Properties of Chromophoric Dyes

The dye molecules that are having substantial dipolar characters, especially those with notable difference between their dipole moments in the ground state and excited state (μ_g and μ_e, respectively), are generally very prone to show large modulations in their absorption and emission features upon alteration in solvent polarity. This is commonly referred as the solvatochromic behavior of the dye molecules. Commercially available BODIPY dyes generally do not show any significant solvatochromic behavior, due to their similar dipolar characteristics either in the ground state or in the excited state (Bañuelos et al. 2012). However, incorporation of suitable substituents

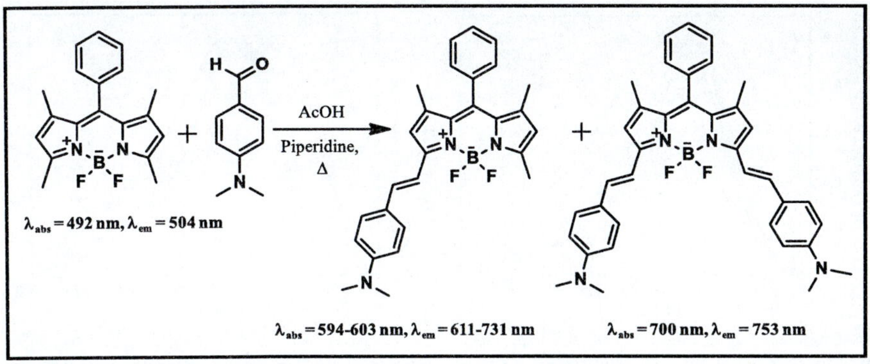

Fig. 14.2 BODIPY dyes with different styryl substituents at the 3 and/or 5 positions of the BODIPY core. Absorption and emission spectral positions are also indicated

in the BODIPY core can make these dyes susceptible to display substantial solvatochromic behavior. In one of the investigations, it has been shown that attachment of the benzimidazole groups at the 2 and/or 6 positions of BODIPY core, as shown in Fig. 14.3, makes the derivatized dyes highly responsive towards solvent polarity as compared to the unsubstituted dye counterpart (Thakare et al. 2018). Detailed investigations have revealed that for benzimidazole substituted BODIPY molecules, namely Dye 1 and Dye 2, as depicted in Fig. 14.4, a large extent of intramolecular charge transfer (ICT) occurs from the benzimidazole unit to the BODIPY core, introducing substantial dipolar character to these dyes. Evidently, thus, both Dye 1 and Dye 2 show significant extent of red shifts in their absorption as well as emission spectra with the increasing polarity of the solvents from non-polar to polar region. Representative emission spectra of Dye 1 and Dye 2, recorded in solvents of varying polarities, are shown in the left and right panels of Fig. 14.4, respectively. Photophysical properties as observed for the two dyes in different solvents are also listed in Table 14.1 for comparisons.

Another class of polyene-based laser dyes, known as hemicyanine dyes, often show negative solvatochromism, a quite uncommon feature for most other organic dyes. This unusual behavior of the hemicyanine dyes is due to their higher ground

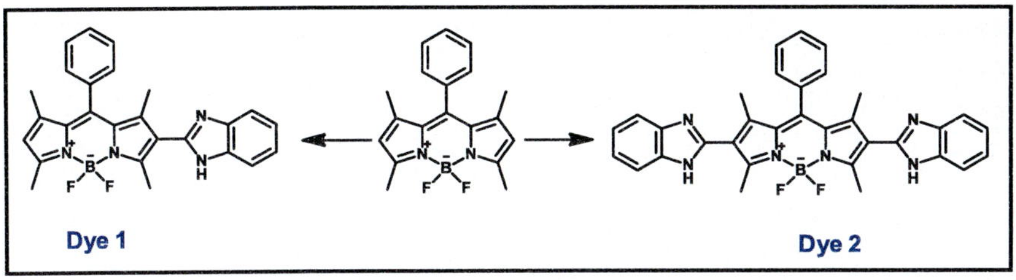

Fig. 14.3 Mono and di-substituted BODIPY dyes with benzimidazole groups at 2 and 6 positions

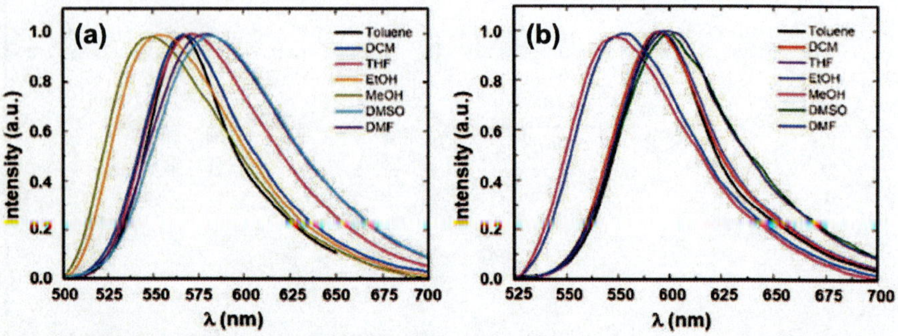

Fig. 14.4 Normalized emission spectra of Dye-1(left) and Dye-2 (right) in different solvents. The spectra show gradual bathochromic shift as the solvent polarity is increased. This figure is replicated with permission obtained from Elsevier (Thakare et al. 2018)

Table 14.1 Photophysical features of Dye-1 and Dye-2, as estimated in different solvents. This table is adapted with permission obtained from Elsevier (Thakare et al. 2018)

Dye/solvent	λ_{abs} (nm)	λ_{em} (nm)	$\Delta\lambda$ (nm)	ε_{max} (10^4 M^{-1} cm^{-1})	Φ	τ (ns)
Dye-1						
Toluene	516	567	51	4.14	0.47	4.19
EtOAc	513	569	56	3.40	0.22	2.96
THF	516	571	55	3.07	0.21	2.93
DCM	511	567	56	4.66	0.30	3.46
EtOH	506	553	47	4.17	0.19	1.99
ACN	508	574	66	3.60	0.09	1.66
DMF	514	579	65	3.51	0.07	1.49
DMSO	514	581	67	3.88	0.08	1.40
Dye-2						
Toluene	540	595	55	0.69	0.33	3.55
EtOAc	535	592	57	3.83	0.27	2.88
THF	538	589	51	3.97	0.21	2.89
DCM	532	594	62	1.78	0.29	2.89
EtOH	522	577	55	5.46	0.15	1.57
ACN	529	586	57	4.13	0.09	1.42
DMF	537	591	54	4.46	0.07	1.40
DMSO	535	598	63	4.94	0.07	1.24

state dipole moment than excited state. Accordingly, these dyes display atypical red shift in their absorption as well as emission spectra on decreasing the solvent polarity (Seth et al. 2009). Sarkar et al. reported this negative solvatochromism for a near-IR hemicyanine dye, LDS-798 (Doan et al. 2017). Detailed insight of such anomalous solvatochromism for hemicyanine dyes containing charged push–pull polyene (CPPPs) systems was explained by Hynes and co-workers (Laage et al. 2003), considering the mixing of the two associated valence bond structures of the dye molecules, grossly represented as the Donor–bridge–Acceptor$^+$ and Donor$^-$–bridge–Acceptor type of the canonical structures. To be mentioned here that even though the solvatochromic responses of the dye molecules is largely modulated by the dielectric constant of the solvents, there are also other solvent parameters like polarizability, acidity, basicity, refractive index, etc., which can also significantly influence the solvent dependent properties of the dyes. Thus, along with the solvent polarity arising through the orientational polarization of the solvents, the other solvent effects are also needed to be considered appropriately during the selection of a solvent system for a dye molecule aiming for its specific application. It is commonly found that the dye molecules which are highly fluorescent in organic solvents, often suffer from dramatic fluorescence quenching in aqueous solutions. This happens mostly due to the participation of dye-water H-bonding process and/or the enhancement in the ICT-mediated non-radiative deexcitation process in the excited dye, assisted by the high polarity of water as the solvent. Both of these processes introduce additional non-radiative deexcitation channels for the excited dye molecules, making them poorly emissive in nature (Sarkar et al. 2010). In a work by Sarkar et al., it is shown that the LDS-698 dye, which is fairly emissive in polar aprotic solvents, suffers from a severe fluorescence quenching in highly polar protic solvent like water (Seth et al. 2009). In this respect, the Rhodamine B dye, which is recognized as an important laser dye, has its fluorescence quantum yield as ~0.7 in ethanol solution, but its quantum yield gets reduced to ~0.32 in aqueous medium, due to both dye-solvent H-bond formation and enhanced ICT process in the dye (Mohanty et al. 2010). Quite similar observations have also been observed for coumarin dyes along with various other chromophoric molecules (Gupta et al. 2012).

14.4 Modulation in the Photophysical Properties Through Supramolecular Interactions

14.4.1 Modulation Through Conventional Host–Guest Complex Formation

One of the most convenient ways of modulating physicochemical features of organic chromophoric molecules is the involvement of supramolecular host–guest interactions, where a dye molecule (guest) interacts non-covalently with a suitable receptor molecule (host) to produce a supramolecular complex with unique architecture and significantly improved physicochemical attributes relative to the free state of the dye in a solution. Number of non-covalent processes, namely hydrophobic interaction, hydrogen bond formation, van der Waals interaction, electrostatic interactions (e.g. dipole–dipole, ion–dipole and ion-ion interactions), and so forth, can easily lead to the construction of supramolecular complexes/architectures with largely modulated characteristics of the guest dyes. The most important aspect of supramolecular modulation is that it does not require any tedious chemical synthesis. Just mixing of the required molecular components together in a solution with appropriate proportions can simply lead to the materialization of the desired supramolecular host–guest complex. Since these complexes involve simple non-covalent interactions, their formation is highly reversible in nature, making them quite susceptible towards different external stimuli, such as pH, temperature, ionic strength, etc. Such reversible nature of the host–guest assemblies has indeed prompted many researchers from all over the globe to study their minute details, which are having direct relevance towards their applications in different areas such as drug formulation, targeted drug delivery, enzymatic assay, development of fluorescent sensors, catalysis, dye laser systems, electronic devices, nano-medicines, functional materials, etc. (Wang et al. 2016; Ma and Zhao 2015; Monti and Manet 2014; Amabilino et al. 2017). All these utilizations necessitate easy availability of the desired hosts that can bind guest dyes suitably with high affinity and substantial selectivity. In this regard, various research groups have designed and synthesized several macrocyclic hosts, which are capable of forming substantially stable host–guest assemblies by inclusion of suitable guest dyes into their cavities. Among different host molecules, the notable ones to mention are cyclodextrins (CDs), calix[n]arenes (CXn), cucurbit[n]urils (CBn), crown ethers, pill[n]arenes, cylophanes, and so on. These host molecules, depending upon their chemical constitution, cavity size, shape, and other specific characteristics, can interact significantly with various organic dyes/drugs, to form supramolecular dye-host complexes with improved fluorescence quantum yield, photostability, solubility, drug efficacy, modulated redox properties, changed chemical reactivity, altered acid-basic properties, and so on (Boraste et al. 2017, 2018; Chakraborty et al. 2021a; Zhang et al. 2021; Guo et al. 2011).

As discussed in the earlier sections, photophysical properties of the chromophoric dyes are influenced largely by the solvent environments around the dyes. Formation of inclusion complex markedly changes the solvent microenvironment around the dyes due to their transfer from bulk solution phase into the host cavity. This causes a significant alteration in the properties of the dyes in their excited states, especially the dynamics of their excited state deexcitation processes. In most fluorescent dyes, the internal conversion (IC) process, which is one of such excited state non-radiative deactivation pathways, is greatly hampered by the formation of host–guest supramolecular complexes. It is well known that the IC process depends very strongly on the solvent polarity around the dye. Decrease in the polarity for the solvent microenvironment around the dye due to its entrapment into the host cavity significantly retards the IC rate, making the dye more fluorescent in its host-bound state. Additionally, structural rigidization that experiences by the dye inside the host cavity also results in a dramatic reduction in vibrational and rotational flexibility for the dye molecule, which eventually impedes the non-radiation excitation processes of the excited dye significantly (Boraste et al. 2017, 2018; Chakraborty et al. 2021b). Furthermore, dye encapsulation inside host cavity makes the dye molecule physically isolated from the bulk solvent environment and thus provides protection to the dye from the reactive species (mainly singlet oxygen) present in the bulk solvent phase, enhancing the overall stability of the dye (Boraste et al. 2018; Mohanty et al. 2007).

A very notable example of host–guest complexation-induced enhancement in the stability of a chromophoric dye is the realization of ultra-stable rhodamine B (RhB) solution in water in the presence of cucurbit[7]uril (CB7) host, as reported by Mohanty and Nau (2005). Encapsulation of RhB into CB7 cavity shields the dye molecule from bulk aqueous medium, leading to a substantial enhancement both in its photo and thermal stability. Taking a lead from these observations, Mohanty et al. have shown that CB7 encapsulated RhB dye in aqueous solution can demonstrate almost comparable lasing efficiency as that of the dye in ethanolic solution (cf. Fig. 14.5) (Mohanty et al. 2010). Moreover, it is also shown by these authors that the intensity distribution in the beam profile of the aqueous RhB-CB7-based dye laser system is far more homogeneous and symmetric as compared to that of the ethanolic RhB-based dye laser system. This particular observation can be attributed to superior thermo-optical characteristics (heat capacity, thermal conductivity, thermal expansion, thermal coefficient of refractive index, etc.) of water as compared to that of ethanol. It is worth mentioning that RhB-CB7 system in water provides very efficient lasing action, which is otherwise not possible for an aqueous solution of RhB, in the absence of the host. It is thus demonstrated that supramolecular dye-host complexation can help in a great way in the development of aqueous dye laser systems with better stability and superior quality as compared to the dye laser systems based on organic solvents (Mohanty et al. 2010).

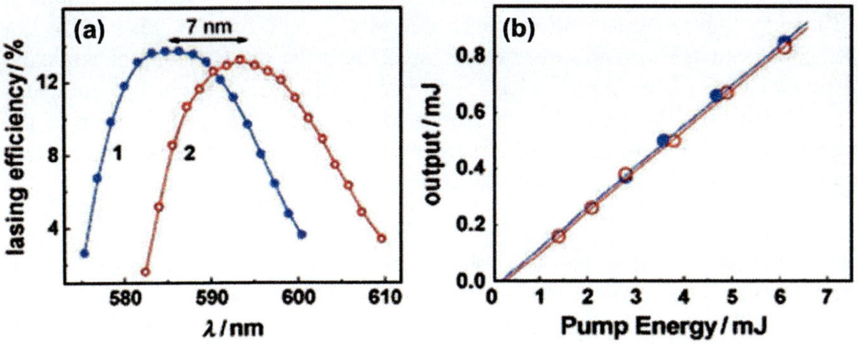

Fig. 14.5 (**a**) Laser tuning curves of (1) RhB in only ethanol and (2) in water containing 200 μM CB7. The pump laser energy (Q-switched Nd-YAG laser, 10 Hz, 532 nm) was 6.3 mJ for both the cases. (**b**) Variations in the output energy for both the dye laser systems with respect to the pump laser energy. The figure is adapted with the permission obtained from Wiley–VCH (Mohanty et al. 2010)

In the same line of RhB-CB7 system, the laser dye coumarin 1 (C1) alone in aqueous solution is seen to have very low quantum yield of fluorescence ($\Phi_f \sim 0.04$). However, upon complexation with CB7 host, the dye registers a huge enhancement in its fluorescence yield ($\Phi_f \sim 0.52$), which effectively becomes quite comparable to its Φ_f value in ethanolic solution ($\Phi \sim 0.54$) (Gupta et al. 2012). Table 14.2 lists different photophysical properties of C1 dye, taken alone in ethanol and water solutions, and also in aqueous solution in the presence of CB7 host. Results in the table clearly indicate substantial changes in the photophysical characteristics of the dye caused by its complexation with the CB7 host.

Megyesi et al. have reported an unusually large enhancement (~500 fold) in the fluorescence intensity of an important alkaloid family of chromophoric molecule, berberine (BBR), upon its encapsulation into CB7 cavity (Megyesi et al. 2008). Since CB7 portal is highly electron rich because of the presence of highly polarizable multiple carbonyl groups, the host can form a very strong supramolecular complex with cationic BBR molecule, involving strong ion–dipole interaction along with the usual hydrophobic interaction arises due to the dye encapsulation into the CB7 cavity. The decrease in the rotational and vibrational flexibility for the dye along with the decrease in its propensity to undergo ICT process inside the hydrophobic host cavity significantly retards the radiationless deexcitation for the excited dye molecule, causing a significant escalation in its fluorescence intensity (Megyesi et al. 2008). Similarly, encapsulation of BBR into homologues cyclodextrin (CD) cavities also causes an increase for the fluorescence intensity of the dye, though in these cases, the enhancements are not as large as it happens upon association of the dye with CB7 host. This is attributed to the relatively weaker binding interaction of BBR with CD hosts because in these cases only hydrophobic interaction is involved for the formation of the dye-host complex in the solution. In a recent study, however, it has been shown that on using a sulfated derivative of β-cyclodextrin (βCD) as a host, which is designated as SCD, the complexation of BBR with the host becomes very strong and results in ~130 fold enhancement in fluorescence intensity for the dye (cf. Fig. 14.6). For BBR-SCD system, since positively-charged BBR experiences strong electrostatic interaction with the negatively-charged sulfate groups of SCD which participates in cooperation with hydrophobic interaction arises due to dye encapsulation by the SCD cavity, the dye-host binding effectively becomes very strong for the present dye-host system (Chakraborty et al. 2021a).

Table 14.2 Photophysical parameters of Coumarin 1 in Ethanol, Water, and Water with the Host CB7. This table is reproduced with permission obtained from the American Chemical Society (Gupta et al. 2012)

Dye/CB7 molar ratio	Solvent	λ_{abs} (nm)	λ_{em} (nm)	Φ_f	τ_f (ns)	τ_r (ps)	k_r (10^8 s − 1)	k_{nr} (10^8 s − 1)
1:0	Ethanol	373	448	0.54	3.09	90	1.75	1.49
1:0	Water	380	468	0.04	0.4	72	1.0	24.0
1:20	Water	396	468	0.52	5.1	290	1.02	0.91

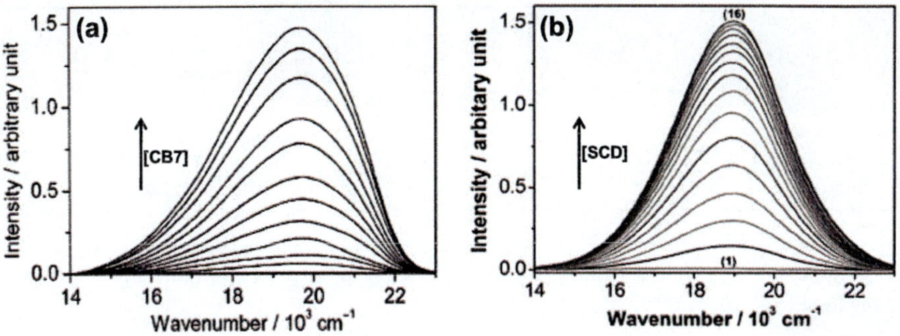

Fig. 14.6 Modulations in emission spectra of BBR with the addition of (**a**) CB7 from 0 to 15.8 μM and (**b**) SCD from 0 to 114.4 μM. The figure in panel (a) is replicated from ref. (Megyesi et al. 2008) and that in panel (b) is replicated from ref. (Chakraborty et al. 2021a) with the permissions obtained from American Chemical Society and Elsevier, respectively

It is to be noted in the present context that encapsulations of guest dyes inside host cavities do not always give an enhancement for the fluorescence intensity for the bound dye molecules. There are in fact examples where dye molecules are seen to experience a reduction in their fluorescence intensity upon binding with some supramolecular hosts (Boraste et al. 2017; Guo et al. 2011). Such quenching in fluorescence intensity is ascribed to the phenomena like ICT between the dye and the host molecules, formation of H-bonds between participating dye and host, and encapsulation of multiple dye molecules into the cavity of a larger size macrocyclic host to form a higher order complex. All these phenomena can cause excited-state dyes to suffer an additional non-radiative deexcitation process to dissipate their excited-state energy, making them to be feebly emissive, or simply non-fluorescent in nature (Boraste et al. 2017; Guo et al. 2011).

Sun and co-workers have shown that binding of methylene blue (MB) dye with CB8 host brings about a large reduction in fluorescence intensity for the dye. Since CB8 cavity is fairly larger, two MB molecules can simultaneously bind strongly into a CB8 cavity, leading to the 2:1 dye-host inclusion complex formation (Sun et al. 2014). Due to non-fluorescent nature of dimeric MB, a substantial quenching in fluorescence is observed for the dye on its dimeric association with CB8 cavity. Nonetheless, a large fluorescence reversal has been observed for the MB-CB8 system upon addition of analyte, paraquat (PQ), a well-known herbicide, into the solution of the MB-CB8 complex. In this case, added PQ competitively binds to the CB8 cavity, releasing the MB molecules from the CB8 cavity into the bulk solution, and thus restoring the fluorescence intensity for the dye, as shown schematically in Fig. 14.7. It is thus suggested that MB-CB8 system can be suitably deployed as a fluorescence 'Off–On' sensor for the detection of PQ in the geological environments as well as in other bio-systems (Sun et al. 2014).

In a study, it has been demonstrated that interaction of the molecule, quinine (QNH^+), a well-known antimalarial drug, with the CB7 host, at pH ~ 7, results in

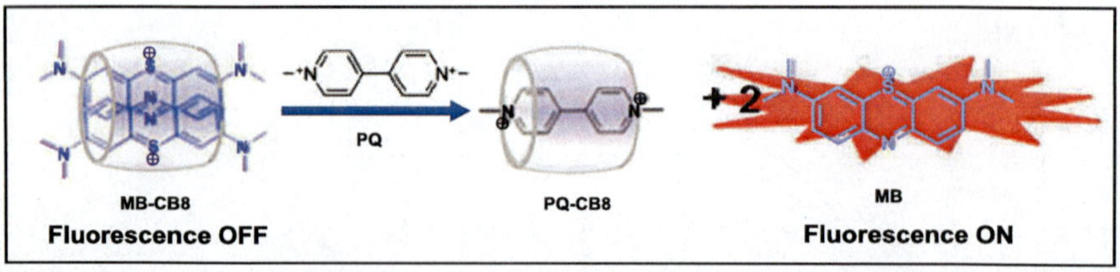

Fig. 14.7 Schematic presentation of the fluorescence 'Off–On' response of MB-CB8 system towards PQ as the analyte. The figure is adapted with permission obtained from the Nature Publishing Group (Sun et al. 2014)

a notable reduction in its fluorescence intensity. Such fluorescence quenching of quinine fluorophore is attributed to charge transfer interaction between electron-donating quinoline moiety of the drug, whose oxidation potential is moderate (~0.99 V against SCE), with the electron-accepting carbonyl groups present at the portals of the CB7 host (Boraste et al. 2017). Similarly, Nau and coworkers have shown that chemiluminescent probe, lucigenin (LCG), binds very strongly with calix[4]arene (CX4) host, with a binding constant (K_a) as ~$2.8 \times 10^7\ M^{-1}$, leading to a large reduction (~140 fold) in fluorescence intensity for LCG dye (Guo et al. 2011). Conversely, addition of protamine, an important neurotransmitter, which is also used as an antidote for heparin, an anti-coagulating agent, to the solution of the LCG-CX4 assembly, leads to fluorescence recovery of the dye. The present system, thus, represents detection of protamine in living cells via indicator displacement assay (Guo et al. 2011). Since protamine is having very high binding affinity towards CX4, its binds very strongly with this macrocyclic host (K_a ~ $1.24 \times 10^9\ M^{-1}$), releasing the LCG dye from the CX4 cavity and thus restoring the dye fluorescence.

14.4.2 Photophysical Modulations Through Host-Assisted Dye Aggregations

Host-assisted nano-aggregate formation for chromophoric dyes is one of the important topics in contemporary supramolecular research fields, and this aspect is very useful for tuning the photophysical features of the dye molecules. Because of their excellent attributes, such as improved repeatability, self-correction and ordered pattern, such dye nano-aggregates are widely used in various areas in physical sciences, chemical sciences, pharmaceuticals, electronics, material sciences, and so on (Lou and Yang 2018; Kwok et al. 2015). For most of the dye molecules, it is generally found that formation of nano-aggregates in most cases leads to a quenching of their fluorescence intensity, due to the detrimental effect towards radiative process, usually referred to be the aggregation-caused-quenching (ACQ). However, in 2001, Tang's group unraveled a new direction to the fluorescence spectroscopy and its applications by their discovery of aggregation induced emission (AIE), a reverse phenomenon of ACQ. The AIE effect is caused by the formation of dye nano-aggregates, which leads to a large restriction for the intramolecular non-radiative processes occur through structural relaxation of the dye molecules, enhancing their fluorescence yield (Kwok et al. 2015). In this perspective, Shi et al. have reported the AIE phenomena for a tetraphenyl ethylene (TPE) derivative, namely $TPESO_3Na$, in aqueous solution, resulting in the enhanced blue emission due to AIE (Shi et al. 2019). While γCD was added to this solution, it led to the disappearance of AIE following the formation of 1:1 $TPESO_3Na$−γCD inclusion complex. Conversely, addition of α-amylase to the latter solution resulted in the cleavage of α-1,4-glycosidic bonds of γCD, restoring the blue emission of the solution due to AIE, caused by the reformation of the $TPESO_3Na$ aggregates again, as schematically shown in Fig. 14.8 (Shi et al. 2019). Thus, the $TPESO_3Na$−γCD system can suitably be used for detection as well as quantification of α-amylase, which is an important biomarker and plays a crucial role towards carbohydrate digestive physiology in human.

In a study, almost a reverse phenomenon than just discussed above has been reported. Thus, interaction of an anionic sulfonated derivative of TPE dye, Su-TPE, with a cationic amino substituted βCD host, AβCD, is seen to cause a significant surge in the fluorescence intensity of Su-TPE, due to the AβCD-assisted formation of dye aggregates (Chakraborty et al. 2021c). The anionic Su-TPE dye at the vicinity of the positively charged portal of AβCD host causes a substantial charge neutralization and thus diminishes the intermolecular electrostatic repulsion between the dye molecules, helping the dye molecules to assemble together to form π-π stacked dye aggregates (cf. Fig. 14.9). Formation of such dye aggregates significantly impedes the radiationless decay channels for the excited dyes, making the dye molecules to

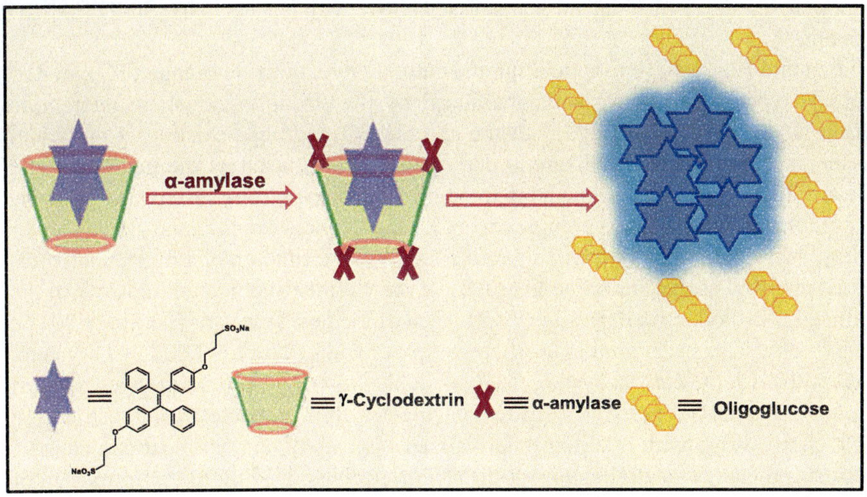

Fig. 14.8 Schematic presentation of the emission response of $TPESO_3Na$ probe in determining α-amylase activity. The figure is redrawn based on the concept obtained from Wiley–VCH (Shi et al. 2019)

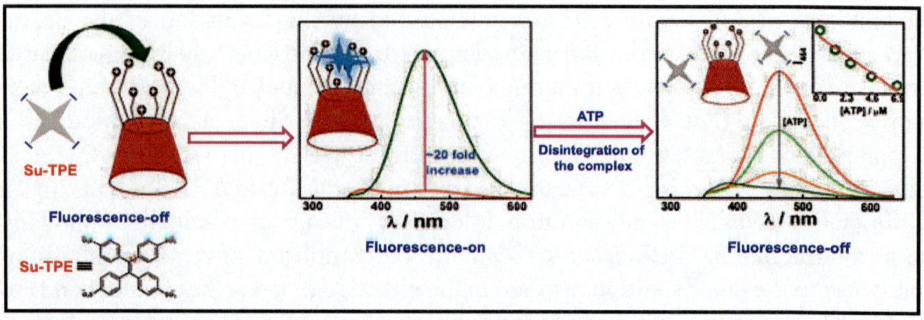

Fig. 14.9 Pictorial presentation of the sensing mechanisms for ATP by Su-TPE-AβCD complex as a fluorescence probe at pH 4.5. In the schematic, the seven $-NH_3^+$ units at the narrower rim of the AβCD host have been denoted by (+) signs. The figure is adapted with the permission obtained from American Chemical Society (Chakraborty et al. 2021c)

be highly emissive in nature (Chakraborty et al. 2021c). Further, it has also been observed that the introduction of adenosine tri-phosphate (ATP) to the same solution drastically reduces the fluorescence of the solution, due to disintegration of the dye aggregates in the presence of ATP, because the multi-anionic ATP has a stronger binding affinity towards cationic AβCD host as compared to the Su-TPE dye. In a similar study, it has also shown the formation of AβCD-assisted aggregation of pyrene tetrasulfonate (PTS) dye, triggering large modulation in its emission profile and lifetime (Chakraborty et al. 2021d). The system is also found to be highly responsive to the external stimuli, such as pH, temperature, and ionic strength. This system has also been utilized for the quantification of ATP using ratiometric fluorescence method, following the same mechanism as discussed above with respect to the Su-TPEAβCD system.

Lau and Heyne have reported that the cationic dye, thiazole orange (TO), undergoes H-type aggregate formation, assisted by the strong electrostatic interaction of the positively-charged dye with the anionic SO_3^- groups existing at the wider terminal of the sulfated calixarene derivative (SCX4), used as the host (Lau and Heyne 2010). The supramolecular interaction in the present system leads to the development of host–guest complex with 3:1 dye-to-host stoichiometry. In a similar study, Awasthi and Singh have reported J-aggregate formation of a molecular rotor-based amyloid probe, auramine-O (AuO), at the vicinity of polyanionic portal of the sulfated β-cyclodextrin derivative, SCD, used as the host (Awasthi and Singh 2017). The mechanism is very similar to what we have elucidated for Su-TPEAβCD system, except that for the current system the dye and the host are just oppositely charged as compared to the Su-TPEAβCD system. Evidently, the strong electrostatic attraction of a charged dye with a suitably derivatized oppositely charged host can cause a substantial charge neutralization of the dye molecules, leading to the formation of the adequately arranged (H-type or J-type) dye aggregates. Since the formation of such aggregates largely reduces the torsional relaxation of the dye molecules, there is a consequent reduction for non-radiative deexcitation processes of the dye excited

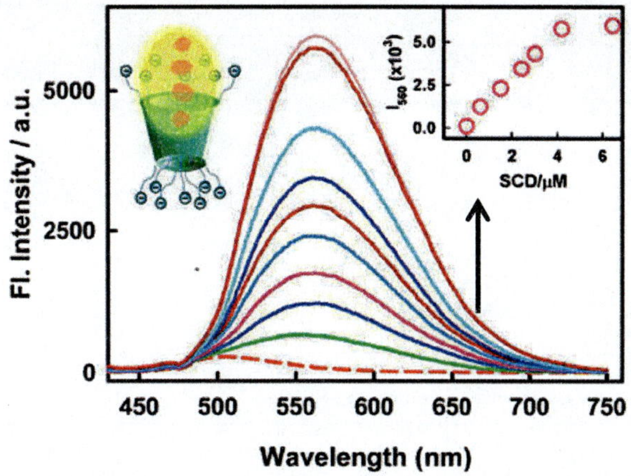

Fig. 14.10 Increase in fluorescence intensity for AuO solution with increasing concentration of SCD host. Inset shows the increase in the fluorescence intensity at 560 nm as a function of SCD concentration. λ_{ex} was 410 nm. The figure is adapted with permission obtained from American Chemical Society (Awasthi and Singh 2017)

state, and accordingly a large increase in the fluorescence intensity of the dye (cf. Fig. 14.10). It is worth mentioning that the AuO—SCD system is also found to display extremely good sensitivity towards external stimuli like ionic-strength, and temperature, making this system highly useful for detection and quantification of various analytes.

Along with the macrocyclic host-assisted nanostructured assemblies of chromophoric dyes, there are many other kinds of aggregated chromophoric species that have many technological advantages as the fluorescence sensors for various bio-analytes like protein, DNA, enzymes, etc. and can be used also in bio-imaging applications as well as in the developments of molecular logic gates, stimuli-responsive functional materials, and so on. In this context, Zhao et al. (2009) have utilized the AIE property of a silole-based cationic probe having quaternary ammonium substitution, for detection as well as quantification of ATP. Substituted silole is non-fluorescent in aqueous buffer solution, because of its highly torsional motions and its tendency to undergo twisted ICT state (TICT state) formation. However with gradual addition of ATP, a negatively charged bio-analyte, there is the formation of silole-ATP adduct, which makes the complex electrically neutral as a whole, leading eventually to the formation of quite strongly emissive silole aggregates in the solution. Formation of the aggregates dramatically reduces the TICT state conversion of the probe and also restricts its molecular torsional motions, making the probe to be highly fluorescent in nature. Therefore, the emission intensity enhancement for the silole derivative with increasing ATP concentration forms the basis for ATP quantification using this probe. To be noted, however, that no significant AIE was witnessed with adenosine monophosphate (AMP) and adenosine diphosphate (ADP) as the analytes, which are the hydrolyzed products from ATP. It is suggested that the high selectivity for the studied silole derivative towards ATP is because of the tetra negative charge in the ATP analyte along with the presence of a nucleoside unit in its structure, a combination of which effectively induces aggregation for the studied silole-based

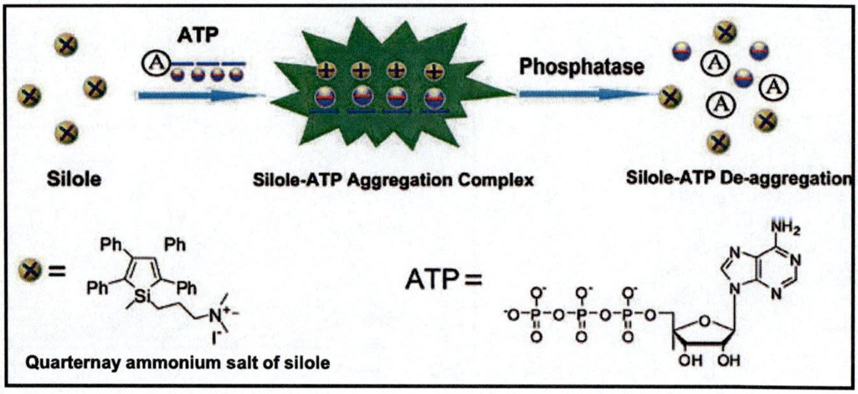

Fig. 14.11 ATP detection and phosphatase assay utilizing aggregation-induced emission property of ammoniated silole derivative. The figure is adopted with permission obtained from American Chemical Society (Zhao et al. 2009)

chromophoric molecule. Further, on addition of phosphatase to the fluorescent silole-ATP complex, there is a substantial quenching in the fluorescence intensity for the concerned solution. Thus, the present system can be used as a useful fluorescence switch with the ability to detect suitably charged bio-analytes (cf. Fig. 14.11).

In another study, Wang et al. (2008) have also used the same silole derivative for the detection of heparin (Hp), the common blood anti-coagulating agent used not only to treat and prevent blood clots under certain medical conditions but also before the procedure of open-heart and bypass surgery, especially, to reduce the risk of blood clots. Overdose of Hp in these cases is known to cause fatal consequences like critical hemorrhage and thrombocytopenia. Thus, careful monitoring of Hp concentration in blood is of utmost necessary under the aforementioned conditions. Since, Hp is a highly negatively charged bio-macromolecule, its addition to the solution of cationic silole derivative induces strong aggregation of the probe, leading to a large enhancement (~90 fold) in the probe fluorescence. Accordingly, following this fluorescence enhancement, detection of Hp is possible with the detection limit as ~23 nM in a buffer (HEPES) solution. In the similar line, Tong and coworkers (Liu et al. 2014) have synthesized a novel salicylaldehyde azine derivative with ammonium group functionalization, namely, N,N′-bis[4-[[3-(trimethylammonio)-thyl]oxy]salicylidene]-ethylenediamine (BTASE), as an AIE probe, for the quantification of Hp (cf. Fig. 14.12). The cationic probe BTASE forms aggregates with incremental addition of Hp in the studied solution. This probe is also found to be useful in detecting Hp even in dilute horse serum sample, which clearly indicates the potential application of this probe for Hp detection even in complex biological matrices.

In an independent study, Tang and coworkers have used number of tetraphenyl ethylene (TPE)-based cationic AIE probes for sensing of bioanalytes like BSA (bovine serum albumin) and ct-DNA (calf thymus DNA) (Tong et al. 2006). In

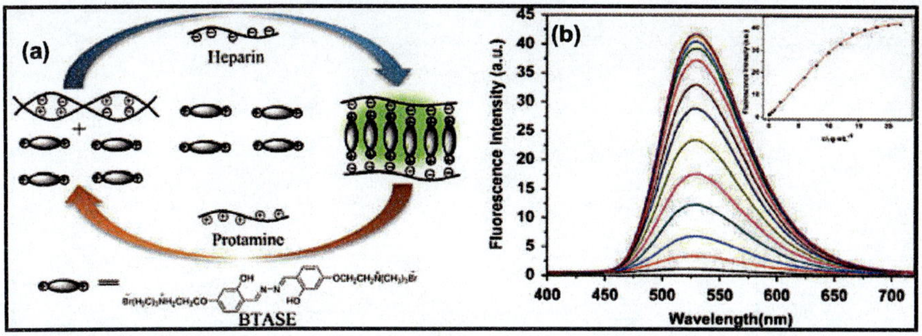

Fig. 14.12 (**a**) Schematic presentation heparin detection using AIE characteristics of BTASE probe. (**b**) Modulation in the fluorescence characteristics of 30 μM BTASE solution with increasing amount of heparin (from 0 to 22 μg/mL) in 10 mM Tris–HCl buffer at pH = 7.0. The λ_{ex} = 391 nm. The figure is adopted with permission obtained from Elsevier (Liu et al. 2014)

aqueous buffer solution, these probes are found to be practically non-emissive in nature, but they display substantially strong fluorescence light-up on binding with protein and DNA molecules, involving hydrophobic and/or electrostatic kinds of non-covalent interactions. At pH ~ 7, since both ct-DNA and BSA are negatively charged species, addition of these bio-analytes to the solutions of the TPE probes induces aggregation for the probe molecules, leading to the large enhancements in their fluorescence intensity (cf. Fig. 14.13).

Hong et al. (2010) and Wang et al. (2012) have reported the use of an anionic TPE derivative, BSPOTPE, and a distyryl antharecene derivative, BSPSA, respectively, as the efficient fluorescence turn-on sensors, with the prospective detection of serum albumins. These dye molecules having large hydrophobic/non-polar groups in their structures tend to aggregate in aqueous medium, in order to reduce their surface area exposed to polar bulk aqueous solvent. When albumin proteins are

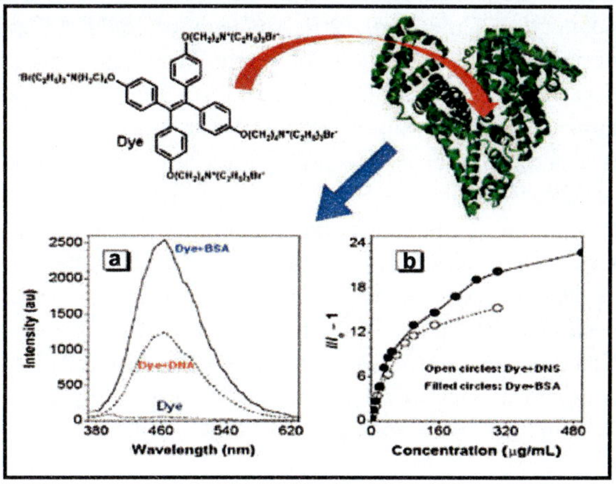

Fig. 14.13 Schematic presentation ct-DNA and bovine serum albumin (BSA) detection using AIE characteristics of the cationic TPE derivatives in a solution at pH 7. The figure is replicated with permission obtained from Royal Society of Chemistry (Tong et al. 2006)

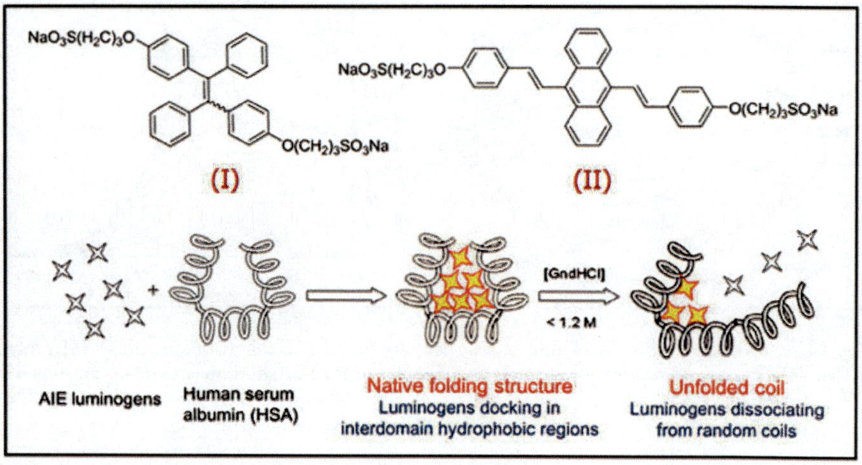

Fig. 14.14 Schematic presentation of the AIE characteristics of anionic TPE derivative (I) and distyryl anthracene derivative (II) inside BSA cavity. The figure is adopted with permission obtained from American Chemical Society and Royal Society of Chemistry, respectively (Hong et al. 2010; Wang et al. 2012)

added to the aqueous solution of these dyes, the dye molecules are readily transferred from bulk solution to the hydrophobic regions of the proteins, due to their strong hydrophobic character. Since volumes inside the conventional binding pockets of the albumin proteins are less in comparison to the dimension of these probe dyes, the dye molecules tend to aggregate at suitable hydrophobic regions of the proteins (cf. Fig. 14.14). Due to such aggregation, there is a dramatic reduction in the intramolecular torsional motions of these dye molecules, triggering a large escalation in the fluorescence intensity.

Apart from the bio-analyte induced dye aggregations, various other kinds of AIE effects for chromophoric dyes have also been described in the literature. In this connection, Chien et al. (2012) have reported the AIE phenomenon for an ammonium derivative of a tetraphenylthiophene (TP)-based dye, TP-NH_3^+, due to its aggregation in the presence of polyelectrolytic polymer, poly(sodium vinylsulfonate), PSV, assisted by long range electrostatic interaction between oppositely charged probe dye and polymer species. The authors have shown that the fluorescence quantum yield for TP-NH_3^+ − PVS system can be increased quite substantially by changing the polymer to dye molar ratios from 1 to 4. Further, an increase in the NaCl concentration to the TP-NH_3^+− PVS solution leads to the disintegration of the dye-polymer aggregates. Accordingly, there is a decrease in the fluorescence yield of the studied system. In the presence of the salt, due to the screening of the electrostatic attraction that operates between polymer species and the dye results in the disintegration of the previously formed TP-NH_3^+ − PVS complex in the solution. Formation and dissociation mechanisms of the TP-NH_3^+PVS complex as revealed in this study are conceptually shown in Fig. 14.15.

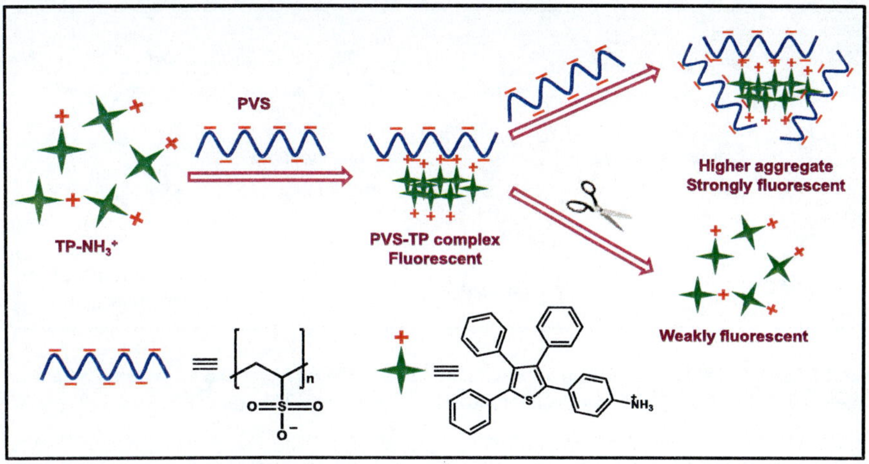

Fig. 14.15 Conceptual presentation of the construction and disintegration of the TP-NH_3^+PVS complex under different conditions. The figure is redrawn based on the concept obtained from Wiley–VCH (Chien et al. 2012)

Quite similar to the above study, Singh and co-workers have also reported aggregation of highly flexible thioflavin T (ThT) dye in the presence of polyanionic polymer, polystyrene sulfonate (PSS), triggering a large augmentation (~80 fold) in fluorescence intensity of the dye (Mudliar et al. 2020). Electrostatic attraction between anionic PSS and cationic ThT leads to the charge nullification of ThT and thus assisting the stacking of the dye molecules in a co-axial fashion, giving rise to the H-type of aggregate formation (cf. Fig. 14.16). Decrease in intramolecular torsional motions of the dye molecules due to such aggregation causes a substantial escalation for the dye fluorescence at the vicinity of PSS. Further, the ThT−PSS system is appeared to be very sensitive to ionic strength of the solution, which enables the authors to use this system as an efficient 'turn off' probe for protamine (Pr) sensing, with the limit of detection as ~18 nM. Protamine being highly positively charged bio-analyte, it disintegrates the ThT−PSS complex due to its competitive binding towards PSS, leading to a large decline in fluorescence intensity for the ThT−PSS system. Singh and co-workers have also reported the aggregation of a cationic pyrene derivative, pyrene methyl ammonium, PMA, in the presence of PSS, resulting in a large ratiometric change in the fluorescence behavior of PMA. Further, they have utilized the PMA–PSS system as an efficient ratiometric probe for the detection of basic amino acids (Pandey and Singh 2020).

Similar to the TP-NH_3^+−PVS complex, as discussed previously, Yao et al. (2011) also documented formation of H-aggregates of the cationic 3,3′-diethylthiacyanine (TC) dye, induced by tetrakis(4-fluorophenyl)borate (TFPB) anion as an additive, in an aqueous solution containing a neutral polymer as the stabilizing agent for the so formed dye aggregates. The authors have shown that along with the modulation in the optical properties, the morphology of the aggregated nanoparticles formed in the

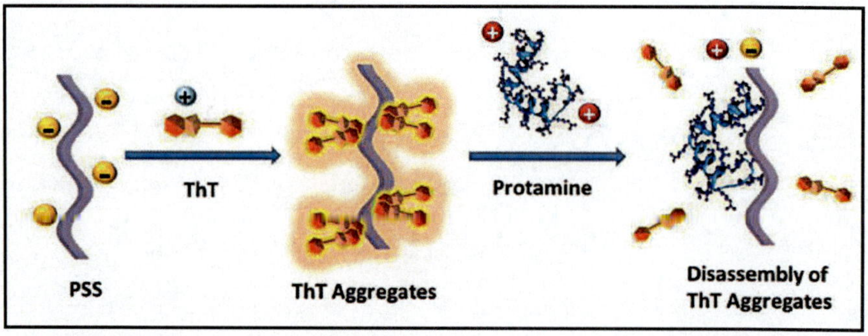

Fig. 14.16 Schematic presentation of the formation ThT-PSS complex and its protamine-induced dissociation. The figure is adopted with permission from Elsevier (Mudliar et al. 2020)

system also depends largely on the molar ratios of the dye to the counter ion used in this study (cf. Fig. 14.17). Thus, with an increase in the dye to counter ion molar ratio (r) from 1 to 4, the distribution of the nanoparticle sizes decreased gradually from 40–90 nm range (r = 1) to 30–60 nm range (r = 2), to 15–40 nm range (r = 4). Such decrease in the nanoparticle sizes with the increasing r value is ascribed to the surplus adsorption of anionic TFPB onto the initially formed nanoparticle surfaces, which thus inhibits the larger growth of these nanoparticles (Yao and Ashiba 2011). It is suggested by the authors that the increase in the surface adsorption of TFPB leads to the surface tension diminution for the nanoparticles, causing the reduction in the particle size.

In another study, Yao et al. (2009) have reported the nanoparticle formations involving the association of well-known cationic fluorescence probe, 2-(4-(dimethylamino)styryl)-1-ethylpyridinium (DASPE), with the species like tetraphenylborate (TPB) or tetrakis(4-fluorophenyl)borate (TFPB) anions, using the stabilizing agent like poly(vinylpyrrolidone). In contrast to previous example of TC-TFPB system, particle size for DASPE-TPB/TFPB systems changes very slightly, though the size distribution of the nanoparticles increases quite significantly with the increase in the molar ratios of the dye to counter ion (TPB or TFPB) present in the solution. The fluorescence quantum yields for the DASPE-TPB/TFPB systems are found to increase substantially with an increase in the particle size, which is attributed to both restricted intramolecular rotations around styryl single bonds of the dye and a reduced polarity effect caused by the TPB or TFPB matrix, decreasing the formation of non-fluorescent TICT state for the dye molecules in their aggregated states (Table 14.3).

Tohnai and coworkers (Hinoue et al. 2012) have investigated the crystal structures of aggregated species of anthracene-1,5-disulfonate dye and revealed that depending upon the nature (size and branching) of the counter ammonium cations, the dye aggregates can adopt different types of crystal arrangements, comprising face-to-face, zig-zag stacked, slipped column, and brick-like motifs, with substantial π-π overlaps between the dye molecules in each of these arrangements. The optical

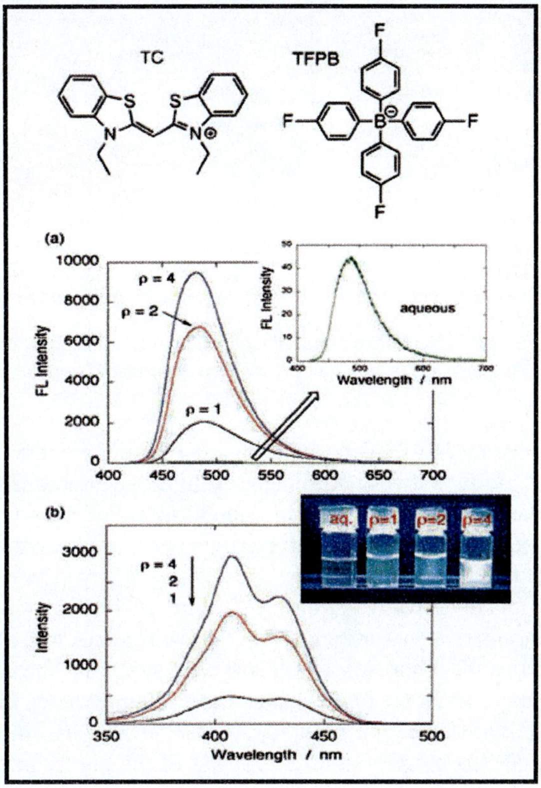

Fig. 14.17 (**a**) Emission spectra of TC nanoparticles formed at r = 1, 2, and 4, along with that of TC only in water (λ_{ex} = 407 nm). The inset represents enlarged emission spectrum of monomeric TC in water. (**b**) Excitation spectra of TC nanoparticle keeping monitoring wavelength at 480 nm. The inset depicts the photographs taken under 365 nm UV light irradiation for the solutions containing nanoparticles formed with r values of 0 to 4. The figure is adopted with permission obtained from Royal Society of Chemistry (Yao and Ashiba 2011)

Table 14.3 Effect of counter ion (TPB or TFPB) on the size and emission quantum yield (Φ) of the formed nanoparticles in the DASPE-TPB and DASPE-TFPB systems. This table is adopted with the permission of the American Chemical Society (Yao et al. 2009)

Samples	Phase	Size (nm)	Φ
DASPE	Aqueous	–	(3–5) × 10^{-3}
DASPE	Chloroform	–	0.036
DASPE-TPB (r = 1)	Nanoparticle	31.4 ± 3.05	0.085
DASPE-TPB (r = 2)	Nanoparticle	72.6 ± 6.46	0.092
DASPE-TFPB (r = 1)	Nanoparticle	46.5 ± 4.09	0.080
DASPE-TFPB (r = 2)	Nanoparticle	98 ± 12.74	0.12

properties (absorption, emission, fluorescence yield, and lifetime) of these crystals are seen to be substantially dependent on the formation and meticulous control of the fluorophore arrangements, which is controlled by the selection of the counter ions and dye-counter ion stoichiometry. It is revealed that instead of the electronic properties of the ammonium cations, the arrangement of the fluorophores in the

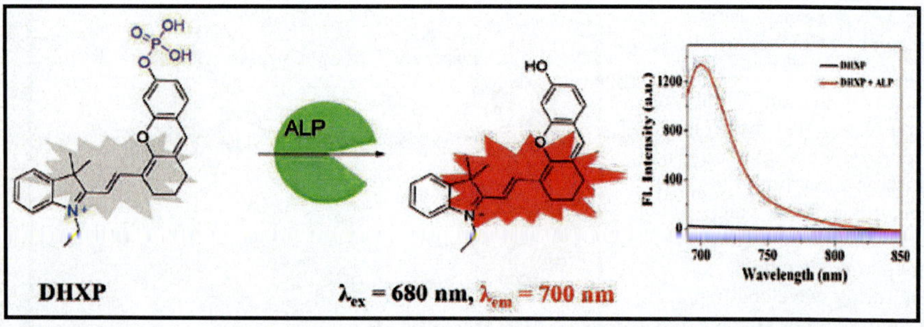

Fig. 14.18 Schematic presentation of action of alkaline phosphate of NIR dye (DHXP). The figure is adopted with permission obtained from American Chemical Society (Tan et al. 2017)

aggregates largely determines the optical properties of the concerned systems. Thus, such study is a very prominent example that demonstrates the correlation between the arrangements of the chromophoric moieties and the concerned photophysical properties, especially involving the dyes representing the organic salts.

Alike the earlier studies, Sun and coworkers (Tan et al. 2017) have synthesized a phosphate derivative of the dye 3-dihydro-1H-xanthene-6-ol dye (DHXP) and used as a sensitive near infrared (NIR) probe for detection of alkaline phosphatase (ALP) activity under both in vitro and in vivo conditions. The authors have shown that DHXP registers about 66-fold enhancement in fluorescence intensity upon incubation with ALP, with detection limit for ALP as ~0.07 U/L. Initially, the probe was soluble in aqueous buffer, but after removal of the highly polar phosphate group from the probe, due to the action of ALP, the probe molecules undergo aggregation among themselves due to their hydrophobic interaction, which significantly suppresses the intramolecular torsional motions for the dyes, causing an increase in their emission intensity. The schematic of this phenomenon is shown in Fig. 14.18.

In the similar manner, Lin et al. (2019) have reported AIE phenomenon involving a TPE-based cyanostilbene dye for ALP detection. They have developed a fluorescent probe introducing a strong electron withdrawing cyano group in the molecule at one of its ethene carbon, and incorporating the phosphatase recognizing biphenyl-based moiety with a phosphate group at the terminal. The synthesized probe dye has a typical donor–acceptor-donor type of structure having a comparatively high degree of conjugation with excellent water solubility. When water solubilizing phosphate group from the probe is cleaved by the action of ALP, the dye undergoes aggregation in the solution and thereby displays a strong AIE feature. The modulations in emission intensity thus monitored against the ALP concentration, form the basis of ALP detection by the developed probe.

14.5 Supramolecular Polymers and Nanostructures as the Exotic Functional Materials

Formation of supramolecular polymers and related other supramolecular nanostructures are the topics of intense investigations in contemporary research areas of chemical and material sciences. These assemblies are considered as the exotic supramolecular functional materials, having promising applications, not only in material sciences but also in different other areas in physics, chemistry, biology, pharmaceutics, electronics, and so on. In this respect, Kim et al. (Kim et al. 2016) have reported a highly emissive supramolecular polymer system comprising of 1-methylpyridium substituted cyanostilbene (Py^+–CNSB–Py^+) derivative as the chromophoric guest and CB8 as the macrocyclic host. The free Py^+–CNSB–Py^+ molecule is almost non-fluorescent in nature due to its large structural flexibility and its tendency to form TICT state, causing very efficient non-radiative deexcitation for excited dyes. Nonetheless, the dye solution registers a substantial escalation in fluorescence upon its binding with CB8 host, primarily because of the formation of linear and significantly rigid polymer assembly. Due to the presence of positive charges on both the 1-methylpyridinium end groups of Py^+-CNSB-Py^+ dye, either of its end groups can undergo very strong complexation with the host, CB8. Further, since CB8 cavity is reasonably larger in size, it can suitably act as an anchor to simultaneously encapsulate two Py^+ end groups of two independent dye molecules, approaching along the two opposite portals of the host cavity. Such mode of binding subsequently continues at both ends of the so-formed complex in a sequential manner, giving the eventual construction of the unique supramolecularly formed polymeric system. It is interestingly observed by these authors that the other stilbene derivative that does not contain the cyano substitution at its ethene unit did not undergo the formation of supramolecular polymer, but results in the formation of 2:2 cyclic CB8–(Py^+-CNSB-Py^+)$_2$–(CB8) species, giving rise to fluorescence quenching instead of fluorescence enhancement, as associated with the supramolecular polymer formation. The fluorescent supramolecular polymeric system, as obtained for Py^+-CNSB-Py^+ derivative in the presence of CB8 host, is predicted to find usages in chemical and biochemical sensing in addition to bioimaging applications. In a similar line, Pang and coworkers (Xu et al. 2011) have also reported that interaction of aqueous thiazole orange (TO) with CB8 induces linear supramolecular dye polymerization. The authors have suggested that the supramolecular polymerization process in the studied system is governed by both hydrophobic and π–π interactions among the building blocks (cf. Fig. 14.19). Several experimental evidences, such as atomic force microscopy, viscosimetry, electrospray ionization mass spectrometry (ESI–MS), nuclear magnetic resonance, and dynamic light scattering were used in this study to obtain the supports for the development of TO–CB8 based linear supramolecular polymer.

Jeon et al. have documented the formation of huge supramolecular vesicles, driven by CT complex formation of two complementary dye molecules inside a CB8 cavity (Jeon et al. 2002). In this study, simultaneous encapsulation of a methyl viologen (MV) derivative, having either 12-carbon-long or 16-carbon-long aliphatic chains

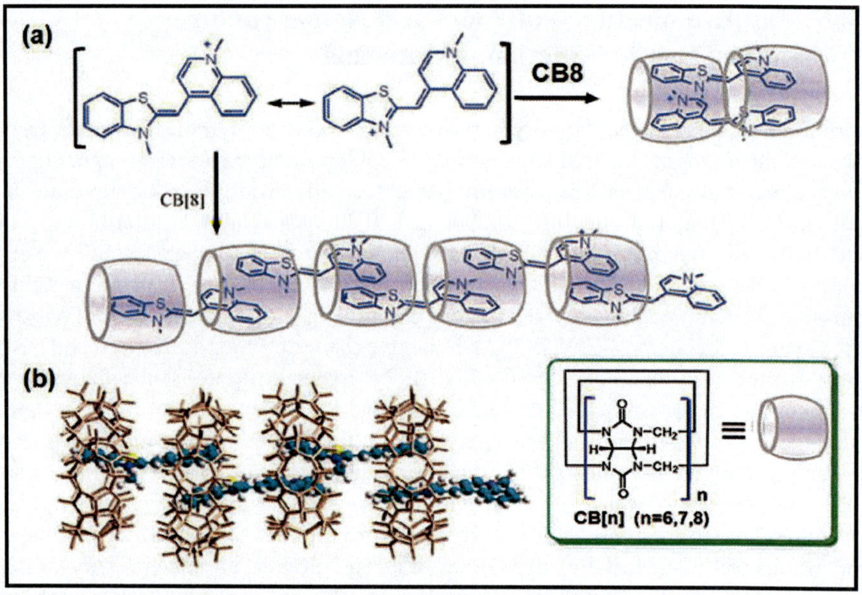

Fig. 14.19 (**a**) Chemical structure of TO dye and schematic for the formation of TO–CB8 assemblies. In the structure of TO above, its counter ion, *p*-toluene sulfonate, has been avoided deliberately for clarity of presentation. (**b**) Molecular modeling of the TO–CB8 polymer, where CB8 structure is displayed by the brown sticks for the clarity of the supramolecular assembly. The figure is replicated with permission obtained from Royal Society of Chemistry (Xu et al. 2011)

(MV-C12 and MV-C16, respectively), along with one 2,6-dihydroxynaphthalene (DHNp) molecule, inside a CB8 cavity, resulted in the formation of MV–DHNp CT complex. Since, the supramolecular complexes thus formed possess a long hydrophobic tail, in the form of C12 and C16 alkyl chains, along with a polar head group consisting of CB8 encapsulated MV–DHNp complex, they show very strong amphiphilic character. The so-formed complex thus undergoes self-association in aqueous solution to produce large vesicles, confirmed through SEM, NMR, mass spectroscopy, and TEM techniques. The mean diameter of the vesicles formed in the case of MV-C12 derivative is found to be ~20 nm. In the case of MV-C16 derivative, however, vesicles are indicated to be formed with much larger size, having ~400 to ~950 nm diameters. The study has clearly indicated that size of these supramolecular vesicles can be tuned suitably by varying the length of the associated alkyl chains.

Differential interactions of a CT type of amphiphilic dye containing pyridinium-functionalized anthracene (AnPy) moiety with CB7 and CB8 hosts have been reported by Assaf et al. (2018). In these cases, formation of the dye-host complexes ultimately led to the assembling of either supramolecular nanoparticles or nanorods, depending upon the size of the host cavities. The dye, AnPy contains a cationic pyridinium group at one terminal and electron-rich hydrophobic anthracene ring at the other terminal, separated from each other by an alkyl unit. For the AnPy-CB7 complex, due

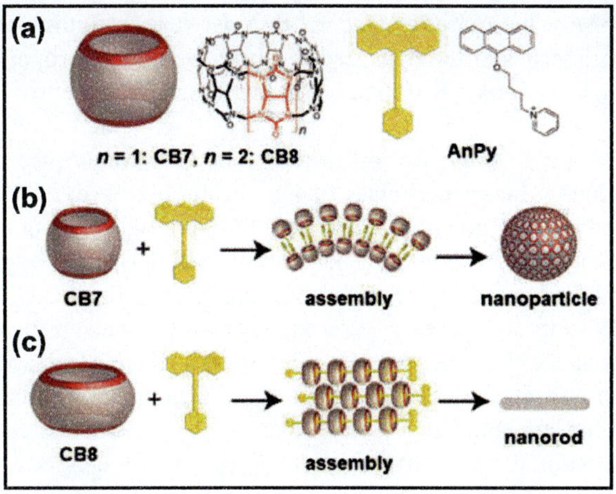

Fig. 14.20 (**a**) Schematic presentations of the structures of CB7 and CB8 hosts and AnPy dye. (**b**, **c**) Pictorial illustrations of the formations of supramolecular assemblies of AnPy derivative assisted by CB7 and CB8 hosts, respectively. The figure is adopted with permission obtained from Royal Society of Chemistry (Assaf et al. 2018)

to smaller cavity size, the CB7 host can engulf only the cationic pyridium group of AnPy dye, leaving its electron rich anthracene group exposed to bulk aqueous phase. These dangling anthracene groups, which are hydrophobic in nature, then interact with each other through π–π stacking, leading to the construction of stable nanoparticles having diameters in range of ~400 to ~1000 nm, confirmed through DLS and TEM (transmission electron microscopy) measurements. Conversely, due to its larger cavity size, the host CB8 can accommodate both pyridium and anthracene groups of two AnPy molecules simultaneously, causing a strong CT interaction between the two encapsulated units, which eventually leads to the construction of supramolecularly formed polymeric chains of AnPy molecules, via simultaneous incorporation of pyridinium and anthracene units of two different AnPy into one CB8 cavity in a successive manner. Hence, the formed polymeric chains can also interact with one another subsequently, resulting in the construction of supramolecular nanorods. Various types of nanostructures formed through the involvement of AnPy–CB7 and AnPy–CB8-based polymeric systems are schematically presented in Fig. 14.20. The formed assemblies show strong stimuli-responsive behavior and can be assembled and stripped off in the presence of suitable external stimulus like temperature, competitive binder, etc., indicating the potential of the studied nanostructured systems for stimulus responsive drug delivery applications, especially.

14.6 Future Scopes

To summarize, it is quite evident to realize that the supramolecular dye–host complexation can lead to the formation of exotic molecular assemblies that can display many interesting properties, quite different than the isolated dye molecules in the solution. Such supramolecular systems, especially the ones having fluorogenic dyes as one of the ingredients, can bring out very intriguing changes in the physiochemical properties of the dyes, especially in the optical properties, offering their widespread applications in different fields like fluorescence sensors, bioimaging, optoelectronics, dye lasers, and many others. Accordingly, such supramolecular studies have occupied a special position in the realm of fluorescence spectroscopy and its applications. The special class of the chromophoric dyes, designated as AIEgens, which are having no emission in their monomeric states, but produce quite large enhancement in their emission intensity by virtue of their aggregation, either on their own, or under the effect of various supramolecular hosts, like, macrocyclic molecules, proteins, enzymes, and so on, have garnered a significant research interests in recent years. Such AIEgen molecules are realized to have applications in various scientific areas ranging from chemical, biological and environmental sciences. Moreover, the physicochemical properties of these supramolecular assemblies along with the developed new photophysical characteristics of the associated dyes and AIEgen systems can also be modulated effectively under the influence of various stimuli like ionic strength, pH, temperature, etc. Therefore, the supramolecular systems that are decorated with various chromophoric dyes can be utilized very efficiently in different fundamental and applied areas, including drug formulation, targeted drug delivery, enzymatic assays, development of fluorescent sensors, catalysis, dye laser systems, electronic devices, nano-medicine, functional materials, etc. In fact, the supramolecularly assisted dye aggregates are presently considered as the futuristic functional materials that can promise many interesting and diverse applications in physics, chemistry, material sciences, biology, and environmental studies.

Acknowledgements Authors are grateful to Dr. Archana Sharma, Director, Beam Technology Group, and Martin Mascarenhas, Head, Laser and Plasma Technology Division, for their generous support during writing this chapter. HP is especially grateful to Department of Atomic Energy for his Raja Ramanna Fellowship. The present work was financially supported by Department of Atomic Energy (DAE), Government of India, under project number R&D-044-2018.

Author Contributions All the authors contributed in writing this manuscript. Further, all the authors have also given consent to the final version of this manuscript.

Conflicts of Interest Authors have no conflict of interest to declare.

References

Amabilino, D.B., Smith, D.K., Steed, J.W.: Supramolecular materials. Chem. Soc. Rev. **46**, 2404–2420 (2017)

Assaf, K.I., Alnajjar, M.A., Nau, W.M.: Supramolecular assemblies through host–guest complexation between cucurbiturils and an amphiphilic guest molecule. Chem. Commun. **54**, 1734–1737 (2018)

Awasthi, A.A., Singh, P.K.: Stimulus-responsive supramolecular aggregate assembly of auramine O templated by sulfated cyclodextrin. J. Phys. Chem. B **121**, 6208–6219 (2017)

Bañuelos, J., Arbeloa, F.L., Arbeloa, T., Martinez, V.M., Arbeloa, I.L.: BODIPY laser dyes applied in sensing and monitoring environmental properties. Appl. Sci. Innovations Pvt. Ltd., 641–677 (2012)

Batat, P., Cantuel, M., Jonusauskas, G., Scarpantonio, L., Palma, A., O'Shea, D.F., McClenaghan, N.D.: BF2-azadipyrromethenes: probing the excited-state dynamics of a NIR fluorophore and photodynamic therapy agent. J. Phys. Chem. A **115**, 14034–14039 (2011)

Boens, N., Leen, V., Dehaen, W.: Fluorescent indicators based on BODIPY. Chem. Soc. Rev. **41**, 1130–1172 (2012)

Boraste, D.R., Chakraborty, G., Ray, A.K., Shankarling, G.S., Pal, H.: PH-responsive interaction of fluorogenic antimalarial drug quinine with macrocyclic Host Cucurbit[7]uril: modulations in photophysical and acid-base properties. ChemistrySelect **2**, 5128–5142 (2017)

Boraste, D.R., Chakraborty, G., Ray, A.K., Shankarling, G.S., Pal, H.: Supramolecular host-guest interaction of antibiotic drug ciprofloxacin with Cucurbit[7]uril macrocycle: modulations in photophysical properties and enhanced photostability. J. Photochem. Photobiol. A **358**, 26–37 (2018)

Chakraborty, G., Pillai, V.S., Chittela, R.K.: Complexation-induced tuning of optical properties of a medically important alkaloid, berberine in the presence of charged cyclodextrin. J. Photochem. Photobiol. A **419**, 113454 (2021a)

Chakraborty, G., Choudhary, M.K., Sundararajan, M., Ray, A.K., Mula, S., Pal, H.: Stimuli responsive confinement of a molecular rotor based BODIPY dye inside a Cucurbit[7]uril nanocavity. J. Phys. Chem. B **125**, 7946–7957 (2021b)

Chakraborty, G., Malegaonkar, J.N., Bhosale, S.V., Singh, P.K., Pal, H.: Host-assisted aggregation-induced emission of a tetraphenylethylene derivative and its responses toward external stimuli. J. Phys. Chem. B **125**, 11122–11133 (2021c)

Chakraborty, G., Singh, P.K., Pal, H.: A cationic cyclodextrin assisted aggregation of an anionic pyrene derivative and its stimuli responsive behavior. J. Mol. Liq. **321**, 114499 (2021d)

Chien, R.-H., Lai, C.-T., Hong, J.-L.: Complexation of tetraphenylthiophene-derived ammonium chloride to poly(sodium vinylsulfonate) polyelectrolytes: aggregation-induced emission enhancement and long-range interaction. Macromol. Chem. Phys. **213**, 666–677 (2012)

Díaz, M.S., Freile, M.L., Gutiíerrez, M.I.: Solvent effect on the Uv/Vis absorption and fluorescence spectroscopic properties of berberine. Photochem. Photobiol. Sci. **8**, 970–974 (2009)

Doan, H., Castillo, M., Bejjani, M., Nurekeyev, Z., Dzyuba, S.V., Gryczynski, I., Gryczynski, Z., Raut, S.: Solvatochromic dye LDS 798 as microviscosity and pH probe. Phys. Chem. Chem. Phys. **19**, 29934–29939 (2017)

Guo, D., Uzunova, V.D., Su, X., Liu, Y., Nau, W.M.: Operational calixarene-based fluorescent sensing systems for choline and acetylcholine and their application to enzymatic reactions. Chem. Sci. **2**, 1722–1734 (2011)

Gupta, M., Maity, D.K., Singh, M.K., Nayak, S.K., Ray, A.K.: Supramolecular interaction of Coumarin 1 dye with Cucurbit[7]uril as host: combined experimental and theoretical study. J. Phys. Chem. B **116**, 5551–5558 (2012)

Hinoue, T., Shigenoi, Y., Sugino, M., Mizobe, Y., Hisaki, I., Miyata, M., Tohnai, N.: Regulation of p-stacked anthracene arrangement for fluorescence modulation of organic solid from monomer to excited oligomer emission. Chem. Eur. J. **18**, 4634–4643 (2012)

Hong, Y., Feng, C., Yu, Y., Liu, J., Lam, J.W.Y., Luo, K.Q., Tang, B.Z.: Quantitation, visualization, and monitoring of conformational transitions of human serum albumin by a tetraphenylethene derivative with aggregation-induced emission characteristics. Anal. Chem. **82**, 7035–7043 (2010)

Jeon, Y.J., Bharadwaj, P.K., Choi, S., Lee, J.W., Kim, K.: Supramolecular amphiphiles: spontaneous formation of vesicles triggered by formation of a charge-transfer complex in a host. Angew. Chem. Int. Ed. **114**, 4654–4656 (2002)

Kim, H.-J., Whang, D.R., Gierschner, J., Park, S.Y.: Highly enhanced fluorescence of supramolecular polymers based on a cyanostilbene derivative and Cucurbit[8]uril in aqueous solution. Angew. Chem. Int. Ed. **55**, 15915–15919 (2016)

Kwok, R.T.K., Leung, C.W.T., Lam, J.W.Y., Tang, B.Z.: Biosensing by luminogens with aggregation-induced emission characteristics. Chem. Soc. Rev. **44**, 4228–4238 (2015)

Laage, D., Thompson, W.H., Blanchard-Desce, M., Hynes, J.T.: Charged push−pull polyenes in solution: anomalous solvatochromism and nonlinear optical properties. J. Phys. Chem. A **107**, 6032–6046 (2003)

Lau, V., Heyne, B.: Calix[4]arene sulfonate as a template for forming fluorescent thiazole orange H-aggregates. Chem. Commun. **46**, 3595–3597 (2010)

Lin, M., Huang, J., Zeng, F., Wu, S.: A fluorescent probe with aggregation-induced emission for detecting alkaline phosphatase and cell imaging. Chem. Asian J. **14**, 802–808 (2019)

Liu, H., Song, P., Wei, R., Li, K., Tong, A.: A facile, sensitive and selective fluorescent probe for heparin based on aggregation-induced emission. Talanta **118**, 348–352 (2014)

Lou, X., Yang, Y.: Manipulating aggregation-induced emission with supramolecular macrocycles. Adv. Opt. Mater. **6**, 1800668 (2018)

Ma, X., Zhao, Y.: Biomedical applications of supramolecular systems based on host−guest interactions. Chem. Rev. **115**, 7794–7839 (2015)

Megyesi, M., Biczók, L., Jablonkai, I.: Highly sensitive fluorescence response to inclusion complex formation of Berberine Alkaloid with Cucurbit[7]uril. J. Phys. Chem. C **112**, 3410–3416 (2008)

Mohanty, J., Nau, W.M.: Ultrastable Rhodamine with cucurbituril. Angew. Chem. Int. Ed. **44**, 3750–3754 (2005)

Mohanty, J., Pal, H., Ray, A.K., Kumar, S., Nau, W.M.: Supramolecular dye laser with Cucurbit[7]uril in water. ChemPhysChem **8**, 54–56 (2007)

Mohanty, J., Jagtap, K., Ray, A.K., Nau, W.M., Pal, H.: Molecular encapsulation of fluorescent dyes affords efficient narrow-band dye laser operation in water. Chem. Phys. Chem. **11**, 3333–3338 (2010)

Monti, S., Manet, I.: Supramolecular photochemistry of drugs in biomolecular environments. Chem. Soc. Rev. **43**, 4051–4067 (2014)

Mudliar, N.H., Pettiwala, A.M., Dongre, P.M., Singh, P.K.: An anionic polyelectrolyte induced aggregate assembly of Thioflavin-T: a prospective platform for Protamine sensing. Int. J. Biol. Macromol. **164**, 1174–1182 (2020)

Pandey, S.P., Singh, P.K.: A polyelectrolyte based ratiometric optical sensor for Arginine and Lysine. Sens. Actuators B **303**, 127182 (2020)

Sarkar, P., Luchowski, R., Raut, S., Sabnis, N., Remaley, A., Lacko, A.G., Thamake, S., Gryczynski, Z., Gryczynski, I.: Studies on solvatochromic properties of aminophenylstyryl-quinolinum dye, LDS 798, and its application in studying submicron lipid based structure. Biophys. Chem. **153**, 61–69 (2010)

Seth, D., Sarkar, S., Pramanik, R., Ghatak, C., Setua, P., Sarkar, N.: Photophysical studies of a hemicyanine dye (LDS-698) in dioxane-water mixture, in different alcohols, and in a room temperature ionic liquid. J. Phys. Chem. B **113**, 6826–6833 (2009)

Shi, J., Deng, Q., Li, Y., Chai, Z., Wan, C., Shangguan, H., Li, L., Tang, B.: An aggregation-induced emission probe based on host-guest inclusion composed of tetraphenylethylene motif and γ-cyclodextrin for the detection of α-amylase. Chem. Asian J. **14**, 847–852 (2019)

Sun, S., Li, F., Liu, F., Wang, J., Peng, X.: Fluorescence detecting of paraquat using host-guest chemistry with cucurbit[8]uril. Sci. Rep. **4**, 3570 (2014)

Tan, Y., Zhang, L., Man, K.-H., Peltier, R., Chen, G., Zhang, H., Zhou, L., Wang, F., Ho, D., Yao, S.Q., Hu, Y., Sun, H.: Reaction-based off–on near-infrared fluorescent probe for imaging alkaline phosphatase activity in living cells and mice. ACS Appl. Mater. Interfaces **9**, 6796–6803 (2017)

Thakare, S.S., Chakraborty, G., More, A.B., Chattopadhyay, S., Mula, S., Ray, A.K., Sekar, N.: Modulation of optical properties of BODIPY fluorophore via intramolecular charge transfer. J. Lumin. **194**, 622–630 (2018)

Tong, H., Hong, Y., Dong, Y., Ha¨ußler, M., Lam, J.W.Y., Li, Z., Guo, Z., Guo, Z., Tang, B.Z.: Fluorescent "light-up" bioprobes based on tetraphenylethylene derivatives with aggregation-induced emission characteristics. Chem. Commun., 3705–3707 (2006)

Wang, M., Zhang, D., Zhang, G., Zhu, D.: The convenient fluorescence turn-on detection of heparin with a silole derivative featuring an ammonium group. Chem. Commun. **37**, 4469–4471 (2008)
Wang, F., Wen, J., Huang, L., Huang, J., Ouyang, J.: A highly sensitive "switch-on" fluorescent probe for protein quantification and visualization based on aggregation-induced emission. Chem. Commun. **48**, 7395–7739 (2012)
Wang, Q., Li, Z., Tao, D., Zhang, Q., Zhang, P., Guo, D., Jiang, Y.: Supramolecular aggregates as sensory ensembles. Chem. Commun. **52**, 12929–12939 (2016)
Xu, Y., Guo, M., Li, X., Malkovskiy, A., Wesdemiotis, C., Pang, Y.: Formation of linear supramolecular polymers that is based on host–guest assembly in water. Chem. Commun. **47**, 8883–8885 (2011)
Yao, H., Ashiba, K.: Highly fluorescent organic nanoparticles of thiacyanine dye: a synergetic effect of intermolecular H-aggregation and restricted intramolecular rotation. RSC Adv. **1**, 834–838 (2011)
Yao, H., Yamashita, M., Kimura, K.: Organic styryl dye nanoparticles: synthesis and unique spectroscopic properties. Langmuir **25**, 1131–1137 (2009)
Zhang, Y., Wang, Z., Yao, X., Zhang, Y., Wei, T., Yao, H., Lin, Q.: Novel tripodal-pillar[5]arene-based chemical sensor for efficient detection and removal paraquat by synergistic effect. Sens. Actuators B **327**, 128885 (2021)
Zhao, M., Wang, M., Liu, H., Liu, D., Zhang, G., Zhang, D., Zhu, D.: Continuous on-site label-free ATP fluorometric assay based on aggregation-induced emission of silole. Langmuir **25**, 676–678 (2009)

Chapter 15
Dimensional Engineering of 2D/3D Perovskite Halides for Efficient and Stable Solar Cells

Arif D. Sheikh and Kiran Kumar K. Sharma

Abbreviations

3D	Three-dimensional
2D	Two-dimensional
1D	One-dimensional
0D	Zero-dimensional
PCE	Power conversion efficiency
PSCs	Perovskite solar cells
MA	Methyl ammonium ($CH_3NH_3^+$) ion
FA	Formamidinium ($HC(NH_2)_2^+$) ion
ETL	Electron transporting layer
HTL	Hole transporting layer
DSSCs	Dye-sensitized solar cells
CZTS	Copper zinc tin sulfide (Cu_2ZnSnS_4)
spiro-OMeTAD	2,2′,7,7′-Tetrakis(N,N-pdimethoxyphenylamino)-9,9′-spirobifluorene
PEDOT	PSS-poly(3,4-ethylenedioxythiophene) polystyrene sulphonate
PTAA	Poly[bis(4-phenyl)(2,4,6-trimethylphenyl)amine]
P3HT	Poly(3-hexylthiophene-2,5-diyl)
ABHB	4-(Aminomethyl)benzoic acid hydrogen bromide
GBL	Gamma butyrolactone

A. D. Sheikh (✉)
Center for Nanoscience and Nanotechnology, Amity University Maharashtra, Bhatan, Panvel, Mumbai 410206, Maharashtra, India
e-mail: adsheikh@mum.amity.edu

K. K. K. Sharma (✉)
School of Nanoscience and Biotechnology, Shivaji University, Kolhapur 416004, Maharashtra, India
e-mail: kks.snst@unishivaji.ac.in

R. S. Ningthoujam and A. K. Tyagi (eds.), *Handbook of Materials Science, Volume 1*, Indian Institute of Metals Series, https://doi.org/10.1007/978-981-99-7145-9_15

DMSO	Dimethyl Sulphoxide
MQWs	Multi-quantum wells
QWs	Quantum wells
LP-VASP	Low-Pressure Vapor-assisted Solution Process
FF	Fill factor
WBH	Wide-bandgap halide
FESEM	Field emission scanning electron microscopy
RH	Relative humidity
V_{oc}	Open circuit voltage
GIXRD	Grazing Incidence X-ray diffraction
FTO	Fluorine-doped tin oxide

15.1 Introduction

In the era of the Internet of Things, artificial intelligence, machine learning, and the rapid development of the world, the use of electronic devices relentlessly demands alternative, renewable, and carbon–neutral energy sources. In recent years, the world storage of coal, crude oil, natural gas, and other non-renewable energy sources are impacting climate change and the environment's global warming issue. With time, the problem of energy shortage is becoming more severe for the development of the global economy. Against this background, most countries continuously focus on harvesting energy using alternative sources like wind, geothermal, photovoltaics, hydrogen generation, solar thermal power plants, etc. All these types of renewable energy sources could effectively promote economic growth, enhance energy security, increase energy access, and reduce carbon footprint (Satapathi 2022). Among different renewable energy sources, solar energy is the largest available source that has the potential for scaling up to meet future energy demands. The solar energy supply to the earth's surface is gigantic: 3×10^{24} J per year (Grätzel 2001). In principle, the complete conversion of solar energy that strikes the surface of the globe in 1 h into electricity can fulfill the current global energy consumption of an entire year required for humanity. It is predicted that harvesting less than 0.02% of the solar energy that falls on the earth's surface would satisfy the present needs (Hammarström and Hammes-Schiffer 2009). Solar cells can straightforwardly transform light into electricity through the photovoltaic effect, and currently, silicon solar cell is the most sophisticated commercialized technology. Nevertheless, because of the complexity of the fabrication technique and the excessive cost of silicon materials, it remained a bottleneck for the absorption of the technology to ground level. Despite this present Si-based photovoltaic technology, finding alternative cost-effective materials and methods to fabricate economical solar cells is necessary. The dye-sensitized solar cells (DSSCs), CZTS-based solar cells, organic solar cells, and quantum dot solar cells are some of the alternative types of solar cells, which are focused on for the last couple of decades. However, even after the success so far achieved with solar cells, their low photo conversion efficiency (PCE) and lack of low-cost solution processing techniques still remain a bottleneck. In the last decade, hybrid perovskite halide-based solar cells have developed as an emerging era of promising low-cost solution-processed photovoltaic technologies.

The most studied inorganic–organic hybrid perovskites like $MAPbI_3$, $FAPbI_3$, and $CsPbI_3$ and their compositions have attracted considerable attention due to their superior electronic and optoelectronic properties resulting in spectacular successes in solar cell applications. Various strategies are used to improve the PCE of PSCs from 3.8% to 25% through the optimization and processing of various device architectures (Cell-Efficiency n.d.). The PSCs are the first category based on ultra-low-cost fabrication via low-temperature solution processing techniques with high efficiency close to commercial Si solar cells of 25% and have a high potential for integrated photovoltaics (Liang et al. 2021; Gangadharan and Ma 2019). Unfortunately, these conventional hybrid perovskite materials are sensitive to humidity, thermal cycling, ultraviolet radiations, and ion migration due to the volatile nature of the organic group (Schileo and Grancini 2020; Sheikh et al. 2015, 2017). This instability of 3D perovskite halides under extreme ambient conditions like moisture, heat, oxygen, and a significant roadblock remains to be the ultraviolet light for the commercialization of perovskite-based photovoltaic technology. Therefore, even after surpassing the photoconversion efficiency of PSCs, issues regarding ambient stability need to be resolved. Over the past decade, the PCE of PSCs performance and stability was improved by numerous strategies employed by the researchers including device encapsulation techniques (Ma et al. 2022), compositional and solvent engineering (Jeon et al. 2015, 2014a), dopant/additive incorporation (Ugur et al. 2017), selection and modification of charge transporting materials/layers (Haque et al. 2017), implementation of novel metal electrodes (Lee et al. 2019), and dimensional engineering of perovskite film (Krishna et al. 2019; Zhang et al. 2018). Among these, the research on engineering of low-dimensional perovskite materials has drawn tremendous interest because of the remarkable progress made in both device efficiency and the stability of PSCs.

This chapter is focused on the fundamental understanding of dimensional engineering of perovskite halides and explored recent advances reported in the literature. In the dimensional engineering strategy, one can modify the size and shape of the material at the nanoscale by tuning the processing parameters and precursors. Further, introduced the basics of 3D perovskite and 2D perovskite materials and elaborated on their use in PSCs. Recent research demonstrated that 2D perovskite materials are more stable as compared to 3D perovskite materials under the influence of oxygen, light, and humidity. Attaining high solar cell efficiency and ambient stability simultaneously with a mixed dimensional strategy necessitates a delicate balance in the composition of 2D and 3D perovskite along with the structural motifs for achieving optimized contribution from each of the 2D and 3D perovskite phases. Two major protocols are envisaged for the combination of 2D and 3D perovskites viz. mixed 2D + 3D perovskite, i.e. incorporation of 2D perovskite in bulk 3D perovskite and bilayer 2D/3D thin films, i.e. surface treatment with 2D perovskite phase on a 3D perovskite film. However, depending on the type of synthesis techniques, segregation of the 2D and 3D perovskite phase in thin films, and type of 2D and 3D perovskite materials, there is a significant influence on the PCE and ambient stability of 2D/3D PSCs.

15.2 Introduction to Perovskite Halides

Gustav Rose in 1839 first coined the term "*perovskite*" for the $CaTiO_3$ mineral, named after the Russian nobleman and mineralogist Count Lev Alekseyevich von Perovski. Victor Goldschmidt in the year 1926, first used as a general term for describing the crystal structure group (Akkerman et al. 2018). The perovskite crystal lattice is defined as a network of corner-sharing BX_6 octahedra that crystallize with a general ABX_3 stoichiometry (where A = $CH_3NH_3^+$, FA^+, and Cs; B = Pb, Sn, and X = I, Br or Cl). The standard depiction of the aristo-type cubic perovskite represented either with all the atoms (up) or the BX_6 octahedral network only and the A atoms (down) as depicted in Fig. 15.1a. The typical lattice structure of perovskite material lies in the 3D-perovskite category (Akkerman and Manna 2020). Hybrid perovskites halide were first reported in 1893, but it took around 100 years for researchers to use this material for light-emitting devices and transistors due to their intriguing optical and electronic properties (Shamsi et al. 2019). In the year 2009, the real potential of these materials was tested and discovered for the first time in the field of solar cells (Kojima et al. 2009). Here afterward, the family of PSCs showed promising PCE in a short period.

15.2.1 Dimension-Dependent Types of Perovskite Halides

The dimensionality of perovskite materials engineered by adding/substituting alternative A-site cations, which can modify the 3D perovskite lattice structure either to 2D (layered), 1D (columnar), or 0D (quantum dots) nanostructures. The dimensional-dependent schematic depiction of the crystal structures of perovskite halides to (i) 3D, (ii) layered (2D), (iii) 1D nanorods/nanotubes, and (iv) 0D colloidal nanoparticles is demonstrated in Fig. 15.1b. Among these various nanostructures, this chapter is focused on 2D perovskite halide structures for solar cell applications. Therefore, it is introduced here about the 2D perovskite in detail for the reader's reference.

Most commonly employed 2D halide perovskite is fabricated by slicing analogous 3D perovskite in the (100) plane, resulting in the formation of 2D perovskites with different numbers of organic layers as depicted in Fig. 15.2. (Ma et al. 2018). Depending upon the type of large chain organic cations selected, 2D perovskites are categorized into three, i.e. Ruddlesden-Popper (RP), Dion-Jacobson (DJ), and alternating cations in interlayer (ACI) (Soe et al. 2018, 2017). Weak van der Waals gaps are observed between each bilayer of monovalent cations and adjacently located two lead halide sheets in the RP-phase 2D perovskites. The difference between the crystal

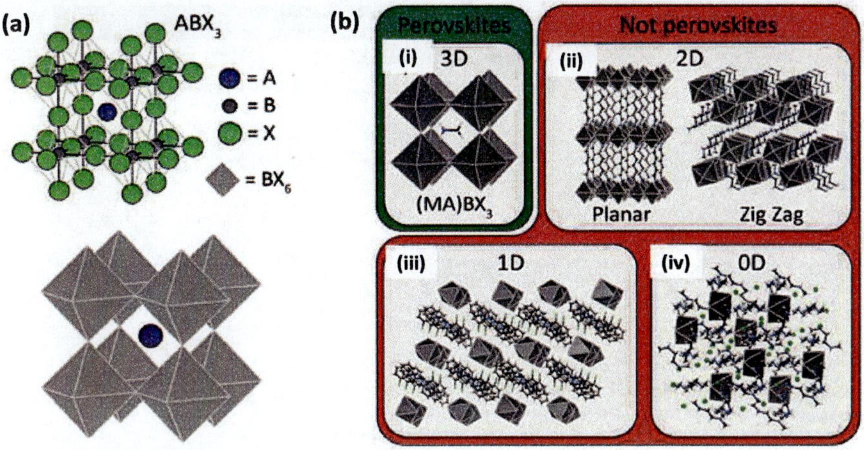

Fig. 15.1 **a** The lattice structure of ABX_3 perovskite structure with two different approaches. **b** Dimensional-dependent classification of perovskite halides into (i) 3D hybrid organic–inorganic metal halides, (ii) Layered (2D) hybrid organic–inorganic metal halides, (iii) 1D metal halides, and (iv) 0D hybrid organic–inorganic metal halides. [reprinted with permission from Akkerman and Manna 2020; Copyright 2020 American Chemical Society)

structures of 3D to quasi-2D to 2D perovskite is shown in Fig. 15.2. The compositions of RP are generally described by $L_2A_{n-1}B_nX_{3n+1}$, (Mao et al. 2018) where L is a long-chain organic cation (spacer) examples include phenyl–ethyl-ammonium (PEA^+) and butylammonium (BA^+), A is a short-chain organic cation that is typically Methyl ammonium ($CH_3NH_3^+$) ion (MA^+), or Formamidinium ($HC(NH_2)_2^+$) ion (FA^+), B is a divalent metal cation like Sn^{2+} or Pb^{2+}, X is a halide anion I^-, Br^-, or Cl^- and n defines the number of inorganic layers held together. Typical PSCs employ 3D ABX_3 structures, where a monovalent "A-site" cation in the cuboctahedral site bonds with the BX_6 octahedra. It is worth noting that when the number of inorganic layers held together $n = \infty$ represents a 3D perovskite, when $n = 1$, it represents the perovskite of pure 2D, and when $n = 2$ to 5, it represents a quasi-2D perovskite. A blend of 3D perovskite and lower-n phases ($n \leq 3$) or higher-n phases ($n = 30$–60), can lead to the formation of a quasi-3D perovskites (Mei et al. 2014). The parameter "n" in the general formula is equal to the number of octahedral layers. There will be one octahedral layer between two organic layers when $n = 1$. Whereas the inorganic layers contained in this structure behave as a quantum well for the charge carriers while the organic layers act as the barrier (Younis et al. 2021). An alternative to 3D perovskites is quasi-2D perovskites for solar cell application. The quasi-2D perovskites address critical issues of stability faced by the 3D PSCs with high PCE performance. The inspiring achievement sprouted from the developments in the scientific understanding of synthetic chemistry of 2D perovskites and the advancements in the fabrication of stable PSC devices.

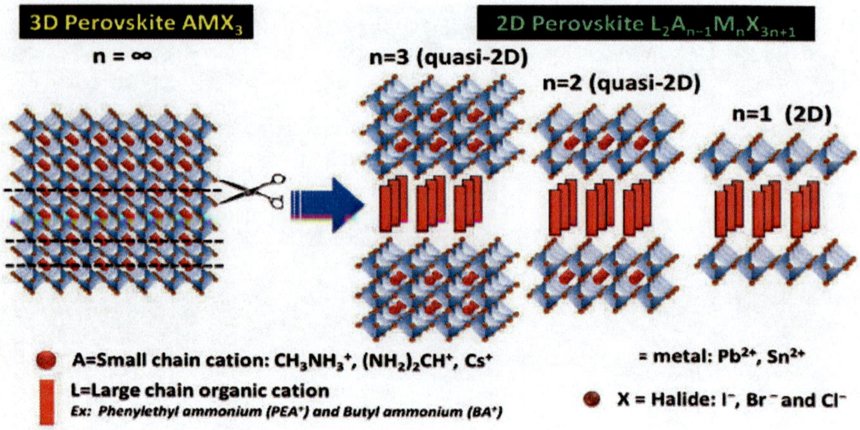

Fig. 15.2 Schematic representation of crystal structure transformation from 3D perovskite to 2D perovskite with the different values of n

15.3 Synthesis of Perovskite Halides and PSCs

Solution-based methods for the deposition of 2D-layered perovskites suffer due to the formation of multiple quantum wells mixtures having a good width (n number) in a random distribution. These factors have substantially hindered the progress in the field of 2D-layered perovskites (Quan et al. 2016; Tsai et al. 2018). Therefore, the deposition of quasi-2D-oriented thin films is still a critical challenge on the thermodynamics of the stability compounds in a solution state.

The perovskite halide materials with 3D, 2D, and mixed dimensional 2D-3D can be synthesized with the most common synthesis process, which is schematically summarized in Fig. 15.3a. They include spin coating, deep coating, screen printing, slot-die coating, and vapor deposition techniques. The hybrid method represents the synthesis method involving intermixing among these techniques. Similarly, the deposition of the electron transporting layer (ETL) and hole transporting layer (HTL) was carried out by deposition techniques including, spin-coating, spray-pyrolysis, hydrothermal, sputtering, and hybrid, which is a combination of these methods. The synthesis methods for the deposition of ETL and HTL are summarized in Fig. 15.3b. In general, perovskite films prepared using vacuum deposition have several advantages such as films with high phase purity, uniformity, pin-hole free and smooth as compared to the films that are processed through solutions. However, this method has disadvantages like a slow deposition process and the requirement of a meticulously controlled vacuum environment. This could be one of the reasons for the few studies reported on the vapor deposition of perovskite and ETL/HTL as compared to the solution processes. As the present chapter is devoted to the engineering of dimensional perovskite halides, the following section covered the synthesis of low-dimensional perovskite in detail. The synthesis protocols for the synthesis of 2D perovskite and

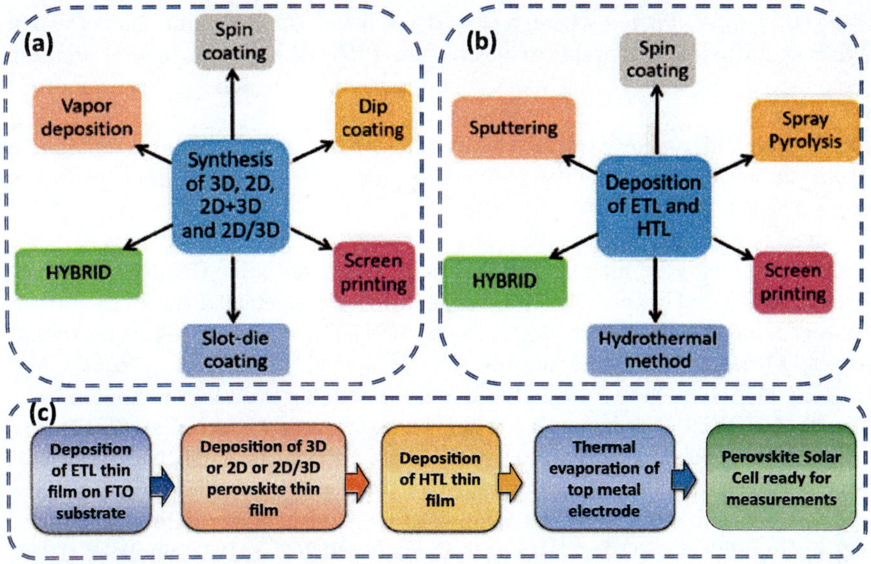

Fig. 15.3 Summary of various synthesis techniques used for the deposition of **a** perovskite halide thin films and **b** electron transport layers (ETL)/hole transport layers (HTL). **c** The important steps involved in the fabrication of perovskite solar cells

different processing strategies for the combination of 2D and or 3D perovskites in PSCs for improved PCE and robust stability are discussed in the following sections.

15.3.1 *Synthesis of 2D Perovskite Thin Films*

There are several processing strategies by which scientists explore 2D perovskite thin films for application in solar cells. Some of the techniques based on literature are discussed herewith. As demonstrated by Tsai et. al. in $(BA)_2(MA)_3Pb_4I_{13}$; where, n = 4; (BA = butan-1-ammonium), the 2D perovskites were deposited by hot-casting method (Tsai et al. 2016). In the described method, the initial hot precursor is deposited using a spin-coater on a pre-heated substrate, which is called the hot casting method. They observed that better crystallinity was achieved by hot-casting film at an optimized temperature of 110 °C, further confirmed by Grazing Incidence X-ray diffraction (GIXRD) patterns. However, the hot-casting method suffers from reproducibility because of the difficulty in maintaining a uniform temperature for the deposition in large-area substrates. The method is further modified by Zuo et. al. (2019) and demonstrated a process of substrate-heated drop-casting to deposit 2D perovskite. They also achieved the deposition of the 2D perovskite film in large area by developing a method of slot-die coating. A slightly modified method was also proposed by Wu et. al. through slower post-annealing process for the deposition of the $(BA)_2(MA)_3Pb_4I_{13}$; where, n = 4 (BA = butan-1-ammonium) 2D perovskite (Wu et al. 2019a). 2D perovskite with structure $(PEA)_2MA_4Pb_5I_{16}$, ($n = 5$) synthesis is reported by Hu et. al. using vacuum poling treatment which resulted in uniform nucleation during crystallization for obtaining the arranged inorganic sheets (Zhang et al. 2019a). The synthesis of $(PEA)_2MA_4Pb_5I_{16}$ perovskite thin films is reported by Gao et. al., which uses a simple spin-coating process. In this method, the growth of highly orientated 2D perovskite films under optimized solvent parameters resulted in improved optoelectronic properties. The optimized solvent composition was (5:5) a mixture of dimethyl formamide (DMF) and dimethyl sulfoxide (DMSO) compared to the film prepared in pure DMF (Gao et al. 2019). Zhang et. al. (Zhang et al. 2017a) explored a single-step spin coating technique for the deposition of a highly crystalline vertically oriented perovskite thin-films of 2D layered $(BA)_2(MA)_{n-1}Pb_nI_{3n+1}$ (BA = $CH_3(CH_2)_3NH_3$, MA = CH_3NH_3, n = 3, 4) with the additive, ammonium thiocyanate (NH_4SCN). Further, as lead (Pb) is a toxic element, Cao et. al. (2017) have reported the synthesis of 2D RP $(CH_3(CH_2)_3NH_3)_2(CH_3NH_3)_{n-1}Sn_nI_{3n+1}$ perovskite using a single step spin-coating method. In brief, the perovskite precursor solution was prepared using 1 mL of DMF over a magnetic stirrer with constant stirring having either 0.2 M of $(BA)_2(MA)_2Sn_3I_{10}$ (0.367 g, 0.2 mmol) or 0.2 M of $(BA)_2(MA)_2Sn_4I_{13}$ (0.474 g, 0.2 mmol) mixed with 0.2 M of SnF_2 (0.031 g, 0.2 mmol) at 70 °C for 1 h. For the triethylphosphine, 5 μL corresponding to 0.034 M was directly added to the perovskite solution mixture of 1 mL DMF. Spin-coating process for the deposition of the perovskite solution on mesoporous TiO_2 substrates, which was preheated at 120 °C at 3000 rpm for 30 s resulted in the 2D films. The films were then annealed at 75 °C for 5 min. and subsequently allowed to cool to room temperature.

15.3.2 *Synthesis of 2D + 3D Mixed Perovskite Layer*

The 2D + 3D mixed perovskite layer is also known as the bulk incorporation method. This method involves the incorporation of the spacer organic cationic compounds in the identical precursor solution containing the components of the 3D perovskite (PbI_2, $PbCl_2$, MAI, FAI, CsI, RbI, etc.). This single precursor containing the cations and anions of both 2D and 3D perovskite is spin-coated on the substrate followed by annealing to get the desired thin film of 2D + 3D perovskite phases. Therefore, several parameters like length of organic cation, solvent, concentration, annealing temperature, and annealing time need to be optimized for the high quality of 3D bulk-incorporated 2D perovskite film. A mixed dimensionality perovskite of $BA_{0.05}(FA_{0.83}Cs_{0.17})_{0.95}Pb(I_{0.8}Br_{0.2})_3$ is reported by Wang et. al. with the addition of BA^+ into $FA_{0.83}Cs_{0.17}Pb(I_yBr_{1-y})_3$ 3D perovskite (Wang et al. 2017).

A low-pressure vapor-assisted solution deposition process also known as LP-VASP method was applied by Li et. al. for the first time. It was used for the preparation of several 2D/3D hybrid perovskite films. The method uses a low-pressure heating oven to spin coat the substrate with phenyl ethyl-ammonium iodide (PEAI)-doped lead iodide (PbI_2) followed by a reaction with methyl-ammonium iodide (MAI) vapor. In this process, the authors investigated PEAI/PbI_2 ratio ranging from 2 to 0 to realize the effect of the ratio of doping PEAI in MAI-vapor-treated perovskites as plays a significant role on the crystallinity, morphology, optical absorption, and the resultant device performance (Li et al. 2018).

In another report, Chen et. al. incorporated 2D PEA_2PbI_4 [PEA = $C_6H_5(CH_2)_2NH_3$] in 3D $MAPbI_3$ [MA = CH_3NH_3] fabrication through the Lewis acid–base adduct method. They termed this mixed dimensionality perovskite as $(PEA_2PbI_4)_{0.017}(MAPbI_3)_{0.983}$ thin films. According to the authors, this thin film contains a structure of nanolaminate, which is made up of layered $MAPbI_3$ nanobricks produced in the presence of 2D PEA_2PbI_4 (Chen et al. 2017). For increasing the stability of α-$CsPbI_3$, $EDAPbI_4$ (EDA = ethylenediamine) was used by Zhang et. al. to obtain 2D perovskite. In this method, authors reported the α-$CsPbI_3$ is stabilized by the ethylenediamine (EDA) cation using a little amount of 2D $EDAPbI_4$ perovskite for avoiding the undesirable formation of the non-perovskite phase (Zhang et al. 2017b). Further, deposition of $(MAPbI_3)_{1-x}[(PEI)_2PbI_4]_x$ 3D/2D mixed perovskite films through spin coating process is also reported by Yao et. al. (2015). In the two-step deposition, PbI_2 was initially deposited along with precursor solution of PEI·HI (polyethyleneimine hydroiodide) followed by the deposition of CH_3NH_3I. The formation of $(PEI)_2PbI_4$ in the first step lowers crystallization rate of the 3D perovskite and promotes the growth of the 2D perovskite phase among the grains of the 3D perovskite phase. A dip coating with two-step approach was proposed by Koh et. al., which lead to the formation of $(IC_2H_4NH_3)_2(CH_3NH_3)_{n-1}Pb_nI_{3n+1}$ (Koh et al. 2016). Briefly, the first step involved spin-coating of a pure 2D perovskite ($n = 1$) with a precursor solution containing PbI_2 and $IC_2H_4NH_3I$ in a particular stoichiometric ratio, which was subsequently followed by the dipping of the substrate in the MAI solution for different time intervals, i.e. varying from 1 min. to 5 min.

Mohite and co-workers in 2016 reported the hot-casting method, which is a new and efficient approach. In the hot casting technique, the substrate consisting of FTO/poly(3,4-ethylene dioxythiophene) polystyrene sulfonate (PEDOT: PSS) was heated at ≈150 °C before the spin coating process of the perovskite. A superior quality thin film having favored out-of-plane alignment of $(BA)_2(MA)_3Pb_4I_{13}$ was achieved (Stoumpos et al. 2017). The authors observed that the PSCs prepared using this method had improved the stability and efficiency of the 2D/3D PSCs.

Grancini et. al. have proposed a method for possible engineering of the interface between the composite of AVA_2PbI_4 2D perovskite and $MAPbI_3$ 3D perovskite halides (Get and al. 2017). Authors prepared 2D-3D perovskite thin films of different compositions through the mixing of different quantities of 1.1 M solution of AVAI:PbI_2 in the ratio of 2:1 and MAI: PbI_2 precursor solution molar ratio of 0, 3, and 5%, respectively. These mixed solution blends are infiltrated in a mesoporous oxide scaffold using a one-step deposition process. The authors observed that the slow drying process allowed the restructuring of the 2D and 3D perovskite phase components in the mixed AVA_2PbI_4/$MAPbI_3$ film before solidification. Finally, according to the authors, a high PCE is obtained for the devices prepared with 3% AVAI perovskite film.

15.3.3 Synthesis of 2D/3D Perovskite Layer

Generally, for the preparation of 2D/3D perovskite bilayers, the two individual precursors containing components of 2D perovskite and 3D perovskite are deposited separately to form individual layers of 2D/3D perovskite thin films. Enhanced stability, a long carrier lifetime, and lower trap density were achieved in 2D perovskite films upon application on the surface of the 3D perovskite film which is further sandwiched to other charge-collecting layers in multi-stacked PSCs (Koh et al. 2018). Spin-coating process of the long-chain alkyl ammonium halides dissolved in isopropanol was used for the deposition at the surface of 3D perovskite film. The presence of water-resistant bulky organic groups at the top of the 3D perovskite films promotes the formation interfacial layer of 2D perovskite for an enhanced stability (Liu et al. 2019a). Deposition of PEA_2PbI_4 2D perovskite film on the top of $Cs_{0.1}FA_{0.74}MA_{0.13}PbI_{2.48}Br_{0.39}$ device was achieved by Nazeeruddin et. al. having an efficiency of 20.1% (Cho et al. 2018). The fabricated device had long-term stability as proven after 800 h under the ambient condition at 50 °C sustaining 85% of the initial efficiency. A post-deposition treatment of 3D perovskite by applying an isopropanol solution of penta-fluorophenyl-ethyl ammonium iodide (FEAI) was achieved by Liu et al. (2019b). In another work, the deposition of 2D perovskite film on the top of 3D perovskite film using a mixed solvent o-dichloro benzene and iso-propanol in the 97:3 v/v ratio with n-hexyl trimethyl ammonium bromide (HTAB) as demonstrated by Jung et. al. (2019a). A transparent 2D perovskite deposition on the top of 3D perovskite film was achieved using the dissolution of precursor strategy with alkylammonium bromide dissolved in chloroform by Yoo et. al. (2019). A synthetic approach for the fabrication of a bilayer structure of the type 2D PEA_2SnI_4 along with 3D $MASnI_3$ perovskite (PEA = phenethylammonium; MA = methylammonium) is reported by Choi et. al. The formation of the bilayer structure was achieved through a sequential vapor process combined with vacuum deposition and followed by vapor reaction. The advantage of the vapor process lies in ensuring the deposition of the uniform upper layer of the 2D perovskite thin film layer without damaging the lower 3D perovskite thin film layer (Choi et al. 2020). But later Joker et. al. (2021) reported a solution processing-based sequential method of deposition using hexafluoro-2-propanol as a solvent for the on the top of the 3D layer deposition of eight bulky ammonium cations to form a 3D/quasi-2D layer preventing the moistures and acting as a protecting layer of the tin perovskite grains.

Yukta et. al. (2022) recently reported the synthesis of 3D $MAPbI_3$ perovskite passivated by NH_4SCN followed by the deposition of the capping layer of 2D perovskite with organic xylylene diammonium iodide cation. For the deposition of 3D perovskite + NH_4SCN, an adequate amount of NH_4SCN was added to the 3D perovskite precursor solution and stirred for 12 h at 60 °C before use. This mixed precursor solution was spin-coated with chlorobenzene drip @ 1000 rpm for 10 s. and continuously at 6000 rpm for 30 s. which was followed by annealing at

100 °C. For fabrication of a 3D/2D heterojunction, 100 μL of p-xylylene diammonium iodide/isopropyl alcohol solution was spin-coated on the 3D perovskite + NH_4SCN perovskite surface at 3000 rpm for 30 s and annealed at 100 °C for 10 min.

The synthesis of 2D-3D graded perovskite was reported by Garai et al. (2021). The 2D-3D graded perovskites use a multifunctional 4-(aminomethyl)benzoic acid hydrogen bromide (ABHB) molecule synthesized strategically for the treatment surface in the fabrication process. Briefly, $MAPbI_3$ precursor solutions consisting of dissolved MAI and PbI_2 in 1:1 ratio in a mixed solvent system containing Gamma butyrolactone (GBL)/Dimethyl Sulphoxide (DMSO) (7:3, v/v) were spin-coated at 750 rpm for 20 s and 4000 rpm for 60 s with 160 ml anhydrous Toluene dripped after 20 s as antisolvent and annealed at 80° C for 10 min. Different concentrations of ABHB solution in isopropanol containing 1, 2, and 3 mg/mL were spin-coated at 4000 rpm for 30 s above the perovskite layer. The ABHB-treated perovskite film was annealed further at 80° C for 5 min. This could effectively form 2D perovskite thin film over $MAPbI_3$ 3D perovskite thin films for solar cell applications.

With the above examples, it can be concluded that different processing approaches are used for only 2D, 2D + 3D mixed, and 3D/2D bilayer perovskite thin film depositions for solar cell applications. However, it has been observed that in all these types of thin films, the growth of 2D and 3D perovskite is different, and an in-depth analysis of growth mechanisms needs to be explored. This is because synthesis techniques play a key role in the formation of perovskite layers with different crystal structures, different lattice orientations, and the distribution of 2D and 3D perovskite material. All these changes in the perovskite thin films affect the charge transport mechanism and in turn, decide the photovoltaic performance of the solar cell devices. It is also important to note that the synthesis is also sensitive to the type of initial precursors like the size of organic cations, additives, solvents, humidity, and substrate temperature.

15.4 Fabrication of PSCs with Different Perovskite Phases

The in-depth procedure for the fabrication of a general perovskite solar cell is elaborated as follows. The important steps involved during the fabrication of perovskite solar cells are schematically depicted in Fig. 15.3c. The step-by-step process for fabricating perovskites solar cells is discussed herewith. Zinc metal powder and 2 M HCl were utilized for etching FTO-coated glass substrates followed by washing with deionized (DI) water. Further, the etched substrate was subjected to ultrasonic cleaning using detergent, DI water, acetone, and ethanol, which was then dried finally using dryer. The hydrophilicity of the substrate surface was enhanced by UV-ozone treatment for a few minutes. The electron transporting material (ETM) can be deposited with any suitable method mentioned in Fig. 15.3. In general, techniques like spin-coating and/or spray-coating followed by annealing of thin films were used to deposit ETM thin films. The popular metal oxide-based electron transporting materials are TiO_2, SnO_2, and ZnO (Haque et al. 2017), whereas (Gangadharan and Ma

2019; Gangadharan and Ma 2019)-phenyl C61 butyric acid methyl ester (PCBM) (Xiao et al. 2014), C60 and their derivatives (Liang et al. 2014) have been used as n-type type charge carriers. In the next step, depending upon the type of perovskite materials, a precursor solution discussed in the previous section needs to be prepared separately. This precursor solution can be deposited on the ETM layer-coated FTO substrates. Here, note that depending upon the type of perovskite thin film, i.e. pristine 2D, pristine 3D, mixed 2D + 3D, or 2D/3D bilayer thin films, the precursor solution can be prepared as discussed in the above sections. Again, the perovskite layer can be deposited with any solution processing techniques or physical vapor deposition methods discussed in the above sections. This perovskite layer is required to deposit in the inert and controlled humidity conditions, which were found in the glove box. The fabrication of high-quality perovskite thin films depends on many processing parameters and hence is highly sensitive to the photovoltaic performance of perovskite solar cells. The various methods investigated in the literature for 2D, 2D + 3D, and 2D/3D perovskite layers have been discussed in this chapter later.

Next, the perovskite layer was covered by the deposition of the HTL, which could consist of either organic or inorganic materials. The most common way of deposition of HTL is again the spin coating technique for all types of organic materials. For inorganic materials, physical vapor deposition techniques were also used. One precaution required to take during the HTL deposition is to protect the perovskite thin film layer. The most efficient and widely used organic hole transport materials are poly(3,4-ethylenedioxythiophene) polystyrene sulphonate (PEDOT:PSS) (Xiao et al. 2014; You et al. 2014), 2,2′,7,7′-tetrakis(N,N-pdimethoxyphenylamino)-9,9′-spirobifluorene (spiro-OMeTAD) (Sheikh et al. 2015, 2017, 2019), poly[bis(4-phenyl)(2,4,6-trimethylphenyl)amine] (PTAA) (Jeon et al. 2014b; Yang et al. 2015), and poly(3-hexylthiophene-2,5-diyl) (P3HT) (Conings et al. 2014) in PSCs. Finally, metals like gold or silver thick layers were deposited using the thermal evaporator on the top of HTL, which acts as the top electrode.

In the early work, the device design of PSCs is based on solid-state DSSCs. The two most promising device architectures include mesoporous electron transporting layer-based PSCs and planar PSCs, which were initially focused on the optimization of the efficiency of the PSCs. In mesoporous perovskite solar cells, the mesoporous (scaffold) layer of inorganic metal oxides like TiO_2, Al_2O_3, SnO_2, $SrTiO_3$, etc. was deposited on a FTO substrate with a compact TiO_2 thin film-coating. The perovskite layer was deposited on this mesoporous layer followed by the HTL layer. In the case of planer perovskite solar cells, the perovskite thin film was deposited directly on the compact TiO_2 thin film-coated FTO substrate followed by deposition of the HTL layer. More details about the device architecture can be found in a recent review article (Ge et al. 2020).

This part is focused on the dimensional engineering of the perovskite layer and its impact on the performance of the device and ambient stability. It was observed from the literature that researchers explored various combinations of 2D and 3D perovskite materials for optimizing the device performance with higher stability. Some of the most common device architectures are schematically represented in Fig. 15.4, which will be discussed in this chapter. As observed apart from 3D perovskite, 2D, 2D

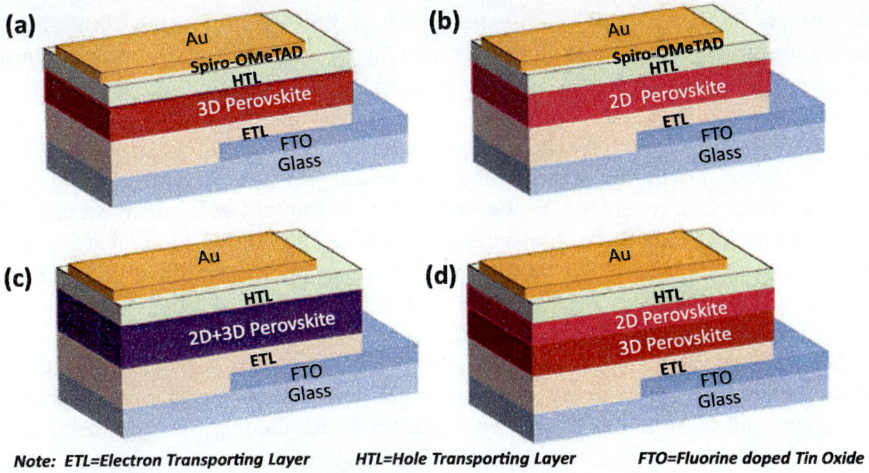

Fig. 15.4 Schematic representation of the device structure for **a** 3D perovskite solar cell, **b** 2D perovskite solar cell, **c** 2D + 3D mixed perovskite solar cells, and **d** 2D/3D layered perovskite solar cells

+ 3D mixed perovskite, and 2D/3D layered perovskite-based solar cells have been fabricated and tested in the literature. As denoted in Fig. 15.4, only the perovskite thin film with its different dimensional phases has been changed and all other layers are required to complete the solar cell device. The photovoltaic properties of the PSCs having a similar structure shown in Fig. 15.4 are discussed in the section below.

15.5 Photovoltaic Properties of 3D, 2D, 2D + 3D, and 3D/2D-Based PSCs

The following section has reviewed the recent work on the key issues that affect enormously the performance of 3D, 2D, 2D + 3D, and 3D/2D perovskite-based solar cells. First, the device configuration structures and properties of 3D, 2D, 2D + 3D, and 3D/2D perovskites will be introduced briefly and the photovoltaic performances are summarized.

15.5.1 3D Perovskite Halide-Based Solar Cells

The journey of hybrid perovskite halide-based solar cells has started with the 3D perovskite halide materials for sunlight harvesting due to their excellent optoelectronic properties. The first report on 3D PSCs has been published by Miyasaka and co-workers in the year 2009 with modest PCE of 3.8% (Kojima et al. 2009). As a breakthrough in the field was observed when the PCE of PSCs was improved from 10 to 20% in a short period (Saliba et al. 2016; Li et al. 2016). Over the past few years, PSCs have undergone a miraculous journey, showing continuous and rapid progress the PCE of PSCs reached a recent record value of 25% (Green et al. 2020). This indicates that the PCE achieved in 3D PSCs, which are competing with the commercial silicon solar cells. However, the PSC technology faces the most crucial issue of ambient long-term atmospheric instability. In addition to solar cells, the perovskite halides have shown promising applications in light-emitting diodes (Kim et al. 2022), photocatalysis (Deshpande et al. 2021), and gas sensors (Sheikh et al. 2022; Bhosale et al. 2023).

Several attempts (Sheikh et al. 2015, 2017; Ugur et al. 2017; Haque et al. 2017) to explore the impact of atmospheric moisture, atmospheric oxygen, atmospheric UV light, and temperature have been made for understanding the degradation of PSCs. Despite the innovations for improving PSC stability under various environmental conditions, the stability of 3D perovskite halide materials against moisture and temperature is still the focus of researchers working in the area of PSC development.

15.5.2 2D Perovskite Halide-Based Solar Cells

According to recent studies, the 2D perovskite halides have higher ambient condition stability as compared to those 3D perovskite halides. The presence of bulky organic spacer in the 2D perovskite structure contributes to more hydrophobicity. Further, highly oriented lattice structures and compact packing density lead to the reduction in the grain boundary density, minimizing the direct contact of water molecules with the 2D perovskite layer. (Ortiz et al. 2019) In addition to this, the layered 2D perovskites structure shows higher thermal stability and larger cations hamper internal ionic motion which further improves the stability (Smith et al. 2014a). However, the wide optical bandgaps and limitations in the charge transport of the 2D perovskites, the 2D perovskite-based solar cell shows limited performance. The PCE of 2D perovskite material is required to be enhanced compared to the performance of 3D perovskite-based solar cells. The multi-functional role of the 2D perovskites in a solar cell is as a capping layer, as a passivating layer, as a prime cell absorber and as an absorber in the hybrid 3D/2D perovskite-based solar cells. Recently, the production by a spatially disconnected 3D structure with a "protective" species, incorporation of emerging 2D-layered perovskites, provides a promising solution to the environmental stability issue of 3D PSCs (Gangadharan and Ma 2019). Therefore, 2D-layered metal-halide perovskites have generated the highest research interest and as one of the best materials for the fabrication of long-term, operationally stable with high-performance PSCs because of their desirable optoelectronic properties (Quintero et al. 2018; Zhou et al. 2017).

The most relevant works of literature on the fabrication of 2D/quasi-2D PSCs are tabulated in Table 1. Some significant points are briefly summarized as follows. The first quasi-2D perovskites ($PEA_2MA_2Pb_3I_{10}$) were demonstrated by Karunadasa et. al. in mesoscopic PSCs with an efficiency of 4.73% in 2014 (Smith et al. 2014b). A hot-casting method by Tsai et. al. was the first method reported for the preparation of 2D perovskite $(BA)_2(MA)_3Pb_4I_{13}$ (n = 4), achieving a PCE = 10% for the first time (Tsai et al. 2016). Zuo et. al. have achieved a PCE of 14.9% for the solar cell based on self-assembled 2D $(BA)_2(MA)_3Pb_4I_{13}$ perovskite films with uniform and oriented, deposited using the drop-casting method on hot substrates (Zuo et al. 2019). A slow post-annealing process was proposed by Wu et. al. for the deposition of $BA_2MA_3Pb_4I_{13}$-based 2D PSCs having PCE of 17.26%, a favorable bandgap energy alignment within the 2D perovskite was achieved owing to the vertical distribution of different perovskite phases in gradient (Wu et al. 2019b). According to the authors, these slow post-annealing processed devices exhibit satisfactory stability with less than 4.5% of degradation after 2000 h in N_2 environment without encapsulation. Similar efforts have been carried out to improve the performance and stability of 2D PSCs (Cheng et al. 2018; Lai et al. 2018; Yao et al. 2016). Another strategy reported by Zhang et. al. uses vacuum poling treatment for arranging variable-n-value nanoplates, enforcing homogenous nucleation in the process of crystallization (Zhang et al. 2019b). This approach yielded $(PEA)_2MA_4Pb_5I_{16}$-based PCE of 18.04%. As observed different reports focused on various growth mechanisms of 2D perovskite and observed different PCE values. Thus, insight mechanistic studies of the crystallization process of hot-casting or solvent processing are emphasized for deeper understanding (Zhang et al. 2020). Soe et. al. (Soe et al. 2018) reported deposition of highly oriented 2D RP perovskite (n-butyl-$NH_3)_2(MeNH_3)_4Pb_5I_{16}$ thin films from dimethyl formamide/dimethyl sulfoxide mixture solvents using the hot-casting method. As per the data, the PSC prepared with this method shows a PCE of 10% with an open-circuit voltage of $\approx$ 1 V. The photovoltaic devices based on $(BA)_2(MA)_2Pb_3I_{10}$ and $(BA)_2(MA)_3Pb_4I_{13}$ perovskite using the precursor's BAI: methylammonium iodide: PbI_2: NH_4SCN molar ratio composition of 2:2:3:1 showed average PCE of 6.82% and 8.79%, respectively, as reported by Zhang et. al. (2017a). Both the devices without shielding performed with excellent stability with a constant PCE during the time span of the storage in a glove box purged with purified N_2. The lead-free 2D perovskite-based PSCs were reported by Cao et. al. (2017), however, the PCE is found to be 2.5% and needs more investigation to further increase the PCE.

According to Liang et. al. and co-workers (Liang et al. 2021), the 2D thin films deposited using the conventional halide spacers result in thin films having a mixture of random well with width (n number) distribution and multiple quantum wells, which are treated as multi-quantum wells (MQWs). They have designed a strategy that can deposit phase-pure metal-halide perovskite quantum wells (QWs), which have highly oriented lattices. The schematic representations of the structure of the multi-phase and the pure-phase 2D perovskite film are shown in Figs. 15.5a, b, respectively. Figure 15.5c, d shows the XRD patterns of the pure phase 2D perovskite film deposited using BAI and BAAc, respectively. In multi-phase 2D perovskite film deposited using BAI, clear periodic 2D crystallographic planes (001) and (010) are observed for n = 1 and n = 2 film, respectively. However, according to the authors, for the largely mixed-phase films with $n \geq 3$, the 2D crystallographic planes disappear. In contrast, all the deposited films having $n = 2, 3, 4, 5$ with BAAc show a clear periodicity of the 2D crystallographic planes (0k0), which confirms the high phase purity. The authors analyzed surface morphologies of the films, which were characterized using SEM as shown in Fig. 15.5e, f. These results indicate to obtain, smooth surface having a controlled large average grain size of ~10 μm a phase-pure QW films should be used, while for smaller grains of ~1–2 μm and recognizable pinholes, the MQW films should be used. A cross-sectional SEM demonstrates the architecture of the device: ITO/SnO_2 (~50 nm)/layered perovskite (~500 nm)/spiro-OMeTAD (~200 nm)/MoO_3 (~5 nm)/Au (~80 nm) is shown in Fig. 15.5g. The J-V characteristics of the best-performing PSCs are shown in Fig. 15.5h. As observed the PCE of the phase pure QW-based PSC and the MQW-based PSC, as 16.25% and 13.81%, respectively. Further authors claim that these solar cells are stable for 4,680 h after keeping at 65 ± 10% humidity. Only less than 10% efficiency degradation was reported after keeping them for 1100 h in a continuous light illumination condition.

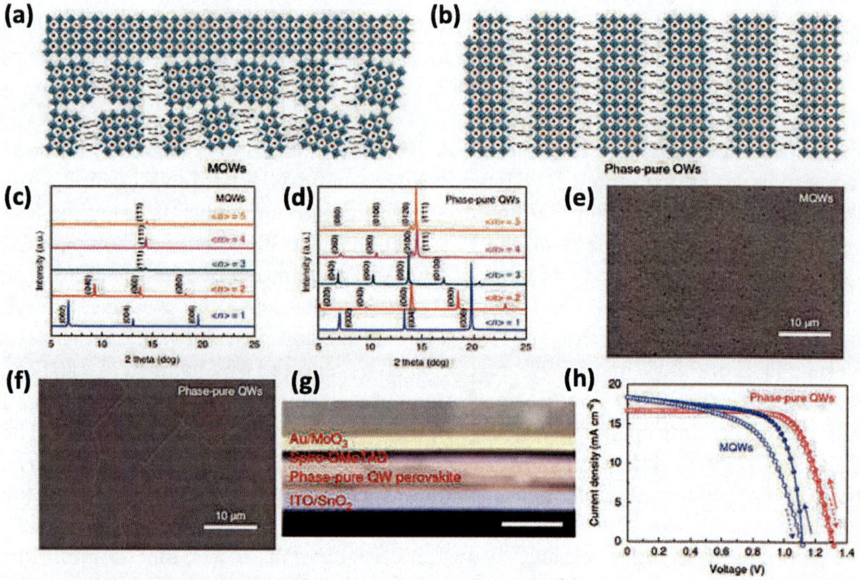

Fig. 15.5 Schematic structure of the multi-phase (**a**) and the pure phase (**b**) 2D perovskite film. The XRD patterns of the multi-phase (**c**) and pure phase (**d**) 2D perovskite films. The SEM images of the multi-phase (**e**) and pure phase (**f**) 2D perovskites thin films and (**g**) cross-section SEM image showing the device architecture. (**h**) Comparative current density vs voltage measurements of the best-performing PSCs with multi-phase and the pure phase 2D perovskite film. (Reprinted with permission from Ref. 79; Copyright 2020 Springer Nature)

15.5.3 Halide-Based 2D/3D Perovskite Solar Cells

The 3D perovskite-based solar cell is more advantageous than the 2D perovskite-based solar cell. The 2D perovskite-based solar cell shows a lower efficiency as a result of hindrance in the charge carrier transport because of the presence of bulky organic cations in the crystal structure. On the other hand, the highly efficient 3D perovskite solar cell suffers from low environmental stability. The incorporation of 2D perovskite into the 3D perovskite structures together can lead to balancing of the problems of a solar cell device, i.e. stability and efficiency (Krishna et al. 2019; Manjunath et al. 2022; Kim et al. 2021). The integration of 2D and 3D perovskite can be done in two main types either by blending/mixing the precursors of 2D and 3D or by layer-by-layer sequential method of deposition to obtain an engineered 2D/3D bilayer structure. The progress of combining 3D perovskite and 2D perovskite strategies reported in the literature has been explored in the following sections.

15.5.3.1 2D + 3D Perovskite-Based Solar Cells

The disadvantage of 2D perovskite as compared to 3D perovskites is the limitation in the performance of pure 2D perovskite-based solar cells. Some of them are mentioned herewith. After the transformation of 3D perovskite to 2D perovskite, the organic spacers used in such solar cells, form thin insulating layer sandwiched between the two slabs of conductive metal halide, leading to a reduction in conductivity of the current in the 2D perovskite films and impedes the extraction of charge as compared to 3D perovskites. Similarly, in comparison to 3D perovskite, 2D perovskite has a relatively higher optical band gap (≈2.5 eV), which requires a shorter spectral range for light absorption. Also, the binding energy of 2D perovskite is higher than that of 3D perovskite. Therefore, to overcome these issues of 2D perovskite, mixed 2D-3D or 2D + 3D perovskite films are preferable for fabricating PSCs, which could offer both high photovoltaic performances from 3D perovskite and superior device stability due to 2D perovskite.

In 2014, mixed 2D/3D multidimensional perovskite-based PSCs were conceptually achieved for the first time by Smith et. al. wherein MA cations were mixed with a large amount of PEA to produce a Ruddlesden–Popper structure with a composition: $(PEA)_2(MA)_2[Pb_3I_{10}]$ (Smith et al. 2014a). Despite the low PCE of 4.73%, the mixed 2D + 3D perovskite of $(PEA)_2(MA)_2[Pb_3I_{10}]$ composition-based device showed a high V_{oc} of 1.18 V. The work also achieved long-term stability of 46 days of air exposure subjected to relative humidity (RH) of 52%, for the 2D/3D mixed perovskite film while the PSCs prepared with $MAPbI_3$ completely degraded. Li et. al. (2018) proposed a device that consists of FTO/C-TiO_2/m-TiO_2/PEAI-$MAPbI_3$/Spiro-OMeTAD/Ag prepared with the LP-VASP method. In this device, according to the authors, it is proposed the $MAPbI_3$ perovskite grain is surrounded by the PEAI-containing perovskite grain, which improves their grain growth. The fabricated device based on ($PEAI/PbI_2 = 0.05$) was able to achieve the PCE of 19.10%

with a J_{sc} of 21.91 mA/cm^2, V_{oc} of 1.08 V along with a excellent fill factor (FF) of 80.36%. However, the authors have not reported their ambient device stability. Similarly, Hu et. al. used an antisolvent engineering method for the PEA incorporation in the 2D/3D Pb–Sn alloyed perovskite solar cell and observed the highest PCE of 15.93% for the device of 2D/3D $PEA_xMA_{1-x}Pb_{0.5}Sn_{0.5}I_3$ (Zhang and Hu 2020).

Koh et. al. reported a perovskite solar cell based on 2D + 3D perovskite with excellent stability in the air at a controlled relative humidity of 70–80% for both encapsulated and non-encapsulated are observed for the perovskite thin films obtained within 5 min. of dipping time. An outstanding 9% (J_{sc} = 14.88 mA/cm^2, V_{oc} = 0.883 V, and fill factor (FF) = 0.69) power conversion efficiency under AM 1.5G condition was achieved (Koh et al. 2016). Stoumpos et. al. reported a solar cell device in which hot casting of $(BA)_2(MA)_3Pb_4I_{13}$ was done on PEDOT: PSS coated FTO substrate (Stoumpos et al. 2017). This work demonstrated a major advancement in the device PCE of 12.5% for the first time. The 2D material incorporation for stability is also validated in this work, as the non-encapsulated device fabricated using a mixed-dimensional perovskite was found to retain 60% of the initial efficiency up to 2250 h at room temperature in a continuous incident light exposure condition. Further, this work also reports an improved tolerance to 65% relative humidity compared to 3D perovskites.

Grancini et. al. reported a highly stable and robust device based on mixed 2D $(HOOC(CH_2)_4NH_3)_2PbI_4$ and 3D $CH_3NH_3PbI_3$ perovskite thin films (Get and al. 2017). A schematic representation of PSC with and without HTM is shown in Fig. 15.6a. As observed authors prepared one device consisting of regular spiro-OMeTAD-based HTL followed by the metal electrode and in the other PSC device the conducting carbon was used, which works as an efficient hole transporting layer and no need for the metal electrode. Figure 15.6b shows the J-V plot and characteristics of 2D/3D perovskite solar cell with an optimized 3% AVAI mixture and Spiro-OMeTAD/Au. The device with HTM shows PCE of 14.6% for mixed 2D/3D perovskite and 15% for 3D perovskite thin film. The comparative time-dependent variation of normalized PCE of the PSC devices with standard 3D and with the mixed 2D/3D perovskite is shown in Fig. 15.6c. The PSCs with mixed 2D/3D perovskite exhibit a better performance in the stability of the device as compared to the pristine 3D $CH_3NH_3PbI_3$ cells as observed in Fig. 15.6c. The stability study carried out by the authors indicated the PCE of mixed 2D/3D perovskite halide achieved 60% of the initial value in argon atmosphere even after 300 h of continuous illumination and is more stable as compared to the standard 3D perovskites. Further, the PCE of PSC prepared with this optimized mixed perovskite in a carbon-based architecture observed a yield of 12.9% efficiency. The authors demonstrated the up-scaling by fabricating a solar module of the dimension 10 × 10 cm^2 using an industrial-scale printable process. The solar module delivered 11.2% efficiency and a stability of 410,000 h. Further, it observed under controlled standard conditions a zero loss in the performances as shown in Fig. 15.6d.

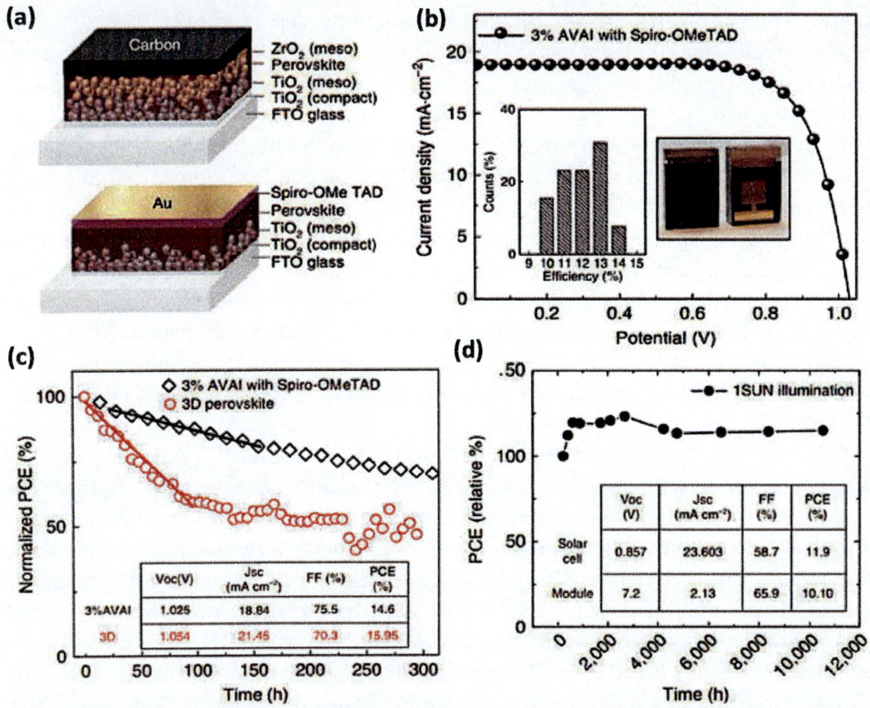

Fig. 15.6 **a** Schematic representation of PSC with and without the HTM **b** J-V curves for the 2D/3D PSCs with 3% $HOOC(CH_2)_4NH_3I$, AVAI hereafter, with device statistics and picture of the cell in the inset. **c** Comparative stability curves of PSCs with the standard 3D and the mixed 2D/3D perovskites. **d** Typical module stability test under 1 sun AM 1.5 G conditions at a stabilized temperature of 55 °C and short circuit conditions. [Reprinted with permissions from Ref. 44; Copyright 2017 Springer Nature under Creative Commons License]

15.5.3.2 2D/3D Bilayer Perovskite-Based Solar Cells

In this strategy, the 2D perovskite layer was deposited over the top of a 3D perovskite layer, which ultimately protects the 3D perovskite from moisture. This section elaborated on recent work so far conducted on 2D/3D bilayer perovskite-based solar cells.

As developed by Cho et. al. (2018), a two-step deposition method was used to fabricate perovskite solar cells of 3D MAPbI3/2D PEI_2PbI_4 bilayers (Grancini and Nazeeruddin 2019). The dynamical spin deposition of isopropanol solution of PEAI over the mixed halide 3D perovskite with excess PbI_2 induced layer-by-layer growth. Segregation of the excess PbI_2 was observed over the top of the 3D perovskite. Thus, the in-situ reaction of PEAI with PbI_2 at the top surface leads to the deposition of a thin layer of 2D on the top of the 3D layer. Thus, the 2D perovskite lies at the interface with the hole transporting material (HTM) forming the top surface which reduces the interfacial charge-carrier recombination leading to the rise PCE over 20%, which is retained even after 800 h exposure at 50 °C using 1 Sun illumination under suitable environmental conditions. Significantly, these results provide encouraging developments toward achieving significantly improved stability without compromising the PCE. Further, the base surface of n–i–p devices architecture can be developed using 2D perovskite anchored on the TiO_2 surface. In a recent development, an ultrathin wide-bandgap halide (WBH) was used to achieve a double-layered halide architecture by Jung et. al., i.e. 2D perovskite halide stacked on a narrow band-gap light absorbing halide layer resulting in a 3D perovskite halide, prior to HTM deposition. The insertion of the WBH layer was found to effectively reduce the charge recombination at the interface of perovskite and P3HT. Therefore, an enhanced PCE of around 23% with long-term (1370 h) operational stability with 1 Sun illumination with 95% of the initial efficiency at room temperature was achieved (Jung et al. 2019b).

Garai et. al. (2021) reported the 2D ABHB-3D $MAPbI_3$ graded perovskite heterostructure-based perovskite solar cells. The authors observed that after precise treatment with ABHB, the surface of $MAPbI_3$ 3D perovskite converted into a superior quality 2D/3D perovskite heterojunction film with a reduced grain boundary and better surface coverage. The ABHB-dependent variation in the J-V characteristic of the solar cell is shown in Fig. 15.7a. A PCE of 15.14% was observed for the device without modification while the device after 2 mg/mL ABHB treatment shows improved efficiency of 21.18%. The reproducibility of the process was checked with 15 fabricated devices using the process and plotting of all the parameters of the device in the form of a box chart. The histogram plot, Fig. 15.7b shows the optimum distribution of the device parameters for 15 individual cells with and without ABHB treatment. The images of the films with and without ABHB treatment captured by FESEM are shown in Fig. 15.7c, d. From the images, it was observed that ABHB treatment helps to improve the average grain size and compactness of the thin film morphology. Therefore, better surface coverage, higher quality perovskite films with a reduction in the grain boundary are achieved by ABHB treatment. These factors are responsible for the increase in the PCE of ABHB-treated films as compared to pristine films. In addition to this, the authors claim the role of carboxylic acid and amino groups for effective reduction in the trap states while the halide vacancies are filled by the bromide ions incorporated into the perovskite lattice of 3D $MAPbI_3$ thin film surface. This type of 2D-3D perovskite-based solar cells is able to minimize effectively the carrier recombination, enhances charge transport and a higher generation rate as compared to pristine 2D or 3D perovskite-based solar cells. As expected, the 2D layer on top devices showed improved stability in ambient conditions owing to better hydrophobicity as shown in Fig. 15.7e. Finally, the device stability was measured by authors for PSCs with and without treatment of ABHB measured at different time spans as shown in Fig. 15.7f. The storage of the devices was maintained at room temperature at a relative humidity range of 40–50%. The improved environmental stability was co-related with the hydrophobicity of the ABHB-treated perovskite.

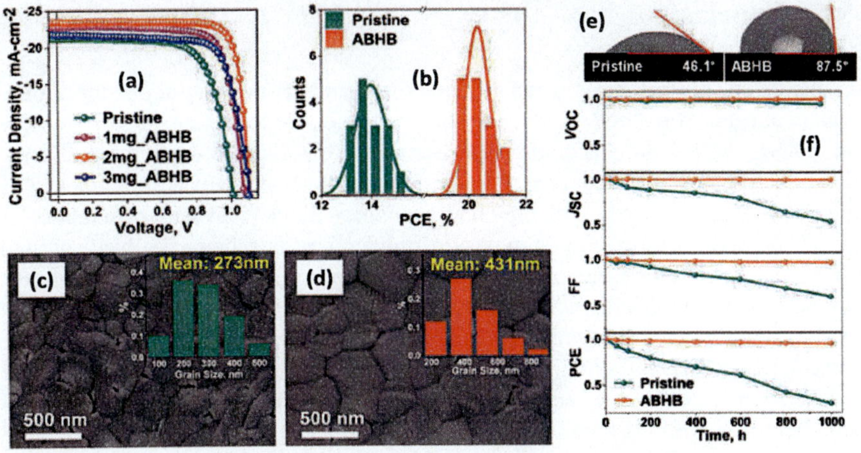

Fig. 15.7 **a** J-V curves, **b** histogram, **c**, **d** SEM images of pristine and varying concentrations of ABHB treated devices, **e** measured contact angle of pristine and ABHB treated films, **f** time-dependent normalized V_{OC}, J_{SC}, FF and efficiency of pristine and ABHB modified devices. (Reprinted with permission from Ref. 87; copyright 2021 The Royal Society of Chemistry)

The NH_4SCN passivated 3D $MAPbI_3$ perovskite active layer with a further 2D perovskite capping layer using xylylene diammonium iodide organic cation thin films was used for the fabrication of perovskite solar cells by Yukta et. al. (2022). The NH_4SCN passivated $MAPbI_3$ perovskite-based solar cell device has shown 19.6% PCE, which is higher as compared to pristine $MAPbI_3$ perovskite solar cells (PCE = 17.18%). Further, the author reported the highest PCE of 20.74% for the 2D/3D perovskite heterojunction-based solar cells. According to the authors, reduced defect density and most significantly inhibition of losses due to nonradiative recombination attributed to the improved PCE of the 2D/3D PSCs. Additionally, the exceptional ambient stability of the as compared to pristine 3D perovskite-based PSCs is because of the hydrophobic 2D capping layer on the 2D/3D heterojunction perovskites. The above discussion identifies that the top 2D perovskite layers improve the surface robustness of the 3D perovskite layer by passivating the surface defects, inducing hydrophobicity, and reducing surface charge recombination (Bouduban et al. 2019). This improvement could ultimately improve the photovoltaic performance of the 2D/3D perovskite-based solar cell devices.

15.6 Conclusions

In conclusion, the present chapter provides in-depth knowledge about the fundamental concepts of not only perovskite solar cells but also to engineer the perovskite materials at mixed dimensionality successfully. This chapter explored the introduction to dimensional-dependent structural properties of perovskite halides. The various synthesis methods explained how to prepare the devices with different solar cell device architectures. The summary of the chapter is schematically illustrated in Fig. 15.8. This figure indicates that the 3D perovskite halide materials are most promising to provide high efficiency but are highly sensitive to environmental parameters. As a result, as far as commercialization of the 3D PSC-based technology, there is no balance of PCE and stability (Fig. 15.8a). Further, by reducing the 3D perovskite to 2D perovskite, the properties of the 2D materials get changed significantly. Moreover, there is a significant increase in the band gap and exciton binding energy due to which pristine 2D-based PSC performance is limited. However, the large organic spacers in the 2D perovskite allow higher hydrophobic surface, which makes them more robust compared to 3D perovskite. This is schematically presented in Fig. 15.8b, where, the stability of the 2D perovskite gained more weight compared to its PCE performance. The mixed halide strategy provides a promising opportunity where the balance between the ambient stability and PCE of the PSC could be obtained. As depicted in Fig. 15.8c, the combination of 2D and 3D perovskite could synergistically provide high efficiency and high stability.

Overall present chapter attracts the attention of readers towards dimensionality tuning of the perovskite in particular the low-dimensional perovskites with greater stability against moisture as compared to their standard three-dimensional native

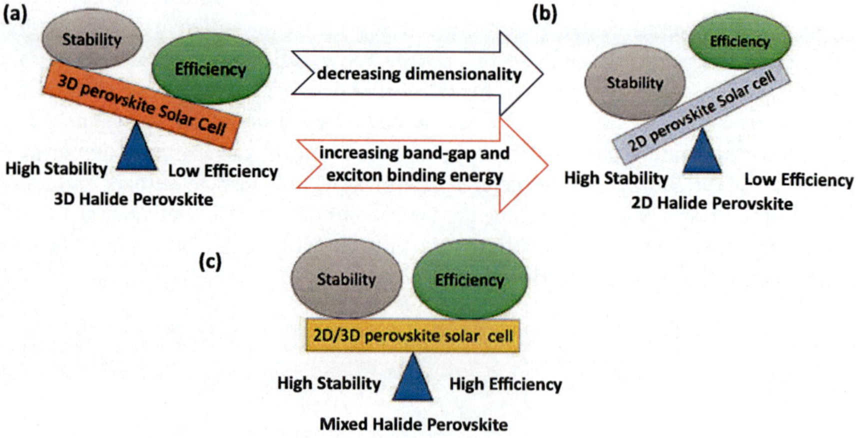

Fig. 15.8 Graphical Illustration of the summary of the PCE and stability of perovskite solar cells prepared with **a** pristine 3D, **b** 2D, and **c** mixed-dimensional 2D/3D perovskites

structures. Thus, a deeper understanding of the fundamental relationship of structure–property of different 2D perovskite phases in combination with 3D perovskite phases is necessary for developing new strategies to design and fabricate a durable, high-performing photovoltaic device. The diversity in the composition of 2D perovskites along with property-dependent choice of novel organic spacer cations provides a huge scope for further scientific exploration. Understanding the mechanism of the organic spacer cations influencing the structures and properties relationship of the 2D perovskites is the necessary step in the advancement of designing better materials for high performance with decisively high stability for optoelectronic applications. The investigation of novel HTLs for inverted PSCs could influence leading to a wider range of new developments with significant improvements in the device performances. Moreover, the proposed studies based on the development and optimizations of novel materials remain important to use cost-effective HTLs for further improvement in the stability of the quasi-2D PSCs.

It has been observed that the robustness of 2D PSCs is higher but has limited PCE. However, the dimensional engineering of perovskite materials plays a multi-functional role in the field of solar cells, like the capping layer, passivating layer, hydrophobic layer, absorbers, and high PCE. This chapter explores recent developments and accomplishments involved in combining 3D and 2D perovskites for improving both the device efficiency along with ambient stability. This approach will provide the way ahead for research in the field of mixed 3D and 2D perovskite-based solar cells and their commercialization.

Acknowledgements The author, ADS, would like to thank the Science and Engineering Research Board (SERB), Department of Science and Technology (DST), New Delhi, India, for the financial support under, INSPIRE Faculty Award Number DST/INSPIRE/04/2015/002601 and SERB-Research Scientist award number SB/SRS/2021-2022/62/PS.

Author Contributions "The manuscript was written through the contributions of all the authors. All authors have given approval to the final version of the manuscript."

References

Akkerman, Q.A., Rainò, G., Kovalenko, M.V., Manna, L.: Nat. Mater. **17**, 394–405 (2018)
Akkerman, Q.A., Manna, L.: ACS Energy Lett. **5**, 604–610 (2020)
Bhosale, M.K., Kazi, A.I., Pawar, K.K., Shingate, R.S., Kadam, A.D., Patil, N.J., Sheikh, A.D.: Nanotechnology **34**, 065501 (2023)
Bouduban, M.E.F., et al.: J. Phys. Chem. Lett. **10**, 5713–5720 (2019)
Cao, D.H., Stoumpos, C.C., Yokoyama, T., Logsdon, J.L., Song, T.B., Farha, O.K., Wasielewski, M.R., Hupp, J.T., Kanatzidis, M.G.: ACS Energy Lett. **2**, 982–990 (2017)
Chen, J., Lee, D., Park, N.G.: ACS Appl. Mater. Interfaces. **9**, 36338–36349 (2017)
Cheng, P., et al.: ACS Energy Lett. **3**, 1975–1982 (2018)
Cho, K.T., et al.: Eng. Env. Sci. **11**, 952–959 (2018)
Cho, K.T. et al.: Nano Lett. **18**, 5467–74 (2018)
Choi, W.G., Park, C.G., Kim, Y., Moon, T.: ACS Energy Lett. **5**, 3461–3467 (2020)

Conings, B., et al.: Adv. Mater. **26**, 2041–2046 (2014)
Deshpande, S.V., Bhiungade, R.A., Deshpande, M.P., Pawar, K.K., Bhat, T.S., Kulkarni, S.K., Sheikh, A.D.: Mater. Res. Bull. **142**, 111433 (2021)
Efat, J., Po, Y.C., Chia, Y.L., Narra, S., Shahbazi, S., Diau, E.W.G.: ACS Energy Lett. **6**, 485–492 (2021)
Gangadharan, D.T., Ma, D.: Energy Environ. Sci. **12**, 2860–2889 (2019)
Gao, L., Zhang, F., Xiao, C., Chen, X., Larson, B.W., Berry, J.J., Zhu, K.: Adv. Fun. Mater. **29**, 1901652 (2019)
Garai, R., Gupta, R.K., Hossain, M., Iyer, P.K.: J. Mater. Chem. A **9**, 26069 (2021)
Ge, C., Xue, Y.Z.B., Li, L., Tang, B., Hu, H.: Front. Mater. **7**, 601179 (2020)
Grancini, G. et al.: Nat. Commun. **8**, 15684 (2017)
Grancini, G., Nazeeruddin, M.K.: Nat. Rev. Mater. **4**, 4–22 (2019)
Grätzel, M.: Nature **414**, 338 (2001)
Green, M.A., Dunlop, E.D., Ebinger, J.H., Yoshita, M., Kopidakis, N., Hao, X.: Prog. Photovoltaics **28**, 629 (2020)
https://www.nrel.govcell-efficiency
Hammarström, L., Hammes-Schiffer, S.: Acc. Chem. Res. **42**, 1859 (2009)
Haque, M.A., Sheikh, A.D., Guan, X., Wu, T.: Adv. Energy Mater. **7**, 1602803 (2017)
Jeon, N.J., Noh, J.H., Kim, Y.C., Yang, W.S., Ryu, S., Seok, S.I.: Nat. Mater. **13**, 897 (2014a)
Jeon, N.J., et al.: Nature Mater. **13**, 897–903 (2014b)
Jeon, N., Noh, J., Yang, W., et al.: Nature **517**, 476–480 (2015)
Jung, E.H., Jeon, N.J., Park, E.Y., Moon, C.S., Shin, T.J., Yang, T.Y., Noh, J.H., Seo, J.: Nature **567**, 511–515 (2019a)
Jung, E.H., et al.: Nature **567**, 511–515 (2019b)
Kim, E.B., Akhtar, M.S., Shin, H.S., Ameen, S., Nazeeruddin, M.K.J.: Photochem. Photobiol. c: Photochem. Rev. **48**, 100405 (2021)
Kim, J.S., Heo, J.M., Park, G.S., Woo, S.J., Cho, C., Yun, H.J., Kim, D.H., Park, J., Lee, S.C., Park, S.H., Yoon, E., Greenham, N.C., Lee, T.W.: Nature **611**, 688–694 (2022)
Koh, T.M., Shanmugam, V., Schlipf, J., Oesinghaus, L., Buschbaum, P.M., Ramakrishnan, N., Swamy, V., Mathews, N., Boix, P.P., Mhaisalkar, S.G.: Adv. Mater. **28**, 3653 (2016)
Koh, T.M., Shanmugam, V., Guo, X., Lim, S.S., Filonik, O., Herzig, E.M., Buschbaum, P.M., Swamy, V., Chien, S.T., Mhaisalkar, S.G., Mathews, N.: J. Mater. Chem. A **6**, 2122–2128 (2018)
Kojima, A., Teshima, K., Shirai, Y., Miyasaka, T.: J. Am. Chem. Soc. **131**, 6050 (2009)
Krishna, A., Gottis, S., Nazeeruddin, M.K., Sauvage, F.: Adv. Funct. Mater. **29**, 1806482 (2019)
Lai, H., et al.: J. Am. Chem. Soc. **140**, 11639–11646 (2018)
Lee, D.G., Kim, M., Wang, S., Kim, B.J., Meng, Y.S., Jung, H.S.: ACS Appl. Mater. Interfaces **51**, 48497–48504 (2019)
Li, X., Bi, D., Yi, C., Décoppet, J.D., Luo, J., Zakeeruddin, S.M., Hagfeldt, A., Grätzel, M.: Science **353**, 58 (2016)
Li, M.H., Yeh, H.H., Chiang, Y.H., Jeng, U.S., Su, C.J., Shiu, H.W.: Adv. Mater. **30**, 1801401 (2018)
Liang, P.W., et al.: Adv. Mater. **26**, 3748–3754 (2014)
Liang, C., Gu, H., Xia, Y., et al.: Nat. Energy **6**, 38–45 (2021)
Liu, B., Long, M., Cai, M., Ding, L., Yang, J.: Nano Energy **59**, 715–720 (2019a)
Liu, Y., Akin, S., Pan, L., Uchida, R., Arora, N., Milić, J.V., Hinderhofer, A., Schreiber, F., Uhl, A.R., Zakeeruddin, S.M., Hagfeldt, A., Dar, M.I., Grätzel, M.: Sci. Adv., 5 eaaw2543 (2019)
Ma, L., Ke, W., Pedesseau, L., Wu, Y., Katan, C., Even, J., Wasielewski, M.R., Stoumpos, C.C., Kanatzidis, M.G.: J. Am. Chem. Soc. **140**, 3775–3783 (2018)
Ma, S., Yuan, G., Zhang, Y., Yang, N., Li, Y., Chen, Q.: Energy Environ. Sci. **15**, 13–55 (2022)
Manjunath, V., Bimli, S., Shaikh, P.A., Ogale, S.B., Devan, R.S.: J. Mater. Chem. C **10**, 15725–15780 (2022)
Mao, L., Ke, W., Pedesseau, L., Wu, Y., Katan, C., Even, J., Wasielewski, M.R., Stoumpos, C.C., Kanatzidis, M.G.: J. Am. Chem. Soc. **140**, 3775–3783 (2018)

Mei, A., Li, X., Liu, L., Ku, Z., Liu, T., Rong, Y., Xu, M., Hu, M., Chen, J., Yang, Y., Grätzel, M., Han, H.: Science **345**, 295–298 (2014)
Ortiz, C., Carmona, M.P., Solis, I.D.: Chem. Sus. Chem. **12**, 1560–1575 (2019)
Quan, L., et al.: J. Am. Chem. Soc. **138**, 2649–2655 (2016)
Quintero, B.R., et al.: Nat. Mater. **17**, 900–907 (2018)
Saliba, M., Matsui, T., Domanski, K., Seo, J.Y., Ummadisingu, A., Zakeeruddin, S.M., Correa-Baena, J.P., Tress, W.R., Abate, A., Hagfeldt, A., Grätzel, M.: Science **354**, 206 (2016)
Satapathi, S.: ACS Energy Lett. **7**, 906–907 (2022)
Schileo, G., Grancini, G.: J. Phys. Energy **2**, 021005 (2020)
Shamsi, J., Urban, A.S., Imran, M., Trizio, L.D., Manna, L.: Chem. Rev. **119**, 3296–3348 (2019)
Sheikh, A.D., Bera, A., Haque, M.A., Rakhi, R.B., Gobbo, S.D., Alshareef, H.N., Wu, T.: Sol. Eng. Mater. Sol. Cells **137**, 6–14 (2015)
Sheikh, A.D., Munir, R., Haque, M.A., Bera, A., Hu, W., Shaikh, P., Amassian, A., Wu, T.: ACS App. Mater. Inter. **9**, 35018–35029 (2017)
Sheikh, A.D., Patil, A.P., Mali, S.S., et al.: J. Mater. Sci. **54**, 10825–10835 (2019)
Sheikh, A.D., Vhanalakar, V.K., Katware, A.S., Pawar, K.K., Kulkarni, S.K.: J. Alloys Compounds **894**, 162388 (2022)
Smith, I.C., Hoke, E.T., Ibarra, D.S., McGehee, M.D., Karunadasa, H.I.: Angew. Chem. **126**, 11414 (2014a)
Smith, I.C., Hoke, E.T., Solis-Ibarra, D., McGehee, M.D., Karunadasa, H.I.: Angewandte Chemie Int. Edi. **53**, 11232–11235 (2014b)
Soe, C.M.M., Stoumpos, C.C., Kepenekian, M., Traoré, B., Tsai, H., Nie, W., Wang, B., Katan, C., Seshadri, R., Mohite, A.D., Even, J., Marks, T.J., Kanatzidis, M.G.: J. Am. Chem. Soc. **139**, 16297–16309 (2017)
Soe, C.M.M., Nie, W., Stoumpos, C.C., Tsai, H., Blancon, J.C., Liu, F., Even, J., Marks, T.J., Mohite, A.D., Kanatzidis, M.G.: Adv. Eng. Mater. **8**, 1700979 (2018)
Stoumpos, C.C., Soe, C.M.M., Tsai, H., Nie, W., Blancon, J.C., Cao, D.H., Liu, F., Traoré, B., Katan, C., Even, J., Mohite, A.D., Kanatzidis, M.G.: Chemistry **2**, 427 (2017)
Tsai, H., Nie, W., Blancon, J.C., Stoumpos, C.C., Asadpour, R., Harutyunyan, B., Neukirch, A.J., Verduzco, R., Crochet, J.J., Tretiak, S., Pedesseau, L., Even, J., Alam, M.A., Gupta, G., Lou, J., Ajayan, P.M., Bedzyk, M.J., Kanatzidis, M.G., Mohite, A.D.: Nature **536**, 312 (2016)
Tsai, H., et al.: Nat. Commun. **9**, 2130 (2018)
Ugur, E., Sheikh, A.D., Munir, R., Khan, J.I., Barrit, D., Amassian, A., Laquai, F.: ACS Energy Lett. **2**, 1960–1968 (2017)
Wang, Z., Lin, Q., Chmiel, F.P., Sakai, N., Herz, L.M., Snaith, H.J.: Nature Eng. **2**, 17135 (2017)
Wu, G., Li, X., Zhou, J., Zhang, J., Zhang, X., Leng, X., Wang, P., Chen, M., Zhang, D., Zhao, K., Liu, S., Zhou, H., Zhang, Y.: Adv. Mater. **31**, 1903889 (2019a)
Wu, G., et al.: Adv. Mater. **31**, 1903889 (2019b)
Xiao, Z.G., et al.: Energy Environ. Sci. **7**, 2619–2623 (2014)
Yang, W.S., et al.: Science **348**, 1234–1237 (2015)
Yao, K., Wang, X., Li, F., Zhou, L.: Chem. Comm. **51**, 15430–15433 (2015)
Yao, K., Wang, X., Xu, Y.X., Li, F., Zhou, L.: Chem. Mater. **28**, 3131–3138 (2016)
Yoo, J.J., Wieghold, S., Sponseller, M.C., Chua, M.R., Bertram, S.N., Hartono, N.T.P., Tresback, J.S., Hansen, E.C., Correa-Baena, J.P., Bulović, V., Buonassisi, T., Shin, S.S.: Bawendi MG. Energy Environ. Sci. **12**, 2192–2199 (2019)
You, J., et al.: Appl. Phys. Lett. **105**, 183902 (2014)
Younis, A., Lin, C.H., Guan, X., Shahrokhi, S., Huang, C.Y., Wang, Y., He, T., Singh, S., Hu, L., Retamal, J.R.D., He, J.H., Wu, T.: Adv. Mater. 2005000 (2021)
Yukta, P.N., Chavan, R.D., Yadav, P., Nazeeruddin, M.K., Satapathi, S.: ACS Appl. Mater. Interfaces **14**(26), 29744–29753 (2022)
Zhang, J., Hu, B.: Nano Energy **76**, 104999 (2020)
Zhang, X., Wu, G., Yang, S., Fu, W., Zhang, Z., Chen, C., Liu, W., Yan, J., Yang, W., Chen, H.: Small **13**, 1700611 (2017a)

Zhang, T., Dar, M.I., Li, G., Xu, F., Guo, N., Grätzel, M., Zhao, Y.: Sci. Adv. **3**, e1700841 (2017b)
Zhang, F., Kim, D., Zhu, K.: Curr. Opin. Electrochem. **11**, 105–113 (2018)
Zhang, J., Qin, J., Wang, M., Bai, Y., Zou, H., Keum, J.K., Tao, R., Xu, H., Yu, H., Haacke, S., Hu, B.: Joule **3**, 3061–3071 (2019b)
Zhang, F., Lu, H., Tong, J., Berry, J.J., Beard, M., Zhu, K.: Energy Environ. Sci. **13**, 1154–1186 (2020)
Zhang, J., Qin, J., Wang, M., Bai, Y., Zou, H., Keum, J.K., Tao, R., Xu, H., Yu, Haacke, S., Hu, B.: Joule **3**, 3061–3071 (2019)
Zhou, N., et al.: J. Am. Chem. Soc. **140**, 459–465 (2017)
Zuo, C., Scully, A.D., Vak, D., Tan, W., Jiao, X., McNeill, C.R., Angmo, D., Ding, L., Gao, M.: Adv. Eng. Mater. **9**, 1803258 (2019)

Chapter 16
Calcium Aluminate-Based Phosphors for Persistent Luminescence and Radiation Dosimetry

Moirangthem Nara Singh and Anurup Gohain Barua

16.1 Introduction

All persistent luminescent materials are known to exhibit thermoluminescence (TL), providing an attractive and rich field for various basic and applied scientific and technological applications. In addition, clinical and environmental dosimetry measurements are examples of TL dosimetry applications (Matsuzawa et al. 1996; Dong et al. 2012; Olko 2010). According to the report of Smet et al. (2015), about 276 Glow-in-the-Dark phosphors have been prepared, and most of those in Chinese laboratories develop the remarkable green persistent luminescent $SrAl_2O_4:Eu^{2+}$, Dy^{3+}. Scientists have been developing better phosphors by heat and trial methods which is evident from several recent reviews. However, the fundamental problems pertaining to the exact process and the optimal trapping levels of the materials remain unsolved. They have been optimized for persistent luminescence by appropriate doping (Smet et al. 2015), followed by bandgap tailoring, which is a recent strategy to prepare new persistent luminescent materials (Wei et al. 2017) which develop such materials by optical absorption, photoconductivity, thermally stimulated luminescence (TSL), and deep-level transient spectroscopy are most the most frequently used techniques to authenticate or identify trapping levels of these materials.

M. N. Singh (✉)
Department of Radiation Oncology, Dr. B. Borooah Cancer Institute, Guwahati 781016, Assam, India
e-mail: naraphysicist@gmail.com

A. G. Barua
Department of Physics, Gauhati University, Guwahati 781014, Assam, India

R. S. Ningthoujam and A. K. Tyagi (eds.), *Handbook of Materials Science, Volume 1*,
Indian Institute of Metals Series, https://doi.org/10.1007/978-981-99-7145-9_16

16.2 Luminescence

The word luminescence was first introduced by the German physicist Eilhardt Wiedemann which was derived from the Latin word lumen meaning light (Wiedemann 1888). During this process, the radiation interacts with a material and a portion of the energy gets reemitted as light with a wavelength different from the incident radiation, like infrared (IR), visible, or ultraviolet (UV) light (Murthy and Virk 2014). The characteristic time (τ_c), called lifetime or decay time, is defined as the time required to decrease the amplitude from the maximum to 1/e of its maximum value. The luminescence phenomenon is divided into two classes based on the characteristic time (τ_c) spent during energy absorption (excitation) and emission.

(a) **Fluorescence** ($\tau_c \sim 10^9$ s)

The excited electron returns to the ground state quickly, or it could take as long as a fraction of a microsecond, accompanied by light emission after getting stimulated by radiation.

(b) **Phosphorescence** ($\tau_c > 10^3$ s)

The electron excited by radiation from the ground state to the excited state makes a delayed return to the ground level, where the delay is caused by its transition to a metastable level.

The mechanism of fluorescence and phosphorescence in phosphors can be explained in simple terms following the fundamental work of Jablonski (1935), which is crucial to provide for eventual works which are reused in research papers and books.

Many experimental results of phosphors which result, e.g., afterglow, can be treated quantitatively by calculations based on this model. The afterglow of common phosphors ranges from milliseconds to minutes. In the case of persistent luminescent phosphors, it is usually much more about in hours. The lifetime (τ) of charge (electron/hole) in the metastable state decides the afterglow duration. Lifetimes (τ) evaluated at the room temperature of 300 K for some commercial phosphors relevant to persistent luminescence are found to be in minutes to hours, depending on the values of the trapping parameters (Mashangva et al. 2011). The phosphorescence has been divided into short and long periods. Thermoluminescence (TL) and Optically Stimulated Luminescence (OSL) phenomena belong to the long-period phosphorescence (McKeever 1985). The overall phenomenon of luminescence can be classified into subclasses according to the nature of the excitation, as shown in Table 16.1.

Table 16.1 Types of luminescence

Designation	Excitation	Trigger	Acronym
Thermoluminescence	Photons, charged particles	Heat	TL
Photoluminescence	UV, visible photons	–	PL
Radioluminescence	Photons, charged particles	UV	RL
Cathodoluminescence	Energetic electrons	–	CL
Electroluminescence	Electric field	–	EL
Mechanoluminescence	Mechanical energy	–	–
Chemiluminescnce	Chemical energy	–	–
Bioluminescence	Biological energy	–	–
Sonoluminescence	Sound waves	–	–
Optically stimulated Luminescence	Photons, charged particles	Visible/IR Photons	OSL

16.3 Luminescence-Based Dosimetry

The principle of luminescence-based dosimetry states that the intensity of light emitted is proportional to the amount of energy deposited by radiation on a phosphor. It is because such phosphors release electrons that may be trapped at defect or activator sites in the crystal lattice (Olko 2010; Karzmark et al. 1965).

Luminescence-based dosimetry is classified into three classes, viz., thermoluminescence dosimetry (TLD), optically stimulated dosimetry (OSLD) and radioluminescence dosimetry (RPLD) (Olko 2010). Luminescence-based dosimeters have numerous benefits, including their small size, high sensitivity, good storage qualities, wide dose range, and approximative tissue equivalence, and they are becoming increasingly popular in dosimetry. Currently, luminescence-based dosimeters have substituted radiographic films, which were frequently employed in the previous century but had limitations in the personal monitoring services provided to radiation workers (Olko et al. 2006).

16.4 Thermoluminescence (TL)

16.4.1 Theory of Thermoluminescence

Thermally stimulated luminescence or thermoluminescence is the emission of light from on/after heating a wide bandgap material (an insulator or semi-insulator) that has previously been exposed to the radiation from which it has stored energy (McKeever 1985). Thermoluminescence is a two-step process. In the first step, valence band (VB) electrons are excited by exposing them to ionizing radiations; these electrons jump to the conduction band (CB) and move to a metastable state. In the second step,

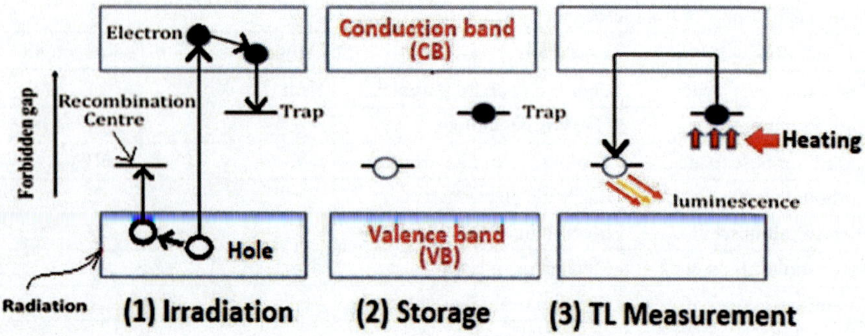

Fig. 16.1 Mechanism of the thermoluminescence process (Redrawn with permission from Elsevier Publisher (Rivera 2012))

trapped electrons are stimulated by heating. Then, electrons and generated holes are combined at the recombination centre, which leads to light emission (Fig. 16.1). The materials exhibiting TL have specific defects in their regular structure in wide bandgap materials, e.g., semi-insulators and insulators. Metals (Conductors) do not exhibit TL properties as they do not have a forbidden gap where electrons could fall into a trap.

During irradiation, some of these free electrons are trapped in the defect centres, and the luminescence emission during thermal stimulation by the trapped charge carriers (electrons and holes) is proportional to the quantity of radiation. Once TL materials are stimulated to emit luminescence, they cannot reemit light by just chilling and reheating. It is possible after re-irradiation to an ionizing radiation again.

16.4.2 Thermoluminescence Analysis

Thermoluminescence equations are complex and contain many parameters, viz. activation energy E, the frequency factor s, the recombination probability A_m ($cm^3 sec^{-1}$), and the retrapping probability A_n ($cm^3 sec^{-1}$). Randall and Wilkins (1945), Garlick and Gibson (1948), and May and Partridge (1964) have been involved in the development of TL equations for first, second, and general orders, respectively, to simplify the complex TL data. Most workers have followed Chen's general order kinetics equation (Chen 1969) and Pagonis et al. (2006) for practical purposes. It can be presented as

$$I(T) = n_0 s'' exp\left(-\frac{E}{kT}\right)\left[1 + \frac{s''(b-1)}{\beta}\int_{T_0}^{T} exp\left(-\frac{E}{kT'}\right)dT'\right]^{-\frac{b}{b-1}} \quad (16.1)$$

where $I(T)$ = intensity at temperature T; n_0 = number of trapped charges at time t = 0 (m^{-3}); E = the activation energy or trap depth (eV); k = Boltzmann's constant (eVK^{-1}); T = the absolute temperature (K); N = the total trap concentration (m^{-3}); b = order of kinetics, ($1 \leq b \leq 2$); s' = effective pre-exponential factor for general order kinetics ($m^{3(b-1)}s^{-1}$); $s'' = s' n_0^{(b-1)}$ = effective frequency factor for general order kinetics (s^{-1}); β = linear heating rate (Ks^{-1}), and T_0 = temperature at time $t = 0$ (s).

16.4.3 Determining Trap Parameters

Conventional classical methods used mainly by new researchers can provide the main kinetics parameters (E, s, and b). However, they have limitations in calculating the lifetime (τ) of the trapped charge in the trap levels. Developments of various methods by different researchers with observations are shown in Table 16.2. The table covers the milestones in the development of the interval of evaluation of electron traps in solids from 1948 till date.

All luminescent materials have their device performance related to the presence and absence of certain definite trapping levels introduced/suppressed by suitable doping of activators, co-activators and elimination of quenching. The large band gap of the semiconductor/insulator serves as the benevolent host. Determination of the trapping levels in phosphors can be traced to the early pioneering works of Randall and Wilkins (1945), Garlick and Gibson (1948), May and Partridge (1964),

Table 16.2 Summary of the development of methods of analysis of TL data

Method	Year	References	Observations
(i) Urbach's formula	1948	Urbach (1930)	It critically assessed the most basic formula for calculating trap depth
(ii) Initial rise method	1948	Garlick and Gibson (1948)	It is quick, simple, and independent of kinetic order, giving a reasonable estimate of trap depth (E)
(iii) Various heating rates	1958	Hoogenstraaten (1958)	Though accurate in well-separated peaks, it can make significant errors in complex TL curves
(iv) Peak shape method	1969	Chen (1969)	For well-separated peaks, it is quick, simple, and reliable. Several peak-shaped methods were developed between 1950 and 1960
(v) Curve fitting	1990-Till	Horowitz and Yossian (1995), Horowitz and Moscovitch (2013), Horowitz (2014)	For practical purposes, it is the most reliable method of determining the main kinetic trapping parameters (E, s, b)

and Hoogenstraaten (1958), who investigated phosphors based on sulphide and silicate host lattice that are relevant even today. These have been optimized for persistent luminescence by suitable doping, followed by band gap tailoring, a technique routinely followed in recent times for novel persistent luminescent materials.

16.4.4 Lifetime of Trapped Charge

The development of the remarkable green persistent luminescent $SrAl_2O_4:Eu^{2+}$, Dy^{3+}, the routes followed have been essentially on a trial-and-error basis till today. Several reviews have appeared over the last decade. However, the fundamental questions of the exact mechanism, as well as the appropriate trapping levels of the materials, have remained unanswered. There seems to be confusion amongst the material scientists in evaluating the trap depth (E) and the critical trapping parameters that decide the lifetime (τ) of the charge, which is related to the equations.

$$\tau = s^{-1}\exp\left(\frac{E}{kT}\right) \tag{16.2i}$$

$$\tau = s^{-1}\frac{\exp(\frac{E}{kT})}{(b-2)} \tag{16.2ii}$$

where E is the trap depth (eV), s is the frequency factor (s^{-1}), and b is the order of kinetics ($1 \leq b \leq 2$).

Equation (16.2i) is applicable for first order ($b = 1$) TL peaks, while Eq. (16.1 (ii) holds for non-first order kinetics ($b \neq 1$). Equation (16.2ii) has been recently derived by Singh and Gartia (2011). Recent papers have discussed its importance in the physical basis of persistent luminescence (Gartia and Chandrasekhar 2014).

16.4.5 Thermoluminescence Experiments

The following are the processes for the TL experimental setup for persistent luminescence and dosimetric applications.

1. Different heating rates (β) were used to get TL data, ranging from 0.01 to 5° Cs^{-1}. Low heating tests have received particular attention since the problem of thermal lag is eliminated in such a setting.
2. Phosphors have been excited by ionizing and non-ionizing radiation, including visible photons in liquid nitrogen and room temperature (RT).
3. All persistent luminescent materials have outstanding TL, which may be tested using simple homemade equipment. Dosimetric experiments have been carried out using typical commercial TL readers.

16.4.6 Implications of TL Peak Position

Conventional classical methods of determination of electron trap parameters failed to calculate the lifetime (τ) of charges trapped in the trap levels. The lifetime (τ) of trapped electrons associated with peak positions in the scintillator, persistence luminescence, dosimetry, and dating are shown in Fig. 16.2. TL peaks having a few years lifetime at temperatures 200–300 °C are helpful in dosimetry. Low-temperature peaks are unstable due to fast fading. High-temperature peaks might contain noise due to black body radiation. The lifetime of trapped electrons within a few minutes to hours is relevant to the persistence of luminescence. The glow peaks at a high temperature above 300 °C are expected to be helpful in dating. In addition to these applications, Gartia (2014) has reported that thermoluminescence could be used for designing scintillators. The depth position of the electron and hole traps in the host lattice are significant for afterglow endurance. Shallow traps generally cause an afterglow that is quite bright at first but rapidly fades. On the other hand, deep traps have the potential to produce an afterglow that is initially much weaker but lasts a longer time than shallow traps. All persistent luminescent materials display outstanding TL that can be measured using simple TL equipment, the pool of TL data on persistent luminescent materials is rather extensive.

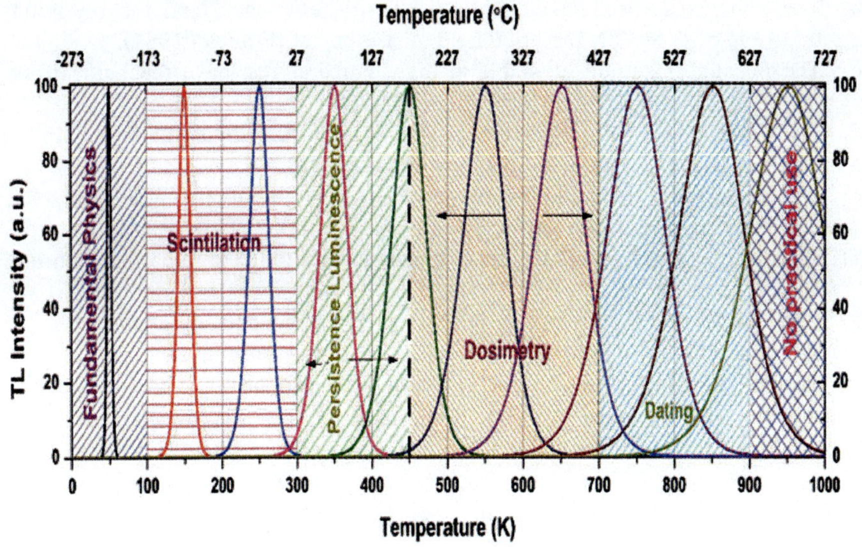

Fig. 16.2 Estimate of lifetime corresponding to glow peak temperature with TL applications

16.4.7 *TL Dosimeter Phosphors*

The most effective TL material among naturally occurring substances is calcium fluoride (CaF_2). Many researchers have worked to find and develop TL phosphors for dosimetry. Phosphors having a wider bandgap enables multiple applications. These phosphors are relevant to radiation dosimetry and persistent luminescence, respectively, because shallow and deep traps are present inside the forbidden gap. The TL phosphors with different dopant types used in TL dosimetry are summarised in Table 16.3 (Christensen et al. 1982; Kortov 2010; Omanwar et al. 2014; Bhatt and Kulkarni 2014).

16.4.7.1 Basic Requirements for Materials Relevant to TLD

The basic required properties of TL materials used in radiation dosimetry (McKeever 1985; Cameron et al. 1968; McKinlay 1981; Horowitz 1984; Mahesh et al. 1989; McKeever et al. 1995; Furetta and Weng 1998; Furetta 2006) are the following.

(a) The phosphor should have a single, sharp, isolated glow peak at temperatures (180–250 °C). However, complex glow curves should have at least the prominent dosimetric peak and be more dominant than the remaining hump peaks.
(b) The phosphors should have high sensitivity, efficiency, and the minimum threshold dose for low radiation field measurement (Pradhan 1981).
(c) The phosphors should be stable at high temperatures, have mechanical and radiation strength, and be chemically inactive.
(d) The TL responses of the phosphors should not affect any climatic conditions and factors such as room temperature and humidity.
(e) It should be non-toxic and tissue-equivalent, essential in clinical and personal dosimetric applications (Kortov 2007).
(f) The dose response should be linear in an extensive range of radiation-absorbed doses.
(g) It should have minimum or negligible energy dependence on radiation.
(h) The phosphor should have good stability, i.e., minimum fading.
(i) The emission spectrum from the TL phosphor during heating should match the sensitivity of the photomultiplier tube (PMT).

16.4.8 *Applications of Thermoluminescence*

16.4.8.1 Persistence Luminescence

In persistence luminescence, when the exciting ionizing or non-ionizing radiation source is removed, the light emission lasts for a long time, from a few seconds to several days. The ancient Chinese Song dynasty (960–1279 A.D.) reported on the

Table 16.3 Dosimetric characteristics of some common TLD phosphors (Gartia 2014; Christensen et al. 1982; Kortov 2010, 2007; Omanwar et al. 2014; Bhatt and Kulkarni 2014; Cameron et al. 1968; McKinlay 1981; Horowitz 1984; Mahesh et al. 1989; McKeever et al. 1995; Furetta and Weng 1998; Furetta 2006; Pradhan 1981; Sun et al. 2018; Lin et al. 2019; Konopka et al. 2017; Aitken 1985; Daniels et al. 1953)

Material	Effective atomic number Z_{eff}	Prominent peak (°C)	Emission maximum (nm)	Relative sensitivity	Fading (at 25 °C kept in the dark)	Linearity dose range
LiF:Mg,Ti	8.14	200	400	1	5%/year	20 μGy–10 Gy
LiF:Mg,Cu,P	8.14	210	368	40	5%/year	0.2 μGy–10 Gy
LiF:Mg,Cu,Si	8.14	240	384	55	Negligible	1 μGy–20 Gy
$Li_2B_4O_7$:Mn	7.30	220	605	0.4	4%/month	0.1 mGy–3 Gy
$Li_2B_4O_7$:Cu	7.30	205	368	8	10%/ 2 months	10 μGy–103 Gy
$Li_2B_4O_7$:Cu,In	7.30	210	368	0.7	6% in 3 months	10–4 Gy to 103 Gy
BeO	7.10	190	330	~1	8%/ 2 months	0.1 mGy–0.5 Gy
CaF_2:Mn	16.3	260	500	5	16%/ 2 weeks	10 μGy–10 Gy
CaF_2 (natural)	16.3	260	380	23	Very slight	10 μGy–50 Gy
Al_2O_3:C	10.2	190	420	60	5%/year	0.1 μGy–10 Gy
MgB_4O_7:Dy/ Tm	8.40	190	490	6–7	4%/ months	5 μGy–50 Gy
Mg_2SiO_4:Tb	11.0	200	380–400	40–53	Very slight	10 μGy–1 Gy
$CaSO_4$:Dy	15.3	220	480–570	30	1%/ 2 months	2 μGy–10 Gy
$CaSO_4$:Tm	15.3	220	452	30	1–2%/ 2 months	2 μGy–10 Gy
CaF_2:Dy	16.3	215	480–570	15	8%/ 2 months	10 μGy–10 Gy
$CaSO_4$:Tm/Dy, Ag	15.3	350	445–570	40	<1% in 1 month	2 μGy to 10 Gy
$SrSO_4$:Eu	30.3	210	380	>40	15% in 30 days	Up to 10^{-2} Gy
$LiCaBO_3$:Tm	18.5	233	455	3	15% in 1 month	1 Gy to 103 Gy
$LiMgBO_3$:Tb	10.2	250	544	4	negligible	1 mGy to 103 Gy
MgB_4O_7:Dy,Na	8.40	190	480, 570	6	8% in 3 months	10^{-4} Gy to 40 Gy

applications of persistent luminescence. However, numerous commercial persistent luminescence materials have been doped with rare earth elements since the 1990s (Matsuzawa et al. 1996).

In the case of persistent luminescent materials, most researchers need to estimate trap levels using TL. It is a technique noted for its simplicity and sensitivity (Mashangva et al. 2011). Furthermore, when the heating rate is 1 °Cs^{-1}, most persistent luminescent materials have displayed strong TL above room temperature (RT) with minor TL above 250 °C (Gartia and Chandrasekhar 2014).

One can notice in public places safety symbols marked as emergency exits for awareness when power fails, which is a unique example of long-lasting persistence luminescence applications.

Many literature reports have been on the applications of persistence luminescence nano phosphors (PLNPs) for deep tissue bioimaging. Scientists have faced critical problems in medical applications of persistence luminescence due to many factors. Phosphors having small and uniform sizes, illuminated initial, and long PL durations have been preferred in medical applications. Most PLNPs have been typically doped with rare earth element ions, the primary concern for health side effects. Finally, excitations of phosphors have also been a significant concern in bioimaging and therapy. The deep sides of tumours need a high penetration mode of radiation like X-rays, which are highly promoted to reactivate PLNPs (Sun et al. 2018; Lin et al. 2019).

16.4.8.2 Radiation Dosimetry

The timelines for the discovery of TL dosimeter phosphors are presented in Fig. 16.3. In the last seven decades, different TLD phosphors have been introduced on large scales but with only a few applications (Konopka et al. 2017). TL phosphors have been used in dosimetry applications in many fields, especially in radiation protection and dose estimation. Technically, TL dosimetry experiments and results differ vastly from persistence luminescence applications. In dosimetry experiments, the heating rates are higher, and the suitable peak position should be in the range of 200–300 °C (Aitken 1985).

In 1953, Farrington Daniels and his research team used LiF phosphor to apply thermoluminescence in radiation dosimetry (Daniels et al. 1953). In 1963, Cameron and his colleagues (Cameron et al. 1963) enhanced the sensitivity of LiF by doping it with Mg and Ti, and Harshaw Chemical Company patented TLD-100, which has been extensively used in radiation dosimetry applications for many years.

In 1978, Nakajima et al. (1978) first reported a thermoluminescence study of LiF:Mg,Cu,P phosphors along with LiF: Mg, Cu, Si, and LiF: Mg, Cu, B. In 1999, LiF:Mg,Cu,Na,Si with better TL sensitivity, about two times that of LiF: Mg, Cu, P was first synthesized by Nam et al. (1999). In 2002, Tang et al. (2002) also described LiF: Mg, Cu, Na, Si phosphor, which is more sensitive, around thirty times higher than TLD-100.

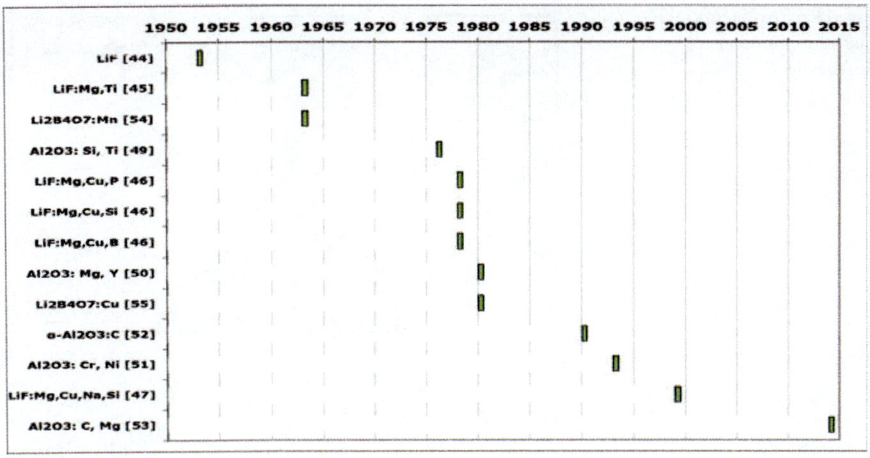

Fig. 16.3 Milestones of TLD phosphors in dosimetry applications (Redrawn and adapted by literature review)

Aluminium oxide (Al_2O_3) has been used as a radiation dosimeter. It is commonly found as a naturally occurring mineral component like Ruby and Sapphire. Many researchers have studied aluminium oxide doped with different elements based on TL phosphors such as Al_2O_3: Si, Ti (Mehta and Sengupta 1976), Al_2O_3: Mg, Y (Osvay and Biró 1980), Al_2O_3: Cr, Ni (Pokorny and Ibarra 1993), α-Al_2O_3:C (Akselrod et al. 1990), Al_2O_3: C, Mg (Saharin et al. 2014).

In 1963, $Li_2B_4O_7$:Mn was first introduced by Schulman and his colleagues (Schulman et al. 1967), a tissue equivalent material phosphor. In 1980, $Li_2B_4O_7$:Cu was prepared by Takenaga et al. (1980), which had better TL sensitivity than Mn-doped lithium tetraborate.

TLDs have been widely used in the personal dosimetry of radiation workers and medical physics (radiotherapy, radiology, and nuclear medicine). Widely explored TLD phosphors include sulphates, borates, fluorides, perovskites, alkali borates and sulphides, alkaline earth borates and sulphides, metal oxides phosphates and halo-phosphates, and different types of glasses (Azorin Nieto 2016).

TLD has been applied in the personal monitoring of radiation workers who work in medical institutions, nuclear reactors, particle accelerators, x-ray machines, and radioactive sources, and the radiation dose delivered to cancer patients has been evaluated.

In India, dysprosium-activated calcium sulphate ($CaSO_4:Dy^{3+}$) phosphors have been used as personal monitoring phosphors for radiation workers in medical institutes, radiation facilities and nuclear reactors. In clinical dosimetry, TL dosimetry has been used to estimate delivered radiation doses to patients and peripheral doses measured in diagnostic radiology and radiotherapy departments. International organizations like International Atomic Energy Agency (IAEA) conduct postal dosimetry worldwide to ensure proper dose delivery of the radiation therapy unit.

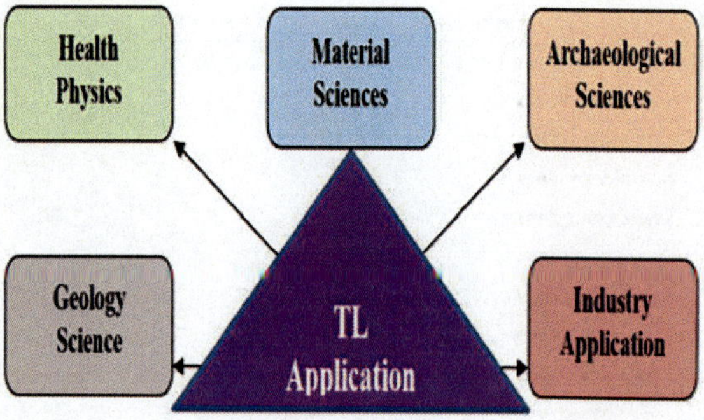

Fig. 16.4 Application of thermoluminescence dosimetry

Thermoluminescence dosimetry applications are used in many fields, such as personal dosimetry (Rivera 2012; Hajek et al. 2004; Vohra et al. 1980), clinical dosimetry (Duggan et al. 2004; Rudén 1976), space dosimetry (Mukherjee 2018), and accidental dosimetry (Wahib et al. 2019; Mesterházy et al. 2012), as shown in Fig. 16.4.

16.5 Thermoluminescence of Calcium Aluminate

Calcium aluminate ($CaAl_2O_4$) is the primary component of cement used in construction industries because of its mechanical strength and chemical stability. Besides, $CaAl_2O_4$-based phosphors are commonly used as blue-emitting long-lasting phosphors, and a few works have reported the possibilities of dosimetry applications. This study reviews the TL studies on undoped and doped calcium aluminate ($CaAl_2O_4$), presented in Table 16.4, which present the possible luminescence applications of calcium aluminate in persistent luminescent and radiation dosimetry.

Many workers have studied the thermoluminescence of calcium aluminate doped with rare earth elements in the last two and half decades. Most of those studies have been on persistent luminescence applications. Madhukumar et al. (2006) have reported the possibility of the dosimetry aspect of calcium aluminate doped with rare earth elements.

16.5.1 Synthesis of $CaAl_2O_4$ Based Phosphors

Over the two decades, various methods have been reported for preparing calcium aluminates for thermoluminescence studies. Many workers have reported various methods of calcium aluminate synthesis, as shown in Fig. 16.5, according to information from Google scholars from 1988 to 2021. The solid-state reaction and

Table 16.4 The literature on thermoluminescence of calcium aluminate-based phosphors

Year	Author	Synthesis	Doping	Paper title	Excitation	Heating rate	Peak Temp. (°C)	References
1997	H. Yamamoto et al.	Solid-state reaction	$Eu^{2+}Nd^{3+}$ $Eu^{2+}Dy^{3+}$	Mechanism of long phosphorescence of $SrA1_2O_4$: Eu^{2+}, Dy^{3+} and $CaAl_2O_4$: Eu^{2+}, Nd^{3+}	365 nm light	10°C/min	77 −113	Yamamoto and Matsuzawa (1997)
2000	T. Zhang et al.	Precipitation	Eu^{2+}, Dy^{3+}	Rare-earth materials for use in the dark	UV lamp	2 °C/s	40	Zhang and Su (2000)
2001	J. Hölsa et al.	Solid-state reaction	Eu^{2+}, R^{3+}	Persistent luminescence of Eu^{2+} doped alkaline earth aluminates, MAl_2O_4:Eu^{2+}	60 W incandescent lamp	VHR	Multiple peaks 90–160	Hölsä et al. (2001)
2001	Y. Lin et al.	Solid-state reaction	Eu^{2+}, Dy^{3+}	The characterization and mechanism of long afterglow in alkaline earth aluminates phosphors co-doped by Eu_2O_3 and Dy_2O_3	Standard lamp	10 °C/min	124	Lin et al. (2001)
2002	D. Jia et al.	Solid state reaction	Tb^{3+}	Electron traps in Tb^{3+}-doped $CaAl_2O_4$	Xe lamp	0.04 °C /s	−5 & 119	Jia et al. (2002a)
2002	D. Jia et al.	LHPG method	Tb^{3+} Ce^{3+}	Green phosphorescence of $CaAl_2O_4$:Tb^{3+}, Ce^{3+} through persistence energy transfer	254 nm UV lamp	9 °C/min	73 & (−87, −64, −40, −1)	Jia et al. (2002b)

(continued)

Table 16.4 (continued)

Year	Author	Synthesis	Doping	Paper title	Excitation	Heating rate	Peak Temp. (°C)	References
2003	D. Jia et al.	Solid-state reaction	Ce^{3+}	Trapping Mechanism Associated with Electron Delocalization and Tunneling of $CaAl_2O_4$:Ce^{3+} A Persistent Phosphor	50 W Hg lamp	9 °C/min	85, −65, −49, −5, 57	Jia and Yen (2003)
2003	Y. Lin et al.	Solid-state reaction	Eu^{2+}, R^{3+}	Influence of co-doping different rare-earth ions on the luminescence of $CaAl_2O_4$-based phosphors	365 nm light	10 °C/min	Multiple peaks 63 to 98	Lin et al. (2003)
2003	T. Aitasalo et al.	Solid-state reaction	$Eu^{2+}R^{3+}$	Persistent luminescence phenomena in materials doped with rare-earth ions	60 W incandescent lamp	VHR	80 °C followed by a tail up to 200 °C	Aitasalo et al. (2003)
2004	T. Aitasalo et al.	Solid-state reaction	$Eu^{2+}R^{3+}$	Low-temperature thermoluminescence properties of Eu^{2+} and R^{3+} doped $CaAl_2O_4$	351 nm Ar^+ ion laser	0.1°C/s	Multiple peaks −88 to 17	Aitasalo et al. (2004a)
2004	T. Aitasalo et al.	Solid-state reaction	$Eu^{2+}R^{3+}$	Annihilation of the persistent luminescence of MAl_2O_4:Eu^{2+} by Sm^{3+} co-doping	20 W incandescent lamp	5 °C/s	Multiple peaks 80 to 130	Aitasalo et al. (2004b)

(continued)

Table 16.4 (continued)

Year	Author	Synthesis	Doping	Paper title	Excitation	Heating rate	Peak Temp. (°C)	References
2004	T. Aitasalo et al.	Pechini method	Eu^{2+}	Eu^{2+} doped calcium aluminates prepared by alternative low-temperature routes	360 & 420 nm UV lamp	5 °C/s	100	Aitasalo et al. (2004c)
2004	J. Hölsa et al.	Solid-state reaction	Eu^{2+}, R^{3+}	Role of defect states in persistent luminescence materials	20 W incandescent lamp	5 °C/s	70°C followed by a tail up to 200 °C	Hölsä et al. (2004)
2005	B. Liu et al.	Solid-state reaction	Dy^{3+}	Potential white-light long-lasting phosphor: Dy^{3+} -doped aluminate	Hg lamps 254 nm & 365 nm	0.2 °C/s	−29, 7, 41, 73	Liu et al. (2005)
2005	T. Aitasalo et al.	dip- and spin-coating methods	Eu^{2+}	Eu^{2+} doped calcium aluminate coatings by sol–gel methods	360 & 420 nm UV lamp	5 °C/s	90	Aitasalo et al. (2005)
2006	T. Aitasalo	Solid-state reaction	Eu^{2+} R^{3+}	Thermoluminescence study of persistent luminescence materials: Eu^{2+}- and R^{3+}-doped calcium aluminates, $CaAl_2O_4$:Eu^{2+}, R^{3+}	20 W incandescent lamp	VHR	Multiple peaks 70 to 320	Aitasalo et al. (2006)
2006	Madhukumar et al.	Solid-state reaction	Dy^{3+}; Sm^{3+}; Ce^{3+}	Thermoluminescence dosimetry of rare-earth-doped calcium aluminate phosphors	γ-rays ^{60}Co source	4	Multiple peaks 110–295	Madhukumar et al. (2006)

(continued)

Table 16.4 (continued)

Year	Author	Synthesis	Doping	Paper title	Excitation	Heating rate	Peak Temp. (°C)	References
2006	C. Chang et al.	Solid-state reaction and co-precipitation	$Eu^{2+}Nd^{3+}$	Luminescence of long-lasting $CaAl_2O_4:Eu^{2+}$, Nd^{3+} phosphor by co-precipitation method	UV lamp	1 °C/s	68 and 64	Chang et al. (2006)
2007	Madhukumar et al.	Solid-state reaction	Ce; Tb	Luminescence studies of rare-earth-doped calcium aluminate phosphor	γ-rays ^{60}Co source	4	140, 190, 295	Madhukumar et al. (2007)
2007	Y. Lin et al.	Solid-state reaction	$Eu^{2+}La^{3+}$	Tunable trap levels observed in La and Eu co-doped $CaAl_2O_4$-based phosphor	UV light	0.21 °C/s	123	Lin et al. (2007)
2007	L. Wang et al.	Solid-state reaction	$Eu^{2+}Nd^{3+}$	Effects of B_2O_3 and SiO_2 on the persistent luminescence property of $CaAl_2O_4:Eu^{2+}$, Nd^{3+}	UV light	1 °C/s	69	Wang and Wang (2007a)
2007	Y. Wang et al.	Solid-state reaction	$Eu^{2+}Nd^{3+}$	Defect states in Nd^{3+} doped $CaAl_2O_4:Eu^{2+}$	Standard artificial light	1 °C/s	46	Wang and Wang (2007b)
2008	V. Singh et al.	Combustion method	Sm^{3+}	Characterization, photoluminescence and correlation between Thermoluminescence and ESR of combustion synthesized $CaAl_2O_4$: Sm^{3+} material	γ-rays	5 °C/s	135, 325	Singh et al. (2008)

(continued)

Table 16.4 (continued)

Year	Author	Synthesis	Doping	Paper title	Excitation	Heating rate	Peak Temp. (°C)	References
2008	X. Teng et al.	Solid-state reaction	$Eu^{2+}Nd^{3+}Re^{3+}$	Influence of La^{3+} and Dy^{3+} on the properties of the long afterglow phosphor $CaAl_2O_4$: Eu^{2+}, Nd^{3+}	Artificial daylight	1 °C/s	Multiple peaks 139–154	Teng et al. (2008)
2008	L. Wang et al.	Solid-state reaction	$Eu^{2+}Nd^{3+}Sr^{2+}$	Effects of Sr^{2+} doping on the persistent luminescence properties of $CaAl_2O_4$:Eu^{2+}, Nd^{3+}	Artificial daylight	1°Cs	127	Wang et al. (2008)
2009	X. Xu et al.	Solid-state reaction	$Eu^{2+}Nd^{3+}Sr^{2+}$	$CaAl_2O_4$:Eu^{2+}, Nd^{3+}, Sr^{2+}: A white light phosphor with yellow-green long afterglow	Artificial daylight	1 °C/s	105	Xu et al. (2009a)
2009	X. Xu et al.	Solid-state reaction	$Eu^{2+}Nd^{3+}Mn^{2+}$	Energy transfer between Eu^{2+} and Mn^{2+} in long-afterglow phosphor $CaAl_2O_4$:Eu^{2+}, Nd^{3+}, & Mn^{2+}	Artificial daylight	1 °C/s	103	Xu et al. (2009b)
2010	X. Xu et al.	Solid-state reaction	Eu^{2+}, $Nd^{3+}Pr^{3+}$	Effects of Pr^{3+} doping on the optical properties of Eu and Nd co-doped $CaAl_2O_4$-based phosphor	Artificial daylight	1°C/s	79	Xu et al. (2010)

(continued)

Table 16.4 (continued)

Year	Author	Synthesis	Doping	Paper title	Excitation	Heating rate	Peak Temp. (°C)	References
2011	W. Xie et al.	Combustion method	$Eu^{2+}Dy^{3+}$	The long afterglow and thermoluminescence properties of MAl_2O_4: Eu^{2+}, Dy^{3+} (M = Ca, Sr, and Ba) phosphors synthesized by combustion technique	UV light	1 °C/s	111	Xie et al. (2011)
2012	L. Wang et al.	Solid-state reaction	Eu^{2+}, Nd^{3+}	Thermal release behavior of carriers in persistent luminescence material $CaAl_2O_4$:Eu^{2+}, Nd^{3+}	Artificial daylight	1 °C/s	Multiple peaks 69–135	Wang et al. (2012)
2012	J. Kaur et al.	Solid-state reaction	Mn^{2+}	TL glow curve study, kinetics, PL and XRD analysis of Mn^{2+} doped $CaAl_2O_4$ phosphors	365 nm UV	6.7 °C/s	358	Kaur et al. (2012)
2013	K V Eeckhout et al.	Sol–gel method	$Eu^{2+}Nd^{3+}$	Revealing trap depth distributions in persistent phosphors	365 nm UV	5°C/s	197	Eeckhout et al. (2013)
2013	A.H. Wako et al.	Combustion method	$Eu^{2+}Nd^{3+}$	Roles of doping ions in afterglow properties of blue $CaAl_2O_4$:Eu^{2+}, Nd^{3+} phosphors	UV light	1°C/s	Multiple peaks 60–170	Wako et al. (2014)

(continued)

Table 16.4 (continued)

Year	Author	Synthesis	Doping	Paper title	Excitation	Heating rate	Peak Temp. (°C)	References
2014	B. Zhang et al.	Solid-state reaction	Eu^{2+} R^{3+}	Long persistent and optically stimulated luminescence behaviors of calcium aluminates with different trap filling processes	354 nm UV	1°C/s	Multiple peaks 40–140	Zhang et al. (2014)
2015	Md. Ziyauddin et al.	Combustion method	Eu^{2+}	Photoluminescence (PL) and thermoluminescence (TL) studies of γ- irradiated $CaAl_2O_4$:Eu^{2+} phosphor synthesized by combustion technique	γ-rays	5 °C/s	394	Ziyauddin et al. (2015)
2015	K. K. Satapathy et al.	Combustion method	Tb^{3+}	Mechanoluminescence and thermoluminescence characterization of Tb^{3+} doped $CaAl_2O_4$: a theoretical and experimental study	γ-rays	5 °C/s	110	Satapathy et al. (2015)
2015	F. Khan et al.	Combustion method	Dy^{3+}	Thermoluminescence characterization of MAl_2O_4 (M = Ba, Ca, Mg) phosphors activated with Dy^{3+}	γ-rays	5 °C/s	110, 270	Khan et al. (2015)

(continued)

Table 16.4 (continued)

Year	Author	Synthesis	Doping	Paper title	Excitation	Heating rate	Peak Temp. (°C)	References
2016	Md. Ziyauddin et al.	Combustion method	Dy^{3+}	Photoluminescence and thermoluminescence studies of $CaAl_2O_4$:Dy^{3+} phosphor	γ-rays	5 °C/s	147	Ziyauddin et al. (2016)
2017	I. Mindru et al.	Precursor route	Tb^{3+}	Tb^{3+}-doped alkaline-earth aluminates: synthesis, characterization and optical properties	UV and x-rays	1 °C/s	87, 313	Mindru et al. (2017)
2017	T. Cui et al.	Solid-state reaction	$Eu^{2+}Nd^{3+}$	Preparation of $CaAl_2O_4$:Eu^{2+}, Nd^{3+} and $SrAl_2O_4$: Eu^{2+}, Dy^{3+} long afterglow luminescent materials using oil shale ash	365 nm UV light	1 °C/s	7	Cui et al. (2017)
2018	Z. Liu et al.	Solid-state reaction	$Eu^{2+}Tm^{3+}$	Effects of the deep traps on the thermal-stability property of $CaAl_2O_4$: Eu^{2+} phosphor	UV source (254 nm and 365 nm)	1 °C/s	102–10[illegible]	Liu et al. (2018)
2018	Y.Z. Halefoglu et al.	Combustion method	$Eu^{2+}Nd^{3+}$	Preparation and photoluminescence properties of aluminate phosphors produced by combustion synthesis	$^{90}Sr/^{90}Y$ beta source	1 °C/s	51, 129–267	Halefoglu et al. (2018)

(continued)

Table 16.4 (continued)

Year	Author	Synthesis	Doping	Paper title	Excitation	Heating rate	Peak Temp. (°C)	References
2019	Md. Ziyauddin et al.	Combustion method	Eu^{2+}	Luminescence properties of calcium aluminate phosphors	254 & 365 nm UV lamp	5 °C/s	121	Ziyauddin et al. (2019)
2019	B. Zhai et al.	Combustion method	Undoped	Blue afterglow of undoped $CaAl_2O_4$ nanocrystals	Artificial daylight	1 °C/s	80	Zhai and Huang (2019)
2019	Nara et al.	Combustion method	Dy^{3+}	Electron life time (τ) in trap levels of Dy^{3+} activated calcium aluminate: implications in TL dosimetry	γ-rays	5 °C/s	135, 410	Moirangthem et al. (2019)
2019	J.M. Kalita et al.	Commercial	$Eu^{2+}Nd^{3+}$	Temperature dependence of persistent luminescence in $CaAl_2O_4$:Eu^{2+}, Nd^{3+} related to beta irradiation and optical excitation	$^{90}Sr/^{90}Y$ beta source	1 °C/s	63	Kalita and Chithambo (2019)
2020	B. Zhai et al.	Combustion method	Undoped	Annealing temperature dependent photoluminescence and afterglow of undoped $CaAl_2O_4$	254 nm UV	1 °C/s	72	Zhai et al. (2020)
2020	R. Ruiz-Torres et al.	Combustion method	$Eu^{2+}Dy^{3+}$	Effect of thermal treatment on luminescence properties of long persistent $CaAl_2O_4$: Eu^{2+} Dy^{3+} synthesized by combustion method	$^{90}Sr/^{90}Y$ beta source	2 °C/s	77	Ruiz-Torres et al. (2020)

(continued)

Table 16.4 (continued)

Year	Author	Synthesis	Doping	Paper title	Excitation	Heating rate	Peak Temp. (°C)	References
2020	Gartia et al.	Combustion method	Undoped	Determination of thermoluminescence parameters in nanocrystalline $CaAl_2O_4$	γ-rays	VHR	Multiple peaks 90–400	Gartia et al. (2020)
2020	Nara et al.	Combustion method	Sm^{3+}	Thermoluminescence of 9 meV electron irradiated $CaAl_2O_4$:Sm^{3+} phosphor and its implication in dosimetry	γ-rays	5 °C/s	190, 400	Singh et al. (2020)
2021	M Zhang et al.	Combustion method	$Eu^{2+}Nd^{3+}Gd^{3+}$	Enhanced afterglow performance of $CaAl_2O_4$:Eu^{2+},Nd^{3+} phosphor by co-doping Gd^{3+}	365 nm UV	1°C/s	80	Zhang et al. (2021a)
2021	M Zhang et al.	Solid-state reaction	$Eu^{2+}Nd^{3+}$	$CaAl_2O_4$:Eu^{2+}, Nd^{3+} anti-corrosive coating and its afterglow-catalytic process	360 nm UV	1 °C/s	72	Zhang et al. (2021b)
2021	Nara et al.	Combustion method	Dy^{3+}, Sm^{3+}, Tm^{3+}	Effects of doping concentration on thermoluminescence parameters of $CaAl_2O_4$:Re^{3+}, ($Re^{3+} = Dy^{3+}$, Sm^{3+}, Tm^{3+}) prepared by combustion method	γ-rays	5 °C/s	Multiple peaks 90–400	Singh et al. (2021a)

(continued)

Table 16.4 (continued)

Year	Author	Synthesis	Doping	Paper title	Excitation	Heating rate	Peak Temp. (°C)	References
2021	Gartia et al.	Combustion method	Sm^{3+}	Deconvolution of thermoluminescence curves and its implications: The case of $CaAl_2O_4$:Sm^{3+}	UV, γ-ray, e-beam and X-ray	5 °C/s	186 and 375	Gartia and Singh (2021)
2021	B. Zhai et al.	Combustion method	Tb^{3+}	Annealing temperature dependent afterglow of Tb^{3+} doped $CaAl_2O_4$	254 nm UV	2 °C/s	54	gai Zhai et al. (2021)
2021	Nara et al.	Combustion method	Tm^{3+}	Thermoluminescence studies of Tm doped nanocrystalline calcium aluminate ($CaAl_2O_4$:Tm^{3+})	γ-ray, e- beam	5 °C/s	190, 299, 409	Singh et al. (2021b)
2021	Nara et al.	Combustion method	Undoped	The physical basis of persistence luminescence in $CaAl_2O_4$ lattice based afterglow phosphors	253 nm UV	1 °C/s	70	Singh and Gartia (2021)

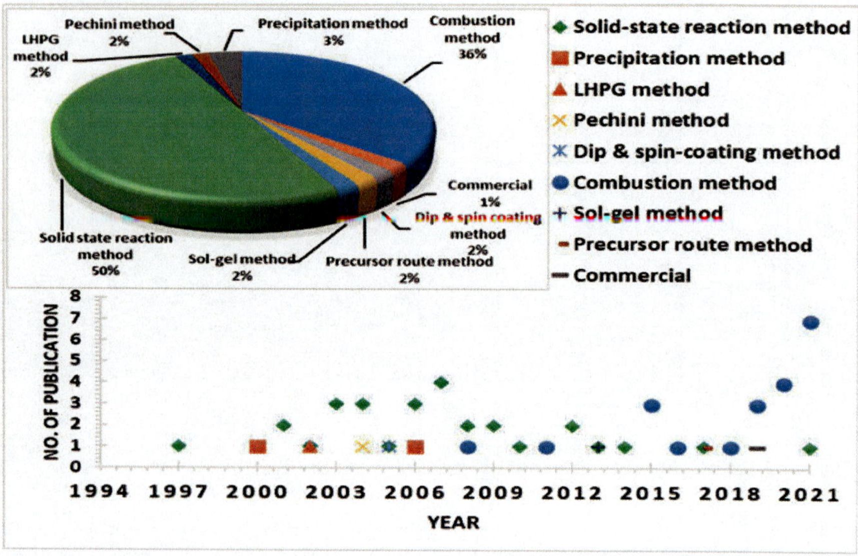

Fig. 16.5 Various preparation methods of calcium aluminate with the publication years 1988 to 2021 (Redrawn and adapted information from Google scholars)

combustion methods have played significant roles in the synthesis process of TL phosphors.

Traditionally, a high-temperature solid-state reaction has been used to prepare the calcium aluminate. Due to numerous drawbacks of the solid-state reaction method, many workers have opted for alternative methods of creating calcium aluminate in recent years. The combustion method has been the most efficient because it is inexpensive, uses conventional techniques that consume less energy and time, does not require a high-temperature furnace for the annealing process, has a high degree of purity, and produces environmentally friendly byproducts (Fumo et al. 1996; Kingsley and Patil 1988). The starting materials of the combustion method are metal nitrates and different organic fuels (Urea, Sucrose). Meanwhile, the solution combustion synthesis process is complex due to mixing starting materials and fuels in stereometric portions.

16.5.2 Structural and Morphological Analysis of $CaAl_2O_4$

Calcium aluminate is an inorganic material belonging to the family of alkaline earth aluminates. Because of its hydration-contributing capabilities, $CaAl_2O_4$ is the major component of calcium alumina cement (Iftekhar et al. 2008). Furthermore, calcium aluminate is employed in optical information storage devices (Zawrah and Khalil 2007). The unit cell of $CaAl_2O_4$ exhibits Ca^{2+} occupying three crystallographic

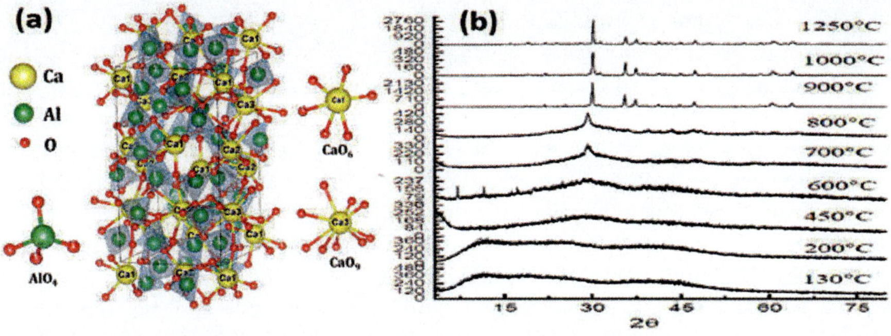

Fig. 16.6 **a** The crystal structure of $CaAl_2O_4$ (Reproduced with permission from Elsevier Publisher. (Kaur and Khanna 2021)) **b** XRD of calcium aluminate (Reproduced with permission from Trans Tech Publications Ltd Publisher (Selyunina and Mishenina 2015))

positions, as shown in Fig. 16.6a. Two Ca ions are octahedrally coordinated with oxygen ions (CaO_6), while the third Ca ion is in nine coordinates with oxygen (CaO_9). Furthermore, Al^{3+} occupies six positions, each having four-coordination with oxygen (AlO_4 units) (Kaur and Khanna 2021).

The effect of heat treatment after synthesis observed with the XRD pattern is shown in Fig. 16.6b. In the XRD diffraction patterns of calcium aluminate, fine and high-intensity peaks were observed along with a small segment of the amorphous phase at an annealed temperature of 900 °C. These imply enough precision in the crystal structure. Verma et al. (2017) have reported the XRD pattern of the $CaAl_2O_4$:Eu^{3+} Gd^{3+} annealed at 1000 °C. They have observed monoclinic phase crystalline behavior unaffected by doping and co-doping in the host lattice's crystal structure. Giilgun et al. (1994) have reported that the phase distribution of calcium aluminate during preparation depends on resin content, time, temperature, and furnace atmosphere during calcination. In particular, the effects of calcination temperature at 800 °C or below and above 800 °C have been amorphous and crystalline, respectively. With increased calcination time at 900 °C, the presence of other phases has been gradually reduced, and after three hours of calcination time, XRD patterns have consisted mainly of the mono calcium aluminate phase.

16.5.3 Applications of Calcium Aluminate Based Phosphors

Calcium aluminate ($CaAl_2O_4$) belongs to the family of alkaline earth aluminate, MAl_2O_4 (M = Ca, Sr, Ba), that are known to be excellent phosphors when doped with Eu^{2+} (Palilla et al. 1968; Abbruscato 1971). They became excellent luminescence phosphors upon suitable R^{3+} co-doping (Sakai et al. 1999; Katsumata et al. 1998, 1997).

16.5.3.1 Persistence Luminescence

Nd^{3+} co-doped $CaAl_2O_4:Eu^{2+}$ has emerged as a benchmark commercial persistent phosphor. Several workers have attempted to improve the persistent luminescence or modify the emission bands in $CaAl_2O_4:Eu^{2+}Nd^{3+}$ afterglow phosphor. Among these have been the works of Xu et al. (2009a, b, 2010) that have not yielded any positive results. However, they have reported TL of $CaAl_2O_4:Eu^{2+}$, Nd^{3+} and Sr^{2+} at 105 °C for a heating rate $\beta = 1$ °C/s. $CaAl_2O_4:Eu^{2+}$, Nd^{3+}, and Mn^{2+} reported a TL peak at 103 °C, while for Pr^{3+} co-doping, it occurs at 79 °C. As routinely argued in many works, Xu et al. have justified persistent luminescence in their phosphors due to the TL peaks around 100 °C. TL peaks have been observed by Jia et al. (2002b), indicating the existence of numerous trapping levels that may cause afterglow at room temperature (300 K), lasting from seconds to hours.

Halefoglu et al. (2018) have reported that $CaAl_2O_4:Eu^{2+}$, Nd^{3+} phosphors exhibit TL peaks above 200 °C, indicating dosimetric possibilities. Shallow peaks have disappeared soon compared with the higher temperature peaks. Liu et al. (2018) have shown changes in the peak temperature with different doping concentrations of Eu^{2+}. Wako et al. (2014) have reported TL of different concentrations of Eu^{2+} and Nd^{3+} doped in the $CaAl_2O_4$ phosphor. They have noticed that the intensity of glow peaks increases with the shifting of peak positions. The effect of the change in the concentration of Nd^{3+} has resulted in a peak deeper than that of Eu^{2+}. Eu^{2+} and Nd^{3+} dopants have enhanced the number of charged particles in shallow and deeper traps.

Calcium aluminate ($CaAl_2O_4$) is an excellent phosphor when doped with Eu^{2+}. It becomes an excellent persistent phosphor when suitably co-doped with R^{3+} (Aitasalo et al. 2004b, 2006). Many workers have studied these materials using thermoluminescence to identify the trapping levels which are relevant to persistent luminescence. Aitasalo et al. (2003, 2004a, b, 2006) have extensively studied the TL of $CaAl_2O_4$: Eu^{2+}, R^{3+} where R = Ce, Dy, Ho, Nd, Tb, Tm, Pr, Er, La, Lu, Gd, Yb, Y, Sm, at below and above the room temperature and documented TL peak temperatures.

Aitasalo and his group (Aitasalo et al. 2003, 2004a, 2004b) have reported a few findings in $CaAl_2O_4$: Eu^{2+}, R^{3+} as persistent luminescent materials, as shown in Fig. 16.7a, b, and it has covered the entire range of rare-earth co-activators in $CaAl_2O_4:Eu^{2+}R^{3+}$. The realistic classification of TL peaks of this phosphor has involved at least three peaks around 80 ± 19, 120 ± 18, and 290 ± 30 °C. Since it has covered all possible R^{3+}, there has been scope for inter-comparisons of global data documented by others using different heating rates, different methods of sample preparation, and composition.

Zhang et al. (2014) have reported TL of $CaAl_2O_4:Eu^{2+}$ & Eu^{2+}, Re^{3+} (Re = Nd, Dy & Tm) after a delay of 48 h (Fig. 16.8). After 48 h gap between excitation and measurement, TL peaks have been observed to decay entirely at the low temperature (<350 K). In the case of higher temperatures (>350 K), TL peaks have sustained a certain amount of intensity. $CaAl_2O_4:Eu^{2+}$ and Tm^{3+} have shown the highest intensity than other doped ones, revealing the maximum storage capacity of the trapped charges.

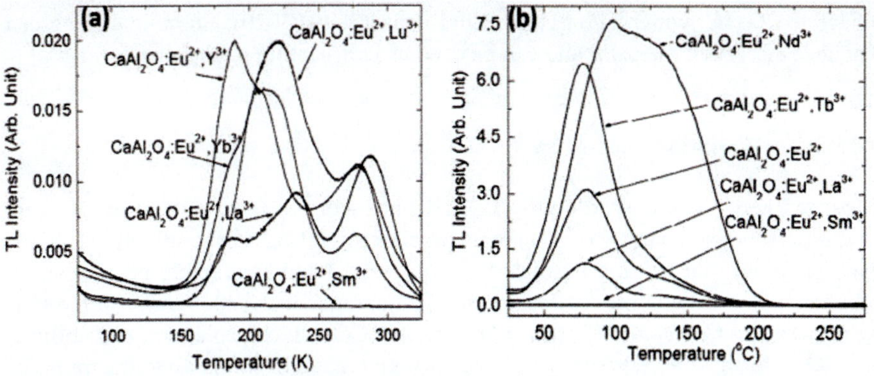

Fig. 16.7 **a** Low and **b** High-temperature TL glow curves of $CaAl_2O_4:Eu^{2+}$, R^{3+} (Reproduced with permission from Elsevier Publisher (Aitasalo et al. 2004b))

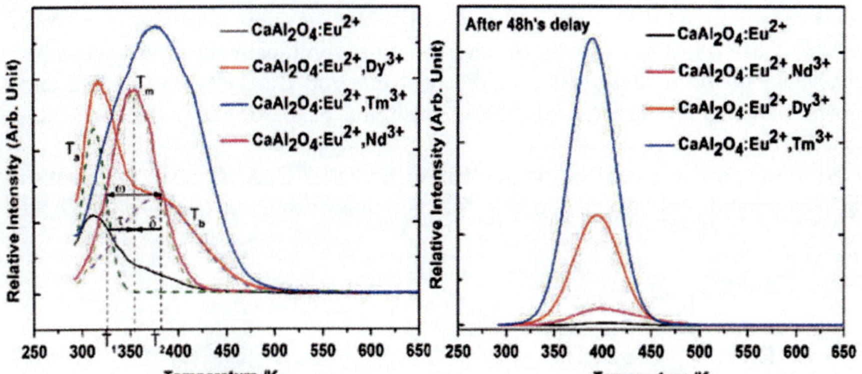

Fig. 16.8 TL of $CaAl_2O_4:Eu^{2+}$ & Eu^{2+}, Re^{3+} (Re = Nd, Dy, Tm), and after a delay of 48 h (reproduce with permission from Elsevier Publisher (Zhang et al. 2014))

Undoped $CaAl_2O_4$ without annealing has been reported by Zhai et al. (2019) to have a blue afterglow, and a broad TL band maximum has been observed at about 82 °C. In addition, Zhai et al. (2020) have reported the effect of annealing on the TL of undoped $CaAl_2O_4$ from 20–200 °C, showing broad TL peaks. TL of undoped $CaAl_2O_4$ before annealing has been shown to have a broad peak at about 72 °C. Annealing undoped $CaAl_2O_4$ has two peaks, at about 63 and 126 °C. Significant changes in the afterglow parameters have been reported for undoped $CaAl_2O_4$ during annealing, extending the length of the blue afterglow to a seven-fold increase in afterglow lifetime.

Nara and Gartia (2021) have reported TL data of undoped calcium aluminate irradiated with UV. The TL peaks have been most suited in the temperature range of RT to 200 °C, which agrees with Zhai et al. (2019, 2020). Suitable doping can enhance afterglow intensity and can be used as long-lasting phosphors.

16.5.3.2 Radiation Dosimetry

Most workers have studied $CaAl_2O_4:Eu^{2+}$ phosphors in the persistence luminescence perspective analysis of TL glow curve up to 200 °C. Ziyauddin et al. (2015) have reported that gamma-irradiated $CaAl_2O_4:Eu^{2+}$ shows a stable peak position and hence found it eligible for dosimetry. Madhukumar et al. (2006) have studied TL glow curves of $CaAl_2O_4:Re^{3+}$ (Re = Dy, Sm, and Ce), which explain the possibilities of $CaAl_2O_4$ phosphor synthesis with solid-state reaction. The peak temperature of $CaAl_2O_4$: Ce^{2+} phosphor at 295 °C has shown better stability than other doping (Dy and Sm).

Thermoluminescence glow curves of $CaAl_2O_4$ phosphors undoped and doped with Dy^{3+}, Sm^{3+}, and Tm^{3+} (Fig. 16.9), prepared by using the combustion method, have been reported by Nara et al. (Singh et al. 2021a). Among these four phosphors, $CaAl_2O_4$: Sm^{3+} has been found to have the most significant sensitivity. However, repairment of the defect centres of the host material $CaAl_2O_4$ by the rare earth dopant elements has suppressed the TL sensitivity of the doped phosphors (Guler et al. 2018).

According to Nara et al. (Moirangthem et al. 2019), $CaAl_2O_4:Dy^{3+}$ TL peaks at high temperatures should be suitable for radiation dosimetry. Gartia et al. (2020)

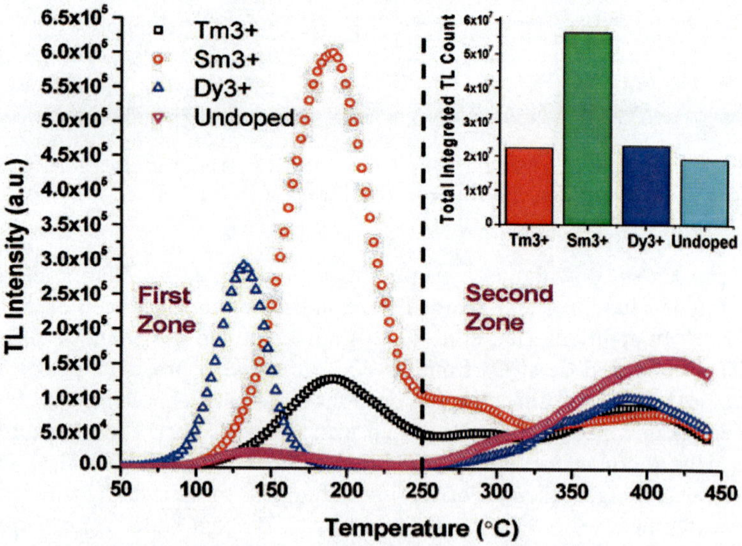

Fig. 16.9 TL glow curves of calcium aluminate undoped and doped with Dy^{3+}, Sm^{3+}, and Tm^{3+} (Reproduced with permission from Elsevier Publisher (Singh et al. 2021a))

have reported multiple trap distributions in undoped calcium aluminate host lattice irradiated with the gamma radiation. $CaAl_2O_4:Mn^{2+}$ irradiated with UV has been reported by Kaur et al. (2012), which shows an intense dosimetric peak at 358 °C.

16.6 Conclusions and Future Status

In the thermoluminescence of calcium aluminate, the combustion synthesis method has gradually replaced the traditional preparation method due to many factors. Post synthesis annealed effect has revealed that crystallinity increases with the annealing temperature. These phosphors have been primarily used as long-lasting phosphors. Commercially, $CaAl_2O_4$: Eu^{2+} Nd^{3+} has been used as a persistent commercial phosphor. The possibility of radiation dosimetry application of calcium aluminate is due to the deep traps in lattices.

Nowadays, TL dosimetry is used in personal monitoring, patient dose estimation, retrospective dosimetry, and high radiation fields. Initially, TLDs were routinely applied to measure doses of low linear energy transfer (LET) radiations like x-rays, gamma-rays, and electron beam radiations in medical applications. Nevertheless, in the last few decades, high LET radiations (protons, carbon ions, etc.) have been increasingly used in cancer therapy. So, the dosimetric studies of TLDs will have to be extended to high LET radiations (Salah 2011). New highly sensitive TL dosimetric phosphors are needed to be developed for expanding the list of TL phosphor materials. Applications in quality control, micro-dosimetry, mixed radiation fields, and neutron dosimetry are the challenges in TL dosimetry, commonly used in nuclear power units and medical-clinical-oncological dosimetry (Horowitz 2014).

Acknowledgements The author (Dr. Moirangthem Nara Singh) expresses his sincere gratitude and thanks to his Guru Emeritus Professor, Dr. R. K. Gartia, for his valuable support and guidance.

Author Contributions The manuscript was written through the contributions of all authors. All authors have given approval to the final version of the manuscript.

References

Abbruscato, V.: J. Electrochem. Soc. **118**, 930–933 (1971)

Aitasalo, T., Dereń, P., Hölsä, J., Jungner, H., Krupa, J.C., Lastusaari, M., Legendziewicz, J., Niittykoski, J., Stręk, W.: J. Solid State Chem. **171**, 114–122 (2003)

Aitasalo, T., Durygin, A., Hölsä, J., Lastusaari, M., Niittykoski, J., Suchocki, A.: J. Alloys Compd. **380**, 4–8 (2004a)

Aitasalo, T., Dereń, P., Hölsä, J., Jungner, H., Lastusaari, M., Niittykoski, J., Stręk, W.: Radiat. Meas. **38**, 515–518 (2004b)

Aitasalo, T., Hölsä, J., Jungner, H., Lastusaari, M., Niittykoski, J., Parkkinen, M., Valtonen, R.: Opt. Mater. **26**, 113–116 (2004c)

Aitasalo, T., Hölsä, J., Jungner, H., Lastusaari, M., Niittykoski, J., Saarinen, J.: Opt. Mater. **27**, 1537–1540 (2005)

Aitasalo, T., Hölsä, J., Jungner, H., Lastusaari, M., Niittykoski, J.: J. Phys. Chem. B **110**, 4589–4598 (2006)
Aitken, M.J.: Thermoluminescence Dating. Academic Press, London (1985)
Akselrod, M., Kortov, V., Kravetsky, D.J., Gotlib, V.I.: Radiat. Prot. Dosimetry **33**, 119–122 (1990)
Azorin Nieto, J.: Appl. Radiat. Isot. **117**, 135–142 (2016)
gai Zhai, B., Xu, H., Huang, Y.M.: Opt. Mater. **112**, 110739–110748 (2021)
Bhatt, B.C., Kulkarni, M.S.: Defect Diffus. Forum **347**, 179–227 (2014)
Binti Wahib, N., Khandaker, M.U., Aqilah binti Mohamad Ramli, N., Sani, S.F.A., Bradley, D.A.: Appl. Radiat. Isot. **148**: 218–224 (2019)
Cameron, J.R., Kenney, G., Zimmerman, D.: Radiat. Res. **19**, 199 (1963)
Cameron, J.R., Suntharalingam, N., Kenney, G.N.: Thermoluminescent Dosimetry. University of Wisconsin Press, Madison (1968)
Chang, C., Xu, J., Jiang, L., Mao, D., Ying, W.: Mater. Chem. Phys. **98**, 509–513 (2006)
Chen, R.: Electrochem. Soc. **116**, 1254–1257 (1969)
Christensen, P., Bøtter-Jensen, L., Majborn, B.: Int. J. Appl. Radiat. Isot. **33**, 1035–1050 (1982)
Cui, T., Ma, P., Sheng, Y., Zheng, K., Zhou, X., Xu, C., Zou, H., Song, Y.: Opt. Mater. **67**, 84–90 (2017)
Daniels, F., Boyd, C.A., Saundersl, D.F.: Science **117**, 343–349 (1953)
Dong, B., Cao, B., He, Y., Liu, Z., Li, Z., Feng, Z.: Adv. Mater. **24**, 1987–1993 (2012)
Duggan, C., Hood, H., Warren-Forward, M., Haque, T.K.: Phys. Med. Biol. **49**, 3831–3845 (2004)
Fumo, D.A., Morelli, M.R., Segadaes, A.M.: Bull. Mater. Sci. **31**, 1243–1255 (1996)
Furetta, C.: Handbook of Thermoluminescence. World Scientific Publishing Co Inc., Singapore (2006)
Furetta, C., Weng, P.-S.: Operational Thermoluminescence Dosimetry. World Scientific Publishing Co Inc., Singapore (1998)
Garlick, G.F.J., F.A. Gibson: Proc. Phys. Soc. Sect. A. **60**, 574 (1948)
Gartia, R.K.: Defect Diffus. Forum **357**, 193–215 (2014)
Gartia, R.K., Chandrasekhar, N.: Defect Diffus. Forum **357**, 171–191 (2014)
Gartia, R.K., Singh, M.N.: Optik. **248**, 168048 (1–11) (2021)
Gartia, R.K., Singh, M.N., Chanu, L.P.. Singh, T.B.: J. Lumin. **219**, 116867 (1–6) (2020)
Guler, I., Isik, M., Ahmedova, F., Guseinov, A., Gasanly, N.: Luminescence **33**, 759–763 (2018)
Gülgün, M.A., Popoola, O.O., Kriven, W.M.: J. Am. Ceram. Soc. **77**, 531–539 (1994)
Hajek, M., Berger, T., Vana, N.: Radiat. Prot. Dosimetry **110**, 337–341 (2004)
Halefoglu, Y.Z., Yüksel, M., Derin, H., Can, N., Topaksu, M., Ozturk, E., Karacaoğlu, E.: Appl. Radiat. Isot. **142**, 46–50 (2018)
Hölsä, J., Jungner, H., Lastusaari, M., Niittykoski, J.: J. Alloys Compd. **323–324**, 326–330 (2001)
Hölsä, J., Aitasalo, T., Jungner, H., Lastusaari, M., Niittykoski, J., Spano, G.: J. Alloys Compd. **374**, 56–59 (2004)
Hoogenstraaten, W.: Phillips Res. Rep. **13**, 515–693 (1958)
Horowitz, Y.S.: Thermoluminescence and Thermoluminescent Dosimetry. CRC Press Taylor & Francis Group, Boca Raton (1984)
Horowitz, Y.S.: Radiat. Meas. **71**, 2–7 (2014)
Horowitz, Y.S., Moscovitch, M.: Radiat. Prot. Dosimetry **153**, 1–22 (2013)
Horowitz, Y.S., Yossian, D.: Radiat. Prot. Dosimetry **60**, 293–295 (1995)
Iftekhar, S., Grins, J., Svensson, G., Lööf, J., Jarmar, T., Botton, G.A., Andrei, C.M., Engqvist, H.: J. Eur. Ceram. Soc. **28**, 747–756 (2008)
JabtoJisk, V.A.: Physik. **94**, 38–46 (1935)
Jia, D., Yen, W.M.: J. Electrochem. Soc. **150**, H61–H65 (2003)
Jia, D., Wang, X.J., Yen, W.M.: Chem. Phys. Lett. **363**, 241–244 (2002a)
Jia, D., Meltzer, R.S., Yen, W.M., Jia, W., Wang, X.: Appl. Phys. Lett. **80**, 1535–1537 (2002b)
Kalita, J.M., Chithambo, M.L.: J. Lumin. **206**, 27–32 (2019)
Karzmark, C.J., Attix, F.H., Wingate, C.L.: Science **150**, 391–394 (1965)

Katsumata, T., Nabae, T., Sasajima, K., Komuro, S., Morikawa, T.: J. Electrochem. Soc. **144**, L243–L245 (1997)
Katsumata, T., Sasajima, K., Nabae, T., Komuro, S., Morikawa, T.: J. Am. Ceram. Soc. **81**, 413–416 (1998)
Kaur, P., Khanna, A.: Ceram. Int. **47**, 14655–14664 (2021)
Kaur, J., Suryanarayana, N.S., Jaykumar, B., Dubey, V.: J. Miner. Mater. Charact. Eng. **11**, 1081–1084 (2012)
Khan, F., Satapathy, K.K., Mishra, G.C.: Int. J. Lumin. Appl. **5**, 26–28 (2015)
Kingsley, J.J., Patil, K.C.: Mater. Lett. **6**, 427–432 (1988)
Konopka, M.N., Bilski, P., Obryk, B., Marczewska, B., Olko, P., Kłosowski, M., Gieszczyk, W.: Nonlinear Opt. Quantum Opt. **48**, 133–146 (2017)
Kortov, V.: Radiat. Meas. **42**, 576–581 (2007)
Kortov, V.S.: Radiat. Meas. **45**, 512–515 (2010)
Lin, Y., Zhang, Z., Tang, Z., Zhang, J., Zheng, Z., Lu, X.: Mater. Chem. Phys. **70**, 156–159 (2001)
Lin, Y., Tang, Z., Zhang, Z., Nan, C.: J. Eur. Ceram. Soc. **23**, 175–178 (2003)
Lin, Y.H., Li, M., Nan, C.W., Zhang, Z.: J. Am. Ceram. Soc. **90**, 2992–2994 (2007)
Lin, Q., Li, Z., Yuan, Q.: Chinese Chem. Lett. **30**, 1547–1556 (2019)
Liu, B., Shi, C., Qi, Z.: Appl. Phys. Lett. **86**, 1–3 (2005)
Liu, Z., Zhao, L., Chen, W., Xin, S., Fan, X., Bian, W., Yu, X., Qiu, J., Xu, X.: J. Am. Ceram. Soc. **101**, 3480–3488 (2018)
Madhukumar, K., Rajendra Babu, K., Ajith Prasad, K.C., James, J., Elias, T.S., Padmanabhan, V., Nair, C.M.K.: Bull. Mater. Sci. **29**, 119–122 (2006)
Madhukumar, K., Babu, K.R., Prasad, K.C.A., James, J., Elias, T.S., Padmanabhan, V., Nair, C.M.K.: Int. J. Mod. Phys. B **21**, 1971–1980 (2007)
Mahesh, K., Weng, P.-S., Furetta, C.: Thermoluminescence in Solids and its applications. Nuclear Technology Publishing, Ashford (1989)
Mashangva, M., Singh, M.N., Singh, T.B.: Indian J. Pure Appl. Phys. **49**, 583–589 (2011)
Matsuzawa, T., Aoki, N.T.Y., Murayama, Y.: J. Electrochem. Soc. **143**, 2670–2673 (1996)
May, C.E., Partridge, J.A.: J. Chem. Phys. **40**, 1401–1409 (1964)
McKeever, S.W.S.: Thermoluminescence of Solids. Cambridge University Press, London (1985)
McKeever, S.W.S., Moscovitch, M., Townsend, P.D.: Thermoluminescence Dosimetry Materials: Properties and Uses. Nuclear Technology Publishing, United Kingdom (1995)
McKinlay, A.F.: Thermoluminescence Dosimetry. Adam Hilger, United Kingdom (1981)
Mehta, S.K., Sengupta, S.: Phys. Med. Biol. **21**, 955–964 (1976)
Mesterházy, D., Osvay, M., Kovács, A., Kelemen, A.: Radiat. Phys. Chem. **81**, 1525–1527 (2012)
Mindru, I., Gingasu, D., Patron, L., Marinescu, G., Calderon-Moreno, J.M., Diamandescu, L., Secu, M., Oprea, O.: Mater. Res. Bull. **85**, 240–248 (2017)
Moirangthem, N.S., Lisham, P.C., Barua, A.G., Gartia, R.K.: J. Phys. Conf. Ser. **1330**, 012009 (2019)
Mukherjee, B.: A LiF and BeO TLD based microdosimeter for space radiation risk assessment of astronauts. J. Instrum. **13**(06), T06002 (2018)
Murthy, K.V.R., Virk, H.S.: Defect Diffus. Forum **347**, 1–34 (2014)
Nam, Y.M., Kim, J.L., Chang, S.Y.: Radiat. Prot. Dosimetry **84**, 231–234 (1999)
Olko, P.: Radiat. Meas. **45**, 506–511 (2010)
Olko, P., Currivan, L., van Dijk, J.W.E., Lopez, M.A., Wernli, C.: Radiat. Prot. Dosimetry **120**, 298–302 (2006)
Omanwar, S.K., Koparkar, K.A., Virk, H.S.: Defect Diffus. Forum **347**, 75–110 (2014)
Osvay, M., Biró, T.: Nucl. Instruments Methods. **175**, 60–61 (1980)
Pagonis, V., Kitis, G., Furetta, C.: Numerical and Practical Exercises in Thermoluminescence. Springer, USA (2006)
Palilla, F.C., Levine, A.K., Tomkus, M.R.: J. Electrochem. Soc. **115**, 642–644 (1968)
Pokorny, P., Ibarra, A.: J. Phys. Condens. Matter **5**, 7387–7396 (1993)
Pradhan, A.S.: Radiat. Prot. Dosimetry **1**, 153–167 (1981)

Randall, J.T., Wilkins, M.H.F.: Proc. R. Soc. London A Math. Phys. Eng. Sci. **184**, 365–389 (1945)
Rivera, T.: Appl. Radiat. Isot. **71**, 30–34 (2012)
Rudén, B.I.: Acta Oncol. **15**, 447–464 (1976)
Ruiz-Torres, R., Chernov, V., Salas-Castillo, P., Zúñiga-Rivera, N.J., Diaz-Torres, L.A., Meléndrez, R., Barboza-Flores, M.: Opt. Mater. **101**. 109763 (1–7) (2020)
Saharin, N.S., Wagiran, H., Tamuri, A.R.: Radiat. Meas. **70**, 11–14 (2014)
Sakai, R., Katsumata, T., Komuro, S., Morikawa, T.: J. Lumin. **85**, 149–154 (1999)
Salah, N.: Radiat. Phys. Chem. **80**, 1–10 (2011)
Satapathy, K.K., Mishra, G.C., Kher, R.S., Dhoble, S.J.: RSC Adv. **5**, 79391–79396 (2015)
Schulman, J.H., Kirk, R.D., West, E.J.: Use of lithium borate for thermoluminescence dosimetry. In: Proceedings of the International Conference on Luminescence Dosimetry (1967)
Selyunina, L.A., Mishenina, L.N.: Adv. Mater. Res. **1085**, 130–133 (2015)
Singh, L.L., Gartia, R.K.: Radiat. Eff. Defects Solids **166**, 297–304 (2011)
Singh, V., Gundu Rao, T.K., Kim, D.K.: Radiat. Meas. **43**, 1198–1203 (2008)
Singh, M.N., Baruah, A.G., Gartia, R.K.: AIP Conf. Proc. **2220**, 020018 (2020)
Singh, M.N., Barua, A.G., Gartia, R.K.: Optik **228**, 166151–166159 (2021b)
Singh, M.N., Gartia, R.K.: AIP Conf. Proc. **2352**, 020087 (2021)
Singh, M.N., Singh, L.R., Barua, A.G.: Radiat. Phys. Chem. **188**, 109631 (1–13) (2021)
Smet, Van den Eeckhout, P.K., De Clercq, O., Poelman, D.: Persistent phosphors. In: Handbook on the Physics and Chemistry of Rare Earths, Including Actinides, vol. 48, Netherlands: Elsevier, 2015, pp. 1–108 (2015)
Sun, S., Wang, H., Yan, X.: Acc. Chem. Res. **51**, 1131–1143 (2018)
Takenaga, M., Yamamoto, O., Yamashita, T.: Nucl. Instrum. Methods **175**, 77–78 (1980)
Tang, K., Zhu, T., Shen, W., Liu, B.: Radiat. Prot. Dosimetry **100**, 239–242 (2002)
Teng, X., Zhuang, W., He, H.: Rare Met. **27**, 335–339 (2008)
Toshiyuki Nakajima, A.K., Murayama, Y., Matsuzawa, T.: Nucl. Instrum. Methods. **157**, 155–162 (1978)
Urbach, F.: Meas. Methods, Meet. Reports Acad. Knowl. Vienna. **139**, 363 (1930)
Van Den Eeckhout, K., Bos, A.J.J., Poelman, D., Smet, P.F.: Phys. Rev. B Condens. Matter Mater. Phys. **87**, 1–11 (2013)
Verma, N., Singh, K.C., Mari, B., Mollar, M., Jindal, J.: Acta Phys. Pol. a. **132**, 1261–1264 (2017)
Vohra, K.G., Bhatt, R.C., Chandra, B., Pradhan, A.S., Lakshmanan, A.R., Shastry, S.S.: Health Phys. **38**, 193–197 (1980)
Wako, A.H., Dejene, B.F., Swart, H.C.: Phys. B Condens. Matter. **439**, 153–159 (2014)
Wang, L., Wang, Y.: Phys. B Condens. Matter. **393**, 147–152 (2007a)
Wang, L., Wang, Y., Xu, X.: J. Appl. Phys. **104**, 1–6 (2008)
Wang, L., Chen, Y., Zhou, R., Xu, Y., Wang, Y.: ECS J. Solid State Sci. Technol. **1**, R72–R75 (2012)
Wang, Y., Wang, L.: J. Appl. Phys. **101**, 053108 (1–5) (2007)
Wei, J., Chen, L., Zhang, W., Yang, Y., Li, Z.: J. Lumin. **181**, 427–432 (2017)
Wiedemann, E.: Ann. Phys. **34**, 446–463 (1888)
Xie, W., Shao, L.X., Wang, Y.H., Zhang, J., Zou, C.W.: Adv. Mater. Res. **197–198**, 318–322 (2011)
Xu, X., Wang, Y., Gong, Y., Li, Y.: Electrochem. Solid-State Lett. **12**, 44–46 (2009a)
Xu, X., Wang, Y., Li, Y., Gong, Y.: J. Appl. Phys. **105**, 11–15 (2009b)
Xu, X., Wang, Y., Li, Y., Gong, Y.: Electrochem. Solid-State Lett. **13**, 126–128 (2010)
Yamamoto, H., Matsuzawa, T.: J. Lumin. **72–74**, 287–289 (1997)
Zawrah, M.F., Khalil, N.M.: Ceram. Int. **33**, 1419–1425 (2007)
Zhai, B., Huang, Y.M.: EPL. **127**, 17001(p1–p7) (2019)
Zhai, B., Xu, H., Zhuo, F., Huang, Y.M.: J. Alloys Compd. **821**, 153563 (1–11) (2020)
Zhang, T., Su, Q.: J. Soc. Inf. Disp. **8**, 27–29 (2000)
Zhang, B., Xu, X., Li, Q., Wu, Y., Qiu, J., Yu, X.: J. Solid State Chem. **217**, 136–141 (2014)
Zhang, M., Li, F., Lin, Y., Li, Y., Shen, Y.: J. Rare Earths **39**, 930–937 (2021a)
Zhang, M., Li, F., Jiang, S., Lin, Y.C., Chen, F., Zhao, X., Shen, Y.: Opt. Mater. **116**, 111049 (1–8) (2021)

Ziyauddin, M., Brahme, N., Bisen, D.P., Kher, R.S.: Int. Res. J. Eng. Technol. **2**, 871–876 (2015)
Ziyauddin, M., Tigga, S., Brahme, N., Bisen, D.P.: Luminescence **31**, 76–80 (2016)
Ziyauddin, M., Brahme, N., Bisen, D.P., Kher, R.S.: Int. J. Lumin. Appl. **9**, 586–589 (2019)

Chapter 17
Silicon-Based Nanocomposites as Photoluminescent Materials: A Perspective and Advantages of the Radiation-Assisted Synthetic Approach

Apurav Guleria and Soumyakanti Adhikari

17.1 Introduction

Silicon (Si) is a non-metallic element and a member of the carbon family (Group 14 [IVa]) of the periodic table. After oxygen, Si is the most abundant element (~27.7%) in the Earth's crust (Croissant et al. 2017; Ghosh and Shirahata 2014; Liang et al. 2019). The name *silicon* is derived from the Latin *silex* or *silicis*, which means "flint" or "hard stone." Si, in addition to being a semiconductor, possesses three important characteristics: natural abundance, low toxicity, and cost-effectiveness. Additionally, Si is although a trace element but belongs to an essential category as it is involved in our metabolic processes. Most people consume Si-based substances, like silica, through their vegetables, grains, and seafood (Croissant et al. 2017; Ghosh and Shirahata 2014; Liang et al. 2019). These characteristics of Si undoubtedly make it one of the most sought-after elements with applications of the utmost significance, particularly in the communication and electronics industries. However, optical applications of Si were not previously considered due to its indirect band gap (bulk crystalline Si band gap ~ 1.1 eV) wherein direct electron–hole pair recombination probability is low (Morozova et al. 2020; Song and He 2019; Cheng et al. 2014; Li and Jin 2017).

A. Guleria (✉)
Radiation and Photochemistry Division, Bhabha Atomic Research Centre, Mumbai 400085, India
e-mail: aguleria@barc.gov.in

S. Adhikari (✉)
Scientific Information Resource Division, Bhabha Atomic Research Centre, Mumbai 400085, India
e-mail: asoumya@barc.gov.in

A. Guleria · S. Adhikari
Homi Bhabha National Institute, Trombay, Mumbai 400094, India

R. S. Ningthoujam and A. K. Tyagi (eds.), *Handbook of Materials Science, Volume 1*, Indian Institute of Metals Series, https://doi.org/10.1007/978-981-99-7145-9_17

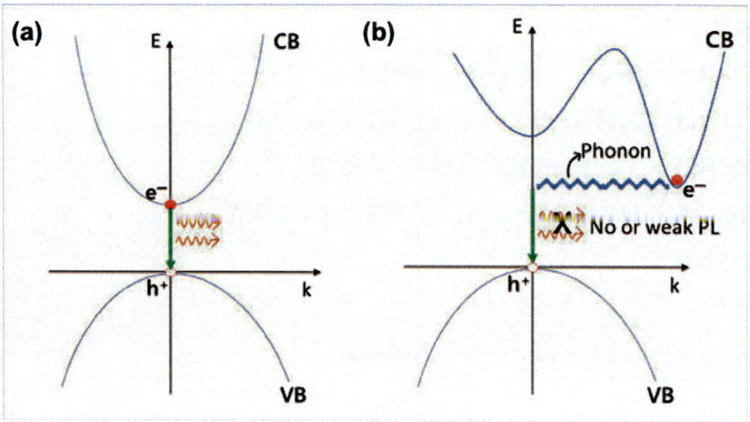

Fig. 17.1 Schematic illustration of energy (*E*) versus wave vector (*k*) plots for direct (**a**) and indirect (**b**) band gap semiconductors

As shown in Fig. 17.1, the minimum of the conduction band and the maximum of the valence band are located at distinct wavevectors in the reciprocal space. Radiative recombination across the band gap therefore requires at least a phonon, rendering it an inefficient 2nd-order process. Some of the common semiconductors used for manufacturing luminescent nanoparticles (NPs) are mentioned in Table 17.1 along with the nature of their band gap and band gap energy values.

Researchers have long desired to gain light from Si. Moreover, taking into account the advantageous properties of Si, there is no doubt that remarkably efficient light-emitting or amplifying materials based on Si will be of enormous value in both commercial applications and fundamental scientific research.

Canham's discovery of luminescence in porous Si in the 1990s provided evidence of alternations in its structure at smaller dimensions (Canham 1990). The significance of quantum confinement to the observed phenomenon was quickly recognized by the scientific establishment. Essentially, nanoscale dimensions alter the indirect character of the band gap, which raises the possibility that electron and hole wave functions will overlap in the Brillouin zone (Li and Jin 2017; Yu et al. 1999; Warner et al. 2005). These effects become noticeable in Si nanomaterials (NMs) when their dimensions approach the exciton Bohr radius of bulk Si (~4.2 nm) (Song and He 2019; Li and Jin 2017). The revelation of such a phenomenon led to a surge in the interest of the researchers to synthesize Si-based NMs with variable morphologies, composition, and surface properties. In the last three decades, researchers have made great efforts to comprehend and control the photoluminescence processes of Si-based NMs in an attempt to provide a substitute for quantum dots (QDs) manufactured of cadmium and lead as well as other hazardous and unstable organic dyes (Ghosh and Shirahata 2014; Cheng et al. 2014; Li and Jin 2017). Furthermore, theoretical predictions indicate that the emission tunability range of Si NPs could be one of the widest among commonly used semiconductors (see Table 17.1), ranging from 1.1 eV in

Table 17.1 Type of band gap and band gap energy values of common semiconductors

Material	Type of band gap	Band gap energy (eV) at 300 K
Si	Indirect	1.12
Ge	Indirect	0.66
ZnS	Direct	3.54
ZnSe	Direct	2.58
ZnTe	Direct	2.25
ZnO	Direct	3.35
CdSe	Direct	1.74
CdS	Direct	2.42
CdTe	Direct	1.5
InP	Direct	1.35
InAs	Direct	0.36
InN	Direct	0.70
GaAs	Direct	1.42
GaN	Direct	3.36
GaP	Indirect	2.26
α-SiC	Indirect	2.99
PbSe	Direct	0.27
PbS	Direct	0.37
PbTe	Direct	0.29

the near-infrared (NIR) to over 3.5 eV in the ultraviolet (UV), as can be observed from Fig. 17.2a. However, experimentally very few studies have demonstrated it, while such deviations from the predicted trend are often related to the presence of defects or strain. Nonetheless, with the advancement in the synthetic methods, the photoluminescence quantum yields (QYs), as well as the tunability exhibited by Si-based NMs, have improved a lot (Ghosh and Shirahata 2014; Cheng et al. 2014; Kang et al. 2007, 2009; Mastronardi et al. 2012a; Dasog et al. 2014; Dohnalová et al. 2014). For instance, there are few reports in which researchers have demonstrated an astonishingly wide range of tunability in the emission colour of Si NMs ranging from near UV to near-IR region (Fig. 17.2b, c) (Kang et al. 2009; Dasog et al. 2014; Dohnalová et al. 2014; Shirahata 2011).

The fabrication of Si NPs with high photoluminescence quantum efficiency (QE) (up to ~90%) and colour-tunable emission has accelerated research and development of a broad spectrum of luminescent Si-based NMs, such as, organosilica, silsesquioxane, silica mixed oxide, mesoporous silica NPs and so on (Croissant et al. 2017; Ghosh and Shirahata 2014; Cheng et al. 2014; Shirahata 2011; Teo and Sun 2007). In recent times, these NMs have been explored as active materials for their applications in sensing, thermoelectronics, light-emitting diodes (LEDs), phototherapeutics, lasers, solar cells, photodetectors, biomedical imaging, and drug delivery (Croissant et al. 2017; Ghosh and Shirahata 2014; Liang et al. 2019; Song

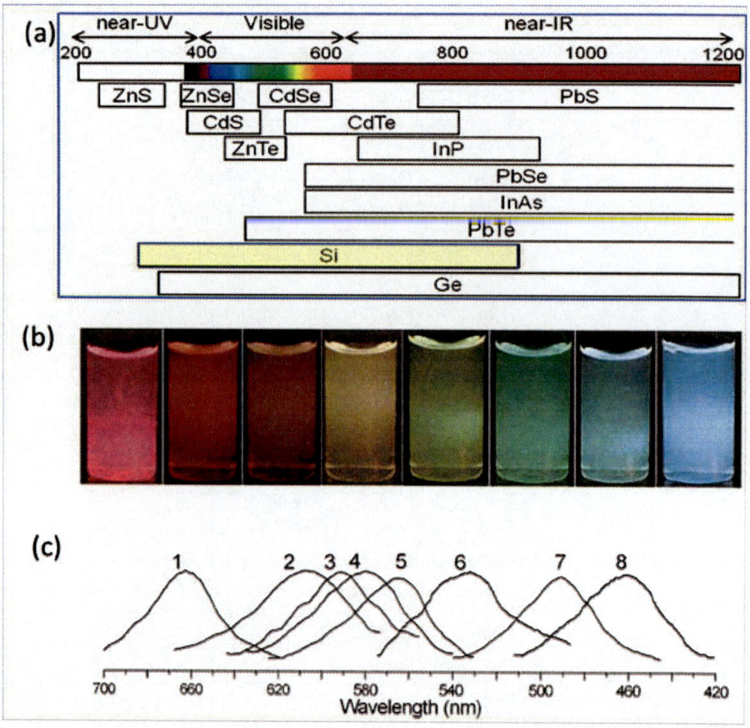

Fig. 17.2 **a** Potential band gap energy tunability for representative semiconductor nanocrystals via size control. Reproduced from Shirahata (2011) with permission from the RSC Publisher. **b** Photograph of H-terminated Si QDs (left, red emission) and seven water-soluble Si QDs under UV light (exhibiting 7 distinct emission colours). **c** Photoluminescence spectra of H-Si QDs (curve 1) and Si QDs after 0.5, 1.5, 3.5, 6, 9, 14, and 24 h oxidization (curves 2–8), respectively (λ_{exc} = 360 nm). Images b and c: Adapted with permission from John Wiley and Sons (Kang et al. 2009)

and He 2019; Cheng et al. 2014; Teo and Sun 2007). In comparison to other conventional semiconductor-based luminescent QDs, the inherent advantages of Si make it a potentially better candidate for such a wide-ranging applications. However, in contrast to conventional QDs, the origin of the optical properties of Si-based NMs is still a matter of contention (Li and Jin 2017; Dasog et al. 2014; Dohnalová et al. 2014). This may be because the properties of Si and its NMs are in general extremely sensitive to the surface chemistry—to an extent not observed in other conventional semiconductors QDs.

In this article, the photoluminescence properties of Si-based NMs along with their origin and mechanism have been discussed. The role of quantum confinement, surface properties, and defects in influencing the photoluminescent behaviour of Si-based NMs has been highlighted. This is followed by a discussion of the various types of Si-based NMs and their synthetic methodologies. The advantages of the radiolytic approach in the preparation of Si-based NMs have been presented, and some of the recent works carried out are briefly discussed. In the subsequent section, the potential applications of Si NMs have been summarized, followed by the challenges and prospects that lie ahead in the field of Nano-Si research.

17.2 Photoluminescence in Si NMs: Size Versus Surface

Typically, conventional QDs are composed of semiconductors with a direct band gap, and follow the effective mass approximation (EMA) model where the variation of the optical band gap (E_g) with the size (d) of the QDs can be fitted by the Brus equation (Ghosh and Shirahata 2014) shown below,

$$E_d = E_g + \frac{(h\pi)^2}{2d^2}\left[\frac{1}{m_e} + \frac{1}{m_h}\right] - \frac{1.786e^2}{\varepsilon_r d} \tag{17.1}$$

where 'd', 'e', and 'ε_r' represent the size/diameter of the QD, the electron charge, and the relative permittivity, respectively. The effective masses of the electron (e) and the hole (h) are represented by m_e and m_h, respectively. However, as mentioned already, unlike the conventional direct band gap semiconductors, bulk Si possesses an indirect band gap. Consequently, bulk Si can hardly emit light. But the scenario changes when the average size of the Si NPs is lesser than the Bohr exciton radius (Li and Jin 2017; Dasog et al. 2014; Dohnalová et al. 2014; Canham 2020; Gupta et al. 2009). Essentially, the delocalization of carriers in k-space caused by geometrical confinement at the nanoscale allows for zero phonon optical transitions, and greatly improves their (i.e., zero phonon transitions) oscillator strength in small Si nanocrystals (probable mechanism illustrated in Fig. 17.3). Not only do the NPs exhibit photoluminescence, but the emission colour could be varied with their size (Ghosh and Shirahata 2014; Dasog et al. 2014; Dohnalová et al. 2014; Canham 2020; Gupta et al. 2009). Various methods have been adopted by the researchers for tuning the emission of Si NMs across the whole NUV–VIS–NIR region, as shown in Fig. 17.4. The mechanism behind this incredible photoluminescent behaviour is still under debate.

For better understanding, the photoluminescence pattern of Si NMs has been categorized into two regions: NUV–blue ($\lambda_{em} = 300$–500 nm) and green–NIR emission ($\lambda_{em} = 500$–950 nm) region. In general, the emission in the green-NIR region has been attributed to the quantum confinement effect, whereas NUV–blue emission is assigned to be originated from oxide-based defects (Ghosh and Shirahata 2014; Dasog et al. 2014; Dohnalová et al. 2014).

According to the quantum confinement model, the band gap of NPs widens as its size shrinks. This can be noticed from the blue shift in the photoluminescence maximum of nanocrystalline Si/SiO_2 nanocomposites with increasing HF etching time (Fig. 17.5a), which is attributed to the quantum confinement effects resulting from decreased particle size (Hessel et al. 2006). Another study by Hessel et al.

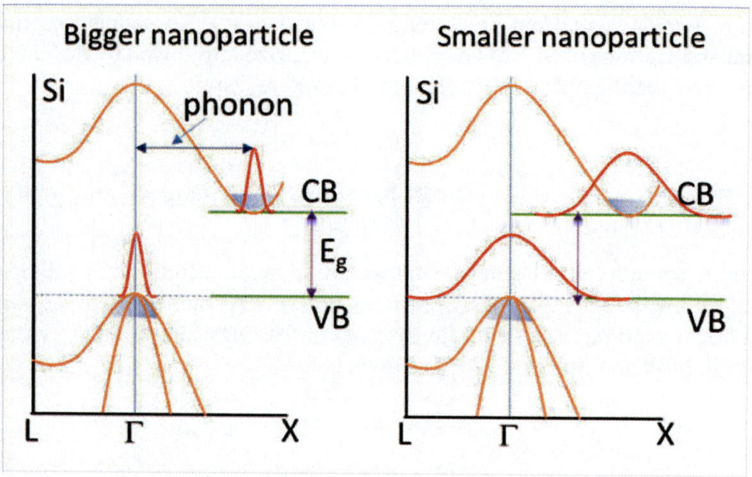

Fig. 17.3 Schematic illustration of the possible emission scenarios according to the quantum confinement model. The valence and conduction bands are denoted as "VB" and "CB", respectively. The 'Gamma (Γ)' point refers to the centre of the Brillouin zone

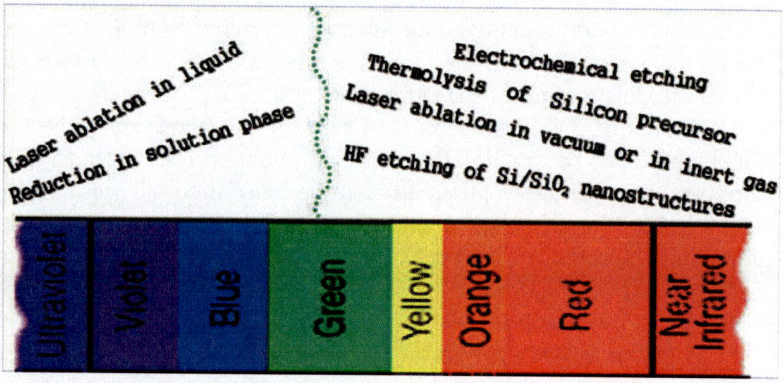

Fig. 17.4 The emission range exhibited by Si QDs prepared by different methods. Reprinted with permission from Taylor & Francis Publisher under Creative Commons Licence (Ghosh and Shirahata 2014)

(2012) showed that as the size of Si NPs decreased from about 10 nm to <2 nm, the photoluminescence peak wavelength blue shifted from 1000 to 600 nm (Fig. 17.5b). Aside from this, the reported influence of size on the photoluminescence QE also provided vital evidence for the existence of quantum confinement effects in Si NPs. Mastronardi et al. (2012b) observed high photoluminescence QE reaching up to ~43% in the near-infrared region for a larger size NPs (about 2.5 nm, as measured from PXRD). However, a monotonic decrease in the QE, as well as a blue shift

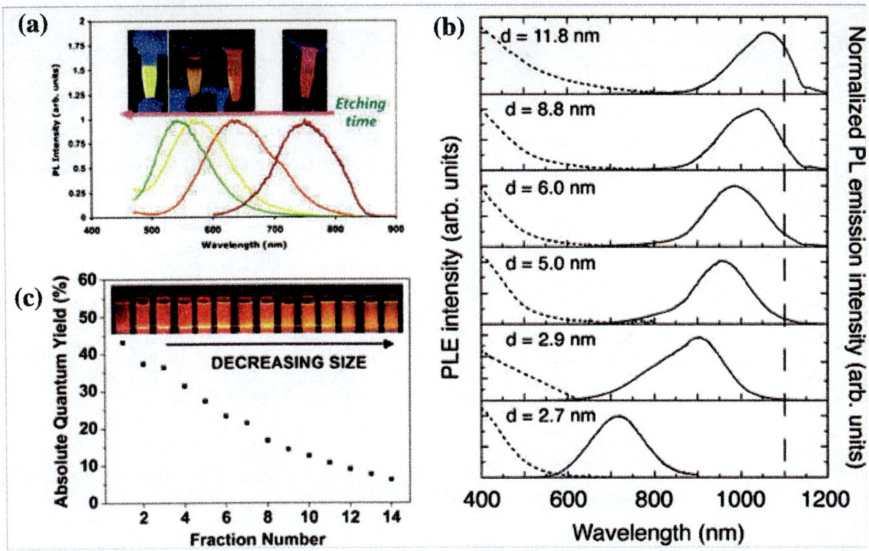

Fig. 17.5 **a** Photoluminescence spectra of pentane solutions of red-, orange-, yellow-and green-emitting Si QDs. Inset: photos of Si QDs pentane suspension in UV light (365 nm). Reprinted (adapted) with permission from Hessel et al. (2006). Copyright (2006) American Chemical Society. **b** Room temperature photoluminescence ($\lambda_{exc} = 420$ nm) and PLE (measured at emission maximum) spectroscopy of Si nanocrystals with different sizes. Reprinted (adapted) with permission from Hessel et al. (2012). Copyright (2012) American Chemical Society. **c** Size-dependent photoluminescence absolute quantum yields of Si NPs with decreasing sizes. Reprinted (adapted) with permission from Mastronardi et al. (2012b). Copyright (2012) American Chemical Society

of photoluminescence to the visible regime, was noticed by these researchers with further reduction in the size of Si NPs (Fig. 17.5c).

However, as mentioned earlier, unlike the case of direct band gap semiconductor QDs, photoluminescence spectral tuning in the case of Si NMs cannot only be discussed based on the quantum confinement effect. Essentially, Si is highly prone to oxidation. Consequently, oxidation of Si NMs results in the formation of oxygen-related surface/interface states within the band gap, which eventually plays a significant role in influencing their photoluminescence properties. For example, Gupta and Wiggers (2011) reported the photoluminescence transformation of freshly etched Si QDs (fabricated by etching) from red to blue region as a result of oxidation (see Fig. 17.6a, b). These researchers concluded from their experiment that the blue emission was caused by defect states at the newly formed Si oxide surface, whereas red emission was connected to the quantum confinement effect. In addition, the optical properties of Si NMs are highly sensitive to the chemical nature of the surface motifs. Several studies have shown that the emission spectrum of Si QDs may shift as a result of surface functionalization. Dasog et al. (2014) reported the preparation of 3–4 nm size colloidal Si nanocrystals as well as strategies for their functionalization with amine, alkyl, acetal, and phosphine groups (Fig. 17.6c, d). The main highlight of this

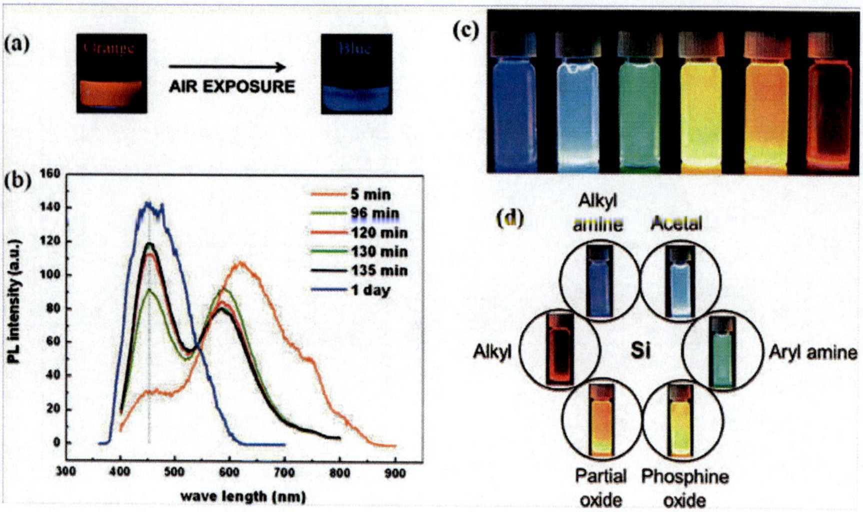

Fig. 17.6 **a** Photo of Si QDs dispersion under UV light: freshly etched (left) and after 1 day of air exposure (right). **b** Photoluminescence spectra of freshly etched Si QDs as a function of air exposure time. Images a and b: Adapted with permission from Gupta and Wiggers (2011). Copyright IOP Publishing, 2011. All rights reserved. **c** Photograph of toluene dispersed Si nanocrystals functionalized with various surface groups under UV light. From left to right, the functionalization of the Si nanocrystals is as follows: blue, dodecylamine; blue-green, acetal; green, diphenylamine; yellow, TOPO; orange, dodecyl (air); red, dodecyl (inert). **d** Photo showing the H-terminated Si NPs with different organic moieties and the corresponding emission colours. Images c and d: Reprinted (adapted) with permission from Dasog et al. (2014). Copyright (2014) American Chemical Society under Creative Commons License

work was the effective tuning of Si nanocrystals photoluminescence across the entire visible spectral range by rational changes in the surface moieties while restricting the variations in the particle size.

Undoubtedly, surface modification of Si QDs with an organic monolayer can suppress long-term oxidation while imparting more stable photoluminescence properties. Additionally, it has been noticed that Si NMs prepared using colloidal solution methods typically exhibit blue-green, whereas red emission is primarily obtained when using high temperature or etching methods (Ghosh and Shirahata 2014; Li and Jin 2017; Dasog et al. 2014; Dohnalová et al. 2014; Hessel et al. 2012; Gupta and Wiggers 2011). Furthermore, a short fluorescence lifetime (on the order of ns) in Si NMs is frequently attributable to the presence of defects or surface sites. In contrast, the quantum confinement state of conventional Si QDs exhibits longer lifetimes on the order of microseconds.

To summarize, the origin of photoluminescence in Si NMs is still controversial, however, there is consensus that photoluminescence is generally determined by two factors: the quantum confinement effect and surface/defect states surrounding the NPs. Nevertheless, the unique intrinsic photoluminescence properties of Si nanostructures, such as, bright emission, long fluorescence lifetime, wavelength-tunable luminescence, and photostability, make them desirable for biomedical imaging, efficient LEDs, optical amplifiers, and sensors.

17.3 Different Types of Luminescent Si-Based NMs

The versatile and promising inherent properties of Si have attracted tremendous attention in the nano-world. Therefore, numerous reports could be found in the literature on the fabrication of Si-based NMs using a range of physical and chemical synthetic techniques (Croissant et al. 2017; Ghosh and Shirahata 2014; Liang et al. 2019; Morozova et al. 2020; Cheng et al. 2014; Dohnalová et al. 2014; Teo and Sun 2007; Hessel et al. 2006; Guleria et al. 2017, 2020, 2021a, b, 2022). Apart from the morphology, and the surface characteristics, the composition of Si-based NMs holds great significance in deciding their applications. For instance, mesoporous silica (SiO_2) NPs gained attention due to their huge potential in the areas of drug delivery, and catalysis (Croissant et al. 2017; Morozova et al. 2020; Song and He 2019). Another category of Si-based NMs is silsesquioxanes, which are basically cage-like or polymeric materials that have Si–O–Si linkages along with tetrahedral Si vertices and organic moieties (Croissant et al. 2017; Morozova et al. 2020; Song and He 2019; Hessel et al. 2006). Silsesquioxanes are generally represented with the general formula, $(RSiO_{1.5})_n$—where R represents an organic group, i.e., alkoxy, aryl, alkyl, or H. These NMs have a range of properties, i.e., high thermal and chemical resistance. Other Si-based NMs include organic–inorganic siliceous NPs that have been designed to acquire desirable properties of the siloxane matrix. Figure 17.7 illustrates various groups of bio-based degradable siliceous NMs (Croissant et al. 2017). Herein, a brief overview summarizing the recent works on the synthesis of different types of photoluminescent Si-based NMs has been presented.

Since the first discovery of photoluminescence from porous Si, researchers have made directed efforts to devise strategies for manufacturing highly luminescent Si-based NMs with variable dimensions and compositions. For instance, Nishimura et al. (2013) developed a novel method based on chemical etching to synthesize photostable fluorescent zero-dimensional (0D) Si NPs for fluorescence imaging and single-molecule tracking. Li et al. (2016) synthesized nitrogen-capped 0D Si NPs with high photoluminescence QE of up to 90% and narrow emission bandwidth. In contrast to 0D NMs, 1D nanostructures (in general) have been observed to possess higher multiexciton generation efficiency, which is a vital requirement for innovating high-performance optical sensors. Taking this into account, 1D Si NMs could be very promising for the development of high-performance optical sensors. Few reports are there on the synthesis of high-quality fluorescent 1D Si nanostructures, which may be due to relatively complicated synthetic procedures. Song et al. (2016) prepared size-tunable fluorescent Si nanorods (NRs) with a relatively high photoluminescence QE of ~15% and excellent photostability through a microwave synthesis route. In addition to 0D and 1D NMs, various other luminescent nanocomposites of Si have been reported by researchers. For example, luminescent Si carbide (SiC) NPs with

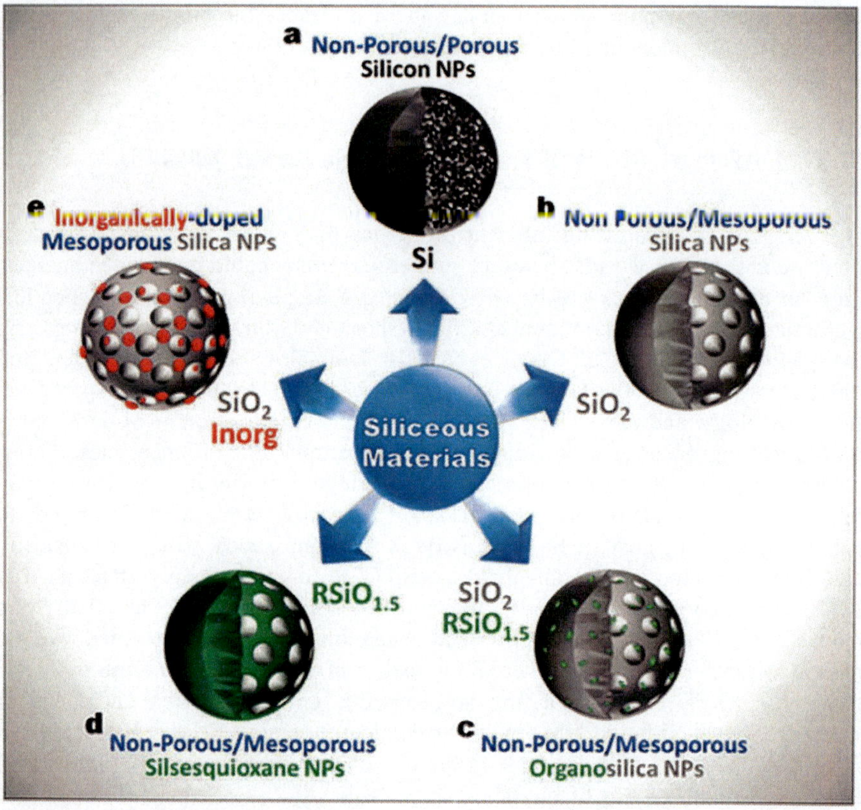

Fig. 17.7 Schematic illustration of various families of bio-related degradable siliceous NMs: **a** silicon, **b** silica, **c** organosilica, **d** silsesquioxane, and **e** inorganically doped mesoporous silica NPs. Reprinted with permission from Croissant et al. (2017) ©2017 Wiley-VCH Verlag GmbH & Co. KGaA, Weinheim

excellent stability, optical properties, and water solubility were synthesized by Yang et al. (2020) using a dielectric barrier discharge (DBD) plasma-based method. Dasog and Veinot (2012) reported the preparation of highly luminescent, freestanding Si nitride nanocrystals via in situ nitridation of magnesium followed by metathesis with silica particles (sol–gel derived). Luminescent polyhedral oligomeric silsesquioxane (POSS) nanostructures have found applications across various fields, especially as sensors with high sensitivity, selectivity, and stability. Zhao et al. (2018) used simple "Azide-acetylene" click chemistry reaction to prepare white-light-emitting POSS-based single molecular nano-hybrid. Chandra et al. (2017) reported one-pot facile preparation of highly fluorescent water-soluble silica NPs and demonstrated their applications in bioimaging. One-step hydrothermal synthesis of blue emitting polymer-like coated organosilica NPs (OSiNPs) (QE ~ 73.3%) was reported by Liang et al. (2021) and illustrated their applications for white-light-emitting diodes and fingerprint recognition.

With this brief overview of photoluminescent Si-based NMs with different dimensions and compositions, their futuristic demand in various areas of energy, healthcare, sensing, and the environment is bound to increase. Therefore, it is imperative to explore and further improvise the existing synthetic protocols for Si-based NMs with more eco-friendly traits and large-scale production facilities. The existing synthetic strategies primarily employed by numerous researchers for the preparation of photoluminescent Si-based NMs have been discussed in the following section with a brief mention of their advantages and disadvantages.

17.4 Synthesis of Si-Based NMs

Numerous studies demonstrating progress in the preparation of Si-based NMs with desirable morphologies and compositions can be found in the literature. The methodologies can generally be categorized into physical and chemical methods. Generally, the majority of physical routes are top-down, involving primarily plasma-assisted synthesis and laser ablation. While both top-down (such as, electrochemical etching and decomposition of precursors) and bottom-up protocols are used in chemical methods (such as, Zintl phases oxidation and reduction of Si halides). These strategies are represented in Fig. 17.8.

In addition to the physical and chemical-based methodologies, the biogenic approach is also being exploited by researchers in recent times. A brief discussion of the strategies mainly used in the synthesis of Si-based NMs has been provided below.

(i) **Laser pyrolysis**: Laser pyrolysis is one of the primary techniques for generating freestanding Si NMs. Herein, laser irradiation leads to a high temperature of up to 1000 °C near the point where the beam intersects the silane gas, resulting in Si nanocrystals formation. Werwa et al. (1994) prepared photoluminescent Si QDs by laser pyrolysis approach wherein Si wafer was ablated at very low pressure, yielding 2–3 nm Si QDs. There are various other reports in which the formation of colour-tunable Si QDs has been shown as a function of their size by varying the parameters of ablation in a gaseous helium environment (Ghosh and Shirahata 2014; Liang et al. 2019; Morozova et al. 2020; Cheng et al. 2014). Nonetheless, despite the initial success of this strategy, Si NMs with strong photoluminescence were formed only in a few cases, especially where the pyrolysis products were further etched with HF. Moreover, the yield of the final product was generally very low.

(ii) **Plasma-assisted synthesis**: In this method, precursor molecules, such as SiH_4, undergo dissociation by plasma electrons at high temperatures. Subsequently, the nucleation, and growth of the particle core through anion-molecule interactions occur. As particle density increases and ion density

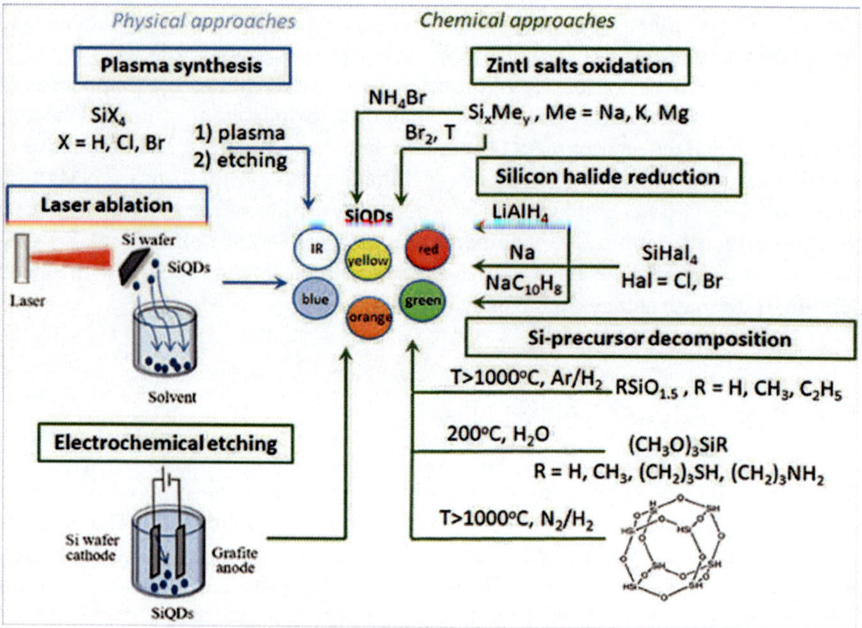

Fig. 17.8 Schematic representation of the typical synthetic strategies of Si NMs. Reprinted with permission from Morozova et al. (2020). Copyright 2020 Frontiers in Chemistry under Creative Commons License

decreases, the growth rate slows. This is a significant distinction from other pyrolysis methods (e.g., laser or thermal) in which there is no mechanism to stop particle growth. Nonetheless, the plasma method has been used to produce both embedded (in thin films) and freestanding Si NPs (Ghosh and Shirahata 2014; Liang et al. 2019; Morozova et al. 2020; Cheng et al. 2014; Shirahata 2011). This approach, however, requires the use of specialized instrumentation and the involvement of stringent experimental conditions.

(iii) **Microwave-assisted synthesis**: Microwaves are the electromagnetic waves with frequencies ranging from 300 MHz to 300 GHz. Although the energy of a microwave photon is low, the corresponding radiations are used for efficient heat transfer through dielectric heating. The microwave-assisted technique has a significant place today in the domain of nanoscience and nanotechnology. Researchers have used this technique for the synthesis of various Si-based NMs (Ghosh and Shirahata 2014; Liang et al. 2019; Morozova et al. 2020; Cheng et al. 2014; Shirahata 2011; Gupta et al. 2009). For example, hollow nano-silica spheres with porous morphology have been reported by using this approach. Further, rice husk was utilized as an inexpensive precursor for producing Si NPs with a QE of ~60% by rapid microwave heating (Ghosh and Shirahata 2014; Liang et al. 2019; Morozova et al. 2020; Cheng et al. 2014; Shirahata 2011). These materials showed strong photostability and storage stability. The microwave-assisted approach has many advantages, i.e., quick reaction times, low energy consumption, and high reaction efficiency. However, it also has some limitations, such as, high costs for dedicated microwave reactors and scale-up issues.

(iv) **Electrochemical etching**: In general, the electrochemical preparation of Si-based NMs employs a Si wafer as the cathode, while graphite is used as an anode (Ghosh and Shirahata 2014; Liang et al. 2019; Morozova et al. 2020; Cheng et al. 2014; Kang et al. 2007; Shirahata 2011). Typically, porous Si is electrochemically etched with the aid of ultrasound and a solution of HF and H_2O_2. Etching leads to the creation of a 'luminescent colloidal suspension' containing Si NPs. Kang et al. (2007) reported another variant of the etching method that involved a graphite rod and Si wafer as anode and cathode, respectively. The novelty in their design was the use of electrolytes generally consisting of aqueous HF with HNO_3 or H_2O_2 and different additives (e.g., polyoxometalates) with a unique ability to simultaneously act as an electron donor and acceptor. By varying the current density, the morphology of the NPs could be controlled with 1–4 nm size range and emission peaks between 450 and 700 nm. Because of its facileness, this method quickly gained attention and is now widely used. However, the use of HF and other harsh conditions poses a high safety risk.

(v) **Zintl salt oxidation**: Herein, Zintl salts (Me_ySi_x, Me = Na, K, Mg, etc.) react with Si halides, ammonium bromide, or gaseous bromine in a boiling glyme solution (Ghosh and Shirahata 2014; Cheng et al. 2014; Shirahata 2011). Although this method offers the benefits of scalability and accessibility, it primarily results in Si nanocrystals formation with photoluminescence in the blue-green region of the spectrum.

(vi) **Reduction of Si halides**: This is a rapid approach to synthesize Si nanocrystals by reducing $SiCl_4$ with agents, such as, lithium aluminum hydride, sodium, sodium naphthalenide, or tetraethylorthosilicate (Ghosh and Shirahata 2014; Liang et al. 2019; Morozova et al. 2020; Cheng et al. 2014; Shirahata 2011). However, one of the main shortcomings of this methodology is the wide particle size distribution. Further, this method yields mainly blue luminescent colloidal Si nanocrystals, but high QE of up to 90% have been reported herein (Li et al. 2016).

(vii) **Decomposition of Si-containing precursors**: This strategy involves the decomposition of Si-containing precursors followed by re-assembly processes, which eventually led to the formation of Si NMs. For example, hydrothermal decomposition of 3-aminopropyl trimethoxysilane (APTMS), or 3-aminopropyl triethoxysilane (APTES) in the presence of reducing chemicals (i.e., $NaBH_4$, sodium citrate, and $LiAlH_4$) leads to Si QDs formation (Ghosh and Shirahata 2014; Cheng et al. 2014; Shirahata 2011). Nanocrystals with high QE of up to 65–85% have been reported by varying the experimental conditions, such as, reaction time, temperature, and the type of reducing agent as well as precursor material. Supercritical fluids have also been used for the thermal degradation of Si precursors into size-monodisperse Si nanocrystals.

Nonetheless, some limitations of this method are the requirement of high temperature, specialized instruments, and the use of toxic chemicals.

(viii) **Sol–gel method**: The sol–gel process, also known as the Stöber method, has been widely used to produce Si-based NMs, especially Silica NPs. It involves hydrolysis and condensation of metal alkoxides, such as, tetraethylorthosilicate (TEOS, $Si(OC_2H_5)_4$) in the presence of a catalyst, which is generally a base or an acid. Various modified versions of the sol–gel method have been explored by researchers to prepare luminescent Si-based NMs (Croissant et al. 2017; Ghosh and Shirahata 2014; Liang et al. 2019; Morozova et al. 2020; Cheng et al. 2014; Shirahata 2011). For example, photoluminescent nanotubes and nanodisks of silica have been synthesized using the reverse micelle sol–gel approach. Nevertheless, long reaction/processing time and scale-up production issues are some of the disadvantages of this method.

(ix) **Biogenic method**: Nature resembles a large "bio-laboratory," with plants, algae, fungi, yeast, and other biomolecule-based organisms. In recent times, these naturally occurring biomolecules have gained a lot of interest in the domain of nanoscience. Many researchers are exploring these biomolecules for the synthesis of NPs with various shapes and sizes, paving the way for innovating greener, safer, and eco-friendly synthetic protocols. The synthesis of strongly luminescent and water-dispersible Si NPs has been reported from silane compounds using green reducing agents, such as, *Citrus limon* (L.), *Fusarium oxysporum, Actinobacter sp. Bacterium*, etc. (Ghosh and Shirahata 2014; Liang et al. 2019; Morozova et al. 2020; Cheng et al. 2014; Shirahata 2011; Tiwari et al. 2019). Although there have been tremendous developments in the field of microorganism-produced NPs, longer synthesis times and poor control over the particle size and morphology are some of the limitations of this methodology.

(x) **Radiation-assisted approach**: The majority of the aforementioned nano-synthetic routes require corrosive, toxic, and/or flammable chemical reagents (such as, stabilizers, reducing agents, and precursors) and harsh experimental conditions (such as, high temperature, inert atmosphere, etc.). However, for the impending application of NMs, it is essential to design and develop sustainable methods that are economical, time-efficient, environment friendly, and scalable. In this context, the synthesis of NMs using ionizing radiations (e.g., e-beam, γ-rays) is a well-established approach that does not require any of the stringent experimental conditions mentioned above (Guleria et al. 2017, 2020, 2021a, b, 2022). Furthermore, by selecting appropriate experimental parameters (such as the dose rate, absorbed dose, and host matrix), the nucleation and growth of NPs can be influenced. Additionally, because the solutions can be irradiated in typical industrial accelerators or γ-ray plants that are typically used to sterilize medical kits and other items, this technique has the potential to produce NMs on a large scale. The mechanism of radiation-assisted formation of NPs is followed by the radiolysis of an aqueous solution, which is represented by the following equation:

$$H_2O \rightsquigarrow e^-_{sol}, H^\bullet, OH^\bullet, H_3O^+, H_2O_2, H_2$$

The solvated electrons, e^-_{solv} and $H^\bullet$ are strong reducing agents with redox potentials of −2.9 and −2.3 V, respectively. The $OH^\bullet$ radicals are highly oxidizing in nature with a redox potential of +2.7 V (in acidic media). In general, NPs formation proceeds via solvated electrons, e^-_{solv} reaction with the precursor ions. Of course, the yield of e^-_{solv} plays a crucial role in the NPs growth kinetics and depends primarily on the reaction medium. To scavenge $H^\bullet$ and $OH^\bullet$ radicals produced during aqueous solution radiolysis, *tert*-butanol is added to the reaction mixture. Their scavenging would allow only the e^-_{solv} to react with precursors, causing the reaction to proceed as follows:

$$CH_3(CH_3)_2COH + OH^\bullet(H^\bullet) \rightarrow {}^\bullet CH_2(CH_3)_2COH + H_2O\ (H_2)$$

Depending on the reaction matrix, other free radicals may react with the precursor. In some cases, even $OH^\bullet$ radicals have been attributed to the formation of NPs. Nonetheless, a diagrammatic depiction representing the general scenario of the radiation-mediated synthesis of NMs is shown in Fig. 17.9.

Owing to the advantages of the radiation-induced approach over conventional methods, many researchers have investigated its potential to produce a wide range of NMs (metals, metal oxides, semiconductors, ceramics, etc.) with distinct morphologies, compositions, and photophysical properties (Guleria et al. 2017). Surprisingly, despite its many advantages, the radiation-assisted approach for the preparation of Si-based NMs in aqueous media is relatively unexplored. Recently, Si-based NMs synthesis in aqueous media using γ-rays and electron-beam irradiation have been reported (Guleria et al. 2017, 2020, 2021a, b, 2022). Interestingly, Si NMs obtained comprised of unique features (such as, amorphous and porous characteristics), which otherwise require meticulous synthetic procedures via conventional methodologies. The highlights of the research work carried out in this regard are presented in brief in the following sections.

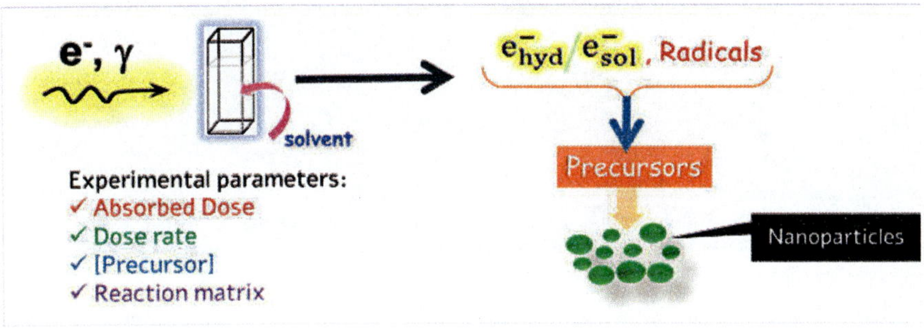

Fig. 17.9 Diagrammatic depiction of a general scenario representing the radiation (e^-, γ) induced synthesis of NMs

Si QDs coated with Si oxides (SiO_x, $0 < x < 2$) shell were obtained by γ-ray irradiation of an aqueous mixture of (3-Aminopropyl) trimethoxysilane (APTS) acting as a Si precursor (Guleria et al. 2017) (Fig. 17.10).

The NMs were found to be porous and amorphous, with an estimated average size of less than 2 nm. By varying the absorbed dose, the QDs size could be conveniently tuned. Further, the QDs exhibited blue photoluminescence under UV light. Comprehensive studies indicated that the blue photoluminescence could be attributed to the defects present in the oxide shell as well as at/or near the Si/SiO_x interface. The QDs were found to be non-cytotoxic and showed robust and highly sensitive thermosensing behaviour. Further, the potential use of these QDs was demonstrated in anticounterfeiting applications.

In another work, photoluminescent organosilicon oxide NPs (OSiNPs) were synthesized in aqueous media using electron beam (sourced from 7 meV LINAC) irradiation approach (see Fig. 17.11A) (Guleria et al. 2020).

This method allows for the commercial-scale production of NPs in timely manner without the use of toxic/hazardous reducing or stabilizing agents. Furthermore, pulse radiolysis studies were carried out to examine the mechanism of OSiNPs formation.

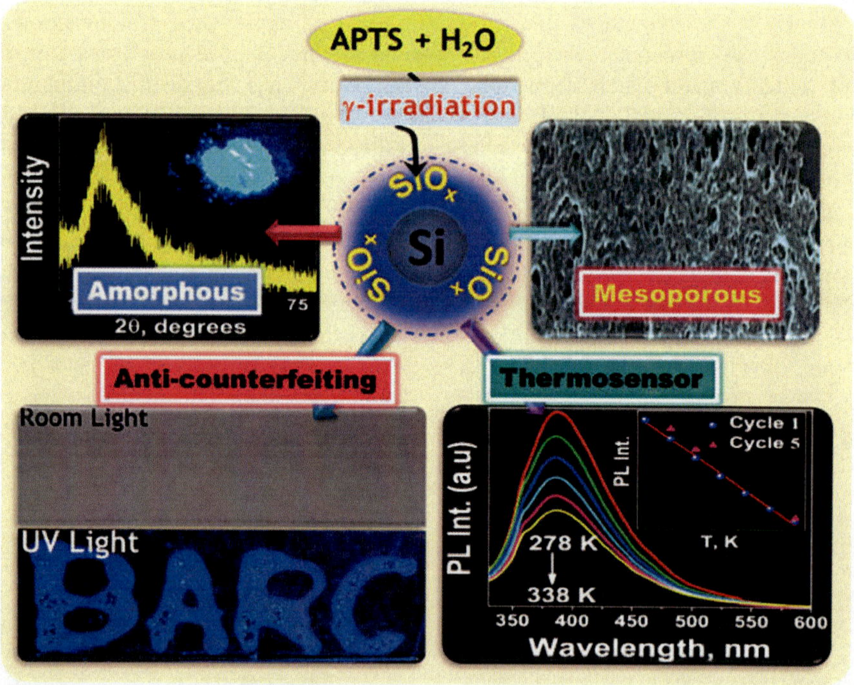

Fig. 17.10 Schematic representation of the γ-ray assisted synthesis of Si QDs (embedded within SiO_x shell, $0 < x < 2$). Other highlights of this work, including blue emission, amorphous and porous characteristics, as well as the potential applications of these NMs, are depicted. Reprinted (adapted) with permission from Guleria et al. (2017). Copyright (2017) American Chemical Society

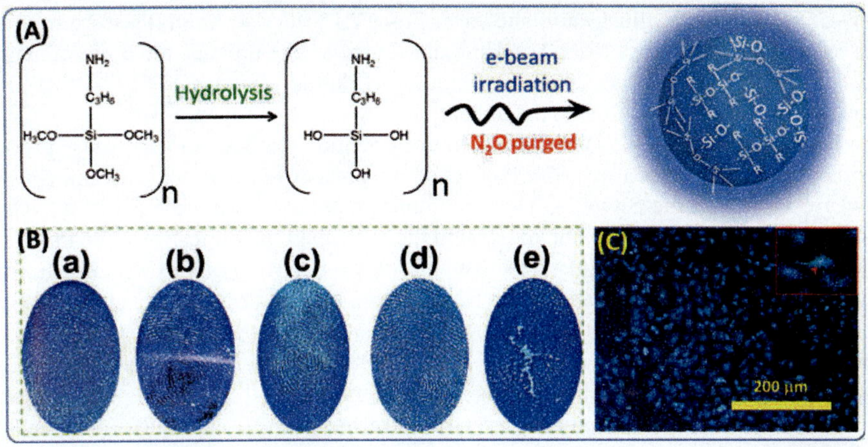

Fig. 17.11 **A** Schematic representation of the formation of OSiNPs by electron beam irradiation. **B** Photographs of fingerprints developed with L-Glu@OSiNPs (under UV light) on various objects: (a) Al foil, (b) glass beaker, (c) CD/DVD, (d) Al metal sheet, and (e) CD/DVD plastic cover. **C** Fluorescence microscopy image of A549 lung cancer cells using a colloidal solution of L-Glu@OSiNPs with a pH ~ 6.5. The inset image illustrates the nuclear localization of L-Glu@OSiNPs. Reprinted (adapted) with permission from Guleria et al. (2020). Copyright (2020) American Chemical Society

It was observed that, contrary to the generally reported solvated electron (e^-_{sol}) driven process (in the case of majority of other NMs), the synthesis of OSiNPs takes place through the reaction of the hydroxyl radical (•OH) with the precursor species. In addition to this, OSiNPs were functionalized with L-Glutathione (L-Glu or GSH), a biocompatible ligand, which significantly improved their emission QE (up to 25%) and colloidal stability. The applications of L-Glu@OSiNPs were demonstrated in fingerprinting (Fig. 17.11B) and cancer cell imaging (Fig. 17.11C). Nonetheless, it indicates their broad range of applications, which can be explored further as a replacement for costly and toxic materials currently in use in the respective areas.

As mentioned earlier, the optical properties of Si NMs are highly sensitive to the nature of ligands present on their surface. Various biocompatible ligands such as cyclodextrin, polyethylene glycol (PEG), and Triton X-100 (TX-100) were used to functionalize the Si NMs. Importantly, functionalization was carried out in situ during the irradiation of the reaction mixture as a one-pot approach. Although the same precursor (APTS) was used in the respective reaction mixtures, the composition and photoluminescence properties of the NMs differed significantly. For instance, SiO_2 NPs were formed in the presence of α-cyclodextrin (α-CD). The photoluminescence properties of as-synthesized α-CD functionalized SiO_2 NPs (i.e., α-CD@SiO_2 NPs) could be tuned from blue to green by merely changing the absorbed dose (see Fig. 17.12a). And, it was ascribed to the creation of a particular type of 'defects' at a proportionately absorbed dose (Guleria et al. 2021b). The maximum QE of α-CD@SiO_2 NPs was determined to be ~21%. Using the TX-100 micellar medium, the photoluminescence QE of radiolytically synthesized OSiNPs enhanced from ~9 to ~55% (schematic illustration shown in Fig. 17.12b). Also, the NPs were found to be comprised of silica (SiO_2) and siloxane/silicone like units as the predominant constituents (Guleria et al. 2021a). In another work, the presence of PEG-10000 in the reaction mixture resulted in the formation of Si oxide nanocomposites (SiONCs) on irradiation (Fig. 17.12c). Multiple characterization studies pointed to the presence of structures, such as, $(Si(CH_3)_2\text{-}O\text{-})_n$ (as present in polysiloxane) and SiO_x (where $x < 2$), in the SiONCs. The maximum photoluminescence QE of SiONCs was determined to be ~49%. In addition to the cell imaging applications, the PEGylated SiONCs exhibited highly sensitive and selective sensing of Cr (VI) ions (limit of detection, LOD ~ 0.74 μM), as can be seen from Fig. 17.12d (Guleria et al. 2022).

In recent times, there has been a lot of focus placed on Si-based NMs in the fields of nanoscience and nanotechnology. However, various challenges are associated with their production, thereby limiting their wider applicability. In this perspective, the radiation-assisted technique shows promise as a simple, time-efficient, environmentally benign, and cost-effective sustainable approach for the production of photoluminescent aqueous-soluble Si-based NMs.

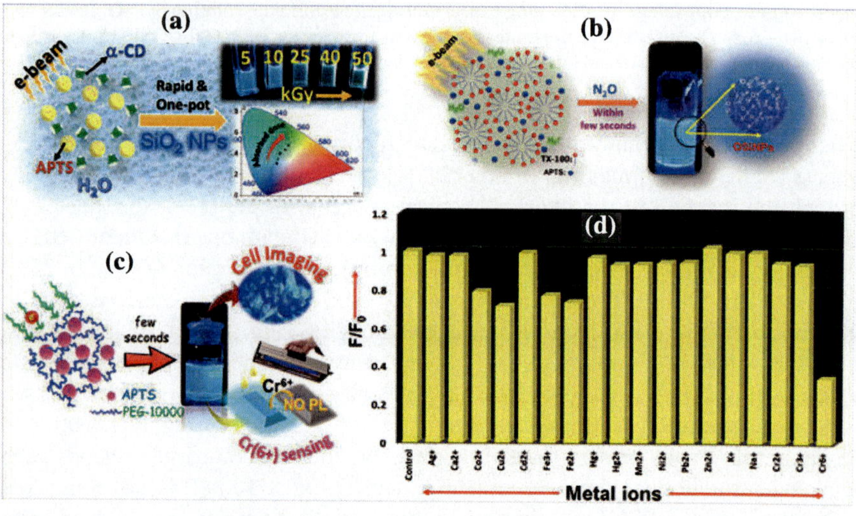

Fig. 17.12 **a** Schematic illustration of the radiation-assisted preparation of α-CD@SiO_2 NPs along with the absorbed dose-dependent photoluminescence tunability displayed by these NPs. Reprinted from Guleria et al. (2021b), Copyright (2020), with permission from Elsevier. **b** Schematic representation of the synthesis of blue photoluminescent OSiNPs in TX-100 micellar medium by electron-beam irradiation approach. Reprinted from Guleria et al. (2021a), Copyright (2021), with permission from Elsevier. **c** Schematic representation of the synthesis of PEGylated SiONCs by irradiation method along with their applications in Cr (VI) sensing and cell imaging. **d** Plot of F/F_0 versus metal ions, showing maximum photoluminescence quenching of PEGylated SiONCs by Cr (VI) ions. Reprinted from Guleria et al. (2022), Copyright (2022), with permission from Elsevier

17.5 Perspective Applications of Si-Based NMs

As previously stated, the NMs of Si have generated tremendous interest in the scientific community owing to their key properties, such as, abundance, less toxicity, cost-effectiveness, and compatibility with CMOS technologies. Si NMs have emerged as a viable alternative to other semiconductor NMs in a variety of healthcare, energy, and environmental applications. Nonetheless, the primary potential applications of Si-based NMs based on their photoluminescence properties are discussed below.

(a) **Light-Emitting Diodes (LEDs)**

Although many QD-based optoelectronic structures, such as LEDs or displays, have been realized over the last two decades using direct band gap materials. However, the majority of these materials are expensive, less abundant (Se, Te, In), and highly toxic (containing elements such as As, Pb, and Cd). All these aspects give an advantageous edge to Si-based NMs (Ghosh and Shirahata 2014; Morozova et al. 2020; Cheng et al. 2014; Mastronardi et al. 2012a; Dohnalová et al. 2014; Teo and Sun 2007). The foremost aspects of LEDs are luminance (L), turn-on voltage (V_T), and external quantum efficiency (EQE), which is linked with the diode efficiency. EQE can be represented by the equation shown below (Morozova et al. 2020):

$$\text{EQE}\ (\%) = \frac{q \times P}{J \times E} \times 100 \qquad (17.2)$$

where the terms q, J, P, and E are the electron charge, current density, optical power density, and photon energy emitted by the LED, respectively. In recent times, some of the best EQE values of red-emitting QD light-emitting diodes (QLEDs) with conventional as well as inverted structures have been reported to be 20.5% and 18.0%, respectively (Ghosh and Shirahata 2014; Morozova et al. 2020; Cheng et al. 2014; Mastronardi et al. 2012a; Dohnalová et al. 2014; Teo and Sun 2007). However, these values are obtained using cadmium-based QDs as optically active (or emission) layers. Compared to these, the work on Si NMs-based LEDs has garnered a lot of interest and improvisation in the last decade. For instance, the fabrication of electroluminescent devices employing size-separated Si nanocrystals (ncSi) as light-emitting material was reported by Maier-Flaig et al. (2013) (see Fig. 17.13). The EQE up to 1.1% was obtained for red-emitting ncSi. There are other reports also with a record value of EQE up to 8.6% for near-infrared emitting Si QDs (Ghosh and Shirahata 2014; Morozova et al. 2020; Cheng et al. 2014; Mastronardi et al. 2012a; Dohnalová et al. 2014; Teo and Sun 2007). Nonetheless, a lot of efforts are being put to increase the EQE by improvising the synthetic protocols with better control over the surface and the morphology of the Si NMs. In addition to that, several other ways are being explored by the researchers for improving LED characteristics, such as maneuvering the thickness of Si QDs, using an inverted structure device instead of a direct one, polyTPD layers (as the hole transport layer), and using aromatic ligands rather than the aliphatic ones (Morozova et al. 2020).

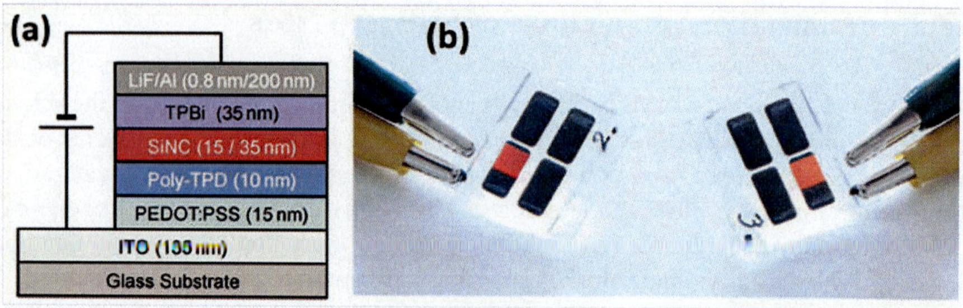

Fig. 17.13 **a** Schematic representation of ncSi-based LED stack. **b** Photographs of two encapsulated ncSi-based LEDs featuring strong red and orange luminescence that can be easily operated under ambient conditions for several hours. Reprinted (adapted) with permission from Maier-Flaig et al. (2013). Copyright (2013) American Chemical Society

(b) **Lasers**

Si-based lasers have the potential to radically transform future microchips, enabling light-speed processing and massive energy savings. However, despite concerted efforts by the scientific community, a satisfactory Si-based laser has not yet been developed. Essentially, the optical gains of Si are low due to its indirect band gap feature. Nonetheless, nanostructuring Si leads to an improvement in optical gain. Recently, Wang et al. (2018) reported the fabrication of an all-Si laser device which was optically pumped. The active medium comprised of hydrogen silsesquioxane (HSQ) with a high spatial density of Si nanocrystals, followed by hydrogen passivation at high pressure for a longer duration. Upon optical pumping, the active layer of the Si nanocrystals displayed very high optical gains. Characteristics of lasing suggested the realization of an optically pumped all-Si laser. However, there are still many challenges associated with the technical feasibility of lasers based on Si NMs. One of the primary obstacles is the broad size distribution, which results in a poor energy resonance between excited Si QDs (Dohnalová et al. 2014). Further, minimizing high scattering losses, and high QE Si NMs are some of the other prerequisites. Nonetheless, the pursuit for an all-Si laser remains an unresolved task and a challenging problem for the scientific community.

(c) **Sensors**

Si NMs provide a lot of scope in the development of efficient and cost-effective sensors. Various reports could be seen in the literature on the biological and chemical sensing applications of Si-based NMs. In addition to that, Si NMs have gained attention in sensing humidity and gas. For instance, the use of Si nanowires in the fabrication of chipless RFID (radio frequency identification) humidity sensor tags has been reported (Ahoulou et al. 2021). Furthermore, highly sensitive sensing of toxic heavy metal ions, such as, Cd^{2+} and Hg^{2+} ions by using field-effect transistors based on single Si nanowire has been shown. Si NMs with photoluminescence have been used as fluorescent probes for detecting low concentration levels of toxic heavy metal ions (e.g., Hg^{2+} detection, Fig. 17.14a), explosives containing nitroaromatic compounds, and various biomolecules (such as, DNA (see Fig. 17.14a), proteins, glucose, etc.) (Xie et al. 2014). Chu et al. (2016) reported the first demonstration for real-time and long-time intracellular measurement of pH in live cells by using fluorescent Si NPs-based sensors. As seen in Fig. 17.14b, pH sensors were developed using modified Si NPs that contained both pH-sensitive dopamine molecules and pH-insensitive rhodamine B isothiocyanate (RBITC) molecules. Wang et al. (2020) demonstrated europium (Eu^{3+})-doped Si NPs (Eu@SiNPs) based nanothermometer for sensing of intracellular temperature with high accuracy in living cells (Fig. 17.14c). Based on these exciting works, it can be stated that Si NMs can provide a cost-effective and reliable analytical platform for sensing/monitoring various compounds of distinctive nature.

(d) **Energy Conversion (Solar Cell)**

Si QDs have a lot of potential in photovoltaics (PV). Additionally, colloidal QDs have many benefits because they can be deposited using low-temperature and low-cost methods like spin coating, stamping, printing, spraying, etc. Various ways are being explored by the scientific community to improve the efficacy of solar cells based on Si NMs (Ghosh and Shirahata 2014; Mastronardi et al. 2012a; Dohnalová et al. 2014; Canham 2020; Pucker et al. 2012). These are briefly discussed as follows. By controlling the size of the QDs, band gap energy can be optimized to maximize the efficiency of solar cells (single-junction based). Meanwhile, efficiency can also be enhanced by arranging or stacking cells with different band gaps into a tandem pattern. Furthermore, to boost the efficiency of the solar cells, photoluminescence downshifters can be used with Si QDs with strong emission in the visible-red portion of the solar spectrum. In fact, Sgrignuoli et al. (2012) have demonstrated a 6% relative enhancement in solar cells efficiency by using Si QDs as spectral down-converters and shifters. In addition to these, it is well known that the relaxation of excited carriers in QDs is slowed down relative to bulk material as the distance between energy states increases (Ghosh and Shirahata 2014; Mastronardi et al. 2012a; Dohnalová et al. 2014; Canham 2020; Pucker et al. 2012). Therefore, solar cell's open-circuit voltage can be increased by extracting the charge carriers in the dot prior to their de-excitation. This concept is explored in the hot carrier solar cell designs. The formation of multiple excitons in Si QDs is another phenomenon that has been investigated for usage in solar cells. If these multiple excitons are effectively retrieved, they would result in an increased current in the solar cell (Pucker et al. 2012). Various reports on solar cells based on NMs derived from Si have been encouraging thus far; however, further in-depth research is required considering the benefits of Si.

(e) **Bioimaging**

Undoubtedly, fluorescent bioimaging is one of the most effective non-invasive methods for a wide range of biological and biomedical studies. For the aforementioned use, high-quality fluorescent probes are required, and they must have favourable biocompatibility, robust photostability, excellent aqueous dispersibility,

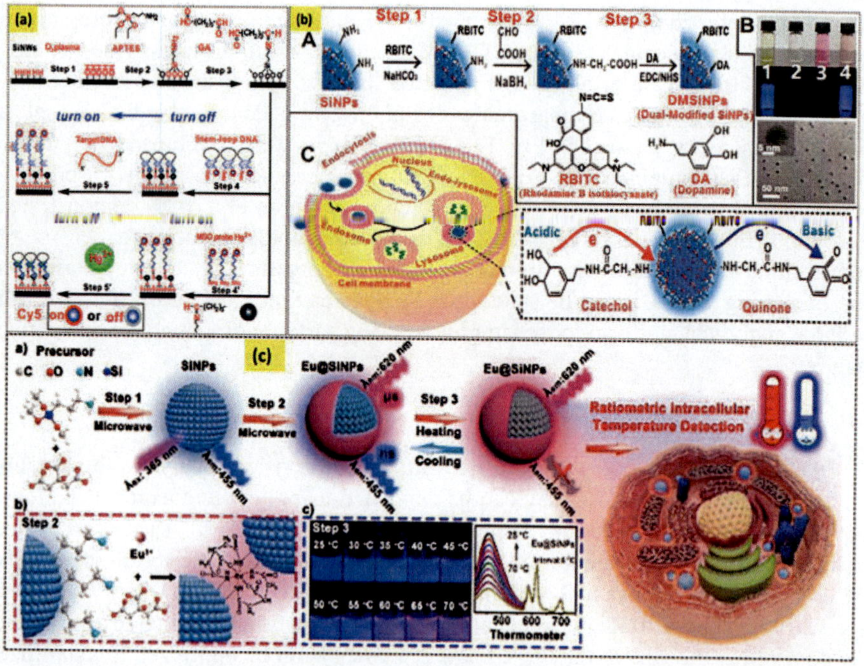

Fig. 17.14 **a** Schematic of the construction of a Si nanowire (NW)-based fluorescent sensor for the detection of DNA and Hg^{2+}. Reproduced from Xie et al. (2014) with permission from the Royal Society of Chemistry. **b** Construction of Si NPs based intracellular pH sensor. (A) Schematic illustration of the synthesis of Si NPs based-sensor. (B) photos of Si NPs (1), dopamine (2), rhodamine B isothiocyanate (RBITC) (3), and modified Si NPs (4) in an ambient environment and under UV irradiation ($\lambda_{ex} = 360$ nm) as well as TEM image of dual-modified Si NPs (DMSiNPs). Inset shows the enlarged HRTEM image of a single DMSiNPs. (C) Schematic illustration of cellular internalizations of DMSiNPs. Inset in part C presents the charge-transfer mechanism of DMSiNPs at acidic or basic conditions. Reprinted (adapted) with permission from Chu et al. (2016). Copyright (2016) American Chemical Society. **c** (a) Schematic illustration of fabricating Eu@SiNPs-based nanothermometer for intracellular temperature detection. The process consists of three typical steps. Step 1: synthesis of Si NPs. Step 2: fabrication of a dual-emission Eu@SiNP-based nanothermometer. (b) Local coordination process between SiNPs and Eu complex. Step 3: temperature response process of Eu@SiNPs and their application in the detection of temperature in a living cell. (c) Photographs under UV irradiation and corresponding photoluminescence spectra of Eu@SiNPs at different temperatures. Reproduced from Wang et al. (2020) with permission from the Royal Society of Chemistry

and high fluorescence quantum yield. The use of organic dyes in biological and biomedical research has generally been extensive, but their poor photobleaching ability makes them unsuitable for long-term and real-time bioimaging. In this pretext, the potential of photoluminescent Si NMs for bioimaging applications has garnered a lot of interest in recent times. For instance, Warner et al. (2005) demonstrated cellular uptake of blue photoluminescent Si QDs with no acute cellular damage detected. He et al. (2009) demonstrated the use of highly luminescent oxidized Si nanospheres

(O-SiNSs)/antibody bioconjugates for immunofluorescent cell imaging. Zhong et al. (2013) demonstrated long-term imaging of cell nuclei with Si NPs (for up to 60 min) with no photobleaching. As shown in Fig. 17.15A, Si NPs-labeled nuclei exhibited stable fluorescent signals over 60-min continuous monitoring, whereas the signals due to fluorescein isothiocyanate (FITC) fluorescent labels almost vanished shortly due to severe photobleaching. In another study, Wu et al. (2015) used *C. elegans* as well-known animal model to illustrate the feasibility of Si NPs-based dual-colour bioimaging in vivo. Figure 17.15B shows the digestive and reproductive systems of *C. elegans* that have been microinjected with blue- and red-emitting Si NPs. Further, the photoluminescence from the blue- and red-emitting Si NPs can be observed from the gut lumen and intestinal cells, and in the uterus of *C. elegans*, respectively. There are various other studies where photoluminescent Si NMs have been used for in vivo bioimaging. For example, Park et al. (2009) demonstrated in vivo imaging by injecting photoluminescent bare porous Si NPs (LPSiNPs) and dextran-capped LPSiNPs (D-LPSiNPs) dispersions into the left and right flanks of a nude mouse. As can be seen in Fig. 17.15C, a considerable amount of bare LPSiNPs was eliminated through renal clearance, while the remaining NPs accumulated in the spleen and liver. In contrast, D-LPSiNPs slowly accumulated and degraded in the liver in comparison to bare LPSiNPs. Nonetheless, based on the literature reports, it can be stated that photoluminescent Si-based NMs can be a great tool for noninvasive high-contrast and high-sensitivity optical imaging. However, there are some limitations also. For example, the majority of the synthetic strategies yielding luminescent Si-based NMs face the issue of poor aqueous dispersibility and thus require laborious procedures of further surface modification with hydrophilic ligands. Besides, scaling issues still exist with these methodologies. Nevertheless, in order to establish Si-based NMs as versatile candidates for bioimaging applications, it is essential to address these fundamental challenges.

17.6 Conclusions and Outlook

Nano-Si with its inherent advantages can offer a multitude of applications in the fields of healthcare, energy, and environment. This chapter provides an overview of Si, its significance, and nanoscale applications. A discussion of the origin of the photoluminescence properties of Nano-Si has been provided along with the role of quantum confinement, its surface properties, and defects. Various types of Si-based NMs have been discussed, especially those having photoluminescence properties. Synthetic strategies mainly developed for luminescent Si-based NMs have been presented with a special focus on the advantages offered by the radiation-assisted approach. The work carried out on the radiation-assisted synthesis of photoluminescent Si-based NMs using different bio-compatible ligands and their applications has been discussed in brief. In the last stage of this overview, various prospective applications of NMs based on Si have been presented. To summarize, an attempt has been made to provide a glimpse of the considerable signs of progress carried out in the field of

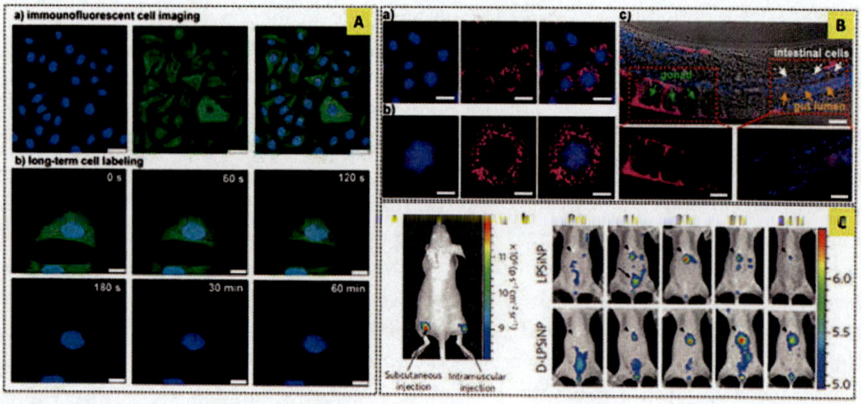

Fig. 17.15 **A** Photos of laser scanning confocal microscopy images of immunofluorescent cells. (a) Left: nuclei are distinctively labeled by Si NPs (excitation: 405 nm, detection window: 450–510 nm); middle: microtubules are distinctively labeled by FITC (excitation: 488 nm, detection window: 515–550 nm); and right: superposition of the two fluorescence images. (b) Time-dependent stability comparison of fluorescence signals of Hela cells labeled by Si NPs (blue) and FITC (green). Scale bar = 5 μm. Reprinted (adapted) with permission from Zhong et al. (2013). Copyright (2013) American Chemical Society. **B** Si NPs-based dual-colour bioimaging in vitro and in vivo. (a) and (b) Photos of dual-colour immunofluorescent cell images. (I: nuclei are distinctively labeled by blue-emitting Si NPs; II: microtubules are distinctively labeled by red-emitting Si NPs; III: superposition of the two fluorescence images. Scale bars of (a) are 20 μm. The scale bars of (b) are 10 μm.) (c) Representative confocal images of the as-prepared Si NPs in *C. elegans* by microinjection. The uterus (left) and gut lumen (right) are stained with red- and blue-emitting Si NPs, respectively (I). II and III are enlarged views of the corresponding to the boxed areas in I; I: scale bars = 20 μm; II and III: scale bars = 15 μm. Reprinted (adapted) with permission from Wu et al. (2015). Copyright (2015) American Chemical Society. **C** (a) In vivo fluorescence image of luminescent porous Si NPs (LPSiNPs) (20 μl of 0.1 mg ml^{-1}) injected subcutaneously and intramuscularly on each flank of a mouse. (b) In vivo images of LPSiNPs and D-LPSiNPs. The mice were imaged at multiple time points after intravenous injection of LPSiNPs and D-LPSiNPs (20 mg kg^{-1}). Arrowheads and arrows with solid lines indicate liver and bladder, respectively. Adapted/translated by permission from (Springer Nature Customer Service Centre GmbH): (Springer Nature) (Nature Materials) (Park et al. 2009), (Copyright) (2009)

Nano-Si ranging from its synthetic methodologies to prospective applications with special emphasis on its optical properties.

Despite advances in Nano-Si research, there are some issues that must be addressed to realize their luminescence-based practical applications. For example, a clear and convincing explanation of the photoluminescence mechanism of Nano-Si, which is still a matter of debate, is necessary in order to provide valuable direction for the synthesis of Si NMs with controllable fluorescence characteristics. Further, the development of eco-friendly and time-efficient methodologies for the large-scale synthesis of monodisperse Si NMs with high photoluminescence QE is highly desirable. In this perspective, the radiation-assisted approach may play a significant role. Moving ahead, the synthesis of novel Si precursors is needed not only for preparing high-quality NMs but also for improving the safety standards of the protocols. For example, most conventional chemical methods employ dangerous materials, such as, Si tetrahydride (silane) gas to create photoluminescent Si NMs. More exploration is required in designing and developing rational surface modifications of Si NMs with well-defined structures and special functionalities.

Undoubtedly, with the intense efforts being put into the preparation and processing techniques of Nano-Si, its commercialization may very well be realized in the near future, with a positive impact on both sustainable development and the global economy.

Acknowledgements The authors thankfully acknowledge the members of the RPCD LINAC facility for their help in the pulse radiolysis and irradiation-based experiments. The authors thank Dr. Awadhesh Kumar, Head, RPCD, and Dr. A. K. Tyagi, Director, Chemistry Group, BARC for their support and encouragement.

Author Contributions The manuscript was written through contributions of all authors. All authors have given approval to the final version of the manuscript.

References

Ahoulou, S., Perret, E., Nedelec, J.M.: Nanomaterials **11**, 999 (2021)
Canham, L.T.: Appl. Phys. Lett. **57**, 1046–1048 (1990)
Canham, L.: Faraday Discuss. **222**, 10–81 (2020)
Chandra, S., Beaune, G., Shirahata, N., Winnik, F.M.: J. Mater. Chem. B **5**, 1363–1370 (2017)
Cheng, X., Lowe, S.B., Reece, P.J., Gooding, J.J.: Chem. Soc. Rev. **43**, 2680–2700 (2014)
Chu, B., Wang, H., Song, B., Peng, F., Su, Y., He, Y.: Anal. Chem. **88**, 9235–9242 (2016)
Croissant, J.G., Fatieiev, Y., Khashab, N.M.: Adv. Mater. **29**, 1604634 (2017)
Dasog, M., Veinot, J.G.C.: Chem. Commun. **48**, 3760–3762 (2012)
Dasog, M., Reyes, G.B.D., Titova, L.V., Hegmann, F.A., Veinot, J.G.C.: ACS Nano **8**, 9636–9648 (2014)
Dohnalová, K., Gregorkiewicz, T., Kůsová, K.: J. Phys. Condens. Matter **26**, 173201 (2014)
Ghosh, B., Shirahata, N.: Sci. Technol. Adv. Mater. **15**, 014207 (2014)
Guleria, A., Neogy, S., Maurya, D.K., Adhikari, S.: J. Phys. Chem. C **121**, 24302–24316 (2017)
Guleria, A., Chavan, A.P., Neogy, S., Gandhi, V.V., Kunwar, A., Debnath, A.K., Adhikari, S.: ACS Appl. Nano Mater. **3**, 5123–5138 (2020)
Guleria, A., Tomy, A., Baby, C.M., Gandhi, V.V., Kunwar, A., Debnath, A.K., Adhikari, S.: J. Mol. Liq. **334**, 116072 (2021a)
Guleria, A., Chavan, A.P., Tomy, A., Baby, C.M., Neogy, S., Debnath, A.K., Adhikari, S.: Ceram. Int. **47**, 2649–2655 (2021b)
Guleria, A., Gandhi, V.V., Kunwar, A., Neogy, S., Debnath, A.K., Adhikari, S.: Colloids Surf. A Physicochem. Eng. **640**, 128483 (2022)
Gupta, A., Wiggers, H.: Nanotechnology **22**, 055707 (2011)
Gupta, A., Swihart, M.T., Wiggers, H.: Adv. Funct. Mater. **19**, 696–703 (2009)
He, Y., Su, Y., Yang, X., Kang, Z., Xu, T., Zhang, R., Fan, C., Lee, S.T.: J. Am. Chem. Soc. **131**, 4434–4438 (2009)
Hessel, C.M., Henderson, E.J., Veinot, J.G.C.: Chem. Mater. **18**, 6139–6146 (2006)
Hessel, C.M., Reid, D., Panthani, M.G., Rasch, M.R., Goodfellow, B.W., Wei, J., Fujii, H., Akhavan, V., Korgel, B.A.: Chem. Mater. **24**, 393–401 (2012)
Kang, Z., Tsang, C.H.A., Zhang, Z., Zhang, M., Wong, N.B., Zapien, J.A., Shan, Y., Lee, S.T.: J. Am. Chem. Soc. **129**, 5326–5327 (2007)

Kang, Z., Liu, Y., Tsang, C.H.A., Ma, D.D.D., Fan, X., Wong, N.B., Lee, S.T.: Adv. Mater. **21**, 661–664 (2009)
Li, Q., Jin, R.: Nanotechnol. Rev. **6**, 601–612 (2017)
Li, Q., Luo, T.Y., Zhou, M., Abroshan, H., Huang, J., Kim, H.J., Rosi, N.L., Shao, Z., Jin, R.: ACS Nano **10**, 8385–8393 (2016)
Liang, J., Huang, C., Gong, X.: ACS Sustain. Chem. Eng. **7**, 18213–18227 (2019)
Liang, J., Wu, Y., Gong, X., Vomiero, A.: J. Mater. Chem. C **9**, 1746–1754 (2021)
Maier-Flaig, F., Rinck, J., Stephan, M., Bocksrocker, T., Bruns, M., Kübel, C., Powell, A.K., Ozin, G.A., Lemmer, U.: Nano Lett. **13**, 475–480 (2013)
Mastronardi, M.L., Henderson, E.J., Puzzo, D.P., Ozin, G.A.: Adv. Mater. **24**, 5890–5898 (2012a)
Mastronardi, M.L., Maier-Flaig, F., Faulkner, D., Henderson, E.J., Kübel, C., Lemmer, U., Ozin, G.A.: Nano Lett. **12**, 337–342 (2012b)
Morozova, S., Alikina, M., Vinogradov, A., Pagliaro, M.: Front. Chem. **8**, 191 (2020)
Nishimura, H., Ritchie, K., Kasai, R.S., Goto, M., Morone, N., Sugimura, H., Tanaka, K., Sase, I., Yoshimura, A., Nakano, Y., Fujiwara, T.K., Kusumi, A.: J. Cell Biol. **202**, 967–983 (2013)
Park, J.H., Gu, L., von Maltzahn, G., Ruoslahti, E., Bhatia, S.N., Sailor, M.J.: Nat. Mater. **8**, 331–336 (2009)
Pucker, G., Serra, E., Jestin, Y.: Silicon quantum dots for photovoltaics: a review. In: Al-Ahmadi, A. (ed.) Quantum Dots—A Variety of New Applications, chap. 4, pp. 59–92. IntechOpen, London (2012)
Sgrignuoli, F., Paternoster, G., Marconi, A., Ingenhoven, P., Anopchenko, A., Pucker, G., Pavesi, L.: J. Appl. Phys. **111**, 034303 (2012)
Shirahata, N.: Phys. Chem. Chem. Phys. **13**, 7284–7294 (2011)
Song, B., He, Y.: Nano Today **26**, 149–163 (2019)
Song, B., Zhong, Y., Wu, S., Chu, B., Su, Y., He, Y.: J. Am. Chem. Soc. **138**, 4824–4831 (2016)
Teo, B.K., Sun, X.H.: Chem. Rev. **107**, 1454–1532 (2007)
Tiwari, A., Sherpa, Y.L., Pathak, A.P., Singh, L.S., Gupta, A., Tripathi, A.: Mater. Today Commun. **19**, 62–67 (2019)
Wang, D.C., Zhang, C., Zeng, P., Zhou, W.J., Ma, L., Wang, H.T., Zhou, Z.Q., Hu, F., Zhang, S.Y., Lu, M., Wu, X.: Sci. Bull. **63**, 75–77 (2018)
Wang, J., Jiang, A., Wang, J., Song, B., He, Y.: Faraday Discuss. **222**, 122 (2020)
Warner, J.H., Hoshino, A., Yamamoto, K., Tilley, R.D.: Angew. Chem. **117**, 4626–4630 (2005)
Werwa, E., Seraphin, A.A., Chiu, L.A., Zhou, C., Kolenbrander, K.D.: Appl. Phys. Lett. **64**, 1821 (1994)
Wu, S., Zhong, Y., Zhou, Y., Song, B., Chu, B., Ji, X., Wu, Y., Su, Y., He, Y.: J. Am. Chem. Soc. **137**, 14726–14732 (2015)
Xie, J., Jiang, X., Zhong, Y., Lu, Y., Wang, S., Wei, X., Su, Y., He, Y.: Nanoscale **6**, 9215 (2014)
Yang, W., Sun, M., Song, H., Su, Y., Lv, Y.: J. Mater. Chem. C **8**, 16949–16956 (2020)
Yu, D.P., Bai, Z.G., Wang, J.J., Zou, Y.H., Qian, W., Fu, J.S., Zhang, H.Z., Ding, Y., Xiong, G.C., You, L.P., Xu, J., Feng, S.Q.: Phys. Rev. B: Condens. Matter Mater. Phys. **59**, 2498–2501 (1999)
Zhao, G., Zhu, Y., Guang, S., Ke, F., Xu, H.: New J. Chem. **42**, 555–563 (2018)
Zhong, Y., Peng, F., Bao, F., Wang, S., Ji, X., Yang, L., Su, Y., Lee, S.T., He, Y.: J. Am. Chem. Soc. **135**, 8350–8356 (2013)

Chapter 18
Graphene and Graphene-Based Nanocomposites: From Synthesis to Applications

Heera Lal Kewat, Rahul Kumar Sharma, Ubaid Sidiqi, and Pushpal Ghosh

18.1 Introduction

Recently, graphene, its derivatives, and graphene-based nanocomposites have drawn huge curiosity in nanoscience and nanotechnology as they are being extensively employed for numerous applications including optoelectronic, magnetic, catalysis, photocatalysis, photothermal therapy, sensing, imaging and solar cells, etc. (Wei et al. 2012; Choi et al. 2010; Ambrosi et al. 2014; Mondal et al. 2021). As the chemical and physical properties can be tuned via soft chemical methods, graphene/graphene-based materials become an interesting area of research (Wei et al. 2012). It is well-known that graphene is a single atom thick, sp^2 hybridized, two-dimensional material with a honeycomb-like structure (Wei et al. 2012; Choi et al. 2010; Ambrosi et al. 2014; Mondal et al. 2021). Graphene has few overwhelming properties as compared to other nanoscale materials. It has high current density, very good thermal conductivity, ballistic transport, and chemical inertness (Choi et al. 2010; Ambrosi et al. 2014; Mondal et al. 2021; Geim and Kim 2008). In addition, it has good optical transmittance and super hydrophobicity at the nanometer regime too (Mondal et al. 2021; Geim and Kim 2008). However, it is highly susceptible for physical and chemical changes. To synthesize the graphene and graphene-based nanocomposites, numerous methods like mechanical exfoliation, chemical oxidation and reduction, CVD, and epitaxial growth on metals, etc. have been explored to date (Ambrosi et al. 2014;

H. L. Kewat and R. K. Sharma are contributed equally.

H. L. Kewat · U. Sidiqi · P. Ghosh (✉)
School of Chemical Sciences and Technology, Department of Chemistry, Dr. Hari Singh Gour University, Sagar 470003, M.P., India
e-mail: pghosh@dhsgsu.edu.in

R. K. Sharma
Department of Chemistry, Government Shyam Sundar Agrawal PG College, Sihora, Jabalpur 483225, M.P., India

R. S. Ningthoujam and A. K. Tyagi (eds.), *Handbook of Materials Science, Volume 1*, Indian Institute of Metals Series, https://doi.org/10.1007/978-981-99-7145-9_18

Jariwala et al. 2011). Though graphene is usually a zero-band gap material; its band gap can be judiciously controlled by altering the size, functionalization onto its surface with various organic and inorganic moieties, etc. (Choi et al. 2010; Ambrosi et al. 2014; Jariwala et al. 2011). For instance, graphene can be turned into graphene oxide (GO) via oxidation in the presence of oxidizing agent. In another case, the incorporation of fluoride ions moiety onto the surface of graphene through fluorination processes is called fluorinated graphene (FG). Subsequently, a new class of nanomaterials such as graphene-based nanocomposites (i.e., semiconductor and rare-earth doped nanomaterials), quantum dots, heteroatoms (N, S, Se, B, P, etc.) doped graphene can be designed (Balaji et al. 2019). In the case of nanocomposites, the blending of two materials results in properties of two materials in a single nanocomposite (Sharma et al. 2022). Recently, several nanocomposites of graphene oxide and semiconductors i.e.,—CdS (Bera et al. 2015), ZnO, CuO, and SnO_2 (Bramhaiah et al. 2017), Fe_3O_4 (Shi et al. 2011), NiO (Xia et al. 2011), Cu–TiO_2/GO, Ag–TiO_2/GO, Pd–TiO_2/GO (Lertthanaphol et al. 2021), etc., have been prepared. When semiconductors are attached onto the surface of graphene, as a result, photocatalytic efficiency, and sensing ability have been significantly enhanced. However, luminescence property of semiconductors is quenched in the presence of graphene (Sharma et al. 2022). Similarly, nanocomposites of graphene/graphene oxide with rare-earth (RE^{3+})-doped lanthanide (Ln)-based oxide and fluorides nanoparticles have also been prepared for several applications (Sharma et al. 2022). In the case of nanocomposites of graphene/GO and RE^{3+}-doped Ln fluorides, two types of lanthanide fluorides are exclusively studied: one is binary (LnF_3/MF_2) fluorides and other is ternary ($MLnF_4$/$MLnF_5$) fluorides (Wei et al. 2012; Sharma et al. 2022; Wang et al. 2013a). In addition, graphene/graphene oxides have also been functionalized with RE^{3+} ion complexes to meet numerous applications including photothermal therapy, optoelectronic, sensing of biomolecules, etc. (Wei et al. 2012; Choi et al. 2010; Ambrosi et al. 2014; Mondal et al. 2021). Moreover, several organic matrices including, Hyperbranched Polyamide-Amine (HPAMAM), microcrystalline cellulose (MCC) (Liu et al. 2022), glucose (Tsujimura et al. 2014), etc. have also been incorporated with graphene to form nanocomposites.

In this chapter, different synthesis methods of graphene, GO, GO/graphene–semiconductor-based nanocomposites, GO/graphene and RE^{3+}-doped binary and ternary fluorides-based nanocomposites, etc. have been emphasized (Fig. 18.1). Finally, various applications of these group of materials are highlighted.

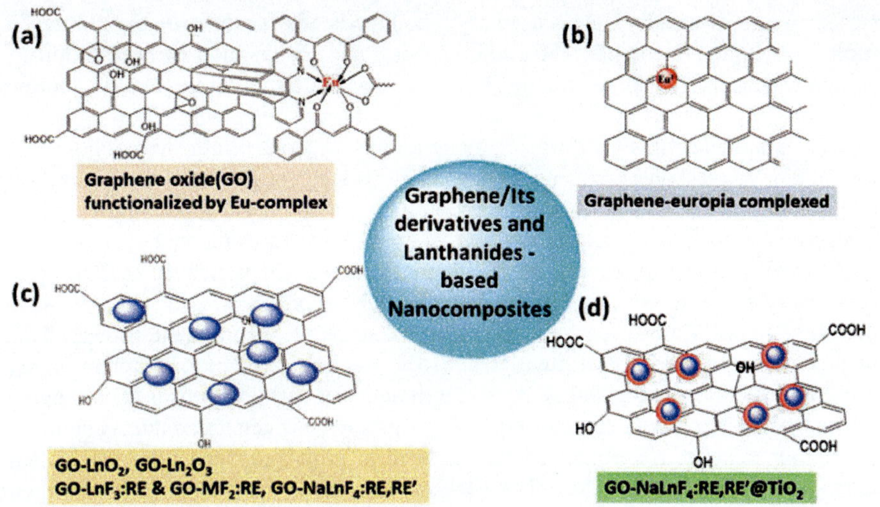

Fig. 18.1 Toolbox of graphene-based nanocomposites

18.2 Synthesis Methods of Graphene and Different Characterization Techniques

18.2.1 Chemical Vapor Deposition Method

Now-a-days, chemical vapor deposition (CVD) is a very important synthesis protocol used in the semiconductor industry to prepare highly pure materials of importance. CVD is a vacuum deposition method and normally a wafer is used as a substrate and is exposed to the volatile precursors which after reaction or decomposition produce the desired product. Graphene or its derivatives are also prepared by these techniques, where carbon species are formed when the gaseous precursor undergoes pyrolysis. Now the deposited carbon species on the surface goes for nucleation and carbon structure of graphene is produced. Sometimes a catalyst may be used to enhance the decomposition of the precursor or to reduce the reaction temperature. Along with thermal CVD, where heating is the important force for the reaction, plasma enhanced CVD or PECVD can also be used as it requires less pyrolysis temperature (Jariwala et al. 2011). Along with these, APCVD (atmospheric pressure CVD) and LPCVD (low-pressure CVD) are also used. The final quality of the as-produced graphene by these methods depends on reaction pressure, temperature, growth time, and on the flow rates too.

Nickel was one of the first substrates on which the first attempt at large growth of graphene by CVD was done (Obraztsov et al. 2007). Polycrystalline nickel films are used and they are annealed at high temperature (900–1000 °C) and in inert (Ar/H_2) atmosphere before synthesizing to increase the grain size and to reduce the surface oxide. Normally CH_4 is used as carbon feedstock for the growth onto the Ni film. As methane decomposes to carbon species and Ni has high carbon solubility, carbon species are dissolved on the Ni film at elevated temperature, and by cooling carbon atoms diffuse to form the Ni-C solid solution. After precipitation on the Ni surface, graphene films are formed. Interestingly, by controlling the cooling rate, multiple or single-layer graphene grows on Ni surface. But graphene layers may not be homogeneously placed as excess carbon may get out at grain boundaries which leads to multiple layer nucleation (Ambrosi et al. 2014; Jariwala et al. 2011).

After Ni, people tried several metal substrates such as Cu, which showed different catalytic effects and carbon solubility. In particular, the polycrystalline Cu films were used by Li et al. (2009) for the high quality of single-layer graphene growth. This method garnered tremendous attention due to its precise control of the graphene layer, low cost, and transferring ability. As lesser quantity of carbon atoms is accumulated on the Cu substrate, only single-/double-layer graphene is generated during cooling. In another method, a hot wall furnace is used to grow graphene films on 25 μm thick Cu foils (Li et al. 2009). The Cu foil was initially annealed in a hydrogen atmosphere at 100 °C and then a mixture of H_2/CH_4 was incorporated to initiate the graphene growth. After continuous graphene growth, the latter was cooled down at room temperature which resulted in the formation of the graphene layer (Li et al. 2009).

18.2.2 Mechanical Exfoliation

Recently, mechanical exfoliation (also called mechanical cleavage) techniques are used for the production of graphene sheets (Choi et al. 2010). It is a useful method of isolating single-layer graphene (Choi et al. 2010). Graphene sheets are comprised together by van der Waals force. The major challenges are to avoid the damaging of layers and avoid the re-agglomeration of the graphene sheets that have been exfoliated during the process (Choi et al. 2010).

This process was discovered by Geim and Novoselov in 2004 and this discovery eventually gave them Nobel Prize (Geim and Kim 2008). This method produces high quality; minimum defects graphene sheets which cause high mobility of electrons. Exfoliation is a repeated peeling technique that is used for the breaking of weak bonds (forces) at the stacked layers and to separate individual sheets. Highly oriented pyrolytic graphite (HOPG) with 1 nm thickness was exfoliated by Viculis et al. (2003). However, single to few-layer graphene in mm size domain is also exfoliated. Solution phase exfoliation using ultrasonication of aqueous suspension and by reduction of hydrazine were proposed by Stankovich et al. (2007). Similarly, Hernadez et al. (2008) and Lotya et al. (2009) also explored the exfoliation of graphene. As there are very weak van der Walls forces between interlayers, exfoliation method can be easily performed for graphene layered structure. Yi et al. have recently discussed the advantages and optimization of the exfoliation method (Yi and Shen 2015). The major advantage of this exfoliation method is cheap, simple, and possibility of large quantity production. However, there are some disadvantages too. The major disadvantage of this method is no control of the size and quality of crystallinity of graphene sheets. In mechanical exfoliation, a normal mechanical force is dominated, that provides a good quality and large-scale graphene flakes. However, this method is extremely time consuming and laborious, and for this reason, it is limitedly used in research laboratory and large-scale industrial production (Yi and Shen 2015).

18.2.3 Chemical Oxidation and Reduction Methods

Chemical oxidation and reduction methods have been used to chemically extract without the exfoliation step from graphite to graphene. Chemical oxidation is a well-known approach to expand few layers using a strong oxidizing agent. Brodie et al. used potassium chlorate in 1859 (Brodie 1859). In the chemical oxidation process, the dissemination of various oxygen functional groups happens on the graphitic layer (Brodie 1859). Oxidation method was improved by several groups such as Staudenmaier (1898), Hoffman and Koning (1937), Hummers and Offeman (1958) etc. On the other hand, reduction methods involve higher temperatures in the range of ~1000 °C, to create gaseous species interacting with the interlayer of the graphene oxide (GO) structure. Here in this process, pressure is increased and O^{2-} (oxygen) functional group are eliminated. During the reduction, small molecules like—CO_2, CO, and H_2O are added to the surface of graphene. Hoffman and König used a hydrazine hydrate reducing agent for the reduction of graphene oxide (Hofmann and König 1937). Some other reducing agents ($NaBH_4$, $LiAlH_4$, etc.) are also employed by several groups (Ambrosi et al. 2014).

18.2.4 Epitaxial Growth Using Metals

Single and multilayer graphene could be prepared by exploiting epitaxial growth on metals. At high range of temperatures, these metals exhibit catalytic activity for hydrocarbon decomposition. The ultrathin growth of graphite films epitaxially has been carried out using this technique (Jariwala et al. 2011). Normally, growth occurs by CVD technique on the catalytic metal surface at high temperature (Jariwala et al. 2011). Generally, high carbon solubility in the metal lattice is noticed, when the lattice is placed not only in high temperature but also in high vacuum too which eventually leads to the segregation-controlled growth. After that, the metal crystal cools slowly and results in a decrease in solubility. Thus, the segregating carbon on the surface grows into the graphene film. The concentration of surface carbon adatoms have a significant effect on the rate of graphene nucleation and growth. There are several limiting factors behind the growth of graphene using this technique. Among them, the attachment of carbon to the graphene islands compared to the surface diffusivity of carbon adatoms is important. Another factor is the orientation and arrangement of carbon atoms at the edges of the island and their attachment with the substrate. Normally, ultra-high vacuum chambers are used to characterize epitaxial graphene films on metallic single crystals. Some techniques to mention are: Low energy electron microscopy (LEEM), angle-resolved photoelectron spectroscopy (ARPES) Raman spectroscopy, etc. (Yazdi et al. 2016). The most widely studied epitaxial growth happens on Silicon carbide (SiC) (Yazdi et al. 2016). In this process, graphene is grown on silicon carbide substrates by heat treatment at Ultra-high vacuum conditions (Jariwala et al. 2011; Balaji et al. 2019; Sharma et al. 2022; Bera et al. 2015; Bramhaiah et al. 2017; Shi et al. 2011; Xia et al. 2011; Lertthanaphol et al. 2021; Wang et al. 2013a; Liu et al. 2022; Tsujimura et al. 2014; Obraztsov et al. 2007; Li et al. 2009; Viculis et al. 2003; Stankovich et al. 2007; Hernandez et al. 2008; Lotya et al. 2009; Yi and Shen 2015; Brodie 1859; Staudenmaier 1898; Hofmann and König 1937; Hummers and Offeman 1958; Yazdi et al. 2016; Emtsev et al. 2009). Here, the silicon atoms sublimate to give an exposed layer of carbon atoms which will further reorganize to generate an epitaxial graphene layer. Judicious tuning of sublimation temperature will result in the formation of the graphite layer, usually a single sheet. Higher temperature (1650 °C) and inert atmosphere (Argon) are prerequisites for coating of monolayer graphene in wafer scale under ultra-high vacuum conditions (Emtsev et al. 2009). The graphene film grown from SiC is generally different from the mechanical exfoliation methods. Substrate-induced corrugations and irregular orientations of the graphene layer are the main responsible for it. However, the transfer of epitaxial graphene on the arbitrary substrate is still a challenge. Thereby, commercialization of this growth procedure is difficult at the present scenario (Jariwala et al. 2011; Yazdi et al. 2016; Emtsev et al. 2009) (Fig. 18.2).

18.2.5 *Different Characterization Techniques*

Different characterization techniques may be used to understand the formation of graphene and graphene-based nanocomposites (Fig. 18.3). For example, using powder X-ray diffraction pattern, difference between graphite, graphite oxide, and graphene oxide can be determined. In addition, nanocomposites formation between GO and other inorganic compounds can also be understood along with their crystallite size determination, etc. (Sharma et al. 2022). Using electron microscopy like Transmission Electron Microscopy (TEM) and Scanning Electron microscopy (SEM), distribution/attachment of nanomaterials on graphene/graphene oxide sheet can be observed. Atomic Force Microscopy (AFM) can be used for depth analysis and surface analysis. Elemental analysis can be evidenced by energy dispersive X-ray (EDX) or X-ray Photoelectron Emission (XPES). Functional group analysis can be done using Fourier Transform Infra-Red (FTIR) spectroscopy.

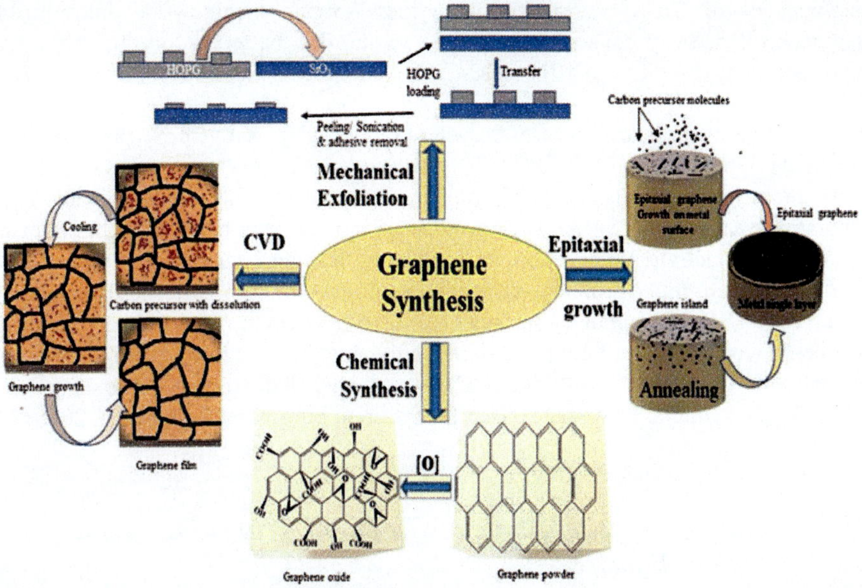

Fig. 18.2 Different synthesis methods for graphene

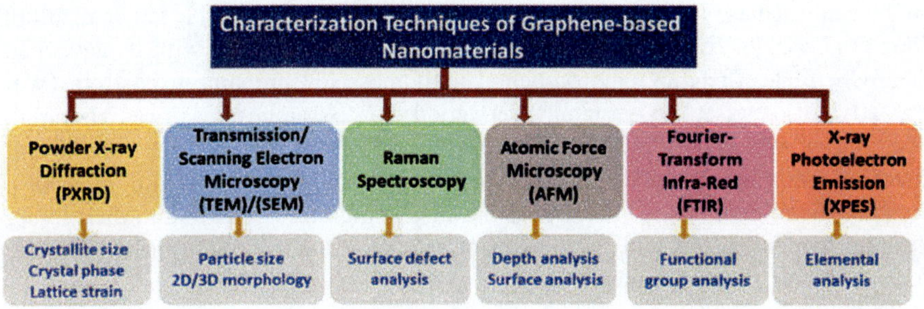

Fig. 18.3 Different characterization techniques used for graphene and graphene-based materials

18.3 Various Nanocomposites Made of Graphene

18.3.1 Graphene/Metal Nanocomposite

Recently, researchers have exploited metals for designing graphene composites and graphene metal nanoclusters (NCs). Different metals like Au, Pt, Pd, Ru, etc. have been inserted into graphene and the resulting composites are used for different applications in sensing and imaging (Liu et al. 2018).

Synthesis of Co_9S_8 nanoflake using metal–organic framework on reduced graphene oxide sheet for making efficient sodium-ion battery anode has recently reported. They revealed that the rGO/Co_9S_8 composite electrode delivers a high discharge capacity of 55/mAhg^{-1} at 0.1 Ag^{+}, a good rate capability of 10 Ag^{-1}, and effective cyclic stability of 500 cycles. Incidentally, rGO/Co_9S_8 promises for practical application in $Na_3V_2(PO_4)_3$K rGO/Co_9S_8 cells too (Huang et al. 2018). A three-dimensional porous patterned graphene-silver nanocomposite is prepared and reported by Xuan et al. (2018). The developed electrode demonstrated high, uniform electrical conductivity. In addition, incorporation of Pt and Au nanoparticles on 3D porous Nitrogen-doped graphene-copper nanocomposite is fabricated and characterized recently, which contains a resistance of 0.16 μ Ω cm at room temperature and is much lower than that of copper. Interestingly, this material has very high thermal conductivity (538 W m^{-1} K^{-1} at 25 °C) even greater than copper. There are many other Graphene/metal composites used for different purposes (Zheng et al. 2018).

18.3.2 Nanocomposites Made of Graphene/Transition Metal Oxide (Hydroxide)

Preparing efficient electrode materials for supercapacitors etc. draws a huge attention and various metal oxides (transition metal) like MnO_2, Mn_3O_4, RuO_2, Co_3O_4, etc. can be exploited as electrode materials for the same (Wei et al. 2011; Hu et al. 2006; Zhou et al. 2011). Just to mention, MnO_2 is a potential material with its advanced electrochemical efficiency, environmental benignness, low preparation cost, etc. (Wei et al. 2011). Likewise, RuO_2 contains very good charge storage capacity as it has high electronic as well as protonic conductivities. In addition, it also contains a significant extent of redox process reversibility over a large potential range (Hu et al. 2006). But pure metal-oxide-based electrodes has also some limitations due to their lesser electrical conductivities and agglomeration tendencies that leads to the incomplete reaction during the redox process. To solve these issues and obtain enhanced electrochemical properties, various transition metal-oxide/carbon composite electrodes are exploited which can exploit complete advantage of double-layer and pseudocapacitance. This leads to enhanced energy density and extended life cycle. In this regard, preparing hybrid films (based on graphene metal oxide nanocomposites) is important. As an example, by reducing permanganate on graphene results in graphene-MnO_2 composite electrodes with a significant specific capacitance (Wei et al. 2011). Graphene-MnO_2 nanocomposites synthesized by electrodeposition method also depicted a high specific capacitance of ≈315 Fg^{-1}. MnO_2 nanowires/graphene, graphene nano-plate/MnO_2, and porous graphene/MnO_2 have also been investigated. As MnO_2 has an insulating nature, the synthesis of graphene/MnO_2 nanocomposites needs higher graphene amounts. This can influence the top density and high electrolyte uptake. To avoid such issues and keeping lower concentration of graphene in the composite, accurate design and synthesis are required. With an alternative to Graphene/MnO_2 nanocomposites (Wei et al. 2011); Graphene/RuO_2 (Hu et al. 2006); Graphene-Co_3O_4 (Zhou et al. 2011); Graphene-Fe_3O_4 (Shi et al. 2011); Graphene-SnO_2 (Bramhaiah et al. 2017); Graphene-NiO (Xia et al. 2011); Graphene-ZnO (Bramhaiah et al. 2017) and many others have also been designed with improved electrochemical capacitance.

Multimetal-oxide contains more than one unit cell structure like perovskite (ABO_3) and pyrochlores ($A_2B_2O_7$) (Jariwala et al. 2011; Gupta and Subramanian 2014). Similarly, a novel perovskite based $CsPbBr_3$ and $CsPbBr_3$/GO nanocomposite materials are reported by Xu et al. (2017). Recently several groups synthesized noble metal-based (Pd, Pt, etc.), rGO/oxide based ternary composite materials using a facile and environmentally benign protocol. Any toxic and hazardous reducing agents such as hydrazine, sodium borohydride, etc. are not used in the synthesis (Liu et al. 2018; Zheng et al. 2018).

18.3.3 Nanocomposites Made of Graphene and Transition Metal Nitrides (or Sulfide)

Along with transition metal oxide (hydroxide) nanocomposites with graphene, transition metal nitrides (e.g., TiN, MoN_x, etc. (Lu et al. 2014; Nandi et al. 2014)) and transition metal sulfides (e.g., CoS, NiS, MoS_2H_2, etc. (Cong et al. 2015)) are also important nanocomposite made with graphene. They can also be used in supercapacitors for their significant specific capacitance, very good electrochemical parameters, and high electrical conductivity (Lu et al. 2014; Nandi et al. 2014; Cong et al. 2015). But both the oxides in their pure form suffer from irreversible oxidation reactions, that cause low chances of cyclability. Thereby metal nitrides or sulfides in conjugation with conductive carbon like graphene, can resist the oxidation reaction and improve the cycle ability of these nanocomposites. Just to mention, poly (vinyl alcohol) (PVA/LiCl) based solid electrolyte, graphene-Fe_2N anode, and graphene-TiN cathode, and generated an irreversible capacitance of about 58 Fg^{-1} at 4 Ag^{-1} (Dong et al. 2020). In addition, this value of capacitance was comparatively stable up to 20,000 cycles. Likewise, very good cycling stability with high capacitance value and retention over 3000 cycles is reported for the graphene/CoS nanocomposites, that is higher than pure CoS (Zhou et al. 2015).

18.3.4 Graphene/RE^{3+}-Doped Binary and Ternary Fluorides Nanocomposites

Nanocomposites made between RE^{3+}-doped binary or ternary fluorides and graphene/graphene oxides have also been explored to further improve the applications in various fields. However, very limited literatures are available in the case of binary or ternary fluorides nanoparticles and GO-based nanocomposites (Wei et al. 2012; Sharma et al. 2022; Wang et al. 2015; Li et al. 2018; Sun et al. 2019; Lingamdinne et al. 2019). For example, to increase the photocatalytic activity of ternary fluorides nanoparticles especially in near infrared (NIR) region of spectrum, β-$NaYF_4$:Yb^{3+}, Tm^{3+} and N-doped P25 nanocomposites are supported onto the surface of graphene. When graphene is attached, the photocatalytic activity of nanocomposites is significantly increased in the NIR region of light. The increased photocatalytic efficiency in the presence of graphene is due to the increasing proximity of nanoparticles, increasing absorption of light, and separation of photogenerated electron–hole pair (Wang et al. 2015). On another side, Wang et al. have synthesized the core–shell structure of α-$NaYF_4$:Yb^{3+}, Tm^{3+}@TiO_2 in a nanoscale regime. Then this core–shell nanostructure are attached onto the surface of reduced GO. As a result, nanocomposites are formed which exhibit photocatalytic activity in the NIR region of light (Wang et al. 2015). Additionally, upconverting nanocrystals are decorated on the surface of GO to enhance the photothermal conversion efficiency of bare graphene oxide. Li et al. have depicted that the GO/$NaYF_4$: Yb, Er nanocomposites prepared by hydrothermal method exhibit significantly enhanced light-to-heat conversion efficiency of 13.9% compared to graphene oxide itself (Li et al. 2018). As a result, graphene oxide/$NaYF_4$: Yb, Er nanocomposites can be potentially used in photothermal and photodynamic therapy. For example, Sun et al. have synthesized β-$NaYF_4$:Yb (25%), Tm(0.3%)@TiO_2@GO nanocomposites for synergistic photothermal and photodynamic therapy (Sun et al. 2019). In addition, graphene oxide-binary fluorides nanocomposites have also been synthesized. Lingamdinne et al. have functionalized the LaF_3 on the graphene oxide surface to obtain the porous flower-like graphene oxide-lanthanum fluoride (GO-LaF_3) nanocomposites (Lingamdinne et al. 2019). Thereafter, the nanocomposite is utilized for adsorptive removal of Arsenic [As (V)]. The adsorption-based removal of As is attributed to mixed processes (i.e., electrostatic, ion-exchange, and surface complexation); however, electrostatic interaction between As and nanocomposites, has played a crucial role in the adsorption of As (Lingamdinne et al. 2019). Furthermore, Sharma et al. have prepared the RE^{3+} (Ce^{3+} or Ce^{3+}/Tb^{3+} or Tb^{3+})-doped binary fluorides–GO nanocomposites using ionic liquid (IL)-assisted hydrothermal method (Sharma et al. 2022). Herein, 1-butly-3-methylimdidazolium tetrafluoroborate [C_4mim][BF_4] IL is employed as a source of fluoride ion, reaction medium, and capping/templating agent. In the synthesis, IL is exploited as a source of fluoride (F^-) ion to form binary fluorides. Numerous binary fluorides-GO nanocomposites such as BaF_2:Ce^{3+}-GO, BaF_2:Ce^{3+}/Tb^{3+}-GO, CeF_3:Tb^{3+}-GO, LaF_3:Ce^{3+}/Tb^{3+}-GO and GdF_3:Tb^{3+}-GO are prepared in this method (Fig. 18.4).

Formation of graphene oxide, BaF_2:Ce/Tb-GO nanocomposites are understood using several characterization techniques like PXRD, SEM, TEM, etc. (Fig. 18.5). During synthesis, GO has played a crucial role in controlling the structural and optical properties of RE^{3+}-doped binary fluorides. In another word, in the presence of GO, the size of RE^{3+}-doped BaF_2 nanoparticles is quite similar. Moreover, it is found that in the presence of GO, excitation and emission spectra of Ce^{3+} ion are significantly

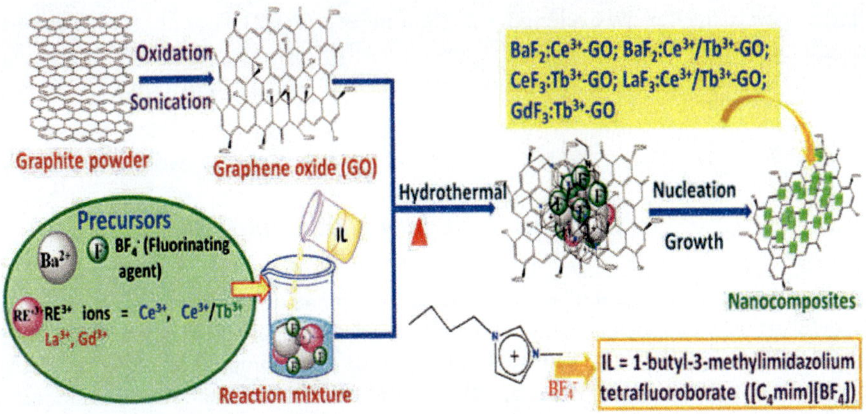

Fig. 18.4 Schematic representation of synthesis of nanocomposites made by rare-earth doped binary fluorides and GO using [C4mim][BF4] IL-assisted via solvothermal method. Part of this is reproduced with permission from ACS publisher under Creative Common Licence (Sharma et al. 2022)

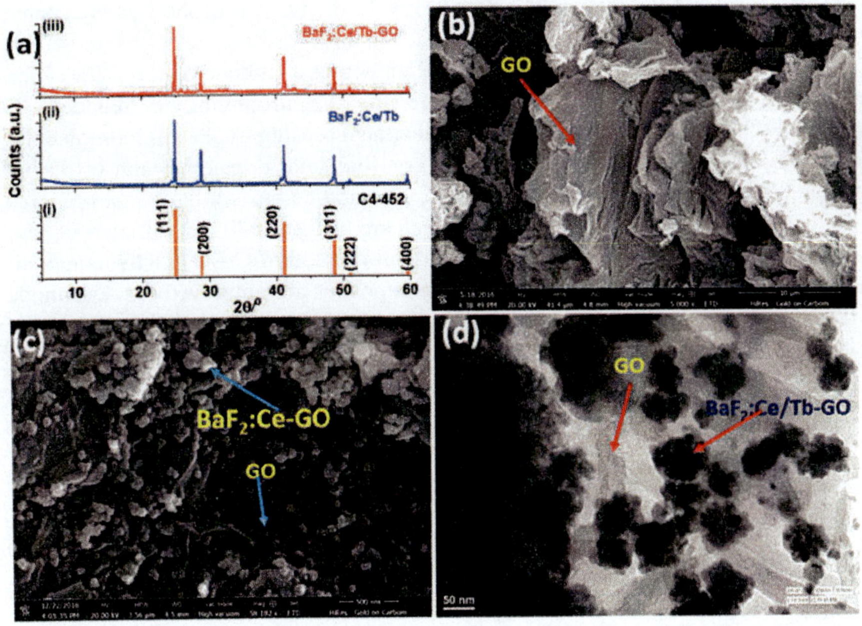

Fig. 18.5 **a** PXRD patterns: (i) BaF_2 standard, (ii) BaF_2:Ce/Tb, (iii) BaF_2:Ce/Tb-GO, **b** low-magnification FESEM image of as-prepared graphene oxide, **c** high magnification FESEM images of as-prepared BaF_2:Ce-GO, **d** low-magnification TEM image of BaF_2:Ce/Tb-GO NCMs formed by one-pot synthesis using the [C4mim][BF4] IL-assisted hydrothermal method

shifted toward the lower wavelength side (Sharma et al. 2022). For enhancing the dye-sensitized solar cell's efficiency, the composite of TiO_2–$NaYF_4$:Er^{3+}/Yb^{3+}–graphene was used as a photoanode to enhance the interfacial electron transfer of FTO/TiO_2 (Li et al. 2012).

18.3.5 Graphene/Lanthanide Oxides Nanocomposites

In addition to transition metal oxides/sulfides, graphene or graphene derivatives can also form nanocomposites with lanthanide oxide nanoparticles especially CeO_2, Eu_2O_3, Gd_2O_3, and so on, and the application of nanocomposites is significantly improved (Bai et al. 2014; Mo et al. 2012). For instance, graphene/CeO_2 nanocomposites are extensively studied for numerous applications in photocatalysis, catalysis, electrode material for supercapacitors, luminescence, etc. (Bai et al. 2014; Nia et al. 2019; Deng et al. 2017). In addition to graphene, a highly stable and methanol tolerant electrocatalyst of nitrogen-doped graphene/nano ceria nanocomposites has also been synthesized for oxygen reduction (Soren et al. 2016). Ju et al. have prepared graphene/Y_2O_3/$LiMn_2O_4$ hybrid materials, where firstly, $LiMn_2O_4$ microspheres were coated with Y_2O_3 and then graphene oxide was added to it for forming the composites. This graphene/Y_2O_3/$LiMn_2O_4$ composites were applied as a cathode material for Li ion battery (Ju et al. 2014). Jafari et al. have prepared the europium oxide/reduced GO (EuONP–rGO) nanocomposite for electrochemical sensing of several biomolecules in the various range of concentrations of ascorbic acid, dopamine, and uric acid, etc. (Jafari et al. 2018). On the other hand, Xue and co-worker have first prepared oleic acid-capped Gd_2O_3 nanoparticles which are further modified with two-photon graphene quantum dots (GQD) leading to the formation of Gd_2O_3/GQD nanocomposites. Thereafter, the Gd_2O_3/GQD nanocomposites are employed for dual-mode MRI contrast and cell labelling agent (Wang et al. 2018).

18.3.6 Graphene Oxide-Rare Earth Metal Complexes

The beauty of graphene oxide is that its surface is enriched with oxygen containing functional group. During the formation of GO-rare earth metal complexes, functional groups assist in the functionalization of the RE^{3+} ions containing complexes leading to the GO-rare earth metal complexes. There are numerous examples which have shown that rare earth metal complexes can be noncovalently functionalized on the surface of graphene oxide (Sharma et al. 2022; Nia et al. 2019; Deng et al. 2017; Jafari et al. 2018). For instance, Cao et al. have synthesized the [Eu(DBM)$_2$(Phen)(SA)](DBM = dibenzoylmethane, Phen = 1,10-phenanthroline, SA = stearic acid) via a noncovalent approach (Cao et al. 2011). According to this preparation method, GO sheets (GOS) were bound to the Eu^{3+}-based complexes via noncovalent interaction. As a result, red luminescent GOS–RE complex hybrids were formed. Employing a similar preparative method, Wang et al. also prepared the GOSs/Eu-acac-phen complexes which emit in the red region (Wang et al. 2013b). Liu et al. have synthesized graphene oxide-La(BTC)$(H_2O)_6$ (H_3BTC = 1,3,5-benzenetricarboxylic acid) metal–organic framework at room temperature. In this composite (LaMOF-GOn), n is varied in the range of 1 to 6 and corresponds to the percentage of GO at 1, 2, 3, 4, 5, and 10%. Thereafter, this composite was employed for selective adsorption of haemoglobin (Hb) (Liu et al. 2014). Multifunctional nanostructures were synthesized by functionalization of rGO with Eu^{3+}-ion complex and vancomycin (Van). On irradiating with near-infrared region of light, the resulting nanostructures were used for detecting and photothermal killing of drug resistant bacteria (Yang et al. 2014). Gupta et al. have complexed the Eu^{3+}-ion with graphene via oxygen containing functional groups. On exciting this complex using 314 nm wavelength, red emission was found at 614 and 618 nm. This complexed graphene exhibits dual functionality wherein the emission of rhodamine B is quenched but shows its own hypersensitive emission in red region (Gupta et al. 2011).

18.4 Applications

18.4.1 Sensors and Bio-sensors

Graphene and graphene-related materials are extensively used to design efficient chemical, biological, and electrochemical sensors (Ambrosi et al. 2014). The unique properties of graphene materials and its ease of functionalizing the graphene surface with the specific elements help their uses in different purposes. Graphene and its derivative can be used for different applications related to sensors and biosensors: (a) Clinical applications e.g., for the detection of glucose, cholesterol, H_2O_2, dopamine, and uric acid; (b) Environmental applications e.g., for the detection of heavy metal ions like Cd^{2+}, etc.; pesticides such as methyl parathion and (c) Food applications e.g., detection of erythromycin, and tryptamine (Krishnan et al. 2019). An electrochemical sensor based on a glassy carbon electrode with copper nanoparticles and graphene is suggested for the detection of glucose (Luo et al. 2012). Recently several graphene-based electrochemical enzymatic biosensors on graphene electrodes were used for the detection of glucose, cholesterol, etc. (Krishnan et al. 2019; Luo et al. 2012).

18.4.2 Glucose Biosensor

Now-a-days, diabetes is a common disease that is affecting millions of people around the world. The high levels of glucose in the blood are considered diabetic. The physiological presence of glucose in blood after ~8 h of fasting confirms the severity

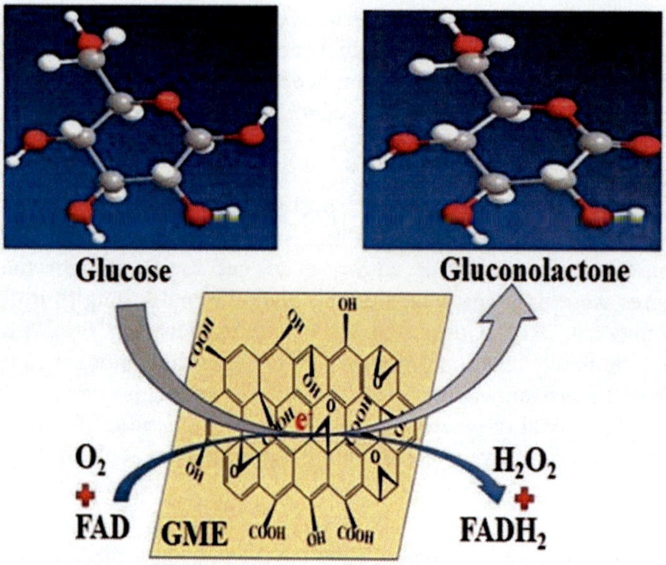

Fig. 18.6 Schematic diagram of graphene-based glucose biosensors

of diabetes. Fasting blood glucose levels in the 70–99 mgdL^{-1} (from 3.9 to 5.4 mmolL^{-1}) range is accounted as normal, while blood glucose levels from 100 to 125 mgdL^{-1} (from 5.6 to 6.9 mmolL^{-1}) are accounted as pre-diabetes, and a blood glucose level of 126 mgdL^{-1} (7.0 mmolL^{-1}) or higher confirms the presence of diabetes (Krishnan et al. 2019). To detect the precise and rapid glucose level in blood is very important for continuous point-of-care. The first report for monitoring blood glucose levels by using glucose oxide (Gox) based enzyme electrodes was done by Clark and Lyons (1962). The electrochemical glucose sensors can be explained by categorizing them into three generations. In the first generation, determination depends on oxygen consumed, specifically the reaction between glucose and the Gox enzyme electrode in the presence of oxygen to produce H_2O_2. After a while, the level of glucose was checked through the amount of enzymatically produced H_2O_2 (Fig. 18.6). The second generation depends on a process called mediated electron transfer (MET). In this process, the usage of electrical mediators to expedite the electron transfer process between the Flavin adenine dinucleotide (FAD) active site of Gox and the electrode's surface is emphasized (Chen et al. 2013a). However, no mediators are used for third-generation enzymatic biosensors to obtain direct electron transfer (DET) between FAD-Gox and the electrode surface (Chen et al. 2013a; Rafighi et al. 2016). In particular, carbon nanotubes (CNT) (Wooten et al. 2014); carbon nanodots (Zhao et al. 2015); porous carbon (Tsujimura et al. 2014), and graphene (Kim et al. 2017) are preferred nanostructures for the efficient immobilization and improved DET characteristics. Among these, graphene is highly advantageous due to its high surface area (2630 m^2g^{-1} for single-layer graphene), excellent electrochemical properties, and biocompatibility. Last but not the least, graphene and graphene oxide (GO) are paying more attention and act as promising electrode material for the electrochemical glucose biosensing and immobilization of enzymes (Luo et al. 2012; Clark and Lyons 1962; Chen et al. 2013a; Rafighi et al. 2016; Wooten et al. 2014; Zhao et al. 2015; Kim et al. 2017).

18.4.3 Photocatalysis

Recently, synergy strategy-based emerging technology has received great attention in controlling water-based pollution (Mondal et al. 2021; Bera et al. 2015; Wang et al. 2013a, 2015; Gupta and Subramanian 2014; Sun et al. 2019; Xiong et al. 2021). The adsorption and photocatalysis of organic pollutants and industrial wastewater are currently an important area of research. Among degrading pollutants, the photocatalytic oxidation treatment via various metal oxides is also considered the most efficient and clean method owing to its non-toxic nature, low cost, excellent eco-friendly nature with high chemical stability and photocatalytic activity (Mondal et al. 2021; Bera et al. 2015; Wang et al. 2013a, 2015; Gupta and Subramanian 2014; Sun et al. 2019; Xiong et al. 2021).

A wide bandgap of TiO_2 (~3.2 eV) shows significantly high photocatalytic efficiency, which absorbs ~5% in the ultra-violet region. In case of doping TiO_2 with other metals or nonmetals, the bandgap is further reduced, and absorption of light is further extended in the visible region (Xiong et al. 2021). Semiconductor nanocrystals can be incorporated into chemically modified GO and rGO to make efficient nanocomposites. Specially, nanocomposites based on semiconductors and transition metal nanoparticles open new ways for the dissociation of various organic dyes into H_2O and CO_2 (Mondal et al. 2021; Bera et al. 2015; Wang et al. 2013a, 2015; Gupta and Subramanian 2014; Sun et al. 2019). In addition, GO, rGO, and metal nanoparticles can efficiently inhibit the recombination rate of electron–hole pairs. It has been proven that the TiO_2/rGO composite has effectively reduced the bandgap and recombination rate, hence it has improved the photocatalytic activity (Xiong et al. 2021). In addition, GO acts as a charge separating on the TiO_2 surface as GO has a conjugated 2D backbone structure, in case of TiO_2/GO nanocomposites. In the presence of a pi (π) electron, GO makes a bond (Ti–O–C) with the Ti atom of TiO_2. Hence, GO acts as an electron acceptor/transporter, resulting in improving catalytic activity (Bramhaiah and Bhattacharyya 2022). Lertthanaphol et al. also studied the photoreduction of CO_2 to ethanol by using TiO_2/GO, Ag-TiO_2/GO, Pd-TiO_2/GO, and Cu-TiO_2/GO-based nanocomposite (Lertthanaphol et al. 2021). Xiang et al. studied the photocatalytic activity of rGO/TiO_2 based on a composite made of GO and TiO_2. They observed 41% hydrogen generation of methanol/water by 1% (weight) rGO loaded with TiO_2 (Xiang et al. 2011). The photocatalytic activity in the visible region is studied by Lee et al. and rGO/TiO_2 shows remarkable degradation of methylene blue (MB) compared to bare TiO_2 and Degusa P_{25} (Lee et al. 2012). Methylene blue

(MB) is a highly soluble organic dye and is widely used in various industries such as textile, dyeing, printing, etc. The removal of MB from wastewater is studied by Nguyen et al., using cellulose nanofiber/GO composite (Nguyen et al. 2022). Similarly, Zhang et al. also studied the degradation of MB using P25-rGO/TiO_2 under UV and visible regions (Zhang et al. 2010). Wang et al. depicted the mechanism of photodegradation of MB dye in the presence of TiO_2/graphene nanocomposite. During photodecomposition of MB dye, graphene plays a crucial role in the separation of photogenerated charge carriers and enhances the adsorption of pollutants onto its surface, leading to enhanced photocatalytic activity of TiO_2 nanosheets (Wang et al. 2012). It is noted that the gaseous oxygen and water participate in the photocatalytic reaction process. Herein, the mechanism of heterogenous photocatalytic activity of nanoparticles, (NPs)-graphene nanocomposites is illustrated in Fig. 18.7. First of all, NPs harvest the incident light resulting in generation of charge carriers (**Step a**), excitation of electron (e^-) to the conduction band (CB) and hole (h^+) in the valance band (VB). When the further recombination of generated charge carriers takes place, the emission occurs. In the photocatalytic reaction, the dissolved gaseous O_2 reacts with electrons of CB leading to the formation of superoxide radical ($O_2{}^-$) (**Step b**). This superoxide radical further reacts with H^+ ions to form HO_2 radical. Thereafter, the formation of H_2O_2 and O_2 are taken place from two HO_2 radicals. On the VB, adsorbed H_2O (as a solvent) or hydroxide ion (OH^-) reacts with hole and then the formation of hydroxyl (OH) radical takes place (**Step c**). Finally, hydroxyl radicals react with organic pollutants to further degrade into non-toxic products (**Step d**) (Wang et al. 2012). Formation of hydroxyl radicals can be confirmed using terephthalic acid, as terephthalic acid serves as hydroxy radical scavenger. As it reacts with hydroxy radical in the presence of light, the formation of hydroxyl terephthalic acid occurs which emits light ($\lambda_{em} = 425$ nm) on exciting at $\lambda_{ex} = 315$ nm. And the emission intensity of hydroxyl terephthalic acid depends on its concentration (Sharma et al. 2017).

Pradhan et al. also found 67% degradation of phenol using $\alpha\sim Fe_2O_3$/rGO nanocomposite (Pradhan et al. 2013). On the other hand, Li et al. synthesized metal-sulfide-based nanocomposites such as CdS/rGO where hydrogenation is seen 5 times faster compared to CdS (Li et al. 2011). Chen et al. prepared rGO/$ZnInS_4$ and have shown the important role played in photocatalysis (Chen et al. 2013b). When $Bi_2Ti_2O_7$ is doped with rGO (1% wt), hydrogenation is increased 3 times and photocatalytic efficiency of 130% compared to $Bi_2Ti_2O_7$ itself (Gupta and Subramanian 2014). Several transition metals like—Fe^{3+}/Cr^{3+}/Mn^{2+}, or $Bi_2Ti_2O_7$, $Sr_2Ti_2O_{7-x}$, etc., based RGO nanocomposite enhanced the photocatalytic activity (Lee et al. 2012; Mukherji et al. 2011). The deposition of ZnO, Cu_2O, and CdS nanoparticles on the rGO surface and graphene sheets are demonstrated (Bramhaiah and Bhattacharyya 2022). The photoreduction approaches on rGO-TiO_2 were well studied by Williams et al. (2008). Also, the photocatalytic reduction approach was studied on ethanol using UV-irradiation. A liquid–liquid interface approach toward the oxidation of rhodamine B (RhB) was applied by Bramhaiah and John (2013). Several groups synthesized various types of nanocomposites including rGO-noble metals; rGO-semiconductor (Fe_2O_3, Ag_2S, ZnO, CuO, and RGO-Ni(OH_2), etc. (Bera

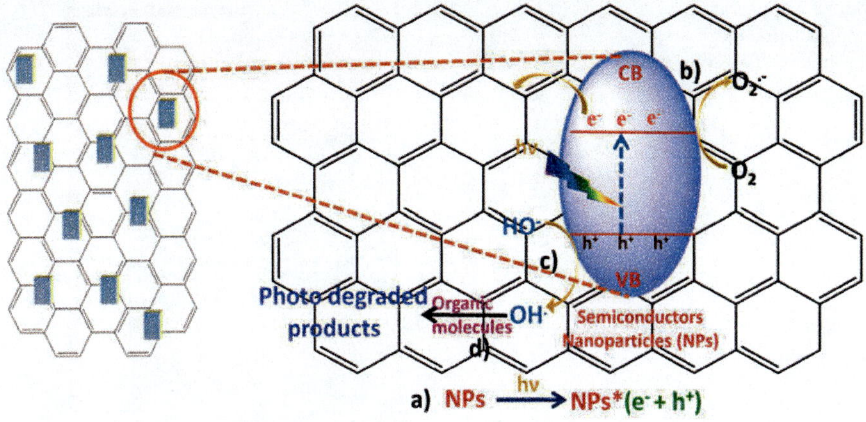

Fig. 18.7 Schematic representation of the mechanism of photocatalytic activity caused by nanoparticles/graphene nanocomposites

et al. 2015; Bramhaiah et al. 2017; Shi et al. 2011; Xia et al. 2011; Pradhan et al. 2013). Acrylamide/GO (Am-GO-SA) hydrogel-based nanocomposite was proposed and successfully employed for the degradation of crystal violet (CV) dye (Pashaei-Fakhri et al. 2021). For the first time, (rGO-Au NP) nanocomposite was successfully photo synthesized by using aqueous plant leaves, Nile banks Persicaria salicifolia (Pluchea salicifolia) which provides higher efficiency than Au NP. In addition, rGO/AuNP act synergistically to promise several unique physiochemical properties (El-Maghrabi et al. 2021).

18.4.4 Solar Cell

In the last decade, Dye-sensitized solar cells (DSSCs) have drawn huge interest as potential alternatives to the widely used, expensive silicon-based solar cells due to their low cost, facile fabrication techniques, and less impact on the environment. However, large-scale and commercial applications of DSSCs are still constrained by their low power conversion efficiencies (PCEs) and poor long range term stability. Since the attributes of the many DSSC components may be tailored by using strategies like the introduction of novel materials and new synthesis techniques, this has sparked a significant interest among the researchers in the realization of high-performance and sustainable devices. Recently graphene-based materials have been used as photoanodes, transparent conducting electrodes, semiconducting layers and dye-sensitizers, or counter electrodes in DSSCs (Fig. 18.8). Graphene and its composite materials have superior optoelectronic, mechanical, thermal, and chemical capabilities, huge specific surface area, flexibility, and excellent stability. Also, these are low cost

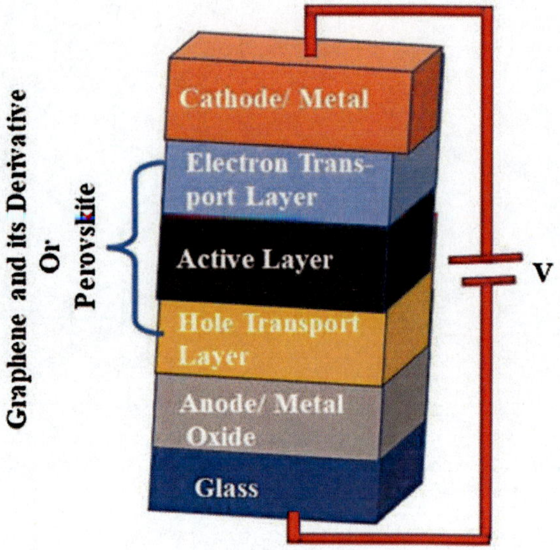

Fig. 18.8 Schematic representation of construction of a solar cell

and non-toxic. All these properties gives plenty of potential to replace the conventional materials that are often employed in the creation of the various DSSC component, which results in a substantive enhancement in power conversion efficiencies (PCE) from ~0.13 to 12% (Muchuweni et al. 2020). Due to its versatile multifunctional properties, Graphene has also been used in quantum dot sensitized solar cells (QDSCs), solid-state solar cells, DSSCs; and in Organic photovoltaics (OPVs), etc. (Muchuweni et al. 2020). Recently perovskite and silicon (ps) solar cell also have gained a great attention in photovoltaic products, with low temperature processing (200 °C), that shows solar cell efficiency 13.56% (Hang et al. 2021).

18.4.5 Photonic and Bio-photonic

Graphene and its derivatives (i.e., graphene quantum dots (GQDs), functionalized graphene, etc.) have shown potential for photonics and bio-photonics applications. The major advantages of GO are tunable fluorescence and its fluorescence quenching capability. In addition, the inherent photophysical properties of GO, and the interaction of electron donor and acceptor molecule with graphene are used to control its optoelectronic properties via charge transfer. The tunable optical properties of GO can also act as excellent candidates for various metals, atoms or ions, and lanthanide metals (Wei et al. 2012; Sharma et al. 2022; Mo et al. 2012; Wang et al. 2018; Gupta et al. 2011). Owing to their intense optical emission and chemical stability, rare-earth-based bio-probes are ubiquitous in bioimaging. On graphene sheets, RE metals such as Eu, Gd, Yb, and Nd show low diffusion barriers and comparatively high adsorption energy (Wei et al. 2012; Sharma et al. 2022; Mo et al. 2012; Wang et al. 2018; Gupta et al. 2011). With the advancement of nanotechnology, miniaturization of devices is important for increasing the myriad of applications in various fields. For instance, graphene quantum dots are luminescent forms of graphene which emit in green or blue regions. The beauty of GQDs is that it can be bound with other molecules, and nanoparticles which turn out to be in nanocomposites (Bacon et al. 2014). As a result, avenue of applications is increased to many folds (Bacon et al. 2014). Along with this, graphene-based nanocomposites can also be used for optical limiting application. Nonlinear optical limiter is required to protect the eye and optical sensor from the powerful pulsed laser. There are basically two roles of an optical limiter: first, it interrupts the transmission of the powerful laser pulse and second, it limits the output in safe level (Hollins 1999). Therefore, a number of graphene/GO-based nanocomposites have been developed as optical materials. For instance, Wei et al. have developed the new GO-based upconverting (GO–$NaYF_4$:Yb/Er) optical limiting materials which work in the NIR region (Wei et al. 2012). In this nanocomposite, GO serves as an optical limiter, and using a 980 nm laser, its optical limiting is focused on the 532 or 1064 nm wavelength. Another group has studied the OL performance of the GO–Fe_3O_4 nanocomposites. In this study, it is found that OL performance of the pristine GO is significantly improved on forming the nanocomposites with Fe_3O_4 nanoparticles. Thus, it can be concluded that GO as well as GO-based nanocomposites can play a crucial role as optical limiting materials (Zhang et al. 2011). Sharma et al. have explored the influence of GO on the luminescent property of RE^{3+}-doped binary fluorides and GO-based nanocomposites (Sharma et al. 2022). The excitation and emission spectra of bare $BaF_2:Ce^{3+}/Tb^{3+}$ and $CeF_3:Tb^{3+}$ NPs are compared to their respective nanocomposites $BaF_2:Ce^{3+}/Tb^{3+}$-GO and $CeF_3:Tb^{3+}$-GO NCMs. The results depict that the emission of bare NPs is quenched in the presence of GO (Fig. 18.9a, b). However, on illuminating the $BaF_2:Ce^{3+}/Tb^{3+}$-GO and $CeF_3:Tb^{3+}$-GO NCMs at 254 nm excitation wavelength using the 6 W lamp, dark green color emission is observed, that can be even seen with bare eyes. This indicates that the NPs are still luminescent even in the presence of GO (Fig. 18.9c, d).

Recently, bio-photonic applications based on graphene and graphene oxide nanocomposites have drawn tremendous attention and played a promising role in both clinical and research areas (Wei et al. 2012; Mo et al. 2012; Wang et al. 2018; Ferrer-Ugalde et al. 2021; Robinson et al. 2011). Better stability, good photoluminescence properties, and high mobility of charge carrier of graphene and its derivatives enable these classes of materials for bio-applications (Fig. 18.10) (Wei et al. 2012; Sharma et al. 2022; Ferrer-Ugalde et al. 2021; Robinson et al. 2011; Gu et al. 2019). Besides these, graphene and graphene-based derivatives play as attractive materials for detecting and recognizing various biological aspects such as antibacterial activity, tissue therapy, therapeutic gene delivery, intracellular mRNA or DNA imaging, live cell imaging (in vivo and in vitro), etc. All these happen due to better stability of graphene quantum dot (GQD) based bio-probe with live cells and tissues (Ambrosi et al. 2014; Wang et al. 2018; Ferrer-Ugalde et al. 2021; Robinson et al. 2011; Gu et al. 2019). Over the last decades, graphene and graphene nanocomposite-based nanomaterials were employed for cytotoxicity assay in human kidney cell line imaging. In

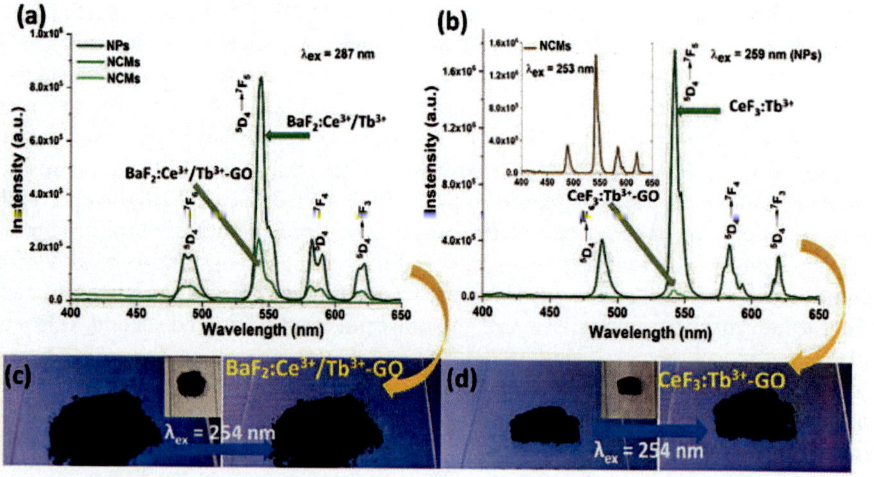

Fig. 18.9 PL spectra: **a** emission spectra of BaF_2:Ce^{3+}/Tb^{3+} NPs and BaF_2:Ce^{3+}/Tb^{3+}-GO NCMs, **b** emission spectra of CeF_3:Tb^{3+} NPs and CeF_3:Tb^{3+}-GO NCMs, **c** and **d** BaF_2:Ce^{3+}/Tb^{3+}-GO and CeF_3:Tb^{3+}-GO are illuminated using UV lamp (6 W). It is reproduced with permission from ACS publisher under Creative Common Licence (Sharma et al. 2022)

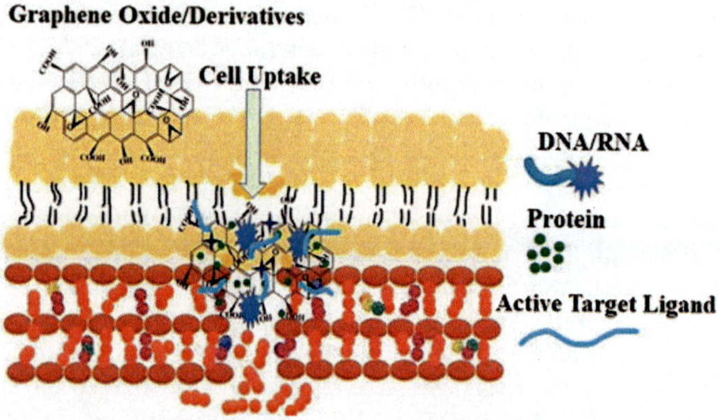

Fig. 18.10 Schematic diagram of targeted attachment of graphene and graphene-based derivatives with DNA/RNA, protein, and other ligands for biocellular imaging purposes

vitro cell imaging by functionalized Gd_2O_3/GQD two-photon probe demonstrated the dual modal imaging agent as magnetic resonance imaging (MRI) and fluorescence probe (Wang et al. 2018). Similarly, Ferrer-Ugalde et al. reported almost 90% in vitro cytotoxicity of HeLa cell using GO-I-COSAN based nanocomposites (Ferrer-Ugalde et al. 2021). The recognition of targeted cancer cell imaging in vivo and in vitro is demonstrated by using PEGylated NGO by Dai and co-workers (2011). Graphene-grafted quantum dot functionalized via poly (L-lactide PLA) and polyethylene glycol

(PEG) are used as simultaneous intracellular micro-RNA (miRNA) imaging and enhance the therapeutic efficiency of gene delivery (Dong et al. 2015). A better inhibition was observed for miRNA and surviving targeted agent. Furthermore, selective recognition of β cell lymphoma is noticed by employing NGO-PEG-Anti-CD20. Similar approach is taken for the recognition of U87 MG cancer cells using peptide target-modified NGO-PEG with Arg-Gly-Asp (Dong et al. 2015). Likewise, Kalluru et al. explored a novel nanocomposite of GO-PEG-FA for the visualization of tumors of the cancer cell (Kalluru et al. 2016). Fan et al. demonstrated the boron-doped GQD used for the detection of biomarkers via cellular imaging (Fan et al. 2014). Similarly, the On–Off mechanism using NIR fluorescence probes for the bioimaging of cancer cells and small animals like mice using transferrin functionalized gold nanoclusters (Tf-Au-NCs/GO) are also studied (Wang et al. 2013c). Antibacterial activity as well as bacteria imaging on nanoscale is performed by using RGO@Eu^{+3} and vancomycin (Eu-Van-RGO) nanocomposite (Yang et al. 2014). In addition, a heteroatom (F/Cl) doped carbon quantum dot-based fluorescence probe is used for cell imaging and antioxidant activity and studied by Marković et al. (2020).

18.5 Conclusion and Future Scopes

In this chapter, different synthesis methods for preparing graphene and its derivatives are highlighted. As these are versatile materials, their chemical and physical properties can be tuned via functionalization with different organic and inorganic moieties. In order to enhance the range of applications, it can be attached with different inorganic nanomaterials and lanthanide complexes too, leading to the formation of novel nanocomposites. The nanocomposites comprise the properties of both graphene and nanoparticles. In this way, graphene and its derivatives can be attached with different nanomaterials like metallic, transition metal oxides, sulfides, nitrides, and RE^{3+}-doped binary/ternary fluorides nanoparticles, etc. to form the nanocomposites with desired properties. In nanocomposites, graphene provides the platform to bring about chemical and physical changes. And thus, these properties of nanocomposites have opened new avenues of applications in different fields such as sensing, photocatalysis, photonics, bio-photonics, solar cells, etc.

Though a significant development has been achieved so far related to graphene and its derivative, still there are challenges which remains in the field. For example, specific attention is required for facile synthesis, functionalization, and high quality with controlled processing. Various aspects of green chemistry must be maintained during the preparation of graphene or its derivative. In addition, a complete understanding of the GO structure at the molecular level, and further exploration of the electronic properties of GO specially after functionalization with different heteroatoms or others is extremely important. Likewise, a clearer understanding of the effect of defects on the GO is necessary for effective and judicious applications in different photonic and bio-photonic domains.

Acknowledgements PG acknowledges Science and Engineering Research Board (SERB), Govt of India [Grant No. CRG/2018/003751], Board of Research in Nuclear Sciences (BRNS) [Grant No. 58/14/22/2022-BRNS/37094] for funding. HLK is grateful to Dr. H. S. Gour University for providing a graduate fellowship.

Author Contributions The manuscript was written through contributions of all authors. All authors have given approval to the final version of the manuscript.

References

Ambrosi, A., Chua, C.K., Bonanni, A., Pumera, M.: Electrochemistry of graphene and related materials. Chem. Rev. **114**, 7150–7188 (2014)

Bacon, M., Bradley, S.J., Nann, T.: Graphene quantum dots. Part. Part. Syst. Charact. **31**, 415–428 (2014)

Bai, G., Wang, J., Yang, Z., Wang, H., Wang, Z., Yang, S.: Self-assembly of ceria/graphene oxide composite films with ultra-long antiwear lifetime under a high applied load. Carbon **94**, 197–206 (2014)

Balaji, M., Mayakrishnan, A., Nithya, P., Muthulakshmi, V., Jegatheeswaran, S., Selvam, S., Sundrarajan, M.: Ornamental morphology of ionic liquid functionalized ternary doped N, P, F and N, B, F-reduced graphene oxide and their prevention activities of bacterial biofilm-associated with orthopedic implantation. Mater. Sci. Eng. C **98**, 1122–1132 (2019)

Bera, R., Kundu, S., Patra, A.: 2D hybrid nanostructure of reduced graphene oxide-CdS nanosheet for enhanced photocatalysis. ACS Appl. Mater. Interfaces **24**, 13251–13259 (2015)

Bramhaiah, K., John, N.S.: Enhanced electrocatalytic activity of reduced graphene oxide-Os nanoparticle hybrid films obtained at a liquid/liquid interface. RSC Adv. **3**, 7765–7773 (2013)

Bramhaiah, K., Bhattacharyya, S.: Challenges and future prospects of graphene-based hybrids for solar fuel generation: moving towards next generation photocatalysts. Mater. Adv. **3**, 142–172 (2022)

Bramhaiah, K., Singh, V.N., Kavitha, C., John, N.S.: Films of reduced graphene oxide with metal oxide nanoparticles formed at a liquid/liquid interface as reusable surface enhanced Raman scattering substrates for dyes. J. Nanosci. Nanotechnol. **17**, 2711–2719 (2017)

Brodie, B.C.: On the atomic weight of graphite. Philos. Trans. R. Soc. A 249–259 (1859)

Cao, Y., Yang, T., Feng, J., Wu, P.: Coating graphene oxide sheets with luminescent rare-earth complexes. Carbon **49**, 1502–1507 (2011)

Chen, C., Xie, Q., Yang, D., Xiao, H., Fu, Y., Tan, Y., Yao, S.: Recent advances in electrochemical glucose biosensors: a review. RSC Adv. **3**, 4473–4491 (2013a)

Chen, Y., Ge, H., Wei, L., Li, Z., Yuan, R., Liu, P., Fu, X.: Reduction degree of reduced graphene oxide (RGO) dependence of photocatalytic hydrogen evolution performance over RGO/$ZnIn_2S_4$ nanocomposites. Catal. Sci. Technol. **3**, 1712–1717 (2013b)

Choi, W., Lahiri, I., Seelaboyina, R., Kang, Y.S.: Synthesis of graphene and its applications: a review. Crit. Rev. Solid State Mater. Sci. **35**, 52–71 (2010)

Clark, L.C., Lyons, C.: Electrode system for continuous monitoring in cardiovascular surgery. N. Y. Acad. Sci. **102**, 29–45 (1962)

Cong, X., Cheng, C., Liao, Y., Ye, Y., Dong, C., Sun, H., Ji, X., Zhang, W., Fang, P., Miao, L., Jiang, J.: Intrinsic charge storage capability of transition metal dichalcogenides as pseudocapacitor electrode. Phys. Chem. C **119**, 20864–20870 (2015)

Deng, D., Chen, N., Xiao, X., Du, S., Wang, Y.: Electrochemical performance of CeO_2 nanoparticle-decorated graphene oxide as an electrode material for supercapacitor. Ionics **23**, 121–129 (2017)

Dong, H., Dai, W., Ju, H., Lu, H., Wang, S., Xu, L., Zhou, S.-F., Zhang, Y., Zhang, X.: Multifunctional PLA-PEG-grafted graphene quantum dots for intracellular microRNA imaging and combined specific-gene-targeting agents delivery for improved therapeutics. ACS Appl. Mater. Interfaces **7**, 11015–11023 (2015)

Dong, Y., Li, Y., Shi, H., Qin, J., Zheng, S., He, R., Wu, Z.-S.: Graphene encapsulated iron nitrides confined in 3D carbon nanosheet frameworks for high-rate lithium-ion batteries. Carbon **159**, 213–220 (2020)

El-Maghrabi, N., El-Borady, O.M., Hosny, M., Fawzy, M.: Catalytic and medical potential of a phyto-functionalized reduced graphene oxide–gold nanocomposite using willow-leaved knotgrass. ACS Omega **6**, 34954—34966 (2021)

Emtsev, K.V., Bostwick, A., Horn, K., Jobst, J., Kellogg, G.L., Ley, L., McChesney, J.L., Ohta, T., Reshanov, S.A., Röhrl, J., Rotenberg, E., Schmid, A.K., Waldmann, D., Weber, H.B., Seyller, T.: Towards wafer-size graphene layers by atmospheric pressure graphitization of silicon carbide. Nat. Mater. **8**, 203–207 (2009)

Fan, Z., Li, Y., Li, X., Fan, L., Zhou, S., Fang, D., Yang, S.: Surrounding media sensitive photoluminescence of boron-doped graphene quantum dots for highly fluorescent dyed crystals, chemical sensing and bioimaging. Carbon **70**, 149–165 (2014)

Ferrer-Ugalde, A., Sandoval, S., Pulagam, K.R., Muñoz-Juan, A., Laromaine, A., Llop, J., Tobias, G., Nunez, R.: Radiolabeled cobaltabis (dicarbollide) anion–graphene oxide nanocomposites for in vivo bioimaging and boron delivery. ACS Appl. Nano Mater. **4**, 1613—1625 (2021)

Geim, A.K., Kim, P.: Carbon wonderland. Sci. Am. **298**, 90–97 (2008)

Gu, H., Tang, H., Xiong, P., Zhou, Z.: Biomarkers-based biosensing and bioimaging with graphene for cancer diagnosis. Nanomaterials **9**, 130 (2019)

Gupta, S., Subramanian, V.R.: Encapsulating $Bi_2Ti_2O_7$ (BTO) with reduced graphene oxide (RGO): an effective strategy to enhance photocatalytic and photo electrocatalytic activity of BTO. ACS Appl. Mater. Interfaces **6**, 18597–18608 (2014)

Gupta, B.K., Thanikaivelan, P., Narayanan, T.N., Song, L., Gao, W., Hayashi, T., Reddy, A.L.M., Saha, A., Shanker, V., Endo, M., Martí, A.A., Ajayan, P.M.: Optical bifunctionality of europium-complexed luminescent graphene nanosheets. Nano Lett. **11**, 5227–5233 (2011)

Hang, P., Cong, J., Li, G., Zuo, L., Kan, C., Li, B., Xie, J., Yao, Y., Wang, Y., Chen, H., Yang, D., Yu, X.: Technoeconomically competitive four-terminal perovskite/graphene-silicon tandem solar cells with over 20% efficiency. J. Energy Chem. **63**, 477–483 (2021)

Hernandez, Y., Nicolosi, V., Lotya, M., Blighe, F.M., Sun, Z., De, S., Mcgovern, I.T., Holland, B., Byrne, M., Gun'ko, Y.K., Boland, J.J., Niraj, P., Duesberg, G., Krishnamurthy, S., Goodhue, R., Hutchison, J., Scardaci, V., Ferrari, A.C., Coleman, J.N.: High-yield production of graphene by liquid-phase exfoliation of graphite. Nat. Nanotechnol. **3**, 563 (2008)

Hofmann, U.V., König, E.: Untersuchungen uber graphitoxyd. Z. Anorg. Allg. Chem. **234**, 311 (1937)

Hollins, R.C.: Materials for optical limiters. Curr. Opin. Solid State Mater. Sci. **4**, 189–196 (1999)

Hu, C.-C., Chang, K.-H., Lin, M.-C., Wu, Y.-T.: Design and tailoring of the nanotubular arrayed architecture of hydrous RuO_2 for next-generation supercapacitors. Nano Lett. **6**, 2690–2695 (2006)

Huang, J., Tang, X., Li, Z., Liu, L.: Metal-organic frameworks derived cobalt sulfide/reduced graphene oxide composites with fast reaction kinetic and excellent structural stability for sodium storage. J. Colloid Interface Sci. **532**, 407–415 (2018)

Hummers, B.W.S., Offeman, R.E.: Preparation of graphitic oxide. J. Am. Chem. Soc. **80**, 1139 (1958)

Jafari, H., Ganjali, M.R., Dezfuli, A.S., Kohan, E.: A platform for electrochemical sensing of biomolecules based on Europia/reduced graphene oxide nanocomposite. J. Mater. Sci. Mater. Electron. **29**, 20639–20649 (2018)

Jariwala, D., Srivastava, A., Ajayan, P.M.: Graphene synthesis and band gap opening. J. Nanosci. Nanotechnol. **11**, 6621–6641 (2011)

Ju, B., Wang, X., Wu, C., Yang, X., Shu, H., Bai, Y., Wen, W., Yi, X.: Electrochemical performance of the graphene/Y_2O_3/$LiMn_2O_4$ hybrid as cathode for lithium-ion battery. J. Alloys Compd. **584**, 454–460 (2014)

Kalluru, P., Vankayala, R., Chiang, C.-S., Hwang, K.C.: Nano-graphene oxide-mediated *In vivo* fluorescence imaging and bimodal photodynamic and photothermal destruction of tumors. Biomaterials **95**, 1–10 (2016)

Kim, J., Park, S.-J., Min, D.-H.: Emerging approaches for graphene oxide biosensor. Anal. Chem. **89**, 232–248 (2017)

Krishnan, S.K., Singh, E., Singh, P., Meyyappan, M., Nalwa, H.S.: A review on graphene-based nanocomposites for electrochemical and fluorescent biosensors. RSC Adv. **9**, 8778 (2019)

Lee, J.S., You, K.H., Park, C.B.: Highly photoactive, low bandgap TiO_2 nanoparticles wrapped by graphene. Adv. Mater. **24**, 1084–1088 (2012)

Lertthanaphol, N., Pienutsa, N., Chusri, K., Sornsuchat, T., Chanthara, P., Seeharaj, P., Lohsoontorn, P.K., Srinives, S.: One-step hydrothermal synthesis of precious metal-doped titanium dioxide–graphene oxide composites for photocatalytic conversion of CO_2 to ethanol. ACS Omega **6**, 35769–35779 (2021)

Li, X., Cai, W., An, J., Kim, S., Nah, J., Yang, D., Piner, R., Velamakanni, A., Jung, I., Tutuc, E., Banerjee, S.K., Colombo, L., Ruoff, R.S.: Large-area synthesis of high-quality and uniform graphene films on copper foils. Science **324**, 1312–1314 (2009)

Li, Q., Guo, B., Yu, J., Ran, J., Zhang, B., Yan, H., Gong, J.R.: Highly efficient visible-light-driven photocatalytic hydrogen production of CdS-cluster-decorated graphene nanosheets. J. Am. Chem. Soc. **133**, 10878–10884 (2011)

Li, Y., Wang, G., Pan, K., Jiang, B., Tian, C., Zhou, W., Fu, H.: $NaYF_4$:Er^{3+}/Yb^{3+} graphene composites: preparation, upconversion luminescence, and application in dye-sensitized solar cells. J. Mater. Chem. **22**, 20381–20386 (2012)

Li, Z., Johnson, O., Huang, J., Feng, T., Yang, C., Liu, Z., Chen, W.: Enhancing the photothermal conversion efficiency of graphene oxide by doping with $NaYF_4$: Yb, Er upconverting luminescent nanocomposites. Mater. Res. Bull. **106**, 365–370 (2018)

Lingamdinne, L.P., Koduru, J.R., Chang, Y.-Y., Kang, S.-H., Yang, J.-K.: Facile synthesis of flowered mesoporous graphene oxide-lanthanum fluoride nanocomposite for adsorptive removal of arsenic. J. Mol. Liq. **279**, 32–42 (2019)

Liu, J.-W., Zhang, Y., Chen, X.-W., Wang, J.-H.: Graphene oxide–rare earth metal–organic framework composites for the selective isolation of hemoglobin. ACS Appl. Mater. Interfaces **6**, 10196–10204 (2014)

Liu, J., Ma, Q., Huang, Z., Liu, G., Zhang, H.: Recent progress in graphene-based noble-metal nanocomposites for electrocatalytic applications. Adv. Mater. **31**, 1800696 (2018)

Liu, Z., Wang, Q., Huang, X., Qian, X.: Surface functionalization of graphene oxide with hyperbranched polyamide-amine and microcrystalline cellulose for efficient adsorption of heavy metal ions. ACS Omega **7**, 10944–10954 (2022)

Lotya, M., Hernandez, Y., King, P.J., Smith, R.J., Nicolosi, V., Karlsson, L.S., Blighe, F.M., De, S., Wang, Z., McGovern, I.T., Duesberg, G., Coleman, J.N.: Liquid phase production of graphene by exfoliation of graphite in surfactant/water solutions. Am. Chem. Soc. **131**, 3611–3620 (2009)

Lu, X., Liu, T., Zhai, T., Wang, G., Yu, M., Xie, S., Ling, Y., Liang, C., Tong, Y., Li, Y.: Improving the cycling stability of metal–nitride supercapacitor electrodes with a thin carbon shell. Adv. Energy Mater. **4**, 1300994 (2014)

Luo, J., Jiang, S., Zhang, H., Jiang, J., Liu, X.: A novel non-enzymatic glucose sensor based on Cu nanoparticle modified graphene sheets electrode. Anal. Chim. Acta **709**, 47–53 (2012)

Marković ZM, Labudova M, Danko M, Matijasevic D, Micusík M, Nadazdy V, Kovacova M, Kleinova A, Spitalsky Z, Pavlovic V, Milivojevic DD, Medic M, Markovic BMT, 2020 Marković, Z.M., Labudova, M., Danko, M., Matijasevic, D., Micusík, M., Nadazdy, V., Kovacova, M., Kleinova, A., Spitalsky, Z., Pavlovic, V., Milivojevic, D.D., Medic, M., Markovic, B.M.T.: Highly efficient antioxidant F- and Cl-doped carbon quantum dots for bioimaging. ACS Sustain. Chem. Eng. **8**(43), 16327–16338 (2020)

Mo, Z., Zhao, Y., Guo, R., Liu, P., Xie, T.: Preparation and characterization of graphene/europium oxide composites. Mater. Manuf. Processes **27**, 494–498 (2012)

Mondal, A., Prabhakaran, A., Gupta, S., Subramanian, V.R.: Boosting photocatalytic activity using reduced graphene oxide (RGO)/semiconductor nanocomposites: issues and future scope. ACS Omega **6**, 8734–8743 (2021)

Muchuweni, E., Martincigh, B.S., Nyamori, V.O.: Recent advances in graphene-based materials for dye-sensitized solar cell fabrication. RSC Adv. **10**, 44453–44469 (2020)

Mukherji, A., Seger, B., Lu, G.Q.M., Wang, L.: Nitrogen doped $Sr_2Ta_2O_7$ coupled with graphene sheets as photocatalysts for increased photocatalytic hydrogen production. ACS Nano **5**, 3483–3492 (2011)

Nandi, D.K., Sen, U.K., Choudhury, D., Mitra, S., Sarkar, S.K.: Atomic layer deposited molybdenum nitride thin film: a promising anode material for Li-ion batteries. ACS Appl. Mater. Interfaces **6**, 6606–6615 (2014)

Nguyen, B.V.T., Ha, L.Q., Nguyen, T.D.L., Ly, P.H., Nguyen, D.M., Hoang, D.-Q.: Nanocellulose and graphene oxide aerogels for adsorption and removal methylene blue from an aqueous environment. ACS Omega **7**, 1003–1013 (2022)

Nia, X., Zhanga, J., Honga, L., Yanga, C., Li, Y.: Reduced graphene oxide @ceria nanocomposite-coated polymer microspheres as a highly active photocatalyst. Colloids Surf. A **567**, 161–170 (2019)

Obraztsov, A.N., Obraztsova, E.A., Tyurnina, A.V., Zolotukhin, A.A.: Chemical vapor deposition of thin graphite films of nanometer thickness. Carbon **45**, 2017–2021 (2007)

Pashaei-Fakhri, S., Peighambardoust, S.J., Foroutan, R., Arsalani, N., Ramavandi, B.: Crystal violet dye sorption over acrylamide/graphene oxide bonded sodium alginate nanocomposite hydrogel. Chemosphere **270**, 129419 (2021)

Pradhan, G.K., Padhi, D., Parida, K.: Fabrication of Fe_2O_3 nanorod/RGO composite: a novel hybrid photocatalyst for phenol degradation. ACS Appl. Mater. Interfaces **5**, 9101–9110 (2013)

Rafighi, P., Tavahodi, M., Haghighi, B.: Fabrication of a third-generation glucose biosensor using graphene polyethyleneimine-gold nanoparticles hybrid. Sens. Actuators B Chem. **03**, 1–31 (2016)

Robinson, J.T., Tabakman, S.M., Liang, Y., Wang, H., Casalongue, H.S., Vinh, D., Dai, H.: Ultra-small reduced graphene oxide with high near-infrared absorbance for photothermal therapy. J. Am. Chem. Soc. **133**, 6825–6831 (2011)

Sharma, R.K., Chouryal, Y.N., Chaudhari, S., Saravanakumar, J., Dey, S.R., Ghosh, P.: Adsorption-driven catalytic and photocatalytic activity of phase tuned In_2S_3 nanocrystals synthesized via ionic liquids. ACS Appl. Mater. Interfaces **9**, 11651–11661 (2017)

Sharma, R.K., Ghora, M., Chouryal, Y.N., Ganguly, T., Acharjee, D., Mondal, D.J., Konar, S., Nigam, S., Ghosh, P.: Multifunctional lanthanide-doped binary fluorides and graphene oxide nanocomposites via a task-specific ionic liquid. ACS Omega **7**, 16906–16916 (2022)

Shi, W., Zhu, J., Sim, D.H., Tay, Y.Y., Lu, Z., Zhang, X., Sharma, Y., Srinivasan, M., Zhang, H., Hoon, H.H., Yan, Q.: Achieving high specific charge capacitances in Fe_3O_4/reduced graphene oxide nanocomposites. J. Mater. Chem. **21**, 3422–3427 (2011)

Soren, S., Mohaptra, B.D., Mishra, S., Debnath, A.K., Aswal, D.K., Varadwaj, K.S.K., Parhi, P.: Nano ceria supported nitrogen doped graphene as highly stable and methanol tolerant electrocatalyst for oxygen reduction. RSC Adv. **6**, 77100–77104 (2016)

Stankovich, S., Dikin, D.A., Piner, R.D., Kohlhaas, K.A., Lleinhammes, A., Jia, Y., Wu, Y.: Synthesis of graphene-based nanosheets via chemical reduction of exfoliated graphite. Carbon **45**, 1558–1565 (2007)

Staudenmaier, B.L.: Verfahren zur darstellung der graphitslure. Dtsch. Chem. Ges. **31**, 1481 (1898)

Sun, J., Song, L., Fan, Y., Tian, L., Luan, S., Niu, S., Ren, L., Ming, W., Zhao, J.: Synergistic photodynamic and photothermal antibacterial nanocomposite membrane triggered by single NIR light source. ACS Appl. Mater. Interfaces **11**, 26581–26589 (2019)

Tsujimura, S., Murata, K., Akatsuka, W.: Exceptionally high glucose current on a hierarchically structured porous carbon electrode with "wired" flavin adenine dinucleotide-dependent glucose dehydrogenase. J. Am. Chem. Soc. **136**, 14432–14437 (2014)

Viculis, L.M., Mack, J.J., Kaner, R.B.: A chemical route to carbon nanoscrolls. Science **299**, 1361 (2003)

Wang, W.-S., Wang, D.-H., Qu, W.-G., Lu, L.-Q., Xu, A.-W.: Large ultrathin anatase TiO_2 nanosheets with exposed 001 facets on graphene for enhanced visible light photocatalytic activity. J. Phys. Chem. C **116**, 19893–19901 (2012)

Wang, W., Huang, W., Ni, Y., Lu, C., Tan, L., Xu, Z.: Graphene supported $NaYF_4$:Yb^{3+}, Tm^{3+} and N doped P_{25} nanocomposite as an advanced NIR and sunlight driven upconversion photocatalyst. Appl. Surf. Sci. **282**, 832–837 (2013a)

Wang, S.-J., Hu, J.-B., Wang, Y.-Y., Luo, F.: Coating graphene oxide sheets with luminescent rare-earth complexes. J. Mater. Sci. **48**, 805–811 (2013b)

Wang, Y., Chen, J.-T., Yan, X.-P.: Fabrication of transferrin functionalized gold nanoclusters/graphene oxide nanocomposite for turn-on near-infrared fluorescent bioimaging of cancer cells and small animals. Anal. Chem. **85**, 2529–2535 (2013c)

Wang, W., Li, Y., Kang, Z., Wang, F., Yu, J.C.: A NIR-driven photocatalyst based on α-$NaYF_4$:Yb, Tm@TiO_2 core-shell structure supported on reduced graphene oxide. Appl. Catal. B: Environ. **182**, 184–192 (2015)

Wang, F.H., Bae, K., Huang, Z.W., Xue, J.M.: Two-photon graphene quantum dots modified Gd_2O_3 nanocomposites for dual-mode MRI contrast agent and cell labelling agent. Nanoscale **10**, 5642–5649 (2018)

Wei, W., Cui, X., Chen, W., Ivey, D.G.: Manganese oxide-based materials as electrochemical supercapacitor electrodes. Chem. Soc. Rev. **40**, 1697–1721 (2011)

Wei, W., He, T., Teng, X., Wu, S., Ma, L., Zhang, H., Ma, J., Yang, Y., Chen, H., Han, Y., Sun, H., Huang, L.: Nanocomposites of graphene oxide and upconversion rare-earth nanocrystals with superior optical limiting performance. Small **8**, 2271–2276 (2012)

Williams, G., Seger, B., Kamat, P.V.: TiO_2-graphene nanocomposites. UV-assisted photocatalytic reduction of graphene oxide. ACS Nano **2**, 1487–1491 (2008)

Wooten, M., Karra, S., Zhang, M., Gorski, W.: On the direct electron transfer, sensing, and enzyme activity in the glucose oxidase/carbon nanotubes system. Anal. Chem. **86**, 752–757 (2014)

Xia, X., Tu, J., Mai, Y., Chen, R., Wang, X., Gu, C., Zhao, X.: Graphene sheet/porous NiO hybrid film for supercapacitor applications. Chem. Eur. J. **17**, 10898–10905 (2011)

Xiang, Q., Yu, J., Jaroniec, M.: Enhanced photocatalytic H_2-production activity of graphene-modified titania nanosheets. Nanoscale **3**, 3670–3678 (2011)

Xiong, P., Xu, S., Yang, T., Jing, K.: Novel silanized graphene oxide/TiO_2 multifunctional nanocomposite photocatalysts: simultaneous removal of Cd^{2+} and photodegradation of phenols under visible light irradiation. ACS Omega **6**, 28813–28827 (2021)

Xu, Y.-F., Yang, M.-Z., Chen, B.-X., Wang, X.-D., Chen, H.-Y., Kuang, D.-B., Su, C.-Y.: $CsPbBr_3$ perovskite quantum dot/graphene oxide composite for photocatalytic CO_2 reduction. J. Am. Chem. Soc. **139**, 5660−5663 (2017)

Xuan, X., Kim, J.Y., Hui, X., Das, P.S., Yoon, H.S., Park, J.-Y.: A highly stretchable and conductive 3D porous graphene metal nanocomposite-based electrochemical-physiological hybrid biosensor. Biosens. Bioelectron. **120**, 160–167 (2018)

Yang, X., Li, Z., Ju, E., Ren, J., Qu, X.: Reduced graphene oxide functionalized with a luminescent rare-earth complex for the tracking and photothermal killing of drug-resistant bacteria. Chem. Eur. J. **20**, 394–398 (2014)

Yazdi, R.G., Iakimov, T., Yakimova, R.: Epitaxial graphene on SiC: a review of growth and characterization. Crystals **6**, 53 (2016)

Yi, M., Shen, Z.: A review on mechanical exfoliation for the scalable production of graphene. Mater. Chem. A **3**, 11700–11715 (2015)

Zhang, Y., Tang, Z.-R., Fu, X., Xu, Y.-J.: TiO_2 graphene nanocomposites for gas-phase photocatalytic degradation of volatile aromatic pollutant: is TiO_2 graphene truly different from other TiO_2 carbon composite materials. ACS Nano **4**, 7303–7314 (2010)

Zhang, X.-L., Zhao, X., Liu, Z.-B., Shi, S., Zhou, W.-Y., Tian, J.-G., Xu, Y.-F., Chen, Y.-S.: Nonlinear optical and optical limiting properties of graphene oxide–Fe_3O_4 hybrid material. J. Opt. **13**, 075202 (2011)

Zhao, M., Gao, Y., Sun, J., Gao, F.: Mediatorless glucose biosensor and direct electron transfer-type glucose/air biofuel cell enabled with carbon nanodots. Anal. Chem. **87**, 2615–2622 (2015)

Zheng, L., Zheng, H., Huo, D., Wu, F., Shao, L., Zheng, P., Jiang, Y., Zheng, X., Qiu, X., Liu, Y., Zhang, Y.: N-doped graphene-based copper nanocomposite with ultralow electrical resistivity and high thermal conductivity. Sci. Rep. **8**, 9248 (2018)

Zhou, W., Liu, J., Chen, T., Tan, S., Jia, X., Luo, Z., Cong, C., Yang, H., Li, C.M., Yu, T.: Fabrication of Co_3O_4-reduced graphene oxide scrolls for high-performance supercapacitor electrodes. Phys. Chem. Chem. Phys. **13**, 14462–14465 (2011)

Zhou, Q., Liu, L., Guo, G., Yan, Z., Tan, J., Huang, Z., Chen, X., Wang, X.: Sandwich-like cobalt sulfide-graphene composite—an anode material of excellent electrochemical performance for sodium ion battery. RSC Adv. **5**, 71644–71651 (2015)

Chapter 19
Quantum Dots as Optical Materials: Small Wonders and Endless Frontiers

Sisir K. Sarkar

19.1 Introduction

The current lifestyle has embraced optics and photonics technologies in various ways spanning from communication, computation, energy sector, medical science, to entertainment. The last decade has witnessed advances in optical fiber communications, which led to a 100-fold increase in the amount of communicable information enabling a society-transforming internet to blossom. Charles Kao pronounced in his 2009 Nobel Prize lecture that our present daily lives have been transformed by fiber communication. Truly such communications have enabled us to visualize a "flat world" and the present internet would not have existed without optics (National Academies Press 2013).

The year 2015 celebrated the International Year of Light (IYL) and light-based technologies for promoting sustainable development to global challenges in energy, education, agriculture, health care, and security. In the last decade, advances in transmitting information, on the one hand, have enabled internet service, and on the other hand, provided the displays on our smartphones and computing devices. This is surely a promising trend for the world's economy.

Optics affecting energy conservation globally is best exemplified by the invention of light-emitting diodes (LEDs) for lighting purposes. Another prominent development is to create white light by combining blue LEDs based on gallium nitride (GaN) with fluorescent phosphors containing rare earth elements like europium and terbium. Synthetic biology has emerged as a new amazing field. This has helped genetically engineered specific optical properties to be incorporated into living organisms and effectively produce optically active materials.

Materials are playing a significant role in the technological evolution catering to new types of solar cells, sensors, and imaging of cellular functions. It is known

S. K. Sarkar (✉)
Bhabha Atomic Research Center, Mumbai 400085, India
e-mail: sisirchhaya@gmail.com

R. S. Ningthoujam and A. K. Tyagi (eds.), *Handbook of Materials Science, Volume 1*, Indian Institute of Metals Series, https://doi.org/10.1007/978-981-99-7145-9_19

from the literature that the pace of material development has always been a slow albeit costly process. But, with the rise of awareness of the importance of certain materials and advent of engineered materials, material research has assumed strategic significance. These days, key optoelectronic materials are playing a significant role in negotiations between countries.

In recent years, low-dimensional materials having strong nonlinear optical properties such as quantum dots (QDs) and core–shell quantum dots (CSQDs), have been the subject of intense research and development. Novel materials like II–VI group semiconductor quantum dots exhibit size and shape-dependent various physicochemical properties, which led to many fascinating application areas in optoelectronics, sensors, and medicines (Tong and Wang 2020; Efros and Brus 2021). Cadmium selenide, CdSe, is a vital member of this group having band gap energy, E_g of 1.75 eV at room temperature. Since this band gap matches well with the solar energy spectrum, the flexibility to tune its optoelectronic properties make them a very useful material (Peng et al. 2005).

The current synthetic methods employ rigorous conditions of temperature, pressure, passive atmosphere, toxic reducing agents, etc. However, a one-step synthesis method employing the electron beam (EB) has provided a simple efficient green method. Herein, the synthesis of CdSe quantum dot (QD) in homogeneous (water) as well as in microheterogeneous systems (water-in-oil microemulsions, ionic liquid) using EB irradiation was reported (Singh et al. 2013a).

Next, advance in this area centered on core–shell architecture, which provided an efficient approach for QDs' surface passivation. In this architecture, the robust shell effectively isolate the core QDs from the adjoining environment and suppress the formation of surface defects/traps. Besides, by appropriately selecting the core and shell materials, it is possible to control the band structure of such core–shell QDs (CSQD). The desired nonlinear optical properties can be achieved by varying the size/shape, and chemical composition of core and shell materials. In this overview, some of the interesting optical properties of QDs, as well as of CSQDs are assessed, such as nonlinear refraction, optical limiting, saturable absorption, reverse saturable absorption, etc. Various tailored optoelectronic devices including solar cells, light-emitting diodes, optical switch/limiter, QD laser, and biological/chemical applications are discussed. These successes have maintained the momentum of this subject area catering to a wide-ranging group of scientists and technologists to work for future prospective applications, which are conversed at the end (Huang et al. 2020).

The expression "optics and photonics" stated above is actually the manifestation of light's dual nature (i) a propagating electromagnetic wave, and (ii) a collection of traveling particles called photons, with similar impact potential like electronics. The best examples of these enablers are laser and associated burgeoning activities. At this point, it may be briefly stated that laser is a source of light that can be (i) coherent, meaning that a group of photons can act as a single unit, and (ii) monochromatic, meaning that the photons can have a well-defined single color. Subsequently, some more often-used terms like optics, electro-optics, optoelectronics, and photonics are described in the literature (Brown and Pike 1995).

Optics, the science of light, is based on several ideas like transmission of light, principle of least time, and well-known law of refraction. These ideas were consequently reinforced in nineteenth century by several path-breaking discoveries such as interference of light (Hooke, Young, Boyle), diffraction (Grimaldi, Fresnel), polarization (Huygens), corpuscular theory (Newton), and, undoubtedly, by Maxwell's theory on electromagnetic fields. End of this century ushered in the age of quantum optics practically ending the field of classical optics. Max Planck introduced the concept of light quanta, marking the beginning of quantum theory. With the invention of the first ruby laser in 1960, many apparently detached principles forwarded by Einstein, Bose, and others were systematized.

The terms "Electro-optics" and "optoelectronics" describe the interaction of light with electrical fields. John Kerr in 1875 established that the refractive index of materials is a function of an electric field. However, the "electro-optics" term became popular in the literature only by the early 1960s and a society called "Lasers and Electro-Optics Society" was formed in 1985 thus legitimizing the name professionally. The field of "optoelectronics" seems to be a subfield of electro-optics studying electronic devices for making source, detection, and control of light. Commonly, "optoelectronics" term refers to the quantum effects of light on semiconductor materials. Electron multiplication in Si and Ge p-n junctions were demonstrated in 1953 and then semiconductors started gaining importance in optics. Subsequently, Neumann indicated the possibility of building a semiconductor laser amplifying the stimulated emission of radiation.

The term "photonics" started appearing in literature around 1981 through reports from Bell Laboratories and Hughes Aircraft Corporation. The renowned International Society for Optics and Photonics (SPIE) organized in 1995 a huge conference in optics and photonics where photonics was showcased as the "engineering applications of light." It was involved in the detection, transmission, storing, process information, display, and generation of energy. However, presently, photonics and optics have become synonymous in the technical literature. Interestingly, based on this development, the "Lasers and Electro-Optics Society" also changed its name to "IEEE Photonics Society" in 2008.

19.2 Back to Basics: Big Excitement by Working Small

The genesis of nanotechnology is associated with two events: first, the famous lecture entitled "There's plenty of room at the bottom" by Nobel Laurate Richard Feynman at the American Physical Society meeting in 1959 (Feynman 1960), and secondly, the talk "On the basic concept of Nanotechnology" by Norio Taniguchi at the International Conference on Production Engineering in Tokyo in 1974 (Taniguchi 1974). Feynman predicted that one day, researchers will have just the right size tools for directly manipulating atoms and molecules.

This was a historic prediction in which Feynman forecasted the creation and modification of materials at a fundamental level besides reworking on nature's products. But, scientists have to wait till the last few decades to start directly working in this strange world. The excitement of working in this nano regime rests on the different behavior of materials at ultra-small levels. It is possible to control the materials' fundamental properties like color, melting temperature, hardness, conductivity, strength, etc., (Montanarella and Kovalenk 2022; Pietryga et al. 2016) at nanometer-scale structures.

Comparing the dimensions of an atom (10^{-10} m), a nanoscale material is essentially an assembly of a few atoms or molecules. Over the years, scientists have explored the properties of micrometer-sized bulk substances. It was shown, of late, that material properties become dependent on its size and shape (1–100 nm). New properties are perceived due to a lack of symmetry at the interface or electron confinement at the nanometer scale. Thus, properties of collections of atoms or molecules at the nanoscale are different from the individual constituents and also the bulk. The total fraction of atoms on the nanoparticle surface increases with decreasing the size of the particle and can be estimated by the relation: $P = 4\,N^{-1/3} \times 100$, where P is the percentage of surface atoms versus N, the total number of atoms in the particle. One can have better insight by taking an example of a 3 nm iron particle, which has 50% of its atoms on the surface, whereas 10 and 30 nm particles have just 20% and 5% atoms on the surface, respectively. Since the properties depend instead on the nature of the material, on the size, a single material can provide a consistent and continuous range of properties. Typically, gold and silver nanostructures have absorption in the visible region, while CdSe quantum dots of different sizes can have emissions covering the whole visible region.

In 1980, Russian Physicist Ekimov working at the Vavilov State Optical Institute discovered QDs in solids (glass crystals) (Ekimov et al. 1980). In late 1982, the same phenomenon was discovered by American Chemist Brus at Bell Laboratories while working with colloidal solutions (Rossetti et al. 1983). As the crystal grew over a period of days, a change in the wavelength of absorbed or emitted light was observed. This phenomenon was conceptualized in the framework of electron confinement ascribing the quantum mechanical nature of the particle.

QDs are mostly used in all kinds of applications requiring precise control of colored light. As an illustration, when a thin filter made of quantum dots fixed on top of an LED lamp, one can obtain light from a redder shade to a blueish color. Quantum dots are now replacing conventional pigments, dyes, and hi-tech reflective paints, and by embedding in another matrix, incoming light of one color can be emitted as entirely different color light.

Valence electrons of bulk phase semiconductors excited to a higher energy conduction band, energy difference being the band gap, relaxes to its ground state by emitting a photon equivalent to band gap energy. A typical semiconductor quantum dot has a core encircled by a layer of organic ligands (Fig. 19.1a) (Shiang et al. 1995). The stacking of atoms in a representative 20 nm QD consisting of 100,000 atoms is presented in Fig. 19.1b (Shiang et al. 1995).

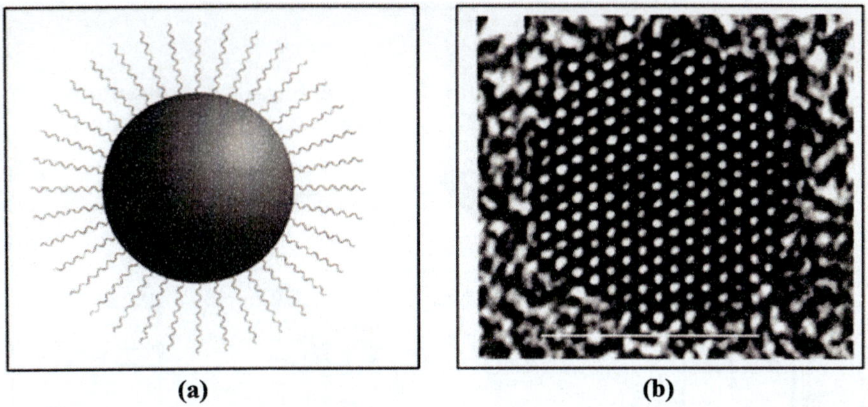

Fig. 19.1 **a** Schematics of a QD with a core (2–20 nm) and a passivating organic ligand shell. **b** High-resolution TEM image of a CdSe nanocrystal showing the array of atoms. Reproduced with permission from ACS Publications (Shiang et al. 1995)

As pointed out already, the optical properties of QDs are heavily dependent on their size. This size dependence was first observed with QD grown by molecular beam epitaxy method. 2D quantum wells having thickness comparable to the Bohr exciton radius, modifies the density of states with fewer band edge states and shifts the bandgap toward blue. Further investigations were continued to make one-dimensional (1D) quantum wires/rods and zero-dimensional (OD) quantum dots. As the QD size gets smaller than the Bohr exciton radius, the photo-excited carriers experience the boundary. This leads to discrete electronic states converting the continuous density of states of the bulk. Figure 19.2 presents the quantum confinement effect in the transition from bulk (3D) materials to quantum dots (0D). Thus, QDs are termed as "artificial atoms" and the quantum dot problem can treated as "particle-in-a-sphere" model. It is shown in the figure that strong confinement results in higher bandgap energy, which in turn shifts the prospective photoluminescence toward blue (Fig. 19.3).

The principle of size-dependent tunability is illustrated in Fig. 19.3a while such tunability of absorption is shown in Fig. 19.3b for the well-studied CdSe quantum dot. It can be seen that the first absorption feature gets red-shifted as the size becomes larger. It is to be noted that the corresponding photoluminescence (PL) always follows the same trend as in emission. Russian researchers used this level-quantization effect in QDs to modify the color of glasses by controlling the size of QDs in silicate glasses. They used a combination of CdSe and CdS QDs put into silicate glasses to produce a range of colors from red to yellow. In the last few decades, synthesis strategies were developed to get precise control of QDs' sizes to make the process commercial, and M/s Quantum Dot Corporation, USA started selling QD products in 1998. Many advanced applications started appearing in the market revolutionizing the industries of display, solar cells, and health care, especially in cancer treatment (Efros and Brus 2021).

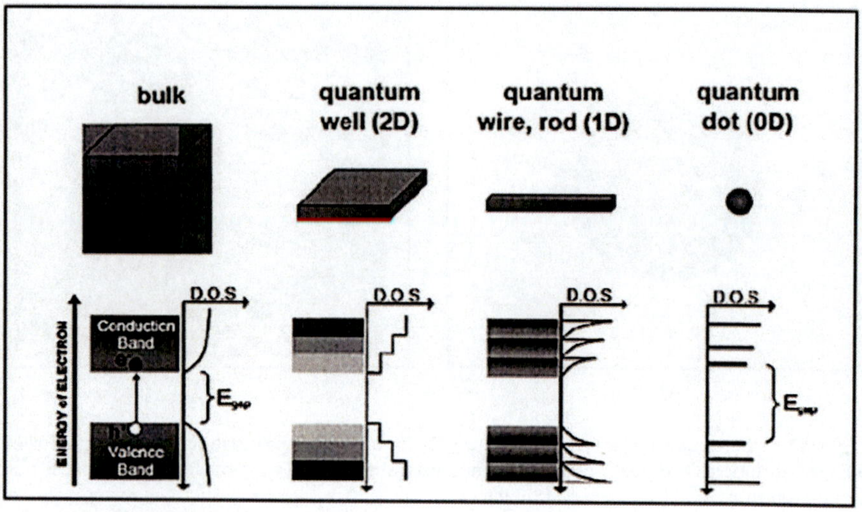

Fig. 19.2 Quantum confinement effect in transition from bulk (3D) materials to quantum dots (0D). Atom-like states are generated owing to confinement in all three dimensions. Reproduced from PhD Thesis (Steckel 2006)

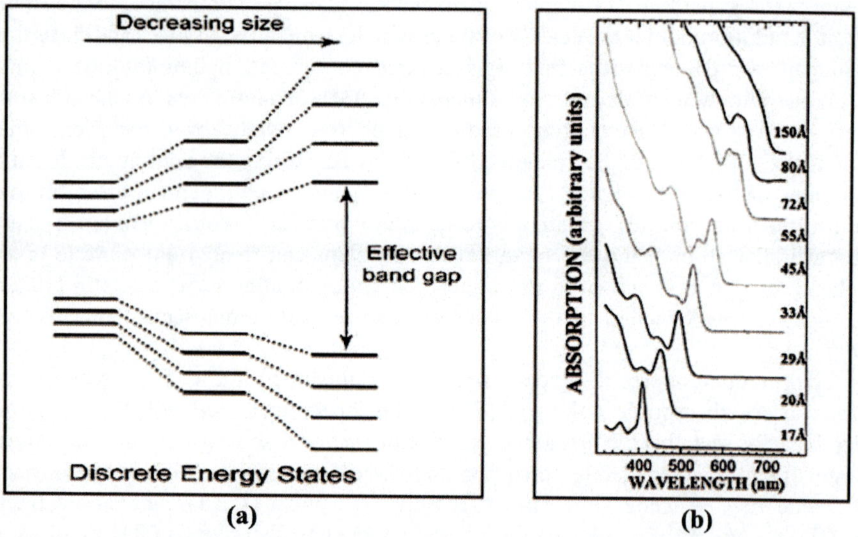

Fig. 19.3 **a** Principle of the size-tunability for QDs. Decrease in size increases the splitting of states and **b** absorption spectra for 2–15 nm CdSe nanocrystals. Reproduced from PhD Thesis (Yen 2007)

M/s Sony advanced the first display in 2013 employing CdSe quantum dots. It improved the backlighting of light-emitting diode (LED) for LCD televisions achieving better color range. Following this, QD-based displays started appearing in the market and M/s Samsung pioneered the upmarket displays acquiring QD Vision in 2016 under the brand "QLED" (using InP instead of CdSe) (Dai et al. 2017). Soon, such displays are expected to infiltrate the market, where quantum dots will be used as color filters/converters. Further, they are also working on the combination of QLED with organic-LED technologies (Palomaki 2022). QDs are expected to play a central role in developing micro-LEDs, flexible displays, and electroluminescent displays (Yang et al. 2021). Besides, researchers are also engaged in developing infrared emitting quantum dots for commercial deployment. M/s Quantum Solutions, currently selling the colloidal materials (i.e., PbS nanocrystals) and M/s Ember ion, has started putting them in infrared cameras and many solid-state devices. Judging from overall rapid deployments, nanotechnology is expected to play a crucial role in the next-generation devices useful in medicine (Shi et al. 2017; Bobo et al. 2016) and photonics (Novotny and Hecht 2006; Mohamed et al. 2021).

With this comprehensive discussion of the basic concept, we now venture into the areas of synthesis, characterization, and applications of QD first and then of CSQD. We will finally explore the future developmental issues and usages of upcoming QD-based optoelectronic devices (Huang et al. 2020).

19.3 Synthesis of QD in Batch Reactors

The motivation for chemical synthesis in solution-phase is to develop a large-scale robust method for producing QDs. It should be pointed out that term "large-scale synthesis" is a comparative concept, where some authors report the production of several grams while others several hundreds of grams. However, a robust large-scale synthesis method should operate under mild conditions with ease of scaling up irrespective of the yield in one batch process.

First, we will take up various batch processes involving hot-injection organometallic synthesis, aqueous synthesis, and biosynthesis approaches. Further, in this context, work is described toward developing a one-step efficient green synthesis method using an electron beam (EB). Using this methodology, CdSe quantum dots (QDs) are synthesized in homogeneous (aqueous) and also in microheterogeneous systems (water-in-oil microemulsions and ionic liquid) and they are characterized by the proven instrumental techniques. The radiolytic processes are shown to proceed through the reactions facilitated by hydrated electrons (e_{aq}^-). Elegant pulse radiolysis technique is used to investigate the formation dynamics of these nanoparticles. Later, a few continuous methods for reproducible large-scale production of QDs are discussed.

19.3.1 Hot-Injection Method

A customary batch reaction pioneered by Murray et al. (1993) for the synthesis of QDs is accomplished by rapidly injecting semiconductor precursors into a hot solvent and organic ligand system. A typical setup is shown in Fig. 19.4 (Murray 1995).

The formation of QDs goes through three stages. First, the reactive monomers are formed by rapid decomposition of the precursors and the monomer concentration continues to grow till a critical supersaturation occurs. This induces energetic nucleation and such nucleation surge lowers the monomer concentration relieving the supersaturation partially. In the intervening time, the existing concentration ensures growth onto the nascent nuclei. When monomer concentration becomes very low due to the continuous reaction depletion, Oswald ripening occurs. Small particles start to dissolve for compensating the growth of large particles and eventually, the growth gets defocused. High-quality, monodisperse QD samples are produced by purposefully controlling the reaction in the growth-focusing region, by adjusting the precursor's amount, the growth time, and multiple injections to keep relatively high monomer concentrations. The relatively high temperature (~300 °C) annealing was found to remove surface trap states following which PL intensity is enhanced. Annealing is believed to first carry out rearranging of semiconductor material and organic ligands at the surface leading to fewer surface trap states. This is also found to passivate the surface and increase the PL efficiency. In subsequent sections, we will have an in-depth discussion on CSQD (Sect. 19.9).

The established technique of solvent/nonsolvent extraction is done repeatedly to obtain the QDs from their original growth solution. Typically, the addition of methanol-induced flocculation of nanocrystals and butanol was added to mix the growth solution with methanol. Precipitated materials are further processed by

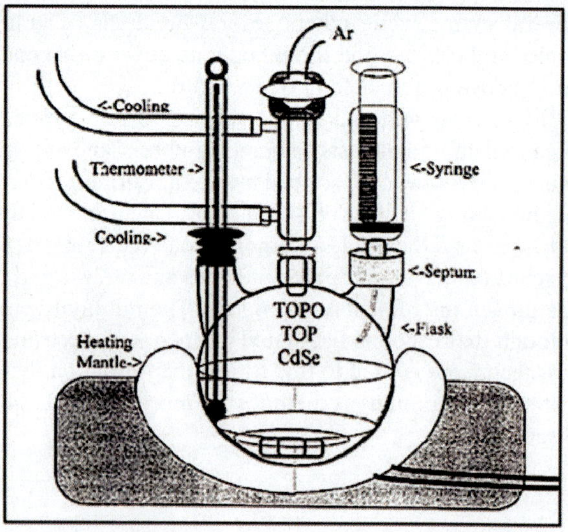

Fig. 19.4 Schematics of a hot-injection organometallic synthesis of CdSe quantum dot. Reproduced with PhD Thesis (Murray 1995)

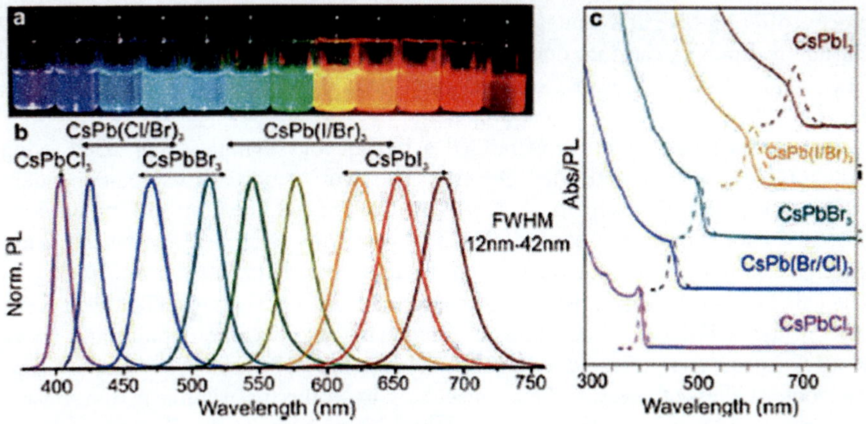

Fig. 19.5 Illustration of narrow and bright emission from perovskite $CsPbX_3$ QDs (X = Cl, Br, I) covering the entire visible region; **a** with UV lamp illumination (λ = 365 nm); **b** fluorescence spectra on excitation with 400 nm, and 350 nm for $CsPbCl_3$; **c** optical absorption and fluorescence spectra. Reproduced with permission from ACS Publications under Creative Common License (Protesescu et al. 2015)

centrifugation and dispersed in hexane. This method is also good for size-selective precipitation, by gently adding a limited amount of non-solvent (methanol).

Working with appropriate sources and synthetic parameters (Abe et al. 2013), this technique has been extended to a large variety of semiconductor QDs, such as PbS, PbSe, InP, Ag_2S, and $Ag_2S_xSe_{1-x}$. Of late, the interesting all inorganic highly luminescent perovskite QDs $CsPbX_3$ (X = Cl, Br, I) has been synthesized by injecting cesium oleate precursor solution into PbX_2 (X = Cl, Br, I) solution (Protesescu et al. 2015). The fluorescence properties (Fig. 19.5) of as-prepared perovskite QDs have comparable high luminescence of CdSe QDs. Further, Wang et al. (2016) have developed a facile solution-phase hot injection method for shape-controlled production of lead-free and stable single-crystalline perovskite derivative Cs_2SnI_6 QDs.

19.3.2 Aqueous Synthesis

The hot-injection organometallic synthesis usually requires additional phase transfer steps in water for producing QDs required in biological and clinical applications. It is worth mentioning that the disposal of a large volume of organic solvents used in large-scale production pose environmental hazards. To alleviate these issues, work was initiated on aqueous synthesis which has proved to be a potential alternative to the organometallic routes. This aqueous synthesis is also considered to be a green process offering several advantages: (1) surface functionalization required for biological applications, can be done during the synthesis step and (2) lower reaction temperatures (usually <100 °C). Incidentally, water-based synthesis was attempted in the early 1980s, when CdS QDs were prepared.

Later, Vossmeyer et al. prepared QDs by aqueous synthesis approach using $Cd(ClO_4)\cdot 6H_2O$, H_2S, and 1-thioglycerol. This work reported the size-dependent UV–visible absorption spectra of CdS QD but did not mention any photoluminescence data (Vossmeyer et al. 1994). The first photoluminescence investigation through this route was published by Weller and coworkers (1996) on CdTe QDs. The CdTe QDs were made by refluxing a mixture of Cd^{2+} with NaHTe at 96 °C for several hours. Following this, aqueous synthesis route became popular and many II–VI QDs, such as CdTe, CdSe, HgTe, CdHgTe, and PbS, were synthesized. Later, this route witnessed several advances such as hydrothermal synthesis, microwave irradiation, and ultrasonic treatment. Recently, Yakoubi et al. (2016) have produced excellent quality CdZnS and Cu-doped CdZnS QDs through a facile and mild aqueous approach. The latter one, a highly emissive QD is made using novel capping agents like 3-mercaptopropionic acid (MPA) or N-acetylcysteine (NAC).

19.3.3 Biosynthesis

Biosynthesis or biomanufacturing is another promising route for the synthesis of metal sulfide QDs like CdS, ZnS, PbS, and Ag_2S (Prasad and Jha 2010) utilizing metal and sulfide ions as precursors and employing the intrinsic enzymatic machinery of microorganism. These ions using the magnesium or manganese transport system first enter into the cell cytoplasm and then get converted into nanocrystals by the intracellular enzymes located in the cytoplasm. Another route was developed known as extracellular synthesis where synthesis of QDs is caused by the enzymes occurring on the cell membrane or excreted to the medium. A single CdSe QD within yeast cells has been produced via genetic engineering of intracellular redox conditions. Further, it has been demonstrated that psychrotolerant, oxidative stress-resistant bacteria from Antarctic microorganisms are able to biosynthesize CdS QDs at low temperatures. Using an engineered strain of stenotrophomonas maltophilia (SMCD1), biosynthesis of CdS and PbS QDs was carried out reproducibly (Yang et al. 2016a). SMCD1 is capable of the direct extracellular biomineralization of CdS QD from a buffered aqueous solution of cadmium acetate and L-cysteine (Fig. 19.6) (Yang et al. 2016a).

In this route, L-cysteine performs both as a sulfur source and the capping agent. The optical properties of the QD can conveniently be controlled by varying the growth time. The progress of this low-temperature, aqueous phase synthesis route clearly indicates its superiority over chemical synthesis. This inherently green synthesis route has the potential to cut down production costs and lead to QDs with new/ improved functionality (Yang et al. 2016a).

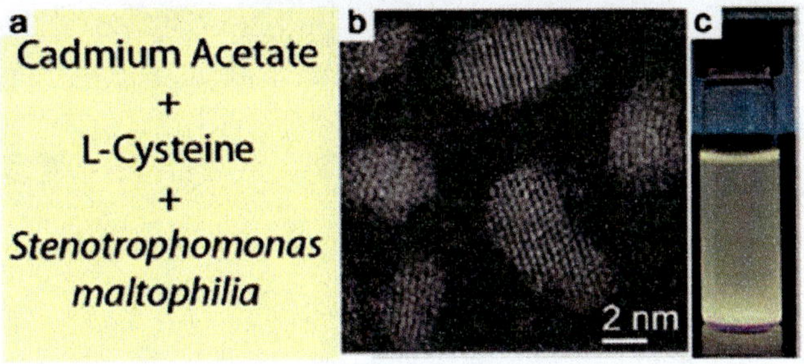

Fig. 19.6 **a** Components for biosynthesis of CdS QDs; **b** its high-resolution TEM image; **c** the visible fluorescence under UV illumination. Reproduced with permission from ACS Publications (Yang et al. 2016a)

19.4 Synthesis in Continuous Reactors

Typically, many reaction parameters were required to be controlled such as temperature, stirring rate, precursor injection position, and mixing to achieve good quality of the QDs. Gram-to-sub-kilogram quantities of QDs have been made in a single batch with proper process control. Recently, Bang et al. (2017) reported the single-batch gram-scale preparation of InP QDs using white phosphorus (P_4) as the precursor. However, in a traditional batch reactor, achieving proper control of the reaction parameters is challenging. To alleviate this, a continuous reactor concept is adopted where the reactants are fed continuously into the reactor and the product is continuously taken out. Since the parameters like temperature, pressure, and residence time are better controlled, a more consistent quality product is obtained.

19.4.1 Synthesis of QDs in Microfluidics

Microfluidic reactor has attracted attention as a promising technology for QDs synthesis since the first report in 2002 (Bang et al. 2010). Microfluidics are essentially continuous reactors integrated with heaters and fluid control elements. This arrangement provides much better control than in conventional macroscale batch-type reactors. Currently, such synthesis in microfluidic reactors is gaining much prominence.

Kikkeri and coworkers (2010) developed the microfluidics method to synthesize surface-functionalized CdSe and CdTe QDs (Fig. 19.7). The chemical reactions are similar to organometallic synthesis of QDs in batch reactors (Sect. 19.3.1). The most striking feature is its operation at a lower temperature (160 °C) than in a batch reactor

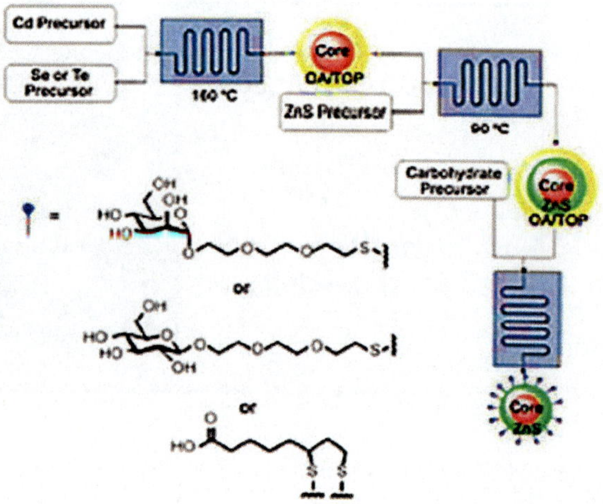

Fig. 19.7 Microfluidic reactor incorporating continuous-flow synthesis of functionalized QDs (OA: oleic acid; TOP: tri-n octylphosphine). Reproduced with permission from Wiley Publications (Kikkeri et al. 2010)

(~300 °C). Both CdSe/CdTe core and ZnS shell can be made in such continuous-flow microchannels. The sizes of the QDs can be tweaked by varying the reaction time in the reactor.

19.4.2 *Thermospray Synthesis*

This synthesis route requires organic ligands which get coated on the nanoparticles' surface preventing the agglomeration of QDs to make them highly dispersed in liquid solutions or solid composites. High-quality semiconductor QDs can be obtained by such spray-based method (Amirav et al. 2005). Using a thermospray nebulizer, aqueous or organic precursor solutions of semiconductor salts are first sprayed (Fig. 19.8) as monodispersed droplets. The precursor reaches over saturation as the droplets move forward and spontaneous condensation occurs due to the solvent evaporation. The ensuing QDs are free-standing, unsupported, and uncoated since each semiconductor dot is produced from a single spray droplet. Nanocrystals obtained using spray-based methods include the II–VI group like CdS, ZnS, and MnS, the IV–VI group PbS, and the metal sulfide such as MoS_2. Thermospray synthetic approach looks very attractive for all applications where films of uncapped and packed QDs are required.

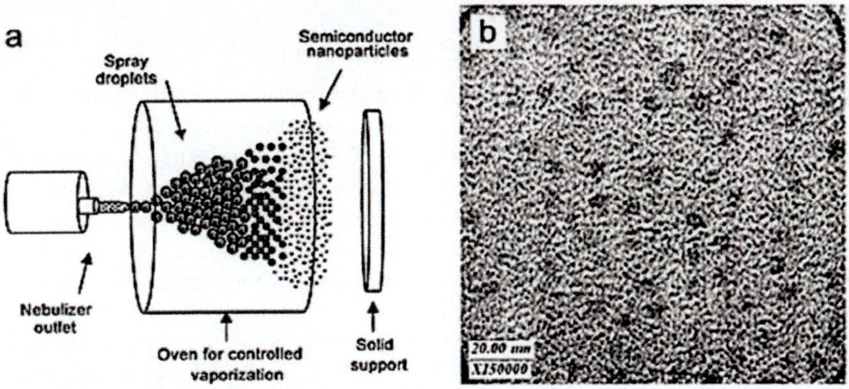

Fig. 19.8 **a** Production of CdS nanocrystals thermospray synthesis; **b** TEM image of CdS QDs (5.1 nm). Reproduced with permission from ACS Publications (Amirav et al. 2005)

19.5 Green Approach Using Electron Beam Irradiation

Radiation-assisted synthesis of nanomaterials is a very promising technique having several advantages: (a) simple, fast, and efficient method under moderate conditions, (b) commercial scale production, (c) since electron being a very strong reducing species ($E_{red} = -2.9$ V) does not require any external reducing agents, and (d) size controlling can be achieved by proper dose and irradiation time.

Sea urchin-like shaped CdSe nanoparticle is produced by this approach in aqueous solutions using citric acid (CA) as a capping agent. The sea urchin-like shaped nanocrystals being 3D in nature exhibit several novel properties, like good conductivity and a larger surface area and useful in several application areas of gas sensors, catalysis, and photovoltaics. Different shapes and sizes of the nanoparticles have been obtained depending on the types of templates or capping agents used during the growth (Liang-Shi et al. 2001). Elias et al. have reported the urchin-like nanostructures of ZnO, which holdgreat promise for photovoltaic applications (Elias et al. 2010). Moreover, compared with the 1D nanomaterials (i.e., nanorods, nanowires, nanotubes, and nanobelts), 3D ordered nanostructural materials can be regarded as the self-assembly of 1D nanomaterials, thereby increasing structural complexity and endowing the possibility of greater functions for some unique applications. Nonetheless, the concentrations of the individual precursors and the CA were optimized keeping in view the desired characteristics of the nanomaterials, i.e., colloidal stability, monodispersity, size tunability, and good PL properties. In the following sections, various techniques are presented which are normally used to characterize QDs. All these investigations along with the transient absorption studies performed by pulse radiolysis technique provided insight into the possible mechanism of such nanoparticle formation (Singh et al. 2013a).

19.5.1 Experimental Details

The precursor solutions were prepared with nanopure water. Sodium selenosulphate (Na_2SeSO_3), precursor for Se was made from 1 g Se powder, 10 g Na_2SO_3 in 50 ml nanopure water by refluxing the solution at 70 °C for 7 h. An appropriate amount of 25% ammonia was added to $CdSO_4$ solution, precursor for Cd till a clear ammoniated solution resulted. Subsequently, to this solution was added first freshly prepared Na_2SeSO_3 solution and then tertiary butanol. The capping agent CA was added to the reaction mixture just before the irradiation. This reaction mixture was irradiated with an electron beam from 7 meV linear electron accelerator (LINAC) at 12 pps rep rate delivering a dose of 140 Gy/pulse. Since the added tertiary butanol scavenged the radiolytically produced OH radicals, the coproduced e_{aq}^- interacts with the precursors inducing the synthetic reaction. The solution turned greenish-yellow or reddish-orange in color on irradiation, demonstrating the formation of the CdSe nanoparticles. The concentrations of CA and the precursor solutions were controlled appropriately to achieve the preferred morphology and the desired optical properties of the nanoparticles.

19.5.2 Pulse Radiolysis Technique

Elegant technique of pulse radiolysis was used to investigate the formation dynamics of the CdSe nanoparticles with a kinetic spectrometer. The aqueous solutions of both precursors containing equimolar (1 mM) amount mixed with 5 mM CA and 1 M tert-butanol. The transient species were monitored using white light from an Xe lamp (450 W) placed orthogonally to the electron beam. The dose measurement was carried out with a potassium thiocyanate, KSCN chemical dosimeter. The time-resolved kinetic traces of the generated intermediate species were recorded (Fig. 19.9) (Singh et al. 2013a).

N_2-purged solutions were used for all the experiments. Figure 19.9 exhibits two distinctive absorption peaks one at 380 nm (band-I) and the other at 510 nm (band-II). It may be noted that the band-I showed weaker absorbance compared to the band-II.

These observations corroborate well with the report of radiolytic formation dynamics in aqueous solution (Singh et al. 2011a). A plausible reaction mechanism for the synthesis of CdSe nanoparticles can be given:

$$e_{aq}^- + [Cd(NH_3)_4]^{2+} \rightarrow [Cd(NH_3)_4]^{\bullet +} \quad (k = 3.1 \times 10^{10}\, M^{-1}\, s^{-1})$$

$$e_{aq}^- + [SeSO_3]^{2-} \rightarrow [SO_3]^{2-} + Se^{\bullet -} \quad (k = 2.3 \times 10^{9}\, M^{-1}\, s^{-1})$$

$$[Cd(NH_3)_4]^{\bullet +} + Se^{\bullet -} \rightarrow (CdSe)_{solvated} \rightarrow (CdSe)_{np}$$

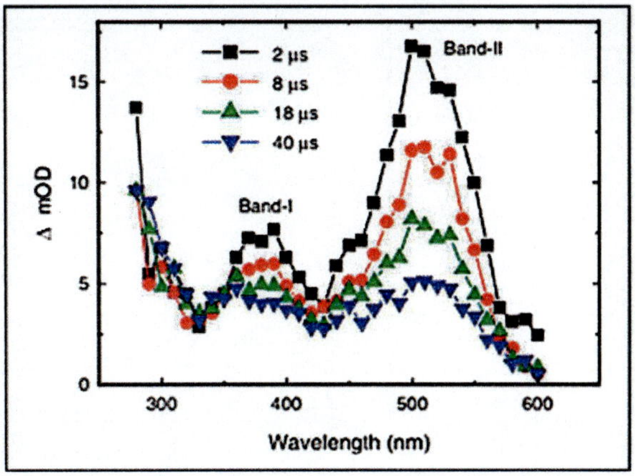

Fig. 19.9 The time-resolved kinetic traces of the generated intermediate species: [precursor] = 1 mM; [CA] = 5 mM and absorbed dose = 25 kGy. Reproduced from American Scientific Publishers as author of this work (Singh et al. 2013a)

The CA molecules facilitate in attaining a sea urchin-like structure by chelating the primary nanoparticles formed which could be due to their tridentate capping nature and short chain length. It may be noted that during the production of nanoparticles, many complex transient species are also generated similar to those reported in Singh et al. (2011a):

$$\left[Cd(NH_3)_4\right]^{2+} : Se^{\bullet -};\ \left[Cd(NH_3)_4\right]^{\bullet +} : [SeSO_3]^{2-};\ \text{and}\ \left[Cd(NH_3)_4\right]^{\bullet +} : Se^{\bullet -}$$

19.6 Characterization of QD

19.6.1 Optical Absorption

The unirradiated solutions containing precursors (1 mM each of Cd and Se precursor) and CA (5 mM) did not show any absorption in the 350–700 nm region (Fig. 19.10). However, the solutions on irradiation showed a clear excitonic peak at around 420 nm. This result confirms that the formation of CdSe nanoparticles takes place by electron beam irradiation. Three important parameters, namely, absorbed dose, precursors' concentrations, and the capping agent, CA were optimized. The solutions irradiated at an absorbed dose of 25 kGy (containing 5 mM CA) changed color from pale yellow to dark orange, as the precursors' concentrations were increased from 1 to 10 mM. These changes are clearly visible in the camera-ready pictures as shown in the top panel (I) of Fig. 19.11 (Singh et al. 2013a).

It may be mentioned that aqueous synthesis of CdSe nanoparticles using equimolar ammoniated $CdSO_4$ and Na_2SeSO_3 solutions, through electron beam irradiation, has been previously reported (Singh et al. 2011b). Therefore, the present investigation expected to produce CdSe nanoparticles was confirmed by XRD measurements. CA

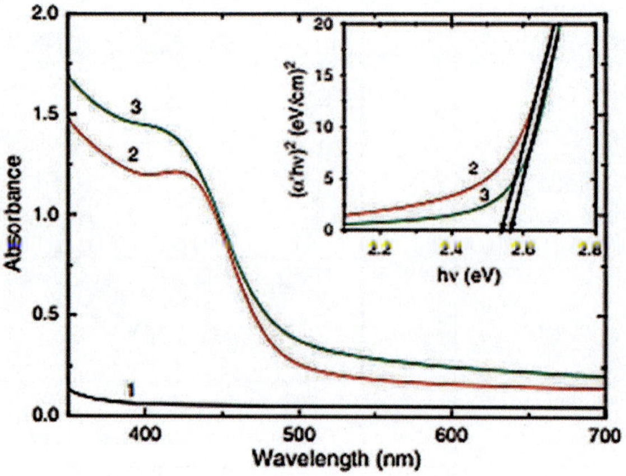

Fig. 19.10 Absorption spectra of unirradiated (1) and irradiated (2, 3) reaction mixtures (25 kGy) containing different equimolar precursor concentrations with a fixed concentration of CA, 5 mM. Precursor concentrations in mM: (1) 0.5 (2) 0.5 (3) 1.0. Inset: Tauc plot of $(\alpha' h\upsilon)^2$ versus $h\upsilon$ for band gap (E_g) determination. Reproduced from American Scientific Publishers as author of this work (Singh et al. 2013a)

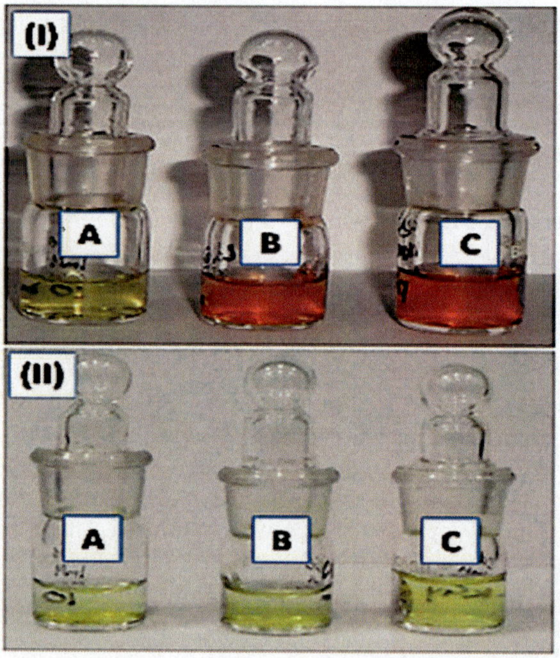

Fig. 19.11 Camera-ready pictures of CdSe prepared by electron beam irradiation. **I** Absorbed dose is kept constant at 25 kGy, [precursors] in mM: (A) 1 (B) 5 (C) 10; **II** fixed [precursors] = 1 mM, fixed [CA] = 5 mM, absorbed doses in kGy: (A) 10 (B) 25 (C) 40. Reproduced from American Scientific Publishers as author of this work (Singh et al. 2013a)

is also known to be a good reducing agent ($E = -1.1$ V vs. Ag/AgCl) and therefore its concentration was optimized in order to avoid any chemical reduction process. It was further observed that the CdSe particles precipitate at a concentration of precursors higher than that of CA, which could be due to the fact that the available concentration of the capping agent was not able to cap the as-grown CdSe nanoparticles efficiently. Therefore, the concentration of the precursors was optimized at 1 mM, where the nanoparticles do not settle down through further agglomeration. Besides, it was noticed that CA with a concentration $\leq$5 mM, produces well-separated smaller spherical nanoparticles. Further, a higher concentration of CA could lead to destabilization effect and induce coalescence, hence was avoided in this study. From these observations, optimized concentrations of precursors and CA were found to be 1 and 5 mM, respectively.

Using the absorption spectra and modified Brus equation the primary CdSe nanoparticles' sizes were calculated and are listed in Table 19.1

$$E_g = E_g(0) + \alpha/d^2$$

where $\alpha = 3.7$ eV nm^2, $E_g(0) = 1.75$ eV, d = particle size in nm, and E_g = band gap in eV. Since this is a direct band gap semiconductor the band gap values were determined from the Tauc plot of $(\alpha' h\upsilon)^2$ versus $h\upsilon$ (inset of Fig. 19.10). The symbol "α" represents the absorption coefficient multiplied with the concentration of the CdSe nanoparticles, calculated from the relation (2.303 A/l), where "A" is the absorbance and "l" represents the cell length (10 cm) and the term "$h\upsilon$" is the photon energy. The estimated values of the particle sizes were between 2.07 and 2.70 nm.

The dependence of absorbed dose on the particle formation was investigated under optimized concentrations of CA and precursors. The camera-ready pictures (showing the dose effect) of the irradiated sol containing the CdSe nanoparticles are shown in the bottom panel (II) in Fig. 19.11. A pale yellow color of the sols with no difference in all three cases is apparent. However, from Fig. 19.10, it is evident that the absorbance increases with the increase of dose (10–25 kGy) and then starts decreasing up to 40 kGy. This was associated with a slight red shift in the excitonic peak position. Correspondingly the particle size increased from 2 to

Table 19.1 Optical absorption measurements yield the CdSe band gap (eV) and nanoparticle size (nm) synthesized with electron beam irradiation (Singh et al. 2013a)

Absorbed dose (kGy)	[Precursor] (mM)	[CA] (mM)	E_g (eV)	Particle size (nm)
25	0.5	5	2.54	2.08
25	1.0	5	2.56	2.07
25	5.0	5	2.25	2.59
25	10.0	5	2.20	2.70
10	1.0	5	2.62	2.00
25	1.0	5	2.56	2.07
40	1.0	5	2.45	2.22

2.22 nm. Instantaneous agglomeration of the particles was observed when the dose exceeded 25 kGy possibly due to radiolytic degradation of CA. Therefore, under the present experimental conditions, 25 kGy seemed to be the optimum dose for the precursors' concentrations of 1 and 5 mM for CA.

19.6.2 TEM, SEM, and XRD

Various characterization techniques TEM, SEM, and XRD were applied for CdSe nanoparticles. The TEM images (Fig. 19.12a) indicated an agglomerated form of CdSe nanoparticles of size about 500 nm, closely resembling sea urchin-like shapes (Singh et al. 2013a). However, the primary nanoparticles were within 3 nm in size. These values were in well agreement with those obtained from the absorption studies. The small angle electron diffraction (SAED) patterns of these nanoparticles show their amorphous nature. The SEM images shown in Fig. 19.12b, also confirm the existence of particles with sea urchin-like shapes and these agglomerates consisted of smaller units of identical patterns (Singh et al. 2013a).

Miller indices for the diffraction patterns are marked in the figure (Fig. 19.13a) and are attributable to those of CdSe (JCPDS Card No. 19-0191) (Singh et al. 2013a). The broadening in the peaks specifies the amorphous nature of synthesized CdSe nanoparticles. The growth process is restricted by the capping agent, CA and so, smaller size (<3 nm) CdSe nanoparticles could be obtained by electron beam irradiation method. However, the primary CdSe nanoparticles being ultra-small and amorphous in nature possess high surface energy, and therefore undergo oriented aggregation of nanoparticle subunits assisted by the CA molecules. In fact, CA is a tridentate chelating ligand having three carboxylic groups and one hydroxyl group, thereby expectedly acting as a structure-directing molecular-coordination stabilizer. Consequently, the

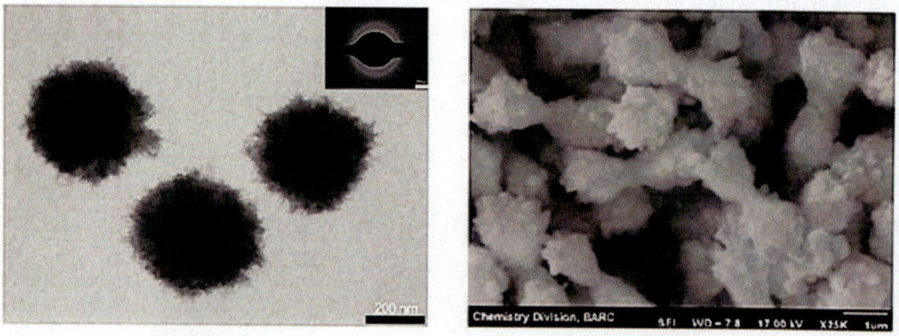

(a) (b)

Fig. 19.12 **a** TEM and **b** SEM image of synthesized CdSe nanoparticles with electron beam irradiation. The parameters are: [precursor] = 1 mM, [CA] = 5 mM; absorbed dose = 25 kGy. Reproduced from American Scientific Publishers as author of this work (Singh et al. 2013a)

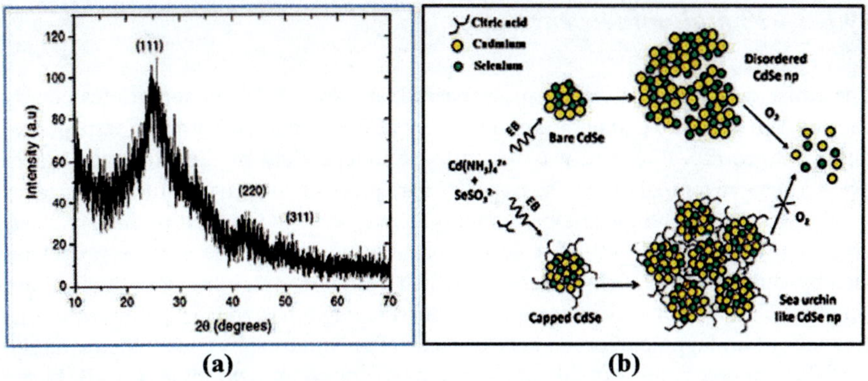

Fig. 19.13 **a** Room temperature XRD pattern of as-grown CdSe nanoparticles. **b** Scheme 1 showing different growth patterns upon EB irradiation in the presence and the absence of capping agent (CA). Reproduced from American Scientific Publishers as author of this work (Singh et al. 2013a)

chelating ability of this molecule for the initially formed amorphous nanoparticles is most likely leading to a 3D anisotropic morphology closely resembling that of sea urchins (Fig. 19.13b; Scheme 1). Basically, seeded growth process along with the kinetically driven mechanism lead to the anisotropic shapes of the nanoparticles. Thus, the phenomenon can be described on the basis of the interaction between the stochastic diffusive forces arising out of nucleation and growth of initially formed primary CdSe nanoparticles into spherical aggregates. They subsequently re-orient into the radially expanding sea urchin-like shapes (see Scheme 1), under the directive forces of CA molecules.

This is further corroborated by the mechanism proposed by Kamat (2008) where it had been shown that initially a few seeds of CdSe nanoparticles are formed, which strongly complex with the citrate anions. It reaches an optimal size by gradual growth of the agglomerates and after reaching that stage further aggregation is prevented by the strong repelling layer of citrate. Subsequently, the Ostwald ripening process produces larger particles. However, in the absence of capping agents, the bare CdSe nanoparticles do not exhibit any preferred morphology, rather settled in disordered agglomerated structures. The uncapped CdSe nanoparticles were found to be unstable upon exposure to air/oxygen (see Fig. 19.13b; Scheme 1), while the capped nanoparticles synthesized in the above studies were found to be quite stable. Therefore, it is recognized that the CA molecules help in achieving the stabilized sea urchin-like shaped nanoparticles and prevent them from reacting with oxygen unlike that observed in the case of bare CdSe nanoparticles (Singh et al. 2010).

19.6.3 Photoluminescence (PL)

The emission spectra of the freshly prepared as-grown CdSe nanoparticles clearly showed 520 nm band gap photoluminescence (BG-PL) and 600 nm trap state photoluminescence (TS-PL), and was found to be independent of the excitation wavelengths at room temperature. The trap states are probably due to unsaturated surfaces involving dangling bonds arising from high dose rate encountered in this synthesis process (Singh et al. 2010). CA is also capable of altering the surface properties thereby reducing the surface defects, which result in obtaining BG-PL in these nanoparticles. However, the uncapped nanoparticles synthesized in aqueous solutions containing similar precursors do not exhibit any PL (Singh et al. 2011b). Interestingly, the CdSe nanoparticles formed in cetyl trimethyl ammonium bromide (CTAB) based water-in-oil microemulsions via electron beam irradiation exhibit strong TS-PL at room temperature (Singh et al. 2013b). However, the CdSe nanoparticles formed through a similar synthesis route in the present study exhibit both BG-PL as well as TS-PL. This demonstrates that CA plays a dual role in stabilizing these nanoparticles under ambient conditions and enhancing the radiative decay pathways for the photoexcited carriers in these nanoparticles.

Using the TCSPC instrument, the time-resolved PL measurements were carried out at room temperature, the excitation and the emission wavelengths, respectively, being 445 and 580 nm (Fig. 19.14).

Assuming tri-exponential decay, the profile was best fitted with the parameters shown (in red) in the figure. The weighted average lifetime was estimated to be 350 ps. The short lifetime component may be due to the BG-PL, whereas the other

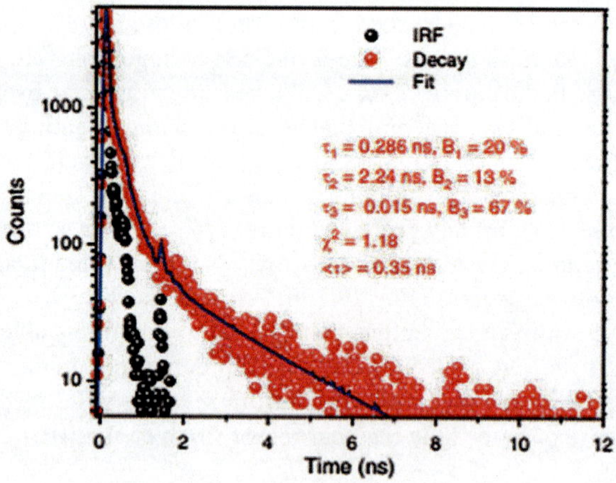

Fig. 19.14 Lifetime emission decay curve (red dots) and IRF (black dots) and the best fitted curve for electron beam synthesized CdSe nanoparticles. The parameters are: [precursor] = 1 mM, [CA] = 5 mM, dose = 25 kGy; and $\lambda_{exc} = 445$ nm, $\lambda_{emi} = 580$ nm. Reproduced from American Scientific Publishers as author of this work (Singh et al. 2013a)

two arise from the trap states. The decay pattern indicates the existence of different deep and shallow trap states in these nanoparticles. A similar trend of PL lifetime values was seen in case of synthesis in CTAB-based water-in-oil microemulsions (Singh et al. 2013b), but the contribution from the BG-PL was quite less compared to the present case. The weighted average lifetime is found substantially longer (4.1 ns) and possibly due to the variance in the surface structure in these two cases.

19.6.4 Raman Spectroscopy

Raman spectroscopy reveals the composition, crystal quality w.r.t. orientation, symmetry, and lattice stress/strain effects of II–VI semiconducting nanomaterials. Room temperature Raman spectra of as-grown CdSe nanoparticles capped with CA (Fig. 19.15) comprises of fundamental longitudinal optical (LO) phonon peak at 206.9 cm^{-1} and its overtone (2LO) peak at 411 cm^{-1}, respectively. An additional shoulder indicating transverse optical (TO) phonon mode is shown near 168 cm^{-1} (Mauro et al. 2008). In addition, the spectrum contains background, which reveals the luminescent nature of as-grown nanoparticles.

A weak wide peak at ~250 cm^{-1} seen in the Raman spectra is due to the small size (2–3 nm, Table 19.1) of these nanoparticles along with their amorphous nature (from XRD spectra and SAED pattern). The appearance of the surface optic (SO) modes near 250 cm^{-1} can be attributed to the extra atoms present on the surface compared to the bulk. There is a red shift observed for LO peak of CdSe nanoparticles from its bulk position at 210–213 cm^{-1}. The size reduction is mostly controlled by quantum confinement but dimensional effects like surface reconstruction and lattice contraction also play some role.

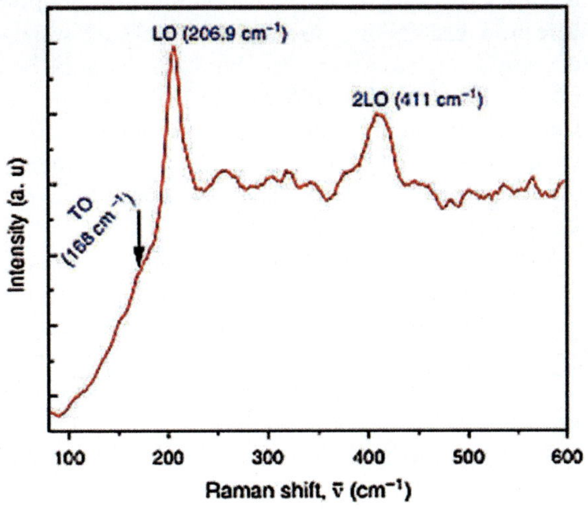

Fig. 19.15 Raman Spectra of electron beam synthesized CdSe nanoparticles. The parameters are: [precursor] = 1 mM, [CA] = 5 mM, and absorbed dose = 25 kGy. Reproduced from American Scientific Publishers as author of this work (Singh et al. 2013a)

The reduced dimensions bring about competition between confinement and strain effects. Consequently, LO peaks experience red shift with respect to their bulk when phonon confinement or strain effect occurs. However, the present observed red shift is due to the phonon confinement effect because the primary nanoparticles (between 2 and 3 nm; Table 19.1) reasonably smaller than the Bohr exciton radius of CdSe (5.6 nm). Further, the LO peak asymmetry with its broadening is sign of the phonon confinement and the structural disorder on the surface. This structural disorder responsible for the rise of surface defects, is corroborated from the broad PL spectra of as-grown CdSe nanoparticles. However, XRD analysis indicated tensile stress leading to lattice contraction. The lattice constant value, "*a*" was found to be 5.97 Å, which is slightly lower than cubic CdSe (6.07 Å). The reason for this lattice contraction is probably due to the surface tension-induced reconstruction on the surface. In effect, the red shift in the LO peak relative to the bulk appears to be the combined result of the phonon confinement and tensile stress. However, the former effect seems to have the dominant role in the observed shift.

19.7 Shape Evolution of Nanomaterials in Microheterogeneous Media

Since QDs' shape is crucial in deciding newer material properties, synthesis of CdSe nanomaterials is investigated in microheterogeneous media (i.e., water-in-oil type microemulsions) using the surfactant, cetyl trimethyl ammonium bromide (CTAB) and cyclohexane as the oil. The central interest is to examine whether the shape evolution of the QDs in such a medium is different compared to an aqueous solution. The investigations were carried out under two different experimental conditions (Singh et al. 2013c): (i) by normal chemical route at room temperature and (ii) by 7 meV electron beam irradiation route. In both the routes, CdSe nanomaterials were produced first in isotropic spherical shape and then evolved to anisotropic rod-like structures. However, the photoluminescence of the CdSe nanomaterials was found to be remarkably different in two cases. In the chemical route, initially, the band gap photoluminescence was observed which changed with aging. After a day, two distinct BG-PL along with TS-PL were recorded. On the other hand, radiation chemical synthesis provided dominant trap state photoluminescence (TS-PL), which persisted even after aging.

In-depth mechanistic investigation on the electron beam synthesis of CdSe nanomaterials was carried out in water-in-oil type microemulsions with less water content, ($w_0 < 20$). It is well established that water undergoes radiolysis and yields three major primary radial species, out of which e_{aq}^- and H are reducing and OH is oxidizing in nature (Spinks and Woods 1976). Tert-butanol (CH_3 $(CH_3)_2COH$) added in the aqueous solution leaves behind hydrated electrons, e_{aq}^- after scavenging the OH/H radicals. This can be termed as a perfectly reducing condition where the reduction is done through the e_{aq}^- species. This reduction of precursor ions produced CdSe nanomaterials (Singh et al. 2011a). Thus, under the above conditions, the steps involved are similar to that described in Sect. 19.5.2:

Radiolysis of water:

$$H_2O \rightsquigarrow e_{aq}^-,\ OH^\bullet,\ H^\bullet,\ H_3O^+,\ H_2,\ H_2O_2$$

Followed by

$$OH\bullet\ (H\bullet) + CH_3(CH_3)_2COH \rightarrow \bullet CH_2(CH_3)_2COH + H_2O\ (H_2)$$

$$e_{aq}^- + \left[Cd\ (NH_3)_4\right]^{2+} \rightarrow \left[Cd\ (NH_3)_4\right]^{\bullet +}$$

$$e_{aq}^- + [SeSO_3]^{2-} \rightarrow [SO_3]^{2-} + Se^{\bullet -}$$

$$Cd^{\bullet +} + [SeSO_3]^{2-} / Se^{\bullet -} \rightarrow (CdSe)_{intermediate} \rightarrow (CdSe)_{np}$$

The radiolytic synthesis turns the colorless microemulsion into a greenish-yellow color. The TEM image of the as-grown CdSe sols showed isotropic spherical nanoparticles (<5 nm) which agglomerated to relatively bigger sizes. The optical absorption spectra showed an intense 450 nm peak. However, the TEM image of the sol recorded after a day's aging, confirmed the rod-shaped CdSe nanomaterials and its absorption peak showed a slight red shift as seen in Fig. 19.16a. It can be seen that the room temperature PL peak at 600 nm (TS-PL) dominates over the hump at 525 nm (BG-PL) (Fig. 19.16b), which gets further augmented with aging. As already shown, this observation was quite different from the result of the chemical route. This is possibly due to the combined effect of internal structure and the defect sites. It is known from the literature that radiation-induced synthesis normally produces nanomaterials with disordered structures with low crystallinity and higher defect density (Singh et al. 2010, 2011b; Kamat 2008). Therefore, the observed photoluminescence behavior may be ascribed to the above factors.

19.8 Shape Evolution in Room Temperature Ionic Liquids (RTIL)

Room temperature ionic liquids (RTIL) have many distinctive properties like very low vapor pressure, good ionic conductivity, large polarity, and stability. These class of compounds have recently been used as media for the nanomaterials synthesis. Though a couple of reviews have already appeared in the literature (Ma et al. 2010), the literature is sparse regarding the role of in-built structure of IL in regulating

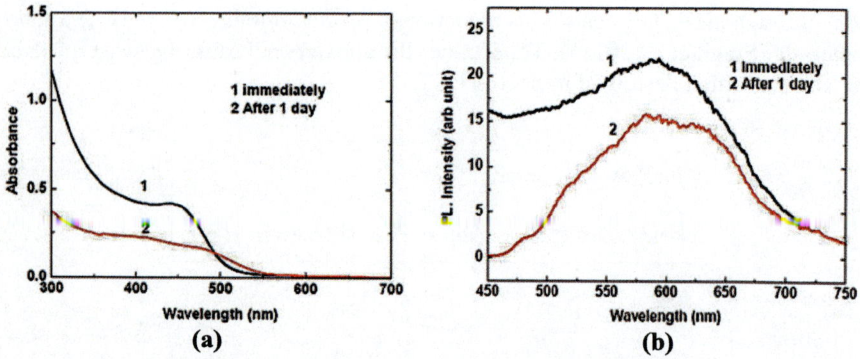

Fig. 19.16 **a** Optical absorption spectra **b** photoluminescence spectra of electron beam synthesized CdSe nanoparticles in water-in-oil type microemulsions with water content ($w_0 = 10$); [precursor] = 10 mM; spectra recorded at different times to check the stability. It is reproduced from the International Association of Advanced Materials as author of this work and under Creative Common License (Singh et al. 2013c)

the nanoparticles' growth and morphology. For a better understanding in this area, the synthesis of CdSe nanoparticles is investigated in a well-studied IL, 1-ethyl-3-methyl imidazolium ethyl sulfate ([EMIM][EtSO4]). The RTIL played three roles in the synthesis such as (i) solvent, (ii) stabilizer, and (iii) shape directing template. The primary nanoparticles produced (2–5 nm range) were characterized by HR-TEM. Then these nanoparticles evolve as nano flake-like units. They further self-assemble and appear as a mixture of 2D sheets and 3D flower-like patterns as revealed by the SEM technique. The stability of these coexisting nano morphologies identifies the inherent microheterogeneity of the RTIL (Guleria et al. 2014).

Figure 19.17 represents the TEM images of the as-grown nanoparticles (Guleria et al. 2014) while its polydispersity (2–5 nm range; average size ~2 nm) is shown in Fig. 19.17b representing the HRTEM image. The most interesting aspect revealed is the self-assembling into some type of superstructures (Fig. 19.17a). Self-assembling was further probed with SEM recorded at different regions (Fig. 19.18) (Guleria et al. 2014). In the inset of Fig. 19.18a, the energy-dispersive X-ray analyzer (EDX) spectrum confirmed the formation of nanoparticles. One unusual peak indicating Si is possibly due to the Si-wafer substrate used in the analysis. Figure 19.18a indicated the globular shape of the primary nanoparticles which later evolve as nano flake-like (NF) structures (arrows in Fig. 19.18b). Later, the formation of a mixture of 2D nanosheets and 3D nanoflower structures from self-assembling of NFs is presented in Fig. 19.18c, d. It is to be noted that the morphologies were quite stable over a long time period. This is indicative of the excellent stabilizing effect of RTIL in preserving the nano morphologies.

SEM images reveal that due to the high surface area of the NFs, they assemble non-homogeneously to form 2D and 3D nano or even micro-organizations. The multilayered 2D sheet structures (10–15 μm long and 1–2 μm wide) (marked in

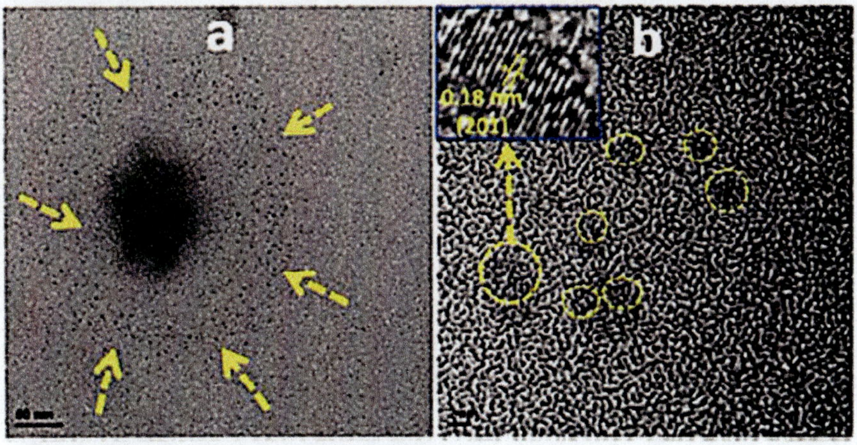

Fig. 19.17 **a** TEM (50 nm scale) and **b** HRTEM image (2 nm scale) of the CdSe nanoparticles made in neat RTIL. The inset of (**b**) presents the lattice fringe pattern (0.18 nm interplanar distance, plane 201) of the CdSe hexagonal phase. Reproduced with permission from RSC Publications as author of this (Guleria et al. 2014)

Fig. 19.18d) originate from the fusion of NFs. It may be worth mentioning that this field is growing at a rapid rate where recent syntheses assisted by RTILs are reported for ZnS nanoparticles, flower-like MoS_2, PbS nanocubes, and CdSe hierarchical dendrites.

Furthermore, FT-IR and Raman spectroscopic investigations indicate π–π stacked aromatic geometry and the hydrogen bond network between the nanoparticles and RTIL cation/anion. Few control experiments were carried out in dilute RTIL with varying water content. The morphology was examined as a function of water content and time. A probable mechanistic scheme is shown in Fig. 19.19 (Guleria et al. 2014) depicting the formation and evolution of the whole series of events employing the RTIL structure and its fluidic aspects.

19.9 Core–Shell Quantum Dots

Two major factors, quantum confinement and surface effects, as discussed in earlier sections influence the QDs' properties. These effects are realized in the development of optoelectronic devices and medical applications. Many basic bulk material properties like melting point, solubility, reactivity, plasticity, magnetism and conductivity, etc., get transformed when QDs are formed. These QDs get easily dispersed in solvents and can be functionalized at their surface. It may be stated again that capping the surface is vital for obtaining luminescent and stable QDs. There are two

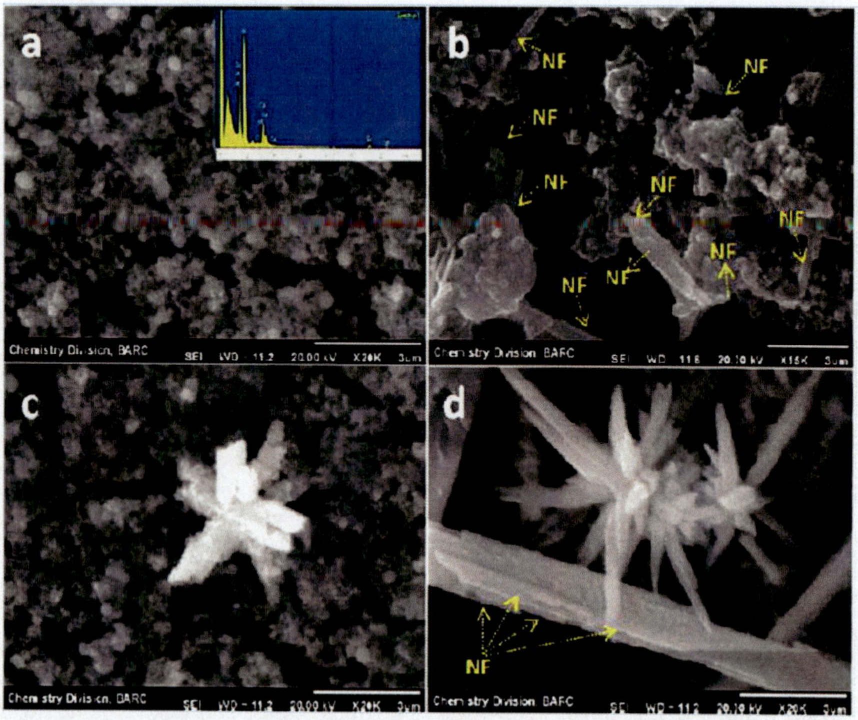

Fig. 19.18 The different transient stages of as-grown CdSe nanostructures recorded with SEM after 5 h of the reaction. They are **a** globular shape **b** NF structure **c** 3D flower-like pattern and **d** 2D nanosheet-like structures (scale bar is 3 μm for (**a**)–(**d**)). Reproduced with permission from RSC Publications as author of this (Guleria et al. 2014)

ways of achieving such capping (i) organic route and (ii) inorganic route (Fig. 19.20) (Bera et al. 2010).

Capping action of organic molecules is provided by binding with surface atoms. The benefits of the above passivation result in monodispersity, colloidal suspension, and bio-conjugation. However, because of large organic capping molecular size, it is not always possible to passivate cation and anion surface sites and achieve full coverage of the surface. But the other strategy namely the inorganic passivation can fully passivate surface trap states by growing another semiconductor layer over the QD. In subsequent sections, the resultant substance called core–shell QDs (CSQD) will be dealt with in detail. In such CSQD, fluorescence QY and photostability are considerably improved. By selecting suitable core and shell materials and their dimensions, the absorption and emission spectra, and other desirable properties can be controlled. However, if there is a considerable mismatch of core and shell lattice structures, the resulting lattice strain produces defect sites. The thicker shell also generates misfit dislocations, which increase the non-radiative processes further and decrease the fluorescence QY.

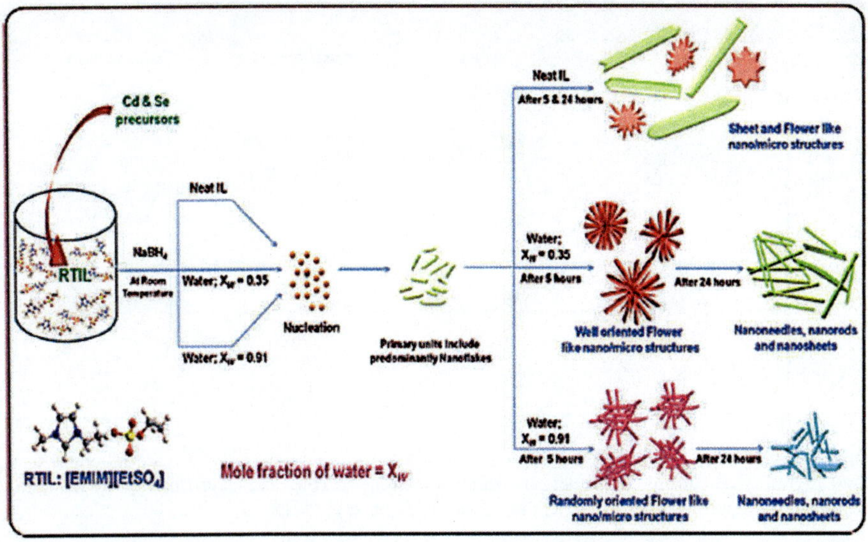

Fig. 19.19 Mechanistic scheme depicting the formation and evolution of the whole series of events in RTIL, ([EMIM][EtSO$_4$]), as a function of time and mole fractions of water. Reproduced with permission from RSC Publications as author of this (Guleria et al. 2014)

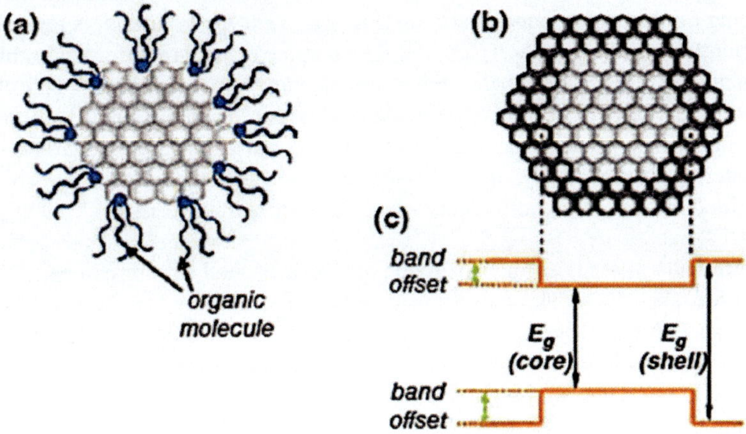

Fig. 19.20 Passivation of QD **a** organic way **b** core–shell QD by inorganic way and **c** energy band offsets of CSQD. Reproduced from MDPI Publications under Creative Common Licenses (Bera et al. 2010)

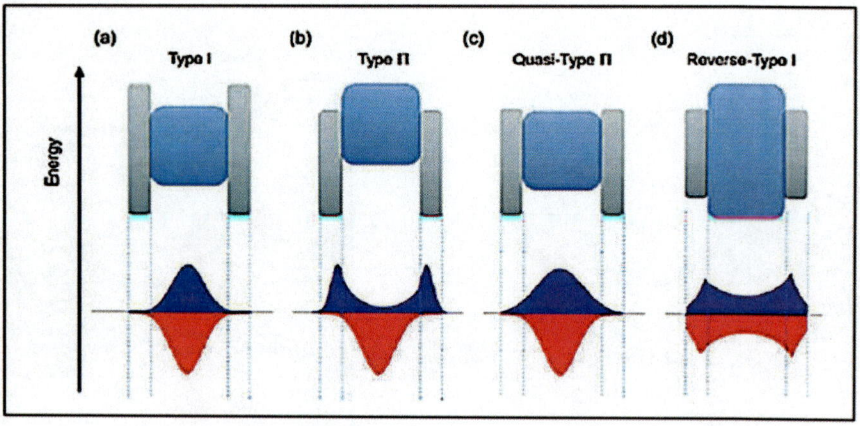

Fig. 19.21 **a** Type I, **b** type II, **c** quasi-type II, and **d** reverse type I CSQD materials: and their band alignment. Electron wave function is denoted by blue and the hole wave function by red color. Reprinted with permission from RSC publications (Zhou et al. 2018)

19.9.1 Classification of Core–Shell Quantum Dots

Three categories of CSQDs, namely, type I, reverse type I, and type II are obtainable depending on the energy bandgap offset (Fig. 19.21). Type I material is represented conveniently where $E_{g,\text{shell}} > E_{g,\text{core}}$ and the charge carriers are enforced within the core. In reverse type I material, the above bandgaps are in reverse order and the charge carriers get partially or completely delocalized within the shell. Finally, in type II, either valance or conduction band edge of shell material lies inside the bandgap of the core material. Consequently, upon photoexcitation, the charge carriers are spatially quarantined in a ditinct region of core–shell heterostructure (Reiss et al. 2009; Zhou et al. 2018).

Additionally, there is an intermediate type of core–shell structure, termed quasi-type II core–shell QDs. The most considered in this category is CdSe/CdS CSQD, although essentially, it is a type I structure. Here the electronic wave function gets delocalized over the entire nanocrystal, while the core material contains the hole wave function (Reiss et al. 2009; Zhou et al. 2018).

19.9.2 Type-I Core–Shell Structures

The major transformation of semiconductor QD is the strongly enhanced band gap fluorescence compared to the bulk phase. When all surface valences are satisfied during synthesis, the surface is termed as well-passivated and QY gets enhanced drastically. In the case of weakly bound capping agents, a dynamic equilibrium of adsorption/desorption results, and they are also sensitive to post-synthetic treatments.

Thus robust method of passivating the surface is needed to produce stable QDs with pronounced quantum yields.

Innovative investigations were conducted where a significant enhancement of QY of CdS QDs obtained by treating with hydroxide ion in the presence of excess Cd^{2+}. A passivating $Cd(OH)_2$ shell formed around the QD prevents the movement of free charge carriers to surface trap states.

Most of the CSQDs are synthesized in high-boiling organic solvents following the procedure of Murray et al. (1993). This organic solvent synthesis provides less polydisperse material compared to the aqueous synthesis route. The first successful shell made of ZnS over CdSe QDs was achieved in 1996 by Hines and Guyot-Sionnest (1996) followed by further work on CdS/CdSe QDs.

The above two CSQDs with high fluorescence QY have been investigated in detail, and are treated as "hydrogen atom" for QDs. Standard core material, CdSe (band gap of 1.7 eV; 728 nm), is chosen for using the entire visible radiation with 1.7–6 nm QD size. Based on two different reasons ZnS and CdS are commonly used as shell materials on CdSe core. Since ZnS has a large band gap (3.7 eV), it is possible to confine the free charge carriers in the core even with a thin shell configuration. But its synthesis via epitaxial growth route is rather difficult due to lattice mismatch in CdSe–ZnS. However, CdS over CdSe offers relatively low lattice mismatch, even with moderately thick shells. But its smaller band gap (2.5 eV) compared to ZnS, leads to less effective electronic passivation for a certain shell thickness.

Figure 19.22 represents the absorption and emission spectra of CdSe/CdS CSQD. Two key inferences can be drawn from the above spectra: (i) with increasing shell thickness, the first absorption maximum gets red-shifted and (ii) the QY of the core–shell configuration is higher compared to the bare QD. These are due to both the charge carriers' relaxation into the CdS shell, while the QY increase is due to surface trap state passivation. For ZnS shell, the observations are similar but the red shift of the absorption maximum is less prominent. It may be noted that the QY reaches a maximum with 1.8 monolayers thick ZnS shell. Such maximum in QY happens due to considerable lattice mismatch. Subsequently, extensive works have been carried out to achieve simple control of the shell thickness for making better devices. Presently, other than the above two, a large variation of type-I CSQDs have been prepared (Hines and Guyot-Sionnest 1996; Li et al. 2003), which include CdS/ZnS, and InP/ZnS.

It may be noted that due to their significant photo and chemical stabilities plus high QY, type-I CSQDs are fast replacing the organic dye molecules in bio-labeling. This kind of in-vitro or in-vivo biolabeling is quite useful in detecting/monitoring the photoluminescence (PL) of the QD (cf. Sect. 19.11.6). In particular, the greatly enhanced photostability along with narrower emission line width compared to the organic dyes, make these CSQDs prospective candidates for even multicolor labeling.

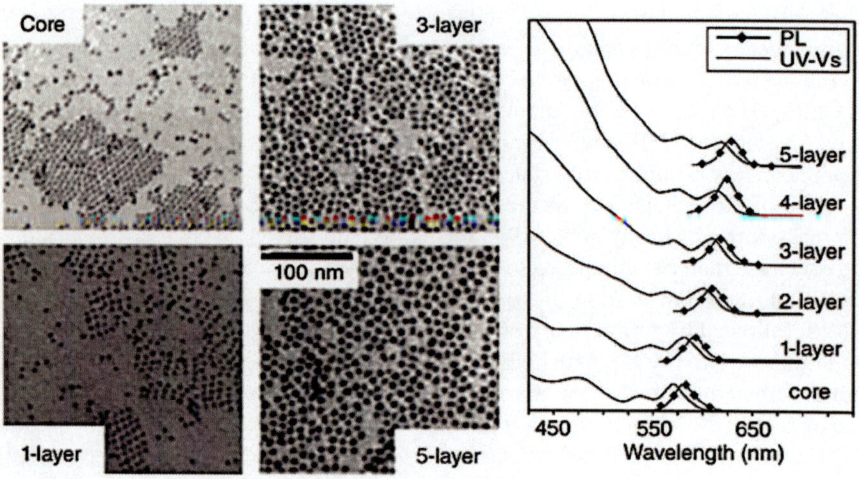

Fig. 19.22 Left: TEM images of bare CdSe and CSQD with CdS shell of 1, 3, and 5 monolayers thick; right: spectra (absorption and emission) of all the configurations. Reprinted with permission from ACS publications (Li et al. 2003)

19.9.3 Type-II Core–Shell Structures

It has been discussed (Sect. 19.9.1) that characteristics of type II core–shell structures arise due to the shell material's valance or conduction band edge lying inside the core material's bandgap. Therefore, the charge carriers produced on photoexcitation are spatially restrained in distinctive regions of the core–shell heterostructure (Hines and Guyot-Sionnest 1996; Li et al. 2003). In effect, the electron and hole being in two different locations undergo remarkably slower recombination rate prolonging the luminescence lifetime.

Usually, type-I nano heterostructures have better performance than core QDs in regard to their fluorescence QY and photostability. But, type-II heterostructures behave differently compared to type-I or core QD material. In this case, due to staggered band alignment, spatially indirect transition occurs at energies lower than both core and shell materials' band gaps. Therefore, the emission of type-II materials occurs at different wavelengths than either type-I or core QDs. Moreover, the luminescence lifetime of type-II heterostructures is observed to be shorter than type-I material.

A promising application of type-II materials namely QD laser has recently been demonstrated. The band alignment and slight charge carrier wave functions' overlap make the exciton-exciton interaction repulsive. Interestingly this behavior is appropriate for building QD laser based on the principle of amplified stimulated emission (ASE). The first such system reported is colloidal type-II (CdTe/CdSe) material where the band offsets allow the electron to be located in the shell CdSe and the hole in the core CdTe. Moreover, the spatially indirect valance (CdTe) to the conduction band

(CdSe) transition affects the absorption and emission wavelengths to be red-shifted compared to pure CdTe. Thus, it is possible even to emit wavelengths not accessible with core and shell materials and the emission was achievable up to 1000 nm.

Klimov et al. studied CdS/ZnSe core–shell nanoparticles in detail and successfully demonstrated their use for ASE (Klimov et al. 2007). The first observation is that shorter emission wavelengths compared to the band gaps of core and shell materials (CdS = 2.5 eV; ZnSe = 2.7 eV). The initial emission spectrum has two components assigned to a very fast decay of doubly excited particles and to a long-lived decay of singly excited particles. The difference between these two transitions is about 100 meV, having a doubly excited transition shifted toward higher energies. Both findings can be correlated to the type-II structure. On the contrary, in type-I nanocrystals where the electrons and holes are constrained in the same space, the exciton-exciton interactions are typically weak and attractive. This provides the doubly excited transition small shift toward lower energies.

The ASE behavior on examination yields some interesting features. Type-I materials can produce ASE at the doubly excited transition wavelength, while the type-II materials produced ASE at the central emission wavelength originating from the singly excited transition. Due to this effect, the type-II systems have a lower pump energy threshold for ASE compared to type-I systems. Consequently, this conduct makes type-II CSQD materials favorable candidates for laser development (Sect. 19.11.5).

19.9.4 Synthesis of Core–Shell QDs

Core–shell QDs can be synthesized involving a two-stage process where the first stage produces core QDs and the second stage makes the shell. We have already dealt in detail the synthesis of core QDs in Sects. 19.3, 19.4, and 19.5. The material and its thickness are two critical parameters for growing the shell. As for the material, generally, semiconductors are chosen for the core and shell with minor lattice mismatch. This mismatch is kept nominal to avoid lattice strain and defect states. These defect states are responsible for trapping the charge carrier and detrimental in lowering PL QY.

Next, the thickness of the shell material decides the effectiveness of the surface passivation. While thin shell material may lead to inadequate passivation, the thicker ones are likely to produce more defect states arising from lattice strain. Using the following simple scheme, the proper shell thickness can be estimated. First, the shell material volume, V_{shell} is calculated for the deposition of "*m*" monolayers of shell thickness:

$$V_{\text{shell}} = \frac{4}{3}\Pi\left[(r_c + m \times d_{\text{ML}})^3 - r_c^3\right]$$

Table 19.2 Important parameters for bulk II–VI semiconductors

Semiconductor	Structure (300 K)	E_{gap} (eV)	Density (g cm^{-3})	Lattice parameter (A^0)
CdTe	Zinc blende	1.43	5.870	6.48
CdSe	Wurtzite	1.74	5.810	4.3/7.01
ZnTe	Zinc blende	2.39	5.636	6.104
ZnSe	Zinc blende	2.69	5.266	5.668
CdS	Wurtzite	2.49	4.820	4.136/6.714
ZnS	Zinc blende	3.61	4.090	5.41

where r_c is the core radius and one monolayer shell thickness is d_{ML} expressed in nm.

Knowing V_{shell}, the material n_{shell} in moles is assessed for depositing "*m*" mL shell

$$n_{shell} = \frac{V_{shell} \times D_{core} \times N_A \times n_{QD}}{MW_{core}}$$

where MW_{core} and D_{core} are the molecular weight and density of the core material, N_A is the Avogadro number, and n_{QD} represents core QDs in solution in moles

After estimating the required shell thickness, successive ionic layer adsorption and reaction (SILAR) technique is employed for deposition. This technique involves careful introduction of anionic and cationic precursors into the core solution and one monolayer is deposited at a time. Repeating the above procedure, the requisite number of monolayers can be deposited on the core QDs' surface. Core–shell QDs having monodispersity and high luminescence yield are obtained by this technique.

A list of important material parameters is given in Table 19.2 for some useful bulk II–VI semiconductors.

19.9.5 *Hydrothermal Process for CdSe–CdTe Core–Shell QDs*

A two-stage hydrothermal process (Ramalingam et al. 2022) is used to prepare CdSe–CdTe QDs where CdSe nanoparticles are prepared first and then added to the precursors of CdTe. The materials required are (i) cadmium nitrate tetrahydrate ($Cd(NO_3)_2{\cdot}4H_2O$), (ii) sodium selenite (Na_2SeO_3), (iii) hydrazine hydrate ($N_2H_4{\cdot}H_2O$), and (iv) ammonia ($NH_3{\cdot}H_2O$).

Aqueous solutions of cadmium nitrate tetrahydrate and sodium selenite were made separately. The Cd^{2+} ion in cadmium nitrate solution was converted into $Cd(NH_3)_4^{2+}$ ion by treating with ammonia. On the other hand, Na_2SeO_3 is dissociated by mixing with hydrazine hydrate to produce hydrogen selenide (H_2Se) where the oxidation state of selenium changed to Se^{2-}. By mixing together these solutions, CdSe cluster is formed which is next added to the CdTe precursor solution prepared by the following method. The key precursor NaHTe is prepared freshly by mixing $NaBH_4$ with Te (2: 1 molar ratio):

$$4NaBH_4 + 2Te + 7H_2O \rightarrow 2NaHTe + Na_2B_4O_7 + 14H_2 \uparrow$$

Stoichiometric amount of $NaBH_4$ is transferred to an ice-chilled flask, followed by the addition of H_2O and Te powder. The black Te powder gets dissolved in sodium borohydride solution after 2 h of vigorous stirring, to produce sodium tetra borate ($Na_2B_4O_7$) as white precipitate and supernatant NaHTe. The pressure build-up from generated hydrogen is released carefully through a vent provided.

This freshly prepared NaHTe solution is used to dissolve cadmium nitrate tetrahydrate ($Cd(NO_3)_2{\cdot}4H_2O$) and then 1.2 g of L-cysteine ($C_3H_7{\cdot}NO_2S$) as a capping agent in 25 ml of deionized water and CdSe is added gradually. The overall reaction is represented as

$$NaHTe + Cd(NO_3)_2 4H_2O + CdSe \xrightarrow{L-Cysteine} CdSe - CdTe \downarrow + NaOH + N_2O \uparrow + 4H_2O + 2O_2 \uparrow$$

The solution is relocated in a teflon-lined autoclave and heated at 180 °C in an oven for six hours. Figure 19.23 illustrates the step-wise hydrothermal synthesis process of CSQD (Ramalingam et al. 2022).

The solution is transferred to a flask from the hydrothermal reactor. Under refluxing conditions (at 110 °C), different durations produced controllable sized

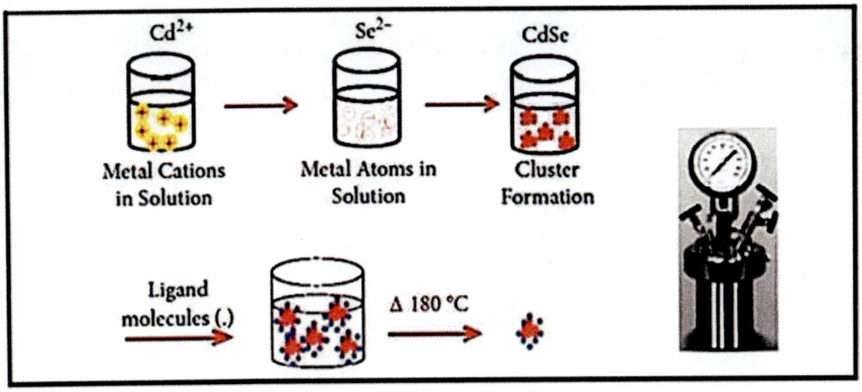

Fig. 19.23 Schematics of CdSe–CdTe quantum dots synthesis. Reproduced from De Gruyter Publications under Creative Common Licenses (Ramalingam et al. 2022)

CSQD. The refluxed solution is subjected to centrifuge and the precipitate is repeatedly washed with deionized water/ethanol. The precipitate was finally isolated using vacuum filtration and exposing it to 110 °C for three hours. The crimson-colored CdSe–CdTe core–shell QD obtained is characterized with XRD, SEM–EDX, TEM, and SAED techniques (Ramalingam et al. 2022). All these techniques are now regularly used for characterization of QDs and CSQDs.

19.10 Characterization of Core–Shell QDs

As-prepared nanocomposites are characterized with powder XRD patterns with nickel-filtered (monochromatic) CuKα radiation (1.5406 Å with a 0.02°/s scanning rate). Morphology of the materials is obtained using a scanning electron microscope (SEM) and energy-dispersive X-ray analyzer (EDX). Microscopic images of nanostructures is obtained through a transmission electron microscope (TEM, 200 kV acceleration voltage). Additionally, selected area electron diffraction pattern (SAED) is recorded with a 120 cm camera. UV–Vis–NIR spectrometer (200–900 nm) is used to characterize the absorption spectrum of the nanocomposite while the emission spectrum is obtained using a spectrometer (300–600 nm) with 380 nm as excitation wavelength.

19.10.1 X-Ray Diffraction (XRD) Studies

Nanocomposites are characterized by powder XRD technique with nickel-filtered (monochromatic) CuKα radiation (Fig. 19.24) (Ramalingam et al. 2022).

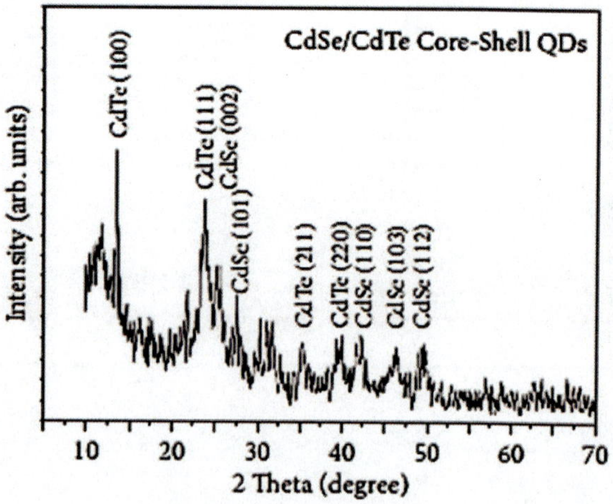

Fig. 19.24 XRD patterns of CdSe–CdTe CSQD. Reproduced from De Gruyter Publications under Creative Common Licenses (Ramalingam et al. 2022)

XRD patterns revealed the synthesis of hexagonal CdSe and cubic CdTe polycrystalline compounds. CdSe hexagonal phase formation is characterized by the peaks at 24.19° (002 plane), 27.62° (101 plane), 42.28° (112 plane), and 56.95° (002 plane). The other recorded peaks at 13.70, 24.19, 35.38, and 39.87° are due to CdTe cubic phase (100, 111, 211, and 220 planes). The diffraction peaks are matched with standard literature corresponding to CdTe and CdSe, respectively. The constrained particle size is reflected in wider signature peaks. Interestingly, even when CdSe is enclosed by CdTe, the diffraction peaks of CdSe cubic structure can be recognized. Williamson-Hall method can be used to find the size (D) and the induced strain (ε):

$$\beta \cos\theta = \frac{K\lambda}{D} + 4\varepsilon \sin\theta,$$

where λ is the wavelength of the X-ray used, β is the intensity FWHM, θ is Bragg's angle, and K is the shape factor. Using this formulation, the estimated 12.6 nm grain size of CdSe/CdTe confirmed the formation of a quantum dot. Further, SEM–EDX and TEM measurements revealed the shape and presence of the compositional elements.

19.10.2 SEM–EDX Analysis

Figure 19.25a presents the SEM image of CdSe–CdTe CSQD (Ramalingam et al. 2022). The monodispersed spherical nanoparticles are found without agglomeration. TEM analysis provides the precise measurement of the QD size and uniform particle size (10–20 nm) was found from the surface morphology. The EDX spectrum (Fig. 19.25b) indicates the presence of elemental Cd, Se, and Te in the QD. The result of the high weight and atomic percentage of Cd (shown in the Fig. 19.25b) indicate that both Se and Te are bonded to Cd.

19.10.3 TEM Analysis

Figure 19.26a–e represent the TEM images of CdSe–CdTe CSQDs (Ramalingam et al. 2022). The TEM images (@50–100 nm scale) (Fig. 19.26a, b) are indicative of vastly monodispersive and spherical nanoparticles. Further magnifications (@20 nm scale) (Fig. 19.26c) show the accurate particle size (~8–10 nm), which corroborates the XRD and SEM observations.

At higher magnifications, the TEM morphology and SEM morphology corroborates each other showing freestanding particle. The employed capping agent L-cysteine aided in size tuning of QDs and also refining the agglomeration. With higher magnifications (@5 nm scales), the surface indicates the unification of two crystal

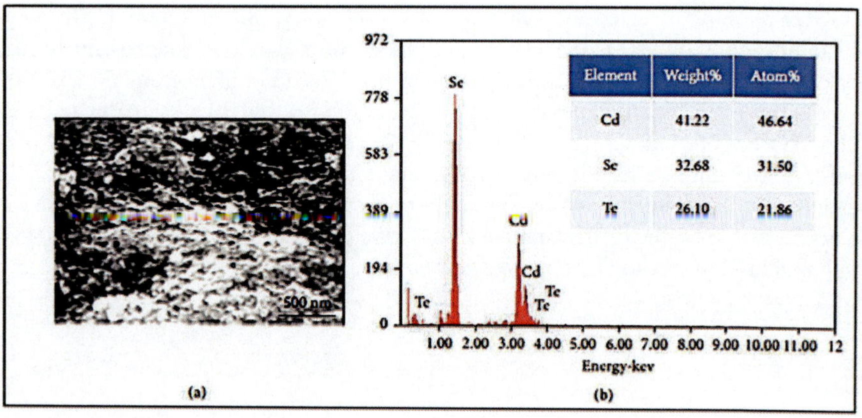

Fig. 19.25 **a** SEM morphology and **b** EDX spectrum of CdSe/CdTe. Reproduced from De Gruyter Publications under Creative Common Licenses (Ramalingam et al. 2022)

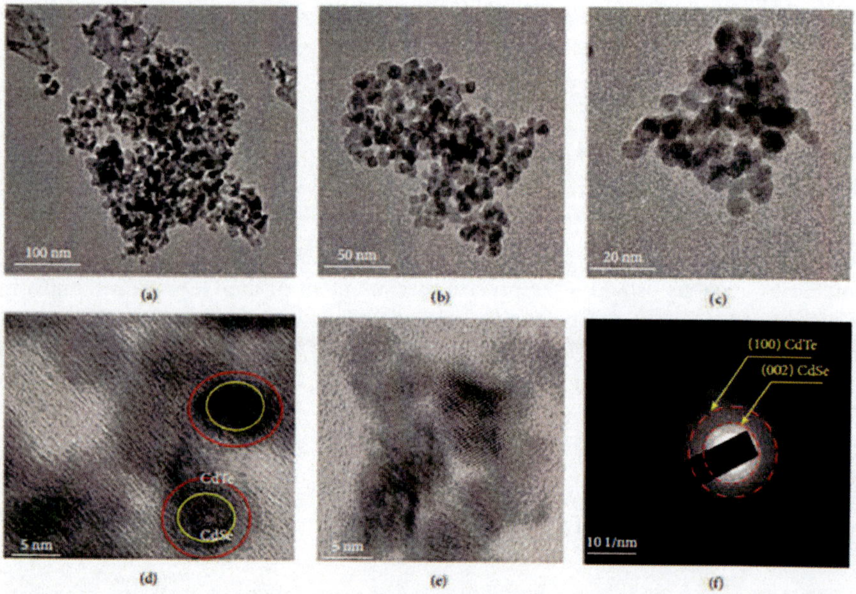

Fig. 19.26 TEM images (**a–e**) and SAED pattern (**f**) of CdSe/CdTe CSQD. Reproduced from De Gruyter Publications under Creative Common Licenses (Ramalingam et al. 2022)

structures probably due to CdSe QDs encircled with CdTe QD patches (Fig. 19.26d, e). In Fig. 19.26d, the CdSe surface and covering compound CdTe are shown by circles.

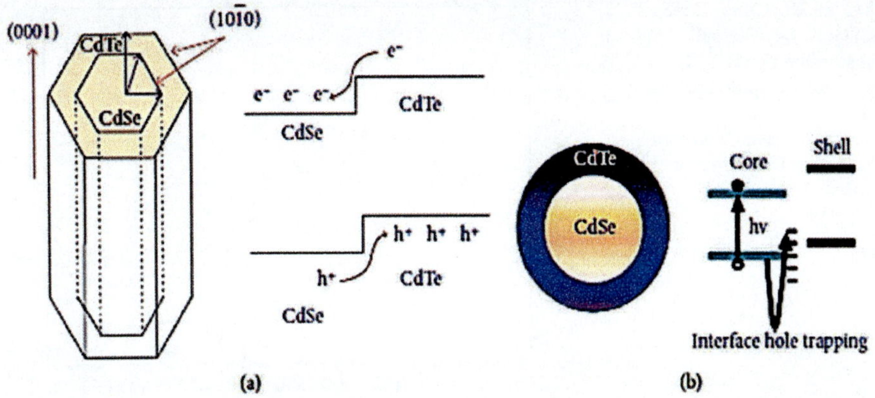

Fig. 19.27 Growth orientation (**a**) and electron–hole trapping (**b**) at the interface of CdSe–CdTe CSQD. Reproduced from De Gruyter Publications under Creative Common Licenses (Ramalingam et al. 2022)

These heterojunction CSQDs have significant advantages when effective exciton production/dissociation is combined with transporting separated charge carriers through discrete channels. It is now feasible to make heterojunction CSQDs with atomic-level control for efficient photovoltaic applications. The core and shell components are recognized by recording the SAED pattern shown in Fig. 19.26f. The inner and outer circular regions are ascribed to the CdSe (002) plane and CdTe (100) plane respectively. Further growth orientation of CdSe–CdTe CSQDs and electron–hole trapping at the interface is schematically presented in Fig. 19.27a, b (Ramalingam et al. 2022). The nonpolar interfaces (1010) of the surface facet are guided toward the wurzite (0001) direction as shown in Fig. 19.27a; the figure also shows the trapping of a fraction of holes at the CS interface (Fig. 19.27b).

19.10.4 Optical Absorption and Photoluminescence (PL)

Figure 19.28 represents the absorption and emission spectra of CdSe–CdTe QDs (Ramalingam et al. 2022). The absorption spectra (200–800 nm) consists of an excitonic peak at 560 nm and photoluminescence peak at λ_{emi} ~ 585 nm. When the PL maximum is nearer to its absorption onset, an increase in emission is observed owing to the direct recombination of HOMO and LUMO charge carriers. The CdSe–CdTe QDs obey type II characteristics and the observed optical absorption together with PL emission conforms the exciton nature of the type II material.

The observed emission in the λ_{emi} ~ 560–590 nm range arises due to several structural defects in the CSQD material. The plausible reason for the emission is due to the significant surface-to-volume (S/V) ratio, which normally creates interstitials and vacancies resulting in trap level. PL intensity is boosted due to low-concentration

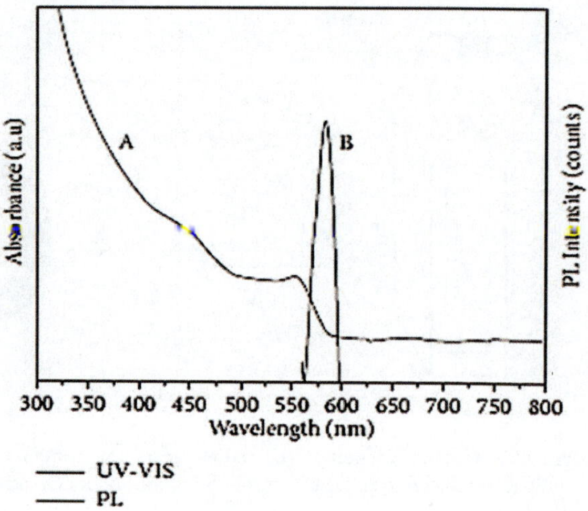

Fig. 19.28 CdSe–CdTe CSQDs: (A) Optical absorption and (B) PL spectrum. Reproduced from De Gruyter Publications under Creative Common Licenses (Ramalingam et al. 2022)

CdTe doping, where radiative recombination takes place at extra centers of CdSe. Three different phenomena are considered based on the optical properties: (i) reduced strain at the graded situations, (ii) lower defect concentration at the interface, and (iii) charge transfer cascading. These factors enhance the charge separation yield and charge carrier mechanism, which are utilized in photovoltaic device to enhance its efficiency (Ramalingam et al. 2022). Of late, some energy storage devices have started incorporating electrodes made with similar uniform spherical CdSe–CdTe CSQDs.

19.11 Applications of Core–Shell QDs

Currently, these CSQDs having excellent properties like size selected bandgap tunability, large PL having narrow FWHM, extended lifetime of excited-state, and good stability are being utilized for numerous applications. Some selected areas like solar cells, light-emitting diodes (LEDs), and biomedical applications will be dealt with in detail in the following sections.

19.11.1 Solar Cells

Quantum dots have revolutionized the area of making efficient solar cells. The traditional solar cell relies on sunlight photon knocking electrons out of a semiconductor into a circuit and producing electric power. However, the process efficiency is quite low, which can be enhanced (typically 10% more efficient) by producing more electrons (or holes) for each solar photon that strikes QD materials. QD devices offer the added advantage of miniaturization in applications like CCDs (charge-coupled devices) and CMOS sensors.

In the initial developments, various QDs such as II–VI (CdSe, etc.) and IV–VI (PbSe, etc.) materials are employed as sensitizers in making the solar cell (Klimov 2006). However, as pointed out earlier, core–shell QDs offer superior qualities of controllable bandgap tunability along with photo-bleaching stability. Presently, type II material provides better cell efficiency though all three types of CSQDs are in use for various solar cell fabrication.

The QD-sensitized solar cell comprised of photo anode, counter electrode, and liquid electrolytes (Fig. 19.29a) (Pan et al. 2018). The photoanode is made of fluorine or indium-doped tin oxide (FTO or ITO) glass on which deposited a large bandgap semiconductor (ZnO, or TiO_2, etc.) material serving as a light absorber. Basically, solar photon excitation produced the charge carriers, i.e., electrons in the conduction and holes in the valance band of the QD. Then, the electrons are quickly introduced into the conduction band of TiO_2 and transported to the counter electrode via the FTO substrate. Meanwhile, reduced species of redox couple in the electrolyte neutralize the oxidized QD while the oxidized species of redox couple get reduced by the electrons from the counter electrode. The resulting photocurrent is measured via an external circuit. With the help of the characteristic J–V curve and the equation below, the power conversion efficiency (PCE) is calculated:

$$\mathrm{PCE} = \frac{I_{\mathrm{mp}} \cdot V_{\mathrm{mp}}}{P_{\mathrm{in}}} = \frac{J_{\mathrm{sc}} \cdot V_{\mathrm{oc}} \cdot \mathrm{FF}}{P_{\mathrm{in}}}$$

where P_{in} is incident power density, I_{mp} and V_{mp} are photocurrent and photovoltage, V_{oc} is the open-circuit voltage, J_{sc} is the short-circuit density, and FF represents the fill factor given by the ratio of specific value P_{max} to V_{oc} and J_{sc} (Fig. 19.29b) (Pan et al. 2018).

First use of core–shell QDs (inverted type I CdS-CdSe) as a sensitizer in solar cell was reported in 2009 (Lee and Lo 2009) where PCE value of 4.22% was obtained. After this, a large number of CSQDs are employed in the fabrication of the solar cell. Following years witnessed a new class of solar cells employing type I CdSe–ZnS CSQDs to sensitize TiO_2, where significantly improved stability of photo anode is reported. Improved PCE to the level of 9.48% was achieved using a novel type I CdSeTe/ZnS CSQDs, which is better than bare CdSeTe QD (8.02%) (Yang et al. 2015).

Further development occurred with the utilization of type II ZnSe/ZnS CSQDs. Due to the better charge separation process, it registered a high PCE efficiency. Synthesis of alloys like $CdSe_xTe_{1-x}$/CdS core–shell QDs by microwave-aided aqueous method produced 5.04% PCE. PCE was enhanced to 6.76% by sensitizing

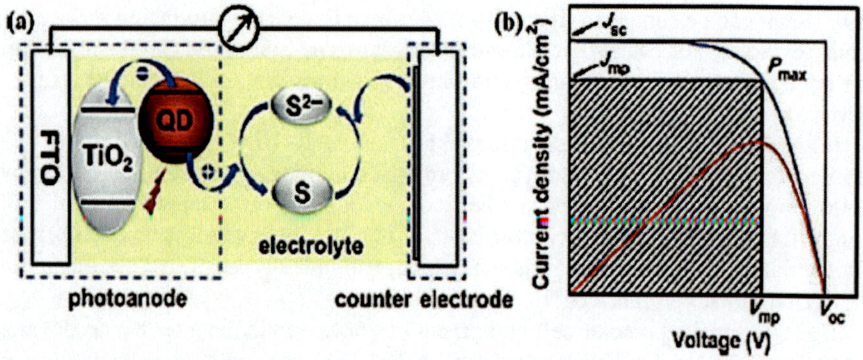

Fig. 19.29 QD-sensitized solar cells (**a**) typical *J–V* graph (**b**). Reprinted with permission from RSC Publications (Pan et al. 2018)

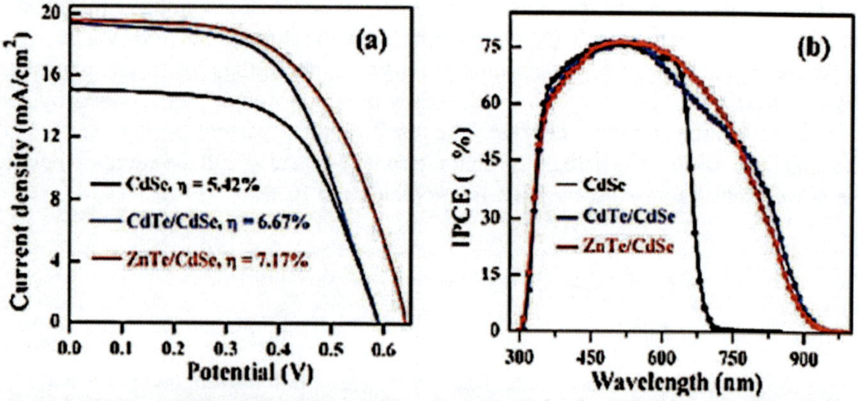

Fig. 19.30 Comparative performances of three types of solar cells based on CdSe QD, ZnTe/CdSe, and CdTe/CdSe CSQD: **a** *J–V* characteristics and **b** IPCE characteristics. Reprinted with permission from RSC Publications (Jiao et al. 2015)

with CdTe/CdSe CSQDs and further to 7.17% by employing ZnTe/CdSe CSQDs, both type II materials (Fig. 19.30) (Jiao et al. 2015).

19.11.2 Light-Emitting Diodes (LEDs)

Quantum dots offering several important advantages, are quickly put to use into computer screens and displays. The typical liquid crystal display (LCD) produces the image made by tiny red, blue, and green crystals. They perform as color filters which can be switched off and on using electronic control and they are also illuminated by a bright backlight. However, QD-based displays look much more realistic and energy efficient since the quantum dots' color can conveniently be tuned and they do not need any backlight. These advantages are important in fabricating portable devices like mobiles where robust battery life is needed. Further quantum dots being much smaller than liquid crystals, they provide a much higher resolution image. It may be mentioned that since the brightness of typical quantum dots is higher compared to organic LEDs (OLEDs), they will possibly replace OLED displays as well. Summing up, QDs can be employed to make brighter TV screens with accurate colors of red, green, and blue pixels than in traditional LCD or competing OLED screens.

Presently, LED based on QDs (QLEDs) have gained much attention for its superior brilliance (~200 kilo cd m^{-2}), color purity (fwhm < 30 nm), and small working voltage (<2 V). The QLEDs are built with components like cathode, anode, QD films, electron/hole transport layers (ETL/HTL) (Fig. 19.31) (Choi et al. 2018). Applying a potential between the electrodes, electrons, and holes are introduced into ETLs and HTLs. Additionally, the charge carriers introduced into the QDs undergo radiative recombination. Above-charge transport materials (ETL/HTL) are crucial for the stability and efficient performance of QLEDs.

QLEDs normally use type I CSQDs because of their broader bandgap shell material confining the exciton and also minimizing surface defect states. This leads to increased quantum efficiency due to better PL QY and stable QD. It may be noted such enhanced PL QY not necessarily indicative of its electroluminescence (EL) performance. Detrimental processes like Augur recombination and inter-particle energy transfer are dominant with the interface structure and thus modification of structure becomes a critical issue (Choi et al. 2018). Better EL performance has been obtained with an enriched intermediate shell of ZnSe or CdS for CdSe/ZnS structure.

Highly stable QD with desired blinking properties can be made either by thick shell or multi-shell coverage on the QDs. This procedure enhanced PL QY and efficient suppression of the charge instability leading to superb device performance.

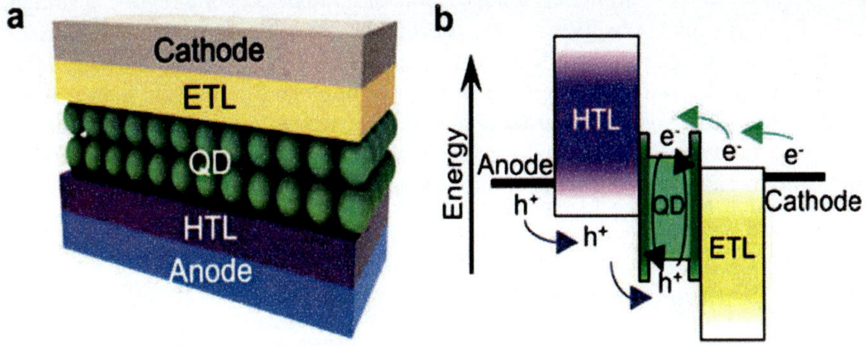

Fig. 19.31 **a** QLEDs device structure **b** energy-level illustration depicting charge introduction scheme. Reproduced from Springer Nature under Creative Common Licenses (Choi et al. 2018)

Recently, excellent green luminescent CdSe/ZnS CSQDs have been made employing a thick ZnS shell covering CdSe (Hao et al. 2019).

19.11.3 Optical Switching and Optical Limiting

It has been pointed out that quantum confinement makes QDs a potential third-order nonlinear optical materials (NLO). In recent times, CdSe QDs, have attracted attention in applications like optical switching and optical limiting. These effects are demonstrated using Z-scan measurement carried out with 532 nm nanosecond laser (Valligatla et al. 2016). It showed that at lower intensities, saturable absorption (SA) dominates while higher intensities recorded reverse saturable absorption (RSA). The SA is essentially ascribed to bleaching of the ground state and RSA arises due to free carrier absorption (FCA) of CdSe QDs.

Figure 19.32 presents the intensity-dependent open-aperture Z-scan curves of CdSe QDs in the input intensities (532 nm) ranging from 90 to 450 MW/cm^2 (Valligatla et al. 2016). It can be seen that in the range of 90–140 MW/cm^2, the SA becomes stronger with a concomitant broad Z-scan curve. Increasing the intensity further, RSA becomes dominant with a decrease in the transmittance. Under such intense excitation, resonant second photon absorption takes place lifting the excited electron in the conduction band to a still higher conduction band level. This process is known as free carrier absorption (FCA), which contributes to RSA. Thus, at higher intensities, SA of lower excited states competes with FCA resulting in RSA over SA.

Thus, the optical limiting behavior in working with nanosecond lasers depends on both FCA and thermal nonlinearities. This performance is utilized in mode-locking laser systems as saturable absorption at lower intensities while based on RSA behavior at high intensities. Therefore, CdSe QDs can have many applications like optical pulse compression, optical switching, and optical pulse narrowing. Extending the open aperture Z-scan studies of nanosecond laser to picosecond regime, a similar shift of SA to RSA is observed. In nanosecond laser excitation, contribution arises from thermal nonlinearities while electronic contribution with picosecond laser excitation. It may be worth pointing out that signs of nonlinearities owe to their origins such as a negative sign for thermal and a positive sign for electronic contribution. It is also found that optical nonlinearities tend to be always higher in nanosecond compared to pico-second laser excitations (Valligatla et al. 2016).

19.11.4 Optical Gain and Lasing

Investigation of optical gain in QDs led to a key development in demonstrating laser action in such a medium. With size-tunable quantum dots, stimulated emission with sufficient gain can be achieved at convenient wavelengths. With continued R&D in this area, QDs are now routinely synthesized comprising 100–10,000 atoms, having

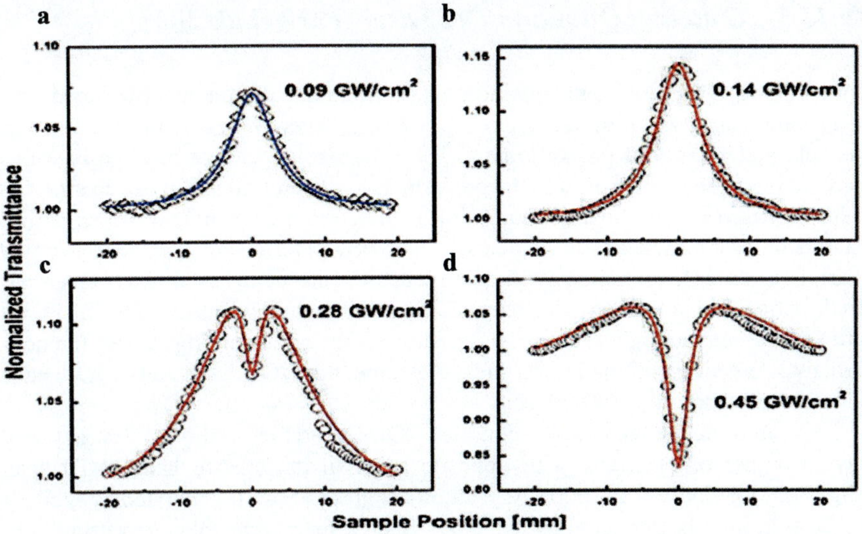

Fig. 19.32 Nonlinear absorption curves of CdSe QDs as a function of intensities of 532 nm nanosecond laser. Reproduced with permission from Elsevier Publications (Valligatla et al. 2016)

various sizes (1–10 nm), shapes (spherical, elongated, hollow, branched, etc.), and compositions (IV, II–VI, I–III–VI, I–II–VI–VI group materials). Such custom-made CSQDs are getting a lot of attention for development of active medium of micro-laser. Critical parameter for the achievement of laser action is to boost the optical gain determined by the difference in rates of absorption and stimulated emission. Recently, extensive investigations are being pursued on colloidal nanocrystals for laser development.

High performance and stable optical gain media have been worked out using tailored CdSe core (spherical; 2.4 nm) and CdS shell (rod; 39 nm) QDs and one- or two-photon pumping schemes. The attractive feature of such a configuration is to obtain tunable laser radiation by controlling the core and shell dimensions. Moreover, the 39 nm long nanorod shell provides a 2–4 times higher two-photon absorption cross-section compared to spherical CSQD. It may be even possible to achieve a still higher two-photon absorption cross-section by optimizing the dimensions and composition of CSQDs. Nanorods are capable of two-photon pumped lasing action even using a spherical optical cavity. Such smart engineering of heterostructures, in turn, offers promising opportunities for application in more areas. With these exciting developments, CSQDs are placed at the center stage of nanotechnology and are eloquently employed for making newer nanodevices using II–VI group compounds, particularly CdSe and CdTe.

19.11.5 Colloidal Quantum Dot Laser (Park et al. 2021)

For spherical QD particles having radius *R*, the quantum confinement-induced shift of its band gap is given by $\Delta E_{\text{conf}} = E_g - E_{g,\text{bulk}}$, where $E_{g,\text{bulk}}$ is the band gap of the bulk material and is proportional to R^{-2}. The size-dependent band gap permits easy tuning of the emission wavelength, which has been utilized in televisions and displays based on colloidal quantum dot (CQD). Proof-of-principle investigations have been carried out about 20 years back when CQD lasing was demonstrated (Park et al. 2021). Presently, CQD nanostructures are being used as commercial-grade optical and optoelectronic materials to build various devices like TVs, displays, and luminescent sunlight collectors. Researchers are now working to beat the non-radiative Auger recombination for the development of CQD laser. Let us now have a look at the working of CQD laser.

Light amplification in strongly confined CQDs resembles a simple 2-level system having a pair of electrons with opposite spins in its valence band (VB) state (Fig. 19.33) (Klimov et al. 2000). An exciton (single e–h pair) is generated by exciting an electron into the conduction band (CB). Such a state is capable of emitting light but cannot amplify the radiation, since the probability of stimulated emission by the CB electron is the same as photon absorption by the remaining electron in the VB.

This situation is similar to "optical transparency" (Fig. 19.33, middle), however, CQD can amplify radiation when the second VB electron is also excited to the CB achieving the population inversion required for lasing action. In effect, CQD starts lasing when an appreciable fraction of the QDs produce biexcitons (two e–h pairs) (Fig. 19.33, right).

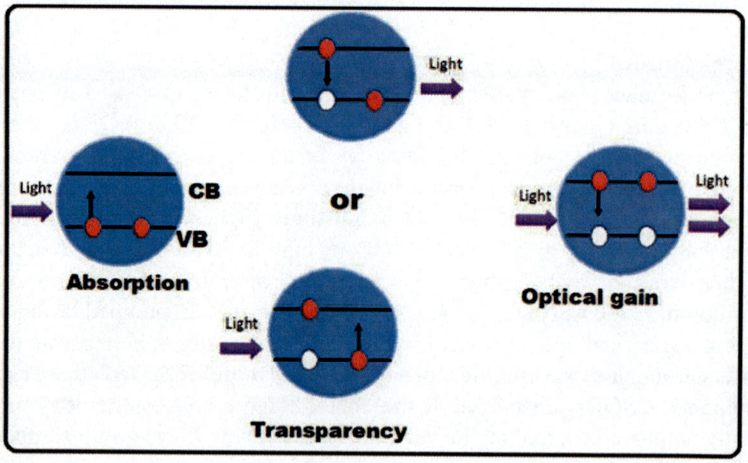

Fig. 19.33 Simple two-level description of CQD laser. When both electrons are excited into a biexcitonic state, the incident photon can induce stimulated emission. This is adapted from Springer Nature (Klimov et al. 2000)

Of late, two schemes have been employed to beat the Auger decay: (i) nonbiexcitonic mechanisms are single-exciton gain, implemented with type II CQDs, and (ii) charged-exciton gain, using photochemical or electrochemical CQD doping. The incentive is to effectively reduce the degeneracy of band-edge states so that biexcitons are not required to achieve optical gain.

Recent investigations have several demonstrations of direct control of Auger decay by making compositionally graded interiors of CQDs. It was implemented with continuously graded CQD material $CdSe/Cd_xZn_{1-x}$ $Se/ZnSe_{0.5}$ $S_{0.5}$ which boosted the biexciton PL QYs to about 45%. Interestingly, Auger decay was heavily suppressed but still preserving large quantum confinement, with ΔE_{conf} of about 10 k_BT at room temperature. Further, R&D in Auger decay engineering led to the demonstration of optical gain with direct-current electrical pumping (Lim et al. 2018) for the first time.

Currently, CQDs are attracting attention as cheap and scalable optical-gain media. Further, recent insight and control of processes impeding light amplification have resulted in several breakthroughs. They are optically pumped cw laser, optical gain with dc-injection, and the demonstration of dual-function electroluminescent devices.

19.11.6 Biological and Chemical Applications

Quantum dots are making rapid strides in various medical fields like cancer treatments. Anti-cancer drugs bound to dots are made so that they get accumulated in particular parts of the body and deliver them. The big improvement over conventional drugs is that they can be precisely targeted at single organs such as liver, kidney, etc. This in effect reduces the undesired side effects associated with traditional chemotherapy.

Quantum dots are fast replacing organic dyes in medical applications as well. They can be used like nanoscopic light sources to illuminate and such color-specific cells can conveniently be studied under a microscope. Chemical and biological warfare agents such as anthrax can be detected using them. Unlike organic dyes, QDs produce a very bright desired color and are also photostable. Based on the knowledge of light dependence on plant growth, experiments have been undertaken using quantum dots to concentrate different colored lights for maximizing plants' growth.

Biological systems are quickly adopting the nanometric approach in their applications. Due to the similarity in size scales, appropriate QDs can be accurately designed and incorporated into many proteins having about tens of nm sizes. It is well known that biological systems being complex, their synthesis, structure, function, etc., are seldom understood in detail. Rational design of QD structures of the same size mimicking biological molecules will also provide a handle to probe and modify biological systems. Rapid entrance of nanostructures is taking place in the field of labeling, sensors, and drug delivery. In subsequent sections, detailed discussion will be made about some of their optical, biological, and chemical applications.

Around 1990, the QDs started playing a crucial role in biomolecular imaging, sensing, gene and drug delivery, and photodynamic therapy (Martynenko et al. 2017). QDs have many attractive features like size, surface chemistry, spectral properties, and finally, their stability. With all these features, they are being utilized for in-vitro and in-vivo detection/imaging in biological systems. While the parameters like emission wavelengths, stokes shift, and large absorption cross section can easily be controlled with changes in their size or composition, but their biocompatibility can be retained.

19.11.6.1 Biosensing

In medical science, with the help of specific biomarkers, it is possible for early detection/therapy of diseases and QDs are finding extensive use in such applications. Biosensing is done by looking at the quenching of QDs' luminescence. For example, investigations are carried out on heme iron absorption and clofazimine-protein interaction (Yang et al. 2016b) by employing core–shell biosensor. For this, Förster resonance energy transfer (FRET) scheme is employed with CdZnSeS/ZnS alloyed CSQD as energy donor. Alongside, another scheme is used by first capping of multifunctional polymer and followed by functionalizing with cyanine 3-labeled human serum albumin (HSA) (Yang et al. 2016b) known as the direct ligand method.

19.11.6.2 Gene and Drug Delivery

The medical applications recently have seen increased demand for gene and drug delivery. Many benefits are recognized by using drug-formulated QDs providing targeted drug delivery, better cellular uptake, and long lifetime. Various in-vivo and in-vitro studies have been made with new bioconjugated QDs (Matea et al. 2017). Recently, the drug paclitaxel (PTX) has been used for human cancer therapy with CdTe/CdS/ZnS core–shell QDs. The PTX drug loaded on ZnSe:Mn/ZnS CSQDs and hybrid silica nanocapsules are used for fluorescence imaging and chemotherapy (Zhao et al. 2017). Recently, antibacterial and anticancer nano-drugs have been developed by quercetin (QE)-loaded CdSe/ZnS CSQDs.

19.11.6.3 Therapy

Both in-vitro and in-vivo photodynamic therapy have seen enormous progress using QDs either alone or with efficient photosensitizers such as phthalocyanines, porphyrins organic dyes, inorganic complexes, etc. It may be noted that photodynamic therapy proceeds through singlet oxygen and various reactive oxygen species (ROS) (Biju et al. 2010). Initially, it was found that QDs have poor efficiency in producing singlet oxygen, however, later ~70% enhancement in efficiency was obtained by conjugating choline e6 with QDs. But, further progress in bio-applications is stalled due to cytotoxicity of cadmium in QDs.

19.12 Summary and Future Prospects

The present overview provides a broad perspective for II–VI semiconductor CSQDs, especially their chemical synthesis and their prospective applications in solar cells, LEDs, biomedical, etc.

It is reasonably tough making forecasts of future technology. It is worth mentioning the way the quality of life improved worldwide with blue GaN LED for solid-state lighting. Further combining this LED with luminescent QDs led to another influential technology. These inexpensive and energy-efficient devices are capable of producing pure colors as well as broad-band white light. They are utilized both in macroscopic and microscopic scale area lighting with white LEDs (Osram) and micro-LEDs.

Future QD materials are best suited for high-quality pictures in TVs and monitors. QD technology started penetrating the TV market from 2015, and by 2018, the sale of QD TVs reached $18.7 million. The forecast indicates (Peach 2015) (Fig. 19.34) that QD technology will be adopted by 2025 for TVs (60%) and monitors (51%). Further, QDs have strategic use in defense or counter-espionage, by integrating QDs into dust to track enemies.

Two break-through application areas with QDs are worth revealing:

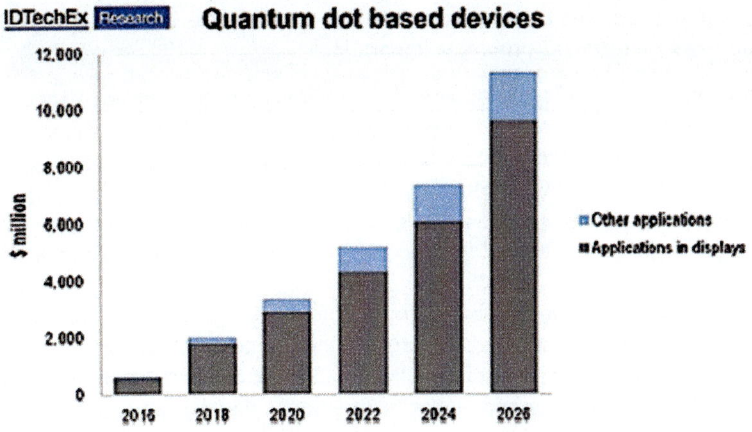

Fig. 19.34 Sale ($ million) forecast for QD devices/components during 2016–2026. Reproduced from Internal (Peach 2015)

Table 19.3 Li-Fi versus Wi-Fi technologies (Kaushik et al. 2017)

Item	Li-Fi	Wi-Fi
Speed	1–3.5 Gbps	54–250 Mbps
Range	10 m	20–100 m
Spectrum range	10,000 times than Wi-Fi	Radio spectrum range
Data transfer medium	Light used as carrier	Radio spectrum
Frequency band	100 THz	2.4 GHz
Safety	Non-hazardous	Hazardous

(a) Li-Fi Technology: With fast capabilities, LEDs, are being utilized in transferring data between computers. Li-Fi (Light Fidelity) technology uses visible light as the carrier in data transmission/networking, whereas Wi-Fi uses radio wavelength. Li-Fi is attractive as one futuristic technology where every bulb can be converted into a whole new Wi-Fi hotspot. With this innovation, information can be networked at a speed as fast as about 10 Gbps. It is contemplated that Li-Fi may be employed soon with its unique properties (Table 19.3) in aircraft, hospitals, and underwater communications (Kaushik et al. 2017).

(b) Quantum Information Science: Currently, the computers are getting faster and smaller with advanced technology, but eventually, the physical limits of materials will impede the activity, till we find completely different technologies. As already discussed one possibility is to store and transmit information with photons in place of electrons, i.e., the technology of photonics. It is knocking at the door to build optical computers using QDs in a similar way the present computers are built upon transistors for memory chips and logic gates. Optical computers are expected to be realized soon depending on the progress made by computer scientists. It may be pointed out that quantum computer works on quantum bits stored by "entangled" atoms, ions, electrons, or photons and they can work on different problems in parallel.

From our previous discussion on CQDs, it may be quite easy to incorporate either single QD or QD arrays. Current insight of such a system is continuously improving about their quantum coherent effects to use them as podiums for quantum information science. CQD designing has the flexibility to achieve control of quantum mechanical properties like charge and spin. Investigations are progressing at a fast rate for optically addressable qubits usable in quantum computation, simulation, and sensing.

Summarizing, R&D on QD is progressing at a great pace. It is working at the cross-roads of chemistry, physics, materials science, and engineering. Development in this area over the last 20 years has parallel in bulk semiconductor technology during mid-twentieth century. In the end, it may be mentioned that the halide perovskite QDs are the latest entrant having many more promising features. Researchers all over the world are hopeful in mitigating the cytotoxicity issue of cadmium, lead, etc.; these small wonders will have endless frontiers.

Acknowledgements The author acknowledges the works of numerous researchers in this field and apologies for omitting many other works. He is indebted to his brother Prof. Salil K. Sarkar, Presidency University, Kolkata for encouragement to work in this subject area. He is also grateful to the Department of Atomic Energy and Bhabha Atomic Research Center for the persistent support provided for his research activities.

References

Abe, S., Capek, R.K., Geyter, B.D., Hens, Z.: Reaction chemistry/nanocrystal property relations in the hot injection synthesis, the role of the solute solubility. ACS Nano **7**, 943–949 (2013)

Amirav, L., Amirav, A., Lifshitz, E.: A spray-based method for the production of semiconductor nanocrystals. J. Phys. Chem. B **109**, 9857–9860 (2005)

Bang, E., Choi, Y., Cho, J., Suh, Y.H., Ban, H.W., Son, J.S., Park, J.: Synthesis of micro and nanostructures in microfluidic systems. Chem. Soc. Rev. **39**, 1183–1202 (2010)

Bang, E., Choi, Y., Cho, J., Suh, Y.H., Ban, H.W., Son, J.S., Park, J.: Large-scale synthesis of highly luminescent InP@ ZnS quantum dots using elemental phosphorus precursor. Chem. Mater. **29**, 4236–4243 (2017)

Bera, D., Qian, L., Tseng, T.K., Holloway, P.H.: Quantum dots and their multimodal applications: a review. Mater. (basel) **3**, 2260–2345 (2010)

Biju, V., Mundayoor, S., Omkumar, R.V., Anas, A., Ishikawa, M.: Bioconjugated quantum dots for cancer research: present status, prospects and remaining issues. Biotechnol. Adv. **28**, 199–213 (2010)

Bobo, D., Robinson, K.J., Islam, J., Thurecht, K.J., Corrie, S.R.: Nanoparticle-based medicines: a review of FDA-approved materials and clinical trials to date. Pharm. Res. **33**, 2373–2387 (2016)

Brown, R.G.W., Pike, E.R.: A history of optical and optoelectronic physics in the twentieth century. In: Brown, L.M., Pais, A., Pippard, B. (eds.) Twentieth Century Physics, vol. III. American Institute of Physics Press, New York, USA (1995)

Choi, M.K., Yang, J., Hyeon, T., Kim, D.H.: Flexible quantum dot light-emitting diodes for next-generation displays. npj Flex. Electron. **2**, 10 (2018). https://doi.org/10.1038/s41528-018-0023-3

Dai, X., Deng, Y., Peng, X., Jin, Y.: Quantum-dot light-emitting diodes for large-area displays: towards the dawn of commercialization. Adv. Mater. **29**, 1607022 (2017)

Efros, A.L., Brus, L.E.: Nanocrystal quantum dots: from discovery to modern development. ACS Nano **15**, 6192–6210 (2021)

Ekimov, A.I., Onushchenko, A.A., Tsekhomski, V.A.: Exciton absorption by copper chloride crystal in glassy matrix. Fiz. Khim. Stekla **6**, 511–512 (1980)

Elias, J., Lévy-Clément, C., Bechelany, M., Michler, J., Wang, G., Wang, Z., Philippe, L.: Hollow urchin-like ZnO thin films by electrochemical deposition. Adv. Mater. **22**, 1607 (2010)

Feynman, R.P.: There's Plenty of Room at the Bottom. California Institute of Technology, Engineering and Science Magazine (1960)

Guleria, A., Singh, A.K., Rath, M.C., Sarkar, S.K., Adhikari, S.: The role of structural and fluidic aspects of room temperature ionic liquids in influencing the morphology of CdSe nano/microstructures grown in situ. Dalton Trans. **43**, 11843–11854 (2014)

Hao, J., Liu, H., Miao, J., Lu, R., Zhou, Z., Zhao, B., Xie, B., Cheng, J., Wang, K., Delville, M.H.: A facile route to synthesize CdSe/ZnS thick-shell quantum dots with precisely controlled green emission properties: towards QDs based LED applications. Sci. Rep. **9**, 12048 (2019)

Hines, M.A., Guyot-Sionnest, P.: Synthesis and characterization of strongly luminescing ZnS-capped CdSe nanocrystals. J. Phys. Chem. **100**, 468–471 (1996)

Huang, X., Tong, X., Wang, Z.: Rational design of colloidal core-shell quantum dots for optoelectronic applications. J. Electron. Sci. Technol. **18**, 100018 (2020)

Jiao, S., Shen, Q., Mora-Sero, I., Wang, J., Pan, Z., Zhao, K., Kuga, Y., Zhong, X., Bisquert, J.: Band engineering in core/shell ZnTe/CdSe for photovoltage and efficiency enhancement in exciplex quantum dot sensitized solar cells. ACS Nano **9**, 908–915 (2015)

Kamat, P.V.: Quantum dot solar cells. Semiconductor nanocrystals as light harvesters. J. Phys. Chem. C **112**, 18737–18753 (2008)

Kaushik, R., Jaiswal, R., Joon, R.: Light fidelity: a new prototype in wireless communication. Int. J. Innov. Res. Comput. Sci. Technol. **5**, 277–280 (2017)

Kikkeri, R., Laurino, P., Odedra, A., Seeberger, P.H.: Synthesis of carbohydrate-functionalized quantum dots in microreactors. Angew. Chem. Int. Ed. **49**, 2054–2057 (2010)

Klimov, V.I.: Mechanisms for photogeneration and recombination of multiexcitons in semiconductor nanocrystals: implications for lasing and solar energy conversion. J. Phys. Chem. B **110**, 16827–16845 (2006)

Klimov, V.I., Mikhailovsky, A.A., Xu, S., Malko, A., Hollingsworth, J.A., Leatherdale, C.A., Eisler, H.-J., Bawendi, M.G.: Optical gain and stimulated emission in nanocrystal quantum dots. Science **290**, 314–317 (2000)

Klimov, V.I., Ivanov, S.A., Nanda, J., Achermann, M., Bezel, I., McGuire, J.A., Piryatinski, A.: Single-exciton optical gain in semiconductor nanocrystals. Nature **447**, 441–446 (2007)

Lee, Y.-L., Lo, Y.-S.: Highly efficient quantum-dot-sensitized solar cell based on cosensitization of CdS/CdSe. Adv. Funct. Mater. **19**, 604–609 (2009)

Li, J.J., Wang, Y.A., Guo, W., Keay, J.C., Mishima, T.D., Johnson, M.B., Xiaogang, P.: Large-scale synthesis of nearly monodisperse CdSe/CdS core/shell nanocrystals using air-stable reagents via successive ion layer adsorption and reaction. J. Am. Chem. Soc. **125**, 12567–12575 (2003)

Liang-Shi, L., Jiangtao, H., Weidong, Y., Alivisatos, A.P.: Band gap variation of size- and shape-controlled colloidal CdSe quantum rods. Nano Lett. **1**, 349–351 (2001)

Lim, J., Park, Y.-S., Klimov, V.I.: Optical gain in colloidal quantum dots achieved with direct-current electrical pumping. Nat. Mater. **17**, 42–48 (2018)

Ma, Z., Yu, J., Dai, S.: Preparation of inorganic materials using ionic liquids. Adv. Mater. **22**, 261–285 (2010)

Martynenko, I.V., Litvin, A.P., Purcell-Milton, F., Baranov, A.V., Fedorov, A.V., Gun'ko, Y.K.: Application of semiconductor quantum dots in bioimaging and biosensing. J. Mater. Chem. B **5**, 6701–6727 (2017)

Matea, C.T., Mocan, T., Tabaran, F., Pop, T., Mosteanu, O., Puia, C., Iancu, C., Mocan, L.: Quantum dots in imaging, drug delivery and sensor applications. Int. J. Nanomedicine **12**, 5421–5431 (2017)

Mauro, E., Eva, P., Jordi, A., Nicolas, S., Thierry, P., Joan, R.M.: Capping ligand effects on the amorphous-to-crystalline transition of CdSe nanoparticles. Langmuir **24**, 11182–11188 (2008)

Mohamed, W.A.A., El-Gawad, H.A., Mekkey, S., Galal, H., Handal, H., Mousa, H., Labib, A.: Quantum dots synthetization and future prospect applications. Nanotechnol. Rev. **10**, 1926–1940 (2021)

Montanarella, F., Kovalenk, M.V.: Three millennia of nanocrystals. ACS Nano **16**, 5085−5102 (2022)

Murray, C.B.: Thesis, Massachusetts Institute of Technology, Cambridge, MA (1995)

Murray, C.B., Norris, D.J., Bawendi, M.G.J.: Synthesis and characterization of nearly monodisperse CdE (E = sulfur, selenium, tellurium) semiconductor nanocrystallites. J. Am. Chem. Soc. **115**, 8706–8715 (1993)

Novotny, L., Hecht, B.: Principles of Nano-Optics. Cambridge University Press, Cambridge, England (2006)

Optics and Photonics: Essential Technologies for Our Nation. National Academies Press, Washington DC, USA (2013)

Palomaki, P.: Quantum dots + OLED = Your next TV: formerly rival technologies will come together in new Samsung displays. IEEE Spectr. **59**, 52–53 (2022)

Pan, Z., Rao, H., Mora-Seró, I., Bisquert, J., Zhong, X.: Quantum dot-sensitized solar cells. Chem. Soc. Rev. **47**, 7659–7702 (2018)

Park, Y.-S., Roh, J., Diroll, B.T., Schaller, R.D., Klimov, V.I.: Colloidal quantum dot lasers. Nat. Rev. Mater. **6**, 382–401 (2021)

Peach, M.: QD Vision, Konka launch quantum dot TVs into China. https://www.optics.org/news/6/4/16 (2015). 2021

Peng, P., Milliron, D.J., Hughes, S.M., Alivisatos, A.P., Saykally, R.: Femtosecond spectroscopy of carrier relaxation dynamics in type II CdSe/CdTe tetrapod heteronanostructures. Nano Lett. **5**, 1809–1813 (2005)

Pietryga, J.M., Park, Y.-S., Lim, J., Fidler, A.F., Bae, W.K., Brovelli, S., Klimov, V.I.: Spectroscopic and device aspects of nanocrystal quantum dots. Chem. Rev. **116**, 10513−10622 (2016)

Prasad, K., Jha, A.K.: Biosynthesis of CdS nanoparticles: an improved green and rapid procedure. J. Colloid Interface Sci. **342**, 68–72 (2010)

Protesescu, L., Yakunin, S., Bodnarchuk, M.I., Krieg, F., Caputo, R., Hendon, C.H., Yang, R.X., Walsh, A., Kovalenko, M.V.: Nanocrystals of cesium lead halide perovskites ($CsPbX_3$, X = Cl, Br, and I): novel optoelectronic materials showing bright emission with wide color gamut. Nano Lett. **15**, 3692–3696 (2015)

Ramalingam, G., Ragupathi, C., Rangasamy, B., Colak, I., Vetrivelan, V., Poudineh, N., Balasubramani, R., Chang, S.W., Gengan, R.M.: Structural and optical properties of CdSe/CdTe core-shell quantum dots. J. Nanomater. 6316716 (2022). https://doi.org/10.1155/2022/6316716

Reiss, P., Protiere, M., Li, L.: Core-shell semiconductor nanocrystals. Small **5**, 154–168 (2009)

Rogach, A.L., Katsikas, L., Kornowski, A., Su, D., Eychmuller, A., Weller, H.: Synthesis and characterization of thiol-stabilized CdTe nanocrystals. Ber. Bunsen-Ges. Phys. Chem. **100**, 1772–1778 (1996)

Rossetti, R., Nakahara, S., Brus, L.E.J.: Quantum size effects in the redox potentials, resonance Raman spectra, and electronic spectra of CdS crystallites in aqueous solution. Chem. Phys. **79**, 1086–1088 (1983)

Shi, J., Kantoff, P.W., Wooster, R., Farokhzad, O.C.: Cancer nanomedicine: progress, challenges and opportunities. Nat. Rev. Cancer **17**, 20–37 (2017)

Shiang, J.J., Kadavanich, A.V., Grubbs, R.K., Alivisatos, A.P.: J. Phys. Chem. **99**, 17417–17422 (1995)

Singh, S., Rath, M.C., Singh, A.K., Sarkar, S.K., Mukherjee, T.: Reversible formation of CdSe nanoparticles in aqueous solutions by high-energy electron beam irradiation. Mater. Chem. Phys. **124**, 6–9 (2010)

Singh, S., Rath, M.C., Sarkar, S.K.: Investigation of dynamics of radiolytic formation of CdSe nanoparticles in aqueous solutions. J. Phys. Chem. A **115**, 13251–13258 (2011a)

Singh, S., Rath, M.C., Singh, A.K., Mukherjee, T., Jayakumar, O.D., Tyagi, A.K., Sarkar, S.K.: CdSe nanoparticles grown via radiolytic methods in aqueous Solutions. Rad. Phys. Chem. **80**, 736–741 (2011b)

Singh, S., Guleria, A., Rath, M.C., Singh, A.K., Adhikari, S., Sarkar, S.K.: Sea urchin like shaped CdSe nanoparticles grown in aqueous solutions via electron beam irradiation. J. Nanosci. Nanotechnol. **13**, 5365–5373 (2013a)

Singh, S., Singh, A.K., Rath, M.C., Adhikari, S., Sarkar, S.K.: Radiolytic synthesis and spectroscopic investigations of cadmium selenide quantum dots grown in cationic surfactant based quaternary water-in-oil microemulsions. J. Colloid Interface Sci. **398**, 112–119 (2013b)

Singh, S., Guleria, A., Rath, M.C., Singh, A.K., Adhikari, S., Sarkar, S.K.: Shape evolution of CdSe nanomaterials in microheterogeneous media. Adv. Mater. Lett. **4**, 449–453 (2013c)

Spinks, J.W.T., Woods, R.J.: An Introduction Radiation Chemistry, 3rd edn. Wiley, New York (1976)

Steckel, J.: Ph.D. Thesis, Massachusetts Institute of Technology, Cambridge, MA (2006)

Taniguchi, N.: On the Basic Concept of Nano-Technology. In: Proceedings of the International Conference on Production Engineering, pp. 5−10. Japan Society of Precision Engineering, Tokyo (1974)

Tong, X., Wang, Z.M. (eds.): Core-Shell Quantum Dots Synthesis, Properties and Devices. Springer Nature Switzerland AG (2020)

Valligatla, S., Haldar, K.K., Patra, A., Desai, N.R.: Nonlinear optical switching and optical limiting in colloidal CdSe quantum dots investigated by nanosecond Z-scan measurement. Opt. Laser Technol. **84**, 87–93 (2016). https://doi.org/10.1016/j.optlastec.2016.05.009

Vossmeyer, T., Katsikas, L., Giersig, M., Popovic, I.G., Diesner, K., Chemseddine, A., Eychmueller, A., Weller, H.J.: CdS nanoclusters: synthesis, characterization, size dependent oscillator strength, temperature shift of the excitonic transition energy, and reversible absorbance shift. J. Phys. Chem. **98**, 7665–7673 (1994)

Wang, A., Yan, X., Zhang, M., Sun, S., Yang, M., Shen, W., Pan, X., Wang, P., Deng, Z.: Controlled synthesis of lead-free and stable perovskite derivative Cs_2SnI_6 nanocrystals via facile hot-injection process. Chem. Mater. **28**, 8132–8140 (2016)

Yakoubi, A., Chaabane, T.B., Aboulaich, A., Mahiou, R., Balan, L., Medjahdi, G., Schneider, R.: Aqueous synthesis of Cu-doped CdZnS quantum dots with controlled and efficient photoluminescence. J. Lumin. **175**, 193–202 (2016)

Yang, J., Wang, J., Zhao, K., Izuishi, T., Li, Y., Shen, Q., Zhong, X.: CdSeTe/CdS type-I core/shell quantum dot sensitized solar cells with efficiency over 9%. J. Phys. Chem. C **119**, 28800–28808 (2015)

Yang, Z., Lu, L., Kiely, C.J., Berger, B.W., Mcintosh, S.: Biomineralized CdS quantum dot nanocrystals: optimizing synthesis conditions and improving functional properties by surface modification. Ind. Eng. Chem. Res. **55**, 11235–11244 (2016a)

Yang, H.Y., Fu, Y., Jang, M.S., Li, Y., Lee, J.H., Chae, H., Lee, D.S.: Multifunctional polymer ligand interface CdZnSeS/ZnS quantum dot/Cy3-labeled protein pairs as sensitive FRET sensors. ACS Appl. Mater. Interfaces **8**, 35021–35032 (2016b)

Yang, J., Choi, M.K., Yang, U.J., Kim, S.Y., Kim, Y.S., Kim, J.H., Kim, D.-H., Hyeon, T.: Toward full-color electroluminescent quantum dot displays. Nano Lett. **21**, 26–33 (2021)

Yen, B.: Ph.D. Thesis, Massachusetts Institute of Technology, Cambridge, MA (2007)

Zhao, T., Liu, X., Li, Y., Zhang, M., He, J., Zhang, X., Liu, H., Wang, X., Gu, H.: Fluorescence and drug loading proprieties of ZnSe:Mn/ZnS-paclitaxel/SiO2 nanocapsules templated by F127 micelles. J. Colloid Interface Sci. **490**, 436–443 (2017)

Zhou, Y., Zhao, H., Ma, D., Rosei, F.: Harnessing the properties of colloidal quantum dots in luminescent solar concentrators. Chem. Soc. Rev. **47**, 5866–5890 (2018)

Chapter 20
XAFS Characterization of Materials—A Realistic Evaluation

Debdutta Lahiri

20.1 Introduction

X-ray Absorption Fine Structure (XAFS) is a synchrotron-based contemporary technique for the simultaneous determination of chemical and structural information for selected element in condensed matter (Koningsberger and Prins 1988). XAFS is essentially a subgroup of x-ray absorption experiments, in which incident x-ray energy is tuned around the binding energy (E_0) of a component element of sample. Representative XAFS spectrum (for Cu foil) is presented in Fig. 20.1a. In the simplest experimental setup (inset of Fig. 20.1a), absorption coefficient $\mu(E)$ is measured by the attenuation of x-rays passing through the sample, governed by Lambert–Beer law: $I_t = I_0 e^{-\mu t}$; I_0 and I_t are incident and transmitted intensities of x-ray and $t =$ sample thickness. XAFS experiments involve measurement of $\mu(E)\ \alpha\ \ln\left(\frac{I_0}{I_t}\right)$, as function of incident x-ray energy E (tuned around E_0). XAFS spectrum (Fig. 20.1a) displays a sharp rise at E_0—the so-called "absorption edge," resembling "resonance" phenomenon. This is followed by a smooth decreasing background, modulated by oscillations, called "fine structure" of x-ray absorption. XAFS oscillations are divided into two sub-regions, marked in Fig. 20.1a: (a) X-Ray Absorption Near Edge Structure (XANES), within $30eV$ of edge and (b) Extended XAFS (EXAFS), beyond $30eV$ past edge. XANES and EXAFS, respectively, contain valence state information of the x-ray excited atom and "local" structural information around this atom (within 10Å). Figure 20.1b depicts the extraction of XAFS oscillations [$\chi(k)$], as function of photoelectron wave-vector (k); Fig. 20.1c depicts the Fourier transform of $\chi(k)$ into radial distribution function $|\chi(R)|$ for extraction of structural information. XAFS technique has dramatically evolved over decades as a unique and versatile tool, for its powerful advantages such as (i) simultaneous information of structure

D. Lahiri (✉)
High Pressure and Synchrotron Radiation Physics Division, Bhabha Atomic Research Centre, Mumbai 400085, India
e-mail: dlahiri@barc.gov.in

R. S. Ningthoujam and A. K. Tyagi (eds.), *Handbook of Materials Science, Volume 1*, Indian Institute of Metals Series, https://doi.org/10.1007/978-981-99-7145-9_20

and chemical state, e.g., in catalysis (Grunwaldt and Frenkel 2009; Timoshenko and Frenkel 2019); (ii) element-resolved information, e.g., for alloys (Frenkel et al. 2000), [*A vs. B*]—sites in ABO_3 perovskites (Lahiri 2010, Lahiri et al. 2019), doped superconductors (Haskel et al. 1999, 2001); (iii) sensitivity to amorphous structure, e.g., metallic glass (Zhang 2019), liquid (Filipponi 2001), soil (Lim et al. 2015); (iv) sensitivity to "local" disorder, e.g., around dopants in crystalline materials (Haskel et al. 1999, 2001); (v) flexibility for diverse sample conditions viz. bulk (surface) information depth, e.g., multilayer (Ramanan et al. 2013), liquid (Zhang 2019), *ppm*-low concentration (Ranieri-Raggi et al. 2003), low temperature (Ramanan et al. 2015a), high temperature (Smith et al. 2019) and high-pressure (Pascarelli and Haskel 2016). These advantages enabled the undertaking of novel research problems and unlocking of under-explored areas such as glass (Cheng et al. 2009; Fujita et al. 2009; Haskel et al. 1999; Lahiri et al. 2014), ultra-small nanosystems (Lahiri 2010), biology (Ascone et al. 2005), environment (O'Loughlin et al. 2010; Suzuki et al. 2002; Thomas et al. 2019), advanced mineralogy (Lahiri et al. 2020), where the scope of diffraction and microscopy techniques are limited.

XAFS has also revolutionized the understanding of fundamental scientific phenomena (e.g., magneto-transport phenomena) by presenting the role of "local disorder" (vis-à-vis long-range-order). Since the length-scale of local disorder $\left(\sim 10\ \mathring{A}\right)$ coincides with that of charge transport, magnetic, and chemical interactions, any perturbation within this distance can significantly modify macroscopic properties. The concept of local disorder has resolved several paradoxes: (i) Anomalous mismatch between crystallographic (from diffraction) and magneto-transport phase boundaries is often observed (Haskel et al. 1996; Shibata et al. 2002a). This anomaly can be resolved with XAFS detection of "hidden" phases, such as locally tilted cuprate octahedra in Cuprate superconductor (Haskel et al. 1996) and nanoscale clustering of Sr in CMR materials (Shibata et al. 2002a).

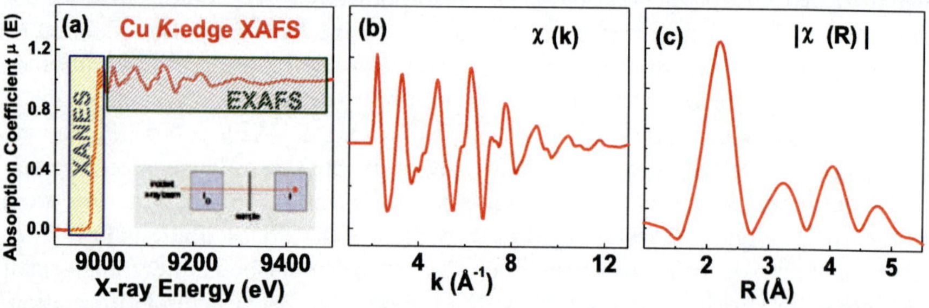

Fig. 20.1 Representative Cu K-edge XAFS spectrum [μ(E)] for Cu foil; XANES and EXAFS regions marked. **a** The simplest (transmission-mode) XAFS setup is shown in inset, in which sample is inserted in x-ray path and attenuation of x-rays through sample is measured; **b** XAFS oscillations [χ(k)] is extracted from μ(E), as function of photoelectron wave-vector (k); **c** χ(R)| = Fourier transform of χ (k), in which different peaks correspond to different bond-lengths around the excited atom (Cu, in this case)

These local deviations are responsible for the (observed) persistence of superconductivity/magnetism even beyond crystallographic transition. (ii) Stabilization of counter-intuitive ground states—(a) metallic ferromagnetic ground state (vis-à-vis antiferromagnetic insulator) was stabilized in manganite nanoparticles (Lahiri et al. 2012), despite the detrimental presence of Jahn–Teller distortion. This paradox was resolved with XAFS detection of inter-octahedral orientational disorder—the latter promotes metallic ferromagnetism by destroying orbital ordering which is the precondition for anti-ferromagnetism. (b) Significantly higher experimental spin polarization (vis-à-vis theoretically predicted value) in spintronic materials could be understood with XAFS detection of the anti-site disorder (Lahiri et al. 2010b). The latter modifies near—neighbor bond parameters and density-of-states favorably, thereby increasing the polarization. These two examples defy the conventional detrimental notion of disorder and present novel beneficial perspective of the disorder. This provides optimism for the development of materials that had been long toiling on minimizing defects. (iii) Paradoxical sensitivity of magnetism to supposedly non-magnetic element (A) in ABO_3 perovskites (Lahiri et al. 2010a, 2019)—(A, B) site-resolved XAFS study resolved this paradox by demonstrating the magnetostructural route viz distortion of $AO_6 \Rightarrow$ rotation of BO_6 [via $A - O - B$ coupling] $\Rightarrow$ modification of $B - O - B$ angle for superexchange magnetic interaction. Some of these works are vividly addressed in a later section. These solutions convey the indispensability of disorder consideration along with basic crystalline structure and thereby, inspired the integration of XAFS probe into fundamental research at large.

Several books besides the basic book (Koningsberger and Prins 1988), review articles and tutorials of XAFS have been dedicated to the development of theoretical and experimental understandings of XAFS (https://xafs.xrayabsorption.org). The present article revisits these concepts with non-XAFS community in mind, who often approach XAFS for supplementary support in their research problems. This presentation is tailored to address their queries, e.g., the suitability of XAFS for their scientific problem, the scale of information that can be extracted with XAFS, selection of samples, importance of good sample and data quality, analytic strategies and the uniqueness of XAFS solution. The ultimate fulfillment of this article lies in inspiring non-XAFS groups to integrate XAFS technique into their mainstream research scheme.

The article is organized into the following sections; each section assumes certain precursory knowledge at the readers' end, as listed along with:

(1) Theory of XAFS—basic theories of quantum mechanics (Schiff 1949,) e.g., wave functions, perturbation, scattering, semiclassical treatment of radiation, interference phenomenon.
(2) XAFS Data Processing—mathematics of Fourier transform (James 2011)
(3) XAFS Data Fitting—principles of data fitting and error analysis (Bevington 1969)
(4) XANES—Bond-Valence-Sum rule of coordination chemistry (Brown and Altermatt 1985), density-of-states (Mermin and Ashcroft 2003).

(5) Experimental Configuration—This section is very briefly covered, since experiments are handled by synchrotron-expert beamline scientists and is beyond the scope of non-XAFS users.
(6) Application Examples.

20.2 Theory of XAFS

XAFS is a subgroup of absorption experiments, in which energy (E) of the incident x-rays is tuned around E_0—the binding energy (BE) of core electron of any element of the sample. Since E_0 is the atomic number (Z)-dependent, i.e. $E_0 \sim Z^2$ (Viswanathan 2018), XAFS probe is element-specific, so that information around different elements (A, B) of sample AB can be separately extracted by exciting (A, B) elements in turn. BE or the so-called "edge energy" (E_0) extends over $5-100$ keV across the periodic table ($Z = 22-92$) (http://www.csrri.iit.edu/periodic-table.html). Depending upon which core electron is excited, XAFS spectra are termed as K-edge ($1s$ electron), L_1, L_2, L_3-edges ($2s, 2p_{\frac{1}{2}}, 2p_{\frac{3}{2}}$ electrons), etc.

Absorption over $5-100$ keV is dominated by photoelectric effect (Veigele 1973). The scheme of photoelectric effect is depicted in Fig. 20.2a. When an isolated atom is excited by x-rays of energy $E(\geq E_0)$, the specific core electron is promoted to unoccupied states of continuum and leaves behind the core hole. The theory of absorption is based on the perception of excited photoelectron as a wave, rather than particle that propagates with kinetic energy ($E - E_0$) through the material (blue circles in Fig. 20.2b). The whole event resembles "resonance" phenomenon of quantum-mechanics that is demonstrated as sharp jump in atomic absorption [$\mu_0(E)$] at the "absorption edge" (E_0), followed by smooth decrease at higher energies.

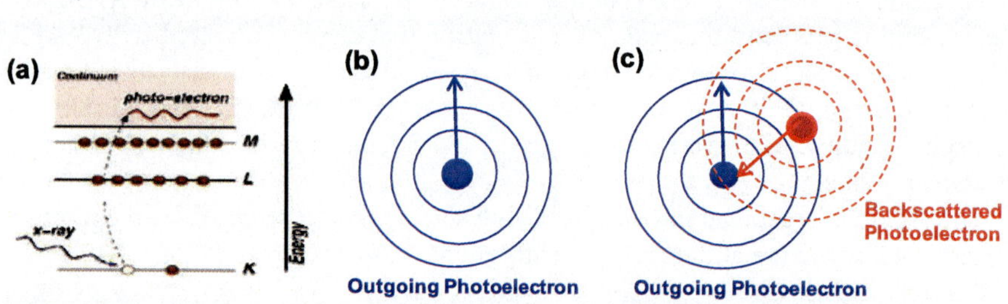

Fig. 20.2 **a** Photoelectric effect—the underlying physics ofXAFS. Core electron (e.g. K-shell) is excited by x-rays of energy equal to binding energy (E0) of the electron and ejected into continuum.The trajectoryof spherical photoelectronwave is depicted for **b** isolated atom: thewave travels outward fromthe absorbingatomand propagates freely (blue solid line); **c** condensedmatter: thewave is backscattered by neighboringatoms and returns to the original atom(red dashed line). Interference of original (blue) and backscattered (red)waves gives rise to XAFS oscillations of finalwave vector. Analysis ofXAFS oscillations reveals information of neighboring atoms: atomic species (Z), radial distance (R), coordination (N), disorder (σ^2)

Quantification of μ_0 is guided by Fermi's Golden Rule (Schiff 1949)—μ_0 is proportional to the absorption matrix:

$$|\langle i_0|H|f_0\rangle|^2 \tag{20.1}$$

$|i_0\rangle$ = Initial state = [x-ray + core electron + no photoelectron],
$|f_0\rangle$ = Final state = [no x-ray + core hole + photoelectron].
H = Electron-photon interaction = $\left(\vec{p} - \frac{e\vec{A}}{mc}\right)^2$ (Schiff 1949);

$\vec{P}$ = electron momentum vector,
$\vec{A}$ = vector potential of electromagnetic field $\sim A_0 e^{i\vec{q}\cdot\vec{r}}\hat{\varepsilon}$, $\vec{q}$ = photon wave vector.

Following certain approximations (not shown), H can be simplified into:

$$\mathrm{H}\,\alpha\,(\hat{\varepsilon}.\,\vec{P}) \tag{20.2}$$

For real materials (vis-à-vis isolated atoms), the photoelectron gets backscattered by neighboring atoms and returns to the original atom (red circles in Fig. 20.2c). The original outgoing $|f_0\rangle$ and backscattered waves $|\Delta f\rangle$ interfere to construct oscillatory final state vector $|f\rangle = |f_0\rangle + |\Delta f\rangle$. This is the origin of XAFS oscillations in absorption spectra [$\mu(E)$] (Fig. 20.1a).

$$\mu(E) = |\langle i \mid H|f\rangle|^2 \approx |\langle i_0 \mid H|f_0\rangle|^2\left[1 + \frac{\langle f_0 \mid H|i_0\rangle^*\langle i_0 \mid H|\Delta f\rangle}{|\langle i_0 \mid H|f_0\rangle|^2} + CC\right]$$
$$= \mu_0(E)[1 + \chi(E)] \tag{20.3}$$

Equation 20.3: CC = complex conjugate; since $|\langle i_0|H|\Delta f\rangle|^2$ the term is very small, it has been neglected; $|i\rangle = |i_0\rangle$, since the core electron is tightly bound to the absorbing atom and is unperturbed by the presence of other atoms. From Eq. 20.3,

$$\chi(E)\alpha\langle i_0|H|\Delta f\rangle \tag{20.4}$$

$\chi(E)$ = XAFS represents the fractional deviation from background atomic absorption.

Although XAFS is an obvious signature of the presence of neighboring atoms (vis-à-vis isolated atom), it took three decades until the late 1970s to convert it into practical structural information. The development of the current XAFS theory is credited to Stern, Sayers and Lytle (Koningsberger and Prins 1988).

20.2.1 Conceptualization of Perturbation Term [$\chi(E)$] in the Absorption Matrix

In the case of isolated atom, e.g., monatomic gas, freely propagating photoelectron may be represented by final state $|f_0\rangle = \psi\left(\sim \frac{e^{ikr}}{kr}\right)$. Origin is set at the absorbing atom. Photoelectron wave-vector (k) is related to the kinetic energy (ΔE) of the photoelectron: $\Delta E = E - E_0 = \frac{\hbar^2 k^2}{2m}$ or alternatively, $k = 0.069\sqrt{\Delta E}$. The outgoing free wave ($|f_0\rangle = \psi$) for isolated atom is depicted by solid blue lines in Fig. 20.2b. For real materials, this wave ($|f_0\rangle = \psi$) travels to neighboring atoms ($\psi \rightarrow \psi_A$), gets backscattered ($\psi_A \rightarrow \psi_B$) and returns to the original atom ($\psi_B \rightarrow \psi_f$) (Fig. 20.2c). The backscattered wave is depicted in red lines in Fig. 20.2c. The perturbation term $|\Delta f\rangle = \psi_f$, so that Eq. 20.4 reduces to:

$$\chi(E)\alpha\langle i_0|H|\Delta f\rangle \sim \langle i_0|H|\psi_f\rangle \tag{20.5}$$

Therefore, the task of $\chi(E)$ determination is essentially reduced to the formulation of ψ_f. Since the absorption matrix is non-zero only in the region where $|i_0\rangle \neq 0$, i.e., at origin, it is sufficient to determine ψ_f at origin.

20.2.2 Assumptions for Derivation of $\chi(E)$

The formulation of ψ_f is simplified by a set of approximations that are valid over the EXAFS regime:

(i) "Single electron approximation"—absorption matrix is calculated only for the excited electron, as the remaining electrons of the atom are assumed to be unperturbed by the transition of core electron.

(ii) "Dipole approximation" (Schiff 1949) i.e., core electron size (d_e) << wavelength of incident photon (λ_p)—this eventually simplifies H term [$\alpha(\hat{\varepsilon}.\vec{P})$] of Eq. 20.2 into ($\hat{\varepsilon}.\vec{r}$)$\sim$ $\cos\theta$ [intermediate steps are not shown]; θ = angle between x-ray polarization ($\hat{\varepsilon}$) and electron position ($\hat{r}$) vectors (Fig. 20.3a).
i.e.,

$$H \sim \cos\theta \tag{20.6}$$

Absorption matrix of Eq. 20.5 reduces to

$$\chi(E)\,\alpha\,\langle i_0 \mid H|\Delta f\rangle \sim \langle i_0 \mid \cos\theta|\psi_f\rangle \tag{20.7}$$

(iii) The next stage of simplification assumes K-edge absorption [i.e., core electron $= 1s$], so that $|i_0\rangle$ = s-wave ($l = 0$) has no angular dependence (l = angular momentum). Since $< \cos\theta >_{avg} = 0$, $\langle i_0 \mid \cos\theta|\psi_f\rangle$ matrix of Eq. 20.7 will

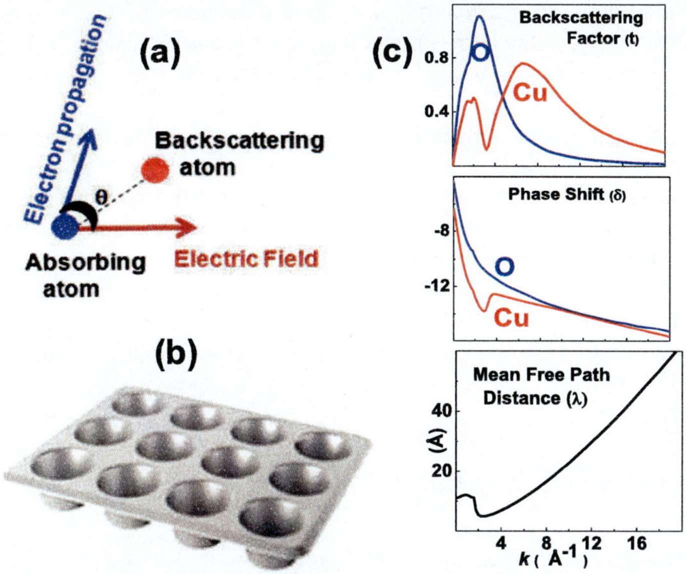

Fig. 20.3 Parameters of XAFS equation: **a** Angle (θ) of photoelectron propagation direction relative to electric field direction; **b** Muffin tin potential– spherical at atomic positions and constant in interstitial region; **c** Backscattering amplitude [t(k)] and phase shift [δ (k)] for high-Z (Cu) vs. low-Z (O) elements and universal λ(k) curve for mean free path of photoelectron

be non-zero only if ψ_f is $\cos\theta$-dependent $[\langle i_0 \mid H|\psi_f\rangle \sim< \cos^2\theta >_{avg} \neq 0]$.

This is satisfied by ψ_f of p $(l = 1)$ symmetry, i.e., $s \to p$ electron transition. Dipole approximation stands valid for all "deep" core electronic shells (e.g., K, L, M) and could be used for derivation of $\chi(E)$ at these edges as well. The approximation breaks down for shallow electronic levels of size $d_e \sim \lambda_p$ (e.g., valence shell), so that XAFS formulation for such edges is ruled out.

(iv) To first order, polycrystalline samples with randomly oriented unit cells are assumed so that site-averaged absorption matrix $\langle i_0 \mid H|\psi_f\rangle \sim \langle\cos^2\theta\rangle = \frac{1}{3}$. This prevents complicated angular-parameterization of $\chi(E)$. On the other hand, polarization-dependent XAFS scheme has been envisaged for judicious exploitation of angular-dependence of XAFS to resolve inter-planar information of anisotropic systems, e.g., layered high-T_c superconductors (Haskel et al. 1999, 2001).

(v) Muffin-tin scattering potential (Fig. 20.3b), which is spherical at atomic positions and constant in the interstitial region (Hartmann and Lauterbach 1998). This warrants that photoelectron is scattered only at atomic sites and travels freely over the interstitial region.

(vi) "Small atom approximation," i.e., atomic diameter $(D) << $ near neighbor distance (R)—this allows the curvature of a spherical wave to be neglected and incoming and scattered waves to be approximated as plane wave (Schiff 1949).
(vii) *De Broglie* wavelength of photoelectron $\left(=\frac{2\pi}{k}\right) <<$ near neighbor distance (R) so that $kR >> 2\pi$.
(viii) Energy (*and* k) of photoelectron is assumed to remain invariant throughout the event. This warrants coherence of states, which is a fundamental precondition for interference phenomenon (explained later).
(ix) Preliminary structural model consists of only *one* neighboring atom at distance R.
(x) *Single* scattering—photoelectron scatters from neighboring atoms and returns to the absorber.

20.2.3 Formulation of Basic $\psi_f \rightarrow \chi(E)$

Successive stages of photoelectron waveform $\left[\psi_i \rightarrow \psi_A \rightarrow \psi_B \rightarrow \psi_f\right]$ are formulated, based on these approximations.

(i) Photoelectron wave, emanating from the absorbing atom at origin: $\psi \sim \frac{e^{ikr}}{kr}$.
(ii) Its value at the neighboring atom position, at distance (R) from the absorbing atom: $\psi_A = \frac{e^{ikR}}{kR}$.
(iii) After backscattering: $\psi_A \rightarrow \psi_B\left(=\psi_A t(k)e^{i\delta(k)}\right) = \frac{t(k)e^{ik(R+\delta(k))}}{kR}[t(k), \delta(k)] =$ backscattering factor and phase shift (Schiff 1949), specific to the species (Z) of backscattering atom and transferable between different samples (Teo and Lee 1979).
Figure 20.3c demonstrates the contrast of $[t(k), \delta(k)]$ between low-Z and high-Z atoms: slow k-dependent decay for low-Z atoms $(Z < 40)$ *vis-à-vis* pronounced features at higher k for high-Z atoms $(Z > 40)$ that is attributed to the Ramsauer–Townsend effect of quantum mechanics (Schiff 1949).
(iv) Backscattered wave (ψ_B) travels distance $\sim R$, back to the absorbing atom. Its value at the origin:

$$\psi_f\left(=\psi_B\frac{e^{ikR}}{kR}\right) = \left[\frac{t(k)e^{ik(2R+\delta(k))}}{(kR)^2} + CC\right]\alpha\frac{t(k)}{(kR)^2}\sin[2kR + \delta(k)] \tag{20.8}$$

Inserting this into the expression for $\chi(E)$ in Eq. 20.7, the simplest form of XAFS can be derived:

$$\chi(E) = \frac{t(k)}{(kR)^2}\sin[2kR + \delta(k)] \tag{20.9}$$

Henceforth, $\chi(E)$ will be referred to as $\chi(k)$. A few facts are imminent from the expression of $\chi(k)$:

(a) The sinusoidal function of $\chi(k)$ is responsible for XAFS oscillations in $\mu(E)$. As E (or k) increases, the *sine* function oscillates between positive and negative, depending on whether $[2kR + \delta(k)]$ is even and odd multiple of π.
(b) The frequency of oscillation ($2k$) is inversely proportional to the near—neighbor distance (R). Slow (fast) oscillations correspond to short (long) bond-lengths. Thus, the frequency of oscillations can be potentially converted into information of discrete bond-lengths, by Fourier transform of $\chi(k) \rightarrow \chi(R)$ = radial distribution function.
(c) The amplitude of $\chi(k)$ is proportional to $[t(k), \delta(k)]$, whose Z-sensitivity can be exploited to identify the species of the neighboring atom.

Thus, basic $\chi(k)$ provides firsthand quantitative theory of the origin of XAFS and its structural implications. The next stage of refinement of $\chi(k)$ is for *real* structure, consisting of *multiple* neighbors of *multiple* species at *multiple* distances.

20.2.4 Refinement and Finalization of $\chi(k)$

(i) Higher coordination of neighbors ($N > 1$) contributes multiplicative coefficient ($\times N$): $\chi(k) \alpha N$.
(ii) At any finite temperature, the positions of the atoms are smeared by the thermal and static disorder. For small disorders, the corresponding bond-length (R) distribution can be assumed to be Gaussian: $\frac{\exp\left(\frac{-(r-R)^2}{2\sigma^2}\right)}{\sqrt{2\pi}\sigma}$; σ^2 = mean-square-displacement for bond-length R. This contributes dampening factor $\left(e^{-k^2\sigma^2}\right)$ into $\chi(k)$. Detailed derivation of Gaussian and non-Gaussian disorder components (Bunker 1984) of $\chi(k)$ is beyond the scope of this chapter.
(iii) Coherence of $|i_0\rangle$ and $|\psi_f\rangle$ were assumed for the preliminary formulation of $\chi(k)$. In reality, the traveling photoelectron loses energy through inelastic scattering, e.g., plasmon, electron–hole pair. Backscattered photoelectron may also fail to return to the absorbing atom before $|i_0\rangle$ changes by filling of core hole through Auger or fluorescence processes in $10^{-15} s$ (Rahkomen and Krause 1974). Such coherence-limiting factors lower the probability of interference (XAFS), which is duly accommodated into $\chi(k)$ with exponential decay factor $\left(e^{-\frac{2R}{\lambda}}\right)$. Mean free path length (λ) represents the distance to which a photoelectron can travel before it scatters inelastically and before core hole is filled. Universal λ(k) curve has been empirically constructed (Fig. 20.3d), based on photoemission and Auger emission results (Powell 1974). Figure 20.3d reveals that $\lambda \leq 30\,\text{Å}$; in other words, $\lambda\left(\sim 30\,\text{Å}\right)$ defines the maximum distance from which XAFS information can be obtained. This is the key factor responsible for transforming XAFS into a *local* probe – its unique strength. It essentially

eliminates the dependence upon long-range order (vis-à-vis diffraction), which makes it uniquely sensitive to amorphous structure. It may be cautioned that core widths can be as large as several eV for the deeper core states of heavy atoms, washing out the fine structure.

(iv) In principle, the initial and final states in the absorption matrix should encompass *all* (n) electrons of the atom: $\langle i^n \mid H | f^n \rangle$. The n-electron states can be disentangled into that of 1 core-electron and $n-1$ passive electrons, so that $\langle i^n \mid H | f^n \rangle = \langle i^{n-1} \mid H | f^{nN-1} \rangle \langle i \mid H | f \rangle$. Under "single electron approximation," $\langle i^{n-1} \mid H | f^{nN-1} \rangle$ was ignored since these $n-1$ "passive" electrons were assumed to be unperturbed by the transition of core electron. In reality, atomic potential changes slightly after the creation of core hole and "passive" electrons adjust to this new potential by relaxing to states different from $| i_0^{n-1} \rangle$ (Meldner and Perez 1971). This is equivalent to the loss of coherence that lowers the probability of XAFS. This limitation is accommodated in $\chi(k)$ through the amplitude reduction factor $S_0^2 = |\langle i^{n-1} \mid H | f^{nN-1} \rangle|^2 = 0.7 - 0.9$. S_0^2 is experimentally determined through the fitting of XAFS spectra of standards (Kelly et al. 2009). The relieving aspect of multielectron excitation is that they do not produce EXAFS. Considering the two-electron excitation mechanism, conservation of energy requires that the sum of the two energies be constant, although the energy of each electron can vary from zero up to the maximum possible energy. The EXAFS from this process will have contributions from electrons with varying energies that will tend to smear out the oscillations. Thus, one expects the main contribution from the single-electron excitation.

(v) Different sets of neighbors exist at different bond-lengths (R_j) with distinct $(N, Z, \sigma^2, \theta)_j$, termed "coordination shells." Nearest neighbors, second-nearest neighbors, etc., are, respectively, termed first shell, second shell, and so forth.

The final XAFS equation, summed over all bonds, stands as:

$$\chi(k) \alpha \sum_j S_0^2 \frac{N_j t_j(2k)}{(kR_j)^2} e^{-k^2 \sigma_j^2} e^{-\frac{2R_j}{\lambda}} \sin[2kR_j + \delta_j(k)] \qquad (20.10)$$

20.2.5 $\chi(k) \rightarrow \chi(R)$

Representative CuK-edge experimental $\mu(E)$ for Cu foil and corresponding $\chi(k)$ are presented in Fig. 20.1a, b. In Fig. 20.1b, $\chi(k)$ contains superimposed structural information, as described in Eq. 20.10. This is decoupled by Fourier transform of $\chi(k)$ into radial distribution function $\chi(R) \left[= \int_{k_{min}}^{k_{max}} \chi(k) k^w e^{i2kR} dk \right]$, presented in Fig. 20.1c.

$\chi(R)$ is complex number, with real (Re[$\chi(R)$]) and imaginary (Im[$\chi(R)$]) components. However, $|\chi(R)| = \left[|\text{Re}(\chi)|^2 + |\text{Im}(\chi)|^2\right]^{\frac{1}{2}}$ is used as the first-hand guide to the atomic distribution around absorbing atoms (Fig. 20.1c).

To the first order, each peak of $|\chi(R)|$ represents a specific bond. Correlation of bond parameters (R, N, σ^2, Z) with [$\chi(k)$, $|\chi(R)|$] is demonstrated in successive Fig. 20.4a–h. [$\chi(k)$,$|\chi(R)|$] were simulated for Cu–O and Cu-Cu bonds, as function of various bond parameters.

(i) Bond-length (R): Figure 20.4a–b presents [$\chi(k)$,$|\chi(R)|$] for Cu–O bond-lengths ($R = 2.5$ Å,$R = 2.7$ Å). In Fig. 20.4a, $\chi(k)$ demonstrates $k\alpha\frac{1}{R}$ dependence, i.e., slow (fast) oscillations correspond to short (long) bond-lengths. Corresponding $|\chi(R)|$ in Fig. 20.4b demonstrates that bond-length (R) is primarily correlated with centroid of the peak. Centroid of $|\chi(R)|$ is shifted (−0.5 Å) relative to R, which is due to linear k-dependence of δ [$\delta \sim -k$] in *sine* term of XAFS equation: $\sin(2kR + \delta) = \sin(2kR - k) = \sin 2k(R - 0.5)$. This generates effective bond-length $R_{eff} = (R - 0.5)$ Å that accounts for the observed displacement of peak. During modeling, the centroid of the peak (+0.5 Å) of experimental $|\chi(R)|$ serves as the preliminary estimate of bond-lengths of the sample. R is secondarily correlated with peak amplitude: $\chi(k)\alpha\frac{1}{R^2}$—for the same coordination, $|\chi(R)|$ for shorter distance has a higher peak amplitude than for longer bond-length.

(ii) Coordination (N): Figure 20.4c, d demonstrate that coordination (N) multiplies [$\chi(k)$,$|\chi(R)|$] by $\times N$ factor.

(iii) Debye–Waller factor (σ^2): Figure 20.4e, f presents [$\chi(k)$,$|\chi(R)|$] as a function of σ^2. Figure 20.4e demonstrates the inhomogeneous dampening and broadening effect of σ^2 on $\chi(k)$, i.e., larger effect at higher k, consistent with $e^{-k^2\sigma^2}$ factor of $\chi(k)$ expression. Dampening and broadening of $\chi(k)$, respectively, lead to reduced amplitude and increased width of peak in $|\chi(R)|$ (Fig. 20.4f); the centroid remains fixed.

(iv) Atomic species (Z): Figure 20.4 g, h presents [$\chi(k)$,$|\chi(R)|$] for $Cu - O$ vis-à-vis $Cu - Cu$ bonds of the same bond-length and coordination. $\chi(k)$ for [$Cu - O$,$Cu - Cu$] in Fig. 20.4g replicate backscattering factors $t(k)$ for (O, Cu), respectively, (Fig. 20.3c). $\chi(k)$ for $Cu - O$ decays fast with increasing k while $\chi(k)$ for $Cu - Cu$ displays beat-like effect. This contrast distinguishes low-Z vis-à-vis high-Z neighboring atoms. Experimental $\chi(k)$ can be used for the identification of near-neighbor species during modeling. The contrast leads to distinct $|\chi(R)|$ peaks for [$Cu - O$,$Cu - Cu$], *wrt* amplitude and centroid (Fig. 20.4 h).

Individual bond parameters are actually retrieved by concomitant fitting of [Re[$\chi(R)$], Im[$\chi(R)$] components of $\chi(R)$. This can be straightforward for simple systems, e.g., binary alloys, simple oxides, but generally non-trivial due to diverse combinations of ($R + N + \sigma^2 + Z$) parameters in the structure of materials.

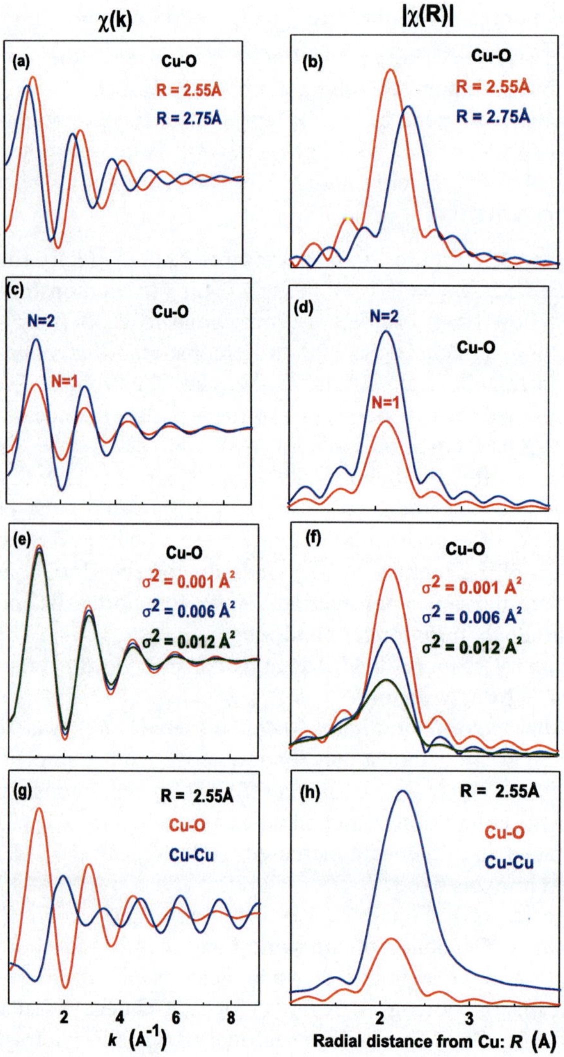

Fig. 20.4 [χ(k), | χ (R)|], as function of bond parameters—atomic species (Z), coordination (N), bond-length (R) and disorder (σ^2). **a**, **b** same (Z, N, σ^2), variable R; **c**, **d** same (Z, R, σ^2), variable N; **e**, **f** same (Z, R, N), variable σ^2; **g**, **h** same (N, R, σ^2), variable Z

20.2.6 Multiple Scattering

The simple formulation of $\chi(k)$ was based on the assumption of single scattering (SS), i.e., photoelectron scatters from one neighboring atom and returns to the absorber. In reality, the photoelectron can be scattered from one neighboring atom (B) toward another neighboring atom (C) before returning to the absorbing atom (A), covering total distance $R_{total} = R_{AB} + R_{BC} + R_{CA}$ (Fig. 20.5a). This phenomenon is called "multiple scattering (MS)." XAFS term for MS:

$$\chi_{MS}(k) = \frac{t(\theta_B)t(\theta_C)}{kR_1R_2R_{BC}} \sin[k(R_{AB} + R_{BC} + R_{CA}) + \delta_{tot}(k)];\ \delta_{total} = \delta_B + \delta_C \tag{20.11}$$

To first order, (A, B) atoms are assumed to be located on the first shell so that $R_{AB} = R_{CA} = R_1$;

$$\chi_{MS}(k) = \frac{t(\theta_B)t(\theta_C)}{kR_1^2R_{BC}} \sin[k(2R_1 + R_{BC}) + \delta_{tot}(k)] \tag{20.12}$$

Two factors from the expression warrant that the first-shell SS peak is unambiguously devoid of MS contribution:

(i) $\sin[k(2R_1 + R_{BC}) + \delta_{total}] \sim \sin[2k(R_1 + 0.5R_{BC}) + \delta_{total}] \Rightarrow$ effective bond-length $R_{MS}(= R_1 + \frac{R_{BC}}{2}) > R_1$

(ii) Relative contribution of MS (vis-à-vis first-shell SS): $\left(\frac{\chi_{MS}}{\chi_{SS}}\right)_I \sim \left(\frac{t}{R_{BC}}\right) < 1$, since $t \leq 1$.

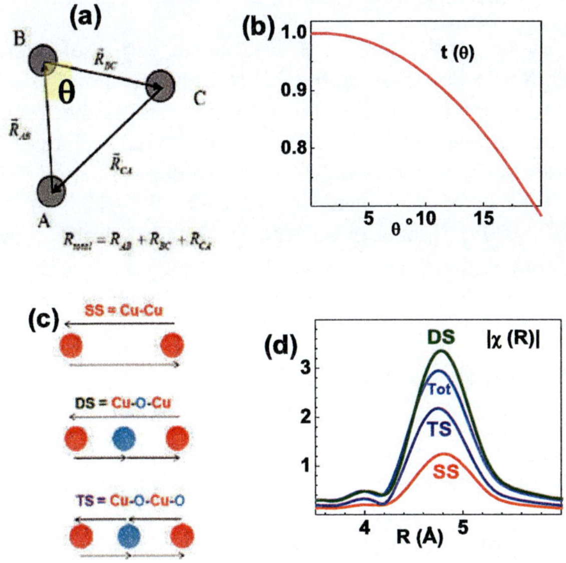

Fig. 20.5 Scheme of Multiple Scattering (MS). **a** MS geometry; **b** backscattering amplitude, as function of MS angle θ; **c** MS paths for linear Cu–O–Cu configuration: Cu-Cu single scattering (SS), Cu-O-Cu double scattering (DS), Cu–O-Cu–O triple scattering (TS); **d** corresponding |χ(R)| for [SS, DS, TS] paths and their cumulative contribution [SS + DS + TS]

However, the same is not obvious for second or higher shells, where $R_{CA}(R_2) = R_{AB}(= R_1) + \Delta$

$$\chi_{MS}(k) = \frac{t(\theta_B)t(\theta_C)}{kR_1R_2R_{BC}} \sin\left[2k\left(R_2 + \frac{R_{BC} - \Delta}{2}\right) + \delta_{tot}(k)\right] \tag{20.13}$$

(i) R_{MS} could be displaced or equal to R_2, depending on $\frac{R_{BC}-\Delta}{2}$. This indicates a finite probability for overlap between second shell (SS, MS) peaks.

(ii) $\left(\frac{\chi_{MS}}{\chi_{SS}}\right)_{II} \sim \left(\frac{t}{R_{BC}} \times \frac{R_2}{R_1}\right)$: Since $R_2 > R_1$, $\left(\frac{\chi_{MS}}{\chi_{SS}}\right)_{II}$ can assume any value, depending on the exact values of parameters. $\chi_{MS}\, \alpha\, t(\theta)$, which is strongly peaked in the forward direction ($\theta = 0^o$) and drops rapidly with increasing θ (Fig. 20.5b). Therefore, MS contribution becomes considerable for near-linear ($\theta \approx 0^o$) arrangement of (A, B, C) atoms (Fig. 20.5c)—the "focusing" effect. The atomic configuration of Fig. 20.5c consists of $Cu - O - Cu$ chain of length $R_2 = 2R_1$. The photoelectron undergoes a plethora of scatterings from the second shell: $Cu - Cu(SS)$, $Cu - O - Cu$ double scattering (DS) and $Cu - O - Cu - O$ triple scattering (TS). Corresponding $|\chi(R)|$ are featured in Fig. 20.5d—their centroids coincide, demonstrating contamination of higher shell SS with MS. Net amplitude $[SS + DS + TS] > SS$, which could deceptively signify SS with higher coordination and must be carefully avoided during fitting. Higher amplitude of MS is due to the intrinsic mathematical degeneracy of scattering paths, e.g., $(Cu - O - Cu) = (Cu - Cu - O)$.

One of the unique applications of MS is the detection of interstitial H atom (Lengeler 1984). XAFS detection of H (SS) is impractical due to its low backscattering coefficient (t). However, if H is the intervening atom B between (A, C) atoms, the focusing effect introduces a detectable increase of XAFS relative to SS of C atom. Further, MS-based analysis is routinely used in the determination of buckling angle along magneto-transport path of materials (Lahiri et al. 2010a, 2012, 2019; Haskel et al. 1996, 1999, 2001; Shibata et al. 2002a). Detailed description of MS-based analysis of buckling angle is presented in the later section.

20.3 XAFS Data Processing

XAFS data analysis is formulated with the ultimate goal of extraction of (R, N, σ^2, Z) fit parameters from experimental data. This involves two main stages (i) Data processing—conversion of raw experimental XAFS data: $\mu(E) \rightarrow \chi(k) \rightarrow \chi(R)$ and (ii) Fitting of $[\chi(k), \chi(R)]$ with (R, N, σ^2, Z) fit parameters.

Data processing is aimed at the extraction of the best and reliable spectra $[\chi(k), \chi(R)]$ for fitting. Data processing algorithm is standardized and operated with ATHENA program (http://cars9.uchicago.edu/~ravel/software/doc/Athena/html). Screenshot of ATHENA is presented in Fig. 20.6a. The steps of data

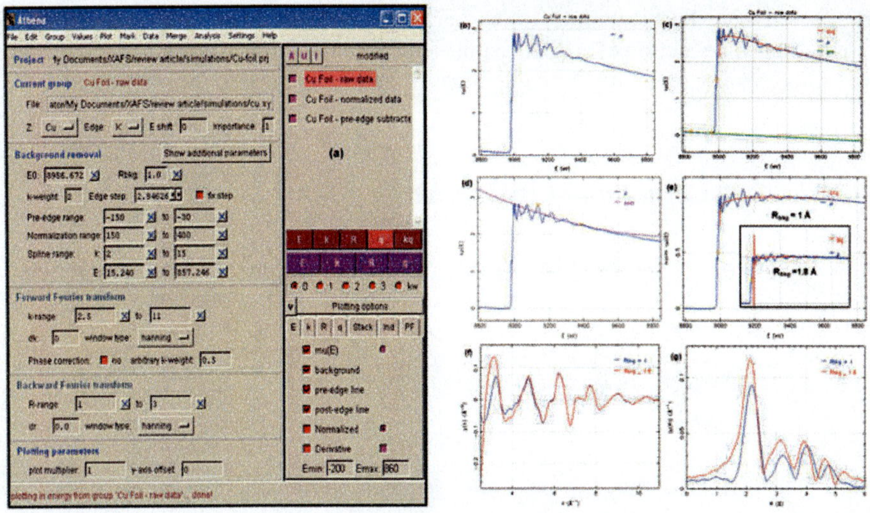

Fig. 20.6 **a** ATHENAprogramfor data processing; **b** RawCu K-edge μ(E) for Cu foil; **c** Edge energy (E0 = 8986.7 eVeV) is set at half of the edge step,marked by small circle. Pre-edge region (E << E0) is fit with linear equation (green line) that is extrapolatedinto post-edge (E << E0) region; **d** After pre-edge subtraction fromthe entire spectrum, baseline ofμ(E) is reset to 0.Normalization range is defined over ΔE = 80–300 eVeV beyond E0. Boundaries of normalizationrange aremarked by orange circles; **e** Normalized μ(E) is shown. Background (smooth) spline fit for Rbkg = 1Åis plotted in orange. (Abrupt) backgroundspline for Rbkg = 1.8 Å is shown in inset. Degeneration of backgrounddue to variation ofRbkg control parameter is demonstrated; **f** Background-subtracted χ **k** and **g** I χ (R)I, correspondingto [Rbkg = 1 Å, Rbkg = 1.8 Å]. **f**, **g** demonstrate spectral distortiondue to incorrect backgroundsubtraction

processing are demonstrated for XAFS of Cu foil (Fig. 20.6b–g). As these subtle steps determine the accuracy of structural solution, data processing could be rigorous, demanding the discretion of experienced analyst and 20 − 30% of total analysis time.

20.3.1 Importing and Pre-processing Raw Data

CuK-edge experimental XAFS data for Cu foil $[\mu_{raw}(E)]$ is imported into ATHENA and presented in Fig. 20.6b. Most datasets are rarely of such high quality; they are susceptible to glitches, statistical noise, and systematic errors due to unfavorable experimental conditions, e.g., low sample concentration, inefficient detectors, higher harmonic content of beam, and glitches from monochromator. Superimposition of such artifacts on XAFS spectra can distort the amplitude and/or period of XAFS oscillations. These artifacts transform into *unreal* peaks in $\chi(R)$ that overlap with the *real* peaks of interest and generate incorrect structural results. Hence, they are eliminated to the best extent with subprograms of ATHENA, e.g., removal of single-point glitches, reduction of statistical noise by averaging multiple reproducible scans and smoothening by interpolative iterative smoothing algorithm.

20.3.2 Determination of Edge Energy (E_0)

E_0 is defined as the energy at half of the edge step (Fig. 20.6c, $E_0 = 8986.67eV$).

20.3.3 Pre-edge Subtraction (User Input = Pre-edge Fit Range)

[$\mu_{raw}(E)$] is divided into two regions by E_0(Fig. 20.6c): (a) *pre-edge* region, with background contribution from Compton scattering, etc., and (b) *post-edge* region, containing XAFS oscillations. ATHENA generates a linear fitting equation [$\mu = A + B * \Delta E$] for pre-edge spectra over user-defined energy range (Fig. 20.6c, $\Delta E = (-150)eV - (-30)eV$). The pre-edge fit range is selected such that it is close to E_0 and free of artifacts. The fit line is extrapolated over the whole energy range and subtracted from $\mu_{raw}(E)$. This exercise resets the baseline of the spectrum to zero, representing $\mu(E)$ with zero background (Fig. 20.6d). $\mu(E)$ = *real* absorption spectrum of interest: $\mu(E) = \mu_0(E)[1 + \chi(E)]$.

20.3.4 Normalization (User Input = Normalization Range)

This step determines the normalization constant [$\mu_0(E)$] of the equation: $\mu(E) = \mu_0(E)[1 + \chi(E)]$. Edge-step normalization approach is adopted in which $\mu_0(E)$ is approximated with $\mu_0(E_0)$—the value of post-edge background line at E_0(Fig. 20.6d). ATHENA generates a linear fitting equation [$\mu = C + D * \Delta E$] for datapoints over user-defined normalization range beyond XANES region (Fig. 20.6d, $\Delta E = 80 - 300eV$). The line is extrapolated to the edge (i.e., $\Delta E = 0$) and corresponding $\mu_0(E_0) = C(= 2.85)$ is the normalization factor (Fig. 20.6d). Normalized spectrum $\mu_{nor}(E)$ is generated by dividing $\mu(E)$ with the normalization factor: $\mu_{nor}(E)\left(= \frac{\mu(E)}{\mu_o(E_0)}\right)$, presented in Fig. 20.6e. Normalization is important, since it scales the post-edge background line of $\mu_{nor}(E)$ to 1 for all datasets. This enables direct visualization of relative variations between XAFS datasets.

20.3.5 Post-Edge Background Subtraction (User Input = Data Range and R_{bkg})

Following normalization, XAFS equation stands as: $\chi(E) = \frac{\mu(E)-\mu_0(E)}{\mu_o(E_0)}$. At this stage, only atomic absorption $\mu_0(E)$ or "background" remains to be determined over the desired data range ($\Delta E_{bkg}/\Delta k_{bkg}$). Homogeneous approximation of $\mu_0(E)$ [$= \mu_0(E_0)$] in Step 4 is justified only for normalization purpose; it does not represent the true slowly varying $\mu_0(E)$ function. Construction of the correct $\mu_0(E)$ is the most difficult and critical step of data processing, since incorrect background subtraction can seriously distort $\chi(E)$. AUTOBK algorithm (Newville et al. 1993) of ATHENA is based on the understanding that $\mu_0(E)$ is the slowest oscillatory component of $\mu(E)$ and corresponds to the shortest distance $R_{bkg}(< R_1)$ in $\chi(R)$, i.e., shorter than first-shell bond-length, since $k\alpha\frac{1}{R}.\mu_0(E)$ is approximated by spline function, with a number of knots (N_{knot}) determined by user-defined [R_{bkg},ΔE_{bkg}] parameters: $N_{knot} = \frac{2R_{bkg}\Delta k_{bkg}}{\pi}$, by Nyquist criterion (Brillouin 1962). The knots are set to be equally spaced in k to ensure a single frequency. Ordinates of the knots are fit to minimize $|\chi(R)|$ of background-subtracted data [$\mu(E) - \mu_0(E)$] over $R \leq R_{bkg}$. The choice of (ΔE_{bkg}, R_{bkg}, k^w) for the best-fit spline often requires the discretion of an experienced analyst. ATHENA allows visual inspection of the spline (red line in Fig. 20.6e). (ΔE_{bkg}, R_{bkg}) are varied until (a) the spline is smooth and non-oscillatory; (b) XAFS oscillations are approximately even about the spline; (c) minimum $|\chi(R)|$ over $R \leq R_{bkg}$. Background variation, via control of $R_{bkg} = 1$Å → 1.8Å, is demonstrated in Fig. 20.6e. $R_{bkg} = 1.8$Å is a poor choice since it encroaches into the tail of the first shell. Spline for $R_{bkg} = 1.8$Å is discontinuous at its lower boundary (inset of Fig. 20.6e). We demonstrate the distortion of data due to incorrect background in Fig. 20.6e –g. Background-subtracted $\chi(E) \rightarrow \chi(k)$ is generated onto an even grid of 0.05Å^{-1} (Fig. 20.6f). Figure 20.6f demonstrates the sensitivity of $\chi(k)$ at lower k to variation of background spline.

20.3.6 $\chi(k) \rightarrow \chi(R)$(User Inputs = Transform Range (k_{min}, k_{max}) and k^w)

Reproducible, clean, and artifact-free portion ($k_{min} - k_{max}$) of $\chi(k)$ is selected from within Δk_{bkg} and Fourier transformed into $\chi(R)\left[= \int_{k_{min}}^{k_{max}} \chi(k)k^w e^{i2kR}dk\right]$. Figure 20.6 g demonstrates the sensitivity of $|\chi(R)|$ to variation of background spline. Incorrect spline contributes significant background to $|\chi(R)|$.

Figure 20.7a–d demonstrates the importance of window k^w in [$\chi(k)$,$|\chi(R)|$], as it selectively emphasizes different regions of k-range.

(i) $\chi(k)k^w$ is presented in Fig. 20.7a, for $w = 0-3$. It is clear that lower (higher) w emphasizes lower (higher) k-region. Thus, k^w can be strategically exercised to (de)emphasize k-regions of (poor) high spectral quality during data processing and fitting.

(ii) $|\chi(R)|$ for $w = 0 - 3$ are presented in Fig. 20.7b; first-shell is commonly rescaled to 1 for all w. Relative intensities and overlap of different higher shell peaks display sensitivity to variation of k^w, which can be strategically exploited to emphasize or resolve the peaks of interest during fitting.

(iii) $[t(k), \delta(k)]$ has been earlier shown to exhibit Z-dependence (Fig. 20.3b)—large $t(k)$ at higher k for high-Z atoms (vis-à-vis low-Z atoms) is routinely exploited to distinguish between (low, high)-Z neighbors. This task becomes more efficient with the application of k^w, since high-Z peak grows significantly for higher w. Figure 20.7c, d demonstrates this for $[Cu-O$ vis-à-vis $Cu-Cu]$, as a contrast of $|\chi(R)|$ becomes significantly more pronounced for $w = 3$ (Fig. 20.7d). Different k^w can be strategically exercised for selective fitting of low-Z (low w) and high-Z (high w) peaks.

Following steps 1–6, the best-processed $[\chi_{expt}(k), \chi_{expt}(R)]$ is ready for fitting.

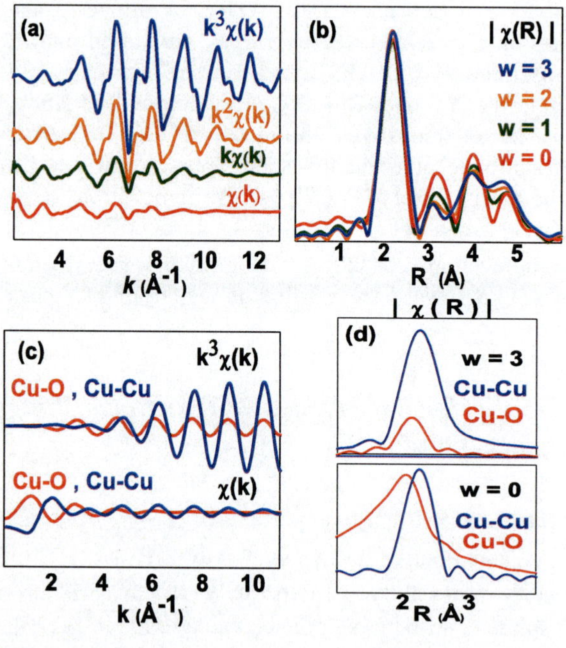

Fig. 20.7 **a, b** [kwχ(k), |χ (R)|] for Cu foil; w = 0–3. k-weights (w) selectively emphasize different regions of χ (k): lower (higher) w ≠ ≠ low (high) k-region. The first peak of |χ(R)| is rescaled to 1. w = 0–3 resolves peaks selectively; **c, d** [χ (k), k^3χ (k)] and corresponding |χ (R)|, for Cu–O and Cu-Cu scattering paths. It is clear that higher k-weight (w = 3) brings out maximum contrast between low- and high-Z backscattering atoms

20.4 XAFS Data Fitting

The flowchart for modeling and fitting programs (Newville 2001) is presented in Fig. 20.8, for the particular example of Cu-foil. The present example elucidates first-shell $(Cu - Cu)$ bond fitting.

20.4.1 *Modeling Program FEFF6.0 Algorithm*

Potential atomic distribution model $(R, N, Z)_j$ is input into *feff.inp* file. This is demonstrated in this example for $Cu - Cu$ path—*absorbing* (Cu) and *backscattering* (Cu) atoms are specified by potential [0, 1], absorption *edge* (K) is specified by HOLE 1.0. FEFF6 program identifies scattering paths, e.g., $Cu - Cu$ and generates their ab initio XAFS scattering contributions: $[\chi_{\mathrm{FEFF}}(\mathrm{k})] = \frac{\mathrm{N_j t_j(2k)}}{(\mathrm{kR_j})^2}\mathrm{e}^{-\frac{2R_j}{\lambda}}\sin\left[2\mathrm{kR_j} + \delta_{\mathrm{j}}(\mathrm{k})\right]$. The program assumes $\left[S_0^2 = 1; \sigma = 0\right]$ and takes $[t(k), \delta(k), \lambda]$ values from its integral database. Output $\chi_{FEFF}(k)$ is stored in *feffxxxx.dat* file, e.g., *feffCuCu.dat*.

20.4.2 *Fitting Program FEFFIT Algorithm*

FEFFIT is programmed for the correction of FEFF6 model and the determination of best-fit values and uncertainties of bond parameters. Program file *feffit.inp* is constructed with.

(a) experimental dataset $\chi_{\exp t}(k)$, with user-defined fit range $[k_{\min} - k_{\max}, k^w, R_{\min} - R_{\max}]$, e.g., $k = 2.5 - 11\ \mathrm{\mathring{A}}^{-1}; k^3; R = 1 - 2.8\mathrm{\mathring{A}}$ (Fig. 20.8).
(b) $\chi_{FEFF}(k)$, assigned a set of correction variables $x[(\delta R, N', \sigma^2, \Delta E_0)]$

δR = Bond-length correction; $[N', \sigma^2]$ = multiplicative factors for coordination and Debye–Waller factor;

ΔE_0 = Energy correction term, to account for possible inaccuracy in the determination of E_0.

FEFFIT allows for refinement of x variables, until the difference (f_i) between experimental $[\chi_{\exp t}(k)]$ and theoretical XAFS $[\chi_{FEFF}(k)]$ is minimized over all "i" data points (Marquardt 1963). The minimal value is defined as $\chi_0^2 = \sum_i f_i^2 = \sum_i \left[\chi_{\exp t}(R) - \chi_{FEFF}(R)\right]_i^2$. Uncertainty (Δx) of each fit parameter is calculated as the tolerable statistical deviation from best-fit value, within $\chi_0^2 + 1$. The figure-of-merit of fit is defined by $R-$ factor $= \frac{\sum_i \{|\mathrm{Re}(f_i)|^2 + |\mathrm{Im}(f_i)|^2\}}{\sum_i \{|\mathrm{Re}(\chi_{\exp t})_i|^2 + |\mathrm{Im}(\chi_{\exp t})_i|^2\}}$. Best-fit solutions $[x \pm \Delta x, R - factor]$ are stored in *feffit.log* file; $|\chi_{fit}(R)|$ is compared with $|\chi_{\exp t}(R)|$.

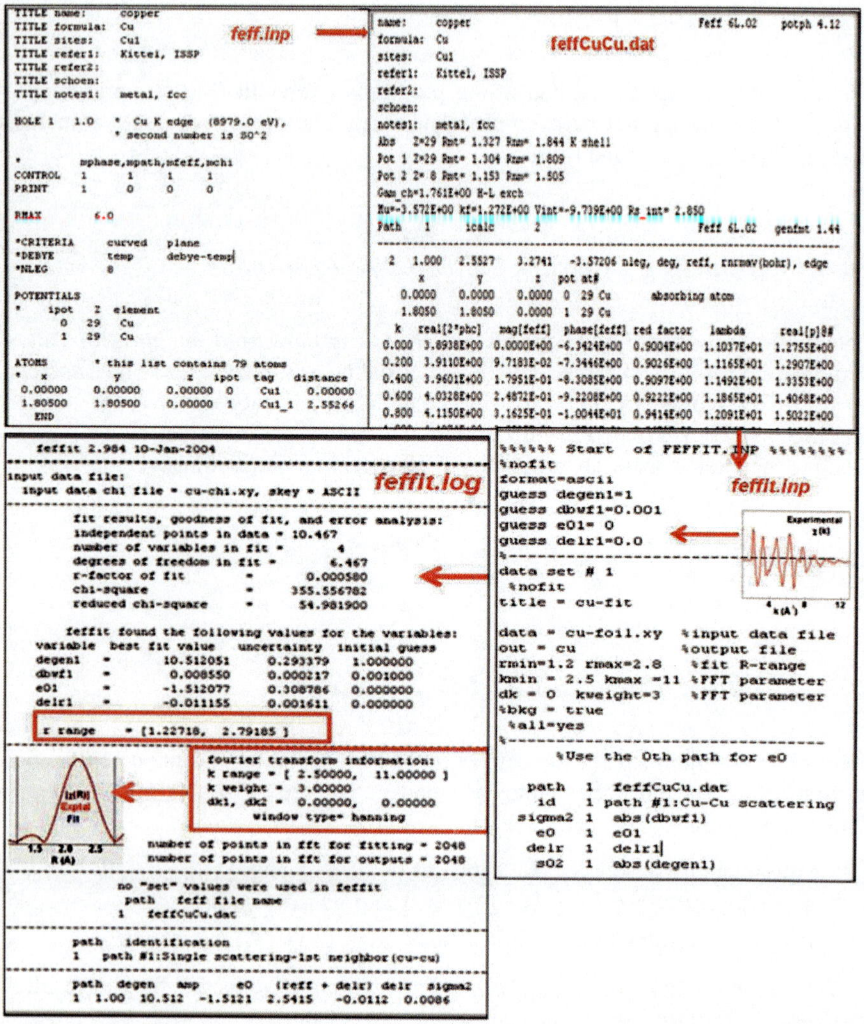

Fig. 20.8 Fitting algorithm for XAFS spectra of Cu foil is outlined schematically with arrow directions. FEFF6 program converts structural model of feff.inp into XAFS oscillation [feffCuCu.dat]. Theoretical XAFS [feffCuCu.dat] and experimental χ(k) are input into feffit.inp file, both marked by red boxes. FEFFIT program minimizes the difference between feffCuCu.dat and experimental χ (k), over user-defined fit range. The best-fit values of path variables (marked by red box) and their uncertainties are recorded in feffit.log file. Experimental and best-fit |χ(R)| are compared over fit-range, defined by vertical lines

In the example of Fig. 20.8, the best-fit values of variables are marked. Note that $N'(= 10.5)$ actually represents S_0^2N; since experimental $S_0^2 = 0.88$, the real coordination $N = N'/S_0^2 = 12$, i.e., reproduces the expected coordination of $fccCu$ foil.

20.4.3 Fitting Conditions

While the above section outlines the basic algorithms for fitting, the success of fit ultimately rests with the discretion of analyst. This comprises (i) selection of reliable data range ($k_{min} - k_{max}$); (ii) distinguishing *real* peaks from artifacts in $|\chi(R)|$; (iii) selection of fit range $[R_{min} - R_{max}]$ over real and analyzable peaks; (iv) choice of k^w for resolution of overlapping peaks; (v) correct starting model (*feffnnnn.dat*), guided by distribution of peaks in $|\chi(R)|$; (vi) reduction of uncertainties (Δx) of fit parameters with strategic fitting; (vii) resolving degeneracy of solution (x) with strategic fitting, physical intuition or theoretical model-based analysis; (viii) realistic assessment of the solution. Fit with R-factor < 0.02 is considered as good but this is a mathematical conclusion, not physical. Therefore, the solution has to be evaluated for actual physical credibility; (ix) parallel consideration of several reasonable models. Uncertainties (Δx) of fit parameters, as defined above, are of mathematical origin and do not represent the physical variation between models. Model-dependent deviation of parameters from the best-fit condition should be incorporated into the error bars that will reveal the robustness of the best-fit model. Some of the intrinsic problems of fitting are described in detail.

(i) Finite information content:

Information content is limited by the number of data points in the spectra (Brillouin 1962): $\left[N_{\inf o} = \frac{2(k_{max}-k_{min})(R_{max}-R_{min})}{\pi} + 2\right]$. This constrains the number of variables (N_{var}) to be lower than $N_{\inf o}$. In the example of Fig. 20.8, $[k = 2.5 - 11\ \text{Å}^{-1}; R = 1 - 2.8\text{Å}]$, i.e., $[(\Delta k \times \Delta R) = 8.5\ \text{Å}^{-1} \times 2.8\text{Å}]$ generates $N_{\inf o} = 17$ data points that can accommodate maximum 17 variables. For fitting the single $Cu - Cu$ peak of Fig. 20.8, 4 variables $\left[= (\delta R, N', \sigma^2, \Delta E_0)\right]$ are sufficient, so that $N_{var} < N_{\inf o}$ condition is satisfied. However, complicated fitting cases, e.g., single peak with multi-atomic contributions, involve larger number of variables $(\mathrm{N_{var}} > \mathrm{N_{info}})$ whereby fitting is disallowed. Collection of reliable spectra extending to high k_{max} is strongly recommended in such cases, in order to increase $N_{\inf o}$. Alternatively, strategies may be exercised to artificially inflate $N_{\inf o}$ or reduce N_{var}: (a) simultaneous fit of $[\chi(k), k\chi(k), k^2\chi(k), k^3\chi(k)]$, with the same variables. This increases $N_{\inf o}(\times 4)$ for the same N_{var}; (b) constraining common bond parameters, e.g., $Cu - Fe$ during simultaneous fitting of $[Cu, Fe]K$-edge $\chi(k)$.

(ii) Peak overlap:

The peak of interest may have contribution from the tail of adjacent peaks which is common for higher shells. The peaks may be decoupled by manipulation of (k-range/k^w) or alternatively, the two adjacent peaks may be jointly fit.

(iii) Intrinsic ambiguity of modeling:

Distribution of peaks in $|\chi(R)|$ is used as the first-hand guide to modeling. However, interpretation of $|\chi(R)|$ has to be carefully based on the understanding that XAFS is essentially an interference phenomenon and not a simple addition of $\chi_j(k)$ contributions from different scattering paths. $\chi_j(k)$ could interfere unpredictably and their net contribution $\chi_{tot}(k)$ could generate a net peak $|\chi_{tot}(R)|$ of unreal magnitude and positions. Two examples are cited to demonstrate that $|\chi_{tot}(R)| \neq |\chi_1(R)|+|\chi_2(R)|$.

Case A (Fig. 20.9a, b)—closely spaced bonds, e.g., *Jahn–Teller* distorted octahedra: $[\chi_1(k),\chi_2(k)]$ in Fig. 20.9a are contributed by $[Cu-O1, Cu-O2]$ bonds at (R_1, R_2); $R_2 - R_1 = 0.2$Å. $[\chi_1(k),\chi_2(k)]$ are slightly out-of-phase; their net oscillation $\chi_{tot}(= \chi_1 + \chi_2)$ generates peaks in $|\chi_{tot}(R)|$ that are of lower magnitude and displaced relative to $|\chi_1(R_1)|$ and $|\chi_2(R_2)|$ (Fig. 20.9b). $|\chi_{tot}(R)|$-based indiscrete modeling of coordination and bond-lengths could generate incorrect fit results.

Case B (Fig. 20.9c, d)—multi-atomic bond, e.g., alloys: $\chi_1(Cu - O)$ and $\chi_2(Cu-Cu)$ in Fig. 20.9c are contributed by distinct neighbors (O, Cu) of the same bond-length. Their net contribution $\chi_{tot}(= \chi_1+\chi_2)$ generates a single $|\chi_{tot}(R)|$ peak, with higher amplitude than $|\chi_1(R)|$ and $|\chi_2(R)|$ (Fig. 20.9d). This practically resembles the peak of *single* elemental component with higher coordination. Modeling of $|\chi_{tot}(R)|$ with a *single* element would be deceptive in such cases.

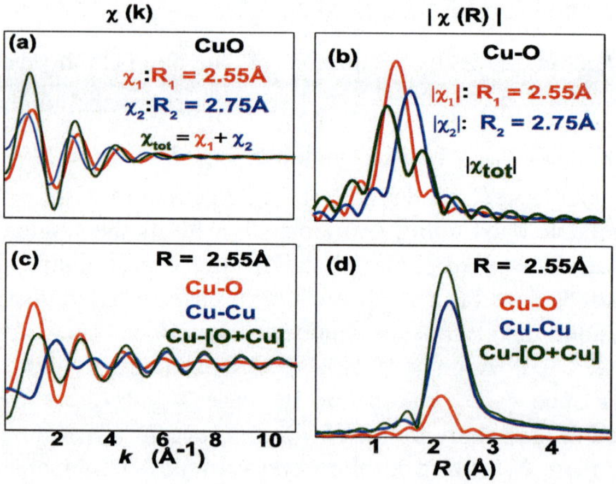

Fig. 20.9 **a, b** [χ(k), |χ(R)|] for closely-spaced [Cu-O1, Cu-O2] bond-lengths.Net | χ tot(R) | fromthe interference of XAFS oscillations of two paths is of lower amplitude and displaced, relative to | χ 1(R)|; **c, d** [χ (k), | χ (R)|] are compared for [Cu–O, Cu–Cu] paths of the same bond-length.Net | χ (R)| fromthe interference of [Cu–O, Cu–Cu] oscillations is distinct fromsimple addition of individual paths

(iv) Correlation between different variables:

$(S_0^2 N e^{-k^2\sigma^2})$ is present as a multiplicative product in the XAFS equation so that several combinations of (S_0^2, N, σ^2) could potentially reproduce the same spectra. Out of these, S_0^2 is pre-determined by fitting $\chi(k)$ of relevant standards (Kelly et al. 2009). For example, Cu foil serves as the standard for CuK-edge XAFS experiments of Cu-based samples. $\chi(k)$ for Cu foil is fit by constraining (N, σ^2, R) to crystallographic values and varying S_0^2. This value of S_0^2 is fixed during the fitting of $\chi(k)$ for the samples of interest at the same edge. In the next step, the correlation between (N, σ^2) is resolved by strategies such as (a) simultaneous (k-weighted) fit of $\chi(k)$, exploiting k^w-sensitivity of σ^2 (vis-à-vis insensitivity of N), (b) simultaneous fit of temperature (T)-dependent datasets with common N but variable σ^2 (since it is sensitive to T).

(v) Local minima:

The computational techniques do not warrant "global" minimum of χ_0^2 but "local" minima based on the initial values of variables (Michalowicz and Vlaic 1998). This could potentially generate non-unique solutions, in worst cases. Such ambiguity is resolved by apt fitting strategies, physical intuition of parameter values, and theoretical model-based analysis.

We finally remark that despite these difficulties, XAFS fit has been able to solve a wide range of complex structures in diverse systems. Some of these case studies will be presented in a later section.

20.5 XANES

XANES shares the same basic physics as XAFS, except that the kinetic energy of photoelectron is low ($\Delta E \leq 30eV; k < 2.5\text{Å}^{-1}$) with the outcome of complex structural interpretation:

(i) *De-Broglie* wavelength of the photoelectron is large $\left(\sim 100\,\text{Å}\right)$ at this low energy (Schiff 1949). The photoelectron extends over several bond-lengths so that it loses resolution to detect local structure.
(ii) Assumptions, e.g., hard-sphere scattering potential, plane-wave approximation, single scattering, for the formulation of simplified solvable XAFS equation, are not valid at low energy (Schiff 1949).
(iii) Photoelectron of low kinetic energy is susceptible to multiple scattering of all orders—infinite in principle—that is ultra-sensitive to the angular distribution of atoms (Koningsberger and Prins 1988). Structural modeling therefore becomes very critical in XANES regime.

(iv) In XANES region ($k < 2.5\,\mathring{A}^{-1}$), photoelectron mean free path increases to $\lambda(k) = 30\,\mathring{A}$ (Fig. 20.3c) and the dampening factor ($e^{-k^2\sigma^2}$) becomes negligible. These factors jointly increase XANES sensitivity to long distances covering $\sim$ 300 atoms, as a result of which simulations become cumbersome.

These problems render the structural formulation of XANES impractical. On the other hand, slow photoelectron becomes advantageously sensitive to the surrounding potential that is defined by the coordination chemistry of the absorbing atom. This enables the detection of the valence state and molecular geometry of the absorbing atom, with widespread applications in catalysis (Grunwaldt and Frenkel 2009; Timoshenko and Frenkel 2019) including in-situ monitoring of redox reactions (Timoshenko 2021), binding of elements in diverse matrices (Thomas et al. 2019), speciation for toxicity (Ramanan et al. 2015b). However, quantitative modeling of XANES could be inconclusive, since it is (i) a complex convolution of structure (coordination, bond-length, symmetry) and charge, and (ii) ultra-sensitive to subtle details of potential. This essentially implies non-converging iterations of model refinement for reproduction of whole XANES spectra. Although recent program FEFF8.0 has successfully reproduced XANES of known metals and oxides, *ab-initio* fitting of XANES spectra for unknown material is still unreported. Hence, a practical approach is adopted by fingerprinting valence state for the sample of interest, relative to standards, based on $2 - 3$ points of XANES spectra viz edge energy (E_0); whiteline (WL) or the first post-edge maxima of $\mu(E)$; pre-edge (PE) peak.

XANES analysis scheme is demonstrated for Fe-doped silicate glass (unpublished work) in Fig. 20.10, based on the reference of $[Fe(0), FeO(+2), Fe_3O_4(+2.+3), Fe_2O_3(+3)]$. FeK-edge XANES spectra for these standards are presented in Fig. 20.10a. Since XANES is an integral part of XAFS spectra, it requires no independent processing. XANES spectra for Fe-doped silicate sample is not shown, since the work is unpublished. We remark that since valence orbital of Fe has d symmetry, FeL_3-edge (p) XANES should be ideally measured to probe (d) valence holes via $p \rightarrow d$ transition, as per the dipole selection rule. This would extract valence state information with the maximum sensitivity. However, inaccessibility of FeL_3-edge X-rays compels one to measure FeK-edge XANES, which probes p-final state instead of the direct valence (d) state. FeK-edge XANES spectra, fortunately, demonstrate valence dependence, with reasonable clarity (Fig. 20.10). Valence-dependence of (E_0, WL, and PE) for the reference spectra is used to calibrate the valence of Fe in silicate (Fig. 20.10b–d); the latter is denoted by "S."

20.5.1 Edge Position (E_0)

Edge position (E_0) is the effective binding energy ($E_0 \alpha Z^2$) of the absorbing atom (Z), defined at half of the edge step or inflection point, i.e., maxima of $\frac{d\mu}{dE}$. E_0 increases (decreases), with increasing (decreasing) valence state (V) (Wong et al.

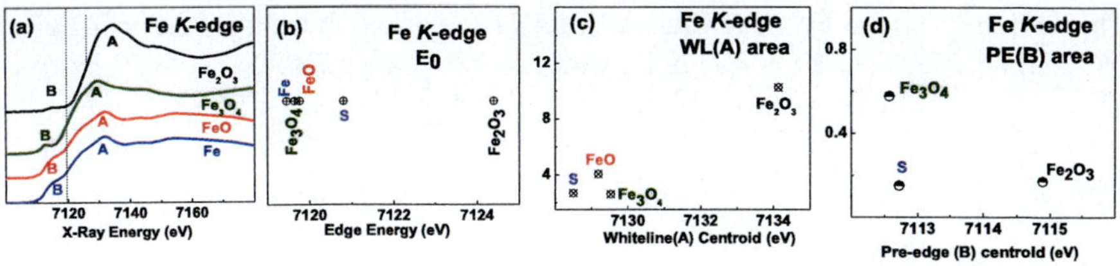

Fig. 20.10 **a** Fe K-edge XANES spectra are presented for standard iron compounds, corresponding to different valence states of Fe. Edge position (E0) is marked by dotted vertical line. Whiteline (WL) and pre-edge (PE) peaks are marked as (A, B) respectively; **b–d** [E0, WL, PE] fit results are derived from the spectra of (**a**). Results for Fe-doped silicate are marked by "S"

1984)—this is conceptually understood to be due to the increase (decrease) of effective nuclear charge $[Z \rightarrow Z_{eff}(= Z \pm \Delta)]$, following the loss (gain) of electron: $E_0 \pm \Delta E_0 \alpha Z_{eff}^2$. In Fig. 20.10a, E_0 for the spectra are marked by vertical dashed lines. From $Fe(0) \rightarrow Fe_2O_3(+3)$, E_0 increases by $4.5eV$ (Fig. 20.10b), i.e., $@\Delta E_0 = 1eV$ per valence state. For "S," E_0 is intermediate between $FeO(+2)$ and $Fe_2O_3(+3)$ that suggests intermediate valence state of Fe in silicate. For fixed valence (V), E_0 can be secondarily correlated with first-shell bond-length (R_1):$E_0 \alpha \frac{1}{R_1^2}$, i.e., E_0 increases (decreases) for shorter (longer) bond-lengths $@|\Delta E_0| \sim 0.3eV/0.01\text{Å}$(empirical). The scheme is consistent with "particle-in-a-box" representation, in which energy levels $\alpha \frac{1}{R_1^2}$;$R_1 =$ box radius (Schiff 1949).

20.5.2 *Whiteline (WL)*

Whiteline (WL) is the first post-edge maxima of $\mu(E)$, representing the density of *empty states* (or hole density ρ_V) of final state $|f\rangle$ (Wong et al. 1984). WL for FeK-edge XANES is marked as A in Fig. 20.10a.

WL is a prominent marker of valence state (V), since hole-density (ρ_V) increases with higher valence (V) due to lower electron content: WL $\alpha \rho_V \alpha V$. "Peak fitting" module of ATHENA allows fitting of WL with Lorentzian, superimposed on the arctangent line shape of the main edge. Fit values for WL [centroid, area] for the standards are plotted in Fig. 20.10c. WL for $Fe(0)$ is non-trivial and therefore, not analyzed. WL area demonstrates $(\times 4)$-fold increase from $FeO(+2) \rightarrow Fe_2O_3(+3)$ $@\Delta WL = 1.3eV$ per valence state (Fig. 20.10c). In principle, WL area for "S" should be intermediate between $[FeO, Fe_2O_3]$, to be consistent with E_0 conclusion. Smaller WL area for "S" than FeO (Fig. 20.10c) could be due to the smearing effect of disorder, consistent with the amorphous character of the sample.

For fixed valence (V), WL could be secondarily correlated with structure. WL broadens for shorter bond-lengths (R_1) due to the broadening of energy levels and increases with ligand coordination (N) via density-of-states (Giachini et al. 2010).

This is demonstrated by FeK-edge WL for $[FeO,Fe_3O_4]$ in Fig. 20.10c—Fe in has comparable valence ($\sim +2$) but distinct coordination geometry in these oxides: $[N = 6,R = 2.14Å]_{FeO}$ vis-à-vis $[N = 4,R = 1.89Å]_{Fe_3O_4}$. WL area for Fe_3O_4 is significantly smaller and broadened, due to lower coordination and shorter bond-length.

20.5.3 *Pre-edge (PE)*

Although the main absorption rise of XANES is dominated by the dipole transition rule ($\Delta l = \pm 1$), weaker pre-edge features for $\Delta l = \pm 2$ could emanate via orbital hybridization. PE peak (B) of Fe K-edge XANES (Fig. 20.10a) represents $s \rightarrow pd$ (hybridized) transition. Hybridization is conditional upon the presence of non-centrosymmetry, so that PE intensity progressively increases from octahedral (FeO) $\rightarrow$ distorted octahedral (Fe_2O_3) $\rightarrow$ tetrahedral (Fe_3O_4) symmetry. Correlation between (centroid, integrated area) of PE is standardized (Wilke et al. 2001), shown in Fig. 20.10d. This is exploited for the calibration of (valence state, coordination geometry), consistent with the Bond Valence Sum rule (Brown and Altermatt 1985). In Fig. 20.10d, PE centroid for "S" coincides with Fe_3O_4 while PE area is smaller than Fe_3O_4 [instead, closer to Fe_2O_3]. This implies that Fe is of mixed valence state as for Fe_3O_4 (which is consistent with E_0 in Fig. 20.10b) and exists in distorted FeO_6 configuration or mixed configuration of tetrahedra and octahedra.

In summary, $[E_0, WL, PE]$ analysis reveals the existence non-equivalent Fe-sites in silicate as expected for amorphous systems, both *wrt* valence (+2, +3) and geometry (FeO_4, FeO_6). For non-amorphous mixed systems, ATHENA allows fitting of XANES (X) with weighted (w_i) linear combination of XANES (X_i) for standards: $X = \sum_i w_i X_i$ (w_i = variable) so that net valence and configuration of the system is their weighted combination. These results demonstrate the unique strength of XANES viz quick fingerprinting of the complete coordination chemistry of atoms (vis-à-vis XPS, EXAFS techniques) to a reasonable extent. The strength is multiplied by its flexibility for diverse sample conditions (vis-à-vis XPS) and ease of measurement due to strong signals. XANES is less sensitive to noise, and therefore, its measurement is less demanding *wrt* alignment and detectors (vis-à-vis EXAFS). These factors jointly contribute to the galloping popularity of XANES -500 publications per year.

20.6 Experimental Configuration

The primary criteria for X-ray source in XAFS experiments are (1) energy tunability for excitation of a wide range of atomic species and (2) high photon flux or ~ 10^3 signal-to-noise ratio for good data quality, both of which are efficiently satisfied by synchrotron source (vis-à-vis laboratory X-ray source). Synchrotrons have therefore emerged as the indispensable seat for modern XAFS experiments (https://lightsources.org). The evolution of the power of XAFS has paralleled that of synchrotrons, as increasing brilliance and decreasing emittance of the source have extended detection limits to ultra-dilute (ppm) samples (Ascone et al. 2005; Suzuki et al. 2002; Thomas et al. 2019; O'Loughlin et al. 2010; Ranieri-Raggi et al. 2003) and supported advanced adaptations of XAFS, e.g., X-ray Magnetic Circular Dichroism (XMCD) for the determination of spin/orbital moments (Laan and Figueroa 2014), polarization-dependent XAFS for layered structures (Haskel et al. 2001, 1999) and High-Pressure XAFS (HPXAFS) for micro-sized samples (Pascarelli and Haskel 2016).

XAFS is measured in either of 2 modes.

20.6.1 *Transmission-Mode XAFS*

This is the standard and simplest XAFS configuration, in which the sample is inserted in the path of X-rays and absorption is measured from the attenuation of X-rays through sample $\alpha \ln\left(\frac{I_0}{I_t}\right)$; (I_0, I_t) are incident and transmitted intensities of X-rays. Logarithmic nature of signal imposes stringent sample conditions such as uniformity, fair concentration, and optimal thickness (http://gbxafs.iit.edu/training/XAFS_sample_prep.pdf). Precision of lengths, e.g., (sample thickness, inhomogeneity) is scaled to $1.5t$; t = distance over which X-ray intensity is 37% attenuated. Distance (t) varies with sample composition and energy edge: $\frac{1}{t}(cm^{-1}) \approx_i = \rho_M(gm/cm^3)\sum \frac{m_i}{M}\sigma_i(cm^2/gm)$ (http://gbxafs.iit.edu/training/XAFS_sample_prep.pdf); σ_i = X-ray absorption cross-section and $\left(\rho_M \frac{m_i}{M}\right)$ = atomic density $\left(= \rho_M \frac{m_i}{M}\right)$ of the constituent atoms; ρ_M = density of the material, $\frac{m_i}{M}$ = atomic mass fraction. Rigorous sample preparation method involves pre-calculation of t and tailoring the particle size by grinding, sample thickness, and sample uniformity to within $1.5t$.

20.6.2 *Fluorescence-Mode XAFS*

This is a flexible but expensive configuration, in which absorption is measured indirectly through detection of fluorescence photons emitted by the excited atom $\alpha \frac{I_f}{I_0}$. Incident and fluorescence detectors are positioned perpendicular to each other and the sample is fixed in an intermediate position at 45^0. Linear signal eliminates the stringent sample conditions and therefore, flexible. However, fluorescence signal within simultaneously thick and concentrated samples is susceptible to "self-absorption" that leads to non-linear distortion of XAFS spectra (Laan and Figueroa 2014). This imposes sample restriction for fluoresence-mode, to this extent. Sample preparation is reasonably flexible within this limit, in contrast to transmission mode. Feasibility of this mode is reliant on the fluorescence yield of the elements which is low for $Z < 30$ and the detection efficiency of fluorescence photons, which mandates expensive detectors, e.g., multi-element Germanium detector. Guidelines for the choice of measurement mode and samples are prescribed in Tables 20.1 and 20.2, respectively.

Description of surface XAFS, e.g., TEY, grazing-angle XAFS is bypassed due to space limitation.

Instrumentation for XAFS experiments is standardized, consisting of the sequence of (a) cryostat-cooled monochromator/polychromator, usually *Si* in the hard X-ray regime, to filter out the required energy range from the entire synchrotron spectrum; (b) harmonic rejection mirror to reject the higher harmonics of the fundamental energy; (c) high-Z material slits to define the beam; (d) translation stage for positioning sample on beam and (e) detectors, e.g., gas-filled ion chamber for

Table 20.1 Transmission versus fluorescence modes of XAFS

	Transmission	Fluorescence
Geometry	Linear	90°
Signal	Logarithmic	Linear
Species of an excited atom	Z < 30	Z > 30
Non-uniform sample	No	Yes
Dilute sample	No	Yes (>PPM)
Concentrated sample	Yes (>5%)	Yes (but thin)
Sample thickness	1.5 absorption lengths	Depends on concentration
Sample preparation	Critical; calculated	Not critical
Detector	Ion chamber	Lytle detector; single- and multi-element solid-state detectors
Expensive	NA	High
Convenience of setup	Yes	No

Table 20.2 Sample considerations for XAFS

Sample composition	AB (in principle, multi-component systems can be measured)
Species of excited atom (A)	$Z > 22$ for hard X-ray beamlines
Elemental contrast of (A, B)	$\Delta Z > 1$ (else, (A, B) edges overlap) $\Delta Z > 5$ (for analysis)
Sample form	Powder (most convenient), liquids, thin ribbon
Sample concentration	>ppm
Sample thickness	$\sim$ 1.5 absorption lengths ($\leq 30\mu$m)
Safety factors	Non-radioactive/non-toxic
Precharacterizations	XRD/neutron diffraction/TEM
Analytic feasibility	Preliminary idea, e.g., whether bond-lengths are resolvable within analytic limits of XAFS

transmission and Lytle detector (Stern and Heald 1979)/solid-state detectors for fluorescence-mode experiments.

Experimental beamtime at synchrotron is allocated through an online proposal submission system. The waiting queue for XAFS beamtime is typically $\sim 1-3$ months. Ultra-sensitivity of XAFS spectra to statistical noise, as explained in Data Analysis Section, mandates precise ($\pm 10\mu m$) alignment at every stage to intercept the best portion of beam. Detailed description of alignment is outside the scope of this article. XAFS alignment consumes $> 80\%$ of the allocated time, leaving $< 20\%$ for data collection. The number and priority of samples should be accordingly planned.

20.7 Application Examples

Most of the users from non-XAFS backgrounds are unaware of the scale of problems that can be adequately solved with XAFS. Several papers in diverse research areas have evaluated the success and limitations of XAFS in solving their specific problems. The present section presents a compilation of representative problems that demand specific analysis strategies. These examples will initiate non-XAFS users into solving such non-trivial problems, besides providing realistic guidelines for defining XAFS problems and sample selection.

20.7.1 Alloys

Empirical Vegard's law (Vegard 1921) has been the standard predictor of alloying behavior for solid solutions viz linear variation of lattice parameters vary with composition. This is based on the assumption that atomic volume is conserved, independent of the extent of the mixing. In reality, breakdown from this assumption is imminent due to *local* chemical ordering, atomic size mismatch, and valence instability (Timoshenko and Frenkel 2019; Tonquesse et al. 2019). These generate diffraction-insensitive *local* distortions, which can be detected with XAFS. Thus, the characterization of atomic-scale alloying or *generic* mixing is one of the unique areas where XAFS technique could be indispensable. The formulation of XAFS analysis for alloys is credited to Frenkel et al. (2000). The criterion for alloy composition is elemental contrast *wrt* atomic number and/or atomic sizes, since XAFS monitors the degree of alloying through fractional composition and/or bond-length of the first coordination shell. For A_xB_{1-x} alloy, consisting of $[AA, AB, BB]$ atomic pairs, (A, B)-edge XAFS can be formulated as the cumulative weighted contribution from different atomic pairs.

$$\begin{aligned} A &: \chi_A = y_{AA}\chi_{AA} + y_{AB}\chi_{AB};\ y_{AA} + y_{AB} = 1 \\ B &: \chi_B = y_{BA}\chi_{BA} + y_{BB}\chi_{BB};\ y_{BA} + y_{BB} = 1 \end{aligned} \tag{20.14}$$

In order to reduce uncertainties of results, simultaneous fit of $[A, B]$-edge XAFS is encouraged with constraints for common bonds: $\left[R_{AB} = R_{BA};\ \sigma^2_{AB} = \sigma^2_{BA};\ \frac{y_{AB}}{y_{BA}} = \frac{1-x}{x}\right]$. These relationships are defined in *feffit.inp* file, following which independent variables are fitted. Detailed description of such formulation is beyond the scope of this chapter. Two extreme cases are presented:

(a) Alloy of metals with contrasting atomic number but similar atomic sizes, e.g., Au_xAg_{1-x}.

For such alloys, coordination composition, rather than bond-length, serves as a marker of alloying.

$$\begin{aligned} Au(L_3) : N_{Au} &= (y_{AuAu} + y_{AuAg})N;\ R_{AuAu} = R_{AuAg} \\ Ag(K) : N_{Ag} &= (y_{AgAg} + y_{AgAu})N;\ R_{AgAu} = R_{AgAg} \end{aligned} \tag{20.15}$$

Independent variables $(y_{AuAu}, N, R_{AuAu}, \sigma^2_{AuAu}, \sigma^2_{AuAg}, \sigma^2_{AgAg})$ are fit, from which the rest of the parameters are solved. In particular, y_{AgAu} result is used to quantify the pairing tendency of (Au, Ag) atoms with $R = \left(\frac{y_{AgAu}}{x}\right)$ = ratio of fractional composition, e.g., y_{AgAu}, relative to stoichiometric abundance (x). $\begin{pmatrix} R = 1 \\ R > 1 \\ R < 1 \end{pmatrix}$, respectively, represent random pairing, affinity, and repulsion of (Au, Ag) atoms.

(b) Alloy of metals with similar atomic number but contrasting atomic sizes, e.g., Pd_xAg_{1-x}.

For such alloys, coordination composition cannot be resolved due to identical backscattering amplitude and phase of (Pd, Ag) atoms. Lattice parameters of (Pd, Ag) are fortunately different [$R_{PdPd} = 2.75$Å vis-à-vis $R_{AgAg} = 2.88$Å]; their alloy (Pd_xAg_{1-x}) can generate conceivable change in lattice parameters [relative to Pd or Ag], which can be exploited as a marker of alloying by Vegard's law.

$$\begin{aligned} Pd(K) : R_{Pd} &= y_{PdPd}R_{PdPd} + y_{PdAg}R_{PdAg} \\ Ag(K) : R_{Ag} &= y_{AgAg}R_{AgAg} + y_{AgPd}R_{AgPd} \end{aligned} \tag{20.16}$$

Independent variables $\left(y_{PdPd}, R_{PdPd}, R_{PdAg}, R_{AgAg}, \sigma^2_{PdPd}, \sigma^2_{PdAg}, \sigma^2_{AgAg}, N\right)$ are fit, from which the rest of the parameters are solved. y_{AgPd} result is used to quantify pairing tendency of (Pd, Ag) atoms with $R = \frac{y_{AgPd}}{x}\begin{pmatrix} = 1 \\ > 1 \\ < 1 \end{pmatrix}$, respectively, represent random pairing, affinity, and repulsion of (Pd, Ag) atoms.

XAFS analysis of alloying has unraveled novel understandings, e.g., bending/stretching force constant-dependent anomalous bond relaxations in ternary alloys (Mayanovic et al. 1990), bond-specific discrete mixing behavior due to asymmetry of pair potential (Frenkel et al. 2000), undermining of chemical pressure in solid solution (Mukherjee et al. 2014), the role of anti-site defects in control of magnetic

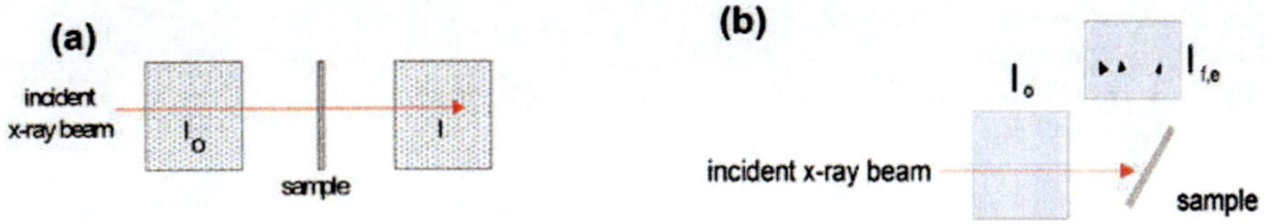

Fig. 20.11 **a** Transmission and **b** Fluorescence-mode of XAFS measurements

polarization (Lahiri et al. 2010b). The schemes of alloying analysis with XAFS are not necessarily limited to bulk but may be generalized for generic mixing phenomena. The following subsection "Nanocomposites" presents the extension of these schemes to decipher complex nano-configurations (Fig. 2.11).

20.7.2 Nanocomposites

The behavior of clusters is markedly different from bulk due to the presence of surface. Surface-free energy, which is comparable with lattice energy, generates anomalously large dynamical fluctuations resulting in the formation of defects, depression of melting point, and faster diffusion rate (Sakai 1996; Mori et al. 1994). Surface energy contribution can induce deviations from bulk phase diagrams of mixing and introduce chemical heterogeneity within clusters, e.g., one of the elements may preferentially segregate at the surface in order to lower surface energy (Wautelet et al. 2000). Such deviations, relative to bulk, become large at smaller cluster sizes (D) due to increasing surface fraction. Nanocomposites of practical interest, e.g., bimetallic, metal–semiconductor, semiconductor–semiconductor could materialize in a plethora of configurations, e.g., core@shell, interfacial alloy, segregated phase (Fig. 20.12). Resolving these atomic-scale details is crucial to their synthesis control and scientific understanding. While HRTEM and electron diffraction are equipped to decipher clusters down to $D \sim 3nm$ size, XAFS becomes indispensible for $D \leq 3nm$.

The generic scheme for mixing analysis (subsection I) can be exercised for nanocomposites, but with due incorporation of modulations due to cluster size and shape. The formulation of XAFS analysis for nanocomposites is credited to Anatoly (Frenkel 2007, 2012) and Lahiri et al. (2014). Cluster size and shape lay their signature on XAFS coordination through the presence of broken bonds or loss of coordination at surface; surface coordination (N_{surf}) is lower than core coordination (N_{core}). Net XAFS coordination for the cluster (N_{eff}) is the weighted average of core (N_{core}) and surface (N_{surf}) contributions: $N_{eff} = \left[(1-x)N_{core} + xN_{surf}\right]$; x = surface fraction of atoms, dependent upon cluster size (D) and shape (AR).

The baseline model for nanocomposite is a spherical cluster (A) of size D (Fig. 20.12a). Assuming fcc symmetry and surface plane = 111, surface fraction (x) of atoms is plotted as a function of cluster size (D) in Fig. 20.13a. Detailed

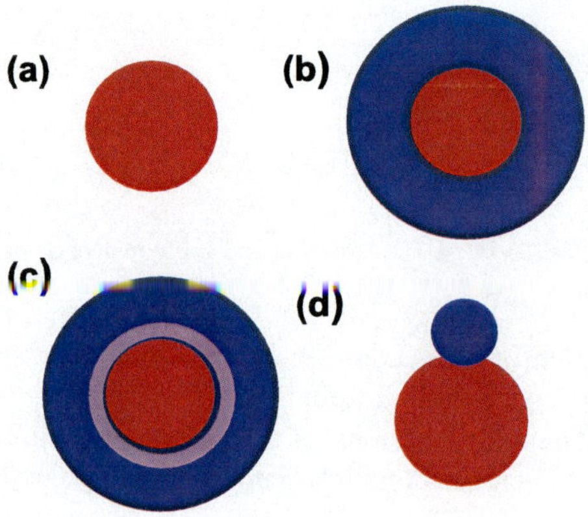

Fig. 20.12 Pt (red)—Ag (blue) nano-composite models. **a** Bare Pt core of size D; **b** Pt-core@Ag-shell [core size = d, shell thickness = t]; **c** Ag atoms intercalating into Pt core. [The interface is shown as light violet ring]; **d** Segregation tendency

derivation of x from D and unit cell parameter is beyond the scope of this article. Assuming $[N_{core} = 12, N_{surf} = 8.3]$ for fcc clusters, cluster coordination (N_{AA}) is plotted as function of cluster size (D) in Fig. 20.13b. N_{AA} increasingly deviates from N_{core} at lower cluster sizes, due to increasing surface fraction (x). In the next stage, A-core@B-shell model (Fig. 20.12b) is considered with A-core size (D) and B-shell thickness (t). Based on $[x_A, x_B]$ = surface atomic fraction of [A-core,B-shell], XAFS coordination can be formulated:

$$\begin{aligned}
\text{A-edge XAFS: } N_{AA} &= [12(1 - x_A) + 8.3x_A],\ N_{AB} = 3.7x_A \\
\text{Total coordination: } N_A &= (N_{AA} + N_{AB}) = 12 \\
\text{B-edge XAFS: } N_{BA} &= 3.7x_B;\ N_{BB} = [12(1 - x_B) + 8.3x_B] \\
\text{Total coordination: } N_B &= (N_{BA} + N_{BB}) < 12
\end{aligned} \tag{20.17}$$

(N_{AA}, N_{AB}) and (N_{BA}, N_{BB}) are plotted as function of cluster size (D) (Fig. 20.13b) and shell thickness (t) (Fig. 20.13c), respectively. These plots reveal maximum surface/interface sensitivity of XAFS at smallest (D, t), which is the size range of maximum scientific interest and at which TEM characterization fails.

For $D \geq 3nm$, core-size (D) and shell-thickness (t) are typically well-characterized by TEM. In these cases, XAFS analysis is projected to resolve the degree of interfacial mixing, undetected by TEM. $[N_{AA}, N_{AB}, N_{BA}, N_{BB}]_{cal}$ are pre-calculated from the plots of Fig. 20.13b, c, based on (D, t) information from TEM. In parallel, $[N_{AA}, N_{AB}, N_{BA}, N_{BB}]_{expt}$ are independently fitted for (A, B)-edge experimental XAFS spectra. Mismatch between experimental and pre-calculated $[N_{AA}, N_{AB}, N_{BA}, N_{BB}]$ is used to identify deviation from the robust core–shell model. Two special configurations are cited:

Intercalation of B atoms into A core (Fig. 20.12c):

$[\left[(N_{AA})_{expt} < (N_{AA})_{cal};\ (N_{AB})_{expt} > (N_{AB})_{cal};\ (N_{BB})_{expt} < (N_{BB})_{cal}\right].$

Fig. 20.13 **a** Surface fraction of atoms, as function of core size (D). Pt-core@Ag-shell: coordination calculations are shown for **b** Pt–Pt and Pt–Ag, as function of core size (D); **c** Ag–Ag, as function of shell thickness (t)

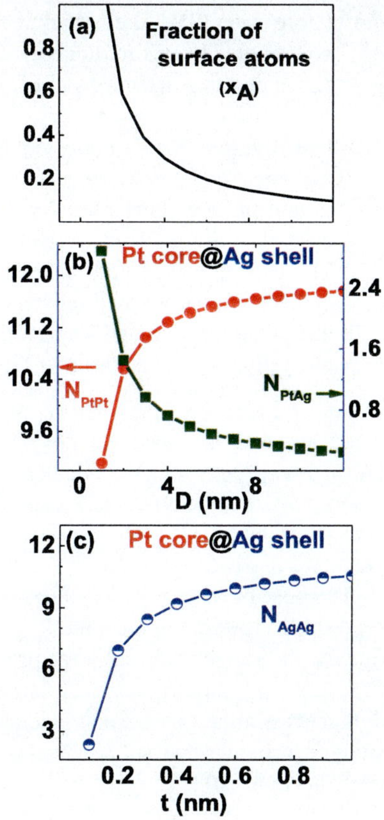

Incomplete B-shell or *partial* wetting (Fig. 20.12d): $[[(N_{AB})_{\exp t} < (N_{AB})_{cal};\ (N_{AA})_{\exp t} = (N_{AA})_{cal};\ (N_{BB})_{\exp t} \neq (N_{BB})_{cal}]$.

The most poorly TEM-resolved core@shell clusters are of $D \leq 3nm$—it is unclear if observed cluster sizes represent D or $(D + t)$. For such systems, XAFS is projected to address:

(a) Speciation of core (A) vis-à-vis shell (B) element via coordination analysis: $[N_A = 12, N_B < 12]$.

In $Pt - Ag$ composite (for example),

$$[\mathrm{N_{Pt}} = 12,\ \mathrm{N_{Ag}} < 12] \Rightarrow \text{Pt core@ Ag shell}$$
$$[\mathrm{N_{Ag}} = 12,\ \mathrm{N_{Pt}} < 12] \Rightarrow \text{Ag core@ Pt shell}$$

This is also true for interfacial mixing (Fig. 20.12c) while both $(N_{Pt}, N_{Ag}) < 12$ for partial wetting (Fig. 20.12d).

(b) [Core size (D), shell thickness (t), degree of mixing (y)] can be reverse-constructed from simultaneous experimental ($N_{AA}, N_{AB}, N_{BA}, N_{BB}$) fit results, based on the correlations of Fig. 20.13b, c. In reality, such direct reconstruction is non-unique due to practical cluster size distributions: $D_i(f_i)$; f_i = respective fraction. XAFS coordination fit results (N) actually represent ensemble-average = $\sum_i f_i N_i$; reverse reconstruction from $N(= \sum_i f_i N_i) \Rightarrow D_i(f_i)$ is non-unique. For example, the distribution $D = 1 - 5nm$ could generate ensemble averaged $N = 10.8$ that corresponds to $D = 2nm$ in Fig. 20.13b. The probability of such discrepancy (e.g., $D = 1 - 5nm$ vis-à-vis $D = 2nm$) is the highest for small clusters ($D \leq 3nm$), since N varies most rapidly over this cluster size range. Therefore, cluster size control with narrow dispersion is crucially important for unambiguous XAFS analysis of small clusters.

An indirect but practical approach toward the solution for such complicated systems is based on $D_i(f_i)$ information from TEM. Ensemble averaged $(N_{AA}, N_{AB}, N_{BA}, N_{BB})_{cal}$ is calculated for this distribution, under the preliminary assumption that the observed clusters represent core size (D)), coupled with possible combinations of [t, y]. Refinement (δ) of core size may be accommodated through $[D - \delta, t + \delta]$.

The best-fit $[D, t, \delta, y]$ combination is selected for which $(N_{AA}, N_{AB}, N_{BA}, N_{BB})_{cal}$ reproduce experimental result $(N_{AA}, N_{AB}, N_{BA}, N_{BB})_{expt}$. In this context, high spectral quality and range is strongly recommended to minimize uncertainties ($\pm 5\%$) of N fit results for unambiguous correlation with $[D, t, \delta, y]$. In difficult cases, molecular dynamics simulations-based modeling may be considered to fit XAFS results. These analytic schemes for spherical clusters may be generalized to any cluster shape (Doudna et al. 2003).

XAFS investigations of nano-composites in numerous publications have unearthed a wealth of exciting science. Very few of these but fundamental ones are being cited due to space limitation, e.g., noble metal in clusters become positively charged at ultra-small sizes with potential applications in drug-binding (Lahiri et al. 2003), defect-mediated room-temperature alloying in clusters (Shibata et al. 2002b), non-stoichiometric recipe of cluster synthesis, for surface segregation-protected stability (Lahiri et al. 2005).

20.7.3 Metallic Glass

In principle, glass or generic amorphous structure offers a unique scope for XAFS application, since most techniques are insensitive. XAFS analysis of the particular subgroup of metallic glass will be discussed. The basic motifs of metallic glass are Voronoi polyhedral clusters. In $Ni - Al - Zr$ glass (for example), Ni-centered cluster may consist of ($Ni - Ni$, $Ni - Al$, $Ni - Zr$) bonds, denoted by "ij"; [$i = Ni$, $j = (Ni/Al/Zr)$]. Cluster parameters are defined as (a) composition (N_{ij})

and (b) total coordination $N_i(=\sum_j N_{ij})$ [fractional composition $f_{ij} = \frac{N_{ij}}{N_i}$]; (c) cluster bond-lengths (R_{ij}), so that mean cluster radius $R_i(=\sum_j f_{ij} R_{ij})$ and cluster distortion = the difference between the shortest and longest bond-lengths; (d) cluster symmetry. The distinguishing feature of glass (vis-à-vis crystalline) is the intrinsic *degeneracy* of cluster parameters. For example, Ni-centers can host a plethora of cluster configurations, so that cluster-averaged results and inter-cluster variance are both important for the representation of glass structure. The role of XAFS is primarily extraction of cluster-averaged parameters around (Ni, Zr) sites with (Ni, Zr) K-edge XAFS, respectively. XAFS is incapable of resolving individual clusters and therefore, unable to detect inter-cluster variance. Fortunately, the variance is small $(\Delta N = \pm 2)$ as per "quasi-equivalent" cluster model, so that XAFS results reasonably approximate the ensemble. Out of the cluster parameters, bond-length and distortion information are directly retrieved from XAFS. On the other hand, interpretation of XAFS coordination and composition is intrinsically ambiguous for glass (Fig. 20.9a, b) as cluster coordination $N = 11 - 12$ could generate XAFS coordination $N = 2 - 3$. Derivation of coordination is therefore indirectly approached by model-based fitting (Lahiri et al. 2014), in which bond-lengths (e.g., R_{NiNi}, R_{NiAl}, R_{NiZr}) and coordination ratio $(x_{NiNi} : y_{NiAl} : z_{NiZr})$ are constrained from model. Common coordination multiplicative factor (F) and Debye–Waller factor (σ^2) for each bond are varied; XAFS fit result for F generates coordination: $N_{NiNi} = F x_{NiNi}$, $N_{NiAl} = F y_{NiAl}$, $N_{NiZr} = F z_{NiZr}$. Thus, XAFS directly or indirectly delivers near-complete information on (Ni, Zr) center-specific clusters viz size, distortion, coordination, and composition. The only gap of information XAFS leaves is cluster symmetry, since XAFS is angle-averaged.

The contribution of XAFS to the understanding of glasses is simply unique, as it provided the first evidence of the existence of short-range order in glass. Site-resolved analysis provided further understanding of the relative roles of atomic size and chemical order upon preferential glass-formation ability (Cheng et al. 2009; Fujita et al. 2009; Lahiri et al. 2014), with implications for composition designing.

20.7.4 Magneto-electronic systems

Representative subgroups of magneto-electronic materials are covered:

(i) Perovskites of formula ABO_3, pervading a wide spectrum of systems, e.g., Ruthenates (Lahiri et al. 2010a), multiferroics (Lahiri et al. 2019), manganites (Shibata et al. 2002a; Lahiri et al. 2012; Booth et al. 1998). Unit cell of ABO_3 consists of A-cation at the body-center and corner-shared BO_6 (Fig. 2.14a); charge transport or magnetic coupling advances via $B - O - B$ path.

(ii) Layered high-T_c superconductors, e.g., Cuprates (Haskel et al. 1996, 1999, 2001a, b; Lahiri 2006), their basic building block consisting of alternate conducting ($B - O - B$) and spacer ($A - O - A$) layers (Fig. 2.14b). Superconductivity is induced (destroyed) by doping spacer (conducting) layers. BO_6 consists of $4O$ atoms in $B - O - B$ plane and 2 bridging O atoms along the vertical axis. Polarization-dependent XAFS measurement can be employed to decouple these in-plane and axial contributions for a clear overview of the transport mechanism (Haskel et al. 1999, 2001a, b; Lahiri 2006).

The crystalline structure of both these groups is well-defined and in principle, readily solvable by diffraction technique. The necessity of XAFS emanates from the presence of *local* defects, e.g., around dopants; since the length-scale of charge transport or magnetic interaction $\leq 2nm$, any perturbation within this distance can significantly modify macroscopic magneto-transport properties. Such deviations generate paradoxical mismatch of crystallographic and magneto-transport phase boundaries (Haskel et al. 1996, Shibata et al. 2002a; Lahiri et al. 2012), which is the driving motivation for XAFS. Since magneto-transport advances along $B - O - B$ path, the structural parameters of interest mainly involve (a) BO_6 unit (Fig. 20.14c) and (b) inter-unit $B - O - B$ coupling (Fig. 20.14d).

(a) BO_6: The most relevant attribute of BO_6 is its distortion, e.g., Jahn-Teller distortion, with implications for polaron formation (Haskel et al. 1999, 2001a; Lahiri et al. 2012). The scheme of undistorted vis-à-vis Jahn-Teller distorted BO_6 is presented in Fig. 20.14c. It has been shown in Fig. 20.9a that interference between XAFS oscillations from closely spaced multiple bond-lengths generates beats in the net XAFS spectra. This phenomenon is exploited in the detection of distortion (or not) of BO_6. XAFS contribution from BO_6 unit $[\chi(q)]$ may be filtered out from the total XAFS spectra by back-transformation $[\chi(R) \rightarrow \chi(q)]$ of $B - O$ peak, using ATHENA. If *no* beat is observed in $\chi(q)$, it implies that BO_6 is undistorted and $\chi(R)$ can be fitted with *single* $B - O$ path of $N = 6$, mean bond-length (R) and (small) spread (σ^2). If beat *exists*, BO_6 is distorted. In that case, the phase derivative method (Piamonteze et al. 2005) can be employed to obtain an approximate estimate of the distortion or bond-length split (Δ) from the phase φ of $\chi(q)$: $\Delta = \frac{\pi}{2k_b}$; k_b = inflection position (k_b) of φ. Subsequent quantification strategy of BO_6 distortion is determined by pre-estimated Δ value.

To the first order, binomial $B - O$ bond-length distribution is assumed to represent distortion: [BO1(R_1),BO2(R_2)]; $\Delta = R_2 - R_1$. For $\Delta \geq 0.2$Å, [BO1,BO2] peaks are well-resolved; these can be fit with $[R_1, \Delta, N_1, \sigma_1^2, \sigma_2^2, \Delta E_0]$ variables and constraints $[R_2 = R_1 + \Delta; N_2 = 6 - N_1]$, i.e., Δ can be directly extracted from fit. For mixed systems such as manganites (Lahiri et al. 2012) with coexistence of distorted and undistorted BO_6 units, determination of the fraction of distorted sites (x_{dist}) becomes an additional task. For such systems, it is preliminarily assumed that $(1 - x_{dist})$ undistorted BO_6 units share the same radius as BO1 (R_1) so that coordination fit constraints for [O1,O2] can be reformulated:

$$N_{O1} = (N_1 \times x_{dist}) + [6 \times (1 - x_{dist})]; \; N_{O2} = [6 - N_1] \times x_{dist} \quad (20.18)$$

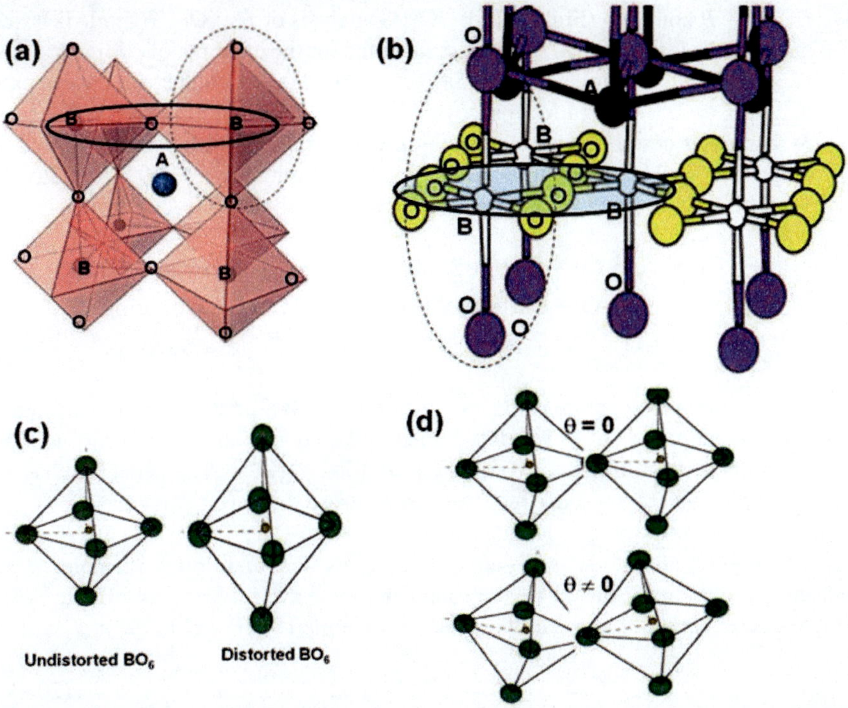

Fig. 20.14 Structures of **a** Perovskites (ABO_3), consisting of inter-connected BO_6 units and **b** Layered superconductors, consisting of alternate conducting (B-O-B) and spacer (A-0-A) layers. BO_6 units and B-O-B compiling are marked by dashed vertical and solid horizontal circles respectively; **c** BO6 units-undistorted vis-à-vis distorted (stretched along vertical axis); **d** Inter-unit coupling B-O-B for collinear ($\theta = 0$) and non-collinear ($\theta \neq 0$) configurations

Similar constraints can be formulated for bond-lengths. We remark that the distorted (or undistorted) sites cannot be correlated with valence state, e.g., $[Mn^{+3}, Mn^{+4}]$ based on only XAFS results, since XAFS provides site-averaged results. This correlation can be drawn from electronic structure considerations and the scope of the crystal field effect.

For $\Delta < 0.2$Å, a large overlap of [BO1,BO2] peaks precludes unambiguous fitting. In such cases, indirect approach may be adopted by truncating k-range below k_b, i.e., bypassing the beats. This strategy diminishes spatial resolution to an extent that [BO1,BO2] become indistinguishable and enables the representation of $B-O$ distribution with *single* $B-O$ path of $N = 6$, mean bond-length (R) and spread$\left(\sigma^2\right)$. The spread $\left(\sigma^2\right)$ then represents distortion of BO_6. With reference fitting of $[\sigma_0^2, \sigma_1^2]$ for end compounds $[x_{dist} = (0, 1)]$, σ^2 can be a marker of x_{dist} in the unknown compound, as $\sigma^2 = (1 - x_{dist})\sigma_0^2 + x_{dist}\sigma_1^2$.

(b) $B-O-B$ coupling (Fig. 20.14d): XAFS analysis of $B-O-B$ angle is based on multiple scattering (MS), as demonstrated for the example of collinear $Cu-O-Cu$ configuration ($\theta = 0^o$) in Fig. 20.5c. The photoelectron covers path length $R \approx 2.5-5$Å in MS process, which defines the fit range for MS fitting. MS analysis could be complicated by the presence of other peaks over this fit range. The analysis for collinear $B-O-B$ configuration ($\theta = 0^o$) is relatively straightforward. The relevant paths are generated by FEFF6 program:

$$\begin{aligned}\text{Cu}-\text{Cu[feffSS00.dat]; SS} &= \text{Single scattering}\\ \text{Cu}-\text{O}-\text{Cu[feffDS00.dat];DS} &= \text{Double scattering}\\ \text{Cu}-\text{O}-\text{Cu}-\text{O[feffTS00.dat];TS} &= \text{Triple scattering}\end{aligned}$$

These are input as separate paths into feffit.inp. As MS parameters are correlated with SS, e.g., amplitude of MS is determined by SS coordination, these correlations are constrained in feffit.inp: $(N_{DS'}, N_{TS'}) = (4,2)N_{SS'}$; $\Delta R_{DS,TS}(0) = \Delta R_{SS}$; $\sigma^2_{DS,TS} = \sigma^2_{SS}$; $\Delta e0_{DS,TS} = \Delta e0_{SS}$. The variables are thereafter solved.

In reality, configuration deviates from collinear (*i.e.*, $\theta \neq 0^o$). MS analysis becomes significantly intriguing, with θ as the additional variable. θ is determined, exploiting the angular dependence of backscattering amplitude $t(\theta)$. Pre-parameterization of angular dependence is credited to Haskel (1998):

$$t(\theta) = t_0 + \theta^2(\gamma t_0) \tag{20.19}$$

As $t = t_0 + \theta^2\gamma t_0$, XAFS from non-collinear configuration may be represented as the sum of θ-independent and θ-dependent components:

$$\begin{aligned}feffDS.dat &= feffDS00.dat + \theta^2 feffDStheta.dat;\ feffDStheta.dat = -\gamma feffDS00.dat\\ feffTS.dat &= feffTS00.dat + \theta^2 feffTStheta.dat;\ feffTStheta.dat = -\gamma feffTS00.dat\end{aligned} \tag{20.20}$$

Detailed derivation is bypassed in this short chapter. During fitting, $[feffDS00.dat, feffTS00.dat]$ and $[feffDStheta.dat, feffTStheta.dat]$ paths are disentangled in feffit.inp, in $1:\theta^2$ degeneracy ratio. Bond-length correction $\Delta R(\theta) = \Delta R + a\theta^2$ is also accommodated. FEFFIT solves θ^2 with an intrinsic error bar $\pm 3^o$, defined by the width of $t(\theta)$ around $\theta = 0^o$ (Fig. 20.5b). Thus, XAFS has poor angular resolution relative to X-ray or neutron diffraction. Since magneto-transport ground states often switch within $\theta < 3^o$, diffraction techniques may be better suited to detect angular correlation than XAFS. The significance of XAFS emerges for heterogeneous, e.g., doped systems, to differentiate the coupling between host and dopant sites (Haskel et al. 2001a). The other significance of XAFS is its sensitivity to rotational disorder of $B-O-B$ path (vis-à-vis diffraction) via

$\sigma^2_{DS,TS} = \sigma^2_{SS}$ (Lahiri et al. 2012). In reality, BO_6 units are rotated relative to each other, unlike the assumed parallel alignment of Fig. 20.14d.

The impact of XAFS in the understanding of magneto-electronic phenomena has been brilliant. It has foremost established the importance of local disorder that has resolved anomalies of phase diagrams (Haskel et al. 1996, 2001a; Shibata et al. 2002a; Lahiri et al. 2012 Booth et al. 1998). Novel scientific perspectives have emerged for (i) perovskites (ABO_3)—active involvement of supposedly non-magnetic $'A'$ ions in magnetism (Lahiri et al. 2010a, 2019), correlation between (A, B)-site distortions (Lahiri et al. 2019), significance of distortions rather than actual bond-length values (Lahiri et al. 2010a, 2012, 2019; Shibata et al. 2002a), role of orientation disorder in phase stabilization (Lahiri et al. 2012); (ii) layered superconductors—spatial correlation of carriers with dopant sites (Haskel et al. 2001a, 1999), distortions specific to superconducting vis-à-vis metal–insulator transitions (Haskel et al. 1999, 2001a, b; Booth et al. 1998; Lahiri 2006), correlation of orbital anisotropy of carriers with phase transition (Booth et al. 1998).

The above coverage over a wide spectrum of systems viz alloys, nanocomposites, metallic glass, perovskites, and layered superconductors constructs a comprehensive review of XAFS analysis strategies. In principle, these strategies may be judiciously exercised to solve XAFS problems for any system. These examples optimistically convey that XAFS problems are challenging at times but solvable to a large extent, with rewarding scientific output.

20.8 Conclusions

The article has presented the theory, experimental scheme, and analytic strategies of XAFS, mainly with the non-XAFS community in mind. The uniqueness, versatility, and vast domain of XAFS (vis-à-vis other techniques) have been duly highlighted. Guidelines for the choice of samples, formulation of XAFS problem and its solution have been prescribed. These should inspire users into undertaking challenging XAFS problems for novel and exciting science.

References

Ascone, I., Fourme, R., Hasnain, S., Hodgson, K.: J. Synchrotron Rad. **12**, 1–3 (2005)

Bevington, P.R.: Data Reduction and Error Analysis for the Physical Sciences. Mcgraw-Hill, USA (1969)

Booth, C.H., Bridges, F., Kwei, G.H., Lawrence, J.M., Cornelius, A.L., Neumeier, J.J.: Phys. Rev. Lett. **80**(4), 853–856 (1998)

Brillouin, L.: Science and Information Theory. Academic Press, New York, USA (1962)

Brown, I.D., Altermatt, D.: Acta Cryst. B **41**, 244–247 (1985)

Brown, M., Peierls, R.E., Stern, E.A.: Phys. Rev. B **15**(2), 738–744 (1977)

Bunker, G.: The EXAFS of disordered systems and the cumulant expansion. In: Hodgson, K.O., Hedman, B., Penner-Han, J.E. (eds.) EXAFS and Near Edge Structure III, pp. 268–272 (1984)

Cheng, Y.Q., Ma, E., Sheng, H.W.: Phys. Rev. Lett. **102**(1–4), 245501 (2009)

Doudna, C.M., Bertino, M.F., Blum, F., Tokuhiro, A., Lahiri, D., Chattopadhyay, S., Terry, J.: J. Phys. Chem. B **107**, 2966–2970 (2003)

Filipponi, A.: Phys. Condens. Matter **13**, R23–R60 (2001)

Frenkel, A.I.: Z. Kristallogr. **222**, 605–611 (2007)

Frenkel, A.I.: Chem. Soc. Rev. **41**(24), 8163–8178 (2012)

Frenkel, A.I., Sh, M.V., Rubshtein, A., Rosenberg, Y., Voronel, A., Stern, E.A.: Phys. Rev. B **62**(14), 9364–9371 (2000)

Fujita, T., Konno, K., Zhang, W., Kumar, V., Matsuura, M., Inoue, A., Sakurai, T., Chen, M.W.: Phys. Rev. Lett. **103** (1–4), 075502 (2009)

Giachini, L., Giachini, L., Veronesi, G., Francia, F., Venturoli, G., Boscherini, F.: J. Synchrotron Rad. **17**, 41–52 (2010)

Grunwaldt, J.-D., Frenkel, A.I.: Synchrotron Radiation News **22**(1), 2–4 (2009)

Hartmann, E., Lauterbach, C.: Adv. Quantum Chem. **29**, 373–387 (1998)

Haskel, D., Stern, E.A., Hinks, D.G., Mitchell, A.W., Jorgensen, J.D., Budnick, J.I.: Phys. Rev. Lett. **76**, 439–442 (1996)

Haskel, D., Stern, E.A., Polinger, V., Dogan, F.: J. Syn. Rad. **6**, 758–760 (1999)

Haskel, D., Stern, E.A., Dogan, F., Moodenbaugh, A.R.: J. Synchrotron Rad. **8**, 186–190 (2001b)

Haskel, D.: Appendix A: measuring buckling angles with XAFS. In: Local Structural Studies of Oriented High Temperature Superconducting Cuprates by Polarized XAFS Spectroscopy. Ph.D. thesis, University of Washington, USA.

Haskel, D., Stern, E.A., Polinger, V., Dogan, F.: Phys. Rev. B **64** (1–10), 104510 (2001)

http://cars9.uchicago.edu/~ravel/software/doc/Athena/html

http://gbxafs.iit.edu/training/XAFS_sample_prep.pdf

http://www.csrri.iit.edu/periodic-table.html

https://lightsources.org

https://xafs.xrayabsorption.org

James, J.F.: A Student's Guide to Fourier Transforms: With Applications in Physics and Engineering (Student's Guides). Cambridge University Press, UK (2011)

Kelly, S.D., Bare, S.R., Greenlay, N., Azevedo, G., Balasubramanian, M., Barton, D., Chattopadhyay, S., Fakra, S., Johannessen, B., Newville, M.: J. Phys. Conf. Ser. **190**(1–4), 012032 (2009)

Koningsberger, D.C., Prins, R.: X-ray Absorption: Principles, Applications and Techniques of EXAFS, SEXAFS and XANES. John Wiley and Sons, USA (1988)

Laan, G.V.D., Figueroa, A.I.: Coord. Chem. Rev. **277–278**, 95–129 (2014)

Lahiri, D.: Physica C **436**, 32–37 (2006)

Lahiri, D.: XAFS Studies of Nanocomposite Systems. Lambert Academic Publishing, Germany (2010)

Lahiri, D., Subramanian, V., Shibata, T., Wolf, E., Bunker, B.A., Kamat, P.V.: J. Appl. Phys. **93**(5), 2575–2582 (2003)

Lahiri, D., Sharma, S.N., Zeena, S., Bunker, B.A.: J. Surf. Sci. Technol. **21**(1–2), 11–22 (2005)

Lahiri, D., Sharma, S.M., Vishwanadh, B., Das, N.C., Dey, G.K., Schumacher, G., Scherb, T., Riesemeier, H., Reinholz, U., Radtke, M., Banerjee, S.: J. Synchrotron. Rad. **21**(6), 1296–1304 (2014)

Lahiri, D., Garg, N., Nayak, B.K.: Aspects Min. Miner. Sci. **5**(2), 563–564 (2020)

Lahiri, D., Shibata, T., Chattopadhyay, S., Kanungo, S., Saha-Dasgupta, T., Singh, R.S., Sharma, S.M., Maiti, K.: Phys. Rev. B **82** (1–6), 094440 (2010a)

Lahiri, D., Khalid, S., Raychaudhuri, P., Dhar, S.K., Sharma, S.M.: Phys. Rev. B **82**(1–6), 134424 (2010b)

Lahiri, D., Khalid, S., Sarkar, T., Raychaudhuri, A.K., Sharma, S.M.: J. Phys. Condens. Matter **24**(1–8), 336001 (2012)

Lahiri, D., Chattopadhyay, S., Kaduk, J., Shibata, T., Shevchenko, E.V., Palkar, V.R.: Mater. Res. Express **6**(1–8), 045005 (2019)
Lengeler, B.: Phys. Rev. Lett. **53**, 74–77 (1984)
Lim, J.E., Moon, D.H., Kim, K.-R., Ok, Y.S.: J. Appl. Biol. Chem. **58**(1), 65–74 (2015)
Marquardt, D.W.: J. Soc. Ind. Appl. Math. **11**, 431–441 (1963)
Mayanovic, R.A., Pong, W.F., Bunker, B.A.: Phys. Rev. B **42**(17), 11174–11182 (1990)
Meldner, H.W., Perez, J.D.: Phys. Rev. A **4**(4), 1388–1396 (1971)
Mermin, A., Ashcroft, N.: Solid State Physics. Cengage Learning, USA (2003)
Michalowicz, A., Vlaic, G.: J. Synchrotron Rad. **5**, 1317–1320 (1998)
Mori, H., Yasuda, H., Kamino, T.: Philos. Mag. Lett. **69**(5), 279–283 (1994)
Mukherjee, S., Nag, A., Kocevski, V., Santra, P.K., Balasubramanian, M., Chattopadhyay, S., Shibata, T., Schaefers, F., Rusz, J., Gerard, C., Eriksson, O., Segre, C.U., Sarma, D.D.: Phys. Rev. B **89**(1–11), 224105 (2014)
Newville, M.: J. Synchrotron Rad. **8**, 96–100 (2001)
Newville, M., Līviņš, P., Yacoby, Y., Rehr, J.J., Stern, E.A.: Phys. Rev. B **47**, 14126–14131 (1993)
O'Loughlin, E.J., Kelly, S.D., Kemner, K.M.: Environ. Sci. Technol. **44**, 1656–1661 (2010)
Pascarelli, S., Haskel, D.: High Pressure Res. **36**(3), 235–236 (2016)
Piamonteze, C., Tolentino, H.C.N., Ramos, A.Y., Massa, N.E., Alonso, J.A., Martínez-Lope, M.J., Casais, M.T.: Phys. Rev. B **71**(1–4), 012104 (2005)
Powell, C.J.: Surf. Sci. **44**(1), 29–46 (1974)
Rahkomen, K., Krause, M.O.: At. Data Nucl. Data Tables **14**, 140 (1974)
Ramanan, N., Roy, S., Lahiri, D., Sharma, S.M., Dev, B.N.: J. Synchrotron Rad. **20**, 137–144 (2013)
Ramanan, N., Rajput, P., Jha, S.N., Lahiri, D.: Nucl. Instrum. Methods Phys. Res. A **782**, 63–68 (2015a)
Ramanan, N., Lahiri, D., Rajput, P., Varma, R.C., Arun, A., Muraleedharan, T.S., Pandey, K.K., Maiti, N., Jha, S.N., Sharma, S.M.: J. Synchrotron Rad. **22**(5), 1233–1241 (2015b)
Ranieri-Raggi, M., Raggi, A., Martini, D., Benvenuti, M., Mangani, S.: J. Synchrotron Rad. **10**, 69–70 (2003)
Sakai, H.: Surf. Sci. **351**, 285–291 (1996)
Schiff, L.I.: Advanced Quantum Mechanics. Mcgraw-Hill Book Company, Inc USA (1949)
Shibata, T., Bunker, B.A., Mitchell, J.F., Schiffer, P.: Phys. Rev. Lett. **88**(20), 207205 (1–4) (2002a)
Shibata, T., Bunker, B.A., Zhang, Z., Meisel, D., Vardeman, C.F., Gezelter, J.D.: J. Am. Chem. Soc. **124**, 11989–11996 (2002b)
Smith, A.L., Verleg, M.N., Vlieland, J., Haas, D., Ocadiz-Flores, J.A., Martin, P., Rothe, J., Dardenne, K., Salanne, M., Gheribi, A.E., Capelli, E., Eijcka, L.V., Konings, R.J.M.: J. Synchrotron Rad. **26**, 124–136 (2019)
Stern, E.A., Heald, S.M.: Rev. Sci. Instrum. **50**, 1579–1583 (1979)
Suzuki, Y., Kelly, S.D., Kemner, K.M., Banfield, J.F.: Nature **419**, 134–134 (2002)
Teo, B.K., Lee, P.A.: J. Am. Chem. Soc. **101**, 2815–2832 (1979)
Thomas, S.A., Mishra, B., Myneni, S.C.B.: J. Phys. Chem. Lett. **10**, 2585–2592 (2019)
Timoshenko, J.: Cuenya B R **121**(2), 882–961 (2021)
Timoshenko, J., Frenkel, A.I.: ACS Catal. **9**(11), 10192–10211 (2019)
Tonquesse, S.L., Pasturel, M., Demange, V., Tayal, A., Solari, P.L., Prestipino, C.: J. Nucl. Mater. **526**(1–8), 151772 (2019)
Vegard, L.: Z. Phys. **5**, 17–26 (1921)
Veigele, W.J.: At. Data Nucl. Data Tables **5**(1), 51–111 (1973)
Viswanathan, C.R.: Introductory Atomic Physics and Quantum Mechanics (1), 10–13 (2018)
Wautelet, M., Dauchot, J.P., Hecq, M.: Nanotechnology **11**(1), 6–9 (2000)
Wilke, M., Farges, F., Petit, P.-E., Brown, G.E., Martin, F.: Am. Miner. **86**, 714–730 (2001)
Wong, J., Lytle, F.W., Messmer, R.P., Maylotte, D.H.: Phys. Rev. B **30**, 5596–5610 (1984)
Zhang, H.A.: Analyzing the short-range order of metallic glass through x-ray absorption fine structure (XAFS) spectroscopy. MSc Thesis, Case Western Reserve University (2019)

Chapter 21
High-Resolution Molecular Secondary Ion Mass Spectrometry for Absolute Quantification of Materials in Low-Dimensional Structures: Foundation, Perception and Challenges

Purushottam Chakraborty

21.1 Introduction

Importance of materials research twigs from increasing applications of advanced materials, low-dimensional structures and smart materials. Rapidly developing areas are thin films for microelectronics, anti-reflective coatings for corrective lenses and camera lens elements, photonic materials for optical data storage, ion-beam modified surfaces with high resistance to wear and corrosion, wood-based materials, composites and plastics, etc. Criterion for these materials demands spatially resolved chemical and structural analysis. For two-dimensional materials, it is important to map lateral and in-depth distributions of chemical composition with high resolution at near-surfaces and interfaces. Sputtering-based micro-sectioning techniques combined with surface analytical methods are frequently applied for the analysis of materials. These methods can have depth resolution down to a few atomic layers over a depth of several microns, thus facilitating the quantitative analysis of thin films, superlattices and quantum-confined structures.

Depending on the domination of physical or chemical effects caused in the material, interaction of ion beam with surfaces may lead to 'physical sputtering' or 'chemical sputtering'. In physical sputtering, the impinging ions collide with target atoms and initiate 'collision cascades' (Thompson 2004) within the material due to "momentum transfer" between ions and target atoms. At higher ion energies, one or more 'collision cascades' are initiated. Some of the cascades recoil back towards the surface. If a part of the 'collision cascade' moving in the upward direction reaches the surface with energy greater than "surface binding energy", an atom or a molecule

P. Chakraborty (✉)
Saha Institute of Nuclear Physics, Kolkata 700064, India
e-mail: purushottam.chakraborty@gmail.com

R. S. Ningthoujam and A. K. Tyagi (eds.), *Handbook of Materials Science, Volume 1*,
Indian Institute of Metals Series, https://doi.org/10.1007/978-981-99-7145-9_21

is ejected from the surface. This phenomenon is known as "sputtering". Three basic points are important in collision cascade-induced sputtering; (1) A collision cascade is confined to a cylindrical volume of ~10 nm diameter and ~10 nm depth from the surface. (2) Average rate of arrival of primary ions (at the highest primary ion-current density, i.e. 1 mA/cm^2) is ~10^{-14} s per ion, which is much greater than the lifetime of a collision cascade. (3) Cascades that are generated by individual ions are independent of one another (linear collision cascade).

The "sputter yield", defined as "number of ejected atoms per incident ion", depends on several parameters; angle of incidence of primary beam, masses and energy of primary ions, masses and surface binding energy of target atoms, etc. If the target is crystalline, the orientations of axes are additional factors of importance. According to the theory of simple collision cascade, sputter yield increases linearly with 'energy density of cascade' deposited to the surface region. In "linear collision cascade" regime, sputter yield is proportional to the number of recoil-atoms, which is also proportional to the energy deposited per unit depth in nuclear process, in which the particle velocity is much less than Bohr velocity (velocity of the electron in the first Bohr orbit of the hydrogen atom; 2.18×10^6 m/s). At low energy densities, the linear dependence matches the experimental data, but at higher energy densities, the results exhibit a sharp transition from linear to nonlinear dependence, indicating the onset of some collective process, known as 'thermal spike' (Davies 1984). In the thermal spike model of sputtering, it is assumed that the energy is deposited instantly in a very small localized region, producing a sharp increase in the temperature of that region, which then spreads and dissipates according to the laws of classical heat conduction in a continuum.

Binary collisions between primary ions and target atoms follow the principles of classical mechanisms and can be applied to describe the sputtering process. Depending on the primary ion energy, elastic or inelastic collisions take place. Elastic collisions govern interactions in the keV range and can be described by nuclear-stopping power expressed as $\left(\frac{dE}{dx}\right)_n$, and defined by the "energy loss of the primary particle per unit path-length". Nuclear stopping predominates at the low ion energies and contributes to the collision cascade causing permanent displacement of the atoms in the solid. These displacements can cause a change in the physical and mechanical properties of the material such as "radiation damage", formation of defect clusters or microstructural changes, amorphization, etc. Ion-tracks are essentially the damage-trails that are created by impinging ions penetrating through solids and are irregular in nuclear stopping. An impinging primary ion whilst traversing through the material loses its energy by three processes; (1) direct collisions with the target atoms, (2) excitations and ionizations of the target atoms and (3) production of phonons due to lattice vibrations. Dominance of one particular process over the others depends on ion-target combination, ion-beam parameters, bombardment conditions, etc. A combination of computer simulations and experimental observations is needed to explain many of the complex processes that take place during ion irradiation. Computer simulations enable the depth of damage, lateral spread and sputtering yields to be predicted fairly accurately. Broadly, the programmes fall into two categories (Harrison et al. 1968): the binary collisions and molecular dynamics approach. The 'molecular dynamics' simulations follow the history of every particle generated in a collision cascade arising from an individual ion-trajectory by a solution of the equations of motion, whereas in 'binary collision approximation', an interaction between two colliding particles is treated as an individual event. Sputter yield Y (atoms/ion) can be estimated from a simple expression: $Y\left(\frac{atoms}{ion}\right) = \frac{0.042S_n(E)}{U}$, where U is "surface binding energy" of the target atom (often equated to the heat of sublimation); α is $\left[f\left(\frac{M_1}{M_2}, \cos(\theta)\right)\right]$—a dimensionless parameter and a function of (M_1/M_2) and cos(θ); S_n(E) is "nuclear stopping cross-section"; θ is the angle of incidence with the surface normal. M_1 and M_2 are respective masses of interacting atoms.

At low energies, sputtering occurs at a threshold of energy, below which the ions transfer insufficient energy to the target to overcome the "surface binding energy". Sputter yield then rises as the incident beam energy increases and eventually at high energies, it decreases as the recoil cascades are formed deeper within the material and thus cannot reach the surface. At very high ion energies (MeV), the inelastic collisions start dominating and the number of collisions increases with the rise of energy. The corresponding parameter to describe this interaction is the **electronic stopping power**, defined as $\left(\frac{dE}{dx}\right)_e$. In inelastic collisions, electrons are removed from the nuclei leading to ionizations along the ion tracks confined to cylindrical damage regions of a few nanometers in diameter. These ion tracks are straight and can be studied by Rutherford backscattering spectrometry (RBS), transmission electron microscopy (TEM), small-angle neutron scattering (SANS), small-angle X-ray scattering (SAXS), etc. As the sputtering causes the removal of atoms from the top monolayers of a solid, 'depth resolution' in the order of monolayer thickness is realizable. However, sputtering results from complex interaction processes between ions and surfaces and may lead to distortions in surface morphology and material composition.

Secondary Ion Mass Spectrometry (SIMS) has placed its exclusive standing amongst all analytical methods. A minute fraction of sputtered particles is ionized positively and/or negatively in the SIMS process and these ions are dubbed as "secondary ions". The effective thickness of a sputtered layer lies in the order of atomic dimensions and can be estimated from the expression of sputter yield involving the parameters like "information depth", "average energy of the sputtered particles" and "nuclear stopping power" (Brundle et al. 1992). The emitted particles originating essentially from the outermost (~1–3) atomic layers provide the information about the composition of the surface and near-surface regions. The sputter-products constitute electrons, atoms, molecules, clusters and molecular fragments. A minute fraction of these sputtered species is excited above respective ground states and subsequently undergoes any of the following processes:

1. De-excitation through radiative electronic transition, leading to photon emission in the ultraviolet to a visible optical range.
2. Non-radiative electronic transition leading to the emission of Auger electrons.
3. Ionization of sputtered species via 'tunnelling of electrons' between these species and the valence band of the solid, both in resonant energy states.

The above dynamic charge-transfer processes can be more complex due to ion-induced electronic and structural perturbations. A fingerprint mass spectrum in SIMS consists of almost all elements present on the top surface of the material. The secondary ions constitute singly- or multiply-charged atomic-, molecular- and cluster-ions. Exact charge-state formation mechanism can be understood in terms of "symmetric and asymmetric collisional effects" on singly- and doubly-charged ion formation in the sputtering process (Mondal et al. 2012). The kinetic energy distributions of secondary ions emitted from the Si surface suggested that Si^+ ions are largely formed in the upper surface layers and Si^{2+} ions are produced owing to the Si 2p shell vacancies created under target–target symmetric collisions, followed by Auger-electron emission (Mondal et al. 2012).

In general, sputtered-ions have large kinetic energy distributions that are mostly peaked at around 2–5 eV. The energy-selected ions are mass analysed and detected using a combination of a suitable mass spectrometer and appropriate ion-collection optics, thus allowing us to get a variety of information about surfaces, sub-surfaces and sample composition. As ion-bombardment goes on, the material is continuously eroded atomic layer by atomic layer, thus providing in-depth quantitative information about material constituents through dynamic SIMS. As a material is irrevocably consumed in this mode of SIMS analysis, it is a consumptive process. However, this problem is principally compromised at the expense of the highest "detection sensitivity", allowing the technique to provide maximum information from infinitesimal volumes of materials. Figure 21.1 represents a typical experimental arrangement for surface and in-depth analysis of materials using ion beams (Baio et al. 2020).

The localized surface chemistry plays a crucial role in 'ionization cross-section' of a sputtered atom or molecule, thus governing the strong dependence of the matrix effect on the intensity of secondary ions of that particular element. For example, an electronegative element like oxygen or fluorine present on the surface of a host matrix increases the emission intensity of positive ions by several orders of magnitude. The stronger is the electronegativity of that influencing element with respect to a matrix element, the higher is the positive ion emission yield of that matrix element. Similarly for a surface exposed under some electropositive alkali element like Cs or Li, an enhanced emission of secondary negative ions and concurrently a decreased emission of secondary positive ions of matrix element is observed. Therefore, the ion yield variation of a particular element in a matrix with varying surface chemistry will not be linear with the concentration of that element in the matrix. This phenomenon is named as "**Matrix Effect**".

Although this 'matrix effect' is an artefact of SIMS, this is often consciously used in the quantification of materials containing elements of extremely low ion yields. Obviously, in order to estimate the exact concentration of a certain species present in a material, "matrix effect" needs to be corrected. The correction is done using "calibration standards" or by the determination of "relative sensitivity factor" (RSF). However, the strong dependence of "ionization probability" on surface chemistry imposes on using acceptable standards in the quantitative analysis. In this scenario, 'ion-implanted standards' have gained pre-eminence, allowing precise quantification (Chakraborty 2002). From an analytical point of view, the motivation towards the

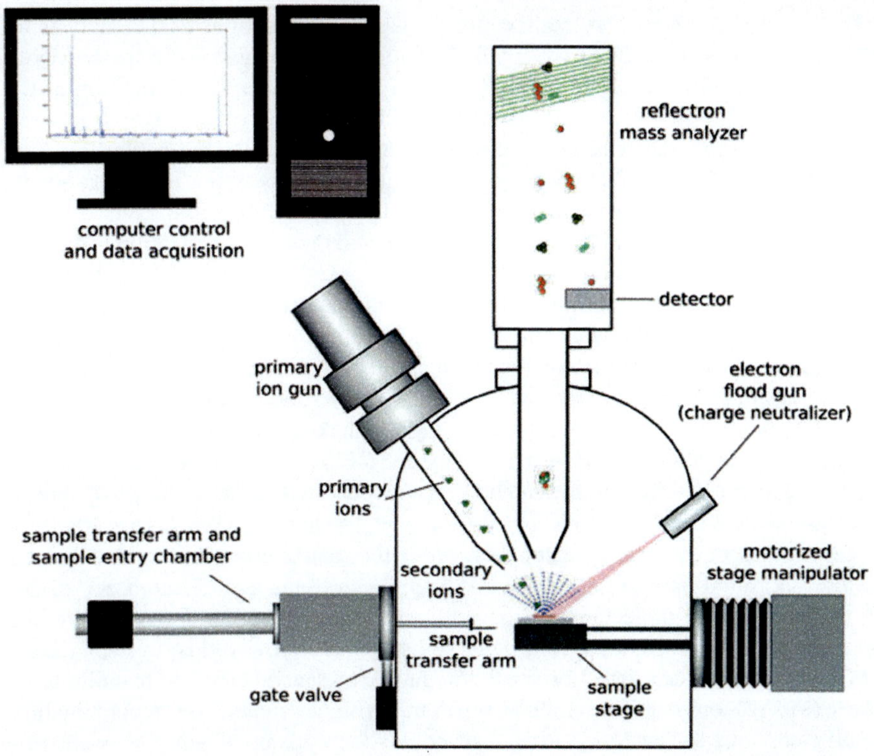

Fig. 21.1 Schematic diagram of the principal arrangement of an instrument used for sputter-depth profile analysis. Reproduced with permission from RSC Publications (Baio et al. 2020)

reduction of matrix effect led to the discovery of "secondary neutral mass spectrometry" (SNMS). Non-dependence of secondary neutral yield on the local surface chemistry makes the SNMS technique directly quantitative without the need for calibration standards (Oechsner and Gerhard 1972). Furthermore, SNMS has the potential to measure depth distributions of elements ignoring the preferential sputtering artefacts, as contrary to the electron spectroscopic techniques. Although SNMS has detection limits and dynamic ranges much poorer to that of SIMS, it has some consequential implications in the routine analysis of thin films. Oechsner and Gerhard (1972) described an experimental arrangement for the SNMS analysis of sputtered neutrals using low-pressure hydrogen fluoride (HF) discharge.

If an alkali element, referred to as "A", exists in the immediacy of the probing matrix element (M), a quasi-molecular ion can form through the attachment of this alkali ion with the probing atom leading to $(MA)^+$ formation in the SIMS process. The formation of this $(MA)^+$ molecular ion is strongly correlated with the 'atomic polarizability' of the element M. Since the emission of resputtered species (M^0) has no connection with $(MA)^+$ ion formation, monitoring of these $(MA)^+$ ions are strongly effective in minimizing the 'matrix effect'. This is very similar to the analysis by SNMS. Caesium, compared to other alkali elements, is strongest in electro-positivity and hence the MCs^+ ions are highly preferred in SIMS analysis. Formation of the MCs^+ molecule can be understood in the light of fundamental attractive force between a Cs^+ ion (resputtered) and the "dipole moment" of the polarized neutral atom of the matrix (M), analogous to Van der Waals interactions between hydrogen bonds. The dipole moment occurs due to the polarization of a neutral atom caused by the electric field of Cs^+ ion in their close proximity, and thus the polarization is directly proportional to the attractive electric field between Cs^+ and M^0. Consequently, the polarizability of an element plays the most significant role in forming the $(MCs)^+$ molecular ion. A detailed understanding of the formation mechanism of this MCs^+ molecular ion can explore its analytical capability in materials quantification. For a large variety of semiconductors, MCs^+ yield has been found to quadratically depend on the atomic polarizability of the element (M) (Chakraborty 2002).

Quantitative SIMS analysis using $(MCs)^+$ and $(XCs)^+$ ions (where M and X denote matrix element and impurity element, respectively) was found to be realizable in the cases where the Cs surface coverage was kept below 5% (Wittmaack 2013). A great success in the complete suppression of the matrix effect has been achieved using "alkali ion-based molecular SIMS" approach (Saha and Chakraborty 2013). Although the MA^+ molecular ions have found their applicability in direct materials quantification in SIMS, these ions generally suffer from low yields. In such a case, $(MA)_2{}^+$ ions provide a detection sensitivity higher by several orders of magnitude, as these $(MA)_2{}^+$ ions have been found to be of much higher yields. The "recombination coefficient" of MA^+ or $MA_2{}^+$ ions, respectively, depends on whether M is electro-positive or electro-negative with respect to A. Apart from the effect of surface binding energy, the changes in 'local surface work function' are seen to have a significant effect on the formation of these molecular ions.

SIMS technique for its extremely high detection sensitivity is enormously powerful to quantify ultra-trace elements with concentrations below parts per billion. It has two operation modes; "Static SIMS" and "Dynamic SIMS". In order to have molecular information of organic or biological materials, an extremely low dose of primary ions is required to strike only a few numbers of surface atoms so that sputter erosion of surfaces can be exceptionally low ($\leq 10^{12}$ ions per cm^2). This mode of operation is referred to as "Static SIMS". On the other hand, "Dynamic SIMS" uses comparatively higher ion doses to make rapid depth profiling of inorganic materials. Use of cluster ions has allowed the profiling of organic and biological materials, as these clusters provide significantly high secondary ion yields with considerably low surface damage. SIMS with "gas cluster ion beams" (GCIBs) have opened possibilities for bio-imaging and molecular depth profiling. Thousands of atoms in a cluster created in supersonic expansion can desorb target molecules with high emission yields and negligible fragmentation (Winograd 2018). The kinetic energy of the cluster consisting of ~1,000 to ~10,000 constituent atoms or molecules is uniformly distributed amongst all constituents, so that the kinetic energies of individual constituents are reduced to energies comparable to chemical bond strengths.

This considerably reduces the probability of fragmentation, surface damage and ion-induced mixing, etc. This is unlike the situation with smaller cluster projectiles, such as Bi_3^+ or C_{60}^+, as kinetic energy per atom in that case ranges from a few hundreds to a few thousands of electron volts (Winograd 2018).

Whilst characterizing polymer or biological materials using static SIMS, the total ion dose accumulated during spectral acquisition is insignificant, yielding an insignificant amount of material consumption. This makes the static SIMS poor in detection sensitivity. As the detection sensitivity scales with the analytical volume consumed in-depth profiling (Haag et al. 1995), dynamic SIMS (primary current density ~ μA/cm^2) obviously has exceedingly high detection sensitivity, a few orders of magnitude higher compared to static SIMS (primary current density ~ pA/cm^2). Dynamic SIMS provides the foremost scientific impact in routine in-depth analysis of low-level dopant concentrations in semiconductor materials. The advantages due to controlled sputter-sectioning and large dynamic ranges have made the dynamic SIMS extremely sensitive for quantification of surfaces, thin films, multilayers, superlattices and low-dimensional structures. However, the main problem in the composition analysis of surface layers with minimal damage is to ensure that a sufficient signal is obtained with minimal incident ion flux. This has been found realizable with a combination of pulsed-ion sources and time-of-flight (TOF) mass spectrometers. TOF–SIMS is significantly resourceful for the acquisition of spectral data with considerably high mass resolution and detection sensitivity. Its applications have increasingly expanded from microelectronics to materials science to biology (Brison et al. 2010). The biological applications of TOF–SIMS have increased with the development of cluster-ion sources. Polyatomic primary ions, such as Bi_n^{q+} (n = 1–7, q = 1 and 2) and C_{60}^{q+} (q = 1–3), can produce a significant increase in the secondary ion yields even in the high mass region of the spectrum.

The present chapter addressed the fundamentals and general perspectives of SIMS, conceptual understanding of MCs_n^+ (n = 1, 2, ...) molecular ion complexes and the phenomenal challenges of "alkali element based molecular ion SIMS" approach in materials quantification without calibration standards. Figure 21.2a and b schematically represent the SIMS process and experimental layout of a quadrupole-based SIMS setup, respectively (Chakraborty 2002). Figure 21.2c gives a schematic representation of a typical quadrupole mass filter, composed of four cylindrical electrodes arranged in a symmetric manner that can filter charged species based on its specific charge, if driven by a mix of constant (DC) and time-varying (RF) potentials. The dynamics of charged particles inside a quadrupole mass filter are governed by the Mathieu's differential equation, represented as $\frac{d_2 y}{dx^2} + (a - 2q\cos(2x))y = 0$; which for a certain charge-to-mass ratio (e/m) defines an infinite number of stability-regions accomplished by combinations of DC voltage, RF amplitude, RF frequency and quadrupole geometry, resulting in stable trajectories of the charged particles inside the quadrupole field.

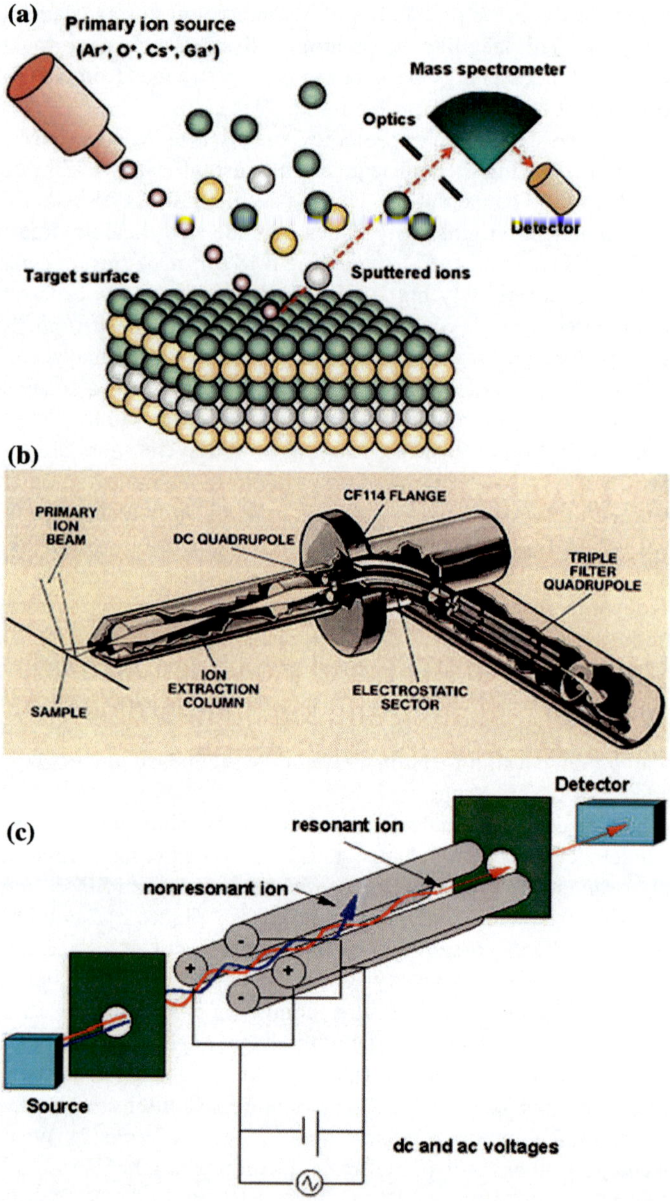

Fig. 21.2 **a** Schematic representation of secondary ion mass spectrometry process. Reproduced with permission from a book published by Nova Science Publisher (Chakraborty 2002). **b** Schematic representation of SIMS setup (Hiden EQS1000). It employs a triple-quadrupole filter and a 45°-sector-field electrostatic energy analyzer with a DC quadrupole input focussing lens. Reproduced with permission from a book published by Nova Science Publisher (Chakraborty 2002). **c** Quadrupole mass filter selects positive or negative ions based on their mass-to-charge ratios by a suitable combination of radio-frequency (RF) and DC potentials. Reproduced from internet link (https://attic.gsfc.nasa.gov/huygensgcms/MS_Analyzer_1.htm)

21.2 Secondary Ion Emission Phenomena: Various Models

Simplistically, ionization potential (I) and electron affinity (A) are two important parameters for positive and negative ionizations, respectively. For a metal target, the domineering parameter is "work function" (ϕ). The quantity (I-ϕ) is the minimum energy required to positively ionize a free atom by transferring an electron from the atom to the surface. Similarly, (ϕ-A) is the minimum energy required by a free atom to form a negative ion by capturing an electron from the surface. Hence, the probability for the formation of a secondary ion, either positive or negative, has a functional relationship with these three quantities; I, A and ϕ. The surface work function ϕ is affected by the modification in the surface chemistry causing thereby significant changes in the secondary ionization probabilities. Although a universal ion emission model is still non-existent, a simple picture can be assumed to be valid for the sputtered ion-flux (n^+ and n^-) that scales approximately as:

$$\begin{aligned} n^+ &= B\exp(\phi - I)/K \\ n^- &= B'\exp(A - \phi)/K' \end{aligned}$$

where B, B′, K and K′ are suitable constants. Andersen and Hinthorne (1973) were the first to make a convincing demonstration of such scaling. They applied a thermodynamic approach for the quantitative interpretation of the intensities of the positive ion signals for various species sputtered from a single matrix. The linearity of the plots with the ionization potentials of the sputtered species was found to be restricted to elements with ionization potentials in the range 5–10 eV. From the similarity of the data to those which could be theoretically obtained from a high-temperature plasma in thermal equilibrium with relative ion populations governed by an equation of the form,

$$\frac{n^+}{n^0} \propto \exp\left(\frac{-I}{kT}\right) \tag{21.1}$$

the sputtering site is proposed to resemble a plasma in the state of "**local thermodynamic equilibrium (LTE)**". Ions, electrons and neutral atoms in the plasma exist in a state of thermal equilibrium with energy distributions following Boltzmann's velocity distributions. Here, k is the Boltzmann constant and T is the plasma temperature. Expression (21.1) essentially gives the secondary ion emission cross-section for positive ions considering the applicability of the LTE model. Andersen and Hinthorne (1973) applied the concept of LTE using the Saha-Langmuir equation for thermal ionization and included the electron density term, n_e, for the calculation of the dissociation constant of the form

$$\frac{n^+ n_e}{n^0} = \frac{u^+(T)}{u^0(T)} 2 \frac{(2\pi mkT)^{3/2}}{h^3} e^{\frac{-(E-\Delta E)}{kT}} \tag{21.2}$$

where n^+, n^0 and n_e are the plasma densities of positive ions, neutral atoms and electrons, respectively. Here, u^+, u^o and 2 are the partition functions of the ions, atoms and electrons, respectively, in the plasma. I is "ionization potential" and (ΔI) is the depression in ionization potential due to plasma. T is equivalent to "temperature". The product kT has the dimension of energy reflecting the average energy of an atom in the collision cascade.

The expression for the positive ionization probability β^+ ($= n^+/n^0$), as derived from Eqs. (21.1) and (21.2), is thus

$$\beta^+ \propto n_e^{-1} \exp\left(\frac{-I}{kT}\right) \tag{21.3}$$

Equation (21.3) explains the increase of positive ionization probability for oxides, thus explaining the matrix effect. The increase in electronic work function for an oxygen-covered surface or for an oxide surface causes a decrease in the electron density n_e in the local plasma, thereby decreasing the probability of neutralization of the emitted positive ions. Similarly, the decrease in work function for a caesium-coated surface causes an increase of n_e giving rise to a decrease in the positive ion emission or an increase in the negative ion emission. Therefore, the LTE model explains the matrix effect quite effectively. However, since this model evolves essentially from a thermodynamic approach, it does not envisage a strong yield dependence on ion emission velocity. Furthermore, the model formidably fails to explain the varying nature of kinetic energy distributions of sputtered atomic-, molecular- and cluster-ion species, in contrast to the Boltzmann distributions.

Another approach is the "**bond-breaking concept**" (Slodzian 1983,) which explains the matrix effect in ionic solids. The concept was found to be applicable for oxides (Williams 1979), where the bonds have partial ionic characters. Considering the sputtering site to be disordered, the electronic nature of the sputtering site was discussed. If the ion is assumed to be initially in the ground state, the probability of ion emission will depend on whether the ion-state remains energetically favoured over the neutral state (Williams 1979). This depends on a critical distance at which the potential energy surfaces of the ion and the neutral cross each other. If this crossing occurs at the surface atom separation of less than 10 Å, there will be a mixing of the ion and neutral states. The charge-exchange will occur at the crossing of the covalent potential curve $M^0 + X^0$ and the ionic potential curve $M^+ + X^-$ at the critical distance (R_c) from the surface, leading to the conversion from covalent to ionic bonding between the metal and the oxygen atoms. The short-ranged covalent force makes the covalent potential energy curve independent of R in the region of crossing. The ionic potential energy curve is given by the Coulomb attraction between M^+ and X^- ions. From the covalent and ionic potential energy curves, it is observed that at infinity the ionic curve lies above the covalent curve by (I-A), where '**I**' denotes the ionization potential of the metal and '**A**' denotes the electron affinity of oxygen-surrounded vacancy. This energy difference is exactly balanced with the Coulombic potential at the crossing point and thus, the "crossing distance" R_c simply becomes equal to $(I\text{-}A)^{-1}$ in atomic units. The probability of transition from covalent to ionic is inversely proportional to the escape velocity (v) with which the system (quasi-molecule) moves through the crossing region, so that the probability of dissociation as ions is $P^{+} \propto \exp(-1/v)$. The transition matrix element H_{12} that essentially dictates the probability of electron-transfer depends on R_c and is expressed as $H_{12} \propto \exp(-R_c)$. The features which influence the crossing distance are (a) ground state energy of the associated species, (b) the difference between ionization potential and electron affinity and (c) the range of electrostatic interaction. Sometimes, gas-phase ion-pair interactions have been used to understand the effects. In SiO_2 system, the level crossing has been seen to occur in the gas phase of SiO_2. Effect of oxidation on ion yield can then be thought in terms of curve-crossing movement further from the surface (Vickerman and Walls 1989).

Ionization through the breaking of chemical bond is qualitatively comparable to ion-pair dissociation of polar molecules according to the Landau–Zener treatment (Zener 1932). Based on this concept, Yu and Mann (1986) have described enhanced emission of metal-positive ions from metal oxides. Sputtering of metal atom M through the breaking of oxide bond creates an oxygen-surrounded cation-vacancy X at the surface. This oxygen-surrounded vacancy, with electron-affinity A, strongly attracts an electron from the ejected particle M, causing enhancement of positive ion yields. Analogous arguments hold for the increased negative ion emission under the effect of Cs. For a few tens of eV, the positive ionization probability β^{+} approximately follows a power-law dependence on ion energy. The model, however, is restricted to positive ion emission, whereas its applicability to negative ions, molecular ions and multiply-charged ions is not established.

Electron tunnelling model (Yu et al. 1991; Weg and Rol 1965) is based on "survival probability of an escaping ion" describing the ionization probability for both positive and negative ions. Figure 21.3 schematically represents the electron tunnelling phenomenon (Yu et al. 1991). Conceptually, this model is framed on the basis of the crossing of the discrete electronic level of the sputtered ion with "continuum states" of the solid. The main feature of this continuum is the Fermi level that separates the occupied and unoccupied states. The Fermi level E_F (Fermi energy) decides the survival probability of the escaping ion above the surface (Chakraborty 2002). A variation in image potential and chemical force of the system causes a variation in ionization energy of the atom E_a with the separation z between the atom and the surface. At the crossing point ($E_a = E_F$), electrons from metal can tunnel out to occupy the atomic level of the ion through "resonant electron tunnelling", leading to the neutralization of the positive ion. If $E_a(z)$ varies rapidly with z and crosses E_F at the crossing distance z_c from the surface with a large slope, the survival probability (P) of an ion has an exponential dependence on both $\Delta(z_c)$ and $v_\perp$:

$$P \cong \exp\left[-\frac{2\Delta(z_c)}{\hbar\gamma\, v_\perp}\right] \tag{21.4}$$

where z_c is the distance of the crossing point from the surface and $\Delta(z_c)$ is atomic level-width at the crossing point. $v_\perp$ is the normal component of ion emission velocity. The functional dependence of P on ion emission velocity v will be according to

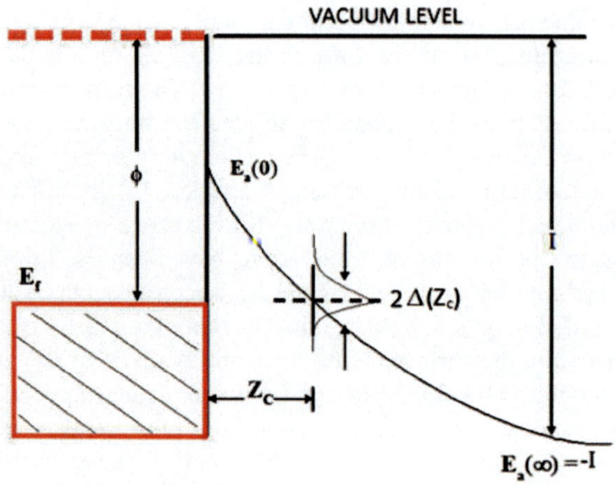

Fig. 21.3 Schematic representation of an ion sputtered from a metal surface. The work function (φ) is the energy difference between the Fermi level ε_F and the vacuum level. The ionization energy of the emitted ion (ε_a) is related to its separation (z) from the surface. For a free atom, at a distance far from the surface, ε_a is the ionization potential **I** of the element. At the crossing point (z_c) where $\varepsilon_a = \varepsilon_F$, the width $2\Delta(z)$ decides the electron tunneling probability. The energy width $2\Delta(z)$ is obvious according to uncertainty principle. Redrawn from a book published by Springer (Yu et al. 1991)

the variation of E_a and $\Delta(z)$ with z. For a small change of surface work function ϕ, the positive and negative ionization probabilities are respectively expressed (Chakraborty 2002) as:

$$\beta^{+} \propto \exp\left[-\frac{I-\phi}{v_{\perp}}\right] \tag{21.5}$$

$$\beta^{-} \propto \exp\left[-\frac{\phi-A}{v_{\perp}}\right] \tag{21.6}$$

Equations (21.5) and (21.6) clearly exhibit the importance of Fermi level in the ionization probabilities. Thus, the electron tunnelling model has convincingly described the role of surface work function and ion emission velocity on ionization efficiency for atoms sputtered from metallic surfaces.

21.3 Quantification

21.3.1 Matrix Effect in SIMS: Theoretical Concepts

Ideally, the number of sputtered ions of a species, irrespective of bombarding conditions, should be linearly related to the fractional concentration of that species in the sample and should be independent of the sample composition. Regardless of the secondary ion formation mechanism, quantification through depth profiling is achieved by converting a secondary ion current I(t) of a given species measured as function of time (t) into the concentration of the same species C(z) measured as function of depth (z). Considering sputtering conditions and erosion rate throughout the measurement to remain unchanged, depth calibration becomes straightforward as the total time of erosion can be directly converted to the eroded depth of the sample. However, in reality, the linearity of time-to-depth conversion appears inoperative, as the ion-induced erosion introduces a spread in the actual depth by rearrangement of target atoms and subsequent changes in surface topography. Considering an equilibrium state in surface composition, secondary ion current of a species (x) in a target matrix (m) can be related to the fractional concentration of the species (x) in the form of the following expression (Chakraborty 2002):

$$I_x^{\pm} = I_p Y_{tot}\, \beta_{x,\,m}^{\pm} f(\Delta E)\, f(\Delta\Omega)\, T_{x,m}\eta_{x,m}C_x \tag{21.7}$$

where $I_x^{\pm}$ (positive or negative) is secondary ion current of isotope-corrected species (i.e. $I_x^{\pm} = I_{xi}^{\pm}/x_i$, $I_{xi}^{\pm}$ is the measured ion current of isotope i of element x, and x_i is its isotopic abundance). I_p is the primary ion current (= $j_p.S$, where j_p is target current density and S is the bombarded area). I_x and I_p are expressed in units of ions/second. The combined subscript (x, m) relative to a certain parameter in Eq. (21.7) indicates a possible dependence of those parameters on the nature of both sputtered entity (x) and target matrix (m). Generally, sputtering from a multi-component target is influenced by preferential sputtering and surface segregation, creating an altered layer whose composition is different from that of the bulk. However, despite the differences in yields of individual species, partial sputtering yield of species (x) is ignored in Eq. (21.7); as in the steady-state (sputter equilibrium), the surface composition is identical to that of the bulk material. This is the basic concept underlying all quantitative analytical techniques that monitor the sputtered flux.

Y_{tot} is global sputtering yield of the target matrix (total number of atoms of all species sputtered from the matrix per incident ion); an estimation of which can be made following any of the standard techniques of sputter yield measurements.

$\beta^{\pm}_{x,\,m}$ is positive or negative "ionization probability" of species (x) sputtered from the matrix (m). The parameter $\beta^{\pm}_{x,\,m}$ depends strongly on the nature of the sputtered entity and local surface chemistry of the sample, thereby causing a great difficulty in SIMS quantification. The product ($Y_{tot}.\beta^{\pm}_{x,m}$) is the ion yield of the species (x) sputtered from the target matrix (m).

T is transmission of instrument (mass spectrometer); f(ΔE), f(Ω) are fractions of ionized particles entering into energy- and solid-angle acceptance-windows of the mass spectrometer; η is detector efficiency and C_x is fractional concentration of element x (atomic density of x atoms/atomic density of the matrix, i.e. n_x/n_m).

The characteristics of SIMS concerns the dramatic variation of ion yields from element to element. Moreover, the secondary ion yield for a certain element present in a matrix depends strongly on the presence of another element in the same matrix. Dependence of ionization probability on local surface chemistry ("**matrix effect**") is established by the fact that an electro-negative species (like oxygen) present on a surface enhances the secondary emission of positive ions of the matrix element significantly (Yu 1987). The enhancement is larger, the stronger is the electro-negativity (Canteri et al. 1991) of the species with respect to the emitted element. For example, fluorine, compared to oxygen, provides a larger positive ion yield enhancement. An equivalent effect occurs for negative ion emission from a surface when exposed to electropositive element, like Cs. A similar enhancement in the positive- or negative-ion emission occurs when the surface is subjected to oxygen or caesium ion bombardment, respectively. The fact that ionization efficiency can abruptly change from matrix to matrix has been evidenced through an estimate of the matrix effect in a two-element Zr-containing alloy by a TOF–SIMS analysis. Zr ionization efficiency was found to strongly vary in Al, Si and Cu matrices (Priebe et al. 2020); higher by more than four times in Si matrix than in Cu matrix and over two times higher when compared to the results obtained in the Al matrix.

The matrix effects, however, constitute serious drawbacks to the analytical capabilities of SIMS method, because they are, in general, not very well characterized and predictable. Whilst determining the interface positions accurately by SIMS depth profiles of organic materials using negative secondary ions, the extent of matrix effect was found to strongly influence the interface positions (Havelund et al. 2018). The shape of the intensity profiles obtained at FMOC/Irganox 1010 and Irganox 1010/FMOC interfaces was seen to be strongly influenced by matrix effects that slightly got reduced by ion-induced surface roughening of long wavelengths, equivalent to a root-mean-square value of 2.5 nm (Havelund et al. 2018). Whilst doing SIMS depth profiling of a delta layer of an organic material, an enhancement or suppression of ion yield due to matrix effect was found to vary widely from element to element and was not related to the energy of the secondary ions (Shea et al. 2016). From the above discussions, it is obvious that 'matrix effect' is detrimental to actual quantification and needs to be appropriately compensated.

21.3.1.1 Compensation of Matrix Effect: Relative Sensitivity Factor

The fractional concentration C_x of the species x (Eq. 21.7) is defined as the ratio n_x/n_m, where n_x is the atomic density for the species (x) in the matrix (m) and n_m is the atomic density of the matrix element itself. Obviously, $C_x = 1$ for a pure single-component matrix. As discussed above, the exact estimation of C_x using Eq. 21.7 is difficult because of the 'matrix effect'. Therefore, the compensation of 'matrix effect' is necessary. Considering only the case of positive ions, Eq. (21.7) can be written as.

$$I_x^+ = \rho_x \cdot C_x \tag{21.8}$$

where $\rho_x \left(= I_p Y_{tot} \beta_x^+ f, \text{with } f = f(\Delta E)\, f(\Delta\Omega)\, T_{x,m} \eta_{x,m}\right)$ is called the "absolute sensitivity factor" (ASF) for the species x. Equation (21.8) gives an estimate of the fractional concentration C_x (or atomic concentration n_x) of the species x in a given matrix m from secondary ion current, provided ρ_x is known.

Whilst undertaking the depth profile of a certain species (x), Eq. (21.8) can be expressed as

$$I_x(t) = \rho' n_x(t) \tag{21.9}$$

where $n_x(t)$ is the atomic concentration of x in the sample at a depth z corresponding to a sputtering time t and ρ' is the 'calibration constant'.

In Eq. (21.7), $f(\Delta E)$ is the fraction of secondary ions with energies in the range E and $E+\Delta E$ entering into the mass spectrometer, $f(\Delta\Omega)$ is the fraction of secondary ions accepted by the mass spectrometer in a solid angle $\Delta\Omega$), $T_{x,m}$ is the transmission factor of the mass spectrometer for ions of mass m of a species x, and $\eta_{x,m}$ is the detection efficiency of the detector or ion-collection optics for ions of mass m of a species x. The instrumental factor f can be taken as constant for a particular mass spectrometer and its associated instrumentation.

The 'absolute sensitivity factor' contains the term $\beta^{\pm}_{x,m}$, an exact evaluation of which is necessary for quantification. In other words, the matrix effect needs to be compensated. The lowest detectable concentration in a given matrix that can be measured by SIMS is called the 'detection limit'. In order to eliminate the fluctuations in I_x^+, $\beta^{\pm}$, Y, f, etc., and to correct for the matrix effect, the secondary ion current is normalized with a reference or matrix current I_m^+ (Chakraborty 2002). From Eq. (8), the fractional concentration $C_x = I_x^+/\rho_x$ or

$$C_x = \frac{I_x^+}{I_p Y \beta_x^+ f} \tag{21.10}$$

Similarly, for the matrix m

$$C_m = \frac{I_m^+}{I_p Y \beta_m^+ f} \tag{21.11}$$

In order to eliminate the fluctuations in I_x^+, $\beta^{\pm}$, Y, f, etc., and to correct for the matrix effect, the secondary ion current is normalized with a reference or matrix current I_m^+ (Chakraborty 2002). Therefore, from Eqs. (21.9) and (21.10),

$$(\mathrm{C_x})_{\min}/\mathrm{C_m} = \left(\frac{(I_x)^+_{\min}}{I^+_m}\right) / \left(\frac{\beta^+_x}{\beta^+_m}\right) \tag{21.12}$$

or

$$(\mathrm{C_x})_{\min} = \left(\frac{(I_x)^+_{\min}}{I^+_m}\right) / \left(\frac{\beta^+_x}{\beta^+_m}\right) (\text{since } \mathrm{C_m} = 1) \tag{21.13}$$

Since I_m or I_{matrix} is related to the material consumption and expressed as $[k \cdot (dz/dt) \cdot S]_{matrix}$, the detection-limit or the minimum detectable fractional concentration $(C_x)_{min}$, as obtained from Eq. (21.12), can be expressed as

$$(C_x)_{min} = \left(\frac{(I^+_x)_{min}}{\left(k.\frac{dz}{dt}S\right)_{matrix}}\right) / \left(\frac{\beta^+_x}{\beta^+_m}\right) \tag{21.14}$$

where (dz/dt) is the erosion-rate or layer-thickness sputtered per second, S is the target area hit by the primary beam and k is a constant depending on the sample and the ion-optics of the mass spectrometer. Therefore, atomic density n_x (atoms/cc) of a certain species (x) can be expressed as

$$n_x = n_m \frac{I^+_x}{I^+_m} / \frac{\beta^+_x}{\beta^+_m} = (RSF)_x \frac{I^+_x}{I^+_m} \tag{21.15}$$

where $(RSF)_x = n_m \frac{\beta^+_m}{\beta^+_x}$, defined as "**Relative Sensitivity Factor**" for a given species (x) in a certain matrix (m). The advantage of RSF over ASF is that the RSF actually takes care of the instantaneous fluctuations in the parameters $I_x{}^+$, $\beta^\pm$, Y, f. As obvious from Eq. (21.14), the lowest detectable concentration is inversely proportional to the erosion-rate or material consumed in SIMS depth profiling, thus signifying dynamic SIMS to be of higher detection sensitivity compared to static SIMS. Equation (21.15) gives an estimation of the atomic concentration of a certain species in a given matrix, provided the value of RSF of that species is known. The value of this factor can be indirectly extracted through a calibration process like 'implantation standard method', which involves accurately preparing a "local standard" by implanting a controlled quantity of a species (x) with a known dose F_x (atoms/cm^2) into another sample of the same matrix. This 'local standard' is then used to calibrate the SIMS signals. The value of $(RSF)_x$ in this method can be calculated by integrating Eq. (21.9) over total sampling time T.

$$\int_0^T I_x(t)dt = \rho' \left[\int_0^Z n_x(z)dz\right] \frac{1}{\dot{z}} \tag{21.16}$$

where, $\dot{z} = \frac{dz}{dt}$ is the sputter erosion rate. From Eq. (21.16),

$$\rho' = \frac{\int_0^T I_x(t)dt}{F_x}\frac{Z}{T} \tag{21.17}$$

where

$$\mathrm{F_x} = \left[\int_0^Z n_x(z)dz\right] \tag{21.18}$$

Here, F_x is the total dose of implanting ions. Z is crater-depth measured after the bombardment time, assuming a constant sputter rate. Equation (21.17) gives the value of ρ', thus measured from the SIMS profile (secondary ion current I_x versus time). As $n_x = I_x/\rho'$ (from Eq. (9)), the atomic density n_x of the species x at any depth of the sample can be measured using I_x and the estimated calibration constant ρ'.

Now, from the measured n_x and using Eq. (21.15), $(RSF)_x$ can be expressed as

$$(RSF)_x = \frac{F_x T I_m}{Z \int_0^T I_x(t)\, dt} \tag{21.19}$$

where I_m is the secondary-ion current of the matrix element (m). Z is the erosion-induced crater-depth and $\int_0^T I_x(t)dt$ is obviously a measure of the total area under the depth-integral curve of the secondary ion-current I_x over a span of the total time of depth-profiling.

There are several other approaches towards quantification without standards. One such is "infinite velocity approach" (Wittmaack 1999), which uses the plots of 'secondary ion intensity' as a function of 'inverse of ion velocity', calculated directly from the energy distributions of secondary ions. The basic idea is that the survival probability of a secondary ion increases exponentially with decreasing 'inverse ion velocity'. Therefore, all sputtered ions in the limit of infinite velocity remain fully ionized. That means, secondary ions with large emission velocities are not affected by any change in local surface chemistry. Or, in other words, the matrix effect is fully removed. However, the difficulty is that the secondary ions have extremely low intensities at the high energy tails of the energy distributions. The kinetic energy distributions of secondary ions, mostly peaked at few eVs, have quite long ranges extending up to several kilo-electron volts. The ion-intensities in such keV regions drop down to the background level and are almost undetectable. Another interesting study was on secondary ion yields and ionization probabilities with the use of slow (few keV/u; u stands for atomic mass unit), highly charged primary ions (such as Xe^{44+} and Au^{69+}) (Schenkel 2002). It was observed that ion emission exceeded by orders of magnitude the values obtained in conventional sputtering with singly-charged ions (Schenkel 2002). Furthermore, the ionization probabilities for secondary positive-ions from UO_2 and GaAs (100) as a function of projectile charge-states showed a strong decoupling of ion production from ionization probability. This established the "highly-charged ion-based SIMS" as an analytical technique free from the matrix effect (Schenkel 2002).

21.3.1.2 Compensation of Matrix Effect: MCs_n^+ (N = 1, 2, ...) Molecular Ions and Their Formation Mechanisms

Although quantification of a material is possible by using 'standards' having composition nearly identical to that of the material, it is difficult to quantify or interpret data across the interfaces of different materials, like metal/semiconductor, metal/polymer, metal/ceramic, etc. In view of this, the alkali element-based molecular MA^+-SIMS approach has been found remarkable for the reduction of the matrix effect (Gao et al. 1995). Caesium has the highest reactivity amongst all alkali elements, as its outermost electron is more loosely bound than the outermost electron of the other alkali metals. Therefore, Cs^+ ions are preferred in MA^+-SIMS approach and the monitoring of MCs^+ ions in SIMS analysis is known as ***MCs^+-SIMS*** technique (Chakraborty 2002). M is referred to "matrix element". Under steady-state condition, MCs^+ ion flux reflects the atomic concentration of element M in the matrix. For a variety of semiconductors, MCs^+ yield quadratically depends on "atomic polarizability" (α) of element M (ion yield $\propto \alpha^n$), where $n \approx 2$ (Gnaser and Oechsner 1994a), establishing the fact that polarizability of M governs the formation of these MCs^+ ions. In this case, the ion and ion-induced dipole moment of neutral species interact with each other. Modelling of the binding strength between M and Cs^+ in terms of polarizability of M was done on the basis of an interaction potential between M and Cs^+ (Gnaser 1999). Kinetic theory was applied to investigate the force of attachment between polarizable neutral atoms and slowly moving ions in a constant field of force (Gnaser 1999). In a study on the influence of isotopic mass difference on the yields of MCs^+ ions through energy distributions of secondary ions of two or more isotopes of a particular element, a pronounced light-isotope enrichment in the flux was observed with atomic ions M^+, whereas the isotopic effects for MCs^+ molecular ions were found to be negligible. The relative insensitivity of MCs^+ signals with respect to sample compositions was evidenced in a study of the oxygen concentration dependence of the intensities of various secondary ions (Ge^+, Cs^+, $GeCs^+$, $GeOCs^+$, etc.) sputtered from $Ge_{1-x}O_x$ samples (Gnaser and Oechsner 1994b). Whilst Ge^+ intensity exhibited a monotonic increase with oxygen concentration, Cs^+ and $GeCs^+$ were found to remain almost invariant. In an additional measurement, the energy distribution of $GeCs^+$ sputtered from a $Ge_{1-x}O_x$ sample was found to be identical to that from a pure Ge sample. The reasonable insensitivity of $GeCs^+$ ion intensity from $Ge_{1-x}O_x$ with varying oxygen concentration and the identical energy spectrum of $GeCs^+$ ions for pure Ge and $Ge_{1-x}O_x$ samples demonstrated the relative inertness of MCs^+ against specimen composition. The MCs^+-SIMS technique was found to be extremely useful whilst analysing a metal/semiconductor interface (Fig. 21.4). An absence of 'matrix effect' was established in MCs^+-SIMS analysis of a thin (0.3 μm) gold film on GaAs substrate (Gao et al. 1995). The appearance of a large hump of Au^+ signal across the interface is a characteristic of interface-peak due to the possible presence of native

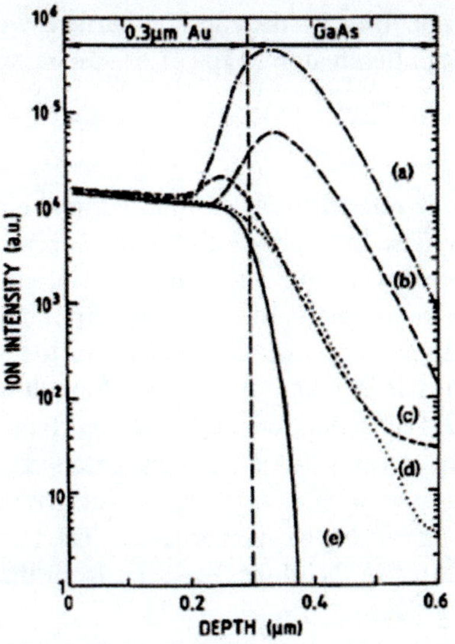

Fig. 21.4 **a** SIMS depth-profiles of gold layer (0.3 μm thick) on GaAs substrate. **a** Au^+ ions under O_2 bombardment, **b** Au^- ions with an off-set of 80 eV in the same bombardment condition, **c** Au^- under Cs bombardment, **d** Au^0 atoms (SNMS) under O_2 bombardment, **e** $AuCs^+$ ions under Cs bombardment. Reproduced with permission from Elsevier Publications (Gao et al. 1995)

oxygen or some other electronegative species at the interface. This interface peak that essentially appeared due to 'matrix effect' was found to be successively reduced under different conditions and completely disappeared whilst $AuCs^+$ signals were monitored (Fig. 21.4e).

MCs^+ ion formation mechanisms and the suitability of MCs^+-SIMS method in analytical applications have been systematically examined (Saha et al. 2008). Taking account of all experimental observations, MCs^+ ions have been predicted to form via recombination of sputtered neutral M atoms and resputtered Cs^+ ions. For such a formation mechanism to be valid, the intensity ratio $I(MCs^+)/I(Cs^+)$ is expected to vary linearly with sputter yield of M and inversely with the mean velocity of M (Gnaser and Oechsner 1994b)

$$I\left(MCs^+\right)/I\left(Cs^+\right) \propto Y_M / < v_M > \tag{21.20}$$

where $<v_M>$ is the mean velocity of sputtered M atoms accounting for the density of M atoms in the combination volume. Normalized MCs^+ intensity has been found to depend exponentially with the change in surface work function (Gnaser 1999). A close agreement between the variation of $AgCs^+$ and Cs^+ intensities with the variation in surface work function indicated a direct correlation between Cs^+ and MCs^+ intensities.

As the ionization probability P^+ depends on surface work function as well as ionization potential I of sputtered atom, Eq. (21.5) can be expressed as

$$P^+ \propto \exp[-(I - \phi)/\varepsilon_0] \tag{21.21}$$

where, the parameter ε_0 is related to the normal component of the velocity of the emitted secondary ions. The normalized Cs^+ counts versus "surface work function changes" ($\Delta\phi$) agreed well with the theoretical prediction for $\varepsilon_0 = 0.29$ eV (Wittmaack 2013). The leading parts of energy distributions of secondary ions are connected with the instantaneous surface chemistry of the sample, and the trailing parts reflect the kinematics of secondary ions. Therefore, the changes in the slopes of the leading edges of energy distributions reflect the surface work function changes that essentially arise due to the variation in surface chemistry (Saha et al. 2008). Figure 21.5 represents a scheme of estimating the surface work-functions from the Cs^+ energy distributions at varying oxygen pressure. The leading parts of the energy distributions have variations in the slopes, as seen from the tangents drawn at the leading edges (Saha et al. 2008).

Under the joint influence of caesium and oxygen on a silver surface, $AgCs^+$ intensity (normalized) was found to be directly proportional to Cs^+ intensity (normalized), and in good agreement with theoretical predictions for all oxygen exposures in the sputter region (Wittmaack 2013). This observation perceptibly authenticated the concept that MCs^+ ions form via the recombination of sputtered neutral M^0 atoms with resputtered Cs^+ ions. The electric field of Cs^+ ion in close proximity of the M-atom polarizes the neutral M^0 atom and creates a dipole moment which subsequently interacts with the Cs^+ ion. The intensity of MCs^+ molecular ions is expressed as

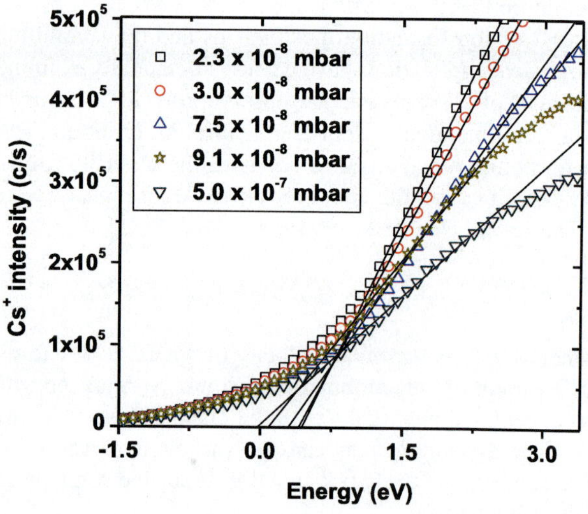

Fig. 21.5 A scheme for surface work functions as estimated from Cs^+ energy distributions. The tangents drawn at the leading edges of ion-energy distributions meet at different points on the energy-axis. The relative changes in the intercepts on the energy-axis provide the estimates of relative changes in surface work function. Reproduced with permission from Elsevier Publications (Saha et al. 2008)

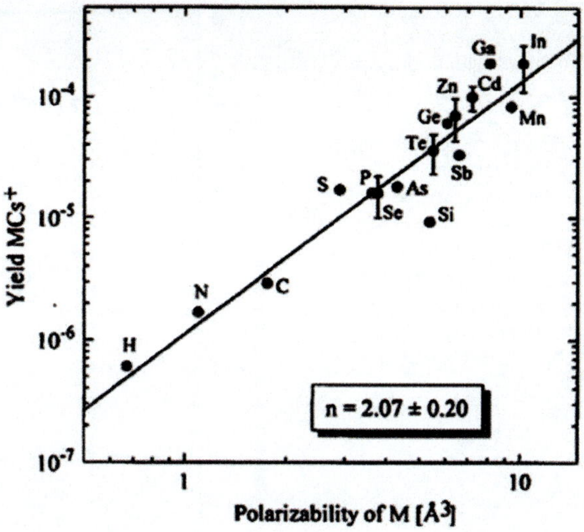

Fig. 21.6 Yields of MCs^+ molecular ions emitted from various elements as a function of atomic polarizability of M. The solid line is a linear least-squares fit with $n = 2.07 \pm 0.20$. Reproduced with permission from Elsevier Publications (Gnaser and Oechsner 1994a)

$$I_{(MCs+)} = \beta_{Cs}(\gamma_{MCs})\, k_M\left(C_{Cs}C_M/C^2_{total}\right)(Y_{total})^2 I_p \tag{21.22}$$

where, β_{Cs} is the ionization cross-section of Cs, γ_{MCs} denotes the recombination coefficient of M^0 with Cs^+; C_{Cs} and C_M are fractional concentrations of Cs and M, respectively; C_{total} is the total concentration; Y_{total} is the global sputtering yield; I_p is primary ion current; k_M is a constant (≈ 1). The recombination coefficient for MCs^+ is found to be greater for elements with higher electro-positivity. The fact that the emission intensity of MCs^+ molecular ions depends linearly with the atomic polarizability of the element M has been established from the measurements of MCs^+ ion yields from silicon-containing various low-concentration elements (Gnaser and Oechsner 1994a). The results showed that the ratio $I(MCs^+)/I(SiCs^+)$ increased monotonically with the atomic polarizability of M. Figure 21.6 shows the yields of MCs^+ ions sputtered from various elements as a function of atomic polarizability.

Although MCs^+-SIMS approach explains the nonappearance of matrix effects, the experimental foundation is still obscure. In an application of MCs^+-SIMS approach under Cs^+ primary ion bombardment to in-depth quantitative compositional analysis of AlGaAs/GaAs multilayer structures, the linear behaviour of the MCs^+ ion yield evidenced the fact that the analysis was independent of the matrix effect. The compositions were determined with an absolute accuracy better than 2% and the depth resolution was smaller than 100 Å (Gao 1988). Most notably, the ion signals for the major elements exactly matched the actual atomic densities of those elements in the layers.

Although MCs^+-SIMS has potential in quantification of materials without the need for calibration standards, it suffers from low yields of MCs^+ ions. In that case, working with MCs_2^+ molecular ions can offer much higher detection sensitivity, as MCs_2^+ ion yields have been found to be much stronger. This happens when the

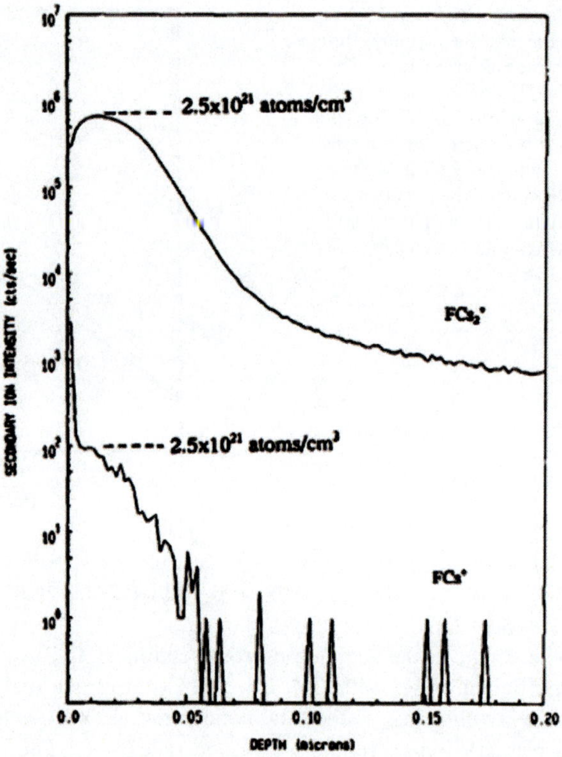

Fig. 21.7 SIMS depth profiles of Si a wafer, implanted with fluorine at fluence of 8×10^{15} atoms/cm^2, monitored with FCs^+ and $FCs_2{}^+$ molecular ions. The implantation profile-peak corresponds to a dopant concentration of 2.5 $\times 10^{21}$ atoms/cm^3, which is almost close to the expected value. Reproduced with permission from Elsevier Publications (Gao et al. 1995)

elements (M) have higher electronegativity with respect to caesium. In MCs^+-SIMS analysis of a fluorine-implanted Si-wafer, $FCs_2{}^+$ ion yield was seen to be significantly higher compared to FCs^+ (Fig. 21.7) (Gao et al. 1995).

Studies on kinetic energy distributions of $MCs_2{}^+$ ions and their constituents are extremely important in the understanding of their formation mechanisms (Sarkar et al. 2004). From the nature of the dependence of energy distributions of $MCs_2{}^+$ molecules on the surface binding energies of the constituents, it was evident that the sputtered neutral M atoms have a strong role on $MCs_2{}^+$ formation (Mootz et al. 1996). The formation mechanisms of these molecular ions depend on the nature of the element M. Two possible mechanisms (Mootz et al. 1996) are

$$M^0 + Cs_2^+ \rightarrow MCs_2^+ \tag{21.23}$$

$$MCs^0 + Cs^+ \rightarrow MCs_2^+ \tag{21.24}$$

While the first mechanism seemed to dominate for electropositive elements M, the second one was important for electronegative elements, thus exhibiting the enhanced yield (Marie et al. 1995) of $MCs_2{}^+$. The formation of MCs^0 molecule

(Eq. 21.24) results from the neutralization of an MCs^+ ion by electron capture (Saha and Chakraborty 2007). According to process (Eq. 21.24), the intensity of $MCs_2{}^+$ molecular ions can be expressed (Saho et al. 2008) as

$$I_{MCs_2^+} \propto Yc_M Y^2 (c_{Cs})^2 P^+ f_{MCs^+} f_{MCs_2^+} \tag{21.25}$$

where, Y is global sputter yield; C_{Cs} is Cs-concentration in the matrix; C_M is concentration of M; P^+ is ionization probability of Cs^+; f_{MCs+} and f_{MCs2+} are formation probabilities of MCs^+ and $MCs_2{}^+$ molecular ions. For a certain element M, c_M and the formation probabilities for MCs^+ and $MCs_2{}^+$ molecular ions can be considered as constant (Saha and Chakraborty 2007). Therefore, Eq. (21.25) can be rewritten as $I_{MCs2}{}^+ \propto P^+$. Thus, intensity of $AgCs_2{}^+$ molecular ions is expected to vary linearly with ionization probability of caesium. This has been evidenced in the exponential dependence of normalized $AgCs_2{}^+$ intensity with the change in surface work function ($\Delta\phi$) (Saha et al. 2008). Those experimental data substantiated the conjecture that $MCs_2{}^+$ molecular ions form via recombination of MCs^0 molecules with Cs^+ ions in the near-surface region (Eq. 21.24). The same argument can be equally applicable for the intensity of MCs^+ ions which varies exponentially with the change in surface work function ($\Delta\phi$) (Saha et al. 2008). The formation probability for $MCs_n{}^+$ (n = 1, 2, …..) is proportional to "spatial and temporal correlation factor" of $MCs_n{}^+$ (n = 1, 2, ….). The constant of proportionality is related to the square of atomic polarizability of M, expressed as (Marie et al. 1995),

$$f_{MCs_n^+} = \gamma_{M^0}^{pol} \Gamma_{MCs_n^+} \tag{21.26}$$

Figure 21.8 shows the special and temporal correlation factors for MCs^+ and $MCs_2{}^+$ molecular ions for different oxygen environments. As presented in the figure, the formation probability for $MCs_2{}^+$ is higher by two orders of magnitude compared to MCs^+ and both remain unchanged under varying oxygen exposure. This observation has also authenticated the higher emission cross-section for $MCs_2{}^+$ molecular ions compared to that of MCs^+ ions.

21.3.2 Composition Analysis of Multilayers, Superlattices and Quantum-Wells

During the analysis of a multilayer structure, special care is taken to obtain the best depth resolution for which the selection of parameters such as high-mass primary ions, low-impact energy and high angle of incidence is necessary so as to minimize the erosion rate. The quantification of matrix constituents in a multilayer structure (e.g. measurement of the value of x in $Al_xGa_{1-x}As$ layers in a $GaAs/Al_xGa_{1-x}As$ multilayer) is complex, as sputter-ion yields vary not only from one matrix to the next but also in an unknown manner in the interfacial regions. It has been shown that for

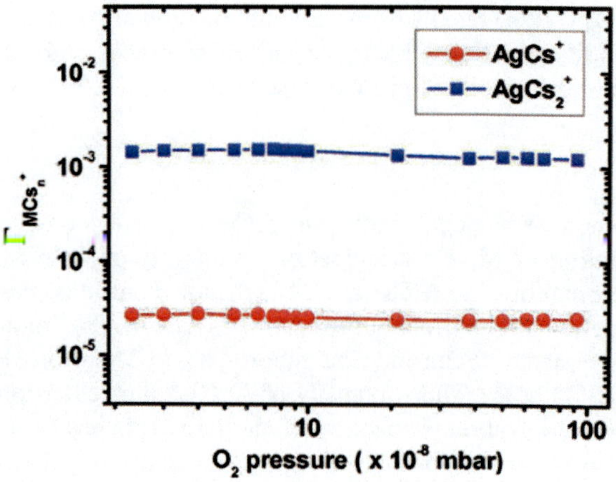

Fig. 21.8 Formation probabilities (spatial and temporal correlation factor) of MCs^+ and MCs_2^+ molecular ions as a function of oxygen pressure in the target chamber. Unpublished data from author

the $Al_xGa_{1-x}As$ system the Al^+ and Ga^+ intensities vary linearly with the aluminium content (up to $x = 0.5$). Based on this observation, the approach is to prepare a series of standard $Al_xGa_{1-x}As$ samples of known compositions and to measure Al^+ and Ga^+ ion intensities from those standards by SIMS. A plot of Al^+/Ga^+ ratios versus (x/1-x) then produces a straight line from which any unknown value of x (Al concentration) and its depth distribution in an $Al_xGa_{1-x}As$ matrix can be determined. This method has many limitations and obviously cannot provide the true compositional profiles of the interfaces in a multilayer structure in general. Depth-resolution can be improved to a great extent by reducing the bombarding-ion energy (which reduces ion-induced collisional mixing) and choosing the correct bombardment conditions, as an additional surface roughening is caused by the oblique incidence of the primary beam (Chakraborty 2002), giving rise to profile broadening. A relatively smooth crater-bottom is seen to be collapsed into a pronounced ripple structure (Chakraborty 2002) at an oblique incidence of the primary beam.

Figure 21.9 represents the SIMS depth profile of an 80-period Si/Fe neutron mirror with layers of 1.8 nm thick using normally incident 1.5 keV O_2^+ primary ions (100 nA). As neutrons are electrically neutral, they have an exceptional ability to penetrate any material and are less reflected. The neutrons are scattered by nuclei and the refractive index of a material for neutrons depends on the number of nuclei and neutron scattering cross-section of the material. 'Scattering length' is the ability of a nucleus to scatter a neutron. Therefore, the neutron refractive index of any material is a function of scattering power, defined as "scattering length density" and the neutron wavelength. Total reflection occurs when neutrons pass from a medium with a higher refractive index to one with a lower refractive index. Since the neutron refractive indices of most of the materials are only slightly less than that of air or vacuum, total external reflection of neutrons normally takes place instead of total internal reflection. This property is very similar to the electromagnetic radiation that falls in the wavelength region of soft X-rays to extreme ultraviolet. Therefore, the design conditions of the Si/Fe multilayer structure are fulfilled in order to reach

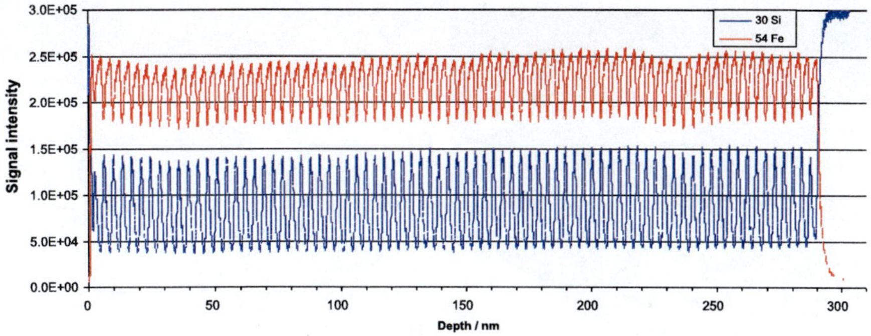

Fig. 21.9 SIMS depth-profile of an 80-period Si/Fe neutron mirror, with layers of 1.8 nm thick, analysed using normally-incident 1.5 keV O_2^+ primary ions (100 nA). Reproduced from internet link (https://www.hidenanalytical.com/applications/surface-analysis/contamination-with-silicone/)

the maximum reflectivity of neutrons at near-normal incidence, similar to that in the layered synthetic microstructures used for X-UV reflecting devices. The SIMS depth profile shows that layer thicknesses are highly consistent with depth and have almost no loss of depth resolution over the entire multilayer stack. Low-impact ion energy and appropriate bombardment conditions have reduced the ion-induced collisional mixing to a great extent thereby increasing the depth resolution substantially.

21.3.3 Metallic Multilayer

Large free energies existing at the interfaces in metallic multilayers are responsible for interfacial mixing and subsequent formation of interfacial alloys, provided those free energies are comparable to or larger than "formation enthalpy". In the case of immiscible metals, mixing is hindered by energy barriers caused by "positive mixing enthalpy". Whereas for miscible metals in multilayer systems, various metastable alloys can form because of negative enthalpies of formation, the values of which can be close to each other. Because of interfacial mixing that essentially arises out of interdiffusion, metastable alloys were found to form across Au/Cu interfaces of an Au/Cu multilayer structure grown by magnetron sputtering. In that case, high quench rate and hyperthermal atoms present in magnetron sputtering systems were effective in the formation of metastable alloy phases across the interfaces (Sarkar et al. 2005). Interfacial confinement in such a structure was the main driving force behind the formation of metastable $D0_{23}$ phase of Cu_3Au alloy across the interfaces (Sarkar et al. 2005). Figure 21.10 summarizes the observations of such a phenomenon. Figure 21.10a shows the high-resolution cross-sectional TEM image of a 10 bi-layered Cu/Au structure grown on glass substrate and Figure 21.10b shows the SIMS depth profile of an identical 2 bi-layered Cu/Au structure grown on Si substrate, indicating Au, Cu and Si signals prominently. The depth resolution

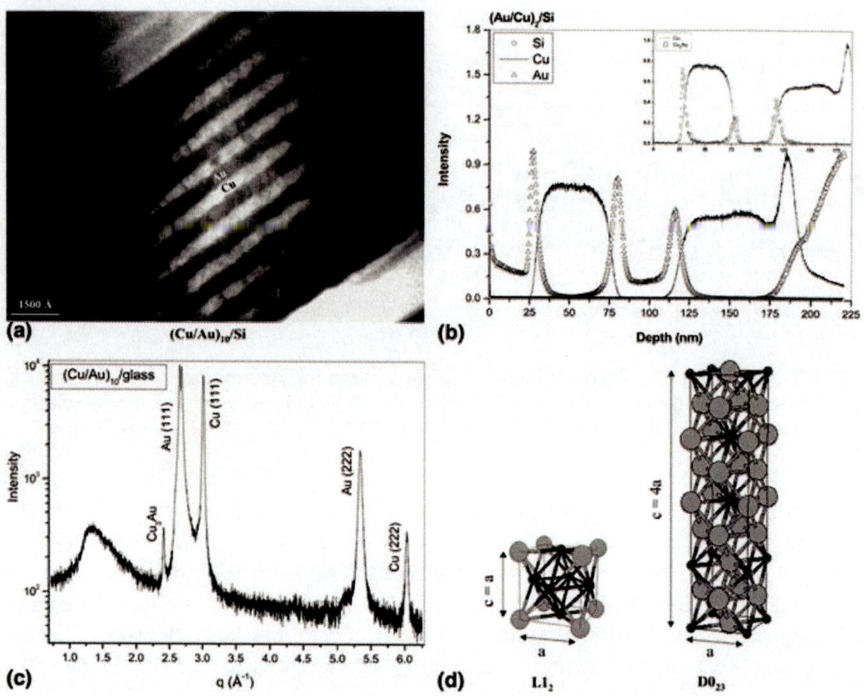

Fig. 21.10 **a** Cross-sectional TEM image of the $(Cu/Au)_{10}/Si$ multilayer, **b** SIMS depth profiles showing Au, Cu and Si signals in a $(Au/Cu)_2/Si$ film, (inset) SIMS depth profiles showing Cu and Cu_2Au alloy fragment signals in the same film, **c** XRD spectrum of the (Au/Cu)10/glass multilayer, and **d** crystal structures of *L1*2 and *D0*23 phases of Cu_3Au. Reproduced with permission from Springer Nature Publications (Sarkar et al. 2005)

of SIMS was good enough to distinguish the various alloy fragments like AuCu, Cu_2Au and Au_2Cu that are formed and spatially separated across each interface. One such alloy fragment like Cu_2Au at the interfaces is indicated in the figure (inset of Fig. 21.10b). Figure 21.10c presents the X-ray diffraction spectrum of (Au/Cu)10 /glass multilayer indicating a small sharp peak in the vicinity of Au (111) peak and that the new peak is a characteristic of the Cu_3Au alloy phase ($D0_{23}$). Figure 21.10d represents the crystal structures of the *L1*2 and *D0*23 phases of Cu3Au.

21.3.4 Semiconductor Quantum-Well

Interdiffusion across interfaces can cause interfacial broadening in a quantum-well structure. The phenomenon was evidenced in an $InP/In_{0.33}Ga_{0.67}As/InP/In_{0.33}Ga_{0.67}As/InP$ double-quantum well grown by metal–organic vapour phase epitaxy (MOVPE) (Sarkar et al. 2000). SIMS and XRD measurements together with

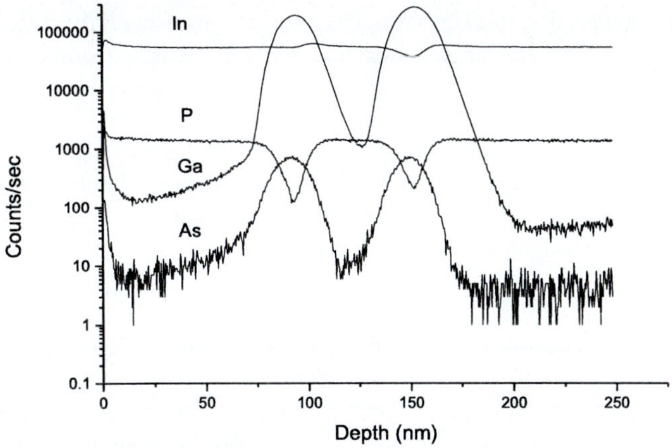

Fig. 21.11 SIMS depth profiles of In, P, Ga and As in an InP/$In_{0.33}Ga_{0.67}As$/InP/$In_{0.33}Ga_{0.67}As$/InP double quantum-well structure. Reproduced with permission from John Wiley and Sons Publications (Sarkar et al. 2000)

a simulation programme (Sarkar et al. 2000) evidenced the interdiffusion of phosphorus into the quantum-wells and the presence of a 10 nm thick intermixed region of In, As and P formed in between the cap-layer and the subsequent quantum well. A combination of these two techniques together with a simulation programme was an innovative approach towards the quantification of in-depth compositional variation in quantum wells. For the indium content present at its maximum in InP layers, the interdiffusion of In from QWs to InP layers was ruled out. Therefore, it was only phosphorus that inter-diffused from InP-layers into the QWs. The shallow valley of phosphorous signal in QW layers supported the phenomenon. Ga and As peaks with equal peak-widths appeared at the same position in the SIMS profile (Fig. 21.11), representing the characteristics of the double quantum-well structure.

A high-resolution X-ray diffraction experiment provided the stoichiometry of the composition in the quantum well layers through the measurement of absolute lattice parameters of the substrate and the phases grown in the deposited layers. A simulation programme, based on the 'kinematic theory of reflection' (Warren 1969), was developed to reproduce the experimental high-resolution diffraction pattern of the double quantum well in the Born approximation scheme (Sarkar et al. 2000). The interplanar spacings were obtained from the measured X-ray data. The scattering amplitude needed for simulation depends on the relative concentration of the elements present in the sample, and thus would change with depth throughout the stack due to the variation of in-depth elemental composition. From the study of high-resolution XRD around the (004) Bragg peak as a function of q_z (= $(4\pi/\lambda)\sin\theta$), interplanar spacings (d-values) in $In_{0.33}Ga_{0.67}As$ and InP layers were determined to be around 1.428 Å and 1.495 Å, respectively. Applying the Vegard's law (Adachi 1982) for a quaternary material like $A_{1-x}B_xC_yD_{1-y}$ constructed from four binary compounds like AC, AD, BC and BD, and using the model based on a linear interpolation scheme, a ‘quaternary material parameter’ was framed that was essentially regarded as the interplanar spacing, which varies almost linearly with the stoichiometry of the material composition. Applying the above law, the desired composition $In_{0.33}Ga_{0.67}As$ of the QWs cannot have the interplanar spacing of 1.428 Å in the (004) direction. On the contrary, assuming the proportions of In and Ga to remain unaltered and an interdiffusion of phosphorus into the QWs, a reconstructed composition, such as $In_{0.33}Ga_{0.67}As_{0.62}P_{0.38}$, was rather found to match well with the estimated interplanar spacing 1.428 Å (Sarkar et al. 2000).

21.3.4.1 Reconstruction of Depth Profiles: MRI Simulation

A general observation in SIMS depth profiling of semiconductor multilayer structures is an asymmetry in ‘leading’ and ‘trailing’ parts. Such an asymmetry in the slopes was observed in depth profile analysis of a delta-layer of AlAs on GaAs and was explained by **MRI** (**M**ixing, **R**oughness and **I**nformation-depth) simulation model (Hofmann 1993). Ignoring the effect of preferential sputtering, the slope of the leading part was determined by considering the quantities like “surface roughness” and “information depth”, whereas the slope of the trailing part was determined by considering the extent of “atomic mixing” (Hofmann 1993). The **MRI** model is a semi-empirical model to explain “depth resolution function” (DRF), which was developed using three fundamental parameters: atomic mixing (**w**), surface roughness ($\boldsymbol{\sigma}$) and information depth ($\boldsymbol{\lambda}$). “Atomic mixing” considers a full atomic redistribution creating a homogeneously mixed region of thickness w. “Surface roughness” is a Gaussian distribution with a standard deviation, σ. “Information depth” is essentially the effective depth of information determined by the mean escape depth (λ) of analytical information. For example, the escape depth in SIMS is the mean information depth of secondary ions and typically ~1–2 monolayer). Therefore, the MRI model is characterized by the “partial resolution functions” of these three parameters (w, σ, λ), such as f(w), f(σ) and f(λ), which act on each other establishing the complete DRF for depth-profile reconstruction. An appropriate choice of these three parameters enables the calculation of DRF, which depends mainly on a detailed understanding of the physicochemical processes during sputtering. By comparing the MRI modelling of depth profiles with experimental data, the calculated DRF enables the deconvolution of the measured profile. By a realistic choice of these three parameters (mixing (w) = 1.0 nm, surface roughness (σ) = 0.6 nm, Auger electron escape-depth (λ) = 0.4 nm) for 600 eV Ar^+ ion sputtering, the MRI model was found to perfectly match the measured AES depth profiles of a GaAs (8.8 nm)/AlAs (9.9 nm) superlattice structure (Hofmann 1994). Figure 21.12 represents the extracted Al-concentration (mole fraction) in a GaAs/$Al_{0.2}Ga_{0.8}As$/GaAs multiple quantum well (MQW) structure through SIMS depth profile using 1.5 keV Cs^+ primary ions (Otomo et al. 2006). As evident, the MRI-convoluted profile perfectly matches the actual SIMS profile. The reconstructed Al profile gives an accurate measure of the quantum-well layer thicknesses, with a difference between the sputtering rates of GaAs and $Al_{0.3}Ga_{0.7}As$ within ~5%.

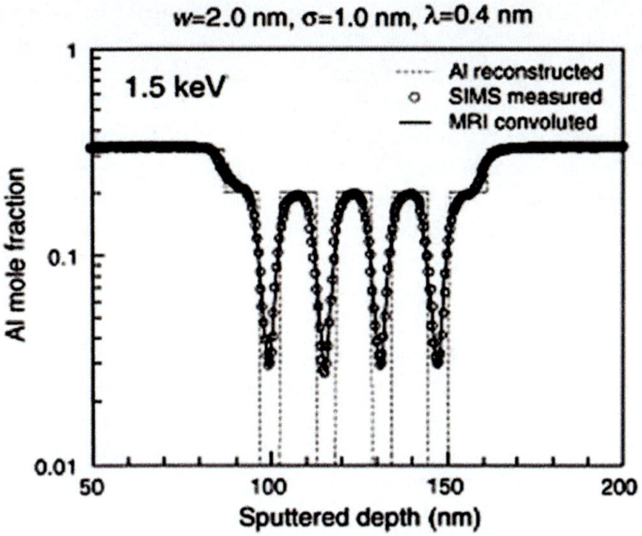

Fig. 21.12 Al-concentration as a function of depth in a GaAs/$Al_{0.2}Ga_{0.8}As$/GaAs multiple quantum well (MQW) structure, fitted with **MRI** (mixing, roughness, information) model. The simulated values: w (atomic mixing) = 2.0 nm, σ (surface roughness) = 1.0 nm, λ (information depth) = 0.4 nm. Reproduced with permission from Elsevier Publications (Otomo et al. 2006)

21.3.5 SiGe Alloy and Si/Ge Superlattice Structures: Composition Analysis Using MCs⁺-SIMS Approach

Silicon–germanium (SiGe) alloy is highly compatible in silicon-based semiconductor technology. Because of poor thermal conductivity compared to silicon and high electron/hole mobility, SiGe is potential in fabricating electronic devices, such as transistors, photodetectors, solar cells, thermoelectric devices and high-speed microelectronic devices. Since this alloy comprises of germanium having a band gap smaller than silicon, the alloy band gap is also smaller than that of silicon, irrespective of silicon-to-germanium content ratios. The narrow band gap of SiGe decreases the thickness of bipolar transistor base without any loss in current gain, thus improving the speed of bipolar transistors. Optimization of such technology is only possible by a precise control of Ge–Si interdiffusion and an accurate determination of Ge concentration in Si_xGe_{1-x} in its complete stoichiometry range. MCs^+-SIMS method was successfully employed for the estimation of Ge concentration up to 23.5% (Prudon et al. 1997) in SiGe alloys. Quantification in the higher range of Ge concentration in SiGe at low sputtering energies was also reported using MCs+-SIMS (Marseilhan et al. 2008a; Holliger et al. 2002; Gavelle et al. 2008). However, for SiGe layers with Ge contents higher than 50%, consistent results were not obtained in the analysis with the MCs^+-SIMS method. Irrespective of incident Cs^+ ion energies, MCs^+-SIMS methodology was found to be a full-proof procedure for exact quantification of Ge concentration in MBE-grown $Si_{1-x}Ge_x$ ($0 < x < 0.72$) alloys, in which the "matrix effect" was completely blocked for all Ge concentrations (Saha et al. 2012). The innovative methodology was thus found to be successful in quantitative composition analysis of Si/Ge multilayer structure (Saha et al. 2012). The lattice parameters extracted from alloy- peaks in X-ray diffraction were utilized to estimate the alloy compositions using Vegard's law. The results were found to be in good agreement with that obtained independently from energy-dispersive X-ray analysis (EDX).

MCs^+ intensity in MCS^+-SIMS analysis for a species M (Gnaser 1995)

$$I_{MCs^+} Y c_M Y c_{Cs} P^+ f_{MCs^+} \tag{21.27}$$

where Y, P^+ and f_{MCs^+} are total sputtering yield, ionization probability of caesium, and formation probability of MCs^+ molecular ions, respectively. c_M and c_{Cs} are the fractional surface concentrations of M (matrix) and caesium, respectively. Considering the formation probability f_{MCs^+} to be constant for both germanium and silicon, Eq. (21.27) can be rewritten as

$$\frac{I_{GeCs^+}}{I_{SiCs^+}} = K\frac{c_{Ge}}{c_{Si}} \tag{21.28}$$

where the constant K can be treated as "relative sensitivity factor" (RSF). Figure 21.13 shows the intensity ratios (I_{GeCs+}/I_{SiCs+}) obtained from MCs^+-SIMS as a function of concentration ratios (c_{Ge}/c_{Si}) obtained from EDX for various impact energies (Saha et al. 2012). For all primary ion energies, the linearity of the graphs, irrespective of germanium and silicon contents in SiGe alloy films, clearly reveals the absence of matrix effects for all possible $Si_{1-x}Ge_x$ compositions; whereas some studies reported the absence of matrix effect only for impact energies lower than 1 keV and for germanium content less than 50% (Gavelle et al. 2008). The complete compensation of matrix effect irrespective of impact ion energy was attributed to a low steady-state surface concentration of caesium due to larger primary impact angle (Saha et al. 2012), signifying the achievement of MCs^+-SIMS technique in quantitative materials analysis without the aid of calibration standards. The linearity of the graph I_{GeCs+}/I_{SiCs+} as a function of c_{Ge}/c_{Si} in the case of $MCs_2{}^+$-SIMS analysis provided a much better perception in respect of its slope, compared to that in the case of MCs^+-SIMS analysis.

In order to understand the motivation of synthesizing Si/Ge superlattice, some basic points are needed to be addressed. Silicon has an indirect bandgap undergoing indirect optical transitions. That means, electrons in conduction band and holes in valence band can recombine only with the help of phonons. Si is an inefficient light emitter, as the interaction probability between electrons and phonons in silicon is low. As "quantum confinement effects" are particularly effective in silicon, light emission from silicon is prominent at ambient condition especially when silicon is in the form of low-dimensional systems like porous silicon, silicon nanocrystals, silicon quantum dots (QD), superlattices, etc. If QD size < exciton Bohr radius, the excitons are squeezed to dominate quantum confinement effects, resulting in controllable

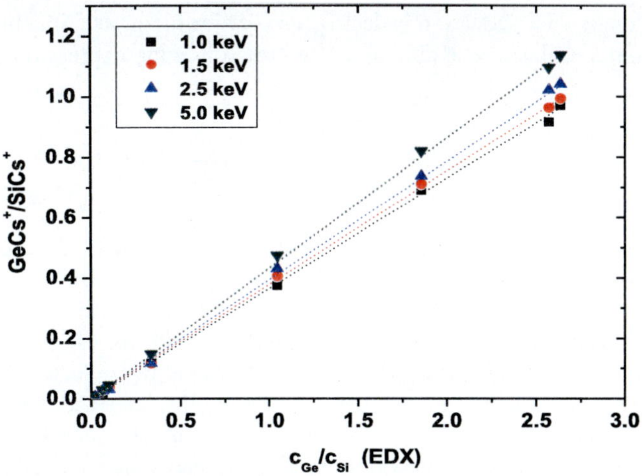

Fig. 21.13 Normalized $GeCs^+$ intensity as a function of germanium concentration ratio. Reproduced with permission from Springer Nature Publications (Saha et al. 2012)

electronic and optical properties. Two fundamental quantum effects occur when Si size is reduced; (1) bandgap increases and the conservation of momentum breaks down because the "localization of charge-carriers in space" causes the "delocalization of those charge-carriers in momentum space". (2) As the emission properties of silicon nanocrystals depend on their size, silicon behaves as "quasi-direct bandgap semiconductor". In that case, phonons are not needed to conserve momentum during recombination of electrons and holes.

An exciting approach is to create a superlattice structure of alternating numbers of extremely thin Si and Ge monolayers by epitaxial method, so that positive holes are confined in the flat Ge islands and electrons are located in the Si spacer layers. Molecular Beam Epitaxy (MBE) is an epitaxial method of thin film deposition of single crystals, in which the growth of a crystalline film is suitably oriented onto a substrate. Therefore, the epitaxial films are thin films with highly ordered atomic arrangements following that of the substrates. In a strained Si/Ge superlattice grown by MBE, a complex structure of minibands can occur due to overlapping of electron wave functions in real and k-space, yielding stronger luminescence with a faster transition. The strain in Si layers corresponds to biaxial tension and the strain in Ge layers corresponds to biaxial compression. Si band-edge dominates in the conduction band and Ge band-edge dominates in the valence band. Such a p-n heterostructure can be used as an infrared emitter at room temperature.

The potential of the MCs^+-SIMS was applied for direct compositional analysis of an MBE-grown Si/Ge superlattice structure, with a nominal thickness of ~10 nm for each Si/Ge bilayer (Sharma et al. 2011). Using Eq. (21.28), the RSFs for Ge were calculated from the slopes of the variation of I_{GeCs+}/I_{SiCs+} (MCs^+-SIMS approach) with c_{Si}/c_{Ge} (EDX) for different ion-impact energies. From the concentration profiles obtained through MCs^+-SIMS, the electron density profile ρ_e as a function of thickness (z) of the multilayer was obtained using the following expression (Saha et al. 2012),

$$\rho_e(z) = N_A \sum_{M=Ge,Si} \frac{Z_M c_M(z)}{A_M} \tag{21.29}$$

where, N_A—Avogadro number of constituent element M (Si or Ge); Z_M—atomic number of constituent element M (Si or Ge); $c_M(z)$ — "concentration versus depth profiles" for Ge and Si; A_M—mass number of constituent element M (Si or Ge). Figure 21.14 represents the electron density profiles (EDP) of the multilayer stack extracted from MCs^+-SIMS data (red circles) and from X-ray reflectivity (blue solid line) (Saha et al. 2012). As obvious from the figure, the interface positions of the multilayer are well matched as reflected by these two measurements. However, the discrepancy between the absolute values of electron densities obtained from X-ray reflectivity and MCs^+-SIMS was attributed to sputter-induced distortional effects across interfaces during SIMS analysis (Saha et al. 2012).

The linearity of intensity ratios (I_{GeCs+}/I_{SiCs+}) as a function of c_{Ge}/c_{Si} in the case of MCs_2^+-SIMS analysis provided a much better perception in respect of its slope, compared to that in the case of *MCs^+-SIMS* analysis. Figure 21.15 presents (I_{GeCs+}/I_{SiCs+}) as a function of (c_{Ge}/c_{Si}) for a set of standard Ge_xSi_{1-x} samples in the *MCs^+-SIMS* and *MCs_2^+-SIMS* analysis modes. As evident, there is a good linear correlation between the ratio of secondary ion intensities $I(GeCs_n^+)/I(SiCs_n^+)$ and the composition ratio [Ge]/[Si], as determined by X-ray diffraction in both the cases. Obviously, there is a perfect one-to-one match between intensity-ratios ($GeCs^+/SiCs^+$) and concentration-ratios (c_{Ge}/c_{Si}) in the case of MCs_2^+-SIMS analysis, thus establishing the superiority of MCs_2^+ molecules over MCs^+ molecules in precise quantification of SiGe alloys of any stoichiometry.

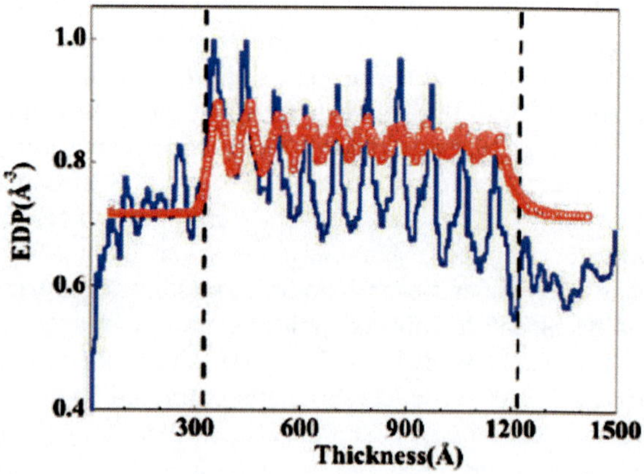

Fig. 21.14 Electron-density profiles of a Si/Ge superlattice, as obtained from X-ray reflectivity (blue solid line) and MCs^+-SIMS (red circles). Two vertical dashed lines indicate the depth positions for expected buffer and cap layers. Reproduced with permission from Springer Nature Publications (Saha et al. 2012)

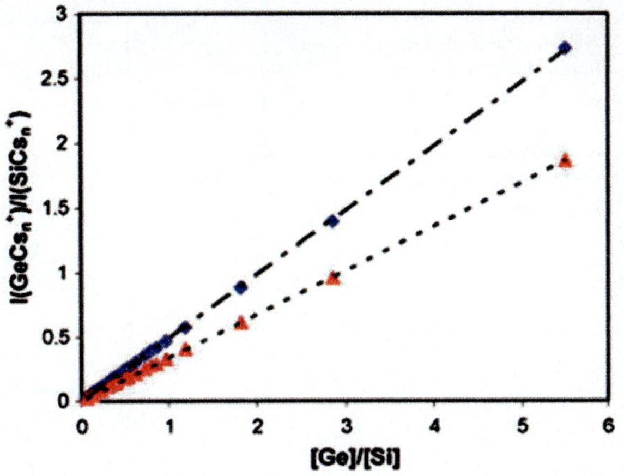

Fig. 21.15 Variation of the intensity ratio $I(GeCs_n^+)/I(SiCs_n^+)$ as a function of the composition ratio [Ge]/[Si] for MCs_2^+ and MCs^+ ions using CAMECA -5F SIMS. Reproduced with permission from Elsevier Publications (Marseilhan et al. 2008b)

21.4 Photoconductive ZnO and ZnS/ZnO Nanowalls: MCs^+-SIMS Analysis

ZnO nanostructures exhibit a range of beneficial optoelectronic properties including good transparency, high electron mobility, wide direct band gap, etc. In bare ZnO nanowalls, sputtering of Zn atoms through breaking of oxide bonds creates oxygen-surrounded vacancies at the surfaces. These oxygen-surrounded vacancies strongly attract the electrons from the ejected Zn^+ ions, enhancing the secondary positive ion yield. Thermal effect of microstructural modifications in self-assembled ZnO nanowalls grown on Al-substrates was reported (Bayan and Chakraborty 2014). X-ray diffraction evidenced the hexagonal wurtzite crystal structure of ZnO and Al_2O_3 layer formation at the interfaces of Al substrates and ZnO nanowalls. Figure 21.16 represents SIMS spectra of unannealed and vacuum-annealed ZnO nanowalls grown on Zn substrates, indicating no considerable changes in Zn^+ intensity under heat treatment (Bayan and Chakraborty 2014). The luminescence features of ZnO nanowalls were seen to depend on surface adsorbates like O_2^- and OH^-, the presence of which was established by SIMS analysis, which also established a large deficiency of lattice-oxygen in vacuum-annealed nanowalls (Marseilhan et al. 2008b). EDX and SIMS study of SnO-coated ZnO needle-like nanostructures (Bayan and Chakraborty 2014) revealed the formation of SnO-phase on ZnO surface. Luminescence response of heterostructured SnO/ZnO system revealed the quenching of defect-related emission through surface passivation of ZnO. The hetero-structured material with lower Sn:Zn molar ratio exhibited higher photoconductive response compared to bare ZnO. Enhancement in photocurrent was explained in the light of carrier separation and carrier multiplication occurring at SnO-ZnO heterojunctions (Bayan and Chakraborty 2014).

The conventional SIMS analysis of bare ZnO- and heterostructured ZnS/ZnO nanowalls indicated ZnS-phase present on ZnO surfaces. MCs^+-SIMS analysis

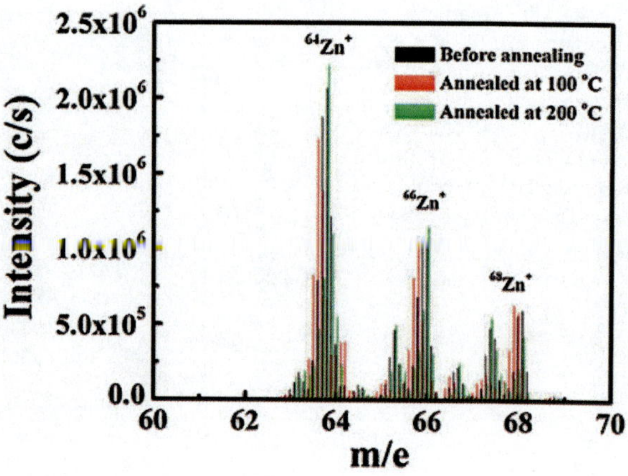

Fig. 21.16 SIMS spectra for Zn^+ ion emission from unannealed and vacuum-annealed ZnO nanowalls. Reproduced with permission from Elsevier Publications (Bayan and Chakraborty 2014)

suppressed the matrix effect and efficiently made a semi-quantitative estimation of Zn and O surface–atom concentrations in both systems (Marseilhan et al. 2008b). Local surface work function was extracted through MCs^+-SIMS and was found lower in the heterostructure supporting the presence of sulphur species of ZnS on top of ZnO nanowalls. Higher electron density on sulphide surface lowers the local surface work function leading to decrease in positive ionization probability, thereby reducing Zn^+ ion intensity in ZnS/ZnO nanowalls. The results authenticated that "Matrix Effect—corrected MCs^+-SIMS" technique provided a comprehensive picture of surface chemistry in ZnS nanoparticle-decorated ZnO nanowalls (ZnS/ZnO hetero-structured systems) (Bayan et al. 2014).

Luminescence responses of ZnS/ZnO heterostructures strongly depend on the degree of ZnS-phase over ZnO. Such heterostructures fabricated with smaller ZnS nanoparticles exhibited higher luminescence responses, explained in terms of a process of charge-carrier transfer from ZnS to ZnO (Bayan et al. 2014). Figure 21.17 represents the SEM images and EDX spectra of bare ZnO nanowalls and ZnS/ZnO hetero-structured nanowalls. The magnified SEM image in Fig. 21.17e clearly depicts the formation of ZnS nanostructures on the top of ZnO nanowalls. Figure 21.18 represents the depth distributions of $ZnCs^+$ to OCs^+ intensity ratios in the MCs^+-SIMS mode. The steady intensities of these molecular ions with depth confirmed the homogeneity of ZnO film. On the contrary, the intensity variation of SCs^+ indicated a momentary existence of ZnS in the heterostructure network. According to MCs^+ mode of analysis, $\frac{I_{ZnCs+}}{I_{OCs+}} \propto \frac{c_{Zn}}{c_O}$. Therefore, the higher intensity ratio I_{ZnCs+}/I_{OCs+} designates the higher concentration ratio c_{Zn}/c_O thus signifying the increase in Zn concentration compared to O due to the ZnS decoration on the top of the ZnO surface.

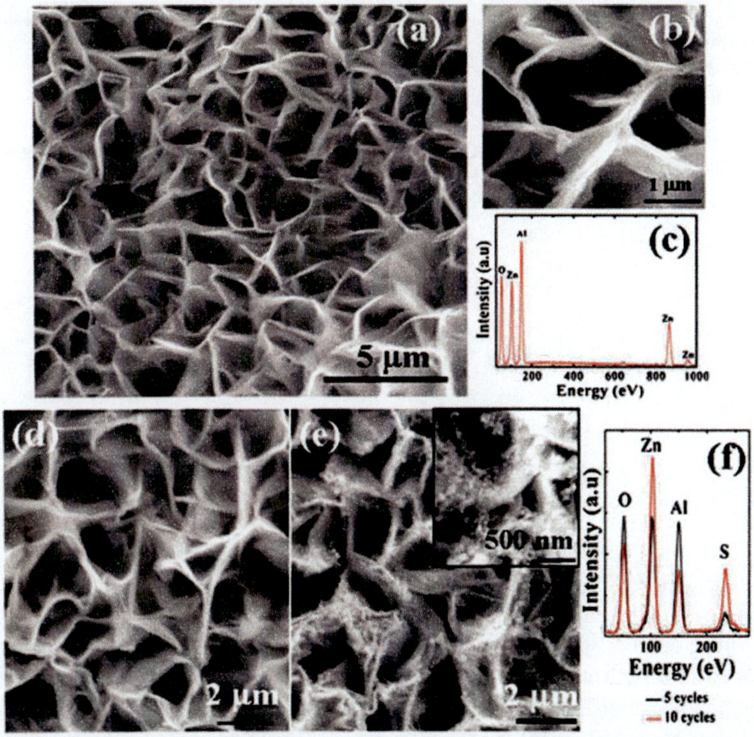

Fig. 21.17 **a** SEM image of the ZnO nanowall film, **b** magnified image of the nanowalls, and **c** EDX spectra of the nanowalls. **d**, **e** SEM images of the nanowalls after ZnS growth in 5 and 10 cycles. Inset of (**e**) depicts the magnified view of ZnS nanostructure formation over the top of ZnO nanowalls. **f** EDX spectra of the ZnS-decorated ZnO nanowalls. Reproduced with permission from John Wiley and Sons Publications (Bayan et al. 2014)

21.5 Photocatalytic Material

Semiconductor-based photocatalysis is an effective method of utilizing the energy of natural sunlight to break water molecules into hydrogen and oxygen. During this process, photons from sunlight excite the electrons of valence band to move to the conduction band, thus creating holes in the valence band. Titanium dioxide (TiO_2) is considered one of the promising photocatalytic materials to eradicate environmental pollution via photocatalysis. TiO_2 is also considered 'green catalyst' as it is environmentally compatible. A substantial improvement in the visible-light photocatalytic activity in oxygen-deficient TiO_2 nanotubes was observed by narrowing the band gap and effective charge-carrier separation (Choudhury et al. 2016). Doping of cerium has extended the photocatalytic activity of TiO_2 nanoparticles to the visible region of illumination (Choudhury et al. 2011). Doping-induced disorder of the TiO_2 surfaces can extend its light absorption in the visible region (Choudhury et al. 2011; Devi

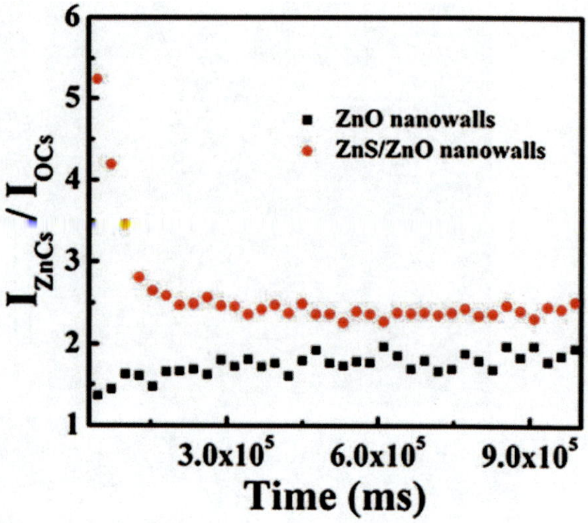

Fig. 21.18 Variation of the intensity ratio of $ZnCs^+$-to-OCs^+ ions with sputtering time. Reproduced with permission from John Wiley and Sons Publications (Bayan et al. 2014)

and Kavitha 2013). Doping of nonmetal or metal ions into TiO_2 introduces new electronic states into the band gap of TiO_2 making it narrower and also lattice-defects into TiO_2. These defects act as "carrier-recombination centres". The oxygen vacancies, introduced into hydrothermally processed TiO_2 nanotubes by vacuum calcination, modify the local coordination in TiO_2, as revealed by Raman spectroscopy and SIMS (Choudhury et al. 2016). Figure 21.19 presents the TEM images of pure (TT) and degraded (DTT) TiO_2 (Choudhury et al. 2016). The high-resolution TEM images of TT and DTT show the inner and outer diameters of the nanotubes.

In such a degraded crystalline system, the sputtering yield is usually higher as the bonds can break more easily. So, the highest Ti^+ emission from disordered TiO_2 nanostructures was due to the increase in total sputtering yield of titanium under vacuum calcination. Similarly, the highest $^{64}TiO^+$ emission (Fig. 21.20) from vacuum-calcined TiO_2 system was attributed to the increased sputter emission of TiO fragments. However, in contrast to the Ti^+ emission, TiO^+ emission from air-calcined sample was slightly lower compared to that in uncalcined sample, confirming the idea that air-calcination leads to the improvement in crystal quality of the TiO_2 nanotubes. The emission of other isotopes of Ti, such as ^{49}Ti and its corresponding oxide ^{65}TiO, followed the similar trend in the SIMS measurements (Choudhury et al. 2016).

The photocatalytic activity of nanostructures essentially originates from the native defect states and is largely influenced by the surface adsorbents. In a recent work, an extensive SIMS study of ZnO nanostructures (nanowires, nanowalls, etc.) has been correlated to their photocatalytic responses (Bayan et al. 2015). The presence of surface-adsorbed H^+, O_2^-, and OH^- species on the surfaces of nanostructures and the relative changes in substrate coverage under varying reaction times have been evidenced through SIMS and substantiated by SEM observation. Compared to nanowires, oxygen adsorption on ZnO surfaces and subsequent oxygen

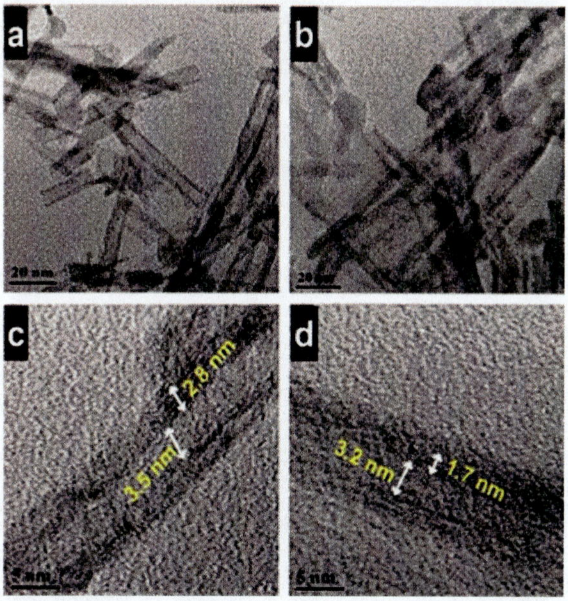

Fig. 21.19 TEM images of **a** pure TiO_2 (TT) and **b** degraded TiO_2 (DTT). High-resolution TEM images of **c** TT and **d** DTT show the inner and outer diameter of the nanotubes. Reproduced with permission from Elsevier Publications (Choudhury et al. 2016)

in-diffusion were found to be more prominent for the nanowall-like structures and highest for the nanowalls which were grown at lower reaction time. Furthermore, the nanowalls were found to exhibit higher photocatalytic activities, which were attributed to higher adsorption of oxygen (Bayan et al. 2015). Photocatalytic decomposition reactions were investigated through TOF–SIMS to understand the details of reaction pathways occurring in nanocrystalline TiO_2 films (Gnaser et al. 2006). Upon irradiation with UV light under atmospheric conditions, the surface composition was found to be pronouncedly changed, an observation ascribed to photocatalytic reactions induced by UV photons: the amount of the parent molecule was diminished, whereas the intermediate reaction products were identified at the TiO_2 surfaces. Eventually, the parent molecule and the characteristic fragment species disappeared completely upon extended exposure to UV light (Gnaser et al. 2006).

21.6 Optical Waveguides

Lithium niobate ($LiNbO_3$), a widely used dielectric material for integrated optics, is characterized by low propagation loss (−0.1 dB/cm), wide transparency band from visible to mid-infrared (0.4–50 μm), minimal optical dispersion in this transparency band, ease in fabrication, etc. Because of its high electro-optic figure-of-merit, it is used for high-speed modulation, switching, and routing of optical signals. The enhancement in the refractive index of $LiNbO_3$, necessary for optical waveguiding, is done by titanium doping locally in the crystal at high temperature or by using

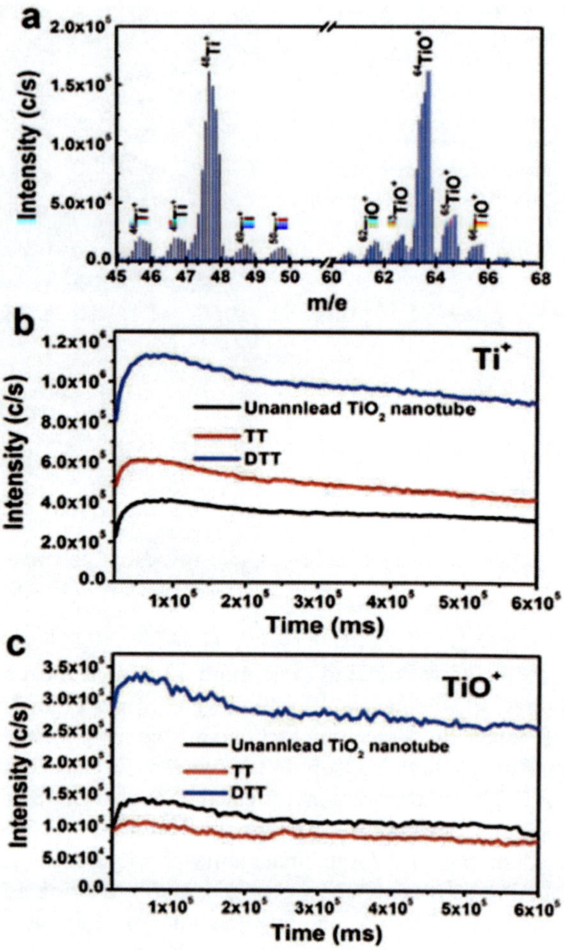

Fig. 21.20 **a** SIMS spectrum of uncalcined TiO_2 nanotube **b** SIMS-depth profile of Ti^+ and **c** TiO^+ for unannealed, air-annealed and vacuum-annealed TiO_2 nanotube. Reproduced with permission from Elsevier Publications (Choudhury et al. 2016)

erbium-doped lithium niobate crystals produced by Er–Li ion-exchange process. More recently, a complete evolution of electro-optical (EO) lithium niobate optical waveguides has been reviewed from Ti-indiffused waveguides to photonic crystals (Courjal et al. 2018).

A combination of SIMS and Near-Field (NF) radiative methods was applied to study Ti in-diffused $LiNbO_3$ optical waveguides (Caccavale et al. 1995a). SIMS was used to measure the in-depth Ti-concentration as a function of diffusion process parameters, where an anisotropy in the diffusion rate for X- and Z-cut crystal directions was observed. The propagating-mode NF method was applied to the refractive index profile reconstruction for single-mode optical channel waveguides. As expected, a sharp change in the refractive index at the air-guide interface was observed. The dependence of refractive index change on Ti concentration was found to be nonlinear, such as quadratic, approximately. Combining electro-optic properties of $LiNbO_3$ with laser-active ions offers the possibility of extra functionality, such as the integration of a high-speed modulator with a waveguide laser to produce an integrated mode-locked laser. Interestingly, Cr is a laser-active ion that has a broad emission spectrum. Combining this property with the electro-optic properties of $LiNbO_3$ offers the possibility of high-speed electro-optic tuning of the laser thereby producing a wavelength-active laser. Furthermore, with Cr ions, it is possible to pump the material at around 670 nm with a laser diode. SIMS measurements provided the depth distribution of Cr concentration. The fluorescence spectrum and upper laser level lifetime of chromium-diffused proton-exchanged and chromium/titanium-diffused lithium niobate waveguides were measured. A simple model was used to estimate the performance of such structures as waveguide-based optical amplifiers and lasers (Almeida et al. 1995). In another work (Caccavale et al. 1995b), the effect of Mg diffusion in pure $LiNbO_3$ and titanium-diffused $LiNbO_3$ optical waveguides was investigated. SIMS provided the in-depth concentration profiles of Mg and Ti in the systems. A numerical study, based on the semi-vectorial finite difference method, determined the coupling loss between single-mode fibers and Ti/Mg: $LiNbO_3$ optical waveguides. The use of Mg was found to reduce the asymmetry-induced coupling loss to almost zero (Caccavale et al. 1995b). Figure 21.21 shows the profile of the optimized Ti/Mg: LiNbO3 refractive index change (Δn). As obvious from the figure, the refractive index change is purely Gaussian in the case of Mg-Ti in-diffused $LiNbO_3$ optical waveguides (Caccavale et al. 1995b). Recently, TOF-SIMS analysis using a low-energy (500 eV) Cs^+ sputter-beam and a 25 keV $Bi_3{}^+$ analysis beam, in combination with Wavelet-Principal Component Analysis (w-PCA) method, has been found extremely efficient in the quantitative depth profiling of organic light-emitting diodes (OLED) (Noël et al. 2019). The w-PCA helped reduce the mass spectra complexity arising out of the characteristic molecular fragments in the organic layers in OLED.

21.7 Photonic Metamaterials: Negative Refractive Index

Photonic metamaterials have emerged as engineered optical materials containing low-dimensional structures, which give extraordinary optical properties. These structures are often made from at least two different materials, involving metals and dielectrics. They are periodic, with the periods being smaller than optical wavelengths. Therefore, the special optical properties do not arise from photonic bandgaps as for certain photonic crystals, but rather from an interaction similar to that of atoms or ions in a normal solid medium. These artificial electromagnetic materials of subwavelength periodicity have a negative refractive index ($n_1 < 0$) that behaves contrary to the conventional "right-handed" interaction of light found in conventional optical materials. Hence, these are often called as left-handed materials or *negative refractive index materials.* All highly anisotropic semiconductor metamaterials provide a straightforward path to negative refraction in the mid-infrared. However, their applicability is restricted by strong frequency dispersion and limited spectral

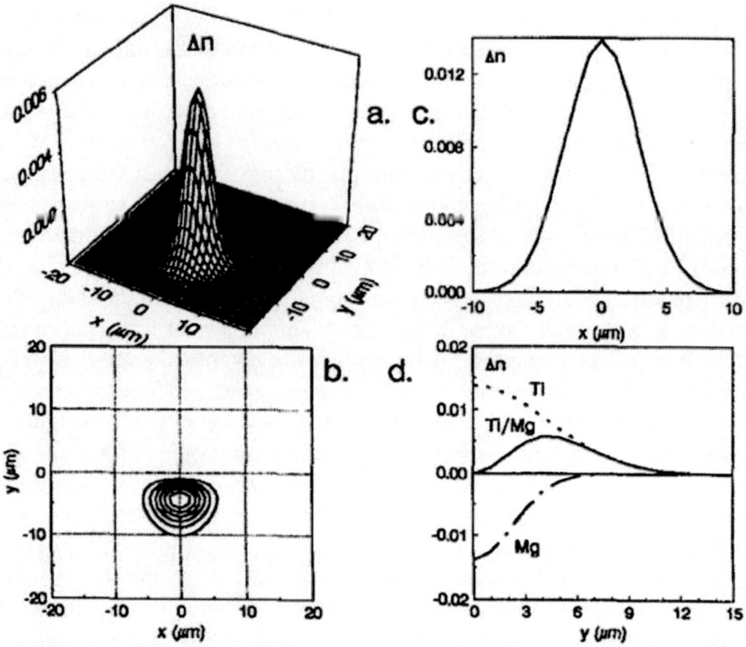

Fig. 21.21 Optimized Ti/Mg: $LiNbO_3$ refractive index change profile; **a** 3D plot, **b** contour plot, **c** horizontal section and **d** vertical section. Reproduced with permission from AIP Publications (Caccavale et al. 1995b)

bandwidth. Enhanced bandwidth and reduced dispersion through stacking multiple optical metamaterials were reported by Escarra et al. (2011). They constructed mid-infrared metamaterials by designing periodic structures formed with alternating layers of highly n-doped InGaAs and intrinsic (undoped) InAlAs, with thickness of each of the layers in the sub-wavelength region, approximately 100 times smaller than the wavelengths. Experiments were conducted on single-stack, two-stack and four-stack metamaterials, grown back-to-back by MBE on undoped InP substrates. Each stack consisted of 80 nm thick highly Si-doped $In_{0.53}Ga_{0.47}As$ layers inter-leaved with 80 nm thick undoped $In_{0.52}Al_{0.48}As$ layers. The InGaAs and InAlAs layer compositions were lattice-matched to the InP substrate. These metamaterials exhibited strongly anisotropic dielectric responses (Escarra et al. 2011).

An accurate free-carrier doping level in each stack of the structure is the most critical parameter to achieve a broadband, flat-response metamaterial. Periodicity in X-ray diffraction (XRD) peaks confirmed the high-quality growth of the epitaxial layers, whereas the SIMS measurements provided precise in-depth quantification of Si, Al, Ga, and In atom concentrations in the growth direction. SIMS depth profiling revealed one, two, and four distinct regions with different Si atom densities in the InGaAs layers, for the one-, two- and four-stack samples, respectively (Escarra et al. 2011). In order to extract free carrier density from the SIMS-measured Si atom density, the conversion efficiency of Si atoms to free electrons is required to be estimated, which still remains a challenge. The calculated beam-shifts of the transmitted light versus wavelength (beam-shift is defined as the distance light translates in the plane of the stacks, as it passes through the sample) are shown in Fig. 21.22 for a p-polarized light (TM mode) for single-stack, two-stack and four-stack metamaterials. These stacked metamaterials have shown enhanced bandwidth of negative refraction and reduced dispersion (Escarra et al. 2011).

Shelby et al. (2001) presented the experimental scattering data at microwave frequencies on a structured metamaterial that exhibits a frequency band where the effective index of refraction is negative. They fabricated a prism using a metamaterial consisting of a two-dimensional array of repeated unit cells of copper strips and split ring resonators on interlocking strips of standard circuit board material and measured the scattering angle of the transmitted beam through the prism. The measured effective refractive index was found to be appropriate to Snell's law. The experiments confirmed the predictions of Maxwell's equations that **n** is given by the negative square root of the product of ε and μ, for the frequencies where both the permittivity (ε) and permeability (μ) were negative (Shelby et al. 2001). Hoffmann et al. (2009) reported a novel class of semiconductor metamaterials that exhibit a strongly anisotropic dielectric function to achieve negative refraction in the mid-infrared region (~8.5–13 μm) of the spectrum. They presented highly doped quantum-well superlattices that are highly anisotropic. Using transmission and reflection measurements and by modelling the highly-doped quantum-well superlattices,

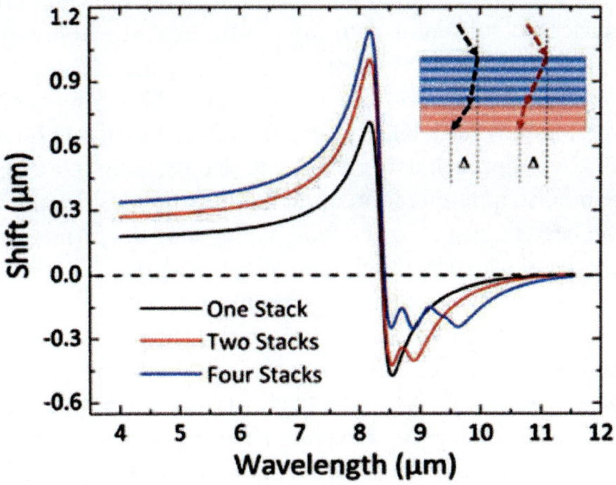

Fig. 21.22 Calculated beam shift of transmitted light versus wavelength for a single stack, two-stack and four-stack metamaterials. These metamaterials have enhanced the bandwidth of negative refraction and reduced dispersion in that region. Inset: Illustration of a two-stack metamaterial, where blue and pink represent the regions with different doping in the InGaAs (darker) layers. The black and red arrows represent two possible beam paths through this material, for different wavelengths, that result in the same beam shift **Δ**. Reproduced from Optica Publications under Creative Commons License (Escarra et al. 2011)

they demonstrated anomalous reflection due to the strong anisotropy of the material (Hoffmann et al. 2009). This new class of semiconductor metamaterials has great potential in materials science, especially for waveguiding and imaging applications in the long-wave infrared region.

21.8 Conclusions

The extremely high detection sensitivity and exceptional depth resolution have made the secondary ion mass spectrometry to be one of the most powerful techniques for the analysis of low-dimensional materials, interfaces of thin films and superlattices. Time-of-Flight (TOF)-SIMS is highly surface sensitive, able to provide both elemental composition and molecular information on a top surface as well as in the depth. It has the potential to provide detailed insight into the 3D chemical composition for a large mass range (0–10,000 amu) for all elements in the periodic table, organics, small molecules, polymers (polymeric films, polymer blends, polymer composites) and biological materials (Mei et al. 2022). On the other hand, dynamic SIMS has the highest detection sensitivity (below parts per billion) and is used to determine the elemental composition and the levels of ultra-trace impurities and dopants in solid materials at any depth.

However, the major limitation of SIMS lies in the materials quantification owing to its strong 'matrix effect'. Although quantification in a given matrix is achieved by using 'calibration standards', it is very difficult to quantify the cross-interfaces composed of matrices of different natures. Continuous attempts have been made towards exploring the possibilities of 'matrix effect compensation' so that accurate quantification is possible without the use of calibration standards. Out of several existing techniques for achieving reduced or 'no matrix' effect, the MCs_n^+-SIMS (n = 1, 2, ...) approach using MCs_n^+ molecular ions is the most useful and beneficial for the precise quantification of quantum structures. In parallel, considerable attempts have been made to conceptually understand the formation mechanisms of these MCs_n^+ molecular ions through theoretical models.

Since SIMS is essentially a sputter depth profiling technique, it is very important to consider the most significant detrimental effects, such as 'surface roughening', 'ion-beam induced atomic mixing' and 'broadening in the information depth' for the correct evaluation and quantification of depth profiles across the interfaces even under optimized experimental conditions. An understanding with accurate estimation of these three fundamental parameters using **MRI** (**M**ixing, **R**oughening, **I**nformation depth) simulation model is extremely useful to construct a depth resolution function (DRF) which has a significant role in depth-profile reconstruction. However, the effects due to nonlinear behaviour such as preferential sputtering, interface segregation, etc. are not yet fully understood so as to make a more refined prediction of DRF.

Interface analysis of multilayers is of continued interest in materials science. In particular, problems in semiconductor technology have been a driving force.

Dynamic SIMS has shown its tremendous applications in various multilayer systems including semiconductor Bragg mirrors and quantum wells. Interfacial diffusion across interfaces is a natural process causing compositional changes across the interfaces in layered structures. In combination with other measurements like X-ray reflectivity, X-ray diffraction, etc., dynamic SIMS has been found to be useful for compositional analysis of the interfaces in semiconductor multilayers. In metallic multilayers, the formation of interfacial alloys can exhibit various crystalline structures. The structure of a particular alloy phase at the interface is determined by many parameters, interfacial energies, composition and ordering conditions. Moreover, the strong confinement effect at the interfaces may give rise to a novel interfacial alloy phase that could have a crystal structure completely different from that in the bulk. SIMS and XRD analyses together can effectively probe such novel 'interfacial alloys' in metallic multilayer systems.

Although there have been enormous instrumental and methodological developments in SIMS for surface and interface analysis in condensed matter systems in general, the futuristic application range of this sputter-induced surface analytical technique is not very clear at this moment. An in-situ SIMS analysis with an extreme level of optimization in-depth resolution during a controlled film growth will be a useful attempt to make a real-time analysis of an interface providing a much better understanding of the interfacial behaviour. Furthermore, adequate flexibility and improvisation in the SIMS instrumentation could possibly lead to an extension of this technique towards new directions, such as analysis of the instantaneous 'layering of liquid' or 'surface melting' states, highly doped quantum-well superlattices for metamaterials, $LiNbO_3$ optical waveguides, etc. Of course, multi-technique approaches are very much essential in which a novel combination of SIMS with other analytical techniques can be useful for the analysis of complex soft condensed matter systems such as colloids, functional materials, supra-molecular organic assemblies, liquid crystals, crystalline polymers, granular media and various sensing materials.

References

Adachi, S.: J. Appl. Phys. **53**, 8775–8792 (1982)

Almeida, J.M., Boyle, G., Leite, A.P., De La Rue, R.M., Ironside, C.N., Caccavale, F., Chakraborty, P., Mansour, I.: J. Appl. Phys. **78**, 2193–2197 (1995)

Andersen, C.A., Hinthorne, J.R.: Anal. Chem. **45**, 1421–1438 (1973)

Baio, J.E., Graham, D.J., Castner, D.G.: Chem. Soc. Rev. **49**, 3278–3296 (2020)

Bayan, S., Chakraborty, P.: Appl. Surf. Sci. **303**, 233–240 (2014)

Bayan, S., Satpati, B., Chakraborty, P.: Surf. Interface Anal. **47**, 37–44 (2014)

Bayan, S., Choudhury, B., Satpati, B., Chakraborty, P., Choudhury, A.: J. Appl. Phys. **117**, 095304–095310 (2015)

Brison, J., Muramoto, S., Castner, D.G.: J. Phys. Chem. C **114**, 5565–5573 (2010)

Brundle, C.R., Evans, C.A., Wilson, S.: In: Brundle, C.R., Evans, C.A., Wilson, S. (eds.) Encyclopaedia of Materials Characterization. Butterworth-Heinemann, MA, USA (1992)

Caccavale, F., Chakraborty, P., Quaranta, A., Mansour, I., Gianello, G., Bosso, S., Corsini, R., Mussi, G.: J. Appl. Phys. **78**, 5345–5350 (1995a)

Caccavale, F., Chakraborty, P., Capobianco, A., Gianelio, G., Mansour, I.: J. Appl. Phys. **78**, 187–193 (1995b)
Canteri, R., Moro, L., Anderle, M.: In: Benninghoven, A. (ed.) Proceedings of SIMS VIII. Wiley, NY, USA (1991)
Chakraborty, P.: Dynamic secondary ion mass spectrometry. In: Chakraborty, P. (ed.) Ion Beam Analysis of Surfaces and Interfaces of Condensed Matter Systems (Chap 7), pp. 217–265. Nova Science Inc., New York (USA) (2002)
Choudhury, B., Borah, B., Choudhury, A.: Photochem. Photobio. **88**, 257–264 (2011)
Choudhury, B., Bayan, S., Choudhury, A., Chakraborty, P.: J. Colloid Interface Sci. **465**, 1–10 (2016)
Courjal, N., Bernal, M.P., Caspar, A., Ulliac, G., Bassignot, F., Manuel, L.G., Suarez, M.: Lithium niobate optical waveguides and microwaveguides. In: You, K.Y. (ed.) Emerging Waveguide Technology, pp. 153–175. Intech Open (2018)
Davies, J.A.: High energy density collision cascades and spike effects. In: Williams, J.S., Poate, J.M. (eds) Ion Implantation and Beam Processing, pp. 81–97 (Chap 4). Elsevier Inc. Amsterdam (1984)
Devi, L.G., Kavitha, R.: Appl. Catal. B Environ. **140**, 559–587 (2013)
Escarra, M.D., Thongrattanasiri, S., Charles, W.O., Hoffman, A.J., Podolskiy, V.A., Claire, G.: Opt. Express **19**, 14990–14998 (2011)
Gao, Y.: J. Appl. Phys. **64**, 3760 (1988)
Gao, Y., Marie, Y., Saldi, F., Migeon, H.N.: Int. J. Mass Spectrom Ion Proc. 11–18 (1995)
Gavelle, M., Bazizi, E.M., Scheid, E., Armand, C.: Mater. Sci. Eng. B **154**, 110–113 (2008)
Gnaser, H.: Surf. Sci. **342**, 319–326 (1995)
Gnaser, H., Oechsner, H.: Surf. Sci. **302**, L289–L292 (1994a)
Gnaser, H., Oechsner, H.: Surf. Interface Anal. **21**, 257–260 (1994b)
Gnaser, H., Orendorz, A., Ziegler, C., Rowlett, E.: Appl. Surf. Sci. **252**, 6996–6999 (2006)
Gnaser, H.: Low-energy ion irradiation of solid surfaces (Springer Tracts in Modern Physics, vol. 146). Springer, Berlin, Germany (1999)
Haag, M., Gnaser, H., Oechsner, H.: Fresenius J. Anal. Chem. **353**, 565–569 (1995)
Harrison, D.E., Levy, N.S., Johnson, J.P., Effron, H.M.: Appl. Phys. **39**, 3742–3761 (1968)
Havelund, R., Seah, M.P., Tiddia, M., Gilmore, I.S.: J. Am. Soc. Mass Spectrom. **29**, 774–785 (2018)
Hoffman, A.J., Sridhar, A., Braun, P.X., Alekseyev, L., Howard, S.S., Franz, K.J., Cheng, L., Choa, F.S., Sivco, D.L., Podolskiy, V.A., Narimanov, E.E., Gmachl, C.: J. Appl. Phys. **105**, 122411–122417 (2009)
Hofmann, S.: Appl. Surf. Sci. **70–71**, 9–19 (1993)
Hofmann, S.: Surf. Interface Anal. **21**, 673–678 (1994)
Holliger, P., Laugier, F., Dupuy, J.C.: Surf. Interface Anal. **34**, 472–476 (2002)
https://attic.gsfc.nasa.gov/huygensgcms/MS_Analyzer_1.htm
https://www.hidenanalytical.com/applications/surface-analysis/contamination-with-silicone/
Marie, Y., Gao, Y., Saldi, F., Migeon, H.N.: Surf. Interface Anal. **23**, 38–43 (1995)
Marseilhan, D., Barnes, J.P., Fillot, F., Hartmann, J.M., Holliger, P.: Appl. Surf. Sci. **255**, 1412–1414 (2008a)
Marseilhan, D., Barnes, J.P., Fillot, F., Hartmann, J.M., Holliger, P.: Appl. Surf. Sci. **255**, 1412 (2008)
Mei, H., Laws, T.S., Terlier, T., Verduzco, R., Stein, G.E.: J. Polym. Sci. **60**, 1174–1198 (2022)
Mondal, S., Gnaser, H., Chakraborty, P.: Eur. Phys. J. D **66**, 197–203 (2012)
Mootz, T., Adriaens, A., Adams, F.: Int. J. Mass Spectrom. Ion Process. **156**, 1–10 (1996)
Noël, C., Tuccitto, N., Busby, Y., Berger, M.A., Licciardello, A., List-Kratochvil, E.J.W., Houssiau, L.: ACS Appl. Polym. Mater. **1**, 1821–1828 (2019)
Oechsner, H., Gerhard, W.: Phys. Lett. A **40**, 211–212 (1972)
Otomo, S., Maruya, H., Seo, S., Iwase, F.: Appl. Surf. Sci. **252**, 7275–7278 (2006)
Priebe, A., Xie, T., Bürki, G., Pethö, L., Michler, J.: J. Anal. Spectrom. **35**, 1156–1166 (2020)

Prudon, G., Gautier, B., Dupuy, J.C., Dubois, C., Bonneau, M., Delmas, J., Vallard, J.P., Bremond, G., Brenier, R.: Thin Solid Films **294**, 54–59 (1997)
Saha, B., Chakraborty, P.: Nucl. Instrum. Methods Phys. Res. B **258**, 218–225 (2007)
Saha, B., Chakraborty, P.: Energy Proc. **41**, 80–109 (2013)
Saha, B., Sarkar, S., Chakraborty, P., Gnaser, H.: Surf. Sci. **602**, 1061–1065 (2008)
Saha, B., Chakraborty, P., Gnaser, H., Sharma, M., Sanyal, M.K.: Appl. Phys. A **108**, 671–677 (2012)
Sarkar, S., Chakraborty, P., Gnaser, H.: Phys. Rev. B **70**, 195427–195431 (2004)
Sarkar, S., Datta, A., Chakraborty, P.: J. Mater. Res. **20**, 2639–2646 (2005)
Sarkar, S., Chakraborty, P., Sanyal, M.K., Caccavale, F., Arora, B.M.: Surf. Interface Anal. **29**, 659–662 (2000)
Schenkel, T.: Surface analysis of slow, highly-charged ions like Au^{69+}: TOF-SIMS and the probing of nano-environments. In: Chakraborty Purushottam (ed) Ion Beam Analysis of Surfaces and Interfaces of Condensed Matter Systems (Chap 8). Nova Science Inc., New York (USA), pp. 267–310 (2002)
Sharma, M., Sanyal, M.K., Mukhopadhyay, M., Bera, M., Saha, B., Chakraborty, P.: J. Appl. Phys. **110**, 102204–102208 (2011)
Shea, M.P., Havelund, R., Gilmore, I.S.: J. Phys. Chem. C **120**(46), 26328–26335 (2016)
Shelby, R.A., Smith, D.R., Schultz, S.: Science **292**, 77–79 (2001)
Slodzian, G.: Phys. Scripta **T6**, 54–66 (1983)
Thompson, M.W.: Phil. Trans. r. Soc. Lond. A **362**, 5–28 (2004)
Van der Weg, W.F., Rol, P.K.: Nucl. Instrum. Meth. **38**, 274–276 (1965)
Vickerman, J.C.: Static secondary ion mass spectrometry. In: Walls, J.M. (ed.) Methods of Surface Analysis, pp. 169–215. Cambridge University Press, Cambridge (1989)
Warren, B.E.: X-ray Diffraction. Addison Wesley Publishing Company Inc., Reading, Massachusetts (USA) (1969)
Williams, P.: Surf. Sci. **90**, 588–634 (1979)
Winograd: Annu. Rev. Anal. Chem. **11**, 29–48 (2018)
Wittmaack, K.: Surf. Sci. **429**, 84–101 (1999)
Wittmaack, K.: Surf. Sci. Rep. **68**, 108–230 (2013)
Yu, M.L.: In: Benninghoven, A. (ed) Proceedings of SIMS VI, Wiley, NY, USA (1987)
Yu, M.L.: Charged and excited states of sputtered atoms. In: Behrisch, R., Wittmaack, K. (eds.) Sputtering by Particle Bombardment III, pp. 91–160. Springer-Verlag, Berlin (1991)
Yu, M.L., Mann, K.: Phys. Rev. Lett. **57**, 1476–1479 (1986)
Zener, C.: Proc Roy Soc A **137**, 696–702 (1932)

Chapter 22
Scintillator Single Crystals

D. S. Sisodiya, S. G. Singh, G. D. Patra, A. K. Singh, S. Pitale, M. Ghosh, M. Sonawane, and Shashwati Sen

22.1 Introduction

Radiation detection and counting is the basis of many applications like nuclear medicine, homeland security, high-energy nuclear physics, nuclear non-proliferation, etc. (Knoll 2010). Radiation detectors have played a major role in research on fundamental physics like in Rutherford's gold foil experiment, God particle (Higgs Boson) experiment, dark matter detection, neutrino oscillations, etc. For the detection of any ionizing radiation several detectors are available like ionization chambers, proportional counters, Geiger Muller (GM) counter, scintillators, etc. Scintillation is the process in which an ionizing radiation such as alpha, beta, gamma, X-ray, neutron, or any high-energy charge particle; interacts with the scintillating materials and converts its energy to visible photons or low-energy photons. This scintillation process is one of the convenient and widely used methods available for the detection and spectroscopy of a wide range of radiations.

The barium platinocyanide ($Ba[Pt(CN_4)]$) one of the first scintillator materials, empowered Rontgen to get the first Nobel prize in Physics in 1901 for the discovery of X-rays, that started a chain of discoveries (Patton 1993). The first scintillator detector was built by Sir William Crookes in 1903 and used ZnS as a scintillating material (Crookes 1874). The former made a device known as spinthariscope which was used to convert the scintillation light into visible light that can be seen by naked eyes. Same technique was used in the famous Rutherford gold foil experiments which led to the discovery of the nucleus and started a new era in modern physics. In 1944, Curran and Baker developed the photomultiplier tube, which totally replaced the naked eye measurement of scintillation light. This was the first step in the era of modern scintillator detectors.

D. S. Sisodiya · S. G. Singh · G. D. Patra · A. K. Singh · S. Pitale · M. Ghosh · M. Sonawane · S. Sen (✉)
Crystal Technology Section, Technical Physics Division, BARC, Mumbai, India
e-mail: shash@barc.gov.in

R. S. Ningthoujam and A. K. Tyagi (eds.), *Handbook of Materials Science, Volume 1*, Indian Institute of Metals Series, https://doi.org/10.1007/978-981-99-7145-9_22

In the present day, scintillators play an important role in our day-to-day life. They find use in X-ray scanning and imaging for both medical and security applications. In the other extreme, they are used as radiation detectors for the detection of particles in high-energy particle accelerators. Also, scintillators find application in modern nuclear medicine and diagnostic applications like Positron Emission Tomography (PET) and single-photon emission computed tomography (SPECT) (Melcher 2000).

There are two main classes of scintillators: inorganic and organic. Organic scintillators are generally the liquid scintillators and plastic scintillators. Inorganic scintillators are halide or oxide single crystals. Inorganic scintillators are generally brighter but with a slower decay time than organic ones. Each scintillator material has its advantages and drawbacks. Thus, no "ideal" scintillator exists and the choice depends on the application.

This chapter will deal with the fundamentals of the scintillation mechanism, a brief description of the different scintillator materials and an in-depth discussion on various single-crystal inorganic scintillators.

22.2 Fundamentals

Before explaining the scintillation process, one has to understand the interaction of radiation with matter. It is well understood that for the detection of any kind of radiation, it must interact with the detector's material. Radiation can be categorized into two types: Charged and neutral. Charged particles interact with the electrons of the material by the Columbic interaction.

Charged particles like alpha, beta, or heavy ions continuously interact with the electrons of the scintillator medium causing atomic excitation or ionization. The rate of energy loss ($-dE/dx$) for charged particles is strongly energy dependent. It is described by the Bethe-Bloch formula for incoming particles in the MeV-GeV range, with atomic shell corrections at lower energy and radiative loss corrections at higher energy.

On the other hand, neutral particle like neutron directly interacts with the nuclei of the scintillating material (Knoll 2010). But to detect neutrons, it has to undergo a direct interaction with the nucleus producing recoil protons or spallation fragments, which will then transfer their energy to the medium in the same way as primary charged particles. For the detection of neutrons using a scintillator, a matrix having ^{6}Li, ^{10}B or ^{136}Gd is used. These isotopes have very high interaction cross-sections with a thermal neutron. The absorption cross-section of ^{6}Li is ~ 940 barns for thermal neutrons. The reaction is given by ^{6}Li (n, α)^{3}H; where a neutron gets absorbed to give an alpha particle and a tritium with 'Q' value energy of 4.8 MeV approx. The released alpha particle deposits their energy-producing electron–hole pairs which is similar to the other scintillation phenomenon. Thus, the absorption of each neutron results in a burst of photons.

The fundamental reactions of neutron are

$$n + {}^{6}Li\ (7.5\%) \rightarrow {}^{3}H\ (2.75\ MeV) + {}^{4}\alpha(2.05\ MeV),\ \sigma = 940\ b \tag{22.1}$$

$$n + {}^{10}B\ (20\%) \rightarrow {}^{7}Li\ (1.0\ MeV) + {}^{4}\alpha(1.8\ MeV)\ BR\ 7\% \tag{22.2}$$

$$\begin{aligned}&\rightarrow {}^{7}Li * (0.83\ MeV) + {}^{4}\alpha(1.47\ MeV) + \gamma(0.48\ MeV),\\ &BR\ 93\%,\ \sigma = 3800\ b\end{aligned} \tag{22.3}$$

(The natural abundance of the given isotope is given in %, interaction cross-sections are σ in the barn at 25 meV), and relevant branching ratios BR).

In case of gamma rays, the interaction is basically through three processes; Compton scattering, photo-electric effect, and pair production. Gamma rays deposit full or partial energy in the detection medium via these interactions. Here, the above three processes of gamma interactions with the matter have been discussed briefly.

22.2.1 Interaction of Gamma Rays with Materials

(i) *Photo Electric Absorption*

In photoelectric absorption, incident X-ray or gamma ray photon gets absorbed by an atom of the absorbing material. The transferred energy during this process in general exceeds the binding energy of the electron in the host atom and as a result electron is ejected with high kinetic energy. After the ejection of electron, there is a vacancy in the electron shell, which is then refilled by a nearby free electron. This filling process again liberates a characteristic X-ray photon, which again gets absorbed by less bounded electrons and produces additional fast electrons. The overall effect of photo electron absorption is the complete deposition of energy of the ionizing radiation. Because, photoelectric effect is the full energy deposition process, it is in some sense an ideal conversion step. The photoelectric absorption probability for a material is proportional to

$$\tau \propto \text{constant} \times \frac{Z^n}{E_\gamma^{3.5}} \tag{22.4}$$

Here, Z is the effective atomic number, and E_γ is the energy of ionizing radiation. The exponent *n* varies between 4 and 5. Photo-electric interaction is the dominant process at low energy (up to a few hundred keV) for high Z materials.

(ii) *Compton Scattering*

In Compton scattering the incoming gamma-ray, photons get deflected by an angle θ with a single free electron in absorbing material. In this process, gamma photons transfer some portion of its energy to the recoiled electron depending upon the deflection angle. The probability of Compton scattering is related to the electron density in the medium and increases linearly with the atomic number of the absorber, favoring, therefore, high Z materials. Since scattering of photon is possible in all possible directions, the recoil electrons are produced with a continuum of energies ranging from zero to maximum which is represented by the backscattering. The energy of a photon or recoil can be calculated by the conservation of energy and momentum. The energy of deflected photon is given by

$$h\nu' = \frac{h\nu}{1 + \frac{h\nu}{m_o c^2}(1 - cos\theta)} \qquad (22.5)$$

Here, ν and ν' are the energy of the photon before and after the scattering and m_o is the rest mass of the recoil electron. At energies above a few hundred keV, Compton scattering becomes predominant depending on the Z_{eff} of the absorbing material.

(iii) *Pair Production*

This process is energetically possible only when the incoming photon energy is above 1.02 meV. In this process, a gamma photon disappears in the presence of a nuclear field of the absorber material and produces an electron–positron pair. The excess energy of gamma photon appears as kinetic energy of electron and positron. Further, the emitted positron participates in a pair annihilation process with the electron of the material and produces two photons.

22.2.2 Scintillation Mechanism

When radiation passes through an ordered material like a single crystal, another mechanism also takes place. When the radiation deposits its energy in the crystal lattice, it leads to the formation of secondary electrons due to the above mentioned three phenomena (Schotanus et al. 1989). These electrons in the keV range, start to interact with the electrons of the atoms of the crystal lattice and excite the electrons from the occupied valence or core bands to different levels in the conduction band. Each of these interactions results in an electron–hole pair formation. If the energy of the electron is higher than the ionization threshold, free carriers are produced, which moves randomly in the crystal until they are trapped by a defect or recombine at a luminescent center. In case the ionization threshold is not reached, the electron and the hole release part of their energy by coupling to the lattice vibration modes until they reach the top of the valence band for the hole and the bottom of the conduction band for the electron. In some materials, they bound and form an exciton whose energy is slightly smaller than the bandgap between the valence and the conduction bands. These excitons lose their energy at luminescent centers through energy or a charge transfer mechanism.

The above described scintillation process can be divided into five main steps (Lecoq et al. 2017):

1. Interaction of ionizing radiation with atoms of the scintillation material and creation of primary charge carrier (electrons and holes).
2. Interaction of charge carrier with the bulk of material. Production of secondary charge carrier.
3. Thermalization of charge carrier via diffusion towards low ionization density region.
4. Energy transfer to the luminescence centers.
5. Light emission by luminescence centers.

In this discussion, the scintillation mechanism is in detail and also elaborates some specific scintillation processes. Scintillation is the collective phenomena of many radiative and non-radiative transition channels that produce the scintillation light. The following paragraphs discuss all the possible channels and elements that cause the scintillation.

For a material to be a scintillator it must contain luminescent centers. The luminescent centers are either extrinsic, generally dopants or co-dopants, or intrinsic like molecular orbitals of the lattice or of defects of the lattice, which possess a radiative transition between an excited and a lower energy state. Moreover, the energy levels involved in the radiative transition must be smaller than the forbidden energy bandgap, in order to avoid re-absorption of the emitted light or photo-ionization of the center.

A simple scheme of scintillation process with respect to electronic transition between bands is represented in Fig. 22.1. This figure shows one core band with maximum energy E_c and bandwidth ΔE_c. The valence band is shown with energy maxima at $E_v = 0$ and bandwidth ΔE_v. The conduction band is shown with energy minima at E_g. The whole scintillation mechanism can be expressed in five steps happening in different time scales.

In the first step, when high energy ($E>E_g$) ionizing radiation interacts with the scintillation medium, it creates primary hot (energy $>2Eg$) charge carriers in the form of holes and electrons in the inner core bands and the conduction bands respectively. In a very short time scale (of the order of 10^{-16}–10^{-14} s) multiplication of the charge starts via inelastic electron–electron scattering and Auger process. Secondary charge carriers are produced in the form of holes in the core and valence band and electrons in the conduction band. This multiplication process ends when all the electrons in the conduction band have energy less than the e-e scattering threshold ($2E_g$) and all the holes occupy the valence band.

The second step is the thermalization of charge carriers which happens in the time scale of 10^{-14}–10^{-12} s. In this process charge carriers lose their kinetic energy due to interaction with phonons of the lattice and the energy of the electrons comes in the minima of the conduction band and holes at the maxima of the valence band.

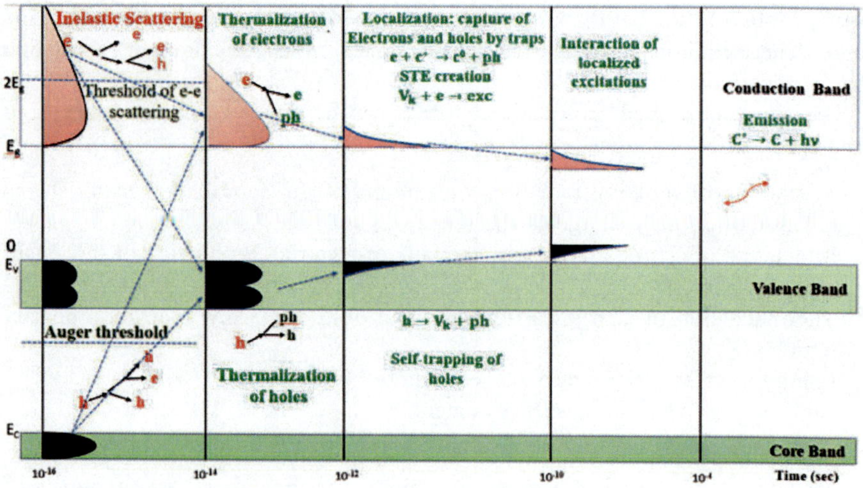

Fig. 22.1 Schematic representation of electronic excitation and relaxation processes in an insulator. In the diagram the different symbols are denoted as follows *e* is electrons, *h* is holes, *ph* is phonons, $h\nu$ is photon energy, V_k is self-trapped holes, *STE* is self-trapped excitons and *cn* is ionic centers with charge *n*. Density of states is represented by red and black areas for electrons and holes, respectively

In the next step, charge carriers get localized through their interaction with lattice defects and impurities of the material and form exciton, self-trapped excitons (STE), self-trapped holes (V_k centers), etc. This process happens in the time scale of 10^{-12}–10^{-10} s.

In the next stage, the relaxed charge carriers migrate in the scintillation medium and recombine radiatively or non-radiatively.

In the final step, the electronic excitations transfer their energy to the luminescence centers and emit photons, which can be detected by photomultiplier tube (PMT).

22.2.3 Scintillator Terminology

Before going into a detailed discussion about scintillation, one has to understand the basic terminologies which are used to characterize a scintillator material.

(i) *Scintillation Yield*

Light yield or scintillation yield of a scintillator (Y) is defined as the number of light photons produced by the scintillator per unit of energy deposited by ionizing radiation in the medium. Production of scintillation light depends on the five-scintillation process discussed in the earlier section, so the light yield can be defined as.

$$Y = \prod y_i = \frac{E_\gamma}{\beta . E_g} S.Q \tag{22.6}$$

Here, y_i is the yield of each scintillation process, β is the conversion factor, E_g is the band gap of material, and E_γ is the absorbed energy. This expression is based on a simple model of scintillation; it becomes complex if energy losses, surface losses, and band structure are taken into account.

(ii) *Non-Proportionality*

Non-proportionality is the nonlinear relationship between scintillator light yield and energy of ionizing radiation. The light yield of any scintillator is not linear to the absorbed photon energy. In the scintillation mechanism, non-proportionality comes into the picture during the production of secondary charge carriers inside the scintillator material. This process is probabilistic and may take place via different channels for the same absorbed energy (Moses et al. 2008). So, the origins of non-proportionality are not well understood.

(iii) *Energy Resolution*

Energy resolution of a scintillator is the ability to distinguish between two incoming radiations with slightly different energy. This is generally applicable in single-crystal inorganic scintillators. This value is given in percentage and defined as.

$$\text{Energy Resolution (E.R.)}\% = \frac{\Delta E}{E} \times 100 \tag{22.7}$$

Here, ΔE is the full width at half the maximum of the photopeak located at the energy E. Good energy resolution (low value of ER) is the major criteria for finding a new scintillator material. Energy resolution of a scintillator depends on various parameters as given by the following equation (Moses et al. 2008):

$$R^2 = R_{np}^2 + R_{inh}^2 + R_{tr}^2 + R_{lim}^2 \tag{22.8}$$

Here, R_{np} is the factor of non-proportionality. R_{inh} is the inhomogeneity factor of the single crystal, which leads to local variation in scintillation light. R_{tr} is related to the efficiency of the light collection by the PMT. R_{lim} is the factor of intrinsic resolution of the detector which can be described by the Poisson law.

$$R_{lim} = 2.35\sqrt{\frac{1 + \nu(PMT)}{N_{phe}}} \tag{22.9}$$

where, ν (PMT) is the variance of photomultiplier tube's gain and N_{phe} is the number of photoelectrons emitted by the PMT.

Therefore, for finding a scintillator with better energy resolution, one has to keep in mind all the above factors that affect the energy resolution. Some factors can be controlled, like the inhomogeneity of crystal, by controlling the single crystal growth process or tuning the growth parameters. Also, for better homogeneity one can choose a material, which has intrinsic luminescence centers like $CeBr_3$ (Lempicki 1995). Light collection efficiency can be improved experimentally by using high-quality reflector and equipments. The major factors that affect the energy resolution are the non-proportionality and light yield (a number of photons emitted by PMT depends on light yield).

(iv) *Decay Time*

Time between the introduction of radiation on the scintillator's surface and the release of light is called the decay time of a scintillator. In simple words, it is the time taken by the scintillator to complete the full scintillation process. Scintillation decay time explicitly depends on the scintillation process (five steps), which has been discussed earlier.

(v) *Pulse Shape Discrimination*

The pulse shape discrimination is the ability of a scintillator material in which it produces different decay pulse shapes due to different scintillation decay kinetics followed by the excitation radiation inside the material. Generally, it is observed in scintillators which have fast and slow decay components. The relative ratio of these components depends on the nature of ionizing particle. Subsequently, the shape of the decay pulse can be used to differentiate the ionizing radiations. This technique has been employed in many materials like CsI(Tl) (Longo et al. 2022).

22.2.4 Important Scintillator Parameters

A perfect universal scintillator material doesn't exist in nature which can fulfill all the requirements of practical applications. Currently, researchers have studied the scintillation mechanism in-depth and developed new concepts for tailoring the scintillation properties according to specific requirements. This section deals with some of the most important scintillator parameters, which are relevant from the user's point of view for the development of a new scintillating material (Knoll 2010).

1. High density and High Z_{eff}

To detect gamma rays efficiently, scintillator should have high density and high effective Z (Z_{eff}). High density reduces the material size required to slow high-energy

γ photon and electrons. High Z_{eff} increases the photoelectric interaction probability, which is used in gamma-ray spectroscopy. Both high density and Z_{eff} offer high stopping power for the incoming radiation in order to have a compact-size detector. Inorganic scintillators can have high density up to 8 g/cm^3.

2. Energy Resolution and Fast Response

The influencing factors for scintillating materials are the high energy resolution and fast response of the detector. High-energy resolution is required for isotope identification whereas fast response is required in advanced nuclear physics measurement of nuclear excitation and for the detection of rare events in particle physics. Inorganic scintillator provides both high energy resolution and fast response simultaneously like $CeBr_3$ (Sisodiya et al. 2022), $LaBr_3$ (Singh et al. 2018a). These features are not available in plastic scintillators and HPGe detectors.

3. Particle identification

Some specific inorganic scintillators have unique features that they have different pulse shapes or responses for different particles, this feature is used in particle identification. This unique feature is used to identify particles and gamma photons in fluxes of mixed products.

4. Radiation Hardness

A radiation detector should not get affected or should have good stability in the presence of intense ionizing radiation. In general, inorganic single crystal-based scintillators don't have this problem. This property of inorganic scintillator is important for high-energy physics experiments where the luminosity is very high.

5. Durability of Physical and Chemical Parameters

For a long life of a radiation detector, the chemical and physical parameters of scintillator must not change with time. So, for long-run experiments, single crystal-based inorganic scintillator detector are required which has high structural quality and have stable chemical composition.

However, experimentally it is impossible to have a scintillator, which can fulfill all requirements simultaneously, one need to tailor the properties according to demand. Commercially, a wide range of scintillator detectors are available that can fulfill several requirements of users. Still there are many challenges (cost, handling, and durability) for the development of the inorganic scintillator material which are unfavorable for users. Researchers of the scintillator community are working with the aim of solving these challenges.

22.3 Classification of Scintillators

From a composition point of view, scintillators can be divided into two types: organic and inorganic (Koshimizu 2022; Weber 2002). Further, they can be divided into solid, liquid, and gas phases. The classification is given in Fig. 22.2 and described in the following section.

22.3.1 *Organic Scintillators*

(i) *Organic single crystals*

Organic scintillators are aromatic hydrocarbon compounds, which contain benzene ring structures interlinked in various ways. Their luminescence typically decays within a few nanoseconds (Koshimizu 2022). In the organic systems, the scintillation arises upon transition of the charge carriers between an excited molecular level and the corresponding electronic ground state. Due to this fact, the decay time is much less in these materials.

The most common organic scintillator crystals are anthracene (Leo 1994) ($C_{14}H_{10}$, decay time ≈30 ns), stilbene (Leo 1994) ($C_{14}H_{12}$, 4.5 ns decay time), and naphthalene ($C_{10}H_8$, few ns decay time). They are very durable, but their response is anisotropic due to anisotropic crystal field (which spoils energy resolution when the source is not collimated). Further, they cannot be easily machined, or can be grown in large sizes; hence, they are not very often used. Anthracene has the highest light output of all organic scintillators and is therefore chosen as a reference. The light outputs of other organic scintillators are sometimes expressed as a percent of anthracene light.

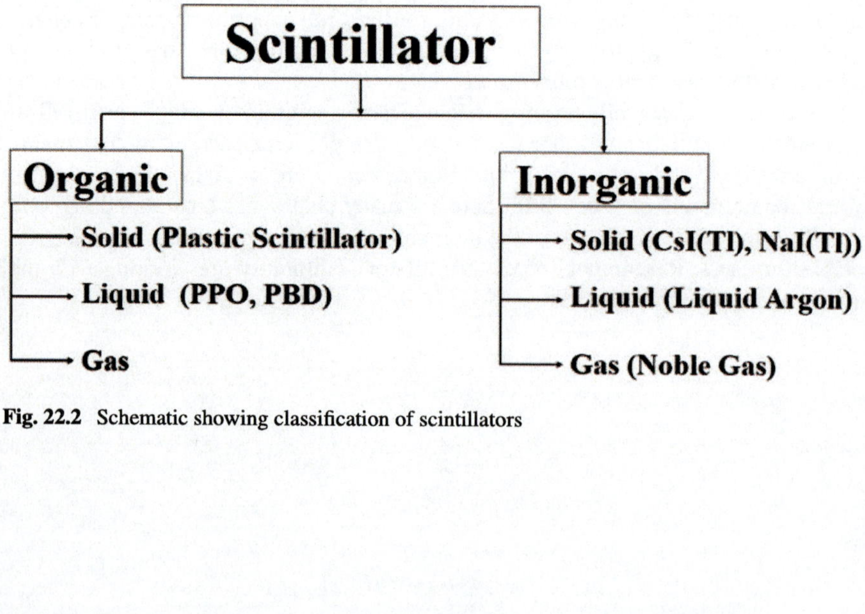

Fig. 22.2 Schematic showing classification of scintillators

(ii) *Organic liquids*

Organic liquids scintillators are liquid solutions of one or more organic scintillators in an organic solvent. The typical solutes are fluors such as p-terphenyl ($C_{18}H_{14}$), $C_{20}H_{14}N_2O$, $C_{24}H_{22}N_2O$, 2,5-Diphenyloxazole commonly known as PPO ($C_{15}H_{11}NO$), and wavelength shifter such as POPOP or 1,4-bis(5-phenyloxazol-2-yl) benzene ($C_{24}H_{16}N_2O$). The most widely used solvents are toluene, xylene, benzene, phenylcyclohexane, triethylbenzene, and decalin. Liquid scintillators are easily loaded with other additives such as wavelength shifters to match the spectral sensitivity range of a particular PMT. These liquids are loaded with ^6Li or ^{10}B to increase the neutron detection efficiency of the scintillation (since these materials have a high interaction cross-section with thermal neutrons).

(iii) *Plastic scintillators*

In a plastic scintillator, the primary fluorescent emitter, called a fluor, is suspended in a solid polymer matrix called the base. This is synthesized by dissolution of the fluor in the monomer matrix followed by bulk polymerization. The biggest advantage of plastic scintillators is the ease of fabrication into very large sizes and any shapes. The other advantages are high light output and a fast decay time of 2–4 ns (Moser et al. 1993).

22.3.2 Inorganic Scintillators

(i) *Gaseous Scintillators*

Gaseous scintillators consist of nitrogen and the noble gases helium, argon, krypton, and xenon. The scintillation process is due to the de-excitation of single atoms excited by the passage of an incoming particle. These scintillators find application in nuclear physics experiments where gaseous detectors have been used to detect fission fragments or heavy charged particles.

(ii) *Glasses*

The most common glass scintillators are cerium-activated lithium or boron silicates. Since both lithium and boron have large neutron cross-sections, glass detectors are particularly well suited to the detection of thermal (slow) neutrons. Lithium is more widely used than boron since it has a greater energy release on capturing a neutron and therefore greater light output. Glass scintillators are, however, sensitive to electrons and γ rays as well (pulse height discrimination can be used for particle identification). Being very robust, they are also well-suited to harsh environmental conditions. Their response time is $\approx$10 ns, their light output is however low thus limiting their application (Knoll 2010).

(iii) *Inorganic Single-crystal scintillator*

The most important type of inorganic scintillator is a single crystal based (Singh et al. 2011). They are used in most of the applications because of their novel detection performance and the rest of the chapter will discuss this category of scintillators in detail. In the inorganic crystals, scintillation arises from thermalized electrons and holes, which move to the bottom of the conduction band or the top of the valence band respectively, by scattering from the initially produced fast charge carriers. Single crystals of inorganic scintillators have very good energy resolution and high density. Thus, they find applications in high-energy radiation detection and spectroscopy.

The most commonly used inorganic scintillator is NaI:Tl (thallium-doped sodium iodide); its scintillation light emission is in the blue region. Other inorganic alkali halide crystals are: CsI:Tl, CsI:Na, CsI, KI:Tl, LiI:Eu, $LaCl_3$:Ce, $LaBr_3$:Ce (Vaněček et al. 2022). Oxide-based scintillators commonly used are $Bi_4Ge_3O_{12}$, CaWO4, CdWO4, $Y_3Al_5O_{12}$:Ce, Gd_2SiO_5:Ce, Lu_2SiO_5:Ce (Kim et al. 2021).

These inorganic scintillators are single crystals with a small amount of dopant which acts as a luminescent center. They are grown in high-temperature furnaces. Before going further it's important to understand why inorganic scintillators are required in single crystal form and how they are grown.

22.4 Why Single Crystal?

Single crystal is the material which has a systematic and periodic arrangement of atoms with no grain boundaries. Crystal structure is made of space lattice and basis, space lattice is defined as imaginary points, which are arranged in space such that each point has identical surroundings. An atom or a group of atoms is called a basis which acts as a structural unit that completes the crystal structure.

Inorganic scintillators are used in many physical forms like single crystal, glass, polycrystalline ceramics, and thin film. But for the effective and quantitative detection of wide energy range of ionizing radiation single crystal-based inorganic scintillator detectors are used (Kim et al. 2021). The physical properties of single crystal-based scintillators which make them a popular choice for radiation detection are:

(i) *Absence of Grain Boundary*

The presence of grain boundaries in the scintillator detector leads to scattering of light photons and decreases the light yield. This leads to losses of information about the incoming radiation. Thus, for high light yield and no loss of information good-quality single crystal are required for the scintillation detection.

(ii) *High Atomic Density*

High atomic density leads to high stopping power for the ionizing radiation. Single crystal is a closely packed system so it has high atomic density compared to power or liquid scintillators so that one can use it in high-energy radiation field.

(iii) *Transparent to own Luminescence*

Generally single crystals are transparent to their luminescence means there is no or less self-absorption of light in the scintillation medium. This leads to a full transmission of information about ionizing radiation. A single crystalline scintillator has a high light yield compared to the same polycrystalline pellet because of transparency and absence of grain boundaries.

(iv) *Uniform Crystal Field Across Activator Ions*

Single crystals are the periodic arrangement of elements, so each incoming radiation sees the uniform crystal field. In activated scintillators, distribution of activators is important, uniform distribution leads to a uniform crystal field across the activator or uniform electronic perturbation in an electronic density of ions. Non-uniform distribution of activator leads to poor photo peak shape and less efficient pulse shape discrimination.

(v) *Durability in Chemical and Physical Parameters*

Single crystals have good chemical stability compared to liquid and powder because of its periodic and closed packing arrangement of atoms. Powder or liquid form can easily get contaminated or reacted, whereas single crystal doesn't have this issue.

22.4.1 Single Crystal Growth

Single crystal growth is the first order-controlled phase transformation from a disordered fluid phase into a highly ordered solid phase (Markov 2016). Phase transformation phenomena take place close to the thermodynamic equilibrium. For crystal growth, the necessary condition is that the change in total Gibbs free energy must be negative. The fundamental entity of the single-crystal growth is the nucleation. The growth starts from a single point which is called a nucleus or small embryo or seed crystal. For the formation of the nucleus of the new phase, it has to pass the thermodynamic energy barrier called the interfacial energy barrier. Interfacial energy which has a positive contribution to the total energy depends on the surface area and volume of the nucleus. At the initial state of nucleation; the surface area is more than the volume so the nucleus has less stability. Once the volume term starts to dominate, contribution of interfacial energy decreases, stable nucleation takes place. There is an intermediate size of nucleus (at $r = r^*$) at which the total free energy of the system is minimum. This is known as the critical size of the nucleus. This effect is known as the Gibbs–Thomson effect. Mathematically one can calculate the critical size of the nucleus, using the fundamentals of thermodynamics (Gunther 2003). The change in Gibbs free energy associated with the formation of a spherical embryo of radius r is given by

$$\Delta G = -\frac{4}{3}\pi r^3 \Delta G_V + 4\pi r^2 \sigma \quad (22.10)$$

Here, σ is the interfacial energy per unit area and ΔG_V is the change in volume energy. Pictorially, this effect can be understood as shown in Fig. 22.3. When the size of the embryo is very small ($r < r^*$) the surface energy term, given by ΔG_s dominates. As ΔG_s is a positive term, ΔG is positive and the embryo is unstable. As r increases further, ΔG_v, the volume energy term starts dominating and ΔG becomes negative and the embryo gets stabilized. The critical radius can then be calculated by differentiating the above Eq. (22.10) with respect to r.

$$r^* = \frac{2\sigma}{\Delta G_V} \quad (22.11)$$

There are two types of nucleation "Homogeneous" and "Heterogeneous." For an efficient radiation detection, large size single crystals are required, so the growth rate must be faster as compared to the nucleation rate. For the prevention of multinucleation sites, several techniques are used like slow cooling of melts, temperature gradients, and introduction of seed crystals. Three basic techniques are used for the growth of single crystals (Pimpinelli and Villain 2010).

1. Growth from Liquid Phase
2. Growth from Solution
3. Growth from Vapour.

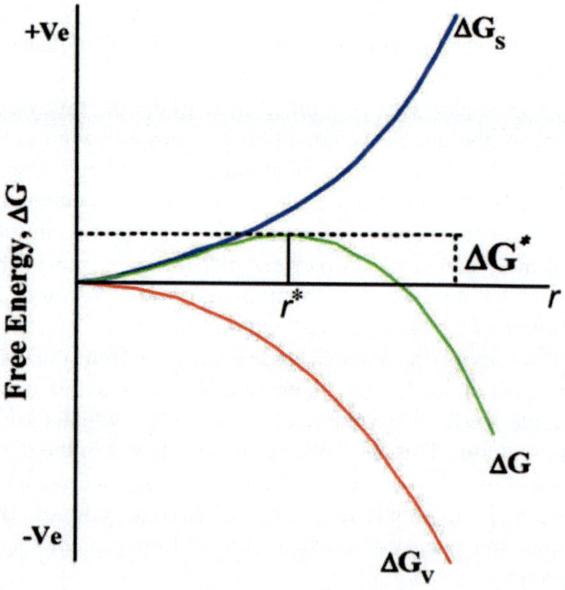

Fig. 22.3 Dependence of the Gibbs free energy as a function of the radius of the nucleation during the crystal growth process

The basic theory for the growth of a single crystal in all techniques is the same. In this chapter, focus on single-crystal growth from melt which is the main technique employed for the growth of scintillator single crystals. In the melt growth technique, the polycrystalline powder of the desired materials is melted by heating to a high temperature and slowly cooling it in a controlled way. Around 80% crystal growth community use the melt growth technique for the growth of single crystals. Melt growth is only applicable to materials that melt congruently and have no destructive phase changes. One of the major advantages of the melt growth technique is the high volume growth rate. Chart of classification of crystal growth techniques is shown in Fig. 22.4.

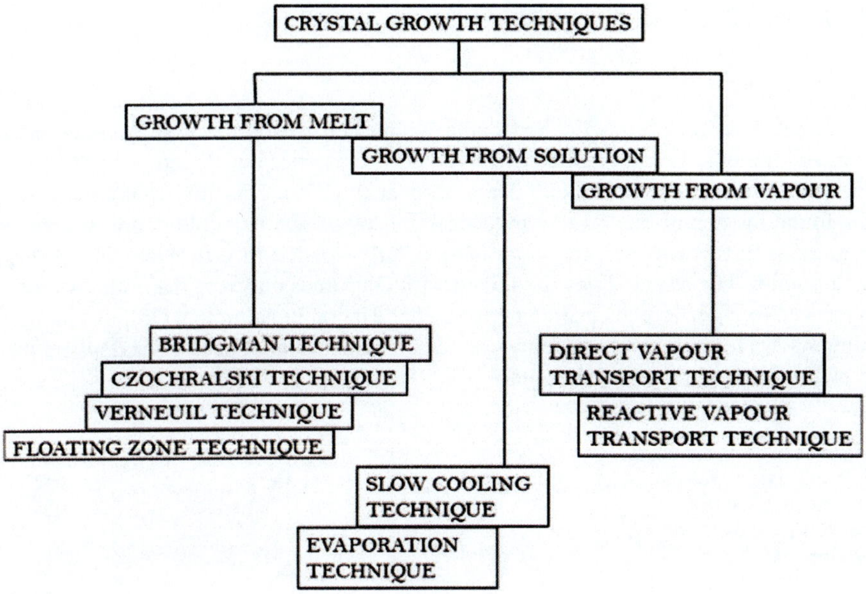

Fig. 22.4 Broad classification of all crystal growth techniques

22.4.2 Growth from the Melt

Melt growth is the process in which the melt of desired material recrystalize via cooling the liquid below its freezing point in a controlled way. This method is commercially used for the growth of technologically important single crystal growth. The growth from melt can be classified into various techniques (Pimpinelli and Villain 2010). (In a small bracket principal of technique is mentioned).

1. Verneuil method ("flame fusion").
2. Kyropoulos technique ("top seeding").
3. Bridgman method ("directional solidification").
4. Czochralski method ("pulling")
5. Floating Zone.
6. Laser-heated pedestal growth.
7. Micro pulling.

Verneuil process is also called flame fusion and it was the first commercially successful method for manufacturing synthetic gemstones developed in 1902 by the French chemist Auguste Verneuil (Barvinschi et al. 1999). This process is considered the founding step of the modern industrial crystal growth technology and shown in Fig. 22.5. In this process, the alumina powder is loaded in a furnace with a very small outlet. The alumina powder mixed with O2 comes out from the outlet where a hydrogen–oxygen torch is maintained to melt the incoming alumina. Drops of molten alumina are formed at this outlet which falls on a pedestal of seed crystal where the growth of a single crystal takes place.

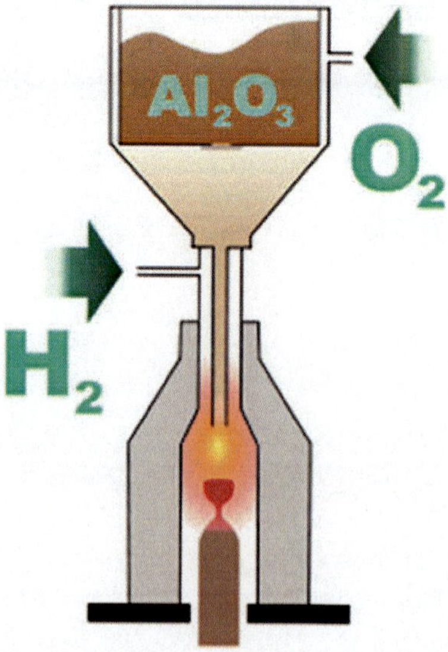

Fig. 22.5 Schematic of Verneuil crystal growth process for the growth of sapphire crystals. (*Source* Wikipedia)

Among the melt growth techniques, most of the scintillator single crystals are grown using the Bridgman, Czochralski, and float zone methods. These techniques are described in the following section.

(i) *Bridgman Method*

This technique was named after its inventor Bridgman in 1925 and Stockbarger in 1938 (Jeffrey 2015). The basic principle of this technique is the directional solidification of melt by translating it from a hot zone to a cold zone of the furnace. Bridgman furnace as shown in Fig. 22.6 has two zones "Cold" and "Hot." The hot zone is maintained above the melting point and the cold zone is maintained below the melting point of the material which is to be grown into a single crystal. In between there is a temperature gradient where the crystal growth takes place. For the growth via the Bridgman technique, the material is sealed and loaded in an ampoule which is loaded in the hot zone of the furnace. The material of the ampoule is chosen so that it should withstand high temperatures and should not react with the material of the crystal. In general quartz, graphite, or nobel metals are used as ampoules. Once the ampoule is heated to a temperature above the melting point of the material, the whole ampoule is translated from a hot zone to a cold zone of furnace. Schematic diagram of the Bridgman furnace has been shown in Fig. 22.6a. A photograph of a typical Bridgman growth furnace which has been designed to growth 2-inch diameter and 3-inch length single crystal is shown in Fig. 22.6b. This technique is used for the growth of mixed halide single crystals like CsI(Tl), LiI(Eu), $CeBr_3$, etc. (Singh et al. 2015). In case of material that has to be grown under inert conditions, the ampoule is evacuated and sealed under a high vacuum ~10^{-5} mbar.

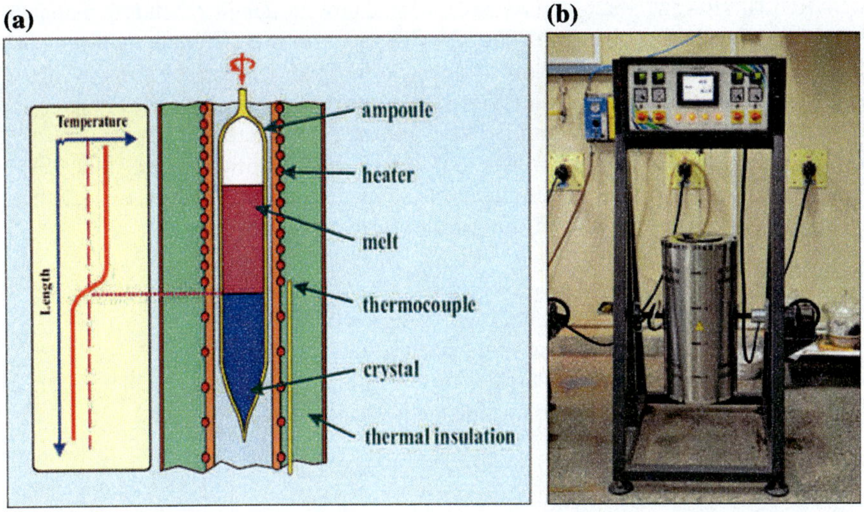

Fig. 22.6 **a** The schematic of a Bridgman crystal growth furnace showing the hot zone and the cold zone **b** Photograph of a typical Bridgman furnace at crystal technology lab TPD BARC

Advantage

1. Growth furnace of the Bridgman technique is less expensive.
2. Vacuum-sealed ampoule permits the growth of hygroscopic material.
3. Selection of the appropriate ampoule can produce a crystal of assigned diameter.

Disadvantages

1. Difficult to retrieve the grown crystal from the ampoule.
2. There is a chance of reaction between the ampoule and the melt.
3. In this technique nucleation starts at the tip of the crucible so there is lesser control on nucleation.
4. As the crystal growth takes place inside the crucible, stress is generated inside the crystal which may deteriorate its quality.

(ii) *Czochralski Method*

Czochralski crystal growth (also called CZ technique) was developed by Jan Czochralski in the early 1900s as a method to solidify desired material in a single crystal boule by pulling it from a melt (Müller 2007). In this method, the charge is melted and maintained slightly above the melting point and a pulling seed crystal is lowered to just touch the melt for starting the nucleation and the crystal is pulled slowly. The pulling rate depends on thermal conductivity, latent heat of fusion, viscosity of melt, and cooling of the pulling crystal. For uniform cylindrical shape of crystal, the seed is rotated during the growth. Shape and size of the crystal can be controlled by the melt temperature, pulling rate and rotation of the seed. Schematic of Czochralski technique and the actual photograph of a CZ puller is shown in Figure 22.7. Figure 22.7b is the photograph of an indigenously developed CZ puller at BARC. This has an induction furnace capable of growing single crystals of diameter 2 inches and melting point upto 2000°C. Also, one can use different environments like ambient gas flow and pressure which will be favorable for the growth of the single crystal. This method is widely used to commercially grow single crystals.

Advancements in crystal pulling have allowed for increased reproducibility, restively rapid growth, and high levels of control of the growth parameters. Scintillation community uses this technique for the oxide scintillator single crystals like Bismuth Germinate (BGO), YAP, tungstate, etc. (Singh et al. 2011).

Advantage

1. Better control over nucleation, shape, and growth rate.
2. Can be used for the growth of large-size crystals.
3. As there is no direct contact between the crucible walls and the crystal, the chance of crystals cracking and stress inside the growing crystal is reduced.

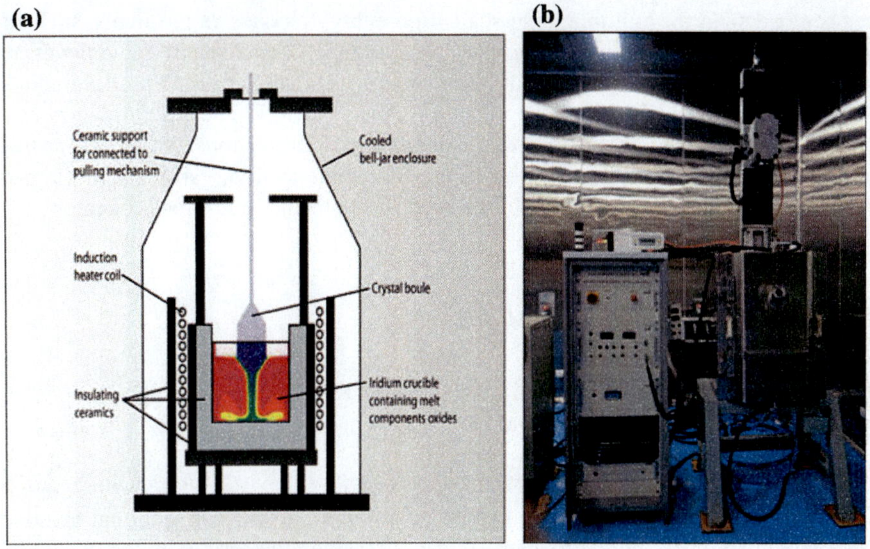

Fig. 22.7 **a** The schematic of a Czochralski crystal growth furnace showing the induction furnace, growing crystal, and the melt. **b** Photograph of a typical Czochralski crystal growth furnace at crystal technology lab TPD BARC

Disadvantages

1. There is a chance of growth in different phases other than the expected one.
2. Special type of atmosphere or running vacuum required during the growth.
3. It is a very expensive and sophisticated technique as compared to Bridgman.
4. There is some probability of evaporation of constituent in many cases.

22.5 Selected Examples of Inorganic Scintillators Single Crystal: Growth and Application

This section discusses the scintillation mechanism and growth technique of a few important inorganic scintillating materials. Most of the inorganic scintillators can be classified into two broad categories: Halide scintillators and Oxide scintillators. Halide scintillators are in general grown by the Bridgman technique but can also be grown by the CZ technique. Few halide scintillators are CsI, NaI, $LaBr_3$, SrI_2, etc. They have low melting points and thus their growth is easy as compared to oxide scintillators. Some examples of oxide scintillators are Bismuth germinate (BGO), $PbWO_4$, Gd3Ga3Al2O12(GGAG), LuYSiO5 (LYSO), etc. Oxide scintillators have high density, mechanical hardness compared to the inorganic ones. $PbWO_4$

has been adapted for multiple large-scale high-energy physics experiments, such as Proton Antiproton Detector Array (PANDA) at GSI (Gesellschaft für Schwerionenforschung) facility in Darmstadt, Germany as well as Compact Muon Solenoid (CMS) and A Toroidal LHC Apparatus (ATLAS) at CERN. These scintillators can be further categorized on the basis of the dopants or activator ions, which acts as the basis of the scintillation mechanism. In the following section, various scintillators based on the dopant ion and thus scintillation mechanism are grouped together.

22.5.1 Tl-Doped Alkali Halides

Tl doped alkali halides like NaI(Tl), CsI(Tl), and KI(Tl) have wide applications in various fields of nuclear physics and these scintillators are commercially available.

(i) ***NaI:Tl***

It is the most widely used single-crystal scintillator. It is a conventional scintillator having application in homeland security like area monitoring, in gamma spectroscopy, in nuclear physics experiments (Suerfu et al. 2020). NaI melts congruently at 661 °C, so a single crystal can be grown by the Bridgman technique and have 3.67 g/cm^3 density. NaI:Tl has emission maxima at 415 nm which is excellently matching with the PMT response function. NaI:Tl has good scintillation performance, it has a decay time around 230 ns, 6.8% energy resolution at 662 keV and has a light yield of 38 Ph/keV. It can be grown in very large sizes. The main advantage of NaI:Tl scintillator is low cost of material and easy growth. NaI:Tl single crystal is grown using the Bridgman technique in a quartz ampoule. However, this is a highly hygroscopic material and highly prone to cracking {cleavage plane <100>}. This makes it very difficult to handle and has to be hermetically sealed for any application. Crystal Technology Section (CTS) of the Technical Physics Division (TPD), Bhabha Atomic Research Center (BARC) has been involved in the growth of NaI:Tl single crystal for various radiation detection application. Figure 22.8a shows the photograph of a grown NaI:Tl single crystal of diameter 50 mm and length 50 mm mounted on PMT. The pulse height spectra of this crystal measured for ^{137}Cs and ^{60}Co gamma source is given in Fig. 22.8b.

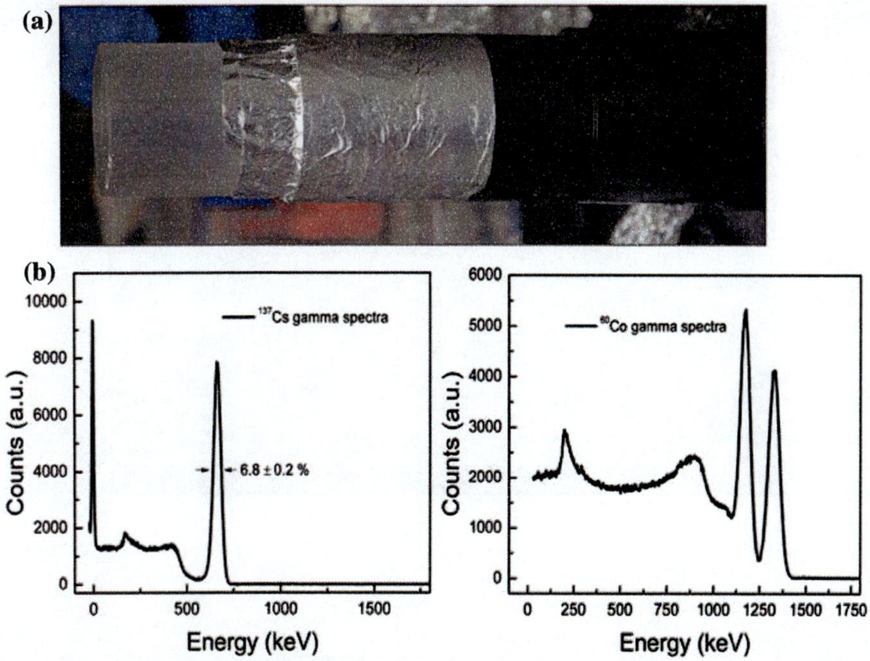

Fig. 22.8 **a** Photograph of NaI:Tl single crystal, diameter 50 mm and length 50 mm mounted on PMT **b** Pulse height spectra measured using a NaI:Tl scintillator detector for gamma source of ^{137}Cs (left) and ^{60}Co (right)

(ii) ***CsI:Tl***

This is one of the best suitable materials for a scintillator detector because of its less hygroscopicity, easy and scalable single crystal growth, ease of detector fabrication and it's machinable nature. CsI has a melting point of 620°C and its high density 4.5 g/cm^3 makes it suitable for efficient radiation detection. CsI:Tl has emission maxima at 550 nm, which is matching with the Si-based photodetectors. In terms of scintillation performance CsI:Tl has a high light yield compared to NaI:Tl 54 (Ph/KeV), 1000 ns decay time, and have 7% energy resolution at 662 keV (Singh et al. 2012).

Figure 22.9a shows the photograph of the CsI:Tl single crystal (size 2″ dia × 2″ length) grown at CTS, TPD using the Bridgman method (Desai et al. 2012; Patra et al. 2020; Tyagi et al. 2015). The growth has been carried out in a Bridgman furnace shown earlier in Fig. 22.6b. The CsI:Tl has an emission at 550 nm thus under UV illumination, it gives yellow color as seen in Fig. 22.9a. To show that the nature of dopant leads to the change in the emission characteristics we have shown the photograph of a CsI:Na crystal under UV illumination. The emission of CsI:Na is 420 nm thus giving blue color.

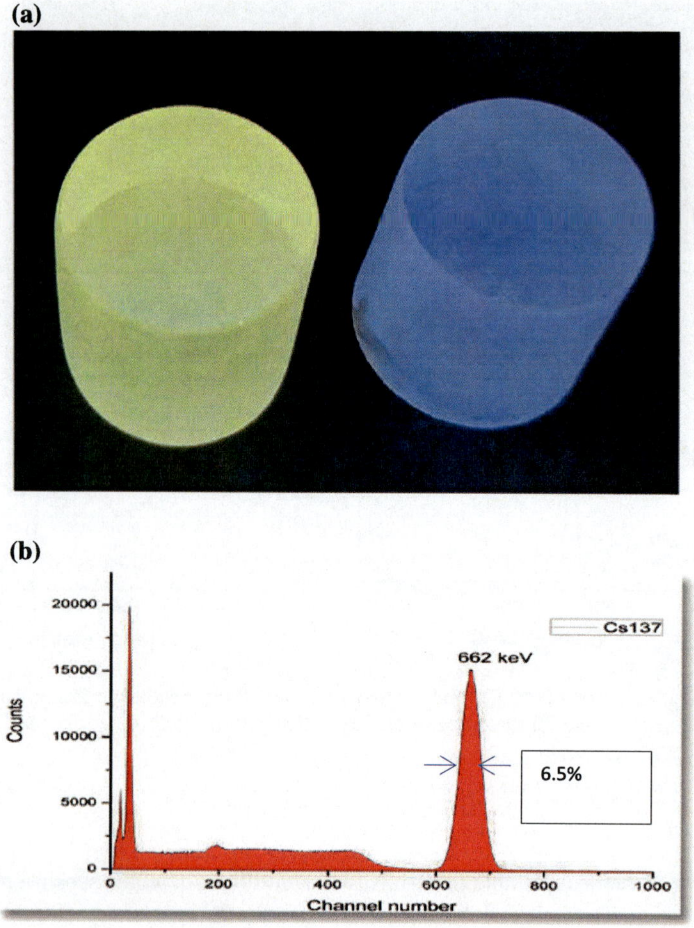

Fig. 22.9 **a** Photograph of processed CsI:Tl (yellow color) and CsI:Na (blue color) single crystal of diameter 50 mm and length 50 mm under UV-illumination. **b** Pulse height spectra measured using a CsI:Tl scintillator detector for gamma source of ^{137}Cs

Further, larger size single crystal of 3″ diameter and 4″ length has been grown in carbon-coated quartz crucible as shown in Fig. 22.10a. Figure 22.10b shows a hand-held gamma spectrometer developed using the 3″ CsI:Tl single crystal for detection of gamma radiation (Tyagi et al. 2015).

The scintillation mechanism of CsI:Tl is very complex and can be briefly explained as follows. In CsI(Tl) scintillator, one Cs^+ ion is replaced by the Tl^+ ion. Tl^+ is a well-known luminescence center and has $5d^{10}$, $6s^2$ electronic configuration which gives rise to $^1s_0$ ground states and $^1p_1$, $^3p_0$, $^3p_1$, $^3p_2$ excited states (Dietrich and Murray 1972). The luminescence occurs due to the $^3p_0$ to $^1s_0$ transition which is spin forbidden. This is why the Tl-doped scintillator has a slow decay profile.

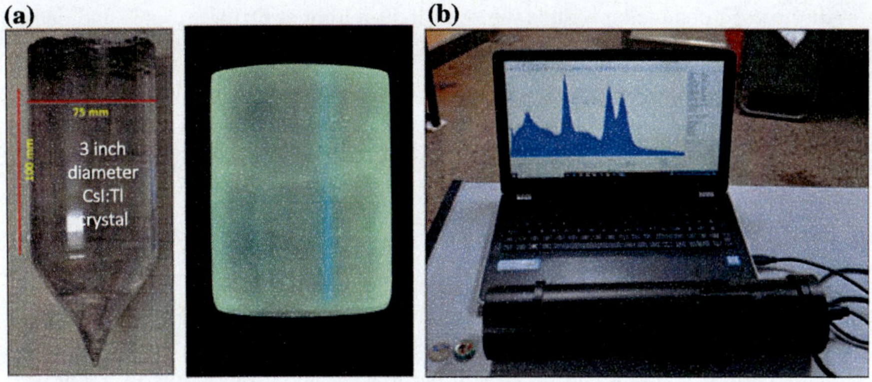

Fig. 22.10 **a** As grown CsI:Tl single crystal of diameter 75 mm and length 100 mm and processed scintillator single crystal **b** A USB-based hand-held gamma spectrometer developed with 3″ diameter CsI:Tl scintillator single crystal

When ionizing radiation interacts with the scintillator matrix and deposits its energy, the formation of *e*–*h* pairs takes place. Here, *e* and *h* are thermalized carriers. The different charge states of Tl ion that exists in the CsI matrix are Tl^0, Tl^+ and Tl^{++}. The lifetime of Tl^0 and Tl^{++} states is longer than the radiative lifetime of Tl^+ (Rogulis et al. 2004). So, apart from other self-trapping mechanisms, there is also a finite possibility that the generated *e* and *h* can be captured by different Tl ions. Trapping mechanism can be expressed as

(i) $e + Tl^+ \rightarrow Tl^0$ $\quad e$ trapped at Tl^+
(ii) $h + Tl^+ \rightarrow Tl^+$ $\quad h$ trapped at Tl^+
(iii) $h \rightarrow V_k$ $\quad h$ self – trapping
(iv) $V_k + Tl^+ \rightarrow Tl^{++}$ $\quad V_k$ dissociation

These trapped charge carriers can recombine in both radiative and non-radiative ways. Also, these processes depend on the temperature. The recombination channels are listed below:

(v) $Tl^0 + h \rightarrow h\nu$ $\quad$ Trapped *e* with free *h*
(vi) $Tl^{++} + e \rightarrow Tl^+ + h\nu$ $\quad$ Trapped *h* with free *e*
(vii) $V_k + e \rightarrow h\nu$ $\quad$ Free *e* with V_k
(viii) $V_k + Tl^0 \rightarrow Tl^+ + h\nu$ $\quad V_k$ with trapped *e*
(ix) $Tl^0 + V_k \rightarrow h\nu$ $\quad$ Trapped *e* with V_k

Some recombination channels like No. 9 are observed only at low temperatures. Other possibilities of scintillation channels are the direct excitation and ionization of Tl^+ by secondary electron and excitation by excitons.

(x) $Tl^+ + e_{sec} \rightarrow (Tl^+)^* \rightarrow h\nu$ $\quad$ Direct excitation
(xi) $Tl^+ + e_{sec} \rightarrow Tl^{++}\ ;\ Tl^{++} + e \rightarrow (Tl^+)^* \rightarrow h\nu$ $\quad$ Direct ionization
(xii) $Tl^+ + ex \rightarrow (Tl^+)^* \rightarrow h\nu$ $\quad$ Excitation by Exciton

All these 12 channels produce the scintillation light at different rates and decide the decay scheme of CsI:Tl scintillator (Mikhailik et al. 2015).

22.5.2 *Intrinsic Fast Inorganic Scintillators*

(i) ***BaF***$_2$

BaF_2 is one of the fastest inorganic scintillator materials with high effective atomic number (Z = 52.7), which makes it attractive for scintillator detectors where high detection efficiency per unit volume and fast response are required. Melting point of BaF_2 is 1368 °C and it can be grown by the vacuum Bridgman and Czochralski technique (Kamada et al. 2005). With the Czochralski technique, BaF_2 can be grown in any crystal orientation and the largest grown crystal by this technique is 8″ ϕin size. BaF_2 has 4.88 g/cc density and has emission maxima at 220 nm (fast component) and 310 nm (slow component) (Diehl et al. 2015). It has 2 decay components 0.8 ns and 630 ns. The energy resolution of BaF_2 is 9.8% at 662 keV and light yield is 8 ph/KeV. Photograph of a 30 mm diameter BaF_2 single crystal grown at CTS, TPD is shown in Fig. 22.11.

Scintillation mechanism in BaF_2 is different from other activated halide scintillators. When ionizing radiation interacts with the BaF_2 single crystal electrons get excited from the lower band (core and valence band) to the conduction band. Hole in the upper core band relax very fast and occupy the upper end of the core band instantly and recombine with an electron from the valence band via a core-to-valence transition. This special kind of intrinsic luminescence is called "cross luminescence" and has a decay time in the order of ps. Cross-luminescence process is dominant when the electron energy is not sufficient for the excitation of Auger electrons from the

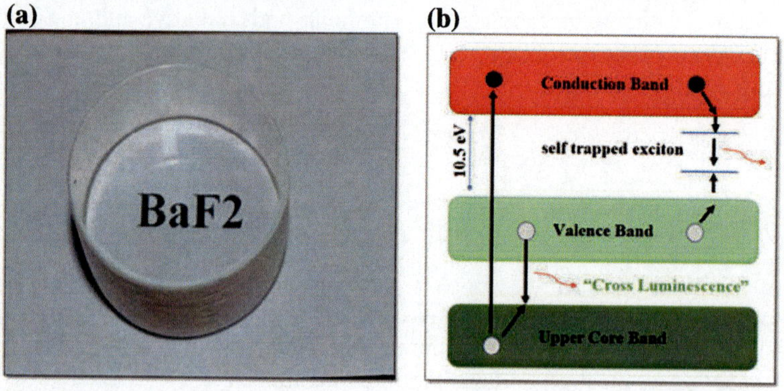

Fig. 22.11 **a** Photograph of a processed BaF_2 single crystal of diameter 30 mm. **b** Schematic for the scintillation mechanism in BaF_2 scintillator

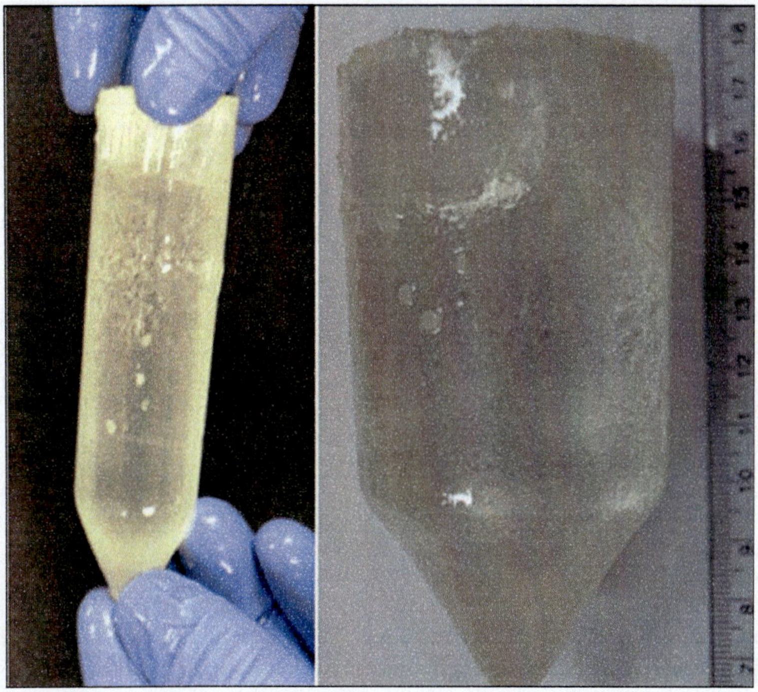

Fig. 22.12 $LaBr_3$:Ce single crystal grown at crystal technology section (left: diameter 25 mm and length 50 mm, right: diameter 50 mm and effective length 50 mm)

valence band. The excited and slowly relaxed electrons in the conduction band can recombine via self-trapped exciton formation with a hole in the valence band. This process is slow compared to the previous one. So, BaF_2 has two decay components. The schematic diagram of the scintillation mechanism in BaF_2 is shown in Fig. 22.12 (Kamada et al. 2005).

(ii) ***Pure CsI***

CsI:Tl has been known as a scintillation for more than half century, which exhibits high light yield at room temperature. Pure CsI is also known as a scintillator but have different properties. It has very fast emission 10 ns but very low light yield at room temperature. At 77 K temperature, the light yield of CsI increases by more than one order of magnitude. This makes CsI an attractive scintillator for detector applications at low temperature (Mikhailik et al. 2015). Low-temperature luminescence studies shows that the main emission features in CsI are due to the radiative decay of self-trapped excitons (STE). Two types of STE exit in CsI: on-center STE and off-center STE. The excited states of these STE have singlet and triplet configurations in thermal equilibrium. Thus, the principal features of CsI luminescence can be interpreted in terms of changing populations of these states with time and temperature (Seema Shinde et al. 2014).

22.5.3 Trivalent Rare-Earth Doped Scintillators (Ce^{3+})

The main criteria for the search for new scintillator material are high energy resolution as well as fast response time. Tl^{+}-doped halide scintillators (NaI:Tl, CsI:Tl) have high light yield but their response is slow (around 1 μs). Other scintillator materials like CsI, BaF_2, and CeF_3 have fast response time (in ns) but light yield is very less compared to Tl doped, which degrades the energy resolution. Rare-earth ions are often used as activating luminescence ions. The main focus of the present day research is to find a bright and fast scintillator can be achieved by using rare-earth ions as a luminescent center. Lanthanides ions as dopants give luminescence due to an allowed *5d–4f* electric dipole transition which is a fast transition. Experimentally, only two rare earth trivalent ions, Ce^{3+} and Pr^{3+}, are acceptable activators with high light output and fast scintillation. Most of the fast scintillators, which have attracted attention in the last few years for their attractive scintillation properties are Ce^{3+} doped both halide and oxide scintillators.

(i) ***$LaBr_3$:Ce***

Newly developed rare-earth doped halide scintillators are lanthanum chloride and lanthanum bromide doped with cerium. They are both very hygroscopic but offer excellent light output and energy resolution. The scintillation yield of $LaBr_3$:Ce is 63 photons/keV for γ radiation and a fast response of 16 ns. It also has excellent linearity, and a very stable light output over a wide range of temperatures. In addition, $LaBr_3$:(Ce offers a higher stopping power for γ rays (density of 5.08 g/cm^3 (Quarati et al. 2007)). Thus $LaBr_3$:Ce is one of the most promising advanced scintillators. However, the main disadvantages of this material are its hygroscopicity, difficult crystal growth, and intrinsic activity which prohibit its use in low background applications.

Figure 22.8 shows the photograph of LaBr3:Ce (Diehl et al. 2015) single crystal grown at the Crystal Technology Section of Technical Physics Division BARC. $LaBr_3$: Ce single crystal has been grown via the Bridgman technique with vibration-free translation and controlled by the PLC system. The furnace is designed so that the temperature gradient between the top and bottom zone is 25 °C/cm. Figure 22.13

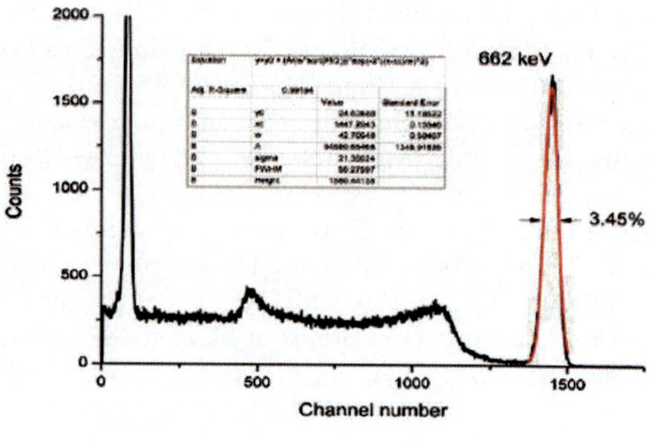

Fig. 22.13 The gamma spectrum for ^{137}Cs source measured using a 2″ diameter $LaBr_3$:Ce scintillator

is the pulse height spectrum recorded using the 2-inch diameter and 2-inch length $LaBr_3$:Ce single crystal grown at CTS. An energy resolution of 3.45% at 662 keV for ^{137}Cs gamma source is achieved.

(ii) ***Cerium Halide***

Cerium halide scintillator is a good choice for the scintillator material because of their relatively high atomic number, fast scintillation response (17 ns) and they are intrinsic scintillators. Scintillation mechanism in Cerium halide is quite simple due to the simple structure of the energy level diagram of Ce^{3+}. $CeBr_3$ scintillator crystal is the best choice for high count rate coincidence measurement because of its high energy resolution 3.6% at 662 keV and fast decay time 17 ns (Markov 2016) also it has an outstanding timing resolution which is better than BaF_2. Growth of $CeBr_3$ single crystal is challenging due to its hygroscopic nature and prone to cracking. These challenges have been addressed using various dopants in the CeBr3. Improvement in hygroscopicity of $CeBr_3$ has been reported using Ba^{2+} doping (Sisodiya et al. 2022). Figure 22.14 shows a single Crystal growth of Pure $CeBr_3$ (Size: 25 mm dia × 35 mm length) grown by employing the Bridgman technique.

Apart from the inorganic scintillators, there are oxide-based materials available, which show scintillation characteristics.

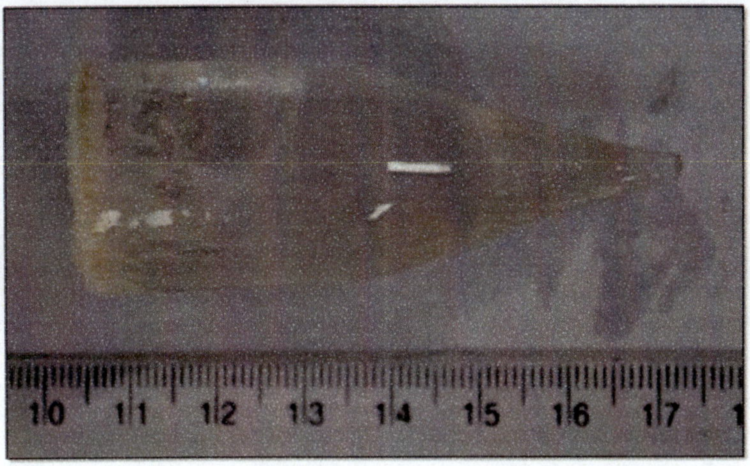

Fig. 22.14 Photograph of a 25 mm diameter and 35 mm length pure $CeBr_3$ scintillator single crystal

(iii) ***YAG:Ce and YAP:Ce***

YAl_5O_{12}:Ce (yttrium aluminum garnet also called YAG) and $YAlO_3$:Ce (yttrium aluminum perovskite also called YAP) scintillator single crystals have many attractive properties (Baccaro et al. 1995). YAG doped with ions such as Nd^{3+}, Er^{3+}, and Yb^{3+} have been used to develop YAG-based laser material. With Ce^{3+} doping YAG and YAP are used as scintillator material. Both these scintillators have high density with YAG:Ce 4.57 g/cc & YAP:Ce of 5.37 g/cm^3. They have good mechanical hardness and high chemical stability. Both the composition melts congruently at a temperature slightly less than 2000 °C and can be grown by the Czochralski technique. YAG:Ce has been grown in both reduction and non-reduction medium using iridium crucible. The growth of YAG is favorable along <100> and <111>. Figure 22.15 shows the photograph of a typically grown YAG:Ce and YAP:Ce single crystal at CTS. The YAP:Ce crystal is grown in an Ir crucible of 50 mm ϕ under Ar ambient. The pull rate was varied between 0.7 and 1 mm/h and the rotation rate 10–15 rpm. After the growth was completed, the crystal was subjected to post-growth cooling: 40–50 °C/h. The dimension of the grown crystal is 30 mm diameter for both YAG:Ce and YAP:Ce.

Fig. 22.15 Photograph of a 30 mm diameter and 50 mm length YAG:Ce (Left) and YAG:Ce (Right) scintillator single crystal grown by Czochralski technique at CTS, TPD

(iv) ***Lu_2SiO_5:Ce***

Lutetium oxyorthosilicate; Lu_2SiO_5 (LSO) activated with cerium has many applications in high energy physics and medical imaging societies because of its high density (7.41 g/cm^3), fast decay 45 ns and high light output (about 25,000 ph/MeV) (Mao et al. 2013). The Emission wavelength is at 420 nm which is well matched with the gain of PMT. The melting point of LSO crystal, however, is 2150 °C. It is also grown by CZ method. The crystal has been grown in CTS in Ir crucible under Ar ambient. The pull rate employed was 1 mm/h and post-growth cooling rate was 20–30 °C/h. Figure 22.16 is the photograph of the as-grown and processed LSO:Ce single crystal under day-light condition as well as UV radiation.

22.5.4 Divalent Rare-Earth Doped Scintillators (Eu^{2+})

Besides, the trivalent rare-earth ions, divalent rare-earth ion Eu^{2+} is also a highly used activator with high light output. Eu^{2+} has relatively slow kinetics with $\tau_r = 1\ \mu$ s (Singh et al. 2012), and gives luminesce due to the $5d^1 \rightarrow 4f^7$ electronic transition. SrI2:Eu and LiI:Eu are two commonly used Eu^{2+} doped halide scintillators.

(i) ***SrI_2:Eu:***

SrI_2:Eu is the best halide-based scintillator in terms of energy resolution, it has 2.8% energy resolution at 662 keV and has 70 ph/KeV light yield which is excellent for gamma-ray spectroscopy. It can be used for low energy detection because SrI_2:Eu doesn't have intrinsic background issues like $LaBr_3$. SrI_2 melts congruently at 518°C. Single crystal growth of SrI_2 is challenging because of its distorted Pbca orthorhombic structure and easily hinder by the presence of foreign elements. The detailed study of nucleation control, material, and crucible purity (Patra et al. 2020) on the growth of SrI_2 has been studied by Patra et al. (2022). The successfully grown SrI_2 2″ ϕsize single crystal after optimizing the crucible design is shown in Fig. 22.17. Figure 22.17 also shows the pulse height spectra of a typical SrI_2:Eu single crystal. 5% Eu doped SrI_2 crystal of diameter 2 inches can be grown using this technique.

(ii) ***LiI:Eu***

LiI:Eu scintillator detector is mainly used for the detection of thermal neutron detection. ^{6}Li has large cross-section (around 1000 barn) for the thermal neutron, thermal neutron interacts with the neutron and produces alpha and tritium with Q value 4.8 MeV as described in Eq. (22.4). Here, Eu works as an activator in LiI matrix. LiI has a melting point of 469 °C and can be grown by the Bridgman technique in a carbon-coated quartz crucible. Figure 22.18 shows a 2-inch diameter and 2-inch length LiI:Eu single crystal. A neutron detector has been fabricated at CTS using the grown crystal and detection of thermal neutron has been demonstrated (Singh et al. 2017, 2018b). Figure 22.18b shows the pulse height spectrum recorded using the LiI:Eu scintillator for thermal neutron and gamma radiation.

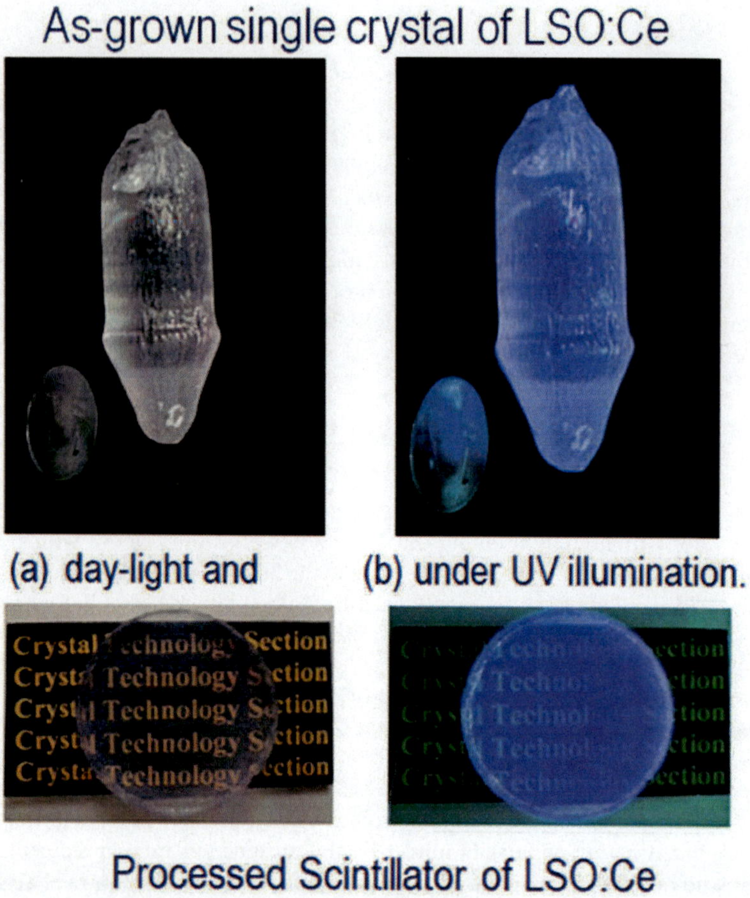

Fig. 22.16 Photograph of a 30mm diameter and 40 mm length Lu_2SiO_5:Ce scintillator single crystal. The top photograph is that of as-grown crystal and the bottom photo is cut and processed crystal. On the left, we have shown the crystal under daylight illumination and the right panel the crystal is photographed under UV illumination

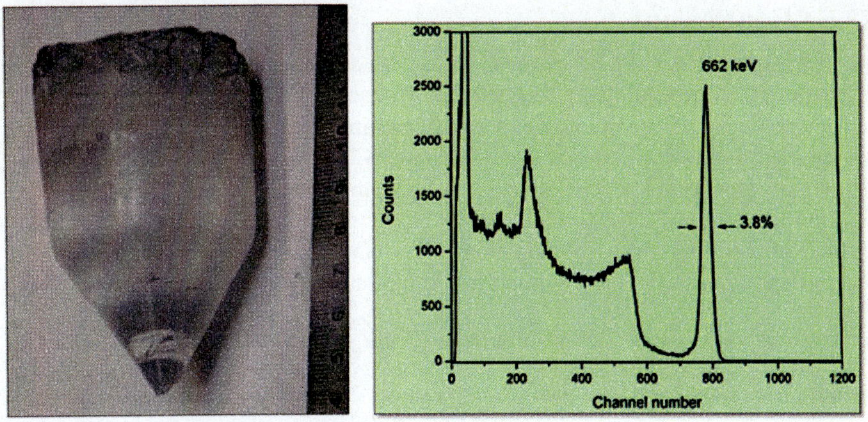

Fig. 22.17 Photograph of a 2″ diameter and 2″ length SrI_2:Eu scintillator single crystal and the gamma spectrum for ^{137}Cs source

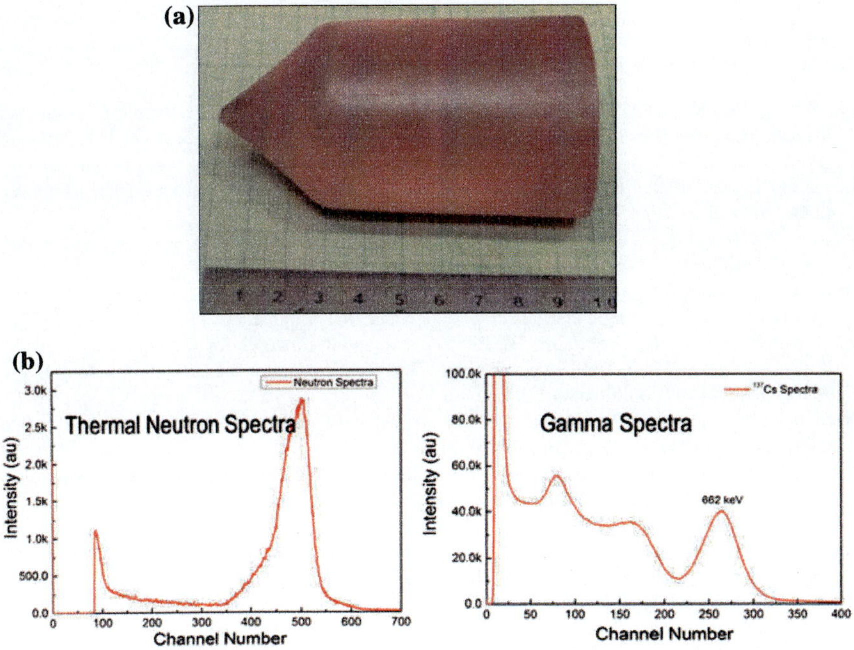

Fig. 22.18 **a** Photograph of a 2″ diameter and 2″ length LiI:Eu scintillator single crystal **b** Energy spectrum recorded using a LiI:Eu single crystal for thermal neutron and gamma radiation

22.6 Conclusions

To summarize, the application of scintillator material is increasing every day with the emergence of new technologies and applications in different fields. With this, the requirement for a scintillator with higher density, high light output, and fast decay is increasing many folds. This chapter has described briefly the physics behind the scintillation process, the characteristics, which make a material a good scintillator. Further emphasis has been given on the growth of inorganic scintillator single crystals and in-depth discussion on few important scintillators has been presented.

Acknowledgements Authors would like to acknowledge all the members of Crystal Technology Lab, TPD BARC for their contribution toward the growth and characterization of scintillator single crystals.

Author Contributions The manuscript was written through contributions of all authors. All the authors have given approval to the final version of the manuscript.

References

Baccaro, S., Blažek, K., de Notaristefani, F., Maly, P., Mares, J.A., Pani, R., Pellegrini, R., Solur, A.: Scintillation properties of YAP:Ce. Nucl. Instrum. Methods Phys. Res. Sect. A **361**, 209–215 (1995)

Barvinschi, F., Santailler, J.-L., Duffar, T., Le Gal, H.: Modelling of Verneuil process for the sapphire crystal growth. J. Cryst. Growth **198–199**, 239–245 (1999)

Crookes, W.: On attraction and repulsion resulting from radiation. Philos. Trans. R. Soc. Lond. **164**; 501–527 (1874)

Desai, D.G., Singh, S.G., Singh, A.K., Sen, S., Shinde, S., Gadkari, S.C.: Scintillation yield uniformity studies on single crystals of Tl doped CsI. AIP Conf. Proc. **1512**, 864–865 (2012)

Diehl, S., Novotny, R.W., Wohlfahrt, B., Beck, R.: Readout concepts for the suppression of the slow component of BaF_2 for the upgrade of the TAPS spectrometer at ELSA. J. Phys. Conf. Ser. **587**(1), 012044. IOP Publishing (2015)

Dietrich, H.B., Murray, R.B.: Kinetics of the diffusion of self-trapped holes in alkali halide scintillators. J. Lumin. **5**(3), 155–170 (1972)

Derby, J.J., Yeckel, A.: Heat transfer analysis and design for bulk crystal growth: perspectives on the Bridgman Method. In: Rudolph, P. (eds). Handbook of Crystal Growth (Second Edition), pp. 793–843. Elsevier (2015)

Gunther, L.: A comprehensive treatment of classical nucleation in a supercooled or superheated fluid. Am. J. Phys. **71**, 351–357 (2003). https://doi.org/10.1119/1.1528914

Kamada, K., Nawata, T., Inui, Y., Yanagi, H., Sato, H., Yoshikawa, A., Nikl, M., Fukuda, T.: Czochralski growth of 8 inch size BaF_2 single crystal for a fast scintillator. Nucl. Instrum. Methods Phys. Res., Sect. A **537**(1–2), 159–162 (2005)

Kim, C., Lee, W., Melis, A., Elmughrabi, A., Lee, K., Park, C., Yeom, J.-Y.: A review of inorganic scintillation crystals for extreme environments. Crystals **11**, 669 (2021)

Knoll, G.F.: Radiation Detection and Measurement, 4th edn. Wiley, Publisher, United Kingdom (2010)

Koshimizu, M.: Organic scintillators. In: Yanagida, T., Koshimizu, M. (eds.) Phosphors for Radiation Detectors (2022)

Lecoq, P., Gektin, A., Korzhik, M.: Scintillation mechanisms in inorganic scintillators. In: Inorganic Scintillators for Detector Systems. Particle Acceleration and Detection. Springer, Cham (2017)

Lempicki, A.: The physics of inorganic scintillators. J. Appl. Spectrosc. **62**(4), 787–802 (1995)
Leo, W.R.: Techniques for Nuclear and Particle Physics Experiments, 2nd edn. Springer (1994)
Longo, S., Khan, M., Dharani, S., von Krosigk, B., Ferber, T.: Energetic neutron identification with pulse shape discrimination in pure CsI crystals. J. Instrum. **17**, P07007 (2022). https://doi.org/10.1088/1748-0221/17/07/P07007
Mao, R., Wu, C., Dai, L.E., Lu, S.: Crystal growth and scintillation properties of LSO and LYSO crystals. J. Cryst. Growthcryst. Growth **368**, 97–100 (2013)
Markov, I.: Crystal Growth for Beginners: Fundamentals of Nucleation, Crystal Growth and Epitaxy, 3rd edn. World Scientific, Singapore (2016)
Melcher, C.: Scintillation crystal for PET. J. Nucl. Med. Offi. Publ. Soc. Nucl. Med. **41**, 1051–1055 (2000)
Mikhailik, V.B., Kapustyanyk, V., Tsybulskyi, V., Rudyk, V., Kraus, H.: Luminescence and scintillation properties of CsI: A potential cryogenic scintillator. Phys. Status Solidi (B), **252**(4), 804–810 (2015)
Moser, S.W., Harder, W.F., Hurlbut, C.R., Kusner, M.R.: Principles and practice of plastic scintillator design. Radiat. Phys. Chem. **41**(1–2), 31–36 (1993)
Moses, W.W., Payne, S.A., Choong, W.S., Hull, G., Reutter, B.W.: Scintillator non-proportionality: present understanding and future challenges. IEEE Trans. Nucl. Sci. **55**, 1049–1053 (2008)
Müller, G.: Review: The Czochralski method-where we are 90 years after Jan Czochralski's invention. Cryst. Res. Technol. **42**, 1150–1161 (2007). https://doi.org/10.1002/crat.200711001
Patra, G.D., Singh, S.G., Desai, D.G., Pitale, S., Ghosh, M., Sen, S.: Effect of OH content in the quartz crucible on the growth and quality of CsI single crystals and remedies. J. Cryst. Growth **54**, 125710 (2020)
Patra, G.D., Sisodiya, D.S., Singh, S.G., Ghosh, M., Pitale, S., Sen, S.: A systematic investigation on the effect of purity of raw material, crucible and crucible design on the single crystal growth of undoped and Eu doped SrI2. J. Cryst. Growth **600**, 126905 (2022)
Patton, D.: Invest. Radiol. **28**(10), 954–961 (1993)
Pimpinelli, A., Villain, J.: Physics of Crystal Growth. Cambridge University Press, Cambridge (2010)
Quarati, F., Bos, A.J.J., Brandenburg, S., Dathy, C., Dorenbos, P., Kraft, S., Owens, A., et al.: X-ray and gamma-ray response of a 2″× 2″ LaBr 3: Ce scintillation detector. Nucl. Instrum. Methods Phys. Res., Sect. A **574**(1), 115–120 (2007)
Rogulis, U., Spaeth, J.-M., Elsts, E., Dolgopolova, A.: Tl-related radiation defects in CsI:Tl. Radiat. Meas. **38**, 389–392 (2004)
Schotanus, P., Dorenbos, P., Eijk, C.W.E., Hollander, R.W.: Recent developments in scintillator research. IEEE Trans. Nucl. Sci. **36**, 132–136 (1989). https://doi.org/10.1109/23.34419
Shinde, S., Singh, S.G., Sen, S., Gadkari, S.C., Gupta, S.K.: Temperature dependent photoluminescence studies in CsI:Tl films with varying thicknesses. Phys. Status Solidi (B) **251**(4), 748–754 (2014)
Singh, S.G., Tyagi, M., Desai, D.G., Singh, A.K., Tiwari, B., Chauhan, A.K., Sen, S., Gadkari, S.C.: Development of technologically important crystals and devices. BARC Newsl. **318**, 16 (2011)
Singh, S.G., Desai, D.G., Singh, A.K., Tyagi, M., Sen, S., Sinha, A.K., Gadkari, S.C., Gupta, S.K.: Growth of CsI:Tl crystals in carbon coated silica crucibles by the gradient freeze technique. J. Cryst. Growth **351**, 88–92 (2012)
Singh, S.G., Desai, D.G., Singh, A.K., Pitale, S., Patra, G.D., Ghosh, M., Tyagi, M., Tiwari, B., Sen, S., Gadkari, S.C.: Development of crystal growth system to grow Tl doped CsI and NaI single crystals for applications in nuclear radiation detection. BARC Newsl. Founder's Day Spec. Issue 288 (2015)
Singh, A.K., Tyagi, M., Patra, G.D., Sen, S., Gadkari, S.C.: Setup of thermal neutron detectors by employing in-house grown LiI:Eu Single crystal. In: Proceedings of the DAE Symposium on Nuclear Physics, vol. 62, p. 1018 (2017)

Singh, S.G., Patra, G.D., Desai, D.G., Sen, S., Gadkari, S.C.: Single crystal growth of Ce doped $LaBr_3$ and its characterization. In: Proceedings of the Materials and Technologies for Energy Conversion and Storage: Book of Abstracts, p. 287 (2018a)

Singh, A.K., Tyagi, M., Desai, D.G., Sen, S., Gadkari, S.C.:Performance stability of in-house developed thermal neutron detector using large size hygroscopic LiI:Eu single crystal. In: Proceedings of the DAE International Symposium on Nuclear Physics, vol. 63, p. 182 (2018b)

Sisodiya, D.S., Singh, S.G., Patra, G.D., Desai, D.G., Sen, S.: Effect of Ba2+ doping on the properties of CeBr3 single crystal. J. Crystal Growth **582**, 126528 (2022)

Suerfu, B., Wada, M., Peloso, W., Souza, M., Calaprice, F., Tower, J., Ciampi, G.: Growth of ultra-high purity NaI (Tl) crystals for dark matter searches. Phys. Rev. Res. **2**(1), 013223 (2020)

Tyagi, M., Singh, S.G., Singh, A.K., Desai, D.G., Tiwari, B., Sen, S., Datta, R., Gadkari, S.C., Gupta, S.K.: Development of $Gd_3Ga_3Al_2O_{12}$:Ce single crystal scintillator and USB based portable gamma-ray spectrometer. BARC Newsl. 33 (2015)

Vaněček, V., Děcká, K., Mihóková, E., Čuba, V., Král, R., Nikl, M.: Advanced halide scintillators: from the bulk to nano. Adv. Photonics Res. **3**, 2200011 (2022)

Weber, M.J.: J. Lumin. **100**, 35–45 (2002)

Chapter 23
Total Reflection X-ray Fluorescence: Suitability and Applications for Material Characterization

Sangita Dhara and N. L. Mishra

23.1 Introduction

X-Ray Fluorescence (XRF) analysis is based on the principle of sample irradiation by a beam of X-rays having sufficient energies and measuring the energies as well as intensities of the X-rays of the individual atoms of the sample, thus excited. It is a near-surface (100 μm) analysis technique, performed without sample consumption or destruction. It is a fast, multi-elemental, versatile, and sensitive method with reasonably good accuracy as well as precision for elemental analysis. For these aforementioned reasons, today XRF has emerged as a very well-established technique worldwide, for the spectrochemical analysis of materials having technological importance in almost all areas of life. It is in fact, a complementary technique to equally versatile phase analysis X-ray methods which include, Ray Diffraction (XRD) and X-Ray Absorption Spectroscopy (XAS) used for material characterization.

Two German scientists, Kirchhoff and Bunsen (1859), a physicist and a chemist, respectively, laid the foundations of spectrochemical analysis of materials, when they made a flame spectroscope. Later, with a series of highly significant discoveries, such as radiowaves by Hertz (1887), X-rays discovered by Roentgen (1895), radioactivity by Becquerel (1896) and both, M. Curie and P. Curie (1898) and cosmic rays by Hess (1910), the level of spectrochemical analysis has advanced drastically. Roentgen's discovery of X-rays was found to have invaluable potentialities in the areas of science, technology, and medicine. The remarkable discovery of this invisible unknown radiation, by Wilhelm Conrad Röntgen, took place after 40 years of the discovery of spectroscopy by Bunsen and Krichhoff (Franks 1997). For this novel discovery, the

S. Dhara (✉) · N. L. Mishra
Fuel Chemistry Division, Bhabha Atomic Research Centre, Mumbai 400085, India
e-mail: sdhara@barc.gov.in

S. Dhara
Homi Bhabha National Institute, Anushakti Nagar, Mumbai 400094, India

R. S. Ningthoujam and A. K. Tyagi (eds.), *Handbook of Materials Science, Volume 1*,
Indian Institute of Metals Series, https://doi.org/10.1007/978-981-99-7145-9_23

first Nobel Prize in physics was conferred to Röntgen, in the year 1901. The foundation of XRF was established by Moseley (1913), using the well-known Moseley's law, which is an empirical law correlating the frequency /wavelength of characteristic X-rays to the emitted atom (Moseley 1913). Later, Barkla (1917) demonstrated that composition of a given sample could be determined by their characteristic X-rays, and for this great research, he was awarded the Nobel Prize in the year 1917. The first XRF spectrometer, which was commercially available consisted of an open X-ray target and a revolving crystal with a photographic plate for detection of X-rays and was built by Siemens and Halske, a German electrical engineering company, in 1924. At present, the number of XRF instruments used worldwide, is more than ~15,000 of which, one fourth operate in energy-dispersive (ED) mode and the rest in wavelength-dispersive (WD). With WD mode, the detection of light elements was optimized using suitable X-ray tubes, detectors with ultrathin windows, and analyzer crystals providing very high resolution. However, at present, EDXRF spectrometers are more frequently built as trends are for miniaturization of spectrometers, fast acquisition of the spectrum, and development of high-performance semiconductor detectors, in order to extend the application towards the low Z elements with improved sensitivity and spectral resolution.

An important advancement in XRF was achieved by the ingenious efforts of Arthur Holly Compton, who discovered the incoherent photon scattering, known as Compton effect, in 1923. He is also responsible for discovering an unusual phenomenon in X-rays known as total external reflection (Compton 1923). This invaluable discovery, in fact, had resulted in the evolution of another very efficacious variant of XRF. Yoneda and Horiuchi, two Japanese scientists, in 1971 (1971) presented another innovative idea of utilizing the principle of total external reflection of X-rays for excitation, in XRF spectrometry. They proposed a method for microanalysis of materials, applied on an optically flat highly reflecting sample support with a glancing angle of the X-ray beam which is incident on the sample being nearly 0.01°. An energy dispersive detector was placed directly over the sample support for measuring the emitted X-rays. This technique subsequently developed as an advanced version of EDXRF called Total Reflection X-Ray Fluorescence (TXRF), which is nothing but a geometrical variation of EDXRF with exceptionally good detection limits (Yoneda and Horiuchi 1971). In contrast to the conventional EDXRF, where the incident beam of X-rays hits the sample at nearly 45° angle, in TXRF the glancing angle of the incident X-rays is less than 0.1° and the shape of the incident X-ray beam is like a paper strip, owing to this grazing incidence. The instrumental arrangement for the conventional EDXRF and TXRF spectrometers, which differs slightly from each other, is illustrated in Fig. 23.1. Since its first reported application for trace element determination in 1971, TXRF has evolved remarkably as a high-performance sophisticated alternative of the traditional EDXRF. It has detection limits at pg to sub-pg level and can analyze all elements irrespective of metals or non-metals, except the very light elements ($Z < 6$). Moreover, TXRF analysis does not encounter the notorious matrix effect, which is quite severe in classical XRF, and the TXRF limits of detection can be compared with the other well-established trace element analysis techniques such as, Electro Thermal Atomic Absorption Spectrometry (ET-AAS), and

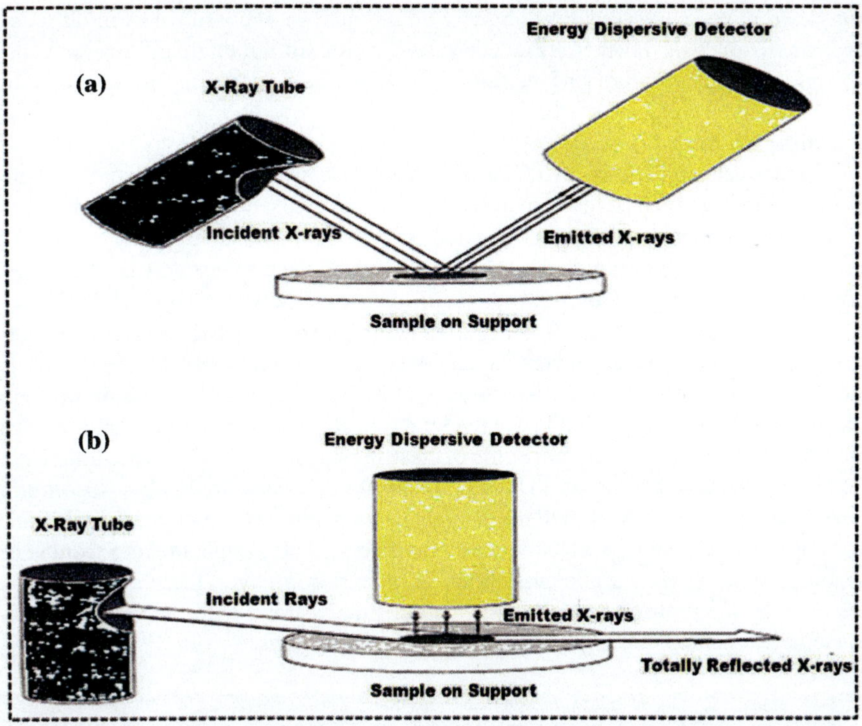

Fig. 23.1 Comparison of the geometries of instrumentation for conventional EDXRF (**a**) and TXRF (**b**). This is redrawn and adapted from John Wiley and Sons Publisher (Klockenkämper and von Bohlen 2015)

the plasma-based techniques such as Inductively Coupled Plasma Optical Emission Spectrometry (ICP-OES) and even, Inductively Coupled Plasma Mass Spectrometry (ICP-MS), which ranks the highest among these competitive techniques. In fact, in the last 50 years, TXRF has acquired a high rank among these competitive trace element determination methods.

Today, TXRF is extensively applied for microanalysis as well as trace elemental analysis of materials in almost every field of science and technology (Dhara and Misra 2019; Pahlke et al. 2001; Moens et al. 1194; Marguí et al. 2010). Samples in small quantities, mostly in the form of solutions or suspensions, are placed on optically flat sample supports.

After evaporation, the residue formed with a thickness of a few microns is excited under a fixed grazing incidence below the critical angle of the support and the X-rays emitted from the individual atoms in the sample are measured by an energy-dispersive semiconductor detector. The high reflectivity of the sample supports and low penetration of the primary beam into the supports eliminate spectral background to almost negligible amount and lowers the detection limits by 4–5 orders of magnitude, in comparison to the classical XRF. Although this technique does not offer an entirely non-destructive analysis, but it definitely offers many new possibilities in ultra and trace micro analysis. Further, its low detection limits, simplified quantification using a single internal standard and negligible matrix effect due to the minute residue or thin sample layers, makes this technique really unique and attractive among the analytical scientists.

Another new application of TXRF was opened up in 1980s by its essential application in the near-surface layer analysis. This was possible due to the dependence of fluorescence intensity of emitted X-rays from different layers, below the substrate's surface, on the angle of incidence, when it is varied in the range at which total reflection of X-rays occurs. This field of spectrometry is known as Grazing Incidence X-Ray Fluorescence (GI-XRF). This effect was applied to investigate the impurities in the surface, near-surface layers, layered wafers, etc. (von Bohlen 2009). With the increase in the diversity of materials and their applications, combinations of different spectroscopies with excitation under the total reflection condition have emerged as a multidisciplinary analytical tool. Some of these hyphenated techniques include, Total Reflection X-Ray Diffraction (TR-XRD) and Total Reflection X-Ray Absorption Fine Structures (TR-XAFS). All these hyphenated methods have greater capabilities and improved sensitivity with a requirement of very small sample making them very useful in areas where sample availability is limited. Moreover, these combinations are capable of structural as well as speciation studies along with the determination of elemental composition for micro-samples and thin films.

23.2 Principle of Total External Reflection of X-rays

The incident, reflected, and refracted rays are coplanar and normal to the interface boundary. The angle of glancing incidence (α_1), which the angle between the incident beam and the interface surface, is always equal to that of the glancing angle of the reflected beam ($\alpha^*{}_1$) as given below:

$$\alpha_1 = \alpha^*{}_1 \tag{23.1}$$

The relation between the angle of glancing of incident as well as the refracted beam (α_2) is established by the well-known Snell's law, which states that:

$$v_2 \cos\alpha_1 = v_1 \cos\alpha_2 \tag{23.2}$$

where, v_1 and v_2 are known as phase velocities of the X-rays in different mediums. Phase velocity is defined as the velocity of the planes of constant phase, of an electromagnetic radiation, propagating in a medium. This velocity is dependent on the specific medium and the wavelength/frequency of the propagating wave. However, in vacuum, the phase velocity is same as the velocity of light (c_0) and is independent of the wavelength/frequency.

Refractive index (n) is defined as the ratio of the phase velocities in vacuum and a specific medium. In vacuum, refractive index has the value of 1. Now, considering that n_1 is the absolute refractive index of medium 1 and n_2 is the same for medium 2, then these indices can be denoted as given below:

$$n_{1/2} = \frac{c_0}{\upsilon_{1/2}} \tag{23.3}$$

So, the above equation (Eq. 23.2) can be rewritten as:

$$n_1 \text{cos}\alpha_1 = n_2 \text{cos}\alpha_2 \tag{23.4}$$

Now, if $n_2 > n_1$, i.e., medium 2 is optically more dense, then the refracted beam in the medium 2 will be deflected away from the interface as shown in Fig. 23.2a. However, if $n_2 < n_1$, i.e., medium 1 is optically more dense, then the refracted beam in medium 2 will be deflected towards the interface as shown in Fig. 23.2b.

The value of refractive index, which is dependent on the phase velocity of the radiation, can be obtained from the Lorentz theory. This theory assumes that the electrons, which are quasi-elastically bound to the atoms, oscillate forcefully in the presence of the radiation field of the incident beam. Hence, the oscillating electron also emits radiation, but with a different phase. So the Phase velocity of primary beam is altered because of the superposition of both these radiations. This superposition causes a slight modification in the values of the refractive index, which now deviates from the value in vacuum, which is 1, by δ and it is called decrement. Hence, the refractive index 'n' now has a complex form having a real (δ) as well as an imaginary part (β) and written as:

$$n = 1 - \delta - i\beta \tag{23.5}$$

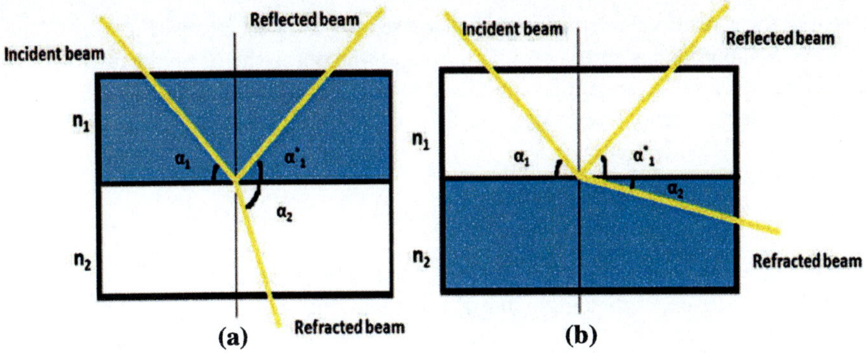

Fig. 23.2 Representation of the path of X-ray beam at the interface of two media with refractive indices n_1 and n_2 **(a)** $n_2 > n_1$ and **(b)** $n_1 > n_2$. This is redrawn and adapted from John Wiley and Sons Publisher (Klockenkämper and von Bohlen 2015)

The real part in the above equation ($n' = 1 - \delta$), in fact, measures the deviation in the refractive index. Since, **n′** is related to the phase velocity (from Eq. 23.3), $\nu \approx c_0/n\prime$, this velocity can exceed velocity of light. Only the velocity by which a real signal is transmitted cannot be greater than c_0, however, the phase velocity can be. The value of δ, for X-rays having energies more than the absorption edge of the elements in medium, can be expressed as:

$$\delta = \frac{N}{2\pi} r \frac{Z}{A} \rho \lambda^2 \tag{23.6}$$

where, **N**: Avogadro's number (6.022×10^{23} atoms/mol), ***r***: classical radius of electron (2.818×10^{-13} cm), **Z**: atomic number of the elements in the medium, **ρ**: density (in g/cm^3), **A** is the atomic mass of the elements in the medium (in g/mol) and **λ**: wavelength of the beam. The value of δ is very small and varies from 0.005 to 0.0000002 (Klockenkämper and von Bohlen 2015).

The imaginary part (**β**), is the value of the attenuation, and written as:

$$\beta = \frac{\lambda}{4\pi} \left(\frac{\mu}{\rho} \right) \rho \tag{23.7}$$

where, **μ/ρ**: mass absorption coefficient (g/cm^2).

For X-rays, air medium is optically more dense than any solid medium ($n < n_{air} \approx 1$). So, as the X-ray beam passes from the air to any solid substrate, then the refracted beam gets deflected toward the interface of the two mediums (Fig. 23.2b), which is in contrast to the visible light which gets deflected away from the interface. Therefore, for X-rays travelling from air to any solid medium, n_1 is $n_{air} \approx 1$, and Eq. 23.4 becomes,

$$\cos\alpha_1 = n_2 \cos\alpha_2 \tag{23.8}$$

Now, for a fixed value of n_2 as α_1 decreases, α_2 also decreases. Further, as the value of α_2 approaches zero, $\cos\alpha_2 \approx 1$, and at this point, the X-ray beam which is refracted, will appear tangentially in the interface. The glancing angle of incidence, at which this phenomenon occurs, is known as the critical angle, $\alpha_1 = \alpha_{\text{crit}}$. At α_{crit} the above equation becomes,

$$\alpha_{crit} = n_2 \tag{23.9}$$

For angles, $\alpha_1 < \alpha_{\text{crit}}$, α_2 does not exist, because the value of cosine cannot be more than 1. Hence, at this angle, the X-ray beam does not penetrate into the solid and the interface between both the medium becomes an ideal mirror, which reflects the entire incident beam of X-rays back into the air medium. This process is known as total external reflection or total reflection of X-rays.

There are three parameters, which define the phenomenon of total external reflection of X-rays in TXRF spectrometry. These are:

(a) Critical Angle
(b) Reflectivity
(c) Penetration depth.

23.2.1 Critical Angle

The critical angle is defined as that glancing angle, having the highest value, where refracted beam becomes non-existent and total reflection of the incident beam occurs. It is calculated from the abovementioned equation (Eq. 23.9), and expressed as

$$\cos\alpha_{crit} \approx 1 - \frac{\alpha_{crit}^2}{2}. \tag{23.10}$$

The above equation can be further simplified with the combination of Eq. 23.5 as

$$\alpha_{crit} \approx \sqrt{2\delta}. \tag{23.11}$$

From Eq. 23.6, the above equation can be further approximated to.

$$\alpha_{crit} \approx \frac{1.65}{E}\sqrt{\frac{Z}{A}\rho} \tag{23.12}$$

where, **Z**: atomic number, **A**: atomic mass of the elements of the medium (g/mol), **ρ**: the density (g/cm^3), **E**: energy of the primary beam (keV) and the value of $\boldsymbol{\alpha_{crit}}$ will be obtained in degrees.

23.2.2 Reflectivity

The ratio of the intensities of the reflected and the incident beam is called Reflectivity (R). The reflectivity depends on the angle between the incident beam and the surface of the medium. For angles, having values more than the critical angle (>than 1°), the reflectivity is in the order of 0.1%. As the glancing angle decreases, the reflectivity increases and at a critical angle, it reaches almost 100%, abruptly. For Si or quartz, the reflectivity shows a distinct steep transition with respect to the glancing angle and for this reason, these materials are mostly used as sample supports in TXRF analysis. A highly reflective sample support is used because the spectral background is proportional to 1 − R.

23.2.3 Penetration Depth

The penetration depth describes that depth, which a beam can penetrate, in a homogeneous material, such that its intensity decays to 37% of the original value. The penetration depth decreases with decreasing glancing angle, with values ranging from 0.01 μm to 10 μm, till it reaches the critical angle. When critical angle is reached, the penetration depth decreases drastically to 2–5 nm and below the critical angle, it attains the same constant value irrespective of the glancing angle.

The effect of total reflection appears only when the substrate or the sample support is optically flat or smooth. Besides optical flatness, high reflectivity is also very essential. The roughness should be of the order of <5 nm with in the area of 1mm^{2}and the waviness should be <0.001° within an area of 1 cm^2. These properties can be checked by a profilometer. Table 23.1 gives the values of these abovementioned parameters, for the commonly used sample supports in TXRF measurements, for different incident energies of X-ray beam.

23.3 Instrumentation

The first commercially available TXRF spectrometer was built in 1980 by Rigaku Corporation, Japan (Model: EXTRAN). Though the instrumentation of TXRF is very similar to traditional EDXRF spectrometers, there is a difference in the geometric arrangement as in TXRF it should provide total external reflection of the incoming beam. Further, it should also ensure that there is interference between the incoming and completely reflected beam, which results in standing waves, just above the surface of the sample support (Bedzy 2013).

In TXRF analysis, the following criteria for instrumentation are required to be followed to ensure the aforementioned conditions:

1. The glancing angle of incident beam needs to be very small, less than the critical angle, to provide a total reflection of the incoming X-ray beam. For primary

Table 23.1 Values of critical angle, reflectivity, and penetration depth for different sample supports and incident X-ray energies commonly used for TXRF measurements (Klockenkämper and von Bohlen 2015)

Medium	Critical angle at photon energies in (°)			Reflectivity at photon energies in (%)			Penetration depth at photon energies in (nm)		
	8 keV	17 keV	35 keV	8 keV	17 keV	35 keV	8 keV	17 keV	35 keV
Plexiglas	0.16	0.08	0.04	87.9	93.2	94.8	132	241	319
Glassy carbon	0.17	0.08	0.04	88.4	93.9	95.0	130	255	311
Quartz glass	0.21	0.10	0.050	73.4	85.5	91.4	42	83	146
Silicon	0.21	0.10	0.051	67.3	81.5	89.5	32	62	115

X-ray beam having energy of 10 keV, the critical angle is nearly 0.1–0.2°and it is further less for higher energies (Table 23.1).
2. The primary X-ray beam should be in the shape of a paper strip, obtained by a line-focus tube. Further, the aperture in the tube should be such that it restricts the beam height to approximately 10–20 μm and the width should be nearly 10 mm.

For the analysis of the sample in the form of granular residues, the measurement is carried out at a fixed glancing angle setting with the following two conditions:

1. The glancing angle should be set at nearly 70% value of critical angle and the divergence of the beam should be restricted to below 0.01°.
2. The bremsstrahlung continuum, which is the high-energy portion of the beam, should be removed, so that total reflection can take place at a smaller angle. Though under a larger glancing angle the high energy X-rays will not be totally reflected, but it would definitely result in an increased undesirable background. If a band pass or low pass or filter is positioned between the source and the sample, it will prevent the background by cutting off the high-energy part of the beam. Use of monochromators can also be extremely helpful as it produces X-ray beams of narrow energy band and this arrangement eliminates the background drastically.

For depth profile measurements using TXRF, the angle-dependent intensity profile has to be recorded. So, instead of a fixed angle measurement, the glancing angle has to be varied and the TXRF spectrum is recorded at each angle. Following are the prerequisite for carrying the measurements:

1. The angle of glancing must be varied between 0 and 2° with very high precision in steps of 0.01°. A very great accuracy of fine-angle control better than 0.005° is necessary for such measurements. This can be obtained by tilting the sample around an axis.
2. The primary beam should be monochromatic in order to ensure an angle-dependent, clear, and distinct intensity profile for thin-layered samples.

The basic design of a TXRF instrument includes a fine-focused X-ray tube having collimator slits, monochromators or low-pass filters, sample holders, and an energy-dispersive detector unit. The charge pulses produced by the detector are processed using electronic registration unit. The commonly used target materials for X-ray tubes are, Mo, Rh, Pd, and W for excitation of elements from Al to U, and for low Z elements, below Ti, Cr and Cu targets are used. Also, dual targets X-ray tubes such as Mo/W, are commercially available. The efficiency of X-ray tubes is very low, as very little of the electric power (0.1%) is actually being converted into X-rays and the rest of it is transformed as heat. Hence, cooling of these tubes is very essential. The X-ray tubes of high power, which operate at 3–5 kW, require continuous cooling with chilled water at the rate of 5 L/min when in operation. The average life span of these tubes is about 4000 operational hours. Rotating anode X-ray sources can operate at very high power (30 kW) with chilled water supply of 15 L/min. These tubes provide high brilliance and intensity many folds higher than the conventional

X-ray tubes. Compact low power table top spectrometers, powered by 50–200 W X-ray tubes, are also gaining importance due to their compactness and simplicity. These spectrometers with peltier-cooled detectors and low-power X-ray tubes do not require any liquid nitrogen cooling for detectors or water chillers for the X-ray tubes, which makes them very compact with negligible maintenance. Moreover, with the introduction of interchangeable X-ray tubes having low power, determination of all the elements above $Z = 6$ (Carbon) is possible using a single XRF instrument (Sanyal et al. 2016). The scientific community, using synchrotron radiation as X-ray source for research, is also growing steadily since the mid-1980.The synchrotron radiation has many features which includes high brilliance which surpasses that of X-ray tubes, by many orders of magnitude, high collimation, variable polarization, broad tunable energy spectrum, etc. (Meirer et al. 2010). Synchrotron radiation-induced Total reflection X-ray Fluorescence (SR-TXRF) offers analytical sensitivity in the range of femtograms, besides applications such as SR-TXRF-XAFS (Synchrotron Radiation-Total Reflection X-ray Fluorescence-X-Ray Absorption Fine Structures) (Meirer et al. 2010). Further, due to pulsed nature of the radiation, it also allows to perform experiments at the timescale of micro and nano-seconds with very high resolution, both lateral and temporal. Use of tunable monochromators for alloy/dual target X-ray tubes e.g. Highly Oriented Pyrolytic Graphite (HOPG) and multilayer monochromators of W-C, W-Si, Ni-C, Mo-B_4C, Pd-B_4C, etc., results in high intensity monochromatic excitation sources of X-rays. There is also a possibility of using evacuated or helium-flushed sample chambers to eliminate air from the chamber, which causes Ar-Kα interferences, as well as loss of the low energy X-rays in the air atmosphere. Initially, lithium drifted silicon, energy dispersive solid state detectors, also known as Si(Li) detectors were used. These detection systems required continuous liquid nitrogen cooling. Recently, detectors have become the most dynamic and vital field for R&D in XRF, in particular, TXRF spectrometry. As a result, the introduction of Silicon Drift Detector (SDD) arrays has improved the sensitivity for low-energy X-rays along with better spectral resolution and no requirement of liquid nitrogen cooling (Rachevski et al. 2016). This has lead to miniaturization of detector size, for compact table top spectrometers. A basic schematic diagram of a TXRF spectrometer showing its operating parts is shown in Fig. 23.3.

23.3.1 Analytical Performance of TXRF for Material Characterization

As TXRF is an advanced version of EDXRF, it shares all convenient features of conventional EDXRF such as the complete spectrum being recorded simultaneously within a few 100 s and all the elements, from Na to U, present in the sample are displayed making it a very fast analytical technique. With advancements in the programming for the acquisition of spectra, automatic peak identification, and fitting

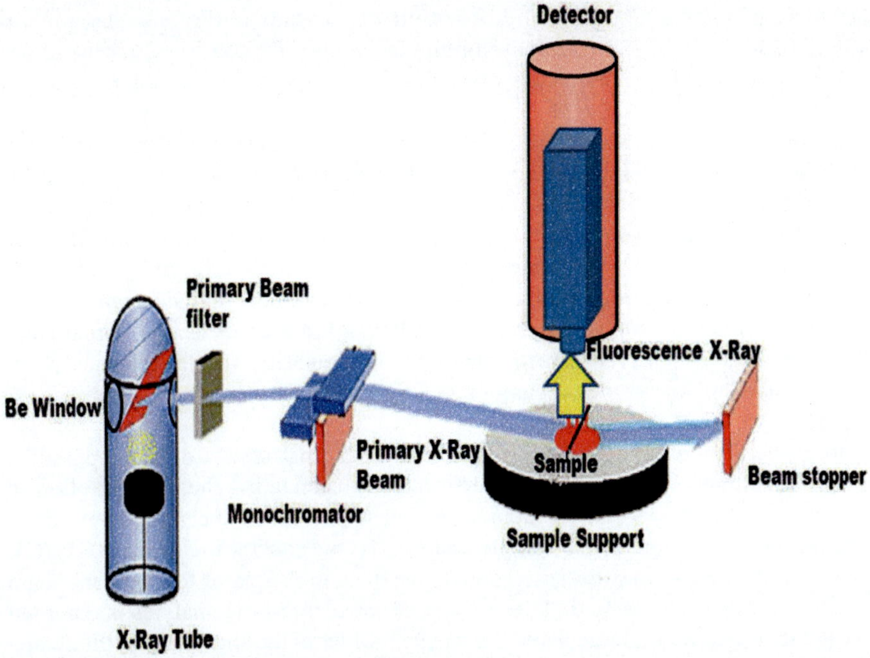

Fig. 23.3 A basic schematic diagram of a total reflection X-ray fluorescence spectrometer. This is redrawn and adapted from John Wiley and Sons Publisher (Klockenkämper and von Bohlen 2015)

of the interfering elements, calculation of the area under the peak is possible very precisely, thereby enhancing the ease of a qualitative as well as quantitative analysis.

Now, with improvements in detection systems and using vacuum sample chamber coupled with ultrathin window detectors, elements with atomic numbers as low as 6 (Carbon) can be detected and analyzed by TXRF. Further, a visual comparison of two or more complete spectra enables a fingerprint analysis of different elements within no time.

TXRF analysis is a method of microanalysis with restricted sample size of 2–100 μL or 200–1000 ng. Consequently, TXRF is now recognized as a method of microanalysis by the International Union of Pure and Applied Chemistry (IUPAC) (Klockenkämper and von Bohlen 2015). Because of the restriction in sample size, it can seldom analyze a complete sample received as it is. Samples in the form of solutions can be analyzed directly, whereas those in the form of suspensions and fine powders require sample processing, prior to analysis. Since a small specimen representative of the bulk sample is analyzed in TXRF, sample must first be homogenized. Further, solid bulk samples need cutting, sawing, grinding to fine powder followed by suspending them in a suitable solvent or digesting them, prior to analysis. Thin films of biological tissues are analyzed by freeze-drying and cutting. Placing of individual particles directly, laser ablation for small area analysis, collection of air particulate matter by impaction, Q-tip method of sample collection and direct contamination control of wafers, are some of the sampling techniques for solids (Klockenkämper and von Bohlen 1999). Sample preparation and presentation on the sample support, prior to analysis, necessitate an extremely clean space for working, mostly a clean bench or room of class100 is sufficient for trace and ultra-trace analysis. Recently, TXRF has gained immense popularity for applications in environmental, geological, chemical, petroleum, and nuclear industries as well as in biological, biochemical, medical, and pharmaceutical (Dhara et al. 2015a; Joshi et al. 2021; Antosz et al. 2012a; Hauser et al. 2022). A complete quantitative analysis is feasible with only a few ng of samples, the troublesome matrix effect is negligible and the quantification can be performed by adding a certified standard, which serves as an internal standard. These features of TXRF are the reason for its popularity. The different analytical steps involved in the sample preparation in TXRF for trace elemental determination, shown in Fig. 23.4.

In case of solid samples, where addition of internal standard is not possible, a semi-quantitative analysis can be performed by calculating the concentration in percentage, with respect to an element which is present in the sample. A direct method for thin or multilayer sample analysis is, by grazing incidence XRF (GI-XRF). In this method, as discussed above, the glancing angle of the incident beam is varied within the critical angle and the peak intensity of the analytes is recorded simultaneously. The angle-dependent intensity profiles of the analytes give the details of elemental concentrations, in different layers of the sample. Since, quantification

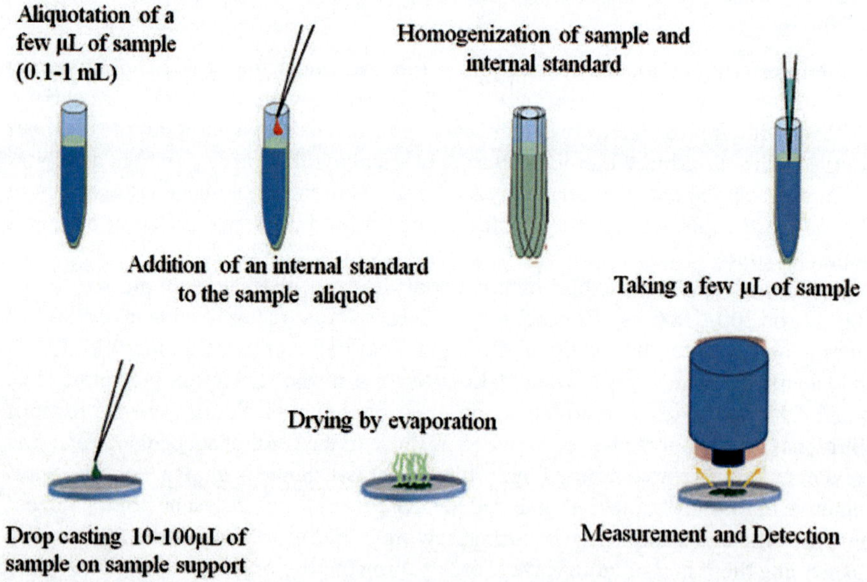

Fig. 23.4 Different steps of sample preparation in TXRF analysis. This is redrawn and adapted from John Wiley and Sons Publisher (Klockenkämper and von Bohlen 2015)

of the elements in such samples cannot be carried out using an internal standard, an algorithm called the fundamental parameter method is used. Fundamental parameter is based on creating mathematical models to calculate the exact peak intensities of characteristic X-rays, while theoretically correcting the matrix effects as well. The fundamental data can be obtained from literature or partly be calculated by use of complex mathematical equations.

23.4 Calibration and Quantification

A qualitative analysis refers to the detection of the elements, which are present in a sample at a concentration level above their detection limits. Using energy dispersive detection systems, it can be performed rapidly and simultaneously, as in EDXRF. XRF analysis gives the details about the elemental composition of a sample, by measuring the energies of the characteristic X-ray peaks and correlating them to the elements. A spectrum of a diluted Merck ICP Multi-Elemental Standard Solution–IV (ICP-MES Solution-IV), measured by a TXRF spectrometer is shown in Fig. 23.5. The figure shows that the elemental X-ray peaks are in compliance with Moseley's law with respect to their energies. Further, the figure shows that the X-ray spectrum has relatively fewer peaks, than the optical spectra, thereby showing the simplicity of XRF spectra. For the determination of the elemental concentrations quantitatively, the net intensities of the analyte X-ray peaks (N) are obtained by calculating the area under the peak, after correcting the area of overlapping peaks and the background, using a suitable peak deconvolution program. In an ideal case, linearity should exist between the net intensities of the analytes and their concentrations (c).

$$N = S * c \tag{23.13}$$

where, S: proportionality constant also known as ***absolute sensitivity***. The plot obtained from this equation is called the ***calibration plot***, and it is linear with slope defined by S. The slope is different for different elements, hence indicating that each element has a different sensitivity value. However, this ideal case is seldom obtained in classical XRF analysis. The discrepancy is due to the presence of matrix, which affects the intensities of analytes and is called the ***matrix effect***.

In TXRF, very small sample residues are analyzed and such minute samples having negligible mass as well as thickness meet the criteria of ideal conditions required for XRF analysis. Since there is very less sample matrix, the matrix effects are negligible, and therefore, for specific excitation source with constant instrumental conditions, the calibration does not change with the composition of the matrix. However, during TXRF analysis it is not possible to fix the sample geometry, because a small sample amount is drop casted or deposited on polished a quartz support and dried. After drying the shape of the sample can differ depending on many factors such as density, drying temperature, mode of drying, etc. So, the actual sample area facing the primary

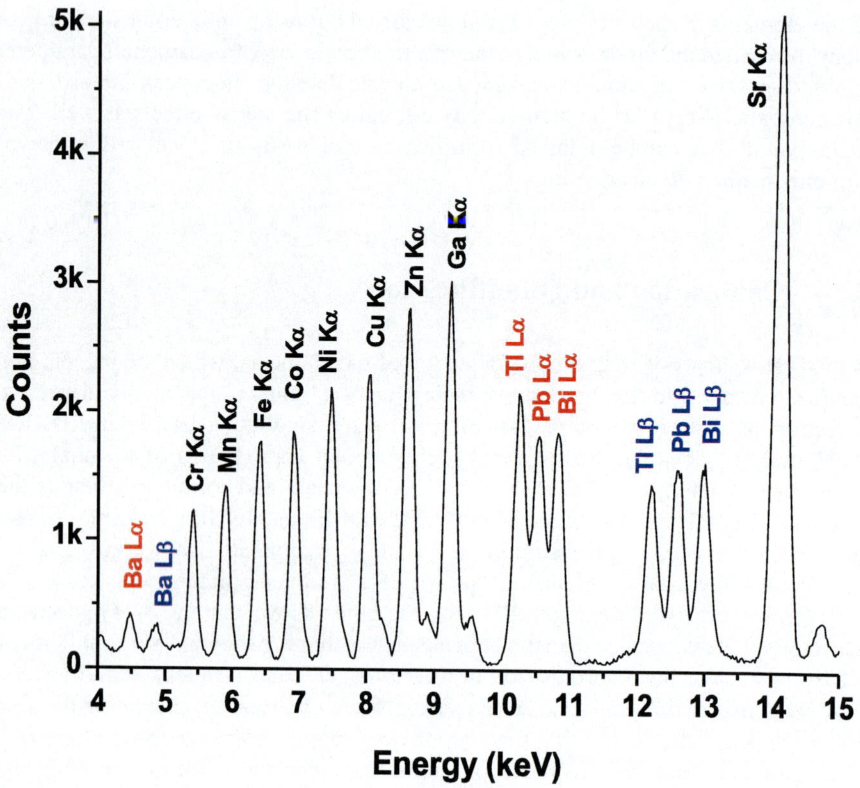

Fig. 23.5 TXRF spectrum of multielement standard solution having elemental concentration of 10 μg/mL (10 μL solution deposited on sample support)

beam may differ in different measurements. This may lead to different X-ray intensities of elements in the same sample, but different specimens. In order to take care of this anomaly a certified standard of an element, is added initially, into the sample (Fig. 23.4). This strategy of adding a standard into the calibration standard as well as sample solution is called internal standardardization, and it makes the procedure for elemental quantification very simple. The requirement for an element to be internal standard is that, the element should be absent in the sample. The ratio of absolute sensitivities of an element and the internal standard is known as ***relative sensitivity*** and it remains constant even if the intensity of the elements changes, due to the exposure of different sample area, in different specimens having same elemental concentrations. For estimation of relative sensitivity values, a certified standard of multielement is used or single elements can be mixed to make a multielement standard and then used. If the concentration of the standard solution containing the internal standard is high (1000–500 μg/mL), then it has to be diluted (100–10 μg/mL) and an aliquot ranging from 2 to 100 μL is drop cast on pre-cleaned support. After complete drying, the supports are finally taken for the measurements. The relative sensitivity values of the elements are determined using the formula given:

$$RS_x = S_x \Big/ S_{IS} = \frac{N_x}{N_{IS}} * \frac{C_{IS}}{C_x} \tag{23.14}$$

where, $\mathbf{RS_x}$: Relative sensitivity of the element (x) with respect to internal standard (IS). $\mathbf{S_x}$: sensitivity of element (x), $\mathbf{N_x}$: net peak area of element (x), and $\mathbf{C_x}$: concentration of element (x) and $\mathbf{S_{IS}}$,: sensitivity of the internal standard, $\mathbf{N_{IS}}$: net peak area of the internal standard and $\mathbf{C_{IS}}$: concentration of for the internal standard.

Figure 23.6 shows the change in the values of relative sensitivity with respect to the atomic numbers. For the evaluation of the relative sensitivity values, K X-ray lines were used for medium Z elements and L X-ray lines were used for high Z elements. The values of relative sensitivity, for the elements, which were absent in the standards, were derived by interpolating this relative sensitivity curve (Fig. 23.6). Moreover, for a definite set of instrumentation such as, mode of excitation, applied current and voltage, sample support and geometry, the values of relative sensitivity remain constant and quantification can be carried out in the samples irrespective of the composition of the matrix. Any modifications in the instrumental conditions will require an entirely new set of calibration and relative sensitivity values. The methodology based on internal standardization, for elemental determinations in TXRF analysis, is quite straightforward and reliable. Mostly, Ga, which is usually absent in the samples, is chosen as internal standard. Further, the elements whose K-lines can be used as analytical lines are preferred as internal standard, because of their higher fluorescence intensities. Light elements have less X-ray energies and low fluorescence yields, which make them unsuitable as internal standard. For determination of elemental concentrations, the net intensities of the analytes as well as the internal standard are calculated from the TXRF spectra, using a suitable peak deconvolution program. Then, by rearranging the Eq. 23.14, the elemental concentrations are determined using the equation given below

$$c_x = \frac{N_x}{N_{IS}} * \frac{C_{IS}}{RS_X} \tag{23.15}$$

For microanalysis of samples in the form of fine powder, pellets, grains, or smears of metals, in which internal standard cannot be added, the quantification is done with respect to an analyte, which is already present in the sample, considering it as an internal standard as discussed above in Sect. 23.4. It is a semi-quantitative analysis, which provides the relative concentration values of the detectable elements, present in the sample (Klockenkämper and von Bohlen 1999). However, there are some conditions and limitations for such analysis. The prerequisite condition is, a very minute sample should be deposited as a thin layer so that the matrix effects are negligible. Further, the specimen thickness, density, particle size, and area get restricted firstly due to the geometry of excitation as shown in Fig. 23.7. The height of the incoming beam is generally 20–30 μm, which is restricted by the slit in the

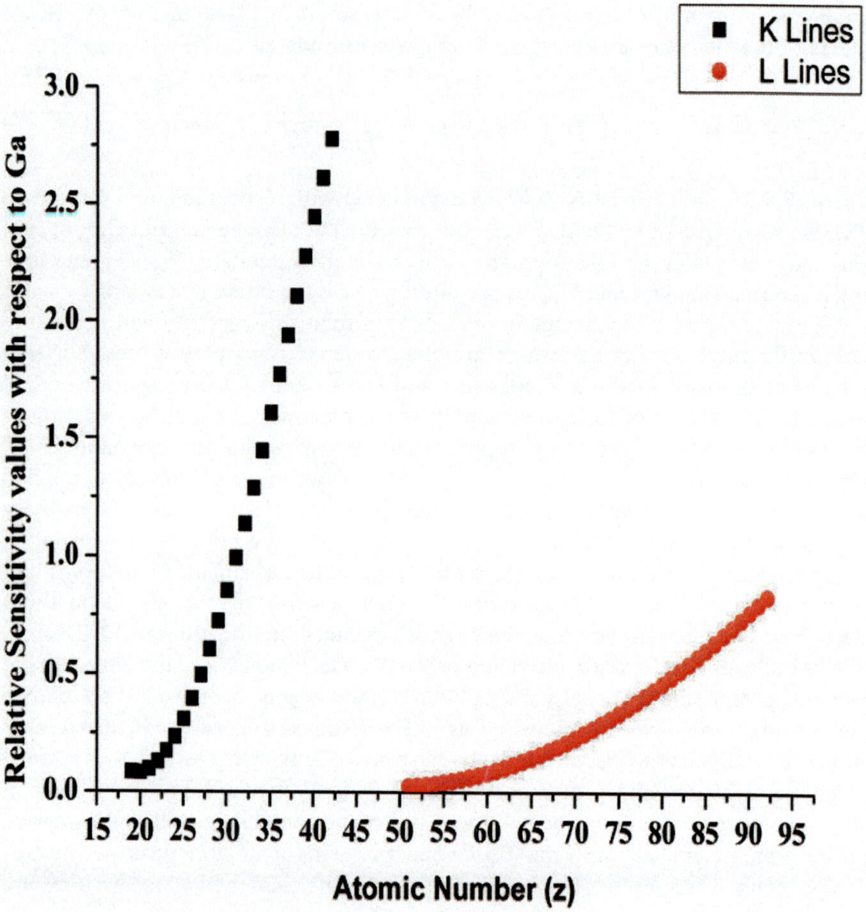

Fig. 23.6 Plot of relative sensitivity values determined with respect to Ga vs. atomic number. (K X-ray lines (for Medium Z elements) and L X-ray lines (for High Z elements)

X-ray tubes. This value is usually 100 μm for synchrotron sources. The height of the triangular region with standing wave, formed due to the intersection of the incident and totally reflected beam, $\mathbf{h_{triangle}}$ is approximately $\mathbf{h_{beam}}$/2. So, the specimen thickness gets restricted to 10–15 μm. The second limitation comes due to the counting capability of the detector. The maximum limit values of sample specimen have been experimentally determined for three typical matrices depending on their densities (ρ). If the approximate densities (ρ) of dried organic tissues is 0.2 g/cm^3, mineral powders it is 1–2 g/cm^3, and metallic smears is 8 g/cm^3 then, the upper limits for mass per unit area (m_{max}) works out to be 250 μg/cm^2 for organic tissues, 140 μg/cm^2 for mineral powders or salts, and 8 μg/cm^2 for metallic smears. This is a rough estimation of the corresponding thickness $\mathbf{d_{max}}$, where.

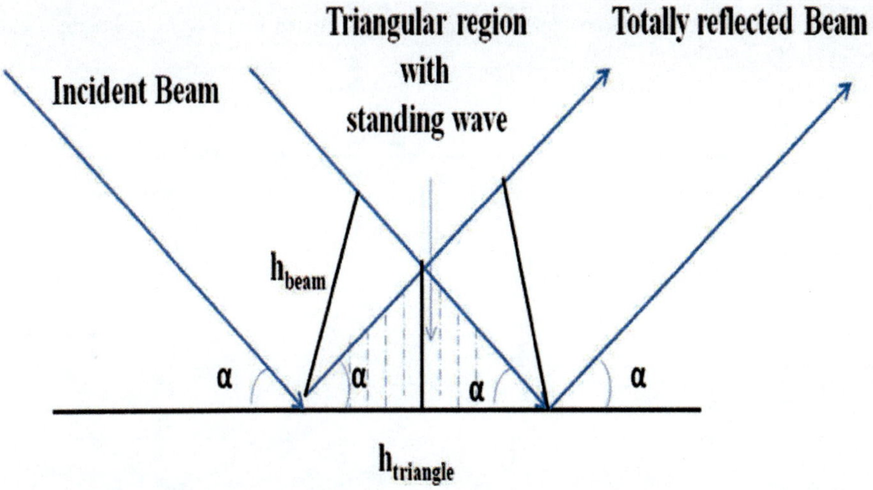

Fig. 23.7 Intersection of the incident and totally reflected beam on a flat substrate. This is redrawn and adapted from John Wiley and Sons Publisher (Klockenkämper and von Bohlen 2015)

$$d_{max} = \frac{m_{max}}{\rho}, \tag{23.16}$$

and this quantity $\mathbf{d_{max}}$ is approximately 4 μm for organic tissues, 50 nm for mineral powder, and 2 nm for metallic smears (Klockenkämper and von Bohlen 1989).

TXRF has gained popularity mainly because of its low detection limits, in comparison to the traditional XRF. It is possible to detect as well as quantify all the elements with atomic numbers greater than 6 (carbon), using different excitation conditions and environment. With special conditions of instrumentation such as detectors with ultrathin windows or windowless detectors, vacuum atmosphere, and low energy excitation source, low Z elements can be analyzed (C to Ti), whereas with high energy excitation, high and medium Z elements can be analyzed (Al to U) (Sanyal et al. 2016). For determination of detection limits (D_{Li}), a diluted MES solution is drop-casted on TXRF sample support and measured for 1000 s. Using the obtained TXRF spectrum, the D_{Li} are determined using the below-given formula:

$$D_{Li} = \frac{\boldsymbol{Concentation} * 3 * \sqrt{\boldsymbol{Background}}}{\boldsymbol{Peak\ Area}}. \tag{23.17}$$

The detection limits of different elements measured using optimized excitation source and atmospheres are tabulated in Table 23.2.

Table 23.2 The TXRF detection limits of the elements excited using optimized source and atmospheres (Sanyal et al. 2016)

Elements	Atomic number (Z)	Excitation source Analytical line Sample atmosphere	Detection limit (Picogram)
C	6	Cr Kα Kα lines Vacuum atmosphere	58,000
F	9		5190
Na	11		1026
Mg	12		753
Al	13		403
S	16		96
K	19		55
Sc	21		29
Ti	22		29
K	19	Rh Kα Kα lines Air atmosphere	3392
Sc	21		1723
Cr	24		615
Co	27		311
Ga	31		133
Se	34		71
Y	39		50
Cd	48	Rh Kα Lα lines Air atmosphere	8756
Gd	64		1737
Ho	67		1170
W	79		1534
Bi	83		285
U	92		202

23.5 Applications

The unique and advantageous analytical features of TXRF have substantially promoted its applications in various fields for different types of sample matrices. Its area of application includes environmental and geological studies of soil, water samples, airborne particulates, etc. These studies help to monitor and prevent the factors, which contribute to environmental pollution. Pure water such as rain and drinking water can be analyzed directly by TXRF, however, water samples with high total dissolved salt (TDS) such as river, pond, and seawater, require some sample preparation prior to analysis. In case of high salt matrix, the salt load in the sample can cause matrix effect, so prior to analysis, the analytes have to be separated either by complexation or chromatographic adsorption followed by elution. The biological and biochemical applications include analysis of plant samples, vegetable and essential oils, milk, tissues, and blood samples. These analyses are required for bio-monitoring of hazardous elements as well as nutrients, which are present in the organic samples. Generally, it is very difficult to analyze these samples directly. But, samples such as tea, juices, and wines can be analyzed directly using TXRF. There are several investigations reported on the determination of heavy metal content in medicines by TXRF. Due to the toxicity of these analytes, its determination in the medicines help to fix the required dose and intake duration of these medicines for proper and effective treatment. Monitoring of elemental contents in blood, urine, hair, etc., shows the signs of heavy metal poisoning and can be helpful in forensic investigations. In fact, recently TXRF has come as a great assistance in such investigations. Because of its versatility, sensitivity, and multi-elemental analysis capability, TXRF has become highly suitable for the analysis of industrial samples as well. Most of the industrial materials require stringent quality control, with sampling and analysis carried out during several stages of its fabrication and synthesis. As a versatile technique, TXRF has many applications in almost all the industries including petroleum, ultra-pure chemicals, and solvents, semiconductor, nuclear, art and forensic. In fact, TXRF has been successfully utilized to solve many important and difficult unconventional problems, and hence it has become competitive with many well-established techniques of ultra trace and trace elemental analysis.

The users of TXRF are distributed all over the world which include a wide variety of research institutes, universities, industries, etc. They use different excitation sources which include synchrotron beam lines as well as laboratory-based excitation sources like X-ray tubes. In a survey which was carried out in 2014, it was observed that nearly 700 individual units of TXRF are present in the world, which is distributed among 57 countries and 6 continents with more than 40% present in Europe itself. A pie chart showing the percentage applications of TXRF spectrometers in different fields for elemental analysis is given in Fig. 23.8 (Klockenkämper and von Bohlen 2014).

23.5.1 Environmental Applications

There are numerous applications of TXRF for environmental samples as shown in Fig. 23.8, carried out to monitor the level of pollution in the environment. TXRF has been applied successfully for ultra-trace and trace analysis of waters, airborne particulate matter, soils, sediments, etc. (Marguí et al. 2010; Kayser et al. 2022; Garg et al. 2022).

Water quality is of high concern to scientists for human health and environmental pollution. TXRF has evolved as a full-fledged analytical technique for the quantitative determination of elemental concentrations in rain, river, and seawater samples. Highly pure water samples such as rain, drinking and ground water (small salt load) can be analyzed directly by adding an internal standard, however, if a troublesome matrix load is present then the trace element has to be separated and pre-concentrated

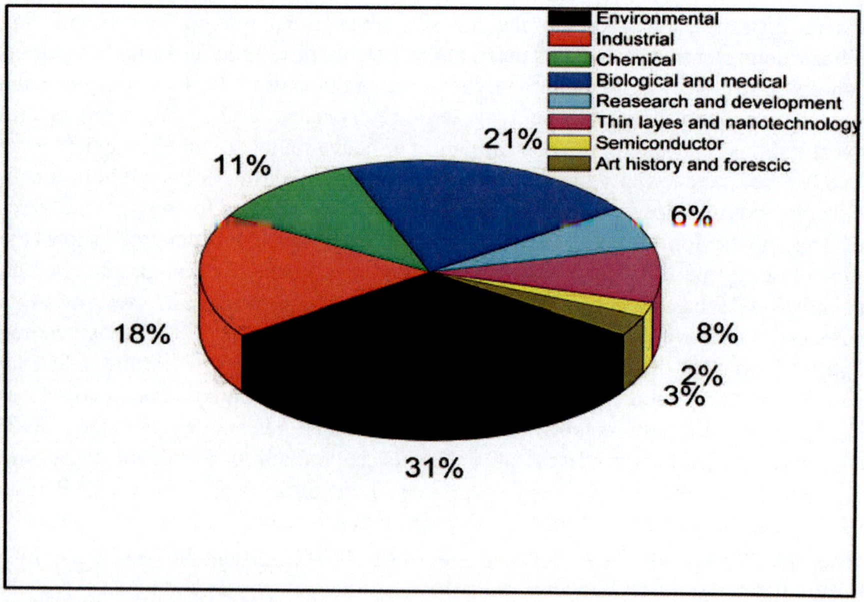

Fig. 23.8 The percentage applications of TXRF spectrometer in different fields for elemental analysis of materials is shown in the pie chart. It is redrawn and adapted from Elsevier Publisher (Klockenkämper and von Bohlen 2014)

by various methodologies like solvent extraction, complexation, solid phase extraction, etc. (Shukla et al. 2020; Sanyal et al. 2020; Maksimova 2022). A unique and simple methodology developed is functionalization of the extractant on the quartz sample support, which has several advantages. The main advantage of this methodology is, the quartz supports functionalized with extractant can be directly put into the solution for equilibration and then removed and directly measured using TXRF (Sanyal et al. 2023).

The elemental analysis of airborne particulate matter is of very high relevance to the modern society as it helps to predict the influence on climatic changes, environmental pollution, and in turn, the human health. The industrial emission of particulate matter cause serious effects on human health, in the form of several diseases such as asthma, respiratory disease, and even lung cancer. Coarse particles are generally obstructed by the hairs in the nose while breathing, and gets collected in the nose itself. But, finer particles enter the bronchi and penetrate and pass through the lung. Different sample preparation methodologies are proposed to collect the particulate matter for analyzing by TXRF. Cascade impactors help in the collection of size-resolved airborne matters, directly on the TXRF sample support (Kayser et al. 2022). Collection and filtration of the airborne particulate on filter paper followed by micro-wave assisted digestion gives the total concentration of all the elements present, irrespective of its particle size (Shaltout et al. 2014).In fact, GI-XRF, where the excitation conditions are varied in a controlled and reliable manner than in TXRF, has been found to be more reliable for analysis of samples of particulate matter. For non-destructive determinations, samples of soil and sediments should be dried, homogenized, and the passed through sieves of nearly 100 μm after grinding. Then, a few mg of the fine powder is suspended in water solution containing a surfactant and after sonication, a few milliliters is drop casted on quartz supports for measurements. The presence of surfactant in the suspensions lowers the sedimentation of particles and increases the homogeneity of the sample solution. This is a unique simple and fast technique for sample screening. In one such studies, the analytical results of two different sample preparation methods: microwave acid digestion and suspension of fine solid particles in 1 wt.% Triton in Milli-Q water, were investigated. A good agreement was obtained between the analytical results of these two different sample preparation methods, especially for the transition elements (Fabjola et al. 2013). This proves that non-destructive, simple sample preparation methodology, which avoids acid handling, can be very successfully used for such analysis.

23.5.2 Biological and Biochemical Applications

Biomonitoring is the study of the uptake of trace elements into the biological ecosystem, along with the understanding of their partition and assimilation, inside the body. Such studies have an impact on both the environmental and nutritional factors. The environmental factors deal with the level of hazardous elements, which are present in the environment, whereas the nutritional factors deal with the uptake and effect of these hazardous elements on the metabolism of an organism. The biochemical applications are generally of medical and pharmaceutical significance. Estimation of the elemental contents, in various organic systems, also helps in investigating the connection between the elemental concentrations and the associated disease or disorder in the body (Mirji et al. 2018). In one such study, the trace elements were determined in the normal as well as malignant tissues and compared in 30–70 years age group, using TXRF (Mirji et al. 2018). The studies revealed that some of the trace elements such as P, S, Cl, K, Fe, and Sr were present in larger concentrations, in malignant tissues of people in the age group upto 60 years, however Pb was found to be more in the age group of 61–70 years old. Though, there are numerous methods available in literature for TXRF determination of biological samples, such as microtomes, slurries, freeze drying, microwave digestion, etc., the suitability of these methods depends on the purpose of the analyses and on the nature of the biological samples (Mishra 2022; Carvalho et al. 2022). Direct sample analysis (DSA) methods by TXRF are interesting as it is fast, relatively simple and avoid possible contamination during sample preparation. However, the main drawback is the sample homogeneity. Another drawback is the difficulty in addition of an internal standard for quantitative determinations. Therefore, using the ratio of net intensity of the analyte and scattered peak (Rayleigh/Compton), the in-homogeneities in density as well as thickness, of the sample, can be taken care of as reported in some literature studies (Wróbel et al. 2017).

The chemical quality control of impurities, in samples of pharmaceutical importance, is very important due to their potential toxic effects. Due to the advantageous features of TXRF methodology described above, it has emerged as an ideal technique for determination of residual metals in the ingredients of pharmaceutical and the intermediate products. In fact, in a study it has been reported that the analytical results of TXRF, were in good agreement with the ICP-MS values and statistical analytical results have indicated that the two technologies are equivalent in terms of analytical parameters at the 95% confidence level for pharmaceutical samples (Antosz et al. 2012b; Mishra and Dhara 2022).

23.5.3 *Industrial and Nuclear Applications*

Nowadays, TXRF is one of the best methods of choice for chemical quality control of several industrial materials in various sectors such as chemical, petroleum, mining, pharmaceutical, electronics, cosmetics, as well as nuclear. In fact, the features of TXRF are highly suitable for nuclear industry as the materials involved are frequently radioactive, and a faster analysis with minimum sample amount is imperative. Some high-purity acids, bases, solvents, etc. can be analyzed nearly directly down to the elemental concentration of level of 0.1 ng/ml using TXRF. In the petroleum industry occurrence of metallic impurities, even at trace level, in the distillates as well as crude oil can detrimentally affect the engines and turbines that process the oil, in addition to the contributions to pollution of environment. Some sample treatment methods, e.g., microwave digestion, sulfate ashing, dilution in various organic solvents, and solvent extraction methods, are used for the separation of matrix in oil-based matrices, for analysis of trace elements. However, these methods suffer many limitations such as poor detection limits, the need of matrix matched standard, undesirably poor ratio for preconcentration, etc. Recently, a TXRF-based methodology has been developed and validated for the determination of trace elements in jet fuel and gasoline samples using surfactant-free microextraction method. This TXRF-based method overcame most of the abovementioned drawbacks (Spanu et al. 2021).

Ceramics are inorganic solids with crystalline structure, having high melting points, chemical and electrical resistance, and hence, used as refractory materials and electrical insulators as well as for applications in the nuclear industry. The presence of trace impurities strongly influences the abovementioned properties of the ceramics. For determination of trace elemental impurities, digestion of these materials is required for analysis by Atomic Absorption Spectrometry (AAS), Inductively Coupled Plasma Mass Spectrometry (ICP-MS) and Inductively Coupled Plasma Optical Emission Spectrometry (ICP-OES). Further, digestion process requires addition of corrosive acids, a long sample preparation time, which in turn can lead to addition of contaminants and also results in the loss of analytes. Direct solid analysis (DSA) using slurry sampling and suspension, are two promising methods of nondestructive determination of trace impurities in refractory materials using TXRF. These methods have been very efficiently utilized for the determination of trace elements in archaeological ceramic shards, and samples of Al_2O_3, SiC, BN, BC, etc., having light element matrix elements. Slurries with good stability were obtained using polyethylenimine at a concentration of 0.1% and a pH of 4. In Table 23.3, a comparison of the results for trace metallic impurities in boron carbide samples, obtained by nondestructive TXRF method and ICP-OES method after dissolution of the sample is given (Ambergera 2009).

Probably, the industry that beneficiated most by TXRF is the semiconductor industry. Silicon wafers are the main product of the semiconductor industry, which requires extremely pure ingots of Si sliced into wafers less than 0.75 mm thick. The total fabrication process for a wafer can take several weeks and is performed in clean rooms of highly specialized plants. But all of these steps are susceptible to infiltration of contaminants, and for that reason contamination control of Si wafer is indispensable. TXRF is a highly suitable method for direct examination of wafer and is capable of detection of 10^7 atoms/cm^2. Most studies reported are concerned with contamination from transition metals on Si-Ge and Ga-As wafers, analyzed by TXRF. In fact, during the past decade, TXRF has become a routine online and inline tool for monitoring impurities on the wafer surface. Other merits of TXRF are, it can be utilized for stratified structures and multilayers, for depth profiling, and thickness measurements below and above the critical angles by scanning the glancing angles (Pahlke et al. 2001). The combination of vapor phase decomposition (VPD), for contaminants preconcentration, and TXRF measurements, is a destructive method but highly sensitive for elements on and in the uppermost layers of silicon wafers (Hellin et al. 2006). A recent development in direct analysis of the whole wafer for contamination is the TXRF measurements recorded in sweeping mode. The wafer is moved step by step automatically and an elemental mapping is obtained (Hellin et al. 2006).

Table 23.3 Analytical results of trace metallic impurities determined using TXRF and ICP-OES techniques in boron carbide samples (Ambergera 2009)

Elements	Detection limits		Concentrations of trace impurities determined	
	TXRF method: direct solid analysis	ICP-OES method: acid digestion	TXRF	ICP-OES
	Concentrations in (μg/g)			
Ca	1.2	0.01	86 ± 2	91 ± 6
Ti	0.6	0.7	206 ± 33	194 ± 22
Cr	0.3	0.3	5.0 ± 0.9	5.2 ± 0.6
Mn	0.3	0.03	4.7 ± 0.5	4.7 ± 0.3
Fe	0.2	0.4	333 ± 18	370 ± 28
Co	0.1	0.5	2.1 ± 0.2	2.4 ± 0.2
Ni	0.1	0.6	14.1 ± 0.9	15.2 ± 1.1
Cu	0.1	0.2	6.20 ± 0.02	6.2 ± 0.6

Nuclear materials are a class of materials having high technological importance, used for power generation and related applications. The materials used have to undergo very stringent chemical as well as physical quality control during the various steps of their production, fabrication as well as after irradiation inside the nuclear reactor called post-irradiation examination. The specific advantage of TXRF analysis such as, the requirement of minute sample for analysis, non-consumption of the sample during analysis, simple and rapid elemental detection and quantification for elements above carbon, insignificant matrix effect, etc. has made this technique very unique as well as highly suitable method (even superior in some respects to the existing methods) for the elemental characterization of radioactive samples (Dhara and Misra 2019; Dhara et al. 2015a). Moreover, management of the radio-analytical waste, produced after radioactive sample analysis, is also a very essential part of the sample analysis. Since the amount of sample required for TXRF analysis is very less and does not require the addition of any reagents (as required in electroanalytical methods), the radio-analytical waste generated during TXRF analysis is also less. In addition, the spectral modifications, such as use of continuum or selective excitation, have been found very helpful in solving many analytical challenges faced in the analysis of nuclear materials. Heavy water (D_2O) is a material of immense importance in the nuclear industry as it is used as a coolant as well as a moderator in pressurized heavy water nuclear reactors. The presence of indium as an impurity in heavy water is undesirable as it gives secondary gamma rays after irradiation in a reactor when used as a coolant. In one such study, for the determination of indium in the sample of heavy water, the Bremsstrahlung continuum from the X-ray tube (Mo tube) was used. A band of energy range, from 25 to 35 keV, was selected and aligned for excitation of the sample using the monochromator. For quantitative determination of indium in D_2O, In Kα (K_{abs} at 27.94 keV) was used as analytical line. Since the monochromatic Mo Kα (Mo Kα = 17.47 keV) X-rays from the Mo target used in the spectrometer could not excite In Kα, the aforementioned continuum strip was used. The desired energy band required to excite indium was obtained by modifying the angle of incidence of the incoming X-ray beam on the surface of the multilayer and the total reflection on quartz support was achieved by changing the inclination of the X-ray tube. After preconcentration of the sample by evaporation and with continuum excitation, the detection limit obtained for indium was as low as 1 ng/mL (Dhara et al. 2015b). The spectrum of a preconcentrated (by evaporation) water sample spiked with a 50 ng/mL multi-elemental standard of Ag, Cd, and In and measured by TXRF, is shown in Fig. 23.9.

23.5.4 *Archeological and Forensic Applications*

Valuable works of historical art and heritage are investigated for their restoration and conservation as well as for their dating and detection of forgeries. Analytical techniques requiring very less sample amount and causing negligible damage to the artifacts are prime requirements for analysis of such materials. In the area

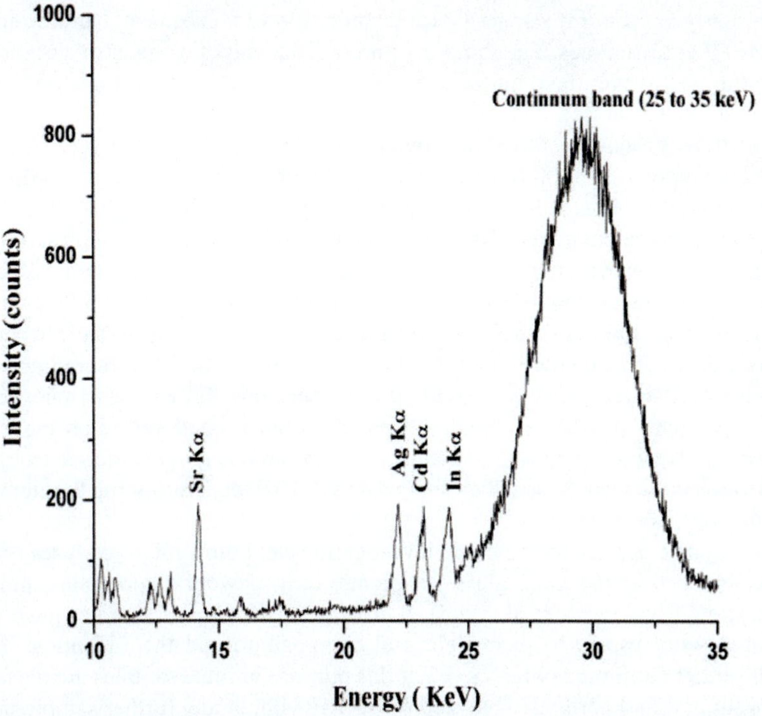

Fig. 23.9 TXRF spectrum showing the X-ray Kα peaks of Ag, Cd, and In having a concentration 50 ng/mL each in a heavy water sample excited using the continuum band

of forensic science, the sample available may be in the range of mg or μg level. Consequently, TXRF as a micro- or ultramicro-analytical tool is highly suitable for the abovementioned applications. Inks and pigments used for painting and writing historical documents, illuminated manuscripts, etc., are valuable cultural heritage. They allow insights of historical and archaeological relevance. The key elements in the color pigments consist of metals in the form of oxide, hydroxide, sulfides, or their mixtures and the metals present in them can be characterized by TXRF. The key element used in the pigments for green color are oxides of Co and Zn, for red Hg and white Pb, in some the cases, of the white color was a mixture of Ba, Ti, and Pb. TXRF is a rapid and convenient systematic screening method applied for the detection of elements present in the coloring pigments, so that the restoration work can be carried out accordingly (Moens et al. 1194). Ink analysis is also important in forensic applications as it can be used to differentiate the forged documents. Atomic spectroscopic techniques such as Inductively Coupled Plasma Mass Spectrometry (ICP-MS), Graphite Furnace Atomic Absorption spectrometry (GF-AAS), Particle-Induced X-ray Emission (PIXE) and X-Ray Fluorescence (XRF) are applied in identifying as well as quantifying metals in historical inks. However, due to the remarkable sampling methods, which is very gentle, causes no visible damage, and requires only micro-samples, the TXRF technique is very unique and preferred one, over the other techniques. The sampling procedure involves using a dry cotton wool bud to rub gently over the required surface containing the paint, without any visible destruction to the surface, and removing a very small amount of the paint. This small paint particle is then transferred on a quartz sample support and analyzed by TXRF (Klockenkämper et al. 2000). The average amount, which gets transferred on the sample support is about 0.1–0.5 μg, which is very small to be weighed, therefore it is impossible to calculate the absolute concentration of the elements present in the paint sample. However, the relative concentrations of the elements with respect to a specific element can be determined by this method. Another method utilizes the analysis of small flakes of varnish that can be taken out by means of a clean scalpel. This is a destructive method in which the sample is weighed and then digested in a mixture containing nitric acid and hydrogen peroxide. Finally, it is analyzed by TXRF. Any hard surface can be simply touched and rubbed gently on the TXRF quartz support which is similar to the ancient and famous nondestructive technique called ***touchstone*** method, and then analyzed by TXRF for determining the elemental compositions, approximately.

The forensic applications include investigations and ultra-microanalyzes of hair, blood, semen traces, paints, glass fragments, drug powders, currencies, gunshot residues, etc. (Ninomiya et al. 1995). Sometimes, two similar samples have to be compared with respect to their elemental composition and their identical nature must be either confirmed or rejected. For this purpose, minute samples are placed on a glass support and normally analyzed by TXRF without any further sample preparation. These TXRF spectral patterns are then used as fingerprints of the samples and compared. In routine examination of the inks of counterfeit currencies, color of inks, the luminescence as well as absorption are evaluated. These methods differentiate the inks in less objective way. However, if it is required to characterization them in a more objective way, elemental determination are required. The elemental determinations in inks can also be used effectively to distinguish the currencies. Rare-earth elements are generally absent in the inks which are commercially available and therefore, spiking of the inks with different rare-earth can be favorably used to distinguish them (Dhara et al. 2010). The TXRF spectra of ink samples tagged as well as untagged with rare-earths of La, Pr, Nd, Sm, and Eu are shown in Fig. 23.10. The sample specimens were prepared by removing the inks from different printed documents, gently with a scalpel in such a way that no damage was done to the printed paper. Then a few micrograms of the removed ink sample which was in the form of powder was transferred on the quartz sample support and analyzed by TXRF. The tagged and untagged inks were clearly distinguished using this method. For determination of the rare earths concentration, nearly 20 mg of printed paper portion containing the rare-earth tagged inks, was digested in a mixture of HNO_3 and $HClO_4$, and after complete digestion, it was analyzed quantitatively using TXRF (Dhara et al. 2010).

Gunshot residues can occur on neighboring surfaces surrounding the discharge of the firearm. The most probable surfaces may include the hand, fingers, nails, face, and also the chest of the person using the firearm. A gunshot residue typically

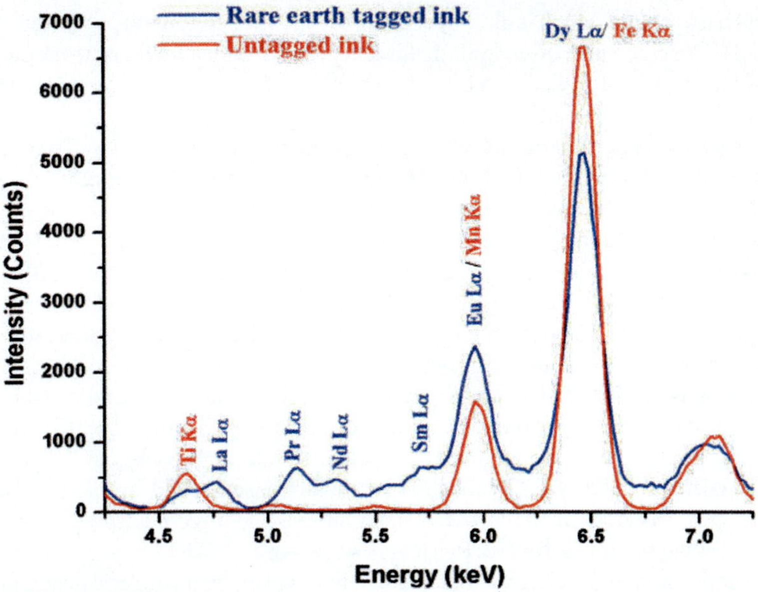

Fig. 23.10 Differentiation of two ink samples by TXRF analysis. It is reproduced with permission from Elsevier Publisher (Dhara et al. 2010)

consists of the elements Pb, Ba, and Sb in it. For TXRF analysis, the traces of samples were taken from the abovementioned body parts of expert shooters. The sampling procedure involved extraction of the gun residue powders using cotton swabs embedded in 5% nitric acid and then its leachate in sub-boiling nitric acid (Sarapura et al. 2019).

23.6 Trends and Future Prospects

TXRF is no doubt a novel variant of EDXRF having numerous applications in wide areas. The principle, instrumentation, analytical performance, and applications have been discussed in the previous sections of this chapter. Now, in this final section, the present trends and future scopes of TXRF will be considered. The trends are mostly developments in the instrumentations in order to extend the analytical capabilities of TXRF. The future prospects mostly include the combination of various X-ray methods performed in total reflection mode aimed to overcome the analytical limitations of TXRF.

Excitation using synchrotron radiation reported in various studies have helped in achieving lowest detection limits which could never have been achieved using laboratory-based X-ray tubes. Apart from high brilliance and flux, synchrotron radiation provides a broad energy range of the X-rays, which can be tuned. In addition, the spectral background which is already reduced by total reflection mode of excitation is further lowered because of the polarization. These features of synchrotron radiation resulted in improved limits of detection, which are almost 3–4 orders of magnitude lower compared to the traditional tube excitation. Recently, a new synchrotron beam line end station has been commissioned at the Elettra Sincrotrone, in Trieste, Italy jointly with the International Atomic Energy Agency (IAEA), for multiple applications of X-ray spectrometry. The special feature of this endstation is that it has an ultrahigh vacuum sample chamber containing a seven-axis motorized sample manipulator. This chamber can be operated at a very low pressure (less than 10^{-8} mbar) (Karydas et al. 2018). The silicon drift detector, having a special window of super light element made of an ultrathin polymer with thickness of nearly a few tens of nm and protected by aluminum layer mounted on Si grid, was used in this facility. The synchrotron beam line delivers X-ray beam of energy in the range of 4 to 14 keV by the means of double crystal monochromator of Si [111]. It has three multi-layers one each for low (0.7–1.8 keV), medium (1.5–8.0 keV) and high energy (3.6–14.0 keV) X-rays (Sanyal et al. 2020). The spectra of a zircaloy disc excited with X-rays of 14.0 and 1.9 keV energies, from the IAEA beamline in TXRF geometry and processed by peak deconvolution software, PyMca, is shown in Fig. 23.11.

The expansion in the range of detectable elements using a single instrumentation with TXRF was a challenging as well as an interesting task. A novel table top TXRF spectrometer with a sample chamber and two easily interchangeable X-ray tubes of

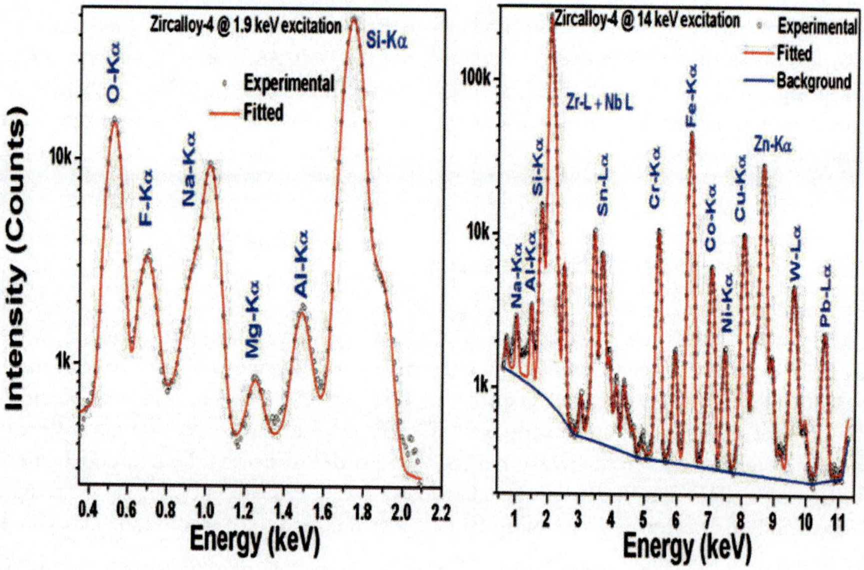

Fig. 23.11 TXRF spectra of a zircaloy sample excited using 14.0 and 1.9 keV energies at IAEA beamline commissioned at Elettra Sincrotrone, Trieste. It is reproduced with permission from Wiley Analytical Science Publisher under Creative Common License (Sanyal et al. 2020)

low power for the elemental detection as well as quantification was developed, which can analyze elements from carbon to uranium using a single TXRF spectrometer (Wobrauschek et al. 2015). The sources of X-rays in this spectrometer were two lightweight, low-power, 35 W air-cooled, X-ray tubes, which can be easily mounted. The targets of the tubes were made of anodes: Cr ($for\ Z = 6$–22) and Rh ($for\ Z =$ 13–92). Further, a small Peltier-cooled detector is placed very near over the sample. These features permits this spectrometer with compact dimensions to be mounted on a tabletop without the requirement of water cooling chiller for the X-ray tubes or cooling of detectors with liquid nitrogen. This spectrometer has the provision to provide vacuum (3 mbar) in the sample chamber for the analysis of low Z elements.

The lowest detection limits achieved using this table top spectrometer was 150, 12, and 3.3 ng for C, F, and Na, respectively, using Cr Kα excitation under vacuum conditions. The lowest achievable detection limits was 90 and 200 pg for Sr and U, respectively, when excited using Rh target, in ambient atmospheric conditions.

Another advanced feature of TXRF mode of analysis is Grazing Exit XRF (GE-XRF), which opens up an entirely a new field for determining the elemental concentrations as a function of depth in samples of multiple layers in the regime of 100 nm to a few nm. In this method the X-rays emitted from the sample at grazing exit are measured, whereas the primary incident beam is made to fall on the sample at an angle of 90° with respect to the sample. GE-XRF can compete very well with the well-established methods of depth-profiling, like Secondary Ionization Mass Spectrometer (SIMS) and Rutherford back scattering spectrometry (RBS). It is nondestructive and does not require ultrahigh vacuum or conducting sample during the analysis unlike the other two methods mentioned above. The feasibility of 3D-XRF spectrometry using 300 μm beam and GE-XRF geometry was also demonstrated using a laboratory setup (Tsujia and Delalieux 2003). Measuring the elemental distribution using micro X-ray beams at different exit angles of 0.8 and 4°, enabled the reconstruction of 3D elemental distributions in the sample. In another study, a simple conical pinhole was successfully used for micro-TXRF analysis of wafers for two-dimensional surface analysis (Tsuji et al. 2006). Further, coupling of TXRF with X-ray Absorption Spectroscopy (XAS) opens up the possibility of speciation studies along with elemental characterization of micro-samples. In TXRF analysis elaborate sample preparation is not required. Samples are slightly rubbed on the quartz support and the TXRF-X-Ray Absorption Near Edge Spectroscopy (TXRF-XANES) analysis of the material is carried out. In one such study carried out to explore the different oxidation states of uranium in various uranates using the U L3 edge, TXRF hyphenated XANES was used (Sanyal et al. 2017). In this study, the oxidation states of uranium in different mixed valent oxide compounds were evaluated. Moreover, their relative amounts in these oxides were also determined in this study. The studies have revealed that inU_3O_8, uranium exists in a mixed valent state consisting of U (+5) and U (+6), in 70 and 30% amounts, respectively, and in U_3O_7U (+5) and U (+6), are present in 66% and 34%, respectively. The obtained results were in concordance with the theoretically calculated values. In Fig. 23.12, normalized spectra for various compounds of uranium, recorded in TXRF-XANES mode, are shown. The chemical shift in the energies of different oxides of uranium is also seen clearly in this figure.

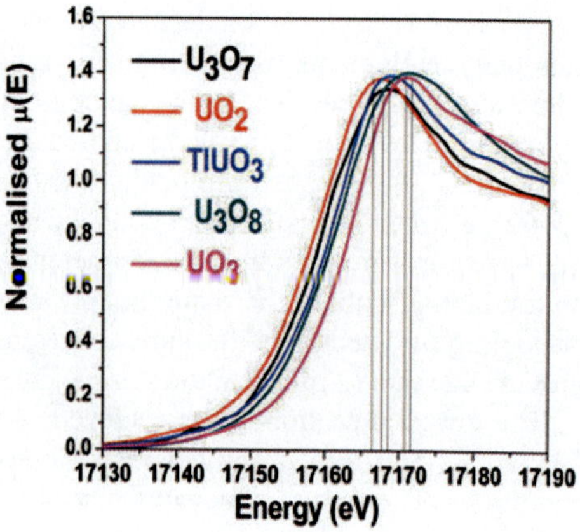

Fig. 23.12 The spectra of various compounds of uranium recorded in TXRF-XANES mode. It is reproduced with permission from ACS Publisher (Sanyal et al. 2017)

Thus, from the abovementioned studies it can be concluded that when excitation of the sample by X-rays in total reflection mode is combined with several other methods of spectroscopy, many information can be obtained for large variety of samples with negligible sample and minimum sample preparation in a very simple manner. Some of the experiments can be performed using lab-based X-ray tubes, while others require synchrotron radiation source. An overview of different combinations of total reflection coupled with different methods of spectroscopy is summarised in Fig. 23.13. Different methods of spectral analysis presented in this figure are based on fluorescence, diffraction, and absorption of X-rays in totally reflected mode. These combinations are applicable to specific tasks, required for some analytical challenges, different from the conventional requirements. So, apart from ultra-trace and trace determinations in micro samples, and their distribution over the sample surface can also be studied using micro-TXRF. GI-XRF and GE-XRF are related to TXRF geometry and used for thin film, surface characterization as well as depth profiling. Using Total reflection X-Ray Reflectrometry (XRR), the absolute analysis of sample thickness, its density and even the roughness in layered materials can be studied. In X-ray Absorption Fine Structure spectroscopy (XAFS), different elemental species i.e., the oxidation state or valence of elements, chemical state and coordination, bond lengths and the structure of molecules can be established. The absorption spectra can be recorded in absorption mode or fluorescence mode. In total reflection mode the TXRF-XAFS is carried out in fluorescence mode as the sample amount used is very less (Sanyal et al. 2017). TR-XRD (Total Reflection X-Ray Diffraction) or GIXRD (Grazing Incidence X-Ray Diffraction) have emerged as a powerful tool for evaluating thin near-surface layers and crystal structures of minute samples owing to the low penetration of the incoming X-ray beam into the sample, at total reflection condition. Similar to GI-XRD, GE-XRD (Grazing Exit X-Ray Diffraction) is used for thickness, density, and surface roughness determination by variation of the exit angle.

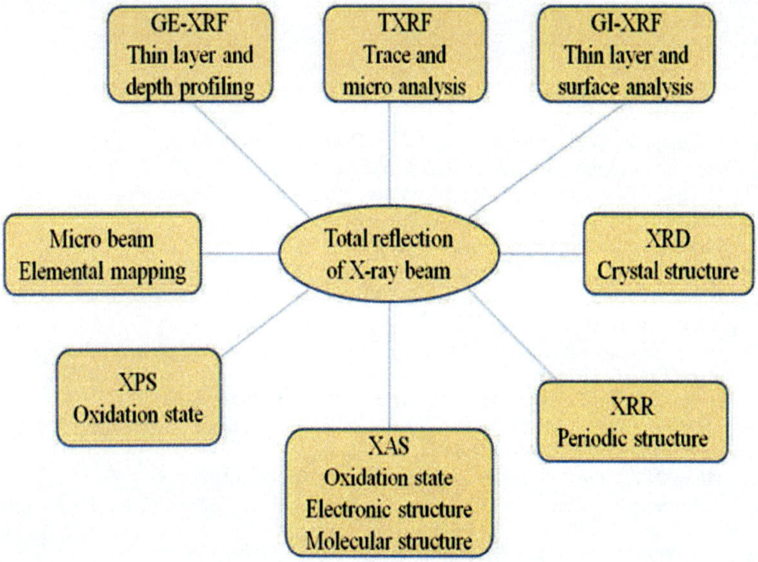

Fig. 23.13 An overview of different combinations of total reflection coupled with different methods of spectroscopy

23.7 Conclusions

Thus, it can be concluded that TXRF is a novel technique for multi-elemental analysis of materials which has inexhaustible applications in various areas of scientific research and development. Due to its non-destructive/non-consumptive capability, high sensitivity toward analysis of samples up to nano gram levels and features of depth profiling, this technique has emerged as a highly competitive method of spectral analysis for characterization of materials. Further, with the introduction of synchrotron radiation source, the achievable detection limits have improved drastically along with the possibility of many other studies which include speciation, structure determination, etc., along with ultra-trace determinations. At present, it is intensively used in the field of semiconductor industries, for process control of Si wafers, but with improvements in the excitation and detection technologies, this technique is gaining its suitability for many other applications of newer scientific areas for material characterization.

Author Contributions "The manuscript was written through contributions of all authors. All authors have given approval to the final version of the manuscript."

References

Ambergera, M.A., Broekaert, J.A.C.: J. Anal. At. Spectrom **24**, 1517–1523 (2009). https://doi.org/10.1039/B906417F

Antosz, F.J., Xiang, Y., Diaz, A.R., Jensen, A.J.: J. Pharm. Biomed. Anal. **62**, 17–22 (2012a)

Antosz, F.J, Xiang, Y., Diaz, A.R., Jensen, A.J.: J. Pharm. Biomed. Anal. **62**, 17–22 (2012)

Bedzy, M.J.: The X-Ray Standing Wave Technique: Principles and Applications. In: Zegenhagen, J., Kazimirov, A. (eds.). World Scientific Publishing Co (2013)

Bohlen, A.: Spectrochim Acta Part B **64**, 821–832 (2009)

Carvalho, P.M., Marguí, E., Kubala-Kukuś, A., Banaś, D., Machado, J., Casal, D., Pais, D., Santos, J.P., Pessanha, S.: Spectrochim Acta Part B **198**, 106548 (2022)

Compton, A.H.: Philos. Mag. 6, **45**, 1121 (1923)

Dhara, S., Misra, N.L., Maind, S.D., Kumar, S.A., Chattopadhyay, N., Aggarwal, S.K.: Spectrochim Acta Part B **65**, 167–170 (2010)

Dhara, S., Parimal, P., Misra, N.L.: Anal. Chem. **87**, 10262–10267 (2015a)

Dhara, S., Sanjay Kumar, S., Misra, N.L., Aggarwal, S.K.: J. Radioanal. Nucl. Chem. **306**, 231–235 (2015b). https://doi.org/10.1007/s10967-015-4180-8

Dhara, S., Misra, N.L.: TrAC, Trends. Anal. Chem. **116**, 31–43 (2019)

Fabjola, B., Borgese, L., Cazzago, D., Zacco, A., Bontempi, E., Guarneri, R., Bernardello, M., Attuati, S., Lazo, P., Depero, L.E.: Environ. Sci. Pollut. Res. **21**, 13208–13214 (2013). https://doi.org/10.1007/s11356-013-2203-y

Franks, A.: X-rays: the first hundred years. In: Michette, A., Pfauntsch, S. (eds.), pp. 1–19. Wiley, New York (1997)

Garg, V.K., Tanta, A., Srivastav, A.L., Tiwari, M.K., Sharma, A., Kanwar, V.S.: Water Environ. Res. **94**, e10759 (2022). https://doi.org/10.1002/wer.10759

Hauser, S., Andres, S., Leopold, K.: Anal. Bioanal. Chem. **414**, 4519–4529 (2022)

Hellin, D., De Gendt, S., Valckx, N., Mertens, P.W., Vinckier, C.: Spectrochim Acta Part B **61**, 496–514 (2006). https://doi.org/10.1016/j.sab.2006.03.008

Joshi, J.M., Garima, P., Sanyal, K., Govalkar, S., Renjith, A.U., Mishra, N.L., Dhara, S.: Spectrochim Acta Part B **182**, 106235 (2021)

Karydas, A.G., Czyzycki, M., Leani, J.J., Migliori, A., Osan, J., Bogovac, M., Wrobel, P., Vakula, N., Padilla-Alvarez, R., Menk, R.H., Gol, M.G., Antonelli, M., Tiwari, M.K., Caliri, C., Vogel-Mikuš, K., Darby, I., Kaiser, R.B.: J Synchrotron Rad **25**, 189–203 (2018). https://doi.org/10.1107/S1600577517016332

Kayser, Y., Osan, J., Honicke, P., Bechoff, B.: Anal Chimica Acta **1192**, 339367 (2022)

Klockenkämper, R., Bohlen, A.: Total-Reflection X-Ray Fluorescence Analysis and Related Methods, 2nd edn. Wiley, Hoboken, New Jersey (2015)

Klockenkämper, R., von Bohlen, A., Moens, L.: X-Ray Spectrom. **29**, 119–129 (2000)

Klockenkämper, R., von Bohlen, A.: Spectrochim Acta Part B **44**, 461–469 (1989)

Klockenkämper, R., von Bohlen, A.: Spectrochim Acta Part B **99**, 133–137 (2014)

Klockenkämper, R., von Bohlen, A. J. Anal. At. Spectrom **14**, 571–576 (1999)

Maksimova, Y.A.: Spectrochim Acta Part B **196**, 106521 (2022)

Marguí, E., Floor, G.H., Hidalgo, M., Kregsamer, P., Román-Ross, G., Streli, C., Queralt, I.: Anal. Chem. **82**, 7744–7751 (2010)

Meirer, F., Singh, A., Pianetta, P., Pepponi, G., Meirer, F., Streli, C., Homma, T.: TrAC Trends Anal. Chem. **29**, 479–496 (2010)

Mirji, S., Badiger, N.M., Sanyal, K., Kulkarni, S.S., Misra, N.L., Gai, P.B.: X-Ray Spectrom. **47**, 432–440 (2018)

Mishra, N.L., Dhara, S.: Total Reflection X-Ray Fluorescence Analysis of some Biological Samples, Total Reflection X-Ray Fluorescence and it's Suitability for Biological Samples, X-Ray Fluorescence in Biological Sciences: Principles, Instrumentation, and Applications, pp. 203–218. Wiley (2022)

Mishra, N.L., Dhara, S.: Total Reflection X-Ray Fluorescence and it's Suitability for Biological Samples, X-Ray Fluorescence in Biological Sciences: Principles, Instrumentation, and Applications, pp. 61–72. Wiley (2022)
Moens, L., Devos, W., Klockenkämper, R., Bohlen, A.: TrAC, Trends Anal. Chem. **13**, 198–205 (1194)
Moseley, H.G.J.: Philos. Mag. 6, **26**, 1024 (1913)
Ninomiya, T., Nomura, S., Taniguchi, K., Ikeda, S.: Analy Sci. **11**, 489–494 (1995). https://doi.org/10.2116/analsci.11.489
Pahlke, S., Fabry, L., Kotz, L., Mantler, C., Ehmann, T.: Spectrochim Acta Part B **56**, 2261–2274 (2001)
Rachevski, A., et al.: Nucl. Instrum. Methods Phys. Res. A **824**, 452–454 (2016)
Sanyal, K., Kanrar, B., Dhara, S., Misra, N.L., Wobrauschek, P., Streli, C.: Adv. X-Ray Anal. **59**, 125–143 (2016)
Sanyal, K., Khooha, A., Das, G., Tiwari, M.K., Misra, N.L.: Anal. Chem. **89**, 871–876 (2017). https://doi.org/10.1021/acs.analchem.6b03945
Sanyal, K., Kanrar, B., Dhara, S., Sibilia, M., Sengupta, A., Karydas, A.G., Mishra, N.L.: J Synchrotron Rad **27**, 1253–1261 (2020)
Sanyal, K., Dhara, S., Gumber, N., Pai, R.V.: Talanta **254**, 124129 (2023). https://doi.org/10.1016/j.talanta.2022.124129
Sanyal, K., Chappa, S., Bahadur, J., Pandey, A.K., Mishra, N.L.: J. Anal. At. Spectrom **35**, 2770–2778 (2020)
Sarapura, P., Gonzalez, M.F., Gonzalez, F., Morzan, E., Cerchietti, L., Custo, G.: Appl. Radiat. Isot. **153**, 108841 (2019)
Shaltout, A.A., Boman, J., Welz, B., Castilho, I.N.B., Al Ashkar, E.A., Gaita, S.M.: Microchem. J. **113**, 4–9 (2014)
Shukla, V.K., Dhara, S., Mishra, N.L.: J. Anal. at. Spectrom. **35**, 1632–1640 (2020)
Spanu, D., Roncoroni, G., Cinosi, A., Furiano, R., Siviero, G., Monticelli, D.: Fuel **310**, 122458 (2021)
Tsuji, K., Kawamata, M., Nishida, Y., Nakano, K., Sasaki, K.: X-Ray Spectrom. **35**, 375–378 (2006)
Tsujia, K., Delalieux, F.: Spectrochim Acta Part B **58**, 2233–2238 (2003)
Wobrauschek, P., Prost, J., Ingerle, D., Kregsamer, P., Misra, N.L., Streli, C.: Rev. Sci. Instrum. **86**, 083105 (2015)
Wróbel, P.M., Bała, S., Czyzycki, M., Golasik, M., Librowski, T., Ostachowicz, B., Piekoszewski, W., Surówka, A., Lankosz, M.: Talanta **162**, 654–659 (2017). https://doi.org/10.1016/j.talanta.2016.10.043
Yoneda, Y., Horiuchi, T.: Rev. Sci. Instrum. **42**, 1069 (1971)